Decimal Equivalents of Common Fractions*

4ths	8ths	16ths	32nds	64ths	To 4 Places	To 3 Places	To 2 Places
				1/64	.0156	.016	.02
			1/32		.0312	.031	.03
				3/64	.0469	.047	.05
		1/16			.0625	.062	.06
				5/64	.0781	.078	.08
			3/32		.0938	.094	.09
				7/64	.1094	.109	.11
	1/8				.1250	.125	.12
				9/64	.1406	.141	.14
			5/32		.1562	.156	.16
				11/64	.1719	.172	.17
		3/16			.1875	.188	.19
				13/64	.2031	.203	.20
			7/32		.2188	.219	.22
				15/64	.2344	.234	.23
1/4					.2500	.250	.25
				17/64	.2656	.266	.27
			9/32		.2812	.281	.28
				19/64	.2969	.297	.30
		5/16			.3125	.312	.31
				21/64	.3281	.328	.33
			11/32		.3438	.344	.34
				23/64	.3594	.359	.36
	3/8				.3750	.375	.38
				25/64	.3906	.391	.39
			13/32		.4062	.406	.41
				27/64	.4219	.422	.42
		7/16			.4375	.438	.44
				29/64	.4531	.453	.45
			15/32		.4688	.469	.47
				31/64	.4844	.484	.48
					.5000	.500	.50
				33/64	.5156	.516	.52
			17/32		.5312	.531	.53
				35/64	.5469	.547	.55
		9/16			.5625	.562	.56
				37/64	.5781	.578	.58
			19/32		.5938	.594	.59
				39/64	.6094	.609	.61
	5/8				.6250	.625	.62
				41/64	.6406	.641	.64
			21/32		.6562	.656	.66
				43/64	.6719	.672	.67
		11/16			.6875	.688	.69
				45/64	.7031	.703	.70
			23/32		.7188	.719	.72
				47/64	.7344	.734	.73
3/4					.7500	.750	.75
				49/64	.7656	.766	.77
			25/32		.7812	.781	.78
				51/64	.7969	.797	.80
		13/16			.8125	.812	.81
				53/64	.8281	.828	.83
			27/32		.8438	.844	.84
				55/64	.8594	.859	.86
	7/8				.8750	.875	.88
				57/64	.8906	.891	.89
			29/32		.9062	.906	.91
				59/64	.9219	.922	.92
		15/16			.9375	.938	.94
				61/64	.9531	.953	.95
			31/32		.9688	.969	.97
				63/64	.9844	.984	.98
					1.0000	1.000	1.00

Military Standards 8C

Drafting
Technology and Practice

Third Edition

William P. Spence

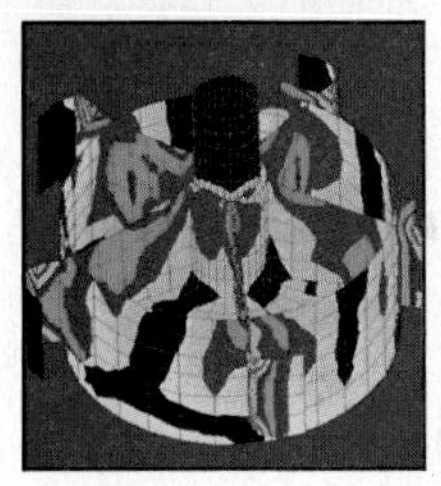

GLENCOE
McGraw-Hill

New York, New York Columbus, Ohio Mission Hills, California Peoria, Illinois

Printed in the United States of America.

Send all inquires to:
Glencoe/McGraw-Hill
3008 W. Willow Knolls Drive
Peoria, IL 61614-1083

ISBN 0-02-676290-0 (Text)

5 6 7 8 9 10 VHJ 99 98 97 96 95

Editing by Lindquist Editorial Services

Credits:
Cover:
General Motors Corp.
Tymet McDonnell Douglas Network Systems Co.
Brent Phelps
Computervision
NASA
Autodesk, Inc.

Title Page:
Ann Garvin
Sverdrup Corp.
Herman Miller, Inc.
NASA
SRAC
IBM RT Systems

Table of Contents:
IBM RT Systems
Hoover Universal
NASA
MacNeil Schwendler
Computervision Corp.

Preface

This book has been designed to serve as a class text and reference book in drafting technology and practice. The content covers a wide range. The text is designed so that it can be used in a beginning course yet includes sufficient additional technical subject matter so that it can serve for advanced classes. It is ideally planned for a broad survey course in which the student is exposed to an extensive range of experiences in applying basic information to a variety of industrial practices.

An important feature of this text is that a large part of it is devoted to specific uses of the graphic language in industrial applications. For example, the student can learn to make a wide variety of *electrical and electronics schematics*, including printed circuits. He or she can explore the much neglected area of *piping diagrams* and drawings. *Technical illustration* is finding ever increasing use in many industries. This area is covered in detail. An introduction to *architectural drawing* is provided to give students the opportunity to have an experience in this area of industry. *Structural drawing* is a big part of architecture. The basic types of structural drawings are explained. Everyone uses *maps* of all kinds, yet students seldom understand their development and how they are made. This text provides a unique experience for students in the area of map design and symbolism. Considerable emphasis is placed on *drawing for production*. Experiences are available in the areas of detail and assembly drawings, gears, cams, and section drawings. The basic introduction to *descriptive geometry* is designed so that the student can grasp the fundamentals and apply these to all types of industrial drawing.

The use of *computer-aided design and drafting* is growing. This text includes a chapter that describes the hardware used in CAD and explains typical CAD commands. In addition, each chapter describes a particular CAD application, such as CAD/CAM or architectural rendering.

A large number of practical drawing problems are included in the end-of-chapter *review sections*. These are carefully selected to give the student a wide range of experiences. The problems selected use objects most students will recognize. This helps relate the subject matter to realistic drawing problems. Also included are a number of problem solving and original design situations. Students will have the opportunity to try their hand at designing and detailing. This affords them a very real chance to apply what they are studying. The study problems range from very simple to very difficult. This enables the instructor to offer a challenge to students of varying abilities.

To provide real-life drawing experience, it is hoped that drafting problems will be drawn on materials used in industry. Students should draw on vellum, tracing paper, and polyester drafting film.

They should run prints of all drawings made. This not only gives them the opportunity to do some reproduction work, but will quickly show the defects in line thickness and darkness. This will enable the teacher to show the students why they need to pay attention to line quality. It is not enough to simply tell them.

Also at the end of each chapter is a section called *Build Your Vocabulary*. Carefully selected technical terms haved been listed for review and definition. The lists will help the student to review the key parts of the chapter and clarify the meaning of items that may not be understood.

Many students have difficulty with basic mathematics. Because a knowledge of math is essential for drafting, math exercises have been included at the end of most chapters. These are labeled *Sharpen Your Math Skills.*

Considerable space has been devoted to the use of *two-color illustrations* to help the student understand the purpose of each illustration and perhaps learn the material more easily.

Many industries are using the metric system. Both the *customary and metric systems* of measurement are explained and used in examples and study problems. Students should be required to produce designs and drawings using both systems of measurement.

To all the individuals and companies who have contributed materials and advice during the preparation of this text, I give my sincere thanks. Certain material from the American National Standards Institute, Inc., is copyrighted by and used with permission of the American Society of Mechanical Engineers, 345 E. 47th St., New York, N.Y.

William P. Spence

Contents

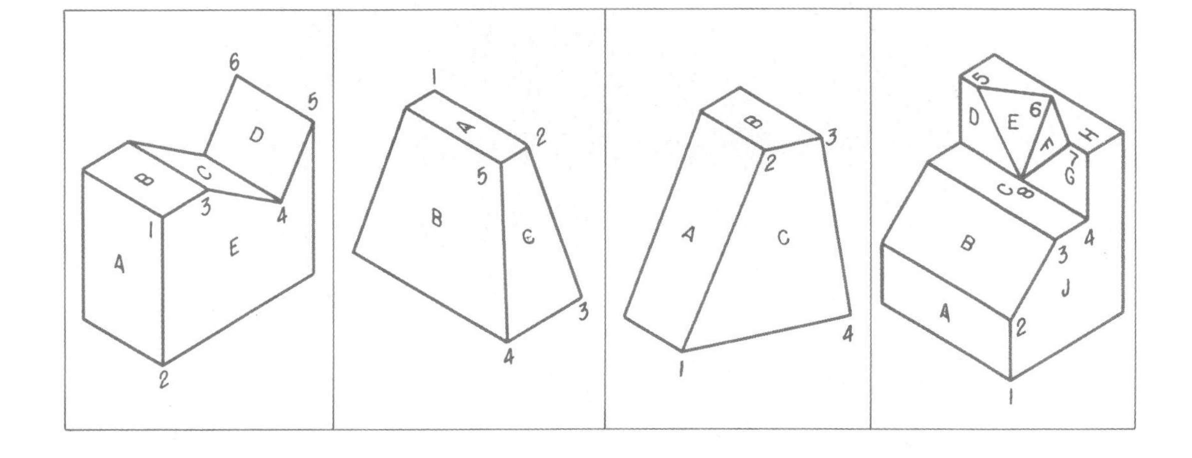
6
5
D
B
C
3
4
1
A
E
2
1
A
2
5
B
C
3
4
B
3
2
A
C
1
4
5
D
E
6
F
H
7
G
C
8
B
4
3
J
A
2
1

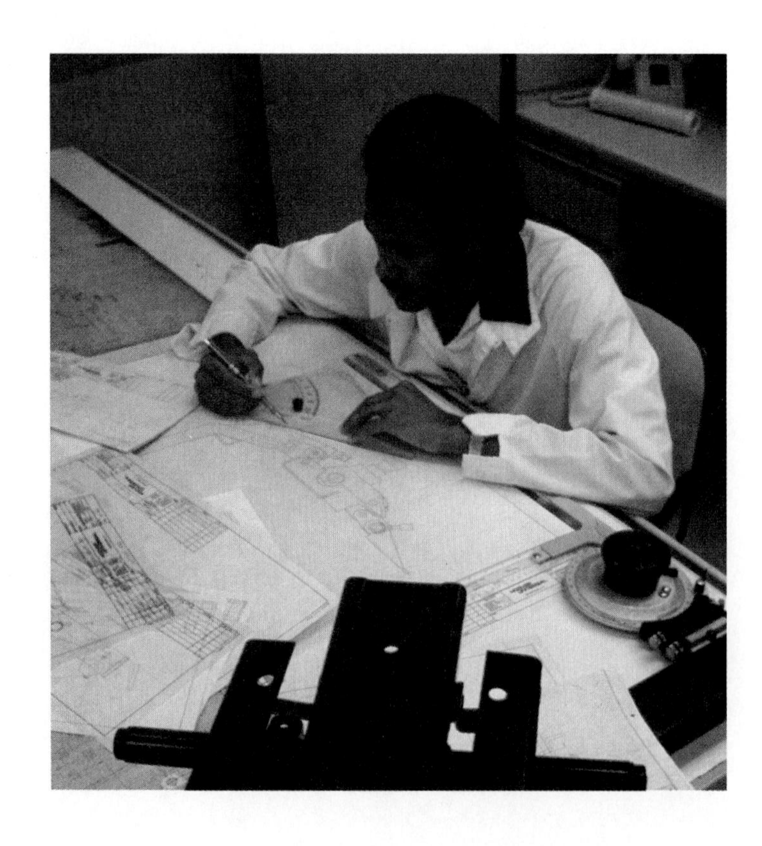

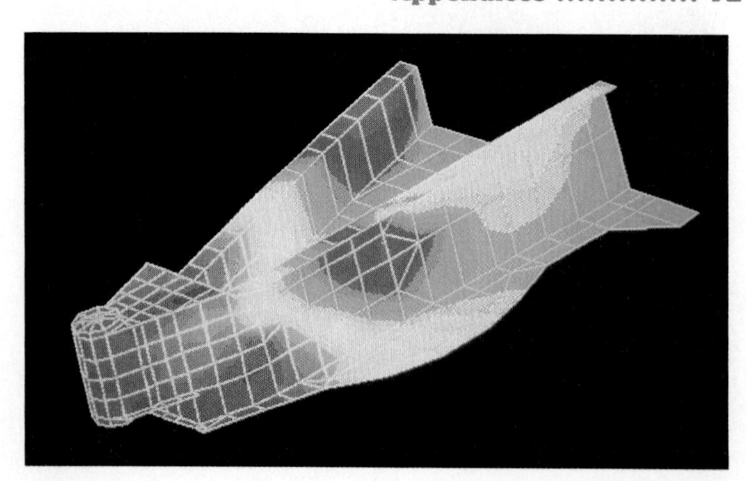

CAD Applications

CHAPTER 1

Drafting in American Industry

After studying this chapter you should be able to do the following:

- Describe graphic communications.
- Understand the range of drafting careers from drafter trainee to design drafting manager and how to qualify for each.
- Understand a small part of the drafting and engineering opportunities that will be available in the future.

The Graphic Language

The earliest records of the human race include many picture drawings. These range from crude drawings on the walls of caves to the more detailed hieroglyphics found on buildings in Egypt. Fig. 1-1. Although people have used drawings to communicate for thousands of years, it has only been since the industrial revolution that people have used the graphic language of technical drawings to the benefit of mankind. **Technical drawings** use lines, symbols, and notes to communicate the information needed to construct an object. The story of the development of the graphic language you are about to study parallels the development of industry and of the products made available to each of us as consumers. As industry and products

Fig. 1—1. Early people recorded events using available tools and materials. (Bettye M. Spence)

have increased in number and complexity, graphic language has changed to keep pace. Today great numbers of technical drawings are made quickly with great detail and accuracy. The process of developing new ways to communicate graphically is vital to the continued success of American industry. The most recent development is the use of computers to assist with the design of a product and to control the plotter which actually produces the drawings. Fig. 1-2.

Graphic Communication

All day long you communicate with other people. Most of the time you talk to them, but often you must communicate in writing. **Graphic communications** are ideas or messages that are written or drawn. Drawings are used in industry to describe things that would be difficult to put into words. Look at the drawing of the simple wedge in Fig. 1-3. The drawing clearly describes this item, including its size. Now try to write a description of the wedge in words. Even if you are fairly successful, the reader may not fully understand what it actually looks like. Consider what it would be like to describe a complicated product in words.

Drafting is the form of communication used in industry. Since drawings are used all over the world, the way drawings are made has been standardized. The lines and symbols used are the same everywhere. Drawings made in Detroit can be read and understood anywhere in this country and abroad.

Who Uses Drawings?

The ability to *read* drawings is important for many jobs, including some that don't involve *making* drawings. For example, a building materials salesperson needs to be able to read drawings to find out what materials must be delivered to a job site. Persons responsible for production supervision in manufacturing plants use drawings constantly. Those who build highways, bridges, and buildings work from drawings. People who service appliances and machines could not do their work properly if they could not read drawings.

On some jobs sketches frequently need to be made. This requires a basic knowledge of drafting because a sketch should be as complete as a mechanical drawing. All of the workers in skilled trades—such as plumbers, carpenters, electricians, machinists, and welders—have to read drawings and often must make technical sketches. Fig. 1-4.

For other jobs, such as drafter or engineer, the person must be skilled at both reading and making drawings. The projects that they are involved with are detailed and complex and may require hundreds or even thousands of drawings. Fig. 1-5.

With some thought, you can see that almost every part of our economy uses drawings. Drafting, therefore, is a very important part of many people's education.

Technological Development

Many of the products we use today did not exist a few years ago. They are the result of the work of engineers and scientists who developed the new materials

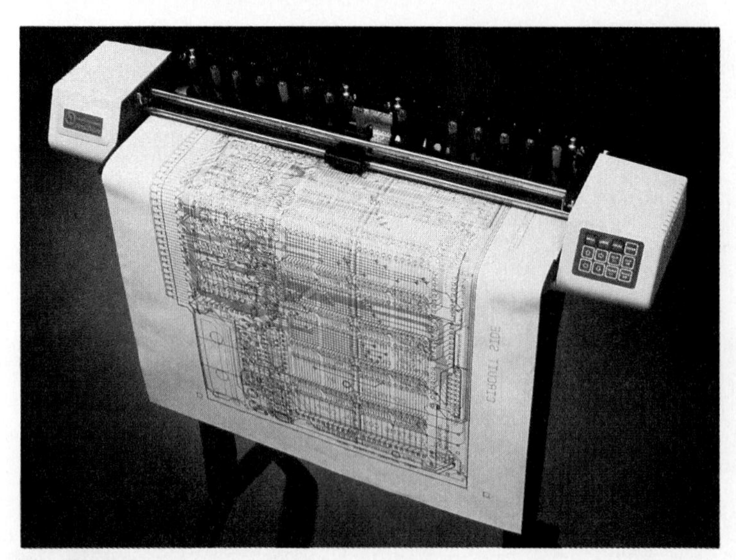

Fig. 1–2. Computer generated drawings are drawn by a plotter. (Houston Instrument)

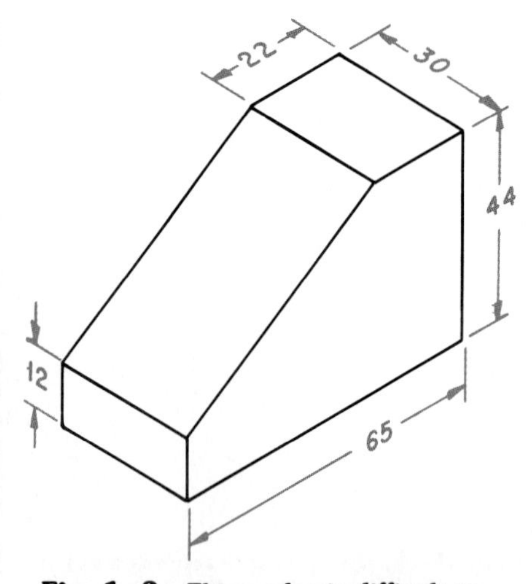

Fig. 1–3. This wedge is difficult to describe in words but is easily described with a simple drawing.

and products. For example, the introduction of plastics has greatly changed the products we use. It has enabled us to not only develop new products but also to change and improve existing products. Plastic pipe and gears are but two examples. Fig. 1-6. These developments mean the drafter and engineer of today must have more technical knowledge than people who held those jobs in the past. They must also continue to learn about new developments throughout their careers.

All of us are consumers of the new products which are the work of the designers, drafters, and engineers of industry. Their work has a direct influence on both where and how we live, work, and play. Fig. 1-7.

The Engineering Design Process

Drafters and engineers become involved in the development of a product and its delivery to a consumer. The exact details and actual involvement vary with the product and the engineering needed for it. In general, however, the **engineering design** process is as follows:

DETERMINE NEED

1. The need for a product is recognized and described. A study is made by the marketing staff to see if there really is sufficient need for this product.

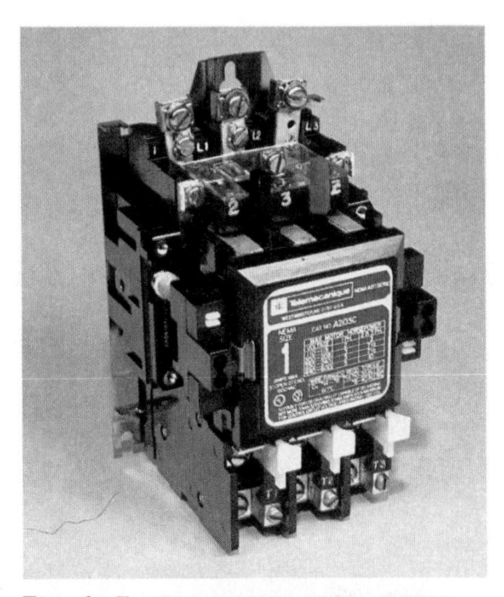

Fig. 1–5. This complex A-line NEMA-rated starter used in industrial control systems required many drawings. (Telemecanique, Inc.)

Fig. 1–4. This technician is operating a single-spindle computer numerically controlled automatic lathe. (New Britain Machine)

Fig. 1–6. These products—electric motor commutators, disk brake pistons, and reactors for automatic transmissions—are typical of the demanding applications for which a plastic called phenolic molding compound is being used. (Rogers Corporation)

Fig. 1–7. **The work of designers, drafters, and engineers is evident in products for our work and play. (National Marine Manufacturing Association)**

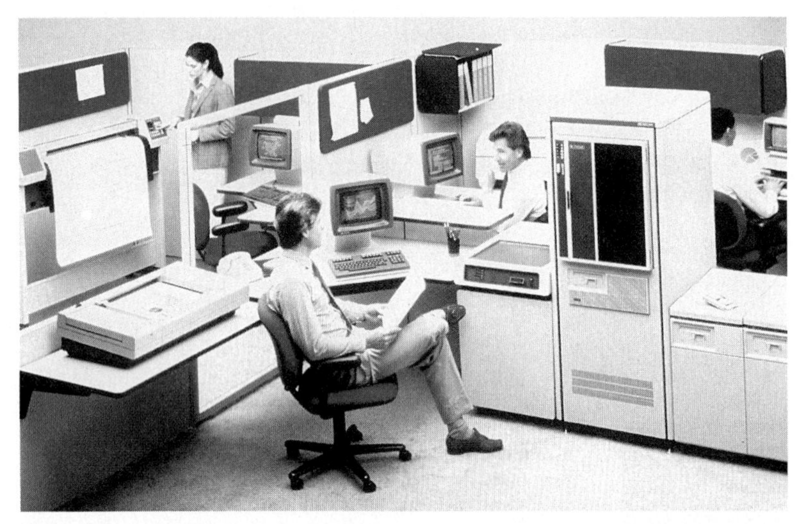

Fig. 1–8. **Engineering design involves many people working together as a team. (Hewlett Packard)**

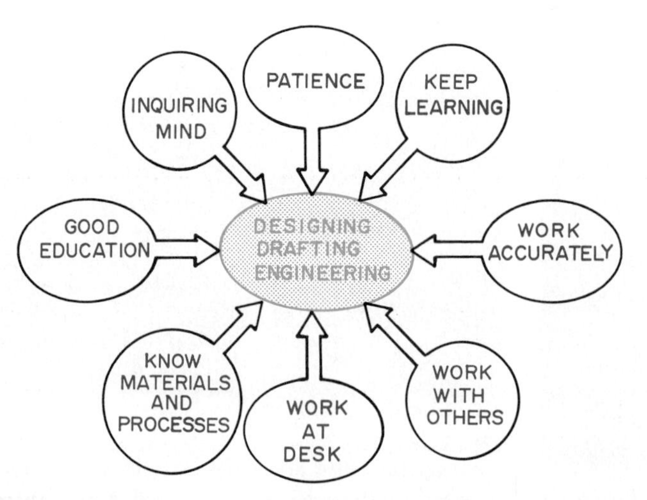

Fig. 1–9. **These traits are required for a successful career in design, drafting, or engineering.**

PREPARE THE PRELIMINARY SKETCHES

2. Preliminary sketches and drawings of the product are proposed by the engineering design staff. Generally several solutions are proposed for study.

CHOOSE FINAL DESIGN

3. After the original ideas have been extensively studied and revisions made, the final design is agreed upon. The specifications for the product are written and funds are provided for the engineering design team to continue their work.

DESIGN THE PRODUCT

4. The team begins to design the product. They bring in any consultants or specialists needed to help with their work. The preliminary engineering drawings are prepared by the drafting staff.

TEST THE DESIGN

5. The product is now tested. Doing so may mean that working models need to be built and used under actual conditions. The results of the tests are given in a final report. Any revisions to the product are made as needed.

PREPARE FINAL ENGINEERING DRAWINGS

6. Final engineering drawings are prepared. They are checked and given final approval. The drawings are released to the production division of the company and the product is made and sold to the consumer.

The Engineering Design Team

The **engineering design team** is made up of people who have varied educational backgrounds and interests as well as being professional drafters and engineers. Fig. 1-8. The drafters will be at different levels of their profession and the engineers may specialize in different areas. For example, many products require the services of several types of engineers such as electrical, mechanical, and agricultural.

The design team needs to interact with other people in the company who have various educational backgrounds and interests. Because the process of developing a new product or an improvement to an existing product usually starts with a study of the needs of the market, people in the marketing department become involved. The product may also include a study of and experimentation with different materials. Technicians and engineers that specialize in those materials would also need to be consulted. Thus a great number of people with different levels of education who are experts in different subjects become involved with the production of a product.

Career Planning

If the idea of becoming a part of a design and engineering team appeals to you, such a career begins in your classes in mechanical and architectural drafting, computing, mathematics, science, and English. You need to succeed in these areas to advance to one of the entry level jobs in design, drafting, and engineering. Employers look for new drafters who have mastered these basic skills. Successful completion of these courses is vital to continuing your education in a technical school, community college, or university.

To work in design, drafting, or engineering, you must have an interest in the work and the personality for the job. Fig. 1-9. Some of the traits required include patience, the ability to work at a desk or drawing board for long periods of time, accuracy in your work, and a good education with an extensive knowledge of materials and processes. Other desirable traits are to have an inquiring mind, the desire to keep learning, and the ability to work cooperatively with others.

The Drafting Field

As you make plans for a career in drafting, consider the level at which you wish to enter and the education required to do so. Then consider the years of experience and education expected for advancement in the field. Entry into the design drafting field is commonly at the drafter trainee or drafter level.

The job classifications and descriptions that follow were developed by the American Design Drafting Association (ADDA). Fig. 1-10.

Drafter Trainee. The minimum education and experience for a **drafter trainee** are as follows:

1. Be a high school graduate and have taken courses that included:
   a. Three years of mechanical drafting.
   b. One year of algebra.
   c. One year of geometry.

Duties and responsibilities of the drafter trainee include:

1. Repair or redraw damaged drawings.
2. Revise drawings by working from instructions, marked prints, or sketches prepared by others.
3. Learn company drafting standards.
4. Learn the fundamentals of drafting.
5. Prepare layouts under close supervision.

Drafter. A **drafter** works under general supervision and has considerable opportunity for individual action. He or she may guide and instruct other drafters or drafter trainees. A drafter exercises no supervisory authority. The minimum level of education and experience for a drafter is one of the following:

1. Two years of post-high school vocational/technical school courses that meet ADDA certification requirements and one year on-the-job experience in a temporary position. (This is the preferred level.)
2. Three years of high school mechanical drawing courses that meet ADDA certification requirements plus three years of drafting experience.
3. A high school diploma with one year each of mechanical drawing, algebra, and geometry plus six years of drafting experience.

Fig. 1-10. Educational Preparation for Positions in Drafting

Job Classification	Recommended Educational Preparation*
Drafter Trainee	High school drafting
Drafter	Two-year post-high vocational school drafting certificate
Designer Drafter	Junior college or technical institute degree
Engineering Designer Design Drafting Supervisor Design Drafting Manager	Four-year college engineering graphics degree

(American Design Drafting Association)

**All of these require on-the-job experience before the classification is awarded.*

CAD Applications:

CAD and Your Future in Drafting

Perhaps one of the most revolutionary changes that have taken place in the drafting field in recent years is the use of computer-aided design (CAD). CAD and CAD systems are used almost as much in manufacturing and engineering as word processors are used in general offices. Today 31 percent of architectural firms and 42 percent of engineering firms in the United States are using CAD. Experts predict that in a few years CAD will be used by 90 percent of all such firms in this country.

CAD brings to drafting both quality and speed. Unlike many drawings done manually, CAD drawings are consistent and clear. The images are precise and can easily be copied, enlarged, or reduced. Although original drawings probably take as long to complete on the computer as they would if done manually, they can be more easily changed. Additions, deletions, and corrections are made much more quickly and with no loss in final drawing quality.

However, CAD is more than a replacement for manual drafting. The real future of CAD in industry seems to be in its ability to compile data and enhance the design process. For example, the same database that contains the plans for a building can be used to produce the bills of materials and cost estimates. And today there are programs that can actually create a design, based on specifications that the CAD operator provides.

Future drafters will find CAD makes many drafting jobs easier. And if the potential of CAD is explored to the fullest, they will find it links them to a larger design process.

Hewlett-Packard Company

Duties and responsibilities of the drafter include:

1. Handle normal drafting assignments under regular supervision.
2. Be completely familiar with drafting standards, symbols, nomenclature, and engineering terms.
3. Know the proper use of materials.
4. Be able to use reference books and catalogs in a specific area of work.
5. Discuss job requirements directly with persons for whom the work is being done.
6. Gather information and data for jobs.
7. Make routine calculations using standard engineering formulae.
8. Take field or shop measurements as required.
9. Perform a limited amount of design under close supervision.
10. Use computer aided design (CAD) facilities as appropriate.
11. Instruct, guide, and check work of other drafters or drafting trainees who may assist on a job.

Design Drafter. A **design drafter** works under general supervision and has considerable opportunity for individual action. He or she often directs the work of drafters and drafter trainees. A design drafter exercises no supervisory authority. The minimum education and experience for a design drafter is one of the following:

1. Two years of a junior college or technical institute curriculum that meets ADDA certification requirements plus one year on-the-job experience. (This is the preferred level.)
2. Two years of vocational/technical school courses that meet ADDA certification requirements for *Drafter* plus two years drafting experience.
3. Three years of high school courses that meet ADDA certification requirements for *Drafter Trainee* plus six years of drafting experience.

Duties and responsibilities for a design drafter include the following:

1. Handle single or multiple design drafting assignments (with assistance from other drafters for multiple assignments).
2. Exercise good judgment in design and layout under minimum supervision.
3. Schedule work or assign projects and report on the progress of projects as required.
4. Make or review calculations needed on projects as required.
5. Perform limited design analysis using engineering computations.
6. Prepare or assist in making material and time estimates, and equipment cost comparisons.
7. Ascertain that designs and drawings conform to engineering and drafting standards and practices adopted by the company.
8. Use computer aided design (CAD) facilities as appropriate.

Engineering Designer. An **engineering designer** directs the work of drafters in lower classifications and serves as supervisor in the absence of the design drafting supervisor. The minimum education and experience for an engineering designer is one of the following:

1. Four years of college training in a program that meets ADDA certification requirements for engineering designer plus one year on-the-job experience. (This is the preferred level.)
2. Two years of a college or technical institute curriculum that meets ADDA certification requirements for *Design Drafter* plus four years of design drafting experience.
3. Two years of vocational/technical school courses that meet ADDA certification requirements for *Drafter* plus six years of design drafting experience.

Duties and responsibilities of an engineering designer include the following:

1. Handle complex design assignments, including multiple assignments, with the assistance of several drafters in lower classifications.
2. Receive assignments directly from supervisor or others requesting work.
3. Check and/or approve all work on projects as delegated including basic layouts, arrangement and design, accuracy of computations, selection of material and equipment in compliance with company standards and safety rules.
4. Prepare studies and reports for estimates, progress, and evaluations.
5. Exercise inventiveness and independent judgment about projects.
6. Regularly review technical literature in his or her field for possible applications to the work.
7. Use computer aided design (CAD) facilities as appropriate.

Successful engineering designers often have the following characteristics:

1. Above average initiative and ability to make the right decisions about the best way to carry out assignments.
2. Exceptional creativity with far-reaching design capabilities.
3. Thorough knowledge of accepted design or method concepts.
4. Working knowledge of basic engineering, design, or other principles related to a specific area of work.

Design Drafting Supervisor. The **design drafting supervisor** reports to the design drafting manager. This person supervises and coordinates the work of the drafting department and delegates duties to subordinates. The minimum level of education and experience for a design drafting supervisor is one of the following:

1. Four years of college training in a curriculum that meets ADDA certification requirements for *Engineering Designer* plus four years of design drafting experience. (This is the preferred level.)
2. Two years of junior college or technical institute curriculum that meets ADDA certification requirements for *Design Drafter* plus eight years of design drafting experience.
3. Two years of vocational/technical school courses that meet ADDA certification requirements for *Drafter* plus ten years of design drafting experience.

Duties and responsibilities of the design drafting supervisor include the following:

1. Supervise and coordinate the activities of a design drafting department.
2. Initiate and develop standards of performance and procedures for subordinates.
3. Recommend standards of design within practical limitations.
4. Review layouts, designs, and calculations to ascertain that the basic intent of each project has been followed and that the end results are practical and within allotted cost estimates or budgeted appropriations.
5. Interview applicants for design drafting positions and make recommendations as to whom to select.
6. Recommend increases in salary, promotions, or separation for subordinates.
7. Handle grievances of his or her department.
8. Regularly review technical literature for possible applications to work.
9. Represent the drafting department at company meetings and participate in technical society meetings.

Successful design drafting supervisors often have the following characteristics:

1. Be capable of working with people smoothly and efficiently.
2. Possess natural qualities of leadership which may be supplemented by training courses, reading, and experience.

Design Drafting Manager. The **design drafting manager** reports to the division or department manager. This person manages a design drafting branch, division, or department and delegates all direct supervision to subordinates. The minimum level of education and experience for a design drafting manager is one of the following:

1. Four years of college training that meets ADDA certification requirements for *Engineering Designer* plus six years design drafting experience. (This is the preferred level.)
2. Two years of college or technical institute training that meets ADDA certification requirements for *Design Drafter* plus ten years of design drafting experience.
3. Two years of vocational/technical school courses that meet ADDA certification requirements for *Drafter* plus 12 years of design drafting experience.

Duties and responsibilities of a design drafting manager include the following:

1. Be responsible for both administrative and technical activities of a design drafting department.
2. Manage and coordinate the work of the department.
3. Determine the priorities of work and approve schedules.
4. Determine the number of employees required.
5. Exercise the right of independent action regarding adjustments in basic personnel requirements of the department to meet changing conditions or to improve procedures.
6. Hire and indoctrinate new employees for the department.
7. Prepare or review performance evaluations of personnel and recommend salary changes and promotions as they are merited.
8. Strive to reduce the cost of operating the department.
9. Administer to the best interest of the company all controllable items of expense incurred by the department and operate within an approved budget which he or she helps to establish.
10. Strive to improve administrative procedures.
11. Enforce compliance with all established rules and procedures of the company and set a high personal standard in this regard.
12. Foster a close liaison with other departments within the company to which work is related.
13. Counsel, guide, and assist personnel in both company and personal problems which relate to work performance.
14. Be responsible for all necessary records, time keeping, and reporting for the department.
15. Represent the department at company meetings and participate in technical societies.

Engineer. **Engineers** receive their education at colleges and universities. The minimum time required to earn a degree in engineering is four years. Some engineering programs require five years of college work.

Engineers must be strong in mathematics and science. Problem solving is their big task. Fig. 1-11. They use drawings of all kinds. Often an engineer will develop the solution to a problem using freehand sketches, or he or she may make instrument drawings instead. An engineer's drawings are not the final drawings needed. Those are made by the drafter.

Engineers typically function in one or more activities such as research, development, design, production, consulting, administration and management, teaching, technical writing, or technical sales and service.

Engineering Technicians. One fast growing career area is engineering technology. **Engineering technicians** complete a bachelor's degree in engineering technology and major in an engineering area such as electronics, civil, mechanical, plastics, or manufacturing. Engineering technicians prepare for positions that require the use of scientific and mathematical theory and the principles of the technical area in which they specialize. They work with engineers and scientists and can make unique contributions as a member of the design team because they learn practical applications of technology in addition to theory. Fig. 1-12. Engineering technicians work extensively with drawings and must be able to produce drawings when needed. They conduct experiments, operate instruments, make calculations, and use the computer.

The Present and Future Outlook

As you read magazines, newspapers, and look at television, the achievements of engineering design teams are very easy to see. Notable among these are the wide range of projects under development and in use by the National Aeronautics and Space Administration (NASA). The better known projects are the Space Shuttle and a space station. Fig. 1-13. To develop these projects it

Design Engineer

Pacific Northwest firm, world leader in computer keyboard technology, has opening for mechanical design engineer with minimum 5 years experience in keyboard switches or similar switch devices. Must have proven record of creativity and resourcefulness. The right applicant will be offered outstanding compensation, comprehensive benefit package, and unsurpassed lifestyle options in Eastern Washington.

Contact:
John Doe
P.O. Box 123
Spokane, Washington
99214
E.O.E. - M - F - H

Fig. 1–11. Considerable education and experience are required for many engineering positions. (Key Tronic)

Fig. 1–12. This technician is using software that makes possible the rapid generation of complex shapes. (Calma Company, a subsidiary of General Electric Co. U.S.A.)

Fig. 1–13. The work performed for the space program involves the highest levels of technology, including the development of materials and processes never tried before. (NASA)

is necessary to do research and development work in materials, fuels, and computer applications. This work involves technologists, engineers, and scientists from many areas of specialization.

The designs of future vehicles for air travel are on the drawing board. Aerospace engineers and related engineering designers and drafters are developing an extensive series of new products. One such proposal is to build an aircraft that will travel at speeds five times the speed of sound and eventually 25 times the speed of sound. In addition to the design of the airframe, a big problem is to develop engines that will make such flight possible. One project team is working on an engine called an air-turbo ramjet. It would burn hydrogen for fuel. Fig. 1-14. These developments will make it possible to fly from New York to Tokyo in two hours.

Work is underway to improve mass transportation through high-speed trains and subway systems. Such trains would run between cities at speeds in excess of 100 miles per hour (160 kilometers per hour). Fig. 1-15.

Land transportation is also undergoing rapid change. Automobiles are more fuel efficient and have aerodynamic bodies. Constant design changes and new materials are producing completely new vehicles for the consumer. The changes are so great that design team members must be experts in many fields. Fig. 1-16.

Plans for more efficient trucks for the future are under development. Aerodynamic exterior designs are producing unique vehicles that slip through the air with little wind resistance. The new trucks will reduce noise both inside the cab and outside on the road. The cabs will have much more room and the trucks will be taller. Since the driver will sit higher than in today's trucks, his

Fig. 1–14. **Some of the greatest engineering advances in the future will be in flight. This aircraft will be able to fly halfway around the world in two hours. (McDonnell Douglas Corp.)**

Fig. 1–15. **Mass transportation is undergoing a revolution through the use of high-speed trains and subways. (The Budd Company)**

Fig. 1–16. **Much research is underway to improve the various systems of land transportation. (Chrysler Motors)**

or her visibility will be greatly improved. The design will also enable the vehicles to handle better than current vehicles, especially when heavy winds hit it from the side. Better brakes and tires are also being developed. Fig. 1-17.

In manufacturing, the use of robots has caused a dramatic change in the way things are made. The design of the robot itself is an engineering achievement. Its applications to the industrial workplace influence design decisions made by design engineers. Fig. 1-18.

Robots are being developed to help people work in space. These robots will allow humans to perform complex tasks safely in zero gravity conditions. They will also perform duties such as duplicate assembly, tightening fasteners, inspection, and other repetitious tasks. They will work under the supervision of someone on site in space or from the central control room of a space station. Monitoring will be through the robot's video system. Fig. 1-19.

The revolution in electronics and computers has changed the way we live. Computers have entered almost every phase of our lives including how we check out of the grocery store and how various functions in our automobiles are controlled. In addition to making it possible to go to the moon or fly aircraft above the speed of sound, computers have had a tremendous impact on the design and development of the products made by industry. The computer has changed how we make and revise designs and how we produce drawings for different products. Fig. 1-20. Computers also control many of the machines that produce the products. The development of powerful personal computers has brought them into schools, homes, and offices. The computer is becoming commonplace at work and at home. Fig. 1-21.

Fig. 1–17. **Equipment to move products long distances rapidly and economically is under development. (Kenworth Trucking Co.)**

Fig. 1–18. **The use of robots in manufacturing and many other applications is producing a revolution in industry. (UNIMATION, Incorporated, A Westinghouse Co. Used with permission.)**

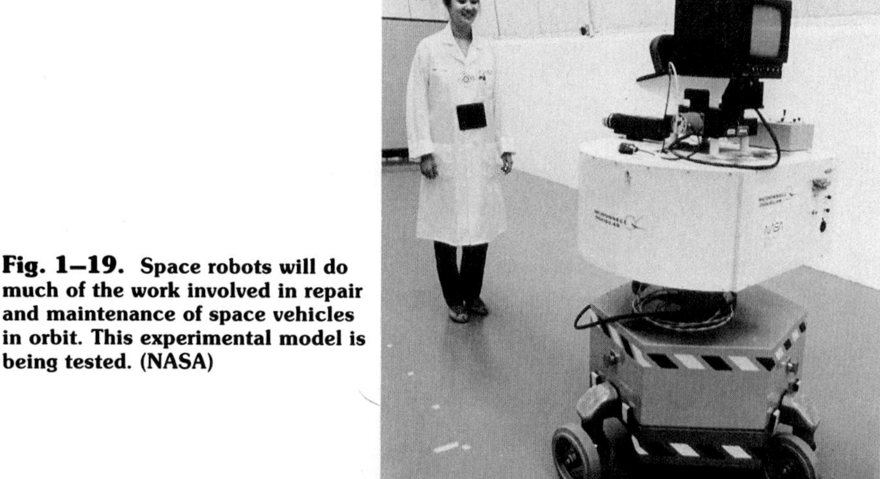

Fig. 1–19. **Space robots will do much of the work involved in repair and maintenance of space vehicles in orbit. This experimental model is being tested. (NASA)**

New developments in communications, especially the use of satellites, have made rapid, world-wide communication a common occurrence. Yet, realize what an enormous engineering achievement such communication is by considering the complexity of the satellite and its ground receiving and transmitting stations, and the engineering problems related in placing a satellite in orbit in space.

World wide communications are a vital necessity today. Various types of satellites have successfully carried the communications burden. In the years ahead, huge communications space stations may be assembled in space. They could serve as a television or telephone communications network, as a huge computer installation, or even to control air traffic on earth. Fig. 1-22. Such space stations would be expected to survive 30 or more years in orbit.

The architect, too, has design possibilities never before available. The development of improved structural systems enables buildings to be made lighter and taller. New construction materials such as a wide range of plastics and advances in concrete technology are but two engineering developments that aid the architect.

Super skyscrapers are a trend in large cities. Some very tall buildings have already been built. Fig. 1-23. The use of computers in the area of structural design makes quite practical techniques of design analysis that would have been too costly just a few years ago. Today's tallest buildings are 110 stories high. Designs are being prepared for a much higher building as seen in Fig. 1-24. Another major factor in high-rise design is the need to efficiently move hundreds of people and supplies in and out each day. Utilities are also a big design problem in such large structures. Architects, engineers, and drafters have a great amount of work to do when large projects are commissioned.

The use of lasers has expanded considerably in recent years and will continue to increase in the future. Fig. 1-25. Lasers are used in many areas, such as for surveying, in medicine, and in manufacturing processes. The engineer and drafter should be aware of the work that can be performed by lasers so that products can be designed to make use of this technology. Industrial laser systems and machining centers perform operations such as cutting, welding, drilling, heat treating, and engraving. Fig. 1-26.

The defense industry also provides extensive design, drafting, and manufacturing challenges. Among these are various launch vehicles and missiles. These have both military and commercial use.

Fig. 1–20. This computer drafting station makes maximum use of the desk space and has a display that is ideal for use in standard drafting office lighting. (Calma Company, a subsidiary of General Electric Co.)

Fig. 1–21. Personal computers are being used in the home, in school, and in industry. (IBM)

Fig. 1–22. The development of government and commercial space launch vehicles requires a vast array of engineering and drafting expertise. (Vitro Corporation)

Fig. 1–23. Architects and engineers have already designed many very tall buildings, which are sometimes referred to as super skyscrapers. Theoretically, it's possible to build a mile-high skyscraper. (Port Authority of New York and New Jersey)

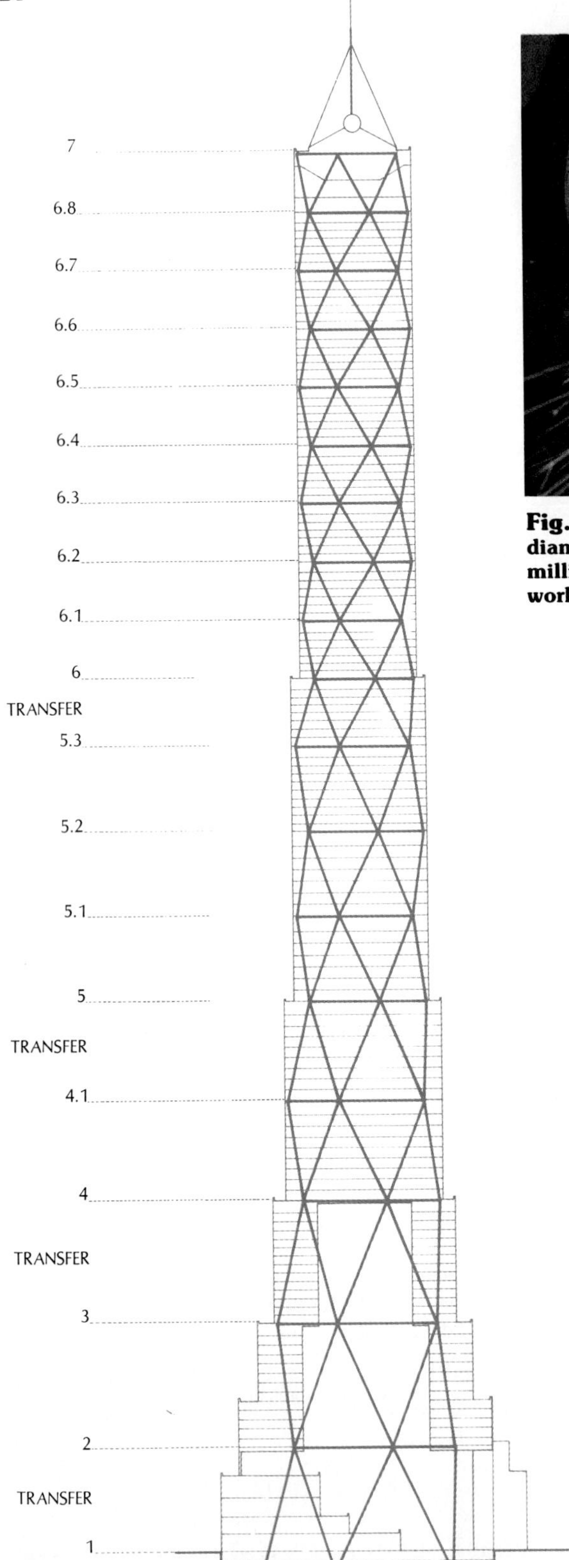

Fig. 1–24. This proposed design is for a 130-story super skyscraper that would be 1535 feet (468 meters) high. (*Popular Science*)

Fig. 1–25. Laser technology can accurately drill holes with diameters of thousandths of an inch (hundredths of a millimeter) and automatically space them properly on the workpiece. (Control Laser Corporation)

Fig. 1–26. This laser is cutting openings in steel plate. (Control Laser Corporation)

One military missile is a small intercontinental ballistic missile (ICBM). This missile will have a single warhead. It will be about 46 feet (14 meters) long and weigh about 30,000 pounds (13,600 kilograms). Its purpose is to serve as a part of our nuclear deterrent defense system. Fig. 1-27.

Another product is the Titan launch vehicle. It is used to launch into orbit satellites that are owned and operated by businesses in the United States and abroad. It has a two-stage liquid-fueled core vehicle and twin solid rocket motors attached at each side of the first stage. It is 14 feet (4.3 meters) in diameter. Fig. 1-28.

Technological developments will continue, no doubt, at a rapid pace in the future. Engineers and scientists will develop new materials, find new uses for existing materials, and develop processes and techniques unknown to us today.

Fig. 1–27. **The development of intercontinental ballistic missiles is of importance to the U.S. land-based strategic deterrent defense forces. (Martin Marietta Corporation)**

Fig. 1–28. **This Titan launch vehicle is used to launch satellites into orbit. (Martin Marietta Corporation)**

Chapter Review

Build Your Vocabulary

You should understand and use the following terms as part of your working vocabulary. Write a brief explanation of what each means.

technical drawings
graphic communications
drafting
engineering design
engineering design team

For each of the following occupations, briefly describe what a person in that occupation does.

drafter trainee
drafter
design drafter
engineering designer
design drafting supervisor
design drafting manager
engineer
engineering technician

CHAPTER 2

Technical Sketching

After studying this chapter you should be able to do the following:

- Sketch the lines used on engineering drawings.
- Sketch various geometric shapes.
- Produce finished engineering sketches.

What Is a Sketch?

A **sketch** is a freehand drawing. When you have an idea, the first thing you may do to preserve it is to make a quick sketch. If someone does not understand something you are explaining, you clarify your idea by making a sketch. **Sketching**, then, is a quick way of making a drawing. Some sketches may be very crude. As you develop and resketch your idea, your drawing begins to take on the characteristics of an engineering drawing. Fig. 2-1.

Sketching is a valuable skill used by people in industry at all levels. Engineers sketch preliminary designs. Drafters work from sketches and often prepare them to show several ways a thing might be drawn. On the manufacturing floor or on the construction site, sketches may be made on a note pad or any other available piece of paper such as the back of an envelope. Engineering change orders often include a sketch to help explain something that is difficult to describe with words alone. The ability to sketch quickly and accurately is needed by the technically trained person.

A sketch takes the same form as any engineering drawing. As you study the different types of drawings in this book, you will see that they could be sketched, made with instruments, or drawn on a computer. Sketches are made up of the same elements as instrument drawings. They are a combination of points, lines, and planes. Pictorial sketches may be shaded to give a more realistic appearance. Fig. 2-2.

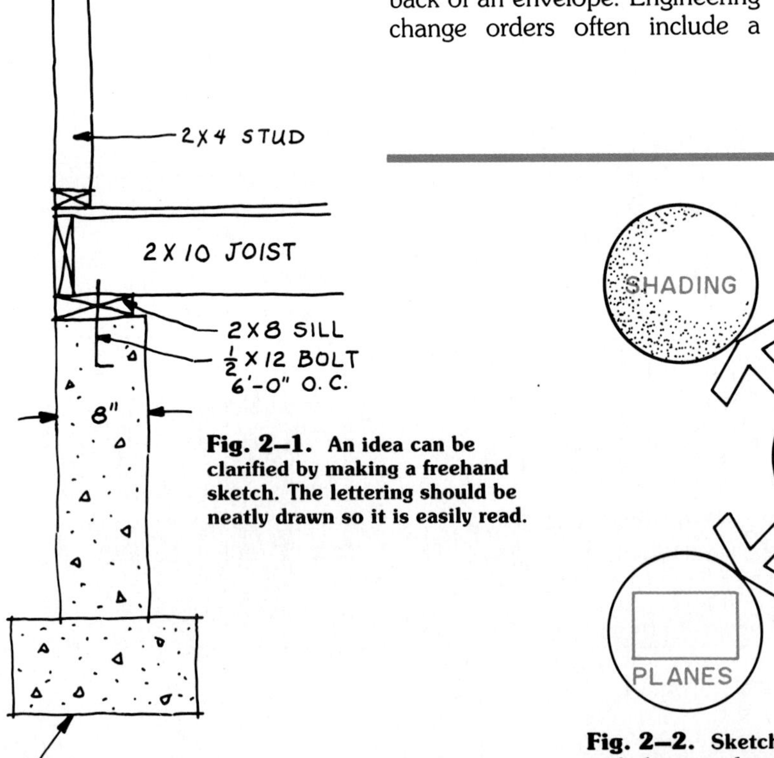

Fig. 2–1. An idea can be clarified by making a freehand sketch. The lettering should be neatly drawn so it is easily read.

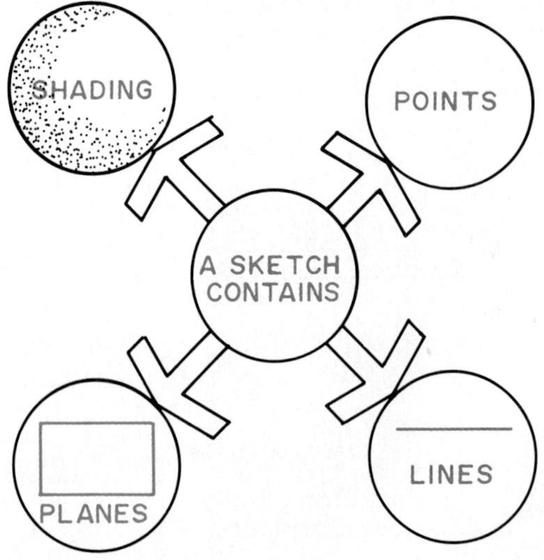

Fig. 2–2. Sketches are made up of points, lines, and planes and are sometimes shaded.

Sketching Tools and Materials

The tools needed for sketching are a pencil, eraser, and paper. Some pencils and some paper are better than others to use but, when needed, any pencil and type of paper will do.

Pencils and Erasers

Any pencil can be used for sketching, but a medium-soft lead such as an F or HB is preferred. An ordinary number two writing pencil is also satisfactory. The medium-soft lead is best because it can produce the light-gray lines needed for preliminary blocking-in of the sketch. The same pencil can be used to produce the darker lines needed on the finished sketch.

Your pencil should be sharpened so that about 3/8 inch (10 millimeters [mm]) of lead is exposed. Fig. 2-3. Three thicknesses of lines are used on sketches—thin, medium, and thick. Fig. 2-4. The following list describes the use of each:

- Sharp points make thin lines that are used for hidden, center, dimension, and extension lines.
- Medium points make medium-thick lines that are used for visible lines.
- Dull points make thick lines that are used for cutting-plane lines and border lines.

Any type of eraser can be used. When drawing on vellum or paper, many drafters like to use a Pink Pearl eraser.

Paper

Sketches can be made on any type of paper. Either plain or notebook paper is satisfactory. Some drafters prefer to use **drafting vellum** when it is available because lines erase from it easily. Also, vellum is translucent, that is, it will let some light through so you can see through it. Lines from a sheet of **graph paper** put beneath a sheet of vellum will show through. The lines from the graph paper help you to sketch straight vertical and horizontal lines.

Other people prefer to sketch directly on graph paper. Graph paper is available with lightly printed squares that measure 4, 8, or 10 divisions to the inch. Metric graph paper has a heavy division every 5 mm with 1-mm divisions in between. Fig. 2-5.

A - WOOD-CASED PENCIL
3/8"
10 mm
B - LEAD HOLDER

Fig. 2–3. Pencils sharpened for sketching.

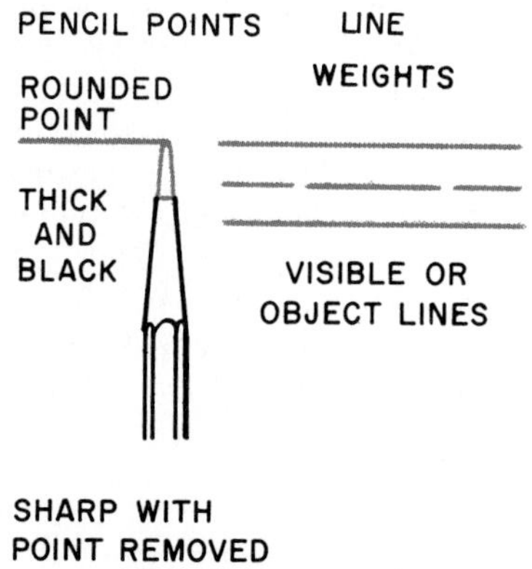

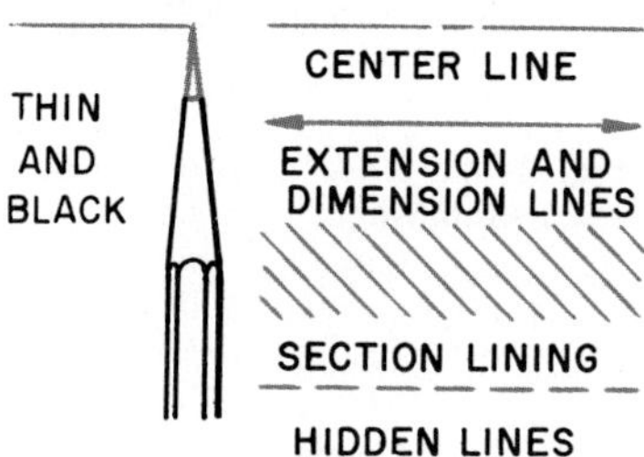

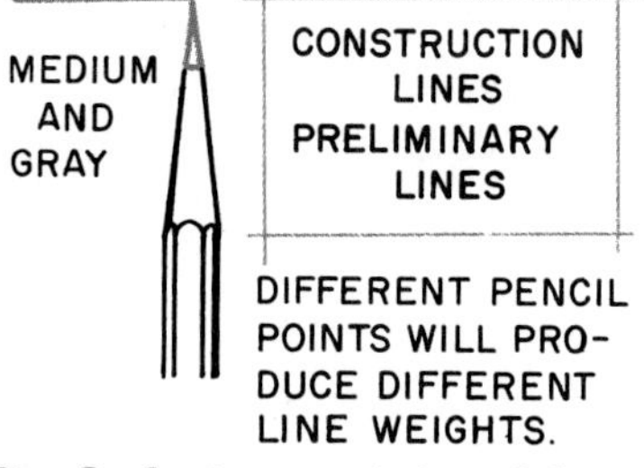

Fig. 2–4. Line symbols and their proper thickness.

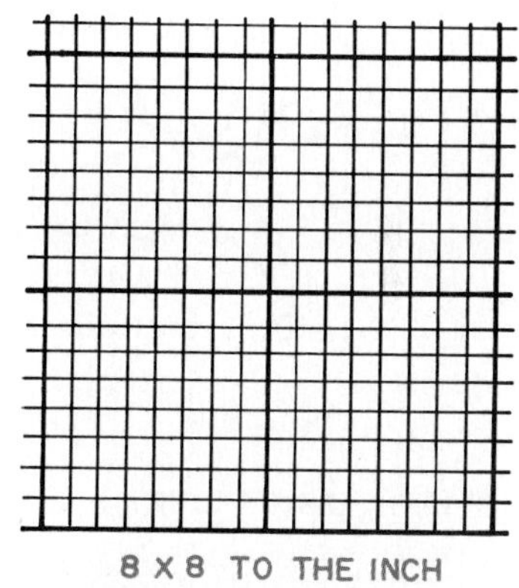

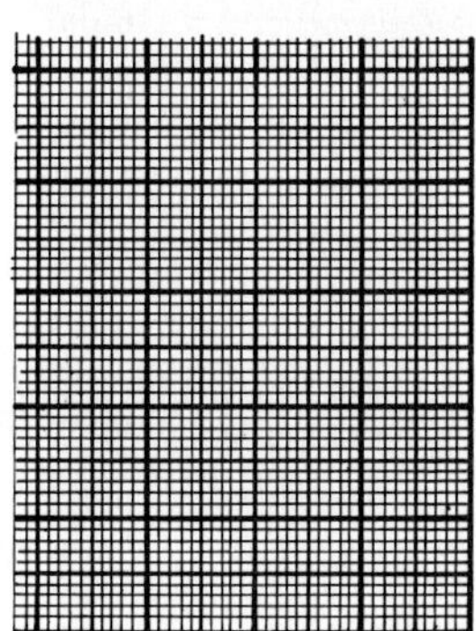

Fig. 2–5. Graph paper used in sketching.

Sketching Techniques

Holding the Pencil

There is no absolute rule for holding a pencil, but do hold it in a natural position. Generally, you should hold a pencil about 1½ inches (38 mm) from the point. Hold it at an angle of 50 to 60 degrees to the paper. Fig. 2-6. As you draw a line, rotate the pencil to keep the point from wearing flat on one side.

Sketching Straight Lines

Methods of sketching straight lines vary with the individual. No line will ever be perfectly straight or uniform, but you will sketch fairly straight lines if you practice the following methods.

Horizontal Lines. Several methods may be used to sketch horizontal lines. The one used most frequently is the point-to-point method. Fig. 2-7. Locate one end of the line on the paper with a pencil mark. Then locate the other end in the same way. Before you draw a line, you may first want to practice by holding the pencil point slightly above the paper and moving it back and forth several times from one mark to the other. Hold the pencil in a natural position and move it from one point to the other with a firm, free motion. As you draw the line, do not watch the pencil but look at the point toward which you are drawing. Right-handed persons should sketch horizontal lines from left to right. Left-handed persons should sketch in the opposite direction.

Some drafters prefer to draw horizontal lines using a series of short strokes. Fig. 2-8. These strokes can have a small gap between them. Another technique is to overlap the ends of each short stroke. Move your hand after each stroke is drawn. Each of these methods is good, so use the one you can do best.

The line you sketch between points will have a tendency to be slightly curved since your arm pivots from the elbow. To compensate for the arc, use a slight finger movement to pull the pencil toward your elbow. Some drafters prefer to compensate for this arc by moving their arms on the fleshy portion that rests on the table top as they draw a line.

Short horizontal lines are drawn with a finger and wrist movement. When you move your fingers and wrist without moving your hand, the line you draw will have a cramped appearance. By combining a hand movement with your finger and wrist movement, the line you draw will have a freer appearance.

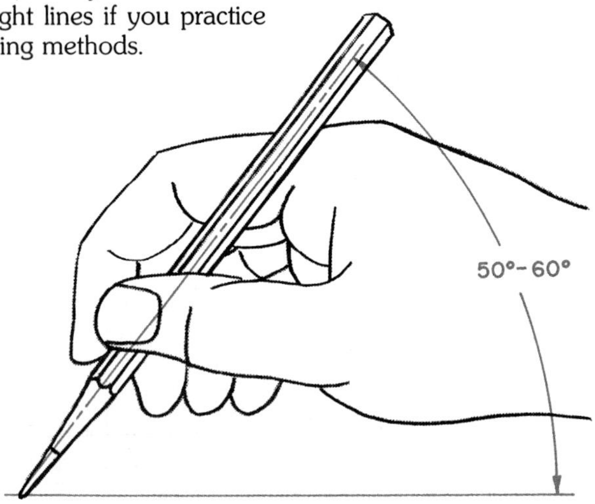

Fig. 2–6. How to hold your pencil when sketching.

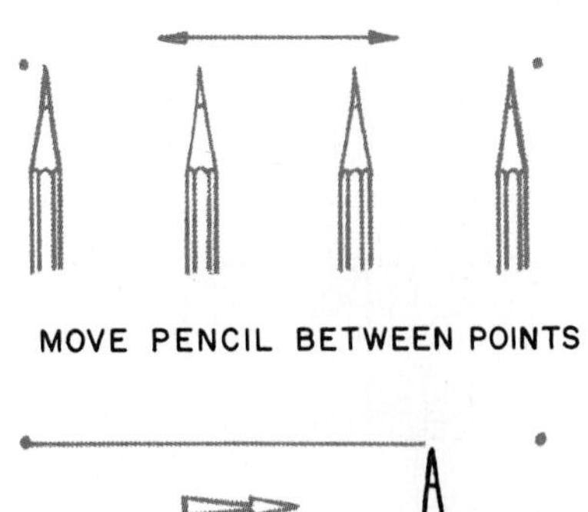

Fig. 2–7. Drawing horizontal lines.

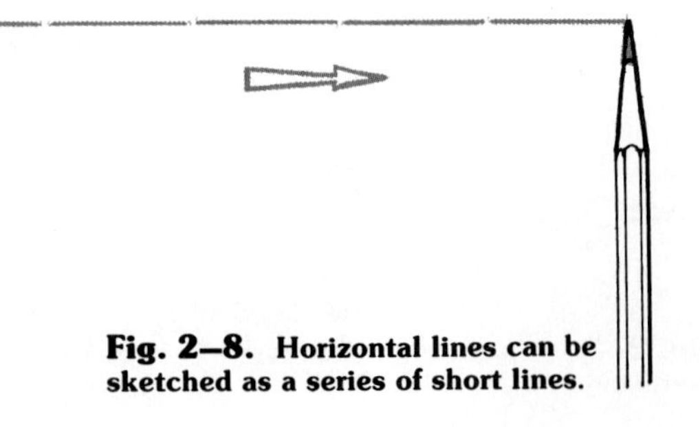

Fig. 2–8. Horizontal lines can be sketched as a series of short lines.

Vertical Lines. Vertical lines are drawn downward, toward your body. Once again, begin practicing by spotting two points on the paper with your pencil. Move your pencil between the two points, keeping your eye on the point closer to the bottom of the paper. Lower your pencil slightly until the point barely touches the paper. In this manner, the direction of the line is established. Now draw the line.

If you need to draw a series of parallel vertical lines, use the "marking gage" method. Fig. 2-9. Hold your pencil in its normal position with the tip of your middle or index finger touching the edge of the sketch pad. Place your pencil on the paper the desired distance from its edge that the line is to be drawn. Then pull the pencil toward you. If another line is to be drawn parallel to the first line, simply move your pencil point to the new distance from the edge of the pad and draw the line.

When sketching on a single sheet of paper, place it next to any straight edge, such as the edge of a table. Use this edge as the guide for your finger. Horizontal lines may be drawn in the same manner as vertical lines by rotating the paper 90 degrees. Lines to be drawn near the center of the sheet pose some problem since your pencil must be extended almost to its full length.

Short vertical lines are drawn by the same method as short horizontal lines. Move your pencil upward with a combined finger, wrist, and hand movement. Remember, you may sketch any vertical line as a horizontal line by simply rotating your paper.

Inclined Lines. Inclined lines that are nearly vertical should be drawn downward, Fig. 2-10 at *A* and *B*. Note in both instances that the pencil is being pulled toward the body. For inclined lines that are nearly horizontal, sketch them as shown in Fig. 2-10 at *C* and *D*. With inclined lines, as with other straight lines, be sure to use two pencil "spots" to aid you.

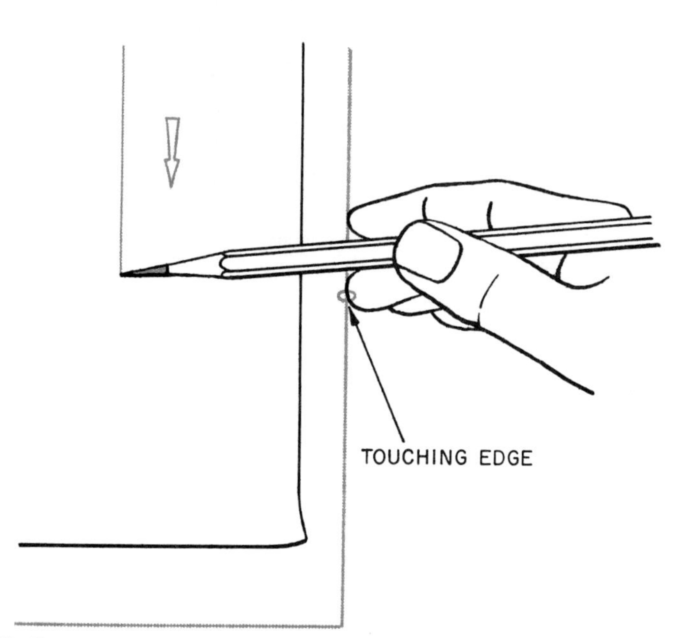

Fig. 2–9. Vertical lines are sketched from the top of the paper to the bottom. The edge of the table can serve as a guide to keep your lines straight.

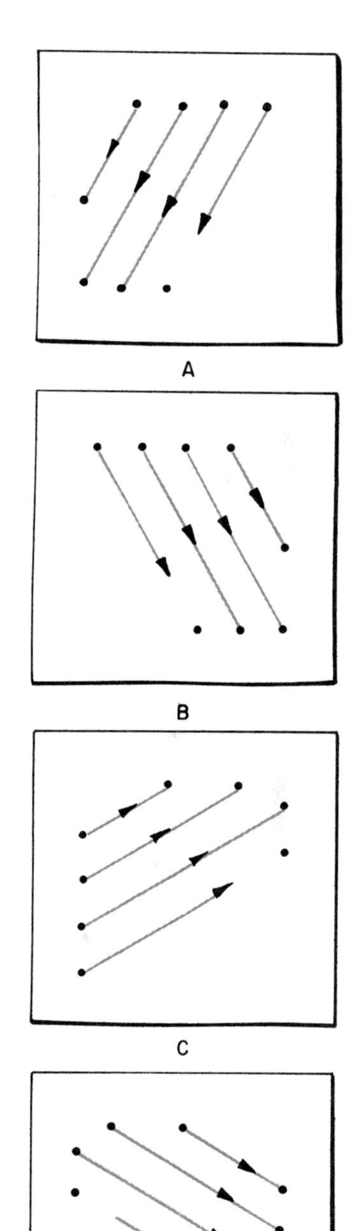

Fig. 2–10. Directions to sketch inclined lines.

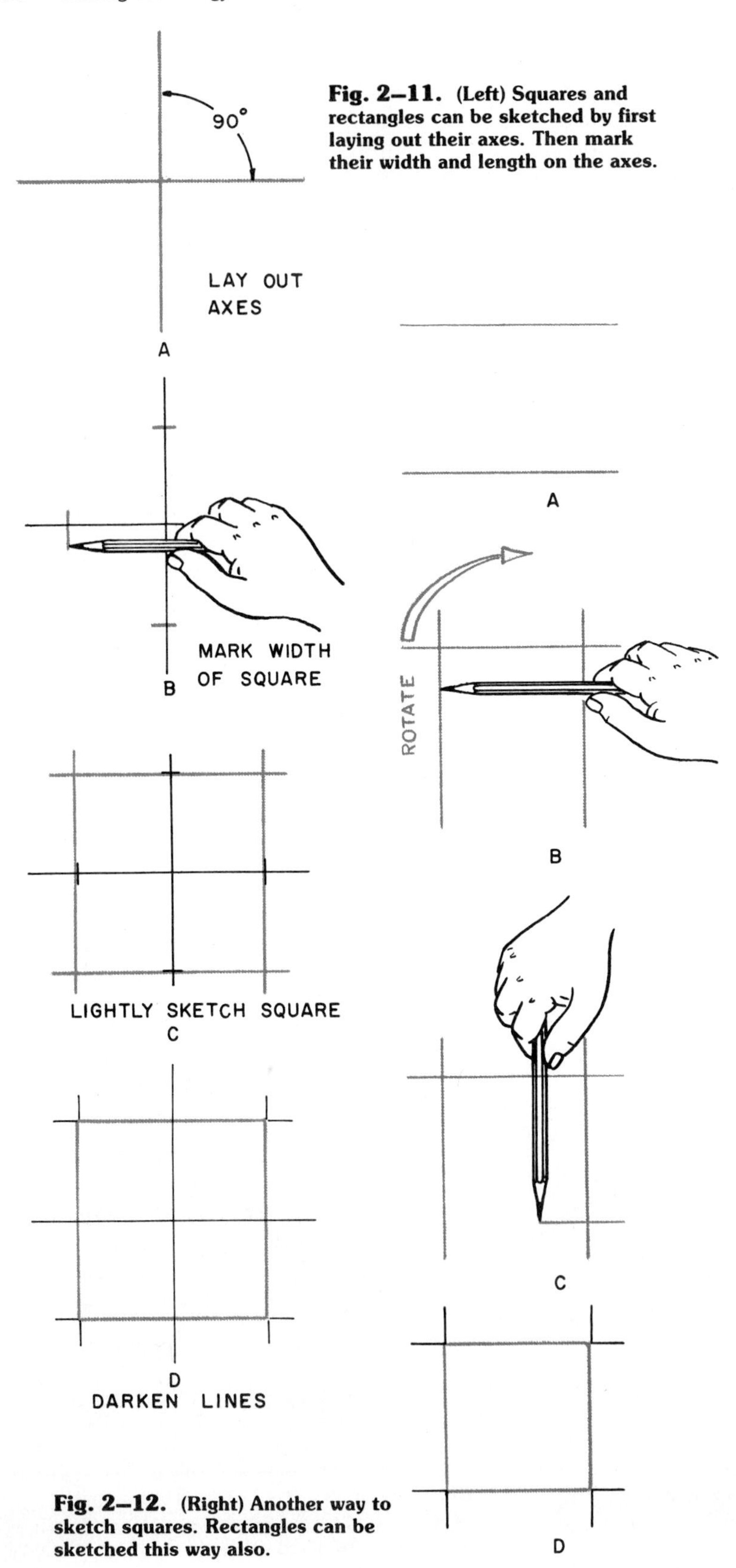

Fig. 2–11. (Left) Squares and rectangles can be sketched by first laying out their axes. Then mark their width and length on the axes.

Fig. 2–12. (Right) Another way to sketch squares. Rectangles can be sketched this way also.

Some students find they can sketch horizontal lines with more freedom and accuracy than inclined lines. If this is your case, rotate your sketch pad so that the inclined lines can be drawn horizontally.

Sketching Geometric Shapes

Pick any nearby object and analyze its basic geometric shapes. Is the object made up of squares, cubes, rectangles, triangles, pyramids, circles, or spheres? All objects, with a few exceptions, may be broken down into basic geometric forms. In technical sketching it is important to learn how to graphically describe each of these shapes. Sketching a straight line is the foundation for sketching most geometric shapes.

Squares and Rectangles. To sketch a square, first lightly lay out two axes at right angles. Fig. 2-11. Then, using the pencil as a marking gage, make a mark on each axis that represents half the width of the square. Sketch light lines through the marks. Then darken the square with firm lines. A rectangle is drawn in the same manner as the square.

Squares may also be drawn by first sketching two parallel lines representing the width of the square. Fig. 2-12 at A. Rotate the paper and sketch a light line representing the third side of the square (B). Use the pencil as a marking gage to transfer the width to one of the other sides with a mark. Lightly sketch the fourth side and darken the figure.

Circles and Arcs. Circles and arcs are easily sketched by first lightly drawing a square with sides equal to the diameter of the circle. Fig. 2-13 at A. Sketch two diagonals on the square, then mark off the radius on the diagonals (B). Now, begin to sketch the circle (C). It is usually easier to draw an arc in one of the quadrants (I, II, III, or IV) than the other three. Where you start depends upon whether you are right- or left-handed. After determining the area in which it is easiest to draw an arc, simply move the drawing sheet around so that the next quadrant is in that first, easy position.

Large circles and arcs may be drawn by using a paper strip. First draw two axes as shown in Fig. 2-14 at A. Mark the radius distance on a strip of scrap paper and move it around the center. Place as many marks as are necessary. Then sketch the circle through these points.

Another method of drawing large circles and arcs is to use your pencil and hand as a compass. Fig. 2-15. Place your little finger on the intersection of the two axes you have drawn. Put the pencil point at the radius and, using your little finger as a pivot, rotate the paper to draw the circle. You must hold your hand rigid in this position as you rotate the paper.

Small circles and arcs, no larger than ½ inch (13 mm) or ⅝ inch (16 mm), may be drawn without an enclosing box. These are made with one complete movement of your fingers and hand.

Large arcs are sketched in the same manner as circles. When sketching an arc, the usual method is to place it in an enclosing square. Fig. 2-16. The side of the square is equal to the radius of the arc. Arcs may also be drawn by either the paper strip method or "pencil and hand as a compass" method.

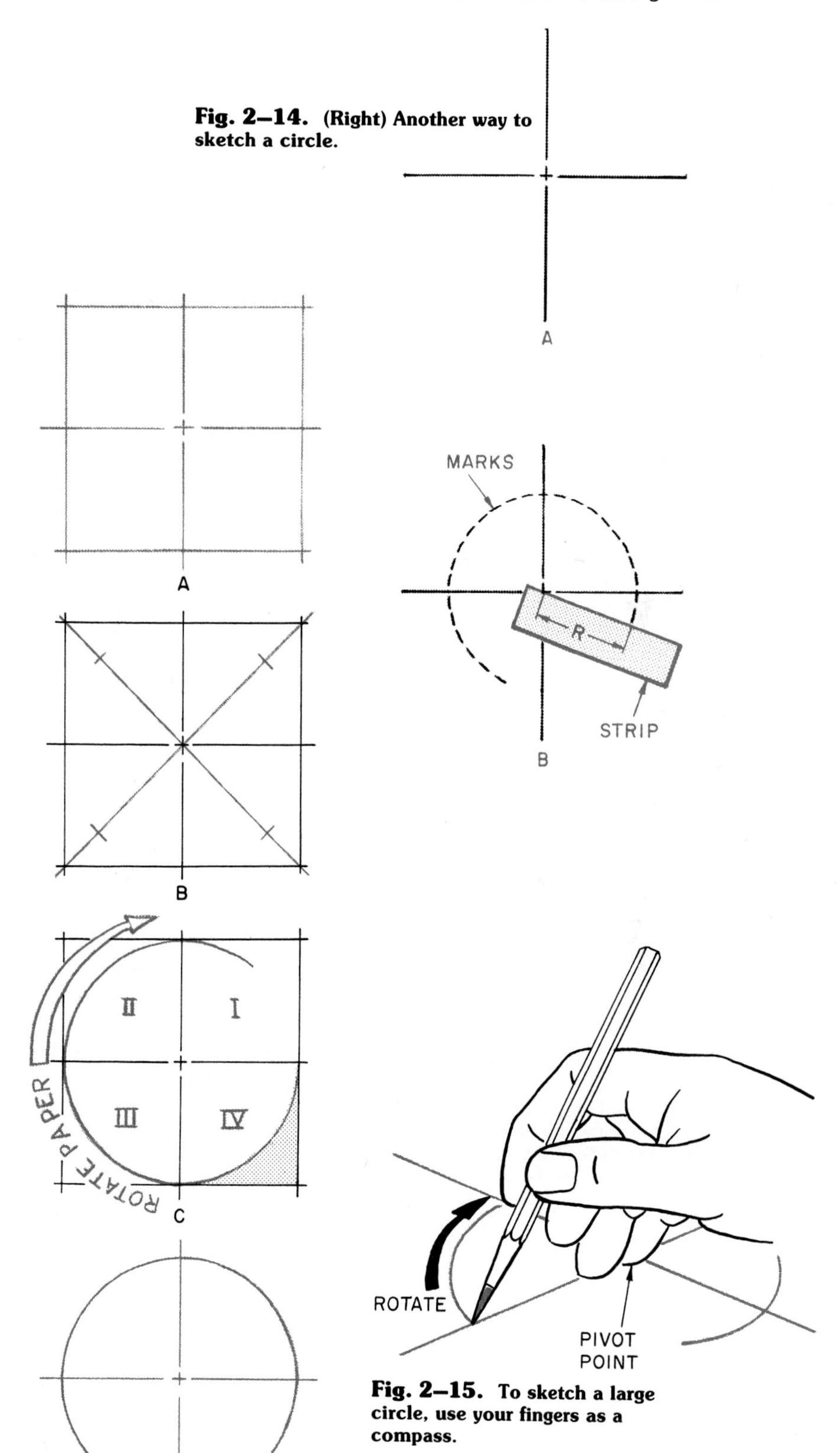

Fig. 2–13. **Circles can be sketched by first laying out a square with sides equal to the diameter of the circle.**

Fig. 2–14. **(Right) Another way to sketch a circle.**

Fig. 2–15. **To sketch a large circle, use your fingers as a compass.**

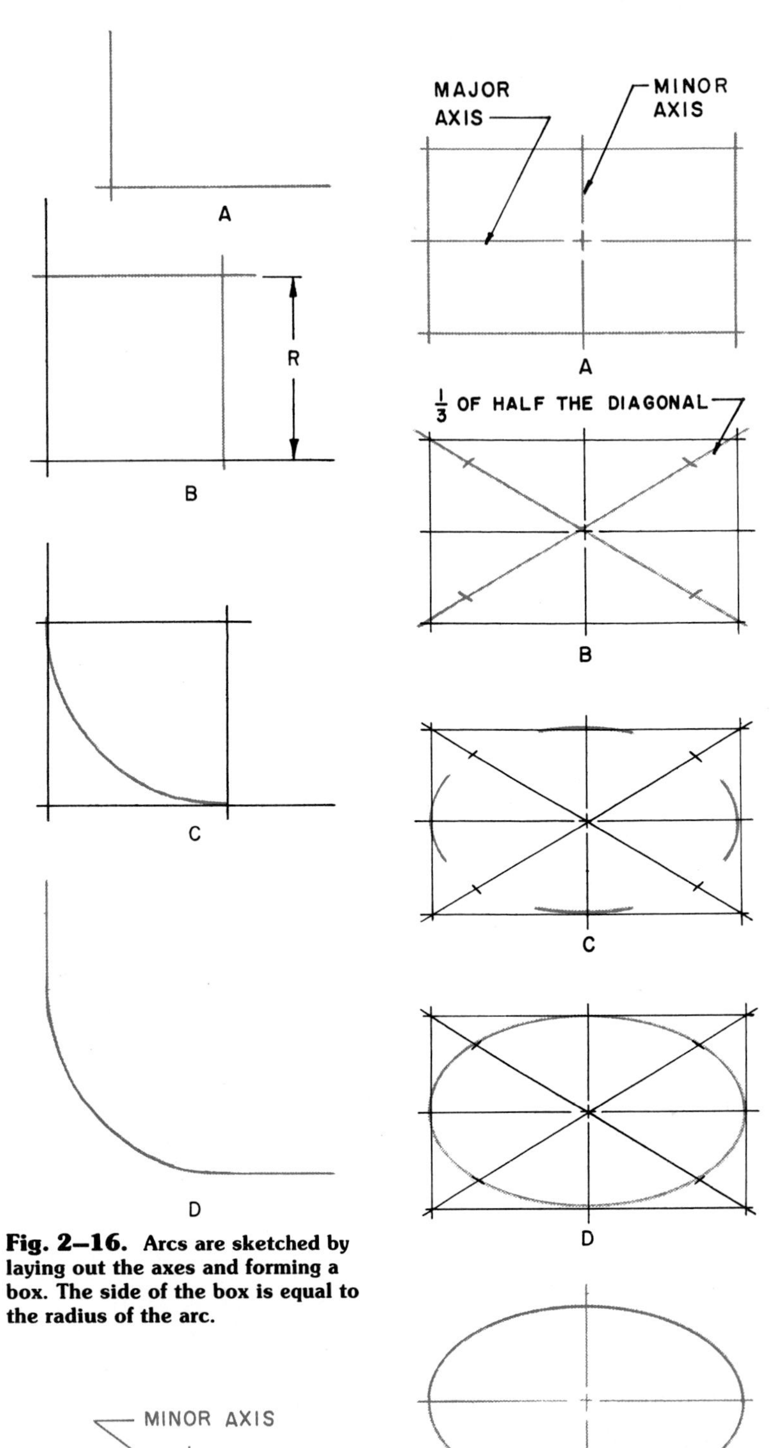

Fig. 2–16. Arcs are sketched by laying out the axes and forming a box. The side of the box is equal to the radius of the arc.

Fig. 2–17. An ellipse has a major and minor axis.

Fig. 2–18. An ellipse can be sketched by drawing a rectangle. One side is equal to the minor axis, while the other is equal to the major axis.

Ellipses. An **ellipse** (oval) has two axes of different lengths. The long axis is the **major axis** and the short axis is the **minor axis.** Fig. 2-17.

The easiest method of sketching an ellipse is to first sketch a rectangle. The length and height are equal to the major and minor axes respectively. Fig. 2-18 at A. Draw diagonals from the corners of the rectangle. Estimate one-third the distance of *half* the diagonal and mark this distance from each corner (B). Then sketch light tangent arcs at the midpoints of the enclosing box (C). Now sketch the ellipse with light lines. When the ellipse has the proper shape, that is, no flat sides or pointed ends, darken it. Fig. 2-19.

Proportions on Sketches

Proportion is one of the key factors in producing a good sketch. The term **proportion** refers to drawing parts of an object in the same size relationship as the object itself. In sketching, this is done by estimating sizes rather than actually measuring them.

Determine the longest dimension of an object and sketch it to the desired length. Then sketch the other sides, keeping them in proportion to the long side. For example, in Fig. 2-20 the long dimension of the file cabinet is the height. The width is about one-half of the height, so the height can be considered 2 units and the width 1 unit. The depth is about 1½ units.

It is important that all parts of a sketch be kept in proportion. The drafter must give careful attention to the size of objects and how they relate to each other.

If a sketch must be very accurate, the object can be measured. This method is most accurate because tools such as a scale, rule, or inside or outside calipers are used to obtain the measurements.

CAD Applications:

Freehand Drawing

Most drawings created on a CAD system are built up of straight lines, arcs, circles, and other geometric features. Some CAD systems also enable the user to do freehand drawings. This is a useful facility when drawing objects with highly irregular outlines, such as maps.

In freehand drawing, lines are entered automatically as the pointing device is moved. It's a little like drawing with a pencil but having the image show up on a computer screen instead of on paper.

There is an important difference, though. When you draw with a pencil, you can create a drawing that combines straight and curved lines in any way you wish. With a CAD system, the freehand drawing is actually a series of many short, straight lines connected end to end. Each line can point in any direction. The shorter these lines, the smoother the image on the screen will appear. Drawings made of very short lines are said to have a "high resolution," while coarser drawings have a "low resolution."

One drawback in using the freehand sketching facility is that it requires a large amount of computer memory. That is especially true of high-resolution sketches. The more lines, the more memory that is required. For this reason, CAD users who do much freehand drawing need to have computer equipment with large capacity storage. In Chapter 4, you will read about computer memory and the ways information is stored in a computer system.

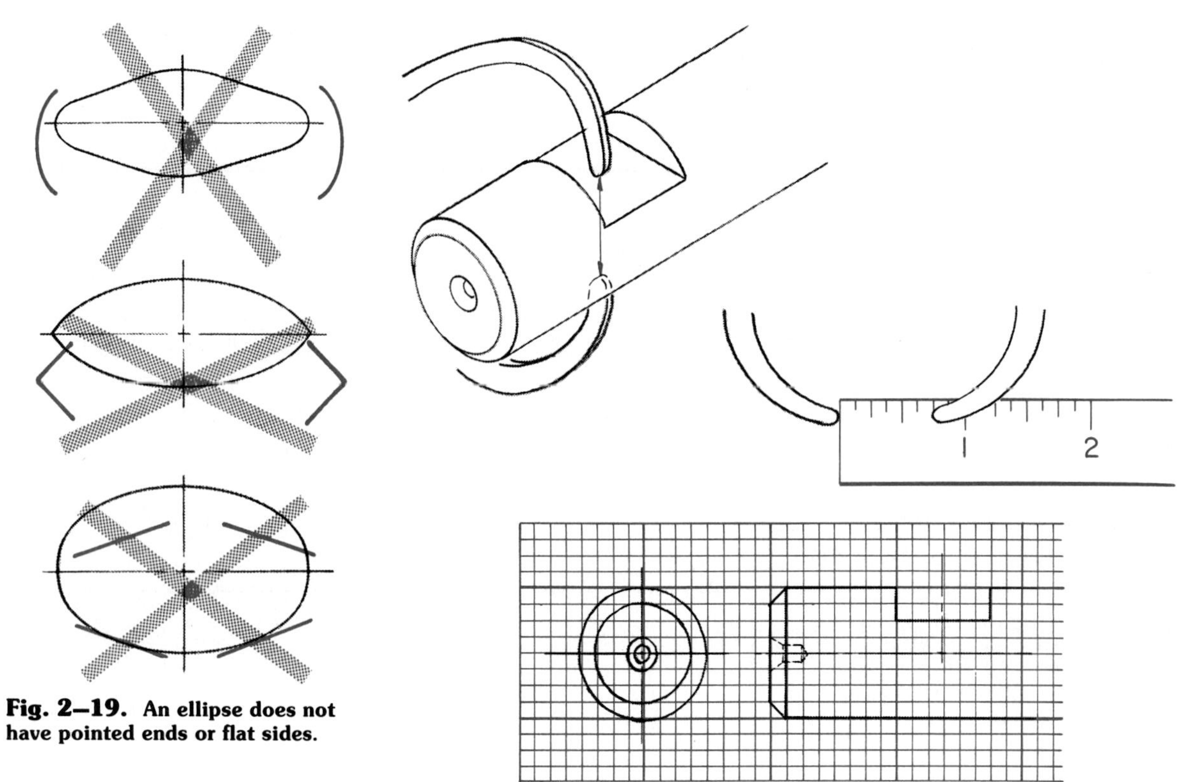

Fig. 2–19. An ellipse does not have pointed ends or flat sides.

Fig. 2–21. The most accurate way to obtain size on a sketch is to lay out the actual measurements on graph paper.

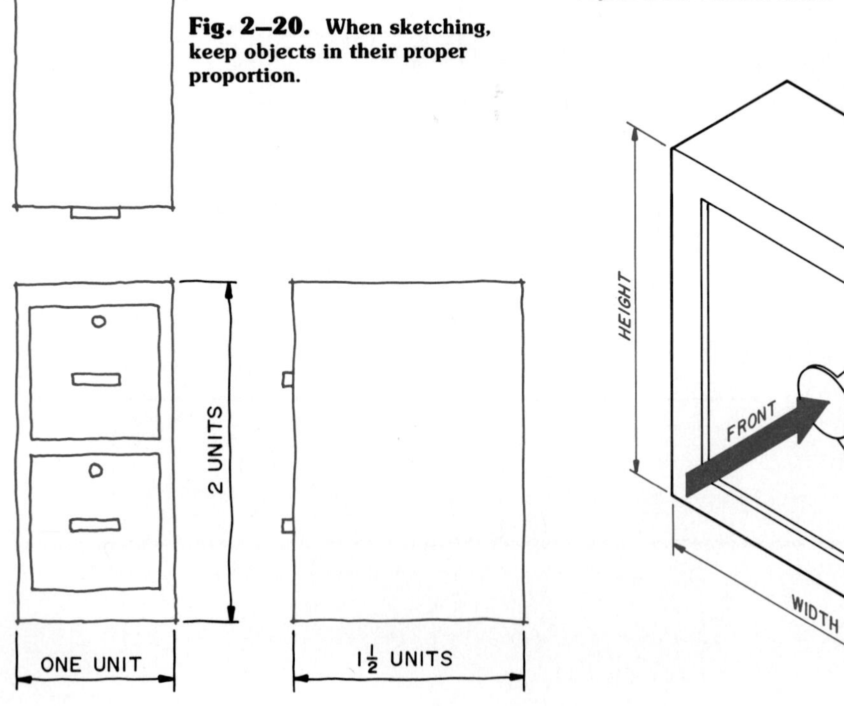

Fig. 2–20. When sketching, keep objects in their proper proportion.

Fig. 2–22. Multiview sketches are made by looking at the object from several sides.

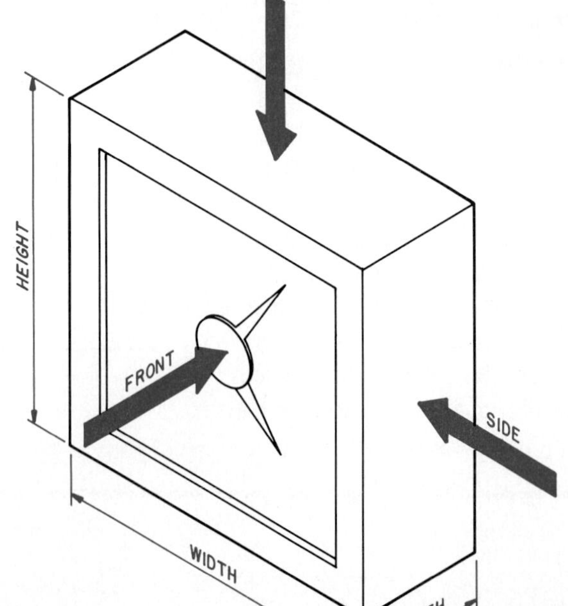

The distances measured are then transferred to a coordinate-ruled sketch sheet. Figure 2-21 shows a measurement taken with a pair of outside calipers and laid out on a ruled sheet (eight divisions to the inch).

When several views compose a technical sketch, all views are developed at the same time. Do not complete one view first. If all views are sketched at the same time, there is a greater probability that they will be in correct proportion.

Multiview Sketches

A **multiview sketch** shows the object from several sides. Fig. 2-22. One-, two-, and three-view sketches are most commonly used. You use only those views that are needed to describe the object.

One-view Sketches

A one-view sketch is possibly the most common kind drawn because many sketches are made to show one thing only, not a complete drawing of an object. Objects that are very thin, such as gaskets, can be shown in one view. Their thickness is given in a note as seen in Fig. 2-23.

Two-view Sketches

Some objects can be completely described with a two-view sketch. This is because views that do not show something different from other views need not be drawn. By studying an object you can determine if it can be described in just two views. The cylindrical object in Fig. 2-24 is a good example of a two-view drawing.

Three-view Sketches

Many objects require three views to describe all of their features. The usual views in a three-view sketch are the top, front, and right-side view. Fig. 2-25. When you study Chapter 6, "Multiview Drawing," you will learn about using other possible views.

Pictorial Sketching

Pictorial (picture-like) sketches can be made to show how the total object would appear. The kind of **pictorial sketches** most commonly used are *oblique* and *isometric.* Additional details about these and other types of pictorial drawing will be learned when you study Chapter 14, "Pictorial Drawing."

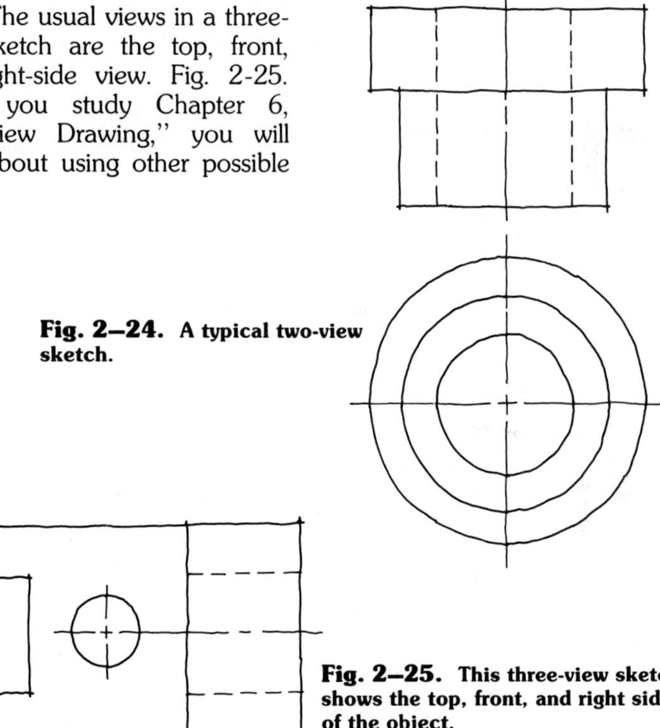

Fig. 2–24. A typical two-view sketch.

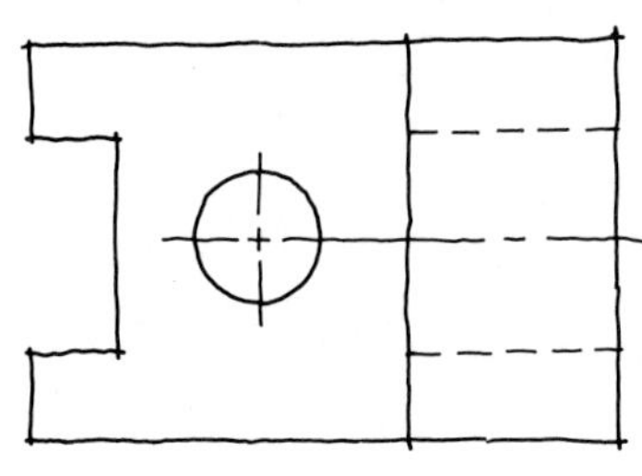

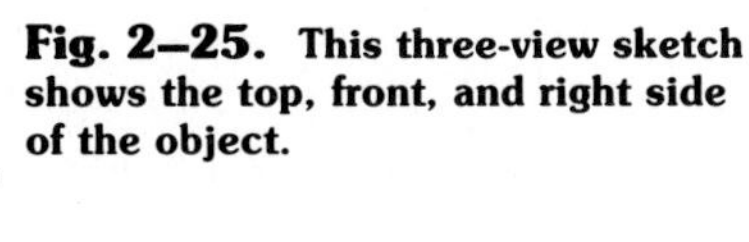

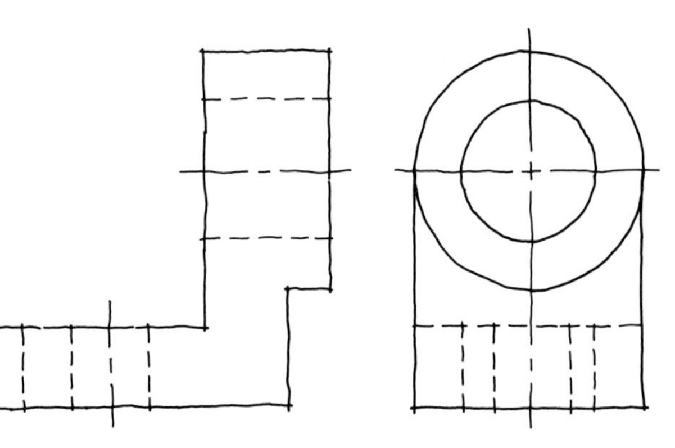

Fig. 2–25. This three-view sketch shows the top, front, and right side of the object.

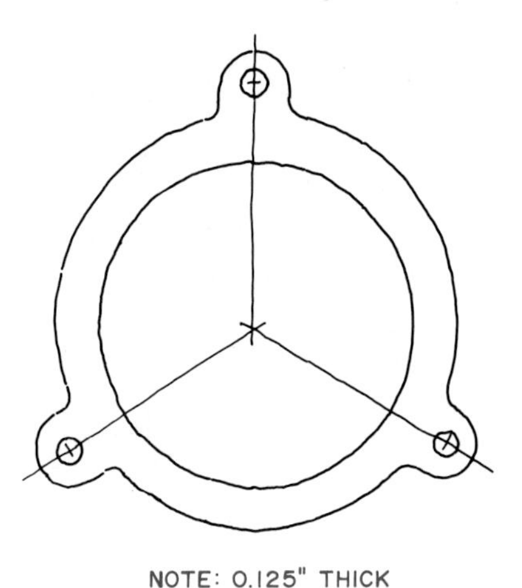

Fig. 2–23. A one-view sketch often shows thickness in a note.

Oblique Sketching

Oblique sketches are made from the front view of an object, with the top and side views lying back at an angle other than 90 degrees. Oblique sketches are often made because they are easy to draw. They are based on three axes. Fig. 2-26. The front view of the object, showing the width and height, is drawn in the same manner as on a multiview drawing. The depth is shown on the third axis, which is drawn on an angle. This type of drawing is useful when sketching a round object. By placing the round surface in the front view, it can be drawn round. On the top or side view, the round surface would be distorted into a different shape.

Making an Oblique Sketch. Graph paper is the best paper to use when you first learn to sketch. To lay out a sketch using graph paper, follow Fig. 2-27 and these steps:

1. Lightly block-in the overall size of the object. Keep it in proportion by counting squares. Any angle can be used for the receding side (depth) but 45 degrees is usually chosen. The 45 degree angle is the diagonal of the squares on the graph paper.
2. Locate all the details of the object with light lines.
3. Remove lines not needed.
4. Darken the remaining visible lines to finish the sketch.

Hidden edges are not usually shown on oblique drawings.

If you choose a large angle such as 60 degrees for the depth, more of the top will be seen. Fig. 2-28. When you draw the depth full size, the drawing is called **cavalier oblique.** A sketch with the depth drawn half size is called **cabinet oblique.** Fig. 2-29.

Circles on the front of an object drawn in oblique appear in their true shape. Those drawn on the top and side appear as elongated circles known as ellipses. The elongation is due to the axis being slanted to one side.

To sketch a circle on the top or side of an oblique drawing, first draw an oblique square with sides equal to the diameter of the circle. Next, locate the center lines of the square. Finally, sketch the ellipse through the points where the center lines touch the sides of the square. Fig. 2-30.

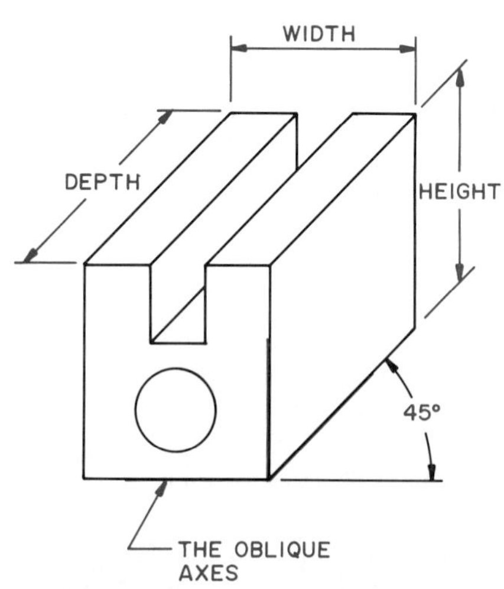
WIDTH
DEPTH
HEIGHT
45°
THE OBLIQUE AXES

Fig. 2–26. These are the three axes commonly used for oblique sketching.

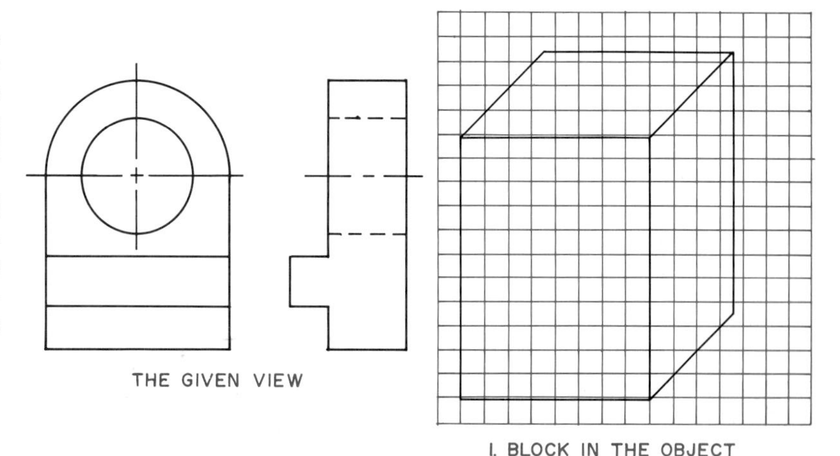
THE GIVEN VIEW
I. BLOCK IN THE OBJECT

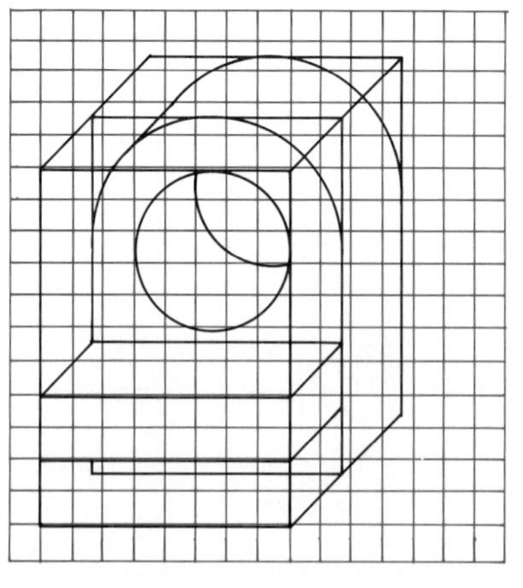
2. LOCATE THE DETAILS

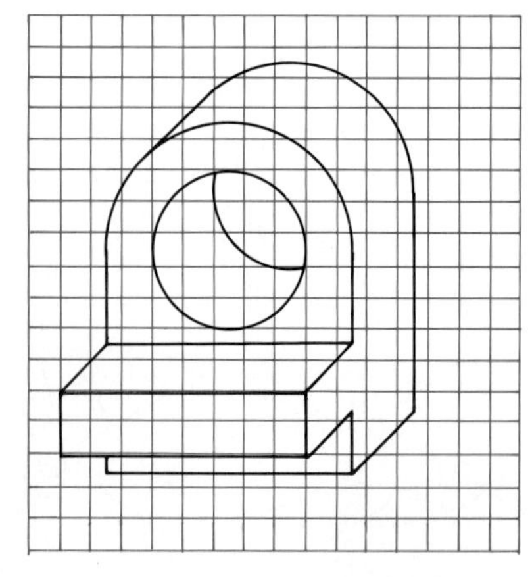
3. DARKEN THE VISIBLE LINES.

Fig. 2–27. How to lay out an oblique sketch.

Isometric Sketching

An **isometric sketch** is based on three axes that are spaced an equal distance (120 degrees) apart. Fig. 2-31. Height and width are measured along two axes, while depth is measured along the third. To lay out an object, a cube for example, measure the actual length of each side along the proper axis. In this example, the height is AB, the width, AC, and the depth, AD. Any lines that are parallel with the isometric axes are called **isometric lines.** True-size distances can be measured along these lines. AB, AC, and AD are each true length and are isometric lines.

The use of isometric grid paper helps when making isometric sketches. However, the angles forming the isometric axes can be estimated if grid paper is not available. Figure 2-32 shows how to lay out a rectangular object made up of all isometric lines on an isometric grid. To lay out the object follow these steps:

1. Block in the overall size.
2. Measure actual sizes along the isometric lines. Remember, lines that are parallel on an object are parallel on the isometric drawing.
3. Connect the visible lines of the object and darken them to finish the drawing.

Sketching Nonisometric Lines. As you just learned, the lines running parallel with the isometric axes are called isometric lines. The actual length of isometric lines can be measured on the drawing. Lines that are *not* parallel with the isometric axes are called **nonisometric lines.** They are drawn by locating each of their ends and then connecting them. This method of drawing nonisometric lines is shown in Fig. 2-33 and described as follows:

1. Block in the view and draw all isometric lines.
2. Locate the end points of the nonisometric lines by measuring a distance from a reference point along the proper isometric lines.
3. Connect the end points of the nonisometric lines with a straight line. Remember, the length is not true size and cannot be measured on the drawing.

Oblique planes are located on isometric drawings by this same method as shown in Fig. 2-34.

Circles and Arcs on Isometric Sketches. Circles and arcs on isometric sketches appear as ellipses. Their position depends upon the plane in which they are located. Fig. 2-35. See Fig. 2-36 and follow these steps to sketch an isometric circle:

1. Lay out an isometric square with sides equal to the diameter of the circle.
2. Sketch the center lines of the square.

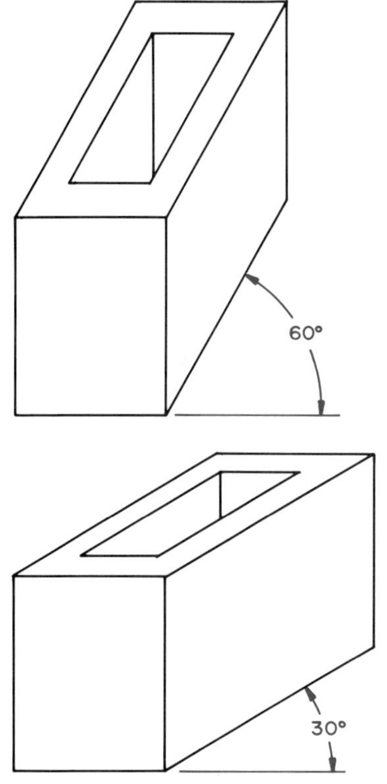

Fig. 2–28. The appearance of the object varies as the angle of the receding axis is changed.

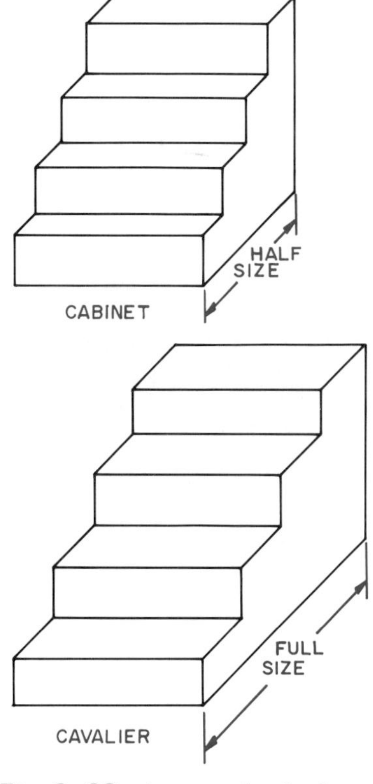

Fig. 2–29. An example of cabinet and cavalier oblique drawings.

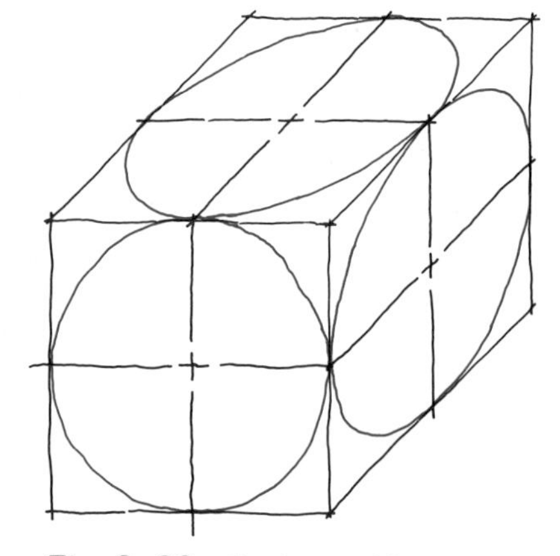

Fig. 2–30. Circles in oblique can be sketched by laying out an oblique box that has sides equal to the diameter of the circle. Then, sketch the circle as shown.

3. Sketch the long arcs tangent to the square at the points where the center line meets the edge of the square.

4. Sketch the short arcs tangent to these same points.

Since an *arc* is part of a circle, use the same technique to sketch an isometric arc. See Fig. 2-37 and follow these steps:

1. Lay out the isometric square including its center lines.

2. Sketch the segment of the circle that is tangent to the surfaces containing the arc.

Cylinders are sketched using these same techniques. Typical layouts for cylinders and cylindrical holes are shown in Fig. 2-38.

Irregular Curves on Isometric Sketches. **Irregular curves** do not have a fixed radius. On an isometric sketch they are laid out using a coordinate system that locates a series of points on the curve. The points are connected to form the curve. The more points used, the more accurate the shape of the curve. To sketch an irregular curve, follow these steps and see Fig. 2-39:

1. On the multiview sketch of the object, draw a series of lines that cross the curve and are parallel with the horizontal surface.

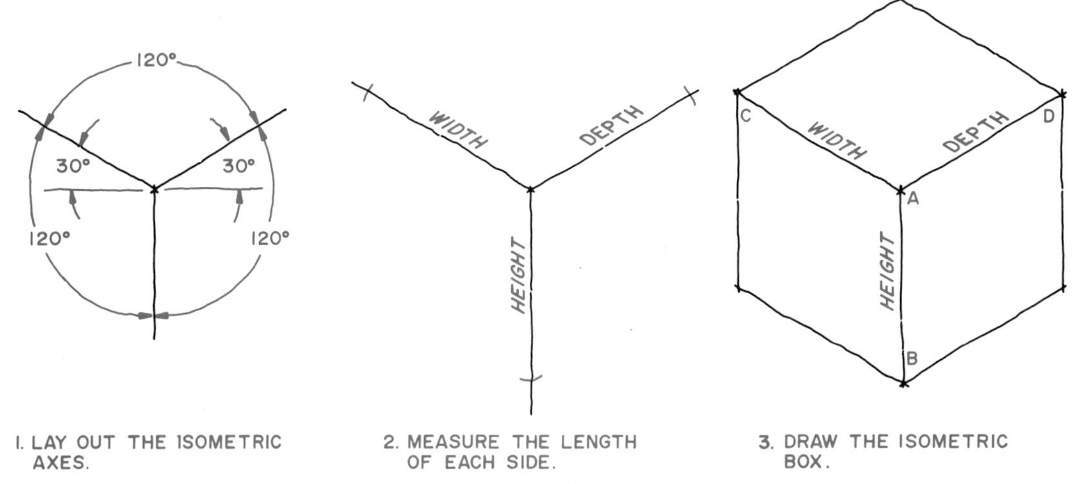

Fig. 2–31. **An isometric sketch is made using axes spaced an equal distance apart.**

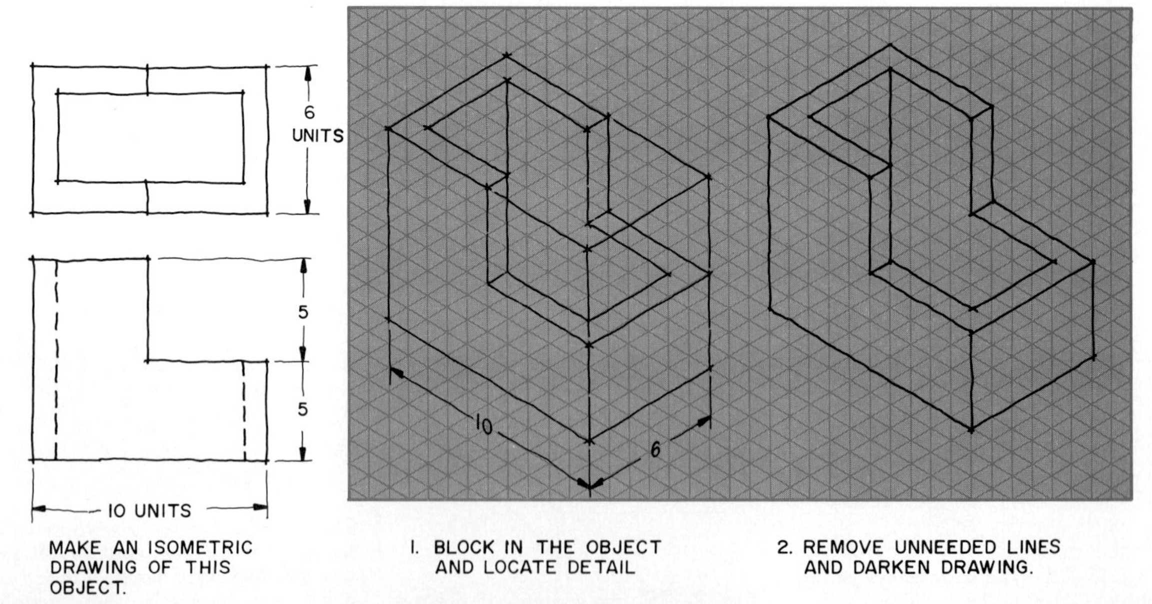

Fig. 2–32. **How to make an isometric sketch using isometric grid paper.**

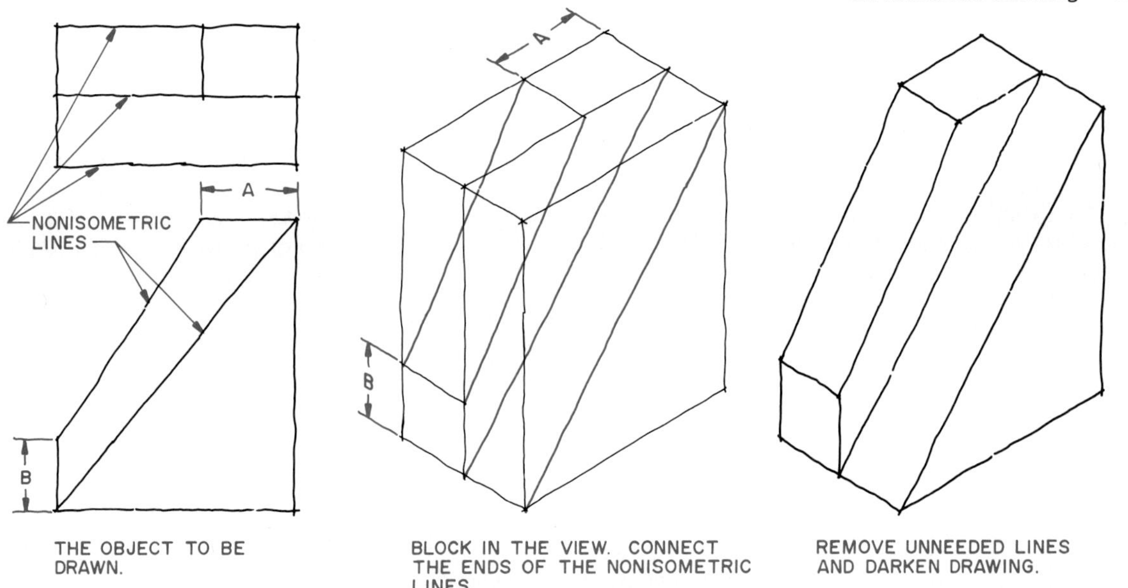
NONISOMETRIC
LINES
A
B
A
B
THE OBJECT TO BE DRAWN.
BLOCK IN THE VIEW. CONNECT THE ENDS OF THE NONISOMETRIC LINES.
REMOVE UNNEEDED LINES AND DARKEN DRAWING.

Fig. 2–33. **To draw nonisometric lines, locate their end points along isometric lines, then connect them.**

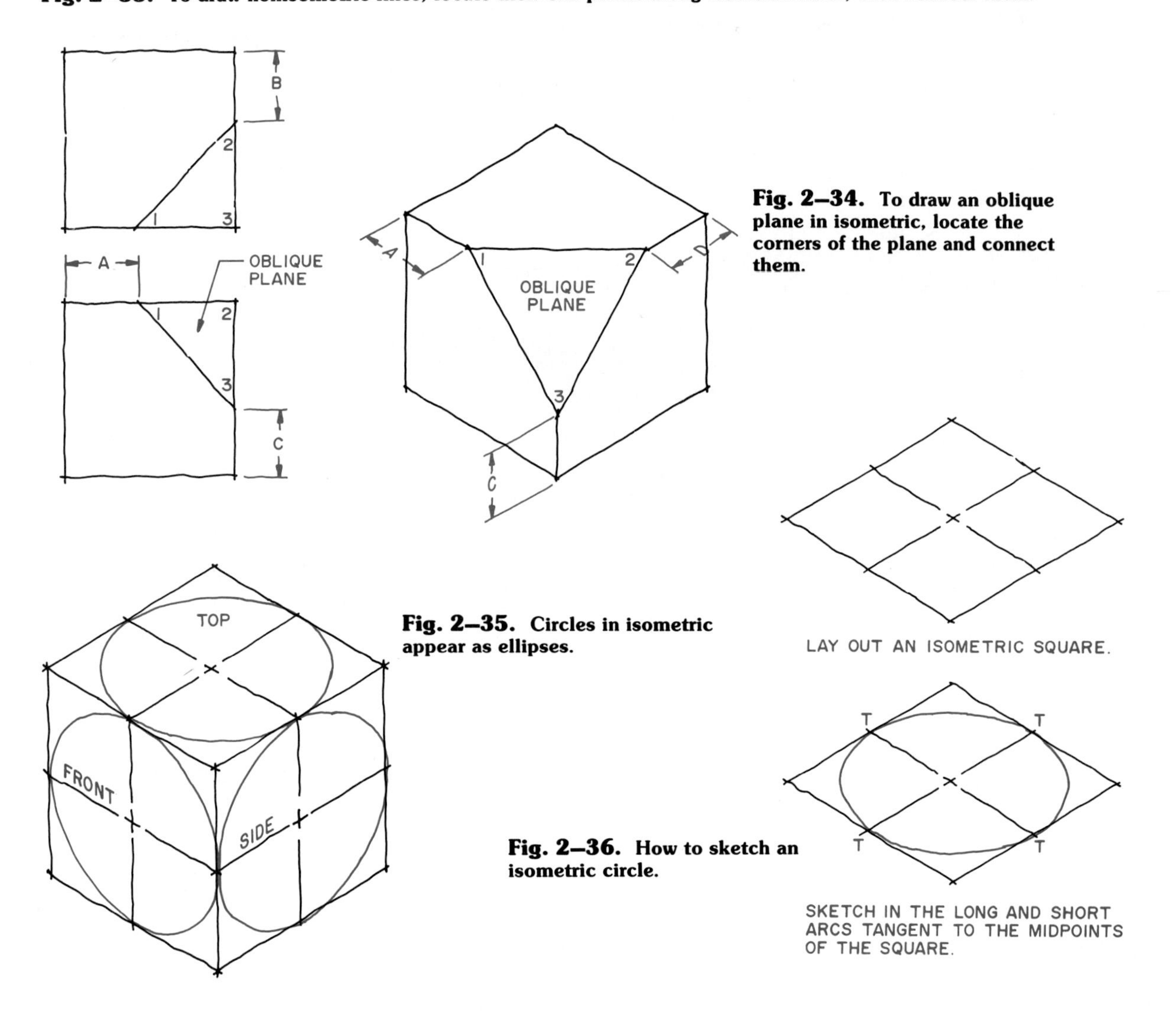
B
2
1
3
A
OBLIQUE
PLANE
1
2
3
C
A
D
1
2
OBLIQUE
PLANE
3
C
TOP
FRONT
SIDE
LAY OUT AN ISOMETRIC SQUARE.
T
T
T
T
SKETCH IN THE LONG AND SHORT ARCS TANGENT TO THE MIDPOINTS OF THE SQUARE.

Fig. 2–34. **To draw an oblique plane in isometric, locate the corners of the plane and connect them.**

Fig. 2–35. **Circles in isometric appear as ellipses.**

Fig. 2–36. **How to sketch an isometric circle.**

Notice that these lines are isometric lines. Lay out the vertical distance from the bottom of the piece to the point where the horizontal lines meet the curve.

2. Block in the isometric view of the object.

3. Draw the coordinate lines on the side of the isometric sketch that is the same as the front view of the multiview sketch. Transfer each coordinate (point) from the multiview sketch to the isometric sketch by measuring accurately.

4. Extend each coordinate across the thickness of the piece to locate points for the curve that forms the back of the object.

5. Connect the points to form the irregular curve.

Other Isometric Axes. Isometric axes can be located in several ways as shown in Fig. 2-40. In all cases, the angles between them are equal (120 degrees). The axes chosen depend upon what parts of the object are to be shown.

Sketching in Perspective

First study the principles behind perspective drawing found in Chapter 14, "Pictorial Drawing." The steps shown in that chapter are also used to make **perspective (picture-like) sketches.** The major difference between perspective drawings and perspective sketches is accuracy. Since perspectives rely upon many projections, a sketch tends to become inaccurate. Even though the perspective sketch may be inaccurate, it can be very descriptive. Figure 2-41 shows a perspective sketch.

Perspective grid paper is a big help when sketching. It has printed lines that will help you project the points needed for the kind of perspective drawing you are making.

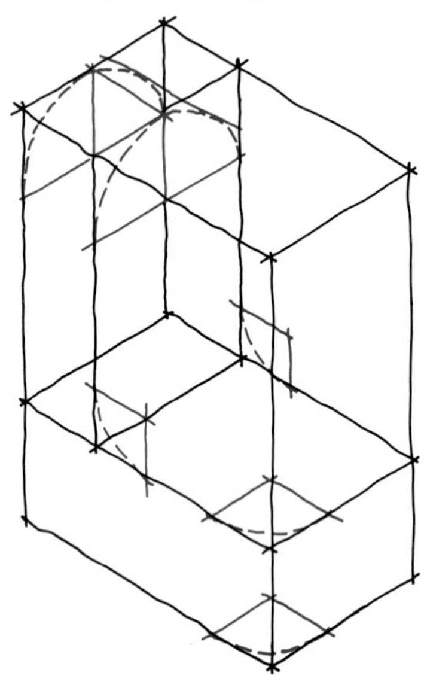
BLOCK IN THE ARCS.

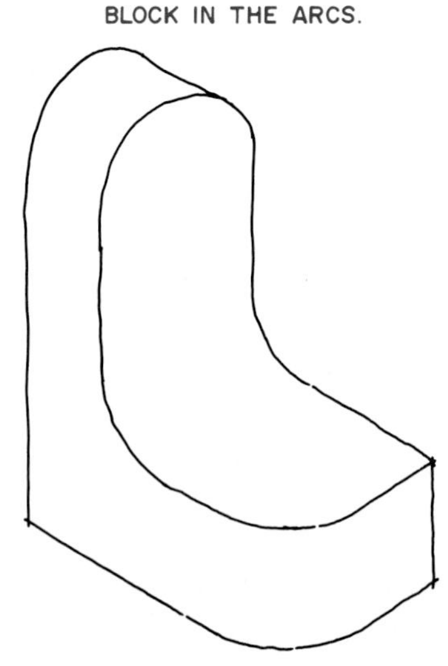
REMOVE UNNEEDED LINES
AND DARKEN OBJECT.

Fig. 2–37. How to lay out arcs on isometric sketches.

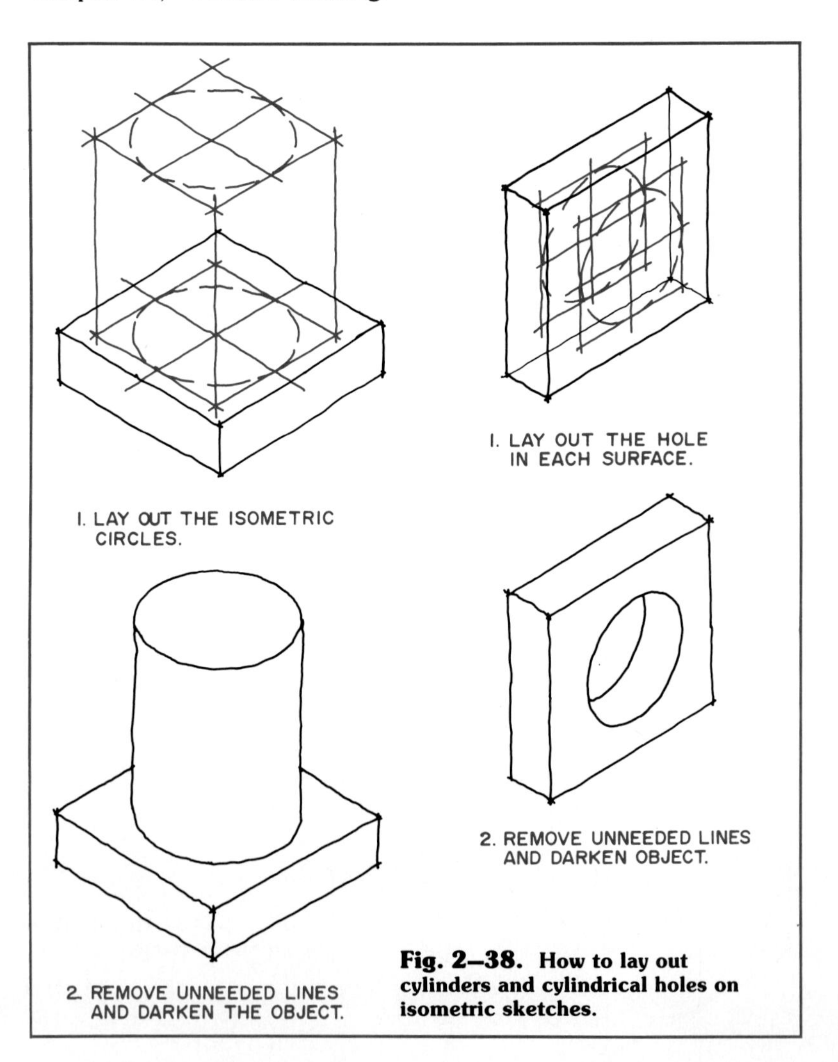
1. LAY OUT THE ISOMETRIC
CIRCLES.
1. LAY OUT THE HOLE
IN EACH SURFACE.
2. REMOVE UNNEEDED LINES
AND DARKEN THE OBJECT.
2. REMOVE UNNEEDED LINES
AND DARKEN OBJECT.

Fig. 2–38. How to lay out cylinders and cylindrical holes on isometric sketches.

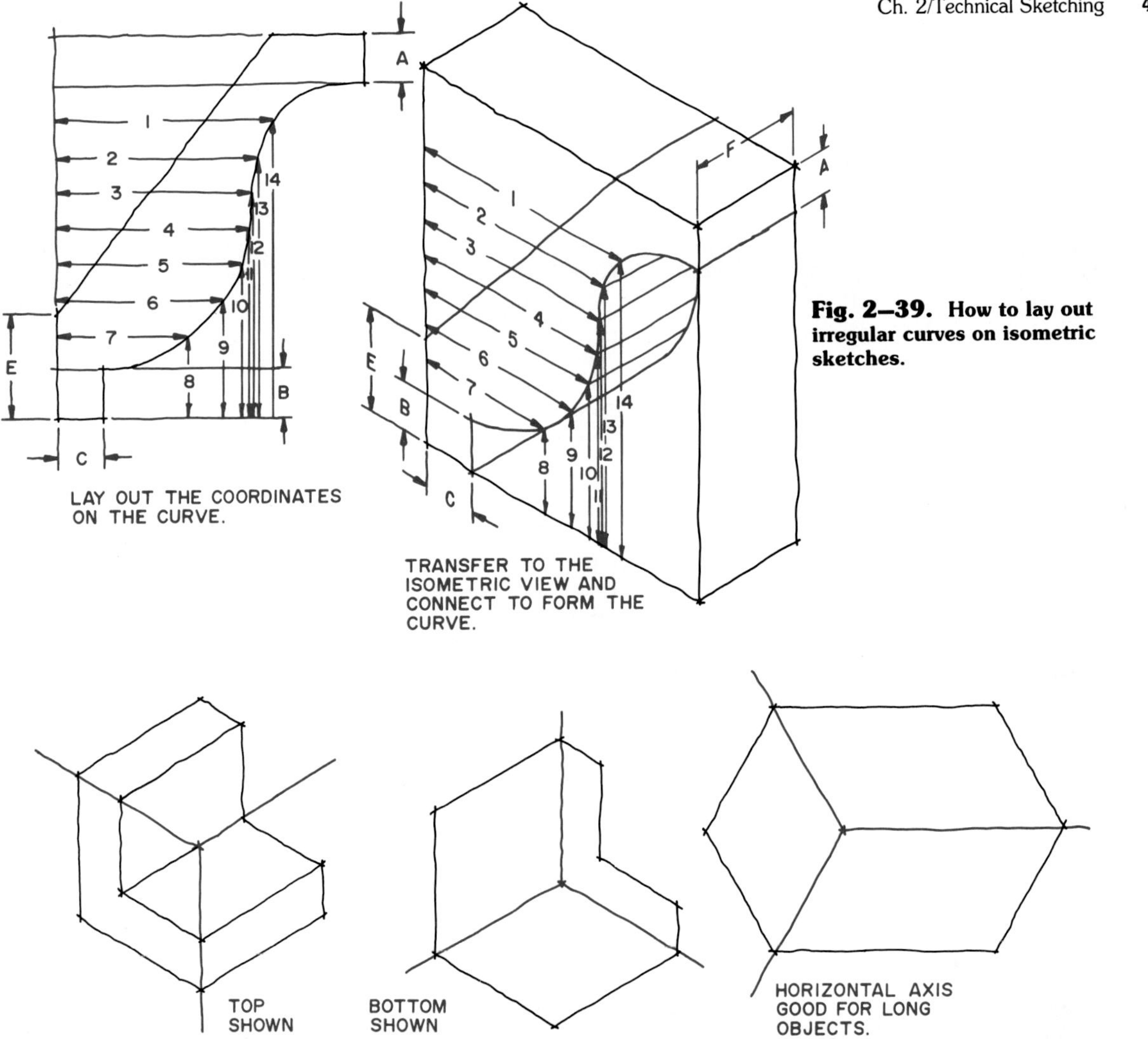

Fig. 2–39. **How to lay out irregular curves on isometric sketches.**

Fig. 2–40. **Other positions for isometric axes that show different parts of an object.**

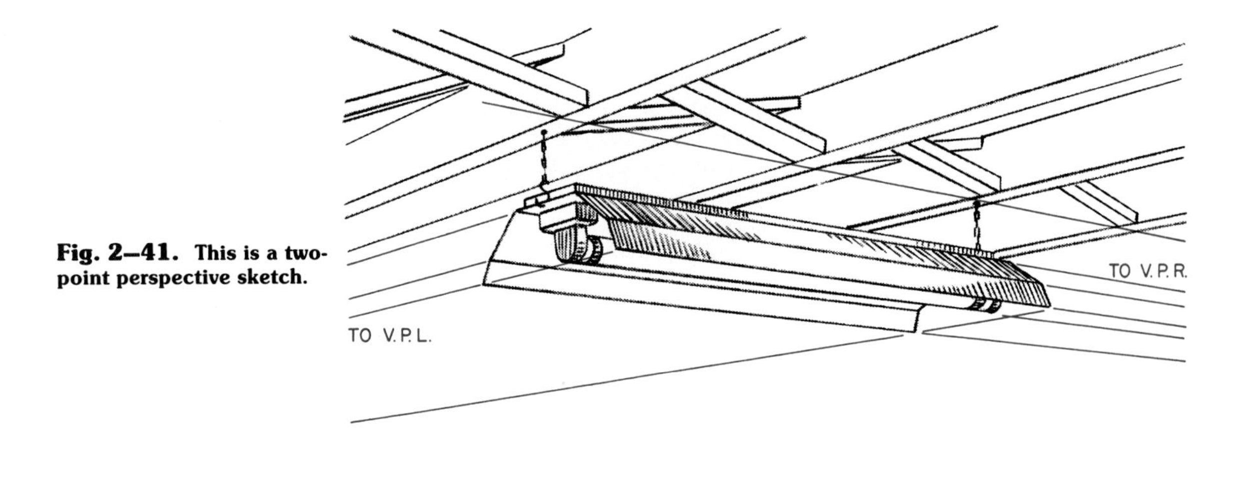

Fig. 2–41. **This is a two-point perspective sketch.**

Chapter Review

Build Your Vocabulary

You should understand and use the following terms as part of your working vocabulary. Write a brief explanation of what each means.

sketch
sketching
drafting vellum
graph paper
ellipse
major axis
minor axis
proportion
multiview sketch
pictorial sketches
oblique sketches
cavalier oblique
cabinet oblique
isometric sketch
isometric lines
nonisometric lines
irregular curves
perspective sketches

Sharpen Your Math Skills

1. Suppose you have graph paper with 8 divisions to the inch. If you have a line of a certain length, you can figure out how many divisions to count off by multiplying the length of the line (in inches) times 8, or the number of divisions per inch on the graph paper. How many divisions should you count off to sketch lines of the following lengths?
a. A line 1¼″ long.
b. A line 2⁵⁄₁₆″ long.
c. A line ¾″ long.

2. You now have graph paper with 10 divisions to the inch. How many divisions high should you count off to sketch a square that is 2.3″ × 2.3″? How many divisions should you count off for the length? The area of a square is the length times the height. What is the area of this square?

3. You have metric graph paper with 1-millimeter divisions and want to sketch a rectangle 5 millimeters long and 7 millimeters high. How many divisions long is the rectangle? How many divisions high is the rectangle? The area of a rectangle, like the square, is the length times the height. What is the area of this rectangle? The perimeter of a rectangle or square is the sum of all four sides. What is the perimeter of this rectangle?

4. The diameter of a circle is the distance from one side to the other, measured through the center. The radius of a circle is ½ the diameter. If you sketch a circle with a diameter of 2½″, what length is the radius? If you have graph paper with 8 divisions to the inch, how many divisions long is the radius?

5. You have a drawing of a rectangle 3″ × 5″. If you redraw the rectangle in proportion at 1½ times the original size, what are the new dimensions?

A CASEMENT WINDOW

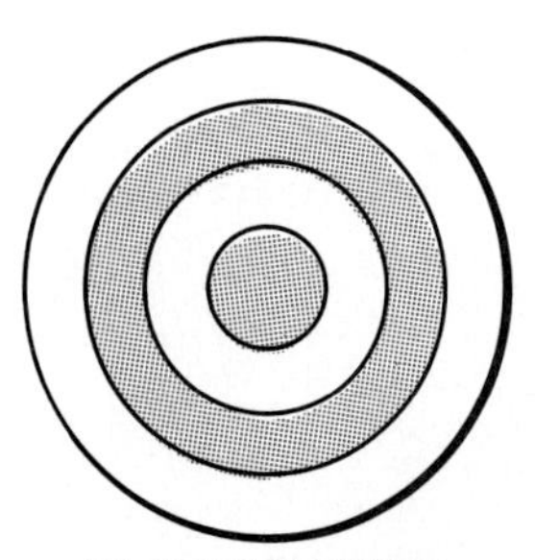
AN ARCHERY TARGET

Study Problems—Directions

The following problems will help develop skill in sketching. As sketches are made, pay close attention to proportion. Sketching skillfully and with correct proportions is important to solving problems graphically. See **Fig. P2-1.**

1. Sketch the archery target twice as large as shown.
2. Sketch the hatbox three times as large as shown.
3. Make a sketch of the casement window. Enlarge it three times the size shown.
4. On graph paper make a full-size sketch of the irregular curve.
5. Enlarge the drawing of the explosive rivet three times.
6. Sketch the model car race track. Make the sketch twice as large as shown.
7. Study the section through the wall of a frame house. Sketch it twice as large as shown.
8. Make a sketch of the doghouse. Make it the same size as shown.
9. Enlarge the drawing of the basketball backboard to twice the size shown.

The following problems can be found in the study problem section of Chapter 6, "Multi-view Drawing":

10. Make an isometric sketch of the isometric drawing of the rubber gasket, **Fig. P6-6.**
11. Make an isometric sketch of the isometric drawing of the asphalt paving block, **Fig. P6-7.**
12. Make an isometric sketch of the isometric drawing of the angle plate, **Fig. P6-13.**
13. Make an oblique sketch of the oblique drawing of the rocker arm, **Fig. P6-55.**
14. Copy the three-view drawing of the tricycle frame, **Fig. P6-26.**

P2–1. Technical sketching study problems.

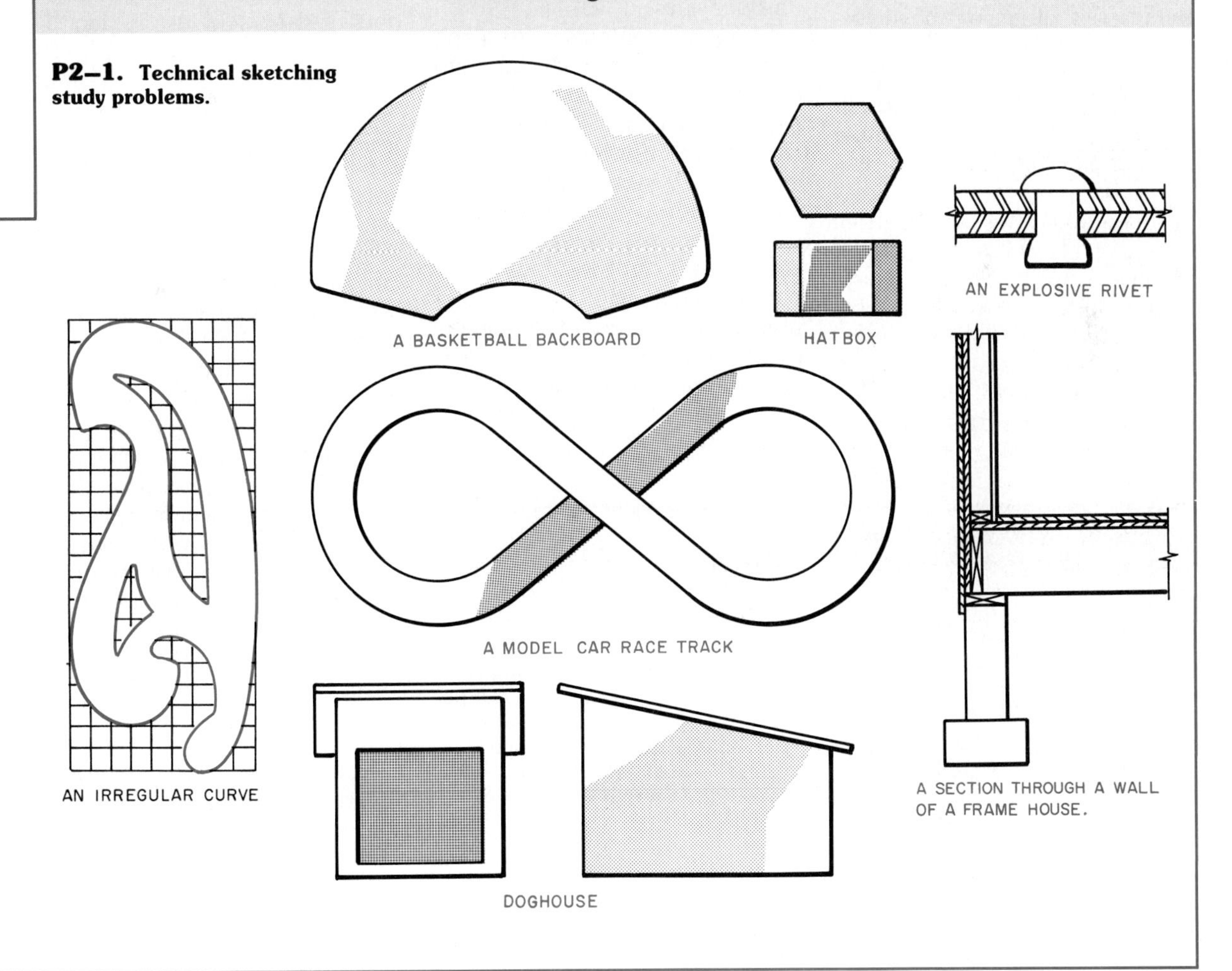

CHAPTER 3

Tools and Techniques of Drafting

After studying this chapter, you should be able to do the following:

- Use basic drafting tools.
- Lay out drawings to scale.
- Select the appropriate drafting paper and vellum.
- Letter drawings in a professional manner.
- Use the standard line symbols on drawings.

Drafting Tables

There are a variety of drafting tables in use. Most have a top that can be raised or lowered to change its distance above the floor and be tilted to change its angle between vertical and horizontal. Fig. 3-1. These adjustments permit a drafter to either stand or sit while drawing as seen in Fig. 3-2.

Table tops are made from wood or a wood composition product with a smooth finish. Many drafters cover their table tops with a vinyl drawing board cover. It provides a smooth surface that is slightly soft. Such a surface allows the drafter to draw dark lines and helps reduce the breaking of pencil leads. These vinyl covers are held to the table top with tape that has adhesive on both sides.

In many companies, drafting tables are being replaced with computer drafting equipment. Drawings are often laid out on a digitizer tablet instead of a drawing board. Lines and points are located with a stylus instead of a pencil. Once the drawing is in the computer, it can be output to paper or vellum with the aid of a plotter. Fig. 3-3. Detailed information about the use of computers in drafting is presented in Chapter 4, "Computer-Aided Design and Drafting."

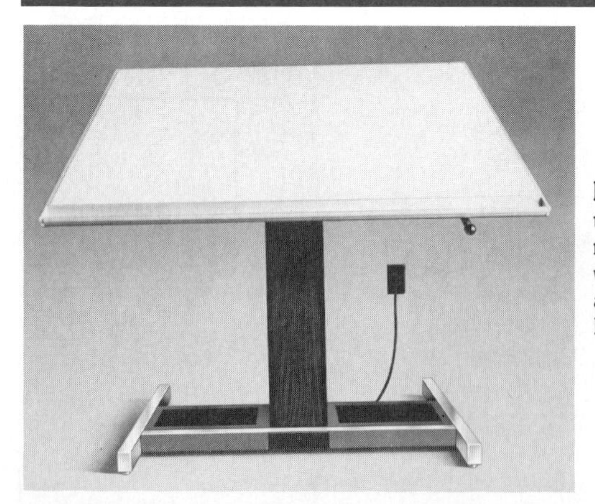

Fig. 3—1. The top of the drafting table can be raised and lowered as well as tilted to any angle. (Hamilton Industries)

Fastening the Drawing Sheets to the Table

Drawing sheets are usually held to the table with **drafting tape** or **dots.** Tape is sold in rolls. Fig. 3-4. Use a piece about ⅜ inch (10 mm) long on each corner of your drawing sheet. Dots are precut circular pieces of tape fastened to a paper backing sheet. Peel the dots from their paper backing and fasten each corner of your drawing sheet to the table. Fig. 3-5. Follow these steps, shown in Fig. 3-6, to fasten your drawing sheet to the table:

1. Line up the top edge of your paper near the upper left corner of the table top using the edge of a T-square, straightedge, or drafting machine scale.
2. Hold the paper in place and carefully slide the tool down the sheet. Place a piece of tape or a dot on each of the upper corners.
3. Smooth the paper toward its bottom edge and tape the lower two corners.

The T-square

The **T-square** is the oldest of the drafting tools used to draw straight horizontal lines. Although out of date with modern industry practice, the T-square is still useful for drafting.

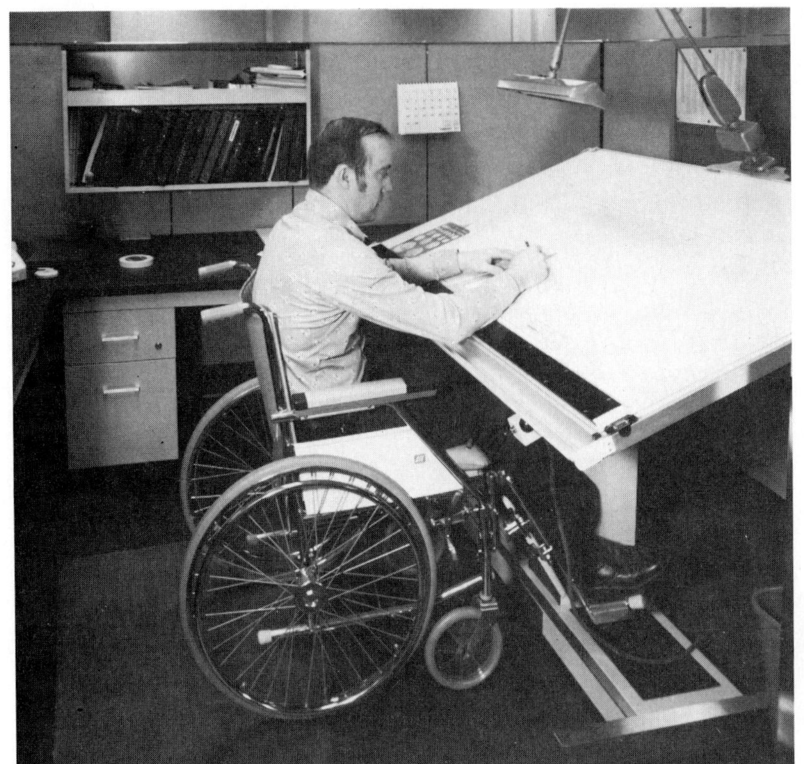

Fig. 3–2. Because of adjustable equipment, drafting is a profession that is open to those people who must sit all day. (Hamilton Industries)

Fig. 3–4. Drafting tape can be used to hold drawings to the board. (3M Company)

Fig. 3–3. Two computer-aided drafting workstations that share a disk drive and plotter are shown here. (Hewlett-Packard)

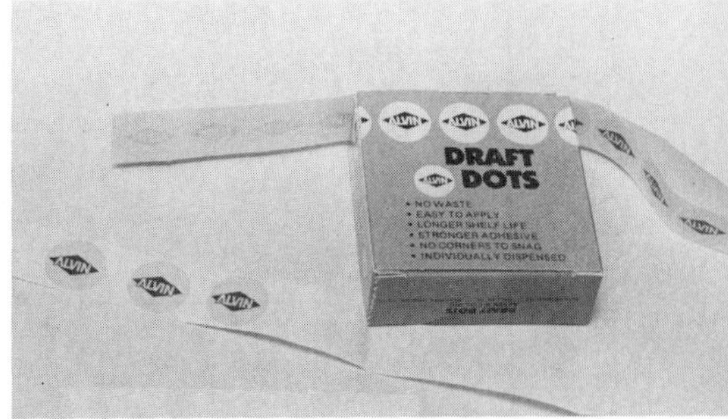

Fig. 3–5. Adhesive-backed pieces of round drafting tape, known as dots, are also used to hold drawings to the table.

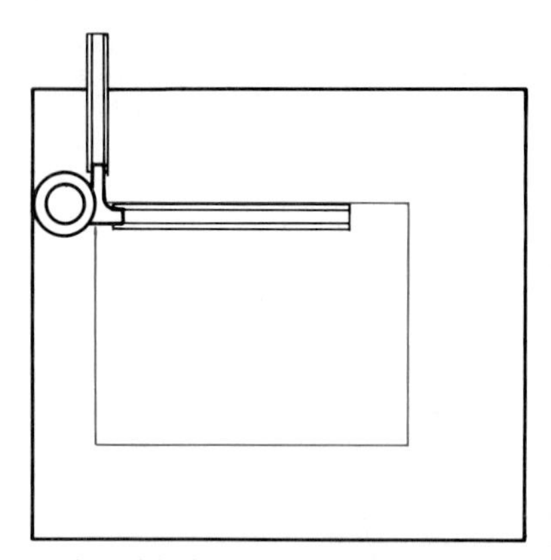

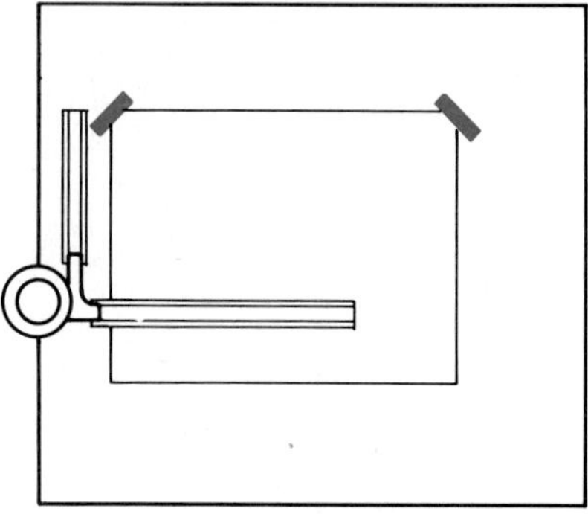

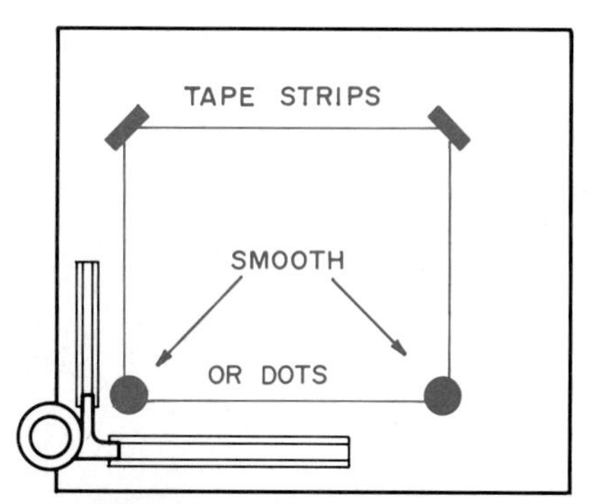

Fig. 3–6. How to fasten the drawing sheet to the drawing surface.

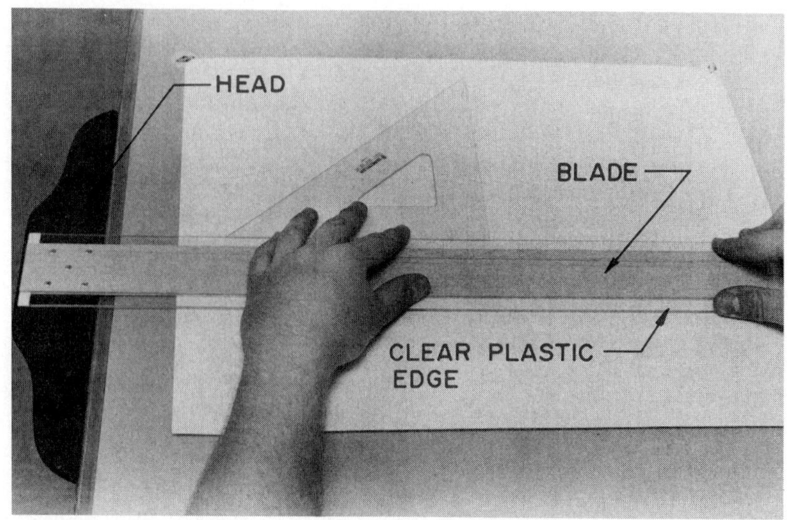

Fig. 3–7. Triangles rest on the T-square.

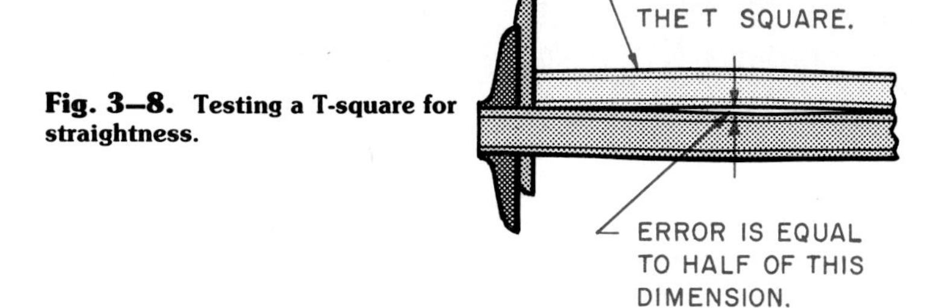

Fig. 3–8. Testing a T-square for straightness.

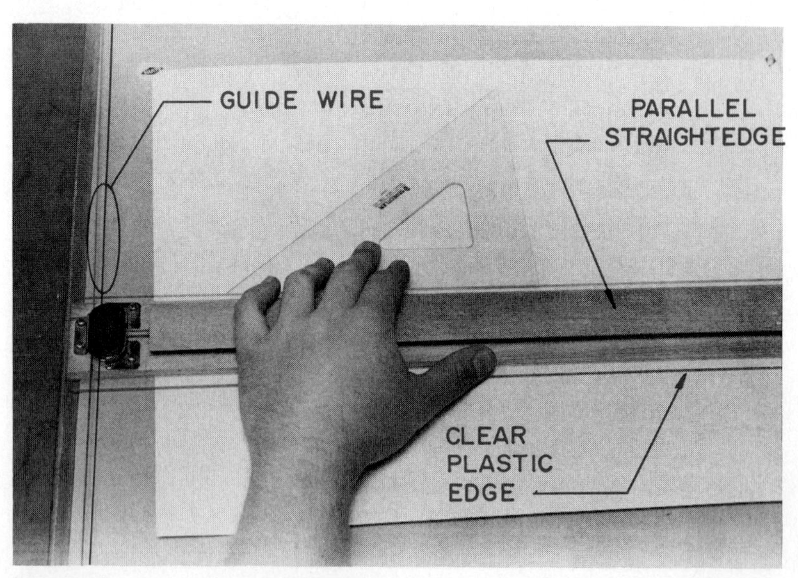

Fig. 3–9. The parallel straightedge runs on wires and serves as a base for triangles.

The T-square has two parts—a head and blade. The blade has clear plastic edges that permit lines beneath them to be seen. Blade lengths are available in sizes ranging from 18 inches (457 mm) to 60 inches (1524 mm). The T-square blade serves as the base for triangles. Fig. 3-7.

Testing a T-square for Straightness

The blade on a T-square must be perfectly straight if you are going to make quality drawings. Test a T-square for straightness by drawing a straight line through two points the length of the blade. Then, turn the T-square over and, drawing on the same edge as before, make a second line between the two points. Any space between these two drawn lines shows that the blade is not straight. The amount of error is equal to one-half the distance between the lines. Fig. 3-8.

The Parallel Straightedge

The **parallel straightedge** takes the place of a T-square. It runs on wires that enable it to move up and down the drafting table. Because of the wires, it always stays parallel and so is more accurate than a T-square. The parallel straightedge is used to draw horizontal lines. It also serves as a base for triangles which are used to draw vertical lines. Fig. 3-9.

Triangles

The two commonly used **triangles** are the 45 degree and 30-60 degree. Fig. 3-10. They are available in a variety of sizes. The size is the distance measured along the longest side of the right angle. They are made of clear or colored transparent plastic. Triangles are easily damaged and should be handled carefully. By combining a 45-degree and 30-60-degree triangle, you can draw angles at 15-degree intervals around a circle.

Adjustable Triangles

Drafters are frequently required to draw angles at increments other than 15 degrees. To save time finding the proper angle, use an **adjustable triangle.** Fig. 3-11. Set the angle you need and draw the line with ease. Most adjustable triangles are graduated in one-half degree increments. The movable portion is held in position by tightening a screw.

Drafting Machines

There are two basic types of **drafting machines.** One is the arm drafting machine shown in Fig. 3-12. The other is the track drafting machine seen in Fig. 3-13. Both types replace straightedges, triangles, scales, and protractors. The main advantage of a drafting machine is that it reduces drafting time since the drafter has most of the tools he or she needs in one machine.

Two scales are fixed at a 90-degree angle to a round head. These arms serve as a straightedge, triangle, and scale. To operate the drafting machine, the drafter holds the head in his or her left hand. He or she can move the machine up, down, and across the drawing at any angle. The round head has a pivot. A release button lets the drafter rotate the arms to any angle desired. A scale of degrees shows on the head. It replaces a protractor.

Drafting machines are made for both right- and left-handed people.

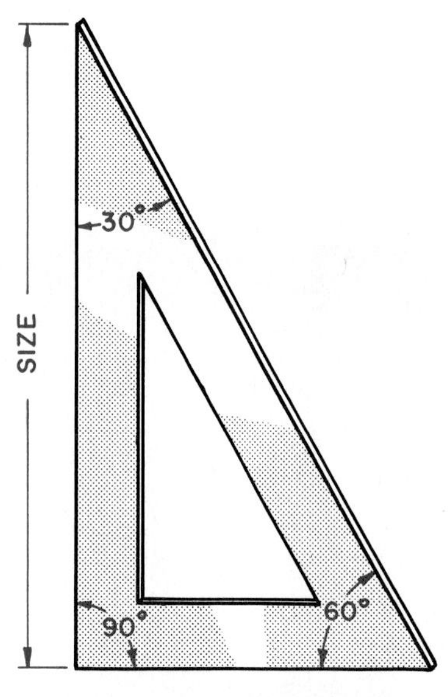

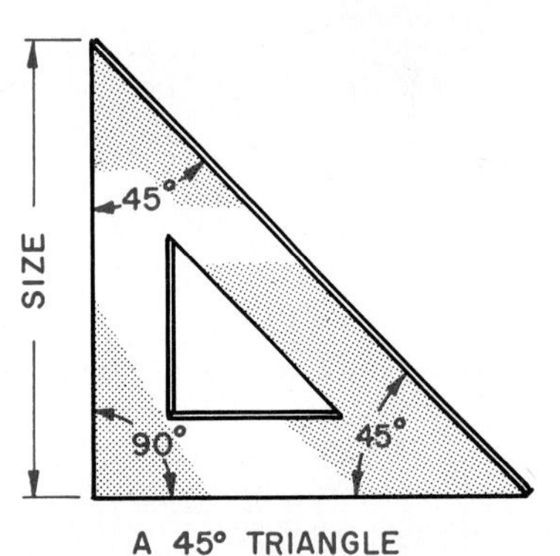

Fig. 3–10. Triangles used in drafting.

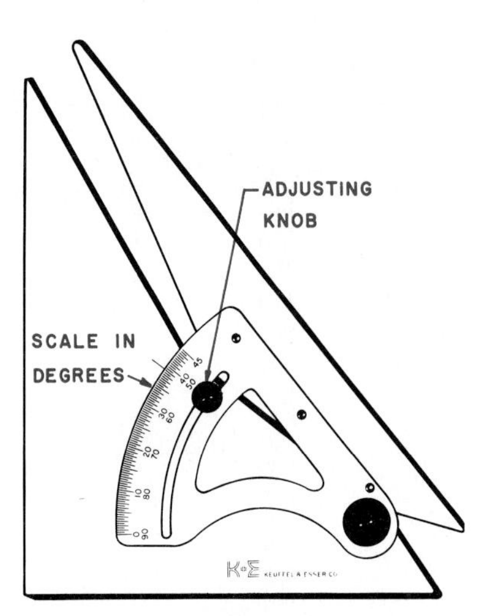

Fig. 3–11. An adjustable triangle. (Keuffel and Esser Co.)

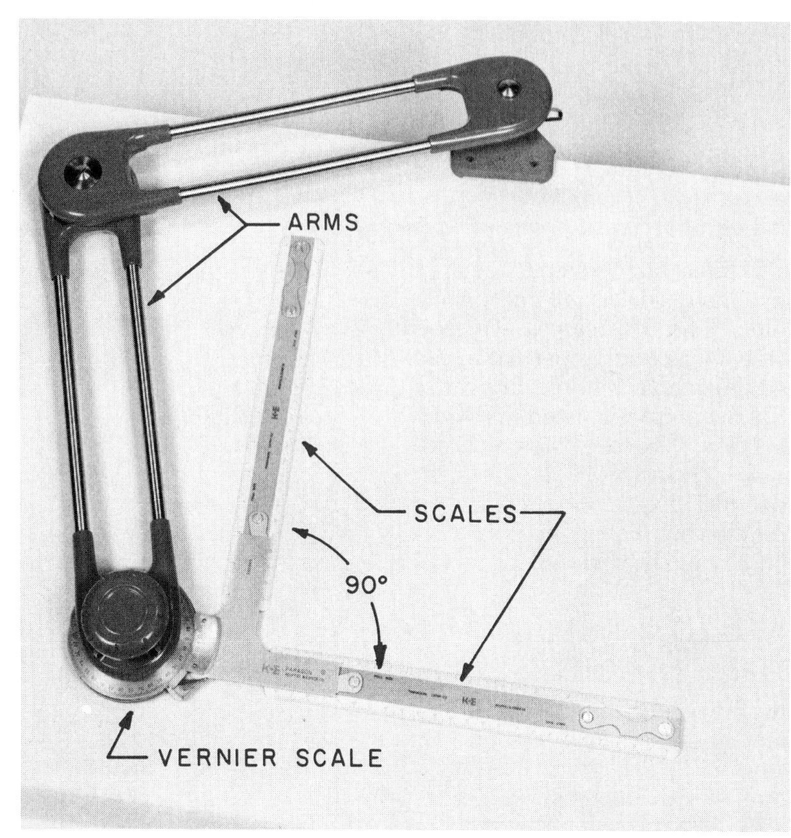

Fig. 3–12. An arm type drafting machine. (Keuffel and Esser Co.)

Fig. 3–13. A track type drafting machine. (Keuffel and Esser Co.)

Drawing Pencils

Wood-cased and mechanical drawing pencils are used in technical drafting. Fig. 3-14. Two types of mechanical pencils are available, standard lead and fine line. The lead of a standard-lead pencil has a large diameter similar to the lead found in wood-cased pencils. The lead of a fine-line pencil, on the other hand, is very fine and is available in 0.3, 0.5, and 0.7 mm diameters. The fine-line pencil does not have to be sharpened because the diameter of the lead is the width of the line drawn.

When drawing on vellum or drafting paper, use a pencil with graphite lead. Graphite leads are made in seventeen degrees of hardness ranging from 6B (very soft) to 9H (very hard) as shown in Fig. 3-15. The hardness is printed on the wooden pencil or on the lead used in standard mechanical pencils. Fine-line pencil leads are made in a more limited range of hardness than graphite leads. In addition, some diameters of fine-line leads have fewer degrees of hardness than others. For example, there are only four choices, B, HB, H, and 2H, for 0.3 mm and 0.7 mm pencils. For the more popular 0.5 mm pencil, however, there are nine choices ranging from 2B to 5H.

The degree of hardness to choose depends upon the thickness and darkness of the line to be drawn. Layout work is very light and you should use a 4H pencil. Finished lines are usually made with a 2H or H lead. Sometimes you may want to use a softer lead such as an F or HB for lettering. Remember, the softer leads are larger in diameter and smear more easily than hard leads. The very hard and very soft leads are not generally used in drafting.

Leads used on polyester drafting film (plastic) are made from special plastic substances. There are three types. One type is called *plastic lead.* It makes dark lines suitable for photographing for microfilm. The other two types are a combination of plastic and graphite. These can be used on film or paper since they produce a dark line and do not smear easily. Drafting film leads are available in five degrees of hardness in wooden and standard mechanical pencils. Pencil manufacturers differ in the way they classify hardness of drafting film leads. In one system E1 is soft; E2, medium; E3, hard; E4, extra hard; and E5, super hard. For fine-line mechanical pencils there are three degrees of hardness.

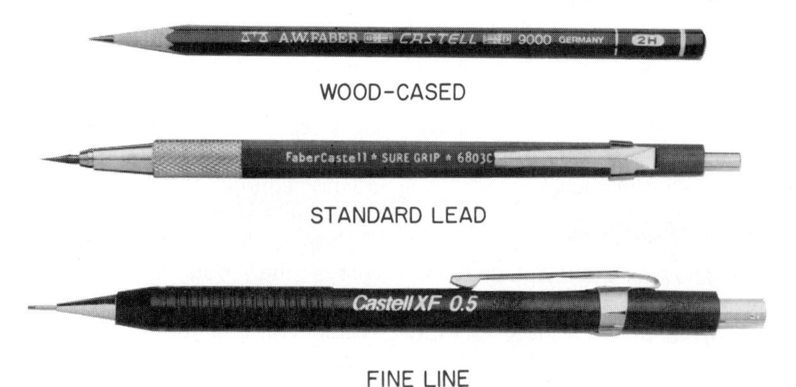

Fig. 3–14. Pencils used for drafting. (Faber-Castell Corporation)

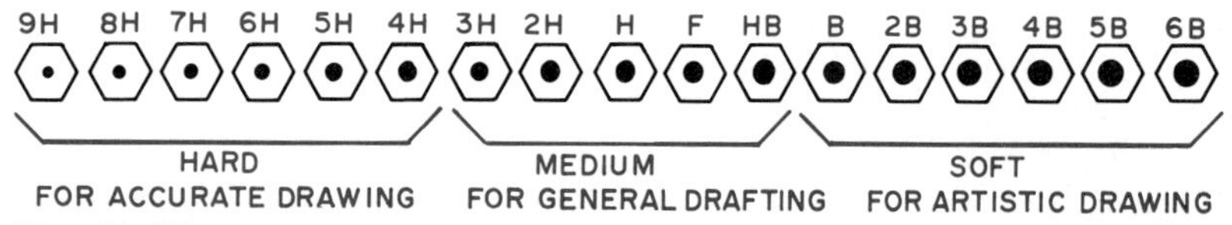

Fig. 3–15. The grades and uses for drawing pencils.

Sharpening a Pencil

Wood-cased pencils are best sharpened with a drafter's pencil sharpener. This pencil sharpener removes the wood only. It leaves about 3/8 inch (10 mm) of lead exposed as shown in Fig. 3-16. Sharpen the end that does *not* show the hardness of the lead. After sharpening, you must put a point on the lead. Figure 3-17 shows both a conical and a wedge point. Most drafting is done with a conical point.

Lead is easily formed to a cone shape in a lead pointer. Fig. 3-18. You place your pencil in the pointer and rotate its top. The lead rubs on an abrasive cylinder inside the pointer.

For some jobs you may want to use a wedge-shaped point. It is stronger than the conical point so it does not wear away as fast. Wedge points are formed with a file or sanding pad. Draw the lead over the file several times on one side. Then turn the pencil over (180 degrees) and repeat the process. Make the wedge long and gradual so that it runs the entire length of the lead.

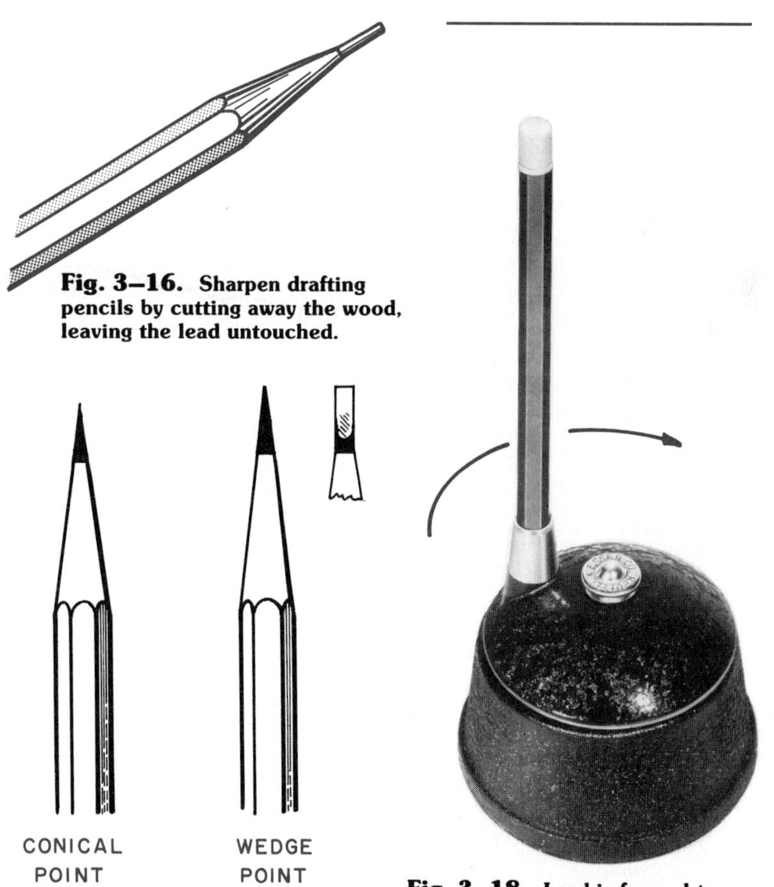

Fig. 3–16. Sharpen drafting pencils by cutting away the wood, leaving the lead untouched.

Fig. 3–17. The two types of pencil points used in drafting.

Fig. 3–18. Lead is formed to a cone shape in a pencil pointer. (Keuffel and Esser Co.)

After the pencil is pointed, wipe the point on a cloth. Or use a special cleaning pad shown in Fig. 3-19. Simply stick the point into the pad and rotate the pencil as you withdraw it.

Pencil Technique

Most drawings are made in pencil. They must be suitable to be reproduced by any of the methods of reproducing drawings. This requires that you follow proper pencil technique.

All pencil lines should be as dark as possible. Your circles and arcs must be as dark as your straight lines. To make them so, you may have to use a softer lead in your compass than in your pencil. By using a template instead of a compass (see Fig. 3-54), you can use your drafting pencil and reduce the darkness problem.

Line symbols should be used properly. The length of dashes in hidden lines should be uniform. Each dash should be dark and clear.

Some people have a tendency to draw extension and dimension lines lightly. These lines must be thin but as dark as all other lines.

Proper line width is essential. The most important lines are thick. This calls attention to them when you read a drawing.

While it is best to draw a line the proper width and darkness with one stroke, you may need to go over a line several times to make it dark. Be certain that you do not change the width as you retrace the line.

Proper pencil technique requires that you sharpen your pencil properly. For instance, a visible line will require a thicker conical point than a center line. Pencils must be sharpened frequently.

Keep your pencil close to the edge of the straightedge or triangle as you draw a line. If it leans in and out as you draw, a wavy line will result. Fig. 3-20.

Select the proper pencil for the job. Wide lines will require a softer pencil than thin lines. Experience will help you decide which hardness is best. The 2H pencil is a good general-purpose pencil for most lines.

Your pencil lead should be soft enough to allow you to draw the line you want without using excess pressure. Too much pressure will form grooves in the paper. This will make it difficult to erase lines to make changes.

When using wood-cased or standard mechanical pencils, slant them on a 60-degree angle in the direction you are drawing the line. When using a fine-line pencil, hold it perpendicular to the paper.

Erasers and Erasing Shields

Pink Pearl, Artgum, and vinyl erasers are soft and used for general erasing and cleaning a drawing. Harder erasers, such as Rubkleen, Ruby, and Emerald, are used for removing pencil and some ink lines. These erasers are coarse and must be used with care. Too much pressure with a coarse eraser may put a hole in your drawing paper.

Vinyl erasers are good for use on polyester drafting film. When used dry, they remove lines drawn with plastic lead. When used on inked lines, these erasers should be moist. There is an eraser made especially for inked lines. It contains an erasing fluid to help dissolve ink. This kind of eraser makes correcting ink drawings very easy.

Electric erasers speed up the work of the drafter. Fig. 3-21. They use a long round strip of eraser held in a chuck. These strip erasers are made of soft or coarse rubber or vinyl.

You use an eraser shield to protect the lines around the area where a correction is being made. Fig. 3-22.

Fig. 3—19. A pencil lead cleaner. (Keuffel and Esser Co.)

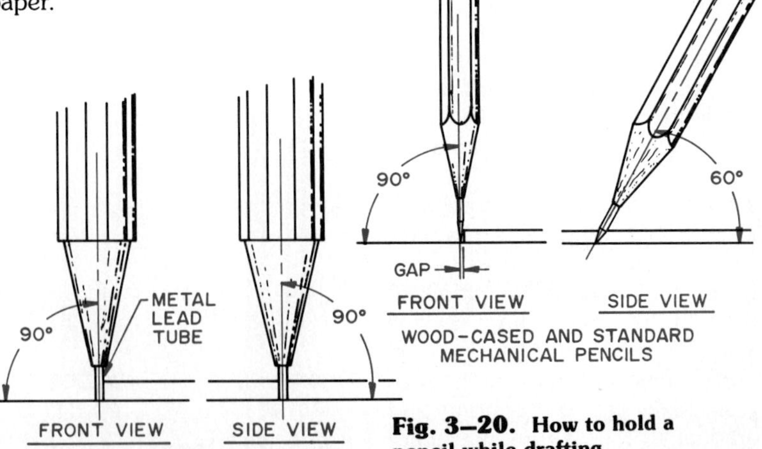

Fig. 3—20. How to hold a pencil while drafting.

The Alphabet of Lines

Line symbols form a part of the graphic language. They have been standardized. The recommendations of the American National Standards Institute (ANSI) for pencil drawings are shown in Fig. 3-23. Two line widths, thick and thin, are used. Thick lines are about 0.04 inches (1.0 mm) wide and thin lines are 0.02 inches (0.5 mm) wide. As a rule of thumb, your thick lines should be twice as wide as your thin lines. All lines should be equally black.

Look at Fig. 3-24 as you read the following descriptions of the line symbols shown in that illustration.

Visible lines are used to show all edges that you can see when looking at the object. These lines are thick and solid.

Hidden lines show surfaces that you cannot see. Hidden lines are made of dashes about 1/8 inch (3 mm) long. The spaces between the dashes are about 1/32 inch (1 mm). The dashes and spaces are often drawn larger if the drawing is large. Estimate these distances as you draw the line. Do not take the time to measure each dash.

Dimension lines are thin, solid lines. They are used to show the extent of the dimension. They contain the dimension number. A leader is one form of dimension line. It is used to connect dimensions or notes to the drawing.

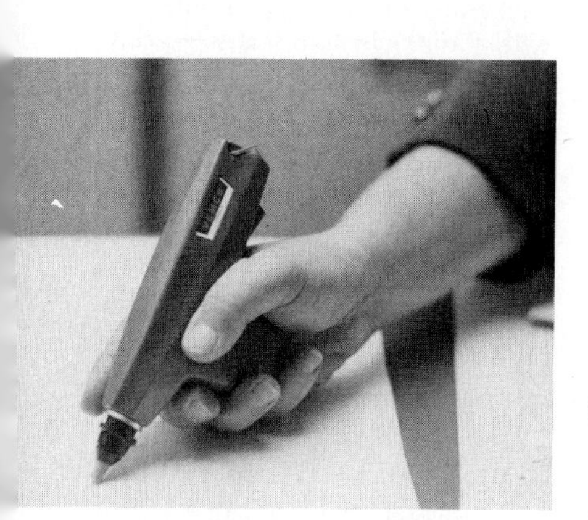

Fig. 3–21. An electric eraser.

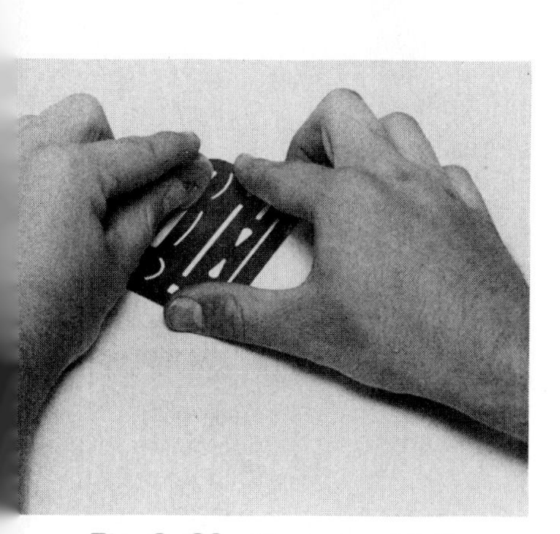

Fig. 3–22. The eraser shield limits the area in which you can erase on a drawing.

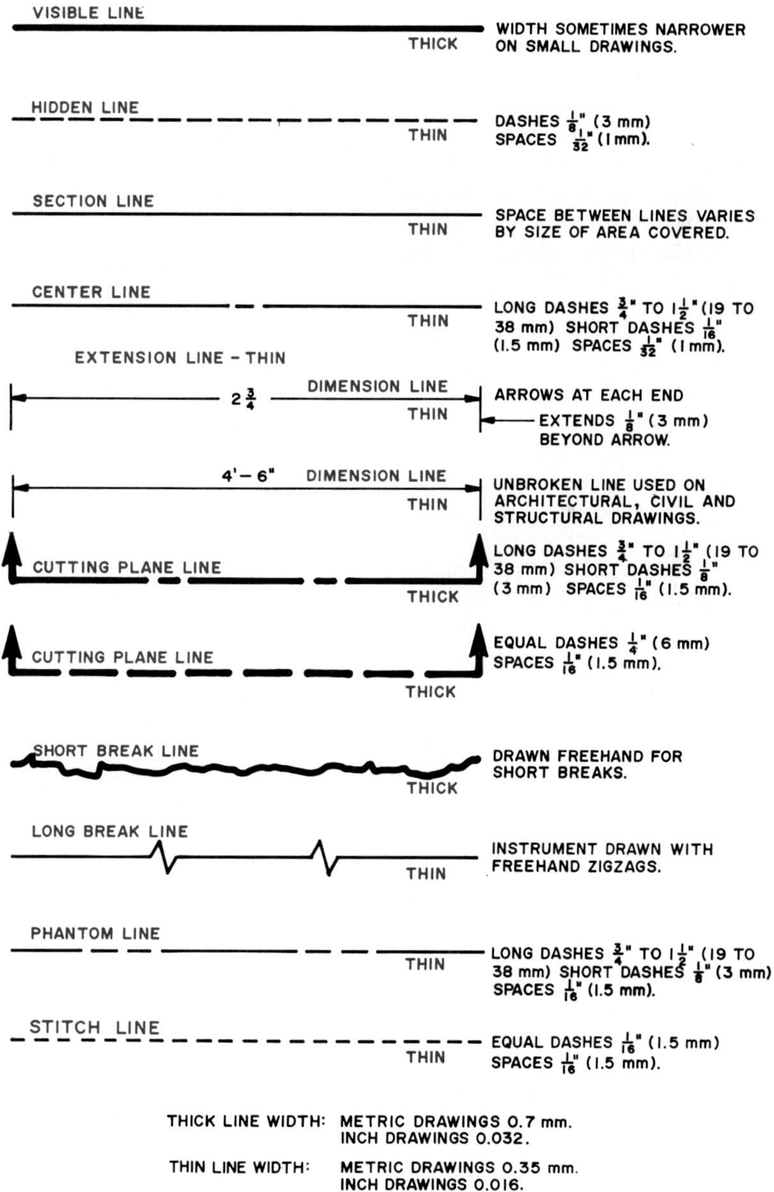

Fig. 3–23. The alphabet of lines and recommended line thicknesses.

Extension lines are thin, solid lines used with dimension lines. They extend to the point on the drawing to which the dimension line refers. Extension lines do not touch the object.

Center lines are used to locate the centers of holes or other circular parts of an object. They are thin lines made of long and short dashes. The long dashes can be from $\frac{3}{4}$ to $1\frac{1}{2}$ inches (19 to 38 mm) long. The length will vary with the drawing and the length of the center line needed. The short dashes are about $\frac{1}{16}$ inch (1.5 mm) long.

Section lines are thin lines used to show a surface that has been cut in a section view. They are drawn parallel and spaced from $\frac{1}{16}$ to $\frac{1}{8}$ inch (1.5 to 3 mm) apart. The spacing varies with the area to be section lined. In large areas you use wide spacing between the lines. The usual angles for drawing section lines are 30, 60, and 45 degrees.

Cutting-plane lines are used to show where a section has been taken. They are thick lines made of either long and short dashes or equal sized dashes. Arrowheads are drawn on their ends to show the direction in which the section was taken.

Break lines are used to show that part of the object has been removed, or broken away. Short break lines are made with a thick solid line drawn freehand. Long break lines are drawn as a thin, solid line with a *Z* symbol inserted in several places along its length.

Phantom lines are used to show alternate positions of parts that can move. If a handle is drawn with visible lines, its rotated position is shown with phantom lines. These thin lines are drawn with one long dash followed by two short dashes.

Stitch lines are used to show sewing or stitching. They are made of short dashes and spaces of equal length.

Drawing Lines

Horizontal Lines

Draw horizontal lines with the top edge of a T-square or parallel straightedge, or the horizontal blade of a drafting machine.

Right-handed people control a T-square with their left hand. Left-handed people control it with their right hand. To use a T-square, hold its head snug against the side of the drawing board and slide it up or down the board as needed. Fig. 3-25.

Right-handed drafters draw horizontal lines from left to right. Left-handed drafters draw in the opposite direction, from right to left.

A
CUTTING PLANE
VISIBLE LINE
A
PHANTOM LINE
LEADER
HIDDEN LINE
Ø 1
SECTION LINING
3
BREAK LINE
CENTER LINE
$2\frac{1}{4}$
SECTION A-A
DIMENSION LINE
EXTENSION LINE

Fig. 3–24. Line symbols used on drawings.

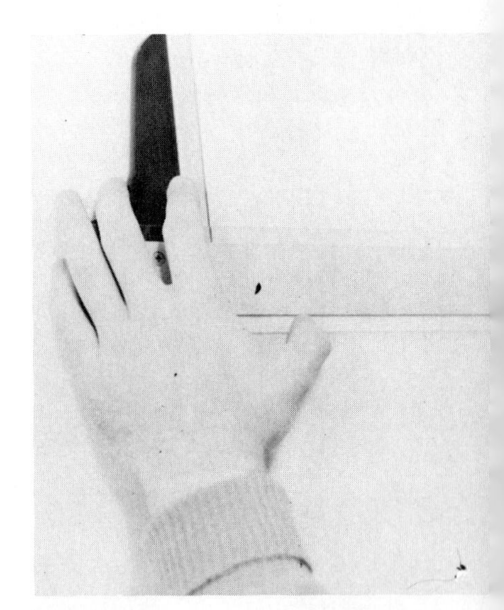

Fig. 3–25. Hold the T-square snug against the edge of the drawing board.

To draw a horizontal line, slant your wood-cased or standard mechanical pencil at an angle of about 60 degrees in the direction the line is to be drawn. Fig. 3-26. As you move your pencil, rotate it between your fingers. Doing so keeps the point conical because the lead wears down evenly on all sides. Draw a horizontal line with your fine-line pencil perpendicular to the paper.

Vertical Lines

You draw vertical lines using triangles. Place a triangle on the top edge of the straightedge. Hold the triangle with your left hand if you are right-handed or with your right hand if you are left-handed.

Draw vertical lines from bottom to top. Fig. 3-27. Left-handed drafters often find it easier to draw from top to bottom. Slant your pencil in the direction the line is to be drawn unless you are using a fine-line mechanical pencil.

Inclined Lines

Draw inclined lines by placing your triangle against the top edge of the straightedge. Right-handed persons draw inclined lines in the directions shown in Figs. 3-28 and 3-29. Left-handed persons often draw in the opposite direction.

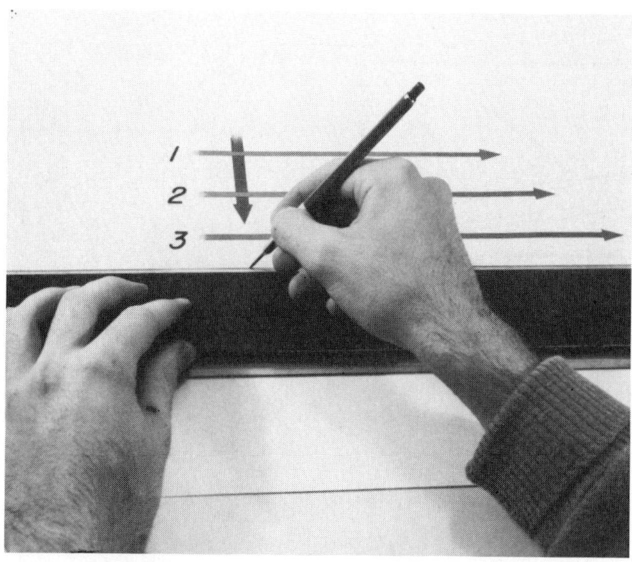

Fig. 3–26. **Draw horizontal lines with your pencil slanted in the direction of the line to be drawn.**

Fig. 3–27. **Draw vertical lines from bottom to top.**

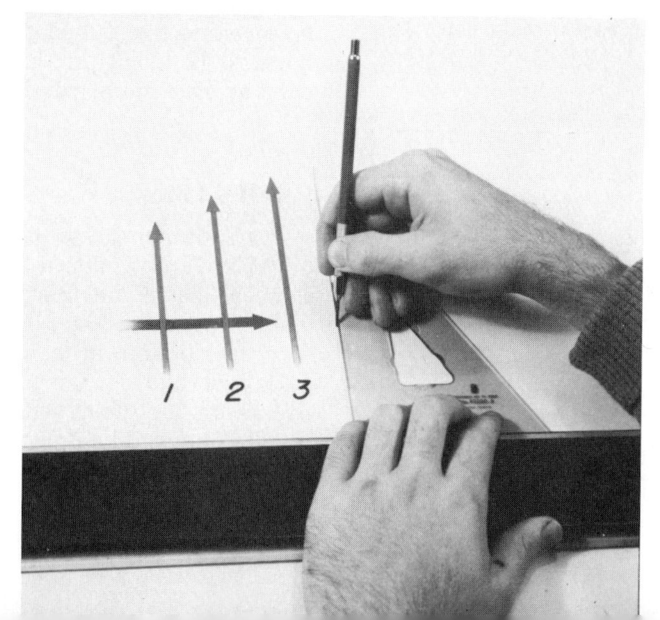

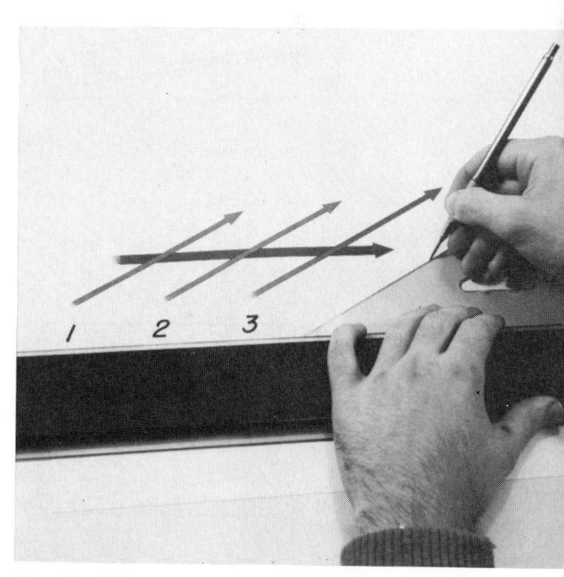

Fig. 3–28. **Draw lines slanting to the right in this direction.**

Fig. 3–29. **Draw lines slanting to the left in this direction.**

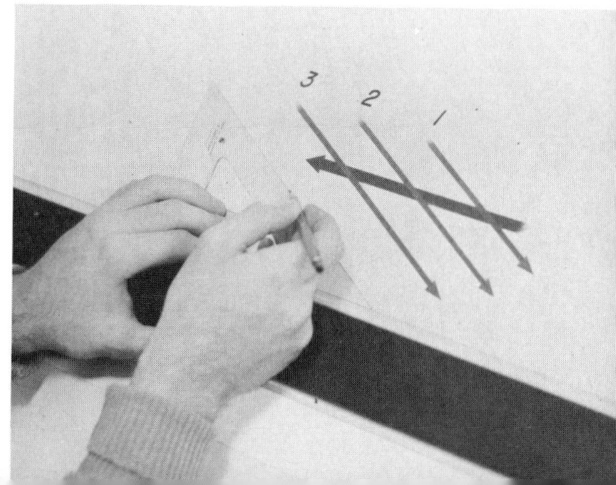

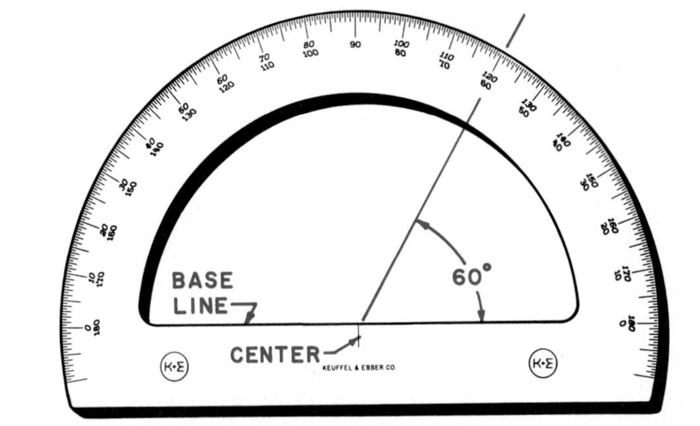

Fig. 3—30. Angles can be measured with a protractor. (Keuffel and Esser Co.)

Angles can be laid out using a protractor. A **protractor** is a flat, semicircular tool that has two sets of degrees marked on it. The degrees run from 0 to 180, as seen in Fig. 3-30. One set is read from right to left and the other from left to right.

To use a protractor, place the center point, found on the bottom edge, at the corner of the angle. Place the base line along one side of the angle. Then read the number of degrees in the angle along the circular scale. Fig. 3-30. Place a mark at the center line and at the degree mark. Use a straightedge to connect these points. Inclined lines can also be drawn using adjustable triangles. See Fig. 3-11.

Perpendicular Lines

To draw a line perpendicular (90 degrees) to a horizontal line, use the edges that meet at a 90-degree angle on a triangle. Refer to Fig. 3-27.

To draw lines perpendicular to inclined lines, use a triangle as shown in Fig. 3-31. Place a triangle with the hypotenuse (long edge) along a drawn line or in position to draw the first line, then draw it. Place the T-square on the bottom of the triangle. Rotate the triangle on the T-square to the adjacent edge. Slide the triangle along the T-square until it reaches the point at which you want to draw the perpendicular.

Parallel Lines

Place a triangle so that one edge is along a drawn line, or draw a line in the position you want. Fig. 3-32. Place a T-square along the bottom of the triangle. Slide the triangle along the T-square to the point where the parallel line is to be drawn. Draw the line.

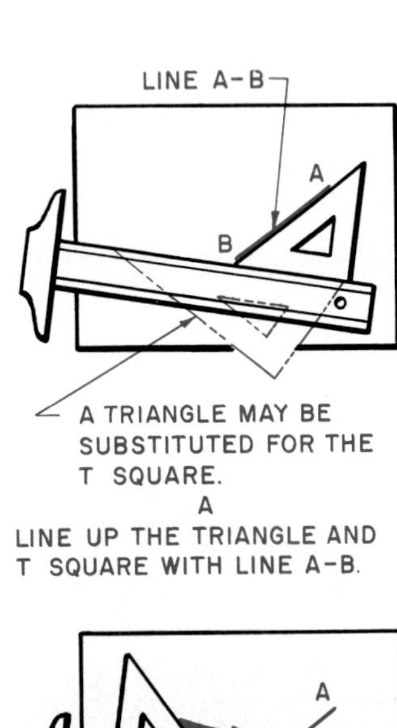

LINE A-B
A
B
A TRIANGLE MAY BE SUBSTITUTED FOR THE T SQUARE.
A
LINE UP THE TRIANGLE AND T SQUARE WITH LINE A-B.

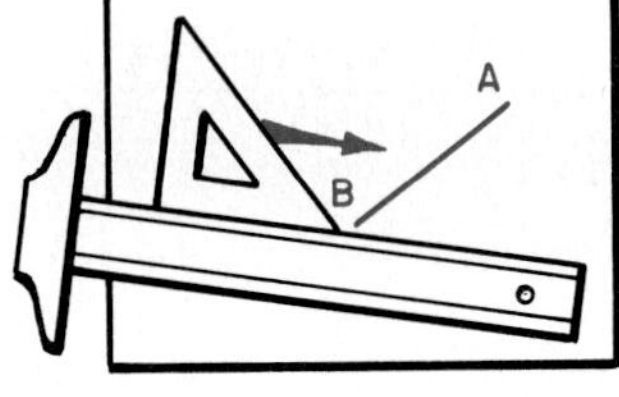

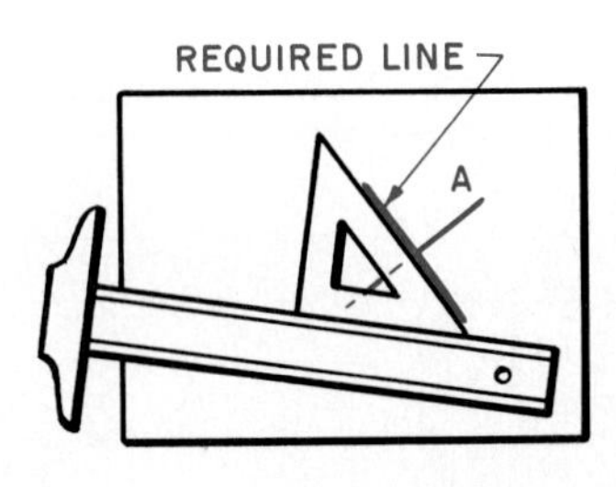

Fig. 3—31. How to draw lines perpendicular to each other.

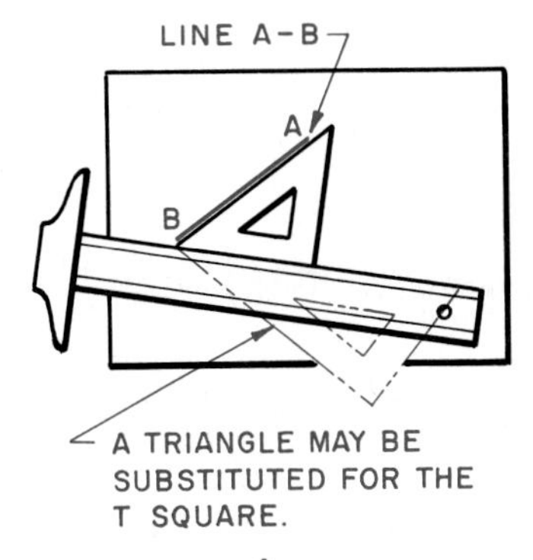

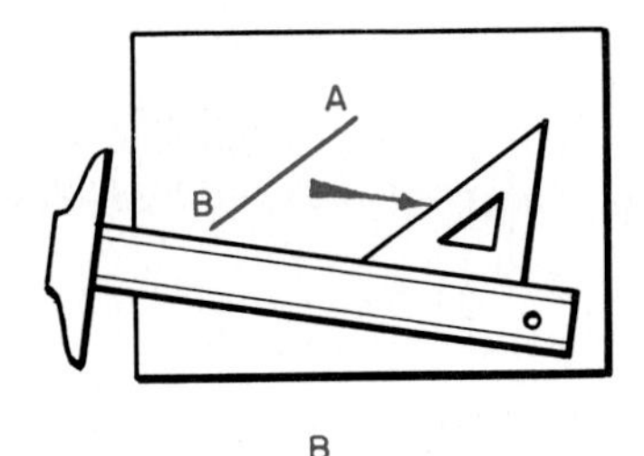

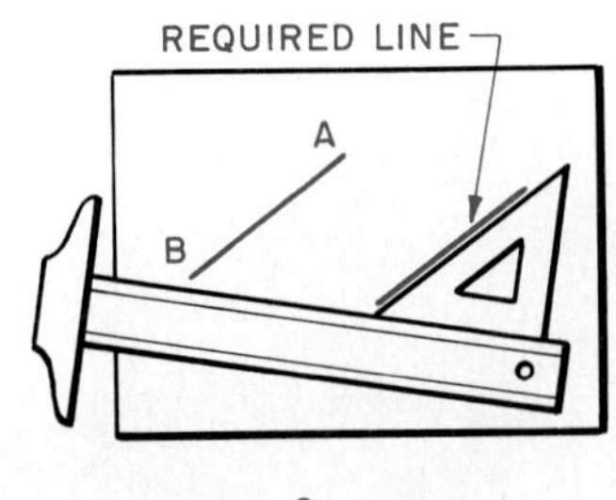

Fig. 3—32. How to draw parallel inclined lines.

Using a Drafting Machine

Right-handed people move a drafting machine head with their left hand. The drafting machine head is moved up, down, and across the board with this hand. Left-handed people use a left-handed drafting machine and hold the head in their right hand.

Use the *horizontal scale* to measure distance and draw horizontal lines. Fig. 3-33A. Use the *vertical scale* to measure distance and draw vertical lines. Fig. 3-33B.

The drafting machine head contains a *vernier scale* marked in degrees to measure inclined lines. Pressing a lever on the head permits the scales to be moved to an inclined position. When the scales are at the correct angle, release the lever to lock the scales in place. A locking lever is available which, when tightened, assures that the angle will not slip. Fig. 3-33C.

To draw lines parallel to each other, lock one scale parallel with the given line. Slide the machine so that the scale is the desired distance away and draw the parallel line. Fig. 3-34.

To draw a line perpendicular to another, rotate the scales so that one is parallel to the given line. Slide the head so that the other scale crosses the given line at the proper point. Draw the perpendicular line. Fig. 3-35.

Keeping Drawings Clean

The finished drawing must be clean. Smudges spoil the appearance, and they will reproduce when copies of the drawing are printed. Follow these steps to help keep your drawings clean:

1. Keep all your tools clean. Your straightedge, triangles, protractor, and templates pick up carbon as they rub over the pencil lines. Wipe your tools fre-

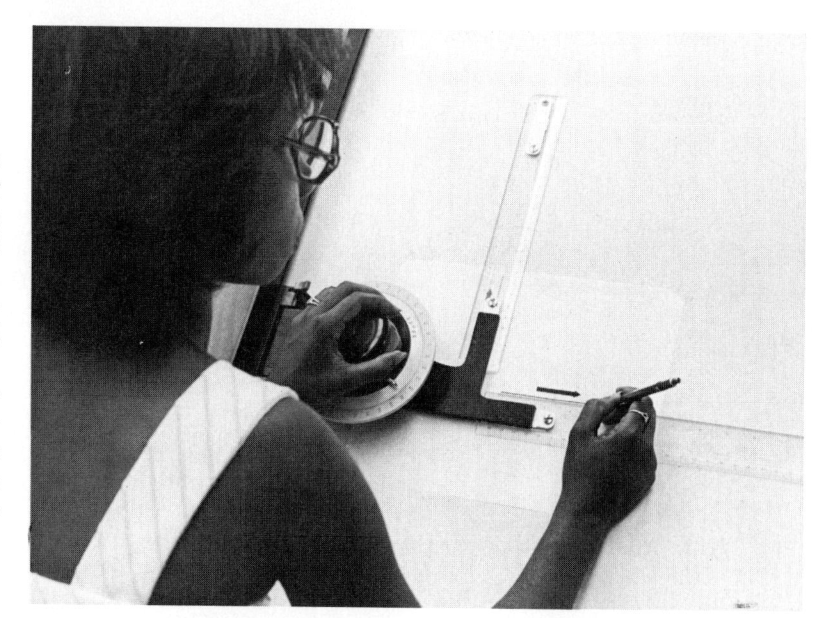

Fig. 3—33A. **Using the horizontal scale on a drafting machine.**

Fig. 3—33B. **Using the vertical scale.**

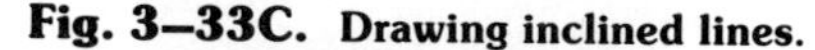

Fig. 3—33C. **Drawing inclined lines.**

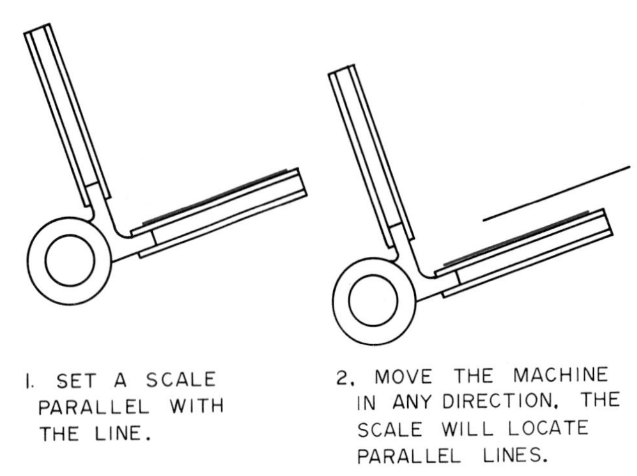

Fig. 3–34. **Drawing parallel lines using a drafting machine.**

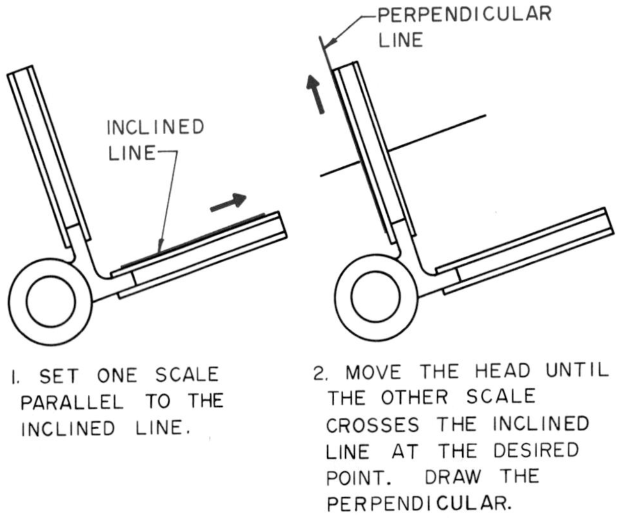

Fig. 3–35. **Drawing a line perpendicular to an inclined line using a drafting machine.**

Fig. 3–36. **A drafter's dusting brush. (Keuffel and Esser Co.)**

Fig. 3–37. **A dry-cleaning pad. (Frederick Post Co.)**

quently with a cloth. If the carbon will not wipe off, wash your tools with a damp cloth.

2. Use a dusting brush to remove eraser crumbs from your drawing. Fig. 3-36. Do not brush away crumbs with your hand because you will smear your pencil lines.

3. To prevent tools from smearing lines that you have drawn, place a clean sheet of paper over those lines.

4. Some drafters use a dry cleaning pad. Fig. 3-37. It is a loosely woven sack filled with eraser crumbs. Rub the pad lightly over the surface of your drawing to remove graphite particles. If you use too much pressure, however, you will lighten your lines. Light lines do not make good prints. Dry cleaning pads must be used with great care.

Scales

A **scale** is used to measure or lay out a line on a drawing in full size, or larger or smaller than full size. The term *scale* also refers to the ratio between the size of a drawing and the actual object. A scale drawing can be larger, smaller, or the same size as the object as long as the proportions are the same. Objects should be drawn full size whenever possible. However, it would be difficult to draw the balance wheel of a watch full size since it is so small. The scale for such a part should be increased greatly. On the other hand, a wheel for a railroad freight car would be drawn smaller than full size. When selecting a scale, remember to allow room for dimensions and notes. This will require that you do some planning and perhaps make a dimensioned, freehand sketch before you begin to draw.

Scales are made in flat and triangular shapes. Figure 3-38 shows the available shapes.

Scales are available that are open divided and fully divided. **Open divided** scales have one fully divided unit at the end. The remainder of the scale's length shows only the main units of measure. The architect's scale, Fig. 3-39, is open divided. A **fully divided** scale has every unit along its length fully subdivided. See the mechanical engineer's scale in Fig. 3-39.

There are several types of scales used in drafting. The mechanical engineer's scale, the architect's scale, and the civil engineer's scale shown in Fig. 3-39 are divided in feet and inches. For metric drawings, use a metric scale.

The Architect's Scale

The architect's scale is used for all types of architectural drawings such as floor plans, foundation plans, elevations, and electrical and mechanical drawings. This scale is also used for other types of drafting including some mechanical drawings. Fig. 3-40.

The architect's scale is divided into units which represent feet and inches. Refer to Fig. 3-39. Each of these units reduces the drawing from full size. The scales used and the ratio of reduction they produce are shown in Fig. 3-41. For example, the scale 3″ = 1′–0″ has a ratio of 1:4 and the drawing is ¼ the true size.

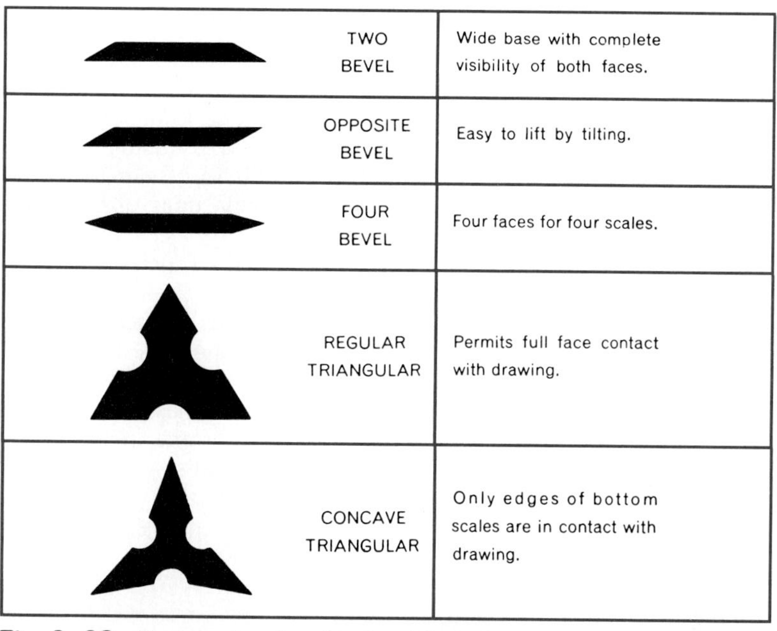

	TWO BEVEL	Wide base with complete visibility of both faces.
	OPPOSITE BEVEL	Easy to lift by tilting.
	FOUR BEVEL	Four faces for four scales.
	REGULAR TRIANGULAR	Permits full face contact with drawing.
	CONCAVE TRIANGULAR	Only edges of bottom scales are in contact with drawing.

Fig. 3–38. Shapes of scales. (Frederick Post Co.)

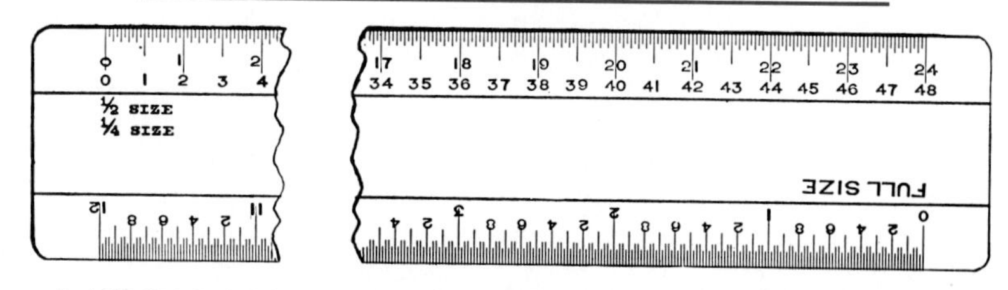

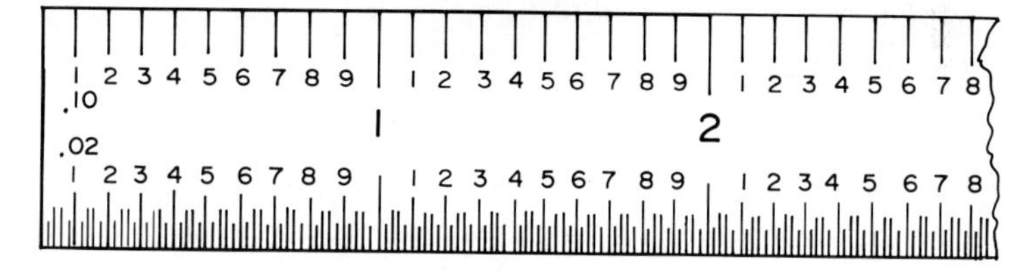

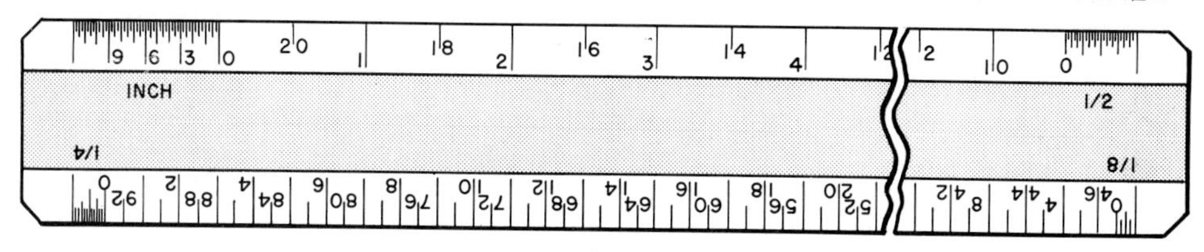

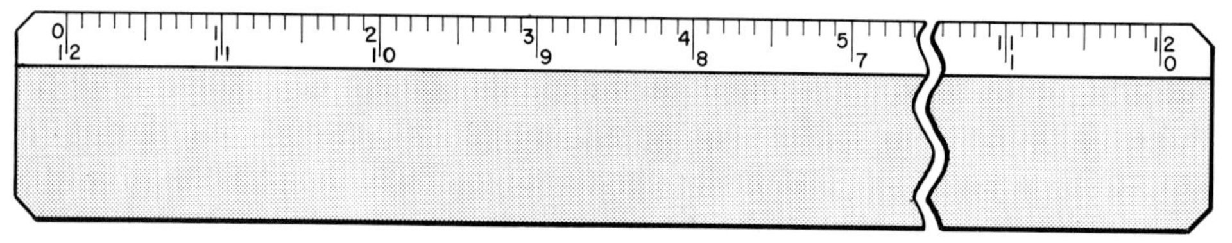

Fig. 3–39. Common types of scales used by drafters. (T. A. Alteneder and Sons)

The distances measured with the architect's scale are in feet and inches. The symbol ′ is used to indicate feet and ″ represents inches. A distance such as four feet, three and one-half inches is lettered 4′–3½″.

When using the architect's scale, the distances in feet are counted on the open divided section beginning at 0. The inches are counted on the fully divided section beginning at 0. In Fig. 3-42, using the ½ scale, you would lay off the distance 3′–5″ by counting 3 feet to the left from the 0 and 5 inches to the right.

When this scale is indicated on a drawing, it is written as SCALE: ½″ = 1′–0″.

The Mechanical Engineer's Scale

The mechanical engineer's scale is divided to make drawings one-eighth size, quarter size, half size, or full size. Refer to Fig. 3-39. The scale units used are the following:

- *Full size* has 1 inch divided into 32nds.
- *Half size* has ½ inch units representing one inch. It is divided into 16ths.
- *Quarter size* has ¼ inch units representing one inch. It is divided into 4ths.
- *One-eighth size* has ⅛ inch units representing one inch. It is divided into 4ths.

These scales use inches and common fractional parts of an inch such as ¼ or ⅛. They are used to draw machine parts and for other general drafting where fractional inches are used.

There are also scales subdivided into decimal parts of an inch such as one-tenth (.1) or one-hundredth (.01). Fig. 3-39. The units used on these scales are:

- *Full size* has 1 inch divided into 50ths (0.02).
- *Half size* has ½ inch divided into 10ths (0.1).
- *Three-eighths size* has ⅜ inch divided into 10ths (0.1).
- *One-quarter size* has ¼ inch divided into 10ths. (0.1).

These scales are used whenever a product is to be drawn and

Fig. 3–40. This architect is laying out a building using a triangular architect's scale.

Fig. 3-41. Scales Available on an Architect's Scale

Scale	Ratio	Size
12″ = 1′–0″	1:1	Full
3″ = 1′–0″	1:4	¼
1½″ = 1′–0″	1:8	⅛
1″ = 1′–0″	1:12	1/12
¾″ = 1′–0″	1:16	1/16
½″ = 1′–0″	1:24	1/24
⅜″ = 1′–0″	1:32	1/32
¼″ = 1′–0″	1:48	1/48
3/16″ = 1′–0″	1:64	1/64
⅛″ = 1′–0″	1:96	1/96
3/32″ = 1′–0″	1:128	1/128

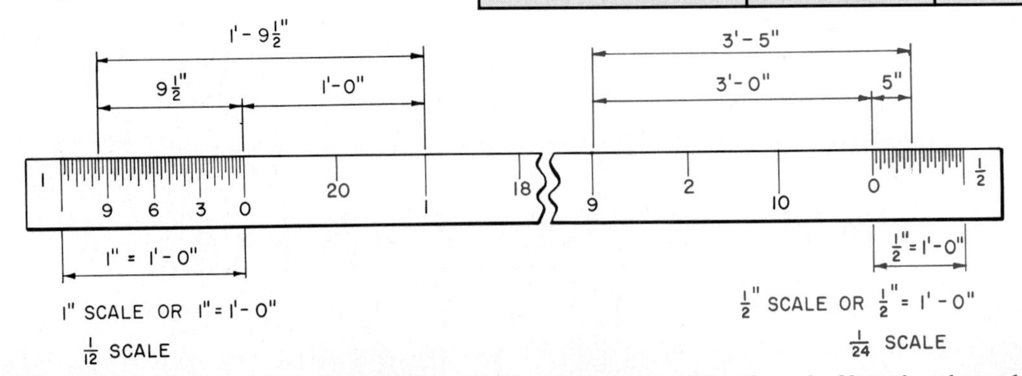

Fig. 3–42. These are the 1″ = 1′ – 0″ and 1/2″ = 1′ –0″ scales on an architect's scale. Note that the scales overlap and that the 9 on the 1″ scale also represents 3 on the 1/2″ scale.

dimensioned in decimal parts of an inch. They are more widely used than the common fraction scale.

When one of these scales is used on a drawing, it is noted as full size, half size, etc. For example, a half-size drawing would have the following note: SCALE: Half Size.

The Civil Engineer's Scale

The civil engineer's scale has inch divisions subdivided into 10, 20, 30, 40, 50, and 60 parts. The scale marked 30 means the inch is divided into 30 parts. Refer to Fig. 3-39.

While this scale is most commonly used to lay out surveys and maps in feet, the one-inch divisions can be used to represent other measurements such as miles or pounds. This scale is also very useful for plotting charts and graphs. Information about scales used on maps is found in Chapter 24, "Mapping."

When showing this scale on a drawing it is written as SCALE: 1″ = 50′, or SCALE: 1″ = 50 lb.

The Metric Scale

The metric scale is divided into millimeters (mm) as shown in Fig. 3-43. Some of the commonly used metric scales are shown in Fig. 3-44. On these scales a distance of one millimeter is used to represent a larger or smaller distance. If one millimeter represents five millimeters, the drawing is reduced at a ratio of 1:5; that is, it will be drawn ⅕ size. Or, if two millimeters on the scale are used to represent one millimeter, the drawing will be enlarged two times and the ratio is 2:1.

CAD Applications:

Voice-Activated CAD

Drafting on a CAD system is usually done with a computer keyboard and one or more hand-held devices, such as a mouse or joystick. Some companies, however, are working on making CAD voice-activated. All a designer then has to do is say a few words into a microphone and the computer does the rest.

Voice recognition CAD works with a special vocabulary. Computer commands hundreds of characters in length can be reduced to one or two words. Each word can have several meanings. All the drafter does is learn the correct vocabulary. Current systems can process vocabularies containing from 250 to 10,000 words.

A voice-activated computer recognizes up to 150 words per minute, which is close to the speed of normal speech. But words cannot be run together as in conversation. Each word must be followed by a short pause.

Voice recognition systems have several disadvantages. They are expensive. The computer can usually recognize only from four to 10 different voices. If more voices are to be recognized, too much space in the computer's memory must be used. Many current systems are not "user friendly." For example, some take as long as 100 hours to set up. Also, words that sound familiar, such as "fine" and "line," are hard for the computer to tell apart.

In spite of these difficulties, voice recognition is worthy of development. Such a system would be helpful to the handicapped. And some drafters find that moving hands and eyes from keyboard, tablet, and other items while trying to concentrate on the screen can be tiring. For such drafters, voice recognition may be just what is needed.

Dragon Systems, Inc.

Several examples of metric scales used to reduce and enlarge drawings are shown in Figs. 3-45 and 3-46. On the 1:2 scale, a distance of one millimeter represents two. On the 1:5 scale, one millimeter represents five millimeters, and so on. When showing the scale on a drawing, it is written as a ratio, such as SCALE: 1:2.

Laying Out a Measurement

To make a measurement, draw a light line in the location of the measurement that is longer than the measurement. Place a scale flat on the paper parallel with and close to the line. Make a short, light dash on the line to mark each end of the measurement. Fig. 3-47.

Drawing Instrument Sets

Figure 3-48 shows a typical set of drawing instruments. There are many different kinds of sets. A basic set will contain a large and small bow compass and a pair of friction dividers.

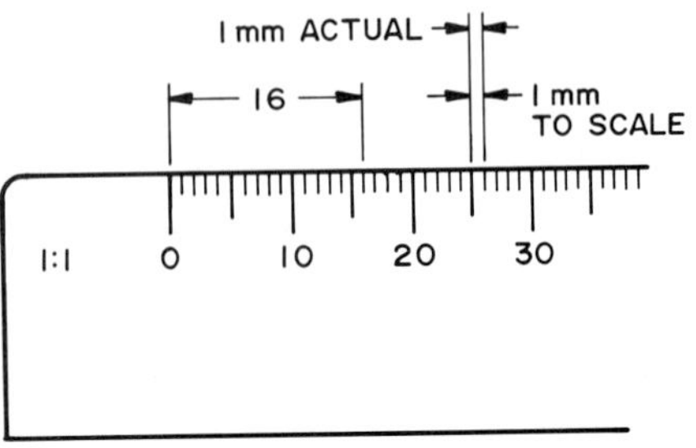

Fig. 3–43. **Divisions on a metric scale.**

Fig. 3-44. Commonly Used Metric Scales

Metric Scale Ratio	Meaning	
1:100	1 mm = 100 mm	(one-hundredth size)
1:50	1 mm = 50 mm	(one-fiftieth size)
1:20	1 mm = 20 mm	(one-twentieth size)
1:10	1 mm = 10 mm	(one-tenth size)
1:5	1 mm = 5 mm	(one-fifth size)
1:2	1 mm = 2 mm	(one-half size)
1:1	1 mm = 1 mm	(full size)
2:1	2 mm = 1 mm	(double size)
5:1	5 mm = 1 mm	(enlarged 5 times)
10:1	10 mm = 1 mm	(enlarged 10 times)
20:1	20 mm = 1 mm	(enlarged 20 times)
50:1	50 mm = 1 mm	(enlarged 50 times)
100:1	100 mm = 1 mm	(enlarged 100 times)

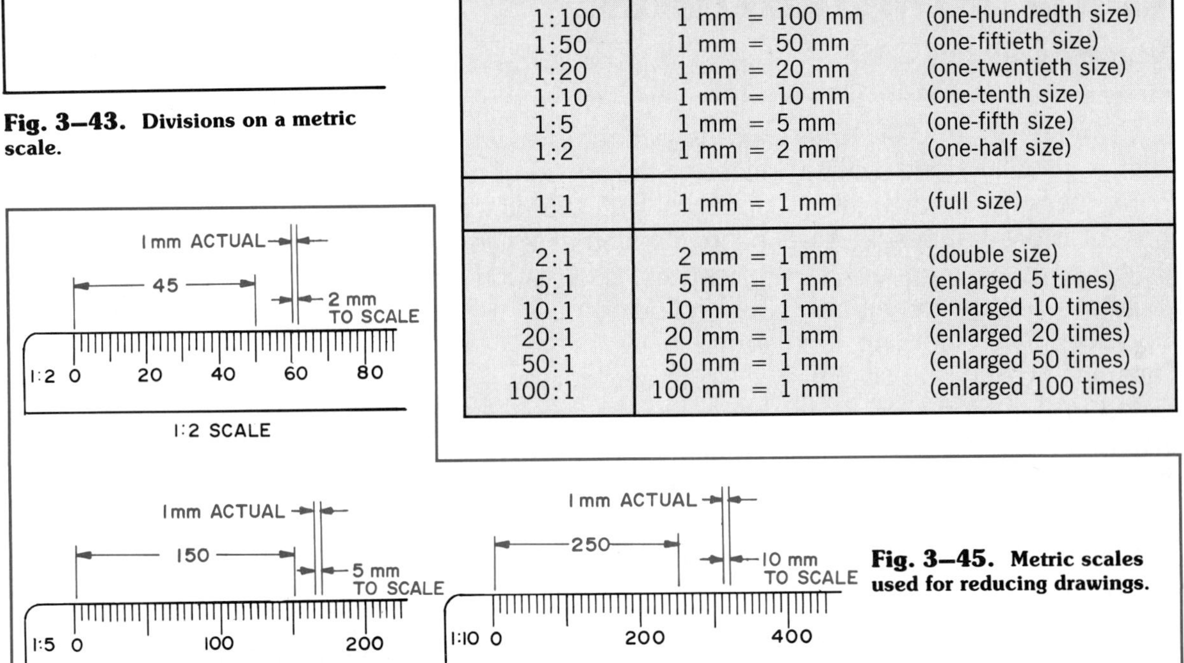

Fig. 3–45. **Metric scales used for reducing drawings.**

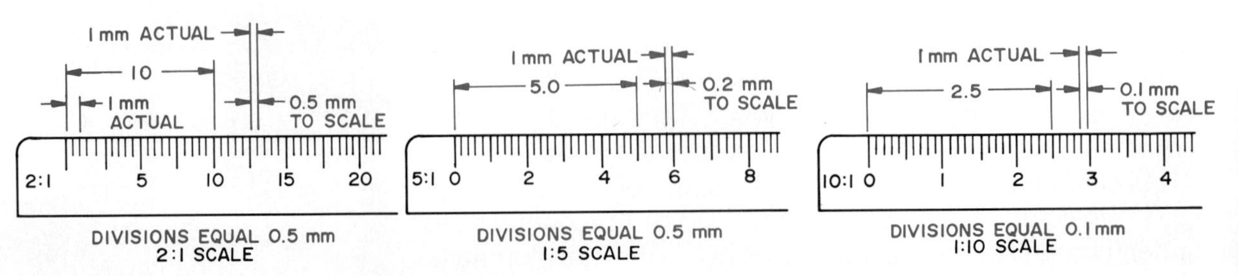

Fig. 3–46. **Metric scales used for enlarging drawings.**

A **compass** has a pin on one leg and attachments that hold either lead or ink on the other leg. It is used to draw circles and arcs. The compass shown in Fig. 3-48 is a bow compass because it has a center wheel adjustment. This wheel is turned to adjust the distance between the compass legs. Compasses without this method of adjustment are called friction compasses.

How to Use a Compass

Measure and mark on your paper the radius of the circle to be drawn. Set the distance between the compass legs to that radius. Place the pin leg on one mark and adjust the other leg until it touches the second mark. Fig. 3-49.

Hold the compass in one hand at its top and turn it clockwise. Tilt it a little in the direction it is moving. Keep enough pressure on the lead point to get a dark line. Fig. 3-50.

Sharpen the lead in your compass on a file or sandpaper pad. Sand the lead on one side only to produce a wedge. Sand the edges lightly to form the final wedge shape. Fig. 3-51.

Fig. 3–47. Using a flat scale. Notice the positions of the pencil when marking the measurement.

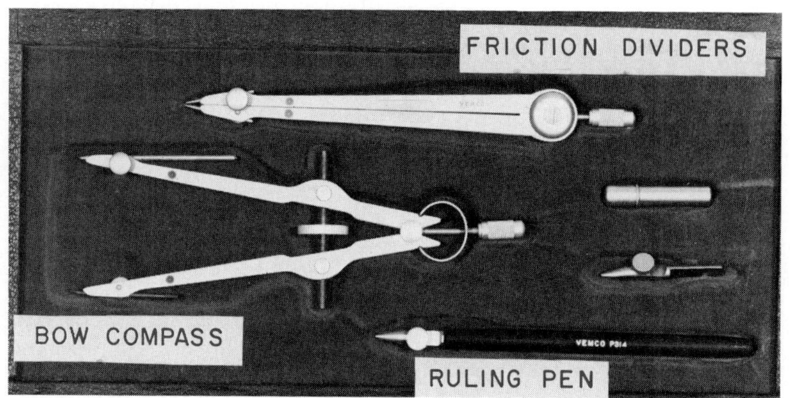

Fig. 3–48. A set of drawing instruments. (V and G Manufacturing Co.)

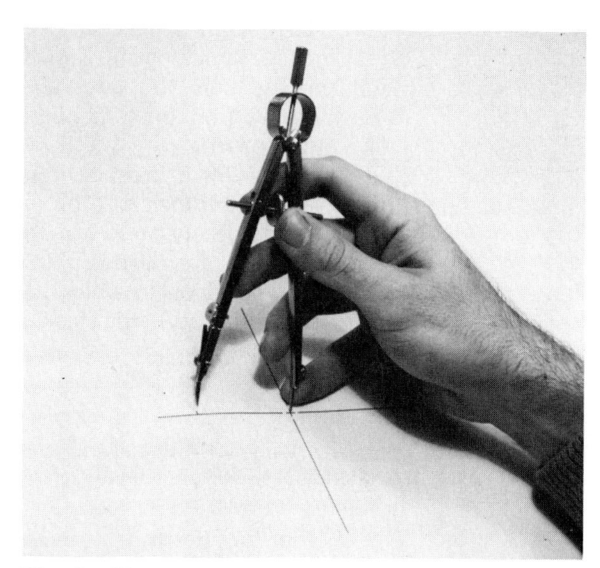

Fig. 3–49. To set the radius on a compass, mark the distance on the drawing. Adjust the compass to this mark.

Fig. 3–50. Swing the compass clockwise. Slant it in the direction it is moving.

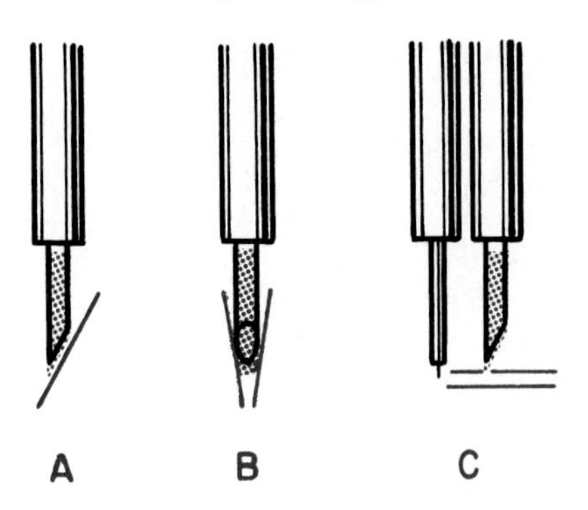

Fig. 3–51. A. Sand a sharp, slanted surface on the lead. B. Sand the edges slightly. C. Set the pin slightly longer than the lead.

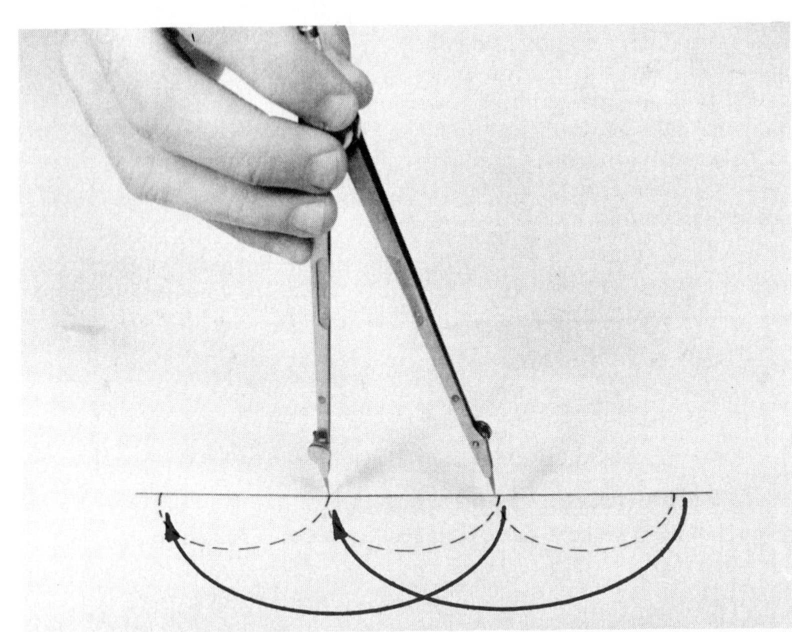

Fig. 3–52. Stepping off equal distances with dividers.

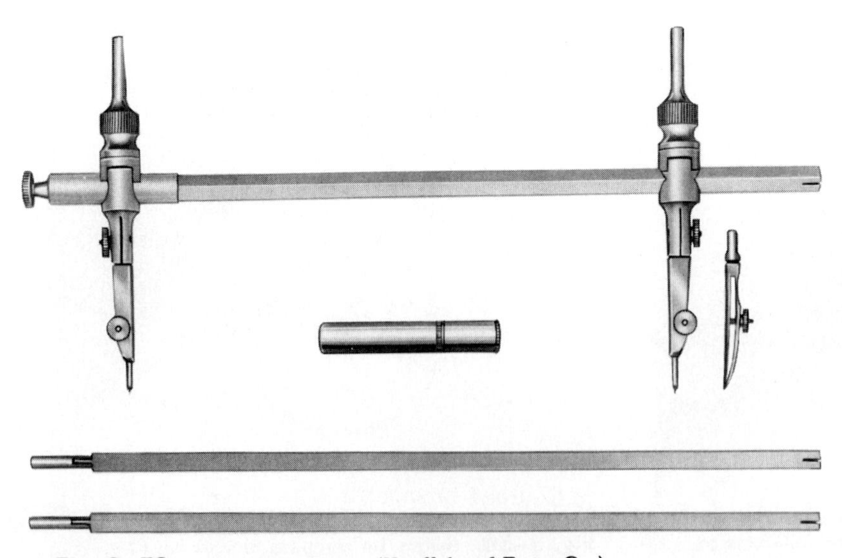

Fig. 3–53. A beam compass. (Keuffel and Esser Co.)

How to Use Dividers

Dividers look much like a compass except that they have pin points in both legs. Dividers are made in both center wheel and friction types. The common uses for dividers are as follows:

1. To mark off equal spaces. This is done by spacing the legs the distance wanted. Place one point at the beginning of the line. Swing the other point until it touches the line and mark a point there. Then, keeping the second leg in place, swing the first leg around and mark a second point. Fig. 3-52. Continue to repeat this stepping process until the desired number of equal spaces has been marked.

2. To divide a line into equal parts. Draw the line to be divided and mark the beginning and end of the line. Estimate the distance wanted. For example, if a line is to be divided into five equal distances, set the dividers at what you think is one-fifth the distance. Place one point at the end of the line. Step along the line five times. If the leg of the dividers does not reach the end of the line, open the divider legs a small amount. If it reaches beyond the end, close them some. Step off the line five times again. Repeat the process until the five steps reach exactly from one end of the line to the other.

3. To transfer a distance from one place to another. Set the dividers to the distance you want to transfer. Place the divider points at the new position to which the distance is to be transferred. Make a mark where the legs touch the new line.

How to Use a Beam Compass

Another tool frequently used is the beam compass. Fig. 3-53. It is used to draw large circles and arcs. One leg has a pin point. The other has lead or inking pen attachments. A pin point can be placed in the second leg to make a large divider. It takes two hands to swing an arc with this compass.

Templates

A *template* is a thin, flat piece of plastic with openings of different shapes cut into it. Fig. 3-54. To use a template, place your pencil inside the opening of the correct shape and size and trace its outline. Most templates are made to allow for the thickness of the pencil lead.

Using templates speeds drafting time considerably. Circle templates have largely replaced using a compass. Ellipse templates have reduced the need to follow the many steps to lay out an ellipse. Lettering templates are especially useful for large letters and titles. A good set of templates is essential for the manual drafter.

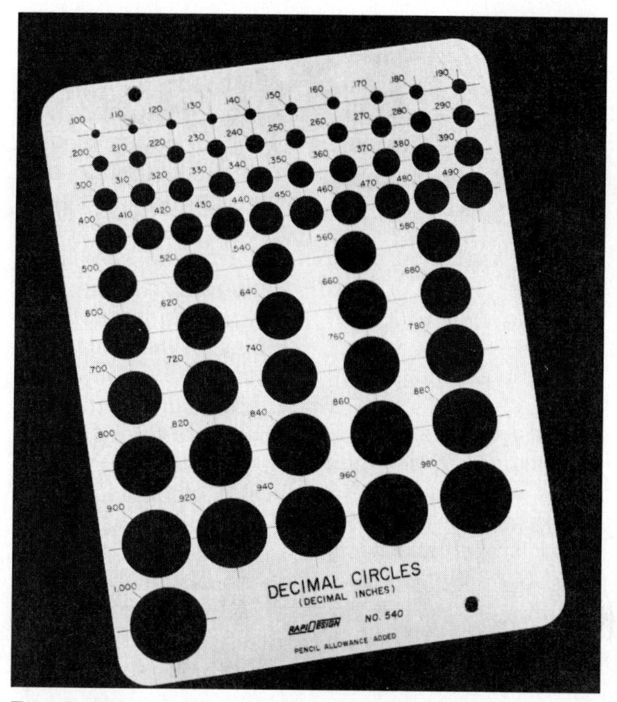

Fig. 3–54. **A circle template. (Rapidesign, Inc.)**

Irregular Curves

Irregular curves are instruments used to draw any noncircular curve. A noncircular curve is one consisting of tangent arcs of varying radii. Such a curve cannot be drawn with a compass. Noncircular curves are found on graphs, charts, involutes (some gear teeth), spirals, and ellipses.

Curves are available in many different shapes and classifications. For example, there are mechanical engineer's curves, rule curves, ship's curves, and body sweeps. Each is used in a special field of drafting, as some of their names imply. Some of the most common irregular curves used in drafting are shown in Fig. 3-55.

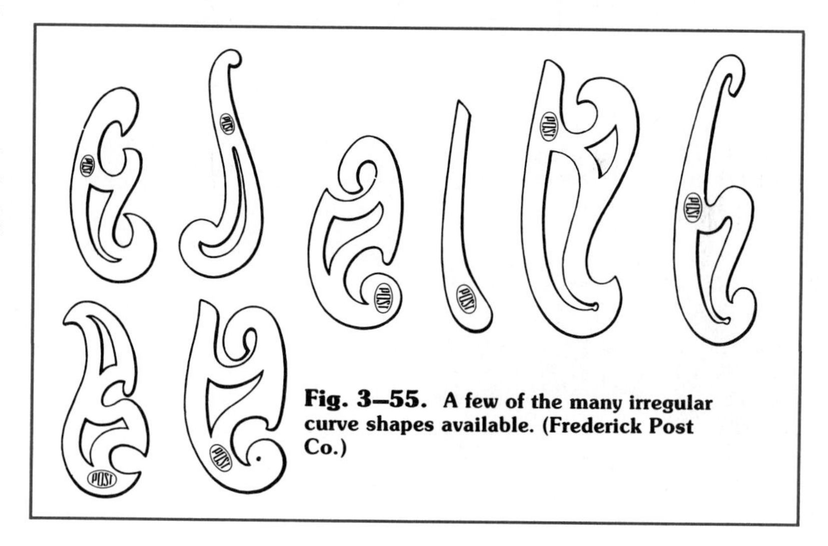

Fig. 3–55. **A few of the many irregular curve shapes available. (Frederick Post Co.)**

Another tool used to draw irregular curves is the *flexible (adjustable) curve.* Fig. 3-56. This tool can be bent to any desired shape. Flexible curves may be made from metal, plastic, or rubber.

Fig. 3–56. **Two types of adjustable curves. (Keuffel and Esser Co.)**

Drawing Irregular Curves

To draw a curve, first plot (locate) points that fall on that curve. Then, move your irregular curve until a section of it lines up through three or more of the points on the curve. Draw that portion of the curve. Move your irregular curve to a new position and join several more points. Fig. 3-57.

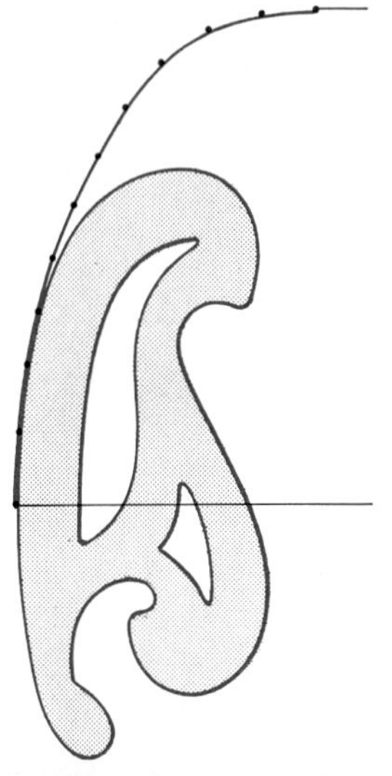

Fig. 3–57. Fit the irregular curve to three or more points at a time and draw that part of the curve.

Inking Drawings

Ink used for drawing on vellum, cloth, and artist's board is a waterproof India ink. A special ink for drawing on plastic film is available. Since ink dries rapidly, keep the lid on its container when not in use. Be sure to cover the points of your inking pens, or they will quickly become clogged with dried ink.

Lettering and lines of all types are drawn with the **technical pen.** It has a removable tip upon which an ink cartridge fits. Fig. 3-58. To fill the pen, unscrew the tip, pull the cartridge away from the tip and fill it with ink. Replace the tip on the cartridge and screw the unit back into the pen handle. Hold the technical pen perpendicular to the drawing surface to draw lines. Fig. 3-59.

You make different width lines by using pens with different size points. The range of point sizes is shown in Fig. 3-60.

When using a technical pen, you will often need to shake it up and down over a piece of scrap paper to start the ink flowing. You will hear a clicking sound as you shake the pen. This is the noise of a flow-regulating device sliding up and down. It is this device that will cause the ink to flow. If you do not hear clicking, the ink has hardened. You must then clean the point by soaking it in a solvent made for that purpose. You could also use an ultrasonic point cleaner if one is available.

Ink will often run under your straightedge. You can prevent this from happening by using a straightedge that has a recessed edge. You might also tape pieces of cardboard to the bottom of your tools to raise them from the drawing surface. Fig. 3-61. If your ink smears, let it dry before trying to erase it. Use an eraser

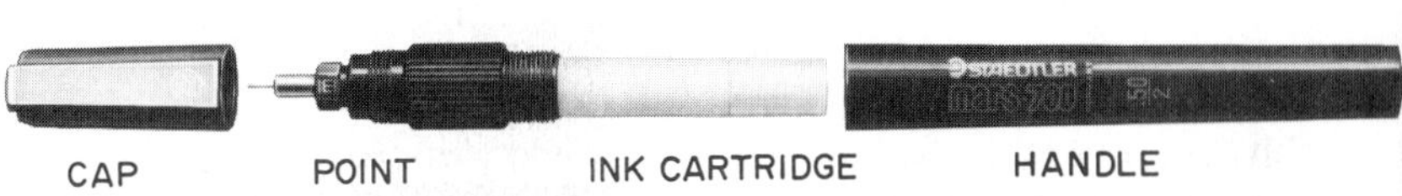

Fig. 3–58. The parts of a technical pen. (J. S. Staedtler, Inc.)

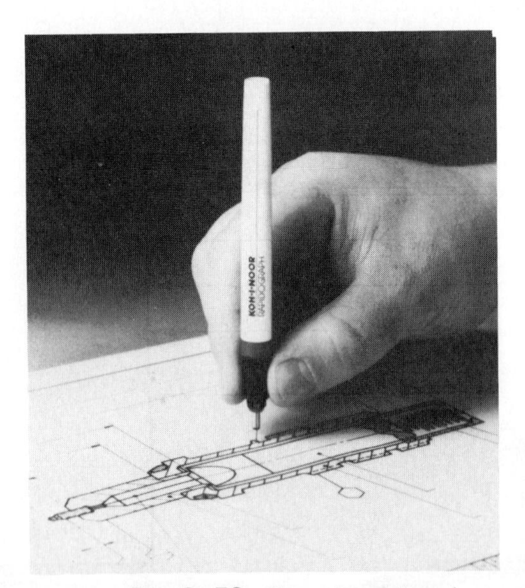

Fig. 3–59. Keep the technical pen perpendicular to the drawing surface. (Koh-I-Noor, Inc.)

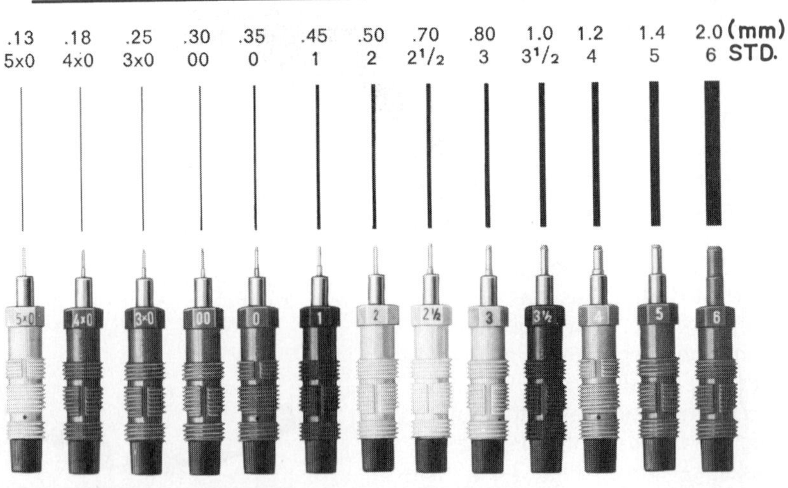

Fig. 3–60. There is a wide range of point sizes available for technical pens. (Koh-I-Noor, Inc.)

designed to remove ink. Do not use an eraser that is coarse or gritty because it will damage your drawing surface.

After a line is inked, it will be shiny until the ink dries. Never try to dry ink by blotting it. Let it dry naturally or, if you must speed up the drying, use an electric hair dryer. Be careful not to blow wet ink across your drawing with a high speed setting. When the inked line appears dull, it is dry.

To obtain the best results when inking a drawing, follow these steps:

1. If you are tracing a pencil line, always center the inked line over the pencil line. Fig. 3-62.
2. Ink center lines first.
3. Next, ink all arcs and circles. Start with the largest; then, reduce the radius of your compass and ink the next largest, and so on.
4. Ink all horizontal visible lines.
5. Ink all vertical visible lines.
6. Ink all inclined visible lines.
7. Ink hidden lines.
8. Ink section lines, extension lines, and dimension lines.
9. Ink arrowheads.
10. Ink all lettering for dimensions, notes, and titles.

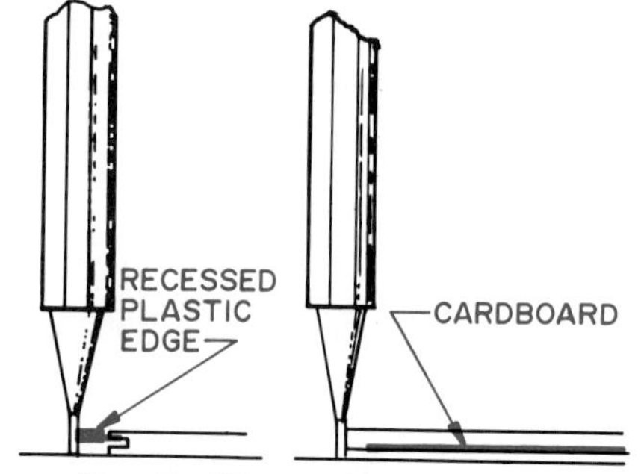

Fig. 3–61. **A recessed edge keeps ink from running under the straightedge.**

Drawing Sheets

The drawing paper you use may be a heavy paper that has a white, cream, or green color. This kind of paper is not used in industry because drawings made on such paper cannot be reproduced by blueprint or whiteprint machines.

Most drawing paper used in industry is smooth on one side and rough on the other. The smooth side is for ink drawings and the rough side for pencil drawings.

Tracing Paper and Vellum

Tracing paper is a thin, untreated, translucent paper. Treated paper has a transparentizing agent applied. Such paper is called tracing **vellum**. Vellums are made from 100 percent pure white rag stock and do not discolor with age. They withstand erasing without leaving marks, so they are good for both pencil and ink drawings. Vellum is tough and can stand a lot of handling.

Tracing Cloth

Tracing cloth is a fabric that has been transparentized. It is tough and strong. Either ink or pencil can be used on cloth. The dull side is used for drawing. Because tracing cloth stretches and shrinks with changes in temperature and humidity, it is not used in industry as much as it used to be.

Polyester Drafting Film

Polyester drafting film is a tough, translucent plastic that is almost impossible to tear. It withstands much erasing. Film changes very little in size as humidity and temperature change. It is therefore especially good for drawings that require a high degree of accuracy. You can use ink or pencil on film and even a typewriter for notes and other lettering. Standard thicknesses for film are .003, .004, .005, and .007 inch.

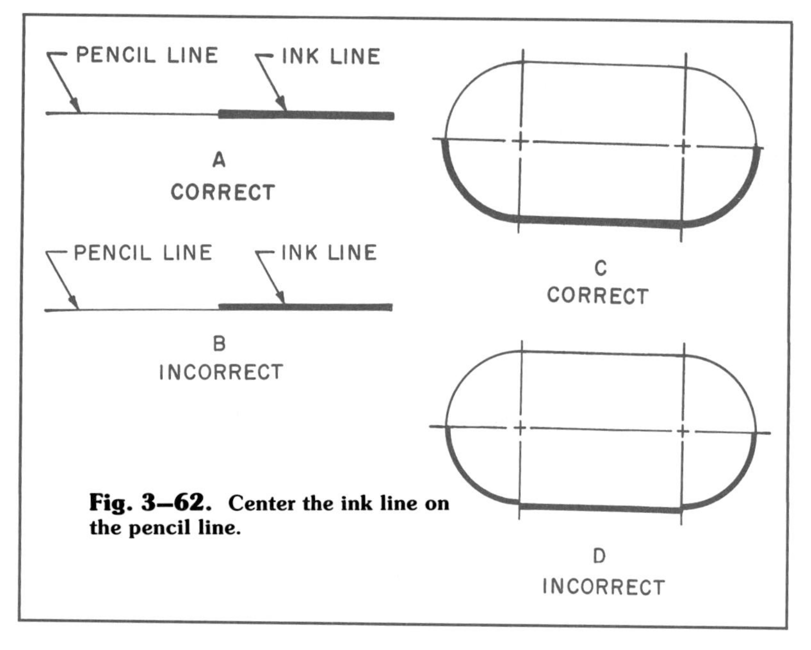

Fig. 3–62. **Center the ink line on the pencil line.**

Sheet Sizes

Drafting paper, tracing paper, vellum, and polyester film are available in both sheets and rolls. Standard sheet sizes are shown in Fig. 3-63. Roll stock varies in width depending upon the material. The most common widths are 30, 36, and 42 inches (762, 914, and 1067 mm). The most common roll lengths are 20 yards (18.29 meters [m]) and 50 yards (45.72 m).

Sheet Layout and Title Blocks

The use of borders and **title blocks** is often set by the drafting standards of a company. As a general practice, borders such as those shown in Fig. 3-64 are recommended. On architectural drawings, a 1-inch border is always drawn on the left side of the sheet. This much space is needed so that the sheets can be bound together as a set.

Fig. 3-63A. U.S. Customary Sheet Sizes

Type	Size in Inches Series One	Series Two
A	8½ × 11	9 × 12
B	11 × 17	12 × 18
C	17 × 22	18 × 24
D	22 × 34	24 × 36
E	34 × 44	36 × 48

Fig. 3-63B. International Standards Organization Sheet Sizes

Type	Size in Millimeters
A4	210 × 297
A3	297 × 420
A2	420 × 594
A1	594 × 841
A0	841 × 1189

The design of the title block depends upon the needs of the company. A typical title block, shown in Fig. 3-65, may contain most or all of the following items:

1. Name and address of the company.
2. Name of the part or parts.
3. Name of the project, machine, or construction.
4. Scale.
5. Tolerances used on the drawing.
6. Names or initials of the drafter, checker, and engineer who worked on the drawing.
7. Date the drawing was made or approved.
8. Drawing number that fits into the numbering system used by the company to file and retrieve the drawing.
9. General notes such as heat treatment or material data.

Fig. 3-64. Recommended Border Sizes

Sheet Size	Border
Metric	
A4, A3	6 mm all sides
A2	10 mm all sides
A1, A0	12 mm all sides
U.S. Customary	
A, B	¼ inch all sides
C	⅜ inch all sides
D, E	½ inch all sides

If the drawing is dimensioned in metric units, that fact is noted directly above or beside the title block.

The title block is generally located in the lower right-hand corner of the drawing. On architectural drawings the title block is often located along the right-hand side. Fig. 3-66.

Assembly drawings include a list of all the parts and identifying data. This parts list is usually located directly above the title block.

Many companies have their title blocks and borders printed on their drawing sheets. Other companies use a printed title block that has an adhesive on one side. These title blocks are fastened to the drawing sheet in the appropriate place.

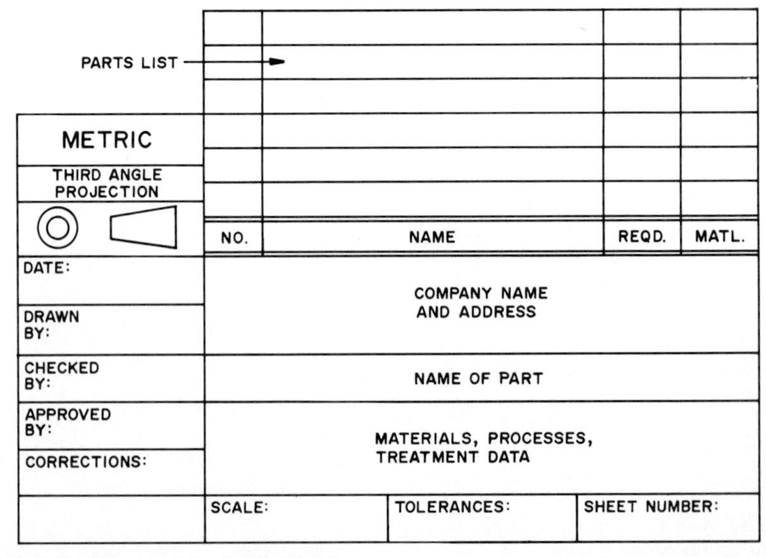

Fig. 3—65. A typical title block.

Lettering

Most drawings contain much lettering in the form of dimensions and notes. These must be easily read and large enough to reproduce. Recommended lettering heights on both U.S. customary (inch) and metric drawings are shown in Fig. 3-67. These sizes are large enough for reproduction by any method in use, including microfilm.

The most commonly used lettering style on engineering drawings is single stroke, uppercase (capital) Gothic. This style of lettering can be either vertical as shown in Fig. 3-68, or inclined, as shown in Fig. 3-69. Gothic lettering is easily done and easy to read. Lowercase letters are seldom used except on maps and some architectural drawings.

Usually a company will use all vertical or all inclined lettering. They are not mixed on a drawing. Vertical letters are easier to read but harder for some drafters to form. The slightest variation from the vertical is easily noticed. You should learn to letter well either way.

Clear lettering, with each letter carefully formed, is essential to an acceptable drawing. The information given in the dimensions and notes is just as important as any other part of the drawing.

Fig. 3-67. Recommended Lettering Heights for Engineering Drawings

U.S. Customary	
Size of Drawing	**Letter Height (inch)**
8.5 × 11 (A)	.125
11 × 17 (B)	.125
17 × 22 (C)	.125
22 × 34 (D)	.156
34 × 44 (E)	.156
Titles	.250
Metric	**Letter Height (mm)**
A4, A3	3.5
A2	4.0
A1, A0	4.5
Titles	7.0

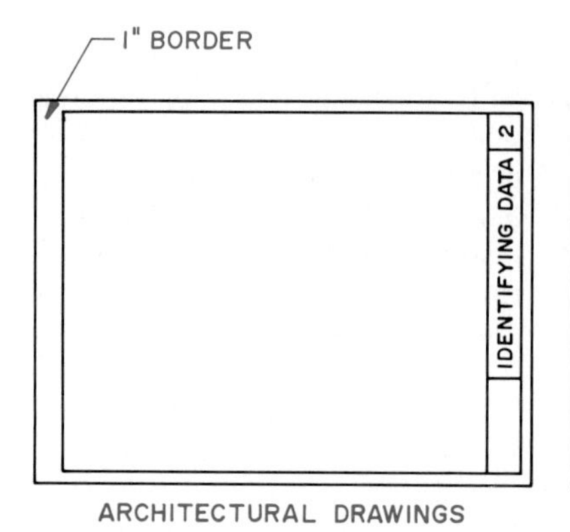

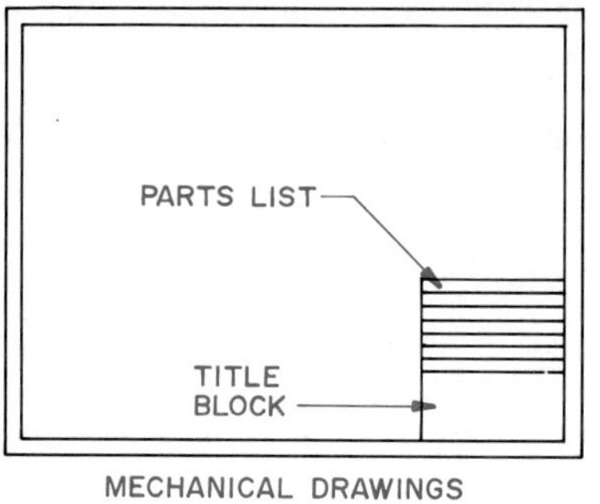

Fig. 3—66. **Typical locations for title blocks.**

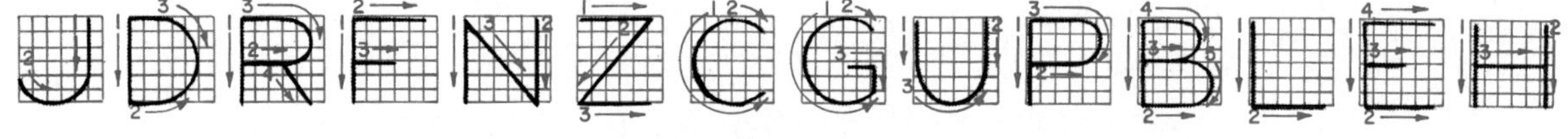

THESE LETTERS ARE 5/6 AS WIDE AS THEY ARE HIGH.

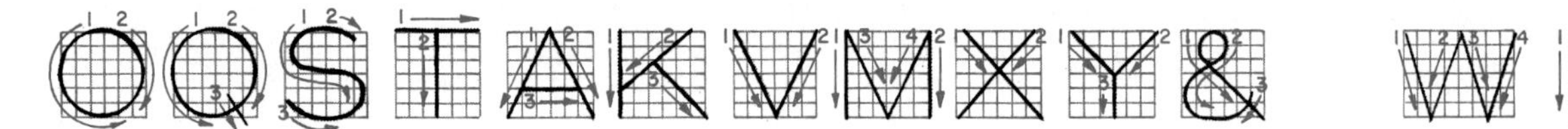

THESE LETTERS ARE AS WIDE AS THEY ARE HIGH.

THIS W IS WIDER THAN IT IS HIGH.

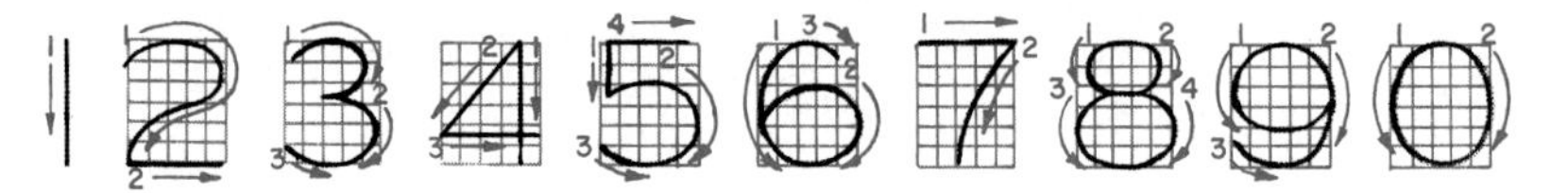

NUMERALS ARE 5/6 AS WIDE AS THEY ARE HIGH.

Fig. 3—68. **These are vertical uppercase Gothic letters and numbers.**

Forming the Letters

Study the letters shown in Figs. 3-68 and 3-69. Notice that some letters are wider than others. Keep this in mind as you form each letter. Some of the letters are 5/6 as wide as they are high. Others are as wide as they are high. One letter is wider than it is high. Which letter is this? No matter what size you make your letters, they are kept in this same proportion as shown. Thus, the higher you make a letter, the wider it must become. Notice that all numerals are 5/6 as wide as they are high.

Figures 3-68 and 3-69 also show the order in which each stroke of each letter is made. These orders have been found to be the easiest way to form the letters while keeping their proportions.

Spacing Lettering

If your lettering is to appear pleasing, the space between each letter must *appear* to be equal. However, the space must not actually be equal. Since letters have different shapes, some must be closer to the next letter than others. Fig. 3-70. The space between letters with vertical strokes, such as H and I, should be larger than the space between letters with open spaces, such as A or V. You judge the amount of space between letters by eye.

The space between words is equal to one letter of the alphabet. Often the letter O is lightly sketched between words to help keep the proper distance. Fig. 3-71.

Guidelines for Lettering

A good drafter always draws horizontal guidelines for lettering. These are spaced according to the height of the letter desired. Fig. 3-72. The letters should touch each guideline without crossing any.

Draw guidelines very lightly. They need only be dark enough for you to see. If made light enough, they need not be erased after you finish lettering. Vertical or inclined guidelines can also be drawn. These are spaced randomly. Fig. 3-72.

Fractions are made as shown in Fig. 3-73. A fraction has the total height of two full numbers. The numbers do not touch the bar between them. This means they are drawn slightly smaller than normal. Draw guidelines for all fractions.

You can use a lettering guide to draw guidelines. Figure 3-74 shows one kind of lettering guide. Notice the numbers two through ten on the bottom of the wheel. They show the height of the lettering in 32nds of an inch. In the illustration, the number eight is touching the vertical mark below the wheel. This means the guidelines will be 8/32 inch (1/4 inch) high. Guides with metric measurements are also available.

To use this kind of lettering guide, place a pencil in one hole and slide the lettering guide along a straightedge to draw a light horizontal guideline. Then choose the next guideline hole and repeat the process.

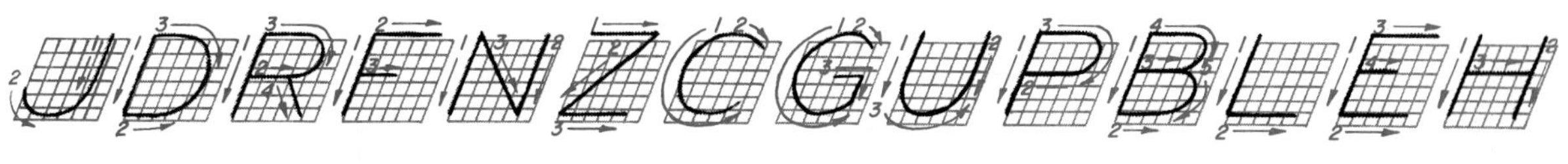

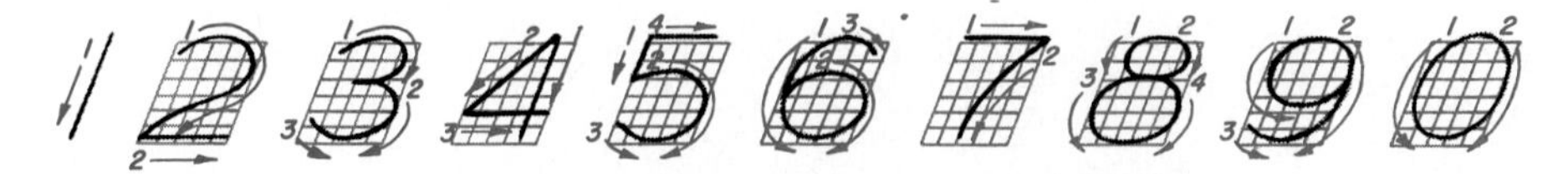

Fig. 3–69. These are inclined uppercase Gothic letters and numbers.

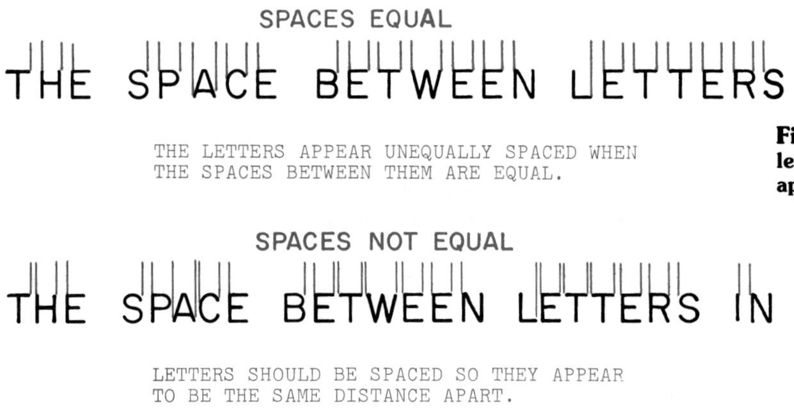

SPACES EQUAL

THE SPACE BETWEEN LETTERS

THE LETTERS APPEAR UNEQUALLY SPACED WHEN THE SPACES BETWEEN THEM ARE EQUAL.

SPACES NOT EQUAL

THE SPACE BETWEEN LETTERS IN

LETTERS SHOULD BE SPACED SO THEY APPEAR TO BE THE SAME DISTANCE APART.

Fig. 3–70. The spacing between letters in a word is important to the appearance of the drawing.

ONE LETTER SPACE BETWEEN WORDS.

THEOSPACEOBETWEENOWORDSOIS

EQUALOTOOONEOLETTER.

Fig. 3–71. How to space between words on a drawing.

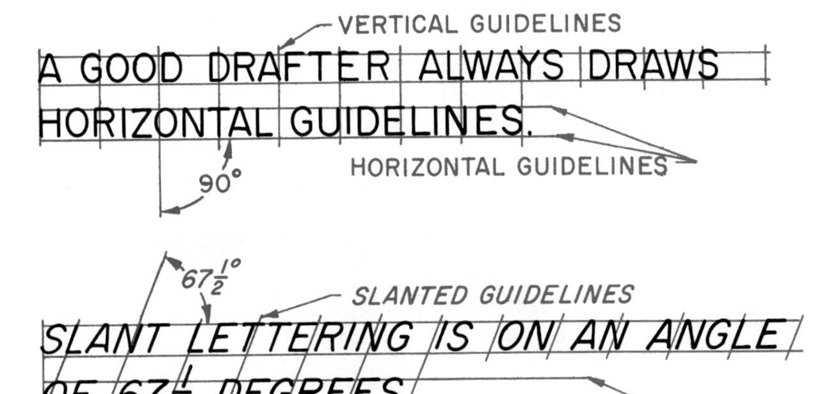

Fig. 3–72. Use guidelines when lettering.

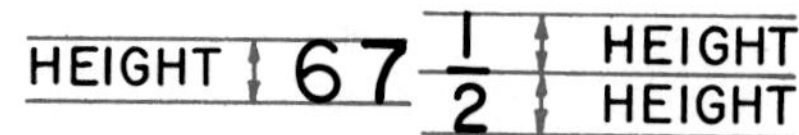

Fig. 3–73. How to letter fractions.

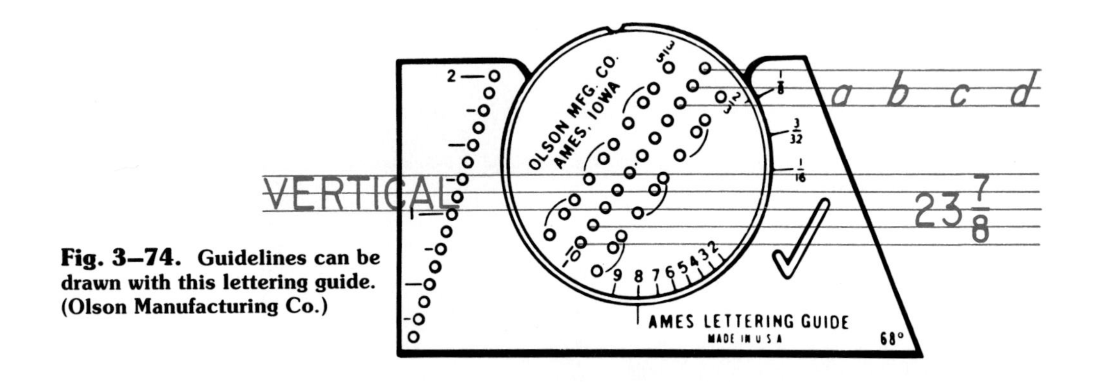

Fig. 3–74. Guidelines can be drawn with this lettering guide. (Olson Manufacturing Co.)

Vertical guidelines can be drawn with the lettering guide. One side has a 90-degree angle and the opposite side has a 68-degree angle. This last angle is correct for inclined lettering. Fig. 3-75.

Lettering Techniques

The height of lettering on most drawings is 1/8 inch (3 mm). That means that fractions are 1/4 inch (6 mm). Some items in the title block, such as the drawing title and number, are lettered larger.

Use a sharp 2H pencil for lettering. Place your arm in a comfortable position on the drawing board. If you draw on a board instead of a drafting table top, you might find it comfortable to turn your board on an angle. With a relaxed grip on your pencil you will make a straighter line.

To prevent smearing your drawing while you letter, place a clean sheet of paper over it. This will keep your hand and arm from rubbing over the drawing surface.

Lettering Instruments

Lettering instruments are a mechanical means of forming letters. There are several types available. One type has a plastic template with the letters sunk into it. Fig. 3-76. A tool called a scriber is used with the template. One leg of the scriber has a metal point that fits into the grooves of the letters. Another leg holds lead or a pen point. As the metal point traces a letter on the template,

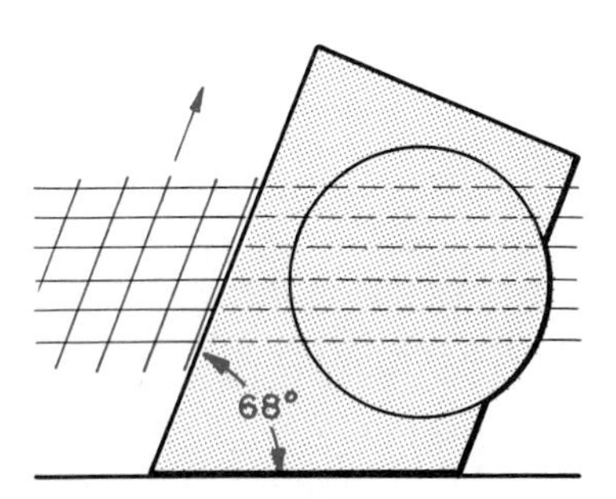

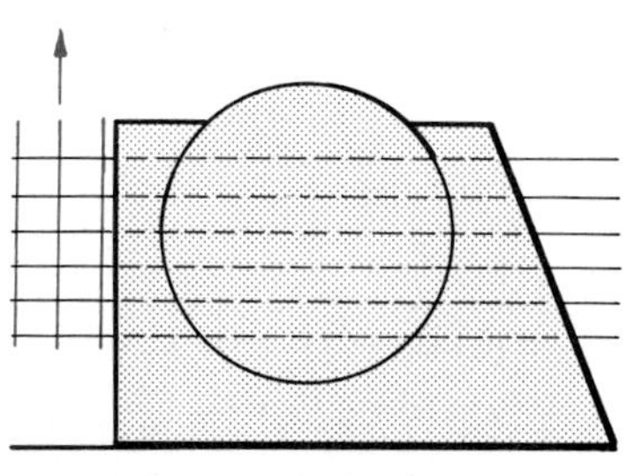

Fig. 3–75. Drawing slanted and vertical guidelines with the Ames lettering instrument.

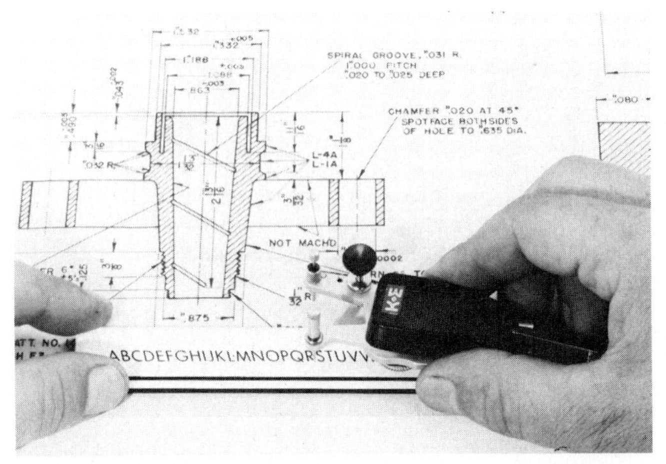

Fig. 3–76. This lettering instrument uses a template to form the letters. It can make vertical and inclined letters. (Keuffel and Esser Co.)

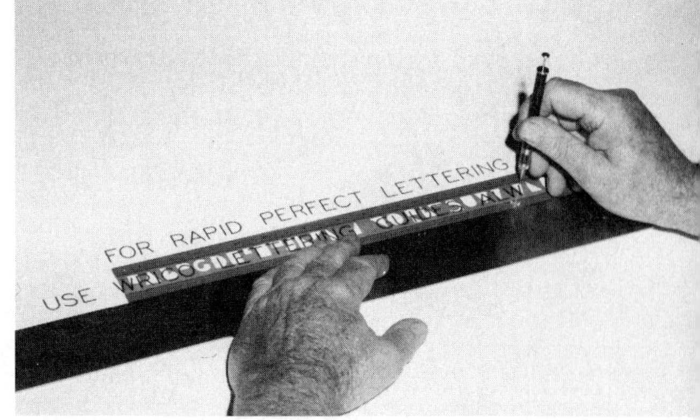

Fig. 3–77. This lettering instrument used by drafters has letters cut out of a plastic template. A technical pen is used to ink the letters. (Wrico Lettering Guide)

Fig. 3–78. This typewriter will handle drawings of any size because of its open-ended carriage. (Diagram Corporation)

the other point draws the letter on your drawing. Holding the template on the top edge of a straightedge keeps the letters straight.

This instrument has several different sizes of templates. The width of the letters depends upon their height. Inclined letters can be drawn by adjusting the legs of the scriber.

Another lettering device is a plastic template with the letters cut out as shown in Fig. 3-77. You trace along the edges of these letters with a pen. To form each word, you slide the template back and forth along the edge of a straightedge.

A typewriter that has a carriage with a long, open end is used for typing information on drawings. Fig. 3-78. A drawing is inserted in the same manner as paper in a regular typewriter. The typewriter provides dark, uniform letters, and a large number of notes and schedules can be completed quickly. Typewritten letters are easier to read than hand-drawn letters.

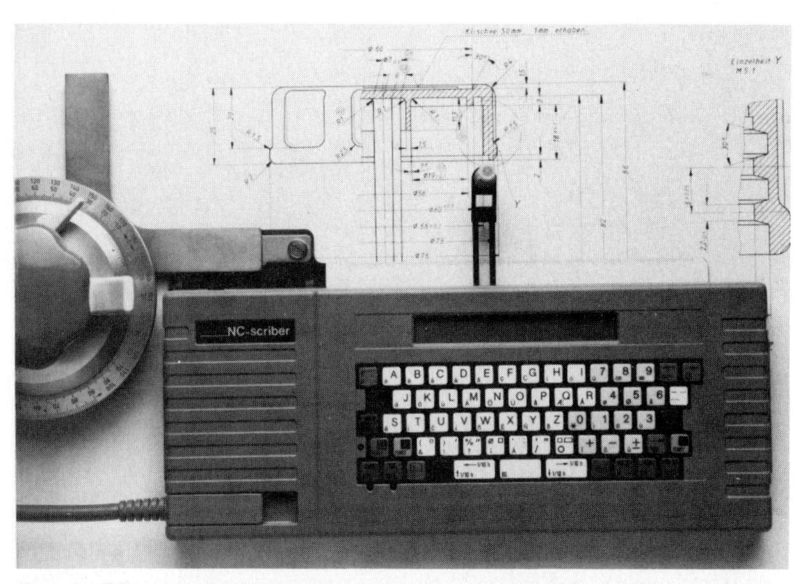

Fig. 3–79. This NC-scriber has memory in an electronic control unit. It uses cassettes to letter notes and dimensions and draw standard symbols used in various engineering fields. (Koh-I-Noor, Inc.)

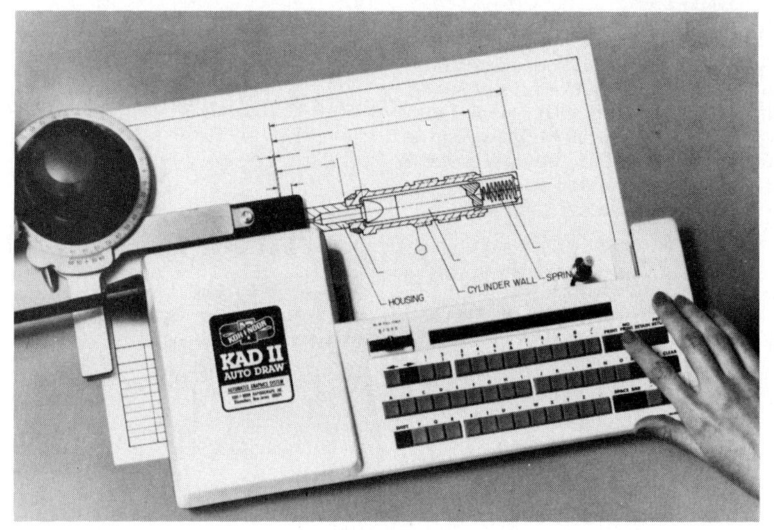

Fig. 3–80. This electronic plotter can be attached to the arm of a drafting machine or used independently. (Koh-I-Noor, Inc.)

Another machine that provides precision lettering, dimensioning, and symbol drawing is shown in Fig. 3-79. It has an electronic memory that enables it to store several thousand characters. The keyboard is mounted on a drafting machine and is connected to a separate electronic control unit. The control unit has twin cassette slots. They make it possible to add both lettering and symbols to drawings. Cassettes are available to make mechanical, electrical, and architectural drawings. This NC-scriber greatly speeds lettering and other routine entries needed on a drawing.

Another electronically controlled letter- and symbol-generating instrument is shown in Fig. 3-80. It is designed for use directly on the drawing board. This unit is able to produce electronically a wide range of lettering styles, symbols, and designs on engineering and architectural drawings. Fig. 3-81. It can also draw symbols and designs for many technical areas using electronic template modules.

Drawings created on a computer may be output on a printer or plotter. In either case, the computer guides the printer/plotter to form the letters and dimensions. Fig. 3-82.

Pressure-Sensitive Letters

Pressure-sensitive letters are available in sheets. The letters have an adhesive on the back. They are placed on a drawing as shown in Fig. 3-83. Examples of several different styles of lettering are shown in Fig. 3-84.

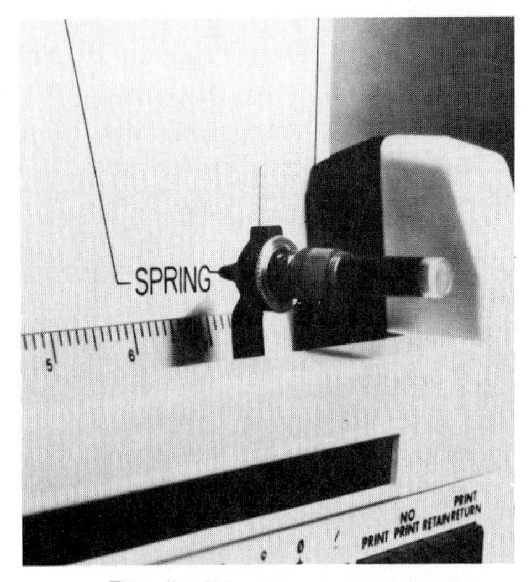

Fig. 3–81. The electronic plotter uses a pen whose cartridge holds a large ink supply. (Koh-I-Noor, Inc.)

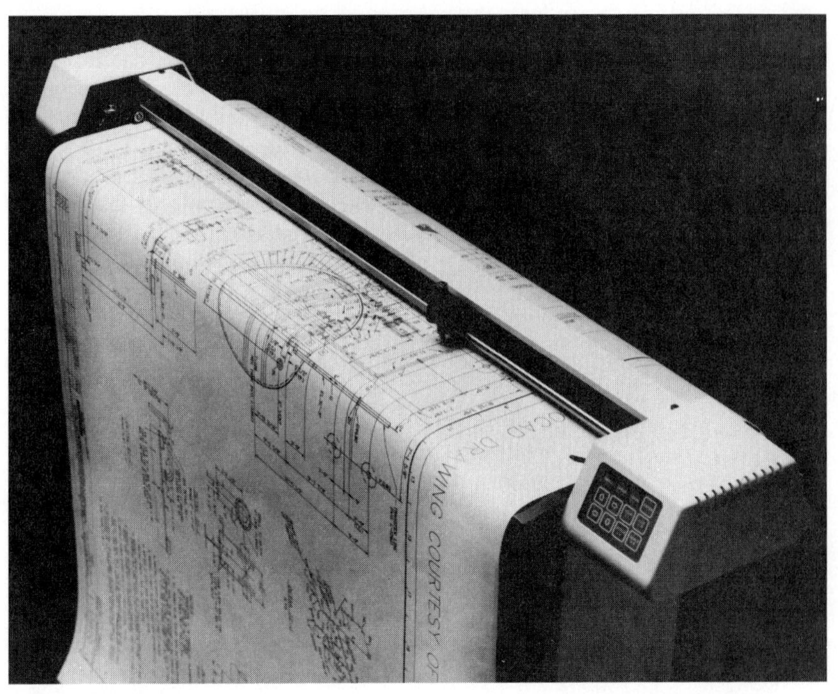

Fig. 3–82. This plotter is producing a computer-generated drawing on paper.

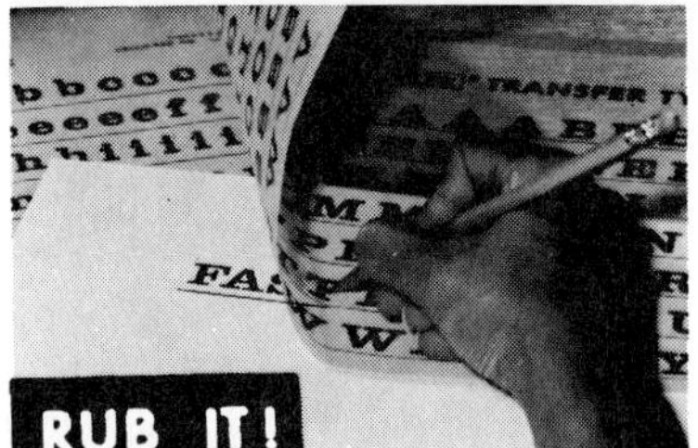

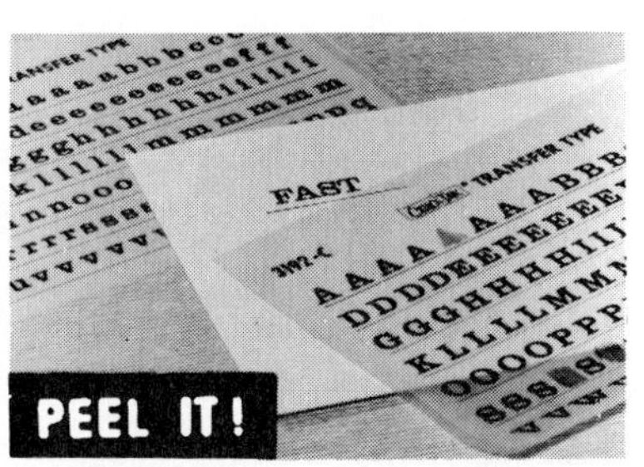

Fig. 3–83. Rule a pencil line. Remove the protective backing and place the desired letter on the pencil line. Rub the letter with a pencil or burnisher. Confine the rubbing to the letter area only. Carefully peel off the transfer sheet, making sure the transfer is complete. (Alvin and Co., Inc.)

Airport Bl

Helvetica Me

Cloister Black

Commercial Scri

Bauhaus He

Caslon Old St

Century Expan

Fig. 3–84. Quick transfer letters are sold in many styles and sizes. (Artype, Inc.)

Chapter Review

Build Your Vocabulary

You should understand and use the following terms as part of your working vocabulary. Write a brief explanation of what each means.

drafting tape
drafting dots
T-square
parallel straightedge
triangles
adjustable triangle
drafting machines
visible lines
hidden lines
dimension lines
extension lines
center lines
section lines
cutting-plane lines
break lines
phantom lines
stitch lines
protractor
scale
open divided scales
fully divided scales
compass
dividers
template
irregular curves
vellum
tracing cloth
polyester drafting film
title block

Sharpen Your Math Skills

1. If the scale on the drafting machine is set on a 15° angle, and you rotate it to a 140° angle, how many degrees did it pass through?

2. If you divide a 90° angle into three equal parts, what size is each part?

3. If you draw a line 10′ long to the scale 1/4″ = 1′–0″, what is the actual length of the line in inches?

4. If you draw a line 500 millimeters long to the scale 1:2, what is the actual length of the line in millimeters?

5. What radius would you use to draw a circle with a 5″ diameter?

6. If you divide a line 9″ long into 4 equal parts, what length is each part? Give the answer in common fractions and decimal fractions.

7. If you cut a D-size sheet (Fig. 3-63) of drawing paper into four equal size sheets, what size are the sheets?

8. If you draw a horizontal and a vertical line in the center of an A2-size sheet (Fig. 3-63) of drawing paper, what size is each of the areas produced?

Study Problems—Directions

The problems that follow will give you the chance to use some of the drafting tools you read about in this chapter. Do the problems that are assigned to you by your instructor.

1. On an A or A4 size drawing sheet, draw a border as recommended in this chapter. Then draw and letter one of the title blocks shown in **Fig. P3-1.**

2. Draw the template shown in **Fig. P3-2**. This problem will help you learn to lay out a drawing on your drawing sheet. Follow these steps to draw it full size on an A or A4 size sheet:

a. Prepare a border on 8½ × 11 inch paper. This will give you a working area of 8 × 10½ inches.

b. Center the template on the sheet.
(1) First find the horizontal location. To do this, subtract the template width, 6 inches, from the working area width, 10½ inches. Divide the resulting 4½ inches by 2 to get 2¼ inches for each side.
(2) Next find the vertical location. Subtract the height of the template, 6 inches, from the height of the working area, 8 inches. This leaves 2 inches. Divide by 2 to get a 1-inch space top and bottom. See **Fig. P3-2** at A.

c. Lightly block in the sides of the template as shown at B. Draw horizontal lines with the horizontal scale of your drafting machine, your parallel straightedge, or your T-square. Draw vertical lines with the vertical scale of your drafting machine or with a triangle.

d. Use your scale to locate the details of the template as shown at C.

e. Block in these as shown at D. Draw very light lines.

f. Erase unneeded lines and darken the remaining visible lines. Since these are visible lines, they must be drawn thick. See E.

g. Letter your name in the lower right corner of the drawing in letters ⅛ inch high.

3. Make instrument drawings of **Figs. P3-3** through **P3-11.** These drawings are dimensioned using fractional inches, decimal inches, feet and inches, and millimeters. Make the drawings to the scale indicated in the captions. Place each drawing on an A or A4 size sheet without a title block. Letter your name in the lower right corner. Make all visible lines thick. Do not dimension the drawing unless directed to do so by your instructor.

4. Make instrument drawings of the one-view drawings in **Figs. P3-12** through **P3-17.** Place each drawing on an A or A4 size sheet and use the scale indicated in the caption. Making these drawings will give you experience working with circles and arcs and drawing thin lines. Locate all center lines. Do not dimension unless directed to do so by your instructor.

5. Make an instrument drawing of the survey shown in **Fig. P3-18.** Letter the distances on each side. Use the civil engineer's scale of 1″ = 40.0′. Use a B or A3 size sheet. Start at closure point A and lay out the drawing clockwise. Record the angles and the length of the closure side.

6. Make instrument drawings of the NASA space station antennas in **Figs. P3-19** and **P3-20**. Draw to a scale of ¼″ = 1′—0″.

Study Problems

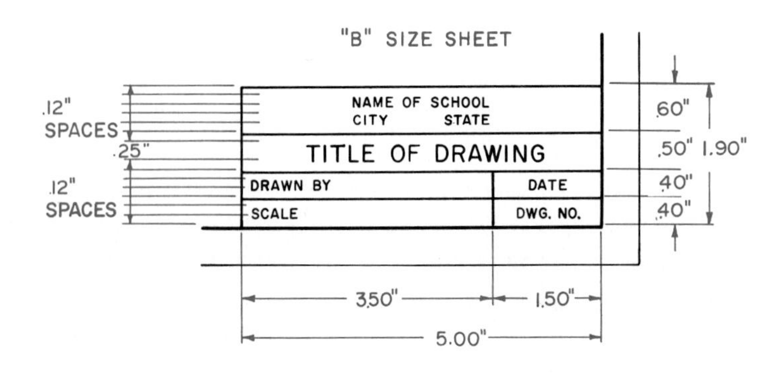

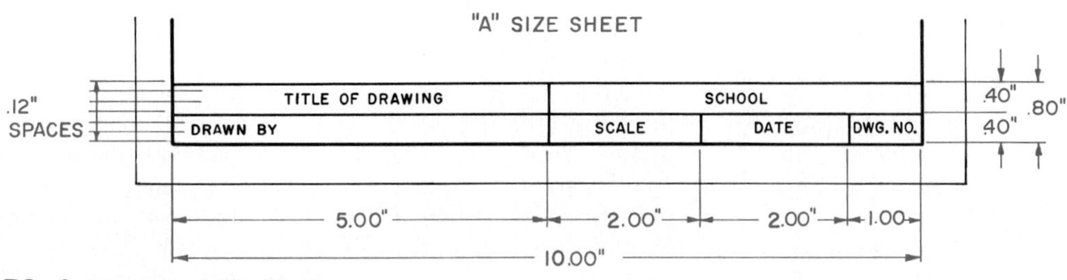

P3–1. Suggested title blocks.

Study Problems

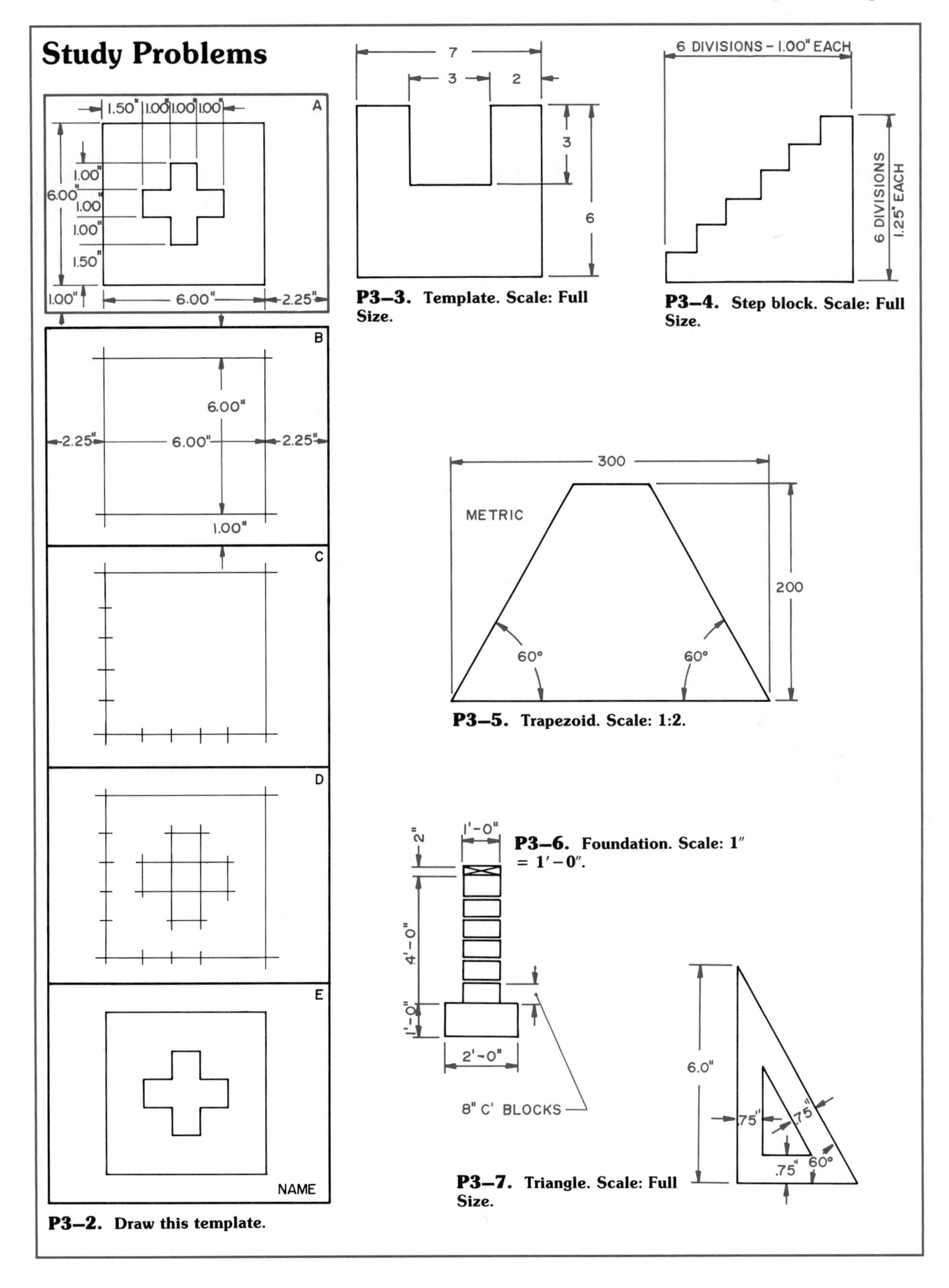

P3–2. Draw this template.

P3–3. Template. Scale: Full Size.

P3–4. Step block. Scale: Full Size.

P3–5. Trapezoid. Scale: 1:2.

P3–6. Foundation. Scale: 1″ = 1′–0″.

P3–7. Triangle. Scale: Full Size.

Study Problems

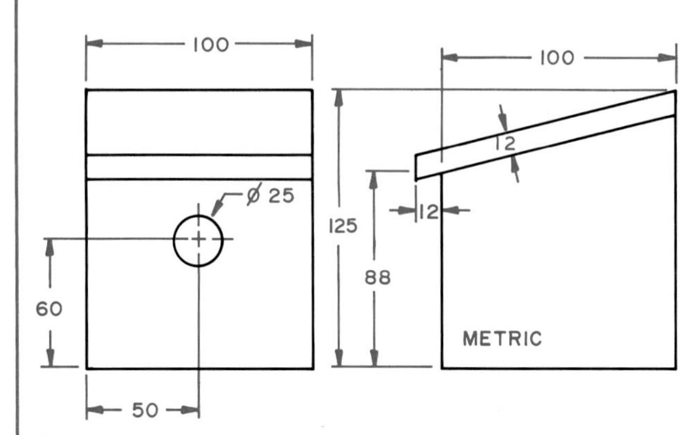

P3–8. Bird house, front and side views. Note the diameter symbol to indicate the diameter of the hole. Scale: 1:2.

P3–9. Louver. Scale: Full Size.

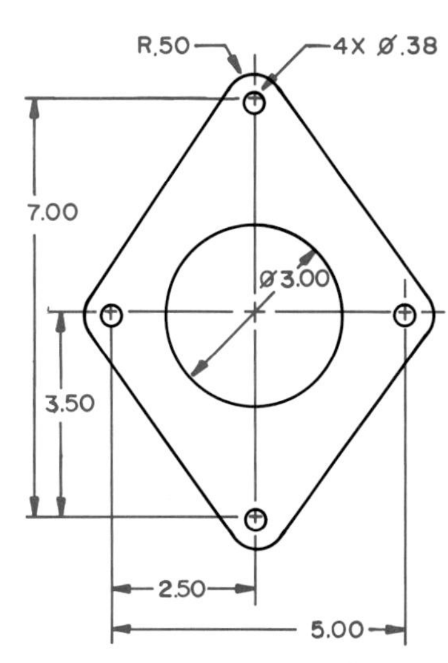

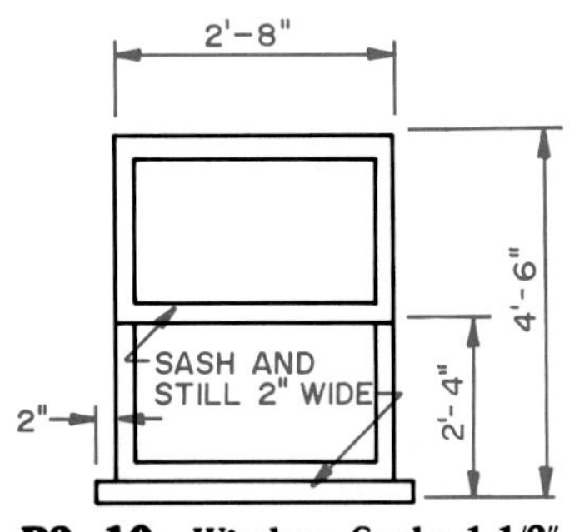

P3–10. Window. Scale: 1 1/2" = 1'–0".

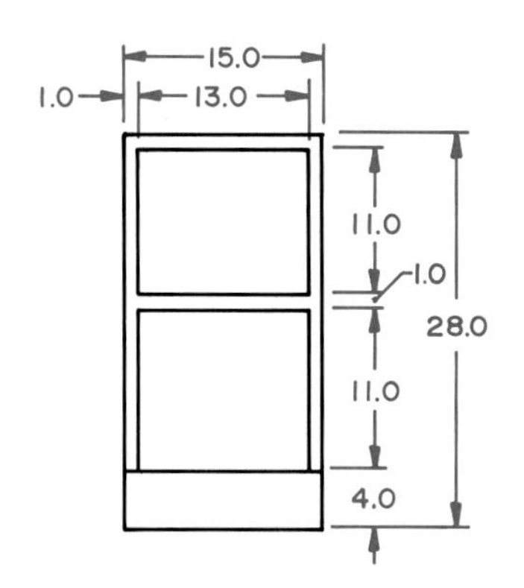

P3–11. File cabinet. Scale: Quarter Size.

P3–12. Gasket. Scale: Full Size.

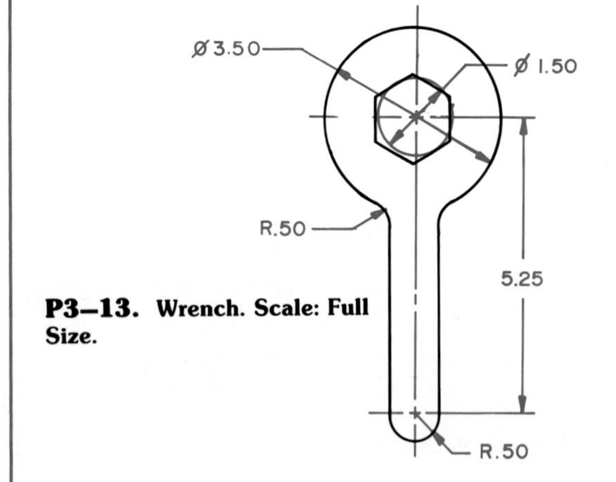

P3–13. Wrench. Scale: Full Size.

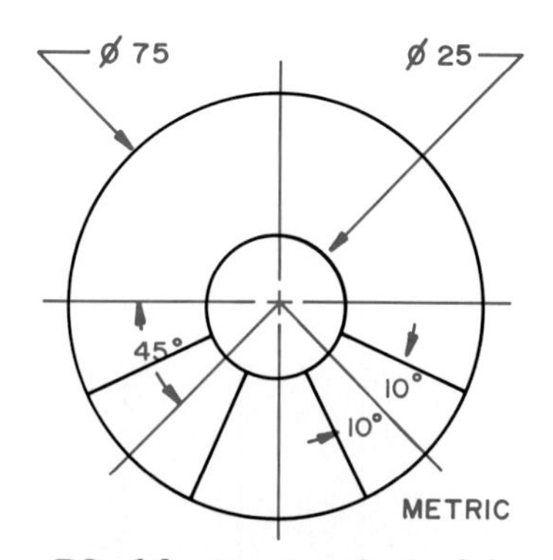

P3–14. Stop lug. Scale: 2:1.

Study Problems

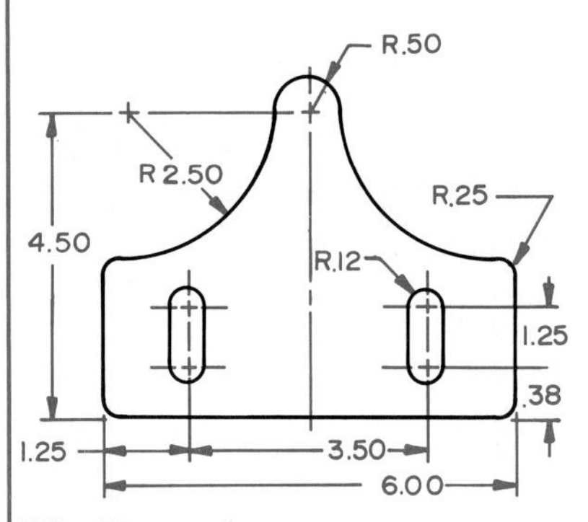

P3–15. Adjustment plate. Scale: Full Size.

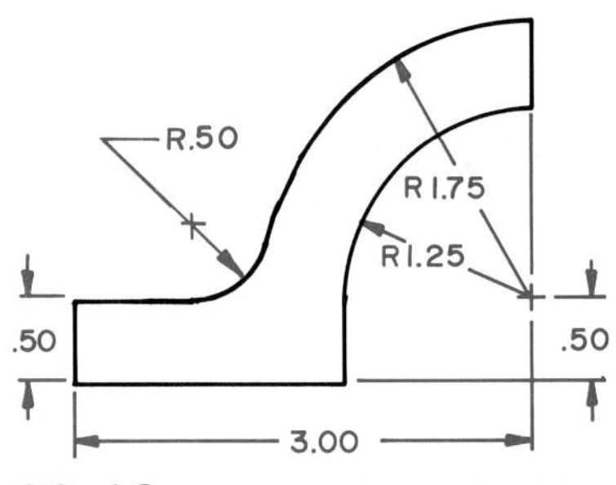

P3–16. Housing. Scale: Double Size.

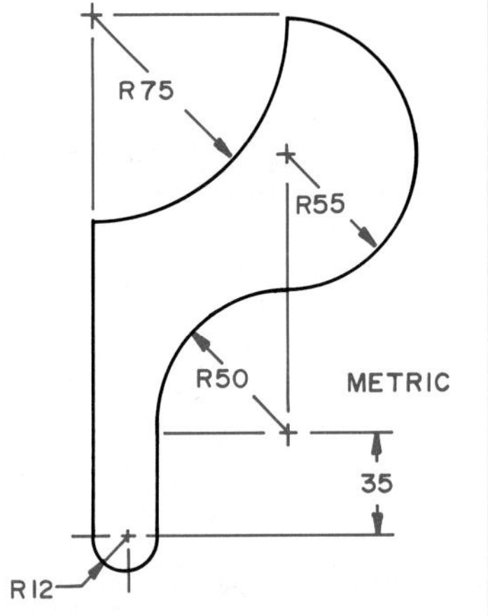

P3–17. Ring compressor. Scale: 1:1.

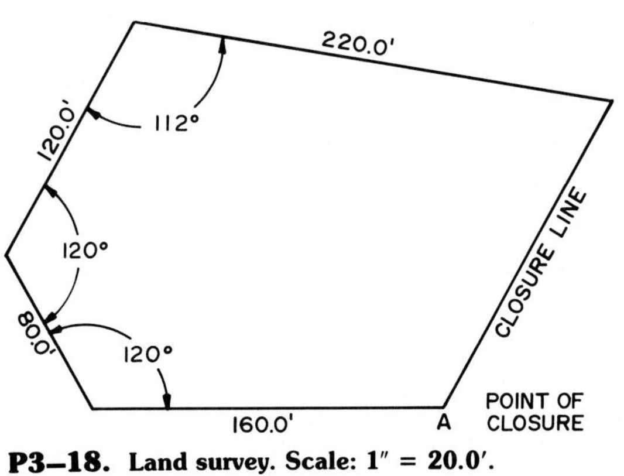

P3–18. Land survey. Scale: 1″ = 20.0′.

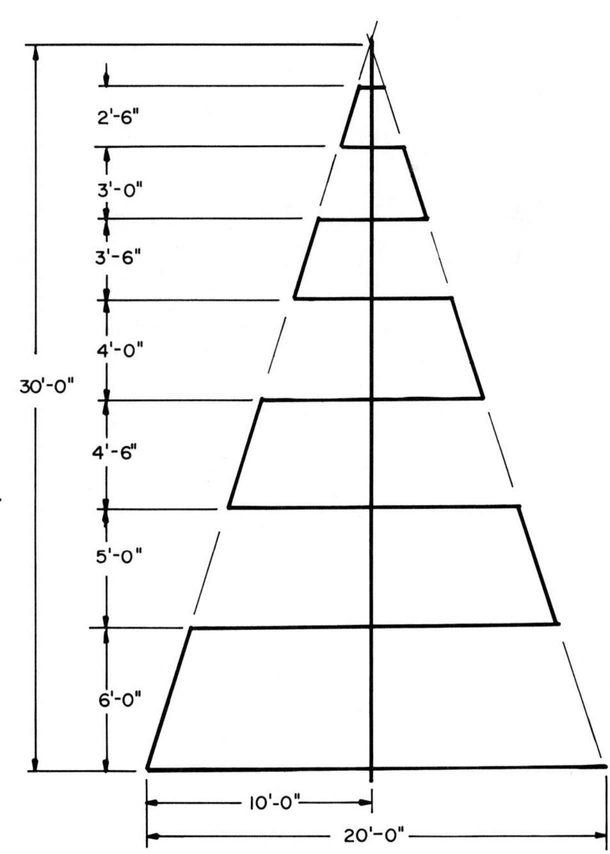

P3–19. NASA space station antenna. Scale: 1/4″ = 1′–0″.

Study Problems

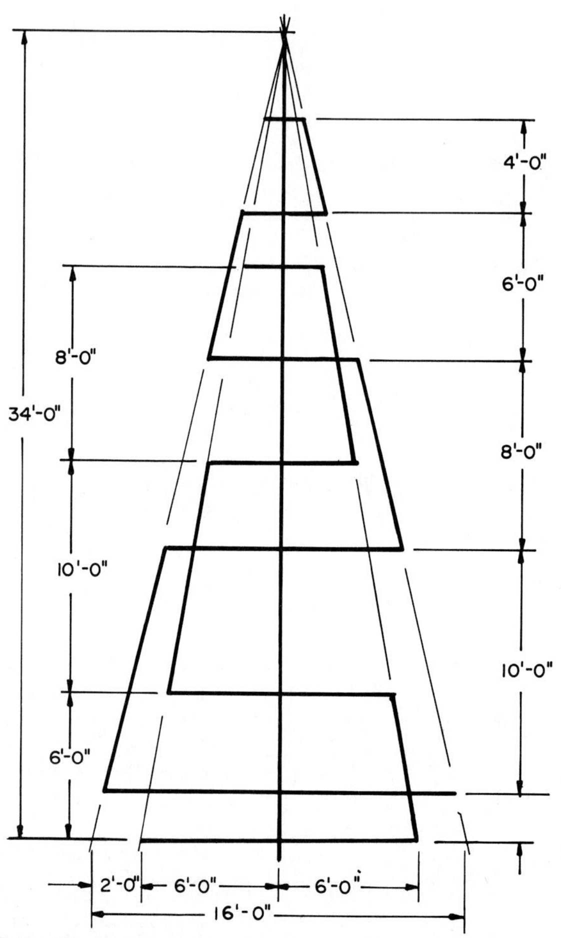

P3–20. NASA space station antenna. Scale: 1/4" = 1'–0".

CHAPTER 4

Computer-Aided Design and Drafting

After studying this chapter you should be able to do the following:

- Explain the advantages of using computers in design and drafting.
- Understand the purpose of each component in a CAD system.
- Understand some basic commands used in CAD systems.
- Explain in a general way how to produce a drawing with CAD.

Advantages of CAD

Drawings have been used to communicate ideas throughout history. Until recently their production has been a manual activity. The development of computers has caused many changes in the way we do things including designing products and producing drawings. The introduction of computer-aided design and drafting (CAD or CADD) has had a great impact on how industry performs these functions. Fig. 4-1.

Computer-aided design and drafting has a number of advantages over traditional drafting methods. These include speed, accuracy, legibility, neatness, consistency, and efficiency.

Speed

Using the memory and speed of operation of the computer, you, the drafter, can rapidly lay out, dimension, and plot a drawing. You can call on the computer to draw any of the many symbols and details in its memory. For example, if a machine screw is in its memory, the computer can draw one in seconds at the location you choose. A CAD system will produce a finished drawing on a machine called a plotter. Plotters run at high speeds and can produce in minutes a drawing that would take a drafter hours to produce manually.

Accuracy

The accuracy of a traditional drawing depends upon the skill and careful work of the drafter. Computer-generated drawings are commonly produced to an accuracy of five to ten decimal places.

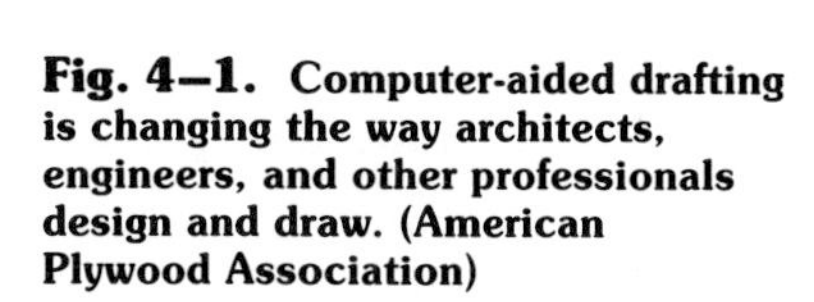

Fig. 4—1. Computer-aided drafting is changing the way architects, engineers, and other professionals design and draw. (American Plywood Association)

Legibility

Traditional drawings rely entirely upon the manual skills of the drafter to produce lines of consistent quality and lettering that is easily read. The computer-driven plotter uses pens to draw lines of high quality, and it letters notes and dimensions with typewriter-like clarity.

Neatness

The manual drafter has a constant problem of keeping pencil lines from smudging and keeping finger marks and perspiration off the drawing. Because no person touches the computer-generated drawing while it is plotted, it is fresh and clean. Since corrections are made on the video monitor, there are no erasure smears on the plotted drawing.

Consistency

Computer-generated drawings are produced in ink by the plotter. The lines and lettering are the same quality regardless of who developed the drawing. Thus, a set of drawings can be developed by several drafters, yet appear consistent when printed.

Efficiency

A drafter using CAD approaches a drawing in a different way than if conventional methods are used. Using recommended techniques and given the speed and accuracy of the system, the drafting process requires a much shorter time from the beginning to the production of a copy of the drawing. The drafter uses his or her knowledge of drafting procedures to think through the task and uses the efficiency of the computer to produce a drawing without having to stop and manually draw the results of his or her intellectual process.

You must understand that the CAD operator must thoroughly know the principles and techniques of drafting before becoming proficient in computer applications. The computer can draw only what it is directed to draw.

CAD Hardware

Hardware is a term used to identify the equipment that makes up a configuration for a CAD system. It includes the following items:

- Computer.
- Keyboard.
- Graphics display, or monitor.
- Graphics tablet, or digitizer.
- Plotter.

Computers

The heart of a CAD system is the computer. There are three main categories of computers: mainframe computers, minicomputers, and microcomputers.

Mainframe computers are large, centrally located computers to which many terminals can be connected so they can be used by many people at the same time. Mainframe computers can process large amounts of data and have substantial memory (storage). They are used by large companies and governmental agencies.

Minicomputers are smaller than mainframe computers and accept several terminals. Thus they can be used by several people at the same time. While they have less capacity than the mainframe, they have more than the microcomputer.

Microcomputers are small desktop computers that are becoming widely used in all areas of business and even in the home. Fig. 4-2. Often called personal computers (PC), they have sufficient capacity to handle a high percentage of the computing tasks in industry. Because microcomputers are less expensive than mainframes, many businesses and schools use them for CAD.

Microcomputer Components

Some of the major components of a microcomputer are the central processing unit, memory, expansion slots, and input/output ports.

The **central processing unit (CPU)** is a microprocessor that is the "brains" of the system. It processes information, performs arithmetic functions, and provides control of the rest of the system. Each instruction fed into the computer is examined and acted upon by the CPU.

The **read only memory (ROM)** holds instructions for the computer that are permanent and cannot be altered. Data can be read from this memory and the memory is retained when the power to the computer is turned off. **Random access memory (RAM)** is used to store programs and data temporarily and to act as work space for the CPU. The larger the RAM, the more work that can be done. Data are written to (stored) or read (retrieved) from the RAM. Data can be added to or changed in the RAM but are lost when the power to the computer is turned off. To save these data they are recorded on a disk or on tape. Data can be reintroduced into RAM when the computer is in operation by copying it from the disk or tape on which it has been stored back into the RAM.

Expansion slots on the CPU enable the memory to be expanded by installing additional CPU/memory cards.

Other devices can be connected to the computer through the **input/output (I/O) ports.** These ports are found at the rear of the computer.

Peripheral Hardware

Peripheral hardware includes the devices added onto the computer. These include various input/output devices such as disk drives, monitors, keyboards, other input devices, printers and plotters.

Disk Drives. A disk drive is the device used to read data from—and write data to—RAM and/or other disks. There are two kinds of disks: floppy disks and hard disks. A **floppy disk** is a circular piece of thin, flexible plastic coated with a magnetic oxide similar to that used in magnetic recording tape. Fig. 4-3. Floppy disk drives may be built into the computer cabinet, or they may be separate units. A **hard disk** is a carefully machined and polished nonmagnetic metal platter that is coated with a magnetic material similar to that used on floppy disks. A hard disk will store the equivalent of several dozen floppy disks, and it allows access to information faster than from floppy disks. Hard disk drives are usually installed inside the computer cabinet. Most hard disks are not removable from their drive unit.

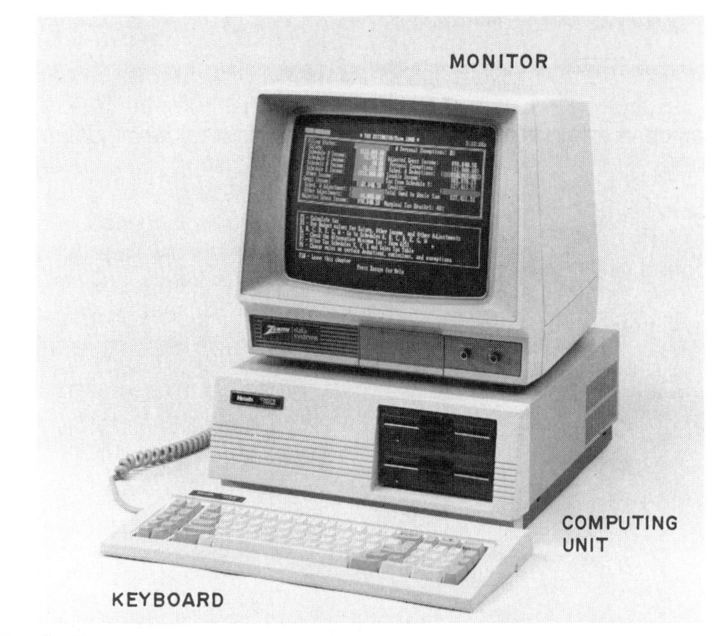

Fig. 4—2. A typical microcomputer that is used in industry. (Heath Company)

Fig. 4—3. A floppy disk upon which data are stored.

Monitors. A **monitor** contains a *cathode ray tube (CRT)* upon which the drawing being developed is displayed. Fig. 4-4. When you make corrections, you see the changes on the CRT. As you draw, you view the monitor and when you are satisfied that the drawing is correct, you then send it to memory where it may be stored or sent on to a plotter to make a hard copy.

Resolution is important when selecting a monitor. Resolution is a measure of the precision and clarity of the display. Like a television screen, the CRT contains many dots, called **pixels,** on its face. The more dots per unit area, the clearer and more uniform the image produced. Low resolution monitors have 200 to 400 pixels horizontally and 100 to 200 vertically. Low resolution screens produce jagged lines. Medium resolution monitors have

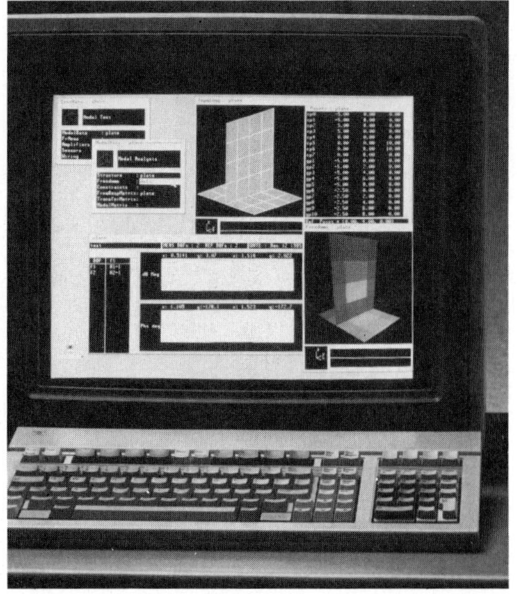

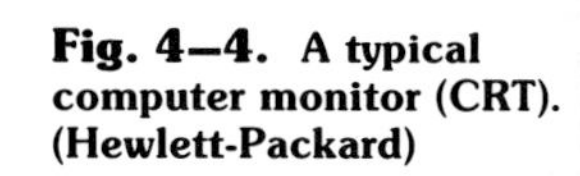

Fig. 4—4. A typical computer monitor (CRT). (Hewlett-Packard)

about 640 by 400 pixels, while high resolution monitors are in the range of 1000 by 800 pixels or higher.

Keyboards. The **keyboard** is an input device and looks similar to one found on a typewriter. However, in addition to letter, number, and punctuation keys it has a number of other keys that provide special commands. Fig. 4-5. A typical keyboard contains alphabetic keys, nonalphabetic keys, function keys, control keys, and a calculator keypad.

The *alphabetic keys* have the standard 26 letters arranged like those on a typewriter. The *nonalphabetic keys* include the numbers 0 through 9, punctuation marks, and special characters. In addition, there are common controls such as a space bar, backspace, tab, and return key.

One group of *function keys* is usually labeled F1 through F10 and is used for special purposes.

Fig. 4–5. A typical computer keyboard. (Texas Instruments)

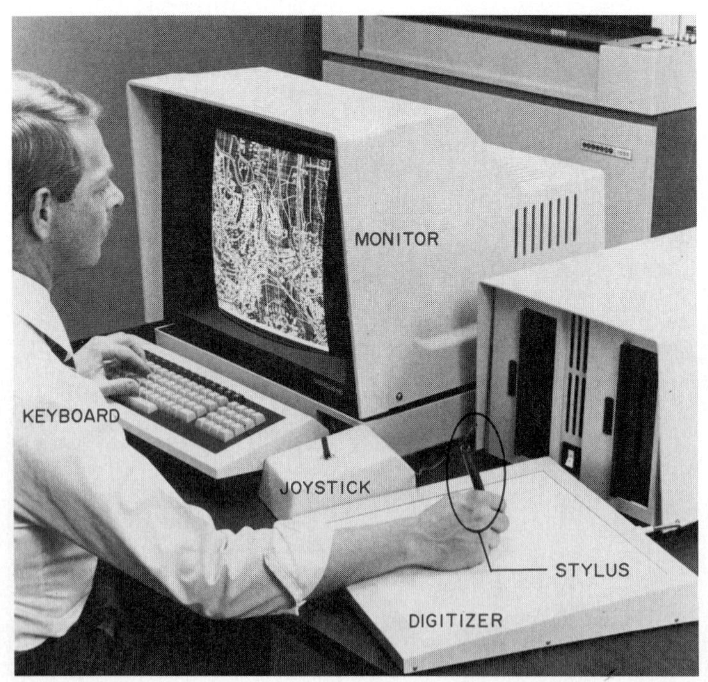

Fig. 4–6. This drafter is drawing with a stylus on a digitizer. (Calcomp/ Sanders)

The purpose is determined by the program or the user. Other function keys perform operations such as moving the cursor or scrolling the screen. *Control keys* are used in combination with other keys to perform a function. These include the Escape, Control, and Alternate keys. The *calculator keypad* keys are used to enter data and functions in the same way as on a calculator.

Other Input Devices. In addition to the keyboard there are a number of other devices used to input data into the computer. These are more important than the keyboard when producing graphics. The devices most frequently used include a digitizer with a stylus or puck, a mouse, a joy-stick, and a light pen.

A tablet **digitizer** is an electronic device that has a flat surface. When a **stylus,** an electronic device that looks like a pen, is held to a point on the digitizer surface and its button pressed, the point is recorded on the drawing. Fig. 4-6. This is called a *cursor point.* Cursor points can also be located on a digitizer pad with a device called a puck. A **puck** is a hand-held device that is flat on the bottom. It has fine cross hairs located on one end that are used to position a point. It has a number of push buttons on the top side. To locate a point on a drawing, position the cross hairs at the desired location and press the appropriate button. Fig. 4-7. Using a stylus or puck is much faster than entering commands using the keyboard.

A digitizer has a fine grid of wires between layers of glass. Fig. 4-8. As the stylus or puck is moved across the tablet, its X and Y coordinates are transmitted to the computer.

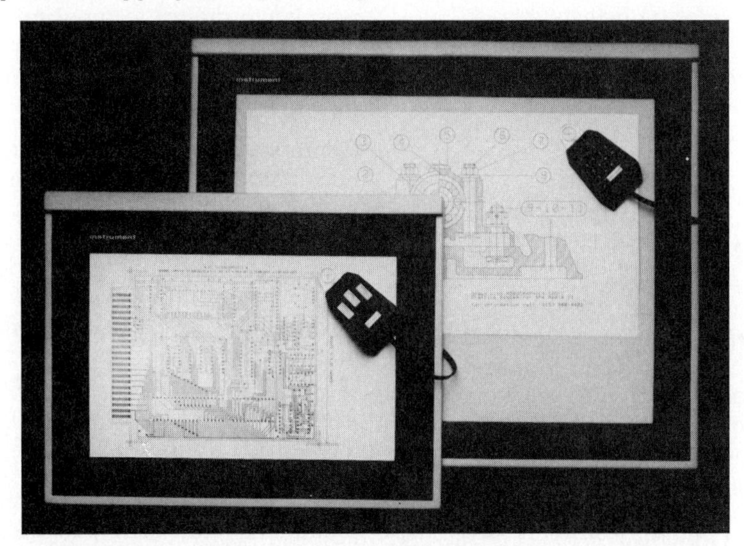

Fig. 4—7. Points on these drawings are being digitized (entered in the computer) with a puck. (Houston Instrument)

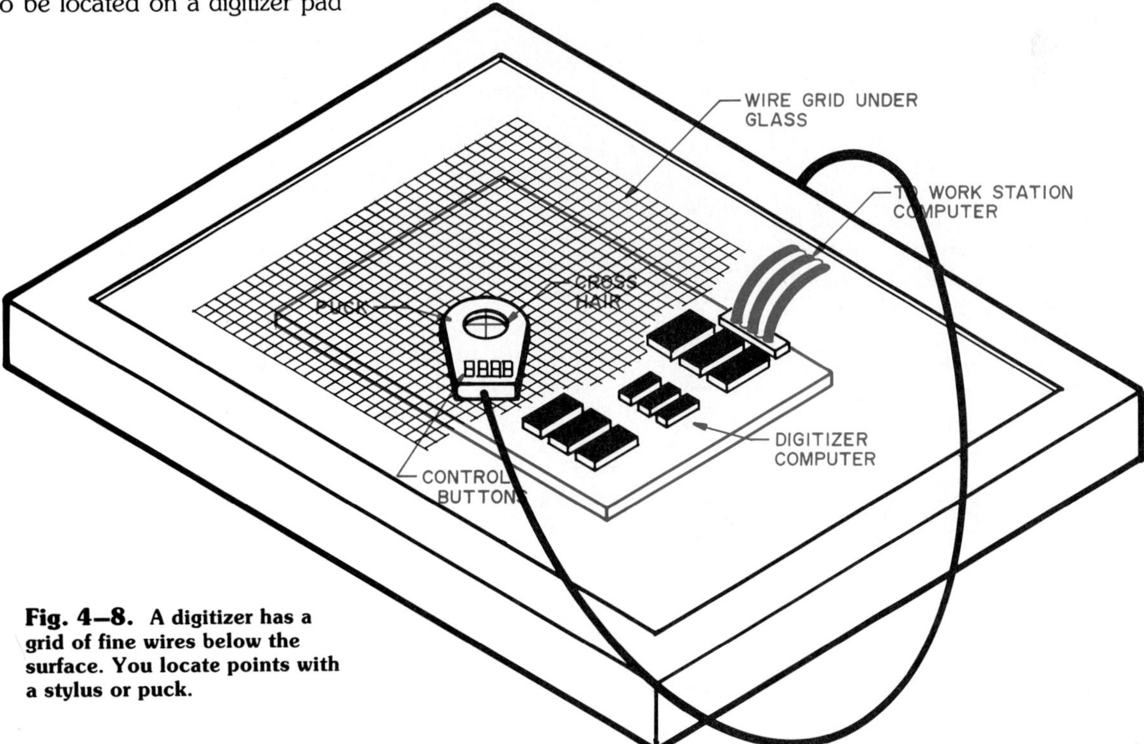

Fig. 4—8. A digitizer has a grid of fine wires below the surface. You locate points with a stylus or puck.

As was seen in Fig. 4-7, it is possible to trace an already completed drawing by first taping a copy of it onto the surface of the digitizer. Then, using the stylus or puck and the proper CAD commands, the line coordinates can be digitized from the drawing. This makes it possible to recreate a drawing quickly and put it on a disk for permanent storage.

A **mouse** is another hand-held device. Like the digitizer, it is used for placing points on the monitor and for selecting items from the menu on the screen. One type has a roller ball on the bottom. As you move the mouse across the table, a cursor moves across the screen. The mouse has several push buttons you use to input commands.

Another type of mouse is optical. It has a small light source that shines upon a pad that contains a metal grid. The contact of the light upon the grid controls the cursor on the screen.

Light pens look much like ordinary pens but are used to enter points on a drawing directly on the CRT. Light pens have a wire connecting them to the computer. They do not actually project a beam of light but detect the presence of light on the screen. The operator controls the light pen by turning on a switch at the tip of the pen. The switch controls a shutter covering the end of the pen. When the shutter is opened, light flows from the CRT into the photoelectric cell in the pen. The light is carried by a fiber-optics conductor to a digital pulse generator which indicates a "hit" and sends this information to the computer. A light pen is difficult to use and requires extensive software to permit it to track across the CRT. However, it does permit the drafter the freedom to work directly with the image displayed.

A **joystick** is a vertical shaft mounted on a base. As it is pulled or pushed, the cursor on the monitor screen moves. The cursor-point movements are accumulated by the digitizer graphics program and their coordinates are stored in the computer. There are also ball and dial controls which do much the same thing.

There are other input devices available such as those which permit the drafter to input information using normal oral commands. These are called *voice data-entry (VDE)* systems.

Printers. A printer is used to produce hard copies of text information as well as for drawing charts, graphs, and other graphics. Three common types of printers are dot matrix, letter quality, and laser.

The **dot matrix printer** produces letters and graphics by striking a ribbon with tiny striker pins. Fig. 4-9. The pins are in a head which shoots the pins out in the proper pattern to form the required image. Dot matrix printers print much faster than letter quality printers but the quality of the image produced is not as good.

The image printed with a **letter quality printer** is equal to that produced by either an electric or electronic typewriter. This type of printer uses an element called a daisy wheel upon which the characters are located. It is slower than a dot matrix printer and while it produces text very well, it is quite limited for the production of graphics.

Laser printers are fast and quiet. They are becoming more and more popular for producing black and white graphics and text. They are also used to produce images when CAD graphics are integrated with computer-aided publishing systems.

Plotters. A plotter is used to produce hard copies of the drawings developed with CAD. The most common type of plotter is the pen plotter. It uses technical pens to draw on vellum, film, or other materials. Some plotters use ball-point or felt-tip pens which produce drawings of lower quality.

After a drawing is completed and stored in the computer, the computer sends commands to the plotter. The pen moves across the paper in response to these commands. The pen is moved by servomotors. They not only control the direction of movement but lift the pen off the paper and put it back down as required.

There are two types of pen plotters: flatbed and drum. The **flatbed plotter** holds the paper stationary on a surface much like a drafting table. Fig. 4-10. The pen is carried in a carriage that moves across (X axis) and up and down (Y axis) this surface. To draw a line, for example, the computer directs the carriage to move the pen to the X and Y coordinates of one end of a line, then signals it to lower the pen and move to the X and Y coordinates of the other end of the line.

The **drum plotter** has a cylindrical drum upon which the drawing is made. Fig. 4-11. The paper is placed over the drum and is rolled back and forth on this cylinder to provide Y-axis movement. The pen is moved left and right on a carriage to provide the X-axis movement. This combination of movements is used to produce the drawing.

Two other kinds of plotters are thermal and electrostatic. **Thermal plotters** use heat to transfer and fuse lines and solid image areas onto the drawing paper. **Electrostatic plotters** reproduce images in the form of an electrostatic charge on paper, vellum, or plastic film. The electrostatic charge attracts a black powder, thus producing an image on the media used. The black powder lines are made permanent by passing it through a fixer.

Fig. 4–9. This high-speed dot-matrix printer is capable of printing near letter quality by switching its printing mode. (Illustration courtesy of Epson America, Inc. All Rights Reserved)

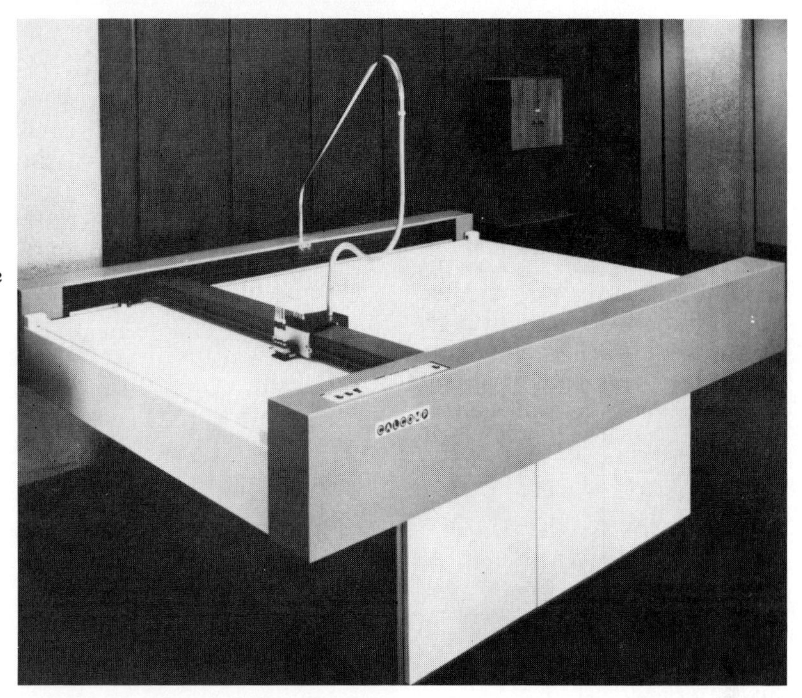

Fig. 4–10. This high-performance flatbed plotter has a pressurized inking system that feeds four pens. (Calcomp Sanders)

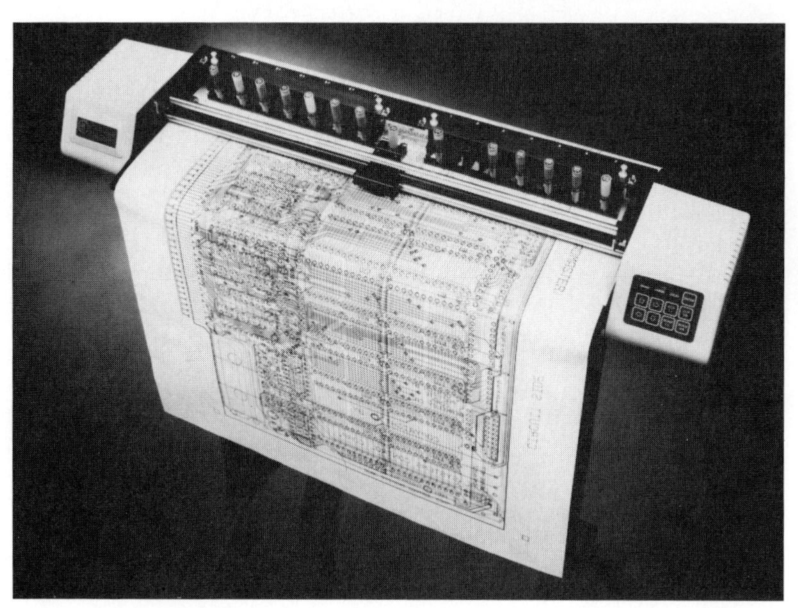

Fig. 4–11. This high-speed drum plotter uses up to 14 pens to produce plots on C- and D-size drawing sheets. (Houston Instrument)

CAD Software

The hardware of a CAD system will not produce a drawing until it is told what to do. The actual drafting functions are contained in a computer program. Computer programs are called **software.** The program is recorded on a floppy or hard disk and is entered into the memory of the computer. A list of commands and functions, such as "LINE" or "EDIT," appears on the monitor's screen. This list is called a **menu.** Fig. 4-12.

Drawings are made up of separate elements such as lines, arcs, circles, and strings of text. These elements are called **entities.** Entities are placed in a drawing with commands. Sometimes it is necessary to specify things such as the radius of an arc to be drawn. As the commands are executed, the results are shown on the screen of the monitor.

CAD Applications:

Skill Olympics

Talented students can now use their CAD skills in competition at the U.S. Skill Olympics. The Olympics, sponsored by the Vocational Industrial Clubs of America (VICA), offer 36 different skill and leadership competitions. Since 1968 students have shown their abilities in such areas as precision machining, electronics, and welding. Then in 1987 CAD was added and became an immediate success.

In one of the recent Olympics, CAD contestants were asked to create fully dimensioned front and right-side views of a valve, based on pictorial drawings. Technical drafting skills and adherence to ANSI standards were tested as well as the ability to use CAD drawing aids and shortcuts. Students had five hours to complete the task. Then their work was plotted. Students surprised judges with their skills in geometric construction and layering, their speed, and their creativity. However, scaling and plotting were common weaknesses.

The Olympics are held each year in June. About 2500 student competitors perform for 800 judges from companies, labor unions, and trade associations. Awards include medals, scholarships, and thousands of dollars in tools and equipment.

The Olympics help cement the partnership between industry and vocational education. Students show what industry can expect from future workers coming out of vocational programs. Industry has a chance to demand quality from those preparing to enter the world of work.

Those interested in more information on the U.S. Skill Olympics should contact VICA, P.O. Box 3000, Leesburg, VA 22075, (703) 777-8810.

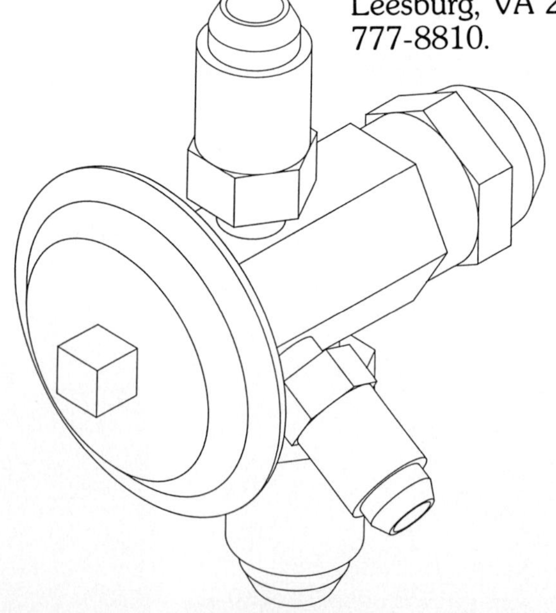

Many CAD programs require the use of a second program that translates information between the drafting program and the computer. This second program is called the disk operating system (DOS). The DOS also permits other functions to be performed, such as copying or erasing files. (Copying or erasing entities is part of the CAD program.) DOS permits disks to be formatted. *Formatting* clears the disk, checks it for defects, and prepares it to receive the input material to be recorded. Copies of programs on disks can be made by using the *diskcopy* command.

Getting Ready to Draw

Before a drawing can be started there are several things that must be set up. These include selecting the units, limits, and scale.

The **units** are used to measure distances. The architect, engineer, and scientist often use different notations for coordinates, distances, and angles. Generally, architects and engineers let one unit equal one inch. For example, 15.50 units would equal 15.5 inches or 1 foot 3½ inches. It is also necessary to choose the fractional denominator to be used, such as 4 (so fractions are ¼, 2/4, etc.) or the number of decimal places.

It is necessary to select the system of angle measurement and the accuracy for it. For example, an angle could be given in decimal degrees, degrees/minutes/seconds, grads, or radians.

The **limits** chosen establish the size of the working area upon which you can make a drawing. The limits are stated as X,Y coordinates. The lower left-hand corner of the working area has a limit of 0,0. The upper right limit should be based on the paper size and the drawing scale. For example, if the paper size is to be 11 by 17 inches, the drawing area inside the border might be 9 by 15 inches. Assume the drawing scale is to be ¼″ = 1′–0″ (or 1″ = 4′–0″). The upper right limit would be 60,36 because each inch represents 4′–0″ (15 × 4 = 60 and 9 × 4 = 36).

Scale has to do with the amount that one unit on the screen represents. You make a drawing in units not feet, inches, meters, etc. One unit could equal an inch or a millimeter. When it is time to plot the drawing it is necessary to specify the amount that one unit is equal to on the drawing. For example, if 1 inch equals 10 units and you decide to count one unit as a foot, then the scale of the plotted drawing is 1″ = 10′.

To relate this to the plotter, assume you are going to plot on 18 by 24 inch paper. With a 1 inch border on all sides the plotting area is 16 by 22 inches. If the scale is 1″ = 10′ and 1 inch of paper contains 10 units, the area inside the border would be 16 × 10 = 160 units and 22 × 10 = 220 units. Fig. 4-13.

Input of Coordinates

Generally the Cartesian coordinate system is used on CAD drawings. This means that points on a drawing can be represented by giving them an X and Y value such as 10,8. Fig. 4-14. The X value is *always given first;* therefore, X = 10 and Y = 8.

Coordinates can be input using the keyboard, a stylus, or a puck. When inputting using a stylus or puck, locate each point by either touching the digitizer with the stylus and pressing the button on it or locating the point on the cross hairs of the puck and pressing the proper button on it. Each point so located has X and Y coordinates.

The X and Y axes intersect at point 0,0. This is usually the lower left corner of the drawing

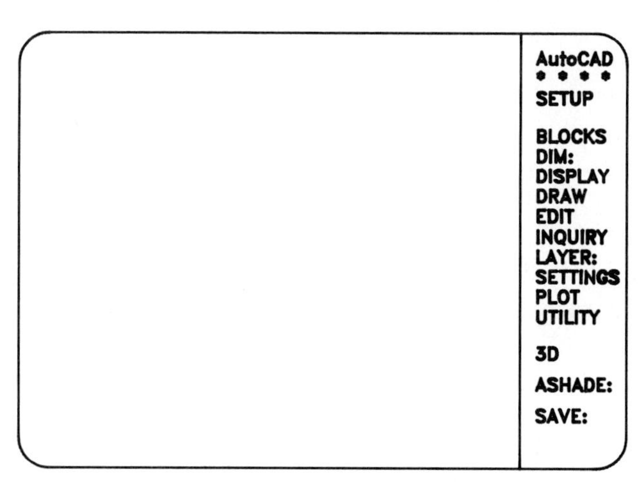

Fig. 4–12. **A typical CAD program menu.**

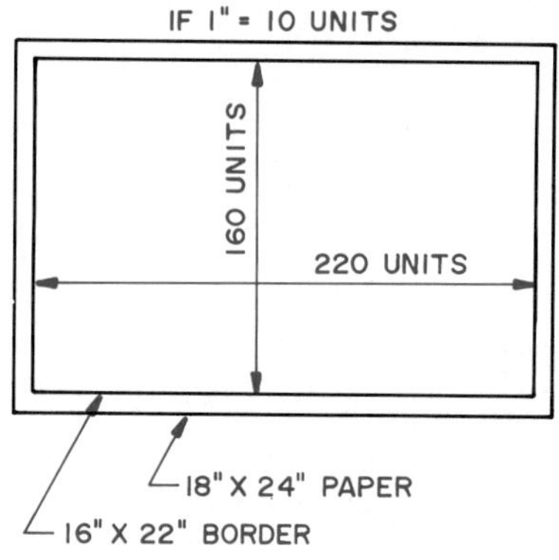

Fig. 4–13. **This is a typical calculation for a plot when 1″ = 10 units.**

area. When the coordinates for a point are given, a small marker appears on the screen. Markers are not a part of the drawing and will not appear on the drawing made by the plotter. Coordinates can be specified by giving absolute coordinates, relative coordinates, or polar coordinates.

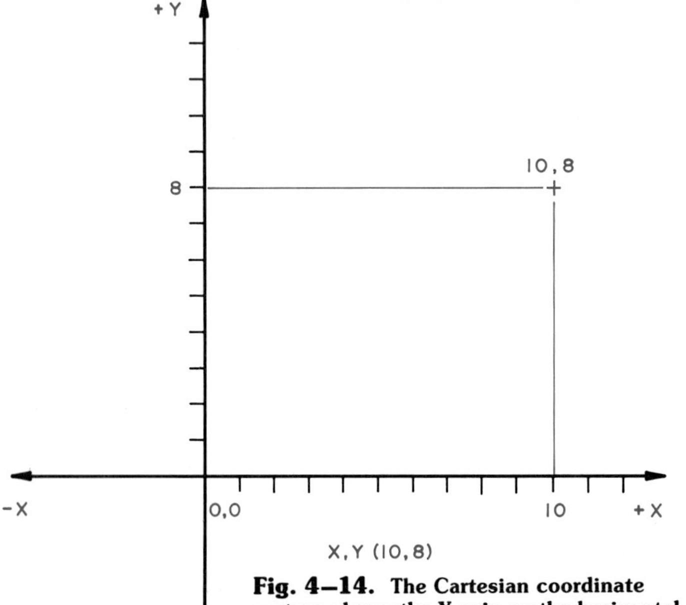

Fig. 4–14. The Cartesian coordinate system places the X axis on the horizontal and the Y axis on the vertical.

Absolute Coordinates. **Absolute coordinates** are the actual X and Y coordinates of each point. For example, a line could be indicated by giving the coordinates of each end, such as 5,2 and 5,9. Fig. 4-15.

Relative Coordinates. **Relative coordinates** are those specified by giving a number of units in the X and Y directions from the last point, which we will call *A*. For example, if the last point known was 3,4, the next point could be given as 6,5. Fig. 4-16. This point would be 6 X-units from point A and 5 Y-units from that same point.

Polar Coordinates. **Polar coordinates** specify the distance and the angle one point is from another. A typical system for locating the coordinates for angle specification is shown in Fig. 4-17. Notice that 0 degrees is on the right horizontal axis. An angle of 45 degrees would put a point above the horizontal axis, while specifying 315 degrees would put it below the horizontal. A polar coordinate such as 10 < 135 would draw a line as shown in Fig. 4-18. The length of the line is 10 units (starting from the last known point) and the angle is 135 degrees.

Typical CAD Commands and Facilities

The following examples of commands and facilities used for creating, changing, viewing, and storing drawings are typical for CAD programs. The names may vary from one program to another, but the actions produced are very similar.

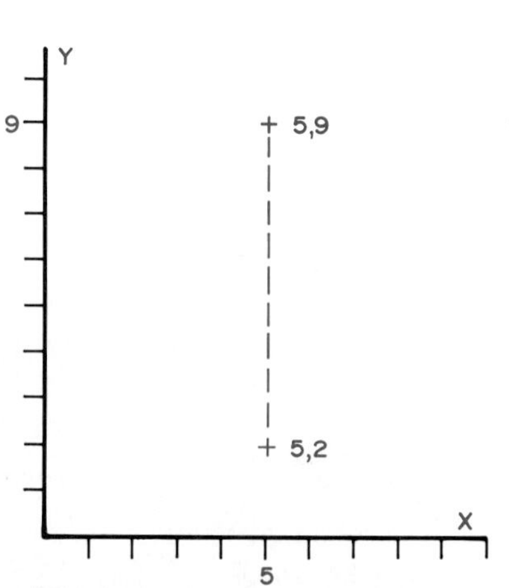

Fig. 4–15. Absolute coordinates are the actual coordinates of a point.

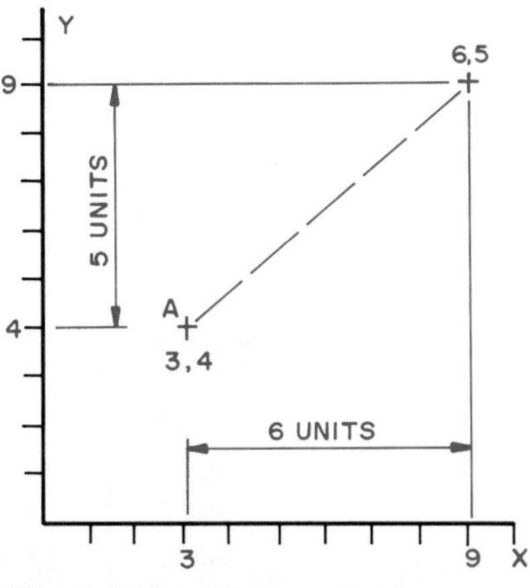

Fig. 4–16. Relative coordinates are measured from the last given point.

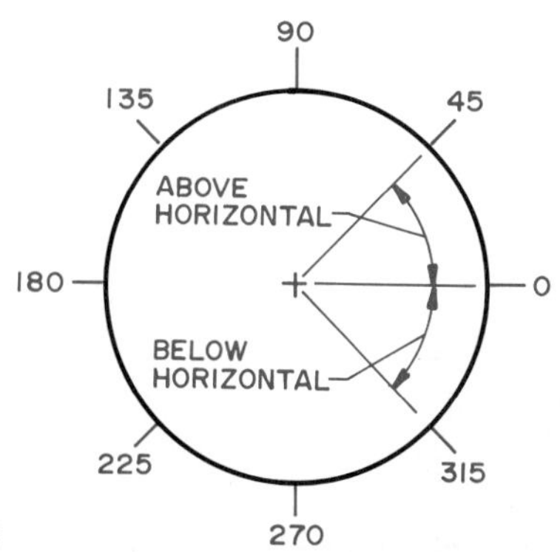

Fig. 4–17. Polar coordinates are located with a distance and an angle.

Commands Used for Drawing

Point Command. The point command is used to draw a point. From the menu select the command "point." Then enter the coordinates of the point's location, such as 10,5. Remember, the X coordinate is first and the Y coordinate second.

Line Command. The line command is used to draw a line. From the menu select the command "line." The following description of the point to point method is one way that a line may be drawn.

First enter the XY coordinates of the first point. Then enter the XY coordinates of the second point. A line will be drawn on the screen connecting these points. As an example, the following operation will draw the box shown in Fig. 4-19. The items shown in **boldface** type represent user input, that is, what you type. The other items represent the computer's messages.

Command: **LINE**
From point: **2,3**
To point: **2,8**
To point: **9,8**
To point: **9,3**
To point: **2,3**

With this input the CAD program will draw a line from point 1 to 2, from 2 to 3, from 3 to 4, and from 4 back to 1.

Circle Command. There are several ways to draw circles with the circle command. Common methods include center and radius, center and diameter, two-point, and three-point.

The *center and radius* method requires that you specify the center point and the radius of the circle. Fig. 4-20. The following is a typical command sequence:

Command:
CIRCLE, CEN RAD
Locate center: **7,7**
Specify radius: **5**

With the *center and diameter* method you specify the center point and the diameter of the circle. Fig. 4-21. The following is a typical command sequence:

Command:
CIRCLE, CEN DIA
Locate center: **7,7**
Specify diameter: **10**

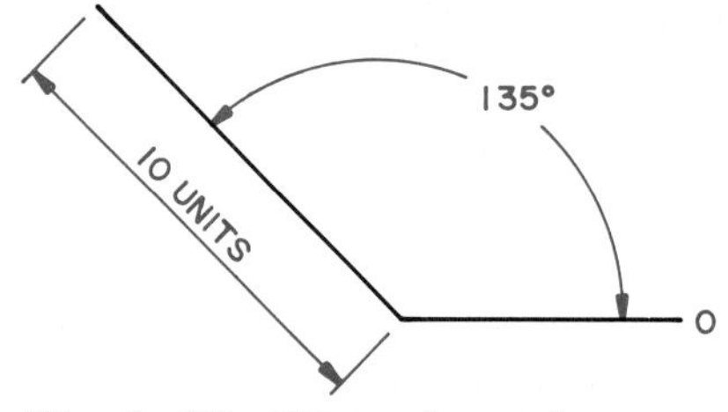

Fig. 4–18. This angle was drawn with polar coordinates, that is, the angle and the length of the line.

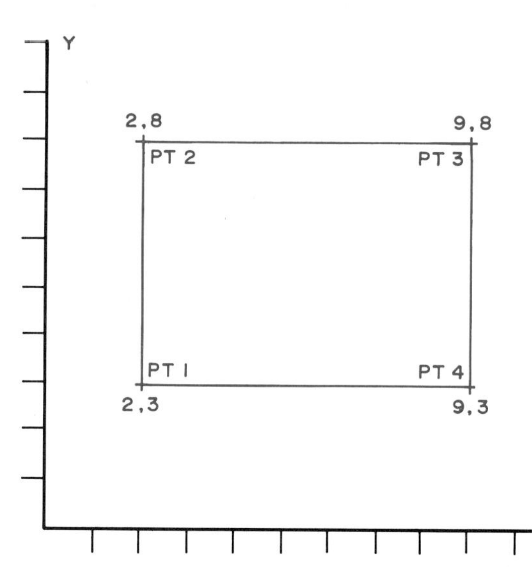

Fig. 4–19. Create a box using the line command by connecting points identified with XY coordinates.

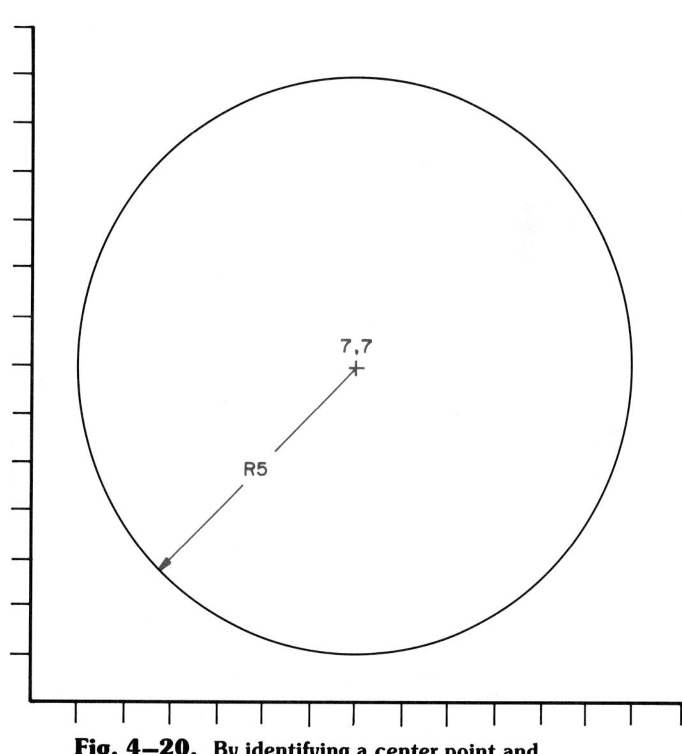

Fig. 4–20. By identifying a center point and specifying a radius, you can draw a circle with the circle command.

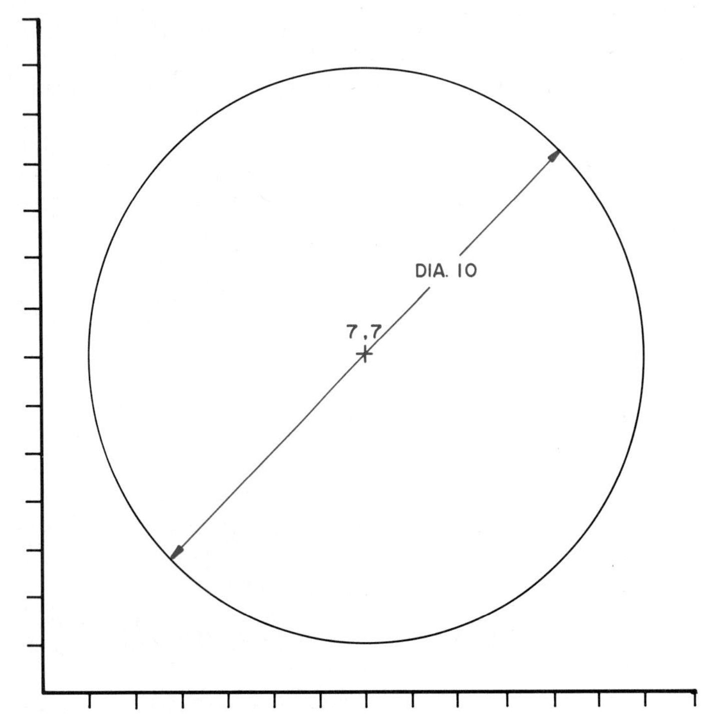

Fig. 4—21. **A circle can also be drawn with a center point and a diameter using the circle command.**

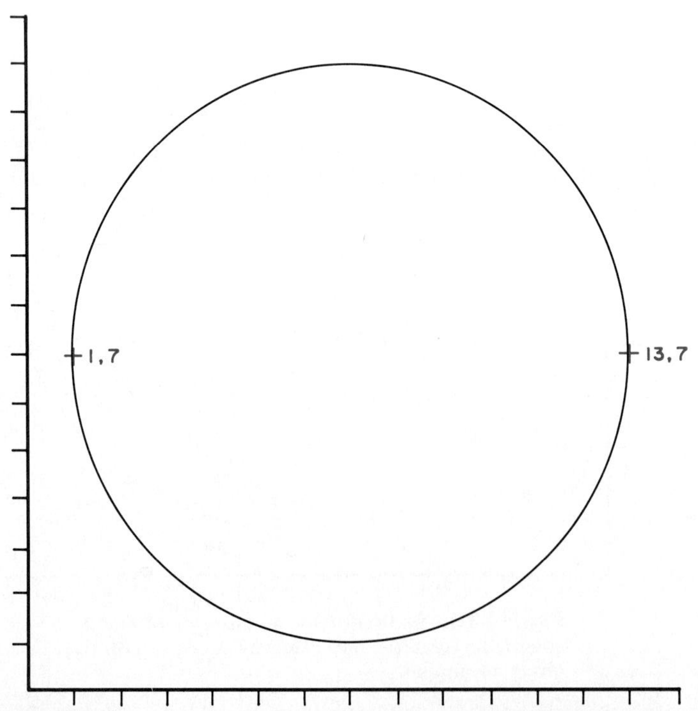

Fig. 4—22. **A circle can be drawn by locating two points on the circumference.**

With the *two-point* method you draw a circle by locating two points on its circumference as shown in Fig. 4-22. These points are actually the ends of a diameter. The following is a typical command sequence:

Command:

CIRCLE, 2 POINT

Enter first point on diameter: **1,7**

Enter second point on diameter: **13,7**

The *three-point* method requires you to locate three points on the circumference of the circle. Fig. 4-23. The following is a typical command sequence:

Command:

CIRCLE, 3 POINT

Enter first point on circumference: **1,7**

Enter second point on circumference: **7,13**

Enter third point on circumference: **13,7**

Arc Command. Using the arc command, there are many ways to draw an arc. The following illustrate only a few of them. Keep in mind that arcs are drawn counterclockwise.

The *start, center, and end* (SCE) method requires that you locate the center of the arc and give the starting and ending points. Fig. 4-24. The radius is equal to the distance from the center to the starting point. If the end point is not the same distance from the center, the arc will not pass through it. The following is a typical command sequence:

Command: **ARC, SCE**

Enter center point: **7,8**

Enter starting point: **2,8**

Enter ending point: **10,4**

The *start, end, radius* (SER) method uses the radius along with the starting and ending points to draw the arc. Fig. 4-25.

The following is a typical command sequence:

Command: **ARC, SER**
Enter radius: **5**
Enter starting point: **9,11**
Enter ending point: **4,3**

The *line/arc continuation* method makes it easy to connect an arc to either a line or another arc that has already been drawn. Fig. 4-26. The starting point for the arc and its direction are taken from the end point and ending direction of the line or arc previously drawn. A typical command sequence follows:

Command: **LINE**
From point: **1,2**
To point: **9,2**
Command: **ARC**
Enter start point: Press **(RETURN)**
Enter end point: **13,6**

Line Type Commands. A variety of line types (hidden, visible, etc.) are used on drawings. The CAD program will have a list of the types of lines available and this can be called up on the screen. This list shows the line and the name given to it. The user's manual for the program will explain how line types are selected from the menu.

Trace Command. On engineering drawings you draw visible lines wider than hidden lines. The CAD system will draw lines to a specified width by using the trace command. When you enter the trace command, you request the desired line width. Then you enter point-to-point coordinates for the lines that are to be widened. The lines are drawn wide on the plotter by having the pen redraw them several times, moving to one side a little more each time.

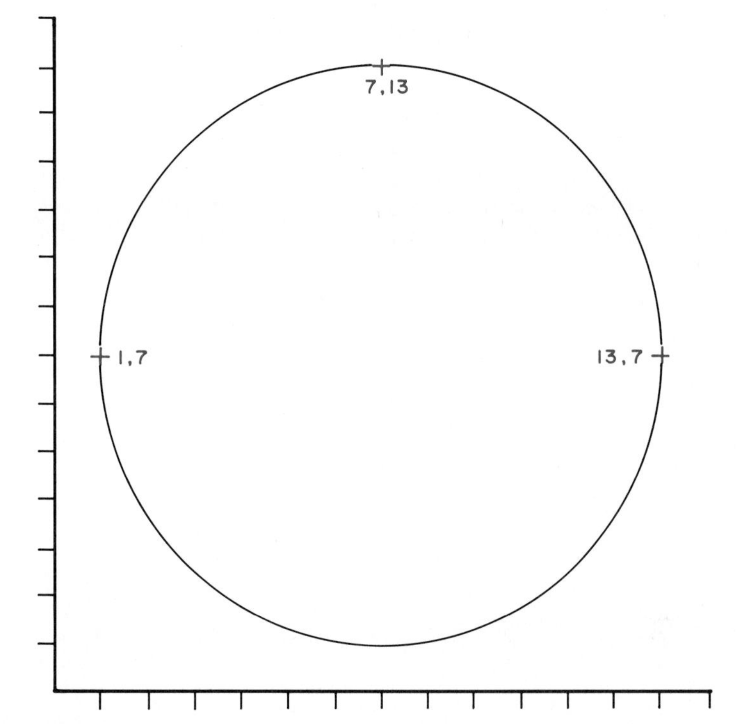

Fig. 4—23. **A circle can be drawn by locating three points on the circumference.**

Fig. 4—24. **An arc can be drawn by giving the coordinates of its starting and ending points plus its center.**

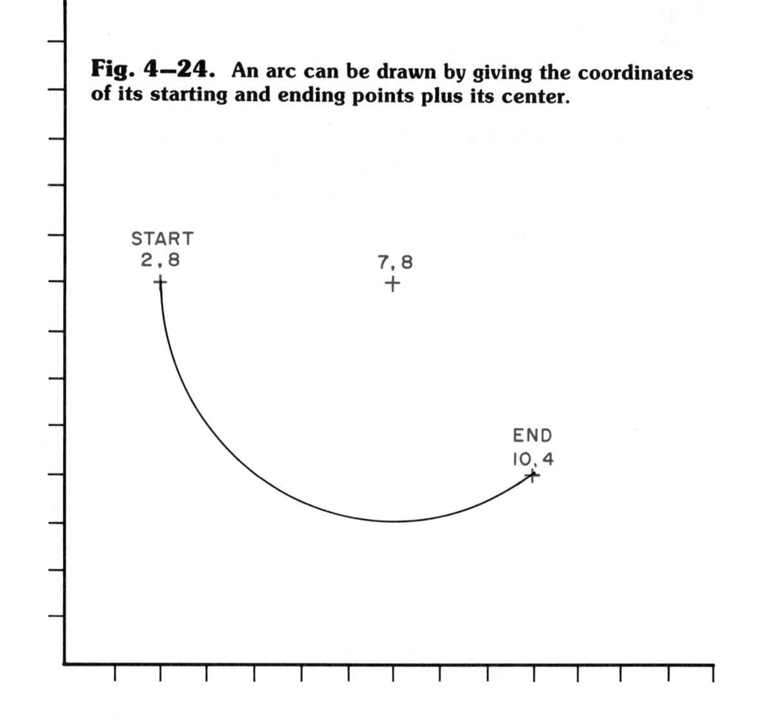

Polygon Command. As you know, a polygon is a closed figure bounded by three or more line segments. The polygon command allows you to draw regular polygons quickly. You simply specify the center point of the polygon, its size, and the number of sides.

Drag Function. The drag function makes it possible to dynamically drag an image on the screen. For example, after you specify the start and center points of an arc, the arc forms on the screen. Moving the pointing device causes the arc to lengthen or shorten. When you achieve the desired length, press the pick button to "set" the image. Drag is usually embedded in other commands and functions, such as the arc command.

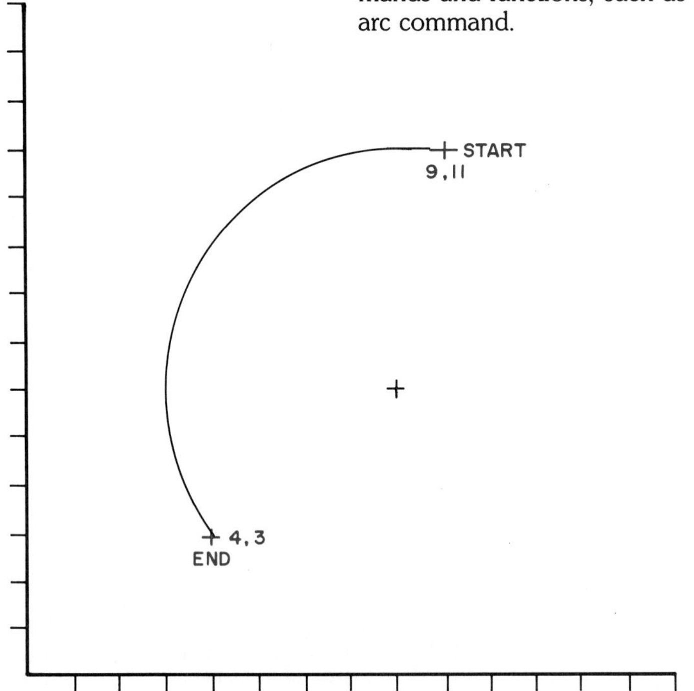

Fig. 4—25. Using the SER method, you give the coordinates of the starting and ending points plus the radius to draw an arc.

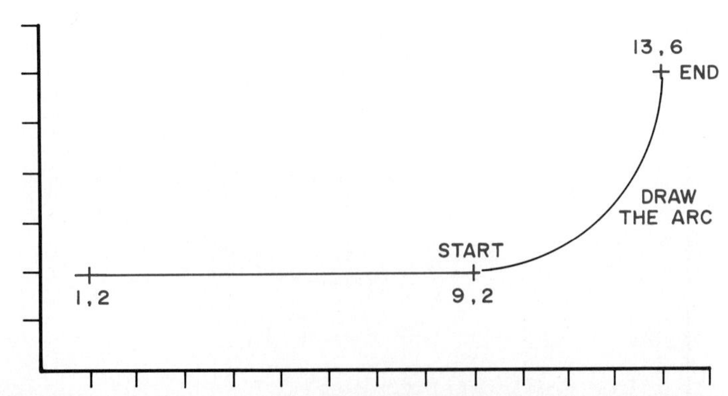

Fig. 4—26. The line/arc continuation method makes it possible to connect an arc to a line or other arc that has already been drawn.

Isometric Drawings. An isometric drawing is one type of pictorial drawing. Some CAD programs have a mode that uses the three drawing planes needed to construct an isometric drawing. Once you establish the three axes, you must indicate whether you are drawing on the right, left, or top plane. The lines are drawn and edited in each plane using the normal commands.

3-D Drawings. Most CAD programs have a three-dimensional (3-D) facility which enables the user to create lines and surfaces in three-dimensional space using X, Y, and Z coordinates. For example, one way to create a tube would be to draw a circle and assign a thickness to it. The circle would be the top view of the tube. When viewed from a different angle, the shape of the tube would become apparent. The base of the tube could be set on any desired elevation so that the tube would appear to rest on a flat surface, float either above or below it, or actually pierce through it.

Symbol Libraries

A **symbol library** is a group of commonly used symbols and details stored on a disk. For example, when making an electrical

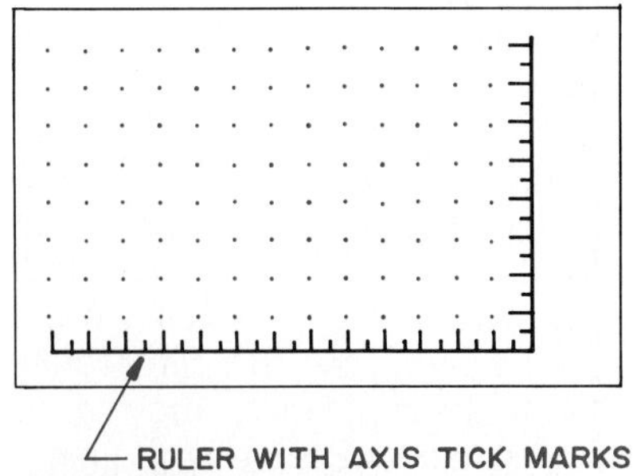

Fig. 4—27. The grid command produces a grid of dots on the screen with spacing that may be specified by the drafter. The axis command places a ruler line along the bottom and right side of the grid. The drafter may specify the spacing of the tick marks.

drawing, you would use the symbol for an electrical outlet many times. Rather than redraw it each time, you can call it up out of the library (storage) and it gets drawn automatically at the location you indicate. CAD programs are available with ready-made libraries for special fields such as architectural drawing. They also provide commands that permit the drafter to add additional items to the existing library.

Drawing Aids

Grid Command. The grid command produces a grid of dots with a specified spacing on the screen. The spacing between dots on the X and Y axes can be different. The grid is used to help lay out a drawing and will not print on the plotter. Fig. 4-27.

Axis Command. The axis command is used to place a ruler line on the lower and right borders of the screen. The spacing of the tick marks can be specified. For example, you may find it helpful if they have the same spacing as the grid dots. Fig. 4-27.

Snap Command. Points entered on the screen can be aligned to an imaginary grid using the snap command. If the drafter attempts to enter a point that is not aligned with a snap point on the grid, the point is automatically forced to the nearest snap point.

Object Snap Facility. The **object snap facility** enables you to draw lines based on specified geometric properties of an existing object. For example, suppose you want to draw a line from the center of a circle to the intersection of two lines. Object snap allows you to do this accurately. Rather than approximating the locations with the cursor, you tell the computer to start the line at the center of the circle and end it at the intersection of the two lines.

Ortho Command. It is difficult to draw lines that are exactly vertical or horizontal. The ortho command can be used to make them exactly vertical or horizontal. When a line is drawn and the end point is not true either vertically or horizontally, the ortho command will force it to the nearest true point. For example, if the end point of a line makes it more nearly horizontal than vertical, it will be forced horizontally.

Dimensioning

When using CAD, a drawing can be dimensioned semiautomatically. This means that the program can construct a dimension line, measure the distance, and enter it in a break in the dimension line.

There are four basic types of dimensioning: linear, angular, diameter, and radius. Each one of these has specific dimensioning commands. Linear dimensions can be horizontal, vertical, aligned with a sloping side, rotated to a specified angle, continued from the baseline (first extension line) of the previous dimension, or continued from the second extension line of the previous dimension (continuous dimension). Fig. 4-28. The angular dimensioning command draws an arc to show the angle between two lines. The diameter command shows the diameter of a circle or an arc. Radius dimensioning commands show the radius of a circle or arc. Fig. 4-29.

The dimensioning menu will list the commands for the four basic types of dimensioning. For example, these might be the words *linear, diameter, angular,* and *radius.* You pick the command for the type of dimension, such as linear-horizontal, and indicate its location on the drawing. The computer calculates the numerical value of the dimension, and the number, dimension line, and required extension lines appear on the monitor.

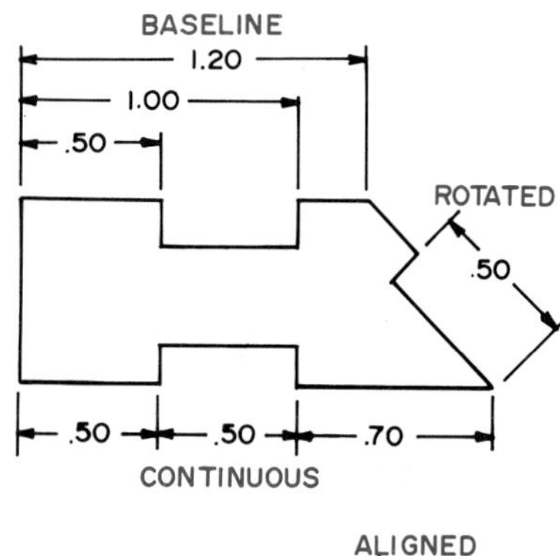

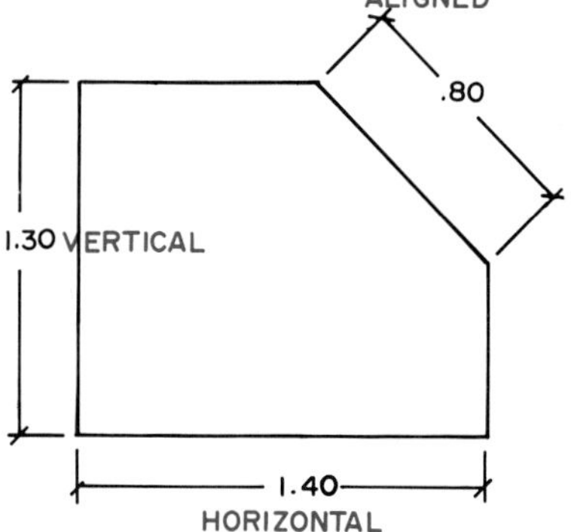

Fig. 4–28. Typical linear dimensions. Note that the dimension lines in the lower drawing have tick marks rather than arrowheads.

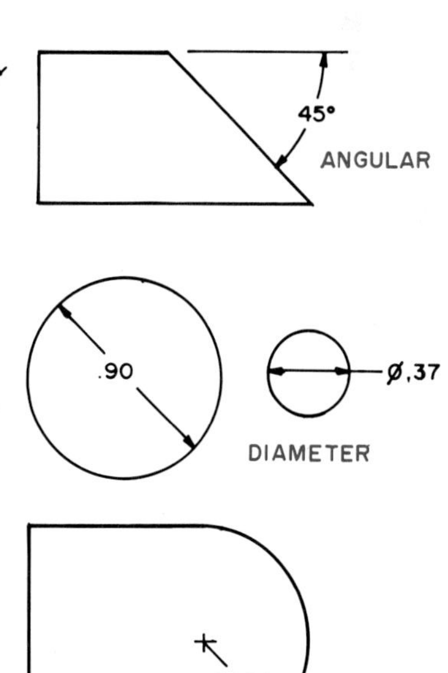

Fig. 4–29. Typical angular, diameter, and radius dimensions.

There are many other features to dimensioning. For example, you can indicate whether you want arrows or tick marks at the ends of dimension lines, specify their size, select the height of the text (words and numerals), and choose other variables such as rounding off values. With **associative dimensioning,** the dimension value is automatically updated if an object's size is altered.

Placing Notes on Drawings

Text Command. Text is placed on a drawing using the text command. A variety of text styles and sizes is often available. Text is placed in several ways. One method is to show the starting point on the screen and type the text. Another will adjust the text size and angle to fit between two designated points. Text can be centered and justified (aligned) on the right side if desired.

Layers and Colors

Layers (also called levels) are similar to the transparent overlays used in manual drafting. A drawing may contain several layers with different data on each. For example, an architectural floor plan could be drawn on layer 1. On layer 2 the electrical system could be drawn. On layer 3 the plumbing system could be shown.

Each layer is given a name or number. The layers can be turned on or off. Layers that are turned off are not displayed on the monitor or plotted, but they remain a part of the drawing file.

Each layer is assigned a *color.* If you have a color monitor the layer will appear in the assigned color. Each layer is also assigned a *line type* (continuous, hidden, etc.). Entities will be drawn in their layer's line type unless otherwise specified.

Commands Used for Editing

As you progress with a drawing it will be necessary to occasionally make changes to it. After deciding what part of a drawing you want to edit, or change, you may have to isolate it. This is done using the object selection process.

Object selection may be done by placing the cursor on the entity you want to change or by drawing a window around it. A window is a box created on the screen by indicating its lower left and upper right corners. Because a window can contain more than one entity, this method of object selection saves time.

Fillet Command. When two lines must be connected with an arc having a specified radius, use the fillet command. You have to indicate which lines are to be connected and give the radius of the fillet. Fig. 4-30.

Chamfer Command. A chamfer is a flat surface made by cutting off the corner of an object. You indicate the locations of the ends of the chamfer on the screen. When the chamfer command is given, the two lines will be trimmed by the specified amount and connected with a straight line. Fig. 4-31.

Erase Command. The erase command is used to remove lines, arcs, and other entities from a drawing. The entities to be erased are selected by placing the cursor on them or by drawing a window around them. This causes the entities to be highlighted. When the ENTER or RETURN key is pressed, the entities will disappear.

Unerase Command. If you erase something and see immediately that it was a mistake to do so, the unerase command will restore the part just erased. This command should be executed before you initiate another command.

Break Command. The break command is used to erase part of a line, circle, or arc. You must select the points for the beginning and end of the break. In Fig. 4-32 a notch is needed in the object. The opening for the notch is located and the line removed with the break command. Then the notch can be drawn using normal techniques.

Redraw Command. As a drawing is executed, a number of markers are used and remain on the screen. If they become a distraction, they can be removed by using the redraw command. This clears the screen of all marks and redraws the entities (lines, etc.).

Move Command. The move command is used to move something on a drawing, such as a hole, to another location. Fig. 4-33.

Copy Command. The copy command makes it possible to copy something on the drawing, such as a hole, in another place on the drawing. The location of the new position is given and the material is copied at this new position. Fig. 4-34.

Mirror Command. The mirror command is used to generate a mirror image of an existing object. A *mirror image* is the reverse of the original image. You place a mirror line on the drawing. The line becomes the axis about which the selected object is mirrored. Fig. 4-35.

Scale Command. The scale command permits the enlargement or reduction of an existing object. For example, if a factor of 2.0 is entered, the object will be enlarged to twice its original size. Fig. 4-36.

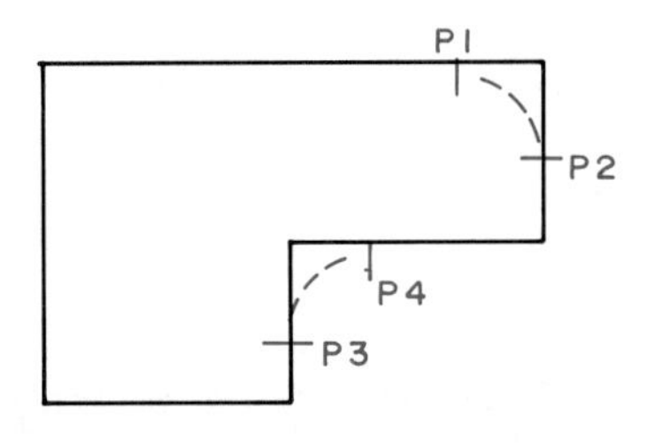
P1
P2
P4
P3
LOCATE THE TWO INTERSECTING LINES AND ENTER THE RADIUS.

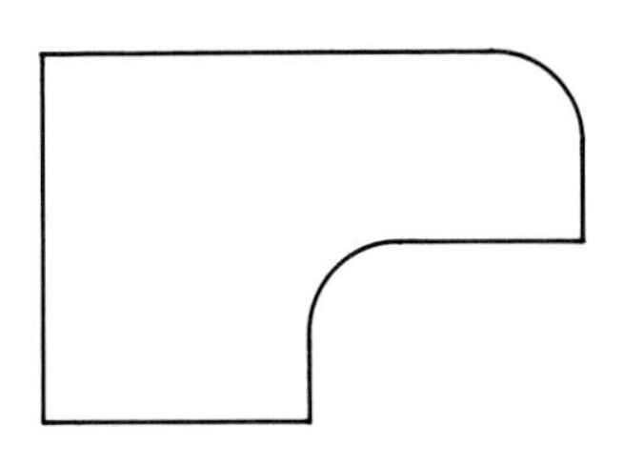
DRAW THE INDICATED FILLETS.

Fig. 4–30. Fillets are drawn by indicating the lines to be connected and giving the radius.

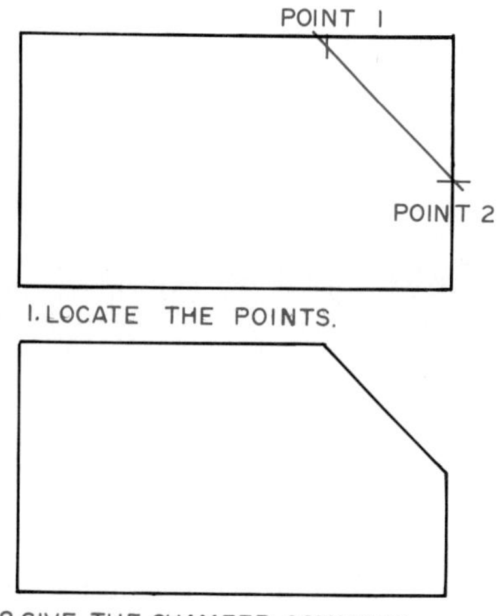
POINT 1
POINT 2
1. LOCATE THE POINTS.
2. GIVE THE CHAMFER COMMAND.

Fig. 4–31. The chamfer command is used to chamfer corners.

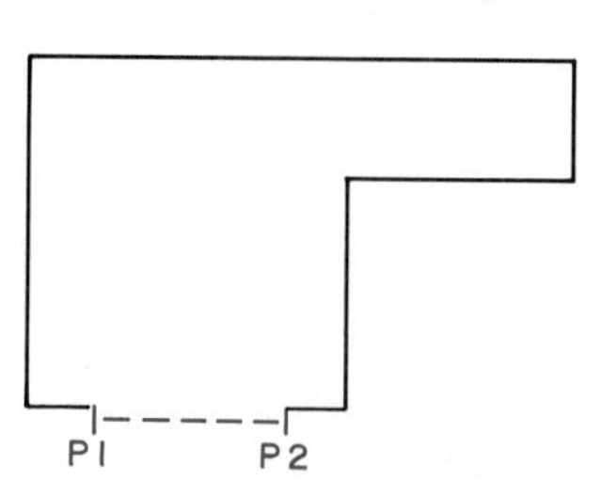
P1
P2
1. LOCATE THE BEGINNING AND END OF SECTION TO BE REMOVED.

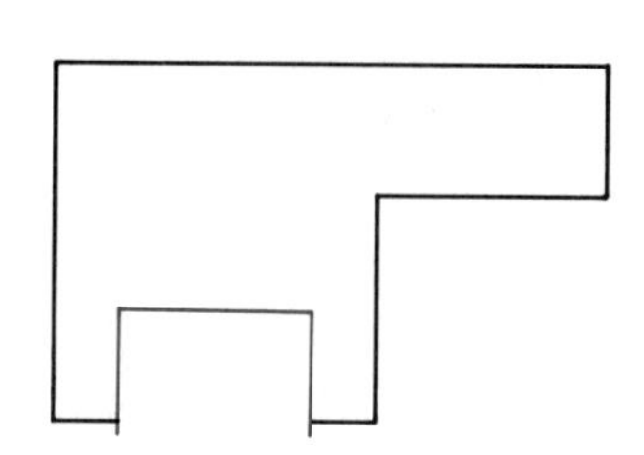
2. USE LINE COMMANDS TO DRAW THE NOTCH.

Fig. 4–32. The break command is used to erase a section of a line or other entity.

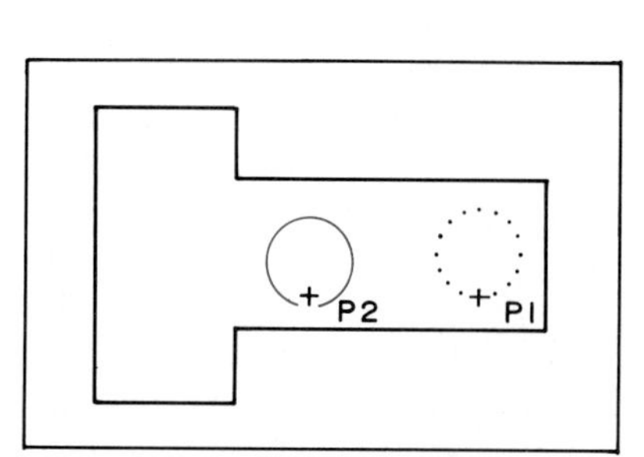
P2
P1

Fig. 4–33. The move command moves a feature such as a hole from P1 to P2.

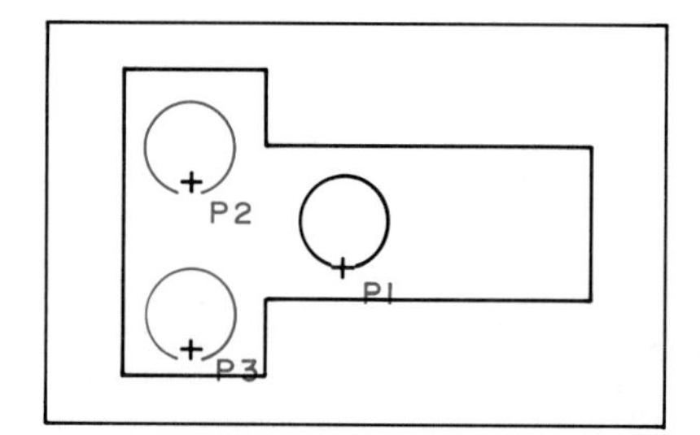
P2
P1
P3

Fig. 4–34. The copy command copies a feature such as a hole at P1 to points P2 and P3.

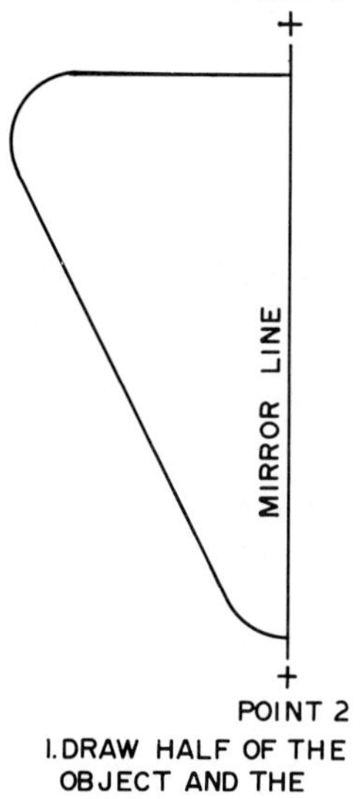
POINT 1
MIRROR LINE
POINT 2
1. DRAW HALF OF THE OBJECT AND THE MIRROR LINE.

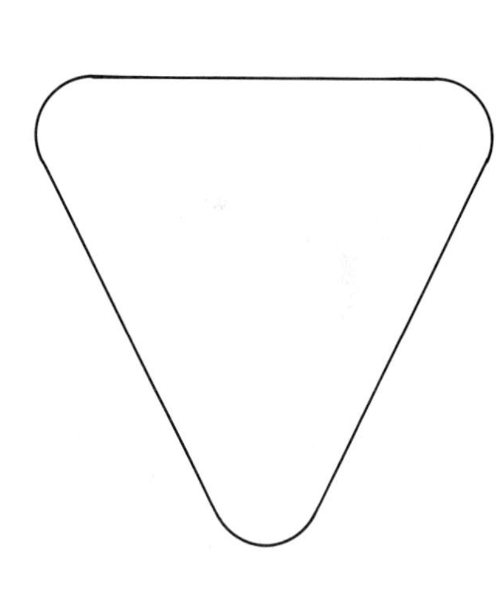
2. DRAW THE OTHER HALF USING THE MIRROR COMMAND.

Fig. 4–35. The mirror command makes it possible to produce the reverse image of the object drawn.

Stretch Command. The stretch command is used to change the size of an object by moving a selected portion. For example, a machine part can be made longer without redrawing. To stretch an object, place a window around the part to be stretched, then select a base point and a destination. Fig. 4-37.

Array Command. An array is a series of items arranged in a pattern. The array command is used to produce multiple copies of one or more objects in a circular or rectangular pattern. Examples of rectangular and circular arrays are shown in Fig. 4-38.

Commands Used to Display Drawings

Zoom Command. The zoom command is used to enlarge or reduce the size of a drawing on the monitor. When a drawing is small and detailed, it can be difficult to draw those details because the lines run together on the screen. With the zoom command the drawing can be enlarged on the screen without changing the size of the actual drawing. The zoom command is followed by a *magnification factor* such as 3. This results in a zoom magnification that shows the drawing three times its original size on the screen. Fig. 4-39. It is also possible to use a zoom window command. This permits only a part of a drawing to be enlarged. You draw a window around the area to be enlarged and give the desired magnification. The enlarged area appears on the screen. Fig. 4-40.

Pan Command. The pan command allows you to see a different portion of a drawing without changing its magnification. Suppose you have zoomed into an area on a drawing. Now you want to move the drawing a short distance so that you can work on a different area. The pan command enables you to do this. Fig. 4-41.

View Command. Views are stored zooms that are identified by a name. Whenever needed, a view can be recalled by using the view command. This is faster than zooming and panning repeatedly.

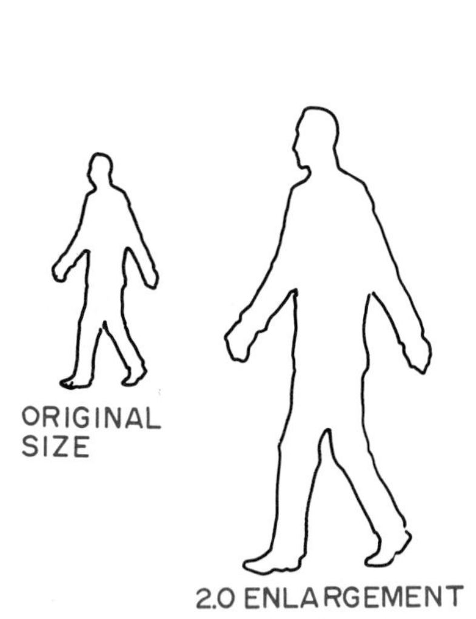

Fig. 4–36. The scale command is used to change the size of an object.

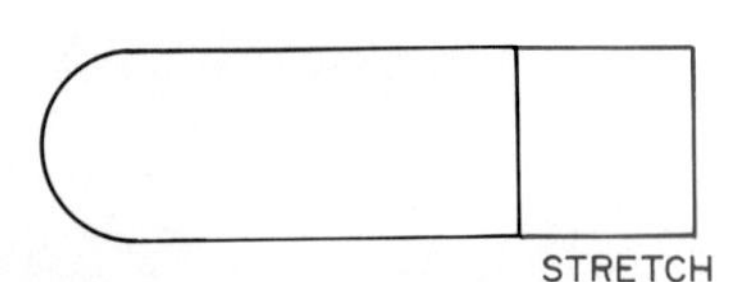

Fig. 4–37. The stretch command was used to change the length of this object.

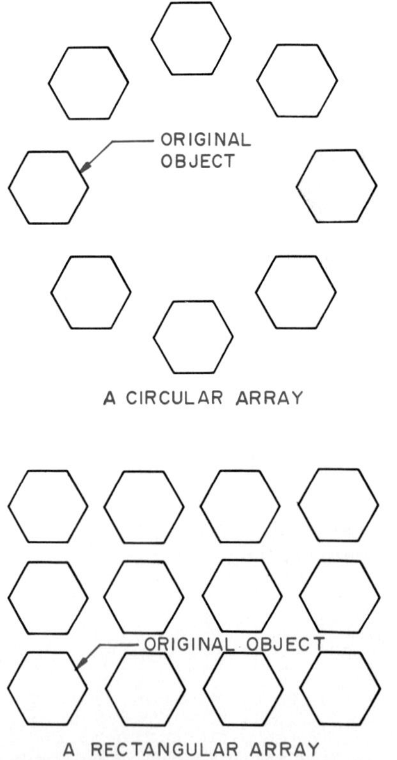

Fig. 4–38. These objects were arranged using the array command.

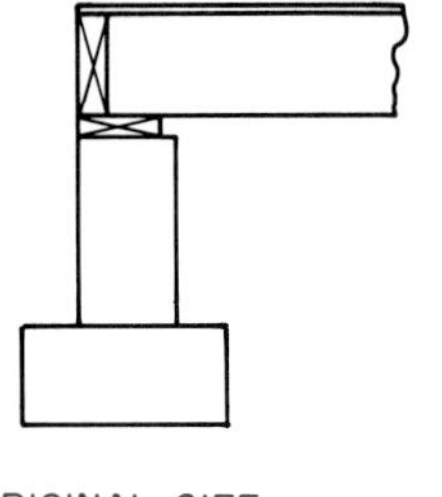

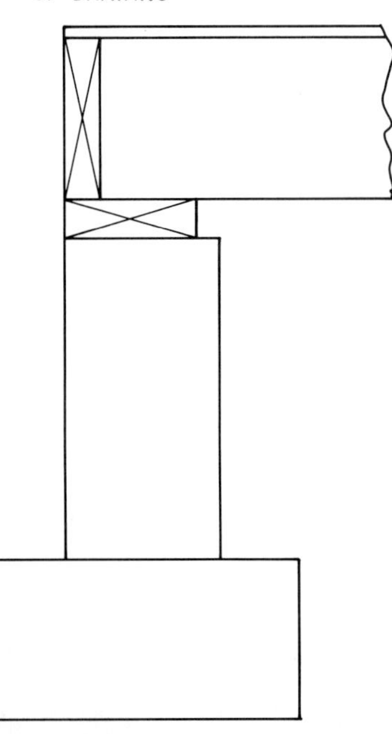

Fig. 4–39. The zoom command is used to enlarge a drawing. (Only part of the right-hand drawing would show on the screen.)

Commands for Saving or Discarding a Drawing

Save Command. The save command permits you to permanently store the work completed on the drawing since it was last saved. For example, suppose you start a drawing and work on it for an hour. Then there is a power failure or you forget and turn off the computer. All of your work will be lost. As you work it is wise to stop and save what you have done so that it will not all be lost accidentally.

End Command. The end command is used when you are finished working on a drawing and wish to save it and exit to the main menu. This command also *saves* the drawing on the disk you have designated by creating a file under the name you called the drawing when you started to work on it.

Quit Command. The quit command discards all the work performed on a drawing since the last save command. If you are working on a drawing and wish to discard everything you added since you last saved it, use the quit command. You will still have your original drawing.

Making a Drawing

The following example uses the AutoCAD® computer drafting program, a product of Autodesk, Inc. While other programs will have somewhat different commands and procedures, this example illustrates a typical sequence for making a drawing. The drawing produced using the following steps is shown in Fig. 4-42.

1. Load the AutoCAD program into the computer. The main menu will appear on the screen.
2. Choose number **1** (begin a new drawing) from the menu and press **RETURN**.
3. Enter the name of the floppy disk drive (or hard disk directory) upon which the drawing will be stored plus the name of the new drawing, such as **B:Example.** Press **RETURN**.

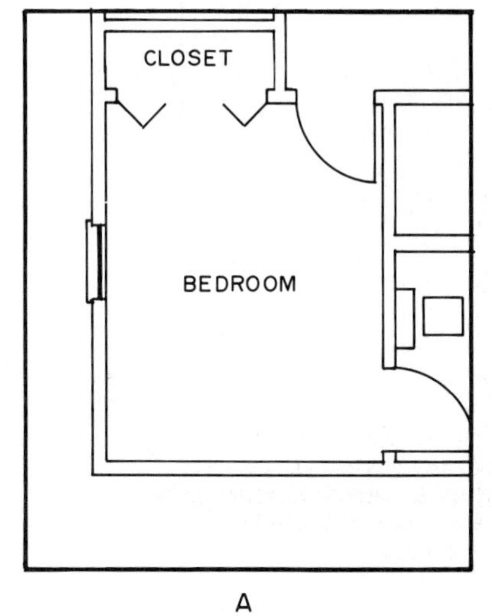

Fig. 4—41A. The zoom command was used to enlarge the bedroom of a floor plan, but now the rest of the floor plan is off the screen.

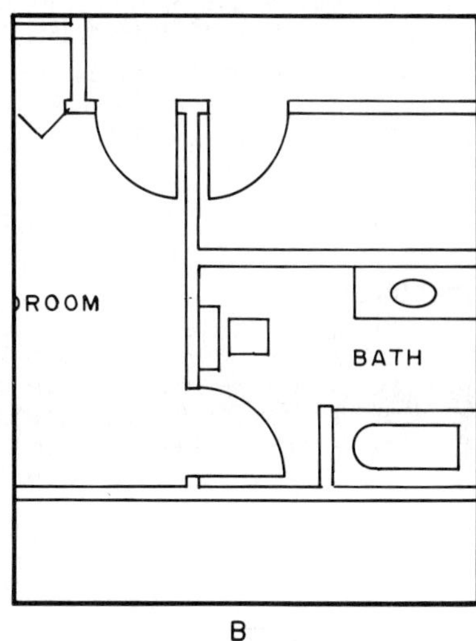

Fig. 4—41B. The pan command was used to move the drawing, bringing the bathroom into view.

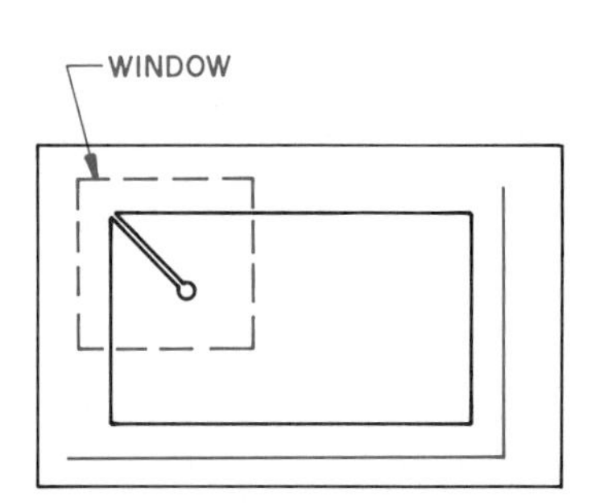

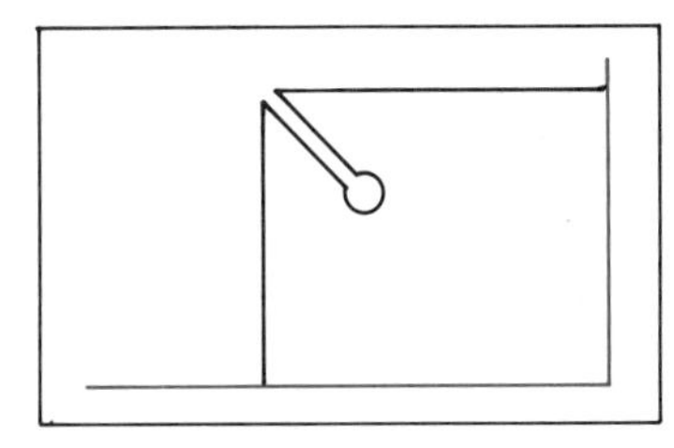

Fig. 4—40. A window area can be enlarged with a zoom window command.

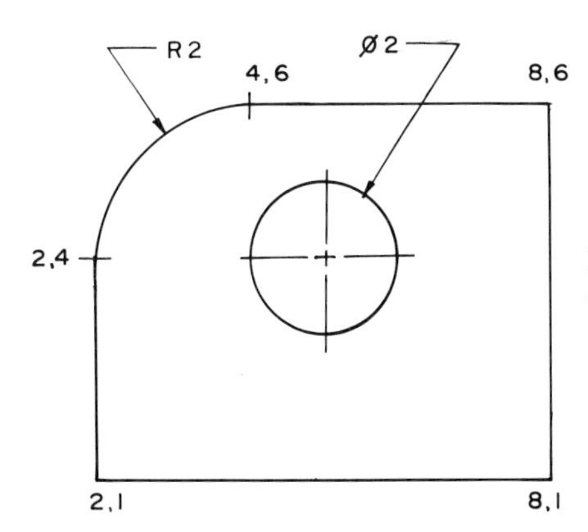

Fig. 4—42. The part to be drawn.

4. The computer will now go into the Drawing Editor mode. The Drawing Editor allows you to create, edit, view, and plot drawings. At this point, you could enter various drawing settings and parameters, such as the limits, units, grid, and snap spacing, etc. However, the AutoCAD program includes a prototype drawing in which all these settings and parameters have already been established. (There are no graphics in the prototype drawing.) Unless the drafter specifies otherwise, these settings and parameters will be in effect whenever a new drawing is begun. For this example, we will allow the prototype drawing to remain in effect.

5. To help with the drawing, place a grid on the screen. Use a spacing of one unit between dots. (In the following examples, the words and symbols preceding the colon show the messages, or prompts, which the computer displays to the drafter. The drafter's input is shown after the colon, in **boldface** type.)

Command: **GRID** (Press **RETURN** key)
Grid spacing (X) or
ON/OFF/SNAP/ASPECT *0*: **1 (RETURN)**

6. To begin drawing, lay out the straight lines using absolute coordinates entered at the keyboard. An alternate method is to pick the points using a stylus, puck, or mouse.

Command: **LINE (RETURN)**
From point: **2,1**
To point: **8,1**
To point: **8,6**
To point: **4,6**
To point: **RETURN**

This gives the layout shown in Fig. 4-43.

7. Next draw the arc starting at 4,6.

Command: **ARC**

From the screen menu, choose **SEA** (start, end, included angle). The program will ask for the start point. Choose **contin** (continue) from the screen menu. The arc will start from the last point entered which was 4,6. The program will ask for the end point, so enter **2,4** and press **RETURN**. Fig. 4-44.

8. Now draw a straight line with 2,4 as the start point and 2,1 as the end point.

Command: **LINE (RETURN)**
From point: **2,4**
To point: **2,1**
To point: **RETURN**

This completes the shell of the object.

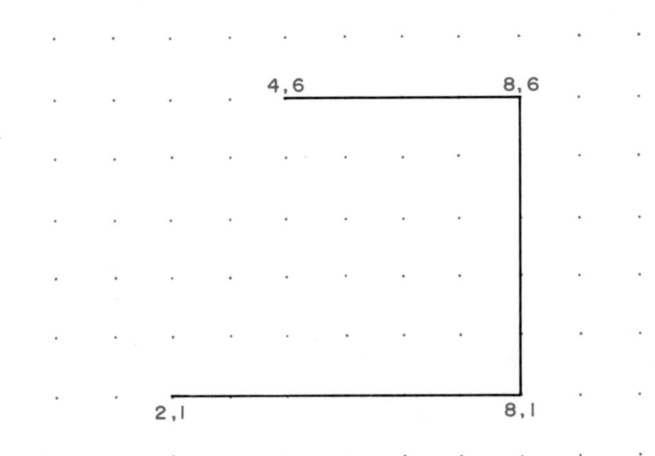

Fig. 4-43. The straight lines are drawn beginning at point 2,1 and ending at point 4,6.

9. Next draw the circle.

Command: **CIRCLE**
3P/2P/TTR/<Center Point>: **5,4**
Diameter/<Radius>: **1**

This draws a circle whose center is at 5,4 and whose diameter is 2. Fig. 4-45.

10. To clean the screen of marks use the REDRAW command.

Command: **REDRAW (RETURN)**

11. Now conclude the work and save the drawing.

Command: **END (RETURN)**

The drawing will be recorded on the disk, and you will be returned to the main menu.

Plotting the Drawing

The drawing can be plotted on a printer or a plotter. Printer plots are less expensive, but the line quality is poor as compared to plotter drawings. Printer plots are an easy way to check a drawing before running it on a plotter.

Types of Plots

There are usually several choices of plots available. These depend upon the software and plotter being used. Following are some of the options:

- A DISPLAY option will plot what is presently on the monitor.
- A LIMITS option uses the drawing limits as the border definition of the drawing.
- A WINDOW option permits you to put a window around part of the drawing and plot the area inside the window.

Operating the Plotter

Exact commands and procedures will vary depending upon the software and equipment you use. You will have to give a command to indicate whether you want a pen plot or printer plot. You must indicate the part of the drawing to be plotted by entering information such as DISPLAY,

LIMITS, or WINDOW. The software will then ask for specifications such as coordinates where the plot begins, plotting area, pen width, and plotting units in inches or millimeters. After all the inquiries are completed, the proper command will start the printer or plotter into operation.

Customizing CAD Systems

Typically, CAD programs are designed for general purpose design and drafting. They can be adapted to meet the needs of many fields such as architecture or structural engineering. These customized systems can be developed by selecting hardware useful to the expanded purpose, purchasing or developing symbol libraries, customizing screen and tablet menus, and adding third-party software to the CAD software.

Third-party software are programs used to enhance the capabilities of the basic CAD program. For example, an architectural program can be linked to the basic CAD program. It will provide an additional library containing architectural symbols, details, and other needed features for architectural drawing. There are many other programs that are used to customize CAD for specific needs such as static and dynamic analyses made by structural engineers as they design structures, preparing part design and tool path information for numerically controlled machine tools, connecting CAD systems within and between companies, and the tie-in of word processors for text editing with the graphic capacity of CAD. Text and extensive notes can then be prepared on a word processor and transferred to a CAD system to rapidly produce text on drawings.

From these few examples you can see that CAD is more than simply making a drawing. It helps users make decisions, provide comparisons, and make engineering analyses. Over the years it will find ever increasing applications in design and drafting.

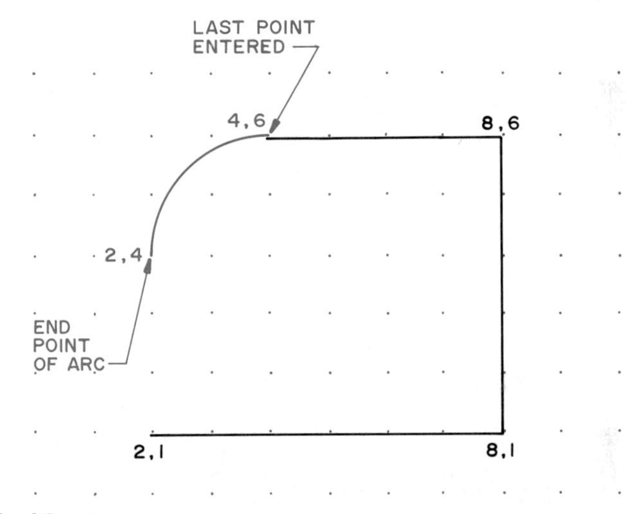

Fig. 4–44. **The arc is drawn starting at point 4,6 and ending at point 2,4.**

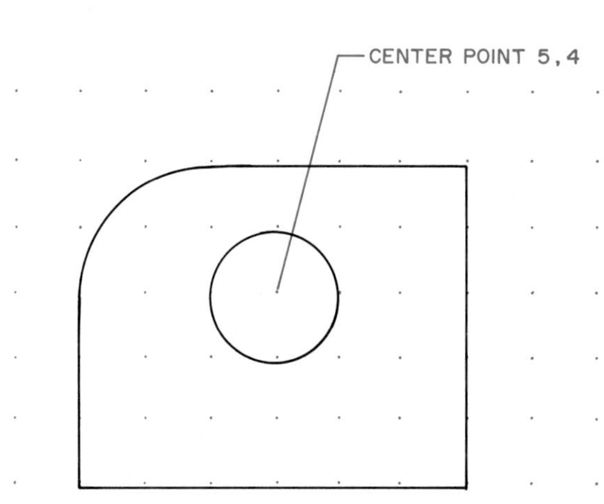

Fig. 4–45. **The circle is drawn using the circle command. The center is located at coordinates 5,4 and a radius of 1 is used.**

Chapter Review

Build Your Vocabulary

You should understand and use the following terms as part of your working vocabulary. Write a brief explanation of what each means.

hardware
central processing unit
read only memory
random access memory
expansion slots
input/output ports
floppy disk
hard disk
monitor
pixels
keyboard
digitizer
stylus
puck
mouse
light pens
joystick
dot matrix printer
letter quality printer
laser printer
flatbed plotter
drum plotter
thermal plotters
electrostatic plotters
software
menu
entities
units
limits
scale
absolute coordinates
relative coordinates
polar coordinates
symbol library
object snap facility
associative dimensioning

Study Problems—Directions

If your class has access to a CAD system, use the study problems in other chapters of this book to gain practice in computer-aided drafting. Keep in mind the CAD features that can simplify and speed up drawing, such as copying, mirroring, semiautomatic dimensioning, and symbol libraries. Plan how you will use these CAD features before you start your drawing. That way, you will make the most efficient use of the system.

To gain practice using the CAD system, try drawing the illustration here. This is an end elevation for a proposed Space Shuttle design.

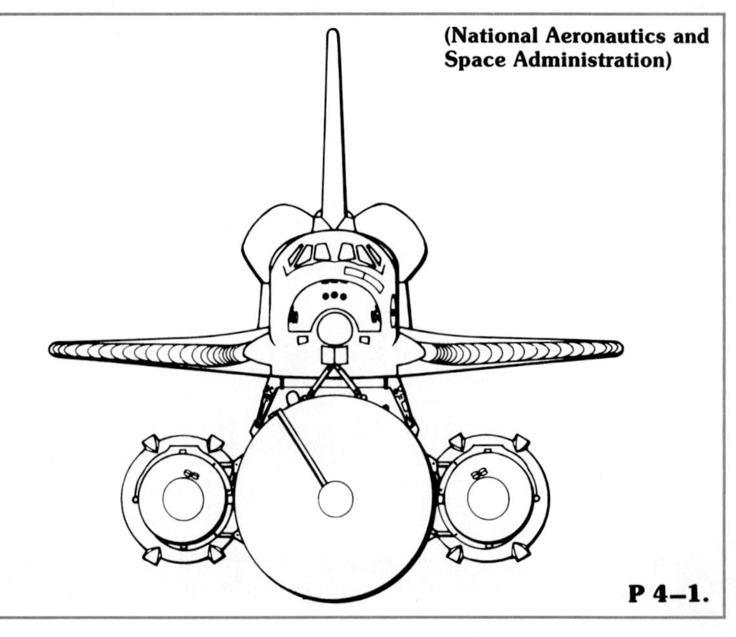

(National Aeronautics and Space Administration)

P 4–1.

CHAPTER 5

Geometric Figures and Constructions

After studying this chapter, you should be able to do the following:

- Lay out two-dimensional geometric shapes.
- Recognize the basic geometric solids.
- Perform basic geometric constructions.
- Locate tangent points.

Geometry in Drafting

Drafters must often draw angles and shapes that represent the intersections of lines and surfaces. Constructing certain basic shapes involves using the principles of geometry. You, as a drafter, must be able to use geometric constructions to make your drawings properly. Geometric constructions are also vital to the work of designers, engineers, surveyors, architects, scientists, and mathematicians. Geometry is a basic knowledge that must be learned for success in any of these fields. Using a computer to produce drawings does not reduce your need to understand geometry. After all, the computer will only draw those geometric figures you tell it to draw. Fig. 5-1.

Geometry is a study of the properties, measurements, and relationships of points, lines, angles, surfaces, and solids. Plane geometric figures that are used often in drafting include squares, rectangles, circles, triangles, and hexagons. Geometric solids that are often drawn include cubes, spheres, triangular prisms, and cylinders. These various geometric constructions are made from lines and points connected in the proper manner.

Figure 5-2 shows many of the geometric figures used by drafters. Study these carefully so you know the meaning of each of their terms. Geometry begins with lines and points. A *line* is the shortest distance between two points. A *point* is an imaginary spot that has no size but is used to locate a position. The intersection of two lines is a point because it is a location. The measurement from one of those intersecting lines to the other is an **angle.** Other geometric terms you should know are the following:

Fig. 5–1. The drafter must understand geometric construction to produce drawings by any method. (Hewlett-Packard)

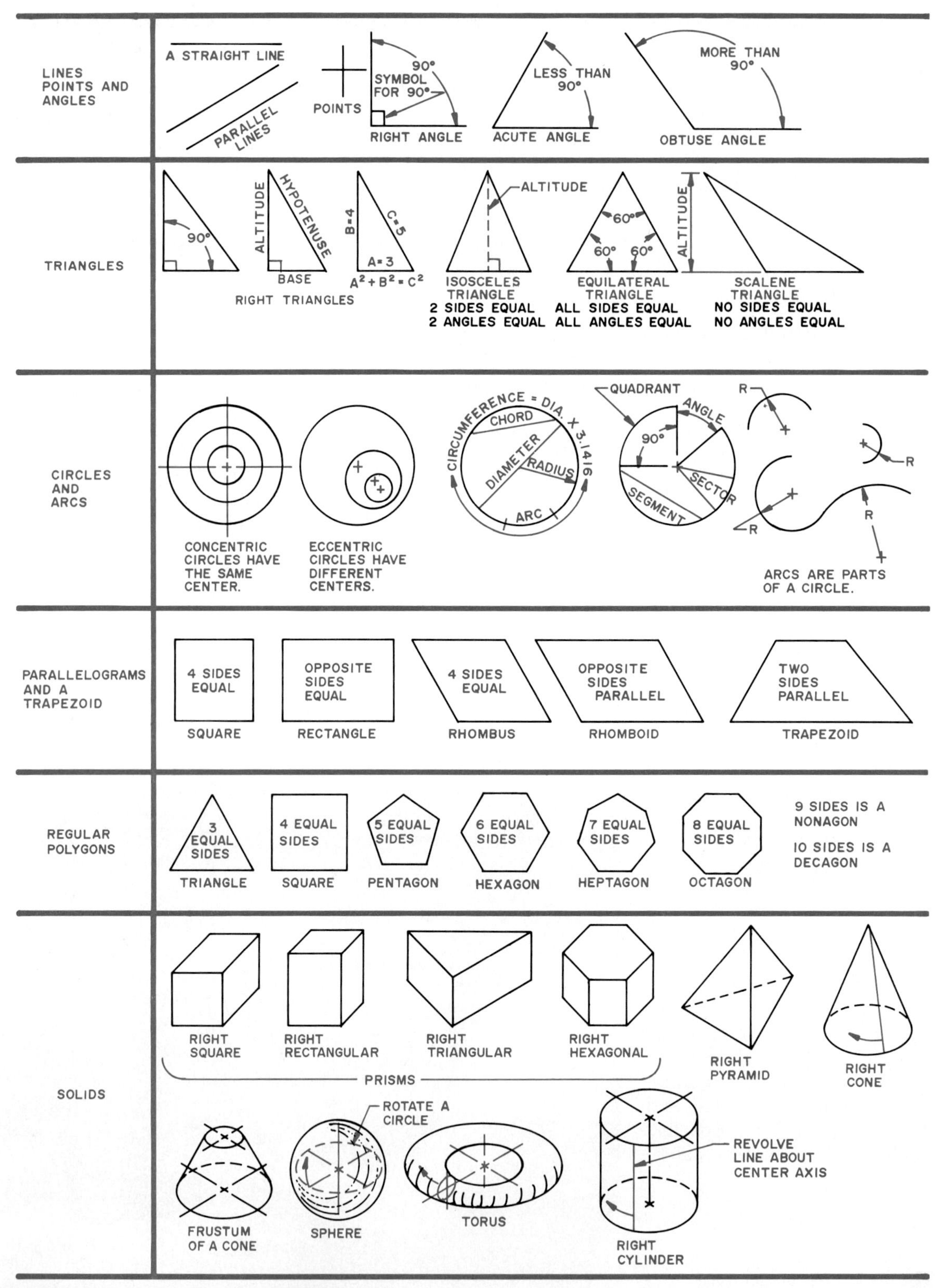

Fig. 5–2. Geometric figures and solids.

- **Parallel** lines are those that are the same distance apart along their entire length.
- A **triangle** is a figure formed by three lines meeting by twos at three points.
- A **circle** is a closed curve that has a fixed **radius** (the distance from the center to the curve). That is, all points on a curve are the same distance from the center.
- An **arc** is a portion of a circle.
- A **parallelogram** is a four-sided figure that has opposite sides parallel.
- A **polygon** is any figure with sides made of straight lines. Figures with sides of equal length and equal angles are called *regular* polygons.
- A **prism** is a geometric solid that has two parallel bases. It may have three or more faces connecting the parallel bases.
- A **cylinder** is formed by moving a line in a circular path around a central axis.
- A **pyramid** has a base in the shape of any polygon, such as a triangle, square, or hexagon. Its sides are triangular and meet at a point.
- A **cone** is formed by moving one end of a straight line in a circle while holding the other end at one point.
- A **sphere** is formed by spinning a circle around its **diameter** (a straight line reaching from one side of a circle to the other and passing through its center).
- A **torus** is formed by moving a circle around a central axis.

Geometric Construction

Bisect a Line

To **bisect** a line means to divide it into two equal parts. The steps to bisect a line are shown in Fig. 5-3.

1. Set your compass with a radius greater than half the length of the line to be bisected (line AB in Fig. 5-3).

2. Put the pin of the compass at one end of the line and swing an arc above and below the line.

3. Place the pin on the other end of the line and swing two more arcs that cross the first arcs drawn. This step forms points C and D.

4. Connect points C and D with a straight line. This line bisects line AB.

Bisect an Angle

To bisect an angle means to divide it into two equal angles. Fig. 5-4.

1. Set your compass to any length radius.

2. Put the pin at point A and swing an arc that crosses both sides of the angle. This locates points B and C.

3. With a new radius longer than half the distance from B to C, put the pin on one of the intersections just found, such as B, and swing an arc. With the same radius, put the pin on the other intersection (C) and swing another arc that crosses the arc just drawn. This locates point D.

4. Connect points A and D with a straight line. This line bisects the angle.

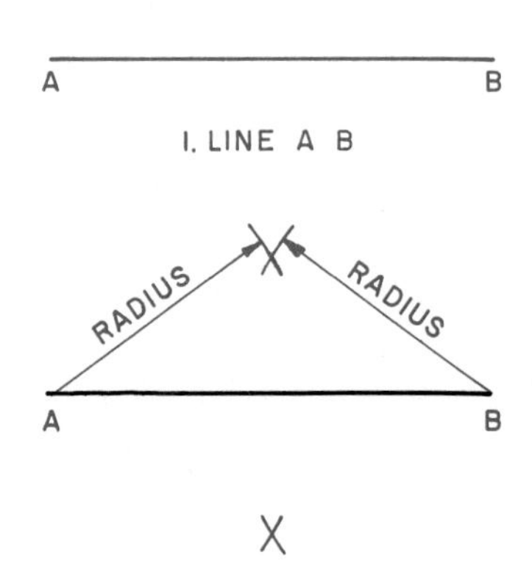
A
B
1. LINE A B
RADIUS
RADIUS
A
B
2. SWING ARCS FROM EACH END OF THE LINE. THE RADIUS MUST BE GREATER THAN HALF THE LENGTH OF THE LINE.

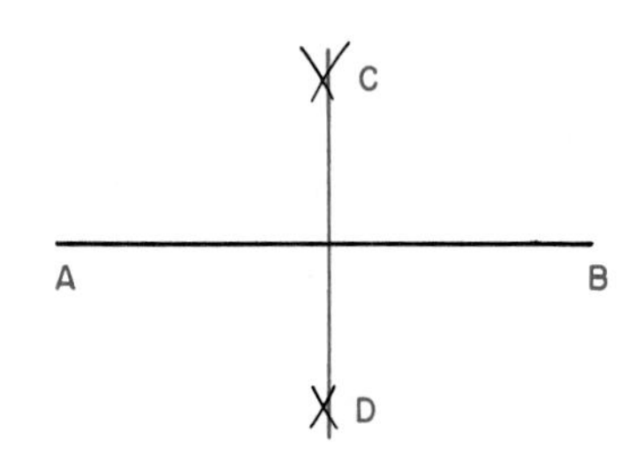
C
A
B
D
3. CONNECT THE POINTS WHERE THE ARCS CROSS. THESE ARE POINTS C AND D.

Fig. 5—3. Bisecting a line using a compass.

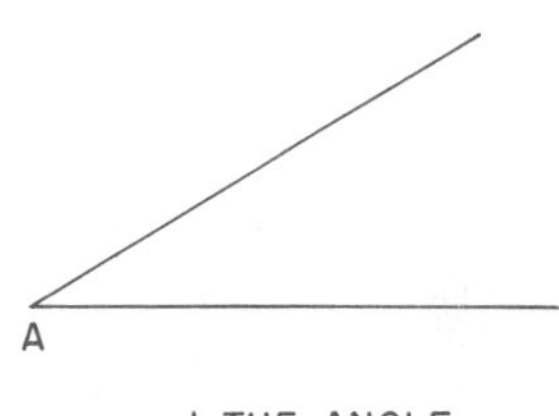
A
1. THE ANGLE

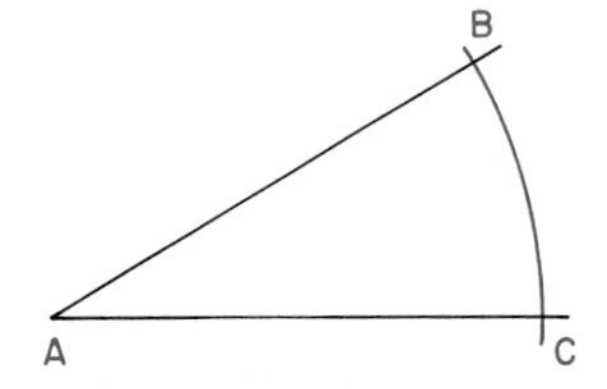
B
A
C
2. SWING AN ARC CUTTING BOTH SIDES OF THE ANGLE.

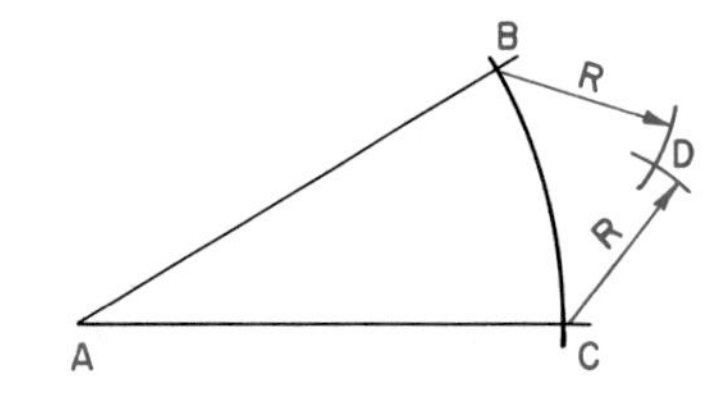
B
R
D
R
A
C
3. USING B AND C AS CENTERS SWING ARCS THAT WILL CROSS.

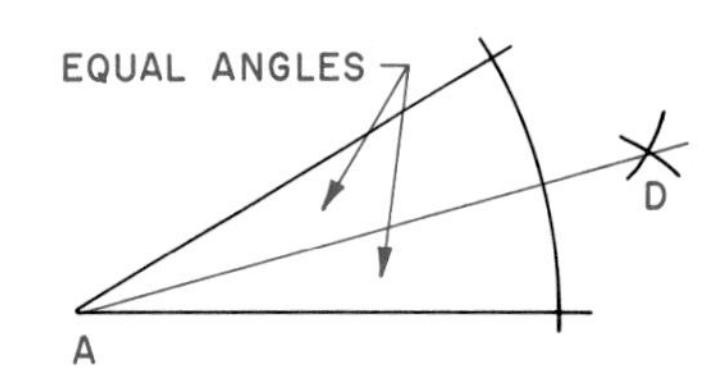
EQUAL ANGLES
D
A
4. DRAW A LINE FROM A TO D. THIS IS THE BISECTOR. THE ANGLES ARE EQUAL.

Fig. 5—4. Bisecting an angle using a compass.

Bisect an Arc

To bisect an arc means to divide an arc into two equal arcs. Fig. 5-5.

1. The arc to be bisected is AB.
2. Connect points A and B with a straight line. This line is called a **chord.**
3. Bisect the chord in the same way you would bisect any line. Doing so also bisects the arc.

Divide a Line into Equal Parts

This construction will work with any number of divisions.

1. Suppose line AB in Fig. 5-6 is to be divided into five equal parts.
2. Draw a line (BC) at any angle to line AB and any length.
3. Along line BC measure and mark five equal lengths using your compass. Call the last point you mark point D.
4. Connect point D with point A. Draw lines parallel with line DA from the other four points on line BD so that they intersect line AB. The place where those lines cross AB marks an equal division (one-fifth) of that line.

Draw a Line Perpendicular to a Point on a Line

A **perpendicular** line meets another line at a right (90-degree) angle. Fig. 5-7.

1. A line is to be drawn perpendicular to AB at point C.
2. From the point (C) where the perpendicular is to meet line AB, swing an arc of any convenient radius so that it crosses line AB in two places. This locates points D and E.
3. From points D and E, swing arcs of the same radius (any length longer than DC or CE) until they cross and form point F. Connect points F and C with a straight line. This line is perpendicular to line AB.

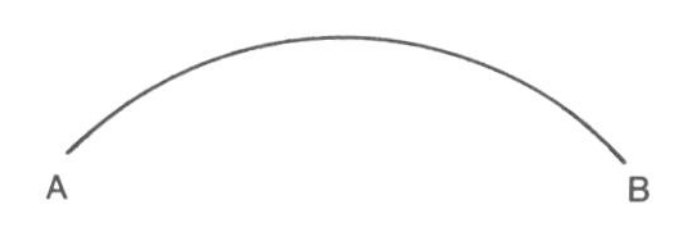

1. THE ARC TO BE BISECTED.

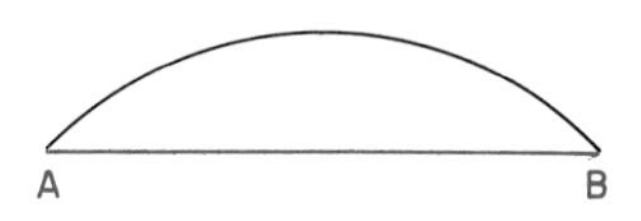

2. DRAW A LINE THROUGH THE ENDS OF THE ARC.

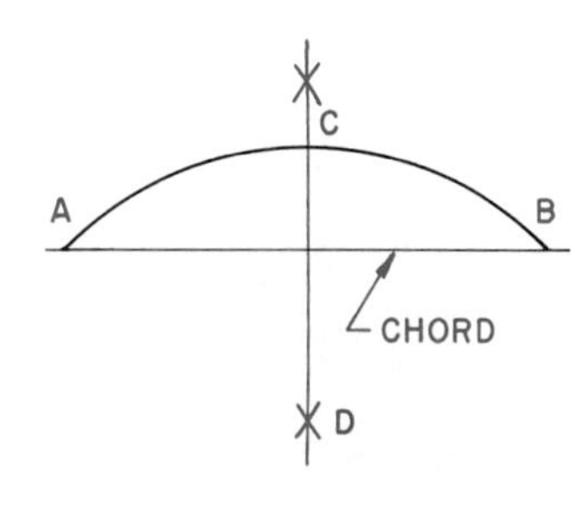

3. BISECT THIS LINE. THE ARC FROM A TO C IS THE SAME LENGTH AS THAT FROM C TO B.

Fig. 5–5. Bisecting an arc using a compass.

1. LINE AB TO BE EQUALLY DIVIDED INTO SEVERAL PARTS.

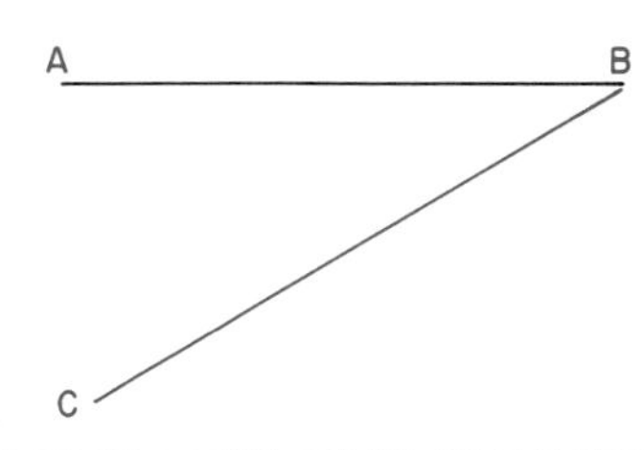

2. DRAW A LINE ON AN ANGLE FROM ONE END OF THE LINE TO BE DIVIDED. THE LINE MAY BE ANY LENGTH AND ANGLE.

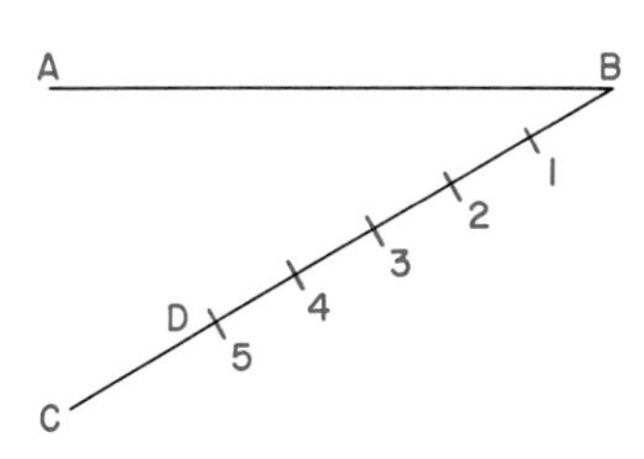

3. MARK OFF EQUAL SPACES, THE SAME NUMBER AS IS DESIRED TO DIVIDE LINE AB (5 IN THIS EXAMPLE.)

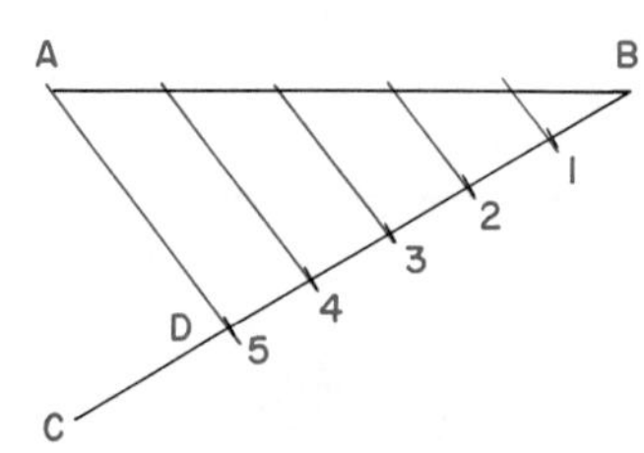

4. CONNECT D TO A. DRAW LINES PARALLEL TO AD THROUGH ALL OTHER POINTS ON BC.

Fig. 5–6. Dividing a line into equal parts.

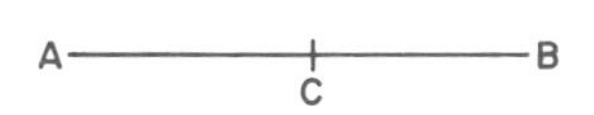

1. A PERPENDICULAR TO LINE AB IS TO BE DRAWN AT C.

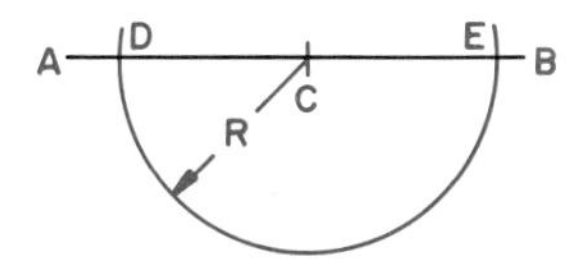

2. SWING AN ARC FROM C, USING ANY CONVENIENT RADIUS, LOCATING POINTS D AND E.

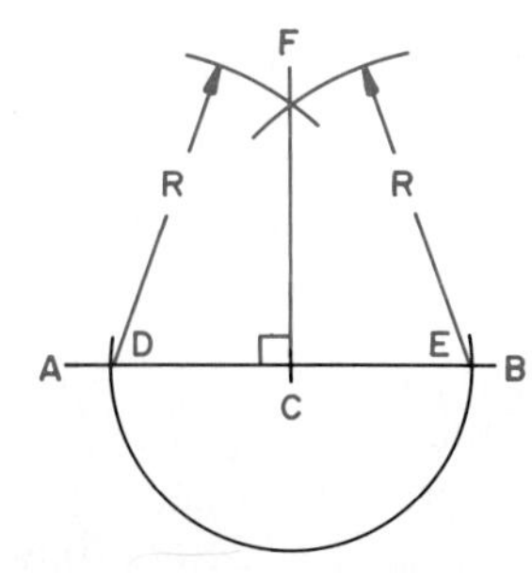

3. WITH D AND E AS CENTERS, SWING ARCS OF ANY RADIUS UNTIL THEY INTERSECT AT F. CONNECT F AND C.

Fig. 5–7. Drawing a line perpendicular to a point on a line.

Draw a Line Perpendicular to Another Line from a Point Not on That Line

This technique is shown in Fig. 5-8.

1. Suppose a line is be drawn perpendicular to AB.
2. From any point (C) not on line AB, swing an arc with a radius long enough to cross AB twice (at points D and E).
3. From points D and E, swing arcs of the same radius (any length longer than half the distance DE) until they cross and form point F. Connect points C and F with a straight line. Line CF is perpendicular to line AB.

Construct an Equilateral Triangle

An **equilateral** triangle is one that has each side the same length. Fig. 5-9.

1. Draw one side, AB, to the correct length.
2. Set a compass to the length of AB. Using that radius, swing arcs with A and B as centers. The point where the arcs intersect is point E.
3. Use straight lines to connect E with A and B to form the equilateral triangle.

Construct an Isosceles Triangle

An **isosceles triangle** has two equal sides and two equal angles. Fig. 5-10.

1. Draw the base, AB, to length.
2. Set a compass with a radius equal to the length of the two equal sides. From points A and B, swing two arcs that intersect at C.
3. Connect AC and BC to form the isosceles triangle.

+C

A B

1. A LINE IS TO BE DRAWN FROM C PERPENDICULAR TO AB.

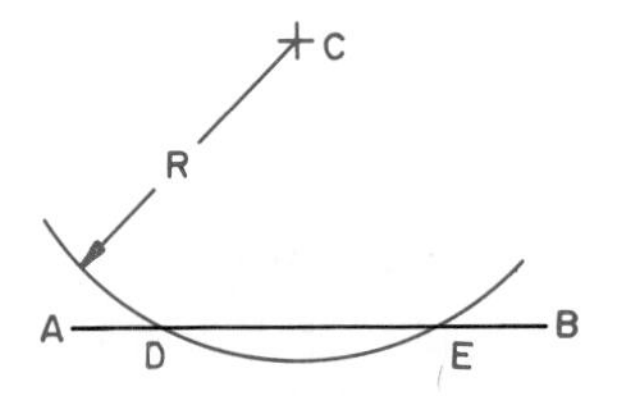

2. SWING AN ARC FROM C WITH A RADIUS LARGE ENOUGH TO INTERSECT AB TO LOCATE D AND E.

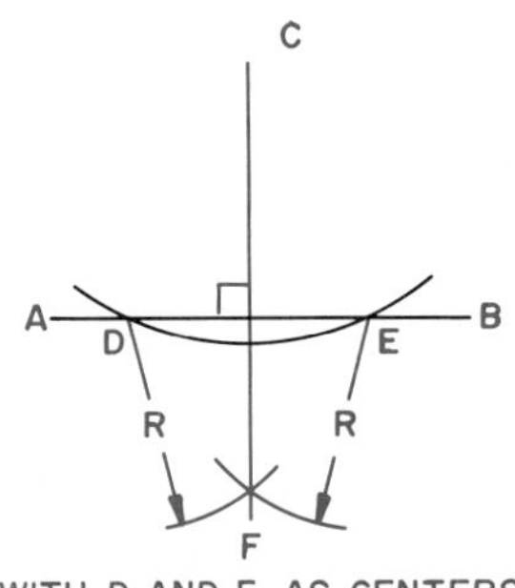

3. WITH D AND E AS CENTERS, SWING ARCS OF ANY RADIUS LONG ENOUGH TO INTERSECT AT F. CONNECT F AND C TO FORM THE PERPENDICULAR LINE.

Fig. 5—8. Drawing a line perpendicular to a line from a point off the line.

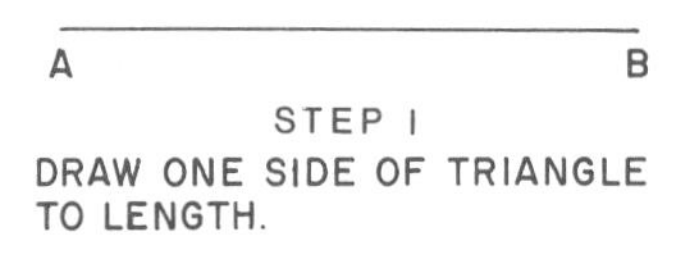

STEP 1
DRAW ONE SIDE OF TRIANGLE TO LENGTH.

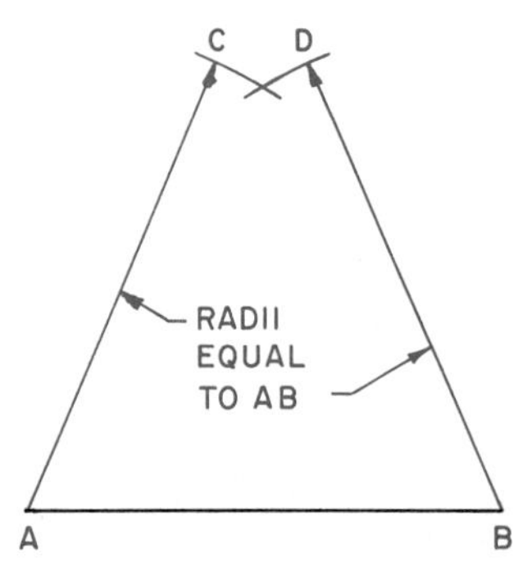

STEP 2
SET A COMPASS TO LENGTH AB. SWING ARCS WITH A AND B AS CENTERS.

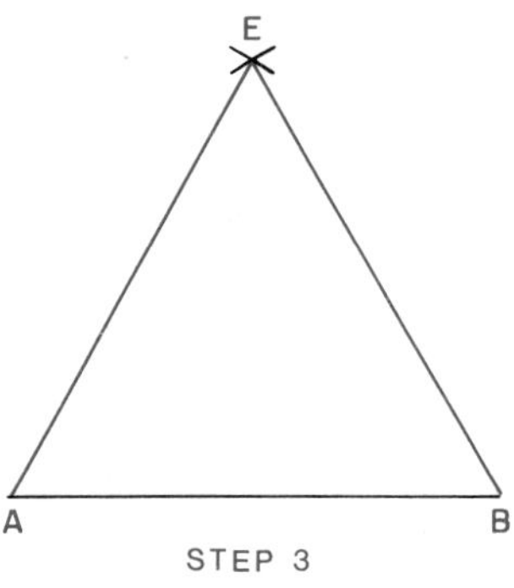

STEP 3
CONNECT POINT E, THE INTERSECTION OF ARCS, WITH A AND B. THIS FORMS THE TRIANGLE.

Fig. 5—9. Drawing an equilateral triangle using a compass.

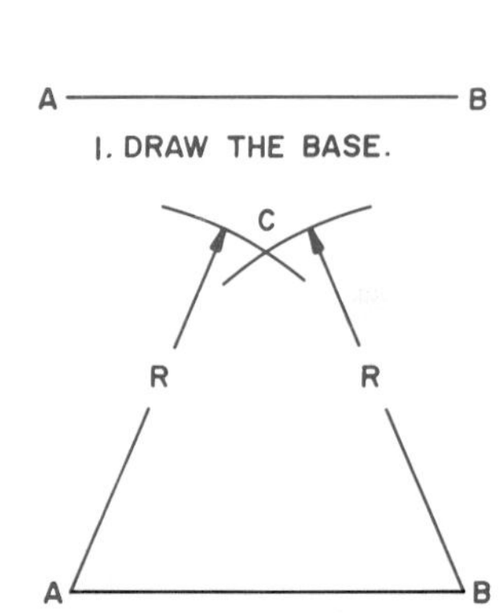

2. SWING ARCS FROM A AND B HAVING A RADIUS EQUAL TO THE LENGTH OF THE SIDES.

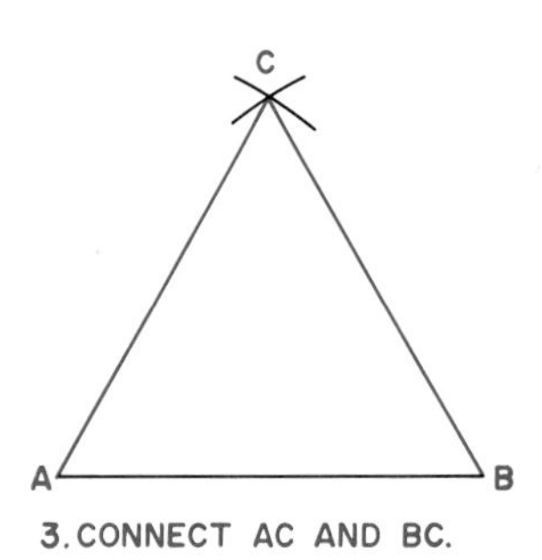

3. CONNECT AC AND BC.

Fig. 5—10. Drawing an isosceles triangle.

Draw a Square When the Length of the Diagonal Is Known

A **square** is a regular polygon made of four equal sides and four equal angles. Fig. 5-11.

1. Draw a circle with a diameter (AB) equal to the diagonal of the square.
2. Draw another diameter at a right angle to AB.
3. With straight lines, connect the points where the diameter lines touch the circle. This forms the square. The square is **inscribed** within the circle; that is, the square is drawn inside the circle so as to touch in as many places as possible.

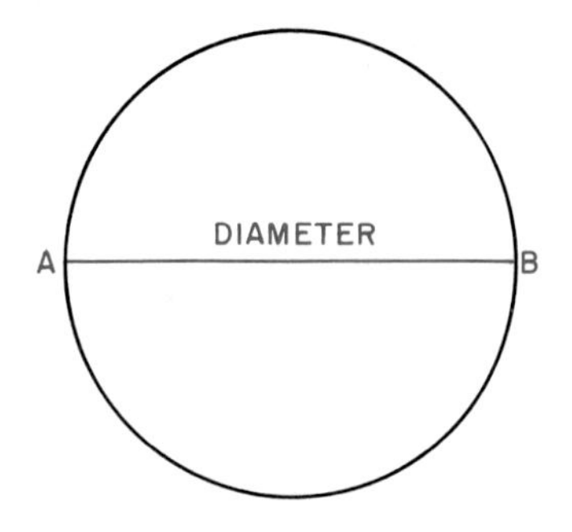

1. DRAW A CIRCLE WITH A DIAMETER EQUAL TO THE DIAGONAL OF THE SQUARE.

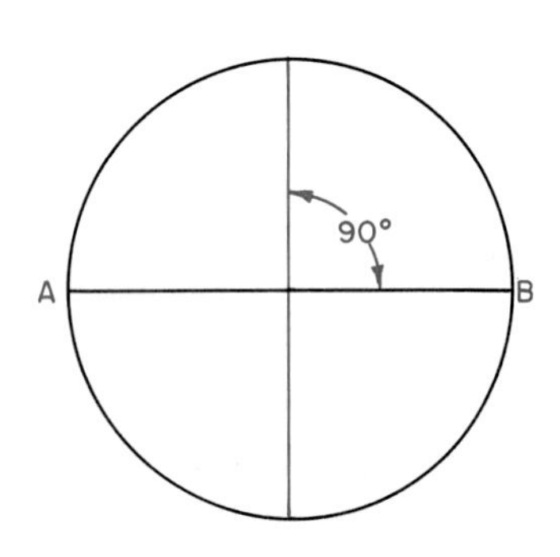

2. DRAW A CENTER LINE AT A RIGHT ANGLE TO THE DIAMETER AB.

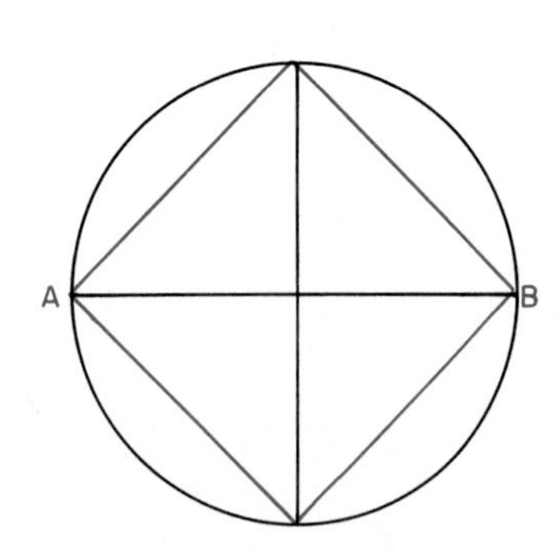

3. CONNECT THE POINTS WHERE THE CENTER LINES TOUCH THE CIRCLE.

Fig. 5–11. Drawing a square when the length of the diagonal is known.

Draw a Square When the Length of the Sides Is Known

When you complete this construction, you will have a square that **circumscribes** (goes around) the circle. Fig. 5-12.

1. Draw a circle with a diameter equal to the length of the side of the square.
2. Draw the center lines of the circle at right angles to each other and longer than the diameter.
3. Draw lines at 45-degree angles from each center line so that they cross the center lines and are **tangent** to (touch but don't cross) the edge of the circle. The 45-degree lines will intersect at the center lines, thus forming the square.

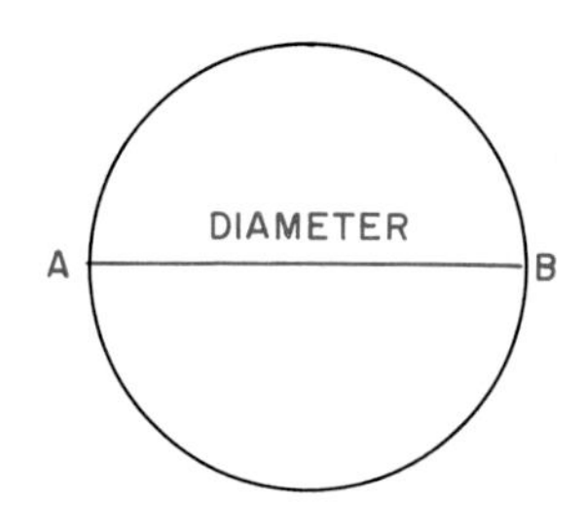

1. DRAW A CIRCLE WITH A DIAMETER EQUAL TO THE LENGTH OF THE SIDE OF THE SQUARE.

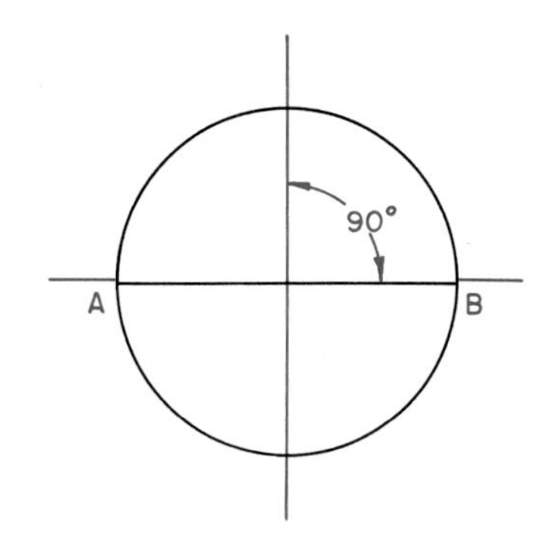

2. DRAW CENTER LINES AT RIGHT ANGLES TO EACH OTHER.

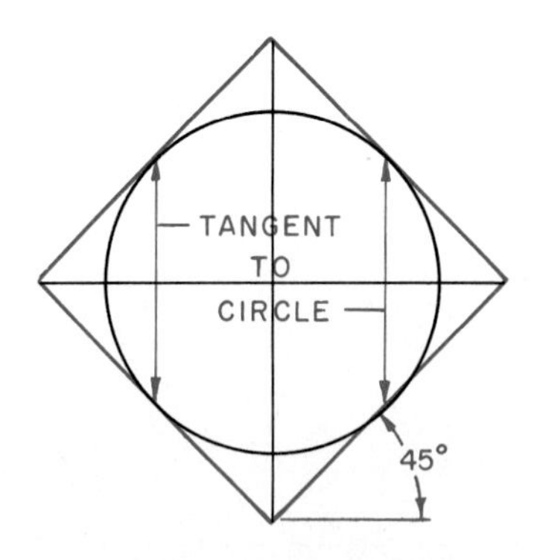

3. DRAW LINES AT 45° WITH THE HORIZONTAL THAT ARE TANGENT TO THE CIRCLE.

Fig. 5–12. Drawing a square when the length of the sides is known.

Draw a Hexagon When the "Across the Corners" Distance Is Known

A **hexagon** is a six-sided polygon. The term *across the corners* means the distance from one corner of the hexagon to the opposite corner. It is measured on a line that goes through the center of the hexagon. Fig. 5-13.

1. Draw a circle with a diameter equal to the distance across the corners of the hexagon. Fig. 5-14.
2. Using the same radius as the circle and A and B as centers, swing arcs until they each cross the circle at two places.
3. Connect each adjacent point of the six points found (including the diameter) with straight lines to form the hexagon. The hexagon is inscribed within the circle.

Draw a Hexagon When the "Across the Flats" Distance Is Known

The distance *across the flats* of a hexagon is the distance from one flat side to the opposite flat side. This distance is measured on a line that goes through the center of the hexagon. Refer to Fig. 5-13.

1. Draw a circle with a diameter equal to the across the flats distance. Draw center lines. Fig. 5-15.
2. Draw horizontal lines on each side of the circle that are tangent to the circle at the vertical center line.
3. Draw the remaining four lines at an angle of 60 degrees to the horizontal and tangent to the circle. The resulting hexagon circumscribes the circle.

Draw an Octagon When the Across the Flats Distance Is Known

An **octagon** is an eight-sided polygon. Fig. 5-16.

1. Draw a circle with a diameter equal to the across the flats distance. Draw center lines.
2. Draw horizontal lines on each side of the circle that are tangent to the circle at the vertical center line.
3. Draw vertical lines on each side of the circle that are tangent to the horizontal center line. Complete the octagon by drawing the remaining four lines at a 45-degree angle so that they are tangent to the circle and intersect the horizontal and vertical lines drawn.

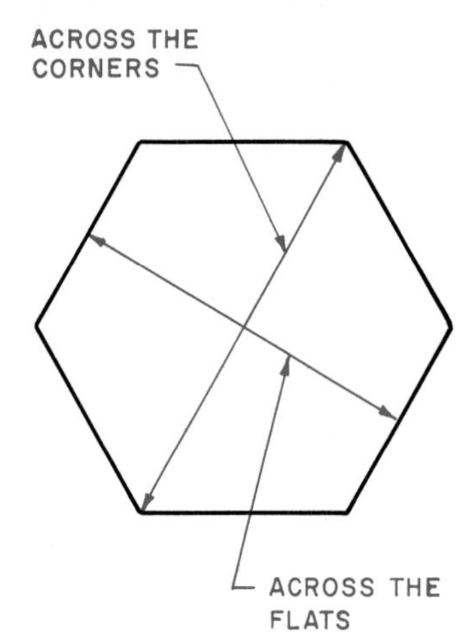

Fig. 5–13. Measurements used with a hexagon.

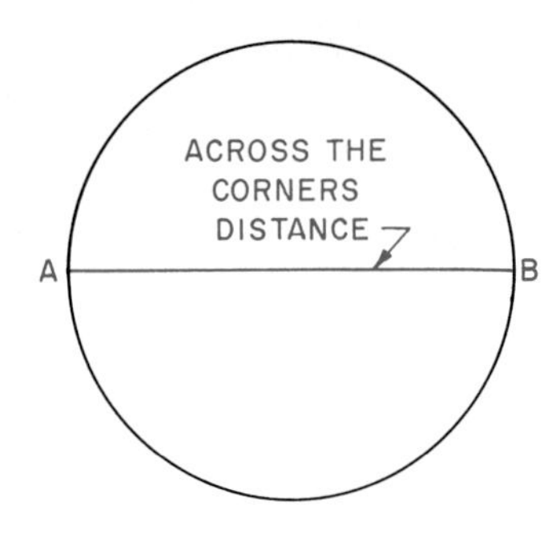

1. DRAW A CIRCLE WITH A DIAMETER EQUAL TO THE ACROSS THE CORNERS DISTANCE OF THE HEXAGON.

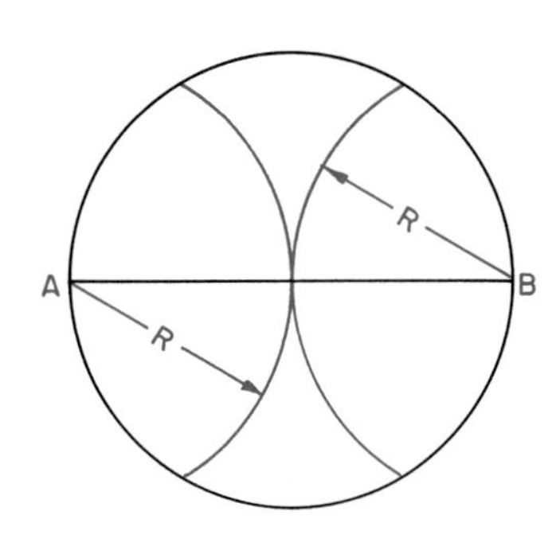

2. SWING ARCS FROM A AND B USING THE RADIUS OF THE CIRCLE.

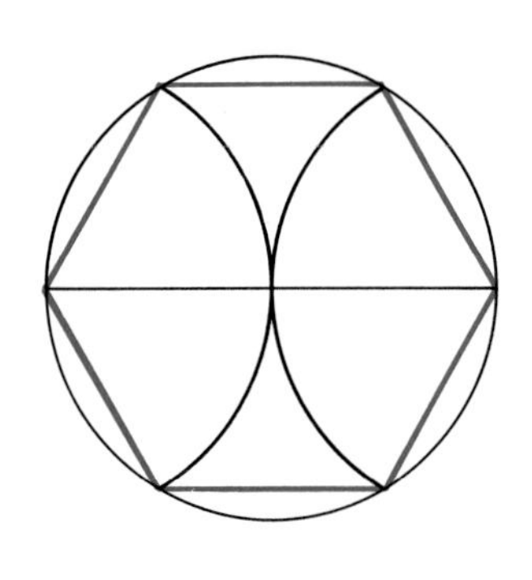

3. CONNECT THE POINTS LOCATED ON THE CIRCLE.

Fig. 5–14. Drawing a hexagon when the "across the corners" distance is known.

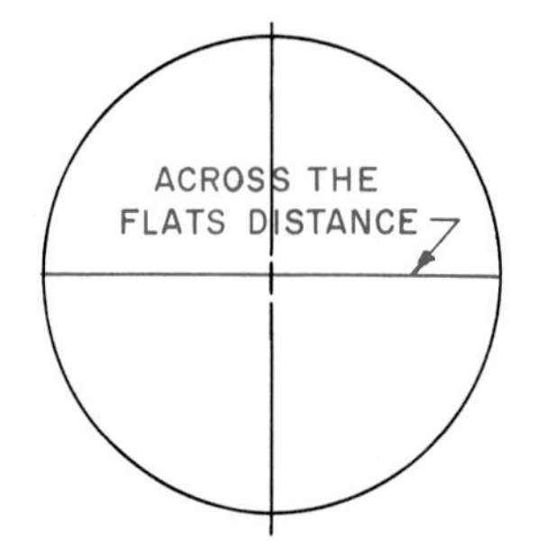

1. DRAW A CIRCLE WITH A DIAMETER EQUAL TO THE ACROSS THE FLATS DISTANCE. DRAW THE CENTER LINES.

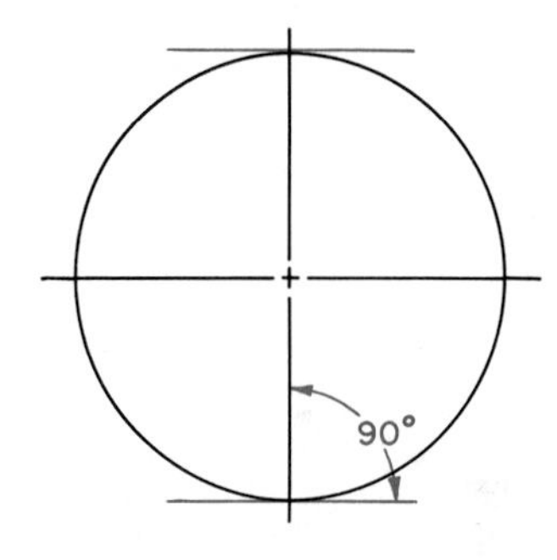

2. DRAW HORIZONTAL LINES TANGENT WITH THE CIRCLE. THEY ARE PERPENDICULAR TO THE VERTICAL CENTER LINE.

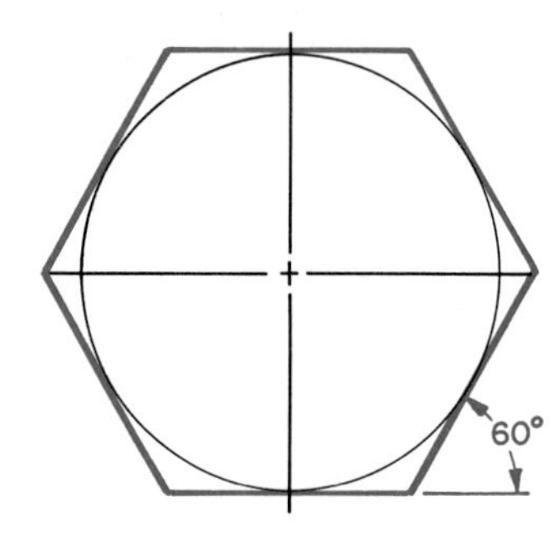

3. DRAW LINES ON AN ANGLE OF 60° TO THE HORIZONTAL, TANGENT TO THE CIRCLE.

Fig. 5–15. Drawing a hexagon when the "across the flats" distance is known.

Construct a Regular Pentagon

A regular **pentagon** has equal sides and equal angles. Fig. 5-17.

1. Draw a circle that will circumscribe the pentagon. Bisect the radius to find point A.

2. With point A as the center, swing radius AB until it crosses the center line at C. With B as the center, swing radius BC to cross the circle at point D.

3. Connect D and B with a straight line to find the length of one side. Set your dividers or compass to distance DB and step off the remaining sides of the pentagon around the circle.

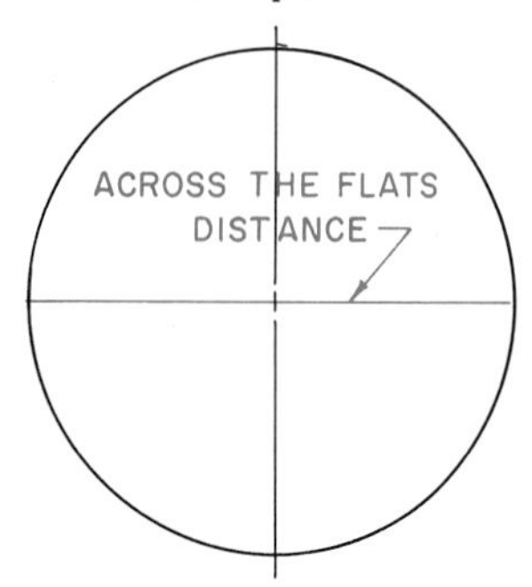

1. DRAW A CIRCLE WITH A DIAMETER EQUAL TO THE ACROSS THE FLATS DISTANCE OF THE OCTAGON. DRAW THE CENTER LINES.

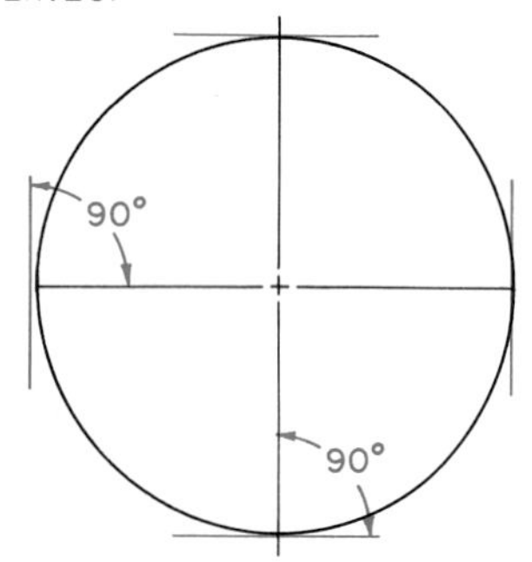

2. DRAW VERTICAL AND HORIZONTAL LINES TANGENT TO THE CIRCLE. THEY ARE PERPENDICULAR TO THE CENTER LINES.

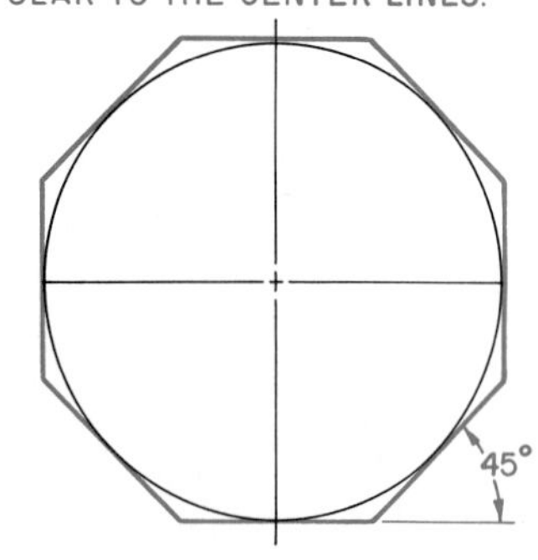

3. DRAW LINES ON AN ANGLE OF 45° WITH THE HORIZONTAL AND TANGENT TO THE CIRCLE.

Fig. 5–16. Drawing an octagon when the "across the flats" distance is known.

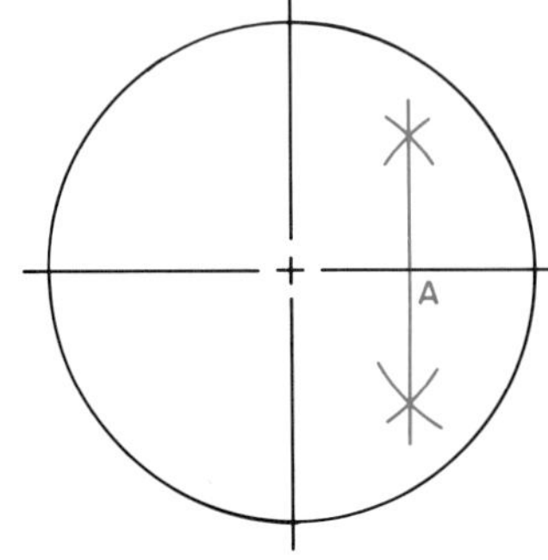

1. DRAW A CIRCLE TO CIRCUMSCRIBE THE PENTAGON. BISECT THE RADIUS.

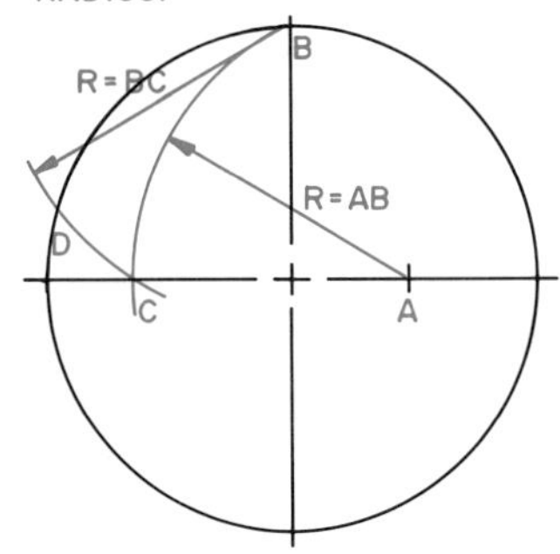

2. SWING AN ARC FROM A WITH RADIUS AB LOCATING C. SWING AN ARC FROM B WITH RADIUS BC LOCATING D.

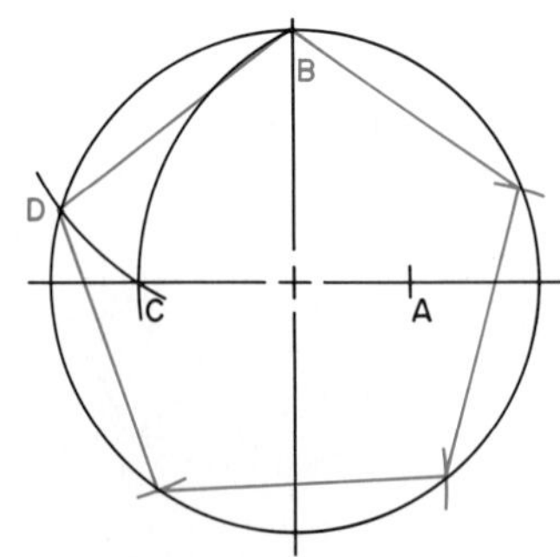

3. STEP OFF DISTANCE DB AROUND THE CIRCLE. CONNECT THE POINTS TO FORM THE POLYGON.

Fig. 5–17. Drawing a pentagon.

Construct an Ellipse

An ellipse is shown in Fig. 5-18. The long axis is called the major axis. The short axis is called the minor axis. A true ellipse is not easy to draw so most drafters draw an approximate ellipse. Although an approximate ellipse is slightly inaccurate, it is accurate enough for most purposes. Follow the steps below and in Fig. 5-19 to draw an approximate ellipse.

1. Draw the major and minor axes of your ellipse.

2. Draw line AC that connects one end of each axis. Set your compass to radius OA and swing an arc to locate point E on an extension of the minor axis.

3. Now set your compass to radius CE and swing an arc to locate point F on line AC. Draw a line that is perpendicular to and bisects line AF. This line crosses line AB at point G, and line CD at point H. With dividers, locate points J and K. Line OG equals OJ and line OH equals OK. Points G, H, J, and K are the centers of the arcs that will form the ellipse.

4. Draw lines KG, KJ, and HJ.

5. Set your compass to radius HC. With H as center, swing an arc until it touches lines GH and HJ. These are the tangent points you will need. Using K as center, repeat this step letting the arc touch lines KG and KJ. Set your compass to radius GA. With G as center, swing the arc to connect the arcs just drawn and form the end of the ellipse. Use J as center and repeat this step on the opposite side to complete the approximate ellipse.

Tangent Points

A tangent point is the exact point at which one of two joining lines stops and the other starts. Fig. 5-20.

The use of tangent points is very important to good drafting.

Drafters always locate tangent points as they make a drawing.

Tangent points must be found when a curved and straight line meet. To find them, first find the center of the curved line. Then draw a perpendicular from the center to the straight line. Where they cross is the tangent point.

When two curved lines meet, the tangent point must be found. This is done by drawing a line from the center of one curve to the other. Where this line crosses the curved lines is the tangent point.

Draw an Arc Tangent to Two Lines at 90 Degrees

This construction is shown in Fig. 5-21.

1. Draw the two lines that meet at a 90-degree angle. These are lines AB and BC. Set your compass to the radius of the corner arc. Put the pin on corner point B and swing an arc that crosses lines AB and BC to locate points D and E. Points D and E are the tangent points needed.

2. With the same radius, set the pin of your compass on points D and E and swing arcs that cross at point F. Point F is the center of the tangent arc wanted.

3. With the same radius, set the pin of your compass on point F and draw the corner arc beginning at D and stopping at E. This is an arc tangent to two lines at 90 degrees.

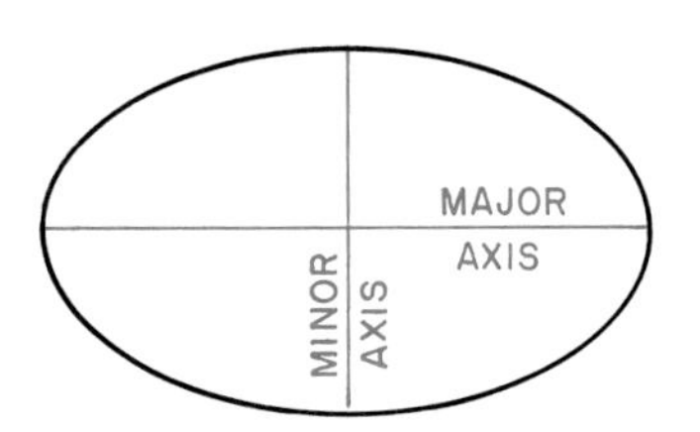
MAJOR
AXIS
MINOR
AXIS

Fig. 5–18. The axes of an ellipse.

Draw an Arc Tangent to Two Lines Not at 90 degrees

This technique is shown in Fig. 5-22.

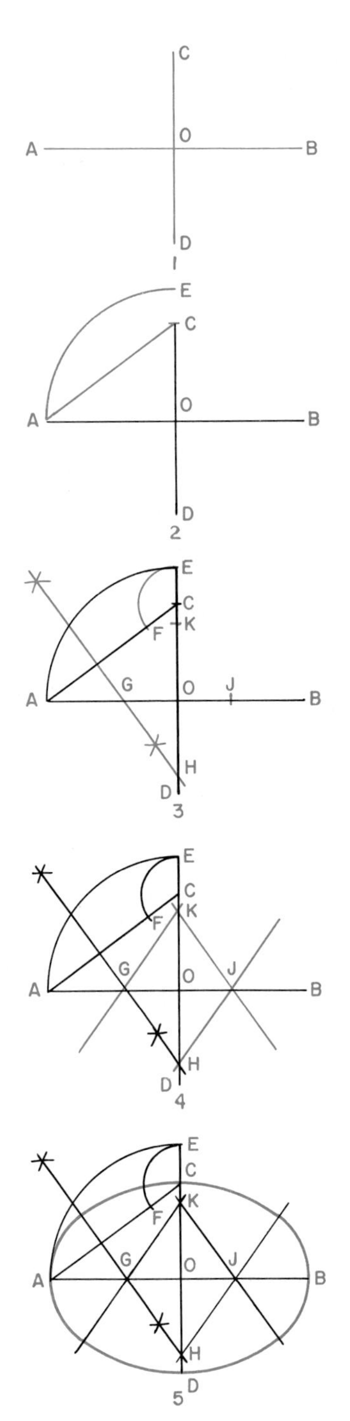
C
O
A
B
D
1
E
2
F
K
G
J
H
3
4
5

Fig. 5–19. Drawing an approximate ellipse.

1. Draw two lines, AB and BC, that meet at an angle other than 90 degrees. Set the compass at the radius of the arc wanted and from any point on AB and BC swing an arc.

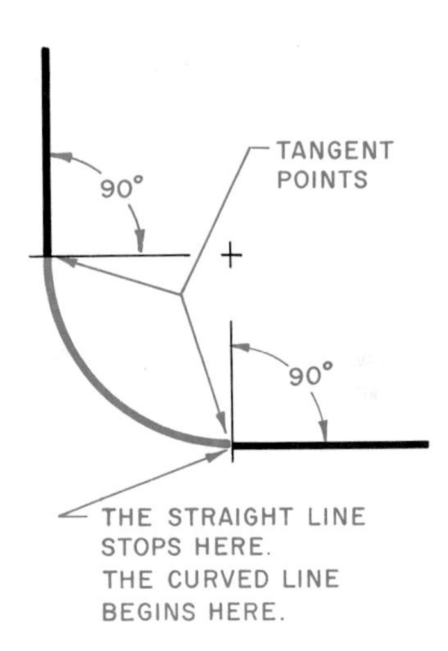
TANGENT
POINTS
90°
90°
THE STRAIGHT LINE
STOPS HERE.
THE CURVED LINE
BEGINS HERE.
THE TANGENT POINT BETWEEN
A CURVED AND A STRAIGHT
LINE.

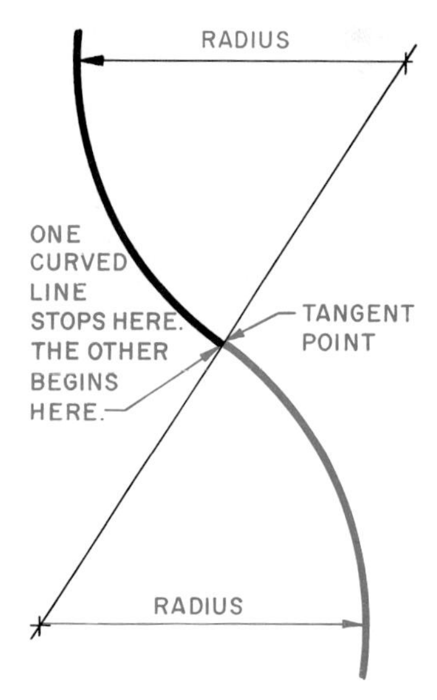
RADIUS
ONE
CURVED
LINE
STOPS HERE.
THE OTHER
BEGINS
HERE.
TANGENT
POINT
RADIUS
THE TANGENT POINT BETWEEN
TWO CURVED LINES.

Fig. 5–20. A tangent point is where one line stops and the one joining it begins.

2. Draw lines tangent to these arcs and parallel with lines AB and BC. These parallel lines are the length of the radius away from AB and BC. Therefore, the intersection of these new lines locates the center, D, of the arc needed.

3. Draw two perpendicular lines from point D. Draw one line perpendicular to AB and the other perpendicular to BC. These lines locate the tangent points.

4. Using D as the center, draw the arc from one tangent point to the other.

Draw an Arc Tangent to a Straight Line and an Arc

1. For this construction you are given arc AB with radius R^1 and line CD. You need to connect these two with an arc of radius R^2. See Fig. 5-23.

2. With the arc and line established, draw line EF parallel to CD at the distance R^2.

CAD Applications:

Geometric Figures

On a CAD system, most geometric figures can be generated by the drafting program. The user simply specifies the figure's characteristics, such as size and number of sides, and its location.

For example, a circle is typically drawn by entering the circle command, selecting a location for the circle's center, and specifying the radius or diameter. Arcs can be drawn by various methods, such as by specifying the start, center, and end or identifying the arc's start, end, and radius.

An ellipse can be difficult to draw manually. On a CAD system, an ellipse can be drawn by specifying the center point, the endpoint of one axis, and the length of the other axis. Another method is to specify the two endpoints of the minor axis and one endpoint of the major axis.

Regular polygons (those having all sides equal in length) can be drawn by specifying the polygon's center, its size, and the number of sides. One program is capable of constructing polygons with up to 1024 sides!

A feature that is particularly useful in geometric construction is object snap. Using the object snap mode causes the cursor to "snap" to predefined points on an object, such as its center or endpoint or the intersection of two lines. The hexagon nut in the following illustration was drawn using the polygon and circle commands and the object snap feature.

Most CAD programs also enable the user to construct tangent lines. Sometimes this is done using an option within the object snap mode. On other programs, it is a separate command.

Geometric construction on a CAD system can be faster and easier than manual drafting. However, it is important to understand the various command options so that the proper ones are selected. For example, selecting one method of drawing an arc may cause the arc to be generated in a clockwise direction. Selecting another may cause it to be drawn counterclockwise. In order to end up with the geometric figure in the desired location and size, you need to know what the options are and how they work.

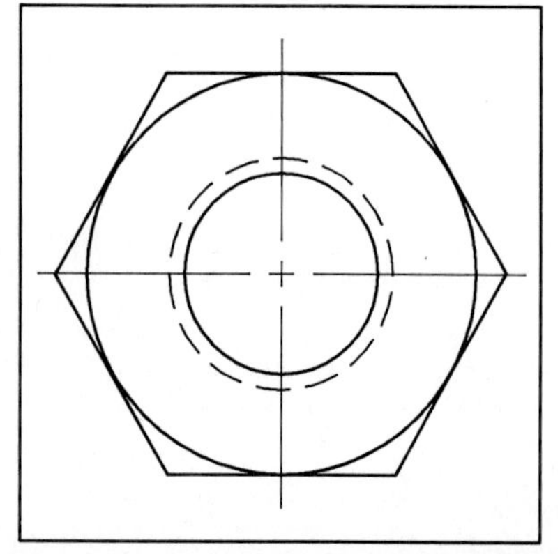

3. Add R^1 and R^2 and set a compass to that length. Swing an arc with radius $R^1 + R^2$ from the center of arc AB. The center of the arc to be drawn, point G, is found where the large arc crosses line EF.

4. From point G, find the tangent points on arc AB and line CD. The tangent point for arc AB is found by drawing a straight line from point G, the center of the new arc, to the center of arc AB. The tangent point for line CD is a perpendicular line drawn from point G.

5. Set your compass to radius R^2 and swing an arc from center point G connecting arc AB and line CD, stopping at each tangent point.

Draw an Arc Tangent to Two Arcs

For this construction you are given arcs AB and CD located as shown in Fig. 5-24 at 1. You are to connect the two given arcs with an arc of radius R^3.

1. Set your compass to the radius $R^1 + R^3$.

2. Swing this arc from the center, E, of arc AB. (Part 2 of Fig. 5-24).

3. Now set your compass to the radius $R^2 + R^3$.

4. Swing this arc from the center, F, of arc CD until it crosses the first arc. This locates center point G. It is the center you will use to draw the arc with radius R^3.

5. Find the tangent points on arcs AB and CD by connecting the centers E and G, and F and G, with straight lines. (Part 3 of Fig. 5-24). The tangent points are located where these lines cross arcs AB and CD.

6. Set your compass to radius R^3 and swing an arc from the center point G. Begin the arc at one tangent point and stop at the other. (Part 4 of Fig. 5-24).

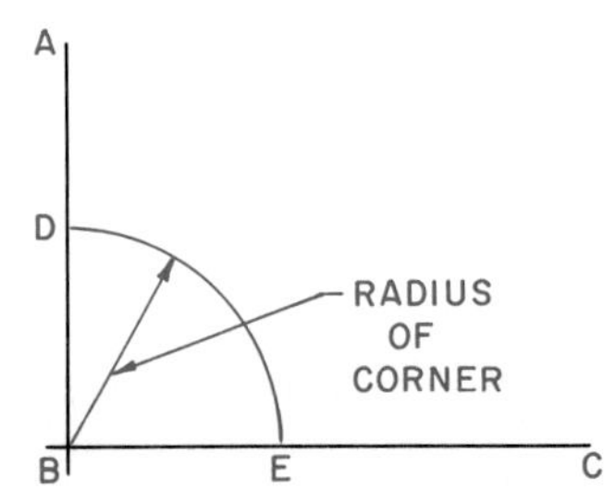

1. SWING AN ARC THAT IS THE RADIUS OF THE ROUND CORNER TO BE DRAWN. THIS LOCATES POINTS D AND E. THEY ARE THE TANGENT POINTS BETWEEN THE STRAIGHT LINES AND THE CORNER TO BE DRAWN.

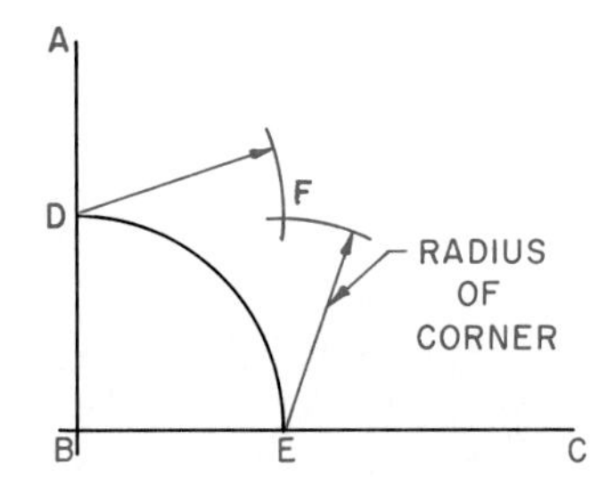

2. SWING ARCS FROM POINTS D AND E USE THE RADIUS OF THE CORNER. THIS LOCATES POINT F. F IS THE CENTER USED TO DRAW THE CORNER.

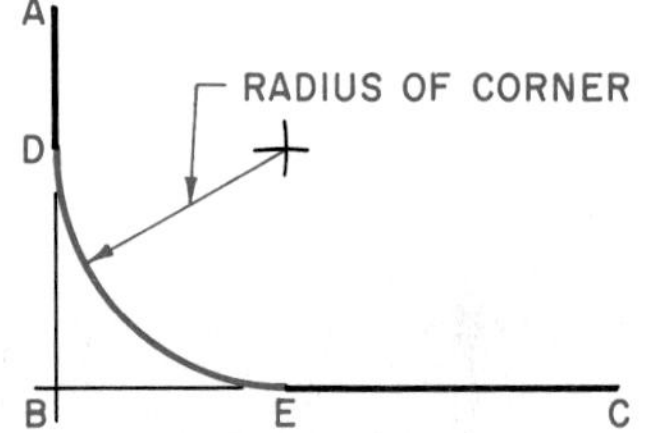

3. USING F AS A CENTER, SWING AN ARC FROM D TO E. D AND E ARE THE TANGENT POINTS. THE ARC ENDS AT THESE TWO POINTS.

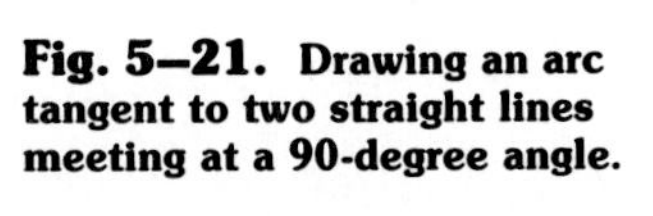

Fig. 5–21. Drawing an arc tangent to two straight lines meeting at a 90-degree angle.

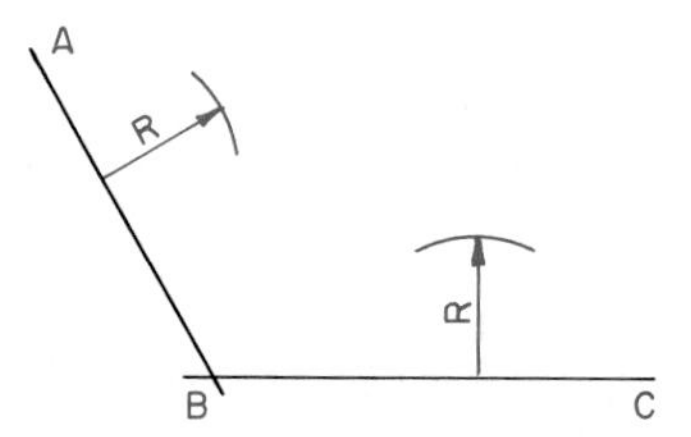

1. SWING ARCS HAVING A RADIUS EQUAL TO THE RADIUS WANTED AT THE CORNER.

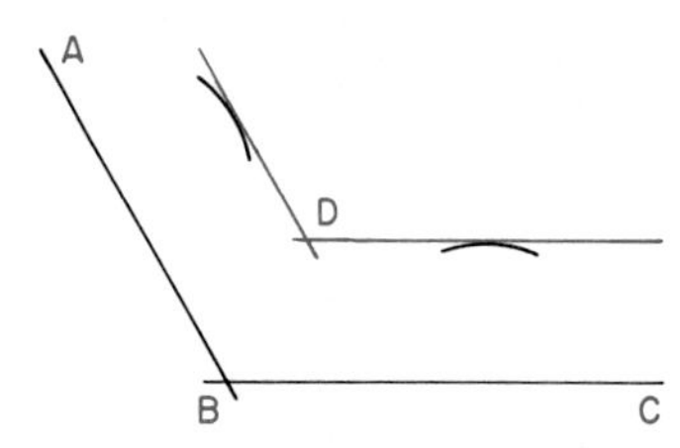

2. DRAW LINES TANGENT TO THESE ARCS AND PARALLEL TO THE LINES. THESE MEET AT D. D IS THE CENTER USED TO DRAW THE CURVED CORNER.

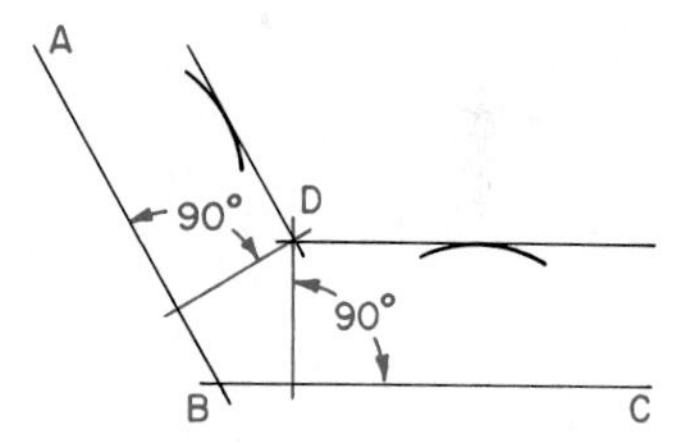

3. DRAW LINES FROM D, PERPENDICULAR TO AB AND BC. THESE LOCATE THE TANGENT POINTS FOR THE CURVED CORNER.

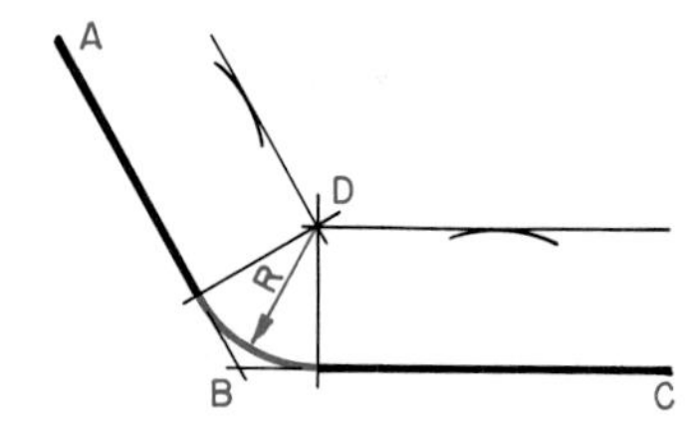

4. USING D AS A CENTER, DRAW AN ARC FROM ONE TANGENT POINT TO THE OTHER.

Fig. 5–22. Drawing an arc tangent to two lines that are not at 90 degrees to each other.

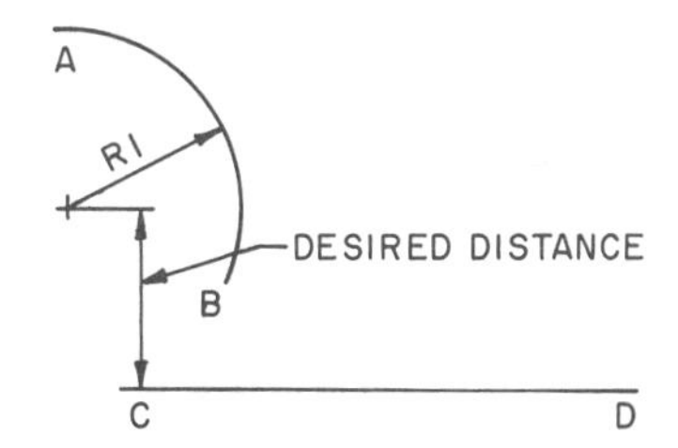

1. DRAW ARC AB (R1). MEASURE THE DESIRED DISTANCE FROM THE CENTER OF AB FOR LINE CD. DRAW LINE CD.

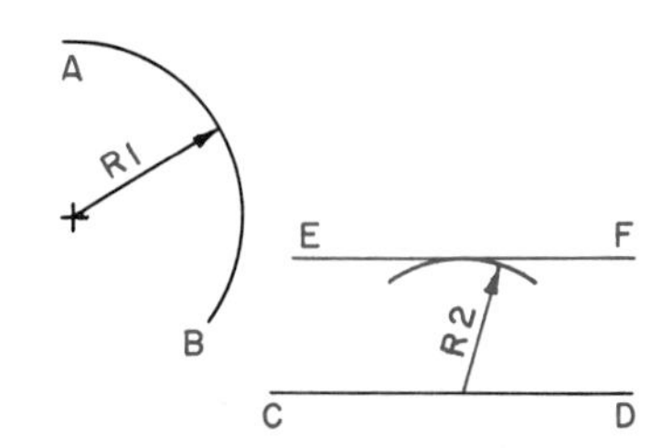

2. DRAW A LINE PARALLEL TO CD AT A DISTANCE EQUAL TO THE RADIUS OF THE ARC TO CONNECT AB AND CD. THE RADIUS IS MARKED R2.

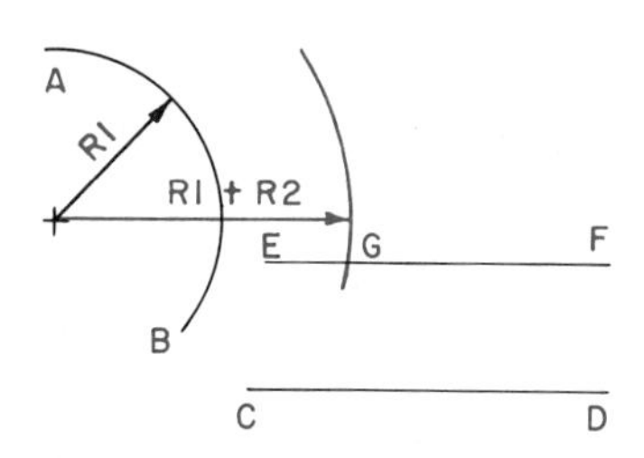

3. SWING AN ARC WITH A RADIUS EQUAL TO THE RADIUS OF ARC AB (R1) PLUS THE RADIUS OF THE CORNER (R2). THIS LOCATES POINT G. G IS THE CENTER USED TO DRAW THE ARC TO CONNECT AB AND CD.

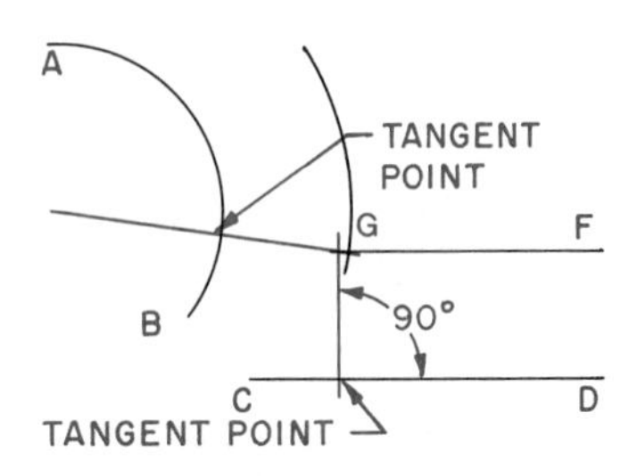

4. TANGENT POINT ON CD IS FOUND BY DRAWING A PERPENDICULAR FROM THE CENTER G TO CD. THE TANGENT POINT ON AB IS FOUND BY DRAWING A LINE FROM THE CENTER G TO THE CENTER OF AB.

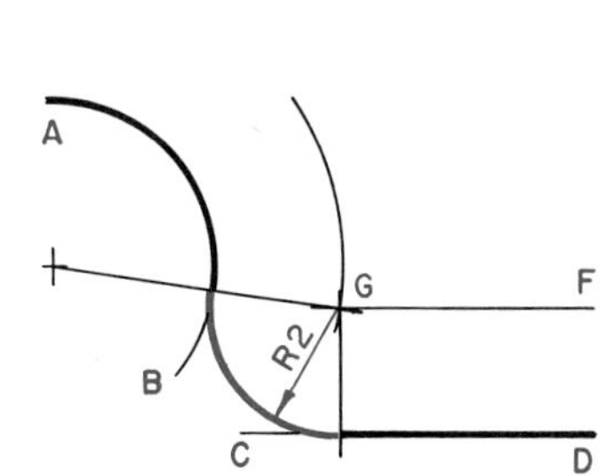

5. DRAW THE ARC WITH RADIUS R2 FROM ONE TANGENT POINT TO THE OTHER.

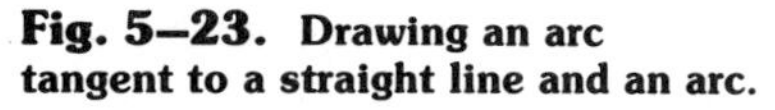
Fig. 5–23. Drawing an arc tangent to a straight line and an arc.

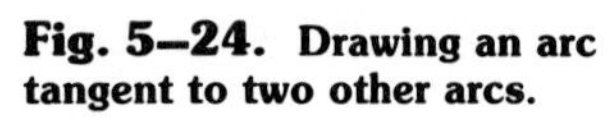
Fig. 5–24. Drawing an arc tangent to two other arcs.

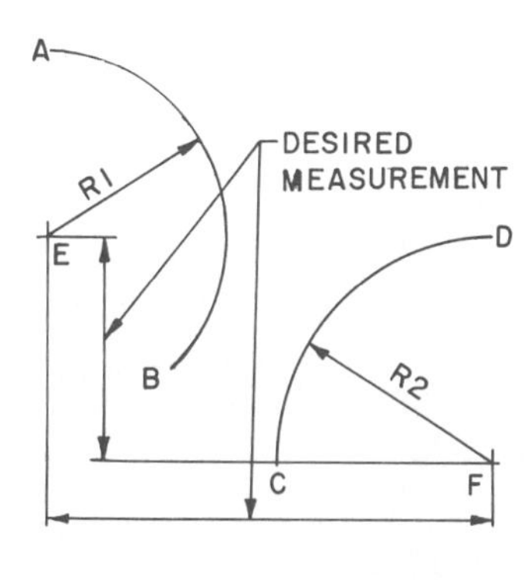

1

DRAW ARC AB (R1) FROM THE CENTER E. MAKE THE TWO DESIRED MEASUREMENTS TO LOCATE CENTER F. DRAW ARC CD (R2).

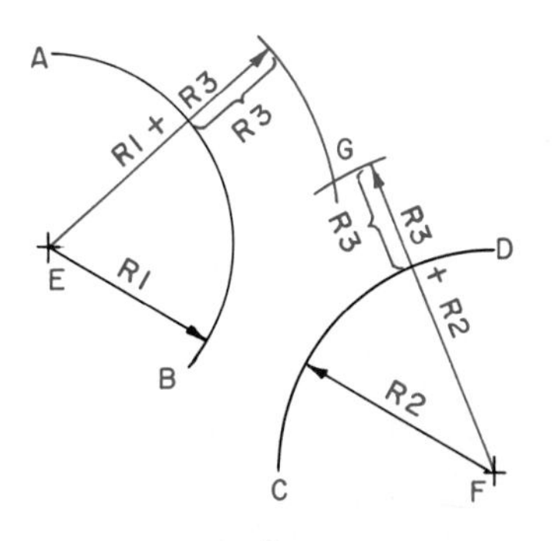

2

ADD THE RADIUS OF THE ARC AB (R1) AND THE RADIUS OF THE ARC TO CONNECT AB AND CD (R3). USING THIS RADIUS (R1 PLUS R3), SWING AN ARC USING E AS THE CENTER. ADD THE RADIUS OF ARC CD (R2) AND R3. SWING AN ARC USING F AS THE CENTER. THIS LOCATES CENTER G. IT IS THE CENTER FOR THE ARC TO CONNECT ARCS AB AND CD.

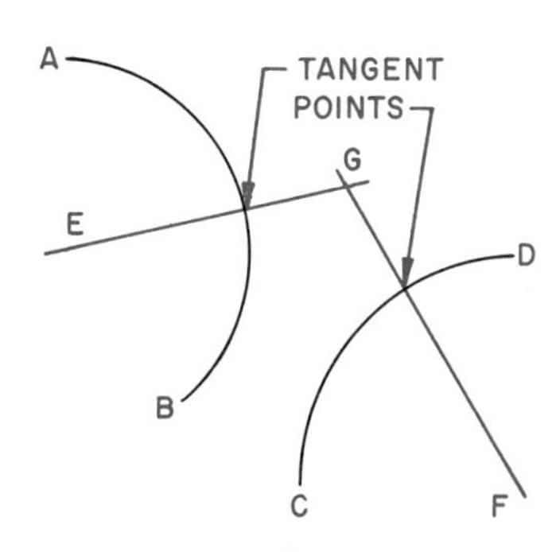

3

CONNECT THE CENTERS EG AND FG. WHERE THESE LINES CUT THE ARCS ARE THE TWO TANGENT POINTS.

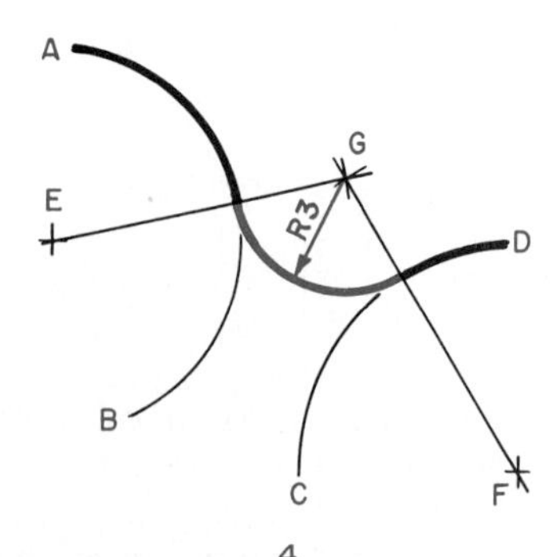

4

SET A COMPASS TO THE RADIUS OF THE ARC TO CONNECT AB AND CD (R3). SWING THE CONNECTING ARC FROM CENTER G. IT RUNS FROM ONE TANGENT POINT TO THE OTHER.

Chapter Review

Build Your Vocabulary

You should understand and use the following terms as part of your working vocabulary. Write a brief explanation of what each means.

geometry
angle
parallel
triangle
circle
radius
arc
parallelogram
polygon
prism
cylinder
pyramid
cone
sphere
diameter
torus
bisect
chord
perpendicular
equilateral triangle
isosceles triangle
square
inscribed
circumscribes
tangent
hexagon
octagon
pentagon

Sharpen Your Math Skills

1. A rectangle measures 4″ × 8″. What is the area of the rectangle?
2. When you bisect a 65° angle, what size is each angle produced?
3. Suppose you draw a line 6 9/16″ long and then draw a line perpendicular to the midpoint. How long is each half of the original line?
4. If an equilateral triangle has a side 4″ long, what is the sum of all three sides?
5. A triangle has two angles, 35° and 47°. The sum of the three angles must equal 180°. What size is the third angle?
6. To find the area of a triangle, use the formula

$$A = \frac{\text{base} \times \text{height}}{2}.$$

What is the area of an isosceles triangle with a base of 3″ and a height of 5″?
7. To find the area of a circle, use the formula $A = 3.14 \times \text{radius}^2$. What is the area of a circle with a diameter of 5″?
8. To find the circumference of a circle, use the formula $C = 3.14 \times \text{diameter}$. What is the circumference of a circle with a radius of 75 millimeters?

Study Problems—Directions

The problems that follow will give you the chance to use some of the drafting constructions you read about in this chapter. Do the problems that are assigned to you by your instructor.

1. Draw an equilateral triangle with each side 3 inches (76 mm) long. Find its center. Inscribe a circle in the triangle.
2. Draw two lines that meet at an angle of 69 degrees. Bisect this angle. Then bisect each of the two angles found. Compare your solution by placing it on top of the others drawn in your class. Do they match?
3. Swing an arc having a radius of 6 inches (152 mm) through 195 degrees. Bisect the arc. Draw chords for each bisected arc and then draw a perpendicular line to each chord.
4. Draw a line 7 13/16 inches (198 mm) long. Divide it into seven equal parts.
5. Draw a hexagon with a distance across the corners of 4 inches (102 mm). Then draw a hexagon with a distance across the flats of 4 inches (102 mm). Compare their sizes. Which is larger and why is it so?
6. Draw an octagon with a distance across the flats of 3½ inches (89 mm).
7. Construct an ellipse having a major axis of 6 inches (152 mm) and a minor axis of 4 inches (102 mm).
8. Draw objects A and B in **Fig. P5-1** and mark the points of tangency. Leave all construction lines on your drawing. Use any convenient size.

Study Problems

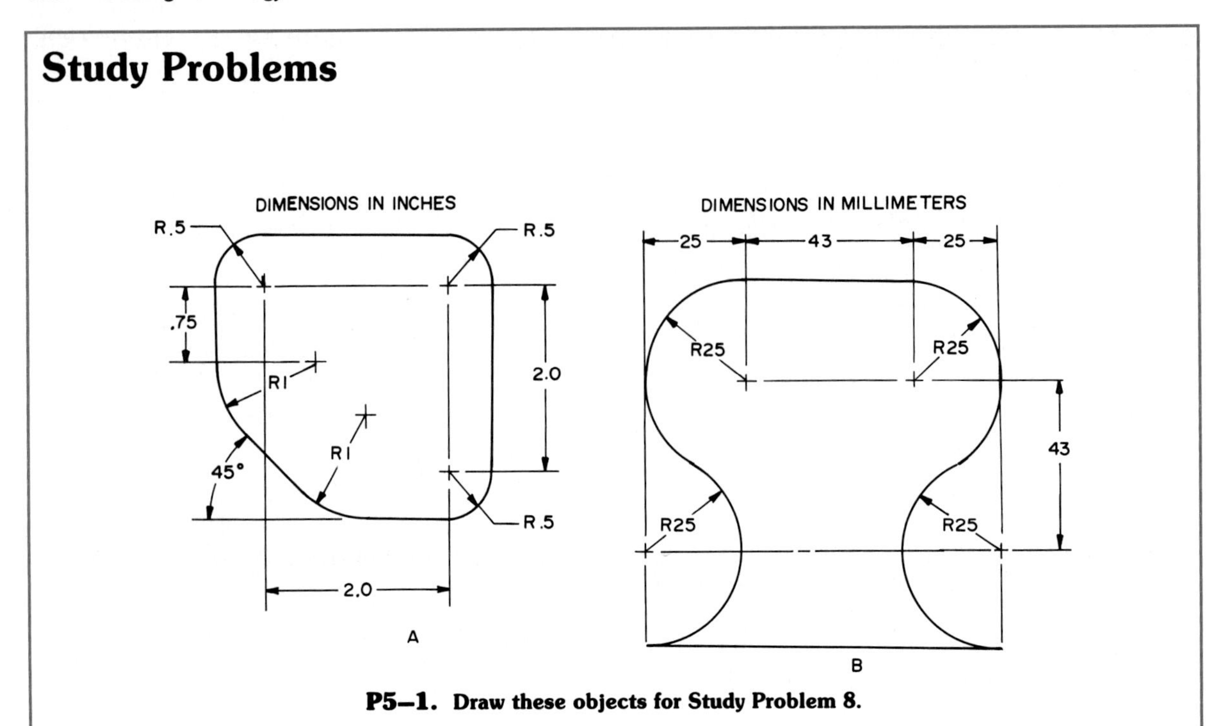

P5–1. Draw these objects for Study Problem 8.

CHAPTER 6

Multiview Drawing

After studying this chapter you should be able to do the following:

- Understand the principles of orthographic projection.
- Lay out multiview drawings.
- Recognize basic lines and planes.
- Use hidden lines.
- Project curved edges.
- Locate intersections between various geometric forms.

The Purpose of Multiview Drawings

Designers and engineers use drawings to present ideas. A pictorial drawing of an object shows a picturelike view that is easy to understand, but it does not present the true shape of the parts. Fig. 6-1. The most frequently used way to present all the details of an object in their true shape is by making a multiview drawing. A **multiview drawing** is a series of separate views of an object arranged so that each view is related to the others.

Figure 6-2 shows a multiview drawing of a machinist's vise. Notice how each separate view on the multiview drawing shows one side of the object. The top view shows the object as it would appear if you were looking directly at it from above. The front view shows the object as it would appear if you were looking directly at the front, and the side view shows it as if you were looking directly at the object's right side. Most objects can be adequately described by showing these three views.

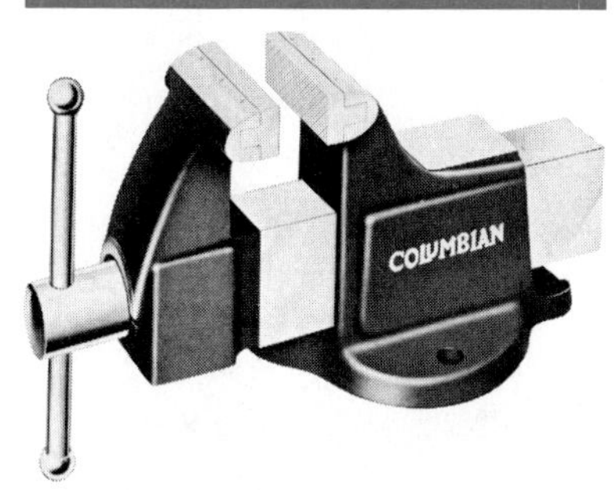

Fig. 6–1. This is a machinist's vise as it appears to the eye. It does not show the true shape of the parts of the vise. (Columbian Vise Mfg. Co.)

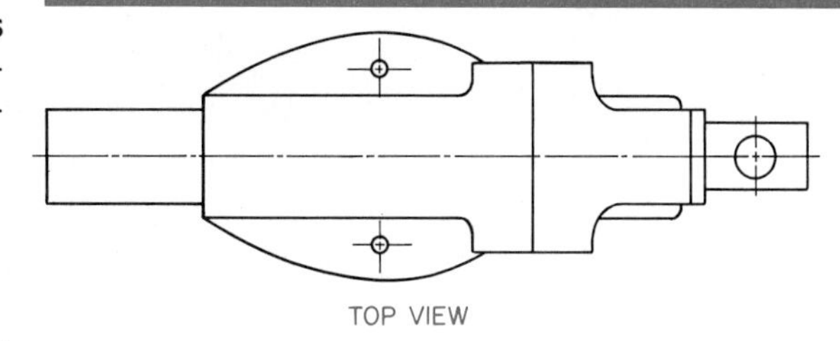

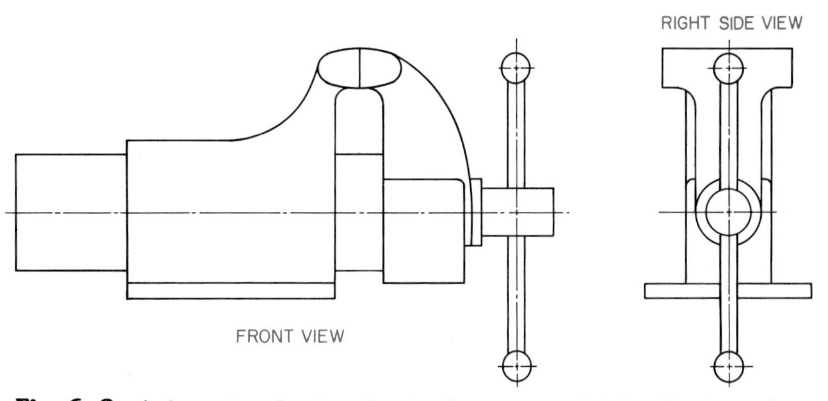

Fig. 6–2. A three-view drawing showing front, top, and right-side views of a machinist's vise.

Orthographic Projection

The views of a multiview drawing are made by projecting them onto planes (flat surfaces). These surfaces are called **planes of projection**. Imagine that an object to be drawn is placed inside a hinged glass box with the sides of the object parallel with the sides of the box. The sides of the box are the planes of projection. You, the drafter, are on the outside looking through the glass at the object. Fig. 6-3.

The glass box has six sides, but for most multiview drawings only three sides (planes of projection) are used. These are the horizontal, profile, and frontal planes. (The profile and frontal planes are both vertical.) All the planes are perpendicular to each other.

The surfaces of the object are projected onto the glass sides at right angles to each side. This *right angle projection* is called **orthographic projection.** All orthographic projections are at right angles to the planes of projection. In order to show the views of the object on a flat surface (such as a sheet of paper), the glass box is unfolded and spread so that it is flat, or in one plane. Fig. 6-4. This is how the orthographic views appear on a multiview drawing.

The top section of the glass box forms the top view on a horizontal plane. It always folds up above the front view. The front view is projected on the frontal plane. The right-side view is projected on the profile plane. It always unfolds to the right of the front view.

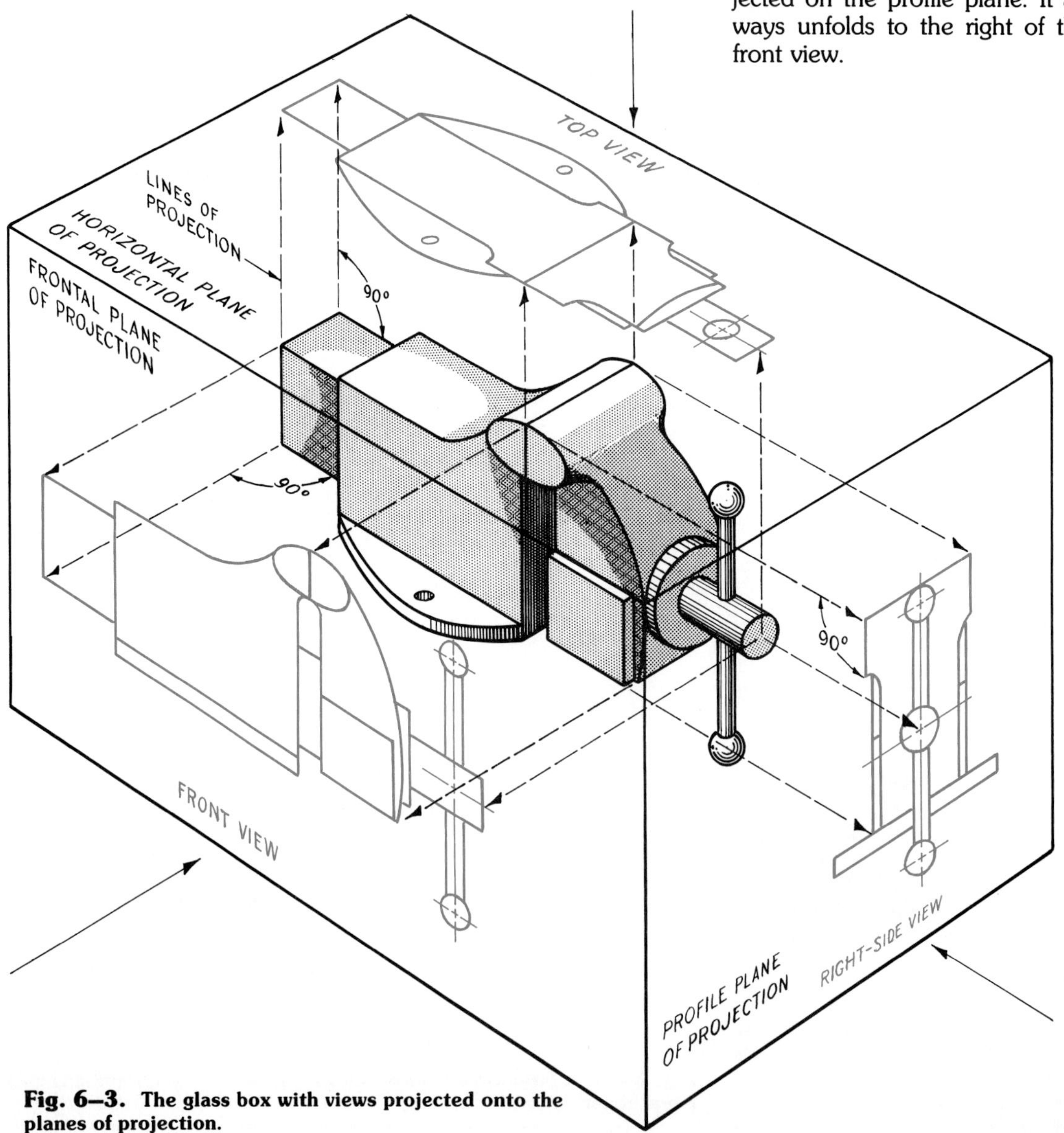

Fig. 6–3. The glass box with views projected onto the planes of projection.

As stated earlier, the glass box has six sides. Views of the object can be projected orthographically (perpendicularly) from any of these sides. Each side has a name to describe the view projected from it. Fig. 6-5.

The left-side view is projected to a profile plane. The bottom view is projected to a horizontal plane. The rear view is projected to a frontal plane.

It is possible to unfold the glass box so that all six sides appear on the same plane. This means that six views of the object can be drawn in this manner. Fig. 6-6. It is good practice to draw only the views that are necessary to describe the object. This could take from one to six views.

Notice how the surfaces of the object line up from one view to another. This projection is used in laying out and drawing the object.

Third-Angle Projection

The three planes shown in Fig. 6-7 form four 90-degree quadrants. These are identified by the numbers 1, 2, 3, and 4. All multiview drawings used in North America place the object in the *third* quadrant. The views are projected to the planes of projection as shown in Fig. 6-8. Since the object is in the third quadrant, or angle, this is called **third-angle projection.** In Europe the object is placed in the first quadrant; therefore, Europeans use first-angle projection. In first-angle projection, the front view is above the top view.

Height, Width, and Depth

You use the terms height, width, and depth when referring to the overall size of an object. *Height* is the perpendicular distance between horizontal planes. *Width* is the perpendicular distance between profile planes. *Depth* is the perpendicular distance between frontal planes. Fig. 6-9.

In the front view, the width and height of the object are shown. Refer to Fig. 6-6. The height projects to the side and rear views and the width to the top and bottom views. The side, bottom, and top views show the depth of the object. The depth is most easily transferred to the top, bottom, and side view using dividers or a scale. Notice how the height, width, and depth are projected to the alternate left, right, and rear views.

Identifying the Edge Views of the Planes of Projection

If you understand the planes of projection, you will understand orthographic projection. In complex problems you may find it necessary to label these planes and their edge views so that you can measure projection distances. Fig. 6-10.

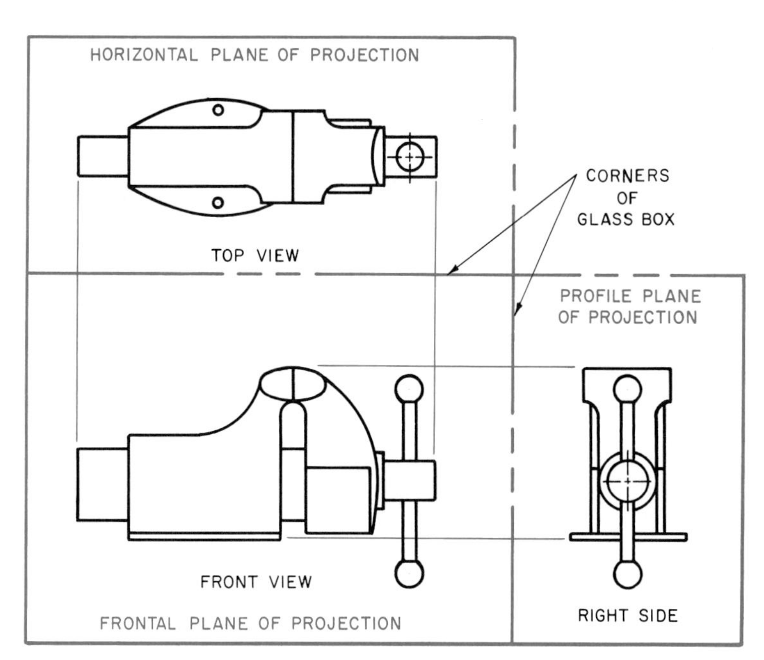

Fig. 6–4. Planes of projection unfolded.

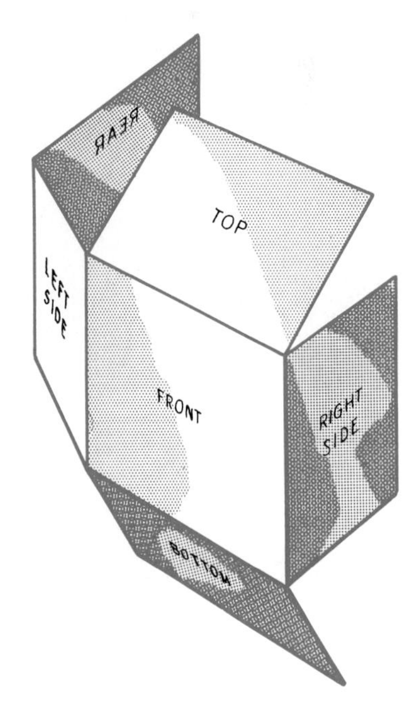

Fig. 6–5. The glass box partly "unfolded" to show the six planes of projection.

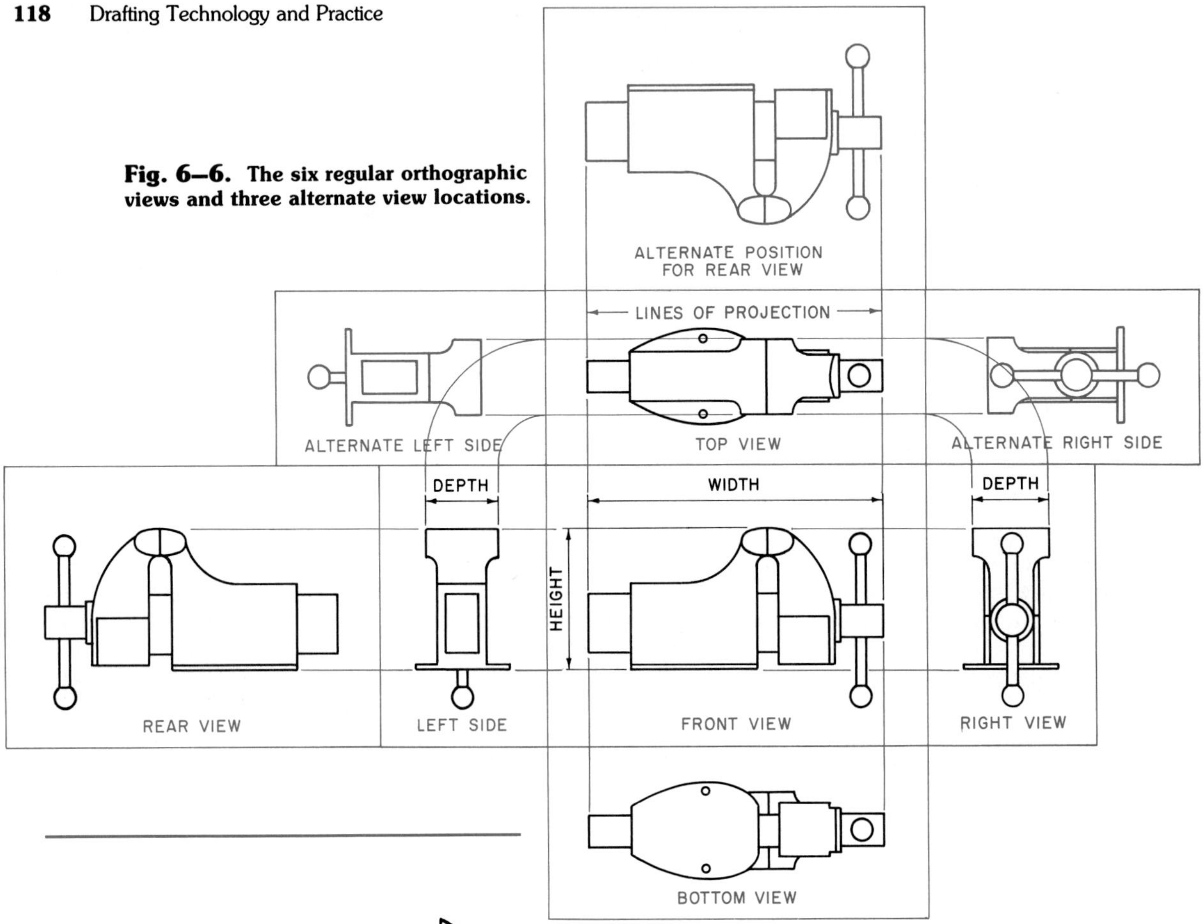

Fig. 6–6. The six regular orthographic views and three alternate view locations.

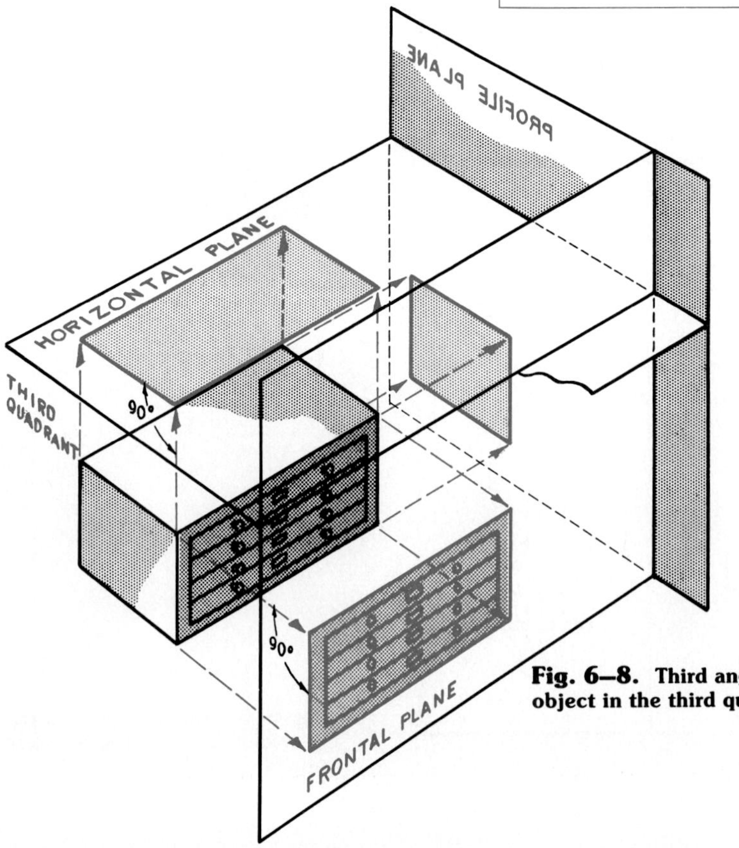

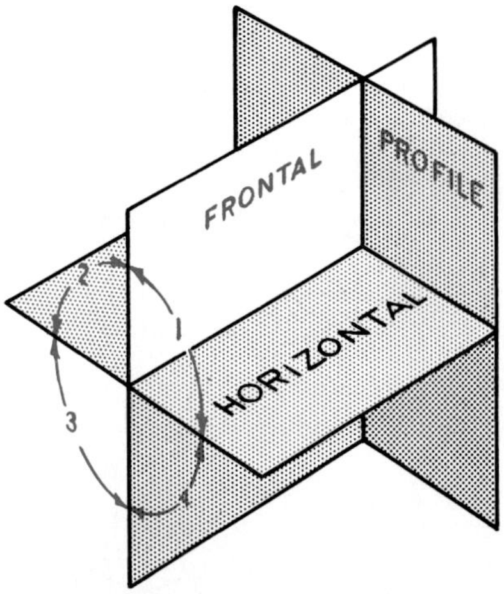

Fig. 6–7. Planes of projection form quadrants.

Fig. 6–8. Third angle projection places the object in the third quadrant.

When your line of sight is perpendicular to the horizontal plane, the profile plane and frontal plane appear in edge view. Figure 6-10 shows these labeled P for profile plane of projection and F for frontal plane of projection.

When your line of sight is perpendicular to the profile plane, the H (horizontal plane of projection) and the F (frontal plane of projection) appear in edge view.

Figure 6-11 shows the three planes placed in a normal position forming the glass projection box. The edge views of the planes are labeled. The planes are unfolded into one plane in Fig. 6-12, which shows how the

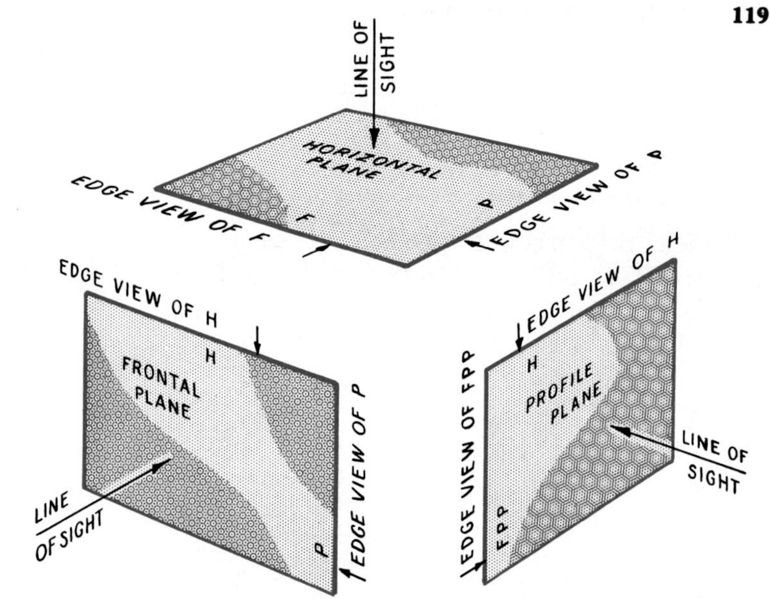

Fig. 6–10. Identifying the edge views of the planes of projection.

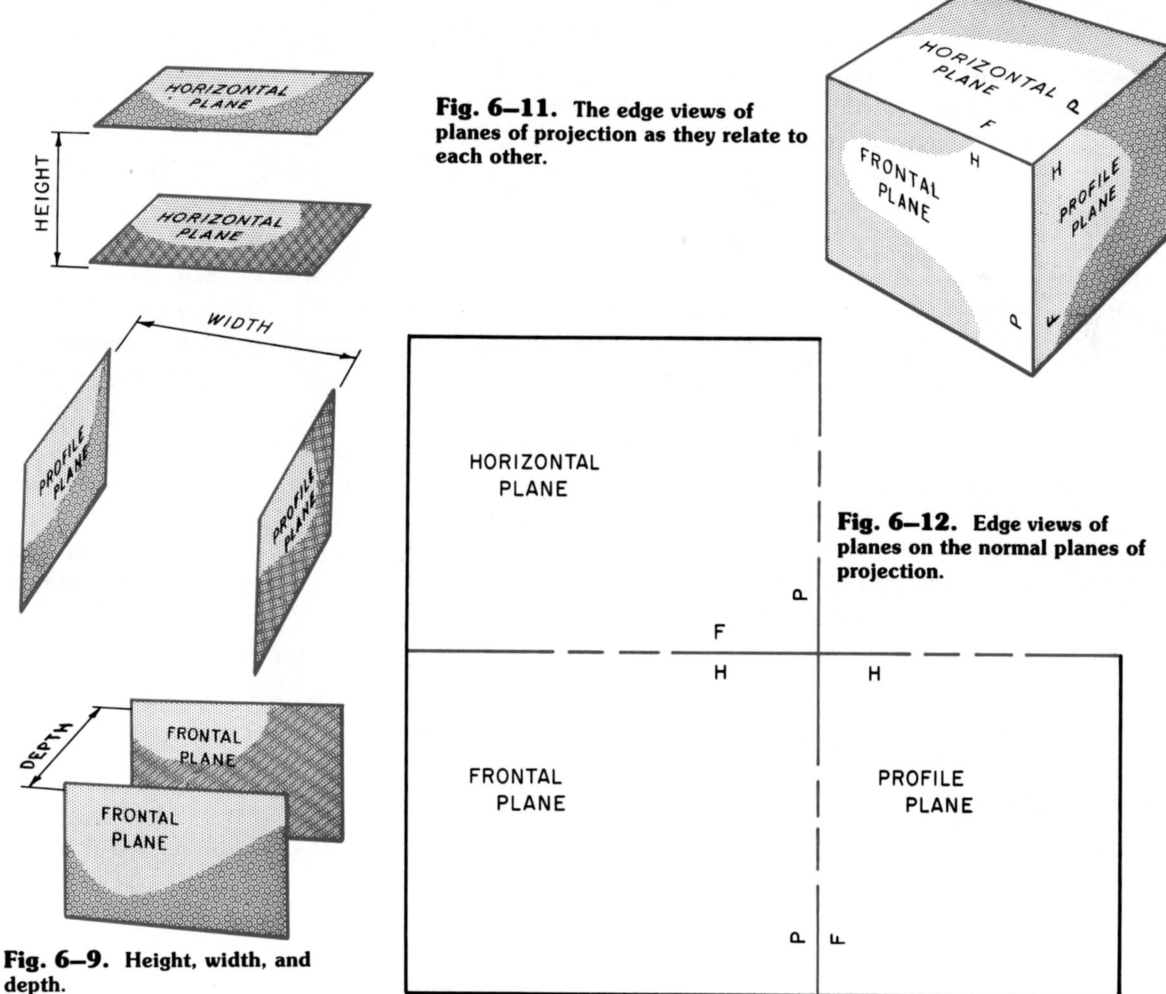

Fig. 6–11. The edge views of planes of projection as they relate to each other.

Fig. 6–12. Edge views of planes on the normal planes of projection.

Fig. 6–9. Height, width, and depth.

CAD Applications:

Multiview Drawings

Since orthographic projection is a universal and basic concept of drafting and design, it is important to understand the CAD system's ability to construct these types of drawings. It is likewise important for the CAD user to know how to generate the various types of lines necessary to accurately compose multiview drawings. These lines include solid, hidden, and center lines. Some CAD systems utilize a linetype command for different types of lines. Others require the loading of different linetypes onto various layers or overlays.

As in manual drafting, proper orientation of views is of prime importance when constructing an orthographic drawing. In a three-view drawing, the front, top, and right side must be properly aligned as illustrated here.

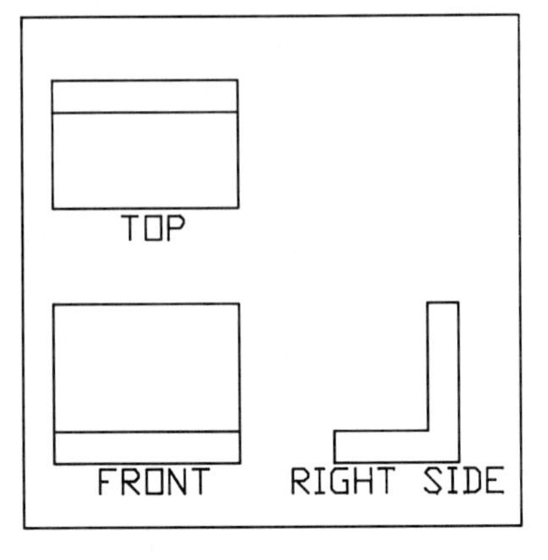

A big advantage CAD has over manual drafting is the ability to move entities and objects, no matter how complex, from one location to another with only a few simple commands. This allows the drafter great flexibility in locating the first view drawn anywhere on the screen and then simply moving that view into its proper location. The next two views are drawn in relation to the first. Any further moving of the views may then be completed. This allows for perfect placement of the three views in proper relation to one another.

The ability to copy, or duplicate, objects is another powerful CAD feature that helps in the development of multiview drawings. This is illustrated in the "before and after" drawings in the next column. The six bolt holes were placed with only one hole actually being drawn. The other five holes were placed in the object using the copy command.

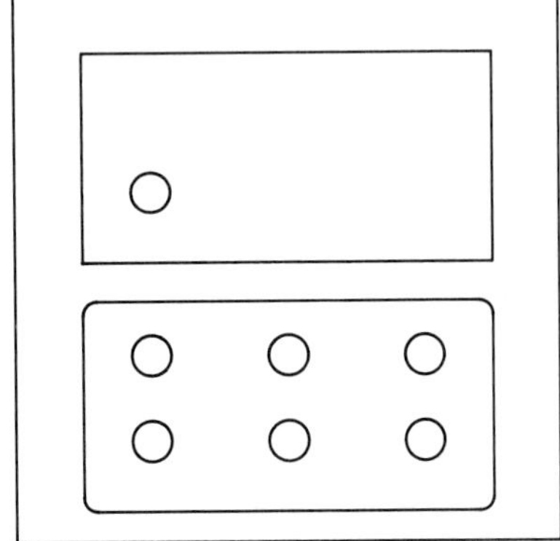

"Fillet" is yet another CAD command that demonstrates the program's great accuracy. The corners of the object were rounded using the fillet command to precisely draw the correct radii. The fillets were drawn more accurately than would be possible using a compass.

These are just some of the functions that help a CAD user develop orthographic, or multiview, drawings. Other items, such as dimensioning, will be covered in subsequent chapters.

edge views of planes are identified on a drawing. After you understand this identification system, you do not need to label the planes on every drawing.

One-View Drawings

Some objects can be fully described by drawing one view and dimensioning it. You must decide if the drawing gives all the information needed to make the article. Sometimes a simple note can remove the need for a view, as shown in Fig. 6-13. How many views are needed to describe the malleable ball handle shown in Fig. 6-14?

Two-View Drawings

Most objects that are shaped like a cone or a pyramid or are **symmetrical** (the same size and shape on both sides of a center line) can be described in two views. The side view of the bearing ring in Fig. 6-15 would be identical to the top view. Therefore the right-side view is not needed. A top view of the adjustment knob in Fig. 6-15 would be identical to the side view and so it is unnecessary.

Three-View Drawings

Most objects require three views for complete shape description. Therefore the three-view drawing is the most common one drawn. In Fig. 6-16, the front and right-side views give all the information needed except the shape of the vertical handle. The top view shows it to be square.

Not all three-view drawings use the top, front, and right-side views. Any of the six regular orthographic views can be used. Figure 6-17 shows a three-view drawing using right and left sides with a front view. A top view would be exactly like the front view and so is unnecessary.

More Than Three Views

Sometimes an object will have a detail that will require that you draw a fourth, fifth, or sixth view. The decision to draw another view is made by the drafter. The additional view or views must present some information about the object that is not shown clearly on the usual views. In Fig. 6-18, the left-side view is needed to clearly show the shape of the hole. It could have been square, circular, or triangular. The clip, Fig. 6-19, uses a rear view to clarify details that are hidden in the front view.

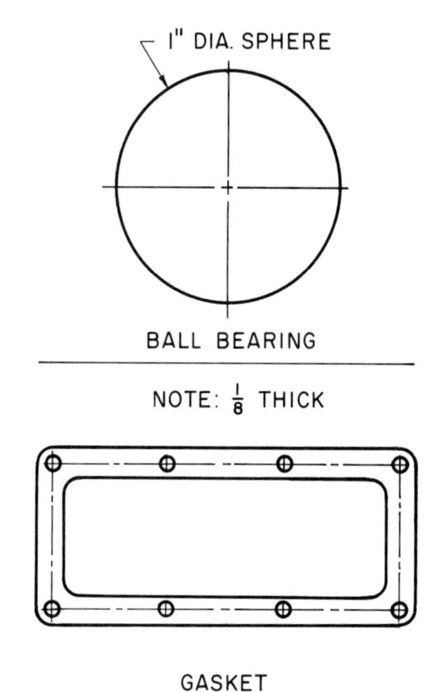

Fig. 6–13. **One-view drawings.**

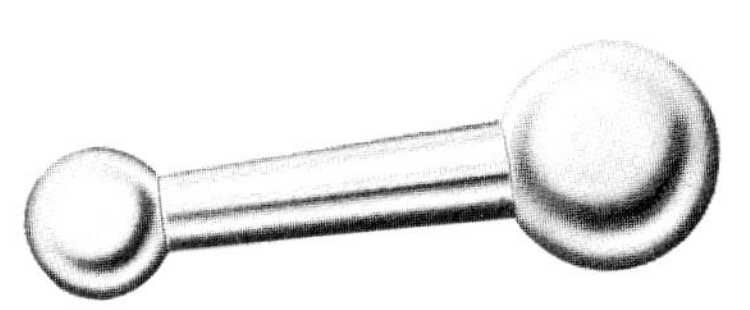

Fig. 6–14. **A malleable ball handle containing two spheres and a conical member. (Reid Tool Supply Co.)**

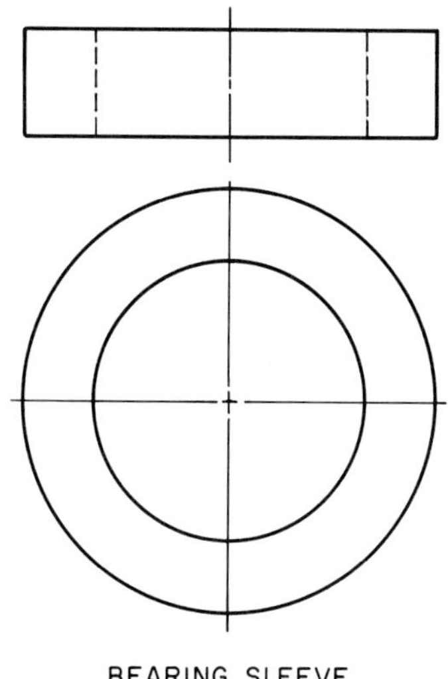

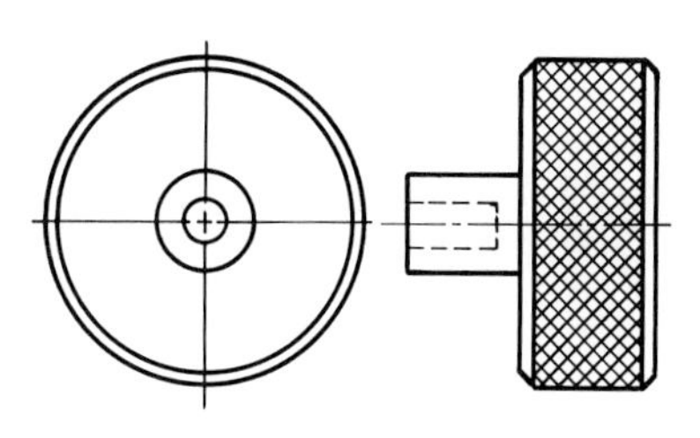

ADJUSTMENT KNOB

Fig. 6–15. **Two-view drawings.**

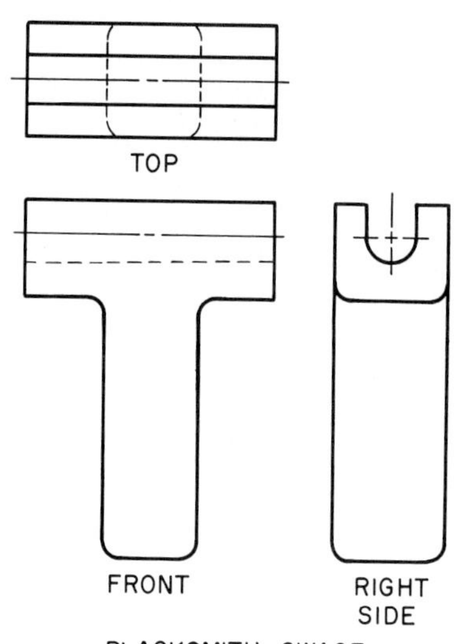

BLACKSMITH SWAGE

Fig. 6–16. **A three-view drawing.**

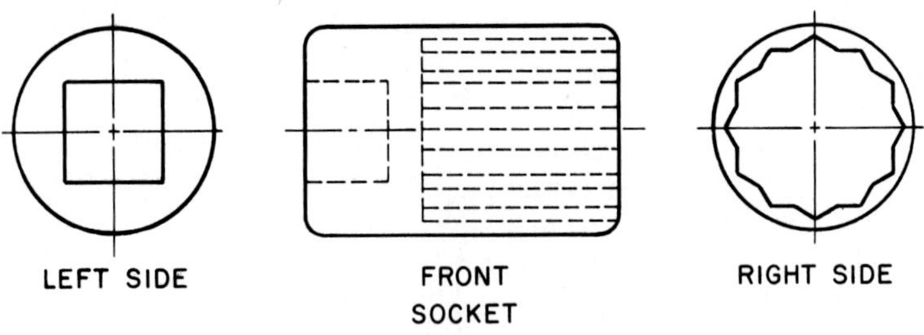

Fig. 6–17. A three-view drawing of a mechanic's socket. The left-side view is necessary to show the shape of the hole in that end.

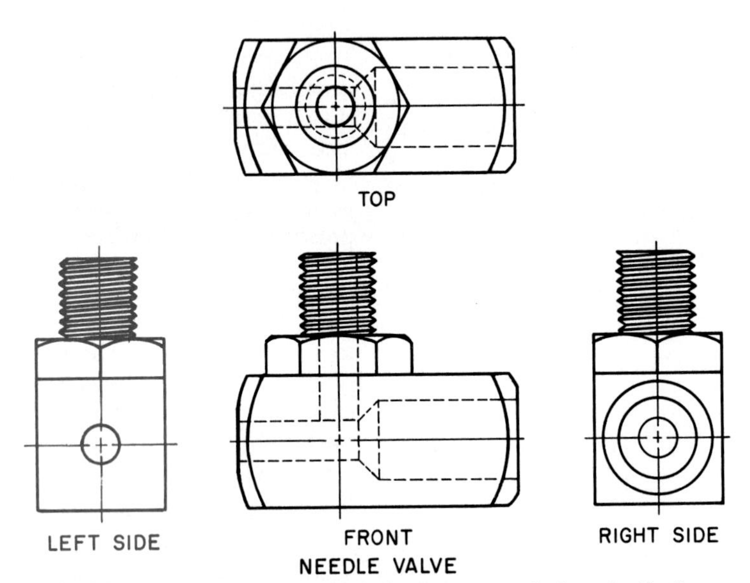

Fig. 6–18. A fourth view, the left-side view, is needed to clarify the details of the needle valve body and hub.

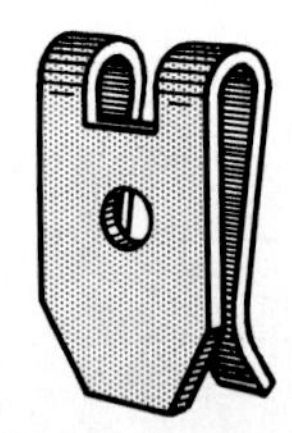

Fig. 6–19. A spring clip requires a rear view to clarify details.

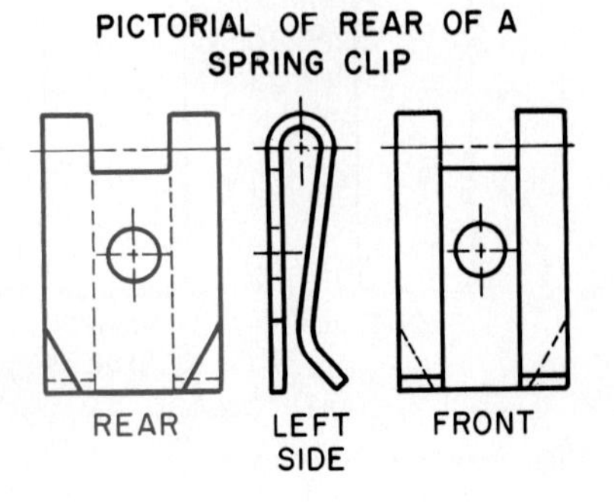

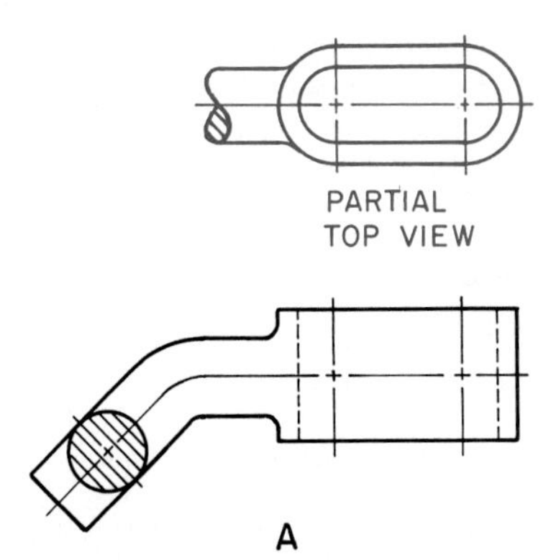

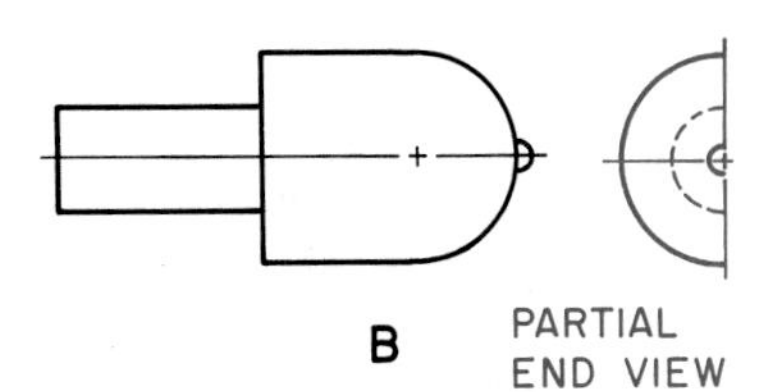

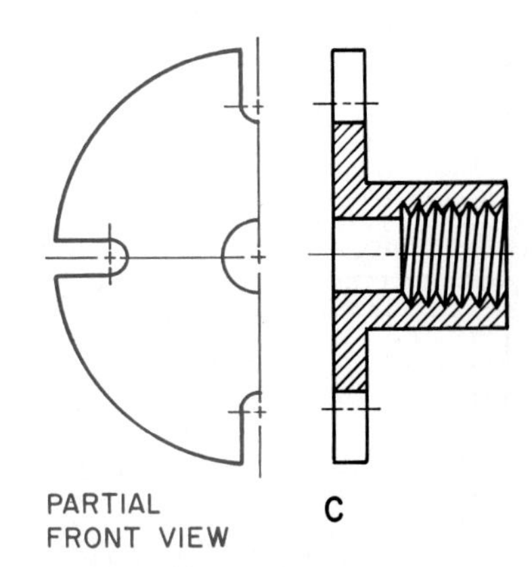

Fig. 6–20. Partial views simplify a drawing.

Partial Views

It is accepted practice to draw views that are not complete but which give enough information to describe the object. These are called **partial views.** They save space and time. They can involve a view of a portion of an object (Fig. 6-20 at A), a half view of a symmetrical object (B), or a partial view with a sectional view (C).

If an object has features on the right and left sides that are different and require both side views, these views need not be complete. Fig. 6-21. Notice that the hidden details on each end are not drawn. That same practice is shown in Fig. 6-17.

Selecting the Views

Before you draw an object, you must decide the number of views needed. If you need two views, usually you will use the top and front or front and right-side views. If you need three views, the top, front, and right-side views are most commonly used.

Select the front view first. The front view is the most important view and usually is the side of the object that shows the shape most clearly. Usually it is also the longest view. Fig. 6-22. The front view should be located so that the least number of hidden lines is needed. Note in Fig. 6-23 that when the motor mounting bracket is drawn with the open side to the front, the edges become visible.

Draw the object in a normal position. If the cup in Fig. 6-24 were drawn in any other position, it would appear quite unnatural.

Alternate Positions of Views

You should draw the object so that it makes good use of the space on the drawing paper and the drawing presents a balanced appearance. In Fig. 6-25 an alternate location for the right-side view was chosen to save space. It

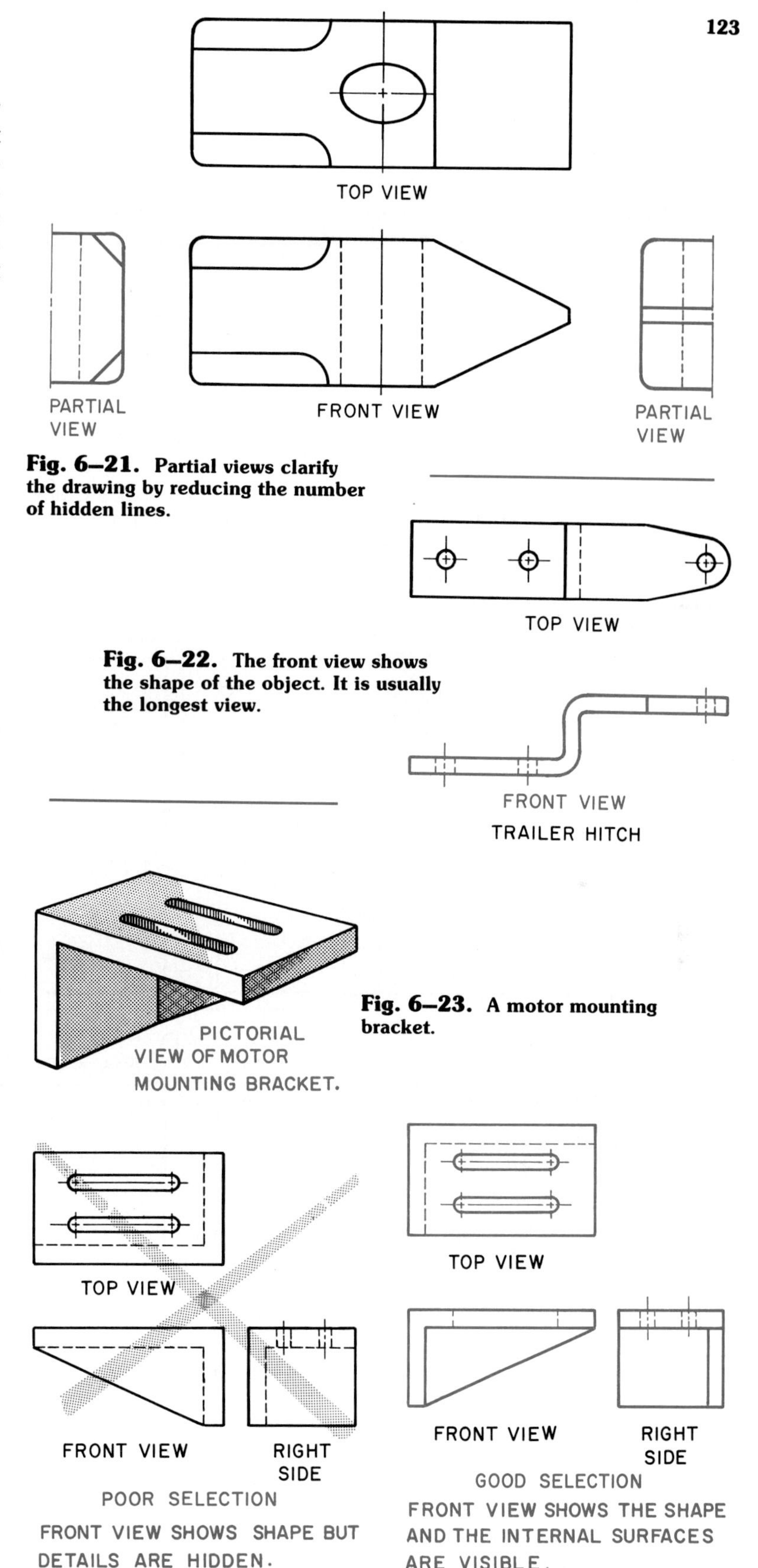

Fig. 6–21. Partial views clarify the drawing by reducing the number of hidden lines.

Fig. 6–22. The front view shows the shape of the object. It is usually the longest view.

Fig. 6–23. A motor mounting bracket.

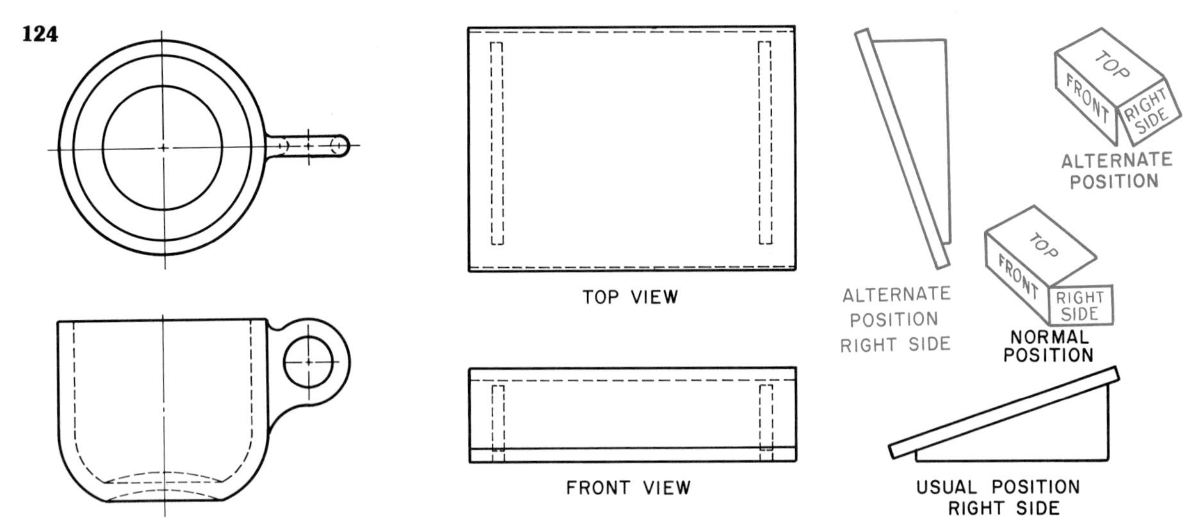

Fig. 6–24. An object should be drawn in its normal position.

Fig. 6–25. The alternate position of the side view of the drawing board balances the drawing and saves space.

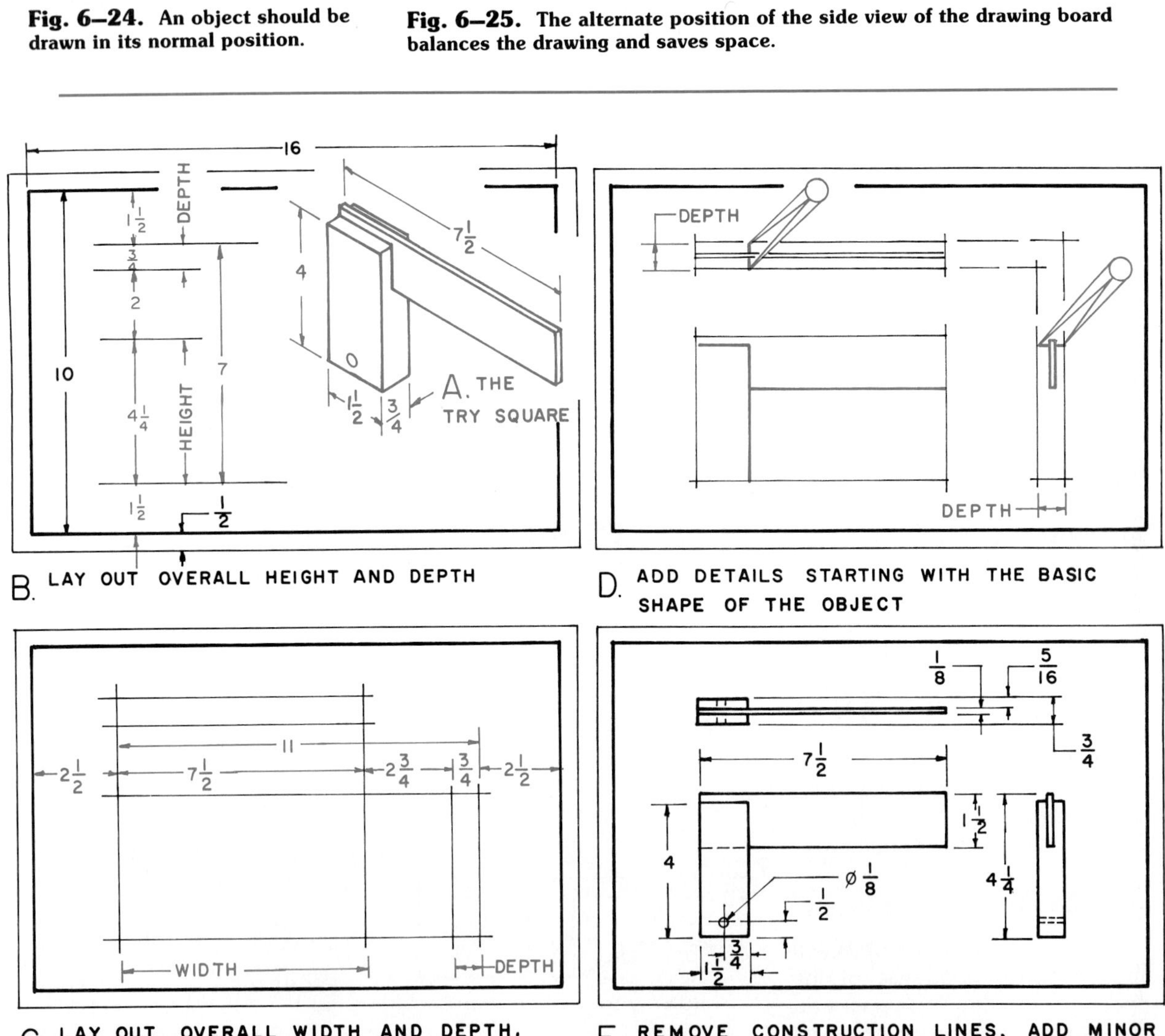

Fig. 6–26. The views should be spaced so as to leave room for the dimensions and allow the drawing to be balanced on the paper.

was projected from the top view. The same thing could be done with a left-side view. Refer to Fig. 6-6 for an alternate location of a rear view. A disadvantage of drawing the rear view off the top view is that the object then appears upside down.

Laying Out the Drawing

Before you can start a drawing, you must make several decisions. You must know the size of the object and decide on the number of views to be used. You must estimate the amount of space needed for dimensions. This is discussed completely in Chapter 7, "Dimensioning." You use these space requirements to decide what size paper you need. Drawing paper is sold in standard sizes as explained in Chapter 3, "Tools and Techniques of Drafting."

The views must be spaced so that they give the drawing a balanced appearance. If the drawing occupies one full sheet, the right and left margins should be equal. Sometimes the bottom margin is slightly larger than the top margin.

Usually you first "block in" the overall dimensions of the object in all views as shown in Fig. 6-26 at C. Use light, thin lines. Next draw the center lines for cylindrical parts.

Draw in the details. Start with the basic shape of the object. Fig. 6-26 at D. Then draw circles and arcs. You draw the minor details next. Carry each detail to all views before starting on a second detail. *Do not complete one view before starting on another.* Carry all views along together to completion.

After all the details are drawn, remove unnecessary lines. The light lines you use to locate details are called construction lines. As you gain experience, you will be able to draw the required lines the proper weight as you lay out the drawing. This is better than going back over the drawing and darkening them. See the alphabet of lines in Chapter 3, "Tools and Techniques of Drafting," for correct line weight.

If the object is cylindrical, locate the center lines in each view. Then draw the circles and arcs followed by the horizontal lines and then all the vertical lines. Fig. 6-27.

Dimension the drawing and letter notes and titles.

One method of spacing views on your drawing paper is shown in Fig. 6-26. To use it, follow this procedure:

1. Choose a sheet size. First look at the overall dimensions of the object: 7½ inches wide, 4½ inches high, and ¾ inch deep. Fig. 6-26 at A. Add the width and depth dimensions of the object to find 8¼ inches. Because you want to draw the object in its natural position, you choose a size B sheet (11 × 17 inches). You do not choose an A-size sheet since you know that it measures 8½ inches wide and the 8¼ inches computed does not allow for space between views nor a border around the sheet. After you draw a ½-inch border on the B-size sheet, you are left with a working space of 10 × 16 inches.

2. The space needed between views for dimensions is 2 inches between the top and front views and 2¾ inches between the front and side views.

3. The sum of the height, depth, and dimensioning space is, therefore, 7 inches. Fig. 6-26 at B.

4. The working space of 10 inches less the distance required by the object, 7 inches, leaves 3 inches to be divided equally above the top view and below the front view of the drawing.

5. The sum of the width, depth, and dimensioning space is 11 inches. Fig. 6-26 at C.

6. The working space of 16 inches less the distance required by the object leaves 5 inches to be divided equally between the right and left margins.

7. Details of the part, Fig. 6-26 at D, do not change the spacing.

8. The completed view, properly spaced with dimensions between views, is shown in Fig. 6-26 at E.

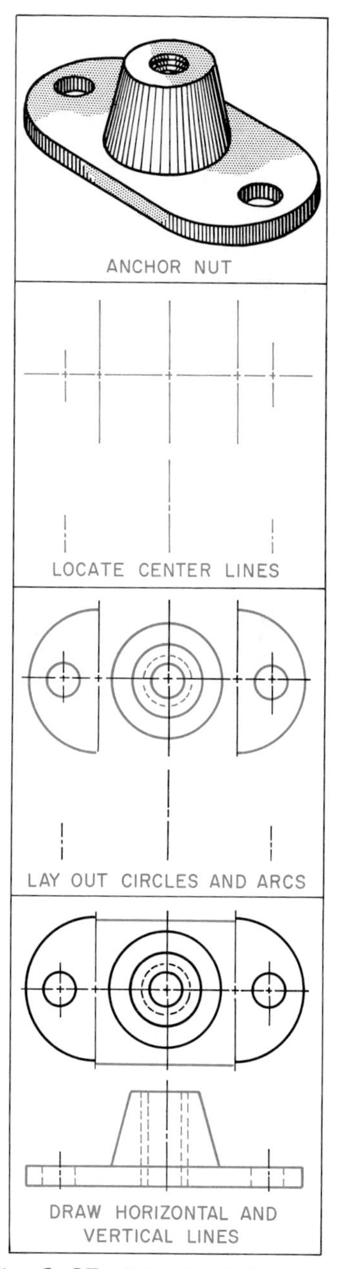

Fig. 6–27. Cylindrical objects can be laid out using center lines.

Hidden Lines

Hidden lines are used to show details that are behind some part of the object. As you recall from Chapter 3, "Tools and Techniques of Drafting," hidden lines are shown on a drawing by a dashed line. The top view of Fig. 6-25 has hidden lines to show the pieces below the top.

Figure 6-28 shows the correct ways to draw hidden lines. Hidden lines should:

- Show a gap whenever a hidden dash would form a continuation of a visible line, detail A.
- End by touching a visible line, detail B.
- Touch to form a corner, detail C.
- Touch when they meet, detail D.
- Cross center lines, detail E.
- Not touch the visible line where they cross, detail F.
- Not touch when representing planes that do not meet, detail G.
- Intersect at corners when they indicate hidden holes, detail H.
- Not touch visible lines because it would make the visible lines appear to extend into the surface of the object, detail I.
- Have their dashes staggered when they are close together and parallel, detail J.
- Touch center lines when they form a corner, detail K.

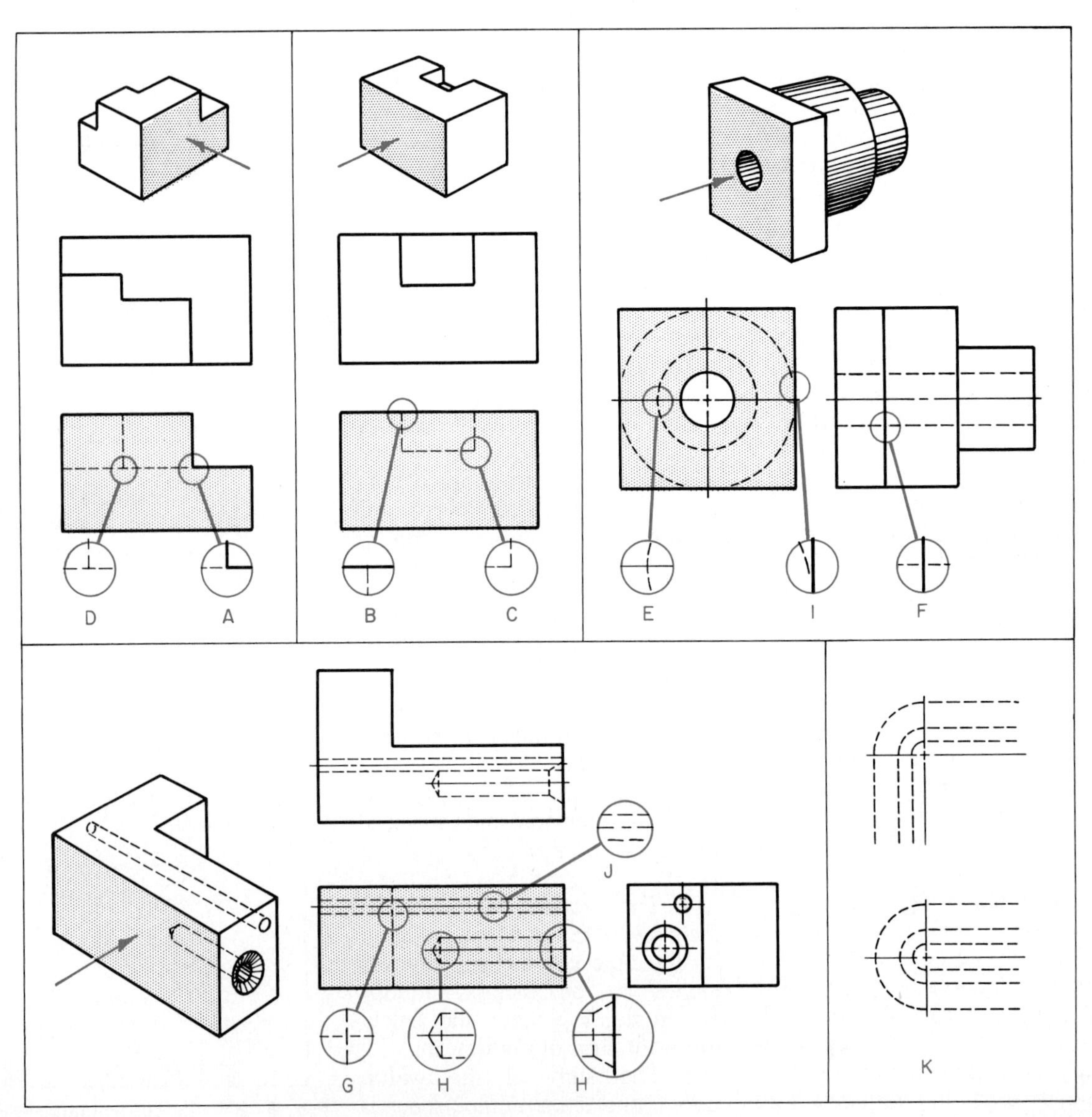

Fig. 6–28. **Correct practices for drawing hidden lines.**

Center Lines

Center lines are used to locate the centers of symmetrical objects and paths of motion. Symmetrical objects, you recall, have the same size and shape on both sides of a center line. Refer to Fig. 6-27. Recall from Chapter 3, "Tools and Techniques of Drafting," that center lines are drawn as a series of long and short dashes. Extend the center line about ¼ inch (6 mm) beyond the outside edge of the object for which it was drawn. It is not used to connect the views.

The center of a symmetrical object is located in Fig. 6-29 at A. The center is usually located by crossing two short dashes. The center lines end with long dashes.

In Fig. 6-29 at B the center line continues across the entire object and serves to locate the holes while it represents the center of the object. In Fig. 6-29 at C the center line is used to describe the circular path of the centers of a series of bolt holes. In Fig. 6-29 at D the path of motion made by the movement of a handle is recorded with a center line.

Center lines are useful in dimensioning. They assist in locating the position of circular parts in relation to the object as a whole. This is covered in more detail in Chapter 7, "Dimensioning."

Precedence of Lines

When two lines in a view fall together, the more important of the two is shown. The order of importance for lines is the following:

1. Visible lines.
2. Hidden lines.
3. Center lines.

When a cutting-plane line falls on a center line, the cutting-plane line is shown.

Projecting Plane Surfaces

Most objects have both plane and curved surfaces. The plane surfaces can be of three types: normal, inclined, and oblique. Fig. 6-30.

A **normal plane** is one that is perpendicular to two regular planes of projection and parallel to the third plane. A normal plane appears true size on the plane to which it is parallel. It appears as a true length line on the planes to which it is perpendicular. Fig. 6-31.

An **inclined plane** is one that is perpendicular to one regular plane of projection and is at an *acute* (less than 90 degrees) angle to the other two planes. It appears true length on any regular plane of projection to which it is perpendicular and foreshortened (not true size) on any regular plane to which it is not perpendicular. Fig. 6-32. Figure 6-33 shows an industrial product that contains both normal and inclined planes.

An **oblique plane** is one that is not perpendicular to any of the regular planes of projection. An oblique plane appears foreshortened on each of the regular planes of projection. Fig. 6-34. Figure 6-35 shows a product containing normal, inclined, oblique, and curved planes.

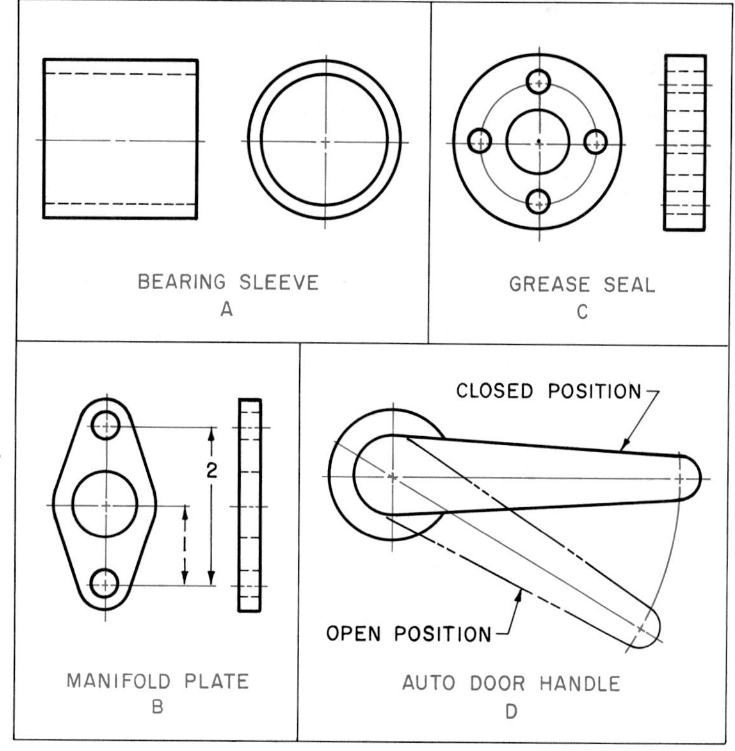

Fig. 6–29. Typical uses of center lines.

Fig. 6–30. **The Aluette II spacecraft was designed using normal, inclined, and oblique surfaces on the exterior. Can you find an example of each surface? (National Aeronautics and Space Administration)**

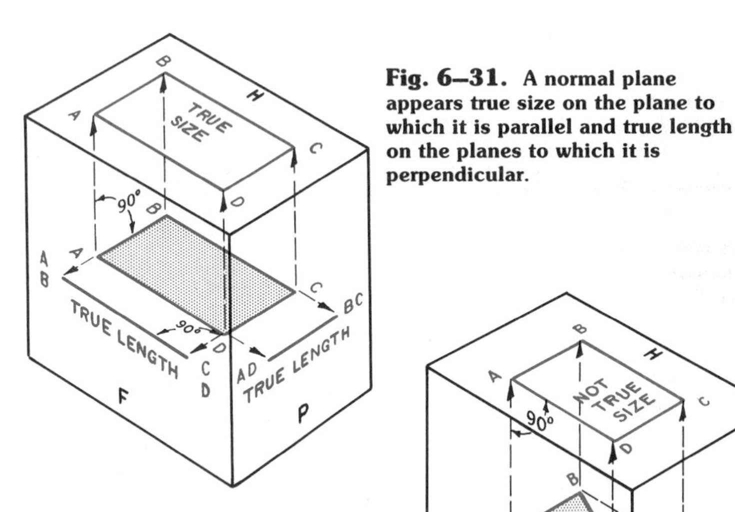

Fig. 6–31. **A normal plane appears true size on the plane to which it is parallel and true length on the planes to which it is perpendicular.**

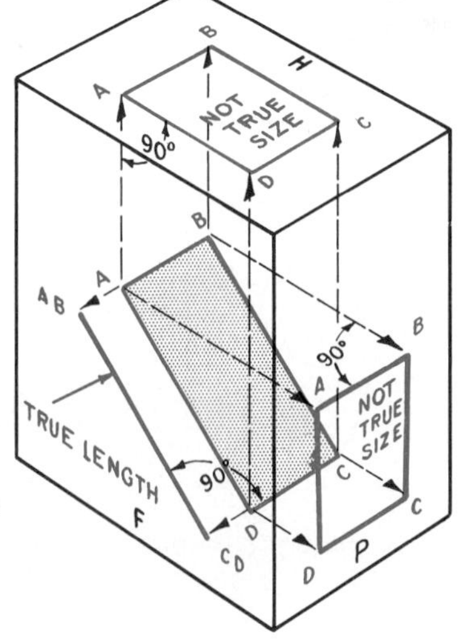

Fig. 6–32. **An inclined plane appears true length on the plane to which it is perpendicular and foreshortened on any plane to which it is not parallel.**

Projecting Lines

Two planes meet at a corner, which appears as a straight line on a drawing. Lines can be of three types: normal, inclined, and oblique.

A **normal line** is one that is parallel to two regular planes of projection and perpendicular to the other plane. It appears true length on the planes to which it is parallel and as a point on the plane to which it is perpendicular. Fig. 6-36.

An **inclined line** is one that is parallel to one regular plane of projection and makes an acute angle to the other two regular planes. It appears true length on the plane to which it is parallel and foreshortened on the other two planes to which it is not parallel. Fig. 6-37.

An **oblique line** is one that is not parallel to any of the regular planes of projection. It appears foreshortened on all regular planes of projection. Fig. 6-38.

Projecting Curved Surfaces

The basic curved surfaces are spherical, cylindrical, and conical. Fig. 6-39. Cylinders and cones appear as curves in views perpendicular with the axis of the object. Fig. 6-40. In other normal views, the cylinder projects as a rectangle and the cone as a triangle. The sphere projects as a circle in all normal views.

Projecting Curved Edges

Simple curves, such as arcs and circles, can be located in each view and easily drawn. Other curved surfaces, such as an irregular curve, must be projected from one view to another. You project a curve by locating a series of points along its edge and then transferring those points to other views. Connect those points with an irregular curve to form the projected curve.

The number of points to use will vary with the curved edge. Usually, the more points you project, the more accurate the projected curve will appear.

Figure 6-41 shows an edge view of a piece of molding drawn actual size in the right-side view. The top view shows the molding cut on a 45-degree angle. Points 1 through 7 are located randomly on the side view and projected to the top and front views. The curve is drawn where those points intersect on the front view.

Figure 6-42 shows a projection when all the surfaces are curved.

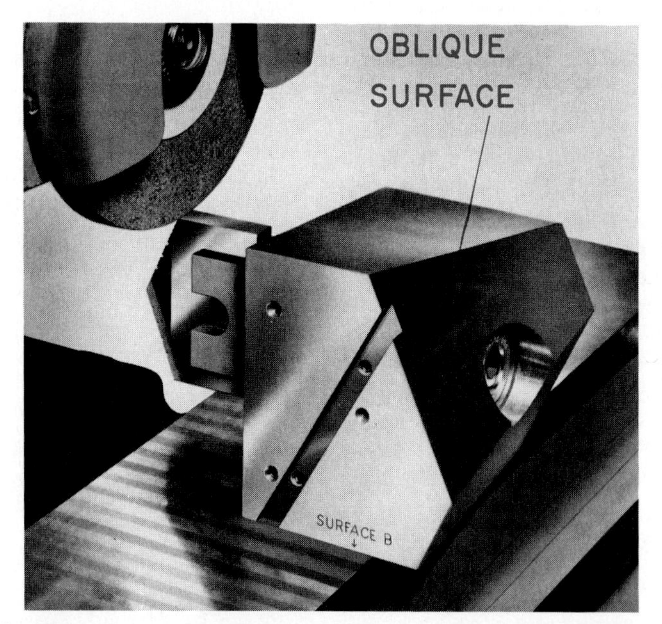

Fig. 6–35. A grinding fixture used to sharpen a spade drill. Can you identify the types of planes on the surfaces of the fixture? (De Vlieg Microbore)

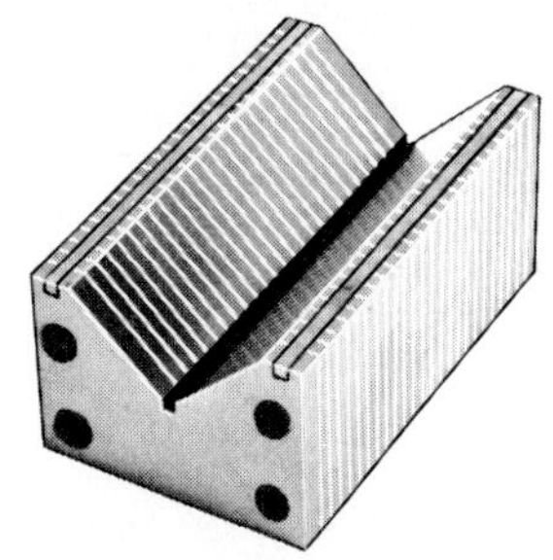

Fig. 6–33. This magnetic V-block contains normal and inclined planes. (Reid Tool Supply Co.)

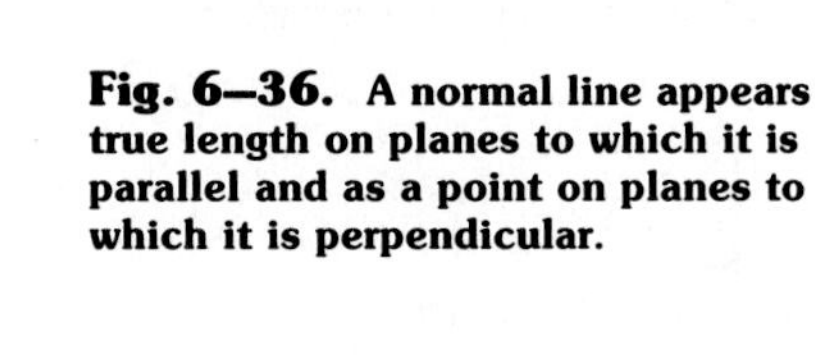

Fig. 6–36. A normal line appears true length on planes to which it is parallel and as a point on planes to which it is perpendicular.

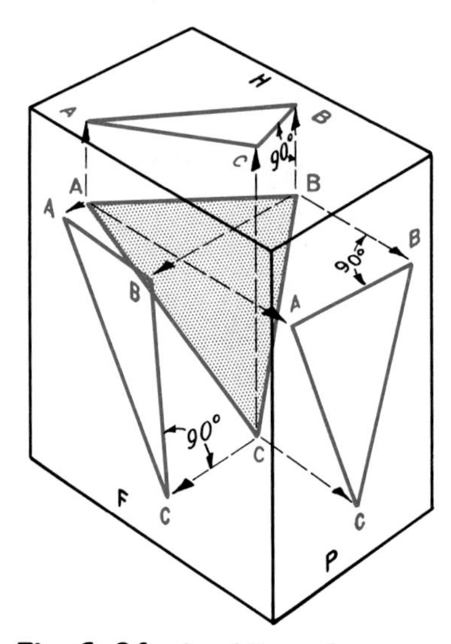

Fig. 6–34. An oblique plane appears foreshortened on all regular planes of projection.

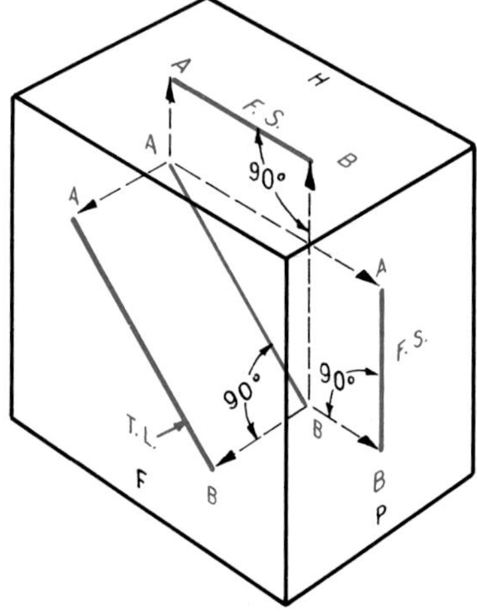

Fig. 6–37. An inclined line appears true length on any plane to which it is parallel.

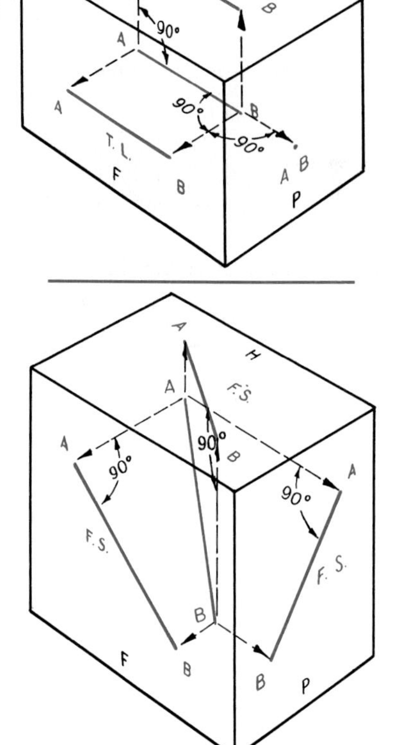

Fig. 6–38. An oblique line appears foreshortened on all regular planes of projection.

Intersections of Planes and Curved Surfaces

Figure 6-43 shows typical intersections between a cylinder and a plane. At A a cylinder is cut by a vertical plane. At B a hollow tube is cut partway through by a vertical plane revealing an interior curved surface. At C the cylinder is cut by an inclined plane, creating an elliptical surface. You can project the elliptical surface by using a system of points as follows:

1. Divide the circular view into several parts.
2. Project the divisions to the other views.
3. Locate the intersection of the projections with light dots.
4. Connect the dots with an irregular curve.

You could also use an ellipse template or construct an approximate ellipse using the method described in Chapter 5, "Geometric Figures and Constructions."

Figure 6-44 illustrates typical intersections between a cone and a plane. At A the cone is cut by a horizontal plane, leaving a circular surface. At B it is cut by a vertical plane, creating a hyperbolic surface. At C the cone is cut by an inclined plane, leaving an elliptical surface. At D the cone is cut with an inclined plane, creating a parabolic surface.

The projection of cylindrical surfaces from view to view is shown in Fig. 6-45. At A notice that a straight section, 1–2, occurs between the points of tangency of the two cylindrical surfaces. This appears as a straight,

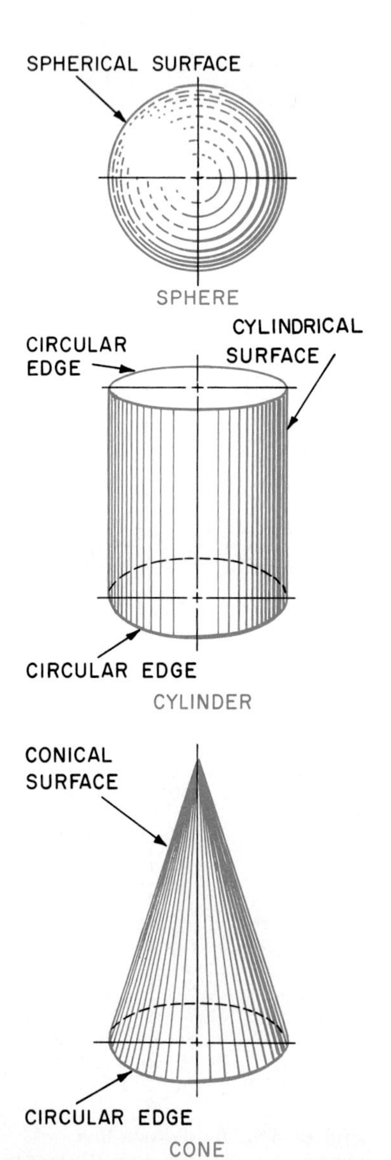

Fig. 6–39. Basic curved surfaces.

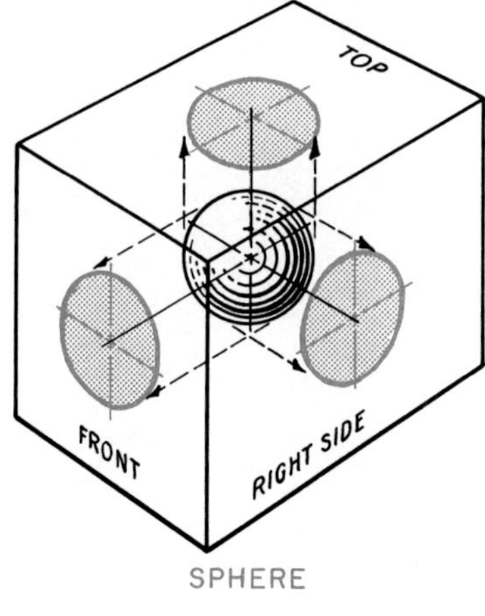

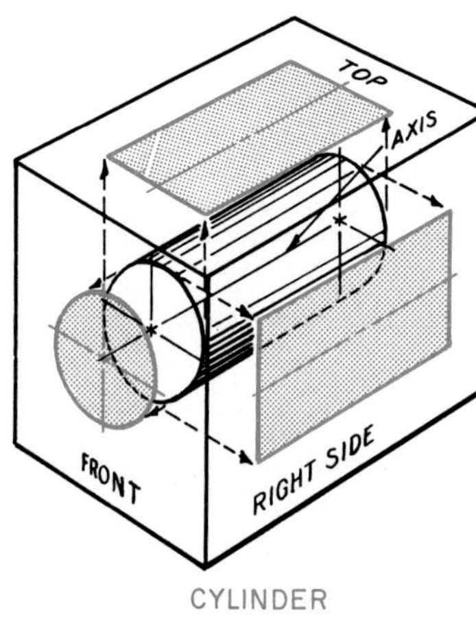

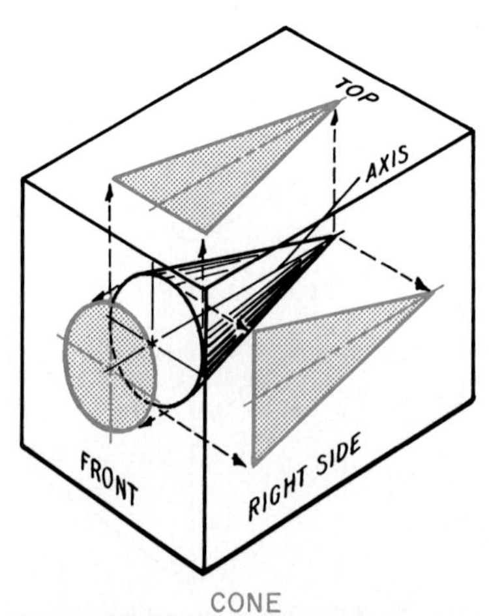

Fig. 6–40. Projections of basic curved surfaces.

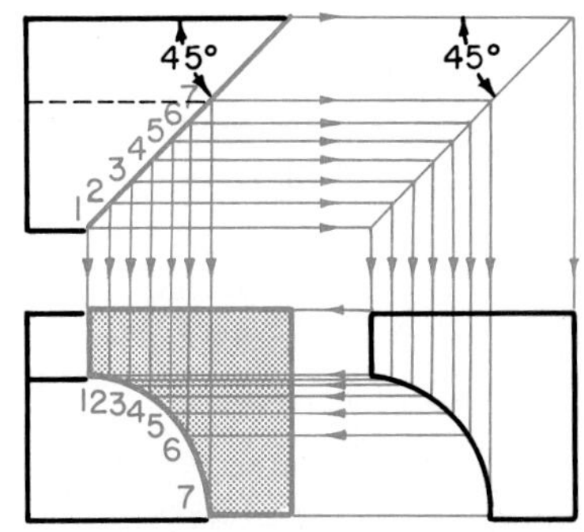

Fig. 6–41. Locate a curved edge by projecting points along the edge.

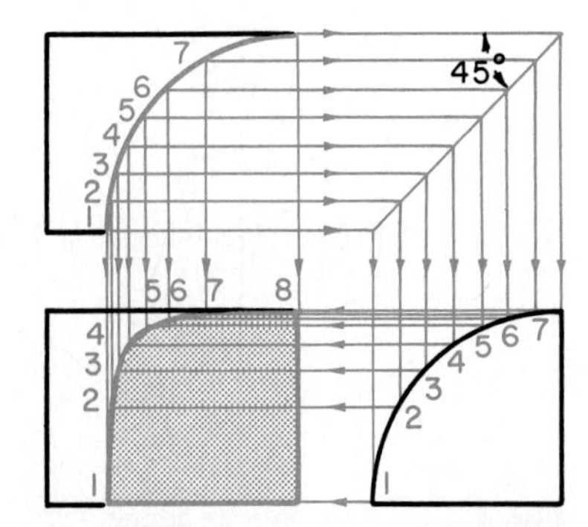

Fig. 6–42. A projection when all surfaces are curved.

visible line in the front view. At point of tangency 3 no edge occurs; therefore no visible line is projected to the front view.

At B the two cylindrical surfaces meet at the point of tangency. This shows on the front view as a straight, vertical line.

At C the surface generated by the two intersecting curves is a continuous slope; therefore no visible line appears in the front view.

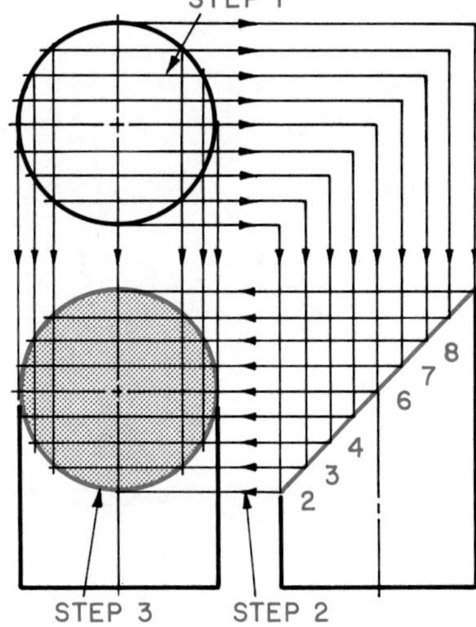

Fig. 6–43. Common intersections between plane and curved surfaces.

Figure 6-46 illustrates the intersections of other cylindrical surfaces and planes.

The intersection of cylinders is shown in Fig. 6-47. At A are two cylinders with the same diameter. They meet in a semi-elliptical curve that projects as a straight line to the normal planes of projection. At B the intersecting cylinder is so small that the curvature of the intersection can be

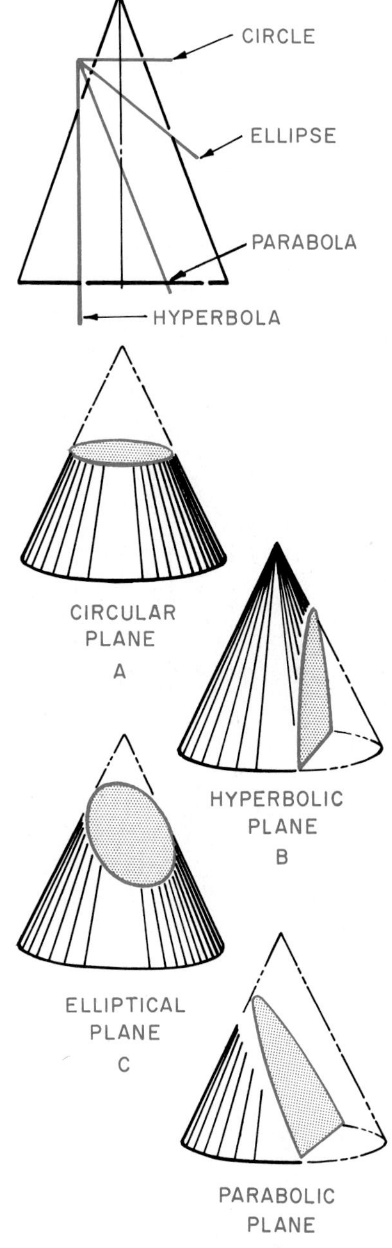

Fig. 6–44. Conical sections are generated by the intersection of a cone and a plane.

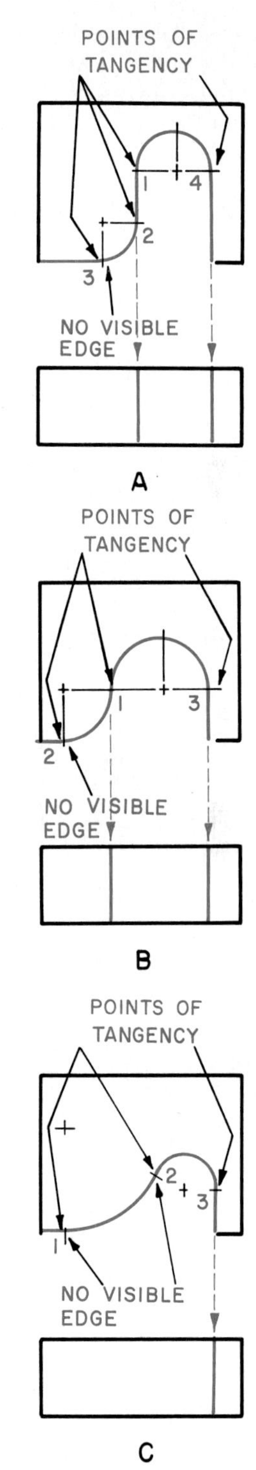

Fig. 6–45. Surface intersections.

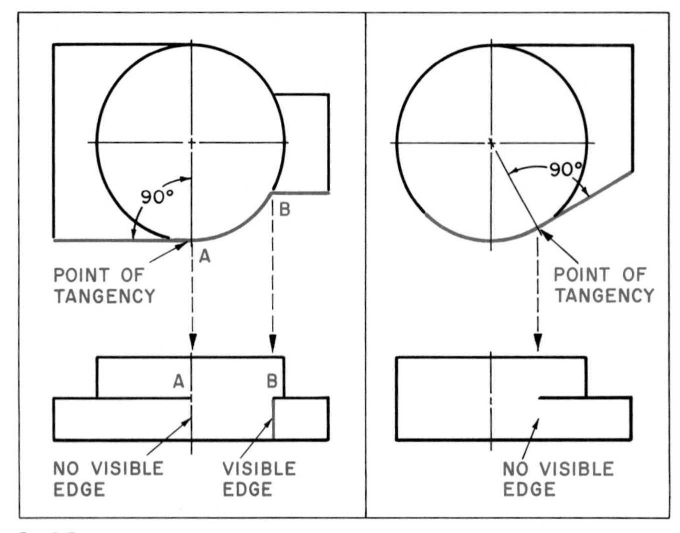

Fig. 6–46. **Common intersections between planes and cylinders.**

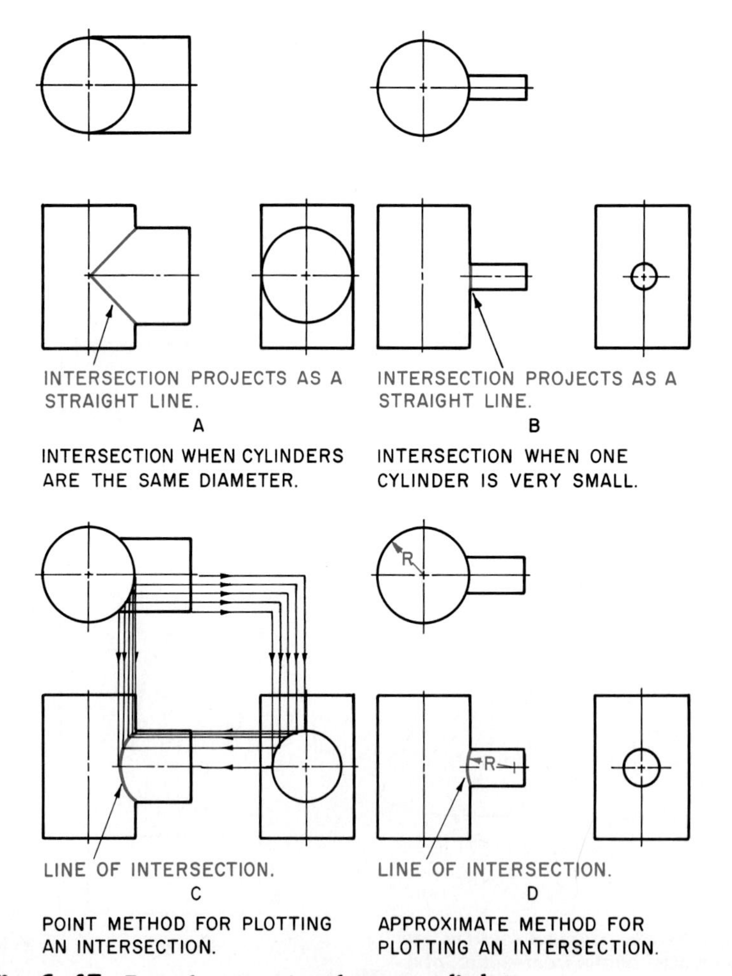

Fig. 6–47. **Typical intersections between cylinders.**

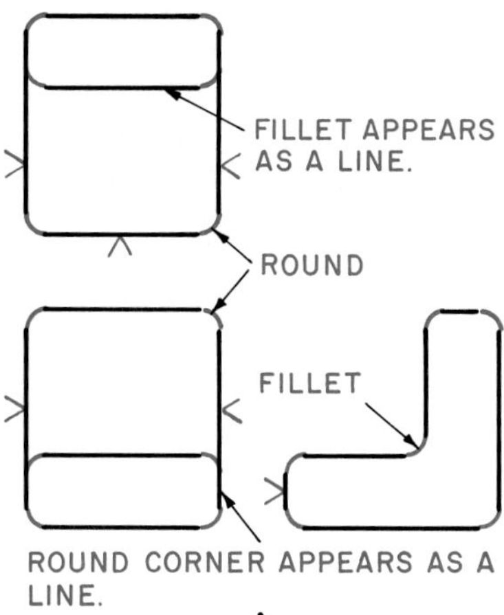

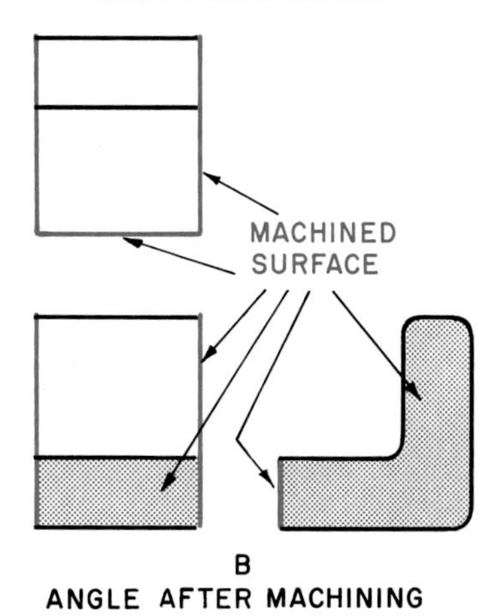

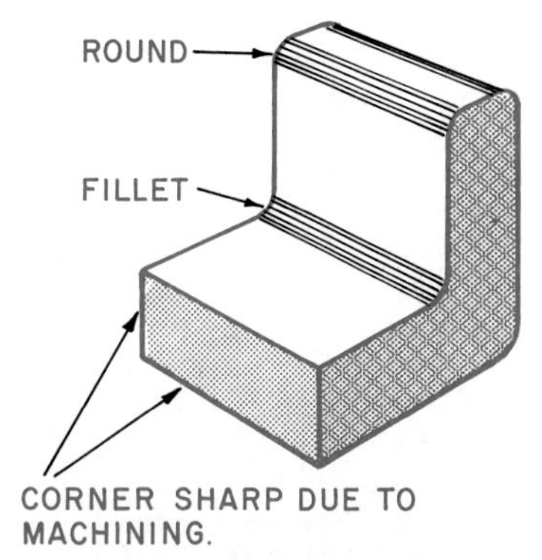

Fig. 6–48. **Methods of indicating rounds and fillets before and after machining.**

ignored. It is projected to the front view as a straight line. At C the intersection is plotted using the point method. This is done whenever the diameters of the intersecting cylinders are large enough to allow plotting the curve. The approximate method, shown at D, may be used when the diameter is small. Small intersecting curves are difficult to plot.

Indicating Rounds and Fillets

External corners on cast and forged parts are usually rounded. Such a corner is called a **round.** Fig. 6-48. The round makes it easier to cast and handle the part. A rounded corner can have any radius the designer decides is necessary.

A rounded interior corner is called a **fillet.** Fillets are necessary to prevent cast parts from fracturing at the corner. Figure 6-48 illustrates the use of rounds and fillets. Notice that even when a corner is a curved surface, it appears as a line in the view to which it is perpendicular.

You usually dimension fillets and rounds with a note such as, "All fillets and rounds ⅛ inch radius unless otherwise specified." If they are not dimensioned, the size is decided by the patternmaker.

Runouts

The intersection of fillets and rounds with other surfaces produces an extension of the curved surface. This extension is called a **runout.** Fig. 6-49. Runouts occur frequently on multiview drawings.

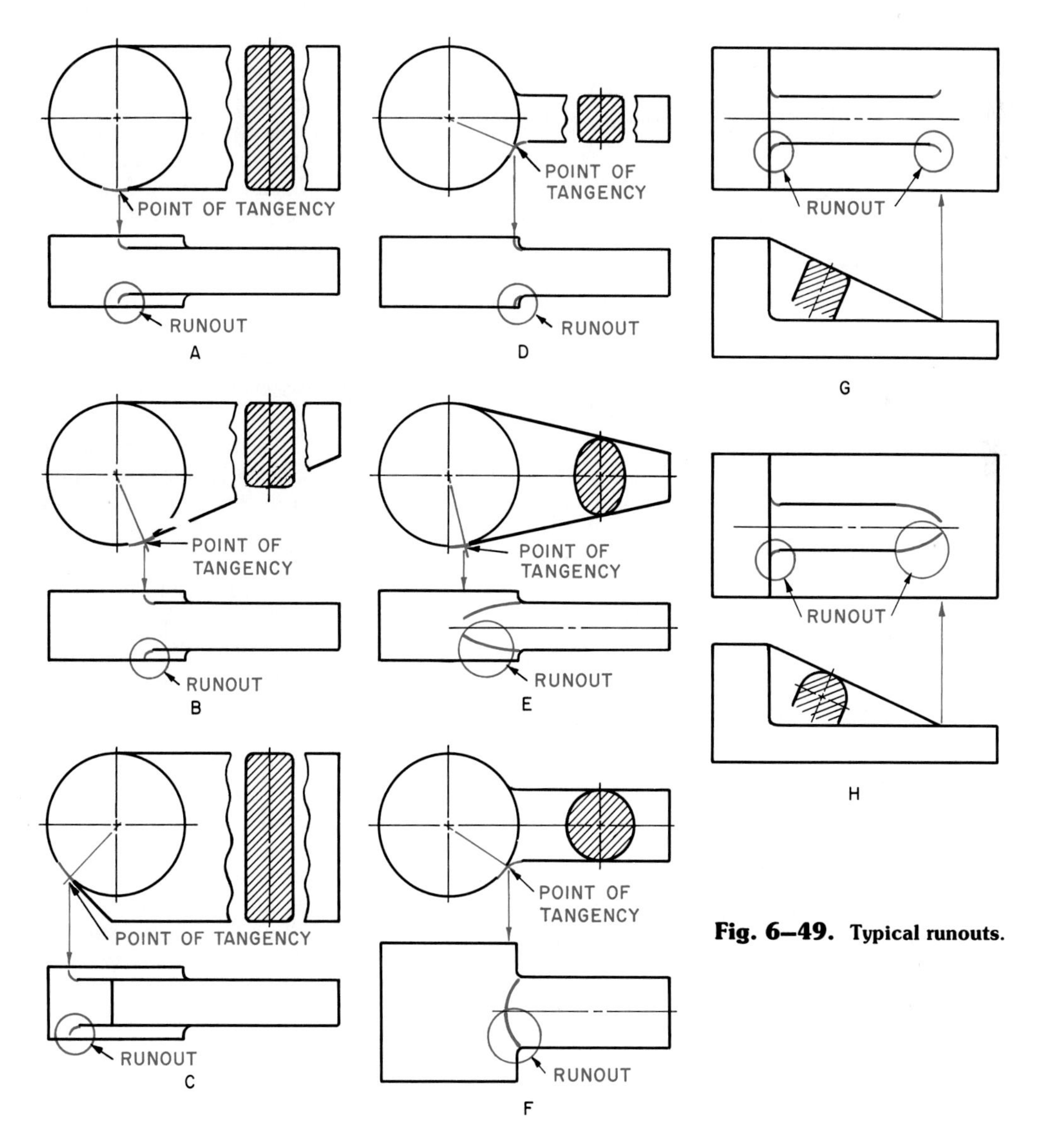

Fig. 6–49. **Typical runouts.**

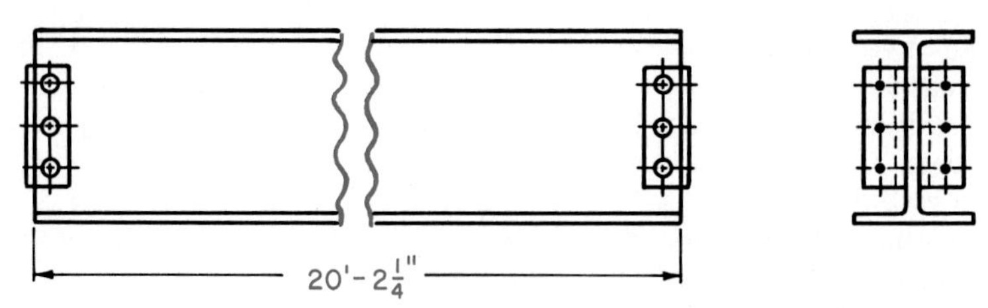

Fig. 6–50. A long object can be shortened by breaking out a section.

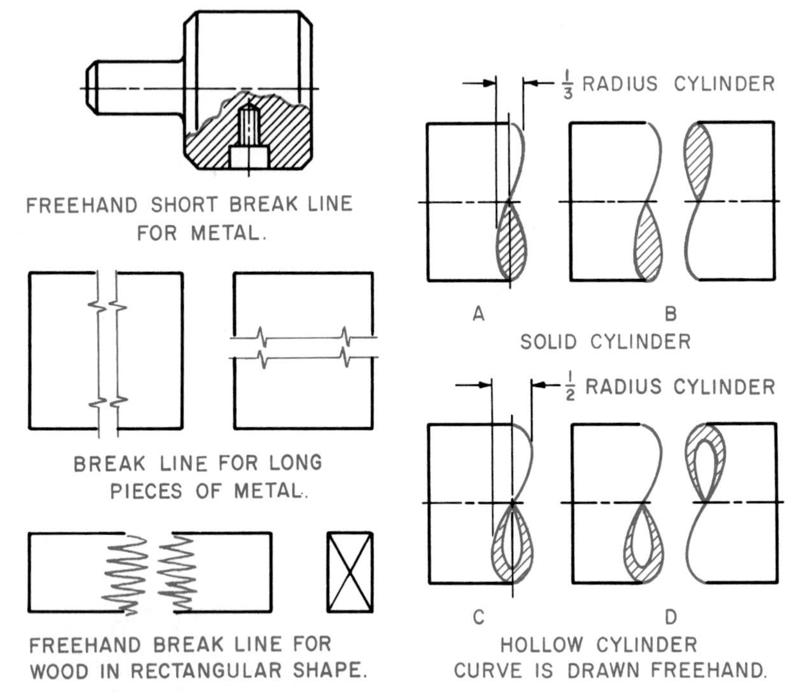

Fig. 6–51. Standard break symbols.

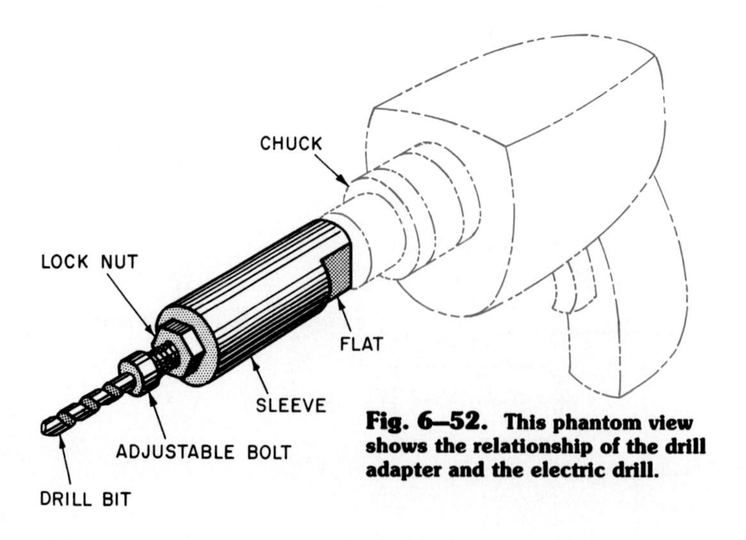

Fig. 6–52. This phantom view shows the relationship of the drill adapter and the electric drill.

A runout arc may turn in or out, depending upon the shape and thickness of the intersecting areas.

Runouts can be drawn with a compass or an irregular curve. Very small intersections can be drawn freehand. It is common practice to draw runouts between small fillets and rounds by using the radius of the fillet. The length of the runout arc is usually one-eighth of a circle.

Figure 6-49 illustrates how common filleted intersections are drawn.

Conventional Breaks

Sometimes on a multiview drawing you need to break away part of an object to show the details inside. You mark the edges where the break occurs with a **break symbol.** Long objects that have no distinctive features for most of their length can be drawn with some of the length removed. This shortens the drawing and permits the object to be drawn to a larger scale. The break lines show that a piece has been removed.

Even though an object is drawn shorter, it is dimensioned true size. Fig. 6-50. Conventional means of indicating breaks are shown in Fig. 6-51.

Short breaks are indicated with a freehand line that is the same thickness as the visible line. If breaks are long, draw a long, thin, ruled line and insert a freehand Z shape at convenient intervals.

You indicate breaks in round shafts by an S-shaped symbol. It may be drawn freehand or with an irregular curve. Figure 6-51 at A shows the proportions to use to draw the symbol for a solid shaft. The symbol to use for a break in a rod is shown at B. The proportions for a break in a hollow tube are shown at C. Notice that the symbol is wider than that for the solid shaft. Estimate the size of the inside loop and draw it freehand. The symbol to use for a break in a tube is shown at D.

Phantom Lines

Phantom lines are used on multiview drawings to show the position of parts next to the object being drawn. In Fig. 6-52 the drill adapter is the important object. The electric drill is drawn with phantom lines to show the position and use of the adapter.

Phantom lines are also used to show the possible positions of moving parts. Figure 6-53 shows the right and left positions of the cabinet lock.

Recall from Chapter 3, "Tools and Techniques of Drafting," that phantom lines are made of a series of one long and two short dashes.

Knurling

A **knurl** is a uniformly roughened surface. It is used on handles and knobs to provide a sure grip. Fig. 6-54. Shafts are also knurled so that they will not slip when forced into a hole. The method for drawing knurls is shown in Fig. 6-55.

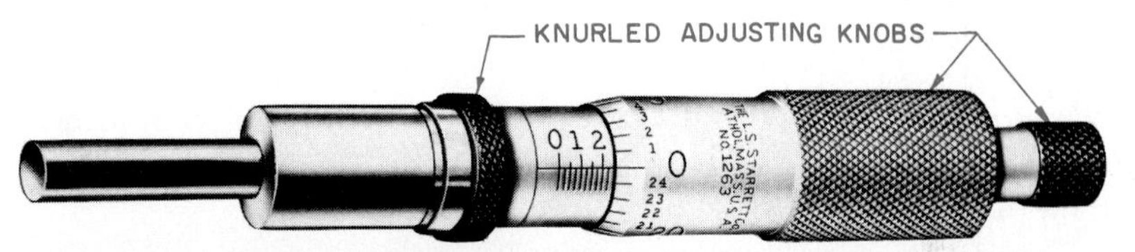

Fig. 6–54. Diamond knurling is used on the adjusting portions of this micrometer caliper. (L. S. Starrett Co.)

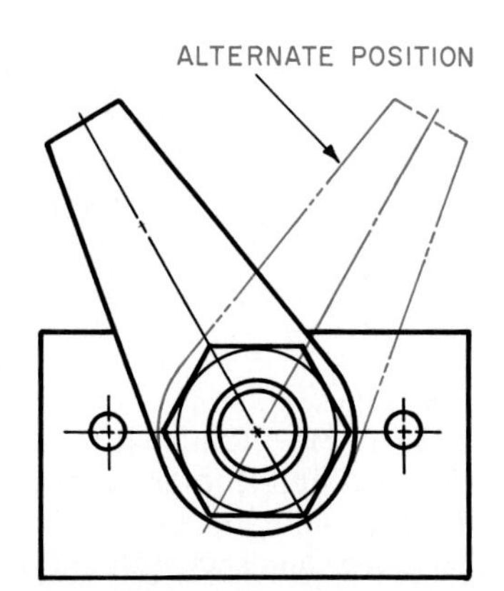

Fig. 6–53. A phantom view of a cabinet lock lever.

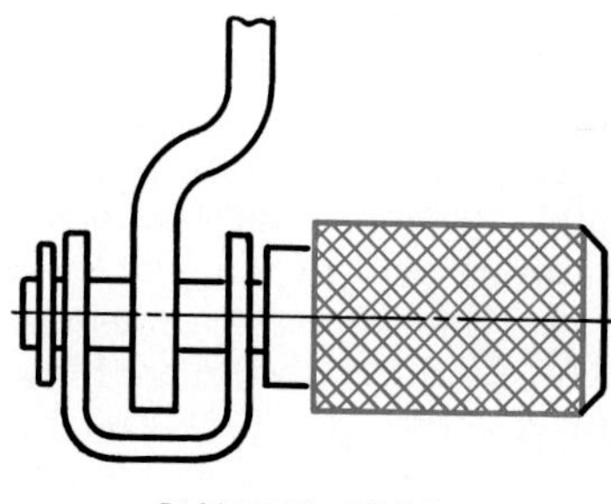

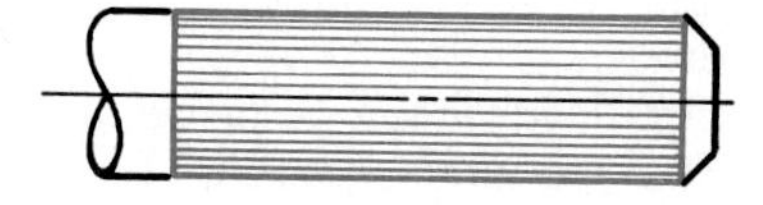

Fig. 6–55. How to show knurling on a drawing.

Chapter Review

Build Your Vocabulary

You should understand and use the following terms as part of your working vocabulary. Write a brief explanation of what each means.

multiview drawing
planes of projection
orthographic projection
third-angle projection
symmetrical
partial views
normal plane
inclined plane
oblique plane
normal line
inclined line
oblique line
round
fillet
runout
break symbol
phantom lines
knurl

Sharpen Your Math Skills

1. If your drawing is 16″ × 22″, how many square inches of work space are available?

2. If you were going to draw a one-view drawing of a 10″-diameter sphere in the center of the sheet in Problem 1, how much clearance would it have on all sides?

3. Dashes used for hidden lines are made 3 millimeters in length and spaced 1.5 millimeters apart. How many dashes would there be in a line 90 millimeters long?

4. If you used the scale ½″ equals 1″, what radius would you use to draw a fillet that is dimensioned with a 3″ radius?

Study Problems—Directions

The problems that follow will give you the chance to use the drafting techniques that you read about in this chapter. Do the problems that are assigned to you by your instructor.

1. Incomplete views. The multiview drawings shown in **Figs. P6-1** and **P6-2** are incomplete. Sketch the given views the same size as shown and supply the missing lines.

2. Surface identification. Some surfaces in **Fig. P6-3** are marked with letters and some edges are marked with numbers. Identify each marked surface and edge as normal, inclined, or oblique.

3. One-view drawings. The products shown in **Figs. P6-4** and **P6-5** can be described with one view. Draw them full size.

4. Two-view drawings. The products shown in **Figs. P6-6** through **P6-12** can be described with two views. Select the two views that give the complete details of the product and make a full-size detail drawing of each.

5. Three-view drawings. The products shown in **Figs. P6-13** through **P6-17** can be described with three views. Select the three views that give complete details of the product and make a full-size detail drawing of each.

6. Intersection problems. The products shown in **Figs. P6-18** through **P6-26** involve the intersection of a plane and curved surface. Select the views needed to describe each product completely. Make a drawing of each. Draw full-size or to a scale assigned by your instructor. Give special care to drawing the intersecting surfaces.

7. Phantom line problem. Draw the necessary views to describe the product in **Fig. P6-27.** Draw it full-size with the lever in the highest position and show the lowest position of the lever with phantom lines.

8. Additional problems. Draw the products shown in **Figs. P6-28** through **P6-56** full-size or to a scale assigned by your instructor. Select which side is to be the front view and decide which other views are needed to completely describe the product.

9. **Fig. P6-57** is a two-view drawing of a proposed design for the Space Shuttle. This is a very detailed and difficult drawing to copy, but see how well you can produce this two-view drawing two or three times as large as printed. Use your dividers to get the sizes needed. (NASA)

Study Problems

P6–1. Incomplete multiview drawings.

Study Problems

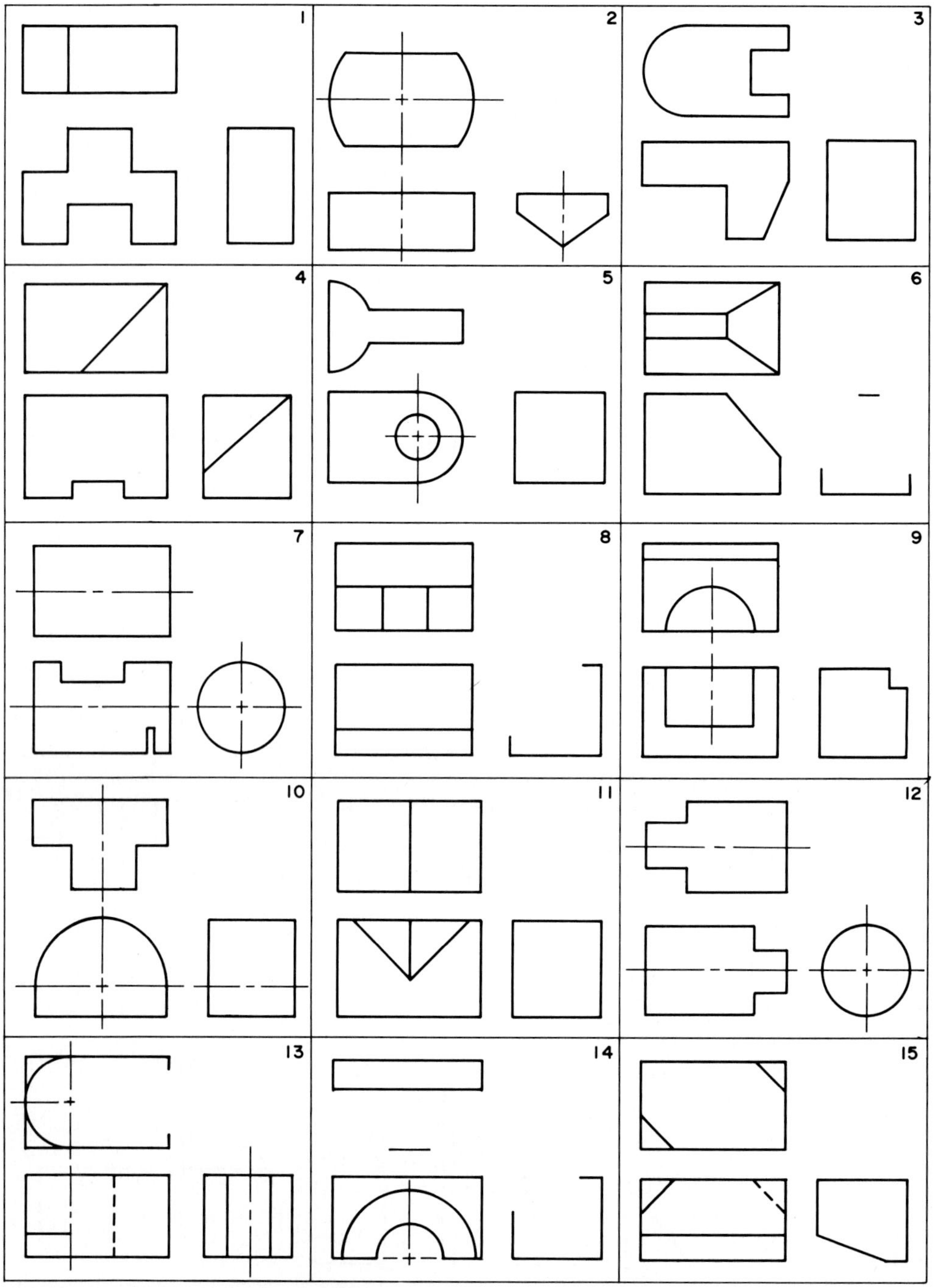

1 2 3 4 5 6 7 8 9 10 11 12 13 14 15

P6–2. Incomplete multiview drawings.

Study Problems

P6–3. Surface identification figures.

P6–4. Protractor.

P6–5. Plastic circle template.

Study Problems

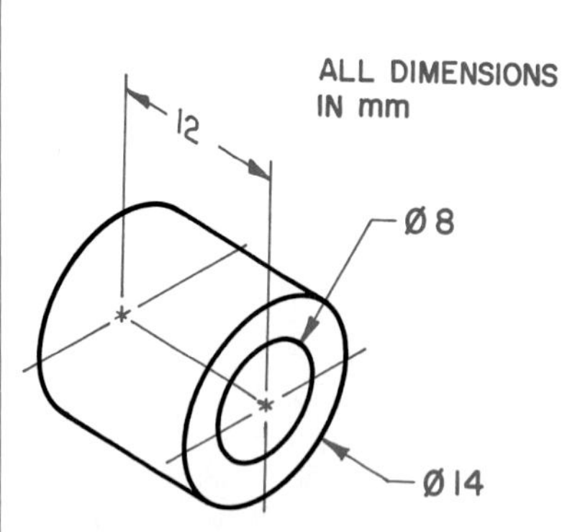

P6–6. Rubber gasket.

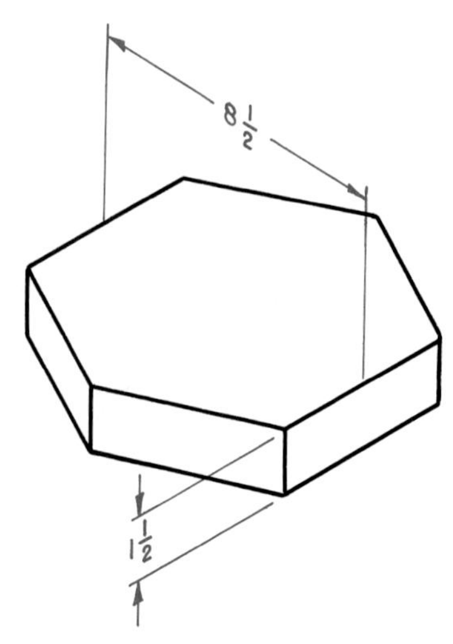

P6–7. Asphalt paving block for industrial floors.

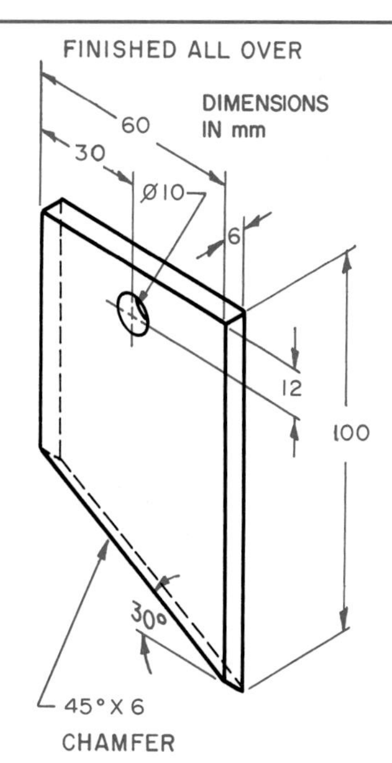

P6–8. Cutting edge for auto glass window tape trimmer.

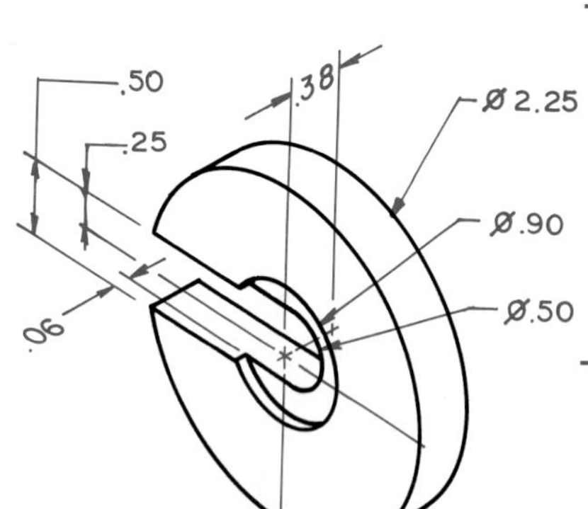

P6–9. C washer.

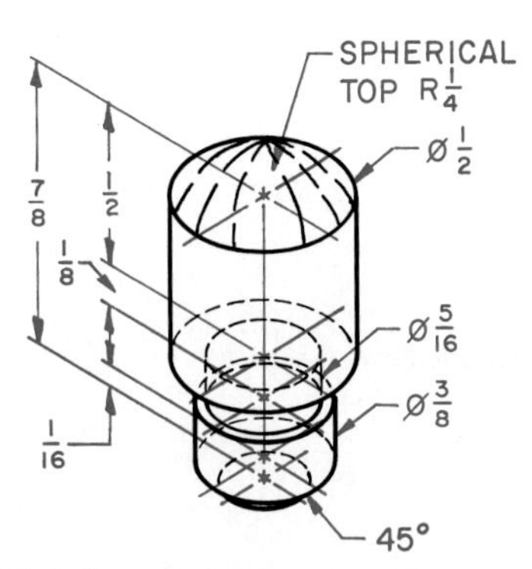

P6–10. Spherical radius locator button.

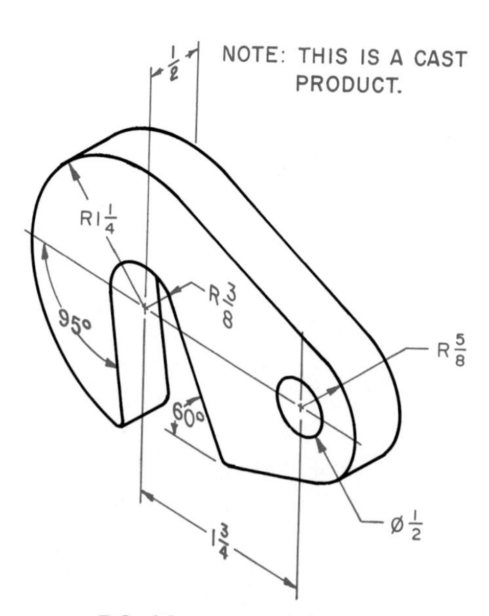

P6–11. Swing washer.

Study Problems

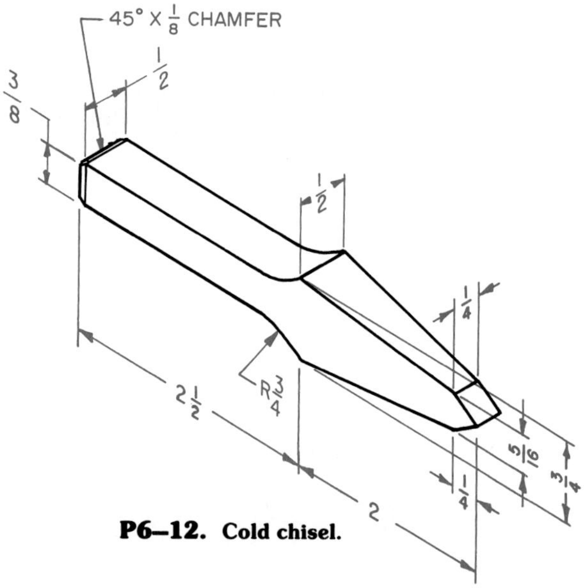
45° X 1/8 CHAMFER
1/2
3/8
1/2
1/4
R3/4
2 1/2
5/16
3/4
1/4
2

P6–12. Cold chisel.

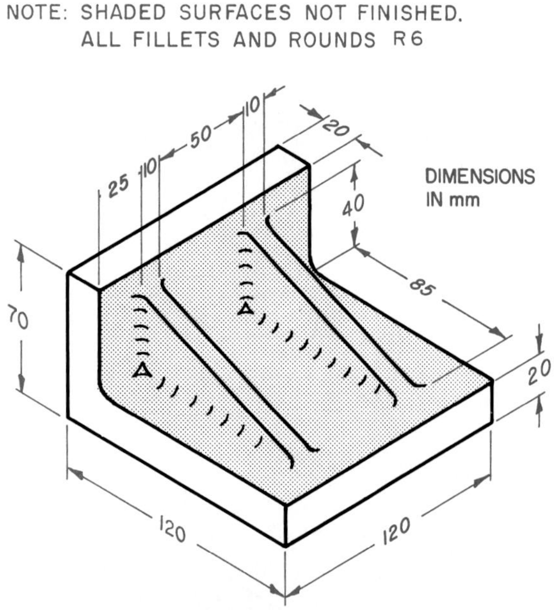
NOTE: SHADED SURFACES NOT FINISHED.
ALL FILLETS AND ROUNDS R6
DIMENSIONS IN mm
25
10
50
10
20
40
85
70
20
120
120

P6–13. Angle plate.

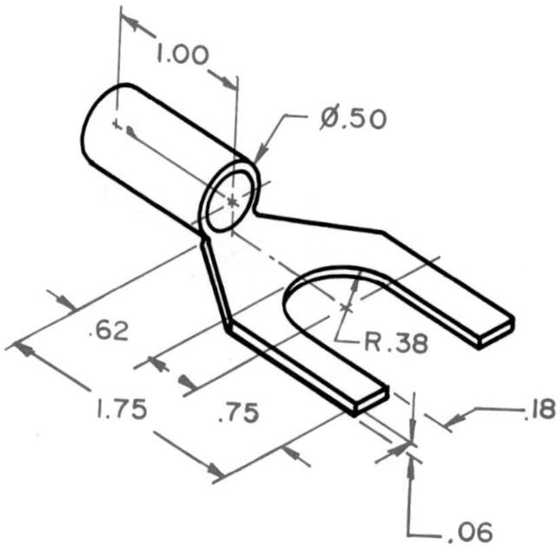
1.00
Ø.50
.62
R.38
1.75
.75
.18
.06

P6–14. Electric cable connection.

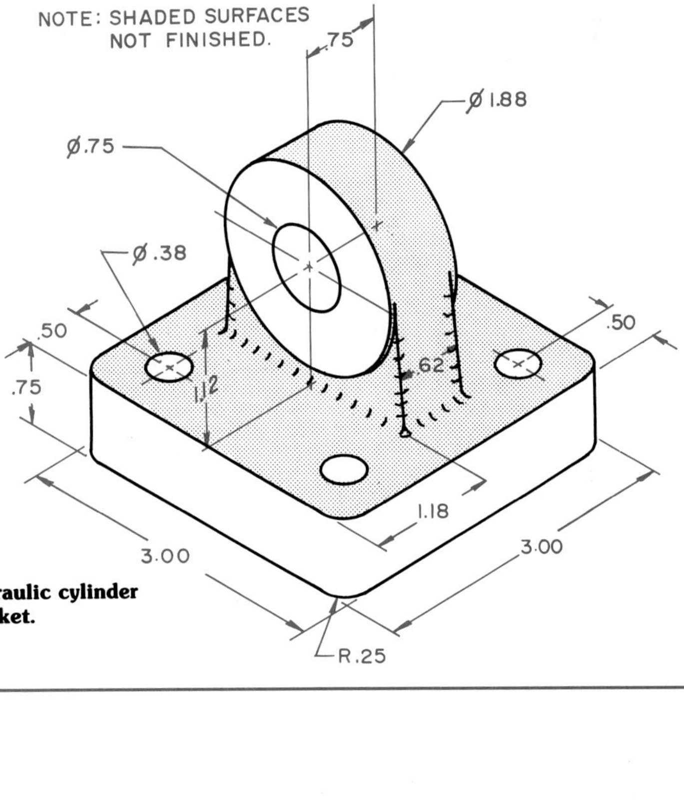
NOTE: SHADED SURFACES NOT FINISHED.
.75
Ø1.88
Ø.75
Ø.38
.50
.75
1.12
.62
.50
1.18
3.00
3.00
R.25

P6–15. Hydraulic cylinder mounting bracket.

Study Problems

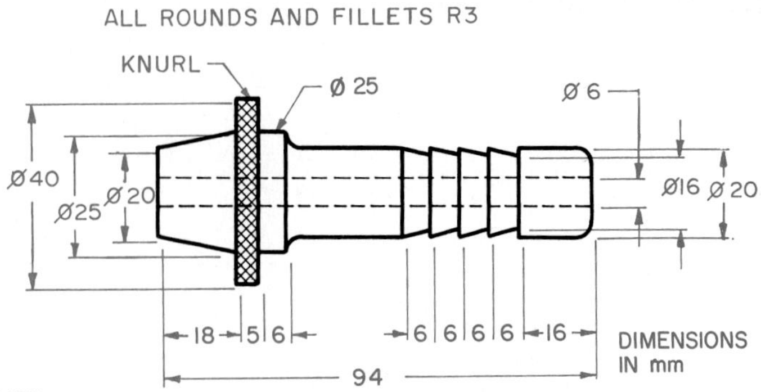
ALL ROUNDS AND FILLETS R3
KNURL
Ø 25
Ø 6
Ø40
Ø25
Ø20
Ø16
Ø 20
18
5
6
6
6
6
6
16
94
DIMENSIONS IN mm

P6–17. Hydraulic hose connection.

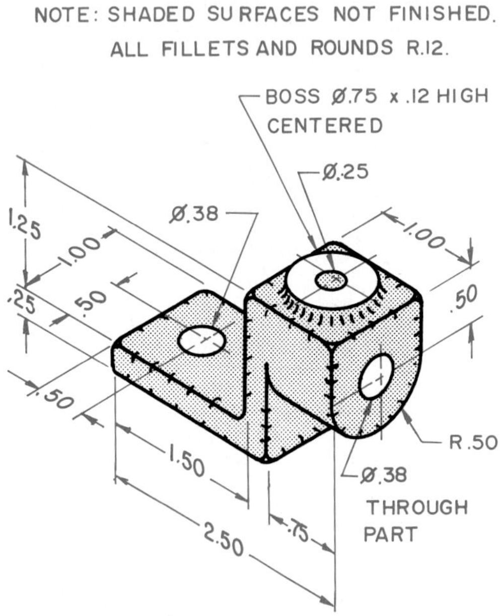
NOTE: SHADED SURFACES NOT FINISHED.
ALL FILLETS AND ROUNDS R.12.
BOSS Ø.75 x .12 HIGH CENTERED
Ø.25
Ø.38
1.00
1.25
1.00
.50
.25
.50
.50
1.50
2.50
.75
R.50
Ø.38 THROUGH PART

P6–16. Rapid release electrical connector.

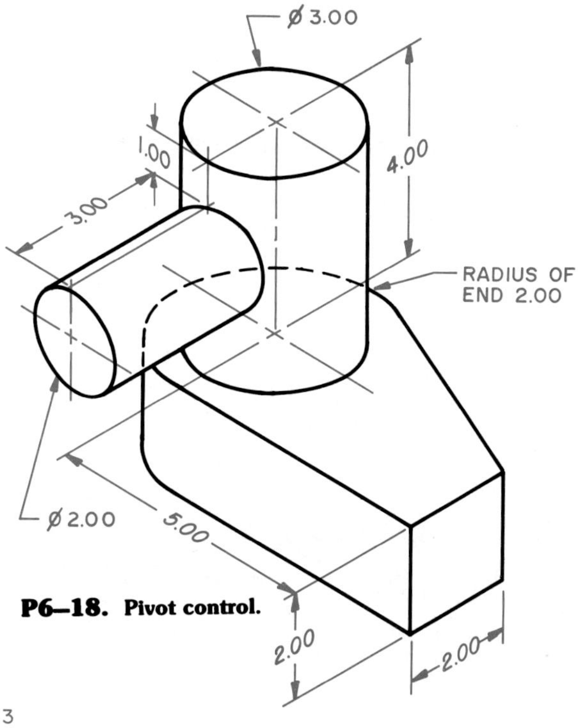
Ø 3.00
1.00
4.00
3.00
RADIUS OF END 2.00
Ø 2.00
5.00
2.00
2.00

P6–18. Pivot control.

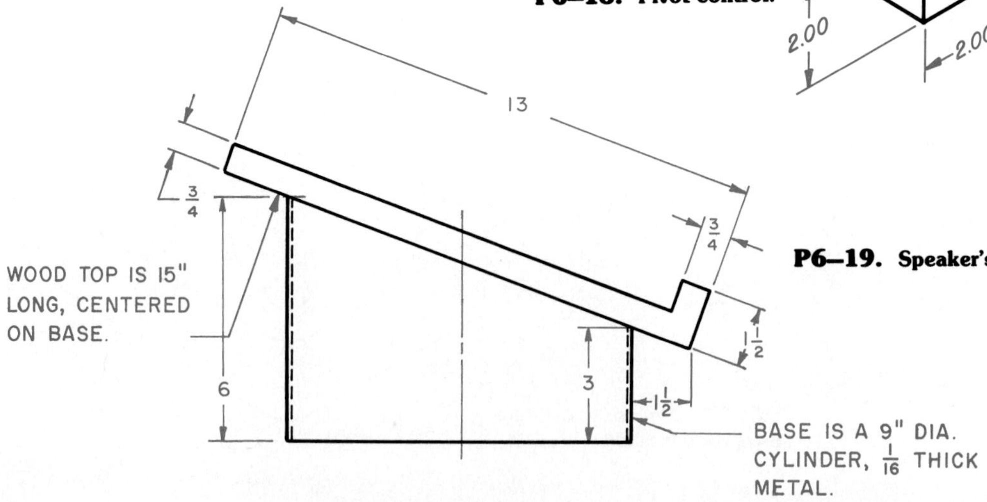
13
3/4
3/4
WOOD TOP IS 15" LONG, CENTERED ON BASE.
6
3
1 1/2
1 1/2
BASE IS A 9" DIA. CYLINDER, 1/16 THICK METAL.

P6–19. Speaker's stand.

Study Problems

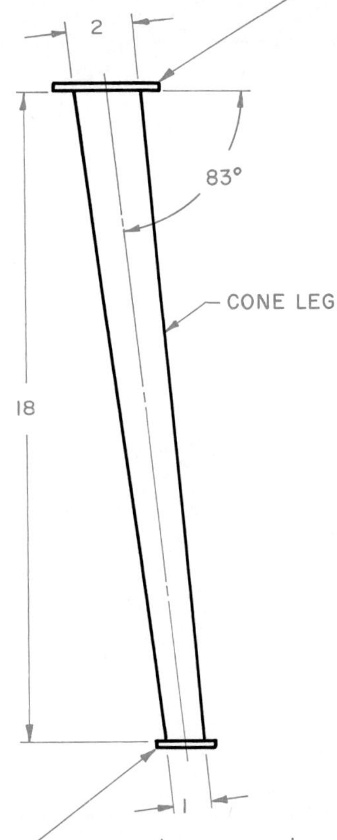

P6–20. Table leg.

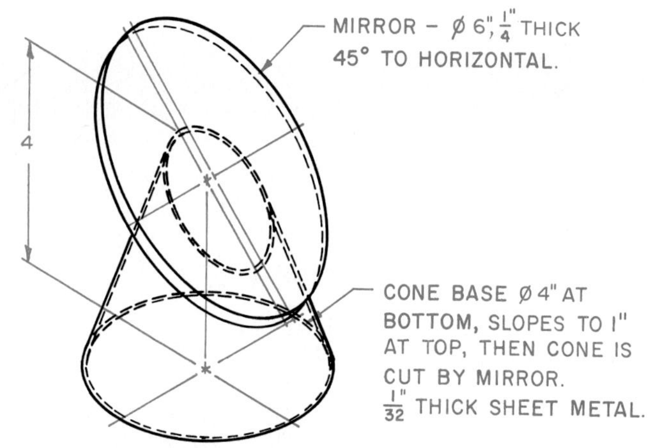

P6–21. Mirror and stand.

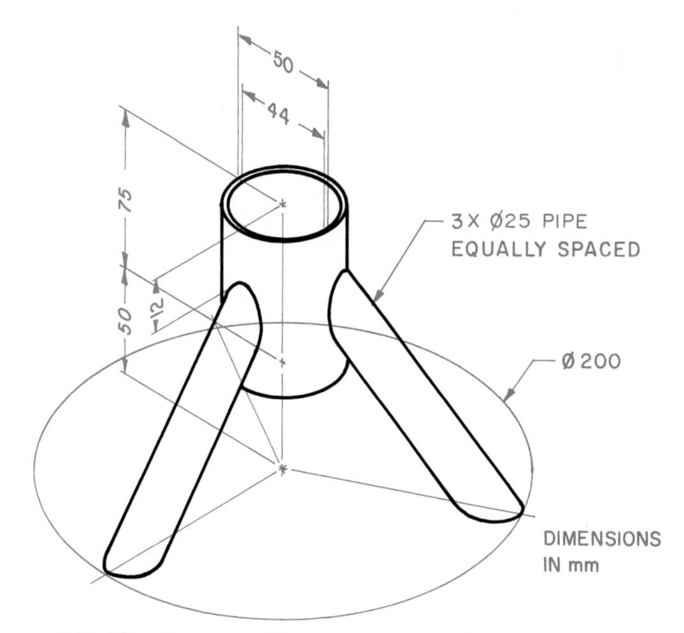

P6–22. Miniature Christmas tree stand.

Study Problems

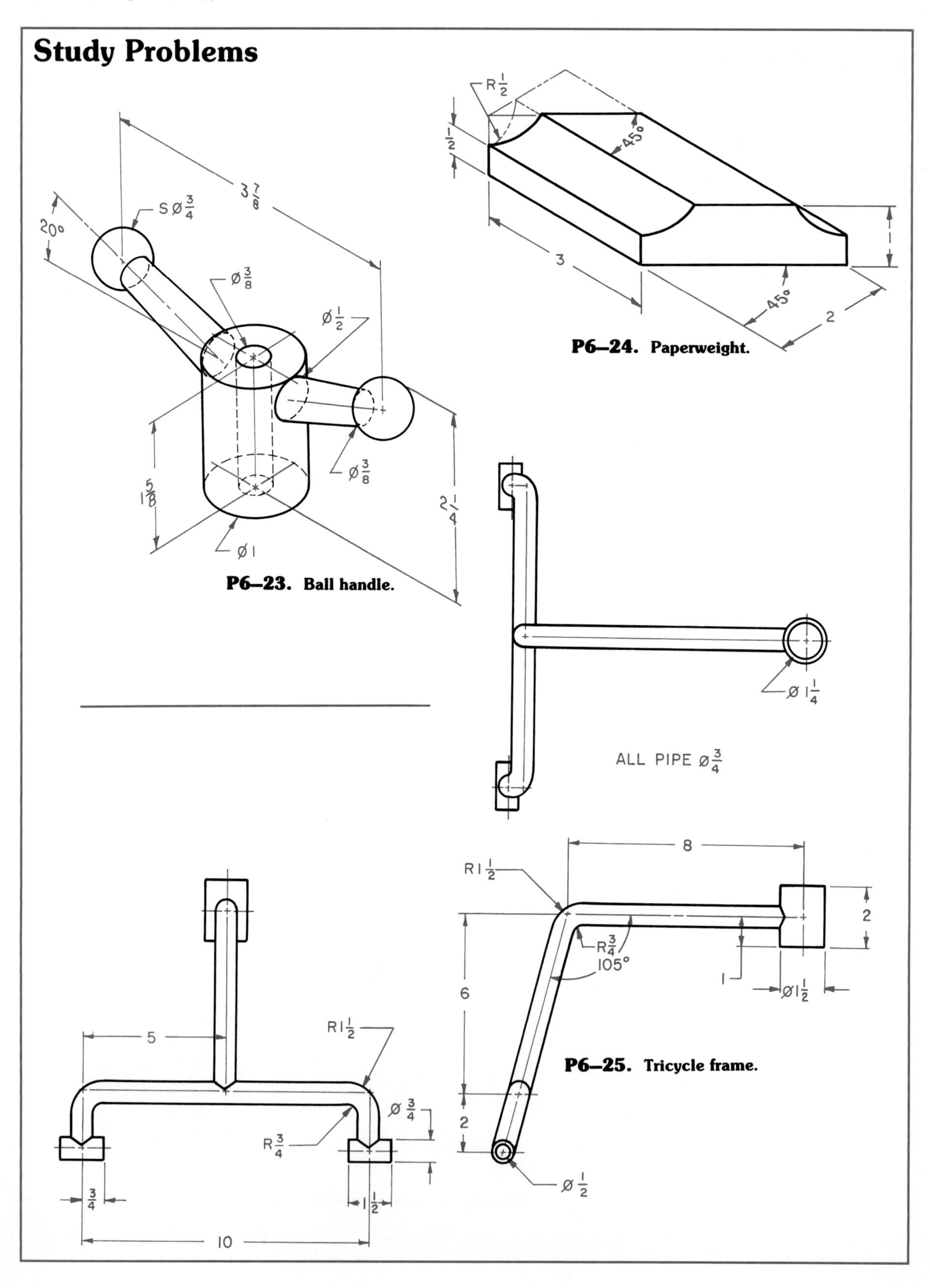

P6–23. **Ball handle.**

P6–24. **Paperweight.**

P6–25. **Tricycle frame.**

Study Problems

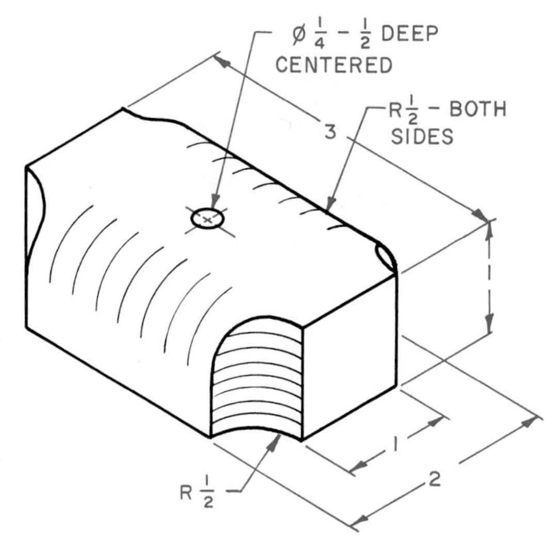

P6–26. **Base for desk penholder.**

R12
M5 x 0.80 FL HD BOLT
COUNTERSUNK FLUSH
HEAD DIA 10
12
5
50
38
R25
DIRECTION OF
MOVEMENT
Ø 10
25
R25
32
33°45'
12
R8
5
DIMENSIONS
IN mm
100

P6–27. **Pressure lever.**

CAST UNIT. NO
FINISHED SURFACES.

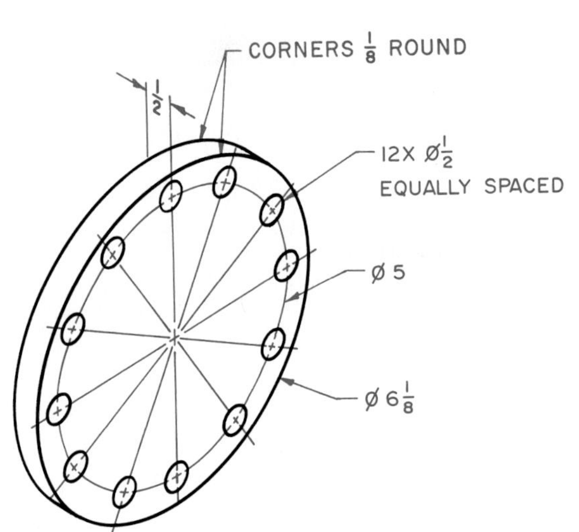

P6–28. **Blind flange.**

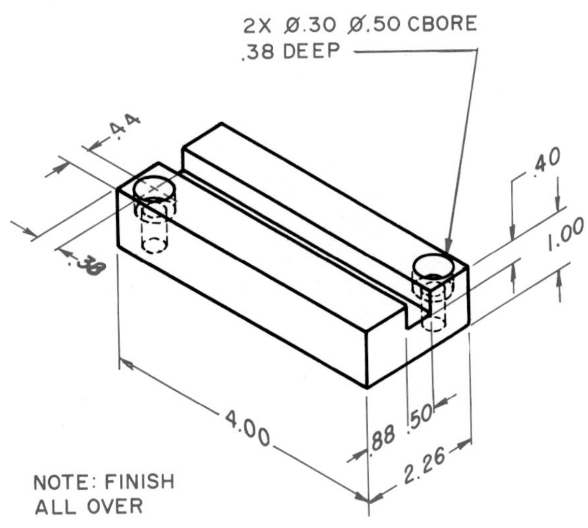

P6–29. **Guide block.**

Study Problems

NOTE: SHADED SURFACES NOT FINISHED.
ALL FILLETS AND ROUNDS R6.

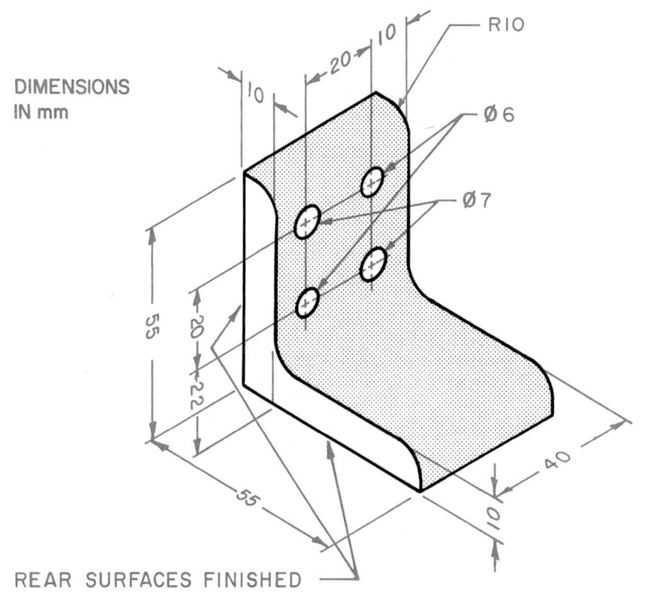

P6–30. Angle bracket. (Carr Lane)

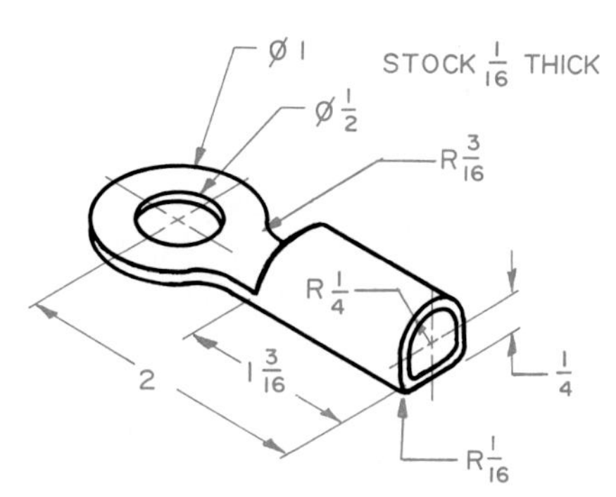

P6–31. Electric cable connector.

SHADED SURFACES NOT FINISHED.

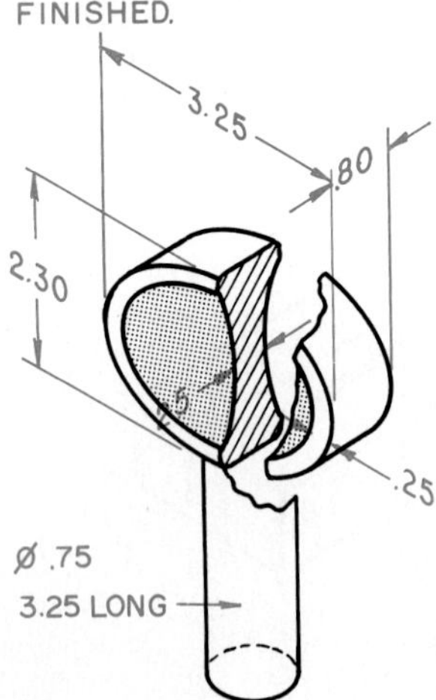

P6–32. Thumb screw blank before threading.

NOTE: SHADED SURFACES NOT FINISHED.
ALL ROUNDS R $\frac{1}{8}$

P6–33. Rivet set.

Study Problems

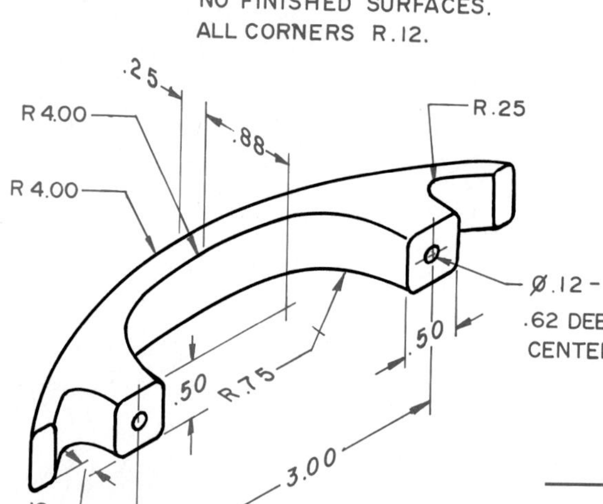

P6–34. Drawer pull.

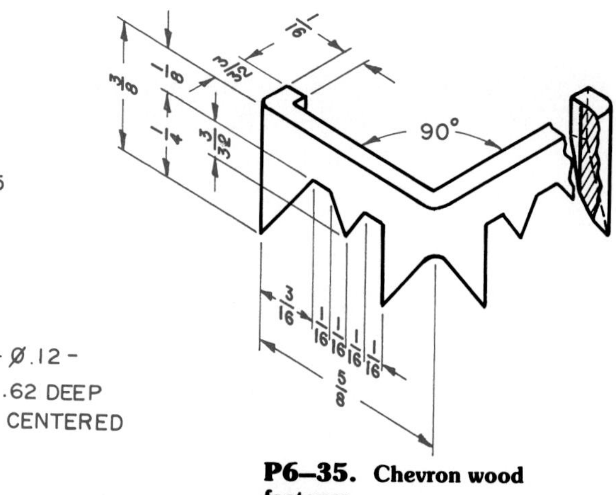

P6–35. Chevron wood fastener.

NOTE: SHADED SURFACES NOT FINISHED.
ALL FILLETS AND ROUNDS R$\frac{1}{8}$.

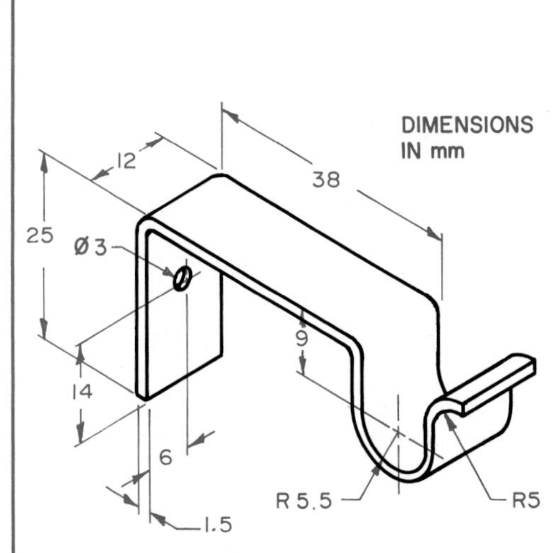

P6–36. Curtain rod hanger.

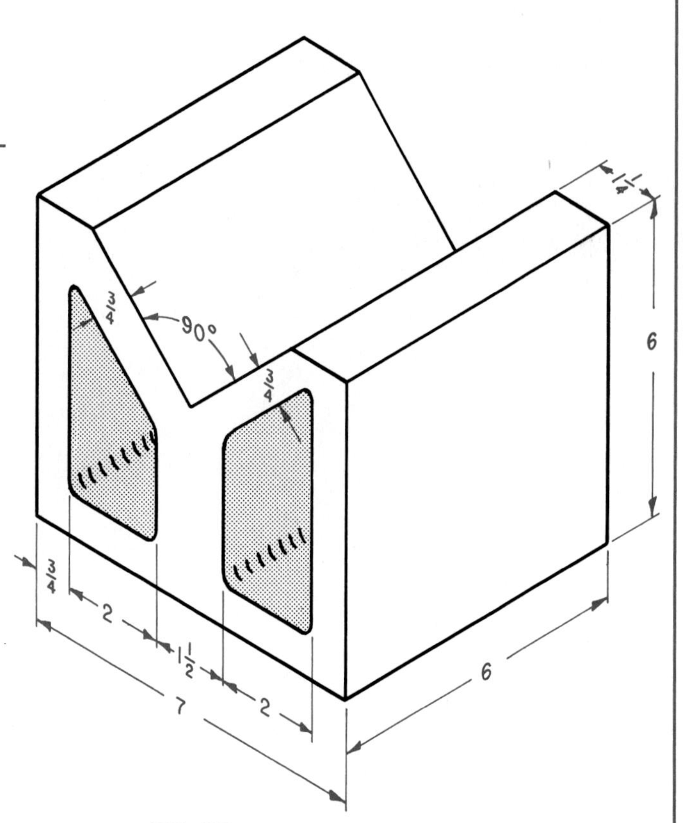

P6–37. Machinist's V-block.

Study Problems

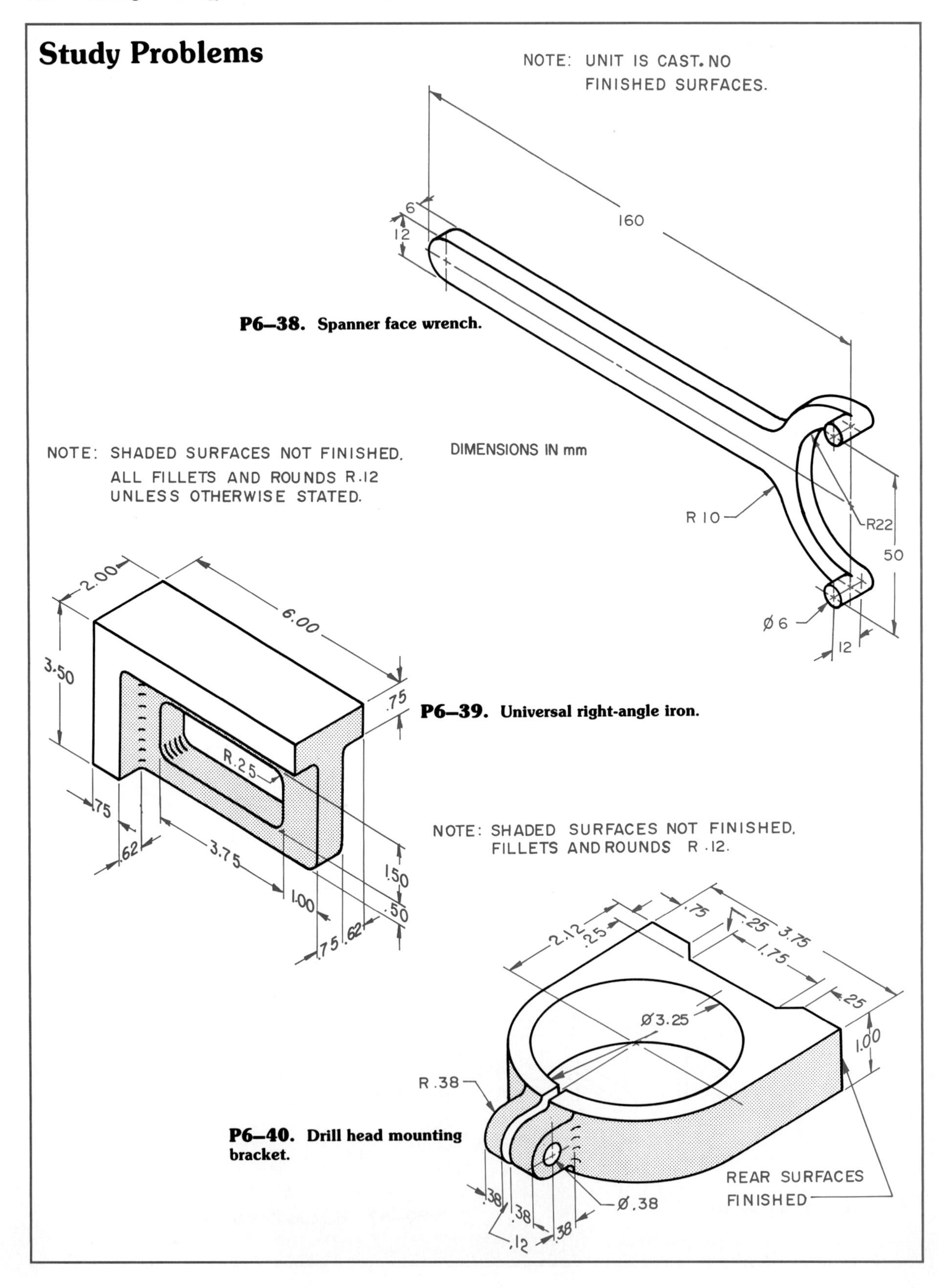

P6–38. Spanner face wrench.

P6–39. Universal right-angle iron.

P6–40. Drill head mounting bracket.

Study Problems

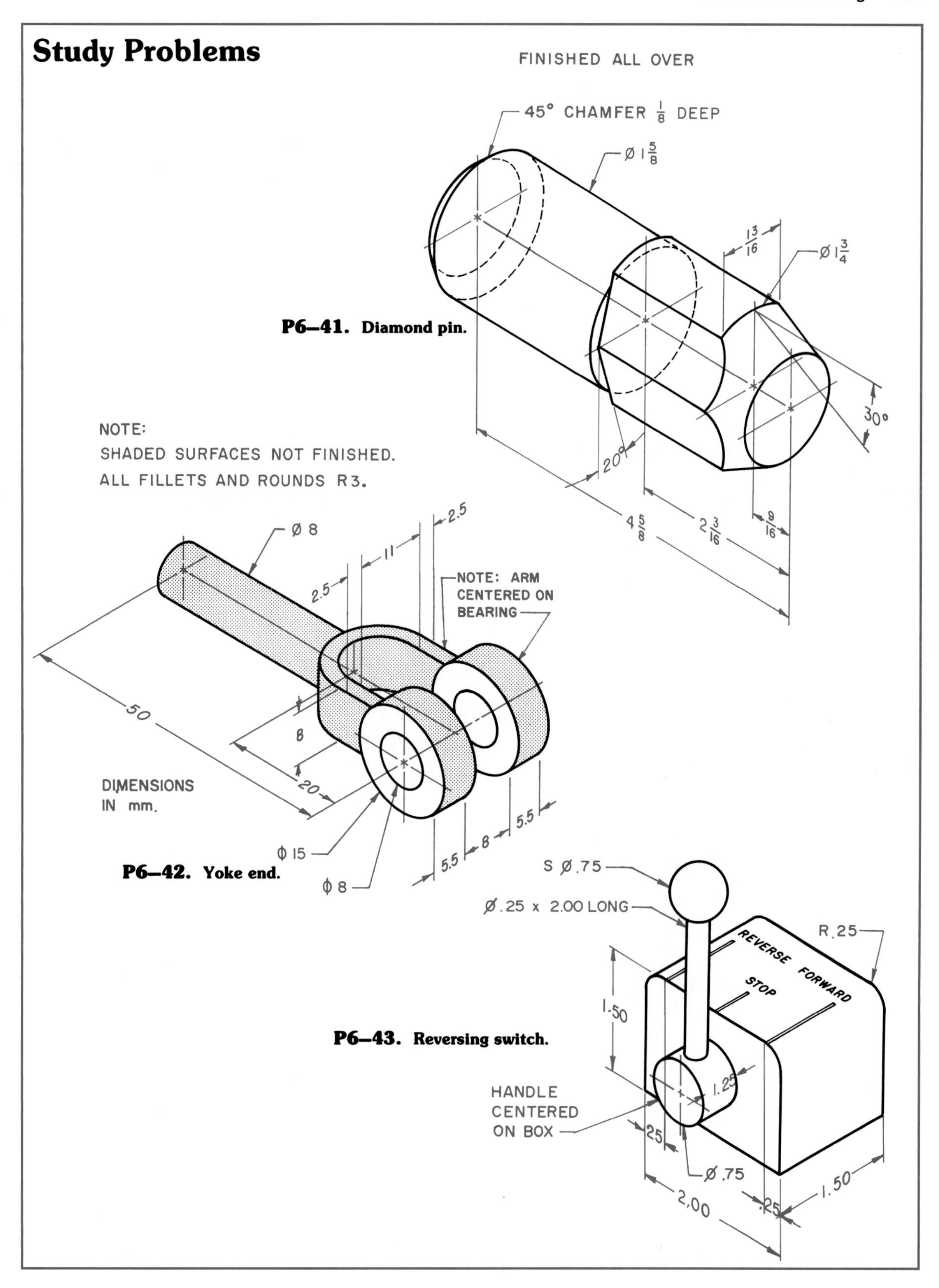

P6–41. Diamond pin.

P6–42. Yoke end.

P6–43. Reversing switch.

Study Problems

NOTE: UNIT IS CAST. NO FINISHED SURFACES. ALL FILLETS AND ROUNDS R $\frac{1}{8}$

$\frac{1}{8}$ DIA WOOD SCREW THREADS

Ø $\frac{15}{16}$ · Ø $\frac{5}{16}$ · □ $\frac{5}{16}$ · Ø $\frac{5}{16}$ · Ø $\frac{1}{2}$ · Ø $\frac{3}{16}$ · R $\frac{3}{8}$ · $\frac{3}{4}$ · $\frac{3}{16}$ · $\frac{5}{16}$ · $\frac{1}{4}$ · $1\frac{1}{2}$ · $\frac{1}{8}$ · $\frac{1}{8}$ · $\frac{1}{8}$ · Ø $\frac{7}{16}$ · Ø $\frac{5}{16}$

P6–44. Door stop.

$\frac{3}{8}$ · Ø $\frac{9}{16}$ · $\frac{1}{8}$ · $\frac{5}{32}$ · $\frac{7}{16}$ · R 1 ON TIP

RUBBER TIP FOR DOOR STOP

NOTE: BOTH PARTS CAST. NO FINISHED SURFACES. ALL FILLETS AND ROUNDS R $\frac{1}{4}$.

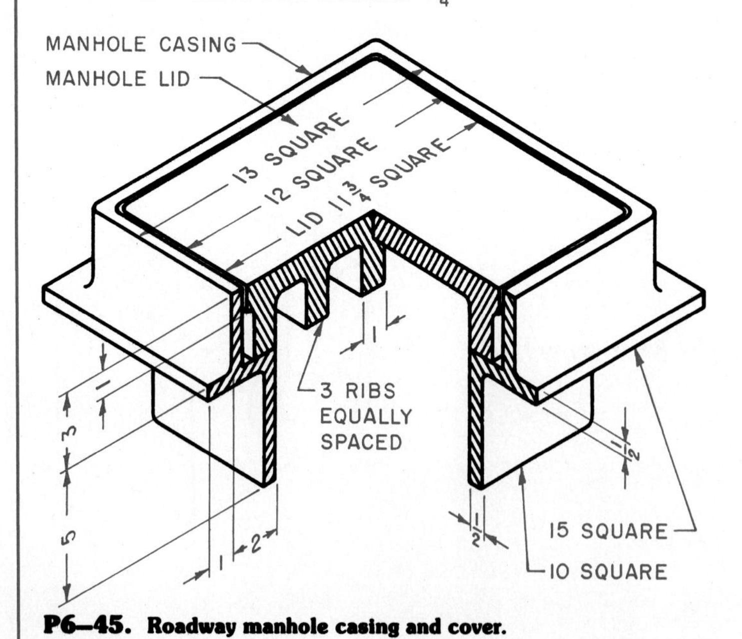

P6–45. Roadway manhole casing and cover.

FINISHED ALL OVER

45° x $\frac{1}{16}$ · Ø $\frac{7}{8}$ · 2X Ø $\frac{1}{4}$ · Ø 1 · $\frac{21}{64}$ · $2\frac{7}{32}$ · $\frac{1}{4}$ · $\frac{11}{64}$ · $\frac{5}{8}$

ANGLE 30° FROM VERTICAL

P6–46. Mill fixture key.

Study Problems

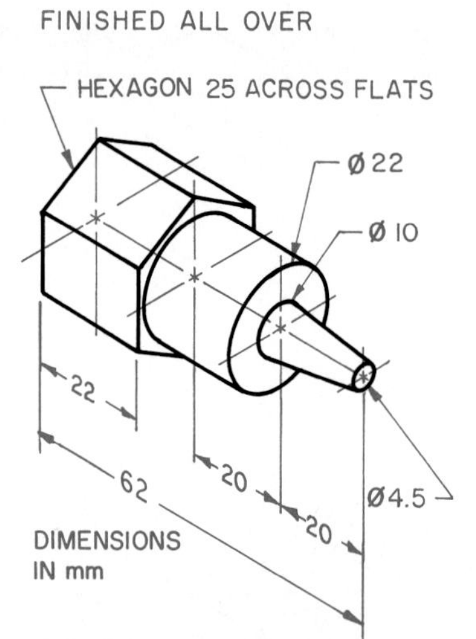

P6–47. **Friction plug.**

Ø $\frac{3}{4}$

WALL THICKNESS OF TUBE $\frac{1}{64}$

$\frac{3}{32}$

$3\frac{3}{8}$

Ø $\frac{1}{2}$

Ø $\frac{5}{32}$

$\frac{3}{16}$

$\frac{5}{64}$ O D

$2\frac{9}{16}$

P6–48. **Water tube for rock drilling machine.**

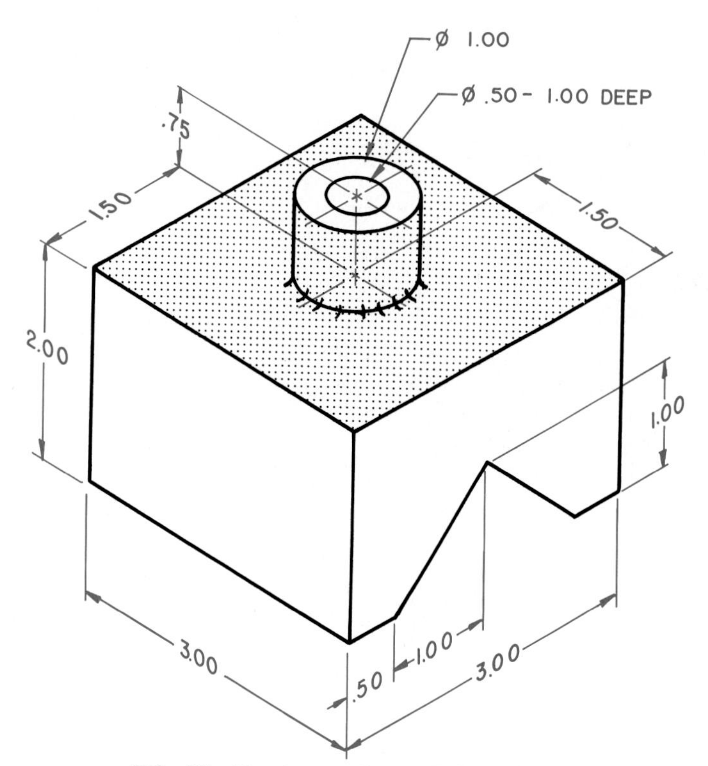

P6–49. **Height gage base. (Ellfeldt Co.)**

Study Problems

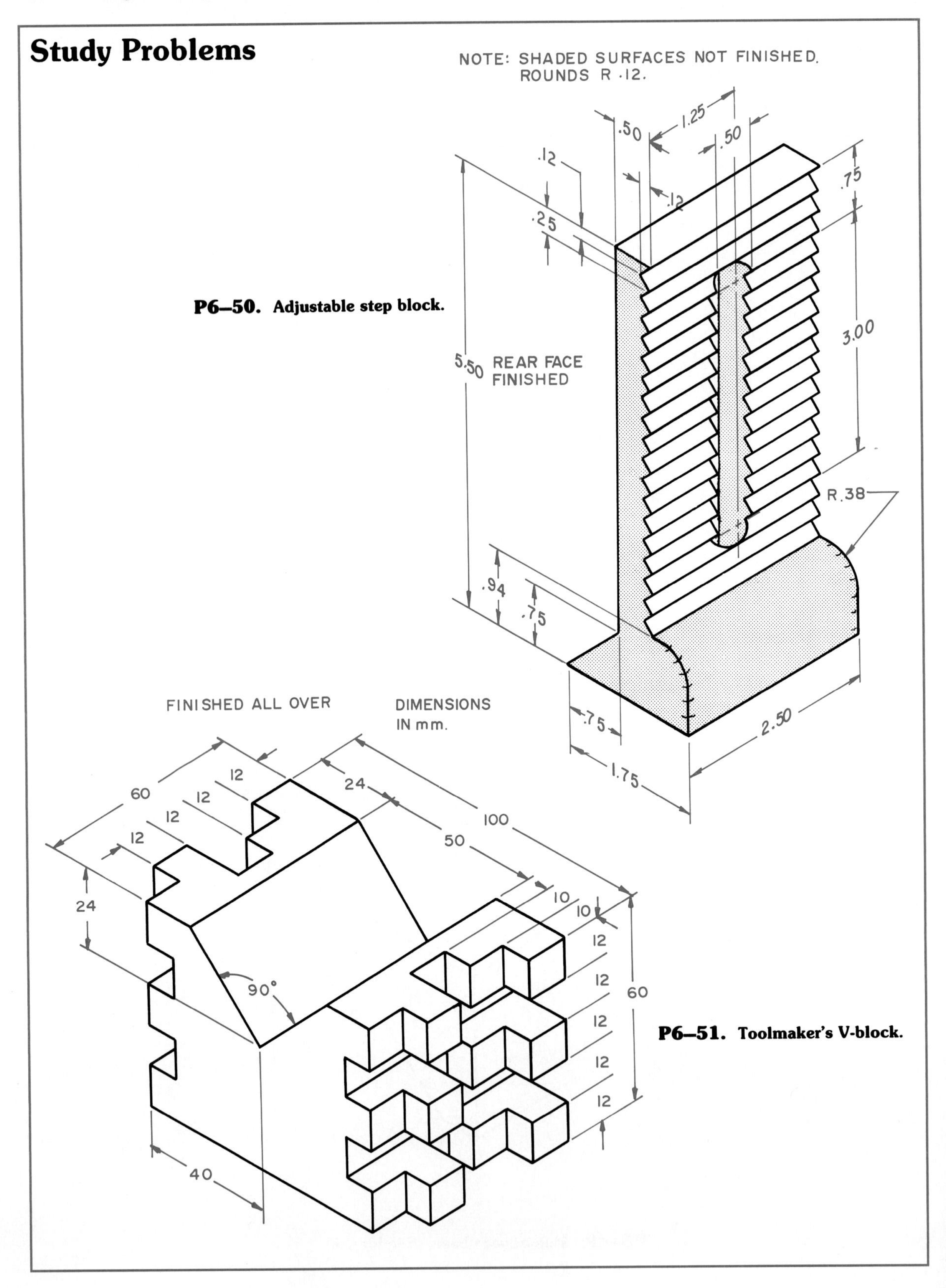

P6–50. **Adjustable step block.**

P6–51. **Toolmaker's V-block.**

Study Problems

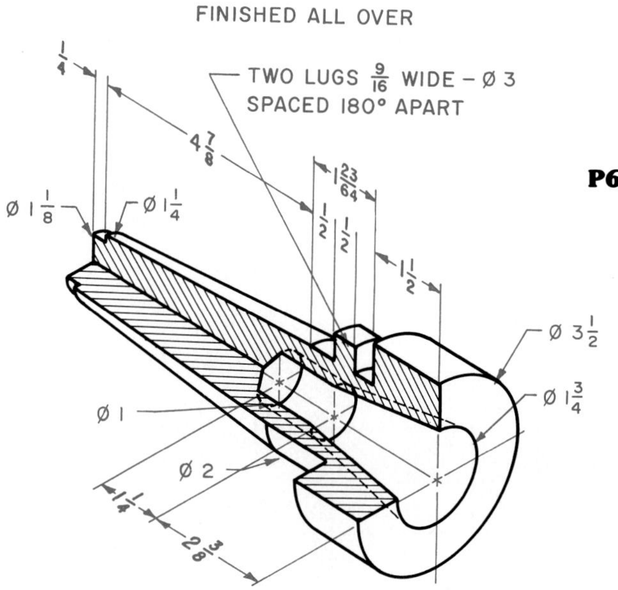

P6–52. Jig borer toolholder.

P6–53. Flanged turret mount.

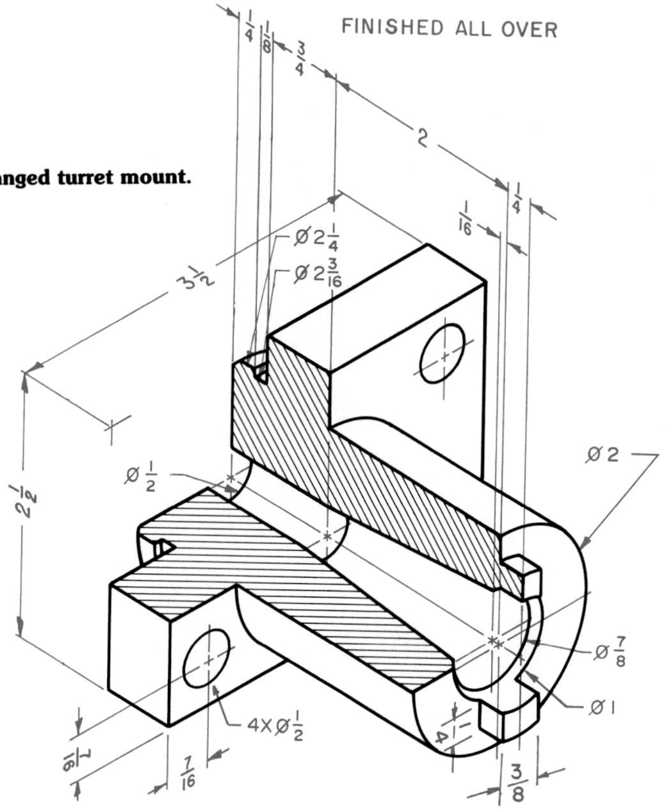

Study Problems

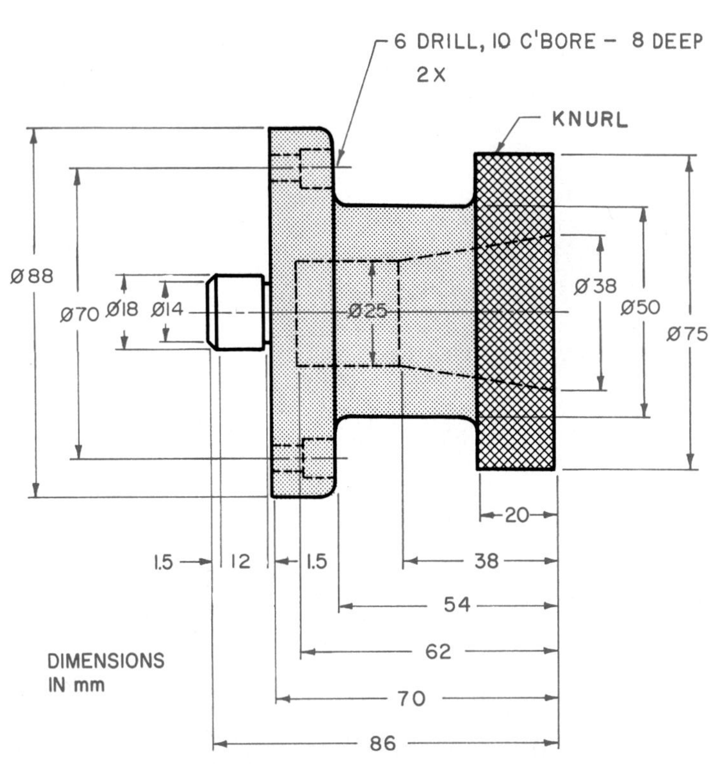

P6–54. Precision boring spindle.

Study Problems

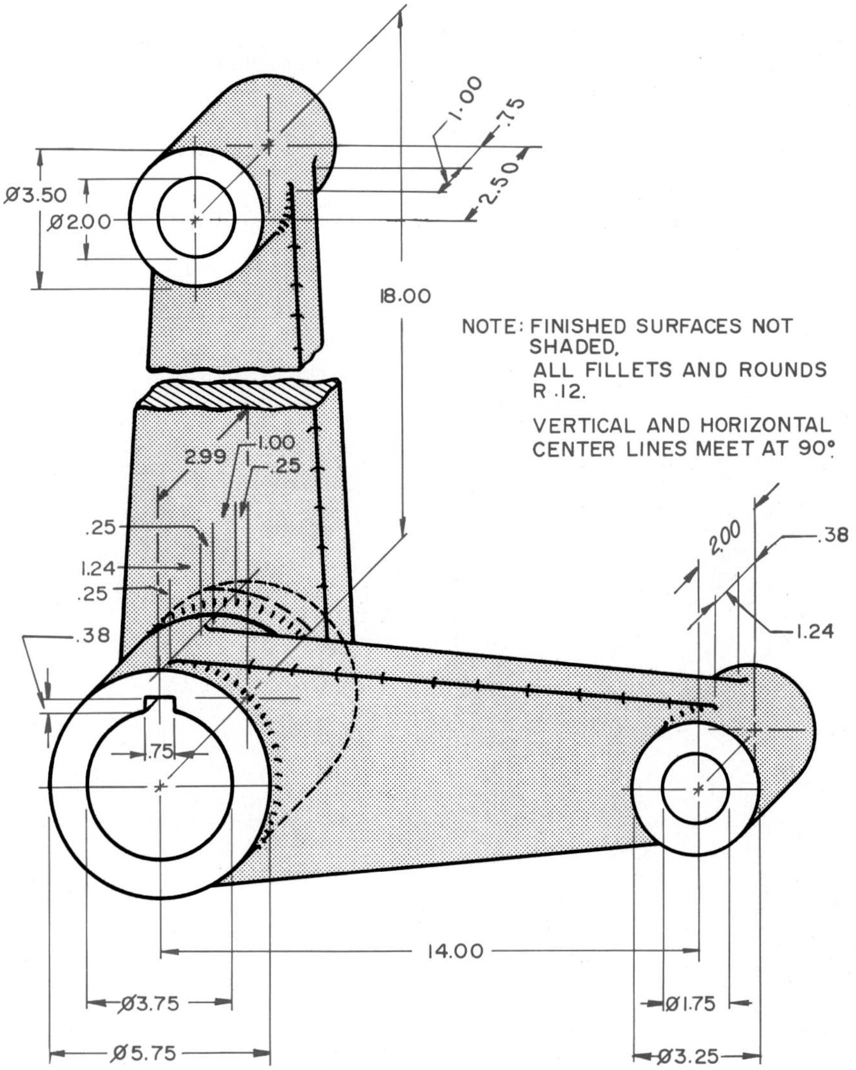

Ø3.50
Ø2.00
1.00
.75
2.50
18.00
NOTE: FINISHED SURFACES NOT SHADED.
ALL FILLETS AND ROUNDS R .12.
VERTICAL AND HORIZONTAL CENTER LINES MEET AT 90°.
2.99
1.00
.25
.25
1.24
.25
.38
2.00
.38
1.24
.75
14.00
Ø3.75
Ø5.75
Ø1.75
Ø3.25

P6–55. Rocker arm.

Study Problems

NOTE: SHADED SURFACES NOT FINISHED.
ALL FILLETS AND ROUNDS R.12 UNLESS NOTED.

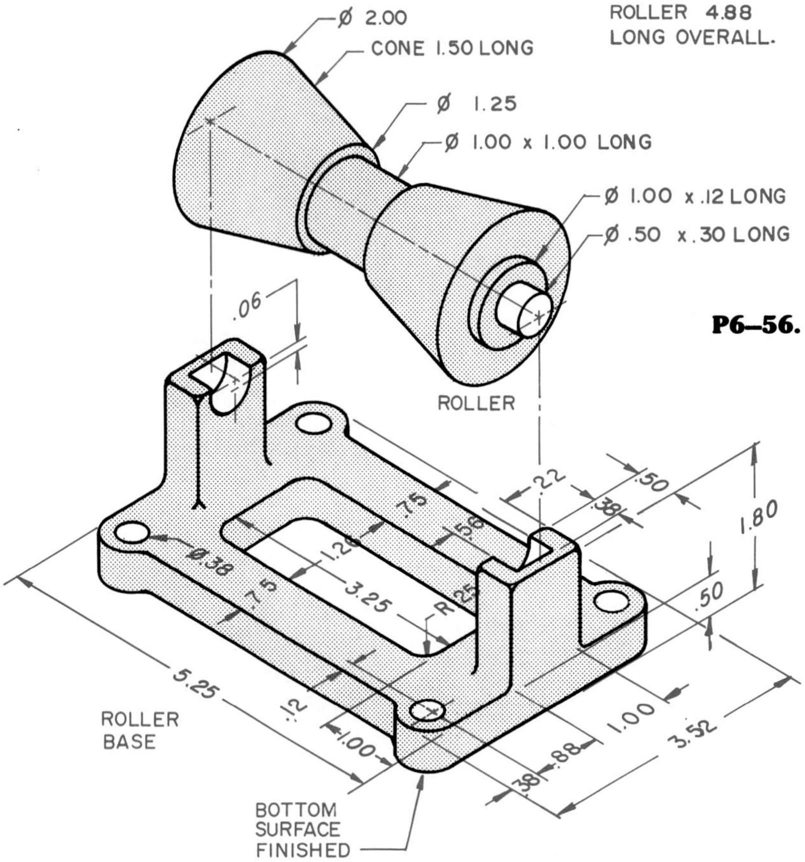

ROLLER 4.88 LONG OVERALL.

P6–56. Pipe roller and base.

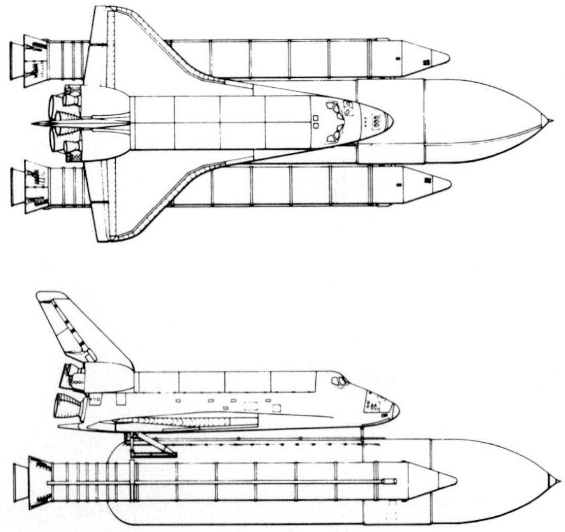

P6–57. Space Shuttle.

CHAPTER 7

Dimensioning

After studying this chapter, you should be able to do the following:

- Use the customary and metric systems of linear measurement.
- Dimension the various geometric shapes of which parts are made.
- Apply both size and location dimensions to a drawing using appropriate dimensioning standards.
- Apply tolerances and limits where needed.
- Apply American Standard or metric fits to cylindrical parts.
- Record surface texture data on a drawing.

Size Description

Any working detail drawing is described in two ways—shape and size. **Shape description** gives only the outline of the object. For example, Fig. 7-1 is a shape description of a cylinder. This drawing could be a balance staff for a watch, a piece of shafting, a boiler for a model steam engine, or any other cylindrical object. For this or any other drawing to be meaningful, a **size description** will give all sizes, or dimensions, general and specific notes, material designations, and specifications about the object. The size description must be complete enough that it may be taken into the shop and produced without any additional explanation.

Dimensions placed on a drawing are the actual dimensions of the finished object. Actual dimensions are used even though the drawing may be smaller or larger than the object. To have a complete working detail drawing, two requirements must be met:

- Sufficient views to describe the object.
- Sufficient sizes and related information to produce the object.

Dimensioning is not difficult to learn if you follow a few simple rules. Perhaps most students encounter difficulty by neglecting to ask, "Could I make this object from this drawing without any additional information?"

Dimensioning Standards

There are several sources of information about dimensioning standards. One is published by the American Society of Mechanical Engineers, United Engineering Center, 345 East 47th St., New York, NY 10017. The standard is *Dimensioning and Tolerancing, ANSI Y14.5M-1982.* Other publications, including those with ISO metric standards, are available from the American National Standards Institute, 1430 Broadway, New York, NY 10018.

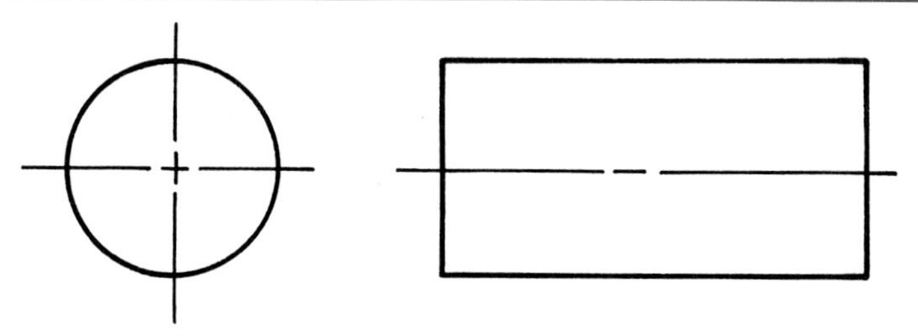

Fig. 7–1. Shape description gives only the outline of the object. Dimensions give the size of the object. Is this a pin for a watch or a boiler for a furnace?

Dimension Lines

Dimension lines are thin, black, solid lines that show where a dimension begins and ends. Draw them parallel to the surface or edge being described. Break the dimension line to allow room for the dimension on engineering drawings. On architectural drawings, draw a continuous dimension line and place the dimension above it. Fig. 7-2.

Draw the dimension line nearest to the view 3/8 inch (10 mm) away from the object line. Space all other dimension lines 1/4 inch (6 mm) away from the first dimension line and each other. Fig. 7-3. Do not use object lines, center lines, or extension lines as dimension lines.

Whenever possible, line up dimension lines to give your drawing an orderly appearance and make it easier to read. Fig. 7-4.

Arrowheads

Arrowheads show the beginning and end of a dimension line. Draw arrowheads freehand using two or three strokes and make them one-third as high as they are long. They may be open or closed. Fig. 7-5. Draw all the arrowheads on a drawing the same size. Make the length of the arrowhead on a small drawing approximately 1/8 inch (3 mm) and up to 3/16 inch (5 mm) long on a large drawing.

Extension Lines

Extension lines are used to extend the lines on a view to show where dimension lines start and end. Place them outside the view beginning about 1/16 inch (1.5 mm) away from the view. Extend them about 1/8 inch (3 mm) past the last dimension. Fig. 7-6. Make extension lines like dimension lines: thin, solid, and black.

It is poor practice to terminate a dimension line at an object line. Fig. 7-7. If an extension line would cause confusion by being too long, you may use one object line as an extension line. Fig. 7-8. However, do not use an object line in this way unless absolutely necessary.

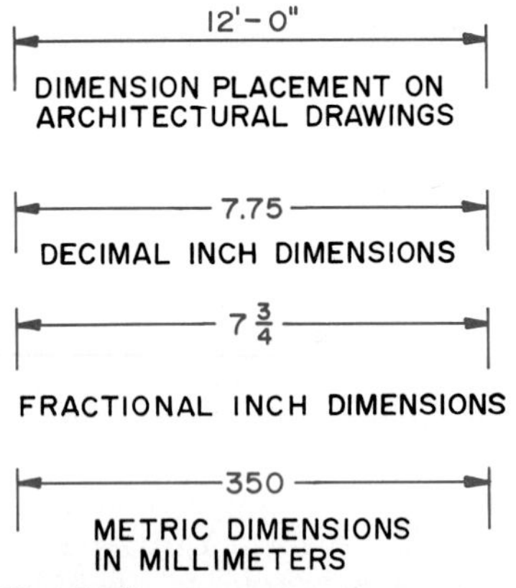

Fig. 7–2. Dimension lines as used on architectural and engineering drawings.

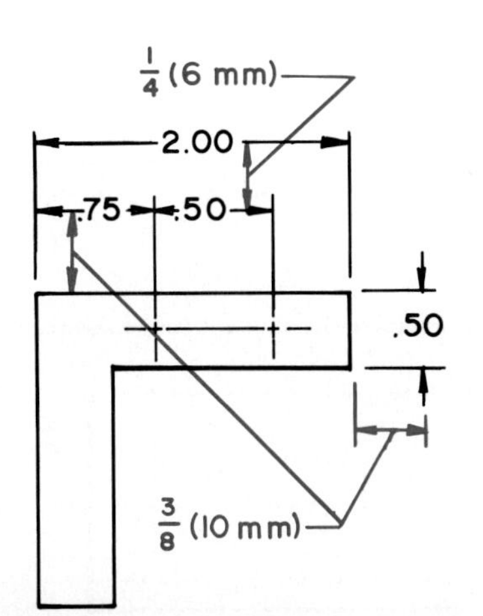

Fig. 7–3. It is important to space dimension lines properly.

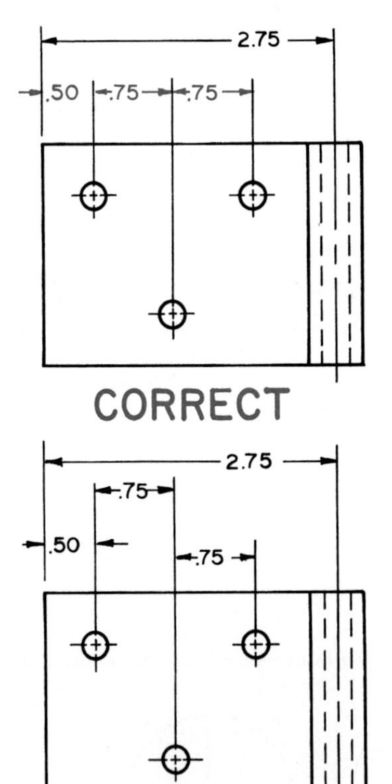

Fig. 7–4. Line up dimension lines.

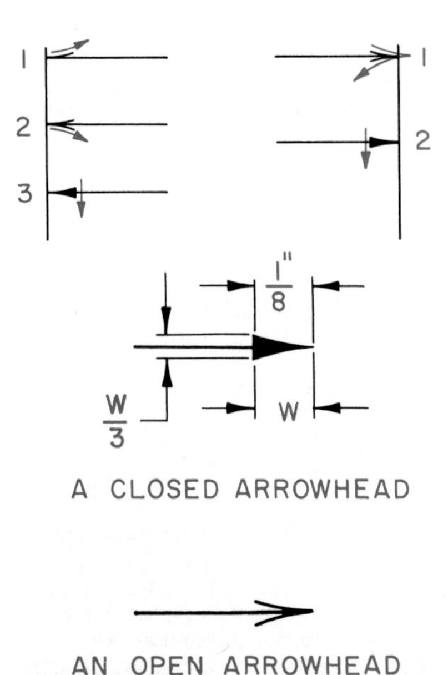

Fig. 7–5. Arrowheads may be open or closed.

Break extension lines when they are crossed by an arrowhead. Fig. 7-9. Do not break extension lines when they cross each other or cross dimension or object lines. Fig. 7-10.

When a point is located by extension lines, the extension line should pass through that point. Fig. 7-11.

Leaders

Leaders are thin lines used to direct dimensions, notes, or symbols to the intended place on a drawing. A leader is the same weight as a dimension line. Use an arrowhead at the end of a leader. Fig. 7-12. If the leader points to a surface, use a dot instead of an arrowhead. Draw a leader as an oblique straight line with a short horizontal shoulder extending to the note. Make the shoulder about ¼ inch (6 mm) long and extend it to the *center* of the first or last letter or number of the note. Never underline the note with the horizontal shoulder. Although you may choose any oblique angle to draw a leader, 60, 45, and 30 degrees in that order are most frequently used. If you must draw two or more leaders to adjacent features, your drawing will be easier to read if you make them parallel.

When a leader is directed to a circle or arc, point the leader to the center. Refer to Fig. 7-12.

Avoid drawing leaders that cross other leaders or are very long, near to horizontal or vertical, or a small angle away from an object line.

Dimension Figures

When using decimal inches, letter the numerals .125 inch high on A, B, and C size drawings and .156 inch high on D and E size drawings. Letter titles .250 inch high regardless of paper size. Fig. 7-13.

On metric drawings, letter the numerals 3.5 mm high on A4 and A3 drawings, 4.0 mm on A2 drawings, 4.5 mm on A1 and A0 drawings. Make all titles 7.0 mm high.

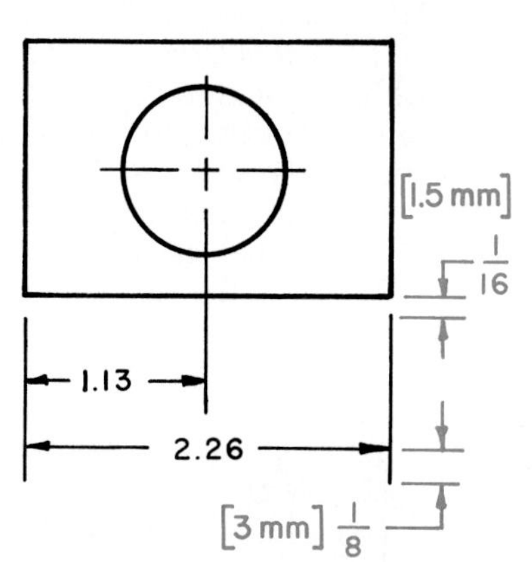

Fig. 7-6. Extension lines should not touch the object. They should extend beyond the dimension line.

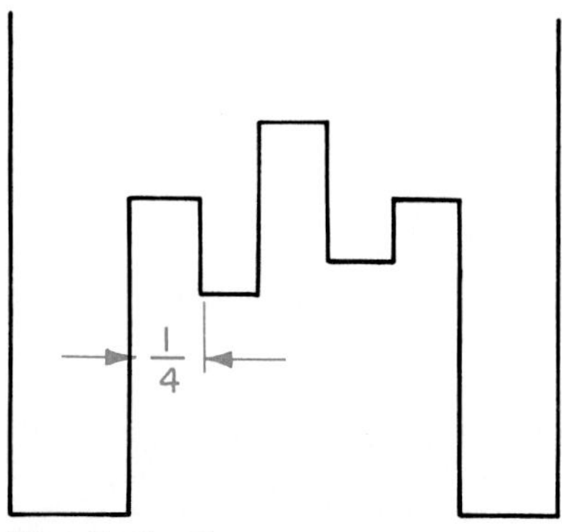

Fig. 7-8. If necessary, one object line can be used as an extension line to avoid using very long extension lines.

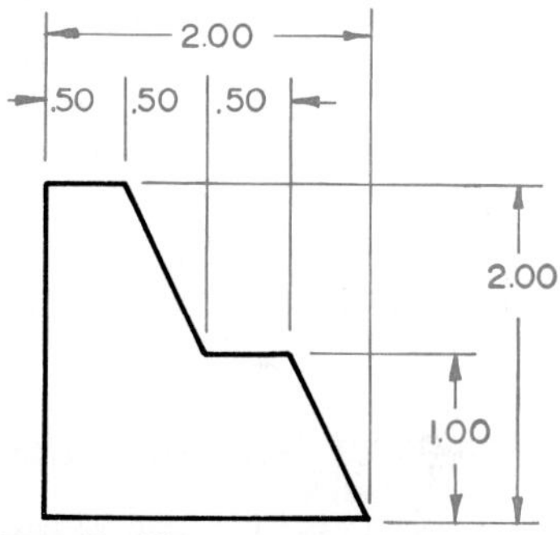

Fig. 7-10. Do not break extension lines when they cross each other.

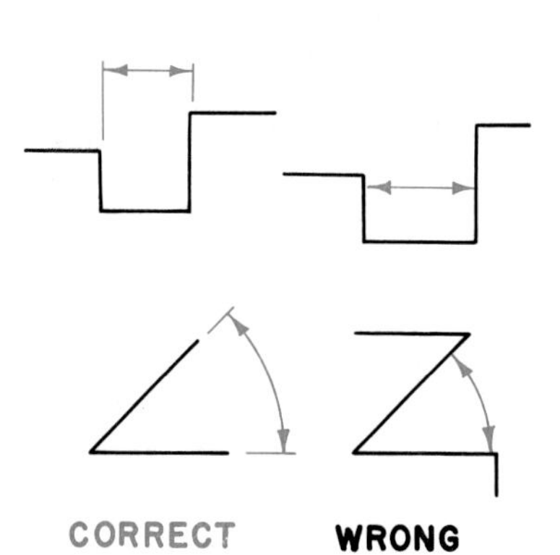

Fig. 7-7. Dimension lines should terminate on extension lines, not on object lines.

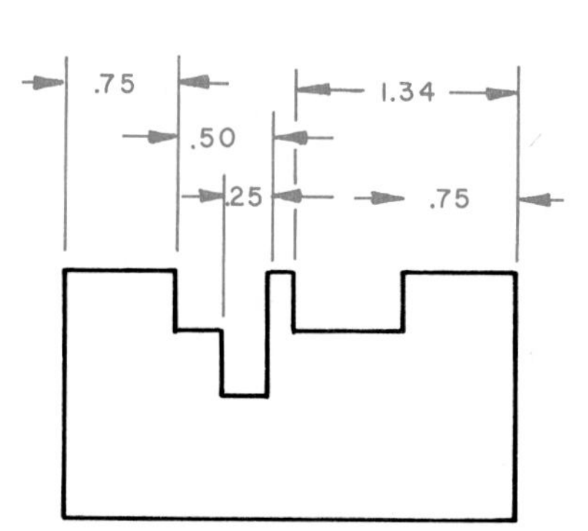

Fig. 7-9. Break extension lines if they cross arrowheads.

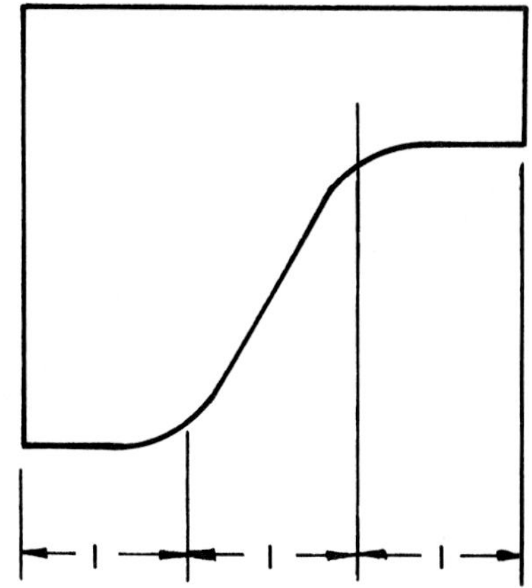

Fig. 7-11. Extension lines pass through points which they locate.

If you use common fractions, letter the numerals ⅛ inch high on A, B, and C size drawings, 5/32 inch high on D and E size drawings, and all titles ¼ inch high. The total height of a common fraction is twice the height of a single numeral. Since there is a bar separating the numerals, letter each numeral slightly smaller than normal to allow space for the bar.

When several dimension lines are parallel, stagger the dimensions to make them easier to read. Fig. 7-14.

Draw very light guidelines when lettering dimensions and notes.

Customary and Metric Linear Measurements

Linear measurement means measurement along a straight line. The United States and a few other countries still mostly use the customary system. The standard unit of measurement for the customary system is the *inch.* Most countries along with many industries in the United States use the International System of Units, commonly called the metric system. The standard unit of measurement for the metric system is the *meter.* A comparison of these two systems is shown in Fig. 7-15.

The Customary System

The basic customary system unit, the inch, is divided into smaller parts that provide a system of fine measurement. The inch can be divided into common fractional parts such as ¾, ⅛, 1/64, and others. It may also be divided into decimal fractions such as 0.1, 0.01, 0.001, and others.

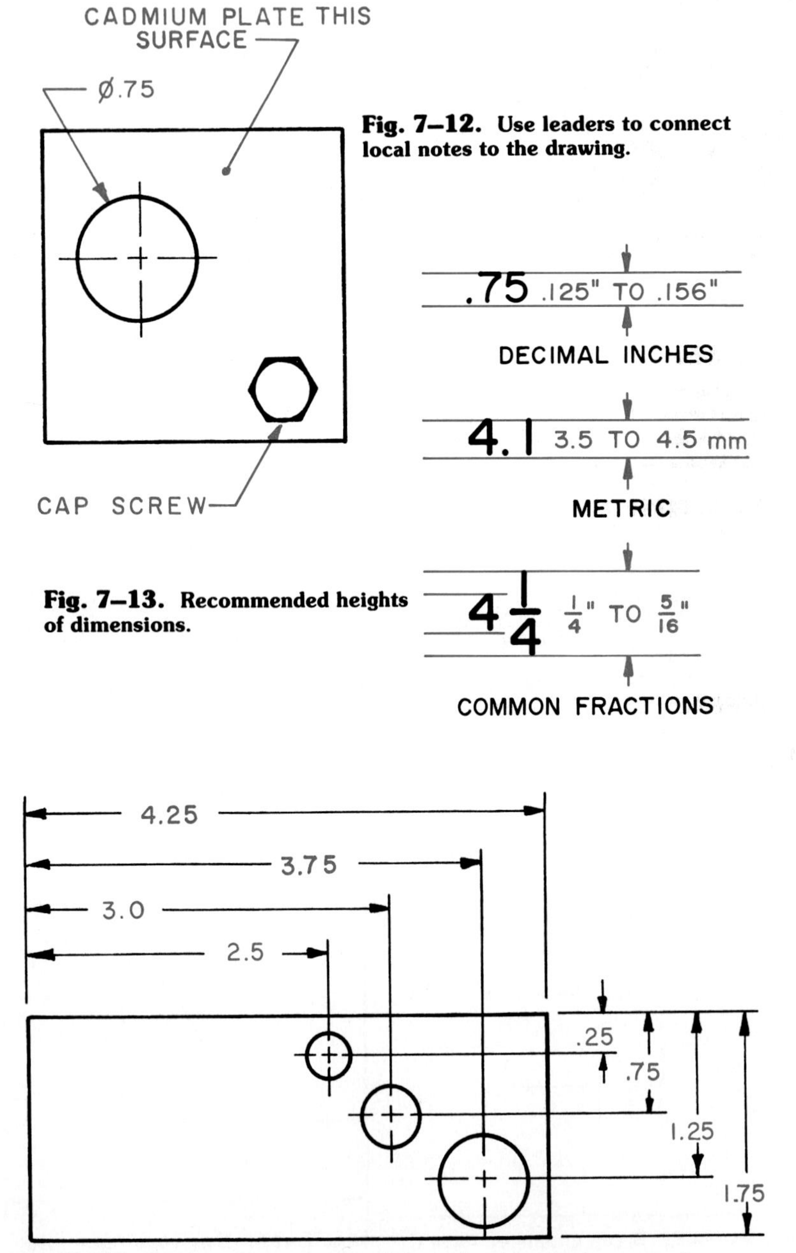

Fig. 7–12. **Use leaders to connect local notes to the drawing.**

Fig. 7–13. **Recommended heights of dimensions.**

Fig. 7–14. **Stagger dimensions to help make them easier to read.**

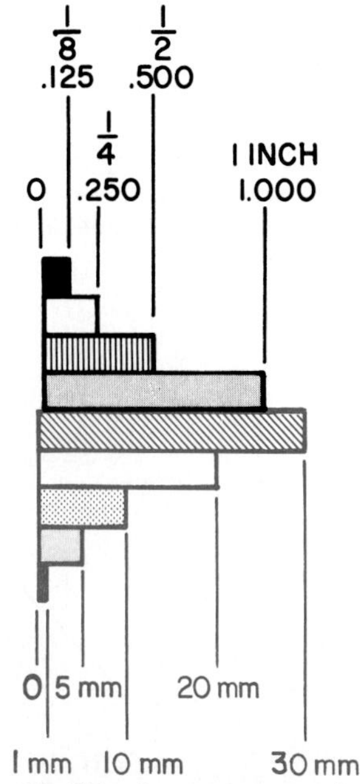

Fig. 7–15. **A comparison of common fractions, decimal fractions, and metric measure.**

The expression of decimal fractions is shown in Fig. 7-16. A table of decimal equivalents of common fractions is also found inside the front cover of this book.

The Metric System

The International System of Units is known as the SI metric system. SI stands for *Système International*, the French name for the system. The SI base units are the *meter* for length, *kilogram* for mass, *second* for time, *ampere* for electric current, *kelvin* for temperature, *candela* for luminous intensity, and *mole* for amount of substance. Fig. 7-17. Note: *Degrees Celsius* is commonly used for temperature, but it is not a base unit.

In addition to the seven base SI units, there are supplementary and derived units. Examples of these are the *ohm* for electric resistance, *newton* for force, and *cubic meter* for volume.

The metric system is a decimal system meaning that the units can be divided or multiplied by ten.

CAD Applications:

Dimensioning

As you learned in Chapter 4, a drawing created on a CAD system can be dimensioned semiautomatically. The user indicates the part to be dimensioned, such as a line, circle, or angle. The CAD program calculates the distance or the angle, constructs the dimension and extension lines, and enters the dimension text (the measurement) on the drawing.

This does not mean that knowledge of dimensioning practices is unimportant to the CAD user. On the contrary, the drafter can make the best use of the system only if he or she understands dimensioning theory and standards.

Before starting to dimension, the CAD user sets the dimensioning parameters, or variables. These variables govern the way dimensions are drawn. For example, the user can specify the size of dimension text, the amount of space to be left between the extension lines and the object, the rounding value for dimensions, the size of arrowheads, and many other variables. Drafters can set these variables to conform with the standards of their industry and company, then save them for repeated use in other drawings.

A useful feature of some CAD programs is associative dimensioning. Suppose an angle has been drawn and dimensioned. Then the drafter decides to make the angle larger. The CAD program automatically changes the dimension of the angle to the new measurement.

Some CAD programs also enable the user to dimension in two systems of measurement at the same time. For example, a drawing done in inches can be dimensioned in inches and millimeters at the same time.

The drawing below was created and dimensioned on a CAD system. Only the shape of the object was actually drawn by the drafter. The dimension text, dimension lines, extension lines, leaders, etc., were all produced by the CAD system according to the parameters set by the drafter.

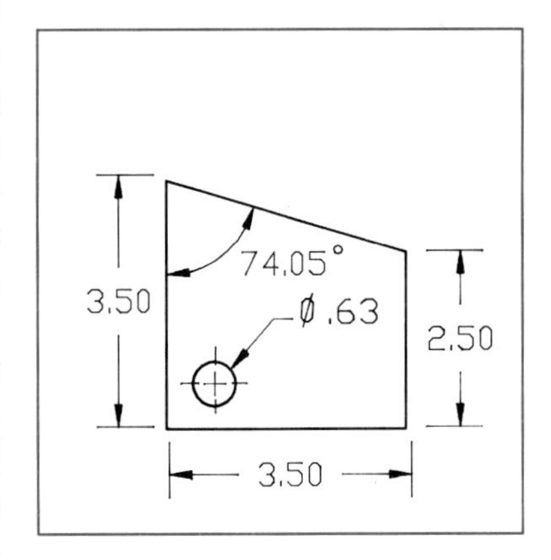

Fig. 7-16. Expressing Fractions in the Customary System

One inch = 1″ = 1.00″
One-tenth inch = 1/10 = 0.1″
One-hundredth inch = 1/100 = 0.01″
One-thousandth inch = 1/1,000 = 0.001″
One ten-thousandth inch = 1/10,000 = 0.0001″
One hundred-thousandth inch = 1/100,000 = 0.00001″
One-millionth inch = 1/1,000,000 = 0.000001″

Fig. 7-17. The Base Units of SI

Quantity	Unit	SI Symbol
Length	meter	m
Mass	kilogram	kg
Time	second	s
Electric Current	ampere	A
Temperature	kelvin	K
Luminous Intensity	candela	cd
Amount of Substance	mole	mol

Fig. 7-18. Metric Decimal Divisions

1 meter = 39.37 inches

1 meter = 10 decimeters (dm)

1 decimeter = 10 centimeters (cm)

1 centimeter = 10 millimeters (mm)

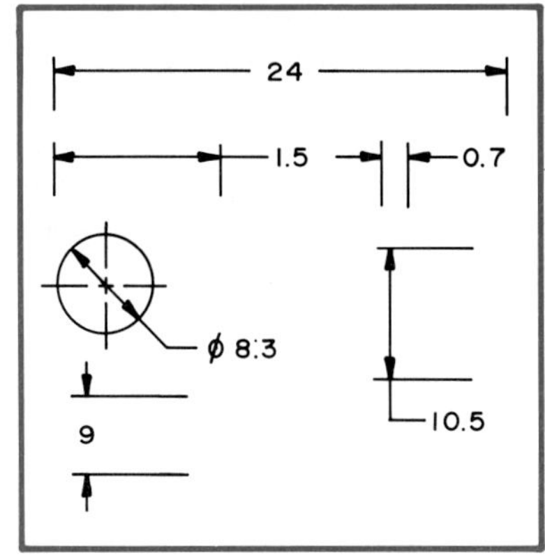

Fig. 7–20. Typical metric dimensions.

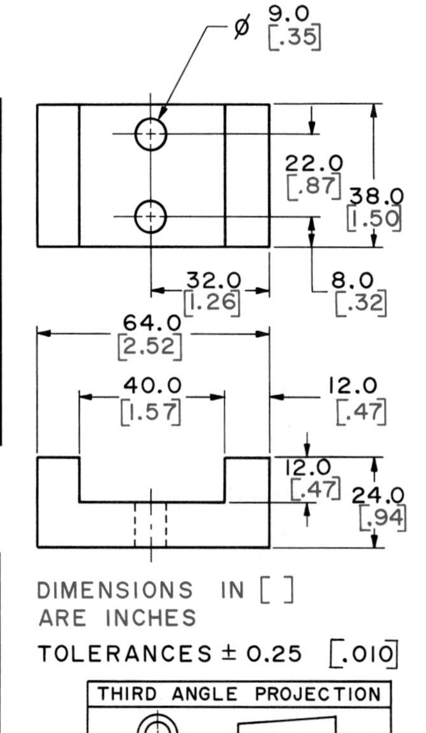

Fig. 7–21. A dual dimensioned drawing.

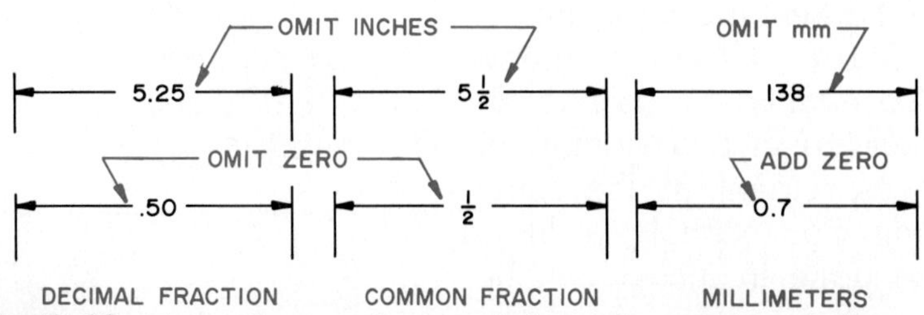

Fig. 7–19. The application of unit symbols and the use of zero before fractions.

For example, the linear base unit, the meter, is divided by ten into tenths called *decimeters,* by ten again into hundredths called *centimeters,* and by ten again into thousandths called *millimeters.* Fig. 7-18. Thus, one meter is equal to 10 decimeters, 100 centimeters, or 1000 millimeters.

Printed inside the back cover of this book are tables for converting common fractions and decimal fractions to metric measure. These tables can also be found in Appendices 56 and 58.

Designing in Metric Units

When a company decides to convert its drawings of existing products from customary measurement to metric measurement without changing any of the actual sizes, it simply finds the metric equivalent of the customary sizes. Such a process is called a *soft* conversion. When a company decides to redesign and resize its products to preferred metric sizes, it is called a *hard* conversion. New products should be designed with metric measurements from their beginning, especially if they are to be sold in other countries. There is a table of preferred metric sizes that should be used when designing with metric measurements. See Appendix 54A.

Metric Dimensioning Practices

When dimensioning a metric drawing, you should know and keep in mind the following:

1. Use correct metric unit names and symbols. See Appendix 55A.

2. The shortened notations of units are symbols, not abbreviations. Therefore, you must not change them such as using capital letters instead of lowercase or vice versa.

3. Unit names are made plural by adding "s" when they are spelled out such as 75 meters or 35.5 newtons. If the unit name represents a value less than one, it is written singularly such as 0.50 meter or 0.75 newton.

4. The unit symbols represent both singular and plural forms such as 0.75 m or 75 m.

5. When a metric dimension is a whole number, such as 15 mm, it is not followed by a decimal point or a zero. Fig. 7-19.

6. The decimal marker is a period, such as 65.3 cm.

7. Unit sizes less than one require a zero before the decimal marker, such as 0.50 m. Fig. 7-19.

8. Large numbers are *not* grouped using commas. The comma is replaced by a space, such as 5 500 100. Four-digit numbers do *not* need a space, such as 1500.

9. When a unit symbol follows a number, leave a space between the number and the symbol. For example, 15 kg or 2.75 m.

10. Because the metric system is a decimal system, only decimal fractions, such as 0.5, are used. Do *not* use common fractions, such as ½.

11. Do *not* mix unit and prefix symbols when recording sizes, such as 30 m 8 dm. Use a single unit, such as 30.8 m.

12. Use millimeters or decimal parts of millimeters to dimension engineering drawings.

13. Use millimeters or decimal parts of millimeters to dimension architectural and construction drawings; however, dimension land surveys in meters and decimal parts of meters.

14. Metric dimensions are usually given to the nearest whole millimeter. If decimal parts are required, one place to the right of the decimal point is most commonly used (tenths of a millimeter). Fig. 7-20.

15. Place a note near the title block of a metric drawing stating, "ALL DIMENSIONS IN MILLIMETERS UNLESS OTHERWISE SPECIFIED."

16. Identify diameters with the symbol "∅."

17. Identify radii with the symbol "R."

Dimensioning Metric Drawings

There are three methods of showing metric dimensions on drawings: dual, chart, and metric only. Dual dimensioned drawings show both inches and millimeters. In Fig. 7-21 the original dimensions (mm) are placed above the converted dimensions (inches). A converted dimension may also be placed to the left of the original dimension and separated from it by a slash.

Using a chart for dual dimensions reduces clutter on a drawing. Fig. 7-22. Dimension the object as usual, then provide the equivalent dimensions in a chart somewhere on the drawing.

Metric only dimensions, of course, show only the metric units. Fig. 7-23.

Customary Decimal Dimensioning Practices

When dimensioning a customary drawing, you should know and keep in mind the following:

1. Dimension all drawings in decimal fractions except those used in industries, such as lumber, where common fractions are customarily used.

2. Identify a diameter with the symbol "∅."

3. Identify radii with the symbol "R."

4. Round off decimal fractions to two places (hundredths of an inch) except when greater accuracy is needed.

5. Express two-place decimals in even hundredths (.02, .10) not odd hundredths (.03, .11). Using even hundredths enables you to divide it by two when necessary, such as for a radius, and still have a two-place decimal. An odd decimal divided by two produces a three-place decimal, which specifies an accuracy greater than intended by the original two-place decimal.

6. Express decimal inches in whole numbers and decimal fractions. Do not place a zero before the decimal point for values less than one. Follow a whole number with a decimal point and zeros to two places. Fig. 7-24.

7. Do not show the inch symbol after each dimension on drawings where all dimensions are in inches. Instead, place the following note on the drawing near the title block: "ALL DIMENSIONS ARE IN INCHES UNLESS OTHERWISE SPECIFIED."

8. Dimension angles in degrees (°), minutes (′), and seconds (″) or decimal parts of a degree. When you specify degrees only, place the symbol (°) after the number. When only minutes and/or seconds are needed, place zero amounts (0° or 0° 0′) in front of the units as necessary.

Rounding Decimals

Round decimals (both inches and millimeters) to a fewer number of places by one of the following methods:

1. When the digit following the last digit to be retained is less than five, do not change the last digit. Fig. 7-25 at A.

2. When the digit following the last digit to be retained is more than five, increase the last digit by one. Fig. 7-25 at B.

3. When the digit following the last digit to be retained is exactly five and the digit to be rounded is even, do not change that digit. Fig. 7-25 at C.

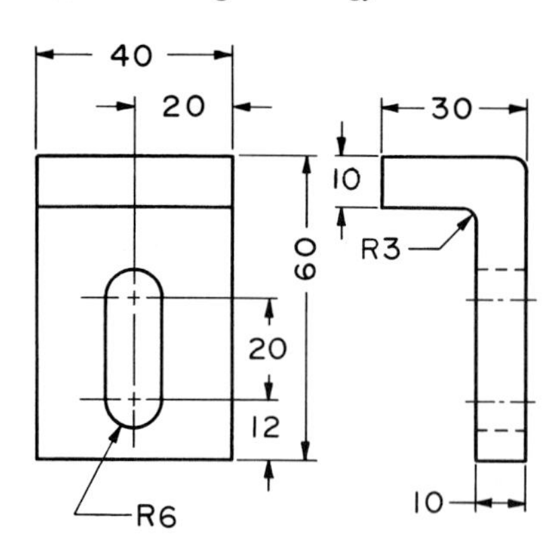

DIMENSIONS IN MILLIMETERS

CONVERSION	
mm	INCH
3	.12
6	.24
10	.39
12	.47
20	.78
30	1.18
40	1.57
60	2.36

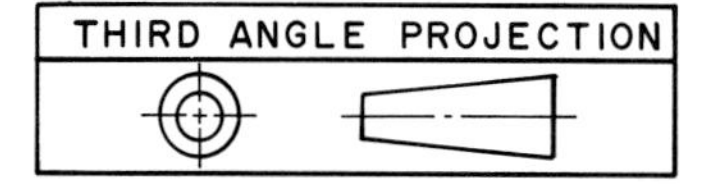

Fig. 7–22. The use of a chart for converting dimensions.

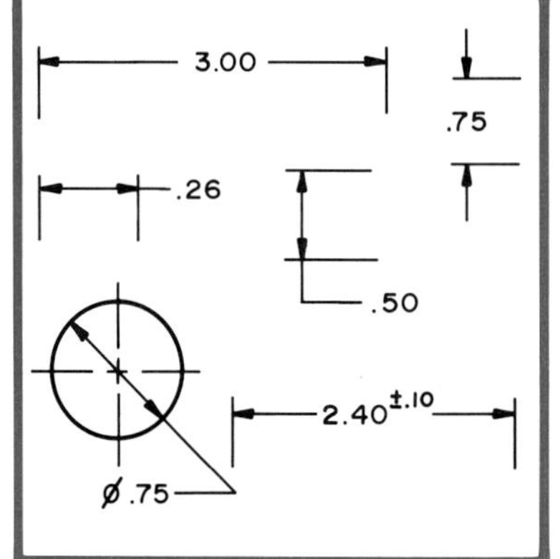

Fig. 7–24. Typical decimal inch dimensions.

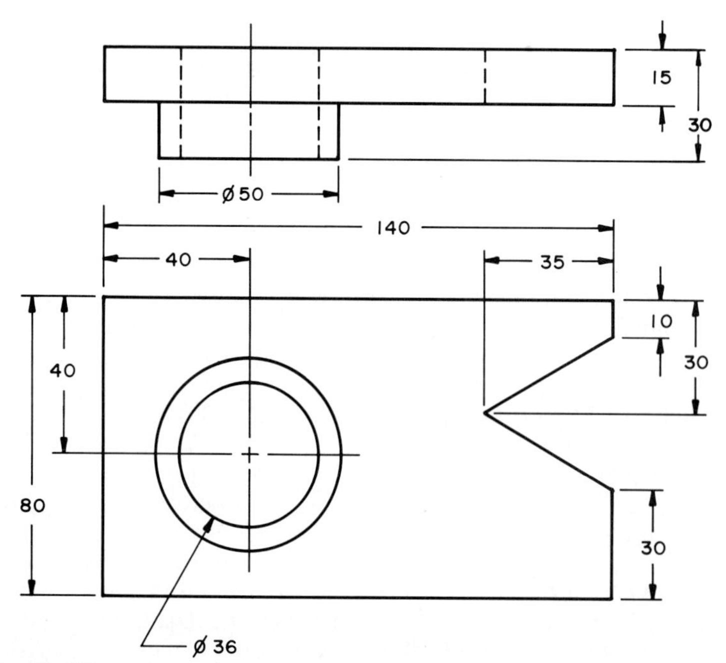

Fig. 7–23. A drawing with metric dimensions.

4. When the digit following the last digit to be retained is exactly five and the digit to be rounded is odd, increase that digit by one. Fig. 7-25 at D.

When rounding diameters of round shafts and holes, use the preferred design sizes shown in Fig. 7-98. When rounding other parts, use the procedure explained above. Remember, if the distance is not critical, even dimensions are preferred.

Notes

Notes are used to give information that is not part of the normal dimensioning system. Notes should be stated as briefly as possible but the meaning must be clear.

Letter notes horizontally and always use guidelines. Form your letters carefully so that they can be read easily. Use as short a leader as possible for notes that are to be related to a specific point on a drawing. Avoid placing notes in crowded areas of the drawing, especially between views.

There are two types of notes you will place on a drawing: general and specific.

General notes apply to the entire drawing. For example, when the note "FINISH ALL OVER" is used, it refers to all the surfaces of the object. Usually you will position general notes in one of two places. One place is above or to the left of the title block. Materials, tolerances, and specifications go there. The other place is near the view to which the note refers. In that case, letter them in an open area so that they are easily seen and not confused with a specific dimension. Place general notes on your drawing after the size and location dimensions and specific notes are in place.

Specific notes, also known as *local notes,* apply to a specific part of the detail. For example, .312-18UNC-2A is a thread note that refers to a specific fastener. Place local notes as close to the detail as possible and keep the leader as short as possible.

Abbreviations are commonly used in notes. You can find common standard abbreviations in Appendix 1.

Dimensioning Systems

All dimensions and notes should be placed so that they are in line with the bottom of the drawing. Fig. 7-26. This placement is called the **unidirectional system** of dimensioning. It is a system that most industries use because it is easy to letter and read. Fig. 7-27. An older system, the **aligned system,** finds little use today. With that system the dimensions are placed parallel to the bottom and right-side edges of the object. Fig. 7-28.

Theory of Dimensioning

All objects are composed of one or more of the following geometric shapes: cylinder, prism, cone, pyramid, or sphere. The principal dimensions of these shapes are shown in Fig. 7-29. Dimensioning an object becomes relatively easy if you first break it down into its component geometric shapes. For example, see how the simple object in Fig. 7-30 is broken into its component geometric shapes.

When dimensioning, think in terms of both positive and negative shapes. A positive shape is one that exists while a negative one is hollow or open but with the same general outline as a positive feature. For example, a drilled hole is a negative cylinder. Dimensioning, therefore, consists of placing sizes on these positive and negative features as well as locating them.

Size dimensions define the sizes of the basic shapes of the object. **Location dimensions** position these shapes relative to each other. Fig. 7-31.

Fig. 7-25. Rounding Off Dimensions

	4 PLACES	3 PLACES	2 PLACES
A	3.1262|4 = 3.1262	.786|2 = .786	5.13|3 = 5.13
B	1.1250|8 = 1.1251	6.187|6 = 6.188	.85|9 = .86
C	.3422|5 = .3422	4.276|5 = 4.276	1.78|5 = 1.78
D	10.7921|5 = 10.7922	2.125|5 = 2.126	4.87|5 = 4.88

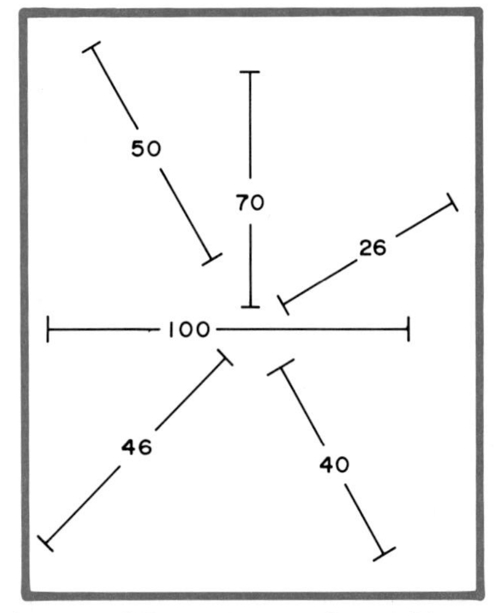

Fig. 7–26. **Unidirectional dimensions are lettered in one direction so that they can be read without turning the drawing.**

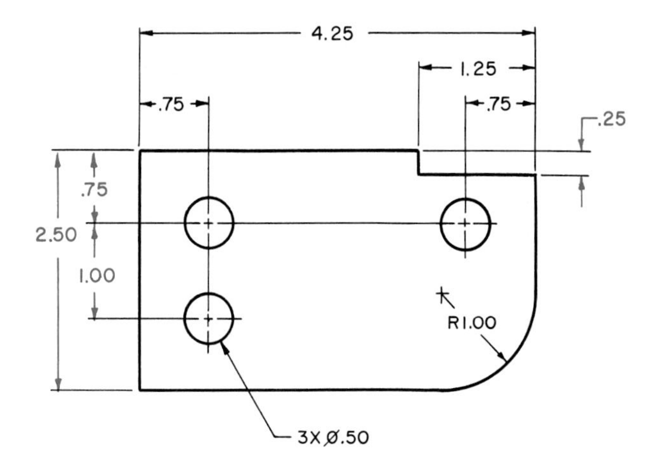

Fig. 7–27. **The unidirectional system of dimensioning.**

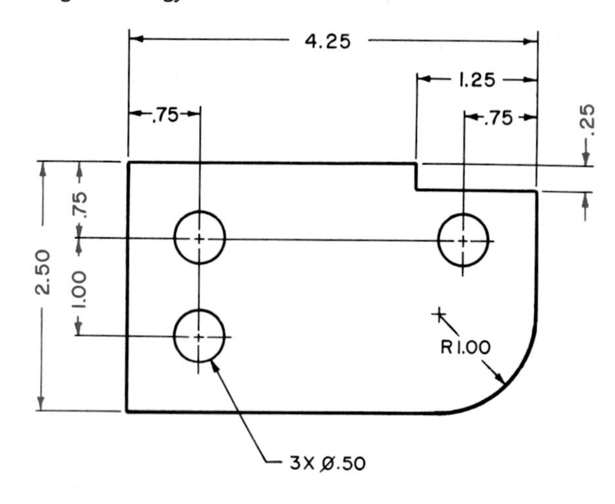

Fig. 7–28. The aligned system of dimensioning.

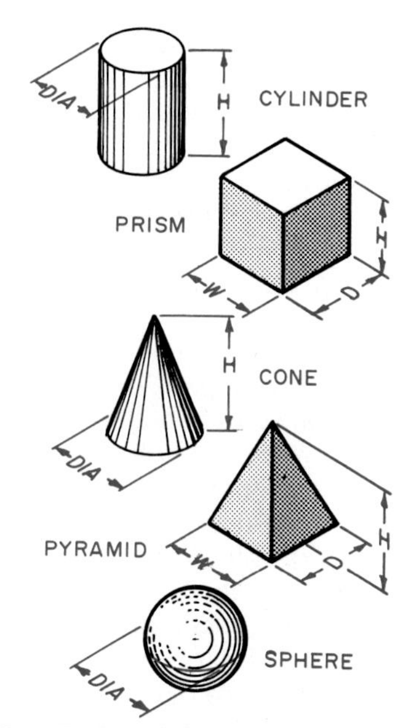

Fig. 7–29. Basic geometric shapes with their principal dimensions.

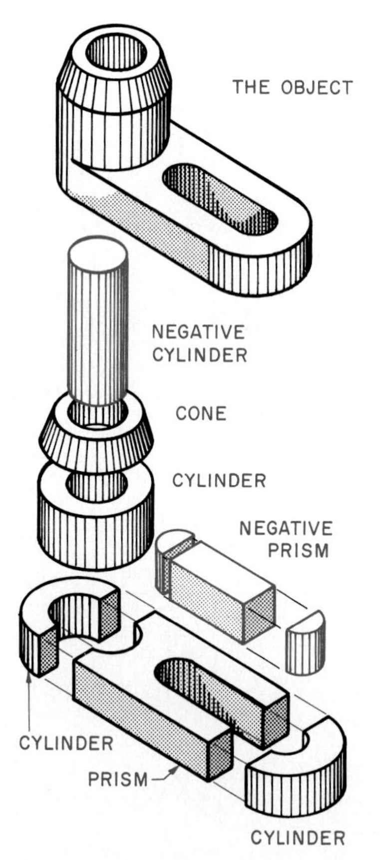

Fig. 7–30. Objects can be broken into various geometric forms.

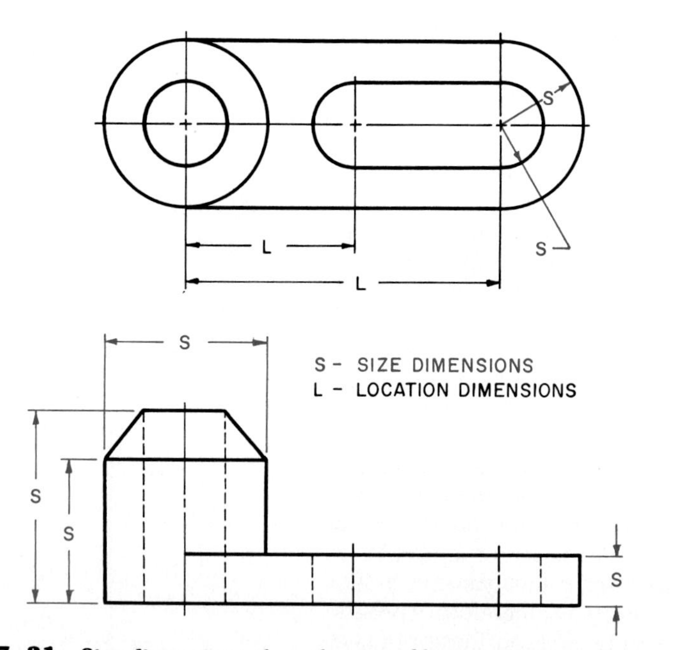

Fig. 7–31. Size dimensions show the size of basic geometric shapes. Location dimensions show the relative positions of these shapes.

Size Dimensioning Geometric Shapes

Prisms

The rectangular prism and variations of it are the most elementary shapes. All rectangular prisms require three principal dimensions: *width, height,* and *depth.*

Three different methods of dimensioning a rectangular prism are shown in Fig. 7-32. Any views which are adjacent (next to each other) will have a common dimension. The dimension which is common to both views is usually placed between the views. For example, in Fig. 7-32 at A the height dimension is common to both the front and right side views. The width is the common dimension in the front and top views as seen at B. When the piece has a uniform depth, it may be dimensioned as shown at C. The depth dimension is given as a note, a practice that can only be done when the piece has a uniform depth.

As a general rule, show the thickness dimension on the end view and all other dimensions on the outline view as shown in Fig. 7-33. The correct method of dimensioning an irregular rectangular prismatic object is illustrated in Figs. 7-34 and 7-35.

Prisms with shapes other than rectangular are dimensioned in a manner similar to a flat rectangular prism as shown in Fig. 7-36. A square prism can be dimensioned by adding a square symbol before the width dimension as seen at A. Hexagonal or octagonal prisms, as seen at B and C, may be dimensioned by using only two dimensions. In some cases, the abbreviation "HEX" or "OCT" may be added to the width dimension. Both prisms may also be dimensioned by giving the distance across corners or across flats. Never give both dimensions. A triangular prism may

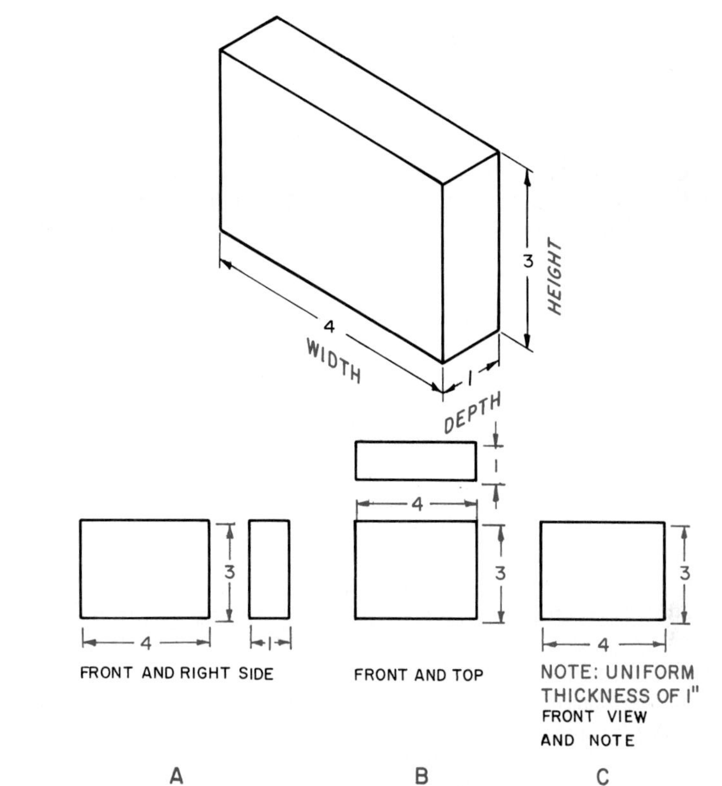

Fig. 7–32. Size dimensions of a rectangular prism.

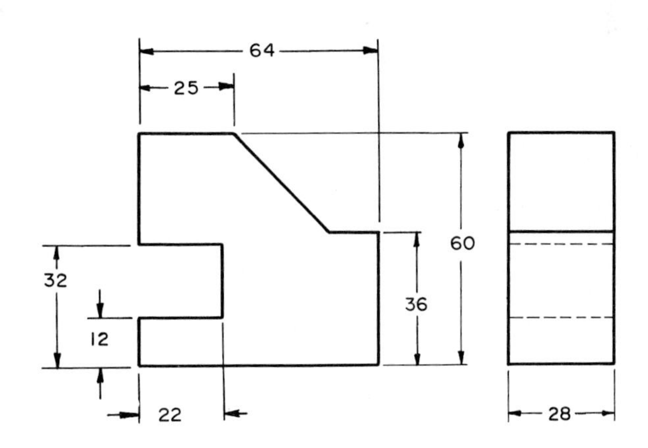

Fig. 7–33. Dimensioning an irregular prism. Notice the use of metric measure.

be dimensioned by giving an angle in degrees or by offset dimensions as shown at D and E. In each case, the dimensions have been placed where they appear with the most *clarity*. If you can follow this rule in dimensioning, you will encounter very few problems.

Cylinders

Cylinders may be positive or negative. The diameter and height of a positive cylinder are given in the rectangular view. Fig. 7-37 shows how positive cylinders can be dimensioned. Notice how some of the dimensions have been placed between the views. Keep extension lines short and diameters as close to the circular view as possible.

Holes are considered negative cylinders. Dimension the size of a hole on the circular view whenever possible. Fig. 7-38 shows common ways to dimension holes. If it is not clear that a hole goes through the part, add the abbreviation "THRU" following the dimension.

The diameter symbol, (∅), is used whenever a diameter is dimensioned. Fig. 7-39.

Arcs

An arc is dimensioned by its radius. The radius is noted by the symbol "R," which is placed before the size figure. Where space permits, draw the radius dimension line from the radius center to the arc. Use an arrowhead to end the dimension line at the arc. Fig. 7-40. Where space is limited, use a leader to place the dimension outside the arc. When space is limited and the complete radius dimension line cannot be shown to scale, the dimension line may be foreshortened as shown in Fig. 7-40.

If an arc does not show its true shape, it may be dimensioned as a true radius. Fig. 7-41. The term "TRUE R" must precede the dimension. This practice may save drawing an auxiliary view of a simple angular portion.

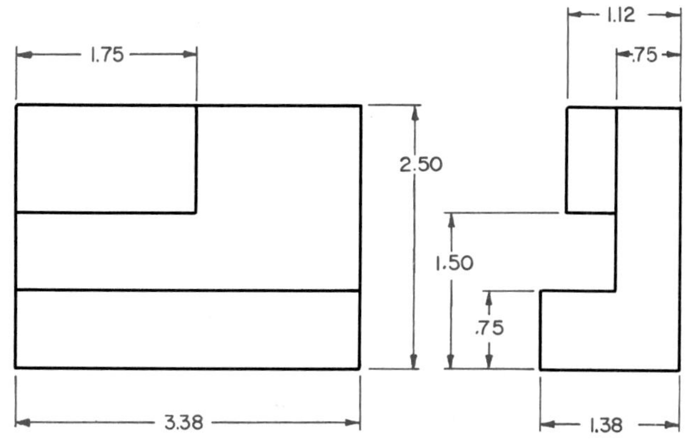

Fig. 7–35. Another example of dimensioning a rectangular prismatic object.

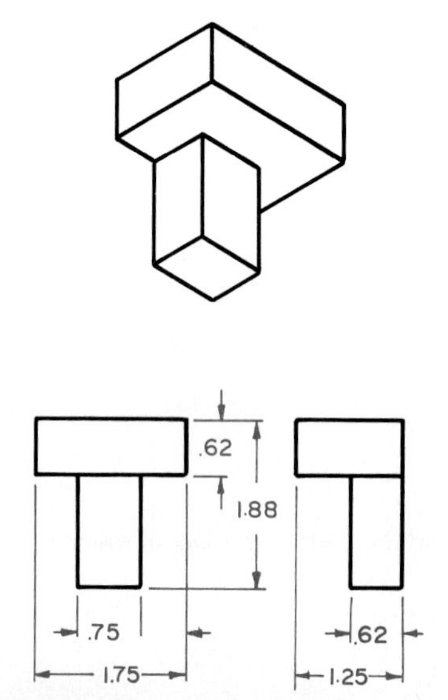

Fig. 7–34. Dimensioning a rectangular prismatic object.

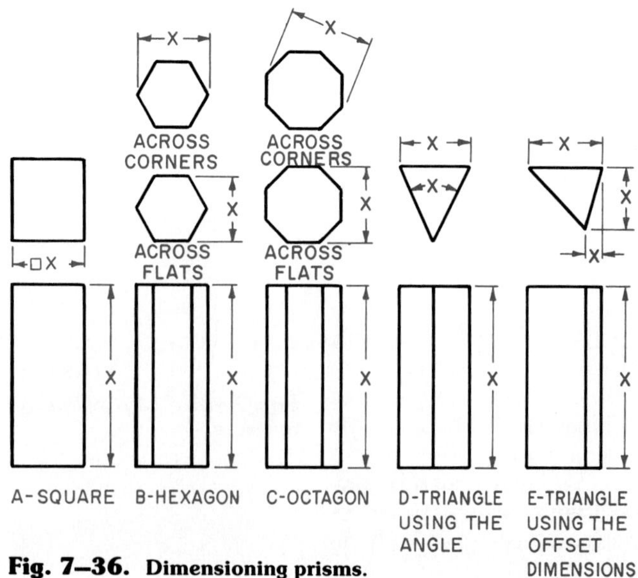

Fig. 7–36. Dimensioning prisms.

Spherical surfaces may be dimensioned as shown in Fig. 7-42. The symbol "SR" precedes the radius of a sphere, and "S∅" precedes a spherical diameter.

Angular Dimensions

Angular dimensions are given in degrees, minutes, and seconds. Some examples of angular dimensions are shown in Fig. 7-43. The symbols used on a drawing are the degree, °, minute, ′, and second, ″. If an angle consists of degrees only, it is dimensioned by the number of degrees and the degree symbol such as 30°. If an angle is less than one degree, place a "0°" notation before the minutes and seconds (if any); for example, 0° 15′. Angles also can be given in decimal form, that is, degrees and decimal parts of a degree, such as 30.5°.

Shapes with Rounded Ends

Parts that have fully rounded ends are dimensioned by giving the diameter and overall length. The radius is indicated but not dimensioned. Fig. 7-44.

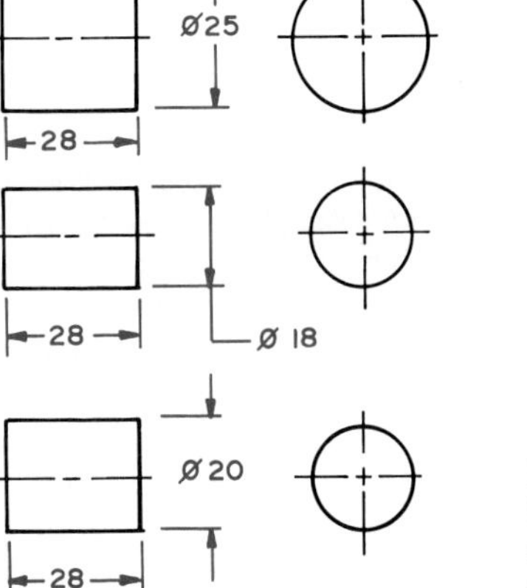

Fig. 7–37. **Dimensioning solid (positive) cylinders.**

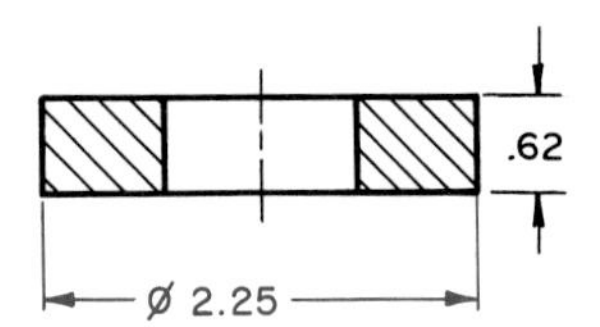

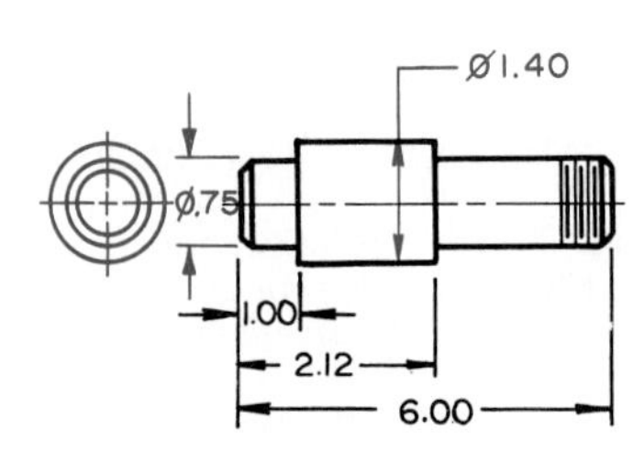

Fig. 7–39. **The circular view may be omitted when the diameter symbol, (∅), accompanies the dimension.**

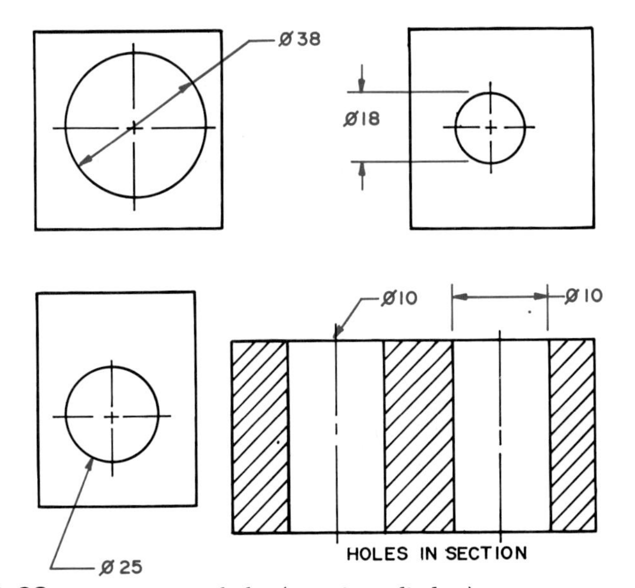

Fig. 7–38. **Dimensioning holes (negative cylinders).**

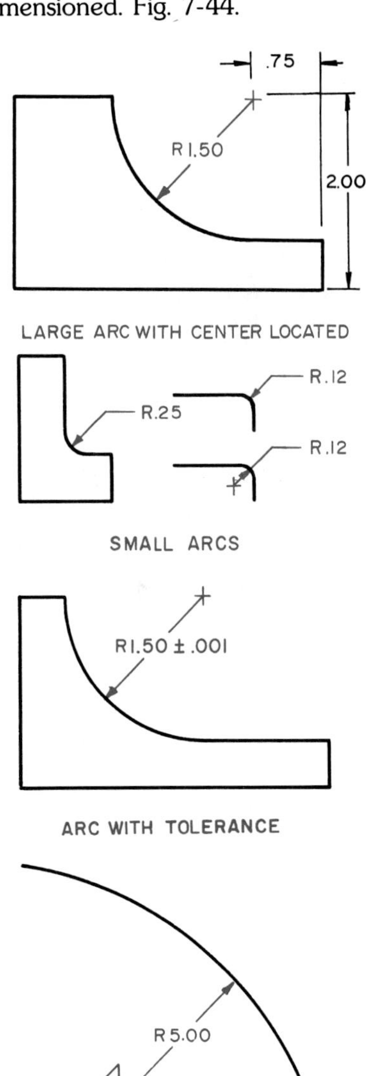

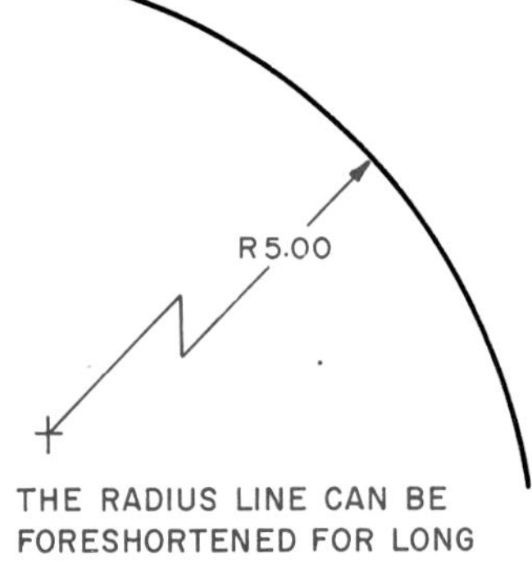

Fig. 7–40. **How to dimension arcs.**

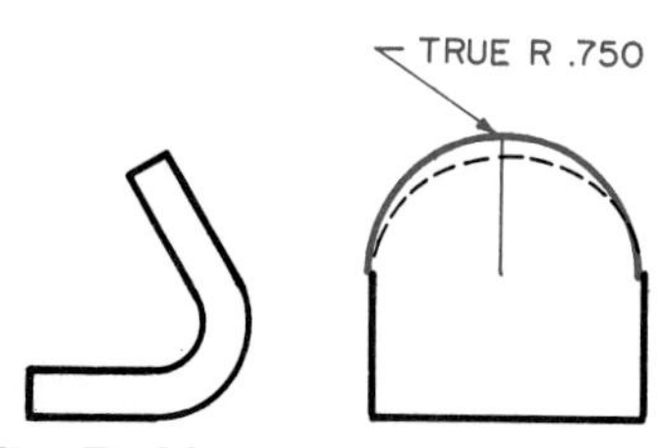

Fig. 7–41. The true radius of foreshortened arcs can be dimensioned in this way. Doing so eliminates the need to draw an auxiliary view.

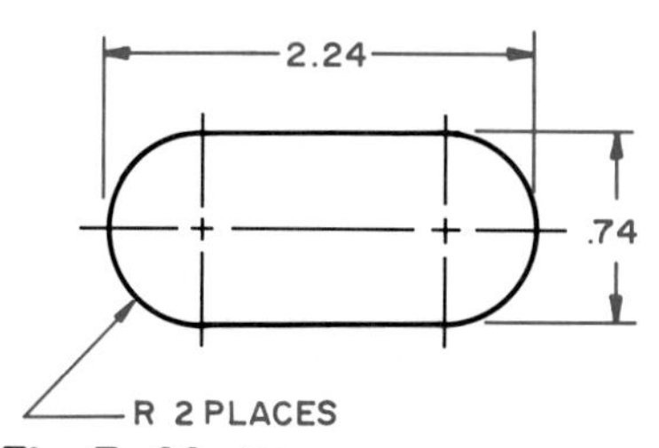

Fig. 7–44. Dimensioning rounded ends. Notice the radius is indicated but not given.

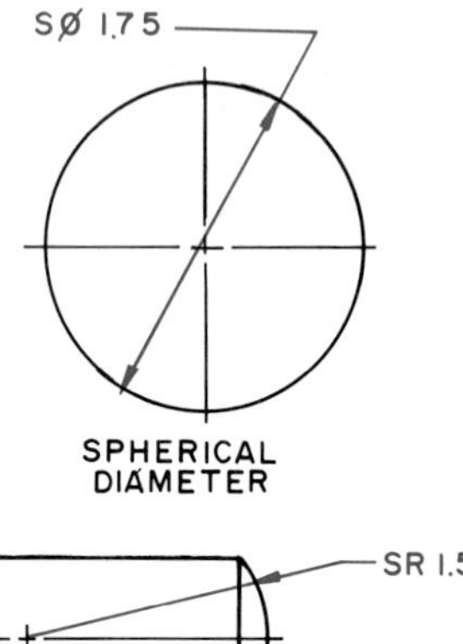

Fig. 7–42. Dimensioning spherical surfaces.

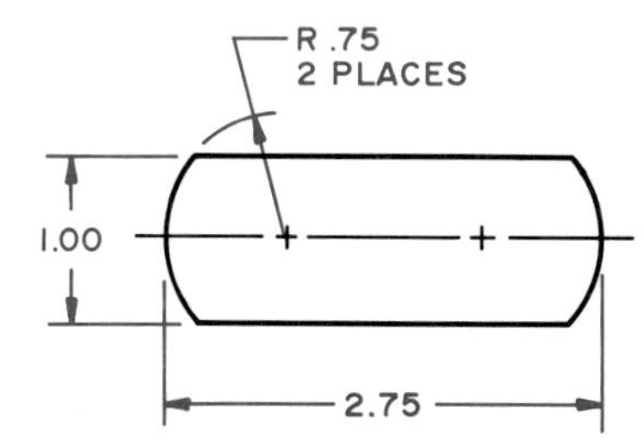

Fig. 7–45. Dimensioning a part that has partially rounded ends. Notice the radii are given.

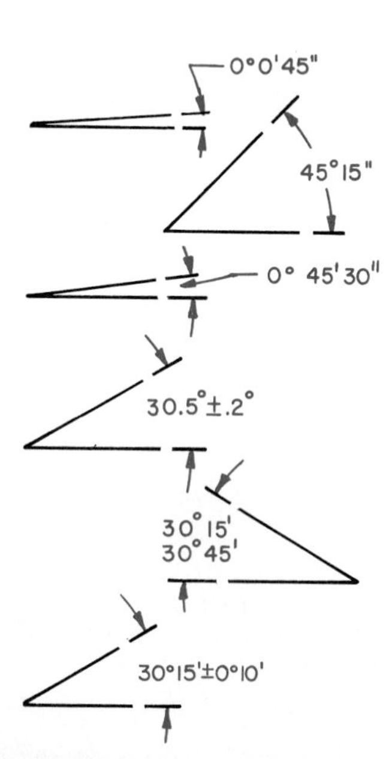

Fig. 7–43. Ways to dimension angular features.

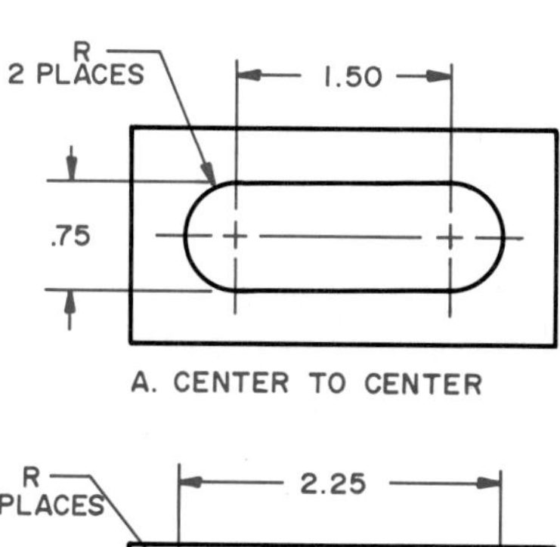

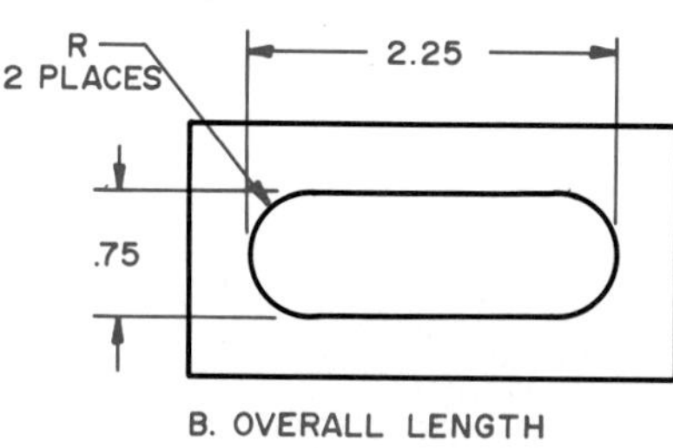

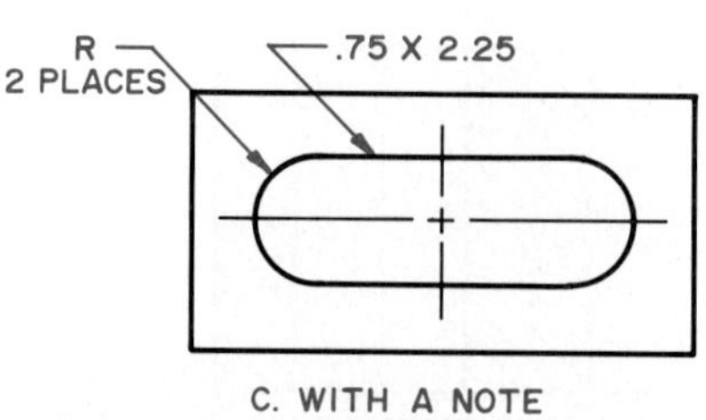

Fig. 7–46. Dimensioning slotted holes.

When a part has a partially rounded end, the width of the part and the radius of arc are given as shown in Fig. 7-45.

Slotted Holes

Slotted holes are dimensioned as shown in Fig. 7-46. When the distance between the centers is important, as shown at A, dimension it. The dimension, consisting of the diameter and overall length, can be given in a note as shown at C. In this case, the radius is noted but not dimensioned since the diameter is given.

Curves

A curved line made up of two or more circular arcs should be dimensioned as shown in Fig. 7-47. The arcs are dimensioned by indicating the radii and locating the centers or points of tangency.

A noncircular, or irregular, curve may be dimensioned by giving the offset, or coordinate, dimensions. These dimensions are always given from a specific line called a datum. Fig. 7-48. Dimensioning from datums is discussed later in this chapter.

Cones, Pyramids, and Spheres

Cones are dimensioned by giving both the diameter and the altitude in the triangular view as shown in Fig. 7-49 at A. In some cases the diameter and angle formed by the sides are specified as seen at B. A frustum (a cut made perpendicular to the axis) of a cone may be dimensioned by showing the altitude and both diameters. Note that the symbol Ø is given before all diametral values. Another method of dimensioning frustums is to give the altitude, one diameter, and the amount of taper per foot. The taper per foot is always specified by note and refers to the difference in diameters in one foot of length.

Cones are usually dimensioned in one view.

Pyramids are dimensioned by giving the altitude dimension in the front view and the dimension of the base in the top view. The dimensions are placed where they are the most meaningful. As with a square prism, the base of a square pyramid may be dimensioned by giving only one dimension preceded by the square symbol as shown at F.

A sphere is dimensioned by giving its diameter as shown at G.

Location Dimensions

Location dimensions are used to position geometric shapes relative to each other or to a particular line. Finished surfaces, center lines, or axes are used in locating geometric shapes. When you are about to dimension a particular view, first plan the size dimensions and then the location dimensions. Usually location dimensions are given in three mutually perpendicular directions. These directions (shown in Fig. 7-50)—bottom to top, front to back, and side to side—are adequate to locate all shapes.

Some knowledge of the function of the piece and its relationship to mating pieces (how it fits with other parts) along with a

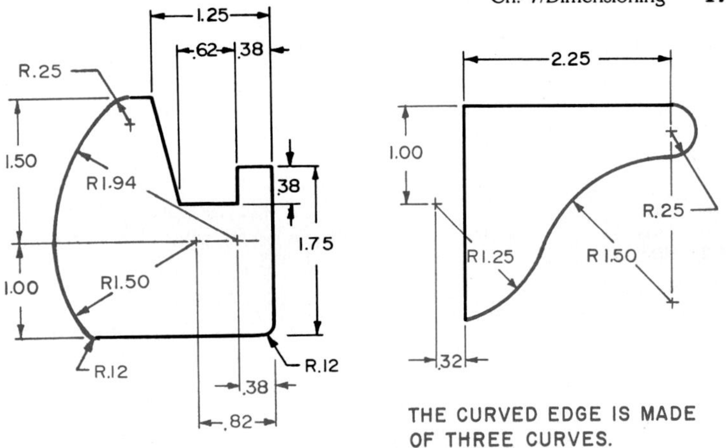

Fig. 7–47. How to dimension curves made of several circular arcs.

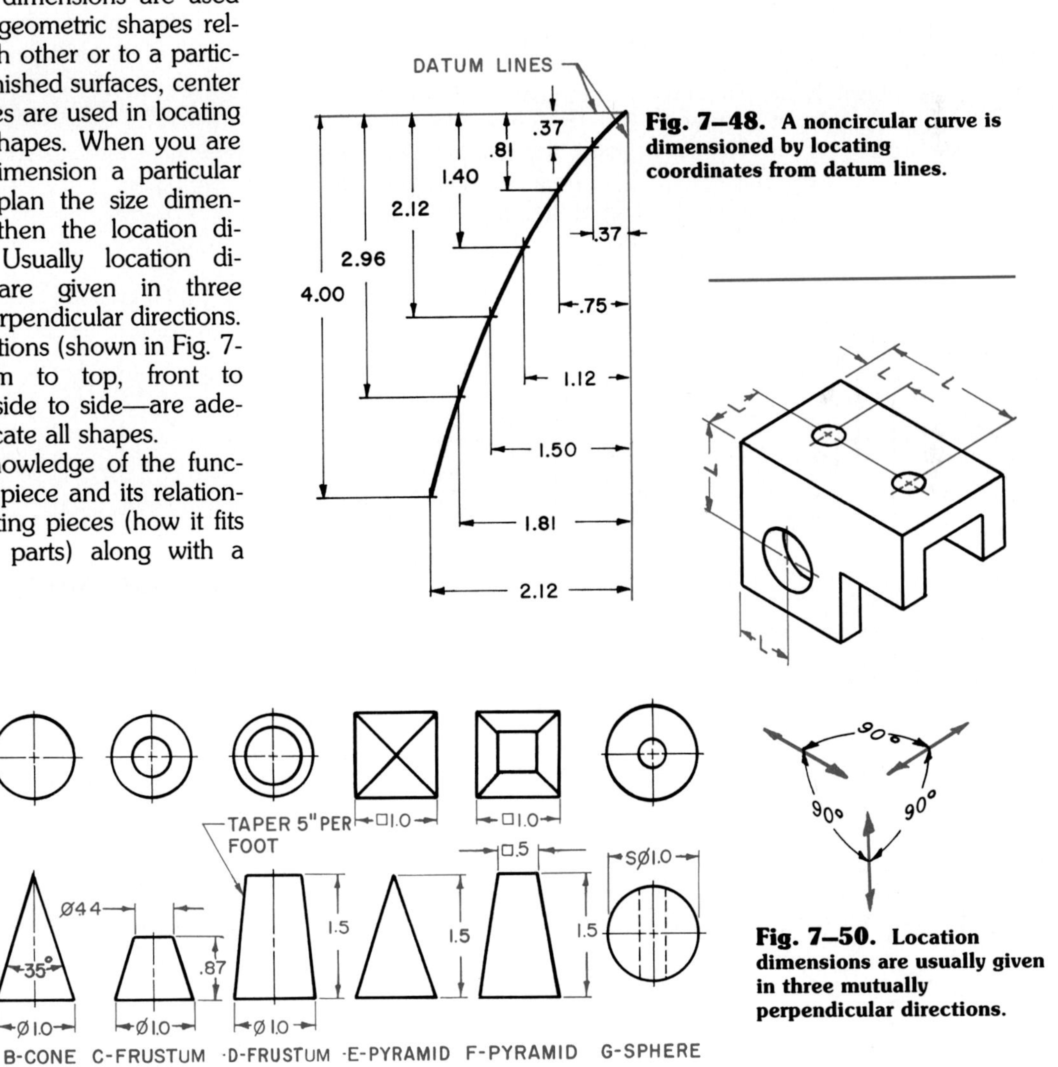

Fig. 7–48. A noncircular curve is dimensioned by locating coordinates from datum lines.

Fig. 7–49. Size dimensions of cones, pyramids, and spheres.

Fig. 7–50. Location dimensions are usually given in three mutually perpendicular directions.

basic knowledge of shop processes is important when dimensioning a part. Understanding these items is valuable in determining how and what location dimensions will be specified.

Prisms

Locate rectangular shapes from surface to surface or from a center line. Fig. 7-51. Dimensions that locate surfaces may be from a datum line. Datums are used as a reference where dimensions may be critical. This is particularly true where one dimension must meet with another. Datums will be described in detail later. Point to point distances are used for describing the location of prisms where the object is simple and when a high degree of accuracy is not required.

Cylinders

Cylinders are located from their center lines. If possible, always give the location dimension in the circular view of the cylinder. Fig. 7-52 shows how cylindrical holes may be located.

When a series of equally-spaced holes in a line is to be located, they may be indicated as shown in Fig. 7-53. Repetitive features or dimensions are specified with an "X" and a number indicating the number of times or places required. A space is left between the X and the diameter.

Holes or cylindrical features when located on a circle or arc may be dimensioned in several ways. Holes may be located by rectangular coordinates. Fig. 7-54. They may also be located by giving the diameter of the center line, hole size data, and the spacing of the holes in degrees. Fig. 7-55.

You may also use angular dimensioning to locate holes along a circular center line. Fig. 7-56.

Distances along Curved Surfaces

Linear distances along curved surfaces may be dimensioned as a chord or an arc. Fig. 7-57. In either case, it should be clear that the dimension line indicates an arc or chord.

Dimensions from Datums

Datums are points, lines, planes, and cylinders which are understood to be exact. These features are used for computation or reference as well as for giving location dimensions. When a datum is specified, Fig. 7-58, all features must be measured from this

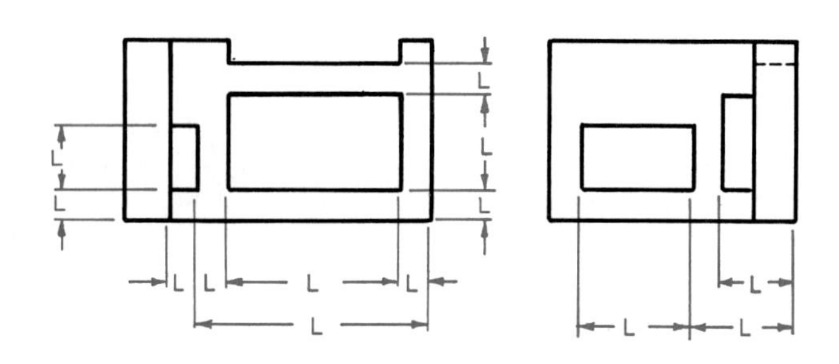

A - LOCATION USING SURFACES

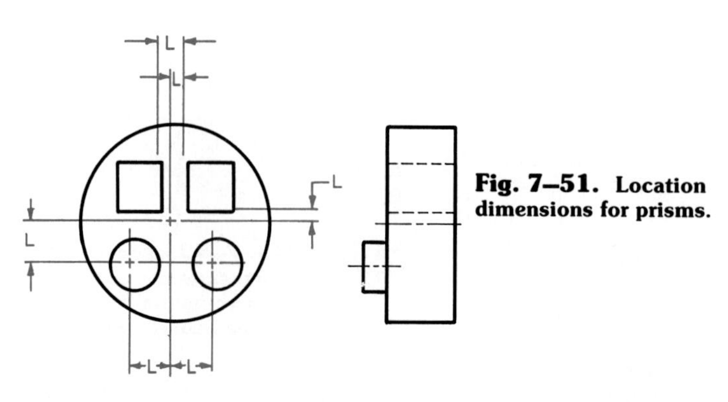

B - LOCATION FROM A CENTER LINE

Fig. 7–51. Location dimensions for prisms.

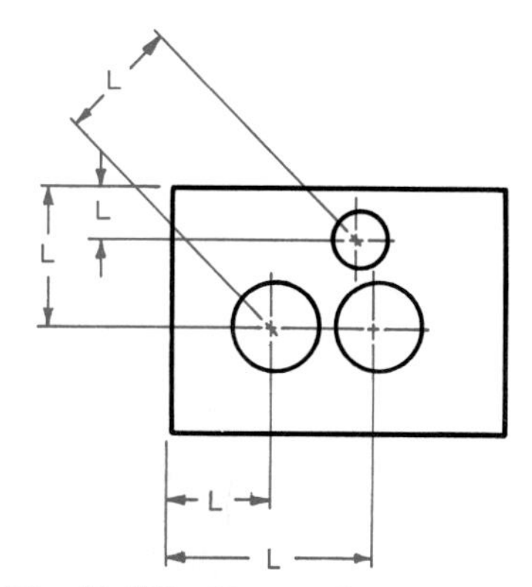

Fig. 7–52. Ways to locate cylindrical holes.

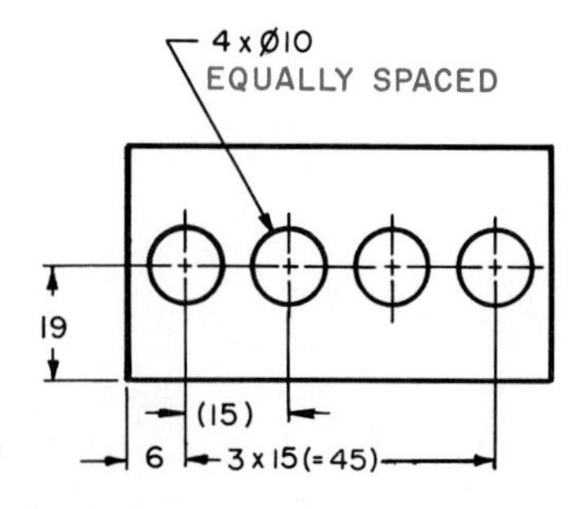

DIMENSIONS IN MILLIMETERS

Fig. 7–53. The note "equally spaced" can be used to locate holes.

datum and with respect to one another. Two and often three datums must be established in locating a feature. Surfaces and center lines that serve as datums must be either easily recognizable or clearly identified. Figs. 7-59 and 7-48 show other applications of datum dimensioning. To be most useful for measuring, the datums on the piece must be accessible during manufacture. Use corresponding features on mating parts as datums to insure proper assembly.

Coordinate Dimensioning. Coordinate dimensioning is a form of rectangular datum dimensioning. All dimensions are measured from two or three mutually perpendicular datum planes. The datum planes are indicated as *zero* coordinates and the dimensions from them are shown on extension lines. Fig. 7-60. Note that the dimensions are shown at the ends of the extension lines. There are no dimension lines or arrowheads. Notice how hole diameters are indicated. Each hole size is given a letter designation and the sizes are shown in a table. This method of dimensioning simplifies a drawing.

Tabular Dimensioning. Dimensions from mutually perpendicular datums should be listed in a table on the drawing instead of on the view. Fig. 7-61 illustrates tabular dimensioning of rectangular coordinates.

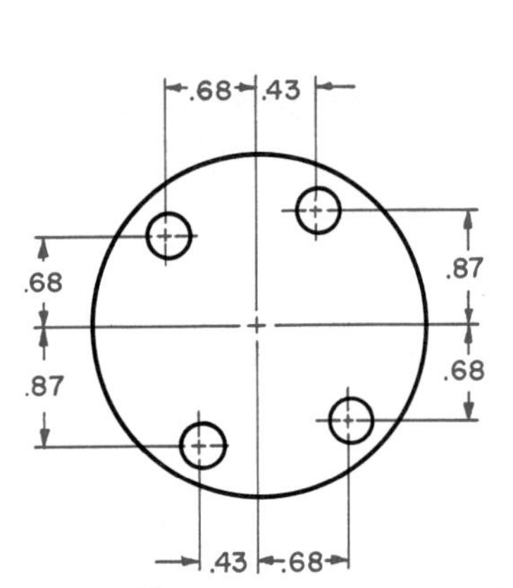

Fig. 7–54. You can locate round holes using coordinates.

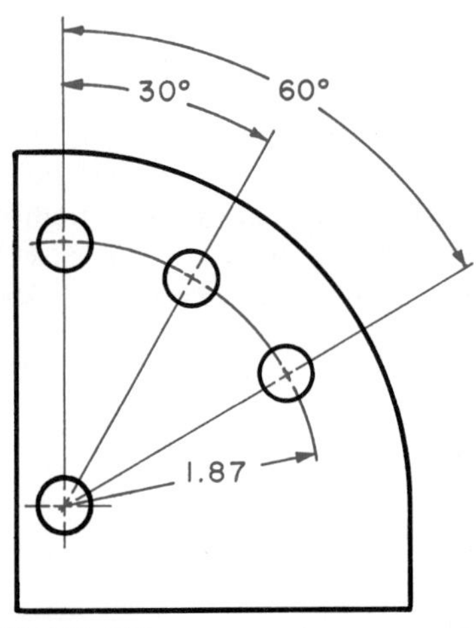

Fig. 7–56. You can locate round holes with a radius and angular dimensions.

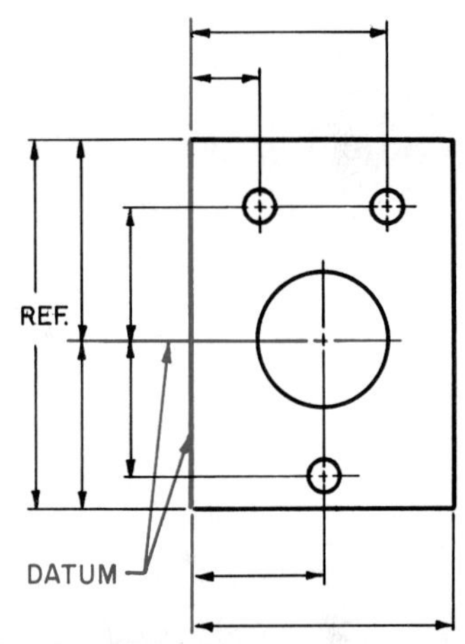

Fig. 7–58. A center line or edge can be used as a datum.

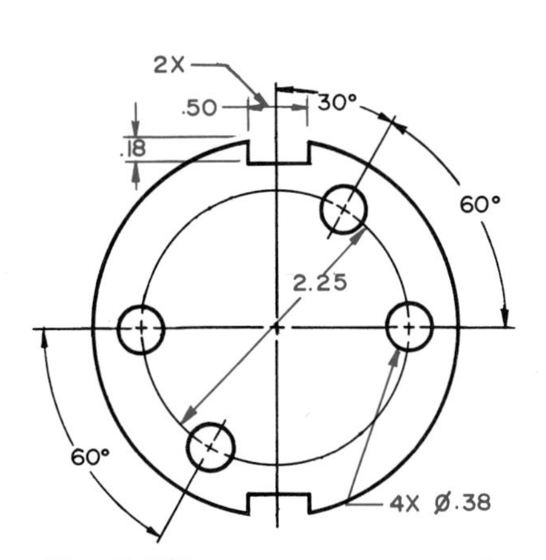

Fig. 7–55. You can locate equally spaced round holes by giving the diameter of their center line, the diameter and number of the holes, and their spacing.

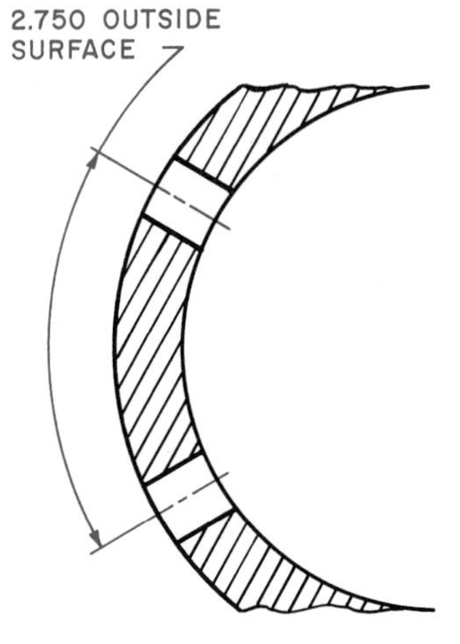

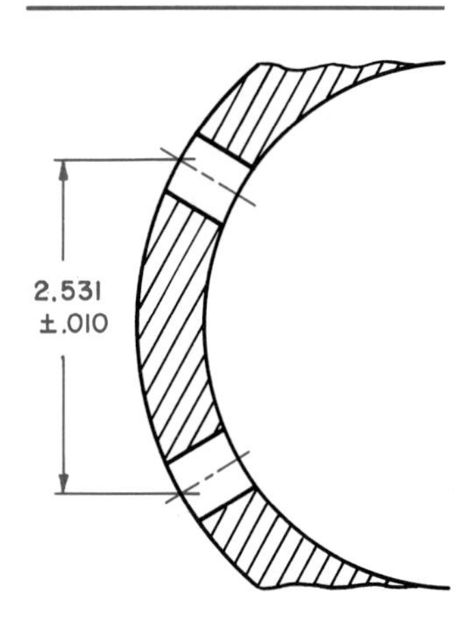

Fig. 7–57. Linear distances on curved surfaces can be dimensioned by giving the chord length of the arc on the surface.

Tabular dimensioning may be used on drawings that require the location of a large number of similarly shaped features. Note in this drawing the baselines from which the coordinates are taken are identified as X, Y, and Z. This method is used a great deal in drawing electronic printed circuit boards.

Reference Dimensions

A **reference dimension** is a dimension *without any specified tolerance*. This type of dimension may be either a size or location dimension and is used for information only. Fig. 7-62. Reference dimensions do not govern any machining or inspection operations. They are indicated on drawings by placing them in parentheses.

Dimensioning Round Holes

Plain Round Holes

Round holes are dimensioned in various ways depending on design requirements and manufacturing methods. The diameter, depth, and number of holes are always specified either by note or dimension. If it is not clear that the hole is to go completely through the piece, the abbreviation for through, "THRU," should be added following the diameter of the hole. A hole that does not go completely through the piece is called a *blind* hole. The depth must be specified. Fig. 7-63. The depth of a hole is identified by its shape rather than by the method of forming. Fig. 7-64 shows methods of dimensioning round holes.

Counterbored Holes

A **counterbored** hole is a cylindrical enlargement of a hole to a specified depth. The bottom of the counterbore is perpendicular to the axis of the hole. A typical use for a counterbored hole is to allow the head of a fillister-head machine screw or cap screw to set beneath the surface of a part. Fig. 7-65. Generally, you dimension a counterbore by giving its diameter and depth. The abbreviation for counterbore is "CBORE." Fig. 7-66 shows the method of dimensioning counterbored holes by a note. Notice that in one case the dimension includes diameter of the hole, and the diameter and depth of the counterbore. In some cases the thickness of the remaining stock

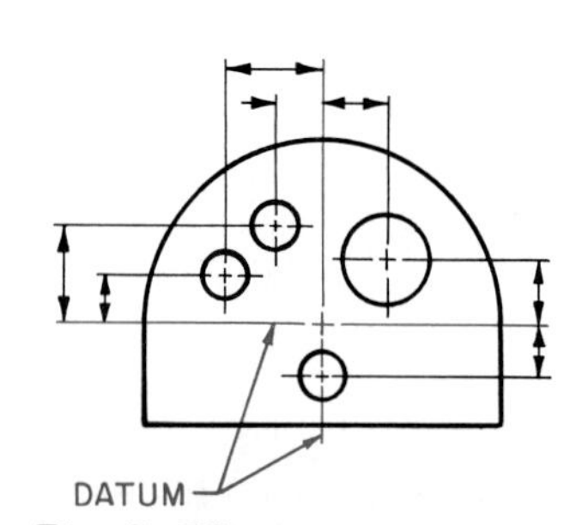

Fig. 7–59. **Datum dimensioning.**

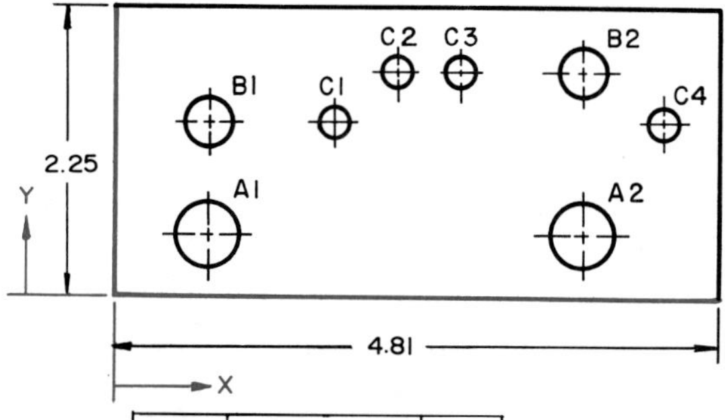

HOLE	DESCRIPTION	QTY
A	Ø12	2
B	Ø10	2
C	Ø7	4

HOLE	FROM	X	Y	Z
A1	X, Y	18	12	THRU
A2	X,Y	95	12	THRU
B1	X,Y	18	34	THRU
B2	X,Y	95	44	THRU
C1	X,Y	45	34	THRU
C2	X,Y	56	43	THRU
C3	X,Y	70	43	THRU
C4	X,Y	111	34	THRU

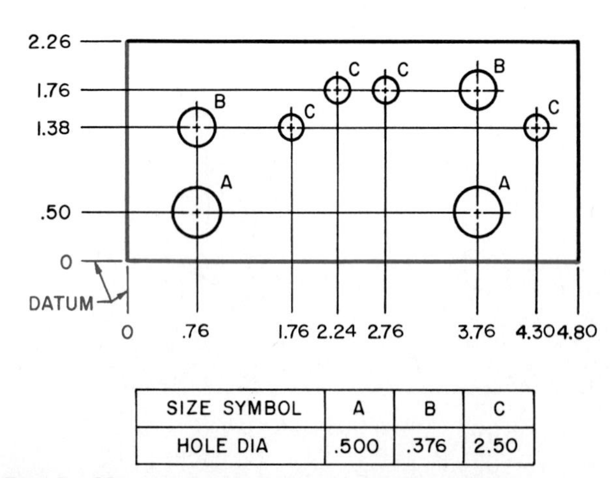

Fig. 7–60. **An example of rectangular coordinate dimensioning without the use of dimension lines.**

Fig. 7–61. **You can locate features with tabular dimensions.**

may be dimensioned rather than the depth of the counterbore. In such a case the thickness is given as a dimension on the view and not in the note. The tool used to produce a counterbore is shown in Fig. 7-67.

Counterdrilled Holes

A **counterdrilled** hole is two holes with different diameters drilled on top of each other. Counterdrilling produces a round recess with a tapered bottom. The information you need to give to dimension a counterdrilled hole includes the diameter of both holes and the depth and angle of the counterdrill. The abbreviation for counterdrill is "CDRILL." Fig. 7-68.

Countersunk Holes

Countersinking is the process of machining the end of a hole to a conical shape. A countersunk hole is used to hold the head of a flathead screw. The tool, called a countersink, that is used to produce the cone-shaped enlargement is shown in Fig. 7-69. Although the countersink tool is made with a variety of angles, draw the countersink at a 90-degree angle. Data given in the note in Fig. 7-70 refer to the diameter of the hole, the diameter of the countersink on the surface of the part, and the angle of the countersink. The abbreviation for countersink is "CSK."

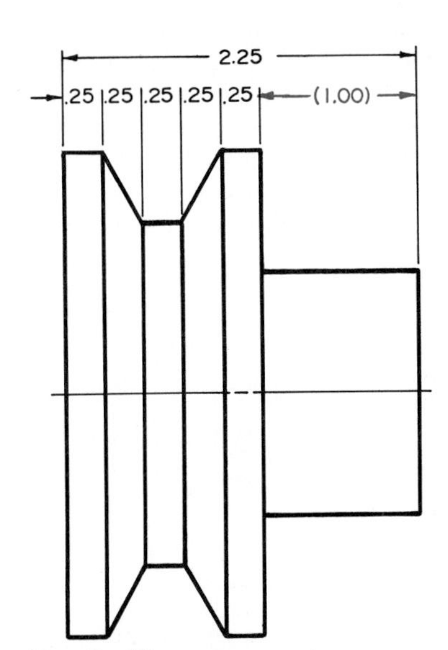

Fig. 7–62. Reference dimensions are used for information only.

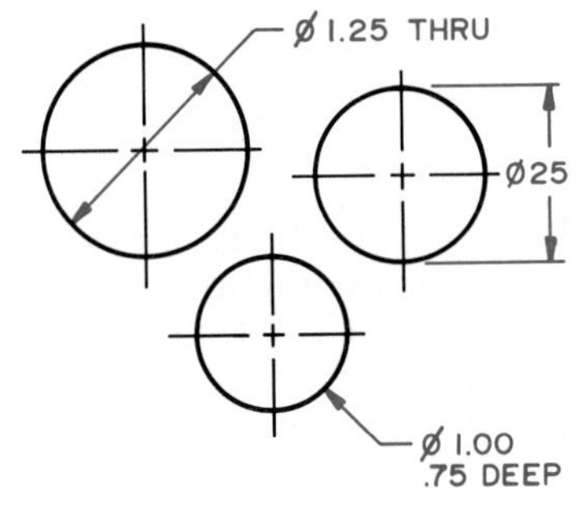

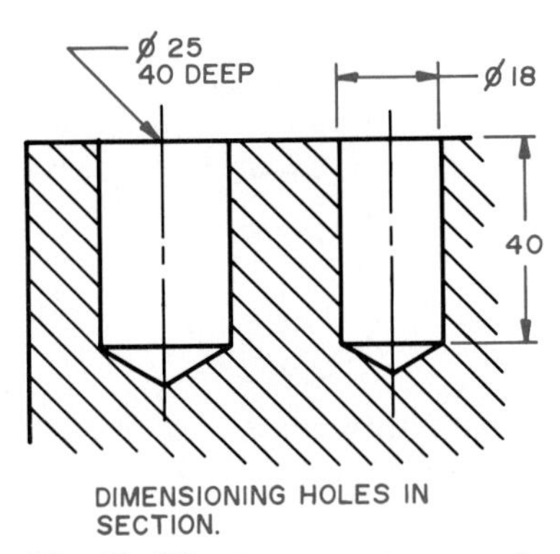

Fig. 7–64. Dimensioning round holes.

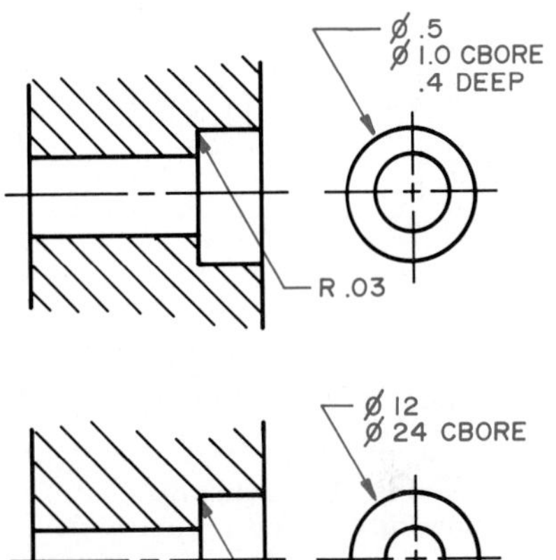

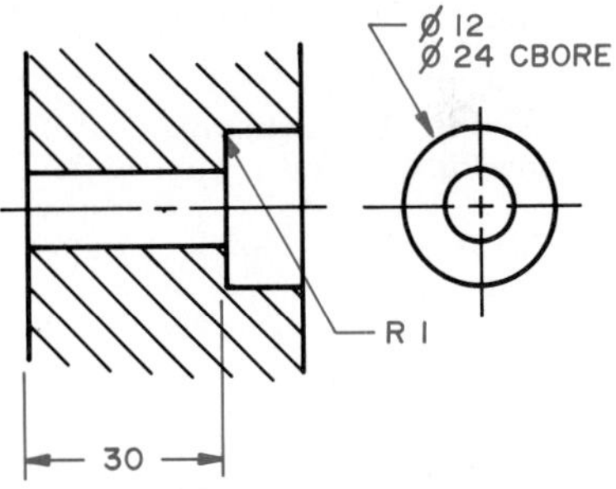

Fig. 7–66. Give the diameter and depth when dimensioning a counterbore.

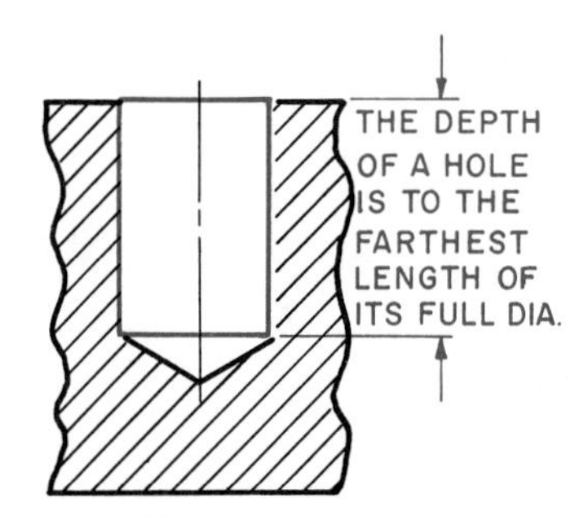

Fig. 7–63. A blind hole does not go all the way through the part.

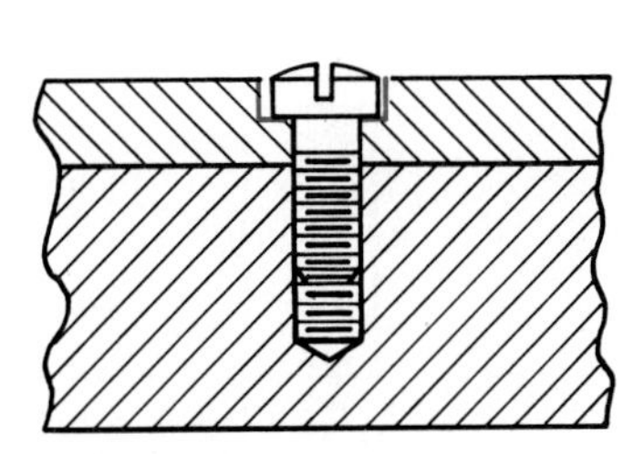

Fig. 7–65. Machine screws are set into counterbores.

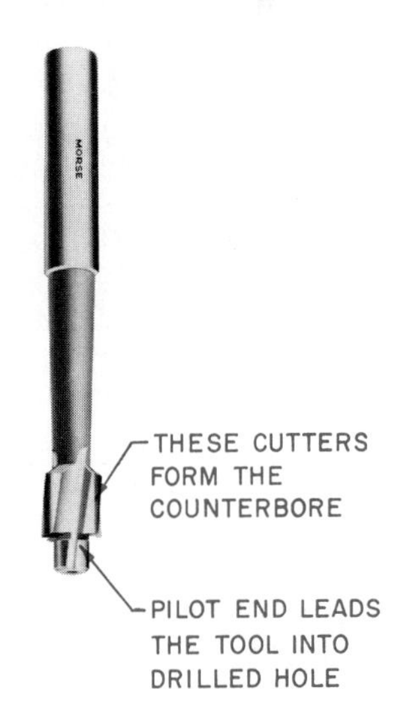

Fig. 7–67. A counterbore tool.

Spot-faced Holes

Spot facing is the process of smoothing and squaring the surface around a hole to seat a washer, nut, or bolt head. The tool used for spot facing is the same one used for counterboring.

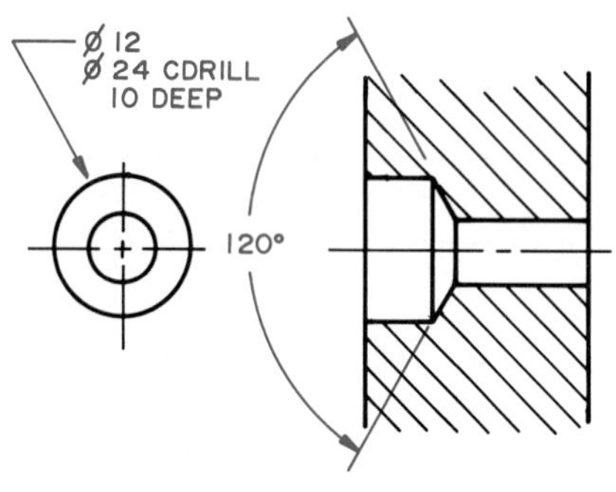

Fig. 7–68. Dimensioning counterdrilled holes.

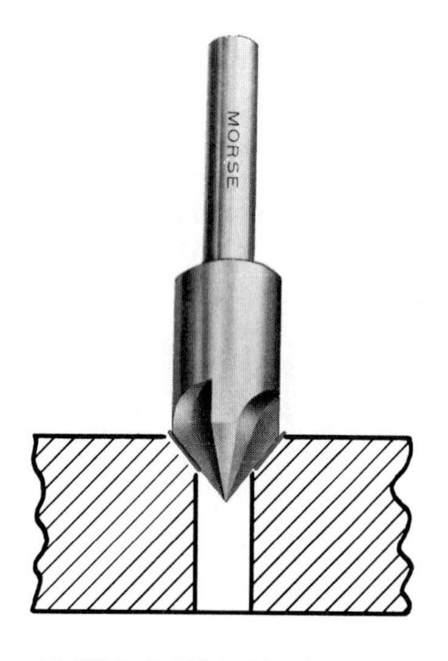

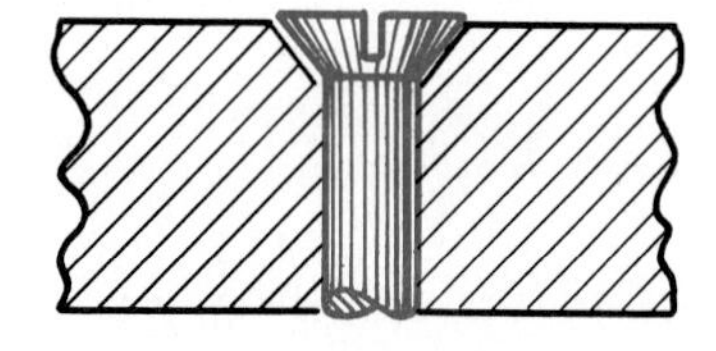

Fig. 7–69. A countersink forms a conical shape at the top of a round hole.

The spot face is indicated by a note as shown in Fig. 7-71. Specify the diameter of the spot face. Either the depth or the thickness of material may be given. Generally, a spot face is not deeper than .06 inch or 2 mm.

Dimensioning Special Features

Chamfers

A narrow, inclined, flat surface along the intersection of two surfaces is called a **chamfer.** Fig. 7-72. The recommended method of dimensioning a chamfer is to give the angle and a linear distance or two linear distances. Fig. 7-73.

Keys

Keys are used to prevent a hub from rotating at a different speed around a rotating shaft, that is, to keep a shaft and hub rotating together. The key is sunk partly into the shaft and extends into the hub. Fig. 7-74 shows a shaft with a key being inserted into a hub. The recess in the shaft that holds the key is called the *keyseat.* Fig. 7-75 illustrates how a keyseat may be dimensioned when stock keys are used. The dimensioning of a Woodruff key and keyseat is shown in Fig. 7-76. See Chapter 12, "Fasteners," for more details.

Knurls

Knurling is the process of pressing a pattern into a metal surface. Fig. 7-77 shows a piece being knurled on a lathe. The pattern of a knurl may be either a diamond or straight as shown in Fig. 7-78. Close tolerances are not necessary for knurls that provide a rough surface for gripping or for decoration. For either of these purposes specify only the pitch and type of knurl and the length of the knurled area. Fig. 7-79. Dimension a knurl for a

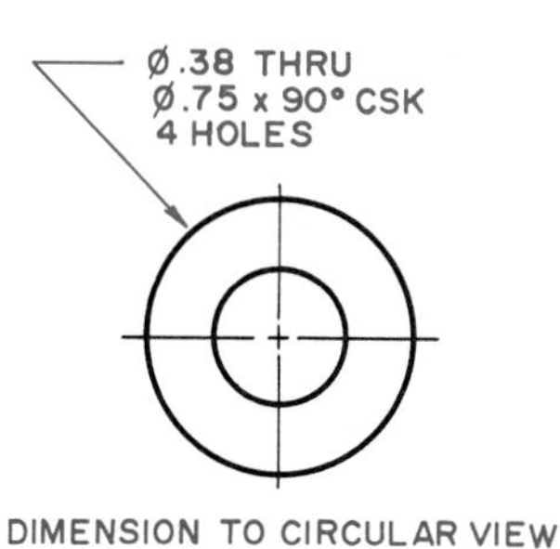

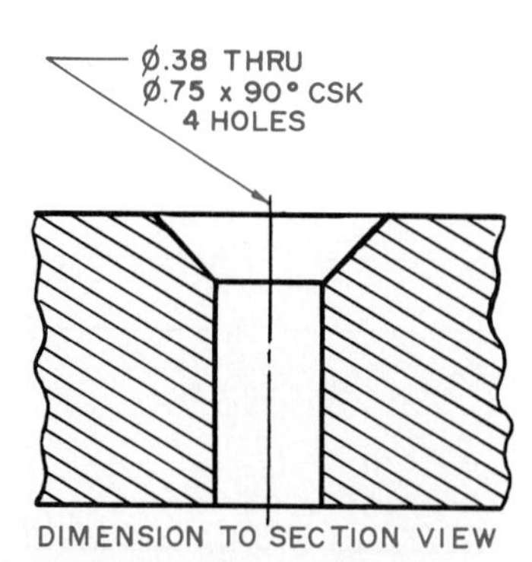

Fig. 7–70. Dimensioning countersunk holes.

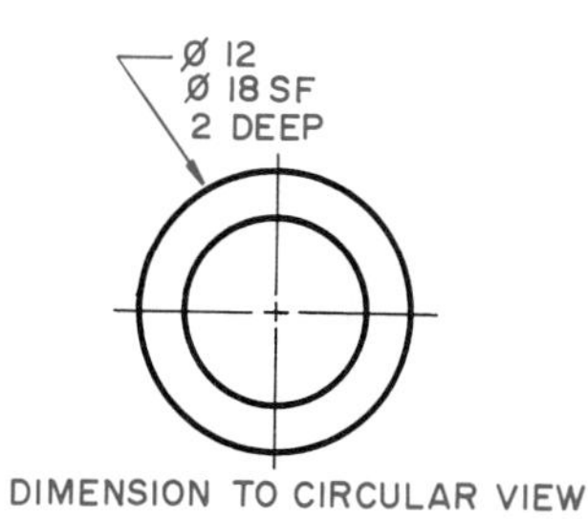

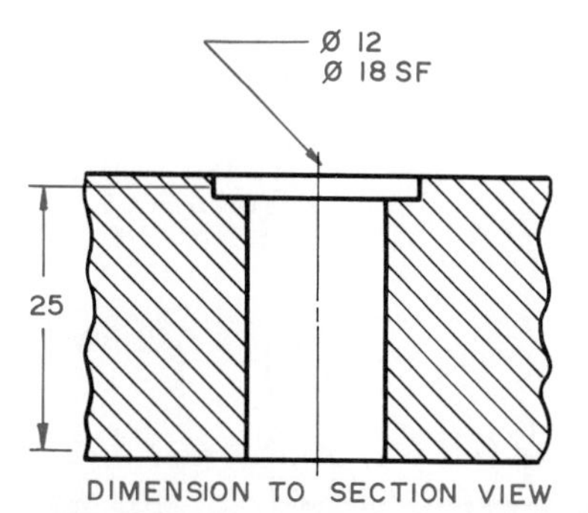

Fig. 7–71. Dimensioning spot-faced holes.

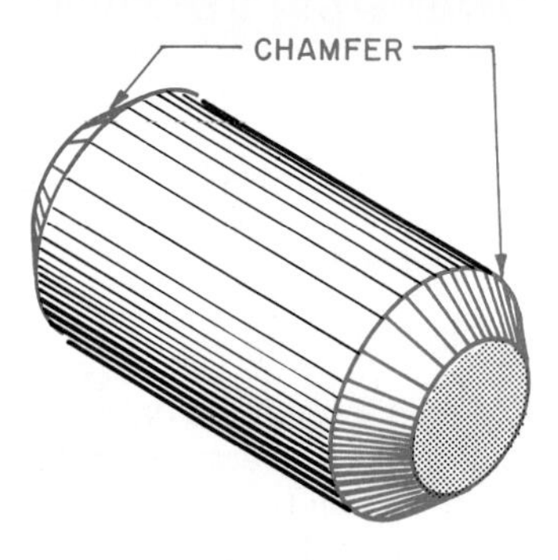

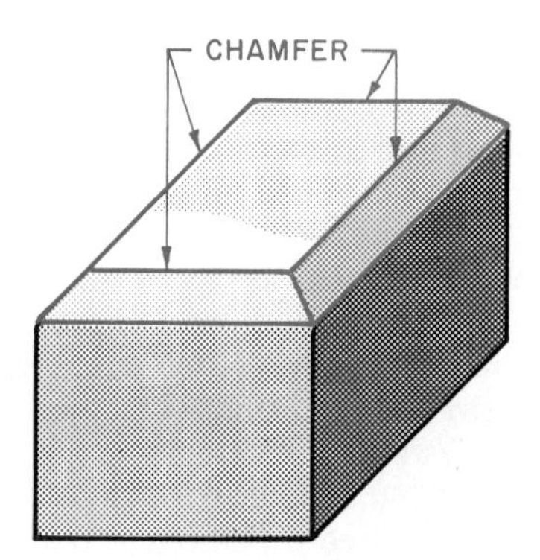

Fig. 7–72. **A chamfer is a narrow, flat, inclined surface which removes the sharp corner where two surfaces meet.**

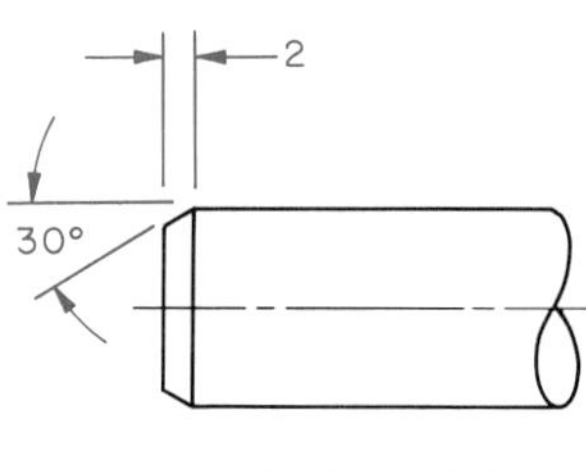

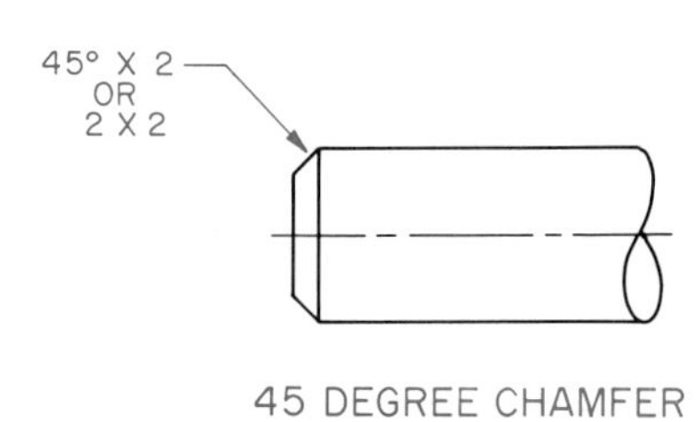

Fig. 7–73. **Dimensioning chamfers. (Reproduced from *ANSI Y14.5M–1982* with permission of the publisher, The American Society of Mechanical Engineers.)**

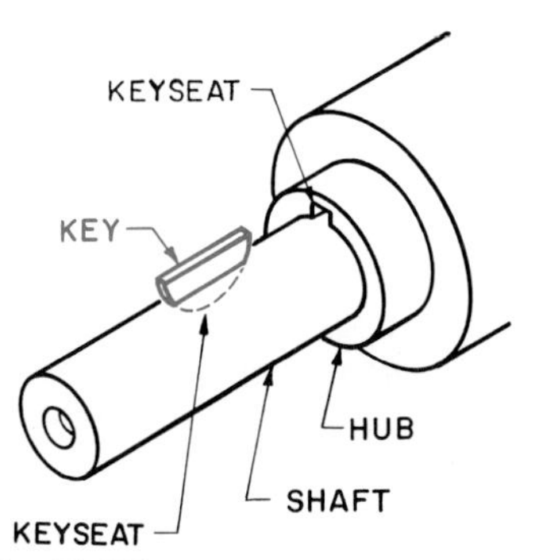

Fig. 7–74. **Keys are used to prevent rotation between a shaft and a hub.**

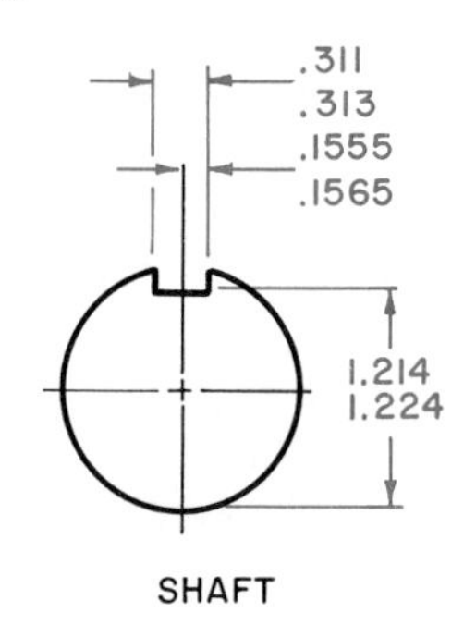

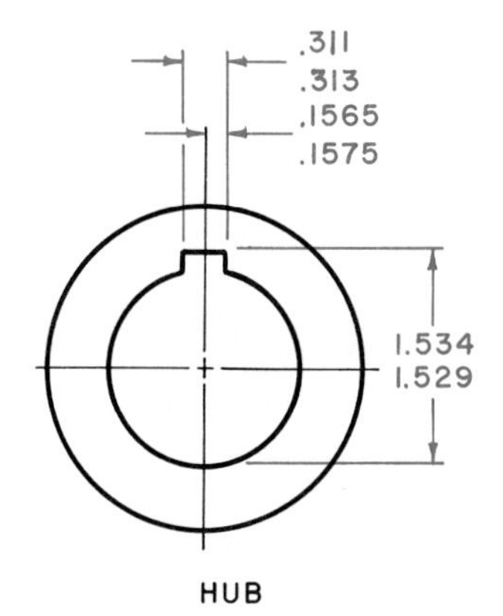

Fig. 7–75. **Dimensioning keyseats for stock keys.**

Fig. 7–77. **Knurling a pattern onto a metal piece.**

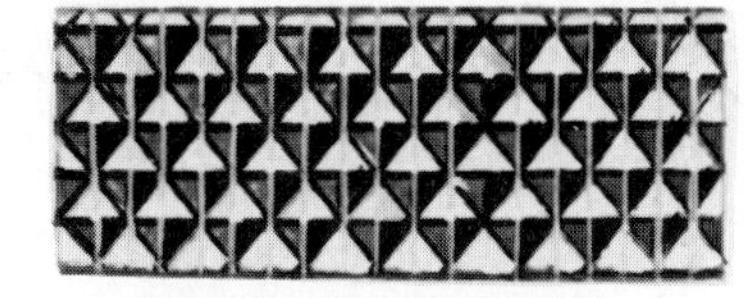

Fig. 7–78. **Common knurl patterns.**

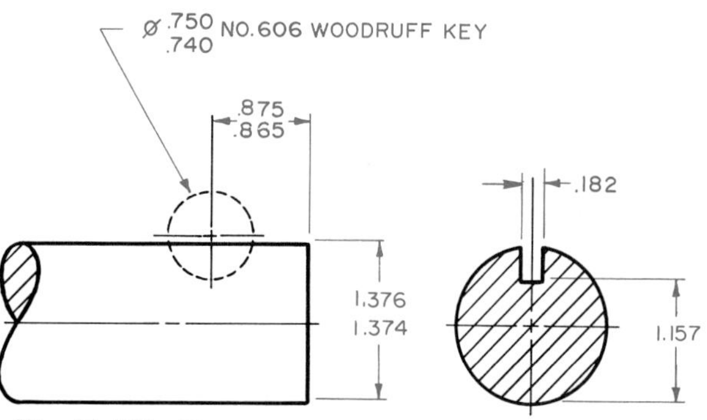

Fig. 7–76. **Dimensioning Woodruff keyseats.**

pressed fit between mating parts by including the following in a note: diameter before knurling, minimum diameter after knurling, pitch and type of knurl, and length of knurled area. Fig. 7-79 shows an example of a straight knurl that is to be used for a pressed fit.

Threads

Both metric and customary threads are used in the United States. Their specifications on drawings are detailed and require careful consideration. Complete information about threads and how to dimension them on drawings is found in Chapter 12, "Fasteners."

Common Dimensioning Errors

A drawing should contain all the dimensions needed to make the object. However, dimensions should not be repeated anywhere on the drawing. Information contained in notes should not be shown again as dimensions. The examples in Fig. 7-80 show common dimensioning mistakes:

- When there are a number of dimensions in a row, as at A, the last dimension is omitted. The overall dimension plus the other dimensions make the last one unnecessary.
- A cylinder is dimensioned on its rectangular view. A hole is dimensioned on its circular view as at B. The distance from the edge of a cylinder to the edge of a hole is not needed. The object is made by measuring from its center line.
- The diameter of a cylinder is properly given on the rectangular view at C.
- The center of the arc, at D, is located by the radius.
- The overall length and width are not needed to make the object shown at E. They may be given as reference dimensions.
- The center of the radius, shown at F, is located from the end of the object. The thickness of the object is established by the radii.
- At G it is clear that several identical features have the same dimension. You do not need to dimension each one.
- General information given in notes should not be repeated on a drawing as shown at H.
- The diameter of the part at I gives its width.
- Hole and thread notes give complete information and do not have to be repeated as shown at J.
- The angle, shown at K, needs only the two outside dimensions *or* the angle and one dimension.

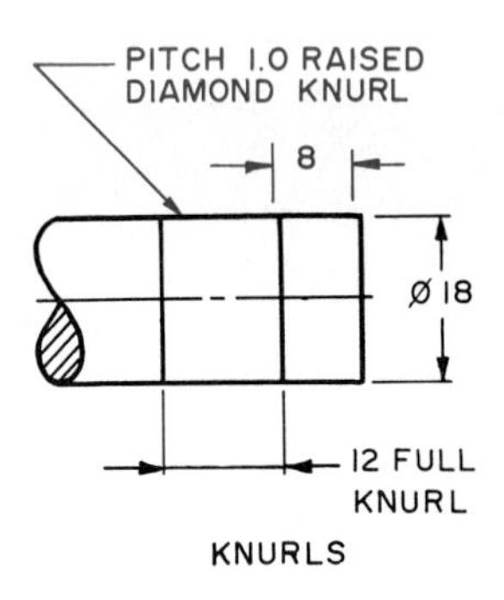

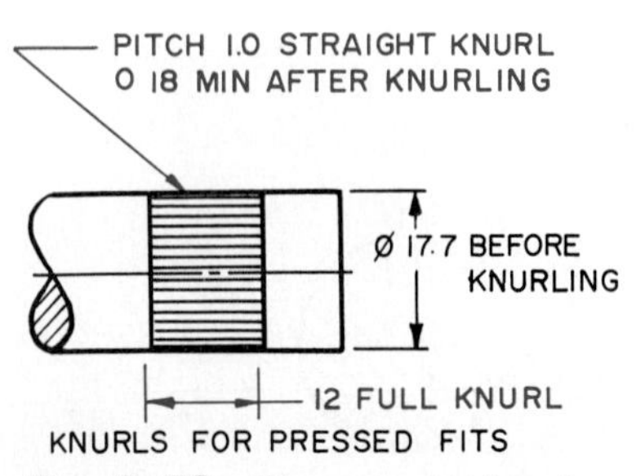

Fig. 7–79. Dimensioning knurls.

Tolerances and Limits

Limits are the maximum and minimum values given for a specific dimension. A **tolerance** represents the total amount by which a specific dimension may vary. Thus, the tolerance is the difference between the limits. Fig. 7-81.

Specifying Tolerances

Tolerances may be *unilateral* or *bilateral.* A unilateral tolerance allows a variation from the design size in one direction only (either plus or minus). A bilateral tolerance permits a variation from the design size in both plus and minus directions. Fig. 7-82.

When using customary dimensions, the limits or plus and minus tolerances must be given to the same number of decimal places as the dimension. Fig. 7-83.

The following rules apply to limits and tolerances used with metric dimensions as shown in Fig. 7-84:

- With unilateral tolerances, show a zero without a plus or minus sign or any decimal values.
- With bilateral tolerances, show the values to the same number of decimal places. Add zeros if necessary.
- With limit dimensioning, show the values to the same number of decimal places. Add zeros if necessary.

When you specify a bilateral tolerance, place the high limit above the low limit. If you show the tolerance on one line, give the low limit first followed by the high limit. Separate the two limits with a dash. Fig. 7-85. This is called limit dimensioning.

Tolerance Accumulation

The three methods of dimensioning, chain, base line, and direct, influence the accuracy of the finished product due to the possibility of the accumulation of tolerances. Fig. 7-86.

Chain dimensioning refers to running dimensions in strings with each one directly related to the one before it. The maximum variation from point X to Y in Fig. 7-86 at A is equal to the sum of

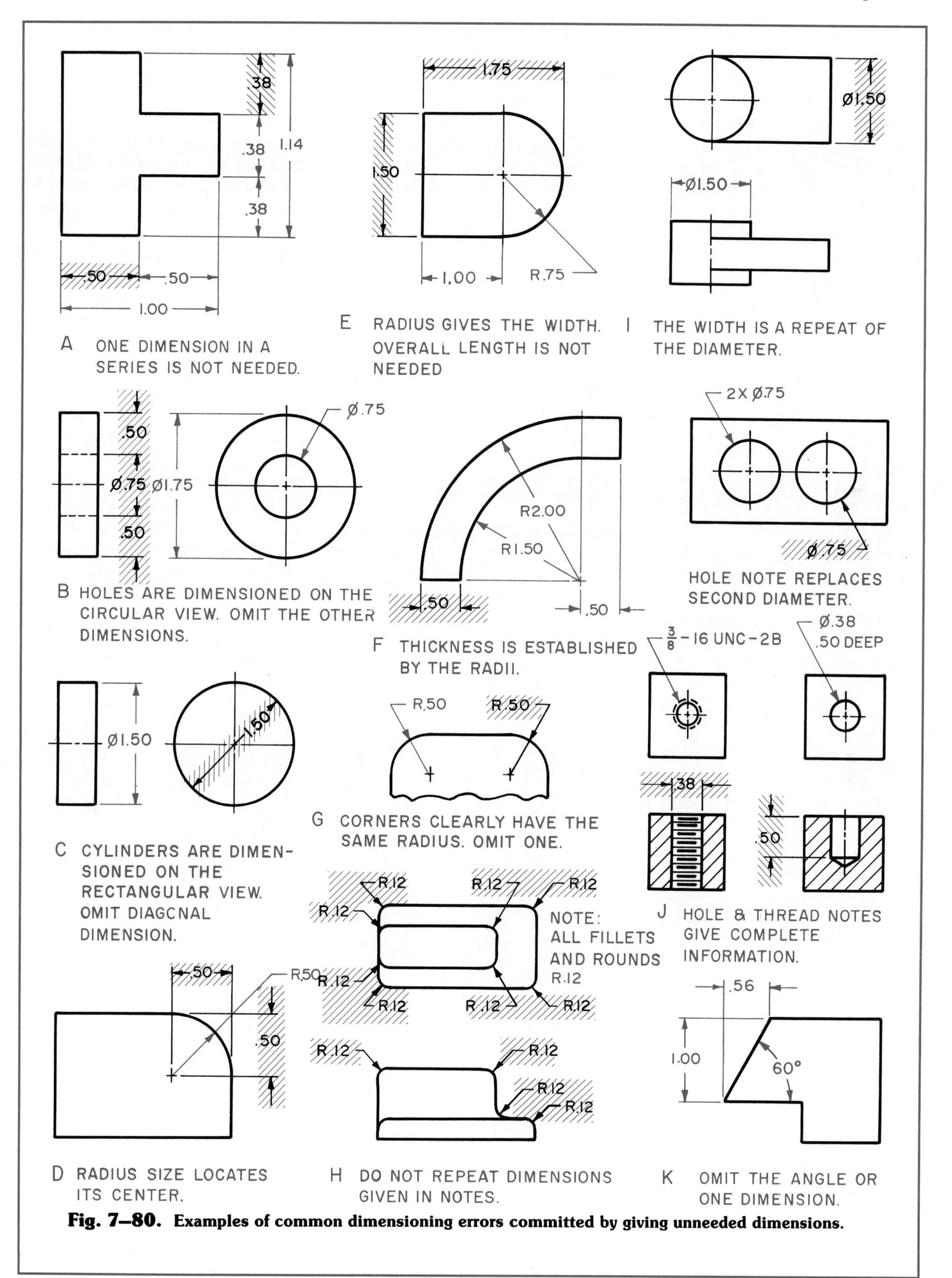

Fig. 7–80. Examples of common dimensioning errors committed by giving unneeded dimensions.

all the plus or all the minus tolerances on each dimension. In this case, the sum is 0.6 mm. This method of dimensioning produces the maximum tolerance accumulation.

With *base line dimensioning,* you show all dimensions from the same surface, called a base line. The maximum variation between points X and Y, shown at B, is equal to the sum of the tolerances of the two dimensions from the base line. In this example, the tolerance from X to Y is 0.4 mm. The X dimension can vary 0.2 mm in one direction and the Y dimension can vary the same amount in the other direction. This method of dimensioning results in less tolerance accumulation.

Direct dimensioning controls the variation between points X and Y by dimensioning and tolerancing the distance between the two points. Notice that the

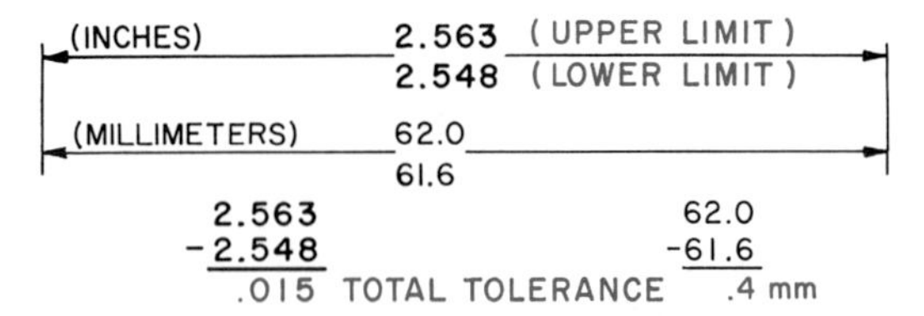

Fig. 7–81. Tolerance is the difference between the upper and lower limits. It is the total amount of variation permitted in a piece.

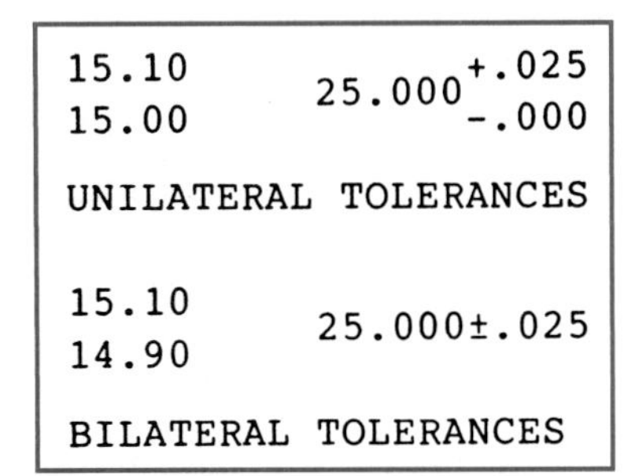

Fig. 7–82. Examples of unilateral and bilateral tolerances.

.650 .648	.75±.01
$.350^{+.005}_{-.000}$	15.0°±.5°

Fig. 7–83. The correct way to specify tolerances in inches.

$40^{\ 0}_{-0.02}$	$40^{+0.02}_{\ 0}$
UNILATERAL	
$30^{+0.15}_{-0.10}$	15.35 15.00
BILATERAL	LIMITS

Fig. 7–84. The correct way to specify tolerances in millimeters.

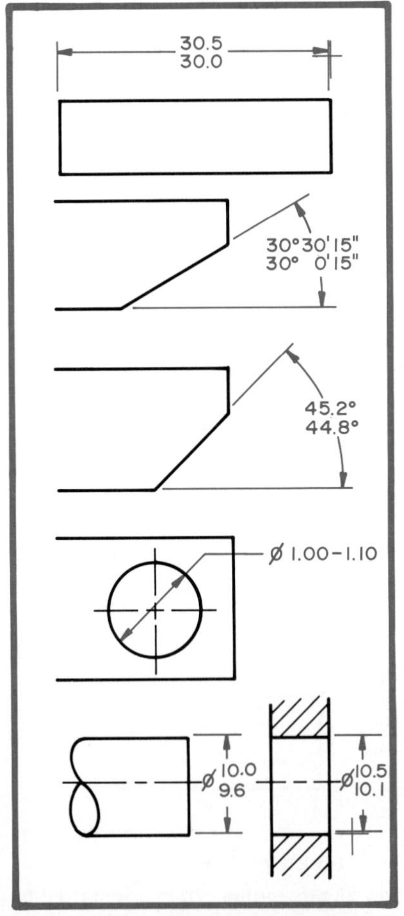

Fig. 7–85. Applications of limit dimensioning.

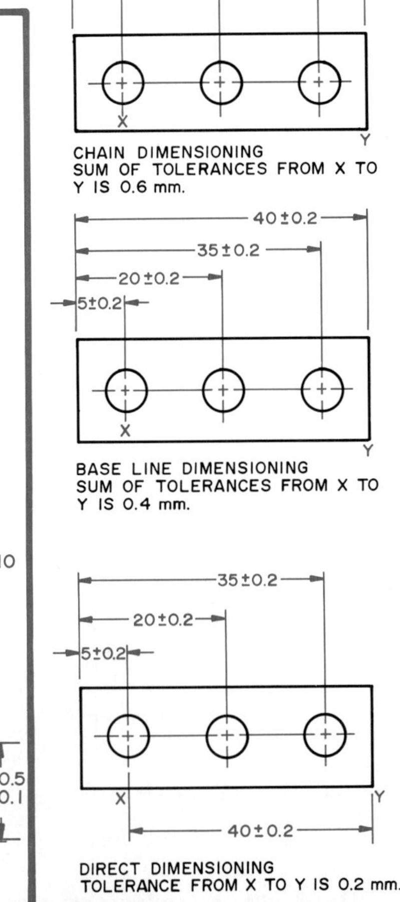

Fig. 7–86. Dimensioning methods influence the accumulation of tolerances.

tolerance of point Y is related to point X only. Any variation of the location of point X does not influence the distance between X and Y.

Tolerances and Tapers

Conical Tapers

A **taper** is the ratio of the difference in the diameter of two sections taken perpendicular to the axis of a cone to the distance between these two sections. For example, if two sections are 1 inch apart and one has a diameter of 2.00 inches while the other has a diameter of 1.90 inches, the taper is .10 inch per inch.

You specify a conical taper by one of the following means:

- A basic taper and a basic diameter.
- A toleranced length and a toleranced diameter at both ends of the taper.
- A size tolerance and a profile of the surface tolerance applied to the taper. This dimensioning technique for a taper is discussed further in Chapter 8, "Geometric Tolerancing."

The basic taper and basic diameter specification is shown in Fig. 7-87. The basic diameter controls the size of the tapered section. It is located from one end of the tapered part. The basic taper is shown as a note. In Fig. 7-87 the taper is .01 inch per inch of length. Notice the symbol used to indicate the conical taper.

Tapers that are not critical are dimensioned by a toleranced diameter at both ends plus a toleranced length dimension. Fig. 7-88.

Flat Tapers

Dimension flat tapers by showing a toleranced taper on its slope and a toleranced height at one end. Fig. 7-89. *Slope* is the ratio of the difference in heights at each end of the inclined surface to the distance between those heights. Notice the use of the flat taper symbol on the note showing the size of the slope.

Tolerances of Angular Surfaces

Angular surfaces can be specified in either of the following ways:

- With a toleranced linear dimension and angle.
- With a basic angle and a toleranced linear dimension.

When a toleranced linear dimension and angle are given, the angular surface must be within a tolerance zone represented by two nonparallel planes. The tolerance zone gets wider the farther it gets from the apex of the angle. Fig. 7-90A.

When a basic angle and a toleranced linear dimension are used, the tolerance zone will be parallel with the boundaries established by the tolerances. Fig. 7-90B.

Limits and Fits for Cylindrical Parts

The designer must specify the type of fit between cylindrical parts and the tolerances on them. The type of fit chosen will depend upon the function of the mating parts. Some mating parts must be very loose with a shaft that is considerably smaller than the hole it fits in. Other mating parts may need to fit so tight that the shaft is slightly larger than the hole and must be forced into it. The following sections will present information about the standard fits for customary and metric sizes.

Definitions

Allowance is the difference between the maximum material limits of mating parts. It is the maximum clearance or interference between mating parts.

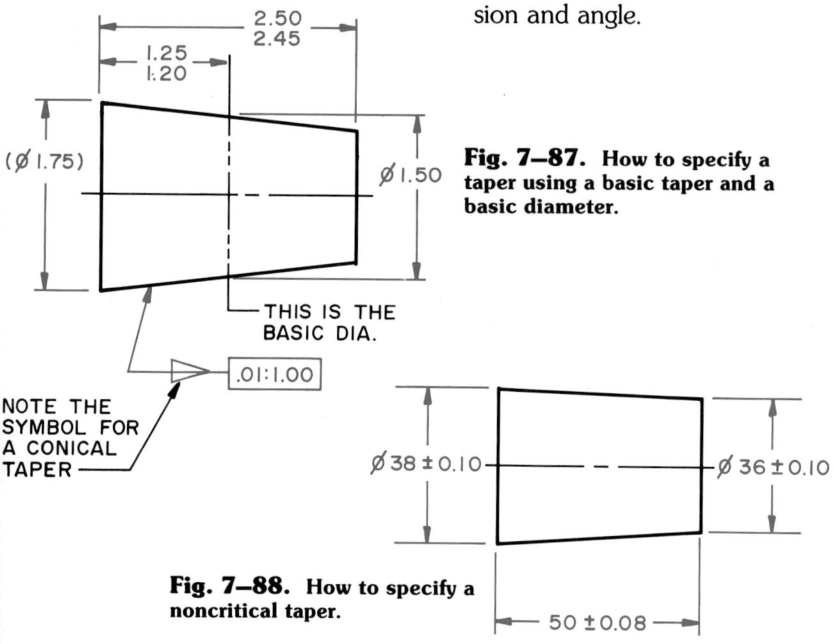

Fig. 7–87. How to specify a taper using a basic taper and a basic diameter.

Fig. 7–88. How to specify a noncritical taper.

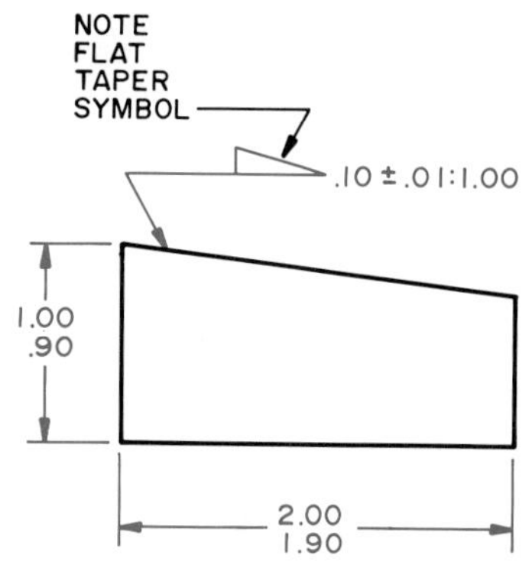

Fig. 7–89. Tolerancing a flat taper.

Maximum Material Condition, MMC, means a feature of a part contains the maximum material possible within the stated limits of size. For example, the largest acceptable shaft diameter is at its maximum material condition. The maximum material condition of a negative feature, such as a hole, is its lowest limit, that is, when it is at its smallest design size.

Least Material Condition, LMC, means a feature of a part contains the least material possible within the stated limits of size. It is the smallest diameter shaft or largest permissible hole.

Basic size is the size from which the limits of size are taken by applying the allowances and tolerances. Fig. 7-91.

Design size is the basic size plus an allowance that may be specified. If there is no allowance, the design size and the basic size are the same.

Clearance fits are those in which an internal member fits into an external member with space between the parts. An example is shown in Fig. 7-92. In it the smallest hole is .5000 and the largest shaft is .4994. The space allowed between them is .0006.

With an *interference fit* the internal member is larger than the external member. Fig. 7-92. In this case the shaft is larger than the hole. When they are brought together, the shaft is forced by pressure into the hole.

A *transitional fit* is one that could have either clearance or interference. When these parts are assembled, some will have clearance and some interference. Fig. 7-92.

Basic Hole System

Tolerance dimensions are usually based on the **basic hole system.** The minimum hole is the basic size. The allowance to be used is subtracted from the basic size to give the largest shaft size. Tolerances are applied to both. For example, if the basic size is 1.000 inch and the allowance is .007, the largest shaft is .993 inch and the smallest hole is 1.000 inch. Tolerances are applied to these sizes to find the largest acceptable hole and the smallest acceptable shaft diameters. The tolerances to be used are found in Appendices 40 through 48.

The basic hole system is the one generally used because many of the standard drills, reamers, and other machine tools are designed to produce standard hole sizes. Shafts are turned to fit standard holes. Fig. 7-93.

Basic Shaft System

Shafts are manufactured in standard diameters so there is an advantage in some cases to apply tolerances using the shaft diameter as the basic size. In the **basic shaft system** the largest diameter of the shaft is used as the basic size to which allowances and tolerances are applied. For example, if the basic size is 1.5 inches, the allowance is added to this amount to get the smallest hole diameter. If the allowance is .004 inch, the smallest hole diameter would be 1.504 inches. Tolerances are applied to these sizes to find the largest acceptable hole and smallest acceptable shaft diameters. Fig. 7-93.

American National Standard Limits and Fits

Standard classes of fits have been designated to insure interchangeability of parts. The type of fit is determined by the requirements of the piece. Standard classes of fits are the following:

- RC Running or sliding
- LC Locational clearance
- LT Locational transitional
- LN Locational interference
- FN Force or shrink

These letter symbols are used with numbers to represent the class of fit. A designation, "FN4," means a Class 4 force fit. When dimensioning, do not put the symbols and numbers on the drawing. Use the limits that you obtain from the tables to size the piece.

The classes of fits are arranged in three general groups: running and sliding, locational, and force.

Running and Sliding Fits (RC)

Running and sliding fits are intended to provide a running performance with suitable lubrication allowance in all ranges of size. The clearances of the first two classes, used chiefly as sliding fits, increase more slowly with diameter than other classes.

RC1—Close sliding fits are intended to locate accurately those parts that, when assembled, have no noticeable movement, or play.

RC2—Sliding fits are intended to locate accurately parts that are to fit, move, and turn easily but not intended to run freely. In larger sizes, this fit may stick fast due to a temperature change.

RC3—Precision running fits are about the closest fits that can be expected to run freely. These are intended for precision work at slow speeds and light journal pressures. The RC3 fit is not suitable when large temperature differences are likely to occur.

RC4—Close running fits are intended chiefly for running fits on accurate machinery with moderate surface speeds and journal pressures. These are used where accurate location and minimum play is desired.

RC5 and RC6—Medium running fits are intended for higher running speeds and/or heavy journal pressures.

RC7—Free running fits are used where accuracy is not essential or where large temperature variations are likely to occur.

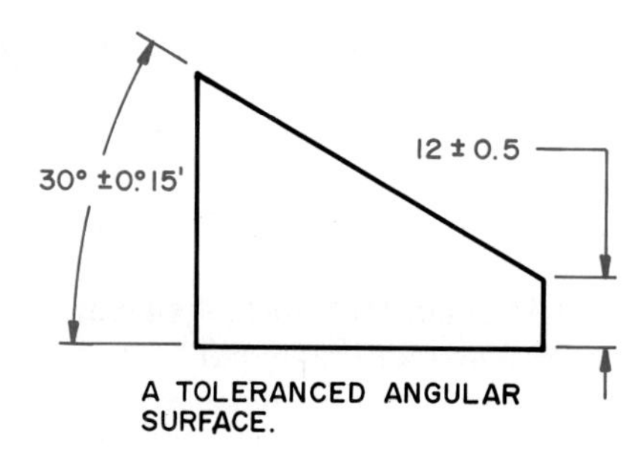

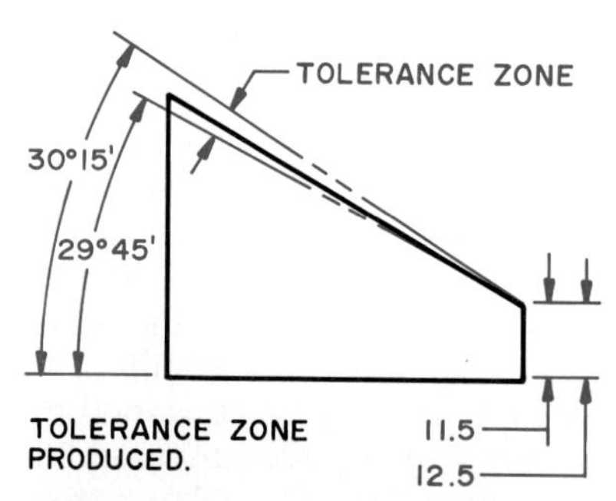

Fig. 7–90A. Tolerancing an angular surface with an angular tolerance zone.

Fig. 7-91. Preferred Customary Basic Sizes for Round Shafts and Holes (Decimal Inches)

0.010	2.00	8.50
0.012	2.20	9.00
0.016	2.40	9.50
0.020	2.60	10.00
0.025	2.80	10.50
0.032	3.00	11.00
0.040	3.20	11.50
0.05	3.40	12.00
0.06	3.60	12.50
0.08	3.80	13.00
0.10	4.00	13.50
0.12	4.20	14.00
0.16	4.40	14.50
0.20	4.60	15.00
0.24	4.80	15.50
0.30	5.00	16.00
0.40	5.20	16.50
0.50	5.40	17.00
0.60	5.60	17.50
0.80	5.80	18.00
1.00	6.00	18.50
1.20	6.50	19.00
1.40	7.00	19.50
1.60	7.50	20.00
1.80	8.00	

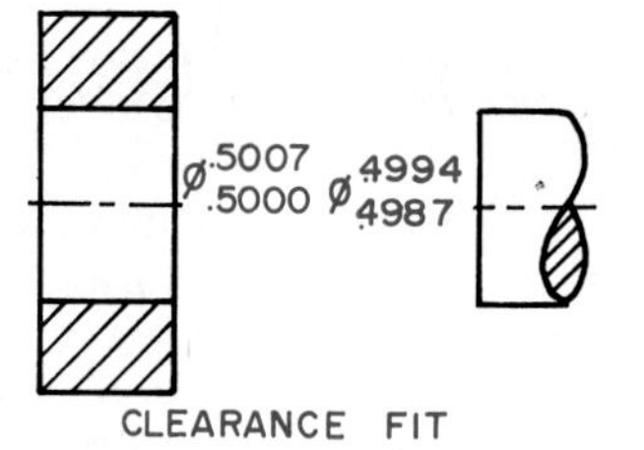

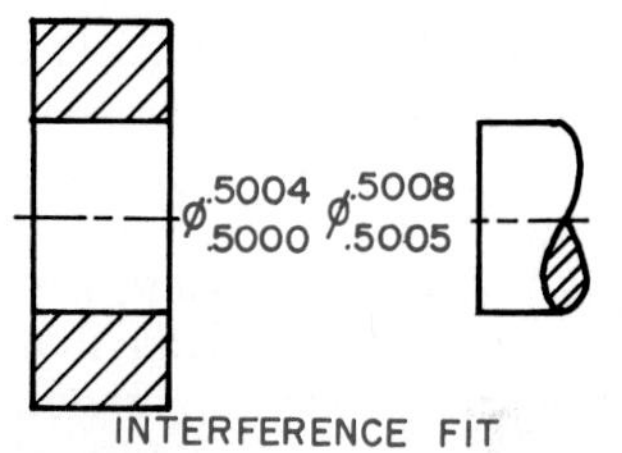

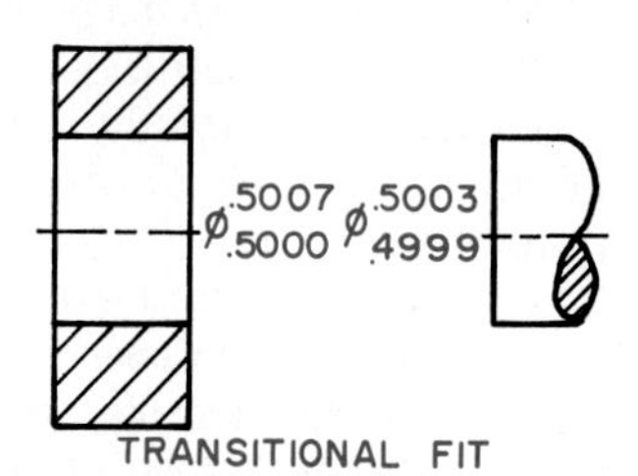

Fig. 7–92. Clearance, interference, and transitional fits.

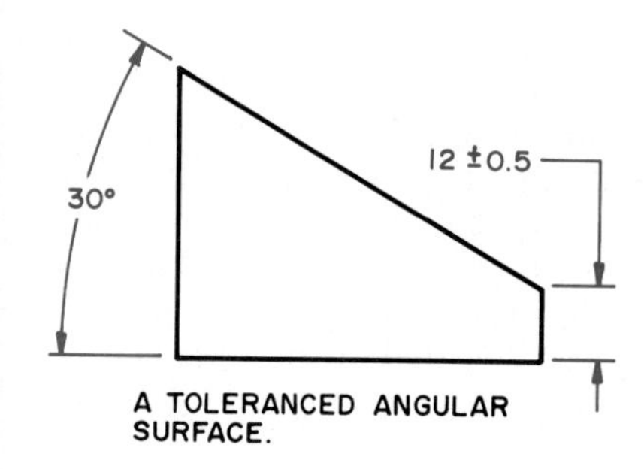

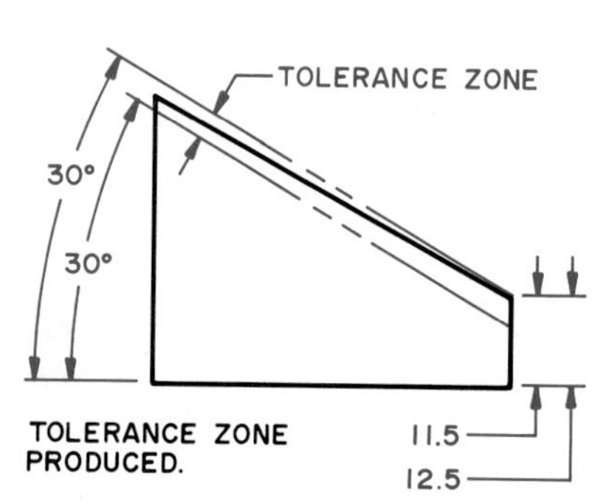

Fig. 7–90B. Tolerancing an angular surface with a parallel tolerance zone.

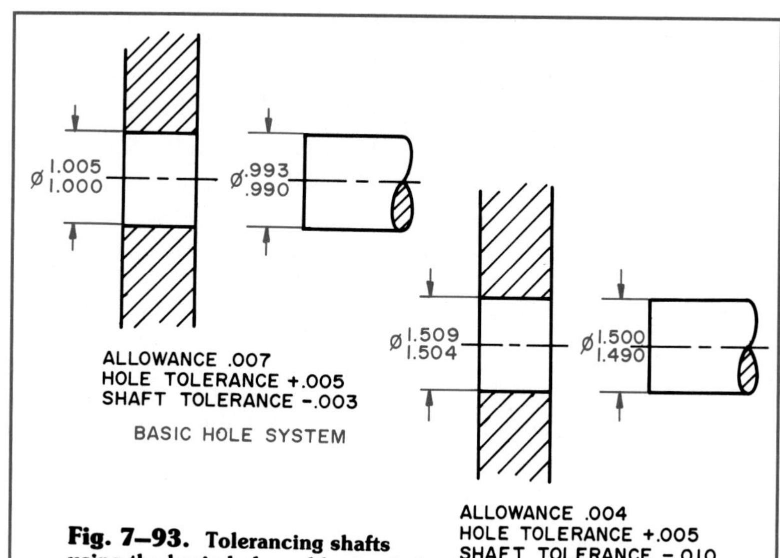

Fig. 7–93. Tolerancing shafts using the basic hole and basic shaft systems.

RC8 and RC9—Loose running fits are used when using materials such as cold rolled shafting and tubing made to commercial standards.

Locational Fits (LC, LT, and LN)

Locational fits are intended to determine only the location of mating parts. They may provide rigid or accurate location, as with interference fits, or provide some freedom of location, as with clearance fits. Locational fits are divided into three groups: clearance (LC), transitional (LT), and interference (LN).

LC—Locational clearance fits are intended for parts that are normally stationary. These fits can be freely assembled or disassembled. They range from snug fits, through medium-clearance fits, to loose fastener fits when freedom of assembly is of prime importance.

LT—Locational transitional fits are a compromise between clearance and interference fits. These are used when accuracy of location is important but either a small amount of clearance or interference is permissible.

LN—Locational interference fits are used where accuracy of location is of prime importance. In addition, these fits are used for parts requiring rigidity and alignment with no requirements for bore pressure. These fits are not intended for parts designed to transmit frictional loads from one part to another. By virtue of tightness of fit, these latter conditions are covered by force fits.

Force Fits (FN)

Force fits or shrink fits are a special type of interference fit. These are characterized by maintenance of constant bore pressure throughout the range of sizes. The interference varies almost directly with the diameter. The difference between these values is small to maintain the resulting pressures within reasonable limits.

FN1—Light drive fits are those that require light assembly pressure and produce more or less permanent assemblies. These are suitable for thin sections, long fits, or cast iron external members.

FN2—Medium drive fits are suitable for ordinary steel parts or for shrink fits on light sections. These are about the tightest fits that can be used with high-grade cast iron external members.

FN3—Heavy drive fits are suitable for heavy steel parts or for short fits in medium sections.

FN4 and FN5—Force fits are suitable for parts that can be highly stressed or for shrink fits when heavy pressing forces required are impractical.

For interchangeable manufacture, the tolerances on dimensions are such that an acceptable fit will result from an assembly of parts having any combination of actual sizes that are within tolerances. Fig. 7-94.

Using American National Tables of Limits and Fits

Tables for American National Limits and Fits are found in Appendices 40 through 44. These tables are based on the basic hole system.

How to Determine the Correct Fit

Most limit dimensions are figured on the basic hole system. The nominal size of the hole is converted to the basic theoretical hole size. The standard limits are then added or subtracted for the hole and shaft size. In the tables of standard fits the nominal size ranges are expressed in inches along the left-hand column. To conserve space, all other values (limits of tolerances and limits for the hole and shaft) are expressed in thousandths of an inch. For example, 2.0 would be read as two-thousandths (.002) of an inch.

The following example will show how to find the limits of a hole with a diameter of 1.7500 and an RC3 fit:

1. Locate the table of running and sliding fits. Fig. 7-95.
2. Find the nominal size range of the hole and shaft. It is the sixth row down since 1.7500 falls between 1.19 and 1.97 and the third major column across headed Class RC3.
3. Find the limits given for the hole (+ 0.6 and 0) and shaft (−1.0 and −1.6) and add these to and subtract them from the basic size. Remember, these figures are given in thousandths of an inch. They may be converted by multiplying by one-thousandth. Therefore: $0.6 \times .001 = .006$.

Hole dimension:

$$\frac{1.7500 + .0006}{1.7500 + .0000} = \frac{1.7506}{1.7500}$$

Shaft dimension:

$$\frac{1.7500 - .0010}{1.7500 - .0016} = \frac{1.7490}{1.7484}$$

Under the heading, "Limits of Clearance," for a 1.7500-inch RC3 fit, the values 1.0 and 2.2 are given. These are the minimum and maximum values of clearance between the hole and shaft. The minimum clearance (smallest hole and largest shaft) is .0010 inch. The maximum clearance (largest hole and smallest shaft) is .0022 inch.

ISO System for Limits and Fits

The ISO system of limits and fits is a standard series of tolerances for mating parts. The tables for selecting these fits are found in Appendices 45 through 48. The tables refer to the "hole" and "shaft." These tolerances are used for mating holes and shafts and to select tolerances for space contained by two parallel faces of any part, such as the width of a keyway or size of a key.

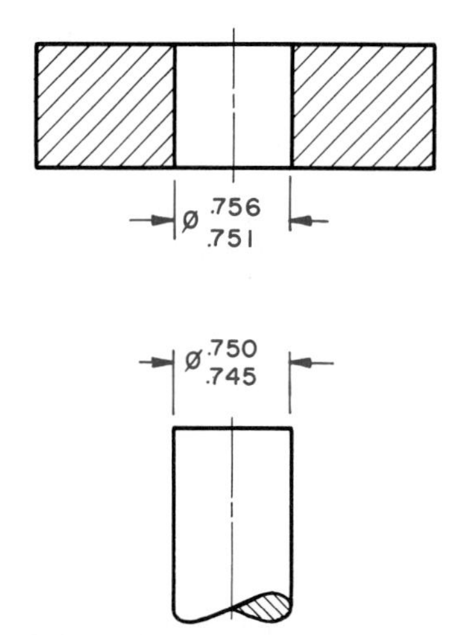

Fig. 7–94. The size of parts designed for interchangeability with a clearance fit. The largest shaft is .750 inch and the smallest hole is .751 inch. This combination gives an allowance of .001 inch. The smallest shaft is .745 inch and the largest hole is .756 inch. This combination gives a maximum clearance of .009 inch.

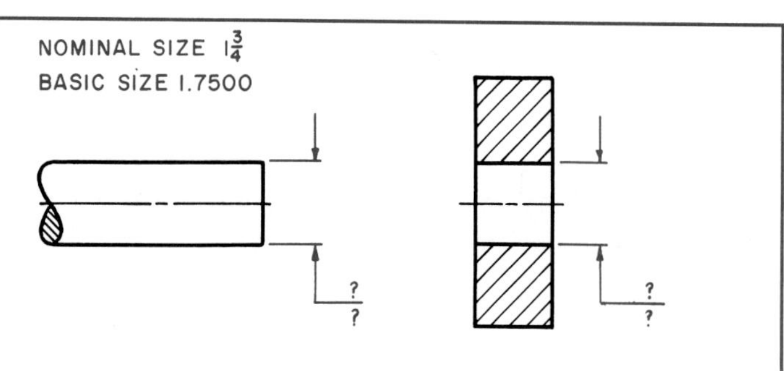

PROBLEM: FIND THE LIMITS IF AN RC-3 FIT IS DESIRED.

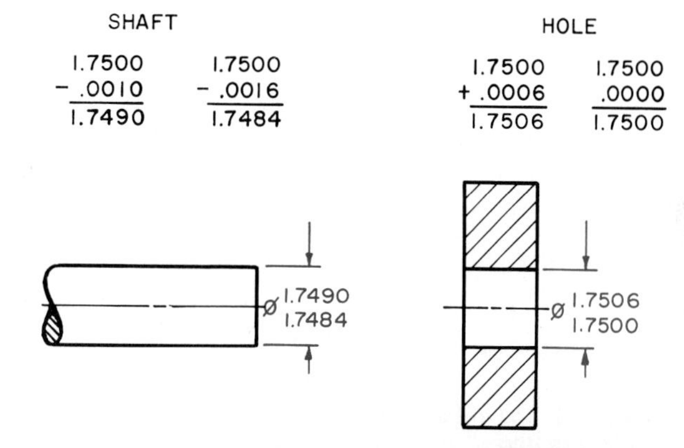

Nominal Size Range Inches	Class RC 1			Class RC 2			Class RC 3		
	Limits of Clearance	Standard Limits		Limits of Clearance	Standard Limits		Limits of Clearance	Standard Limits	
Over To		Hole H5	Shaft g4		Hole H6	Shaft g5		Hole H6	Shaft f6
0.04–0.12	0.1 0.45	+0.2 0	−0.1 −0.25	0.1 0.55	+0.25 0	−0.1 −0.3	0.3 0.8	+0.25 0	−0.3 −0.55
0.12–0.24	0.15 0.5	+0.2 0	−0.15 −0.3	0.15 0.65	+0.3 0	−0.15 −0.35	0.4 1.0	+0.3 0	−0.4 −0.7
0.24–0.40	0.2 0.6	+0.25 0	−0.2 −0.35	0.2 0.85	+0.4 0	−0.2 −0.45	0.5 1.3	+0.4 0	−0.5 −0.9
0.40–0.71	0.25 0.75	+0.3 0	−0.25 −0.45	0.25 0.95	+0.4 0	−0.25 −0.55	0.6 1.4	+0.4 0	−0.6 −1.0
0.71–1.19	0.3 0.95	+0.4 0	−0.3 −0.55	0.3 1.2	+0.5 0	−0.3 −0.7	0.8 1.8	+0.5 0	−0.8 −1.3
1.19–1.97	0.4 1.1	+0.4 0	−0.4 −0.7	0.4 1.4	+0.6 0	−0.4 −0.6	1.0 2.2	+0.6 0	−1.0 −1.6

Fig. 7–95. How to find limits using the standard tables. (The table shown here is an excerpt from Appendix 40.)

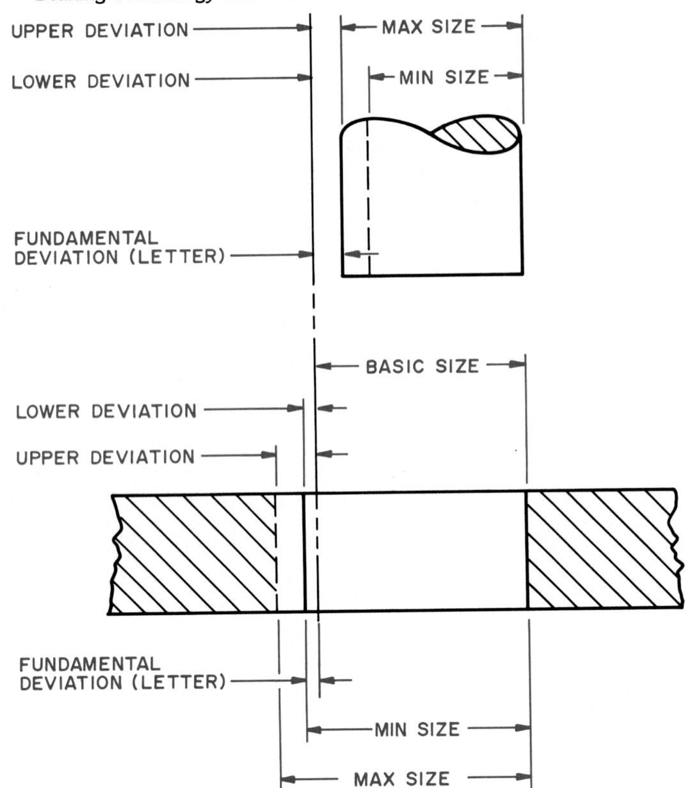

Fig. 7–96. Definitions of the ISO system of limits and fits.

Definitions

The following definitions are illustrated in Fig. 7-96:

Basic size is the size to which limits or deviations are applied. The basic size is the same for both mating members. It is the number "50" in the tolerance symbol "50H11."

Deviation is the difference between a specified size and its basic size.

Upper deviation is the difference between the maximum limit size and the basic size.

Lower deviation is the difference between the minimum limit size and the basic size.

Fundamental deviation is the upper or lower deviation that is closest to the basic size. It is specified by the letter "H" in the tolerance symbol "50H11."

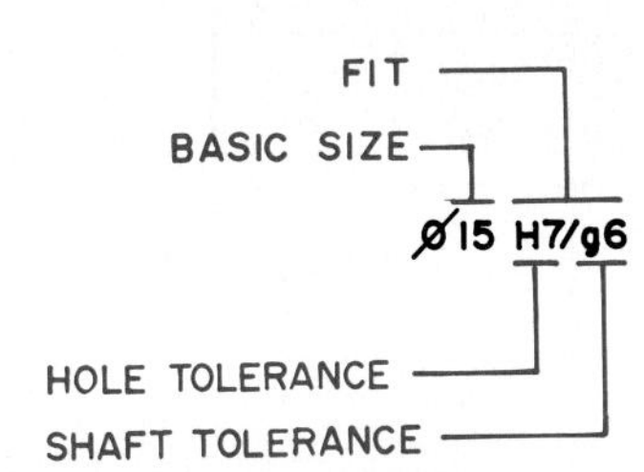

Fig. 7–97. A metric tolerance symbol.

Fig. 7-98. Preferred Metric Basic Sizes (Millimeters) for Round Shafts and Holes

FIRST CHOICE	SECOND CHOICE	FIRST CHOICE	SECOND CHOICE	FIRST CHOICE	SECOND CHOICE
1	1.1	10	11	100	110
1.2	1.4	12	14	120	140
1.6	1.8	16	18	160	180
2	2.2	20	22	200	220
2.5	2.8	25	28	250	280
3	3.5	30	35	300	350
4	4.5	40	45	400	450
5	5.5	50	55	500	550
6	7	60	70	600	700
8	9	80	90	800	900
				1000	

(ANSI B4.2–1978)

Hole basis is a system of fits where the minimum hole is the basic size. The fundamental deviation for the hole basis system is "H."

Shaft basis is a system of fits where the maximum shaft size is basic. The fundamental deviation for the shaft basis system is "h."

Clearance fit means assembled parts have a clearance under all tolerance conditions.

Interference fit means assembled parts have an interference under all tolerance conditions.

Transition fit means assembled parts may have either a clearance or interference fit depending upon the tolerance conditions of the mating parts.

Tolerance Symbols

Tolerance is expressed using a tolerance symbol. It specifies the basic size that is common to both parts and the fit of the hole and the shaft. The first fit symbol represents the internal part (hole) and the second represents the external part (shaft). Fig. 7-97.

Preferred Basic Sizes

When selecting the sizes of mating parts, choose from the table of Preferred Basic Sizes. Fig. 7-98.

Preferred Metric Fits

Fig. 7-99 shows the ten preferred hole basis and shaft basis fits. These range from a loose running fit to a force fit. Notice that with the hole basis fit the position letter is a capital H. This is the basic size. With the shaft basis fit the position letter is a lowercase h. This is the basic size.

Using Preferred Metric Limits and Fits

The tables of Preferred Metric Limits and Fits are found in Appendices 45 through 48. The following example will show how to determine the limits for shaft and a hole that are to mate using the

Fig. 7–99. Preferred Fits in ISO Metric System of Limits and Fits

	ISO SYMBOL			
	Hole Basis	Shaft[1] Basis	DESCRIPTION	
Clearance Fits	H11/c11	C11/h11	Loose running fit for wide commercial tolerances or allowances on external members.	More Clearance
Clearance Fits	H9/d9	D9/h9	Free running fit not for use where accuracy is essential, but good for large temperature variations, high running speeds, or heavy journal pressures.	More Clearance
Clearance Fits	H8/f7	F8/h7	Close running fit for running on accurate machines and for accurate location at moderate speeds and journal pressures.	More Clearance
Clearance Fits	H7/g6	G7/h6	Sliding fit not intended to run freely, but to move and turn freely and locate accurately.	More Clearance
Transition Fits	H7/h6	H7/h6	Locational clearance fit provides snug fit for locating stationary parts; but can be freely assembled and disassembled.	
Transition Fits	H7/k6	K7/h6	Locational transition fit for accurate location, a compromise between clearance and interference.	
Transition Fits	H7/n6	N7/h6	Locational transition fit for more accurate location where greater interference is permissible.	More Interference
Interference Fits	H7/p6	P7/h6	Locational interference fit for parts requiring rigidity and alignment with prime accuracy of location but without special bore pressure requirements.	More Interference
Interference Fits	H7/s6	S7/h6	Medium drive fit for ordinary steel parts or shrink fits on light sections, the tightest fit usable with cast iron.	More Interference
Interference Fits	H7/u6	U7/h6	Force fit suitable for parts which can be highly stressed or for shrink fits where the heavy pressing forces required are impractical.	More Interference

(ANSI B4.2-1978)

[1]The transition and interference shaft basis fits shown do not convert to exactly the same hole basis fit conditions for basic sizes in range from 0 through 3 mm. Interference fit P7/h6 converts to a transition fit H7/p6 in the above size range.

hole basis system. The hole has a diameter of 16 mm and the fit is to be loose running.

Part of Appendix 45 showing the preferred hole basis limits and fits is shown in Fig. 7-100. Find the row into which the basic size, 16, falls. Follow this row to the right to the "Loose Running" column and read the limits. The upper limit for the hole is 16.110. The lower limit is the basic size, 16.000. Next move to the "shaft" column to find the upper shaft limit of 15.905 and the lower shaft limit of 15.795. The fit has a maximum clearance of 0.325 and a minimum clearance of 0.095. These clearances are shown in the "Fit" column to the right of the shaft data.

The tolerance symbol for this hole is written ∅16H11/c11.

Finished Surfaces

There are many times when a part must have a smooth surface in order to fit tightly against another surface. To get the surface smooth, some material must be removed as the surface is finished. Finishing the surface can be done by any of several processes including grinding, shaping, milling, lapping, and turning.

On a drawing, place a **finish mark** on the edge view of every surface that is to be finished. There are two finish marks commonly used and each is a variation of the letter "V." A simple V mark can be used for general purposes. Fig. 7-101. It does not indicate the surface finish desired but simply means that the surface produced by normal manufacturing methods is satisfactory. Using this mark alone is not a good practice. A maximum value for roughness should be specified with a note along with the V.

If an object is to be finished on all surfaces, letter the note "FINISH ALL OVER" on the drawing. Since this general note does not give any indication about the quality of the surface finish, include a maximum acceptable roughness value.

The condition of the finished surface wanted is usually indicated by a surface texture symbol. This symbol is a V with the right leg longer than the left.

Surface Texture

When you must state the quality of the finish of a part, you must use an exact method of giving the information. In general, the quality of finish depends upon the height, width, and direction of surface irregularities. The difference between a rough and smooth surface is determined by the grade of abrasive material used to produce the finish. A coarse abrasive material will produce a rough surface texture because the larger size of the abrasive particles make relatively deep cuts that are widely spaced. By comparison, a fine finish is composed of more surface irregularities that are not as deep. Fig. 7-102. The fine finish is obtained by using finer abrasive particles.

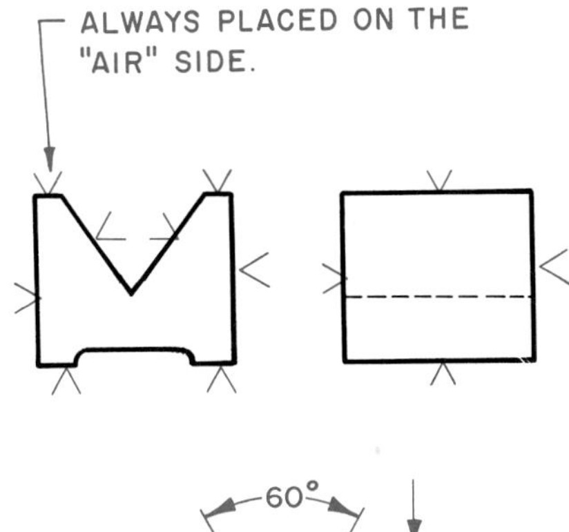

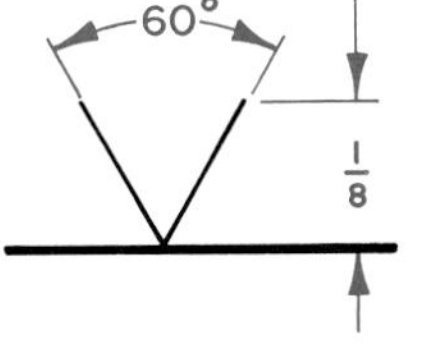

NOTE:
THE FINISH MARK IS DRAWN MECHANICALLY WITH A 30-60° TRIANGLE.

Fig. 7-101. The general purpose finish mark.

ROUGH

FINE

Fig. 7-102. A comparison of a rough and a fine finish, greatly magnified.

Fig. 7-100. Preferred Hole Basis Clearance Fits

BASIC SIZE		LOOSE RUNNING Hole H11	Shaft c11	Fit
12	MAX	12.110	11.905	0.315
	MIN	12.000	11.795	0.095
16	MAX	16.110	15.905	0.315
	MIN	16.000	15.795	0.095
20	MAX	20.130	19.890	0.370
	MIN	20.000	19.760	0.110

(ANSI B4.2–1978)

Use a **surface texture symbol** to identify the surface to be finished and give the information about what is to be done to finish it. Fig. 7-103. The basic surface texture symbol is used to specify roughness. If other texture values are needed, draw the symbol with a horizontal bar on top of the right side of the V. Close the V with a bar if material removal by machining is required. Show a material removal allowance to the left of the symbol. If you put a circle inside the V, it means that material removal is prohibited.

The size of the symbol relates to the height of your lettering. The symbol height should be three times the height of your letters as shown in Fig. 7-104.

Place the point of the symbol on the edge view of the surface to be finished. You may place it on an extension line of the surface to be finished or direct it to the surface with a leader. Draw the long leg to the right as the drawing is read. Fig. 7-105. Place a surface texture symbol for any surface in one place only. Do not repeat it in another view or on a section.

When using the symbol, indicate only those values that are needed to specify the required surface texture characteristics. Those characteristics include roughness, waviness, and lay. Fig. 7-106.

Roughness is the height rating of the relatively fine surface irregularities that are produced by the manufacturing process. The man-

Fig. 7-103. ANSI Surface Texture Symbols

Symbol	Meaning
(a)	Basic Surface Texture Symbol. Surface may be produced by any method except when the bar or circle (Figure b or d) is specified.
(b)	Material Removal By Machining Is Required. The horizontal bar indicates that material removal by machining is required to produce the surface and that material must be provided for that purpose.
(c) 4.5	Material Removal Allowance. The number indicates the amount of stock to be removed by machining in millimeters (or inches). Tolerances may be added to the basic value shown or in a general note.
(d)	Material Removal Prohibited. The circle in the vee indicates that the surface must be produced by processes such as casting, forging, hot finishing, cold finishing, die casting, power metallurgy or injection molding without subsequent removal of material.
(e)	Surface Texture Symbol. To be used when any surface characteristics are specified above the horizontal line or to the right of the symbol. Surface may be produced by any method except when the bar or circle (Figures b and d) is specified.

(ANSI Y 14.36—1978)

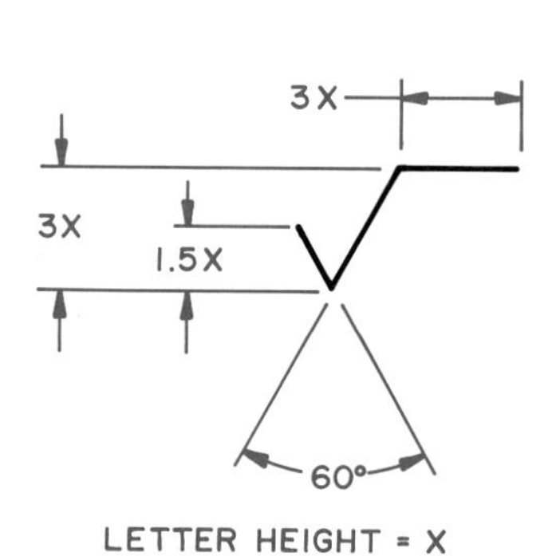

Fig. 7—104. **A standard surface texture symbol. (*ANSI Y14.36—1978*)**

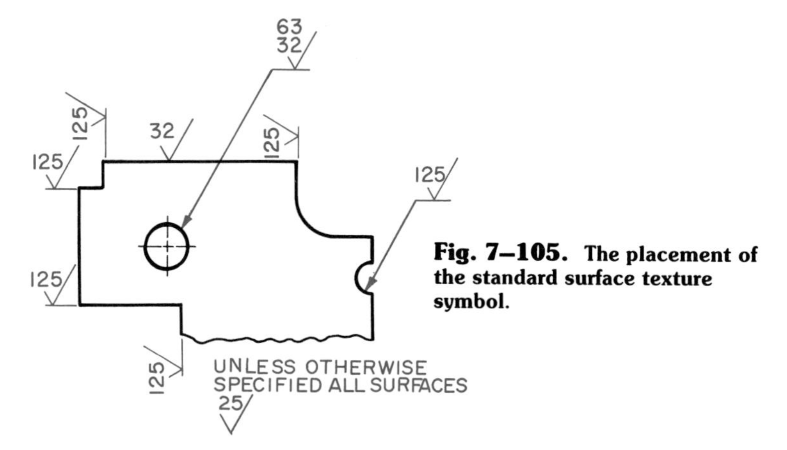

Fig. 7—105. **The placement of the standard surface texture symbol.**

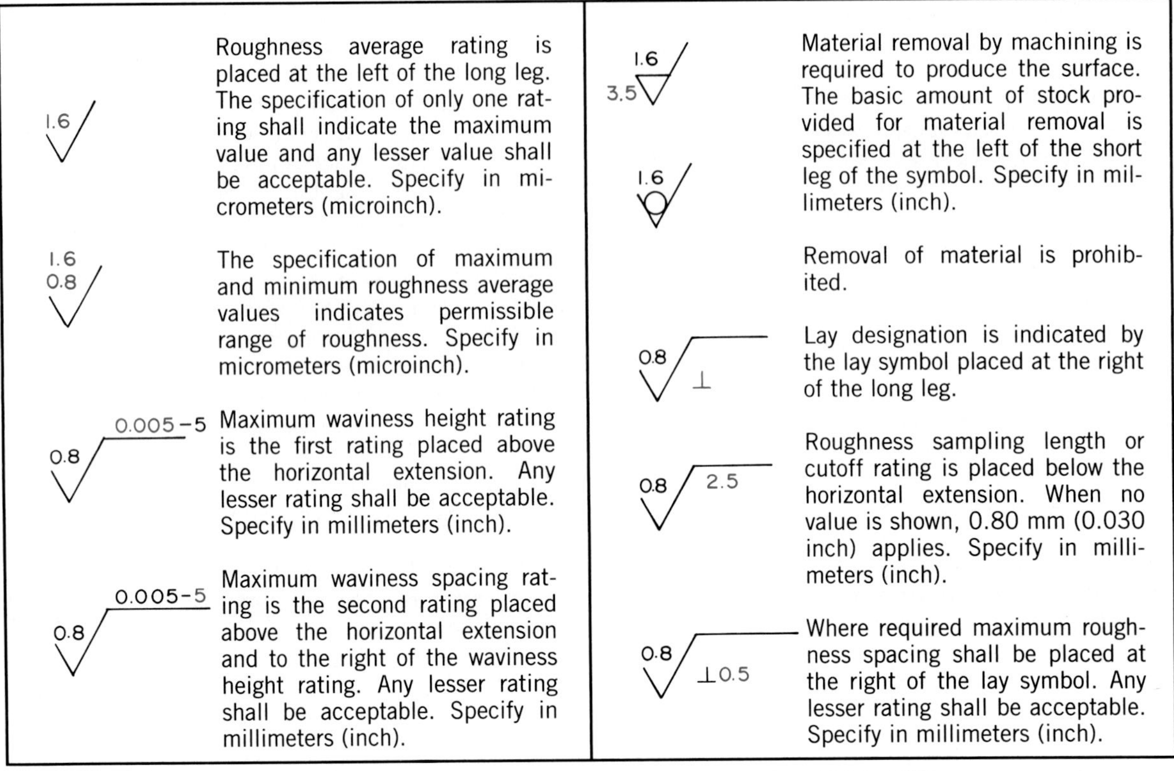

Fig. 7—106. **The application of surface texture values to the symbol. (*ANSI Y14.36–1978*)**

Fig. 7-107. Preferred Series Roughness Average Values (R_a)

Micrometers—µm		Microinches—µin	
µm	**µin**	**µm**	**µin**
0.025	1	1.60	63
0.050	2	3.2	125
0.10	4	6.3	250
0.20	8	12.5	500
0.40	16	25	1000
0.80	32		

(ANSI Y14.36–1978)

Fig. 7-108. Preferred Series Maximum Waviness Height Values

mm	in.	mm	in.	mm	in.
0.0005	0.00002	0.008	0.0003	0.12	0.005
0.0008	0.00003	0.012	0.0005	0.20	0.008
0.0012	0.00005	0.020	0.0008	0.25	0.010
0.0020	0.00008	0.025	0.001	0.38	0.015
0.0025	0.0001	0.05	0.002	0.50	0.020
0.005	0.0002	0.08	0.003	0.80	0.030

(ANSI Y14.36–1978)

ufacturing process may include casting, forging, rolling, coating, molding, machining, grinding, buffing, burnishing, extruding, or a combination of several of these. The average roughness height is measured in microinches or micrometers. A microinch is one-millionth (0.000,001) of an inch. A micrometer is one-millionth (0.000 001) of a meter. Preferred roughness values are shown in Fig. 7-107.

Waviness refers to the irregularities which are spaced farther apart than roughness peaks. Waviness may result from imperfections or vibrations of machine tools, deflections, or warping. Surface roughness may be thought of as being placed upon a wavy surface. Waviness height and width are measured in inches or millimeters. Recommended values are shown in Fig. 7-108.

Lay is the direction of the predominate surface pattern produced by the tool marks. This may also be the result of the surface grain. Lay is usually dependent upon the production method used. Lay symbols are shown in Fig. 7-109.

Production Methods and Surface Roughness

The terms smooth and rough as used thus far in this section have been relative. What is smooth for one specific purpose may be rough for another application. Roughness, then, depends upon the function or purpose of the surface.

What determines the quality of a surface? Several factors influence the choice of surface roughness. The main consideration is what processes can be used to obtain the necessary dimensional characteristics. Another factor is the availability of machinery to perform that process. It is false economy to specify surfaces that have a higher quality than necessary.

Working surfaces on bearings, pistons, and gears are typical of surfaces that require control. *Nonworking surfaces* seldom require any surface control. About the only exception is if the appearance of the part demands a high quality finish. Therefore, do not try to control surface roughness unless such control is essential to performance or appearance. Greater restriction of the finish will increase production costs.

Roughness and smoothness are largely the result of the method of processing a part. A machined or ground part may be rough or smooth for the purpose intended. The ideal surface characteristics for working surfaces may involve such operating conditions as the following:

- Area in contact.
- Load.
- Speed.
- Direction of motion.
- Vibration.
- Type and amount of lubricant used.
- Temperature.
- Material and physical properties of parts.

Various machining processes will obviously produce a range of surface roughness. Fig. 7-110 shows some typical processes and their range of surface roughness values. Remember, many factors contribute to the quality of a finish produced by a specific process. For example, a drilled hole may vary between 63 and 250 microinches in roughness height due to one or more of the following:

- Type of drill used.
- Speed of the drill.
- Feed of the drill.
- Condition of the drill.
- Material being drilled.
- Coolant used, if any.

Variations in any one of these factors can affect the surface quality. Each process listed in Fig. 7-110 has a range of roughness. Within each range is an indication of the less frequent application and the average application. When you specify surface roughness, use the recommended roughness height values. Do not specify a value that is between the recommended values. For example, specify a roughness height of 250, 63, 32, or 16 for boring or turning; do not select 200 or 75.

Fig. 7–109. Lay Symbols

Lay Symbol	Meaning	Example Showing Direction of Tool Marks
=	Lay approximately parallel to the line representing the surface to which the symbol is applied.	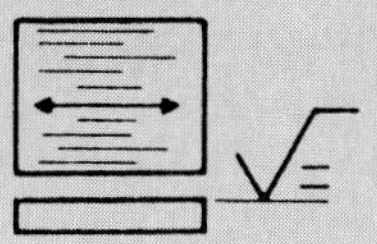
⊥	Lay approximately perpendicular to the line representing the surface to which the symbol is applied.	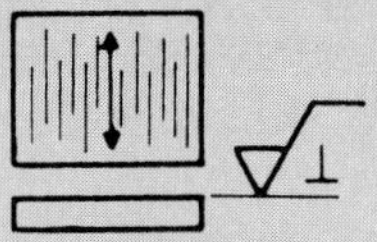
X	Lay angular in both directions to line representing the surface to which the symbol is applied.	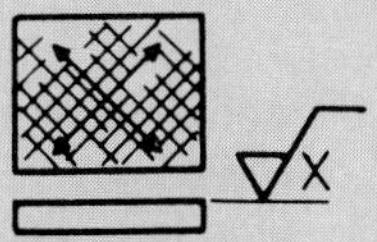
M	Lay multidirectional.	
C	Lay approximately circular relative to the center of the surface to which the symbol is applied.	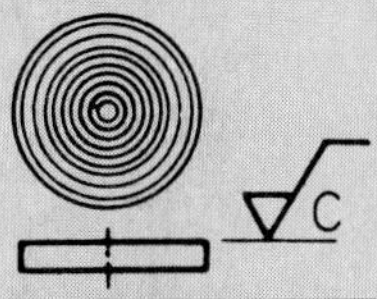
R	Lay approximately radial relative to the center of the surface to which the symbol is applied.	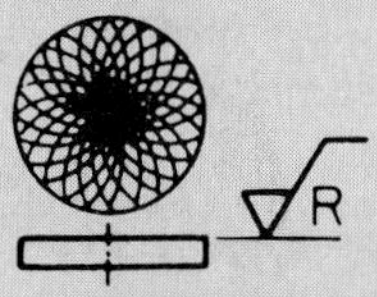
P[3]	Lay particulate, non-directional, or protuberant.	

(ANSI Y14.36–1978)

[3]The "P" symbol is not currently shown in ISO Standards. American National Standards Committee B46 (Surface Texture) has proposed its inclusion in ISO 1302–"Methods of indicating surface texture on drawings."

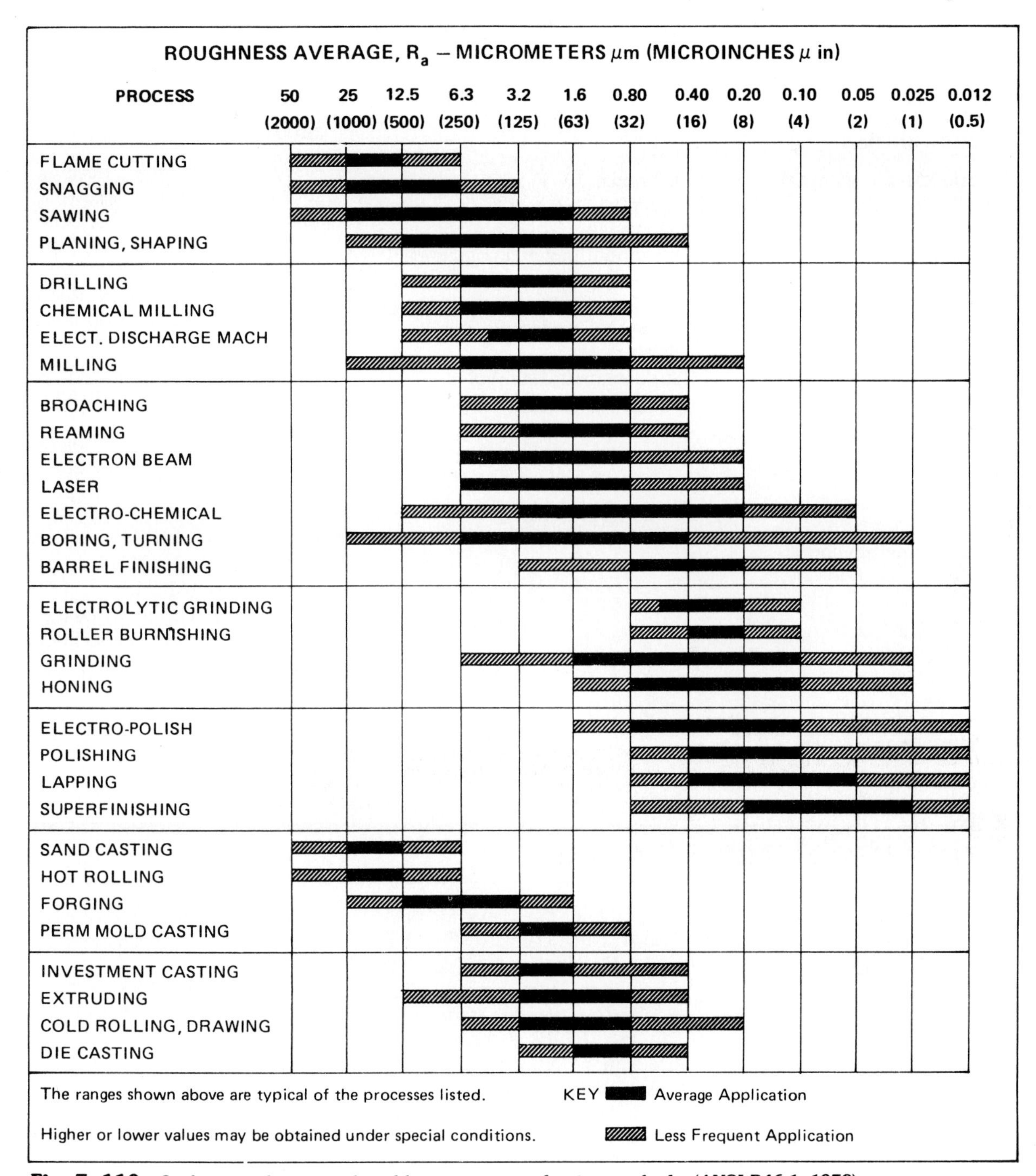

Fig. 7–110. Surface roughness produced by common production methods. (*ANSI B46.1–1978*)

Chapter Review

Build Your Vocabulary

You should understand and use the following terms as part of your working vocabulary. Write a brief explanation of what each means.

shape description
size description
dimension lines
extension lines
leaders
unidirectional system
aligned system
size dimensions
location dimensions
datums
tabular dimensioning
reference dimension
counterbored
counterdrilled
countersinking
spot facing
chamfer
keys
knurling
limits
tolerance
taper
allowance
maximum material condition
least material condition
basic hole system
basic shaft system
finish mark
surface texture symbol
roughness
waviness
lay

Sharpen Your Math Skills

1. What is the decimal equivalent of the following common inch fractions: 3⁄16, 7⁄32, 16⁄64? Round off the answers to 3 decimal places.

2. Add the following decimal inch fractions: 1.050 + .100 + .357 + .250.

3. Convert the following inch dimensions to millimeters: 1.5, 3.7, 5¼.

4. How many centimeters are there in 2500 millimeters?

5. When measuring angles, 60′ are equal to 1°, and 60″ are equal to 1′. Add the following angles: 30°15′ + 15°21′15″ + 8°10′45″.

6. Round off the following decimal inches to 1 decimal place: 3.57, 5.61, 10.19, 3.44, 4.35, 6.65.

Study Problems—Directions

The problems that follow will give you the chance to use some of the drafting knowledge that you learned in this chapter. These study problems are dimensioned actual size. Since some are rather small, you may want to draw them twice their actual size; this will make drawing and dimensioning easier. Do the problems assigned by your instructor at the size he or she specifies.

1. For **Figs. P7-1** through **P7-10**, make a freehand sketch of the views shown. Make your sketch on graph paper. Dimension each problem.

2. Draw the views for each part of the clamp in **Fig. P7-11**. Find tolerances for the hole in the screw and handle.

3. The marking gage head in **Fig. P7-12** is formed with four radii. Draw the views and dimension completely.

4. Use a datum to locate the centers of the ball diameter gage in **Fig. P7-13**. Dimension using two-place decimals.

5. Dimension the lawn mower drive wheel in **Fig. P7-14**. Make the tolerance of the hole ±.01.

6. A plastic boat mooring buoy is shown in **Fig. P7-15**. Each square equals 50 mm. Draw the buoy half size and dimension using millimeters as the basic unit.

7. The plumb bob in **Fig. P7-16** can be dimensioned using millimeters. It has no critical size or location dimensions.

8. The lathe center in **Fig. P7-17** has a tapered body. Draw and dimension it in millimeters.

9. The cutting edge of the hollow punch in **Fig. P7-18** must be drawn with a tolerance as shown. Dimension the other parts with two-place dimensions. Notice the knurl.

10. The bed rail plate in **Fig. P7-19** is a cast object. Di-

mension with the decimal fractions. Notice the countersunk screw holes.

11. A complex catch for a steel cabinet door lock is shown in **Fig. P7-20**. Dimension using two-place decimals. All dimensions are subject to a ±.01 inch tolerance.

12. The garden gate catch in **Fig. P7-21** is stamped from a steel plate and bent to a 90-degree angle. Dimension in millimeters.

13. The drawer pull in **Fig. P7-22** is cast in solid brass. Dimension using two-place decimals. Notice the threaded hole. Select a suitable thread type from the thread tables.

14. The outside dimensions of the bearing journal in **Fig. P7-23** can be dimensioned with two-place decimals. Locate the distance between the bolt holes with a tolerance of ±.001 inch or ±0.025 mm.

15. The width of the clamp strap slot in **Fig. P7-24** must be accurate. Dimension with a tolerance of ±.005 inch. Dimension the other parts with two-place decimals.

16. The C-washer in **Fig. P7-25** does not need to be made very accurately. Use dimensions that permit low-cost manufacturing.

17. The microscope base in **Fig. P7-26** is cast metal. Dimension it completely. The holes have a tolerance of ±.001 inch or ±0.025 mm. Notice the counterbore.

18. The pulley shown in **Fig. P7-27** is used in the clutch on a motorcycle. The outside diameter must have a tolerance of ±.010 inch or ±0.25 mm. It is machined, so show the finish marks. The hole for the shaft has a tolerance of ±.001 inch or ±0.025 mm. Notice the keyseat for a square key.

19. The fixture jaw in **Fig. P7-28** is machined on all surfaces. Show the finish marks. Dimension with two-place decimal inches or one-place millimeters.

20. All surfaces of the spherical washers in **Fig. P7-29** are machined. Draw and dimension each separately. The spherical surfaces have a tolerance of ±.001 inch or ± 0.025 mm. The other surfaces are ±.01 inch or ±0.25 mm.

21. The drill jig in **Fig. P7-30** is used to locate holes accurately when the holes are to be drilled on a production job. Locate each hole from a datum. Dimension with three-place decimals. The tolerance is ±.003 inch.

22. In **Fig. P7-31**, the center and the knurled ring of the adjustable center have an RC2 fit. Dimension using two-place decimals. Notice the threaded hole for a set screw.

23. Draw and dimension the machine handle in **Fig. P7-32**. It is cast with a tolerance of ±.015 inch. The small end is machined to a tolerance of + .001 inch. Notice the chamfer.

24. Locate the hole centers on the printed circuit board in **Fig. P7-33** using tabular dimensions. Dimension to three decimal places. All location dimensions have a tolerance of ±.001 inch or ±0.025 mm. Size dimensions have a tolerance of ±.05 inch or ±1.0 mm.

25. Plugs A and B on the adjustable snap gage in **Fig. P7-34** are machined to the tolerances shown. All other dimensions have a tolerance of ±.001 inch or ±0.025 mm. Draw and dimension each part separately. The cylindrical surface of both plugs has a roughness height value of 80 microinches.

26. Draw and dimension separately each part of the spring jack assembly in **Fig. P7-35**. Carefully dimension the tolerance of each part so that they have the proper fit. The tolerance on the outer surfaces of the body and cap is ±.02 inch or ±0.5 mm. The tolerance on the inside hole is ±.01 inch or ±0.25 mm. All surfaces of the shaft are machined.

27. Draw and dimension separately each part of the revolving clamp assembly in **Fig. P7-36**. The body is a cast unit. The pin and lock are machined on all surfaces. The washer is stamped. Notice the tolerance on the distance between the holes.

28. Solve the tolerance problems shown in **Fig. P7-37**.

29. Draw and dimension separately each part of the circle cutter in **Fig. P7-38**. Notice the tolerances on the machined bar.

Study Problems

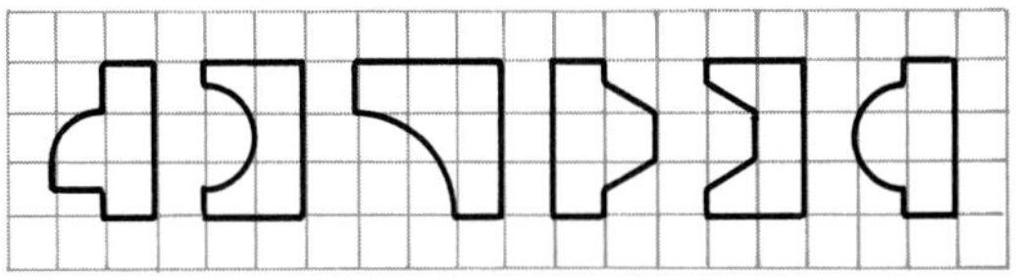

EACH SQUARE EQUALS .25" OR 6 mm

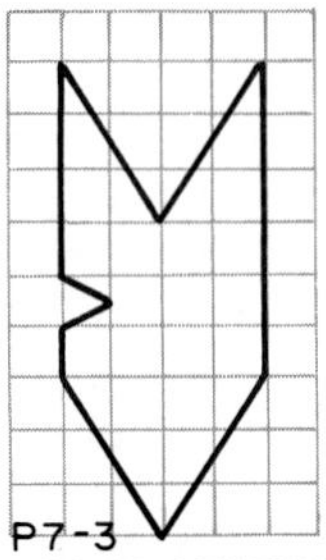

P7-1 MOLDING SHAPES

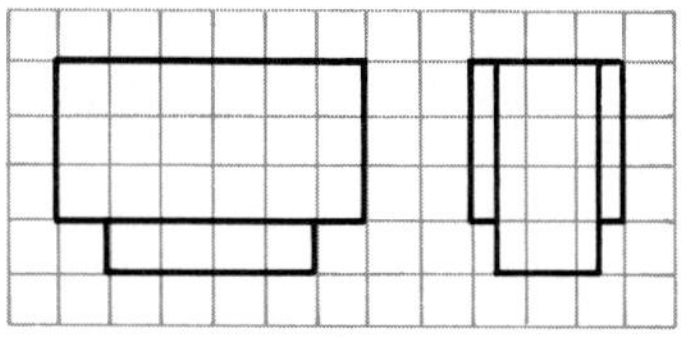

P7-2 DOOR MAGNET

P7-3 LATHE CENTER GAGE 1.5 THICK

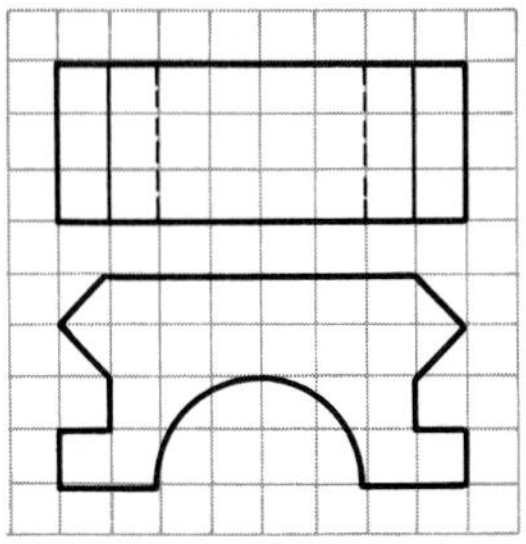

FIXTURE MAGNET P7-4

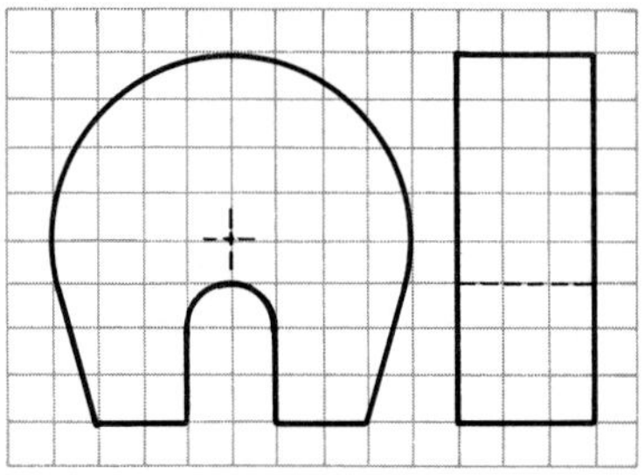

HORSESHOE MAGNET P7-5

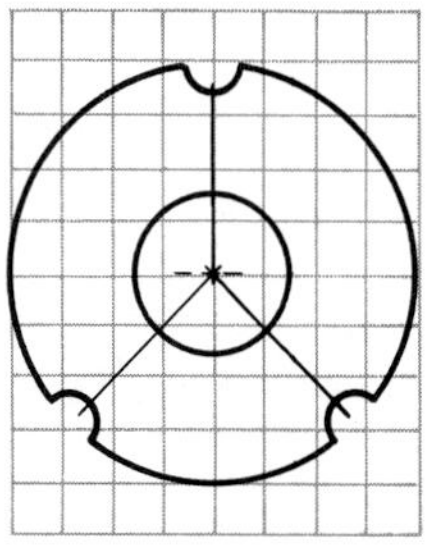

P7-6 CIRCLE MAGNET .12" THICK

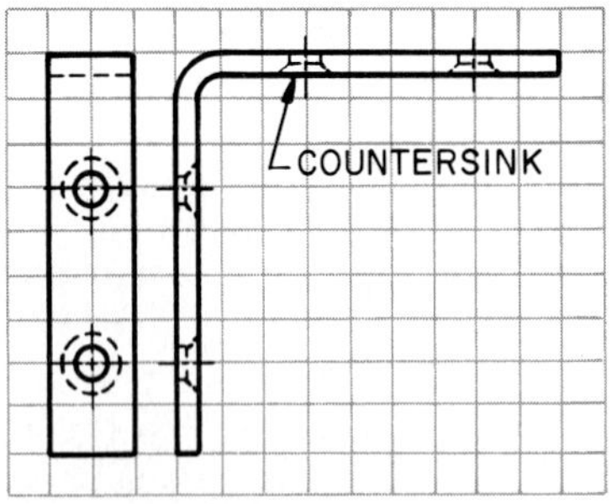

COUNTERSINK

P7-7 SHELF ANGLE

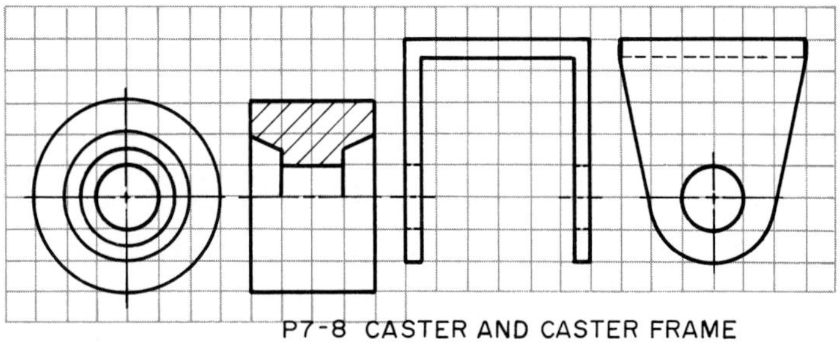

P7-8 CASTER AND CASTER FRAME

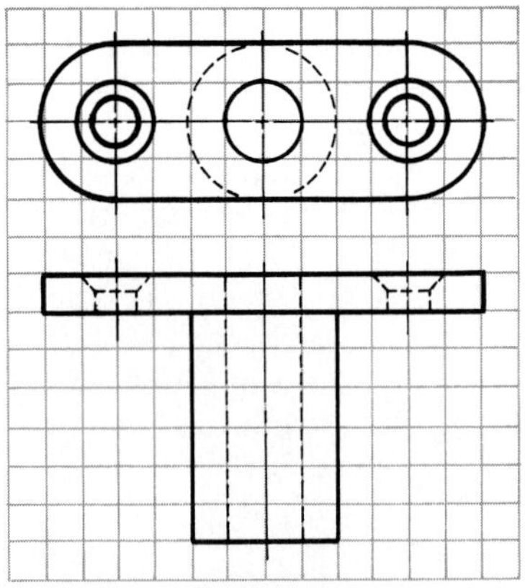

OAR SOCKET P7-9

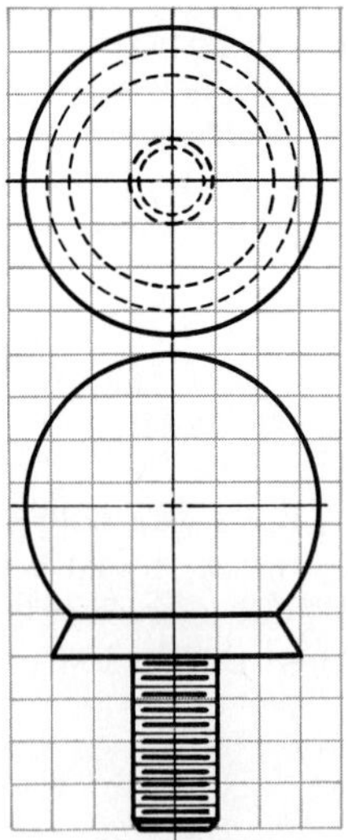

TRAILER HITCH BALL P7-10

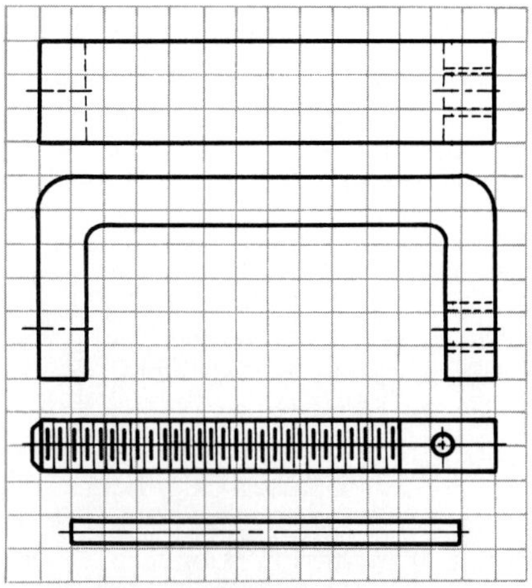

HANDLE & HOLE HAVE AN FN 1 FIT.
C-CLAMP FRAME, SCREW AND HANDLE
P7-11

Study Problems

EACH SQUARE EQUALS .25" OR 6 mm EXCEPT AS NOTED

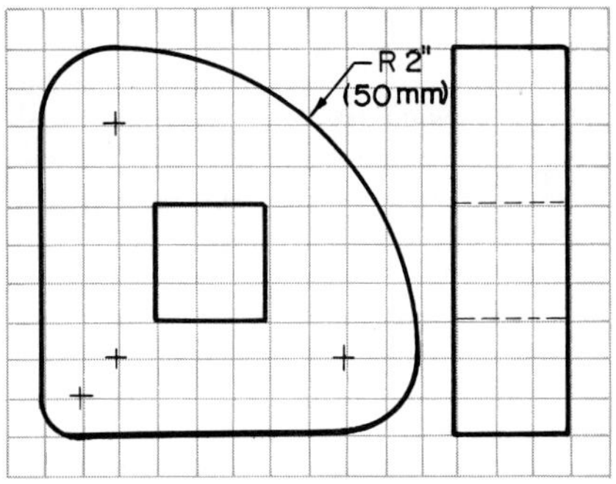

R 2" (50 mm)

MARKING GAGE HEAD
P7-12

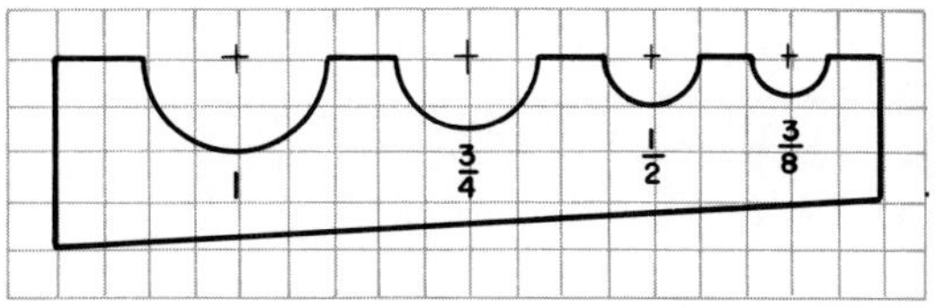

1 3/4 1/2 3/8

BALL DIAMETER GAGE .06"(1.5 mm) THICK
P7-13

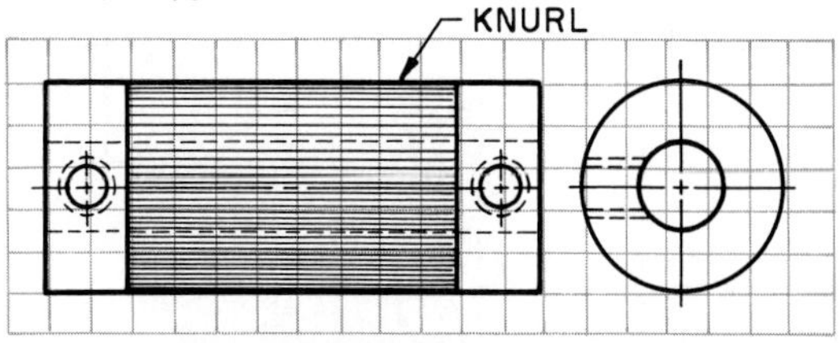

KNURL

LAWN MOWER DRIVE WHEEL
P7-14

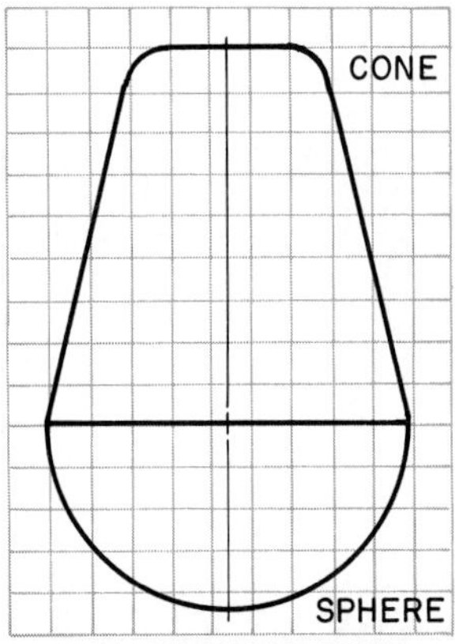

CONE

SPHERE

BOAT MOORING BUOY
EACH SQUARE EQUALS 50 mm
P7-15

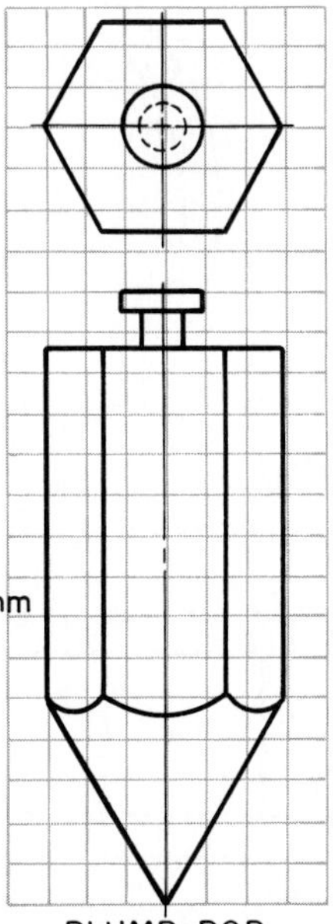

PLUMB BOB
P7-16

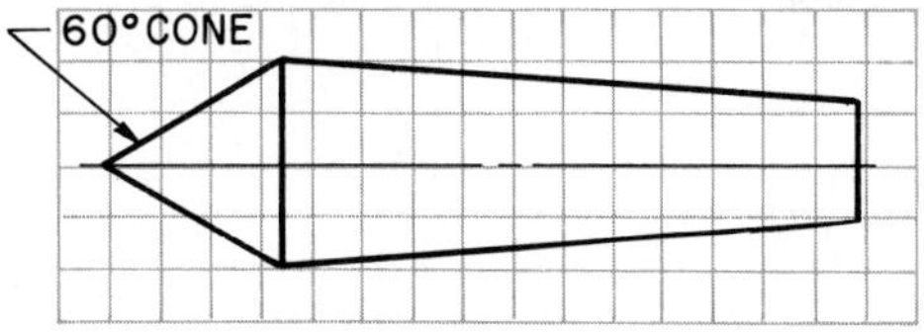

60° CONE

LATHE CENTER
P7-17

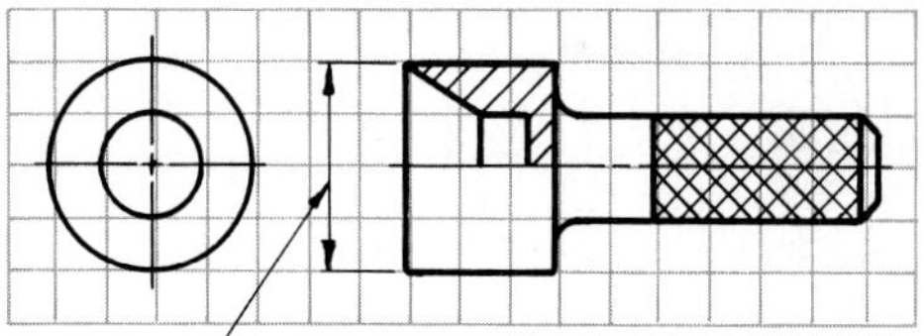

TOLERANCE ± .001" OR ±0.02 mm
HOLLOW PUNCH
P7-18

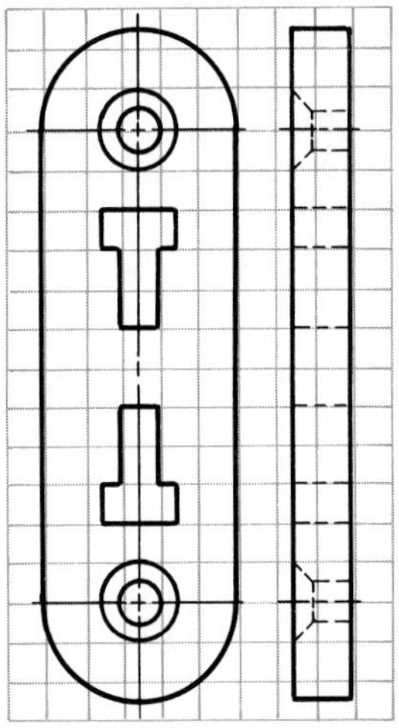

BED RAIL PLATE
P7-19

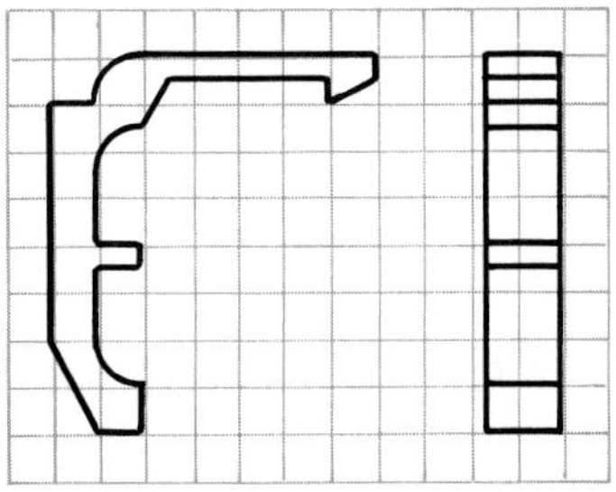

DOOR CATCH
P7-20

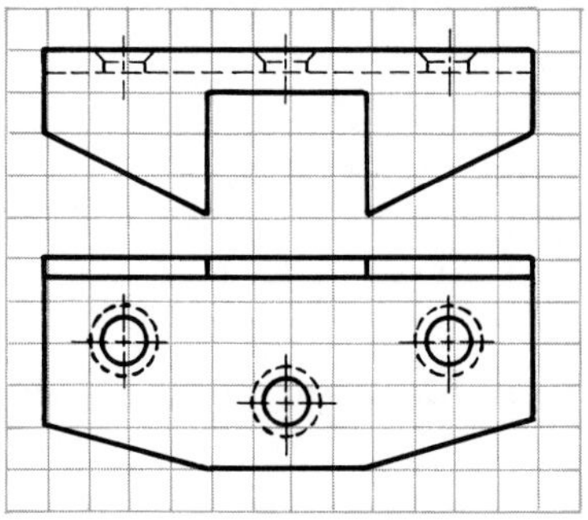

GARDEN GATE CATCH
P7-21

Study Problems

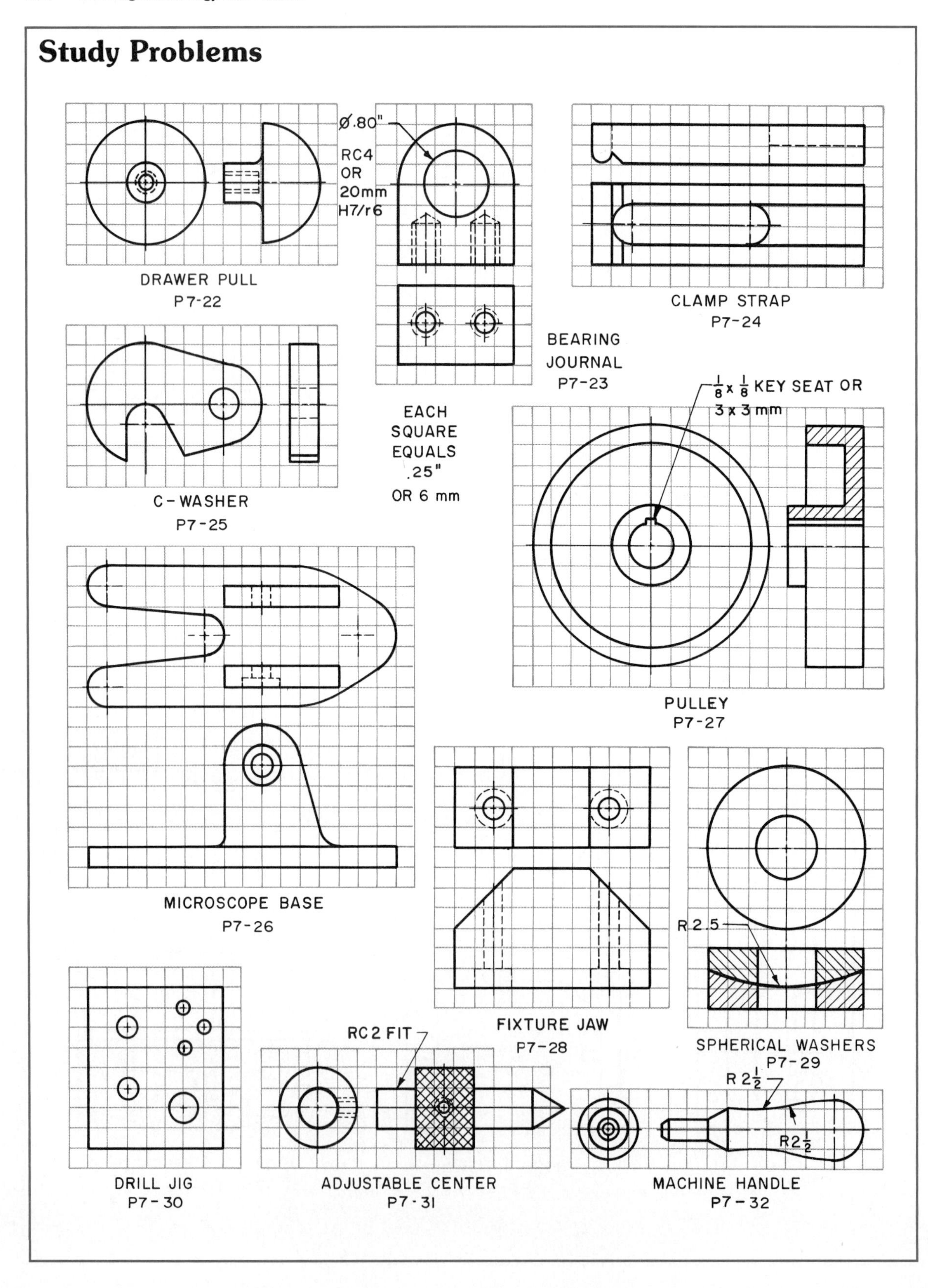

Study Problems

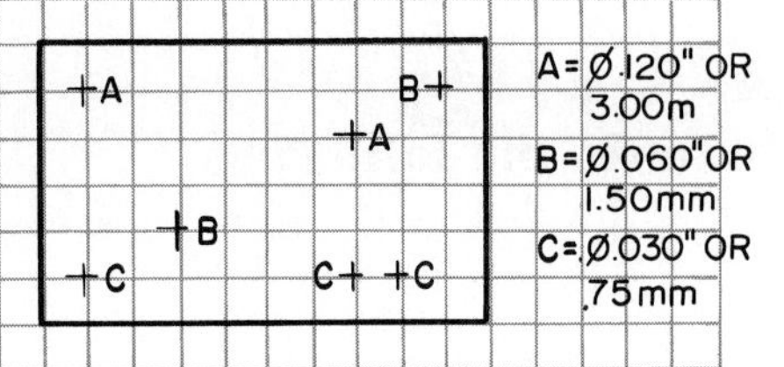

PRINTED CIRCUIT BOARD
P7-33

EACH SQUARE EQUALS .25" OR 6 mm

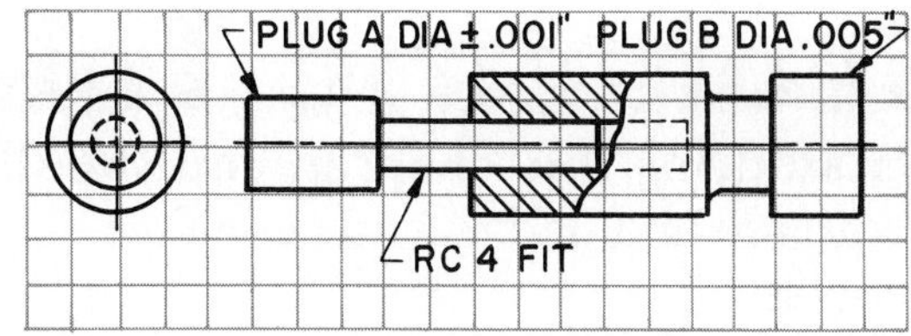

ADJUSTABLE SNAP GAGE
P7-34

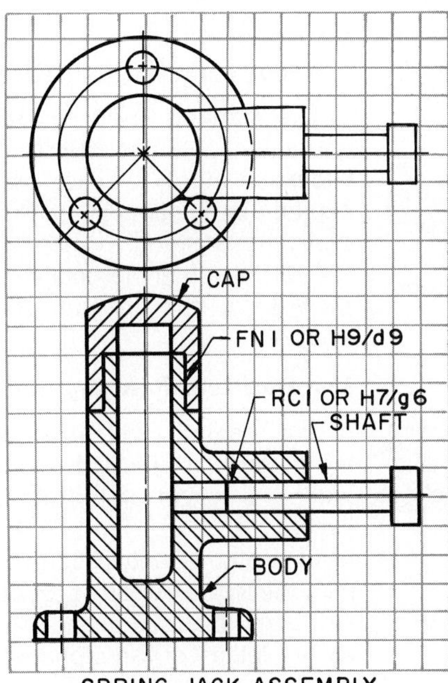

SPRING JACK ASSEMBLY
P7-35

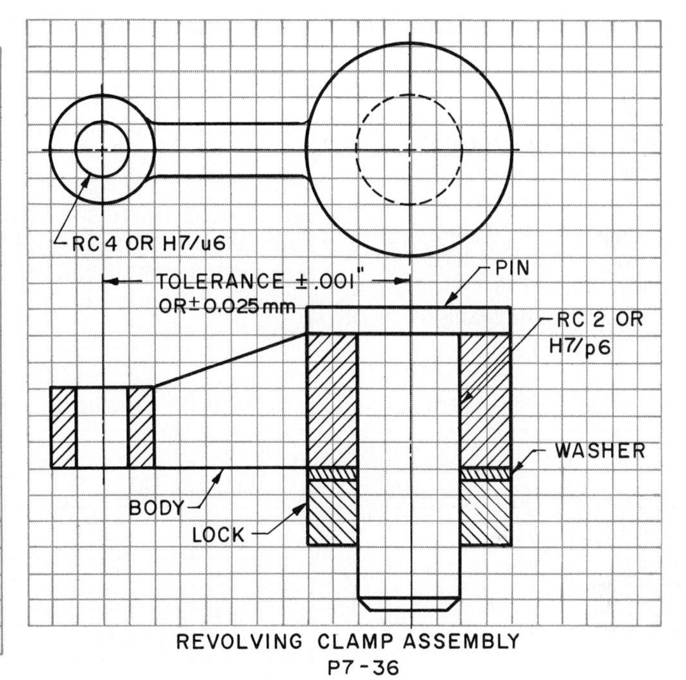

REVOLVING CLAMP ASSEMBLY
P7-36

A NOMINAL DIA 1½ RC 4
BASIC HOLE

B NOMINAL DIA 2" FN 2
BASIC HOLE

P7-37

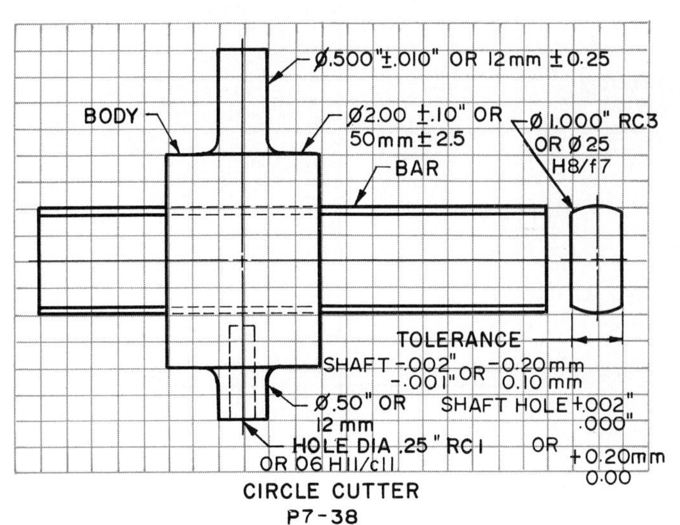

CIRCLE CUTTER
P7-38

CHAPTER 8

Geometric Tolerancing

After studying this chapter, you should be able to do the following:

- Recognize and understand geometric characteristic symbols.
- Dimension drawings using principles of geometric tolerancing.

The Use of Geometric Tolerancing

Geometric tolerancing is used along with size dimensions to increase the accuracy of parts and, therefore, make them more easily interchangeable. This kind of tolerancing is used to specify the shape, position, or relationship of features. ANSI standards specify SI metric units for geometric tolerancing on engineering drawings.

Geometric tolerances are widely used in industry and have application in the area of computer-aided manufacturing. Engineers and drafters should understand their use.

Fig. 8–1. Geometric Characteristic Symbols

	TYPE OF TOLERANCE	CHARACTERISTIC	SYMBOL
For Individual Features	Form	Straightness	—
		Flatness	▱
		Circularity (Roundness)	○
		Cylindricity	⌭
For Individual or Related Features	Profile	Profile of a Line	⌒
		Profile of a Surface	⌓
For Related Features	Orientation	Angularity	∠
		Perpendicularity	⊥
		Parallelism	//
	Location	Position	⌖
		Concentricity	◎
	Runout	Circular Runout	↗*
		Total Runout	⌰*
*Arrowhead(s) may be filled in.			

Reproduced from *ANSI Y14.5M–1982* with permission of the publisher. The American Society of Mechanical Engineers.

Geometric Characteristic Symbols

The symbols used to denote geometric characteristics are shown in Fig. 8-1. They are accepted internationally and are easily understood. When combined with other descriptive data they form the feature control symbol.

Datum Identifying Symbol

Each datum feature on a drawing is given a different reference letter. All letters of the alphabet except "I," "O," and "Q" are used. Datum identifications begin with the letter "A" and proceed in alphabetical order as needed. If all the letters of the alphabet are used but more are needed, double letters such as "AA," "BB," and so on are used. The datum identifying symbol is formed by placing the letter in a box. The letter is preceded and followed by a dash as shown in Fig. 8-2.

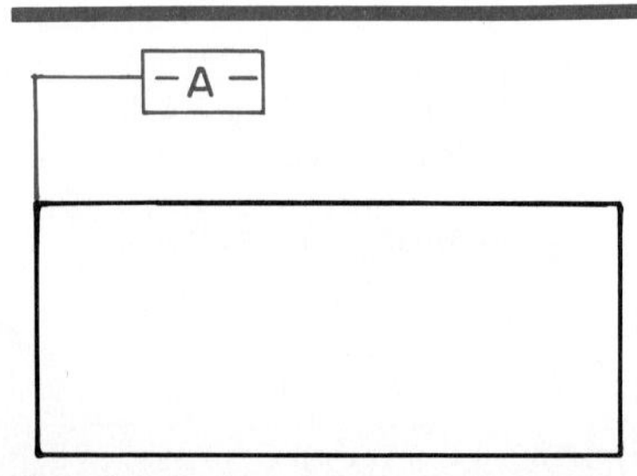

Fig. 8–2. Identifying the datum.

Basic Dimension Symbol

The **basic dimension** is noted by enclosing it in a box as shown in Fig. 8-3.

Modifying Symbols

Fig. 8-4 shows modifying symbols that are used as a part of the feature control symbol. Included are the maximum material condition, least material condition, and regardless of feature size designations.

Maximum material condition (MMC) occurs when a feature contains the most amount of material. Two mating parts will have their tightest fit when both are at their maximum material condition. A hole is at its maximum material condition when it is at the smallest size permitted by its tolerance limits. A shaft is at its maximum material condition when it is at the largest size permitted by its tolerance limits.

Least material condition (LMC) occurs when a feature contains the smallest amount of material possible within its stated limits of size. Two mating parts will have their loosest fit when both are at their least material condition. It is the smallest shaft or the largest hole permitted by tolerance limits.

Regardless of feature size (RFS) means that tolerance limits are applied to the feature regardless of its size.

The abbreviations MMC, LMC, and RFS are used in notes but not in the feature control symbol.

Feature Control Symbols

Form and position tolerances are stated with feature control symbols. Enclose the characteristic symbol, tolerance, data identification, and other information in a frame. Separate the symbols and tolerances by a vertical line. Place the geometric characteristic symbol in the first compartment and the tolerance in the second. Fig. 8-5 at A. When needed, place a diameter symbol before the tolerance and a material condition symbol after. When a geometric tolerance is related to a datum, place the datum reference letter in a compartment following the tolerance. Fig. 8-5 at B. When two datum features are used to establish a single datum, put both datum reference letters

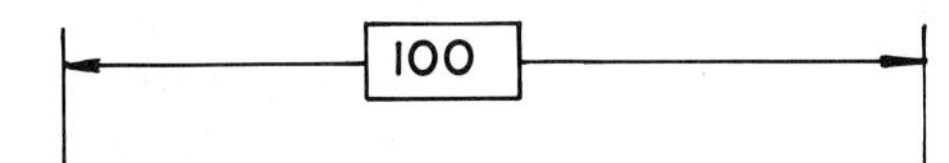

Fig. 8–3. A basic dimension is noted by enclosing it in a rectangle.

Fig. 8–4. Modifying Symbols

TERM	SYMBOL
At Maximum Material Condition	Ⓜ
Regardless of Feature Size	Ⓢ
At Least Material Condition	Ⓛ
Projected Tolerance Zone	Ⓟ
Diameter	Ø
Spherical Diameter	SØ
Radius	R
Spherical Radius	SR
Reference	( )
Arc Length	⌒

Reproduced from *ANSI Y14.5M–1982* with permission of the publisher, The American Society of Mechanical Engineers.

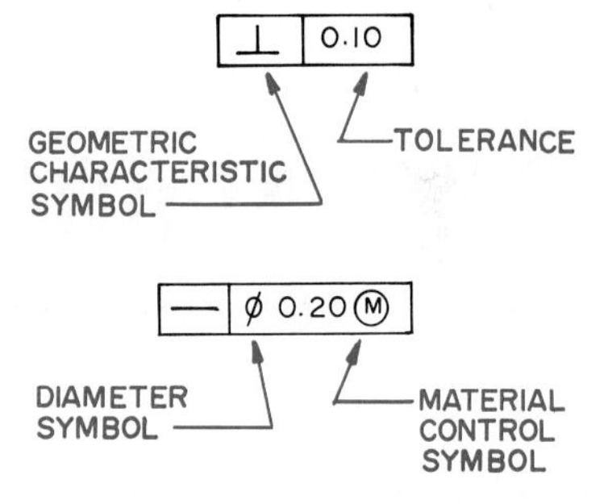

FEATURE CONTROL FRAME

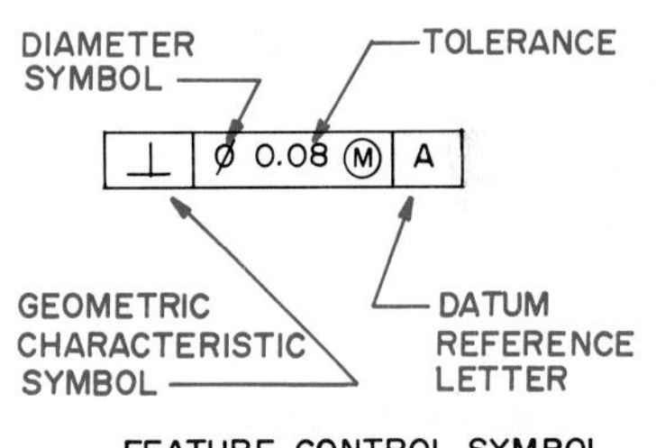

FEATURE CONTROL SYMBOL WITH A DATUM REFERENCE

// 0.16 C-D

MULTIPLE DATUM AS THE PRIMARY

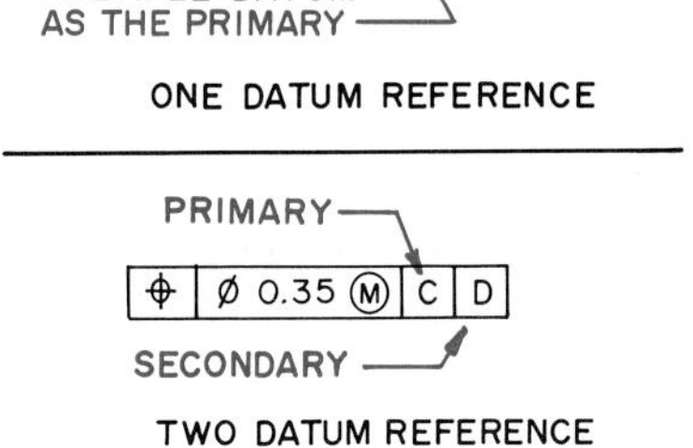

TWO DATUM REFERENCE

Fig. 8–5. Examples of the feature control symbol.

in a single compartment and separate them by a dash. Fig. 8-5 at C. When more than one datum is required, put the datum reference letters in separate compartments in order of precedence from left to right. Datum reference letters do not need to be in alphabetical order in the feature control frame. Fig. 8-5 at D.

The feature control system is related to the feature, as shown in Fig. 8-6, by the following guidelines:

1. Place the symbol box adjacent to or below the dimension or note connected to the controlling feature.

2. Attach a side or end of the symbol box to an extension line or dimension line connected to the controlling feature.

3. Attach a leader from the symbol box to the feature.

Tolerances of Position

Positional tolerances are those used with dimensions that locate a feature, such as a hole, in relation to other features or datums. The three basic types of positional tolerances are true position, concentricity, and symmetry.

The traditional method of locating a feature, such as a hole, is the coordinate method. This method produces a square tolerance zone as seen in Fig. 8-7. While the sides of the square are within the specified tolerance, the diagonal is greater; therefore, the location of the center of the hole could exceed the specified tolerance. This error is not possible using true position tolerancing because it uses a circular tolerance zone.

CAD Applications:

Geometric Tolerancing Symbols

As you've learned in this chapter, geometric tolerancing uses symbols to specify the characteristics or location of a part. When creating a drawing on a CAD system, you could draw each of the symbols individually each time, but there are better ways.

You can draw the symbols once and save them on your computer disk. Then you can insert them into subsequent drawings as needed. Such a collection of symbols is called a library. Here is a simple way to create a library using the AutoCAD® program. Similar procedures would be used for other CAD programs.

1. Give your drawing file a name, such as GEOTOL.

2. On layer 0, draw the symbols for parallelism, cylindricity, etc. (You'll need to determine correct sizes.)

3. On a new layer called NAMES, create a name and insertion base point for each of the symbols.

4. Return to layer 0 and freeze layer NAMES.

5. Store each symbol as a Block, using the names and insertion base points created in Step 3. As you do this, the symbols will disappear from the screen.

6. Now thaw layer NAMES and insert the Blocks into their proper locations, as indicated by the insertion base points.

7. Use the END command to save your drawing.

Now that you've created and saved the symbol library, you can add any of the symbols to any drawing. Here is the procedure. (The following assumes you have a drawing on the screen.)

1. Enter the INSERT command.

2. In response to "Block name" enter GEOTOL.

3. In response to "Insertion point" cancel by pressing Ctrl C.

4. Enter the INSERT command again and type the question mark (?) to obtain a listing of Blocks.

5. Now insert any of the symbols you need in your drawing.

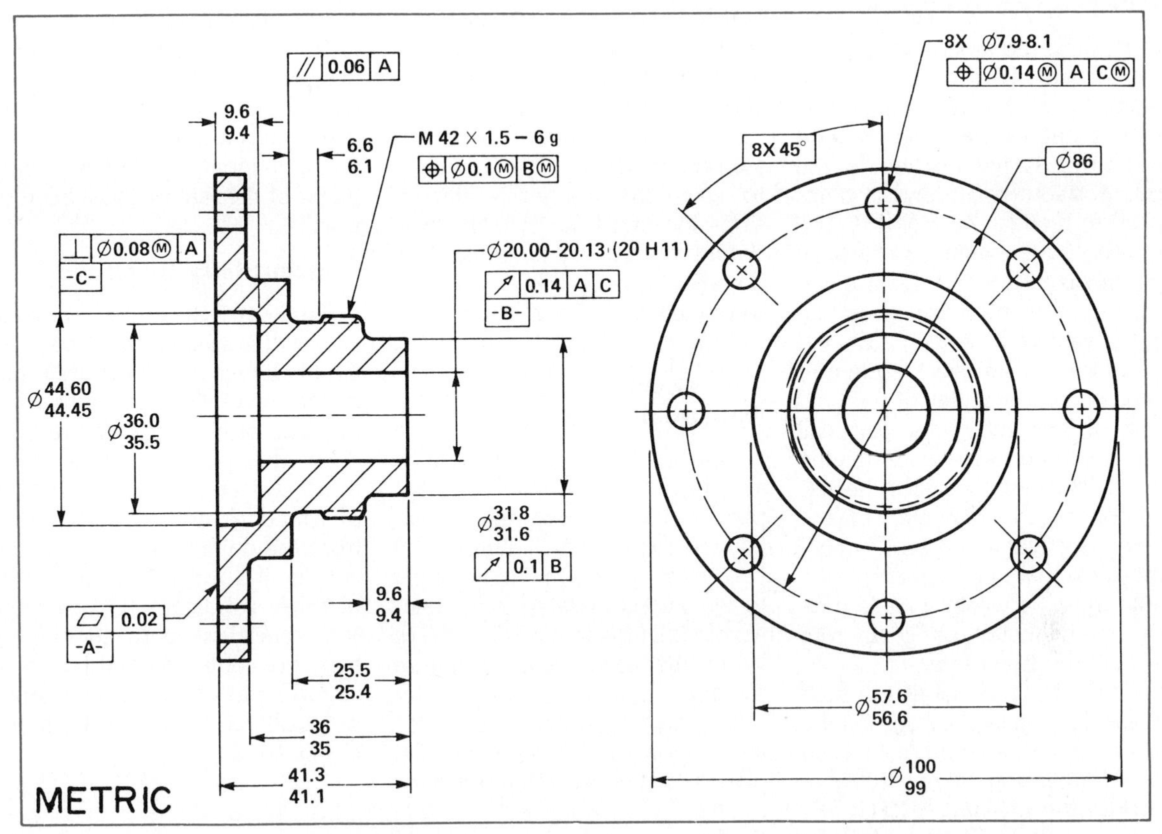

Fig. 8–6. Applications of the feature control symbol. (Reproduced from *ANSI Y14.5M-1982* with permission of the publisher, The American Society of Mechanical Engineers.)

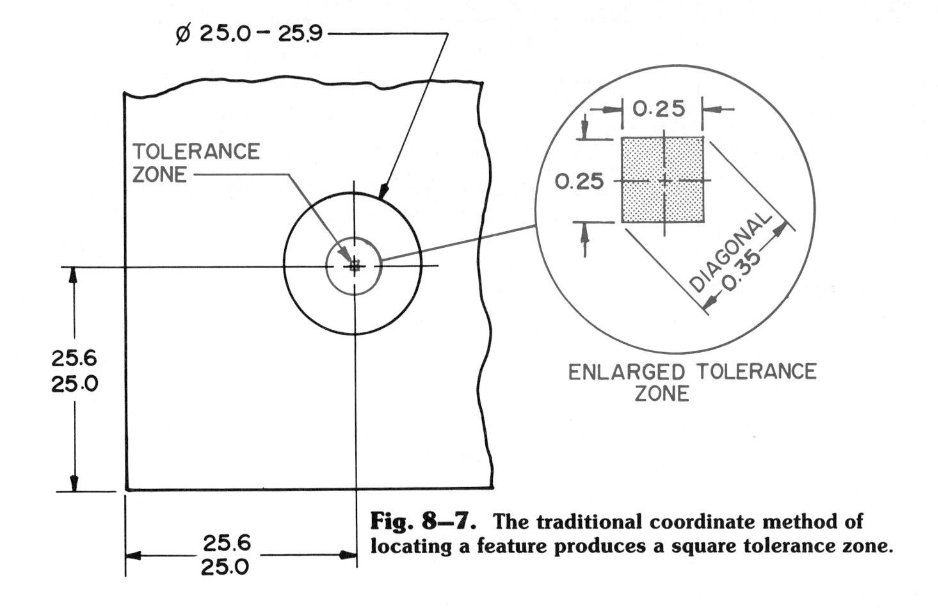

Fig. 8–7. The traditional coordinate method of locating a feature produces a square tolerance zone.

True Position Tolerance

The true position gives the exact location of a point, line, or plane in relation to another feature or datum. The **true position tolerance** is not applied to the true position dimension but is applied to the feature. Fig. 8-8. The true position tolerance is the total permissible amount a feature can vary from its true position. In Fig. 8-8 a circular zone with a diameter of 0.12 is specified. If the true position of the center of the hole falls in this zone, the part will be interchangeable with mating parts. The note specifies the tolerance in the diameter of the hole. The feature control symbol specifies that the center of the hole must be within a circle having a diameter of 0.12 mm around the true center.

The location of a feature, such as a hole, is given by applying the basic dimension symbol to each basic dimension or by adding a note to the drawing: UNTOLERANCED DIMENSIONS LOCATING TRUE POSITION ARE BASIC. The feature control frame is used to specify the size and number of features. Fig. 8–9 shows various types of feature pattern dimensioning. To identify features on a part, it is necessary to establish datums for dimensions locating true positions. The datum features are identified with datum feature symbols and datum references in the feature control frame.

True Position Tolerance for Noncircular Features

Noncircular features, such as a slot, have position tolerances applied in the same manner as explained for circular features. The positional tolerances may apply to the surfaces related to the center plane of the feature. Fig. 8-10. When the feature is at maximum material condition, the center plane must be within a tolerance zone defined by two parallel planes located an equal distance away from the true position. They must have a width equal to the positional tolerance. This tolerance zone also defines the limits within which the variations in the attitude (squareness) of the center plane of the feature must be confined.

Concentricity Tolerance

Concentricity is the condition where the axes of all cross-sectional elements of a feature's surface of revolution are common to the axis of a datum feature. Fig. 8-11. It is possible for the cylindrical surface to be bowed or out of round as well as being offset from the datum feature. It is difficult to find the axis under either of these conditions. Unless there is a special need to control the axis, the control should be specified in terms of a runout tolerance or positional tolerance as shown in Fig. 8-11.

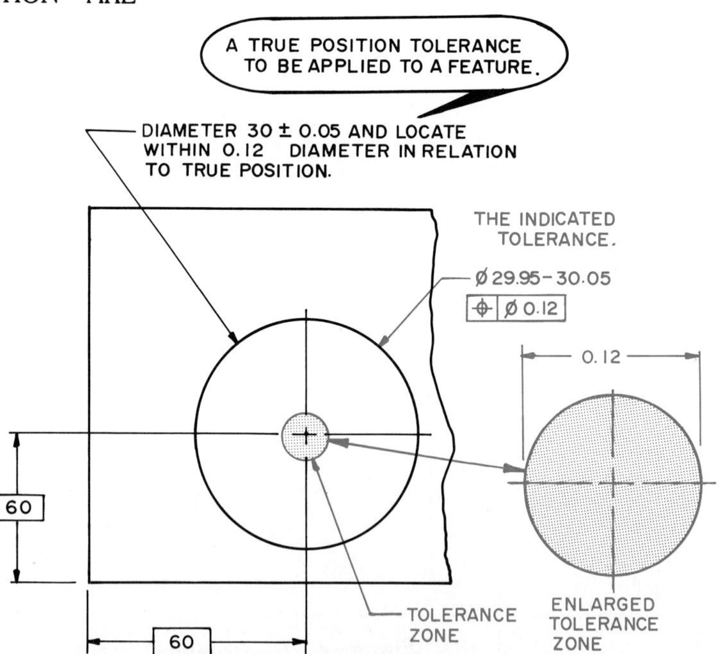

Fig. 8–8. Basic dimensions locate the true center of the circle. In this example the circle has a specified circular tolerance zone of 0.12 mm. The tolerance zone therefore has equal variations in all directions from the true axis of the hole. (Reproduced from *ANSI Y14.5M-1982* with permission of the publisher, The American Society of Mechanical Engineers.)

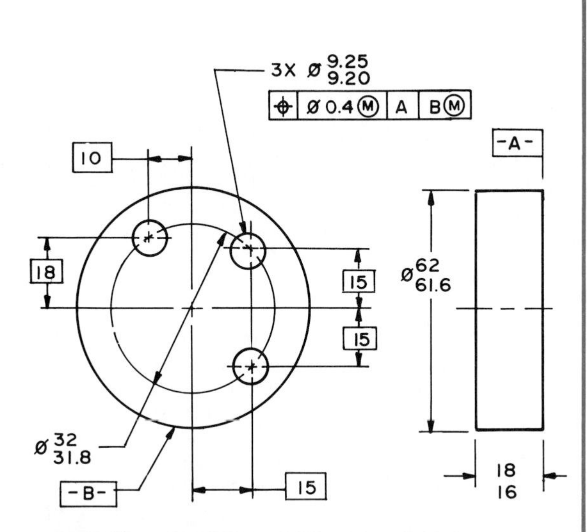

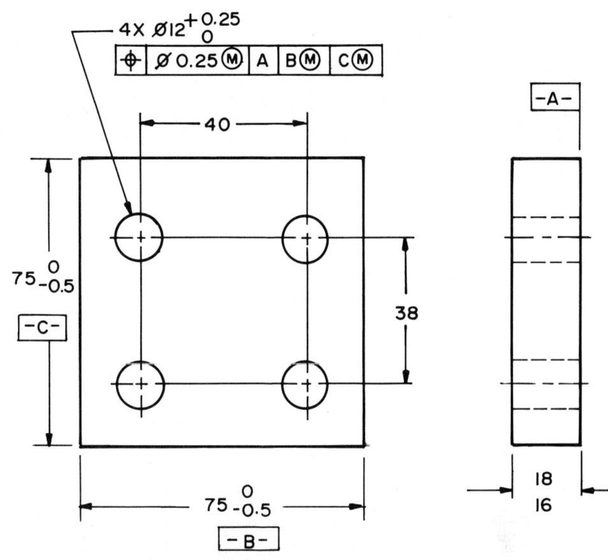

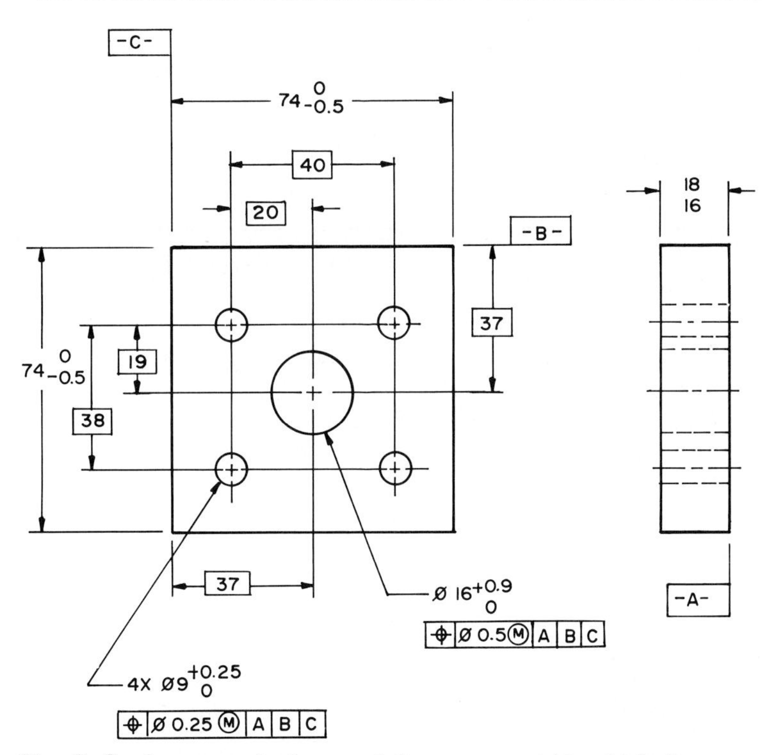

Fig. 8–9. A positional tolerance defines a zone within which the center, axis, or center plane of a feature of size can vary from the true position. (Reproduced from *ANSI Y14.5M-1982* with permission of the publisher, The American Society of Mechanical Engineers.)

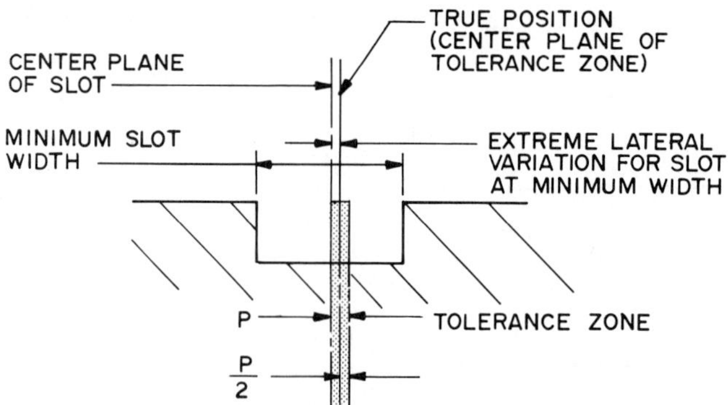

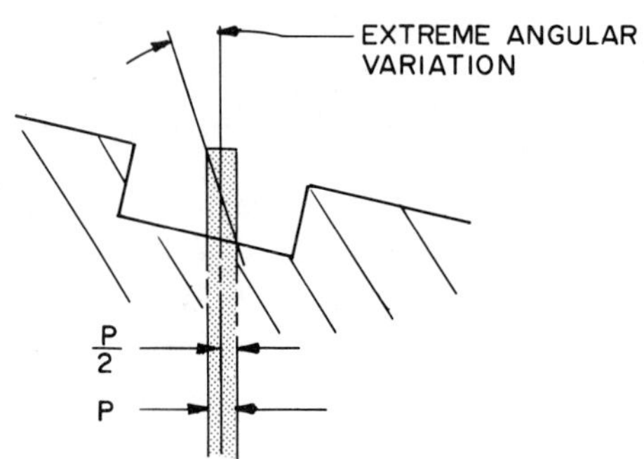

Fig. 8–10. **The true position for a noncircular feature may be applied to the center plane of the feature at MMC. (Reproduced from *ANSI Y14.5M-1982* with permission of the publisher, The American Society of Mechanical Engineers.)**

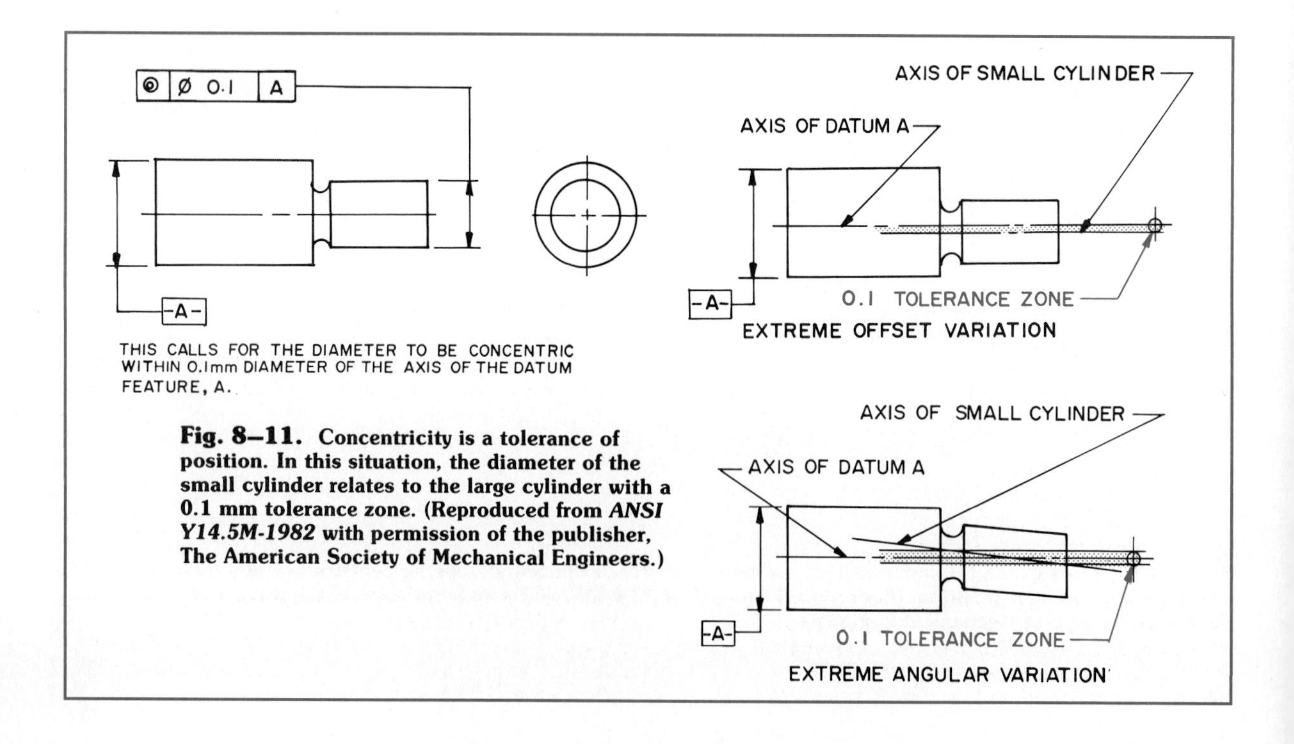

Fig. 8–11. **Concentricity is a tolerance of position. In this situation, the diameter of the small cylinder relates to the large cylinder with a 0.1 mm tolerance zone. (Reproduced from *ANSI Y14.5M-1982* with permission of the publisher, The American Society of Mechanical Engineers.)**

Symmetry Tolerance

Symmetry means that a feature has a corresponding size and shape on opposite sides of a central plane of a datum feature. Symmetry tolerance can be expressed with a note or a feature control symbol. Fig. 8-12. The tolerance may be expressed on either a maximum material condition or regardless of feature size basis.

Tolerance of Form and Runout

Form tolerances control conditions of straightness, flatness, roundness, cylindricity, profile, angularity, parallelism, perpendicularity, and runout. They specify the tolerance zone in which the feature must fall. The forms into which material is fabricated are defined by the use of geometric terms such as a plane, cylinder, cone, square, or hexagon. Since the perfect geometric form specified cannot be produced, the variation must be controlled to permit parts to be interchangeable.

Straightness Tolerance

Straightness of a surface or line is a condition where an element on the surface or an axis is a straight line. The straightness tolerance specifies a tolerance zone of uniform width within which all points on the element or the axis must lie. The tolerance is applied in the view where the elements involved are shown as a straight line. All elements of the feature shown in Fig. 8-13 should lie within the 0.02 mm tolerance zone.

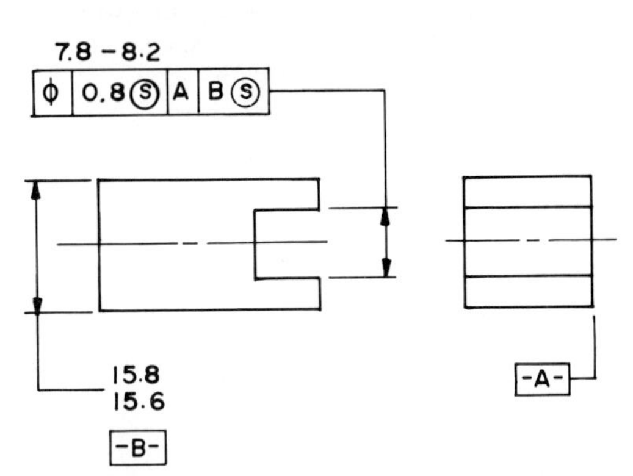
7.8 – 8.2
ϕ 0.8Ⓢ A BⓈ
15.8
15.6
-B-
-A-
THIS CALLS FOR THE CENTER PLANE OF THE SLOT TO LIE BETWEEN TWO PARALLEL PLANES 0.8 mm APART, REGARDLESS OF FEATURE SIZE, WHICH ARE EQUALLY DISPOSED ABOUT THE CENTER PLANE OF DATUM B.

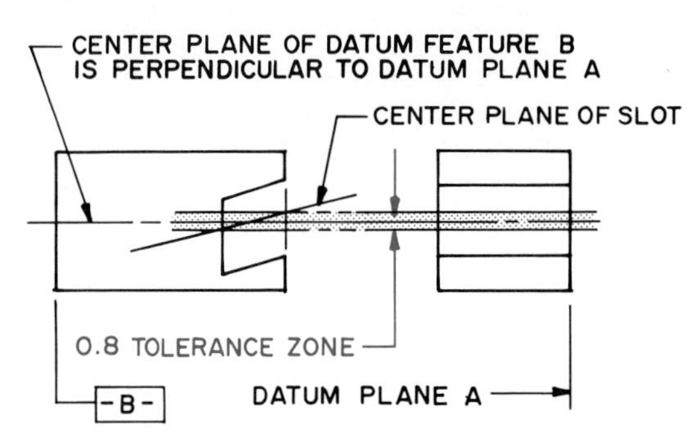
CENTER PLANE OF DATUM FEATURE B IS PERPENDICULAR TO DATUM PLANE A
CENTER PLANE OF SLOT
0.8 TOLERANCE ZONE
-B-
DATUM PLANE A
CENTER PLANE OF SLOT LIES WITHIN THE 0.8 mm TOLERANCE ZONE.

Fig. 8–12. **Symmetry is a tolerance of position. The object's feature (a slot) is symmetrical about the center line of another feature on the same object with an 0.8 mm tolerance zone. (Reproduced from *ANSI Y14.5M-1982* with permission of the publisher, The American Society of Mechanical Engineers.)**

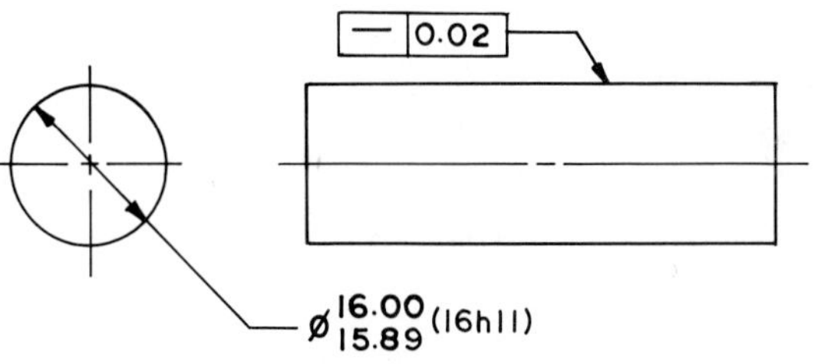
— 0.02
Ø 16.00 15.89 (16h11)
THIS CALLS FOR THE FEATURE TO BE WITHIN THE TOLERANCE ZONE 0.02 mm.

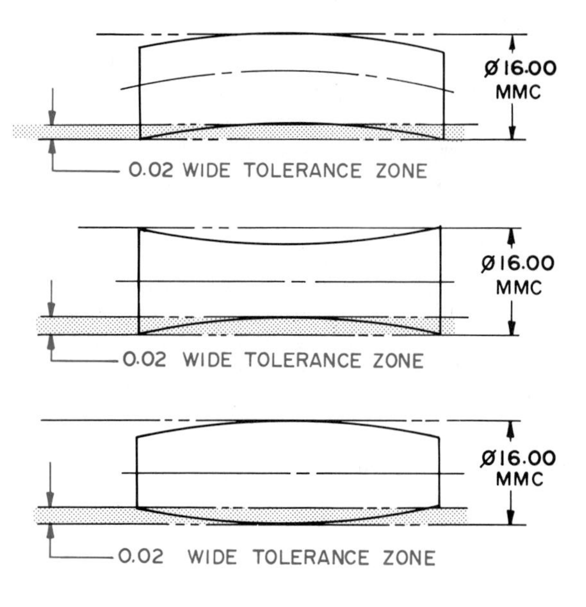
Ø16.00 MMC
0.02 WIDE TOLERANCE ZONE
Ø16.00 MMC
0.02 WIDE TOLERANCE ZONE
Ø16.00 MMC
0.02 WIDE TOLERANCE ZONE
EACH LONGITUDINAL ELEMENT OF THE SURFACE MUST LIE BETWEEN TWO PARALLEL LINES (0.02 APART) WHERE THE TWO LINES AND THE NOMINAL AXIS OF THE PART SHARE A COMMON PLANE.

Fig. 8–13. **Straightness is a tolerance of form in which the elements of a surface are straight lines. The elements of the surface of this object lie within a 0.02 mm tolerance zone. (Reproduced from *ANSI Y14.5M-1982* with permission of the publisher, The American Society of Mechanical Engineers.)**

Flatness Tolerance

Flatness of a surface is a condition where all elements on the surface are in one plane. The flatness tolerance specifies a tolerance zone within which the surface must lie. The symbol is connected to the surface with a leader or an extension line of the surface. The symbol is the place in the view where the surface to be controlled appears as a line. All elements on the surface shown in Fig. 8-14 should lie within the 0.25 mm tolerance zone.

Roundness Tolerance

Roundness of a surface of revolution is a condition where, for a cone or cylinder, all points of the surface intersected by any plane perpendicular to a common axis are equidistant from that axis. For a sphere, roundness means all points of the surface intersected by any plane passing through a common center are equidistant from that center. The roundness tolerance specifies a tolerance

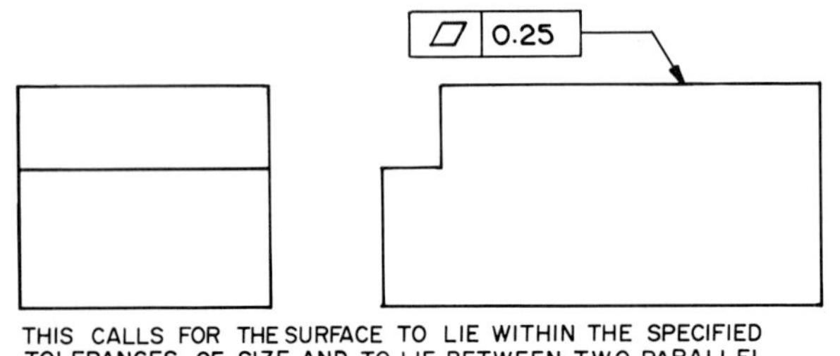

THIS CALLS FOR THE SURFACE TO LIE WITHIN THE SPECIFIED TOLERANCES OF SIZE AND TO LIE BETWEEN TWO PARALLEL PLANES 0.25 mm APART.

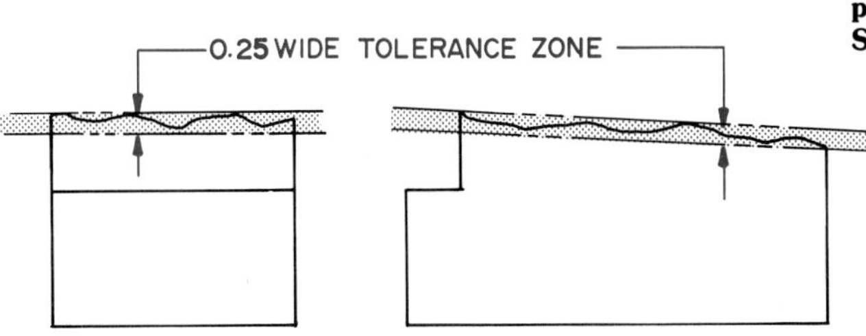

Fig. 8-14. Flatness is a tolerance of form where all the elements on the surface fall within a specified tolerance zone. (Reproduced from *ANSI Y14.5M-1982* with permission of the publisher, The American Society of Mechanical Engineers.)

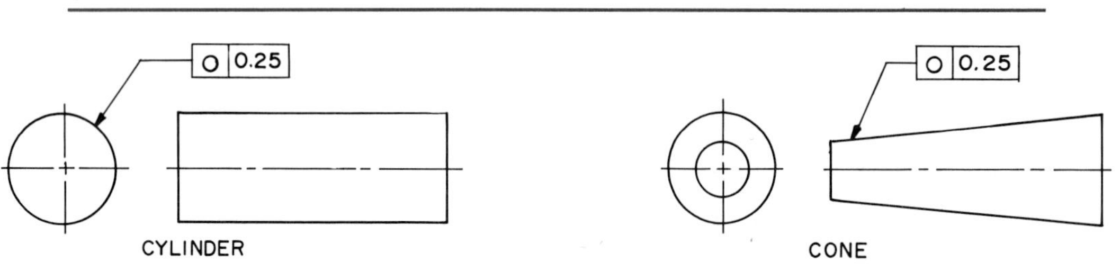

THIS CALLS FOR EACH CIRCULAR ELEMENT OF THE SURFACE IN ANY PLANE PERPENDICULAR TO A COMMON AXIS TO BE WITHIN THE SPECIFIED TOLERANCE OF SIZE AND TO LIE BETWEEN TWO CONCENTRIC CIRCLES (ONE HAVING A RADIUS 0.25 mm LARGER THAN THE OTHER).

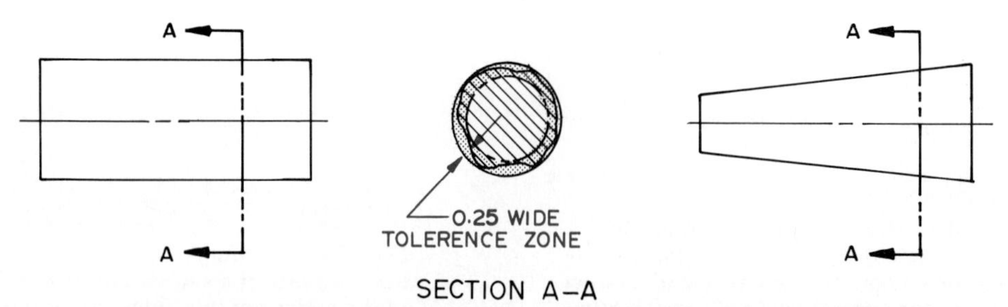

Fig. 8-15. Roundness is a tolerance of form that specifies a tolerance zone defined by two concentric circles. (Reproduced from *ANSI Y14.5M-1982* with permission of the publisher, The American Society of Mechanical Engineers.)

zone defined by two concentric circles. Each circular element of the surface must fall within that zone. Roundness tolerance for a cylinder and a cone is shown in Fig. 8-15, and a sphere in 8-16.

Cylindricity Tolerance

Cylindricity of a surface of revolution is a condition in which all points of the surface are equidistant from a common axis. It is actually a composite control of a form that includes roundness, straightness, and taper of a cylindrical feature. The cylindricity tolerance specifies a tolerance zone composed of two concentric cylinders within which the surface must fall. The tolerance is given as the radius of the circular zone. The tolerance applies simultaneously to the circular and longitudinal elements of the entire surface. The leader from the feature control symbol can be directed to the cylindrical or straight line view. Fig. 8-17.

Profile Tolerance

A **profile** is the outline of an object. Two dimensional objects are projected to a plane. Three dimensional objects are projected to a plane or cut by cross sections revealing the profile at that point. A profile can be composed of arcs, curves, and straight lines. The true profile is defined as basic untoleranced dimensions. The profile tolerance specifies a uniform boundary along the true profile. The elements of the surface must fall within this tolerance zone. Fig. 8-18. The zone may be *unilateral* (fall on one side of the true profile) or *bilateral* (fall on both sides of the true profile). A dimensioned drawing showing the application of the feature control symbol is shown in Fig. 8-19.

Angularity Tolerance

Angularity involves controlling the specified angle that a surface or axis makes with a datum plane or axis. The angularity tolerance specifies the tolerance zone defined by two parallel planes within which the surface or axis of the feature must fall. Figs. 8-20 and 8-21 (p. 212).

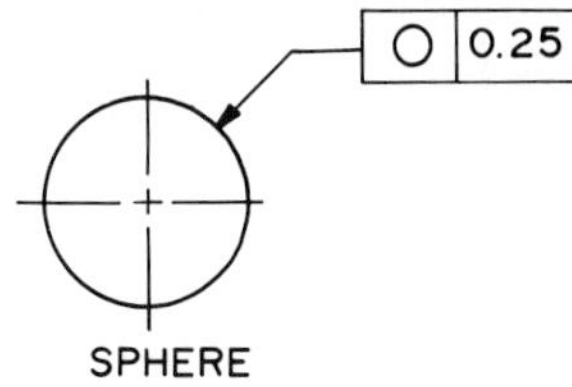

Fig. 8-16. Roundness of a sphere means that any section taken through it will be round and fall within the specified tolerance zone. (Reproduced from *ANSI Y14.5M-1982* with permission of the publisher, The American Society of Mechanical Engineers.)

THIS CALLS FOR EACH CIRCULAR ELEMENT OF THE SURFACE IN A PLANE PASSING THROUGH A COMMON CENTER TO BE WITHIN THE SPECIFIED TOLERANCE OF SIZE AND LIE BETWEEN TWO CONCENTRIC CIRCLES (ONE HAVING A RADIUS 0.25 mm LARGER THAN THE OTHER).

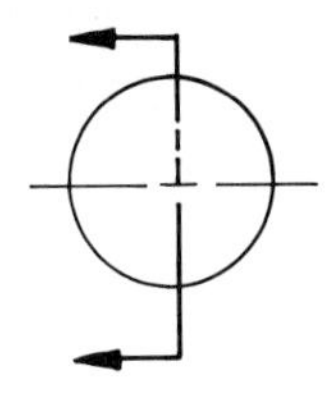

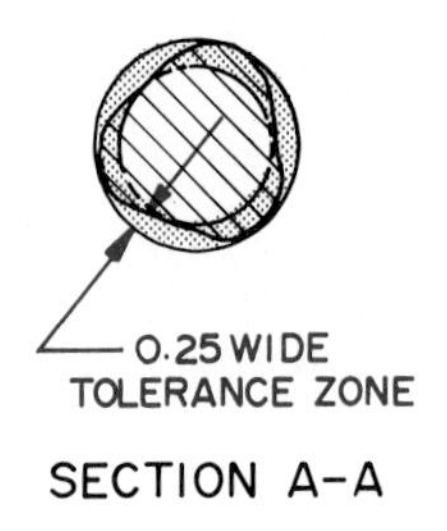

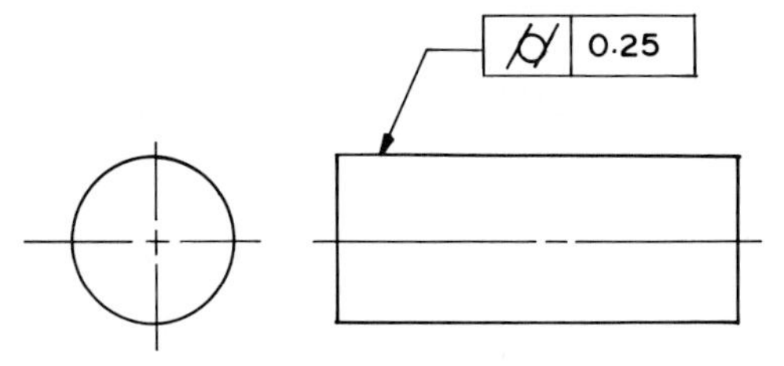

THIS CALLS FOR THE CYLINDRICAL SURFACE TO BE WITHIN THE SPECIFIED TOLERANCE OF SIZE AND TO LIE BETWEEN TWO CONCENTRIC CYLINDERS (ONE HAVING A RADIUS 0.25 mm LARGER THAN THE OTHER).

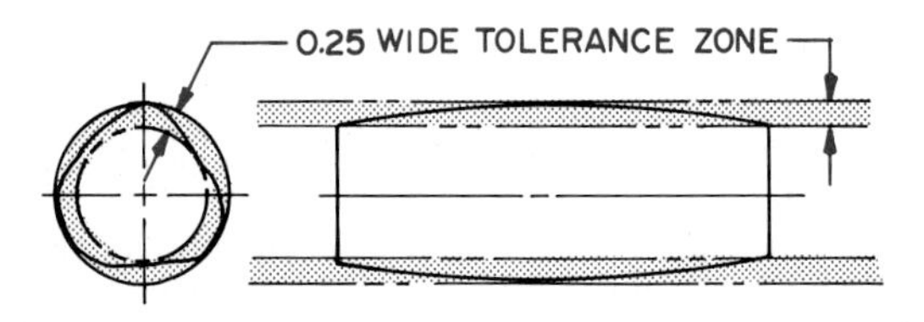

Fig. 8-17. Cylindricity is a tolerance of form specifying that the surface of a cylinder lies between a tolerance zone formed by two concentric cylinders. (Reproduced from *ANSI Y14.5M-1982* with permission of the publisher, The American Society of Mechanical Engineers.)

Parallelism Tolerance

Parallelism means that a surface or line is positioned so that it is equidistant at all points from a datum plane or axis. The parallelism tolerance is specified for a plane surface by a tolerance zone defined by two planes or lines parallel to a datum or axis within which the surface or axis of the feature must fall. The parallelism tolerance is specified for a cylindrical surface by a tolerance zone parallel to a datum axis within which the axis of the feature must fall. Figs. 8-22 and 8-23.

Perpendicularity Tolerance

Perpendicularity refers to a surface or line that is positioned so that it is at a right angle to a datum plane or axis. The perpendicularity tolerance zone can be specified in any of four ways:

1. A tolerance zone defined by two parallel planes perpendicular to a datum plane or axis and at a right angle to a datum plane or axis within which the surface or median plane of the feature must fall. Fig. 8-24.
2. A tolerance zone defined by two parallel planes perpendicular to a datum axis within which the axis of the feature must fall. Fig. 8-25.
3. A cylindrical tolerance zone perpendicular to a datum plane within which the axis of the feature must fall. Fig. 8-26 (p. 214).
4. A tolerance zone defined by two parallel lines perpendicular to a datum plane or axis within which an element of the surface must fall. Fig. 8-27 (p. 214).

Runout Tolerance

A **runout tolerance** is a composite tolerance used to control features of a part having a common axis. The features controlled are either surfaces constructed around a datum axis or those constructed at right angles to a datum axis. Fig. 8-28 (p. 214).

Runout tolerances take into account roundness, straightness, flatness, and parallelism. These are shown in Fig. 8-29 (p. 215). Runout tolerances are checked by rotating the part and checking the surface with a dial indicator.

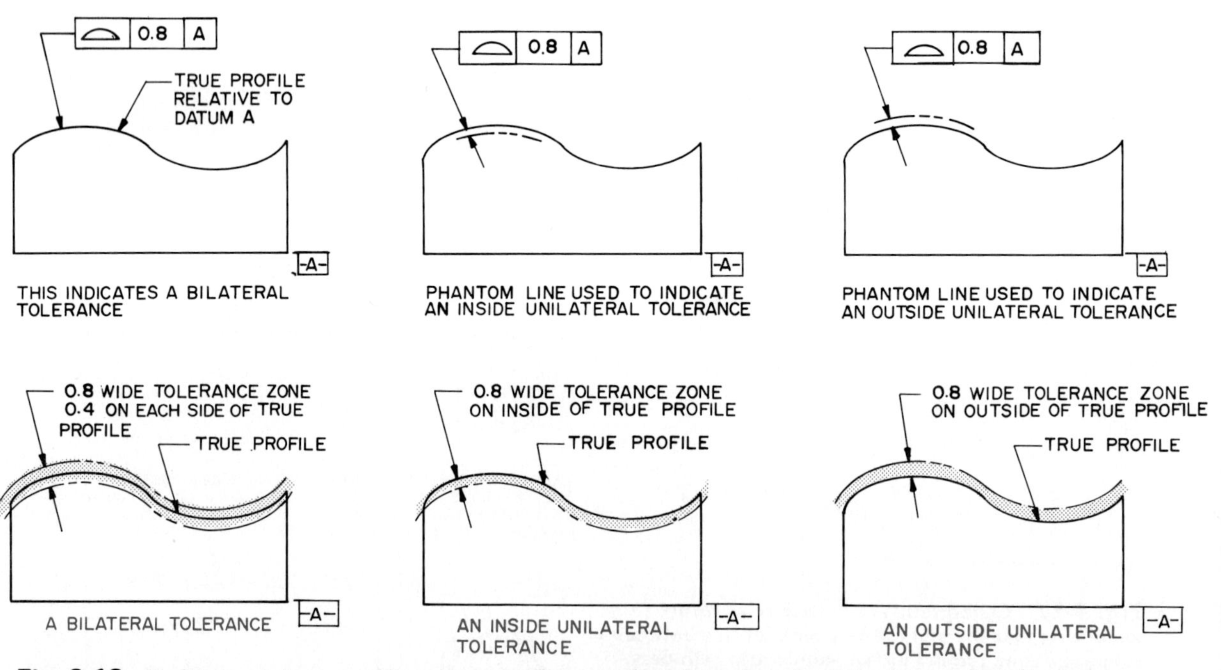

Fig. 8-18. Profile is a tolerance of form that specifies a uniform boundary along the true profile of irregular curves. Elements of the curve must fall within the specified tolerance zone. (Reproduced from *ANSI Y14.5M-1982* with permission of the publisher, The American Society of Mechanical Engineers.)

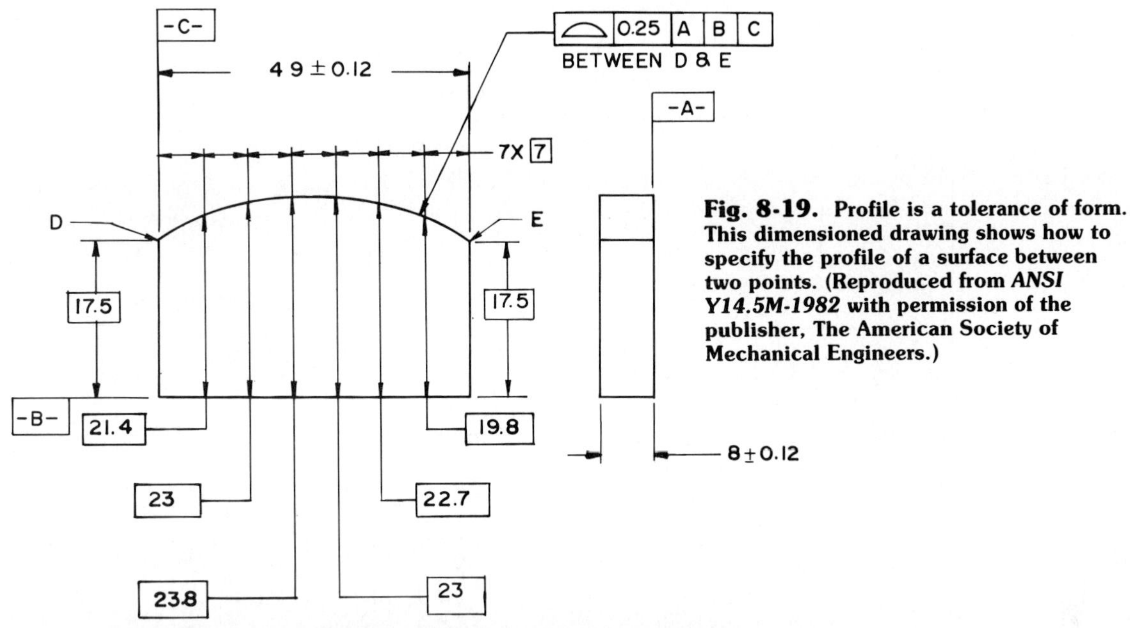
-C-
49 ± 0.12
0.25 A B C
BETWEEN D & E
-A-
7X 7
D
E
17.5
17.5
-B-
21.4
19.8
8 ± 0.12
23
22.7
23.8
23
THE CURVED SURFACE MUST LIE BETWEEN TWO PROFILE BOUNDARIES THAT ARE 0.25mm APART, EQUALLY DISPOSED ABOUT THE TRUE PROFILE, WHICH IS PERPENDICULAR TO DATUM A AND POSITIONED WITH RESPECT TO DATUM PLANES B AND C.

Fig. 8-19. **Profile is a tolerance of form. This dimensioned drawing shows how to specify the profile of a surface between two points. (Reproduced from *ANSI Y14.5M-1982* with permission of the publisher, The American Society of Mechanical Engineers.)**

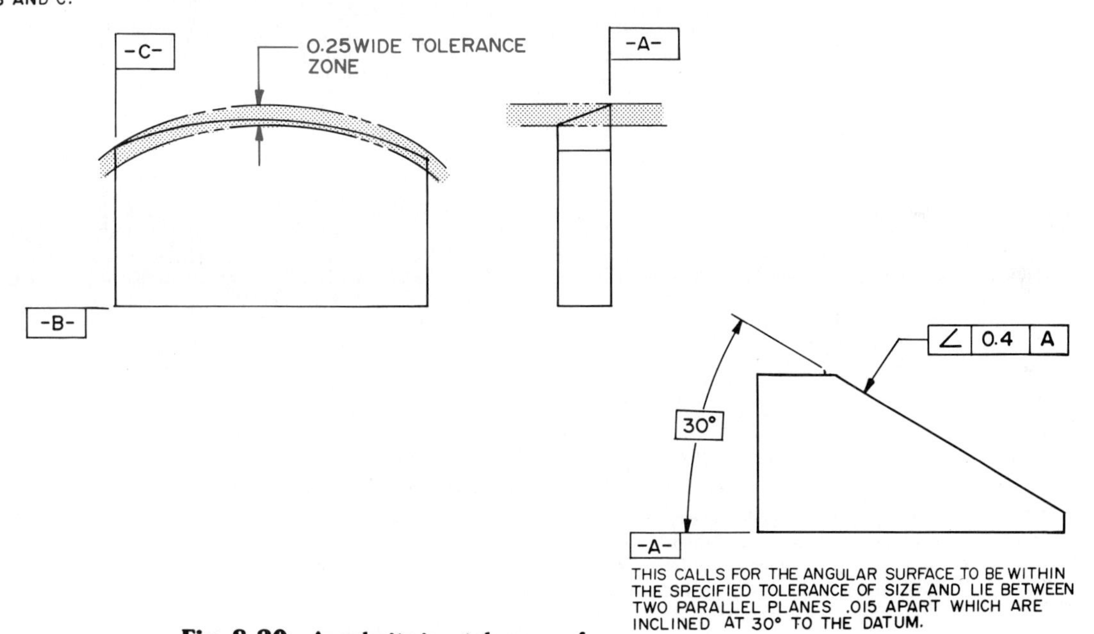
-C-
0.25 WIDE TOLERANCE ZONE
-A-
-B-
0.4 A
30°
-A-
THIS CALLS FOR THE ANGULAR SURFACE TO BE WITHIN THE SPECIFIED TOLERANCE OF SIZE AND LIE BETWEEN TWO PARALLEL PLANES .015 APART WHICH ARE INCLINED AT 30° TO THE DATUM.

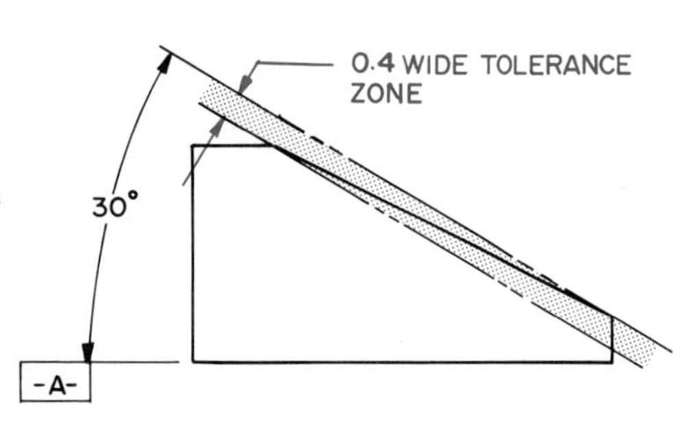
0.4 WIDE TOLERANCE ZONE
30°
-A-

Fig. 8-20. **Angularity is a tolerance of form that controls the angle a surface makes with a datum plane. In this example the 30 degree angle is the true angle, a basic angle. The tolerance specified is 0.4 mm from the basic angle. (Reproduced from *ANSI Y14.5M-1982* with permission of the publisher, The American Society of Mechanical Engineers.)**

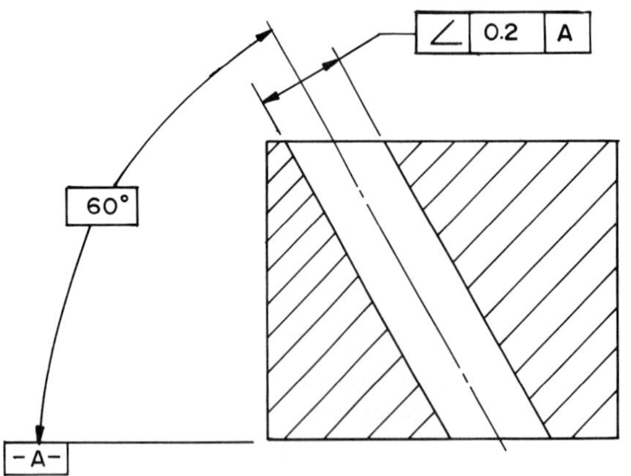

THIS CALLS FOR THE FEATURE AXIS TO BE WITHIN THE SPECIFIED TOLERANCE OF LOCATION AND TO LIE BETWEEN TWO PARALLEL PLANES 0.2 mm APART THAT ARE INCLINED AT 60° TO THE DATUM PLANE.

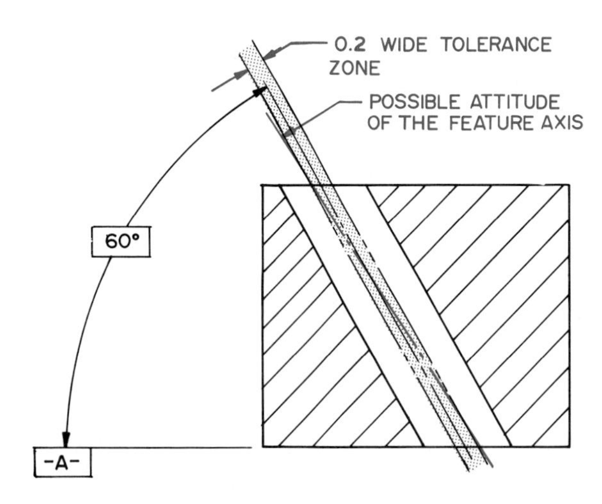

Fig. 8-21. This example shows how the angle of an axis is specified regardless of feature size. (Reproduced from *ANSI Y14.5M-1982* with permission of the publisher, The American Society of Mechanical Engineers.)

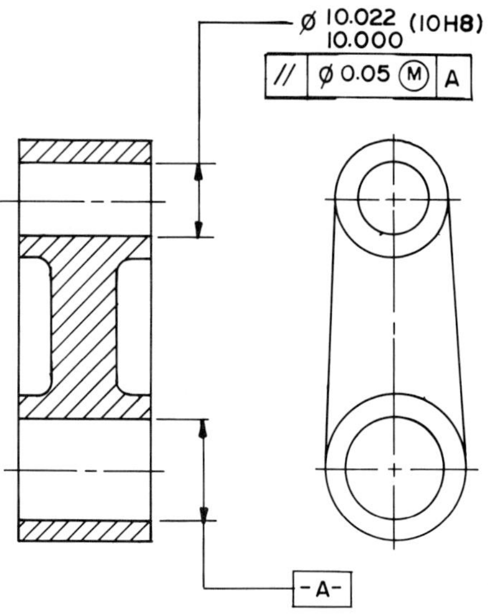

THIS CALLS FOR THE FEATURE AXIS TO BE WITHIN THE SPECIFIED TOLERANCE OF LOCATION. WHERE THE FEATURE IS AT MAXIMUM MATERIAL CONDITION (10.00) THE MAXIMUM PARALLELISM TOLERANCE IS 0.05 DIAMETER. WHERE THE FEATURE DEPARTS FROM ITS MMC SIZE, AN INCREASE IN THE PARALLELISM TOLERANCE IS ALLOWED WHICH IS EQUAL TO THE AMOUNT OF SUCH DEPARTURE.

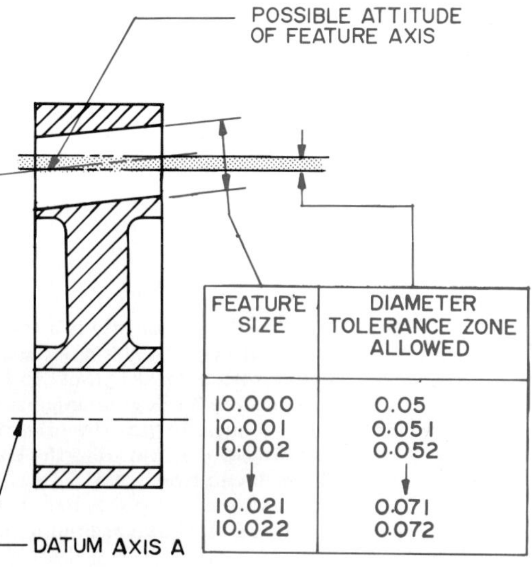

FEATURE SIZE	DIAMETER TOLERANCE ZONE ALLOWED
10.000	0.05
10.001	0.051
10.002	0.052
↓	↓
10.021	0.071
10.022	0.072

Fig. 8–22. This is an example of tolerancing when the features are at MMC and the datum feature is RFS. The upper hole must be parallel to the lower hole (used as a datum) within 0.05 mm diameter. As the hole diameter increases to its maximum size (1.022 mm), the tolerance zone increases to 0.072 mm. (Reproduced from *ANSI Y14.5M-1982* with permission of the publisher, The American Society of Mechanical Engineers.)

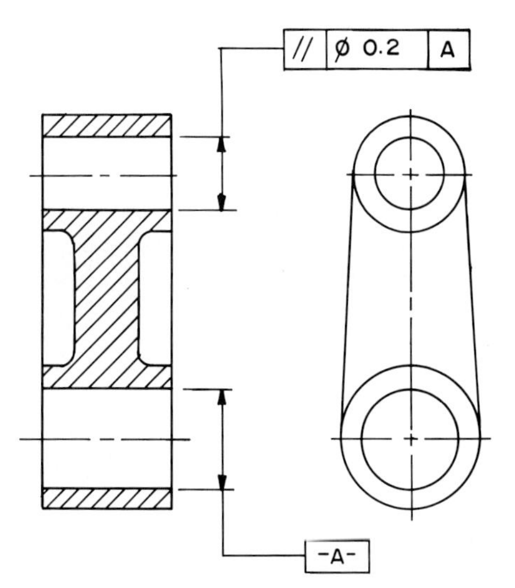
// ∅ 0.2 A
-A-
THIS CALLS FOR THE FEATURE AXIS TO BE WITHIN
THE SPECIFIED TOLERANCE OF LOCATION AND LIE
WITHIN A CYLINDRICAL ZONE 0.2mm IN DIAMETER
REGARDLESS OF FEATURE SIZE, WHICH IS PARALLEL
TO THE DATUM AXIS.

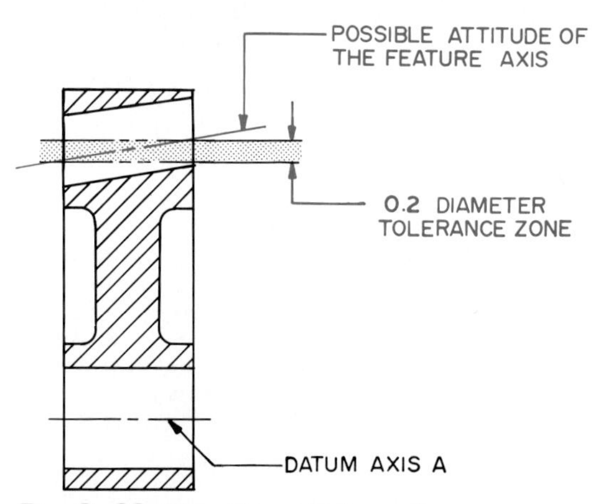
POSSIBLE ATTITUDE OF
THE FEATURE AXIS
0.2 DIAMETER
TOLERANCE ZONE
DATUM AXIS A

Fig. 8–23. **Specify parallelism of one center to another by using one center as a datum and applying the specified tolerance to the other center. (Reproduced from *ANSI Y14.5M-1982* with permission of the publisher, The American Society of Mechanical Engineers.)**

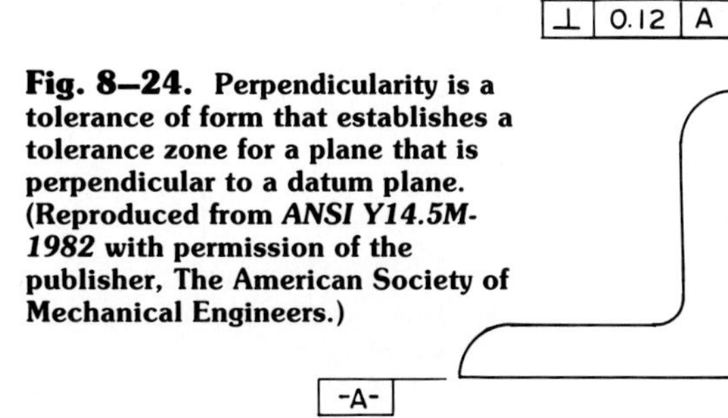
⊥ 0.12 A
-A-

Fig. 8–24. **Perpendicularity is a tolerance of form that establishes a tolerance zone for a plane that is perpendicular to a datum plane. (Reproduced from *ANSI Y14.5M-1982* with permission of the publisher, The American Society of Mechanical Engineers.)**

THIS CALLS FOR THE VERTICAL SURFACE
TO BE WITHIN THE SPECIFIED TOLERANCE
OF SIZE AND LIE BETWEEN TWO PARALLEL
PLANES 0.12 mm APART WHICH ARE
PERPENDICULAR TO THE DATUM PLANE.

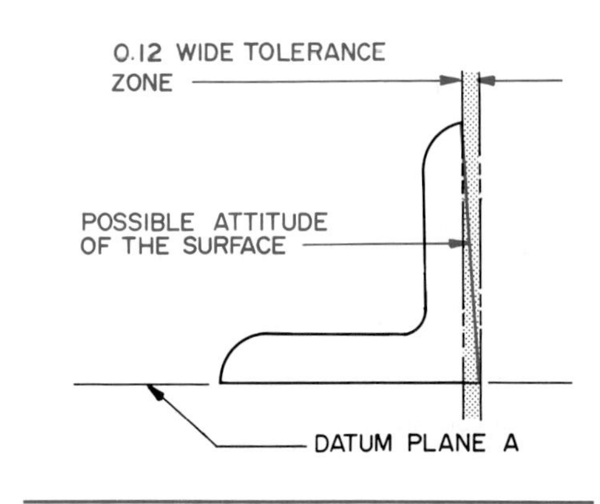
0.12 WIDE TOLERANCE
ZONE
POSSIBLE ATTITUDE
OF THE SURFACE
DATUM PLANE A

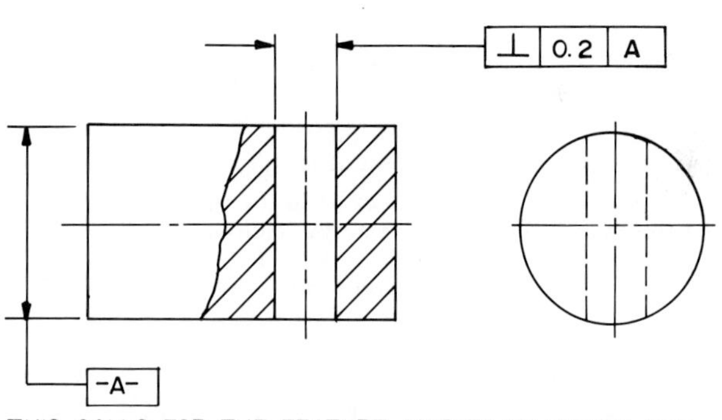
⊥ 0.2 A
-A-
THIS CALLS FOR THE FEATURE AXIS TO BE WITHIN THE
SPECIFIED TOLERANCE OF LOCATION AND LIE BETWEEN
TWO PLANES 0.2mm APART, REGARDLESS OF FEATURE SIZE,
WHICH ARE PERPENDICULAR TO THE DATUM AXIS. THIS
TOLERANCE APPLIES ONLY TO THE VIEW ON WHICH IT IS
SPECIFIED, NOT THE END VIEW.

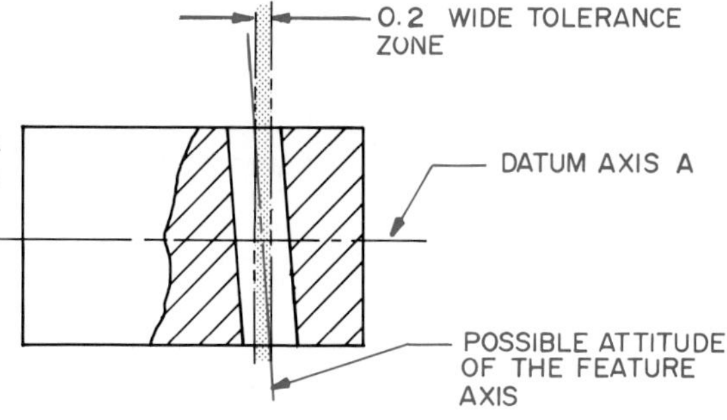
0.2 WIDE TOLERANCE
ZONE
DATUM AXIS A
POSSIBLE ATTITUDE
OF THE FEATURE
AXIS

Fig. 8–25. **Perpendicularity of an axis can be specified by two planes perpendicular to a datum plane. (Reproduced from *ANSI Y14.5M-1982* with permission of the publisher, The American Society of Mechanical Engineers.)**

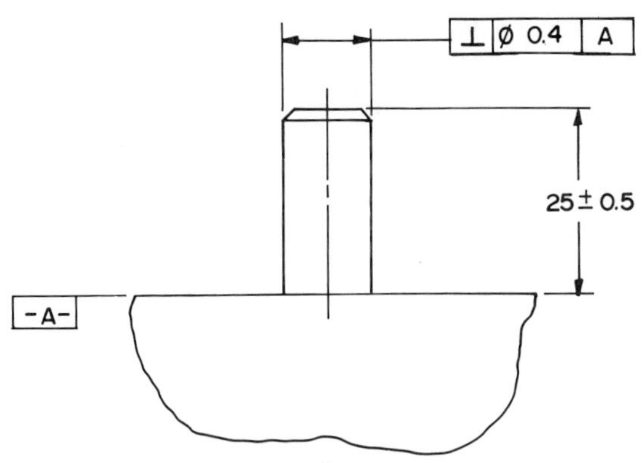
⊥ Ø 0.4 A
25 ± 0.5
-A-
THIS CALLS FOR THE FEATURE AXIS TO BE WITHIN THE SPECIFIED TOLERANCE OF LOCATION AND LIE WITHIN A CYLINDRICAL ZONE 0.4mm IN DIAMETER REGARDLESS OF FEATURE SIZE, WHICH IS PERPENDICULAR TO AND PROJECTS FROM THE DATUM PLANE FOR THE FEATURE HEIGHT.

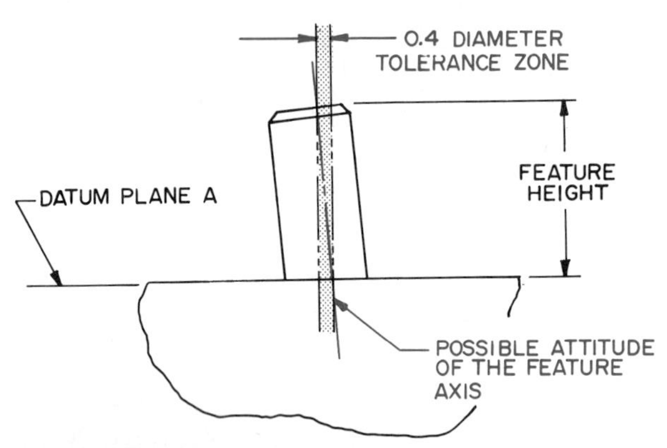
0.4 DIAMETER TOLERANCE ZONE
DATUM PLANE A
FEATURE HEIGHT
POSSIBLE ATTITUDE OF THE FEATURE AXIS

Fig. 8–26. Perpendicularity of an axis can be specified by a cylindrical tolerance zone perpendicular to a datum plane. (Reproduced from *ANSI Y14.5M-1982* with permission of the publisher, The American Society of Mechanical Engineers.)

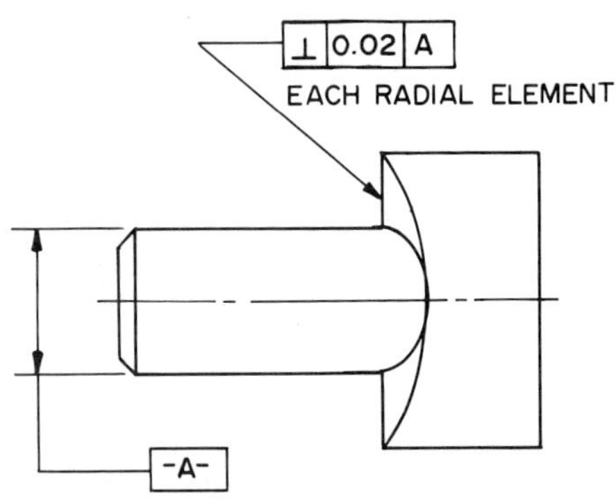
⊥ 0.02 A
EACH RADIAL ELEMENT
-A-
THIS CALLS FOR EACH RADIAL ELEMENT TO BE WITHIN THE SPECIFIED TOLERANCE OF SIZE AND IT MUST LIE BETWEEN TWO PARALLEL LINES 0.02 mm APART WHICH ARE PERPENDICULAR TO THE DATUM AXIS.

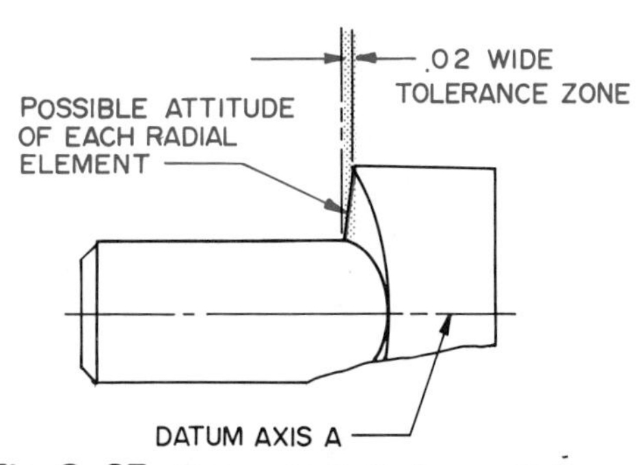
.02 WIDE TOLERANCE ZONE
POSSIBLE ATTITUDE OF EACH RADIAL ELEMENT
DATUM AXIS A

Fig. 8–27. Perpendicularity for a radial element can be specified by two parallel lines perpendicular to the datum plane. (Reproduced from *ANSI Y14.5M-1982* with permission of the publisher, The American Society of Mechanical Engineers.)

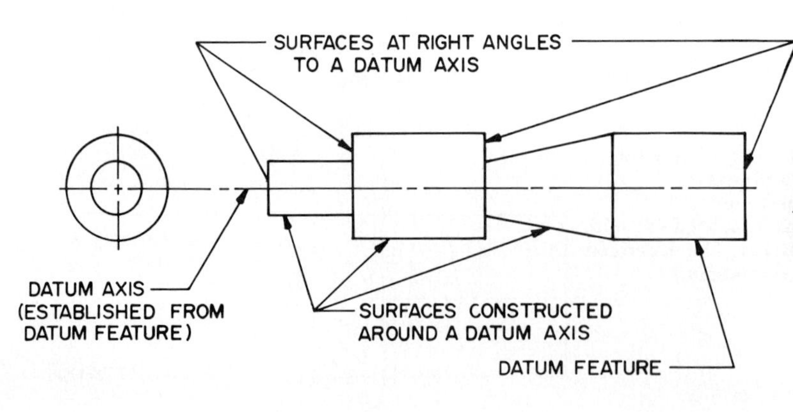
SURFACES AT RIGHT ANGLES TO A DATUM AXIS
DATUM AXIS (ESTABLISHED FROM DATUM FEATURE)
SURFACES CONSTRUCTED AROUND A DATUM AXIS
DATUM FEATURE

Fig. 8–28. Features applicable to runout tolerancing. (Reproduced from *ANSI Y14.5M-1982* with permission of the publisher, The American Society of Mechanical Engineers.)

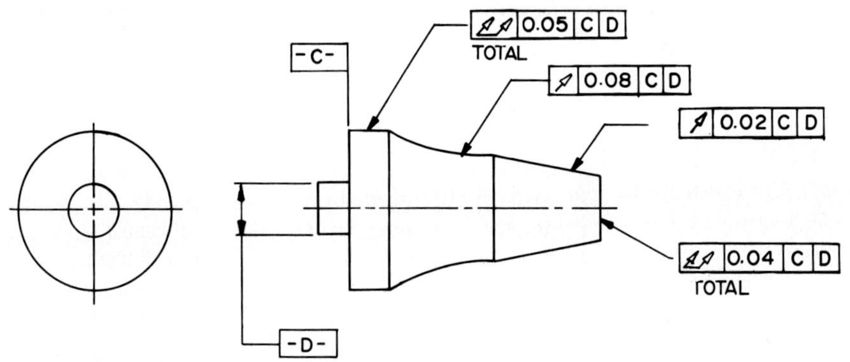

AT ANY MEASURING POSITION, EACH CIRCULAR ELEMENT (FOR CIRCULAR RUNOUT) AND EACH SURFACE (FOR TOTAL RUNOUT) MUST BE WITHIN THE SPECIFIED TOLERANCE WHEN THE PART IS MOUNTED ON DATUM SURFACE C AND ROTATED 360° ABOUT DATUM AXIS D.

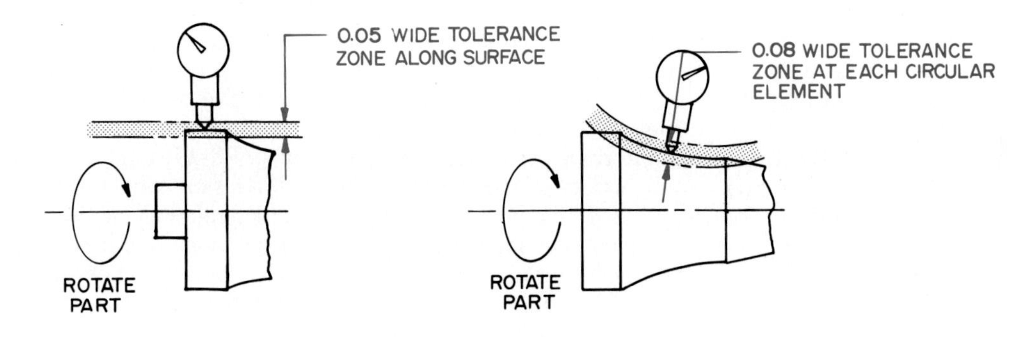

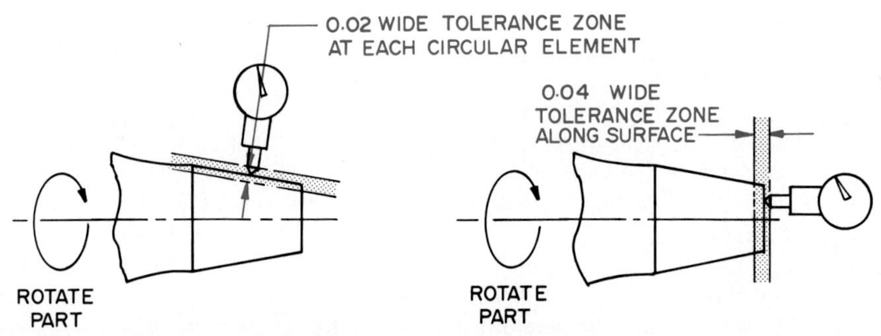

Fig. 8–29. Runout tolerances are specified relative to a datum surface and a diameter. As the object rotates, it is gauged to see if it conforms to the specified tolerance zone. (Reproduced from *ANSI Y14.5M-1982* with permission of the publisher, The American Society of Mechanical Engineers.)

Chapter Review

Build Your Vocabulary

You should understand and use the following terms as part of your working vocabulary. Write a brief explanation of what each means.

geometric tolerancing
basic dimension
maximum material condition
least material condition
regardless of feature size
positional tolerances
true position tolerance
concentricity
symmetry
straightness
flatness
roundness
cylindricity
profile
angularity
parallelism
perpendicularity
runout tolerance

Study Problems—Directions

Note: Each square on the grid used with these study problems is equal to 5 mm or .20 inches.

1. Draw the part shown in **Fig. P8-1**. Locate the two holes with a size tolerance of 0.5 mm and a true-position tolerance of 0.50 DIA. Dimension the drawing and show the proper symbols.
2. Draw the part shown in **Fig. P8-2**. Locate the four holes with a size tolerance of 0.80 mm and a true-position tolerance of 0.50 DIA. Dimension the drawing and show the proper symbols.
3. Use true-position dimensioning to locate the holes in the part in **Fig. P8-3**. Dimension the drawing showing a size tolerance of 1.00 mm and a locational tolerance of 0.80 DIA.
4. Dimension the cylinder in **Fig. P8-4** using the necessary dimensions and a feature control symbol to indicate that the cylindricity of the object is 0.70 mm.
5. Dimension the object in **Fig. P8-5** using the necessary dimensions and a feature control symbol to indicate that the angularity tolerance of the inclined plane is 0.50 from the datum plane (the bottom of the object).
6. Dimension the object in **Fig. P8-6** using the necessary dimensions and a feature control symbol to indicate that the profile of the irregular curved surface falls within a 0.60 mm tolerance zone. Show this for bilateral, unilateral inside, and unilateral outside conditions.
7. Dimension the concentric cylinders in **Fig. P8-7** using necessary dimensions and a feature control symbol to indicate that the small cylinder is concentric with the large cylinder (which is the datum cylinder) within a tolerance of 0.60 mm.
8. Dimension the cylinder in **Fig. P8-8** using the necessary dimensions and a feature control symbol to indicate that the elements of the cylinder are straight with a tolerance of 0.30 mm.
9. Dimension the rectangular object in **Fig. P8-9** using the necessary dimensions and a feature control symbol to indicate that the top surface is flat within a tolerance of 0.90 mm.
10. Dimension the cylinder, cone, and sphere in **Fig. P8-10** using the necessary dimensions and feature control symbols to indicate that the cross section of each is round within a tolerance of 0.50 mm.
11. Dimension the object in **Fig. P8-11** using the necessary dimensions and a feature control symbol to indicate that the top surface of the object is parallel with the bottom surface (the datum) within 0.80 mm.
12. Dimension the object in **Fig. P8-12** using the necessary dimensions and a feature control symbol to indicate that the top hole is parallel with the bottom hole (the datum) within a tolerance of 0.90 mm.
13. Dimension the object in **Fig. P8-13** using the necessary dimensions and a feature control symbol to indicate that vertical surface A is perpendicular to horizontal surface B (the datum) within a tolerance of 0.40 mm.
14. Dimension the object in **Fig. P8-14** using the necessary dimensions and a feature control symbol to indicate that the hole is perpendicular to the base (the datum) within a tolerance of 0.07 mm.
15. Use a feature control symbol to indicate the conical surface of the object in **Fig. P8-15** has a runout of 1.00 mm.

Study Problems

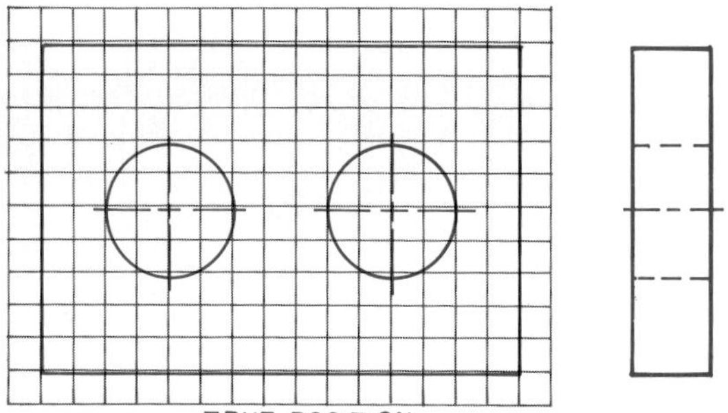

TRUE POSITION

P8–1

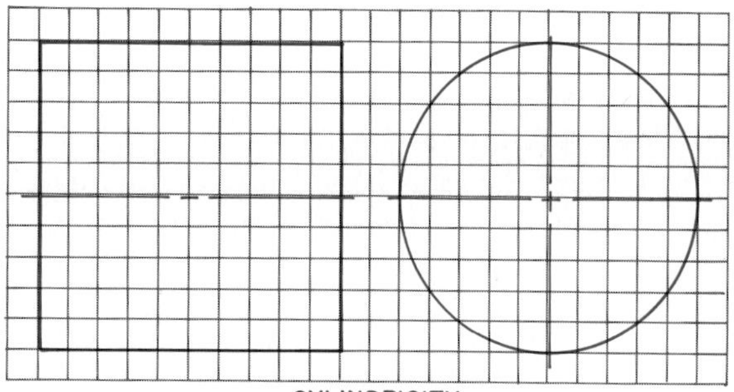

CYLINDRICITY

P8–4

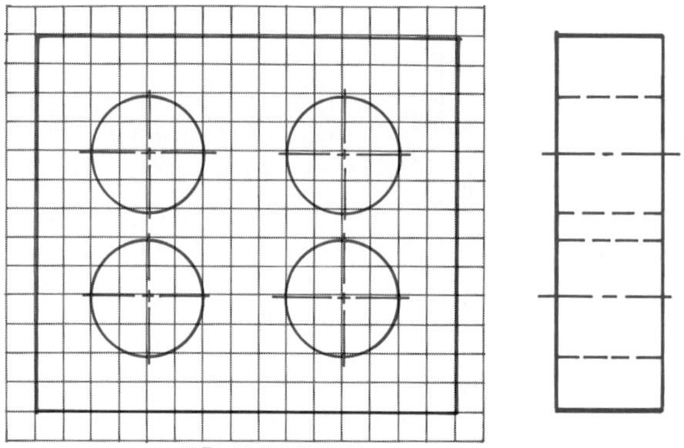

TRUE POSITION

P8–2

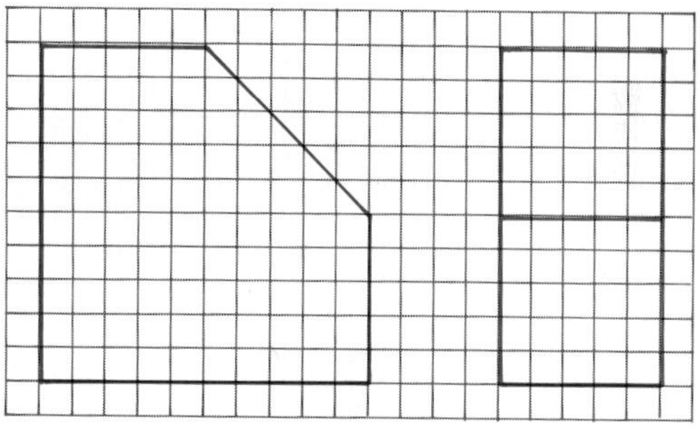

ANGULARITY

P8–5

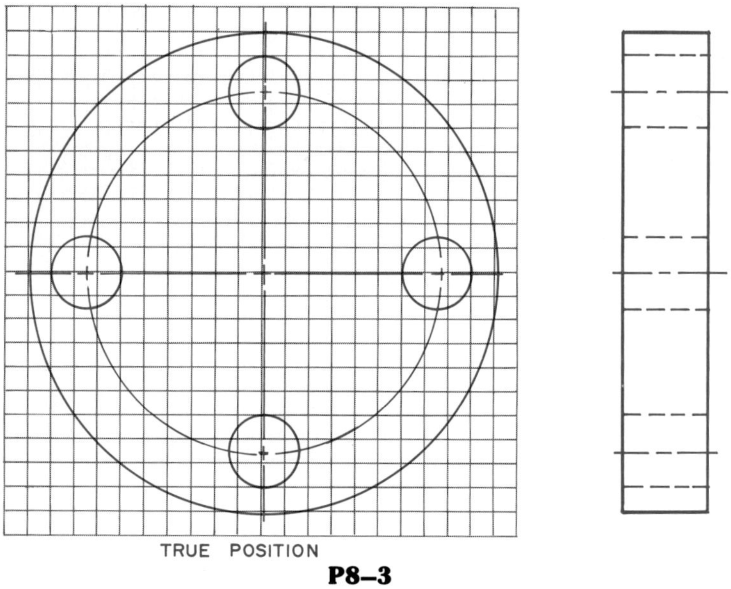

TRUE POSITION

P8–3

Study Problems

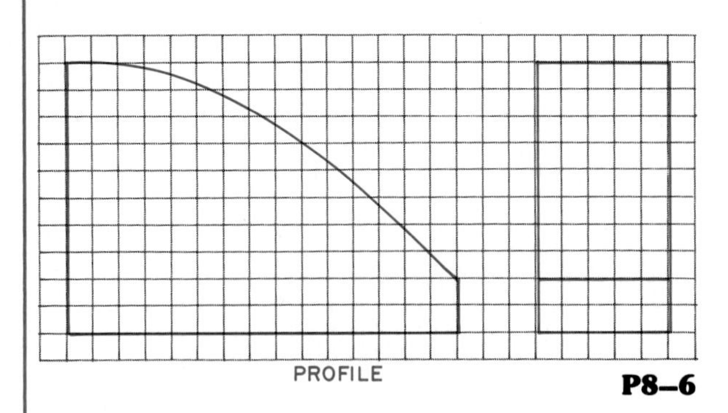

P8–6

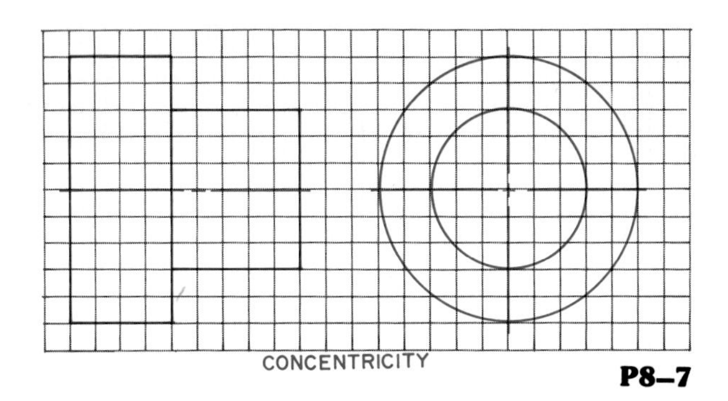

P8–7

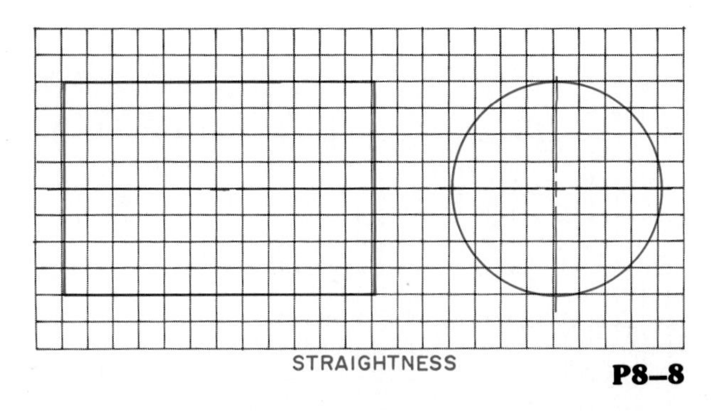

P8–8

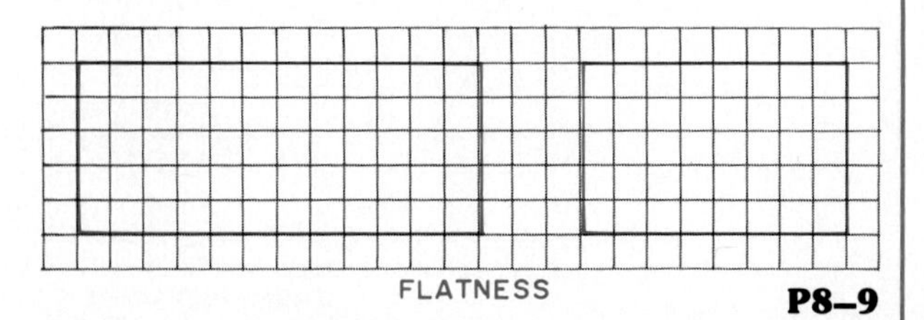

P8–9

Study Problems

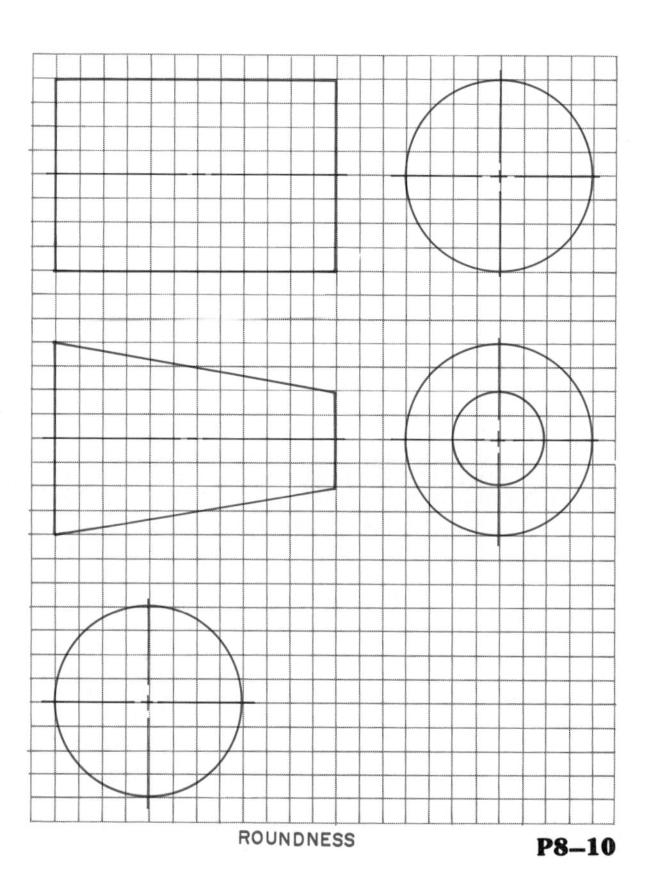

P8–10

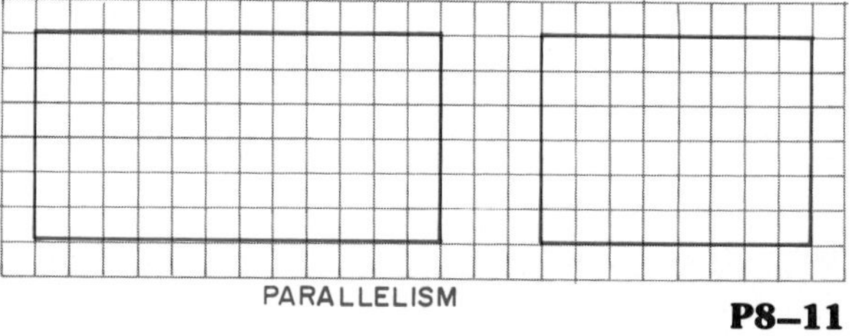

P8–11

Study Problems

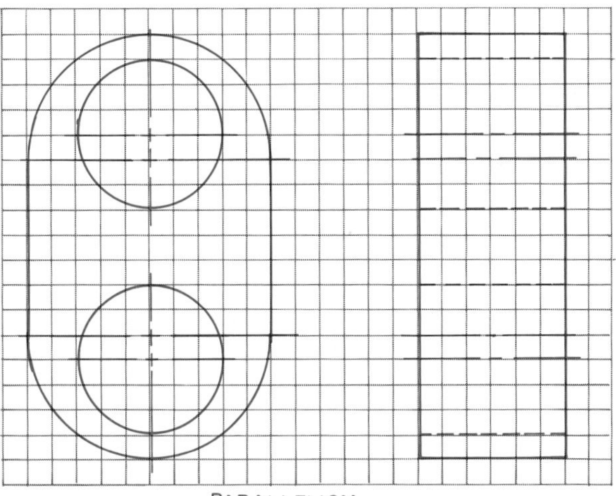

P8–12

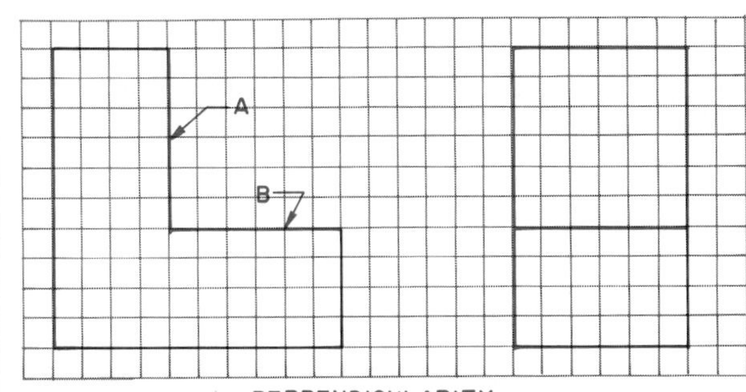

P8–13

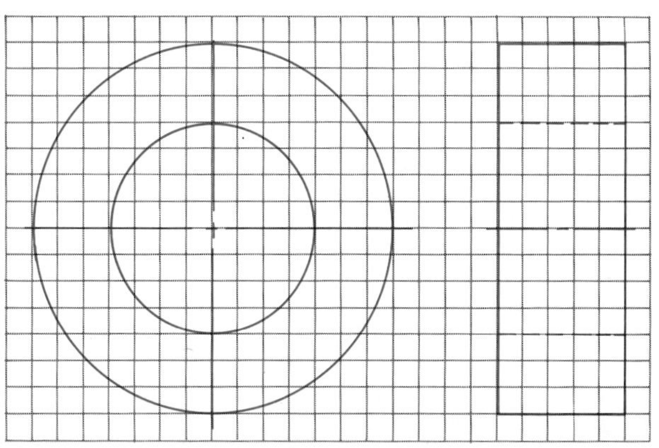

P8–14

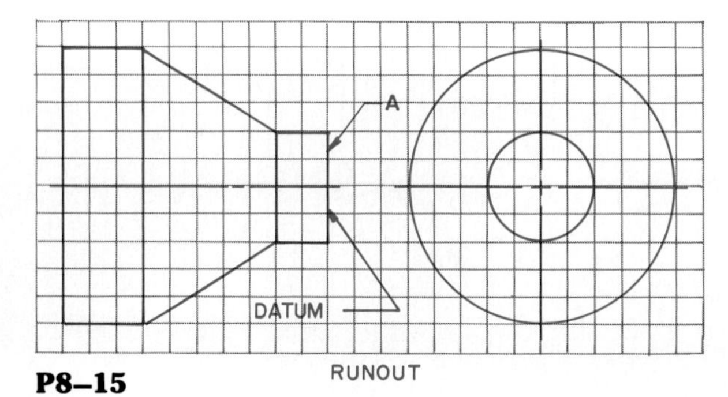

P8–15

CHAPTER 9

Auxiliary Views and Revolutions

After studying this chapter, you should be able to do the following:

- Draw primary and secondary auxiliary views.
- Draw full and partial auxiliary views.
- Project curved surfaces onto an auxiliary plane.
- Use auxiliary views to find unknown quantities.
- Apply the principles of revolution to draw new views of the object.

Auxiliary Views

Some objects have surfaces that are either *inclined* or *oblique* to regular planes of projection. An *inclined surface* is one that is perpendicular to *one* of the regular planes of projection. The face of the voltmeter shown in Fig. 9-1 is on an inclined surface. An *oblique surface* is not perpendicular to any of the regular planes of projection. (Review the section in Chapter 6, "Multiview Drawing," that discusses projection of planes and lines to regular planes of projection.)

Inclined and oblique surfaces do not appear true size when projected to regular planes of projection. In order to show the true size and shape of these surfaces, you must draw auxiliary views. An **auxiliary view** is an orthographic view drawn on a plane of projection that is parallel with the inclined surface. The line of sight is perpendicular to the inclined surface.

Auxiliary views differ from regular views in one way. They are projected upon a plane that is not one of the regular frontal, horizontal, or profile planes. The plane of projection is called an **auxiliary plane.**

The principles of orthographic projection apply to auxiliary views in the same way they apply to the regular views. All lines of projection are perpendicular to the auxiliary plane. An auxiliary view is simply another orthographic view.

Fig. 9-2 shows a drawing of the voltmeter with its surfaces projected onto parallel surfaces of a glass box. Note the inclined surface. To form this inclined surface, an auxiliary plane is passed through the box parallel to the inclined surface of the voltmeter. The inclined surface of the voltmeter is projected onto the auxiliary plane. Figure 9-3 shows the glass box unfolded.

Fig. 9–1. A voltmeter with a face mounted on an inclined surface. (Triplett Electrical Instrument Co.)

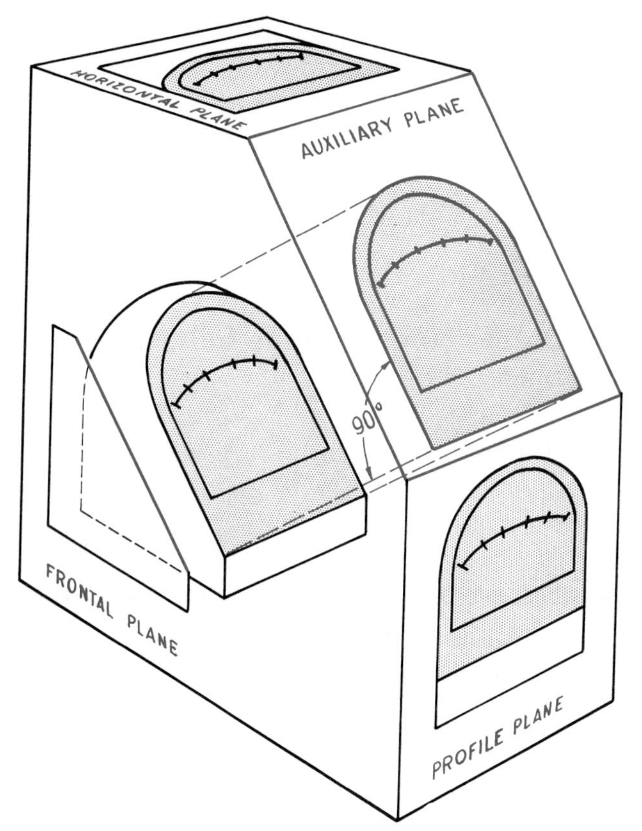

Fig. 9–2. The auxiliary plane is parallel to the inclined plane, and the line of sight is perpendicular to it.

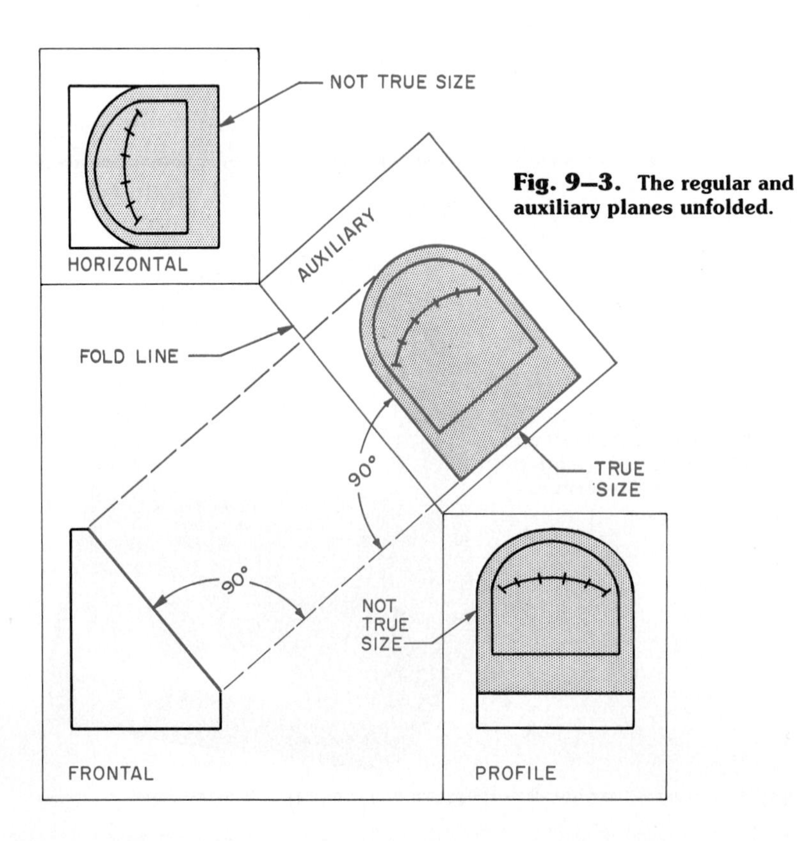

Fig. 9–3. The regular and auxiliary planes unfolded.

Primary Auxiliary Views

The three principal orthographic planes are the horizontal, frontal, and profile. These terms are used to identify the sides of a glass box surrounding the object to be drawn. Auxiliary planes are planes that fold from these principal planes. Fig. 9-4. They may fold from the horizontal plane (top view), frontal plane (front view), or profile plane (side view). The principal planes are identified by the letters H, F, and P. The **primary auxiliary view** is the first auxiliary view projected from a principal plane. It is identified by the number 1 (for first, or primary, auxiliary view).

The auxiliary plane folds from the view in which the inclined surface appears as an edge.

Auxiliary Views Projected from the Top View

When you draw an auxiliary view projected from the top view, notice that the inclined surface is perpendicular to the horizontal plane and appears as an edge in the top view. The steps to draw the auxiliary view are described below and shown in Fig. 9-5.

1. Draw a fold line at any convenient distance from the edge view of the inclined surface and parallel with it. The fold line represents the edge of the glass box. Label it H on the horizontal side and 1 on the primary auxiliary side.
2. Project the edge view to the auxiliary view. The line of sight is perpendicular to the edge view.
3. From the front view, where plane H appears as an edge, measure the distances (A and B) to the top and bottom of the object. Transfer these from the H-1 fold line onto the auxiliary view to form the true-size view of the inclined plane.

Auxiliary Views Projected from the Front View

Remember, when you project an auxiliary view from the front view, the inclined surface is perpendicular to the frontal plane and appears as an edge in the front view. The steps to draw the auxiliary view are described below and shown in Fig. 9-6.

1. Draw a fold line at any convenient distance from the edge view of the inclined surface and parallel with it. Label it F on the frontal side and 1 on the primary auxiliary side.
2. Project the edge view to the auxiliary view. The line of sight is perpendicular to the edge view.
3. From the top view, where the F plane appears as an edge, measure the distances (A and B) to the front and back of the object. Transfer these from the F-1 fold line onto the auxiliary view to form the true-size view of the inclined plane.

Auxiliary Views Projected from the Side View

With an auxiliary view projected from the side view, the inclined surface is perpendicular to the profile plane and appears as an edge in the side view. Follow these steps and Fig. 9-7 to draw the auxiliary view:

1. Draw a fold line at any convenient distance from the edge view of the inclined surface and parallel with it. Label it P on the profile side and 1 on the primary auxiliary side.
2. Project the edge view to the auxiliary view. The line of sight is perpendicular to the edge view.
3. From the front view, measure the distances (A and B) to the right and left sides of the object. Transfer these from the P-1 fold line onto the auxiliary view to form the true-size view of the inclined plane.

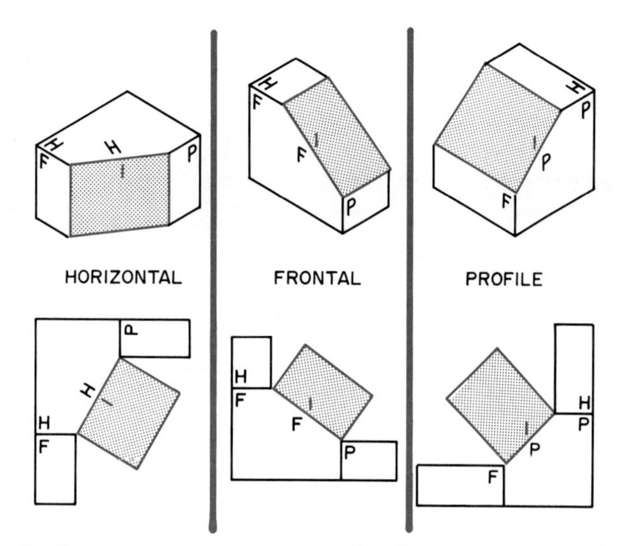
HORIZONTAL
FRONTAL
PROFILE

Fig. 9–4. Primary auxiliary views are taken from the principal planes.

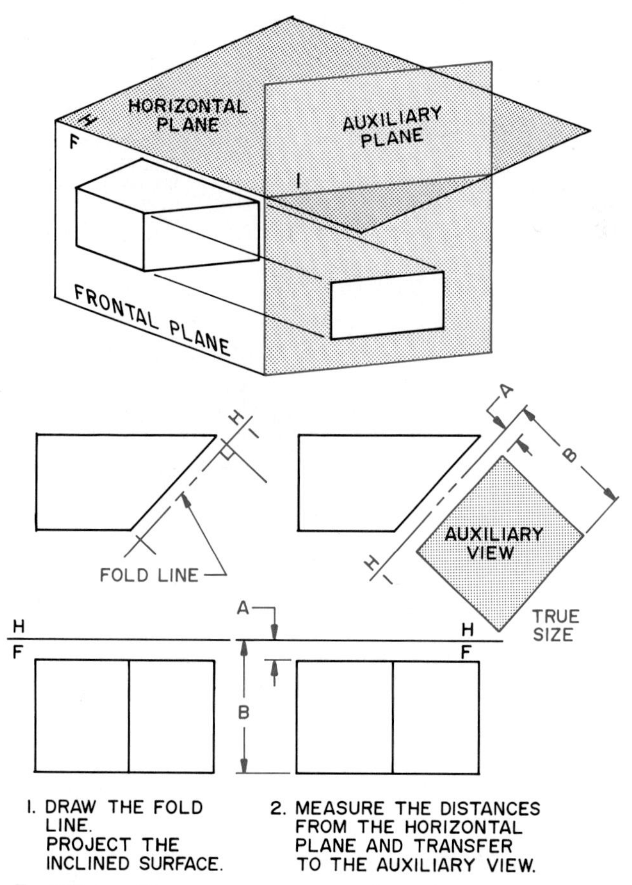
HORIZONTAL PLANE
AUXILIARY PLANE
FRONTAL PLANE
FOLD LINE
AUXILIARY VIEW
TRUE SIZE
1. DRAW THE FOLD LINE. PROJECT THE INCLINED SURFACE.
2. MEASURE THE DISTANCES FROM THE HORIZONTAL PLANE AND TRANSFER TO THE AUXILIARY VIEW.

Fig. 9–5. This auxiliary view is projected from the top view.

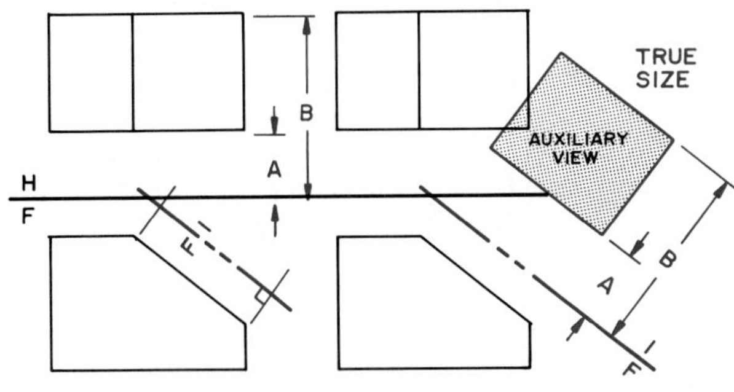

Fig. 9–6. This auxiliary view is projected from the front view.

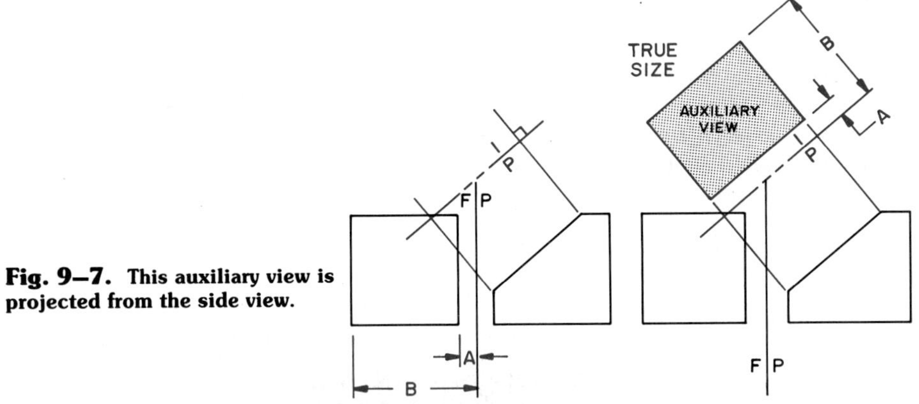

Fig. 9–7. This auxiliary view is projected from the side view.

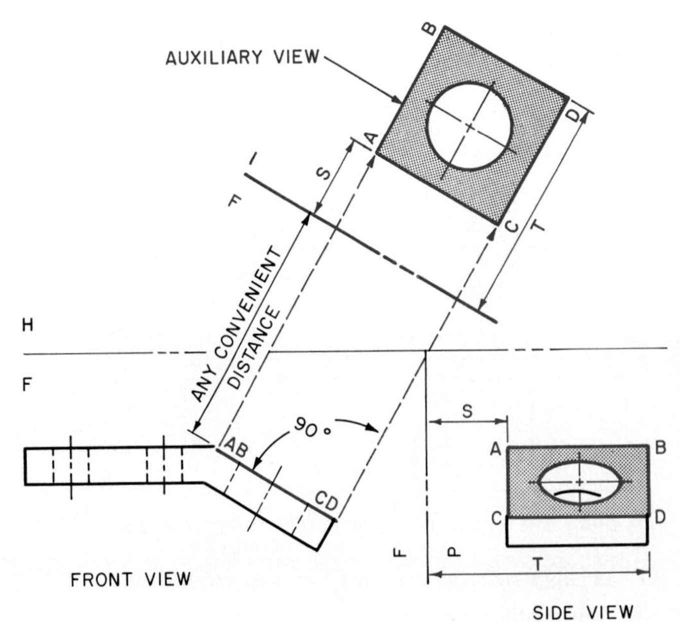

Fig. 9–8. This auxiliary view shows only the true size of the inclined surface. Notice how the corners of the plane are located by measuring their distance from the frontal plane.

Full Auxiliary Views

When drawing an auxiliary view, the inclined surface is the only surface drawn. Fig. 9-8. Since other parts of the object appear true size in the regular views, they are not projected to the auxiliary view. Figure 9-9 illustrates an auxiliary view of the entire object. Notice that all surfaces except the inclined surface are foreshortened. These added surfaces do not make the drawing clear or easy to understand. Since the purpose of a drawing is to show the object clearly and completely, the auxiliary view should show only the inclined surface in its true size. Thus, it is not usually necessary to draw full auxiliary views of an object.

Symmetrical and Nonsymmetrical Auxiliary Views

If a surface to be drawn in an auxiliary view is symmetrical, it is common practice to draw the auxiliary view around a reference plane that passes through the center of the auxiliary view and the regular view. The plane is drawn parallel to the edge view of the surface. The view is laid out to the right and left of the reference plane auxiliary. Fig. 9-10 shows a steel I-beam. Since it is symmetrical, the reference plane is placed through its center. The distances to the right and left of the center line are indicated by the letters X and Y. Distance X is equal to distance Y.

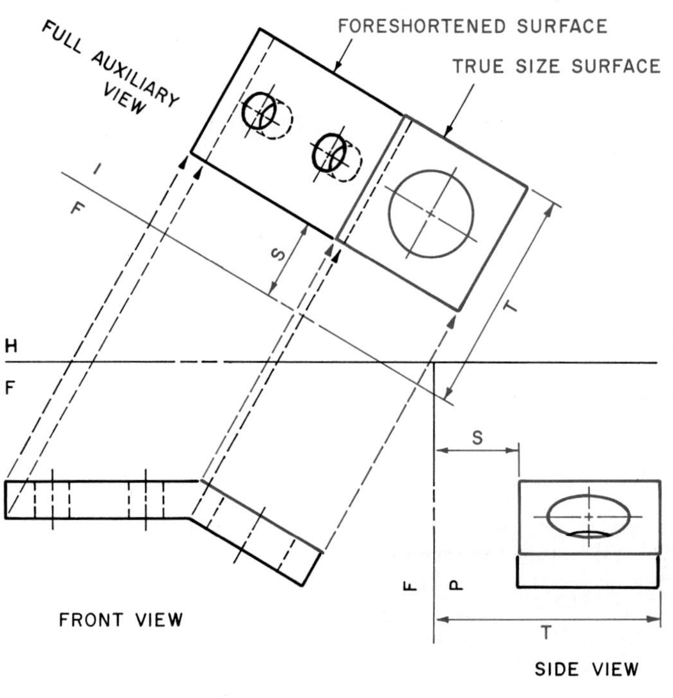

Fig. 9–9. **This is an auxiliary view of the entire object. Notice that everything except the inclined surface is foreshortened.**

Fig. 9–10. **A symmetrical auxiliary view can be constructed around a central reference plane.**

CAD Applications:

Converting Manually Produced Drawings

Companies that switch to CAD usually find it increases productivity. One reason is that frequently used shapes and symbols need to be drawn only once. They can then be inserted into any drawing as needed. (You can read more about this capability in Chapter 4, under "Symbol Libraries).'' Another reason is that changes are usually easier to make. The drawing is called up on the computer and changed as needed. Then the revised drawing is output on a plotter.

Realizing these advantages with new, CAD-produced drawings, some companies have decided that their older, manually produced drawings should be on a CAD system as well. However, re-creating each old drawing "from scratch" on a CAD system can be very costly, both in terms of time and labor. Fortunately, there are faster ways.

One method is called digitizing. It requires a digitizing tablet (see Chapter 4) with a pointing device. The drawing is secured to the surface of the digitizing tablet. By using the appropriate CAD commands and moving the pointing device to certain areas of the drawing, the drawing can be entered into the computer. For example, the user might enter the line command and then point to the beginning and end of a straight line. In this way the line is entered into the computer, at the right length and location.

Another method uses a device called a scanner. A scanner works like a plotter in reverse. (In fact, some are available as plotter accessories.) Instead of outputting a drawing from a computer to paper, a scanner reads a paper drawing and sends the data to a computer. The picture below shows one such device. It is attached to a regular plotter and can automatically input drawings on A-size through E-size paper, vellum, or film.

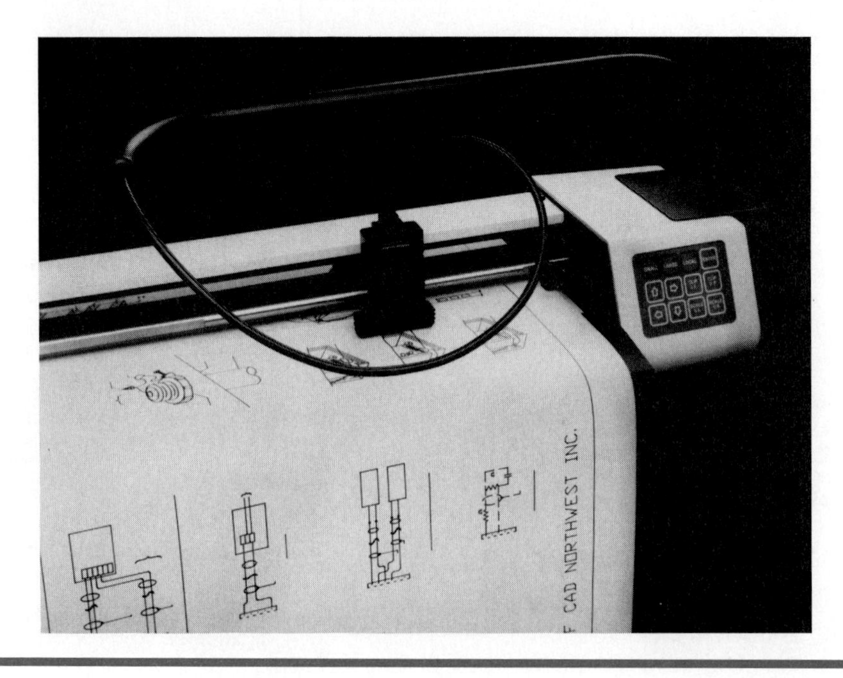

Auxiliary views that are nonsymmetrical can also be drawn using a reference plane. If such a view is **unilateral,** it is drawn entirely on one side of the reference plane. A *unilateral object* is a one-sided part that has all projections falling in one direction. If it is **bilateral,** it is projected right and left of the reference plane. A *bilateral object* is one that has projections relating to two sides. See the examples in Fig. 9-11.

Partial Auxiliary Views

As you can see, auxiliary views are extra views that are not intended to replace the regular orthographic views. They usually show details about the inclined surface, and the rest of the object is omitted. The omitted details are found in the regular orthographic views.

Occasionally only half of a symmetrical auxiliary view is drawn to save time and space on the drawing sheet. Fig. 9-12. This half view is known as a **partial auxiliary view.** Since both halves of the flange are the same, only one-half of it needs to be drawn. Actually, the side view adds nothing to the clarity of the drawing and could be omitted.

Some objects can be clearly described by one regular view and partial views including a partial auxiliary. Fig. 9-13 gives such a case. The full regular front view

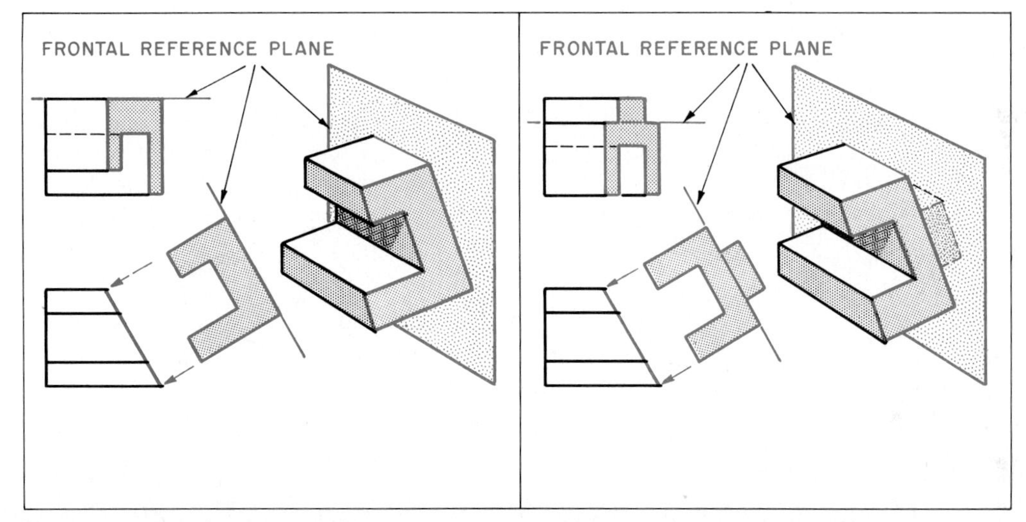

Fig. 9–11. An auxiliary view of a unilateral object is drawn with the reference plane on one side of the object. An auxiliary view of a bilateral object is drawn with projections on both sides of the reference plane.

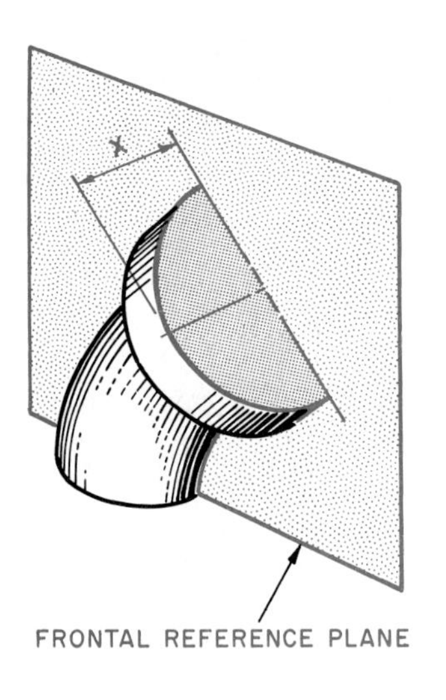

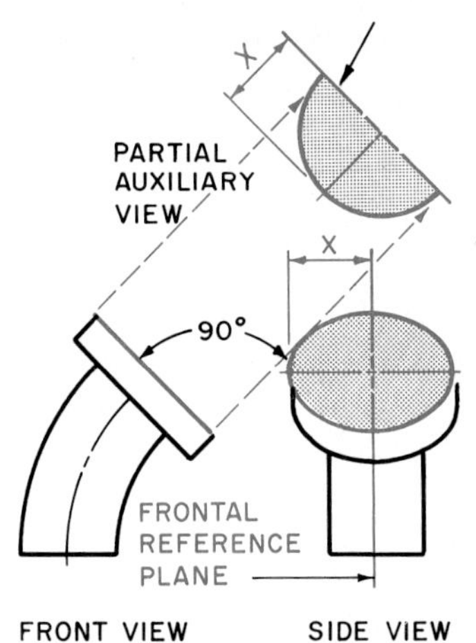

Fig. 9–12. A half auxiliary view can be used to describe some symmetrical objects.

of the tack puller would be distorted and unclear because of the bend. The top view is completely unnecessary since this is obviously a two-view object. An alert drafter would be quick to see the advantage of drawing only the side view, a partial front view, and the partial auxiliary.

Dimensioning Auxiliary Views

Auxiliary views can be dimensioned the same as regular views. Normally dimensions are not given on foreshortened views of inclined faces found on regular orthographic views. Therefore the dimensions should be placed on the auxiliary view. Fig. 9-14.

Auxiliary Sections

Another use for auxiliary views is to show a section through an object. Fig. 9-15 shows the section through the hammer that is not parallel with a normal plane. The cutting-plane line is drawn through the place at which the section is desired. The auxiliary view is projected perpendicular to the cutting-plane line. Actually the cutting-plane line serves as the edge view of the auxiliary plane.

Curved Surfaces in Auxiliary Views

If a curved surface is perpendicular to the inclined surface, like the hole shown in Fig. 9-16, no special problem occurs because the curved surface appears true shape in the auxiliary view.

If the curved surface meets the inclined plane on an angle or is not circular, it must be projected to the auxiliary view using the point method. This method is discussed in Chapter 6, "Multiview Drawing." The curve to be projected should be divided into

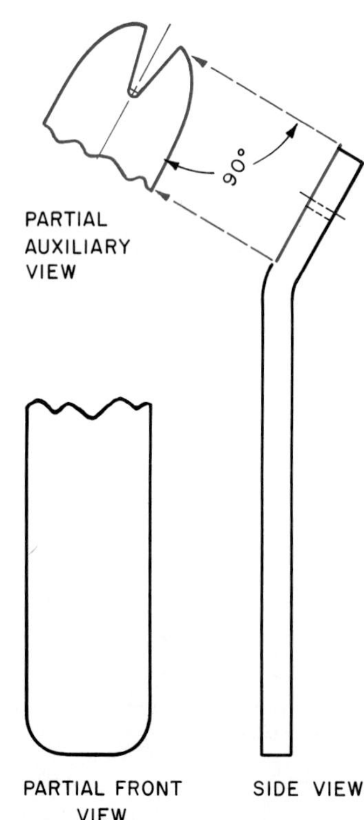

Fig. 9–13. **This drawing of a thumbtack puller uses a partial auxiliary view to replace the regular view.**

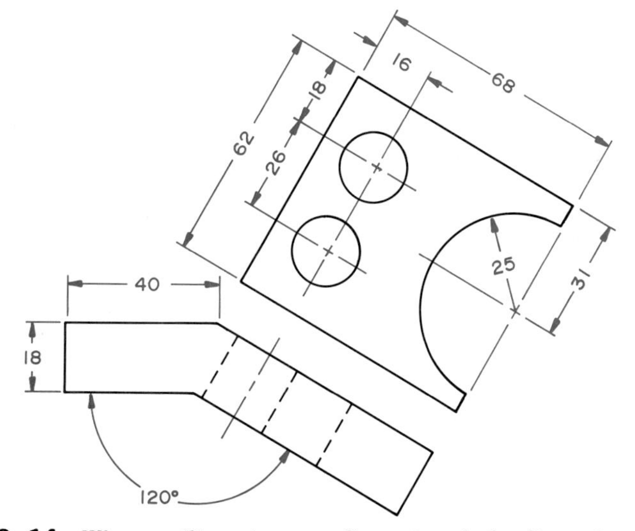

Fig. 9–14. **When auxiliary views are dimensioned, the dimensions are placed parallel with the sides of the views.**

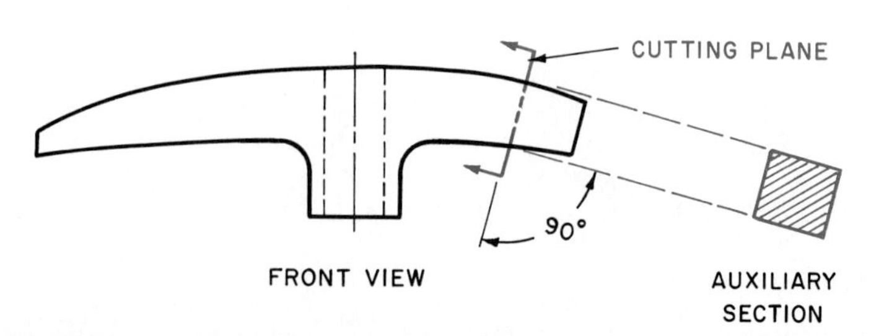

Fig. 9–15. **An upholsterer's hammer with an auxiliary section through one side.**

parts as shown in Fig. 9-17. Any number of points can be used. The more points used, the more accurate the curve will be.

In Fig. 9-17 eleven numbered points are projected onto the view in which the curve appears as an edge. This is the front view of the spotlight reflector shell.

Next, the points are projected from the edge view of the curved surface to the auxiliary view. The distance of each point from the reference plane (distance X) is measured in the same manner as locating any part of an auxiliary view. Only half the surface needs to be measured because it is symmetrical. After the points are laid out, they are connected with an irregular curve.

An irregular curve was plotted in its true shape in the auxiliary view shown in Fig. 9-18 using these steps.

First, points were located on the curve in the top view. The points were then projected to the front view and to the auxiliary view. In this example the curve showed as an edge in the front view, so the auxiliary was taken from the frontal plane. The distance of each point from the frontal plane was found in the top view and marked on the auxiliary view. The points were then connected to complete the true-shape curve.

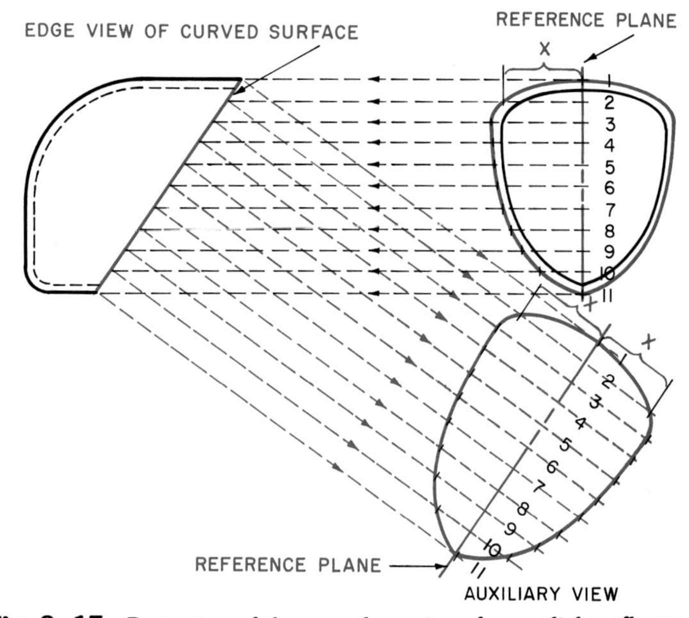

Fig. 9–17. Projection of the curved opening of a spotlight reflector shell.

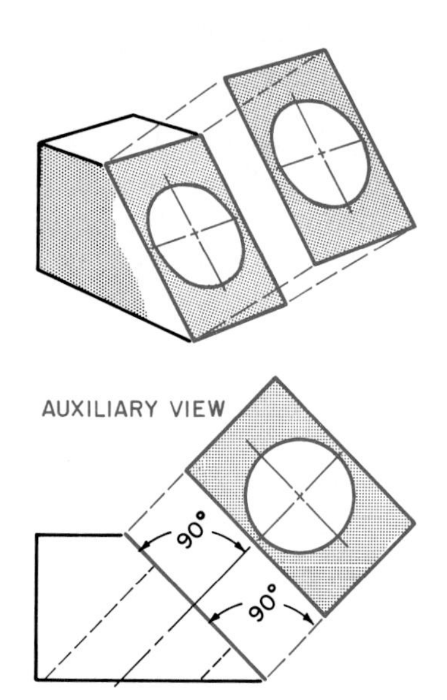

Fig. 9–16. Curved surfaces perpendicular to an inclined surface appear true size on the auxiliary view of the surface.

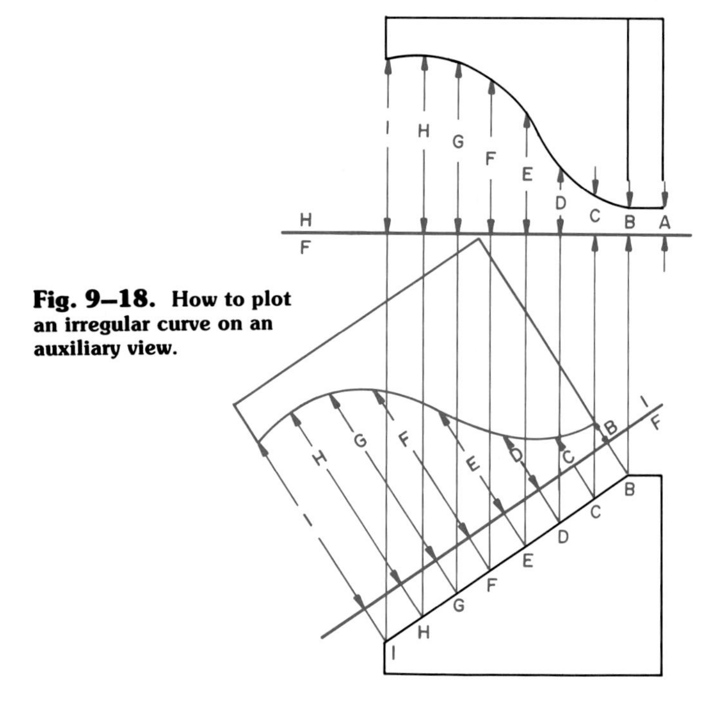

Fig. 9–18. How to plot an irregular curve on an auxiliary view.

Secondary Auxiliary Views

Some surfaces are in such a position that they cannot be shown true size by projecting them upon normal planes or primary auxiliary planes. An oblique surface is a common example. Sometimes you may find it necessary to view an object from a line of sight that is oblique to the normal planes. To show the oblique surface in its true size, you will have to draw a secondary auxiliary view. A **secondary auxiliary view** is one that is projected from a primary auxiliary.

The procedure for drawing a secondary auxiliary view is the same as that for drawing a primary auxiliary. *The views needed are the two preceding views.* In the example in Fig. 9-19, the views that precede the secondary auxiliary are the primary auxiliary and the front view.

The secondary auxiliary view is identified by the number 2.

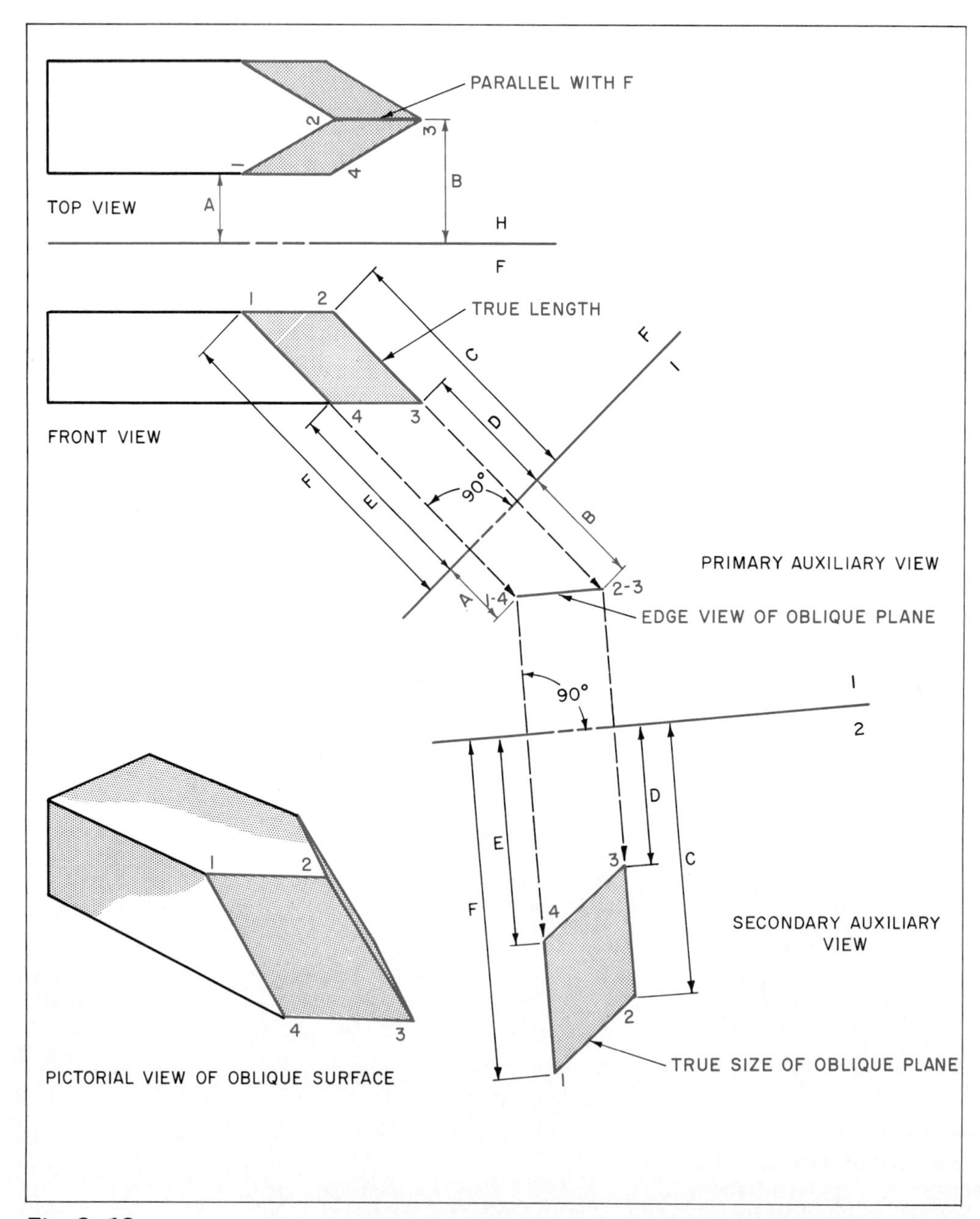

Fig. 9—19. **The true size of an oblique plane is found in the secondary auxiliary view.**

True Size of Oblique Planes

Fig. 9-19 shows a machine lathe tool. The end is ground so that it contains two identical oblique surfaces. To find the true size of these surfaces, work with one of them. In this example, surface 1-2-3-4 was chosen. The general procedure is to first find the edge view of the oblique surface by making a primary auxiliary view. Then, make a secondary auxiliary from the primary auxiliary. The line of sight you take is perpendicular to the edge view that is found in the primary auxiliary. You need to show the edge view of the oblique plane because the only way to find the true size of a surface is to look perpendicular to it when it appears as an edge.

The steps to find the true size of an oblique plane (1-2-3-4 in Fig. 9-19) are as follows:

1. Find the oblique surface in its *edge view*. This is done by drawing a primary auxiliary reference line, F-1, perpendicular to a *true length line,* 2-3, on the oblique plane. In plane 1-2-3-4, edges 2-3 and 1-4 are true length in the front view because they are parallel with the frontal plane of projection. It is necessary to draw the auxiliary reference line, F-1, perpendicular to a true length line because in this way a plane will appear as an edge. In other words, when you look parallel to a true length line on a plane, you see the plane as an edge.
2. Corners 1 and 4 are distance A from the frontal reference plane in the top view. Therefore corners 1 and 4 are distance A from the frontal reference plane in the primary auxiliary view.
3. Corners 2 and 3 are distance B from the frontal reference plane. See the top view.
4. Locate distances A and B in the primary auxiliary. Doing so locates the plane in its edge view.
5. To see the plane true size, draw a secondary auxiliary reference plane parallel to the edge view of the plane. This is auxiliary plane 1-2.
6. In the front view, measure the distances of the corners of the oblique plane from the primary auxiliary plane of projection. Corner 1 is distance F, 2 is distance C, 3 is distance D, and 4 is distance E from the primary reference plane, 1.
7. Measure these distances from the primary auxiliary reference plane in the secondary auxiliary view.
8. Connect the corners of the plane; it appears true size.

Fig. 9-20 shows a secondary auxiliary view of a control cable support. The purpose of making the secondary auxiliary view is to see the oblique surface true size. In this way the true size of the angle can be found.

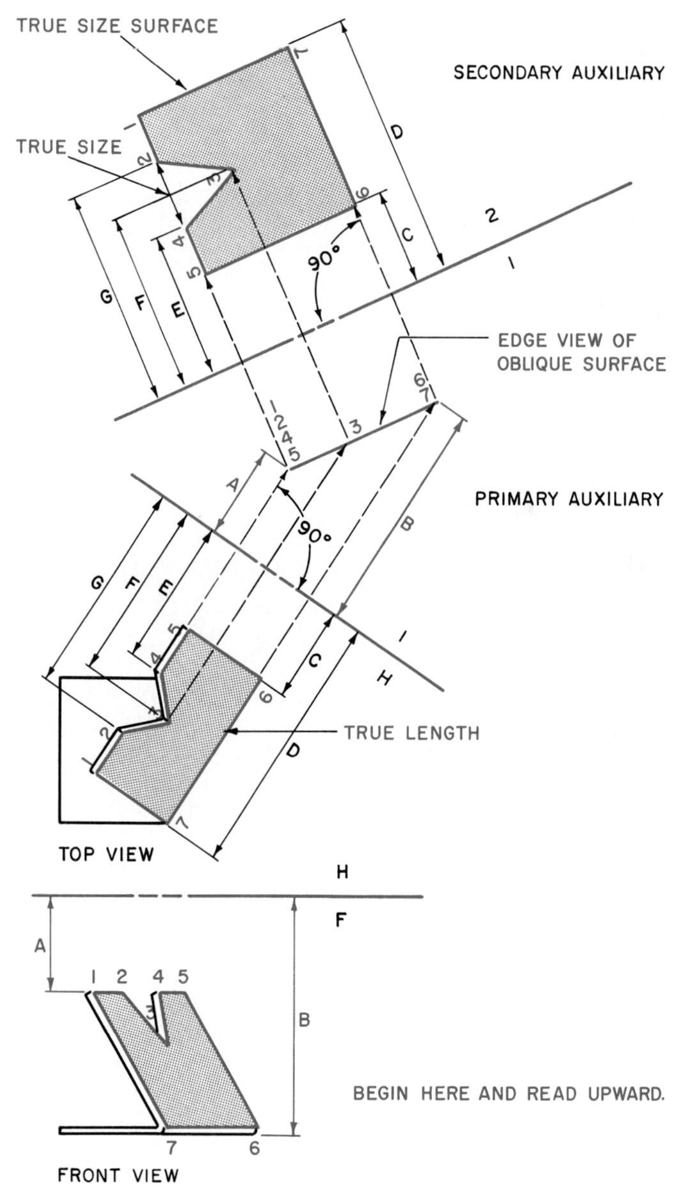

Fig. 9–20. The true size of the angle in the control cable support is found in the secondary auxiliary view.

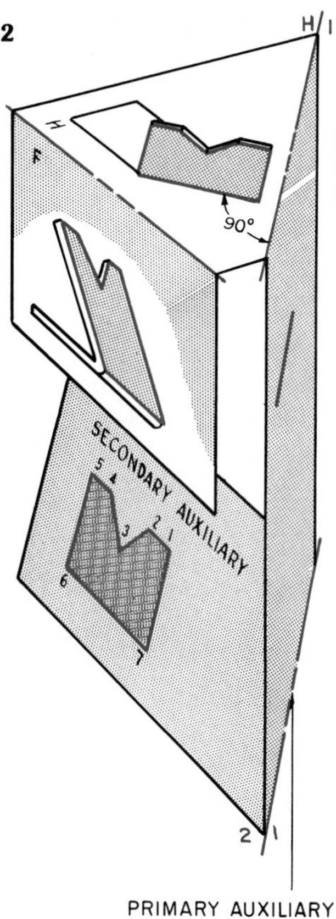

Fig. 9–21. Pictorial drawing of the planes of projection used in Fig. 9–20.

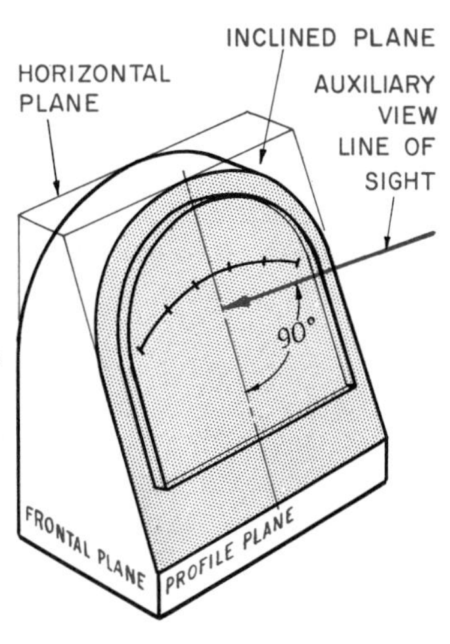

Fig. 9–22. An auxiliary view is one drawn with the line of sight perpendicular to the surface to be described.

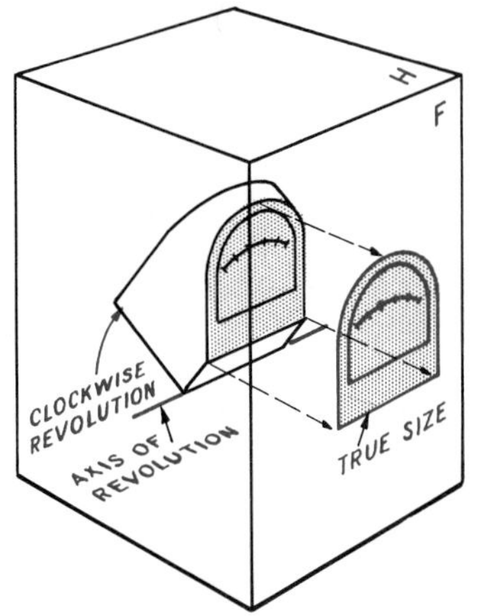

Fig. 9–23. The object is revolved until the inclined surface is parallel to a normal plane of projection. It appears true size when projected to the normal plane.

Fig. 9–24. The Baker-Nunn tracking satellite camera rotates on an axis of revolution.

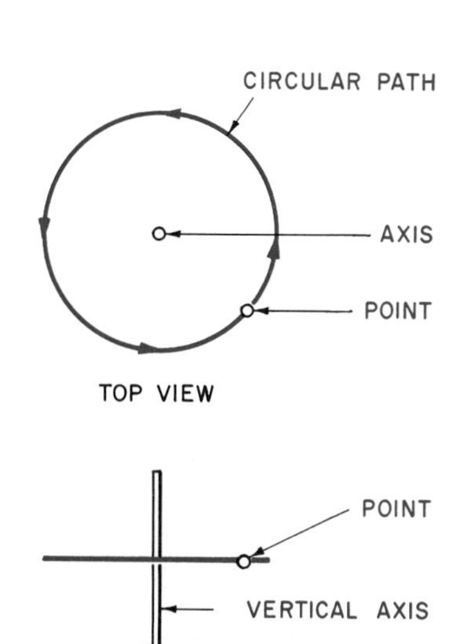

Fig. 9–25. The path of a point revolving around an axis is a circle when the axis appears as a point. The path of a point revolving around an axis is a straight line when the axis appears as a true length line.

The steps to find the true size of the angle are as follows:

1. Draw the top and front views.

2. Find a true length line in one of the views. Edge 6-7 in the top view is chosen.

3. Construct an auxiliary reference plane perpendicular to the true length line. This is H-1.

4. Draw the primary auxiliary view. The horizontal plane appears as an edge in the auxiliary view. It is the plane from which to measure. This plane also appears as an edge in the front view. Measure the distances (A and B) between each point of the plane and the horizontal plane, H-F. Locate these points in the primary auxiliary.

5. Connect the points (A and B) in the primary auxiliary view. The oblique surface appears as a line.

6. Draw a secondary auxiliary plane of projection parallel with the edge view of the oblique surface.

7. Project each point on the oblique surface perpendicular to the secondary auxiliary plane, 1-2.

8. The primary auxiliary plane appears as an edge in the secondary auxiliary. It also appears as an edge in the second view back, the top view. Measure the distance between each point in the top view (C through G) and the primary auxiliary plane, H-1.

9. Locate those points in the secondary auxiliary and connect them. The resulting view is true size including its angle.

The relationship between the planes is shown in Fig. 9-21.

Revolution

In the discussion on auxiliary views, the object is assumed to be in a fixed position. The drafter moves to a different position (line of sight) if he or she wishes to view the object from some position other than a normal one. For example, to see an inclined plane true size, the drafter moves until he or she is in a position to look perpendicular at the inclined plane. Then he or she draws what is seen on an auxiliary plane as shown in Fig. 9-22.

Another way to change the position from which an object is viewed is the **revolution** method. This method allows the drafter to remain in a fixed position while the object is revolved around an axis until it is parallel to a normal plane of projection as shown in Fig. 9-23. The inclined surface appears true size when projected to the normal plane because it is parallel to the plane.

When an object is revolved, it turns around an axis. The tracking satellite camera, Fig. 9-24, has two axes of revolution.

Principles of Revolution

You must learn the following principles before you can completely understand revolution problems. These principles are illustrated in Figs. 9-25 and 9-26.

1. A point revolving around a line (an axis) forms a circular path. The axis is the center of the circle.

a. When the axis appears as a point, the path of the revolving point appears as a circle. See the top view, Fig. 9-25.

b. When the axis is a straight line perpendicular to the line of sight, the path of the revolving point appears as a straight line. This line appears at right angles to the axis, and its length is equal to the diameter of the circular path of revolution. See the front view, Fig. 9-25.

2. The path of revolution of a point around an axis appears to the eye to be an ellipse when the axis is inclined to any of the normal planes of projection. Fig. 9-26.

The most commonly used revolutions involve rotating the object around a horizontal or vertical axis with the axis perpendicular to either the horizontal, frontal, or profile plane.

Revolution with the Horizontal Axis Perpendicular to the Frontal Plane

Revolution around a horizontal axis is shown in Fig. 9-27. The surface of the meter is inclined and foreshortened in all normal views, as seen at A. If you locate an axis at a lower edge and the object is rotated clockwise until the inclined surface is parallel to the profile plane, it will project true size on that plane, as seen at B. You may locate the axis at any convenient point. Then rotate the object either clockwise or counterclockwise—whichever seems best for the particular situation.

The size and shape of the object do not change. Since this is orthographic projection, all points project from view to view exactly the same as in any multiview problem.

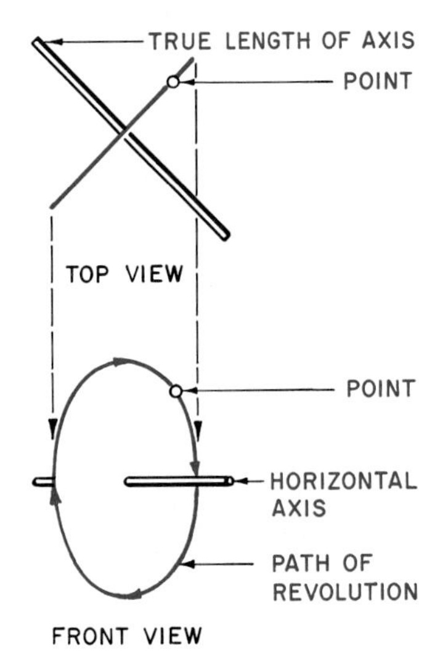

Fig. 9–26. A revolving point appears to form an elliptical path when revolved around an axis inclined to a normal plane.

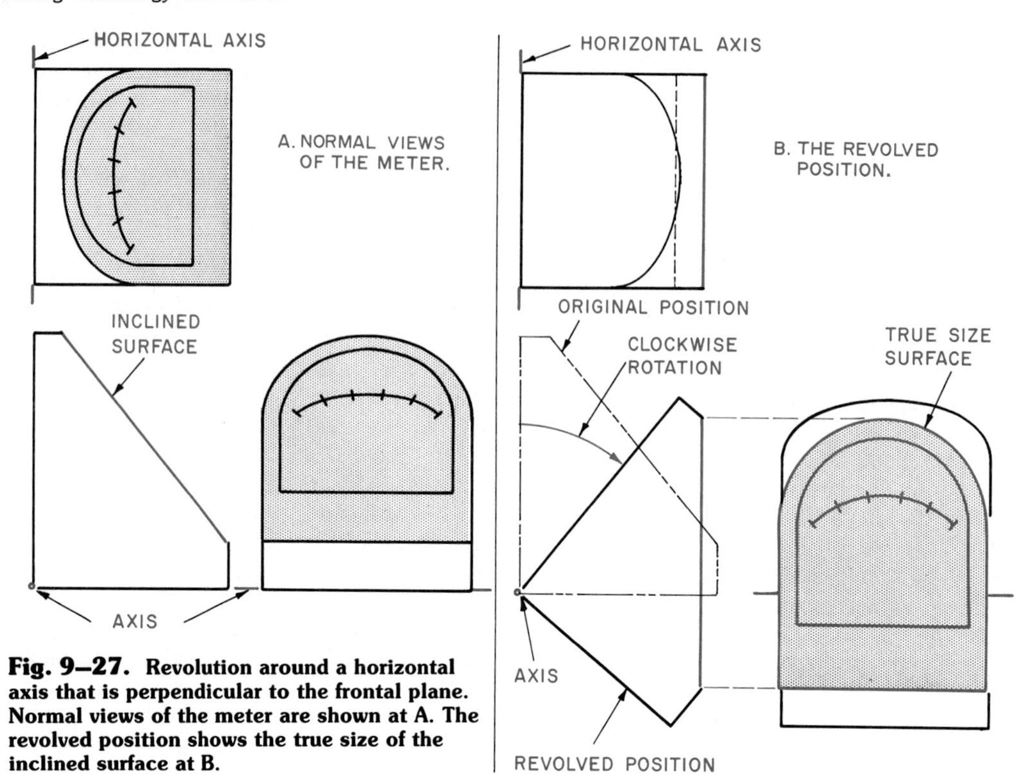

Fig. 9–27. Revolution around a horizontal axis that is perpendicular to the frontal plane. Normal views of the meter are shown at A. The revolved position shows the true size of the inclined surface at B.

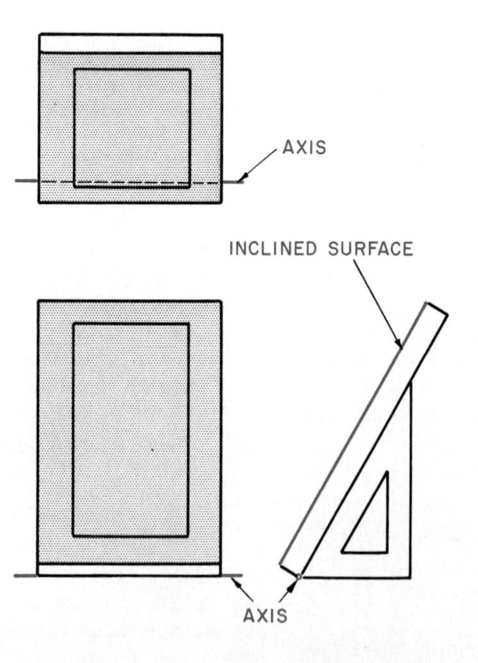

Fig. 9–28. Revolution around a horizontal axis that is perpendicular to the profile plane.

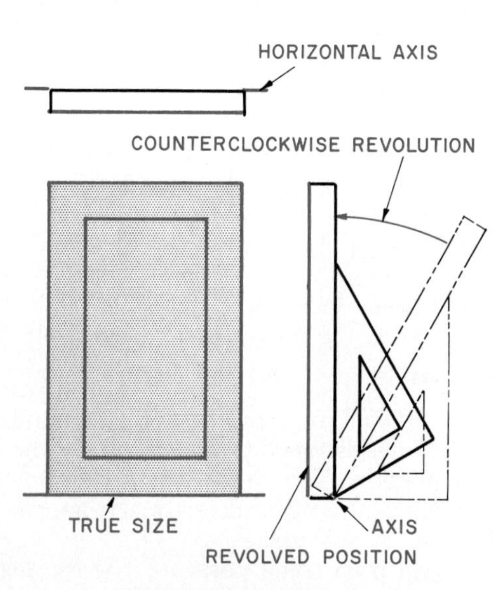

Revolution with the Horizontal Axis Perpendicular to the Profile Plane

A revolution around an axis perpendicular to the profile plane is shown in Fig. 9-28. Place the axis at a lower corner, and rotate the object in a counterclockwise direction until the inclined surface is parallel with the frontal plane, where it appears true size. Then draw the revolved profile view first, and project the other views from it.

Revolution with the Vertical Axis Perpendicular to the Horizontal Plane

Revolution around a vertical axis is shown in Fig. 9-29. Notice that the axis is placed through the center of the object in this illustration. It could have been located at any other convenient place. By choosing the central location, you limit the space needed to revolve the object. Draw the revolved top view first, and project the other views from it.

Finding True Length of a Line by Revolution

Although the true length of a line can be found by making an auxiliary view, using revolution is easier and faster since it does not require drawing another view.

Fig. 9-30 shows a table leg having an inclined plane. Edge AB is an oblique edge. To find the true length of that edge, assume a vertical axis at A in the top view. In the top view revolve AB clockwise until it is parallel to the frontal plane. An edge will appear true length in any plane to which it is parallel. Project revolved point B to the front view. Project B′ in the front view to the side view until it intersects B (revolved). Connect A′ in the front view with B (revolved) in the front view. This is the true length of edge AB.

Additional information about revolution is found in Chapter 10, "Introduction to Descriptive Geometry."

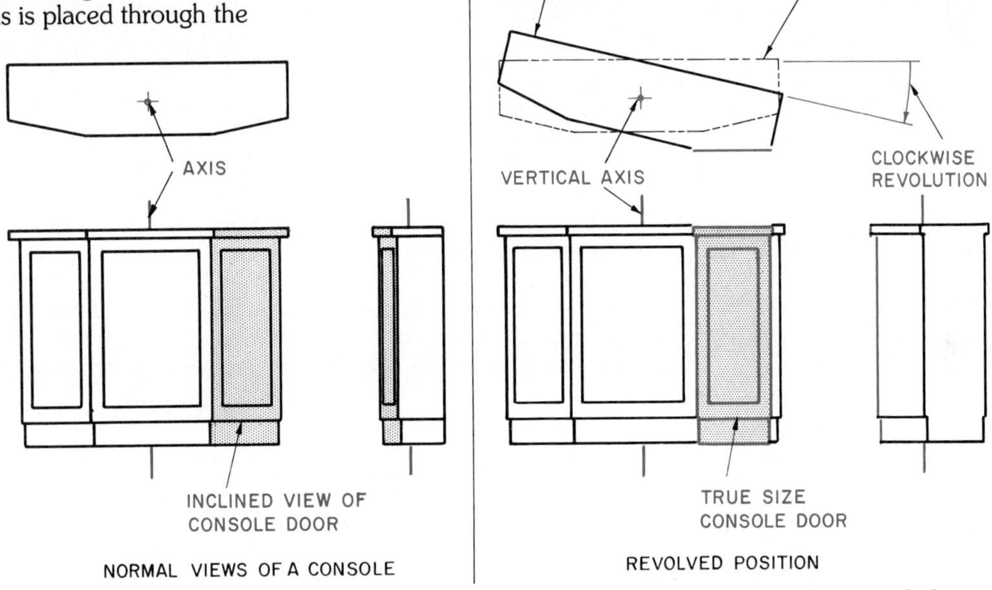

Fig. 9–29. Revolution around a vertical axis that is perpendicular to the horizontal plane.

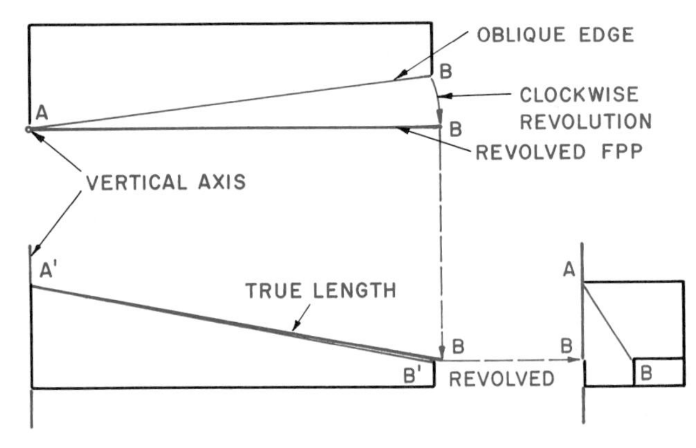

Fig. 9–30. The true length of an oblique line can be found by revolution.

Chapter Review

Build Your Vocabulary

You should understand and use the following terms as part of your working vocabulary. Write a brief explanation of what each means.

auxiliary view
auxiliary plane
primary auxiliary view
unilateral
bilateral
partial auxiliary view
secondary auxiliary view
revolution

Study Problems—Directions

The problems that follow will give you the chance to use some of the drafting knowledge that you learned in this chapter. Do the problems that are assigned to you by your instructor.

Primary Auxiliary Views

1. For **Figs. P9-1** through **P9-20**, find the true size of the inclined surface. This can be done by constructing a primary auxiliary view. Draw the necessary regular views plus the needed primary auxiliary.

2. Find the true size of the angles at each corner of the inclined surface in **Fig. P9-21**.

3. Find the true size of the inclined circular surface in **Fig. P9-22**.

4. For **Fig. P9-23**, find the true size of the inclined surface. Draw the necessary regular views plus the auxiliary view and dimension.

5. For **Fig. P9-24**, draw the front and side views and find the true size of surfaces A and B. Draw a full auxiliary view looking perpendicular to surface A if assigned by your instructor.

6. Find the true size of the two inclined surfaces in **Fig. P9-25**.

7. Draw the true shape of the front and bottom surfaces of the helmet in **Fig. P9-26**.

Secondary Auxiliary Views

The problems shown in Figs. P9-27 through P9-30 can be solved by constructing a secondary auxiliary view. Draw the necessary regular views plus the needed auxiliary views.

8. Find the true size of each piece of glass forming the windshield in **Fig. P9-27**. Record the size of each angle on each piece. Draw your solution at a scale of 1:20.

9. Find the true shape of the two oblique surfaces on the spade drill grinding fixture in **Fig. P9-28**. Draw your solution at a scale of 1:2.

10. Find the true sizes of surfaces A and B on the spade drill in **Fig. P9-29**.

11. Find the true size of oblique surface A in **Fig. P9-30**. If required by the instructor, draw the full object rather than just the oblique surface.

Revolution

12. Solve the problems in **Figs. P9-1** through **P9-26** by revolution.

Study Problems

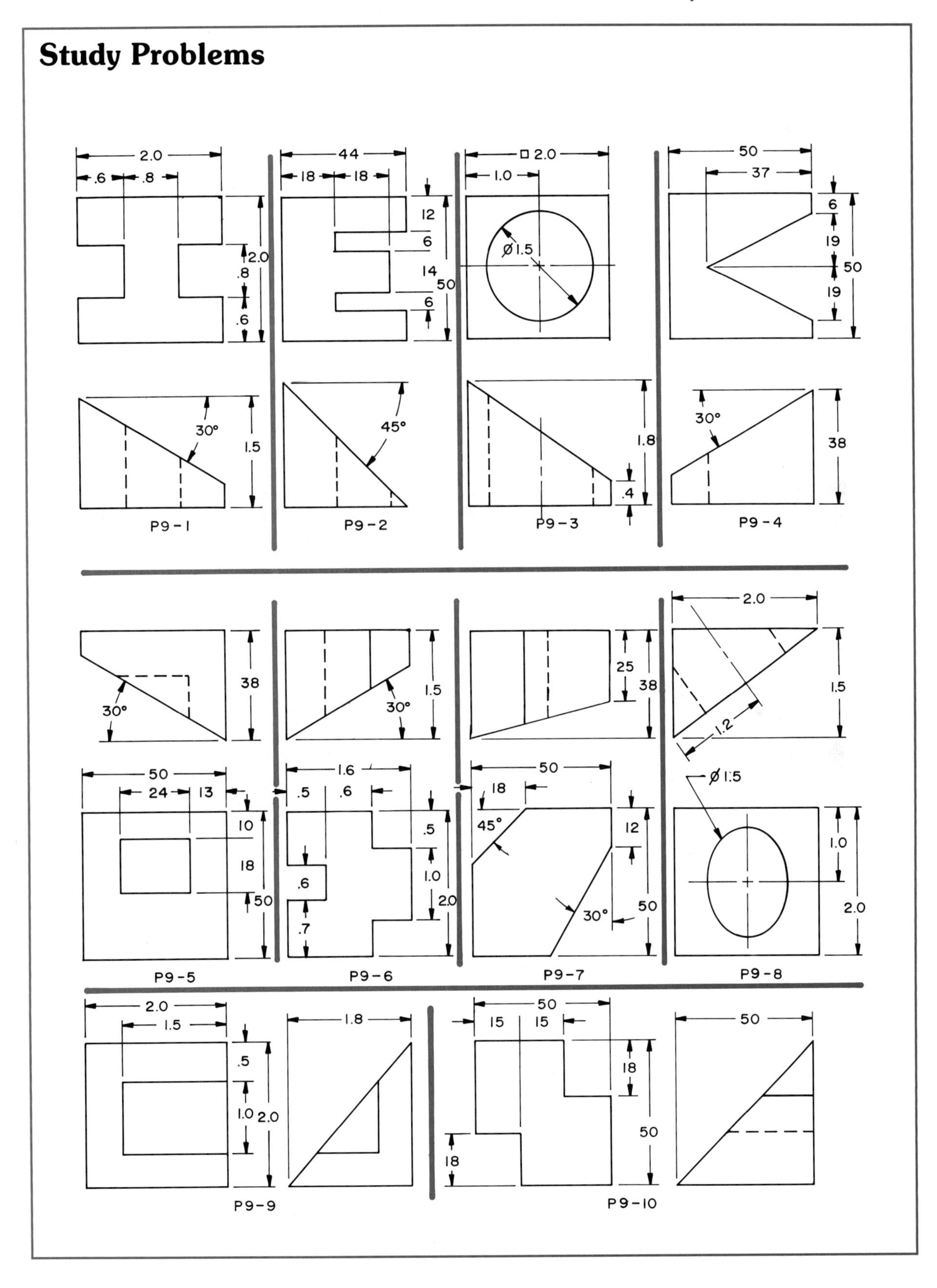

Study Problems

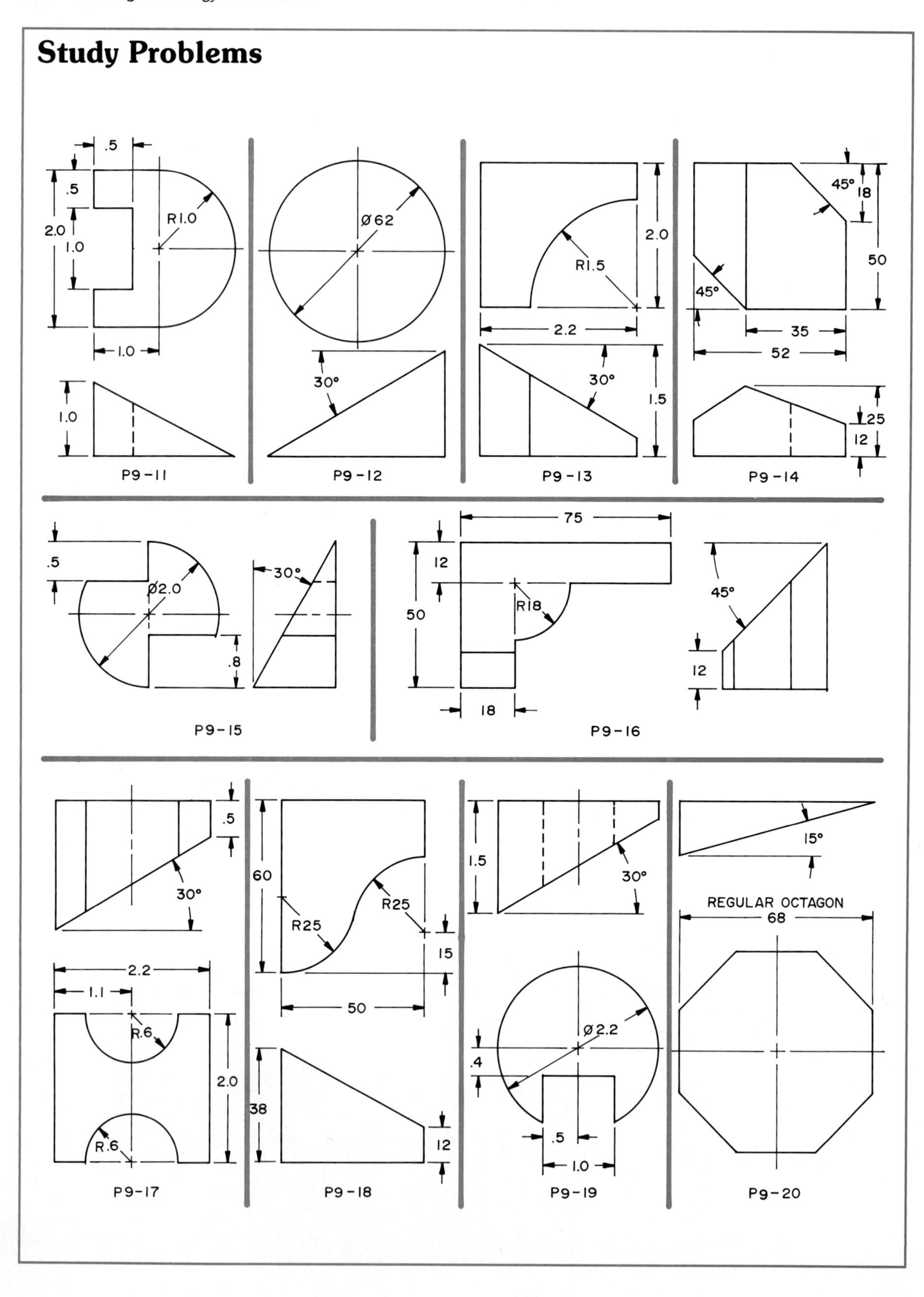

Study Problems

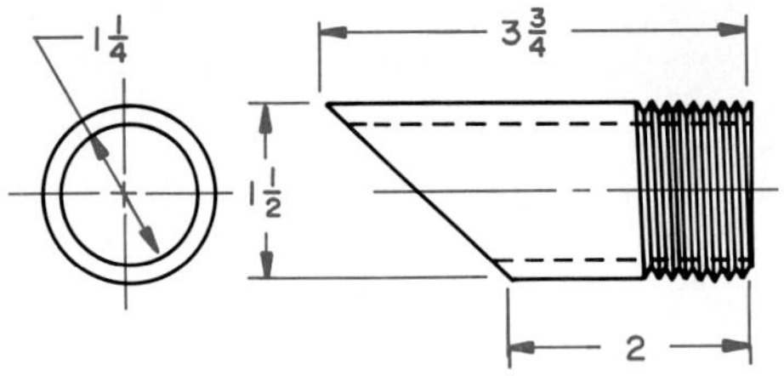

P9-22

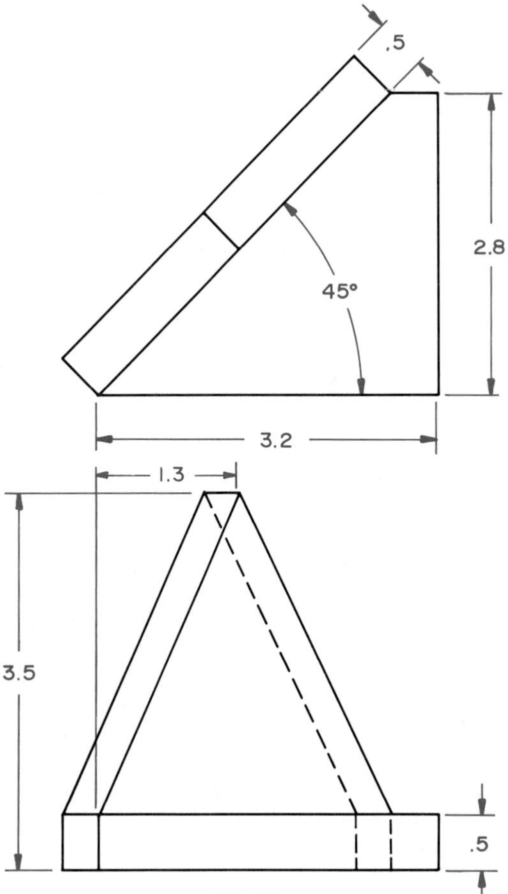

P9-21

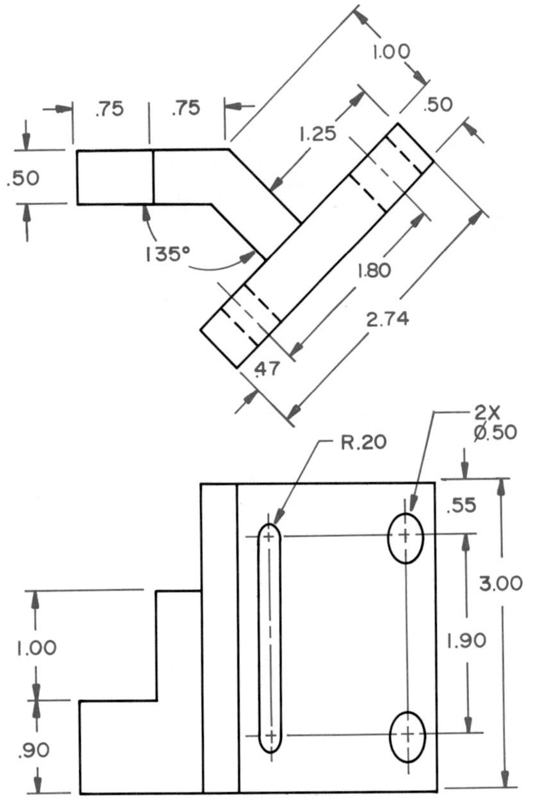

P9-23

Study Problems

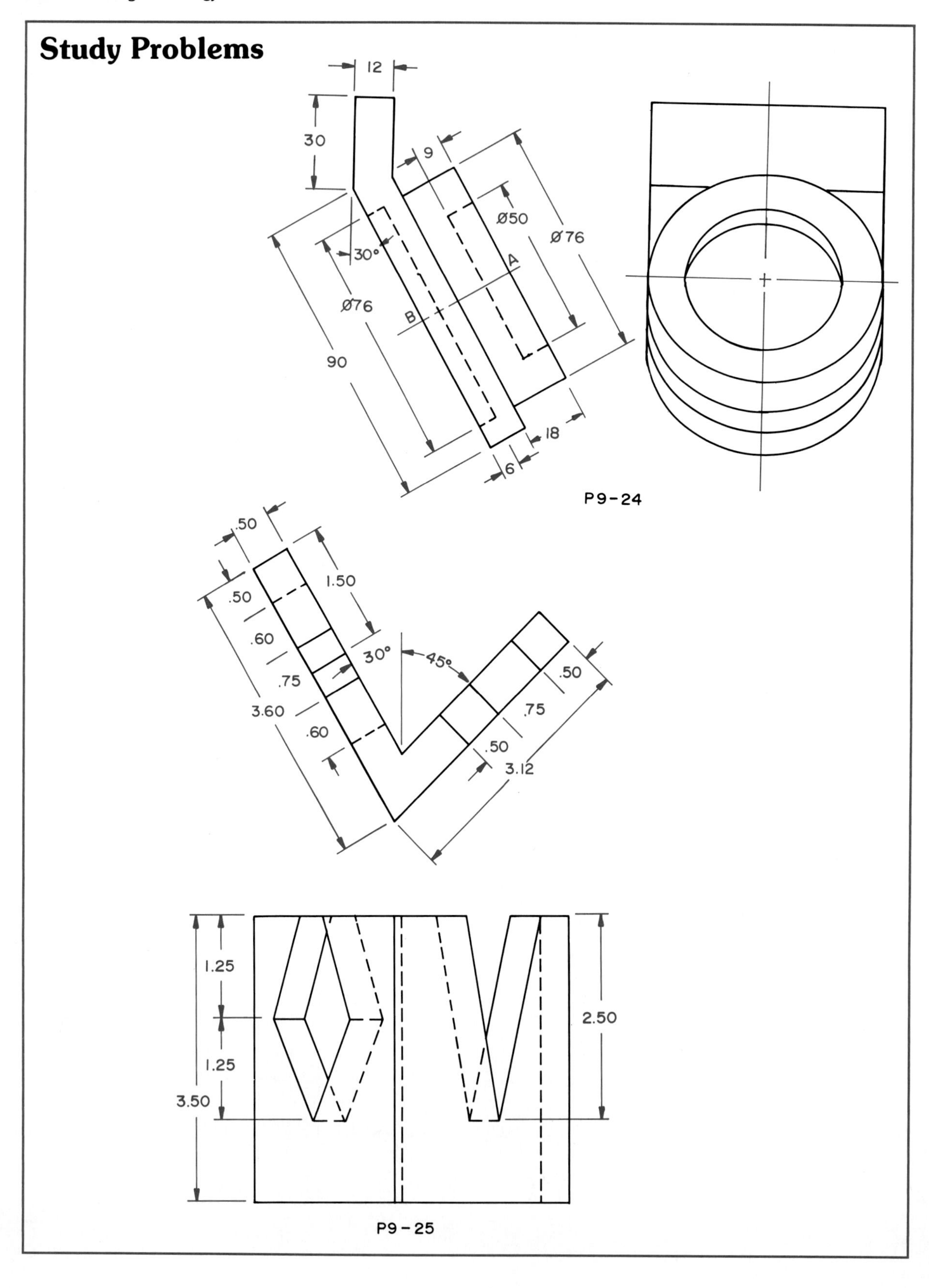

Study Problems

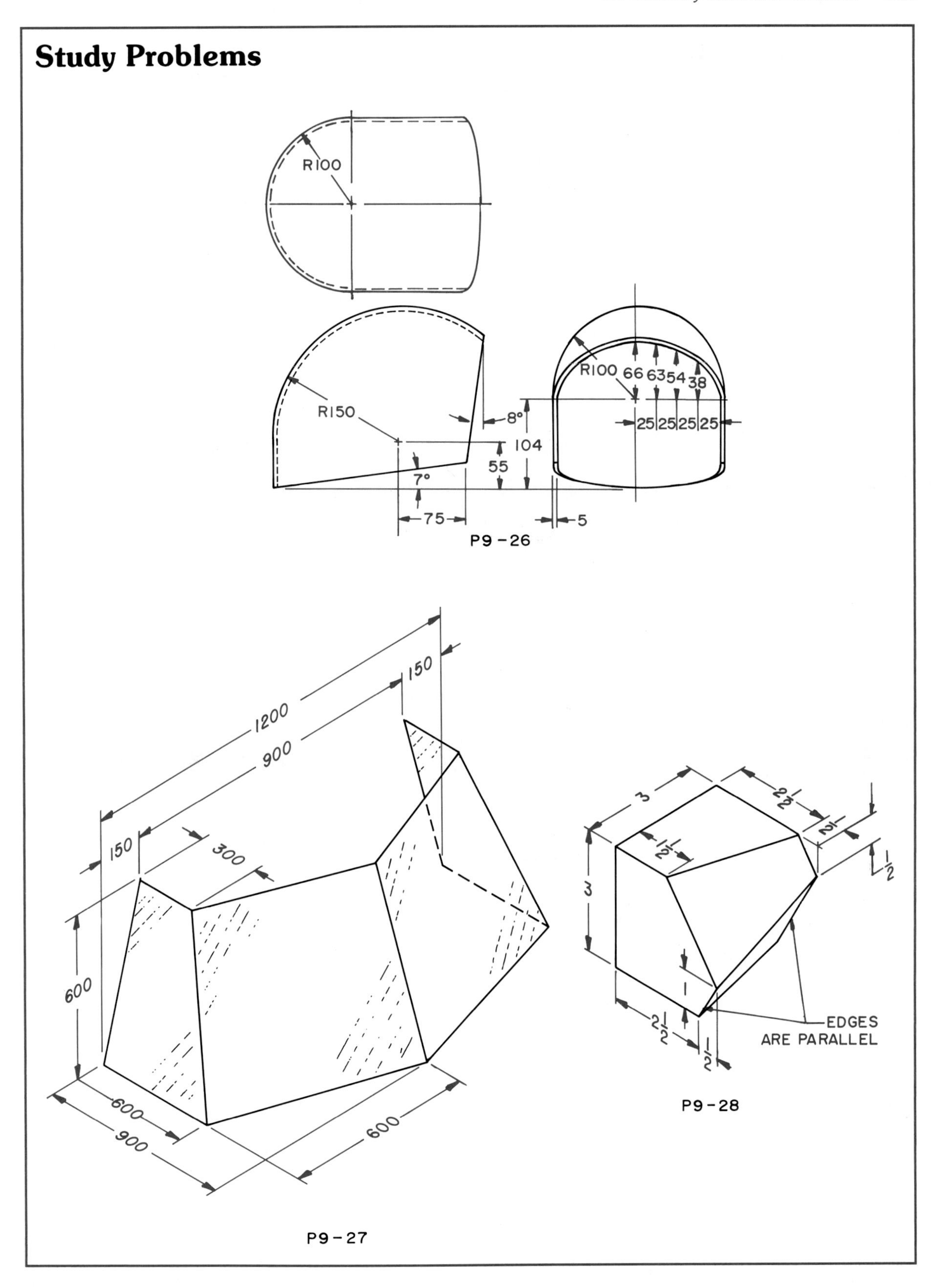

Study Problems

P9-29

P9-30

CHAPTER 10

Introduction to Descriptive Geometry

After studying this chapter, you should be able to do the following:

- Know the relationships between points, lines, and planes.
- Find the shortest distance between points, lines, and planes.
- Find the intersections between lines and planes.
- Find the angles between planes.
- Find the visibility of geometric shapes as they relate to one another.

Geometry

The engineering designer works with three-dimensional objects. Fig. 10-1. These objects are made up of the **geometric elements**—points, lines, and planes—which form three-dimensional solids. These elements are basic to the study of geometry. **Geometry** is the branch of mathematics that deals with points, lines, planes, and solids and examines their properties, measurements, and relationships to each other in space.

Three-dimensional problems can be studied both mathematically and graphically. The graphic study of three-dimensional problems on a plane surface (the drawing paper) is called descriptive geometry. It is the subject of this chapter. The use of the techniques of descriptive geometry enable the engineering designer—and you—to find accurate graphic solutions to three-dimensional problems.

Fig. 10–1. Examine the structure of this tower crane. Notice that it is made of various geometric elements forming its shape. (American Pecco Corporation)

Planes of Projection

You solve descriptive geometry problems using the same planes and projection principles discussed in Chapter 6, "Multiview Drawing." The planes are identified by the letters H (horizontal), F (frontal), and P (profile). The planes are separated with a phantom line. Each phantom line represents a 90° bend between the adjacent projection planes. Fig. 10-2.

Points

A **point** is a position in space that has no size or shape. It locates a specific position. On a drawing a point can be located with either a small dot or a cross.

A point can be projected from its location in space to the planes of projection. Fig. 10-3. In order to establish a position in space, the point must be projected to at least two planes. The point is located by measuring from the fold lines on the planes. Fig. 10-4.

Lines

A **line** is the path generated by a moving point. If it moves in a fixed direction, a straight line is formed. A straight line is also the distance between two points.

Types of Lines

When a line is parallel with two of the principal reference planes, it is called a **principal line**. It appears as its true size on the planes with which it is parallel and as a point on the plane with which it is perpendicular. Fig. 10-5.

If a principal line is parallel with the horizontal plane, it is called a *horizontal line*. It appears true length on the horizontal plane. If it is parallel with the frontal plane, it is called a *frontal line*. It appears true length on the frontal plane. A line parallel with the profile plane is called a *profile line*. It appears true length on the profile plane.

When a line is parallel with one principal reference plane and is neither parallel nor perpendicular to the other two, it is called an **inclined line**. Fig. 10-6. When a line is not parallel with any of the principal reference planes, it is called an **oblique line**. Fig. 10-7.

Terms Used to Describe Lines

In engineering, navigation, mining, and geology, certain lines are described with special terms.

Slope

Slope is the angle that a line makes with the horizontal. When the slope is described as running up from a point, it is given a + (plus) sign. When it runs down from a point, it is given a − (minus) sign. Slope can be given in degrees, as a percent, or as a ratio. Fig. 10-8. In architectural work, roof slope is given in terms of inches of rise per 12 inches run or millimeters of rise per 300 mm run.

Grade. **Grade** is the term used to describe slope on highway, railroad, and sewer projects. Grade is the ratio of the rise (vertical) divided by the run (horizontal) and is expressed as a percent. For example, a rise of 5 feet divided by a run of 100 feet gives a grade of 5 percent. Fig. 10-9.

Batter. **Batter** is the term used to describe the slope that recedes from the bottom to the top. It is used to describe the face of a retaining wall. Batter is given as a ratio of run to rise. Fig. 10-9.

Bevel. **Bevel** is the angle, or slope, that a structural member makes with the horizontal. It is given in inches of rise to 12 inches of run. Fig. 10-9.

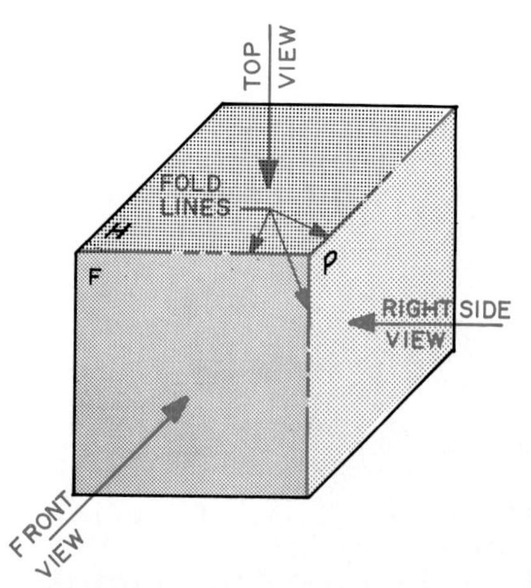

Fig. 10–2. These are the planes upon which the normal views are projected.

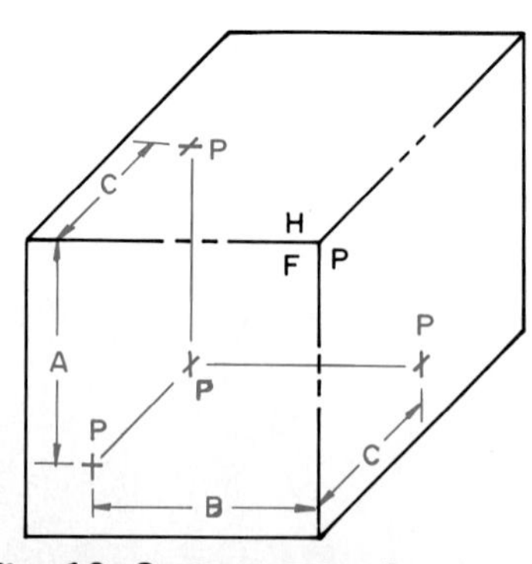

Fig. 10–3. Point P has been projected from its position in space to the three principal planes of projection.

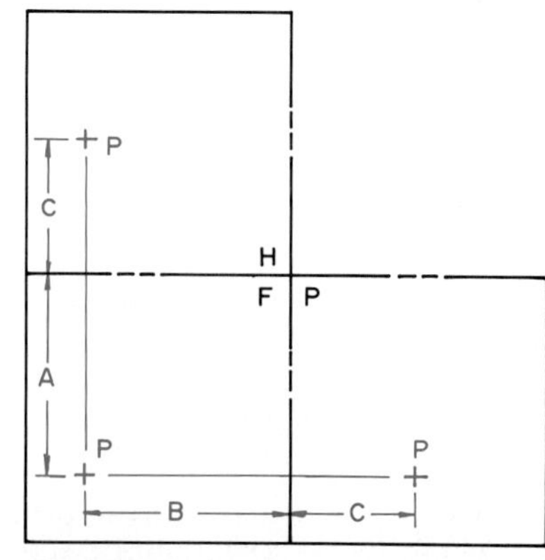

Fig. 10–4. A point is located on the principal planes by measuring from the fold lines.

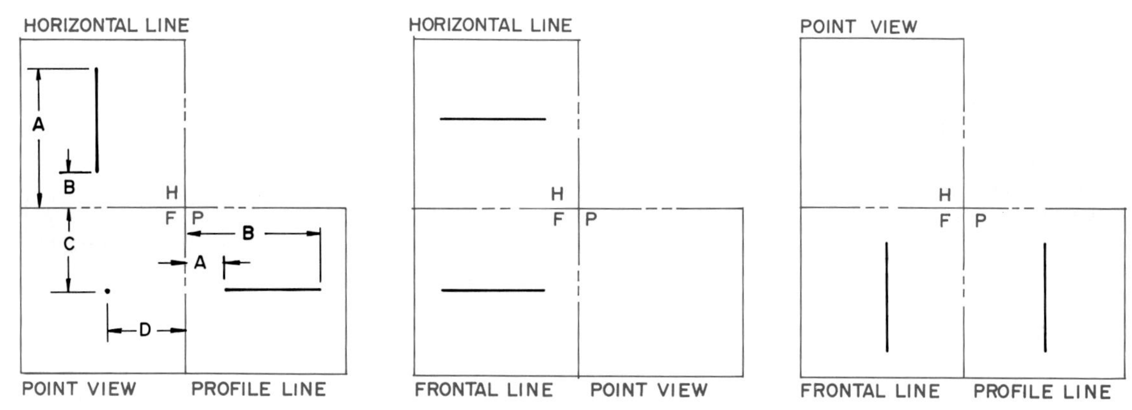

Fig. 10–5. Horizontal, frontal, and profile lines are principal lines.

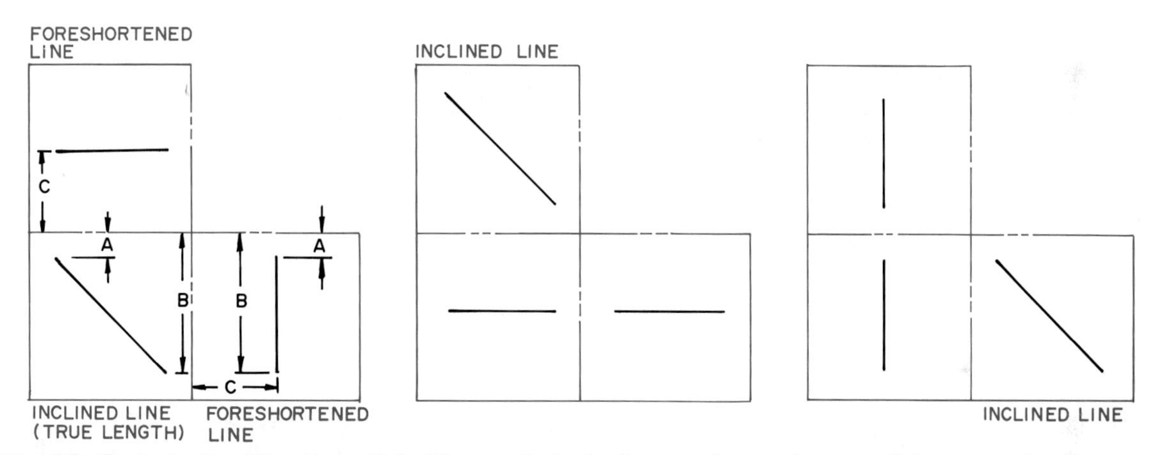

Fig. 10–6. An inclined line is parallel with one principal reference plane and not parallel or perpendicular to the other two.

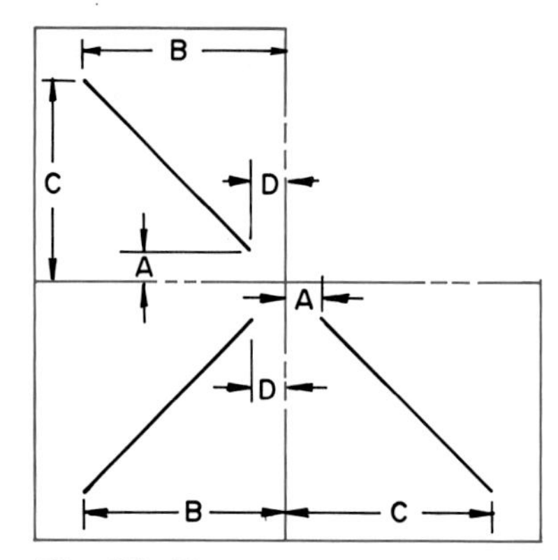

Fig. 10–7. An oblique line is not parallel with any of the principal planes.

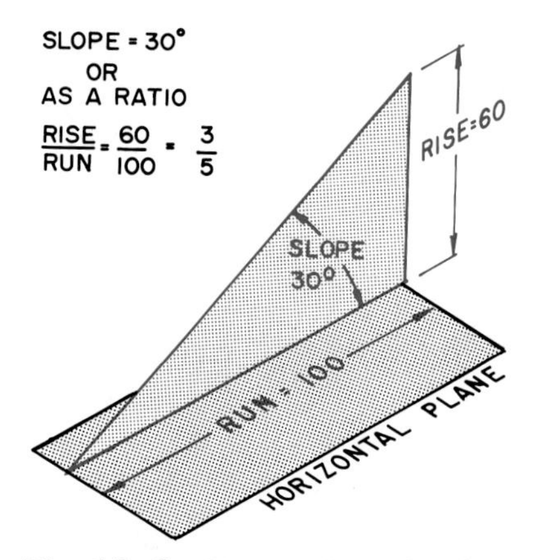

Fig. 10–8. Slope is the angle a line makes with a horizontal plane.

Azimuth

Azimuth is a means of giving direction on a horizontal plane. It is measured from the north pole of the compass in a clockwise direction. All measurements are in degrees. Fig. 10-10.

Bearing

Bearing is a means used by surveyors to record directions in relation to compass bearings. They are read from the north or south. North is to the top of the drawing sheet. Measurements are made on the horizontal plane and measured in degrees. The reading can be east or west of the north-south axis. Fig. 10-10.

Locating a Point on a Line

When you establish the location of a point on a line in one view, it can be projected to all other views. Project it perpendicular to the reference planes until it crosses the line in each view. Fig. 10-11.

Locating a Line on a Plane

A line lies in a plane if the line intersects two lines forming the plane. It also is known to be in the plane if it is parallel with one edge of the plane and intersects one edge.

In Fig. 10-12 line 1-2 is drawn in the top view so that it crosses two sides of the plane. It can be drawn in the other views by projecting to them the points where the line touches the horizontal plane. Connect those points of intersection to form the line.

Locating a Point on a Plane

When you establish the location of a point on a plane in one view, it can be located in the plane of another view. First, pass a line through the point so that it crosses the edges of the plane. Next, project that line to the other view as shown in Fig. 10-13. Then project the point until it touches the line. This locates the point on the other plane.

Intersecting Lines

When lines appear to cross on various views, they may or may not intersect. To determine if they intersect, project the point where they cross to each of the views. If you can project the point to each view, the lines do intersect. Fig. 10-14.

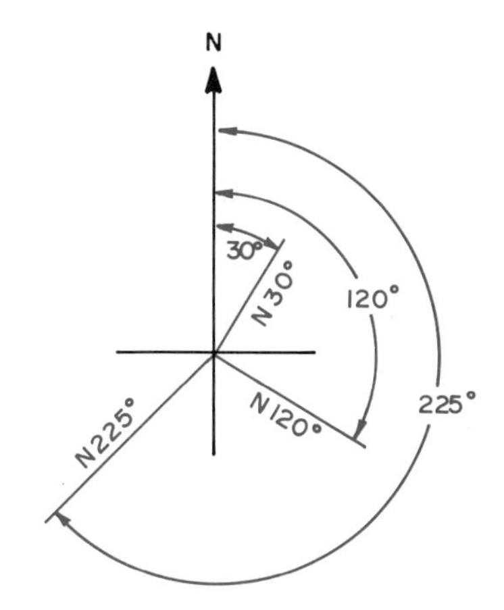

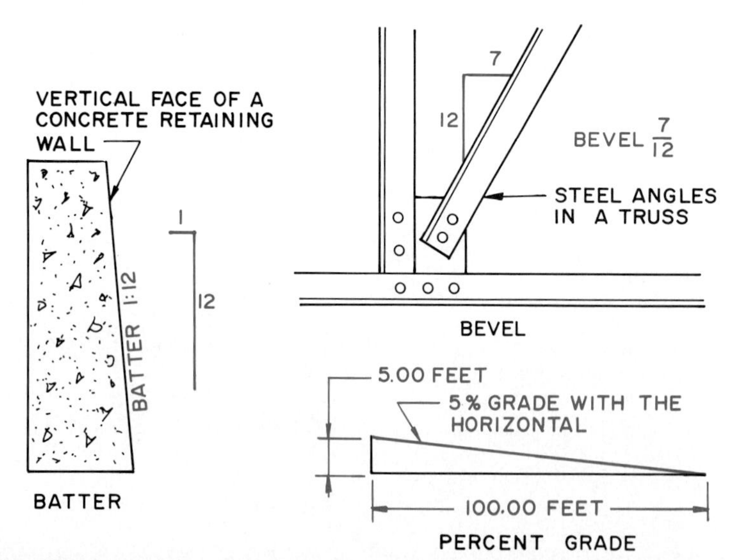

Fig. 10-9. Grade, bevel, and batter are other ways of expressing the slope of a member or surface.

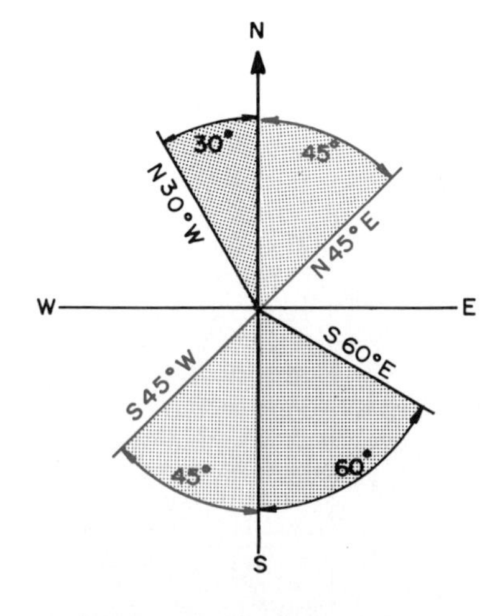

Fig. 10-10. Azimuth and bearing are two other ways to measure angles.

Finding the True Length of an Oblique Line

An oblique line is one that is not parallel with any of the normal planes of projection.

The true length of an oblique line can be found by projecting it perpendicular to an auxiliary plane of projection that is constructed parallel with the oblique line.

To find the true length of an oblique line, follow these steps:

1. Draw a primary auxiliary plane of projection parallel with the line as shown in Fig. 10-15. In this example the auxiliary view was drawn from the front view. Either of the other two views could have been used.
2. Project the ends of the line, C-D, onto the primary auxiliary view.
3. Measure distances S and T from the fold line F-P, and transfer them onto the auxiliary plane. These distances locate the ends of oblique line, C-D. Connect these points to draw the true length line.

Finding the Point View of an Oblique Line

As explained earlier, when a line is perpendicular to a plane, it appears as a point on that plane. An oblique line is not perpendicular to any of the principal reference planes. To find it in point view, a secondary auxiliary must be drawn. First find the line in true length in a primary auxiliary view. Then draw a second auxiliary view perpendicular to the true length line. Follow Fig. 10-16 and these steps:

1. Draw a primary auxiliary parallel with the line in one of the views. In this case the top view was chosen.
2. Project the line onto the primary auxiliary plane.
3. Draw a secondary auxiliary plane perpendicular to the true length line.
4. Project the line to the second auxiliary view. Here it appears as a point.

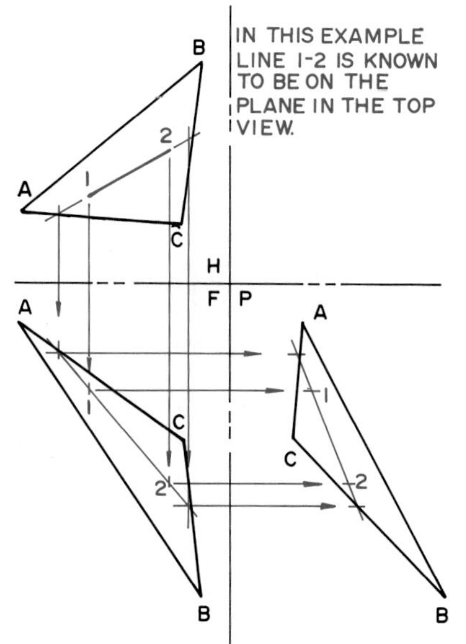

Fig. 10–12. How to locate a line on another plane.

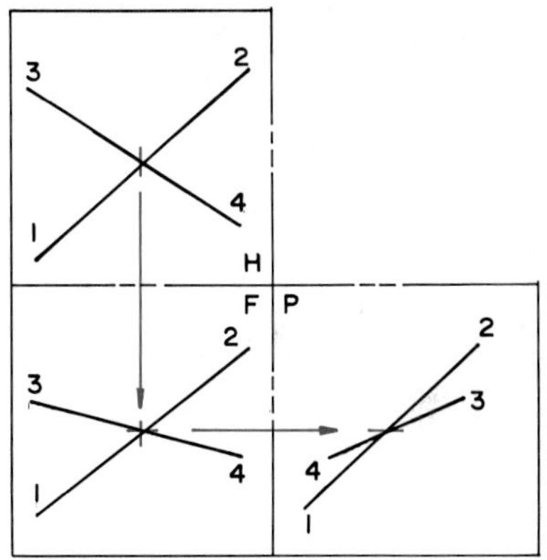

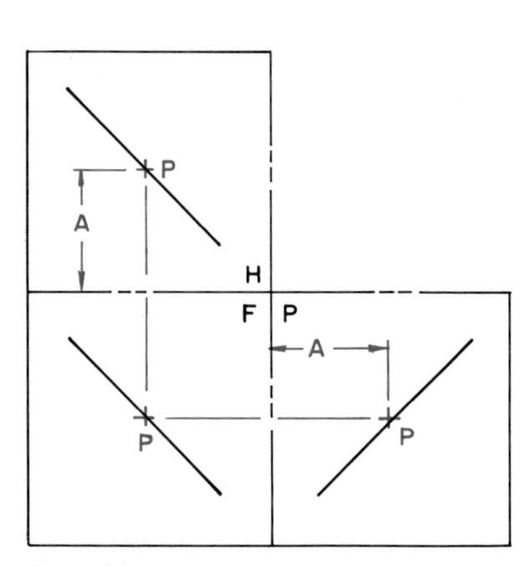

Fig. 10–11. A point on a line in one view can be projected to that line in all other views.

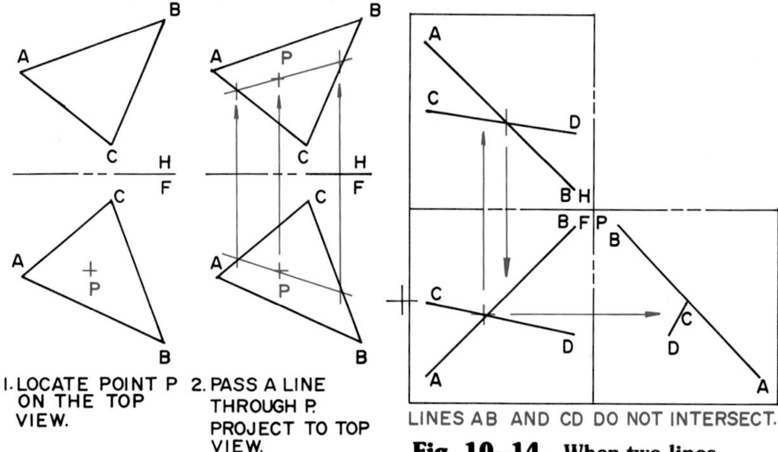

Fig. 10–13. How to locate a point on another plane.

Fig. 10–14. When two lines intersect, the point of intersection will project to all views.

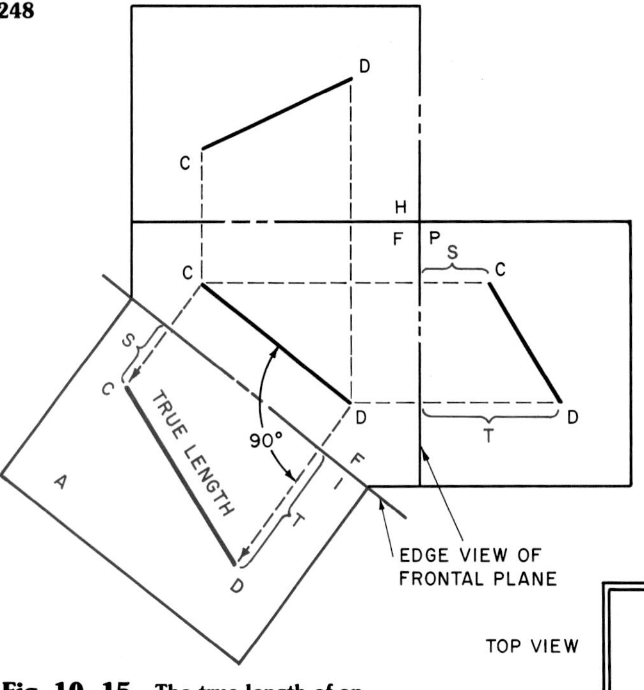

Fig. 10–15. The true length of an oblique line can be found by projecting it to an auxiliary plane.

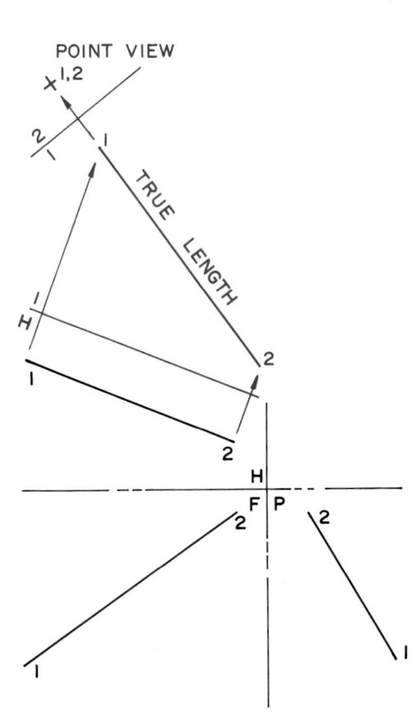

Fig. 10–16. The point view of an oblique line is found in the secondary auxiliary view.

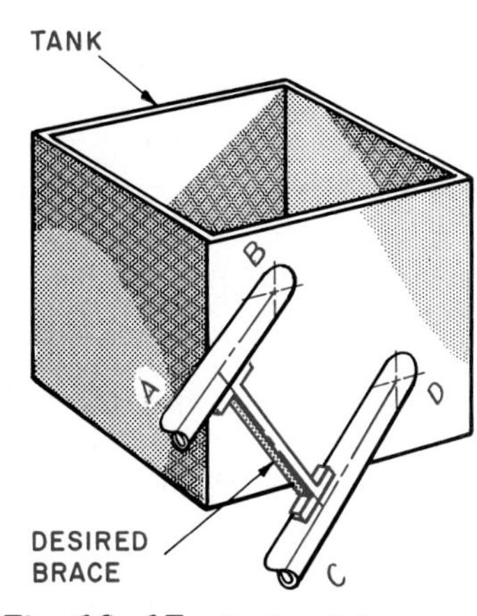

Fig. 10–17. A pictorial illustration of a practical problem that requires finding the shortest distance between parallel lines.

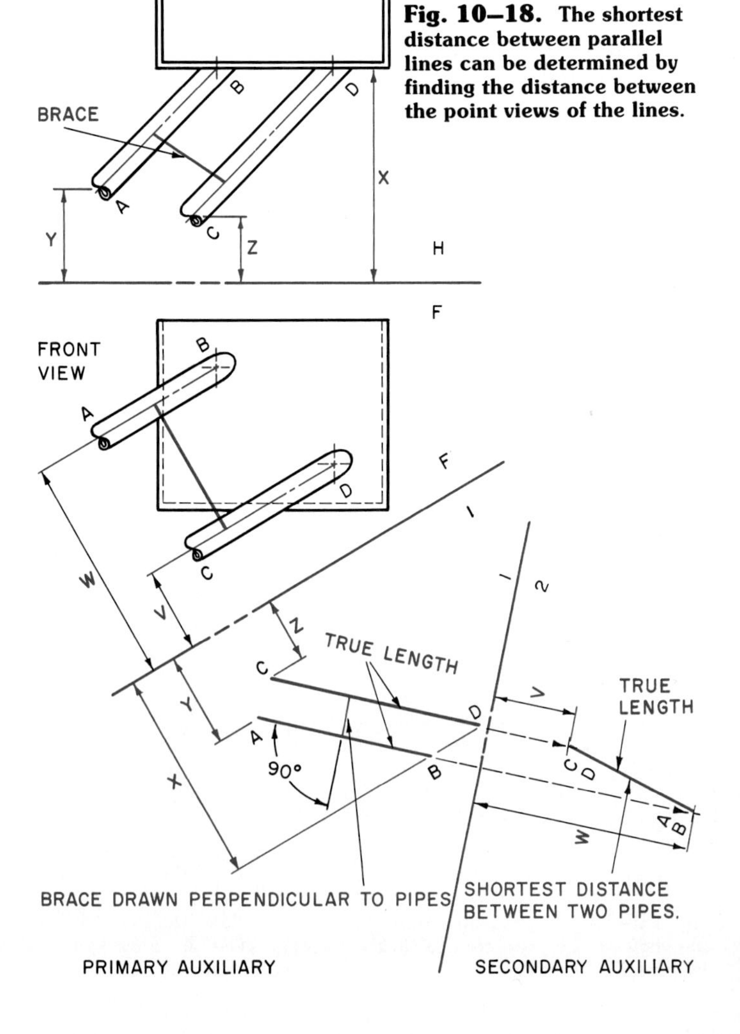

Fig. 10–18. The shortest distance between parallel lines can be determined by finding the distance between the point views of the lines.

Shortest Distance between Parallel Lines

Lines or edges that are parallel appear parallel in all views. Two parallel pipes run from a tank in Fig. 10-17. Suppose you need to find the shortest distance between these pipes so a brace can be built.

The *shortest distance between parallel lines* is a line perpendicular to those lines.

The shortest distance between parallel lines can be found by measuring the distance between the lines when they appear in point view.

To determine the shortest distance between parallel lines, follow these steps:

1. Find the true length of the lines, Fig. 10-18, by constructing a primary auxiliary view parallel to the center lines. Use the center lines of the pipe for measurements.

2. Construct a secondary auxiliary view perpendicular to the true length view of the lines. This projection presents the lines in their point view.

3. Measure the true distance between the point view of the two lines in the secondary auxiliary view. After the two pipes are found true length in the primary auxiliary, the brace can be drawn perpendicular to them. See the line representing the brace in Fig. 10-18. The brace can then be projected back to the front and top views.

Shortest Distance from a Point to a Line

Occasionally in solving design problems you need to find the true distance between a point on an object and an edge of another object. A practical problem is presented in Fig. 10-19: how to transfer a fluid from the tank at point A to the pipe BC using the shortest connecting pipe possible.

The shortest distance between a point and a line is a line from the point perpendicular to the line. Find the line (in the problem, the pipe BC) in its point view. A straight line between the point (A) and the point view of the line (BC) is the shortest distance and can be measured on a drawing.

To determine shortest distance, see Fig. 10-20 and follow these steps:

1. Find the true length of the section of pipe BC by constructing a primary auxiliary plane parallel to it. Use the center line of the pipe for measurements.

A line will appear true length on any plane to which it is parallel. The connecting pipe is drawn perpendicular to pipe BC in this view because BC is true length. The connecting pipe is not true length in the primary auxiliary plane.

2. Construct a secondary auxiliary plane perpendicular to the true length of pipe BC. This will make the center line of the pipe appear as a point. Whenever you look perpendicular to a true length line, it will appear as a point.

3. Carry point A into the secondary auxiliary view.

4. The distance between point A and the point view of pipe BC is the shortest distance. The connecting pipe is true length in this view; it can be measured on the drawing.

Finding the True Angle between Two Lines

If two lines both appear true size, the angle between them can be measured. Figure 10-21 shows two intersecting lines that are in an inclined plane. To find the true size of the angle between them, follow these steps:

1. Draw a primary auxiliary plane parallel to the lines where they appear as a single line (as an edge view of a plane).

2. Project the lines onto the auxiliary plane. Here they are both true size. Measure the true size of the angle.

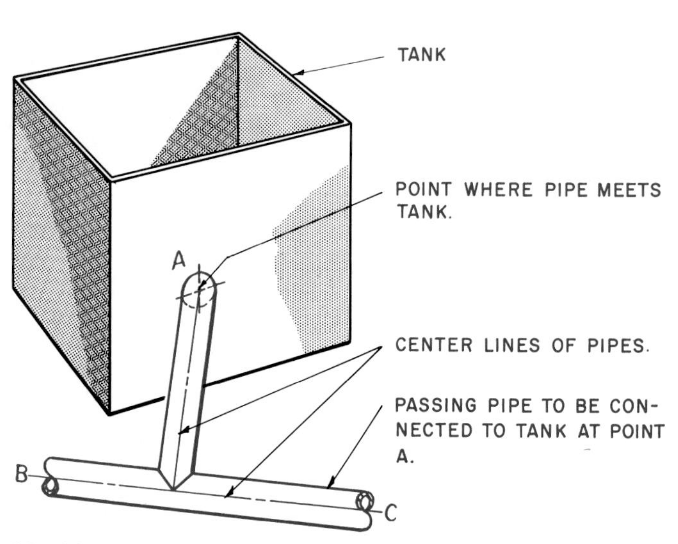

Fig. 10–19. A pictorial illustration of a practical problem that requires finding the shortest distance from a point to a line.

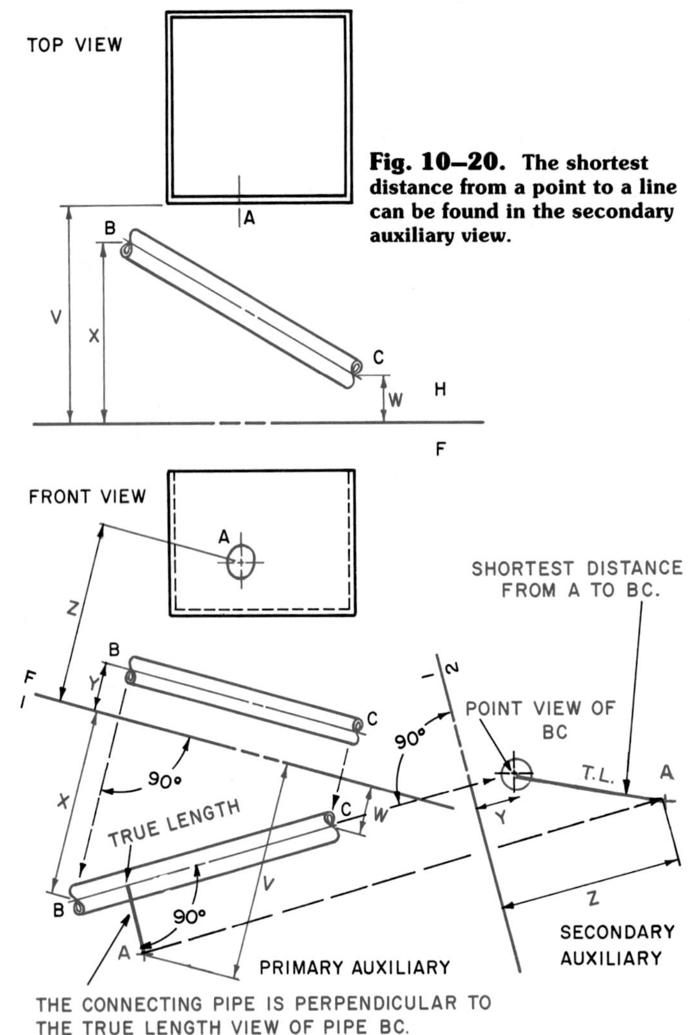

Fig. 10–20. **The shortest distance from a point to a line can be found in the secondary auxiliary view.**

If the two lines are in an oblique plane, you will need to draw the plane in edge view in a primary auxiliary view and then find its true size in a secondary auxiliary view. Follow these steps and Fig. 10-22:

1. Connect the ends of the lines to form a plane. Draw a line across the plane parallel to one of the principal reference planes. In this example the line was made parallel to the frontal plane.
2. Project the plane onto the primary auxiliary view. Here it appears as an edge.
3. Draw a secondary auxiliary plane parallel with the edge view.
4. Project the plane onto the secondary auxiliary plane. Since the plane is true size, you can measure the angle.

Planes

A **plane** is a flat surface produced when a line moves through space. Planes can be represented on drawings several ways. These include three points not in a straight line, two intersecting lines, a point and a line, and two parallel lines. Fig. 10-23.

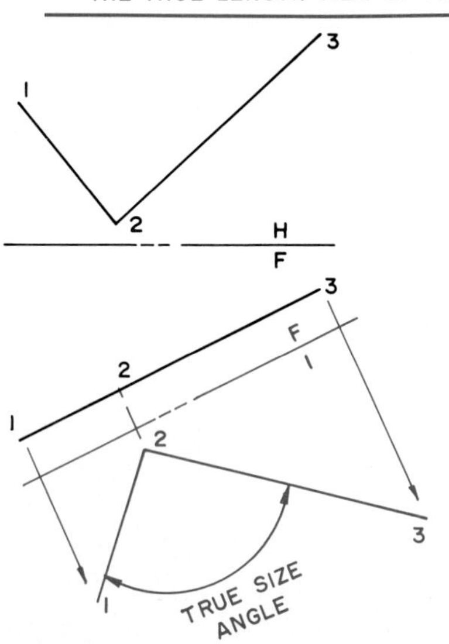

Fig. 10–21. **How to find the true size of the angle between two intersecting lines that are in an inclined plane.**

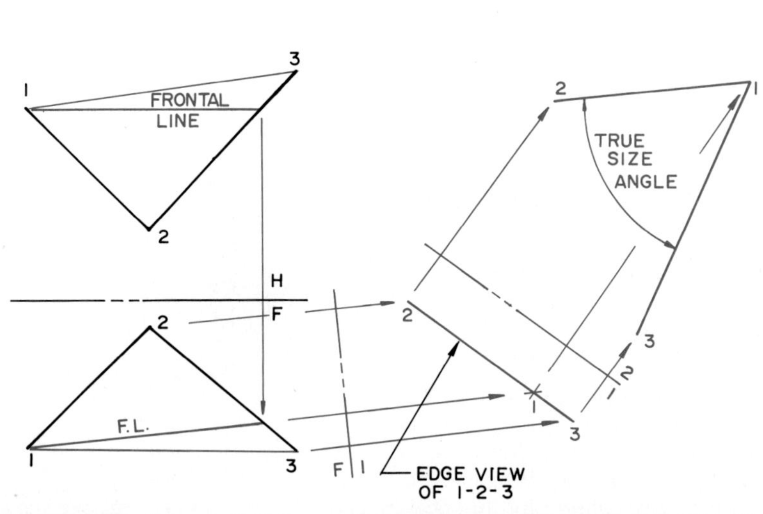

Fig. 10–22. **How to find the angle between two intersecting lines that lie on an oblique plane.**

Types of Planes

Planes parallel with the principal reference planes are also called *principal planes*. Fig. 10-24. If they are parallel with the horizontal plane they are called a *horizontal plane* and appear true size. If parallel with a frontal plane, they are called *frontal planes* and appear true size. When parallel with the profile plane, they are referred to as *profile planes* and appear true size.

If a plane is inclined to two principal planes and perpendicular to one, it is called an *inclined plane*. It appears foreshortened on two planes and as an edge on the one to which it is perpendicular. Fig. 10-25.

If a plane is not perpendicular to any of the principal reference planes, it is called an *oblique plane*. It appears foreshortened on all three principal planes. Fig. 10-26.

Finding the True Size of an Inclined Plane

Figure 10-27 shows how to find the true size of an inclined plane using a primary auxiliary view. To do so, follow these steps:

1. Draw a primary auxiliary plane parallel with the edge view of the plane.
2. Project the plane onto the primary auxiliary plane.
3. Connect the corners to complete the plane.

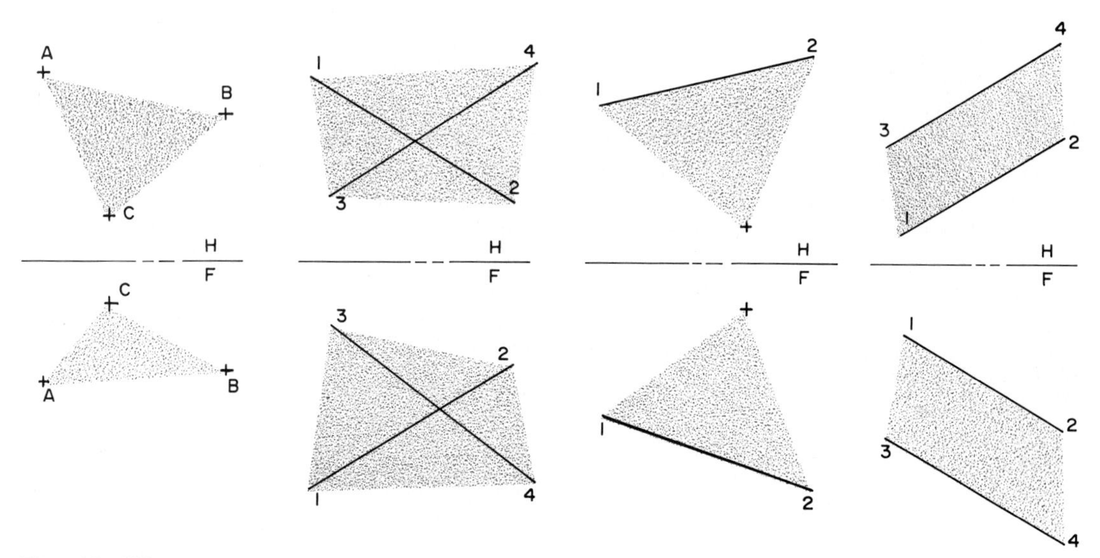

Fig. 10–23. These are the ways planes may be established.

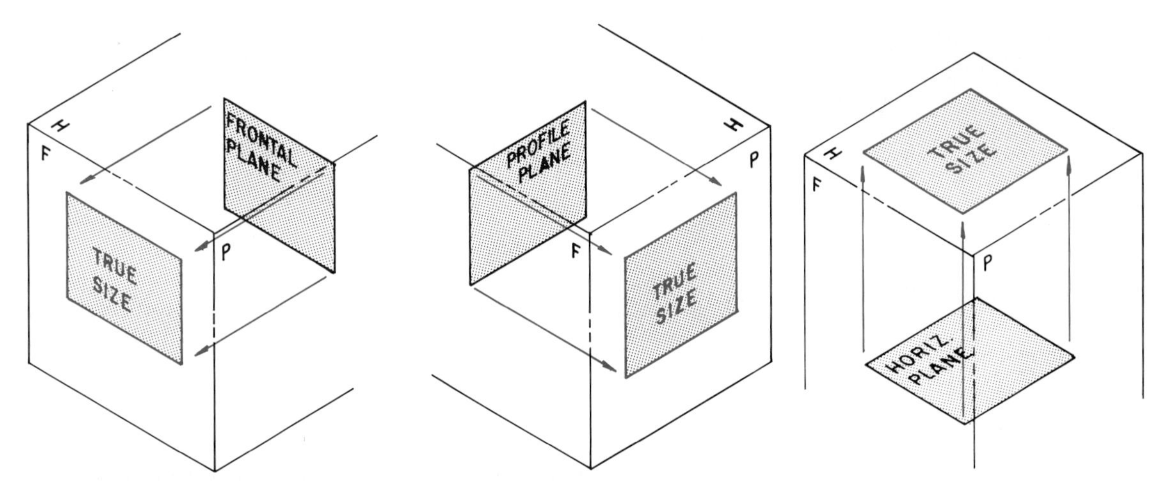

Fig. 10–24. These are the principal planes of projection.

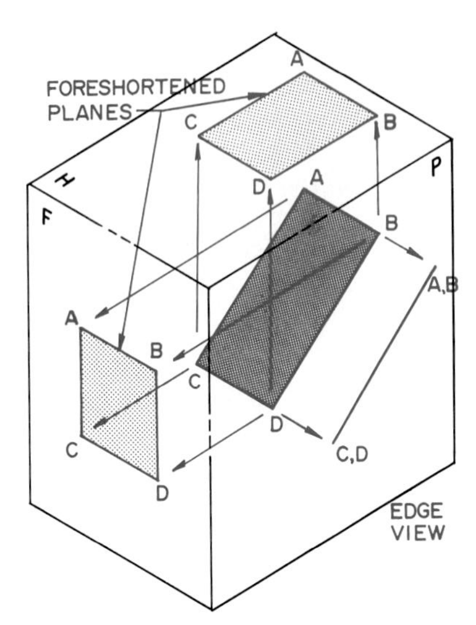

Fig. 10–25. An inclined plane is perpendicular to one principal plane and at an acute angle to the other two.

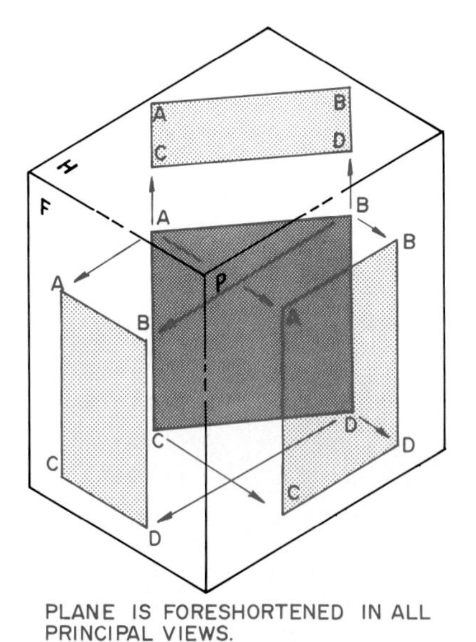

Fig. 10–26. An oblique plane is not parallel with or perpendicular to any of the principal planes of projection.

Finding the True Size of an Oblique Plane

To find the true size of an oblique plane, you will need to construct two auxiliary views. Follow these steps in Fig. 10-28:

1. Draw a line across the oblique plane that is parallel with one of the principal reference planes. In this example it is drawn parallel with the frontal plane. Project this line to the frontal plane where it appears true length.
2. Draw a primary auxiliary reference plane perpendicular to this true length line.
3. Project the oblique plane onto the primary auxiliary where it will appear as an edge.
4. Draw a second auxiliary reference plane parallel with the edge view. Project the oblique plane onto the secondary auxiliary. Here it appears true size.

Finding the Angle between Two Planes

Some objects have two planes intersecting at an unknown angle. This angle is called a *dihedral angle*. Fig. 10-29. The size of the dihedral angle is found by locating the line of intersection between the two planes in its point view. Then the angle can be measured on the drawing. Fig. 10-30.

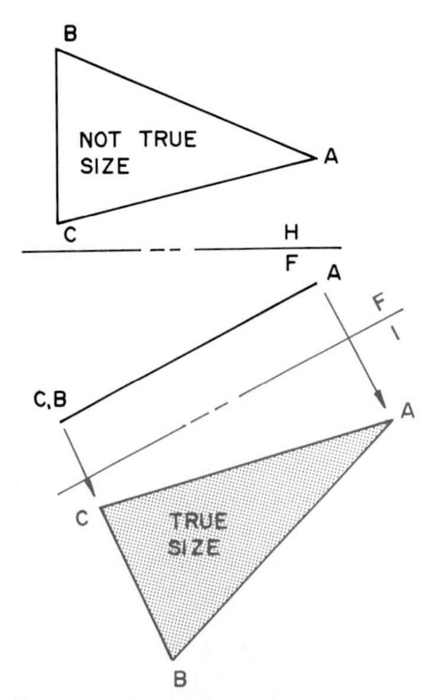

Fig. 10–27. The true size of an inclined plane can be found in a primary auxiliary view.

To find the dihedral angle, follow these steps:

1. The planes intersect at line CE. Draw a primary auxiliary view to find the true length of this line. This is done by establishing an auxiliary plane parallel to CE and projecting the object into the primary auxiliary view.
2. Draw a secondary auxiliary line perpendicular to the true length of CE. This gives a point view of line CE.
3. When the planes BCE and ACE are projected onto the secondary auxiliary view, they appear as edge views, and the true angle between them can be measured.

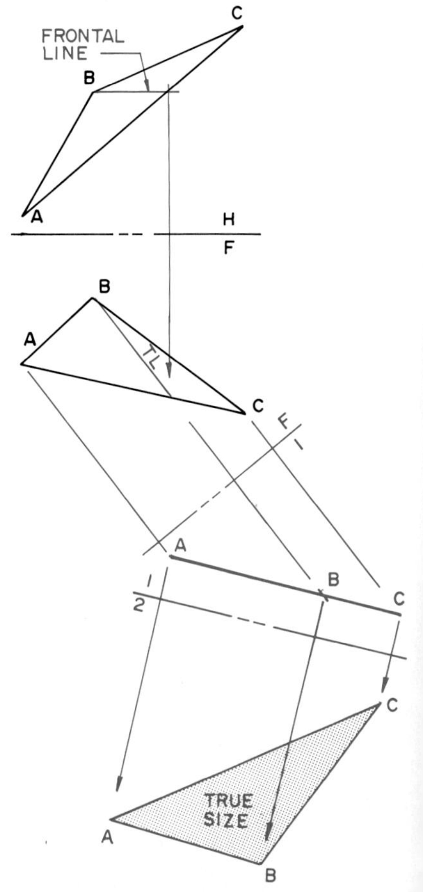

Fig. 10–28. The true size of an oblique plane can be found in the secondary auxiliary view.

Finding the Intersection of Planes

In design problems, two planes on an object often intersect. To describe the object, you need to locate the line of intersection between these planes.

The line of intersection of two planes is found by locating one plane in its edge view and noting where the second plane is cut by the edge view. To do this, follow these steps:

1. Select one of the planes, and show it in edge view in a primary auxiliary view. In Fig. 10-31 plane WXYZ is selected.
2. Locate or construct a true length line on plane WXYZ. Line WV is drawn in the front view parallel to the frontal plane.
3. Project line WV to the top view. Here it is true length.
4. Draw an auxiliary plane perpendicular to the true length line, WV.
5. Project both planes to the auxiliary plane.
6. Where plane ABC crosses WXYZ is the line of intersection. It is labeled 1-2.
7. Project points 1-2 to the regular views. This locates the line of intersection. Notice that point 1 intersects side CB and point 2 intersects side AC.
8. Ascertain the visibility of the planes. This is explained in the following section.

Fig. 10–29. A pictorial illustration of two intersecting planes.

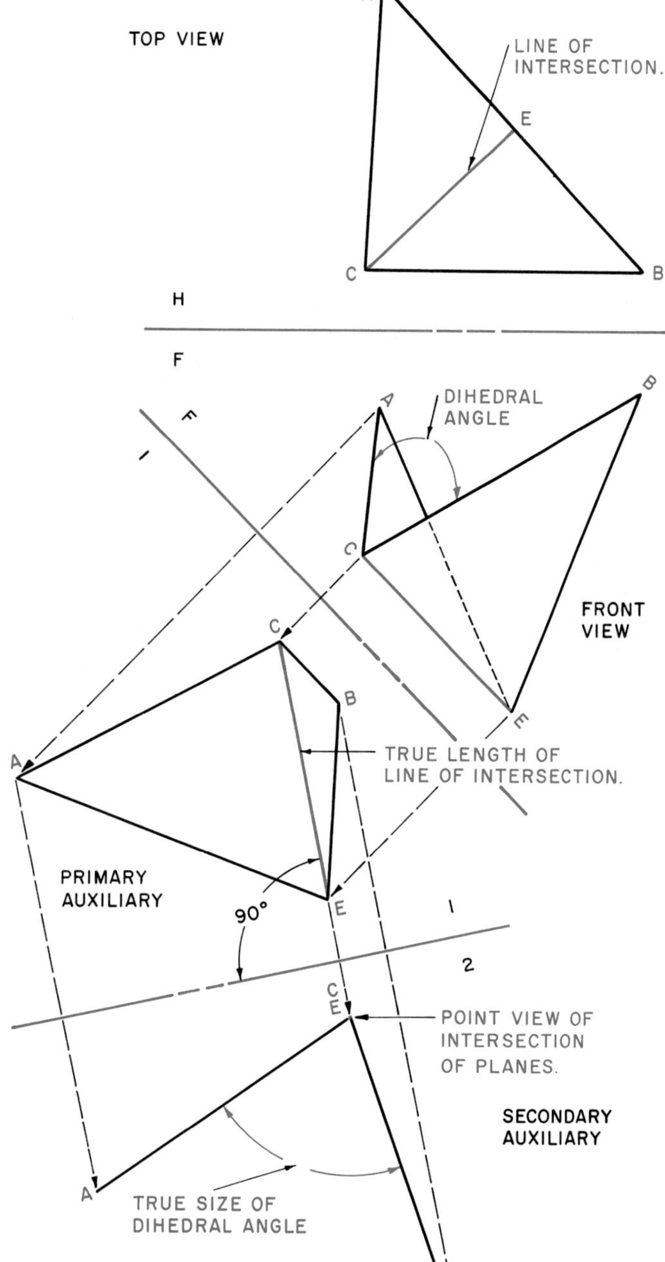

Fig. 10–30. The true size of the dihedral angle is determined by finding the line of intersection in the point view.

Visibility

It is essential when drawing orthographic views to indicate which of several crossing lines is on top and, therefore, visible in the view under construction. The outline of a view will be visible, but it is sometimes difficult to tell whether lines within the outline are hidden or visible.

Many times visibility can be determined by inspecting the given views. Fig. 10-32. The outline is clearly visible in the front view. But the visibility of edges DH and BF must be established. The line of sight in the top view is directed toward the front view of the object. Since edge DH is nearest to the reference line, it is visible. Corner BF is farthest away from the reference line and behind the face CDHG; therefore it is hidden.

Occasionally the visibility of nonintersecting lines cannot be determined by simple inspection, as illustrated in Fig. 10-33. Two hydraulic lines controlling the blade of a bulldozer are shown. To complete accurately a drawing of the piping installation, you need to show which pipe is visible (on top) in each view. The center lines of the pipes are used for drawing purposes. To ascertain visibility, follow these steps:

1. In the top and front views of the pipes, there are two possible points of intersection. These are indicated by X and Y.
2. Project point X from the top view to the front view. It intersects pipe AB at X′ and CD at X″. Since it intersects AB first in the front view, this pipe is visible in the top view.
3. Project point Y from the front view to the top view. It intersects pipe CD at Y′ and AB at Y″. Since it intersects CD first in the top view, CD is visible in the front view.
4. Draw the visible and hidden lines.

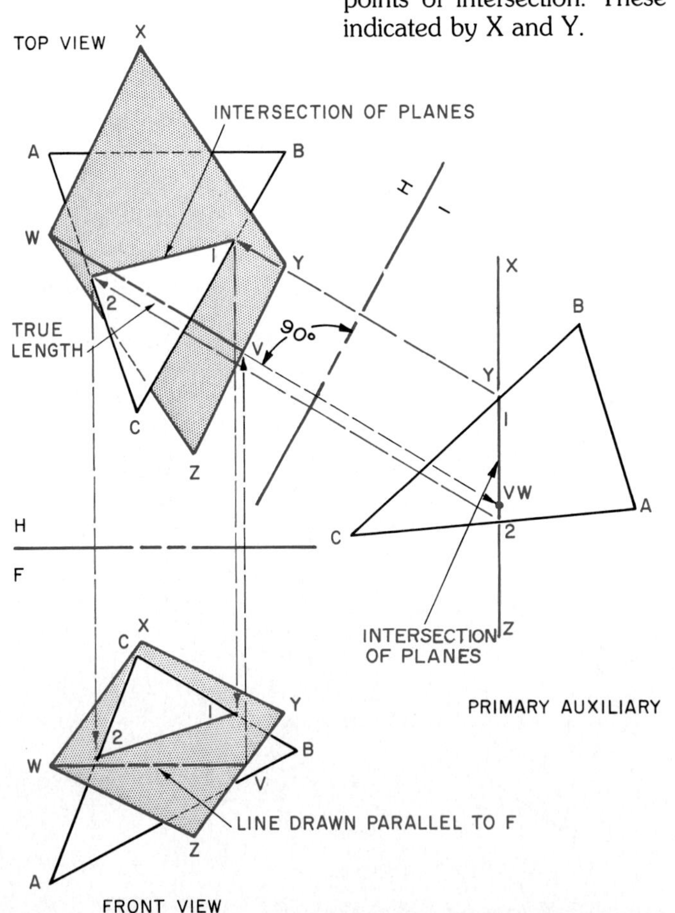

Fig. 10–31. The intersection of two planes can be determined by finding one plane in edge view.

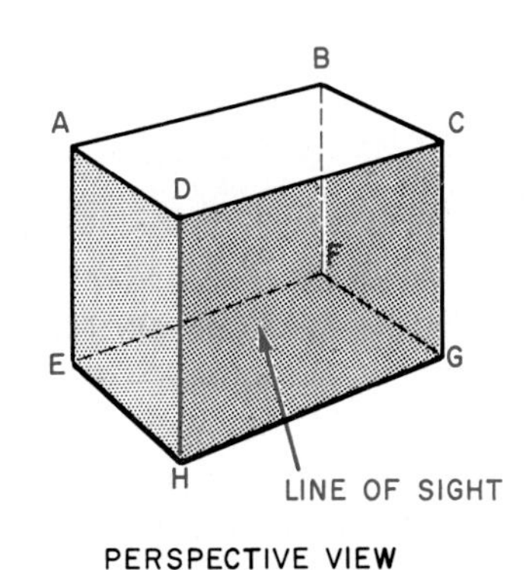

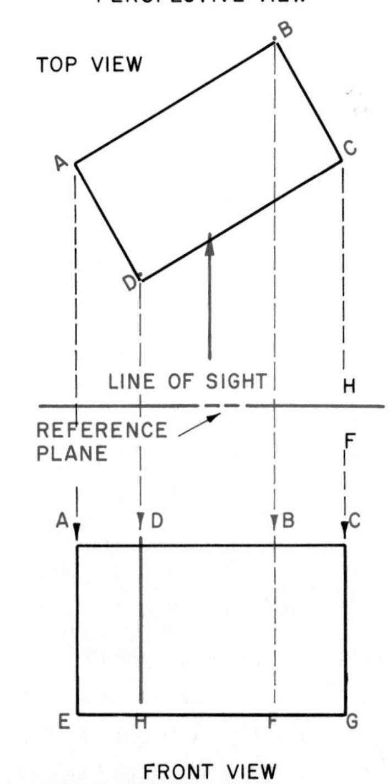

Fig. 10–32. Visibility may be found by inspection.

CAD Applications:

Translating CAD Drawings

In Chapter 9, you read about converting manually produced drawings into CAD drawings. Another challenge many companies face is how to make drawings that were produced on one CAD system usable on another. For example, a manufacturer may receive CAD drawings on disk from a design firm. If the manufacturer's CAD system is different from the design firm's, it may not be able to display or plot the drawings. To solve such problems, methods have been developed for translating drawings from one system to another.

One way is to convert a drawing first into a "neutral" graphics file. The neutral file is not specific to any one system; many systems can accept data in the neutral file format and convert it into their own format. One such neutral file is the Initial Graphics Exchange Specification, or IGES. The IGES was developed by volunteers from over one hundred firms and organizations, and it is now an ANSI standard. It is used to translate drawings from microcomputer CAD systems to mainframe CAD systems and vice versa. Another neutral file is DXF, the Drawing Interchange File. The DXF file was developed by Autodesk, makers of the AutoCAD® drafting program. DXF is used to translate drawings from one microcomputer CAD system to another.

Neutral graphics files such as IGES and DXF have drawbacks, however. The translation is never complete. The untranslated features must be manually added to the drawing. Another disadvantage is that these translations are time-consuming. A complex drawing may take several hours to translate.

In contrast, direct conversion is faster and more accurate. With this method, drawings are converted directly from one system to another. There is no intermediate conversion to a neutral file. However, the software for direct conversion is specific to the systems for which it was designed. In other words, the direct conversion program that translated drawings from system A to system B will not be able to translate drawings from system A to system C. A different conversion program will be needed for that.

Perhaps someday a CAD drawing produced on one system will be usable on all other systems. Until then, however, drafters will have to know how to translate from one system to another.

Octal AutoCAD internal conversion softward by Octal, Inc.

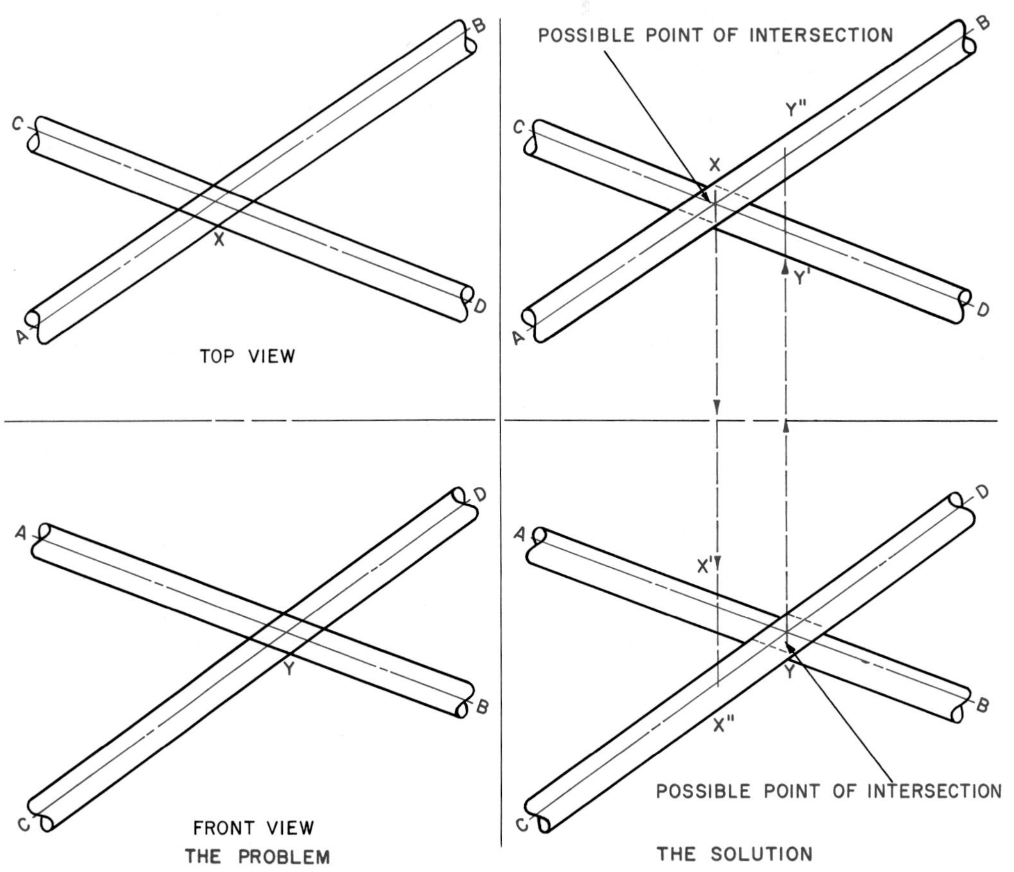

Fig. 10–33. **The visibility of nonintersecting lines is found by projection.**

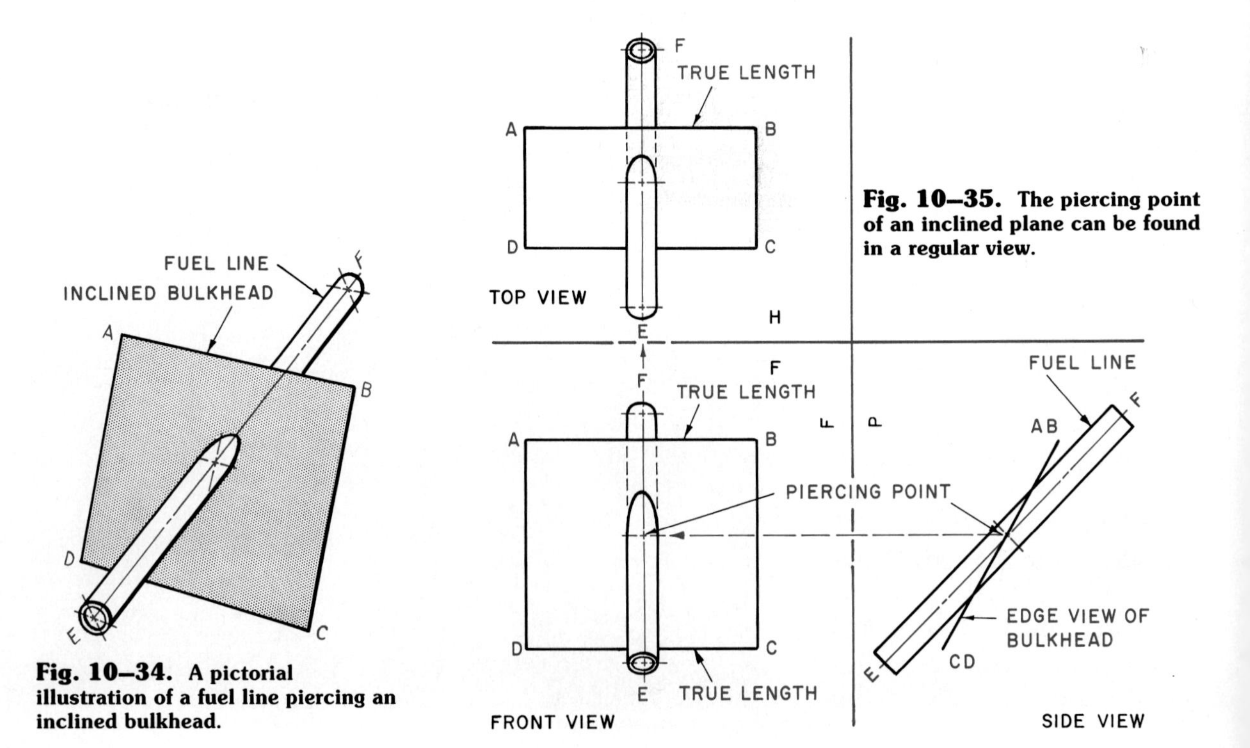

Fig. 10–34. **A pictorial illustration of a fuel line piercing an inclined bulkhead.**

Fig. 10–35. **The piercing point of an inclined plane can be found in a regular view.**

Locating a Piercing Point

Sometimes a line, such as a fuel line in an aircraft, must pass through a bulkhead. In order to locate exactly where the line will pierce the bulkhead, the plane must be located in its edge view. Fig. 10-34 shows an inclined plane (the bulkhead) with the fuel line piercing it. The **piercing point** (intersection of a line and plane) of the inclined plane can be found in the regular side view, as shown in Fig. 10-35. It is then projected back to the other views to locate the point where the pipe pierces the plane. An auxiliary view is not needed because the edges of the inclined plane are true length in the regular views. The inclined plane appears as an edge in the side view.

If the bulkhead is in an oblique position, as shown in Fig. 10-36, the bulkhead must be found in its edge view. This requires a primary auxiliary view. To find the piercing point of a line in an oblique plane, follow these steps:

1. Locate or construct a true length line on plane ABCD. In Fig. 10-37 the oblique plane shown has no true length lines in either the top or front view. It was necessary to construct such a line. In the top view line AG is drawn parallel to the frontal plane. When this line is projected onto the front view, it appears true length.
2. Project the plane ABCD onto the auxiliary plane to view it as an edge view.
3. Project the center line of the fuel line, EF, onto the auxiliary plane. Where it crosses the edge view of the plane is the piercing point.
4. Locate the piercing point in the regular views by projecting it back to them from the auxiliary view.

Fig. 10–36. **A pictorial illustration of a fuel line piercing an oblique bulkhead.**

Revolution

Simple applications for using revolution were shown in Chapter 9. Descriptive geometry problems can also be solved using revolution instead of auxiliary views. Some examples follow.

True Length of an Oblique Line by Revolution

In Fig. 10-38 is oblique line AB. Its true size can be found by revolving it until it is parallel with a principal reference plane. In this example it was revolved in the top view. Hold either end and revolve the other. Project the revolved end to the front view. Connect the revolved end with the held end to form a true length line.

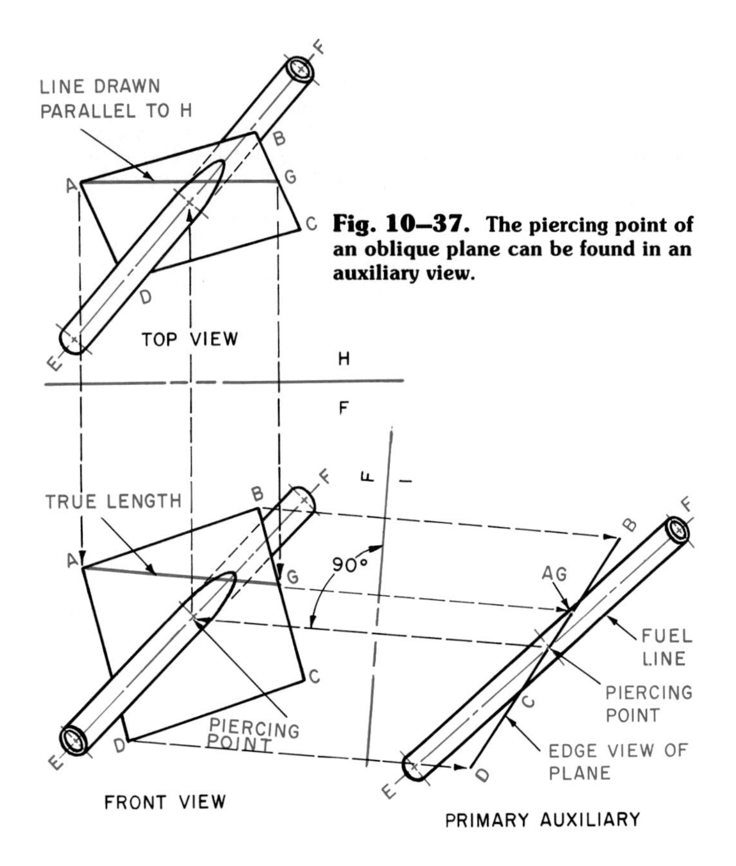

Fig. 10–37. **The piercing point of an oblique plane can be found in an auxiliary view.**

True Size of an Oblique Plane by Revolution

The true size of an oblique plane can be found by using two revolutions. Fig. 10-39. First find the edge view of the plane. Then revolve the edge view to find the true size of the plane. Follow these steps:

1. Draw a line across the plane parallel to a principal reference plane. In this example it was drawn parallel to the frontal plane.
2. Revolve the plane so that the true length line appears as a point in a principal view.
3. Project the corners of the plane to the revolved position.
4. Then, using the true length line as an axis, revolve the plane so that it is parallel with the frontal reference plane in the top view.
5. Project the revolved corners of the plane to the front view to show the plane in its true size.

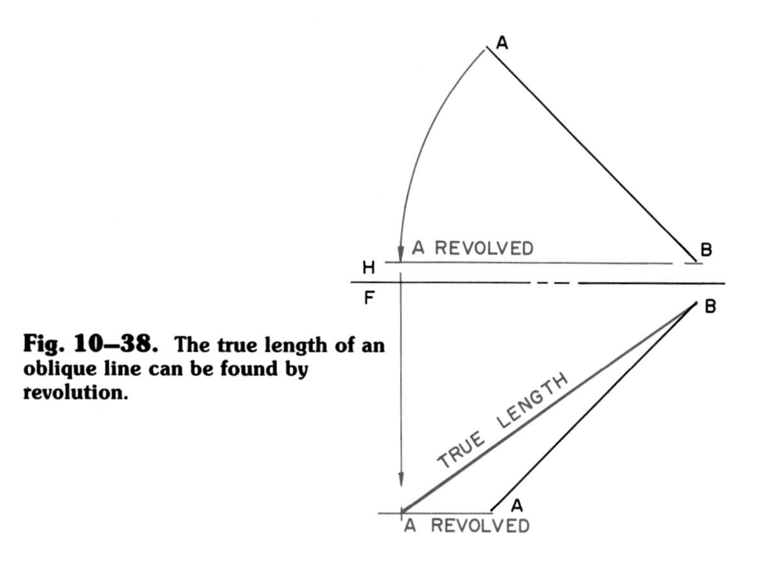

Fig. 10–38. The true length of an oblique line can be found by revolution.

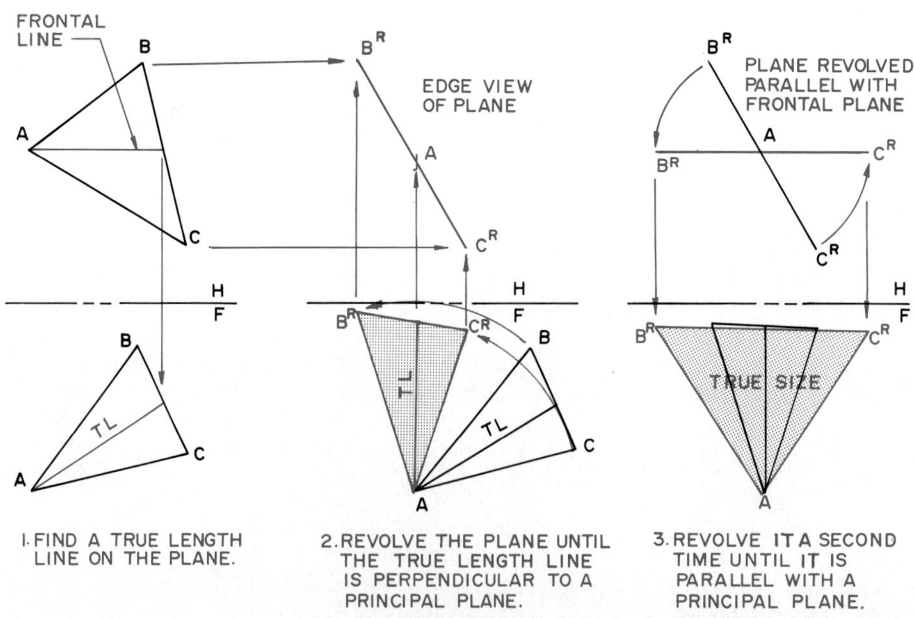

Fig. 10–39. The true size of an oblique plane can be found by revolving the plane twice.

Chapter Review

Build Your Vocabulary

You should understand and use the following terms as part of your working vocabulary. Write a brief explanation of what each means.

geometric elements
geometry
point
line
principal line
inclined line
oblique line
slope
grade
batter
bevel
azimuth
plane
piercing point

Sharpen Your Math Skills

1. If a roof has a slope of 12 to 12, what is its angle in degrees? What is the ratio of the slope reduced to its lowest terms?
2. If a highway has a rise of 10′ per 100′ of run, what is its grade in degrees? What is the ratio of the grade reduced to its lowest terms? What is the percent of the grade?
3. The batter of a retaining wall has a run of 1′ and a rise of 7′. What is the ratio of this batter?
4. What is the azimuth of a point that is 200° from north?
5. What is the bearing of a point with an azimuth of N210°?

Study Problems—Directions

The problems that follow will give you the chance to use some of the drafting knowledge that you learned in this chapter. Do the problems that are assigned to you by your instructor. On the grids, let each square represent .5″ or 12 mm.

1. On **Fig. P10-1**, locate line 1-2 and point P on all views.
2. On **Fig. P10-2**, plot the following azimuth data: 120°, 210°, 350°.
3. Find the point view of line AB on **Fig. P10-3** by projecting from the front, top, and side views.
4. Find the shortest distance from point A to line 1-2 in **Fig. P10-4**.
5. Find the shortest distance between lines AB and CD shown in **Fig. P10-5**.
6. Determine the visibility of the nonintersecting pipes shown in **Fig. P10-6.**
7. On **Fig. P10-7**, find the true size of the angle between intersecting lines 1-2 and 2-3.
8. Find the true size of the angle between intersecting lines AB and BC in **Fig. P10-8**.
9. On **Fig. P10-9**, find the true size of plane ABC.
10. Find the true size of plane 1-2-3 shown in **Fig. P10-10**.
11. Find the true size of the angle between planes 1-2-3 and 1-3-4 in **Fig. P10-11**.
12. In **Fig. P10-12**, find the line of intersection between planes 1-2-3 and ABC.
13. In **Fig. P10-13**, find the point where line AB pierces plane 1-2-3-4.
14. Find the point where line AB pierces oblique plane 1-2-3 in **Fig. P10-14**.
15. In **Fig. P10-15**, find the true length of line AB by revolution.

16. Find the true size of plane 1-2-3-4 shown in **Fig. P10-16** by revolution.

17. In **Fig. P10-17**, find the length of the guy wires from the flag pole.

18. In **Fig. P10-18**, find the length of the cables making the sling and the angle they make with the top surface of the crate.

19. In **Fig. P10-19**, construct the guy wires from the top of the flag pole to points A, B, and C. Find the actual length of each wire.

20. Construct the shortest connecting pipe from point C on the hopper to pipe AB shown in **Fig. P10-20**. Then construct the shortest brace possible between pipes AB and DE.

21. Line EF in **Fig. P10-21** represents a hole to be drilled. Locate the point where the drill would pierce surface ABCD.

22. How long is the mine shaft shown in **Fig. P10-22**? What angle does it make with the horizontal?

23. In **Fig. P10-23**, construct the shortest possible brace from point A on the fuselage to point C on the landing gear strut. What is the true length of the brace?

24. Find the point where line EF in **Fig. P10-24** pierces plane ABCD on the front and top views.

Study Problems

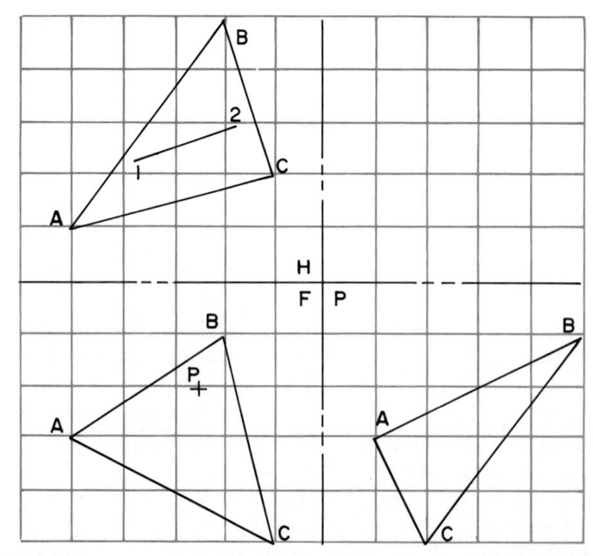

P10–1

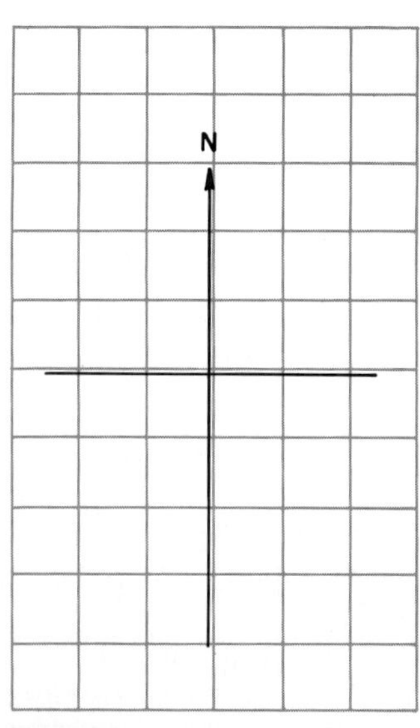

P10–2

Study Problems

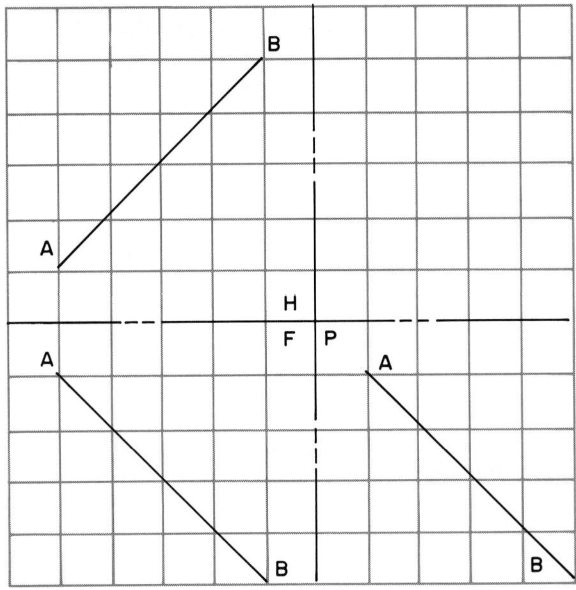

P10–3

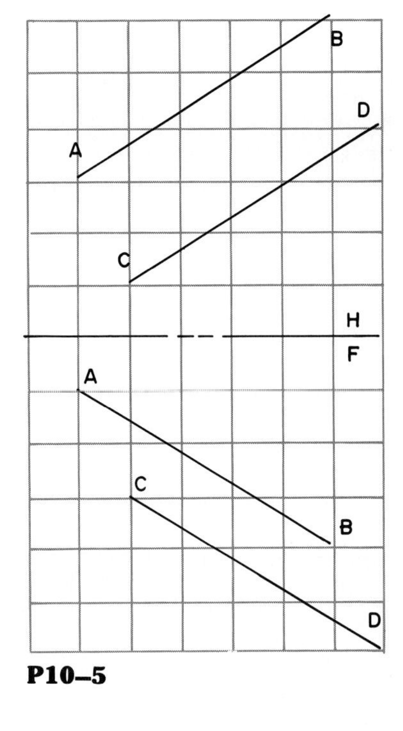

P10–5

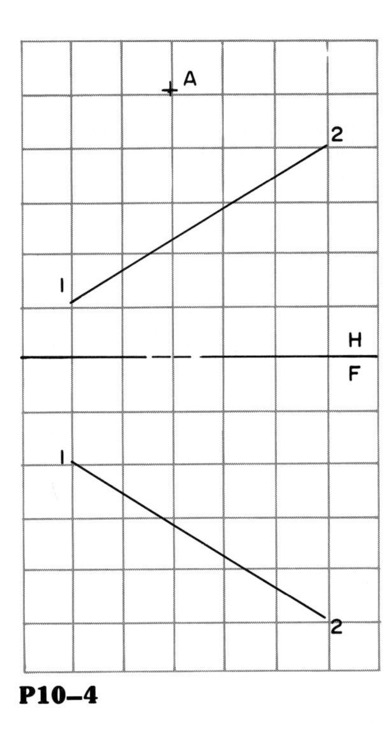

P10–4

P10–6

Study Problems

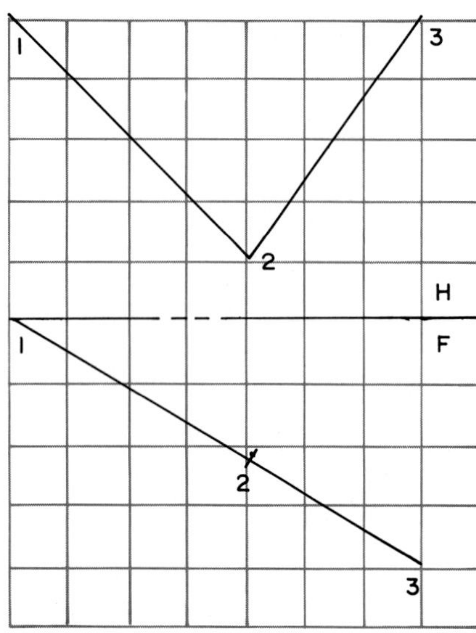

P10–7

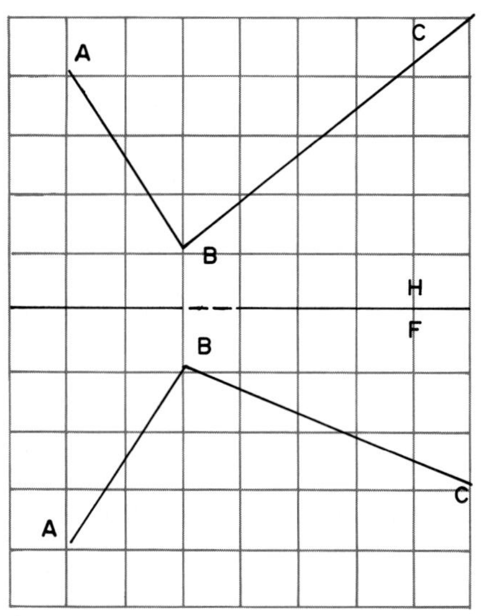

P10–8

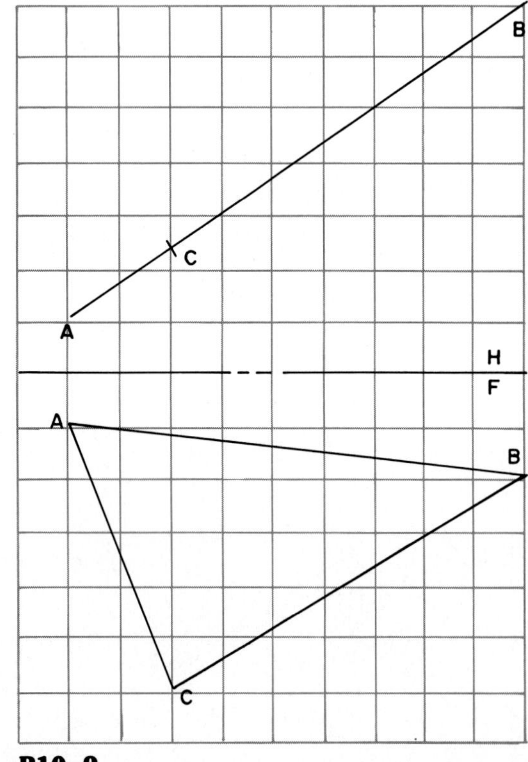

P10–9

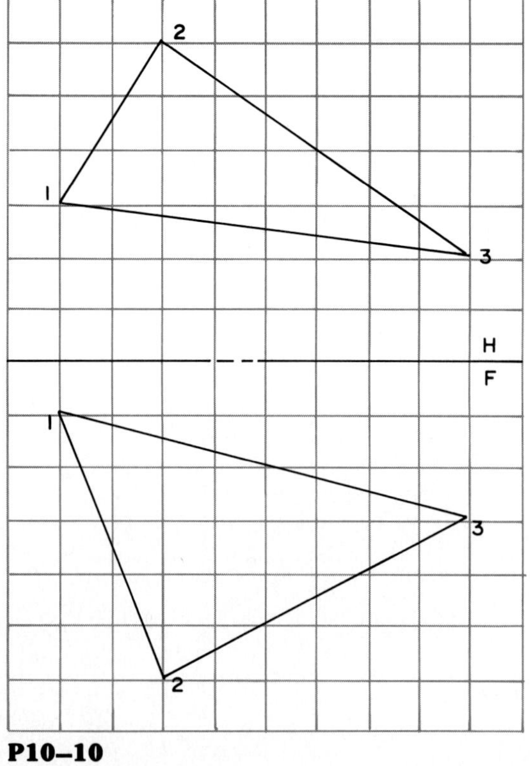

P10–10

Study Problems

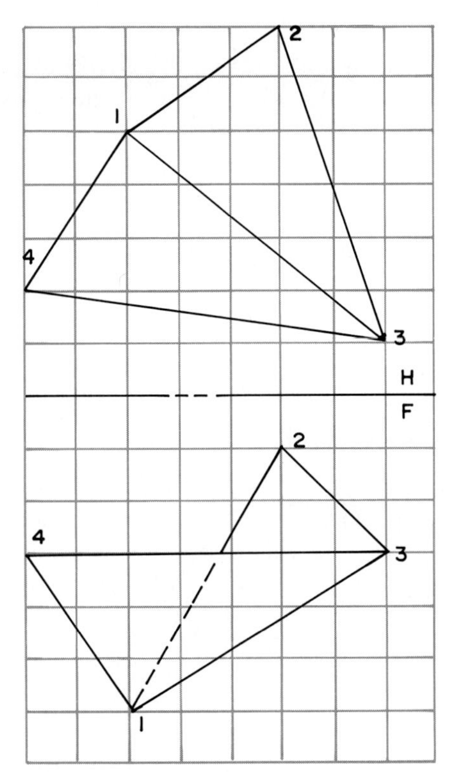

P10–11

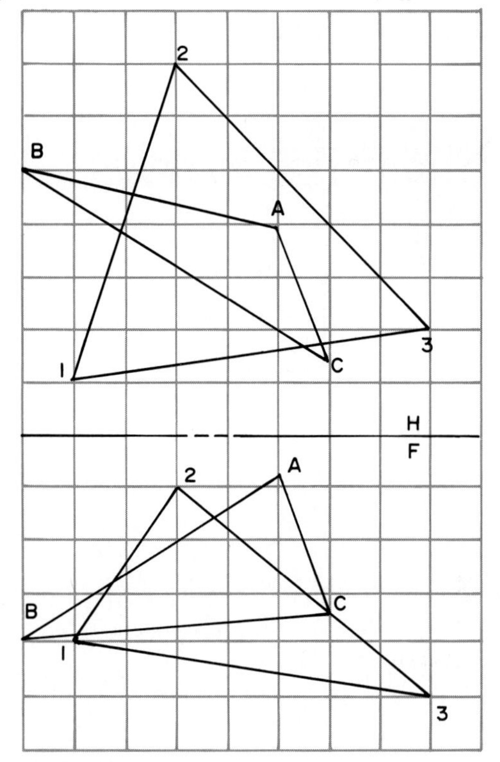

P10–12

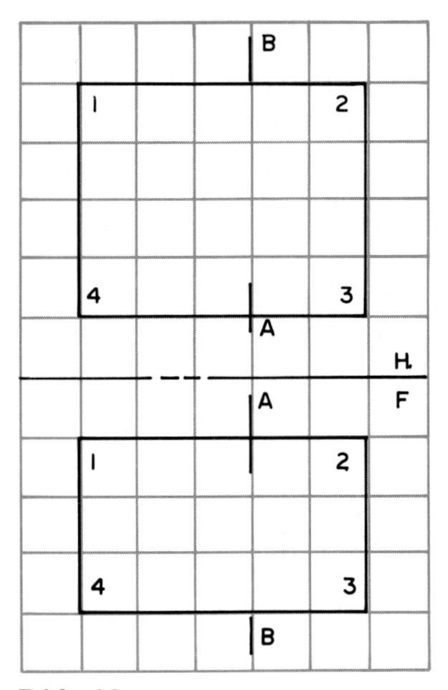

P10–13

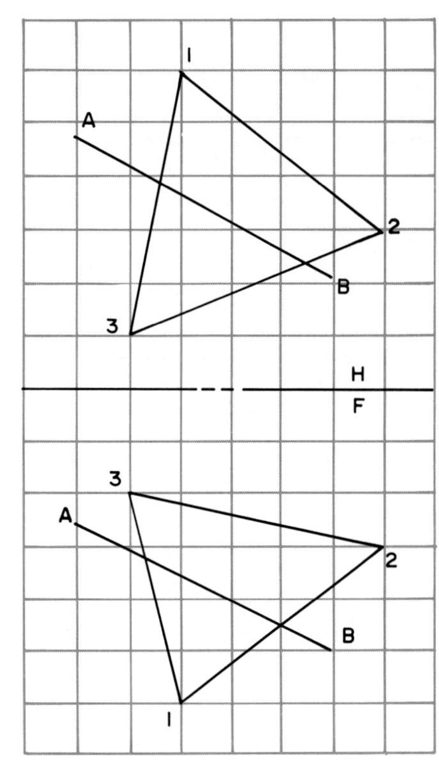

P10–14

Study Problems

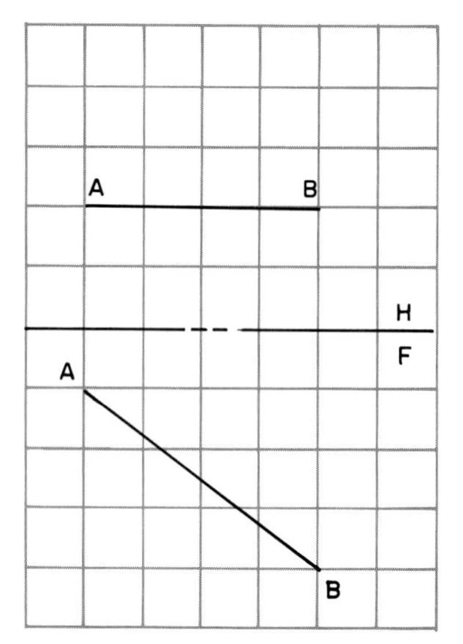

P10–15

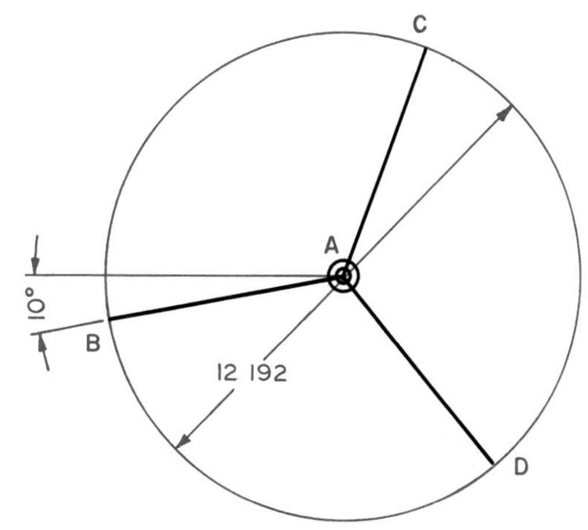

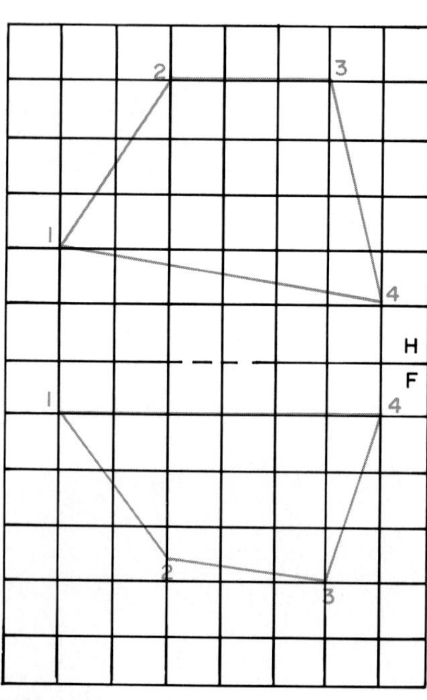

P10–16

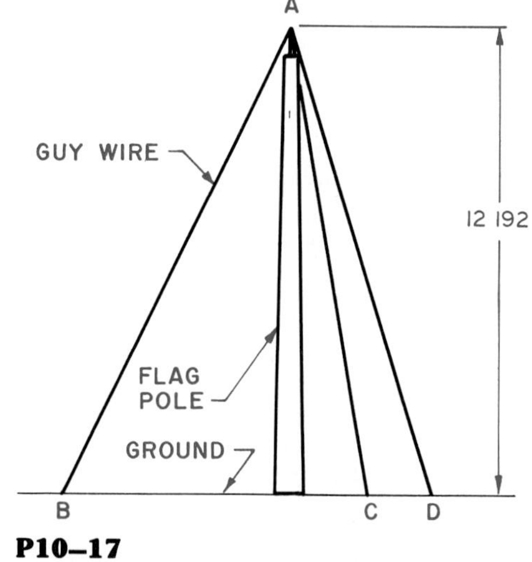

P10–17

Study Problems

PULLEY
CABLE SLING
CRATE

CABLE SLING
CRATE
914
1524

DIMENSIONS IN mm

PULLEY
1220
CABLE SLING
914
CRATE

P10–18

A
D
20'-0"
21'-0"
32'-0"
C
58°
B
28'-0"

D
66'-0"
A
B
C
23'-0"
16'-0"

P10–19

Study Problems

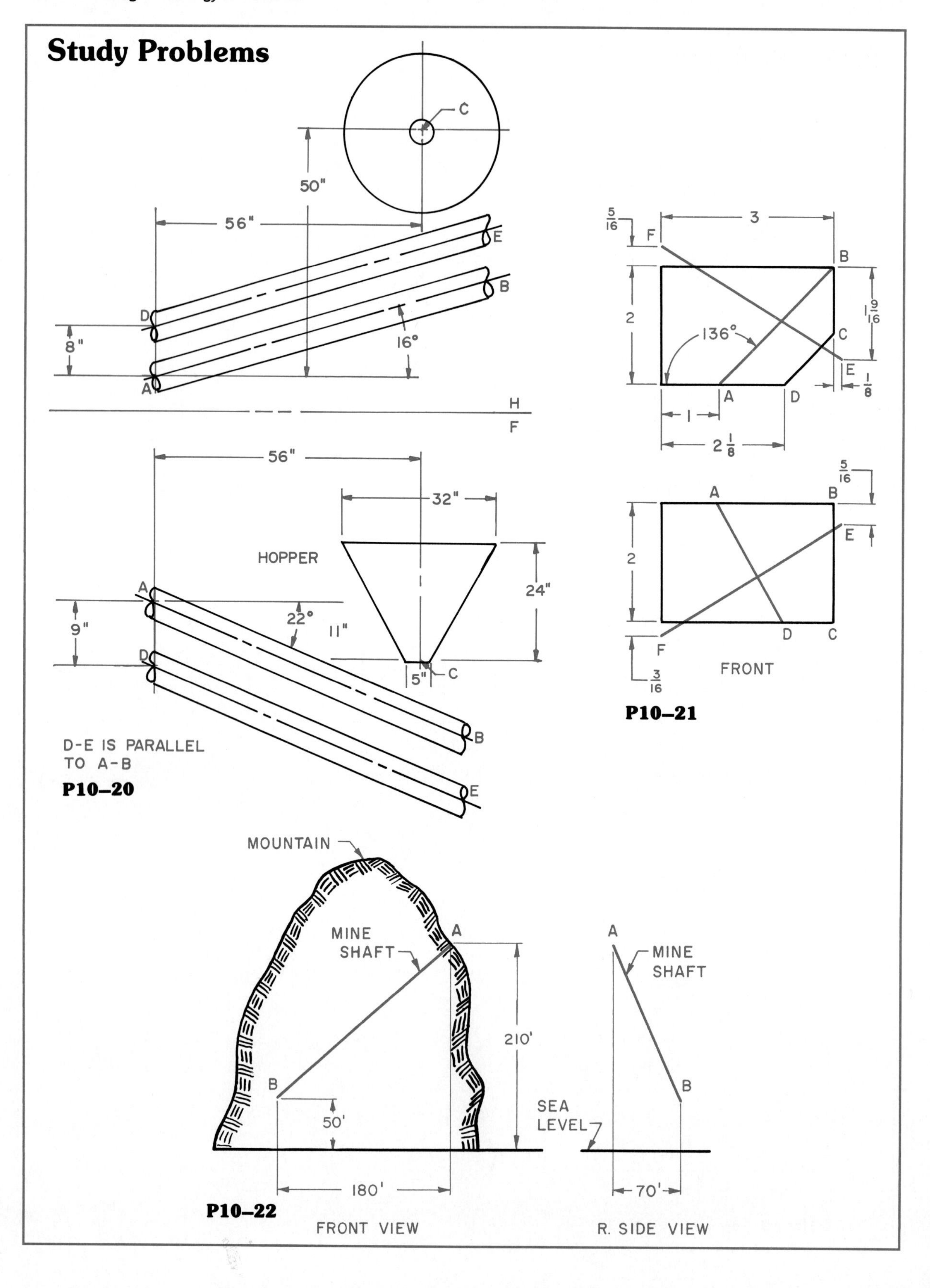

P10–20

P10–21

P10–22

Study Problems

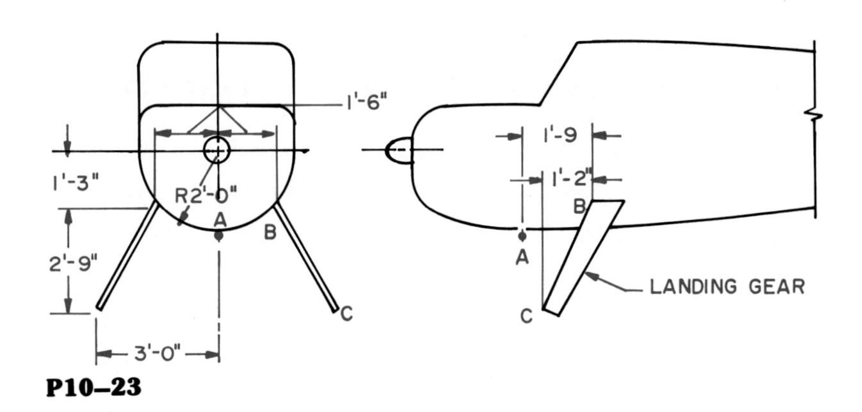

P10–23

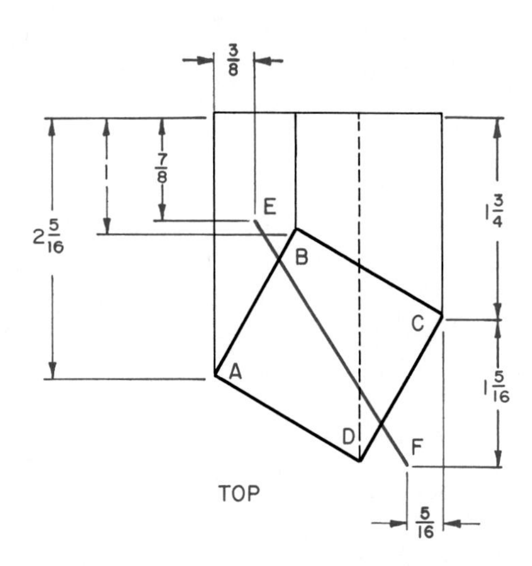

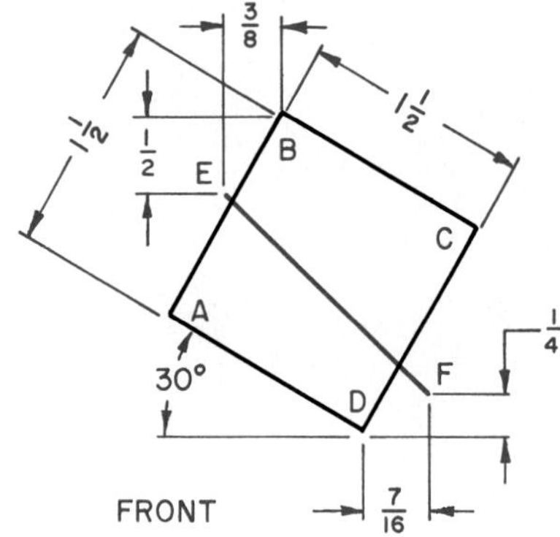

P10–24

CHAPTER 11

Sectional Views

After studying this chapter, you should be able to do the following:

- Draw various types of sectional views.
- Recognize standard material symbols.

The Need for Sectional Views

Some objects have interior details that cannot be seen in the exterior views on a multiview drawing. If these interior details are simple, they are shown with hidden lines. If the hidden lines do not show the interior details clearly, you need to draw a sectional view. Fig. 11-1.

A **sectional view** is made by passing an imaginary cutting plane through the object. Fig. 11-2A. One part is removed, leaving the interior details exposed. Fig. 11-2B. Drawing a sectional view is like sawing through the object and viewing the inside.

In Fig. 11-2C, a three-view drawing, the interior details of the valve are not clear. To clearly show the interior detail, a cutting plane is passed though the object as shown at Fig. 11-2D. The section view is drawn. In this case the section view was drawn as the front view. The exposed surfaces are indicated using **section lining.**

Cutting-Plane Line

The imaginary **cutting plane** is shown on a drawing by symbols. Fig. 11-3. Both types of cutting-plane symbols are in common use. The one you use depends upon the standard practice of your drafting room.

The cutting-plane line is placed on the view in which it appears in edge view. It is from this view that the section is projected. Refer to Fig. 11-2D in which the section is projected from the top view. If it is obvious where the section was taken, as with a full section, the cutting-plane line can be omitted.

Usually the cutting-plane line has arrows at each end. These are drawn at right angles to the cutting-plane line. The arrows point in the direction of the line of sight used to make the section.

The section is identified by placing the same capital letter at each end of the cutting-plane line near the arrowhead. Refer to Fig. 11-3. If only one section is taken and its location is obvious, the letters are not used. If more than one section is made, it's best to use letters to identify the sections.

If the section is made along the center line of a symmetrical object, the cutting-plane line replaces the center line.

Sectional views are projected in the same manner as regular orthographic views. The lines of sight are perpendicular to the cutting-plane line.

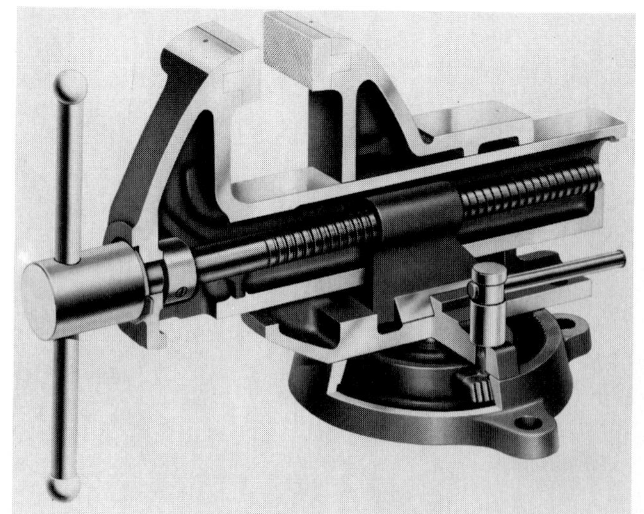

Fig. 11–1. A sectional view is drawn as if part of the object were cut away, leaving the interior exposed. This is a technical illustration prepared for advertising purposes. (The Columbian Vise and Manufacturing Co.)

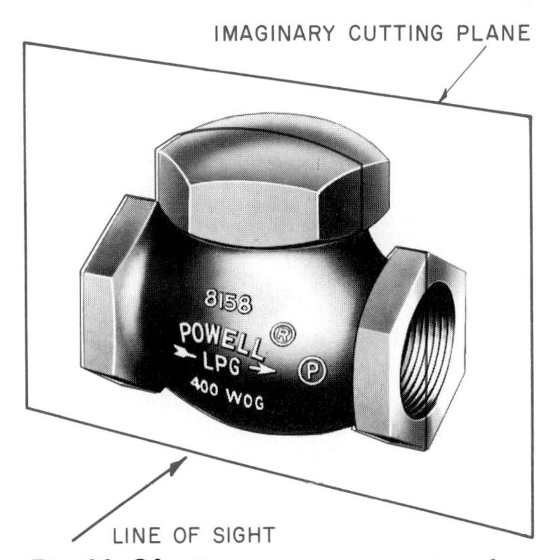

Fig. 11–2A. Here, an imaginary cutting plane is passed through a lift valve parallel to the frontal plane. (The William Powell Company)

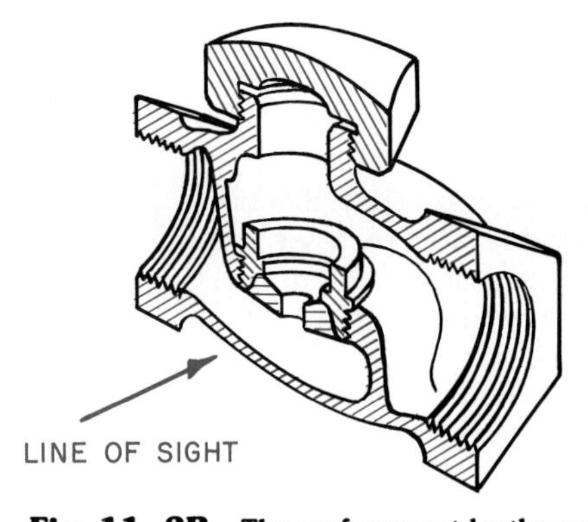

Fig. 11–2B. The surfaces cut by the cutting plane are exposed when the front part is removed.

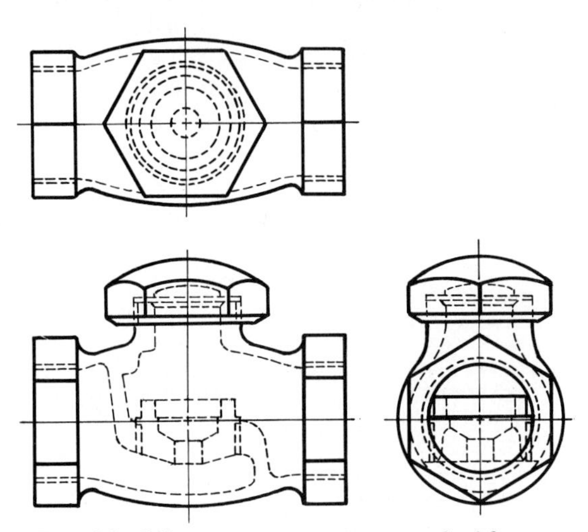

Fig. 11–2C. A three-view drawing of a lift check valve.

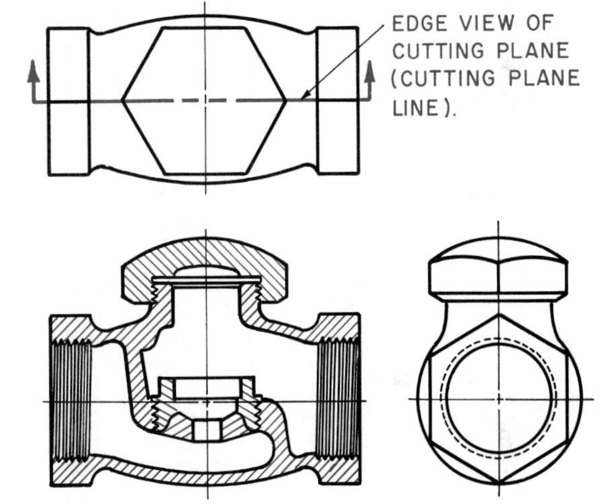

Fig. 11–2D. The surfaces cut by the cutting plane.

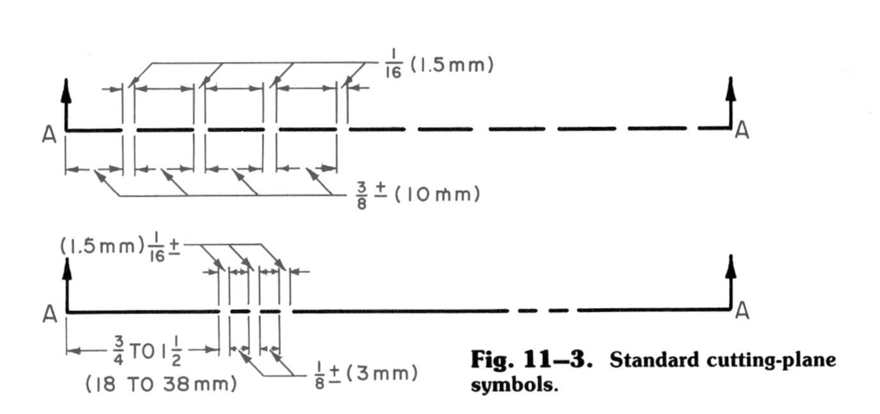

Fig. 11–3. Standard cutting-plane symbols.

Section Lining

Indicate the surface exposed by the cutting plane by drawing very thin lines, called section lines, on the cut surface.

Symbols have been developed to identify the type of material in the sectioned part. The most common **material symbols** used are shown in Fig. 11-4. Since materials are made in many different types, the symbols only generally identify any particular material. For example, there are many different types of steel, and the symbol cannot identify each of them. Therefore, the symbol identifies the material as steel, and the particular type of steel is indicated by a note.

In many drafting rooms it is standard practice to use only one general symbol (the same symbol as cast iron) for all materials. This saves the drafter from having to remember or look up the different kinds of material symbols. If specific material symbols are used, they are usually placed on a drawing where it is necessary to point out the material differences. In Fig. 11-5 the lathe foot is made of cast iron, the leveling pad of steel, and the blocking of wood.

When drawing the general purpose symbol, space the section lines from 1⁄16 to 1⁄8 inch (1.5 to 3 mm) apart. The actual spacing you use depends upon the drawing. On small areas, space the lines closer together than on large areas. Fig. 11-6. Spacing on most drawings should be about 3⁄32 inch (2 mm).

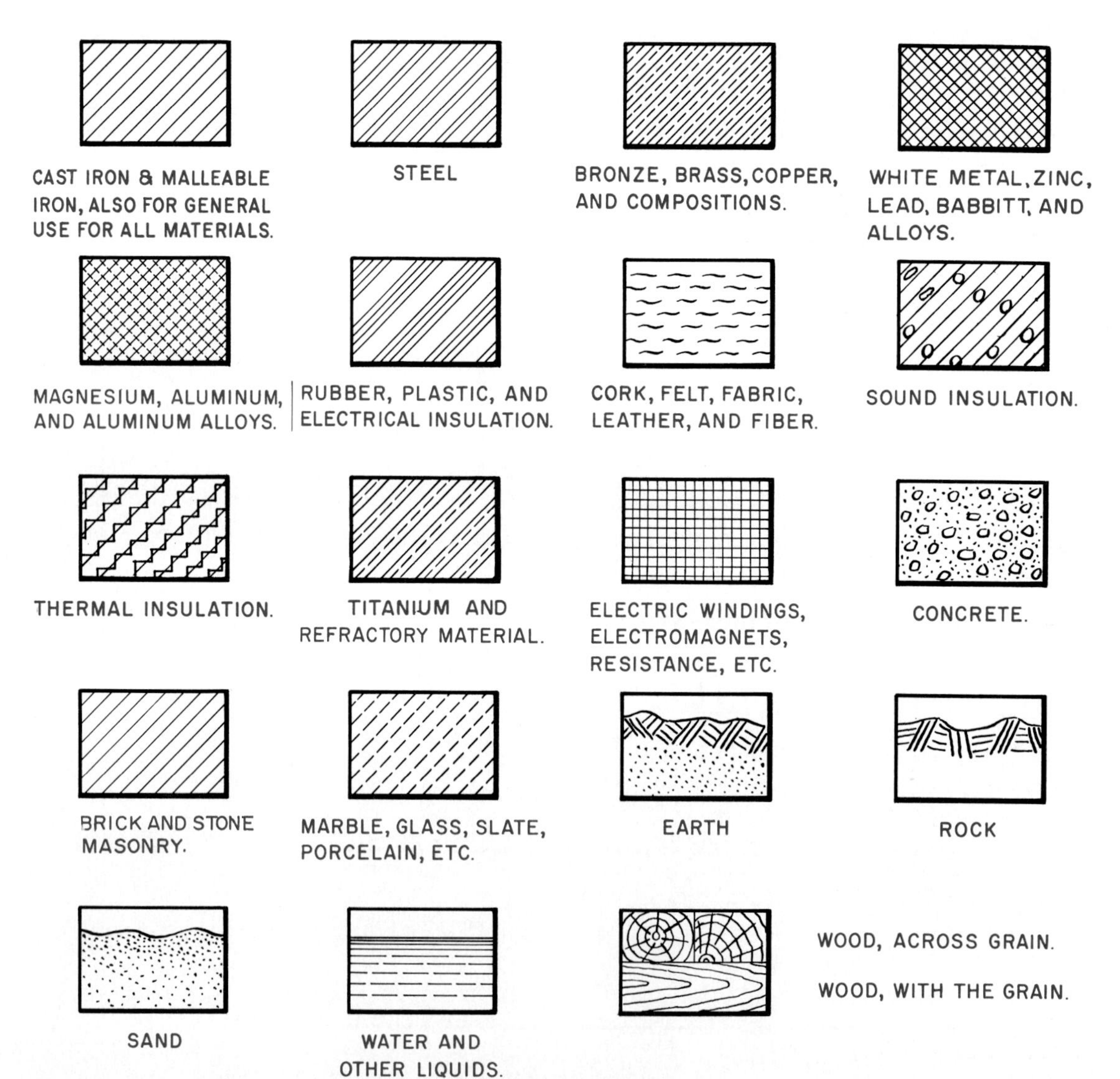

Fig. 11–4. Standard symbols for indicating materials in section.

Very thin parts, such as sheet metal or gaskets, are sectioned solid black. If several thin parts are touching, leave a small white space between the solid sections. Fig. 11-7.

Space section lines by eye. You can make a spacing guide from graph paper. Place the graph paper under the drawing sheet, and use the printed lines that show through as a guide to space section lines.

Draw section lines on a 45-degree angle. Use other angles if the drawing has several parts. Fig. 11-8. Draw the angled section lines of adjacent parts in opposite directions. Do not draw section lines parallel to the sides of the area to be sectioned. Using different angles and directions makes each sectioned part stand out clearly.

If a single piece is cut two or more times by the cutting plane, draw the section lining in the same direction and at the same angle at each cut. Refer to Fig. 11-6.

You may section line a very large area as shown in Fig. 11-9. This section lining is called *outline sectioning* since only the outline of the area has lines drawn upon it.

Visible, Hidden, and Center Lines

Sectional views are drawn to clarify a hidden interior detail. Therefore, omit hidden lines because they would cause confusion in the sectional view. Use hidden lines *only* if they are essential to make the view clear.

Visible edges that occur behind the cut surface are usually drawn as shown in Fig. 11-10. In special cases, such as a removed or auxiliary section, visible features may be omitted. These sections are described later in this chapter.

Place center lines on sectional views. The cutting-plane line takes precedence over a center line. The edges of the cut surfaces are always drawn with visible lines.

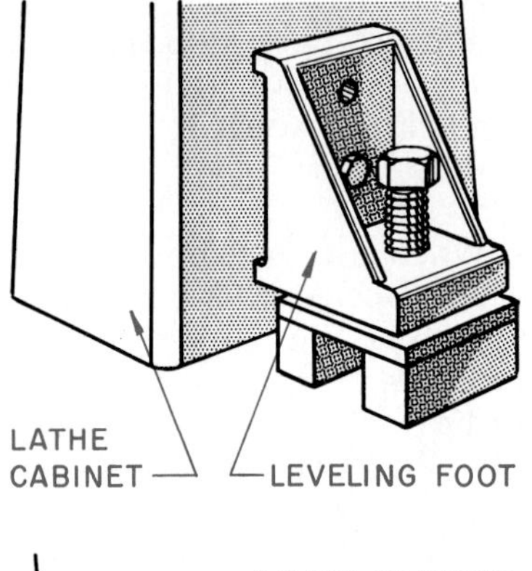

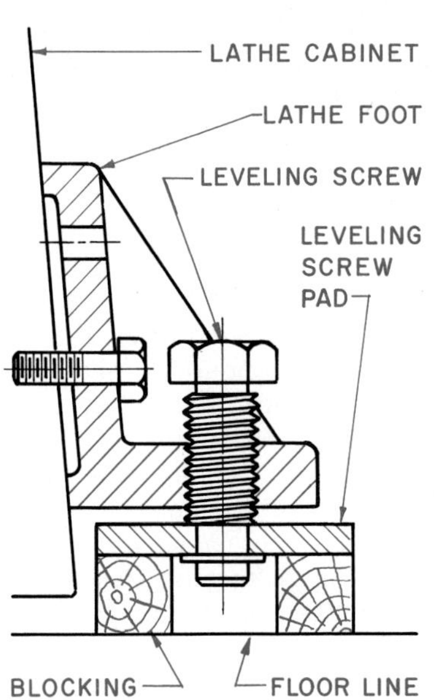

Fig. 11–5. Leveling foot for a lathe. The use of material symbols helps to clarify the drawing. (Atlas Press Co.)

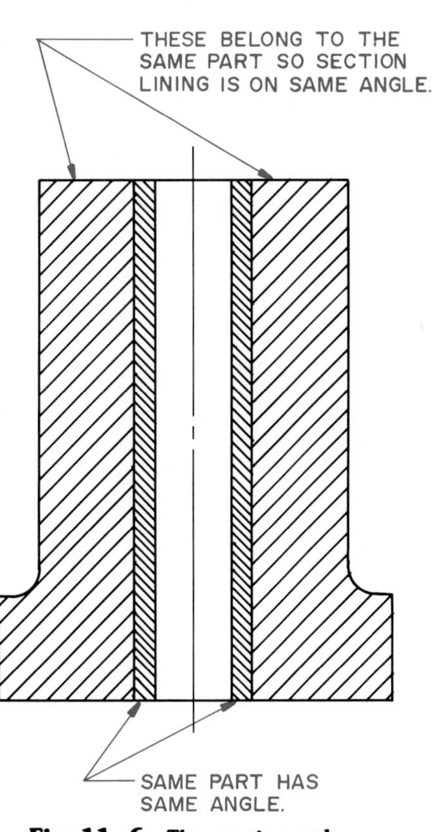

Fig. 11–6. The spacing and direction of section lining.

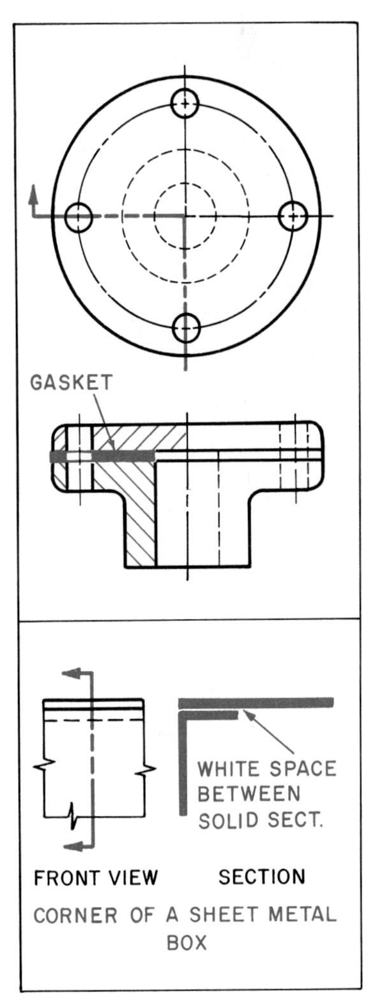

Fig. 11–7. Thin parts are sectioned solid.

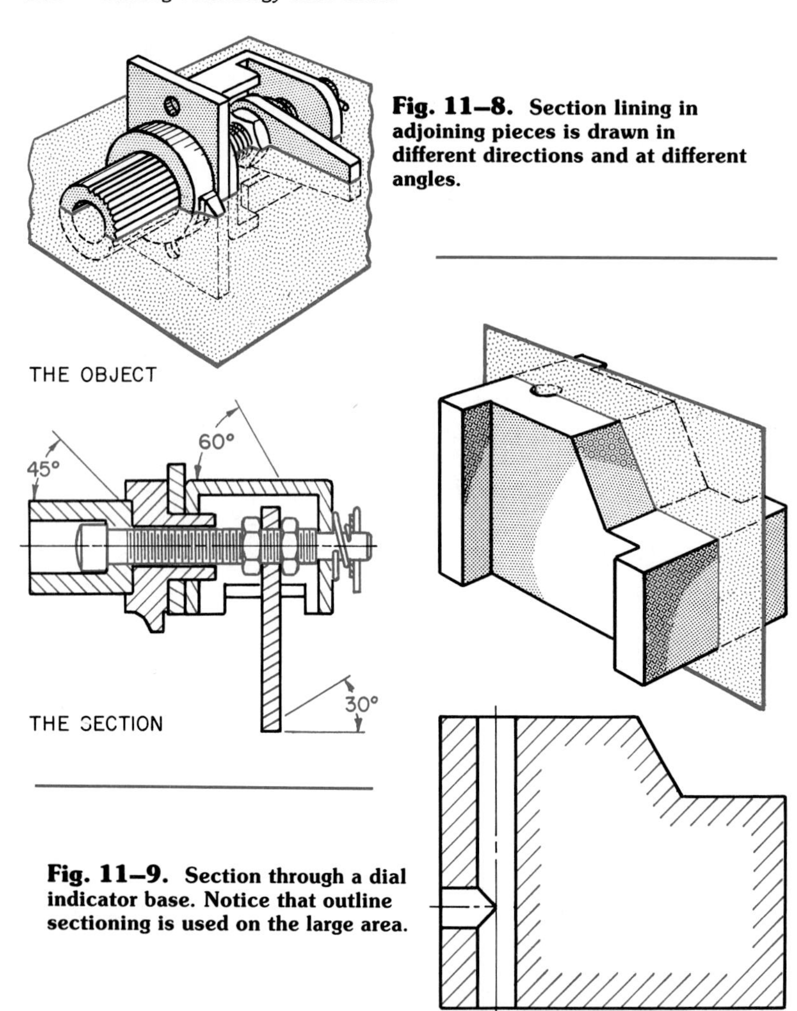

Fig. 11–8. Section lining in adjoining pieces is drawn in different directions and at different angles.

Fig. 11–9. Section through a dial indicator base. Notice that outline sectioning is used on the large area.

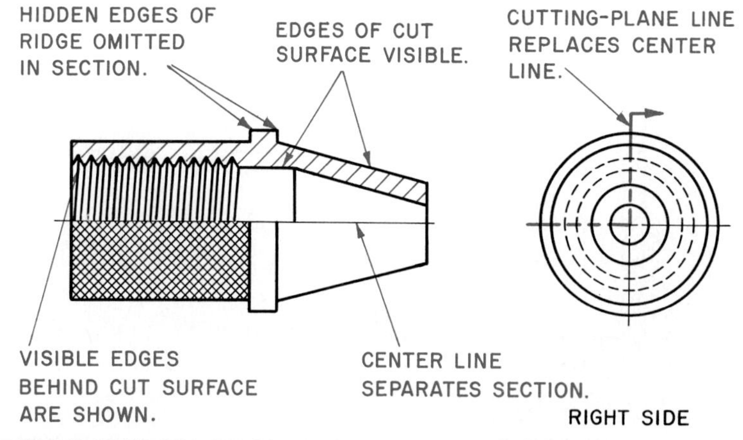

Fig. 11–10. The use of hidden lines, visible lines, and center lines on sectional views.

Full Sections

A **full section** is one in which the cutting plane extends *fully* through the entire object, cutting it into two halves. The full section cutting plane can be parallel to the frontal, horizontal, or profile planes. Fig. 11-11. The cutting-plane line is often omitted on full sections because its location is obvious.

Half Sections

If an object is symmetrical, its interior detail can be fully described by cutting halfway through the object. Since the cutting-plane line passes through one-half of the object, this view is called a **half section.** Fig. 11-12. The half of the object that is not removed shows the exterior. Thus, in one view you see both internal detail and external features. Fig. 11-13.

When a cutting plane falls on top of a center line, omit the cutting-plane line, arrows, and section letters. Fig. 11-14. The half section is commonly separated from the exterior half by a center line.

Usually hidden lines are omitted on both sides of a half section. They are unnecessary since the section shows the hidden details. However, use hidden lines for dimensions or clarity when needed.

Offset Sections

The **offset section** is usually a full section. An offset section is one in which the cutting plane changes direction from the main axis of the object to show details not in a straight line. The cutting-plane line always changes direction at right angles. The change of direction is not shown in the sectional view. Notice in the drawing of the drill press base, Fig. 11-15, that more detail is revealed by the offset section than would be possible by a single full section.

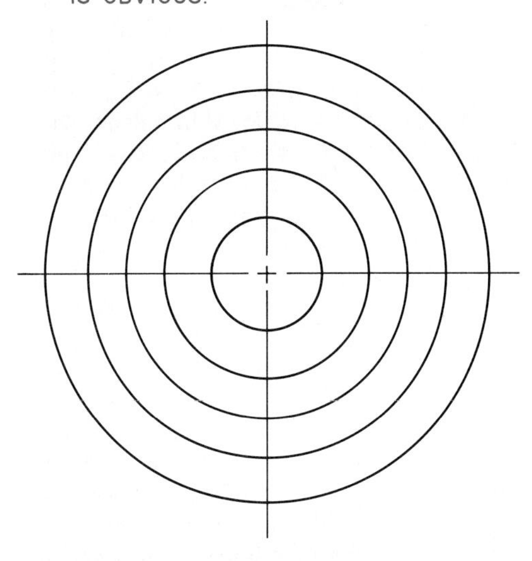

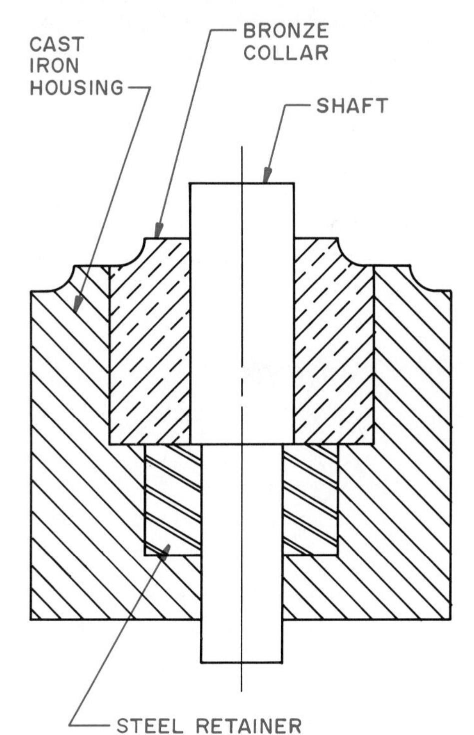

Fig. 11–11. This cutting plane is parallel to the frontal plane.

Fig. 11–12. A half section of a turret tool post for a metal lathe. (Enco Manufacturing Co.)

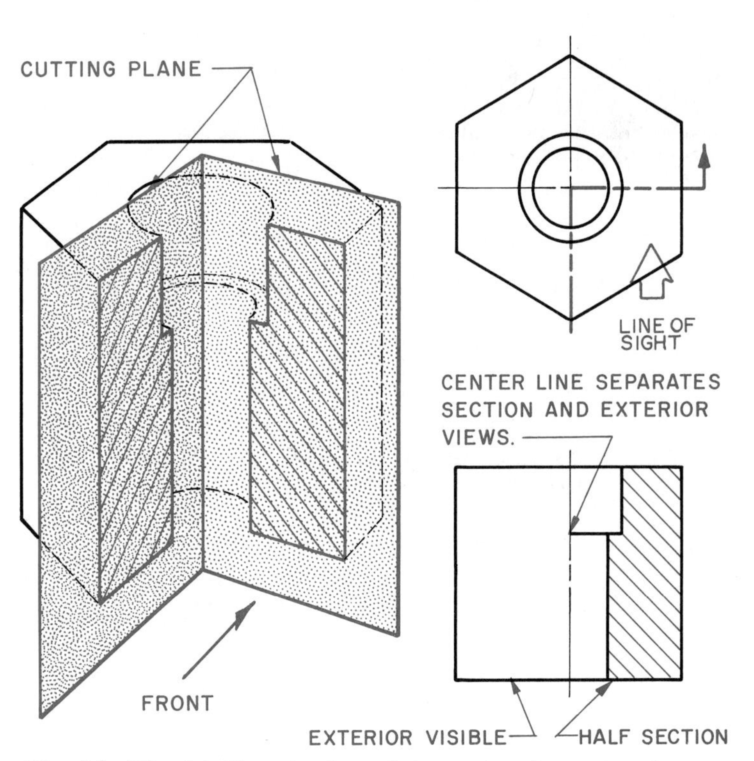

Fig. 11–13. A half section is made by passing the cutting plane halfway through the object.

Aligned Sections

When features are located so that the change in direction of the cutting plane is greater than 90 degrees but less than 180 degrees, the section view is drawn by revolving the angled part into a plane parallel with the line of sight of the sectional view. Such a view is called an aligned section. Fig. 11-16.

Broken-out Sections

Occasionally you need to show the interior detail of a small part of the object. Rather than draw an entire full or half section, you can draw a partial section, called a **broken-out section**. A cutting-plane line is assumed to pass through the feature even though it is not drawn. A freehand break line limits the sectioned area. Fig. 11-17.

Revolved Sections

A **revolved section** is used to describe the cross section of an object, such as a spoke, rib, bar, or other detail, at one special point. To make a revolved section, pass a cutting plane perpendicular to the object's length through the point where the section is to be shown. Then, revolve the section and draw it parallel to the plane of projection. In Fig. 11-18 the section is shown as a slice taken through the handle of a ratchet wrench. At B the proposed section is located on the top view of the handle, and the direction of rotation is shown. At C the rotation is started, and at D it is shown parallel to the frontal plane. Here the section is in position to be projected to the front view where it will appear true size.

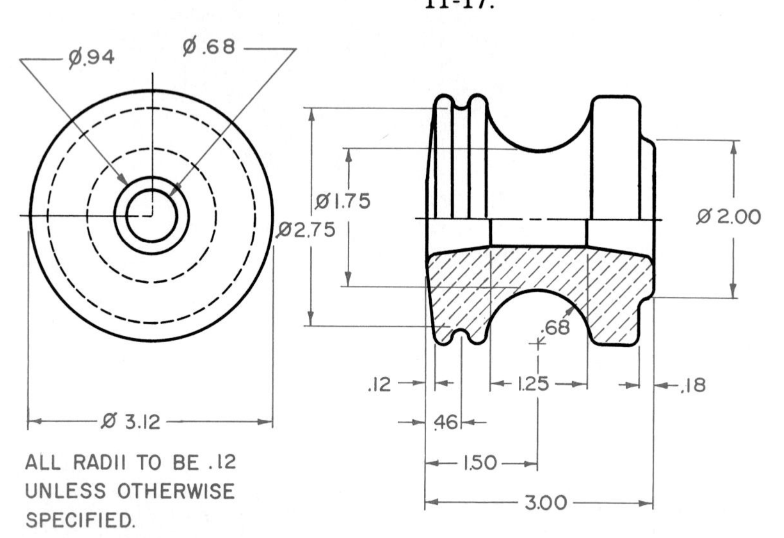

Fig. 11–14. A half section through a ceramic electrical spool insulator. Notice the cutting-plane line has been omitted.

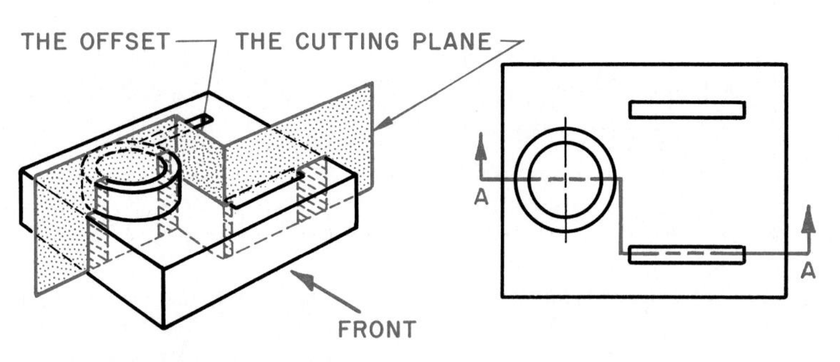

DRILL PRESS BASE WITH OFFSET CUTTING PLANE.

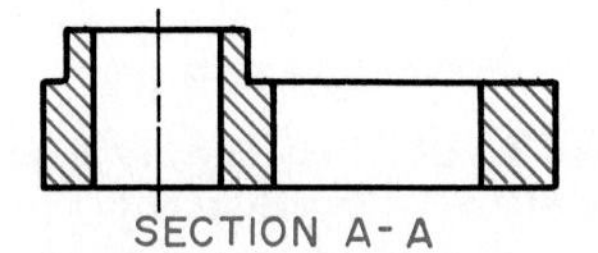

NOTICE THE OFFSET DOES NOT PROJECT TO THE SECTION VIEW.

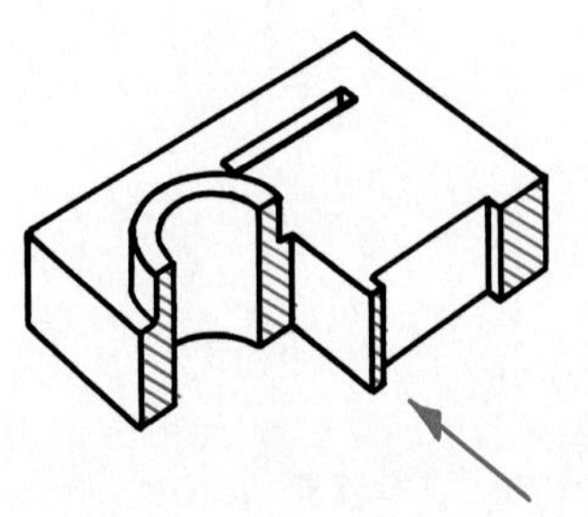

DRILL PRESS BASE WITH PORTION IN FRONT OF CUTTING PLANE REMOVED.

Fig. 11–15. The cutting plane for an offset section changes direction at 90-degree angles.

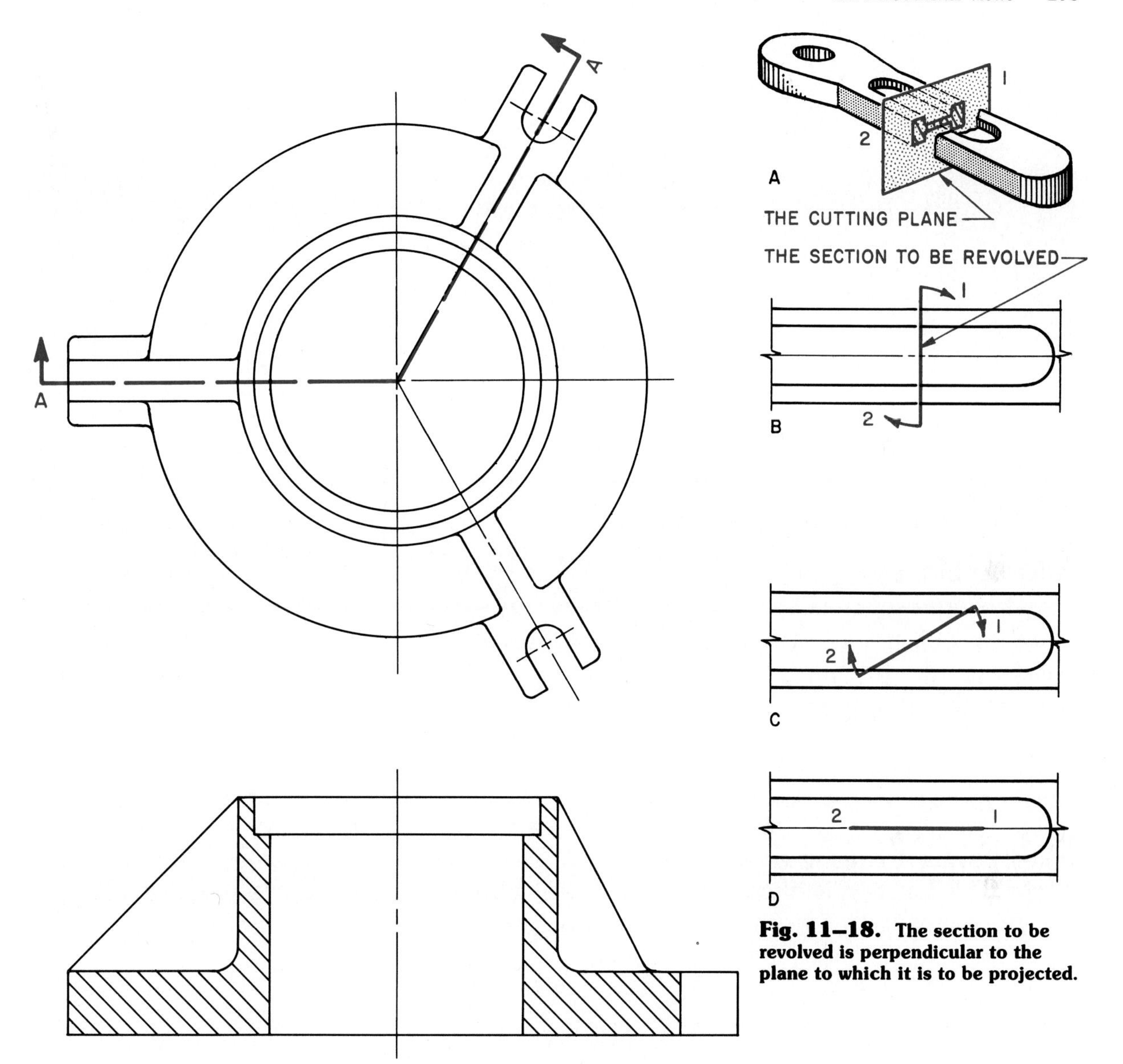

Fig. 11–18. The section to be revolved is perpendicular to the plane to which it is to be projected.

Fig. 11–16. Revolve the angled section to a plane parallel with the line of sight.

Fig. 11–17. Hidden details can be shown by a broken-out section.

FREEHAND BREAK LINE

In Fig. 11-19 a two-view drawing of the ratchet wrench is shown. The section is revolved and drawn on the front view. Notice that the section is drawn directly on the front view in true size. The center line, which serves as the axis of revolution, is always shown.

Usually the view is broken on each side of the revolved section. However, it can be drawn without breaks if desired.

A frequent mistake made in projecting the revolved section is shown in Fig. 11-20. Remember to project the section in its true size and shape regardless of the contour of the object.

Removed Sections

Another means of clarifying the shape of an object with a sectional view is to locate the sectional view away from the normal views. Such a section is called a **removed section.** These sections are usually used when the space on the exterior view is not large enough to draw the section as a revolved section or if the sectional view is to be dimensioned. Frequently more than one section is needed. The views are usually placed on the same sheet as the exterior view.

Notice in Fig. 11-21 how the cutting-plane line and section view are labeled so they can be

CAD Applications:

Section Lining

When drawing sectional views, cut or broken surfaces of objects are often indicated by section lining. A CAD system can automatically produce section lined areas by filling them with equally spaced lines at any angle. This function is often called hatching. Dozens of patterns are available for CAD.

Section lining requires several steps. For example, the process may start with the drafter telling the program the specific area to be section lined. The area can be defined by pointing to what borders it—lines, rectangles, circles, text, and so on. As the objects are identified, they are highlighted on the screen. Then the drafter tells the program whether all or part of the object is involved.

Next, the program actually computes the boundaries of the section lined area. The user must confirm that all the objects selected are correct.

Lastly, the area is actually section lined. The program asks questions about pattern, spacing, angle, and scale. After the questions have been answered, the section lining pattern is added to the drawing.

Section lining is only one of the many CAD functions that can improve the quality of drawings without using a lot of costly drawing time.

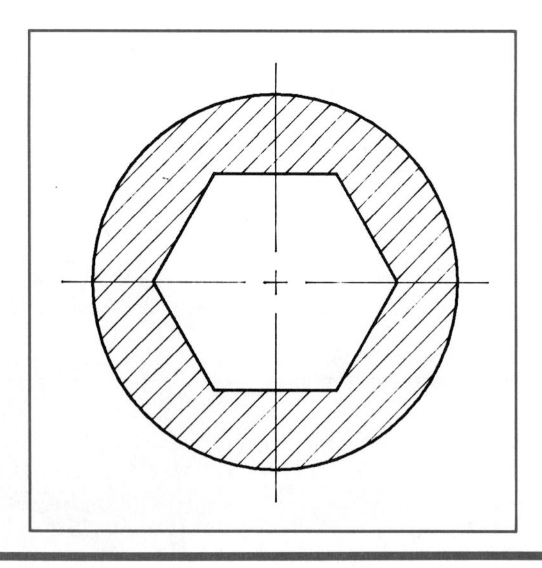

identified. If the removed section is drawn to a larger scale than the exterior view, the scale of the section must be indicated.

The section views should be arranged in alphabetical order from left to right on the drawing or in a vertical position with the first section on top.

Whenever possible, the center line of the removed sections should be parallel to its corresponding line in the normal projected position. Rotation of the view would make the drawing more difficult to read.

Auxiliary Sections

Another type of section is the **auxiliary section.** Some objects have angular elements and require an auxiliary view. If a section is to be taken through the angled portion of the object in order to show the true size and shape at the point to be cut, the cutting plane must be perpendicular to the axis of the section. In this position the angular element will not show true size on the normal reference planes and must be projected to an auxiliary reference plane. The projection is the same as a normal auxiliary view. Fig. 11-22.

Ribs and Spokes in Section

When sectioning circular objects, any element that is not solid around the axis of the object is not drawn in section. For example, if the cutting plane passes through and is parallel to the face of a web, rib, gear tooth, spoke, or some other noncontinuous element, the element is not sectioned. If the noncontinuous element contained section lining, a person reading the drawing would mistakenly think that the

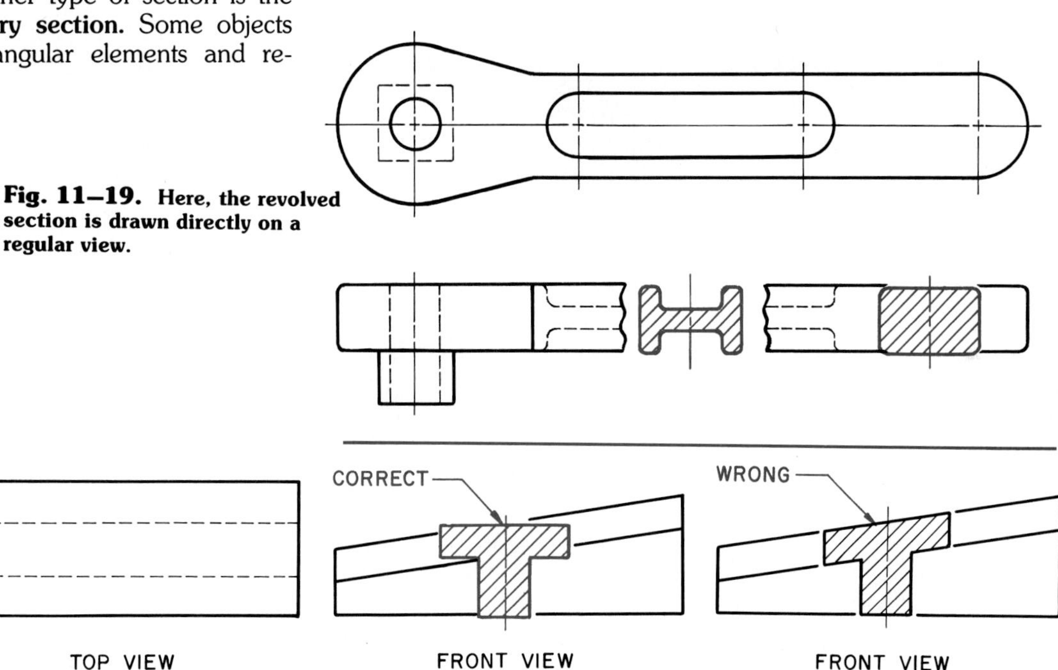

Fig. 11–19. Here, the revolved section is drawn directly on a regular view.

Fig. 11–20. The revolved section should be true in size and shape.

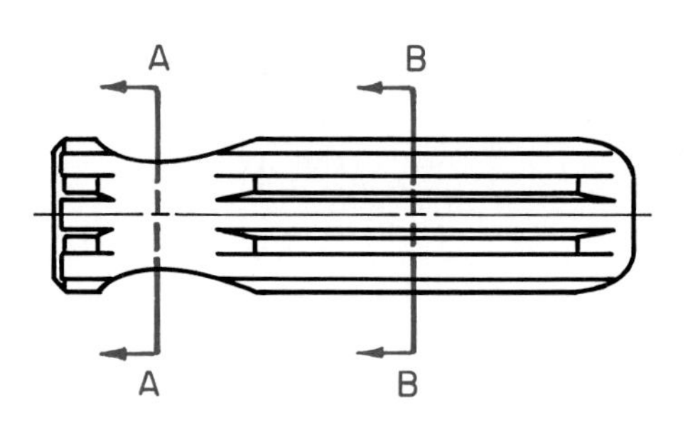

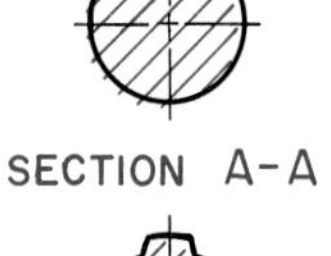

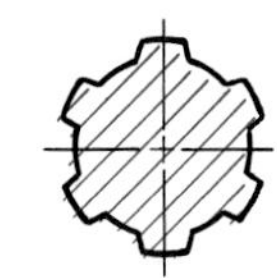

Fig. 11–21. Removed sections are drawn off the regular views.

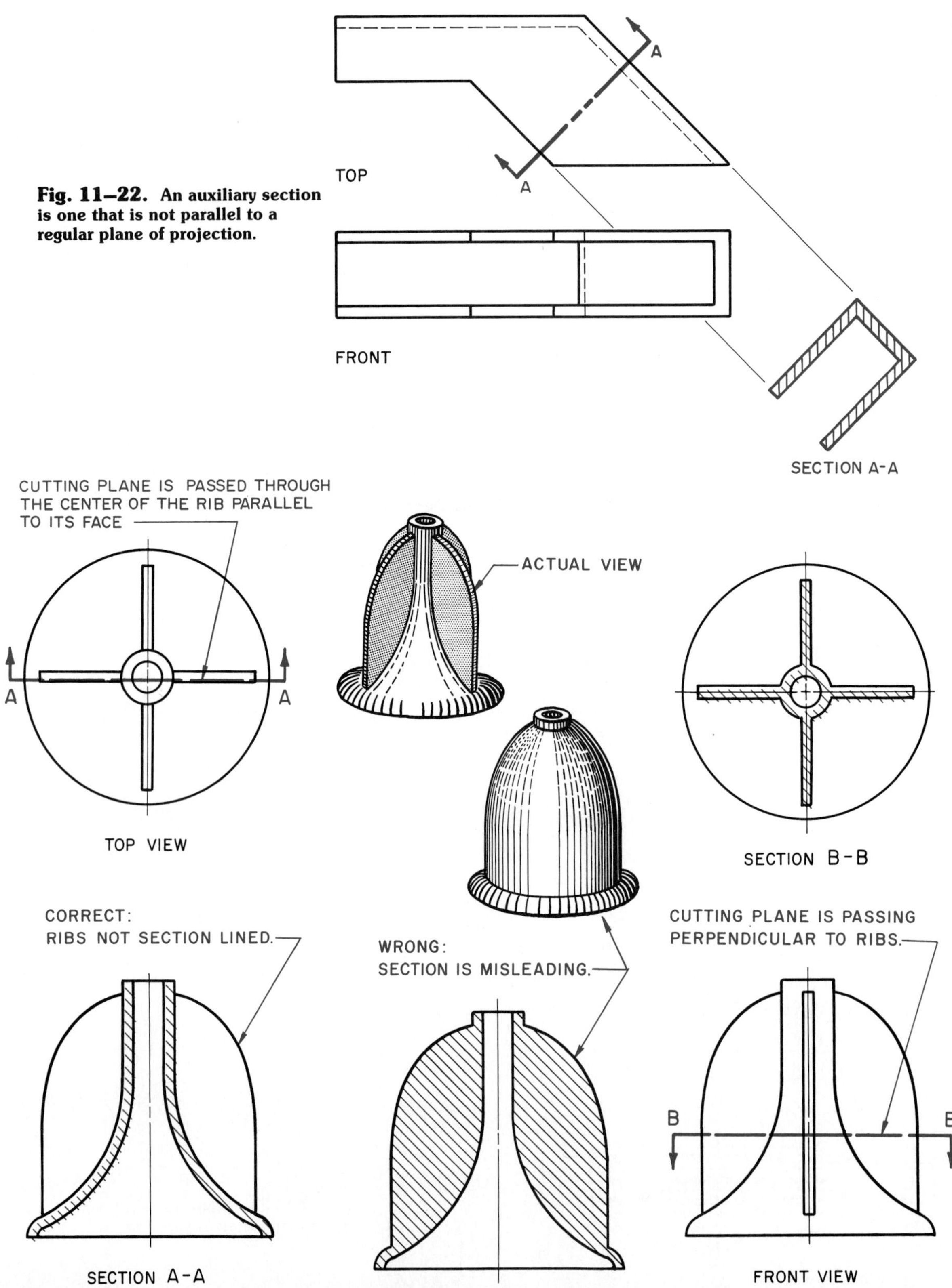

Fig. 11–22. An auxiliary section is one that is not parallel to a regular plane of projection.

Fig. 11–23. When a cutting plane passes through the flat face of noncontinuous elements, the element is not section lined.

web, rib, or spoke is a solid element. Cutting plane A-A in Fig. 11-23 passes through the ribs of a washing machine agitator. If the ribs were drawn with section lining, you would think the ribs were a solid unit.

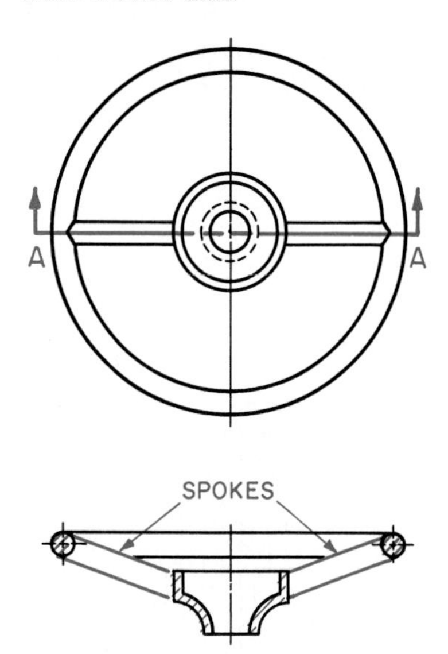

Fig. 11–24. **Spokes are not sectioned when the cutting plane runs parallel through them.**

If the cutting plane passes perpendicular to a rib, section it in the usual manner. Section B-B in Fig. 11-23 is such a section.

A steering wheel is shown in Fig. 11-24. It presents a sectioning problem similar to the washing machine agitator. The spokes are not section lined. If they were, the wheel would appear solid rather than spoked.

A section through a caster having a solid web is illustrated in Fig. 11-25. Since the web is solid, it is section lined.

Revolved Features

Some objects needing a section have features that are shown more clearly when that feature is sectioned in a revolved position. Some typical features that need such treatment are ribs, slots, holes, and spokes. Fig. 11-26 shows a steel cap that is held in place with bolts placed in three slots. If the lower slot was not revolved to a position parallel with a normal plane, the section would be confusing. In its revolved position, it appears true size in the section view.

Fig. 11-27 shows a plastic dial. One of the lower ribs has been revolved. A band saw wheel is shown in Fig. 11-28. The section labeled "correct" shows the spoke in a revolved position. The section labeled "wrong" shows the spoke as it normally appears in an unrevolved position. Such a view is confusing and does not clearly reveal the desired information.

Alternate Section Lining

In some objects, if a rib or similar part is not section lined when a cutting plane passes through it parallel to its face, it is difficult to tell if the rib exists. The rib in section, shown in Fig. 11-29, is not section lined at A. You cannot tell from the drawing if a rib appears here or if the space is open. The use of alternate section lining shows that it is a rib.

Space the alternate section lining over the rib by extending every other section line of the normal section lining. Notice that the shape of the rib is shown with hidden lines.

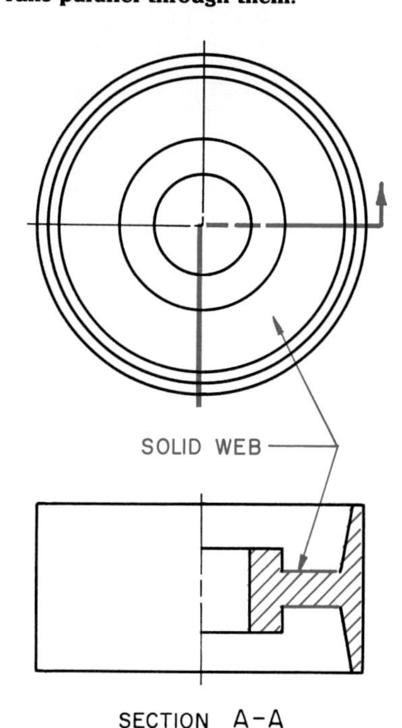

Fig. 11–25. **A solid web is section lined.**

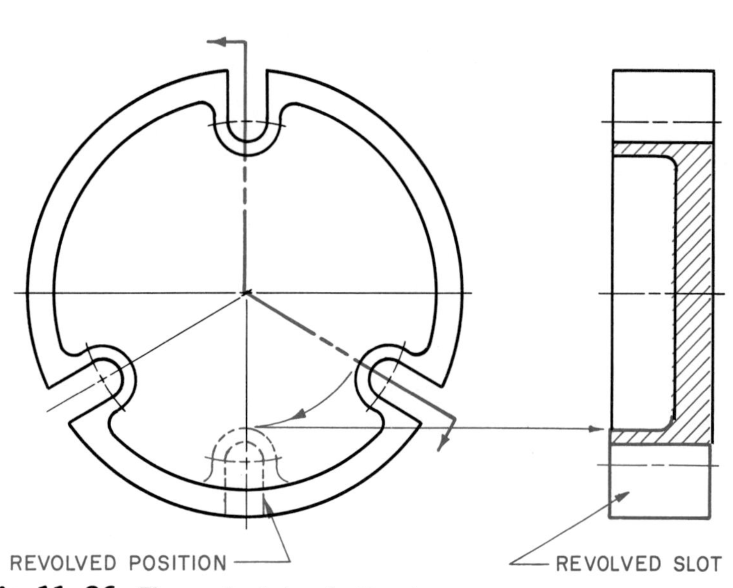

Fig. 11–26. **The revolved slot clarifies the view.**

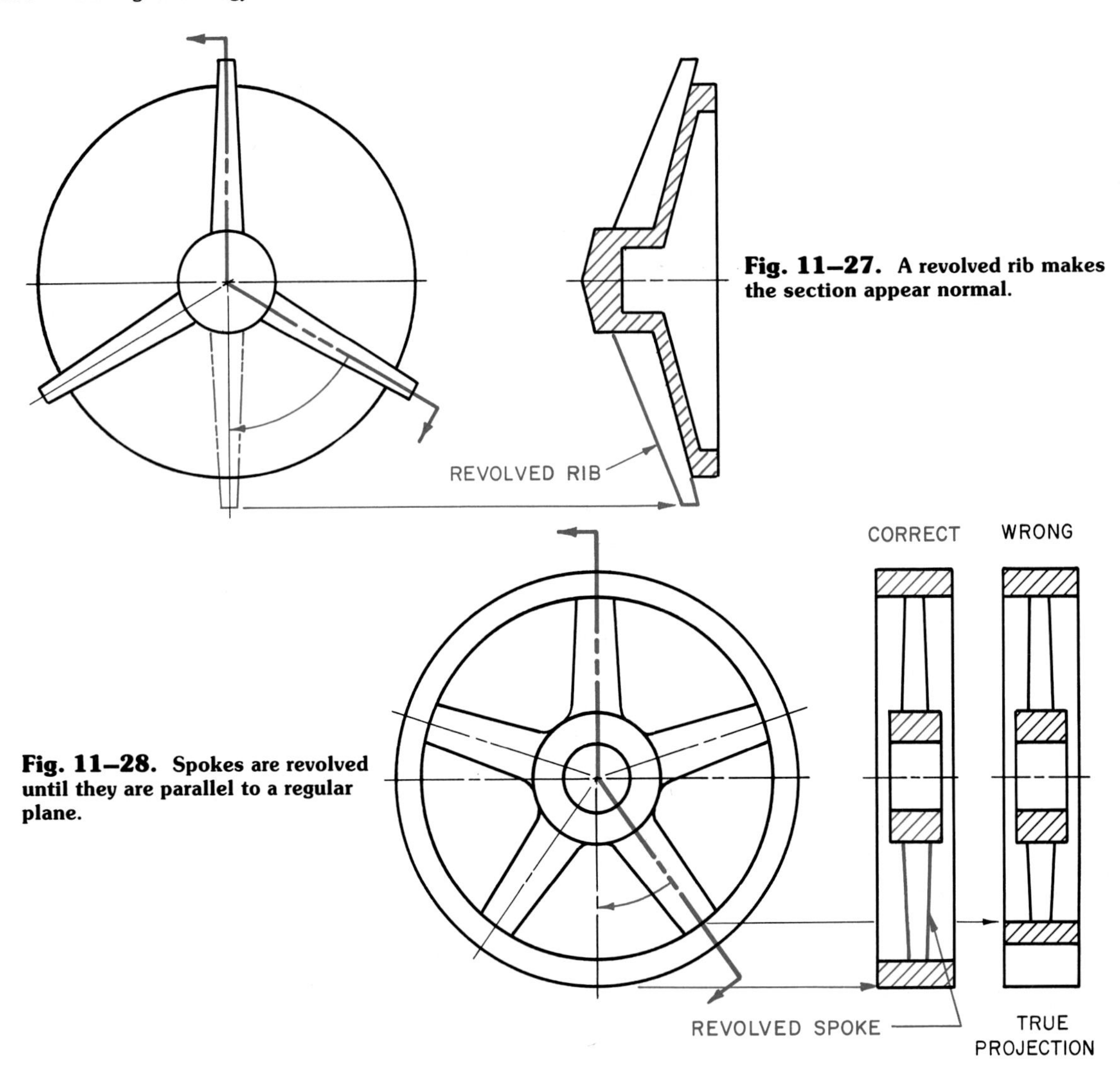

Fig. 11–27. **A revolved rib makes the section appear normal.**

Fig. 11–28. **Spokes are revolved until they are parallel to a regular plane.**

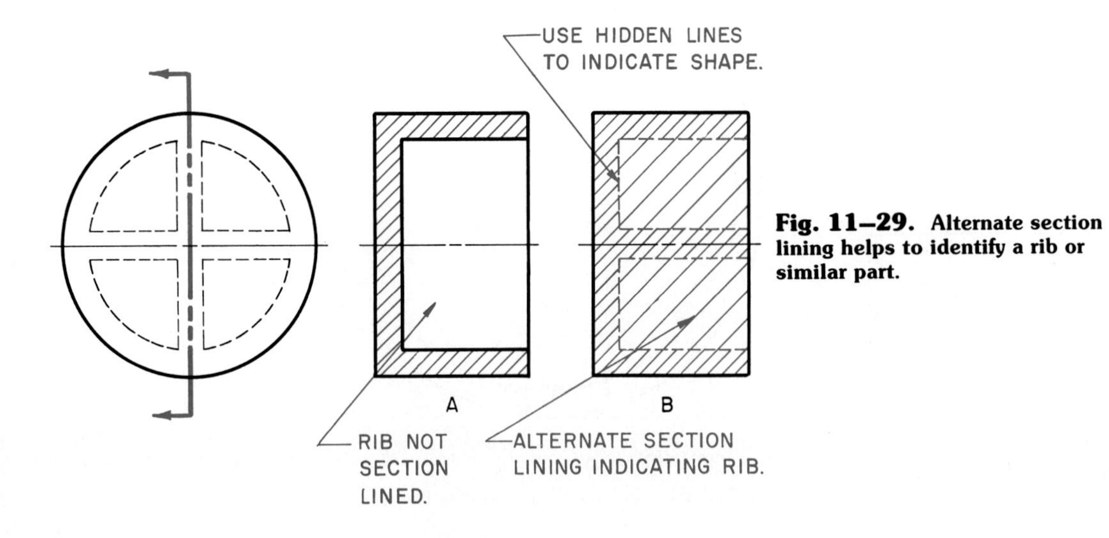

Fig. 11–29. **Alternate section lining helps to identify a rib or similar part.**

Phantom Sections

Draw a **phantom section** on top of a regular view. Use it to show interior shapes without drawing a separate sectional view. The section lining symbol used is like that for general material except that it is broken into short dashes. Fig. 11-30 shows a phantom section through a piston.

Shafts and Fastening Devices in Section

If the cutting plane passes through a shaft, bearing, or the length of a fastening device, these items are not sectioned. Fig. 11-31. Typical fastening devices are pins, keys, rivets, bolts, and set screws. Sectioning these items would serve no purpose because they have no interior details to show. Actually, such sections could be confusing to the reader of the drawing. If these parts are cut across their axes, they are section lined. Typical examples are shown in Fig. 11-32.

Assemblies in Section

An important use of section drawings is to show how a device operates or how its parts are assembled. Sectional views also are used to show the relationships between parts. Sections made through an assembled device are called *assembly sections*.

Fig. 11-33 shows a pump. The assembly section shows the complex interior details. Assemblies in section are discussed in greater detail in Chapter 13, "Drawing for Production: Detail and Assembly Drawing."

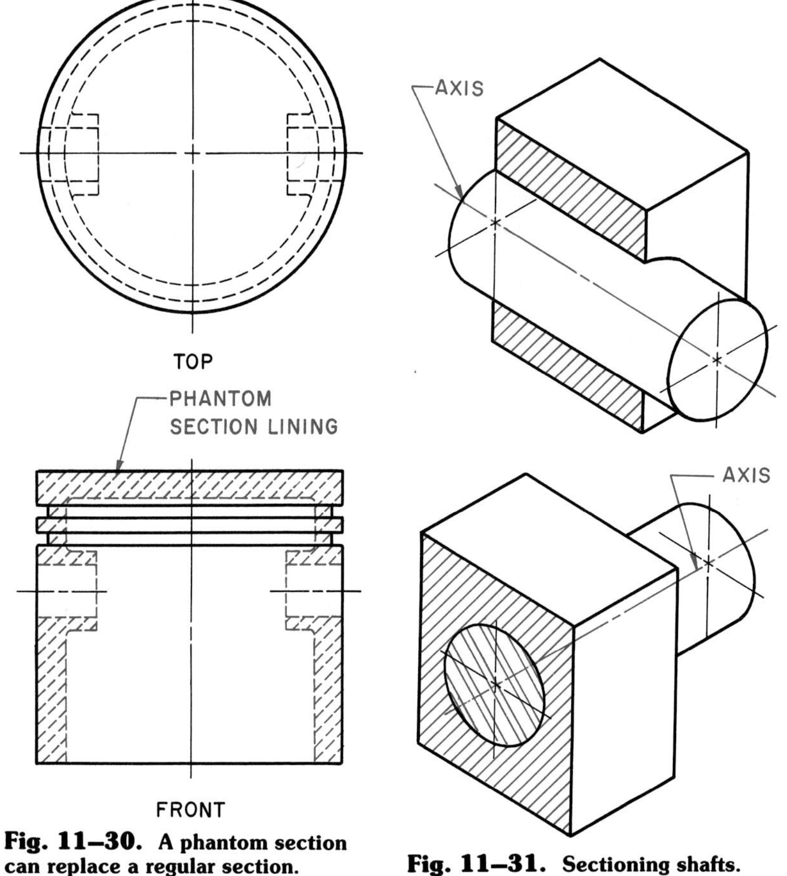

Fig. 11–30. A phantom section can replace a regular section.

Fig. 11–31. Sectioning shafts.

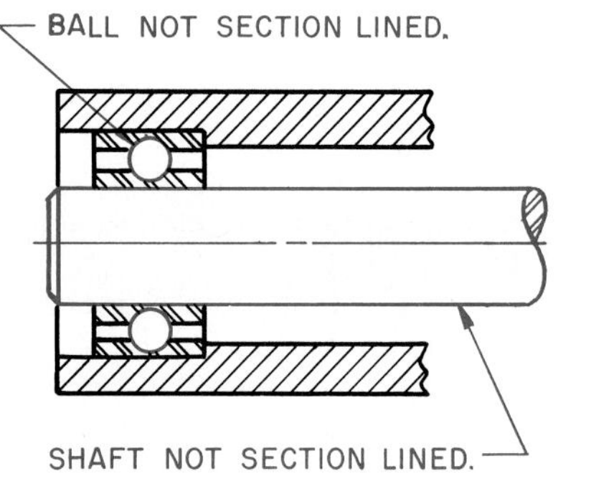

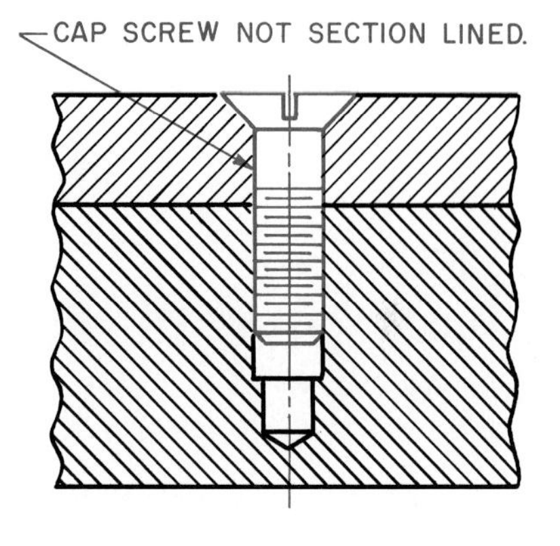

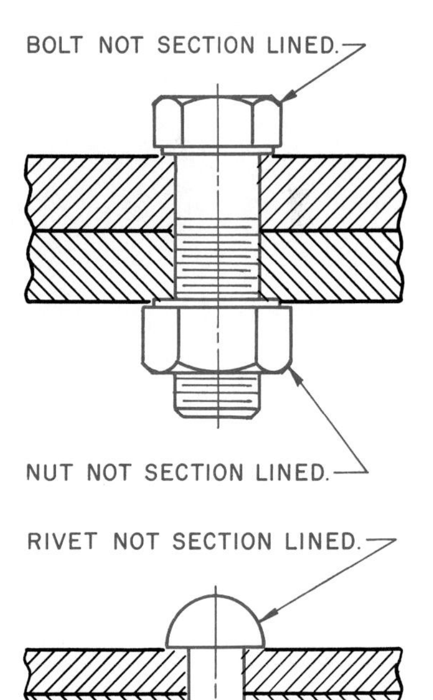

Fig. 11–32. Typical sections through fasteners.

Pictorial Sections

Sectioning is used on pictorial drawings of single pieces and pictorial assemblies. Fig. 11-34 is an example of an isometric sectional view.

Fig. 11-35 shows a pictorial assembly made for use in a sales catalog. This type of drawing is discussed in Chapter 20, "Technical Illustration."

Intersections in Section

When a section is drawn through an intersection, such as where two cylinders meet, and the intersection is small, the true projection is disregarded. Fig 11-36 at A and B. When the intersection is large, the true projection is drawn. When arcs are required, they are approximated. Fig. 11-36 at C.

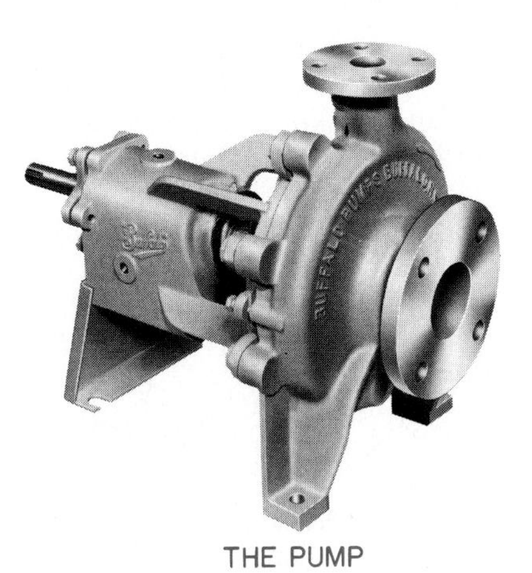

THE PUMP

THE ASSEMBLY DRAWING

Fig. 11—33. The assembly drawing reveals the complex interior of the pump (Buffalo Forge Co.)

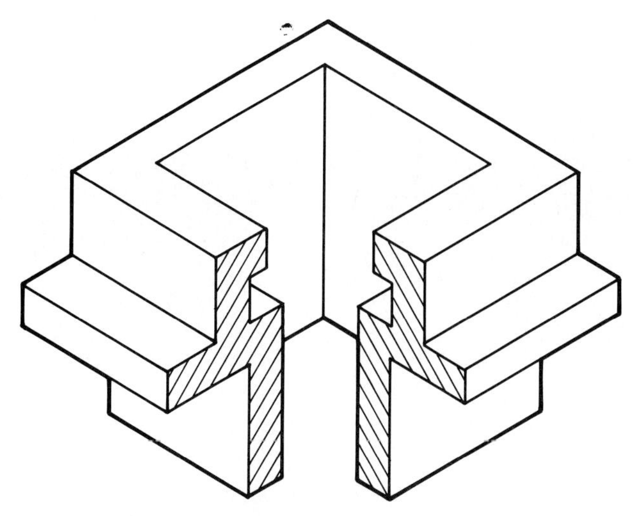

Fig. 11–34. An isometric section.

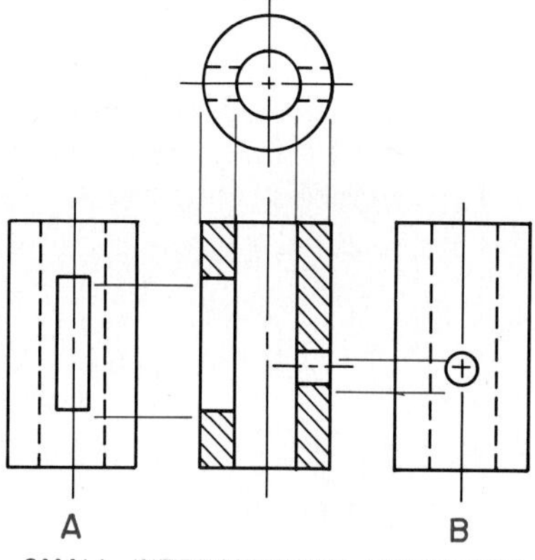
A
B
SMALL INTERSECTIONS DISREGARD TRUE PROJECTION.

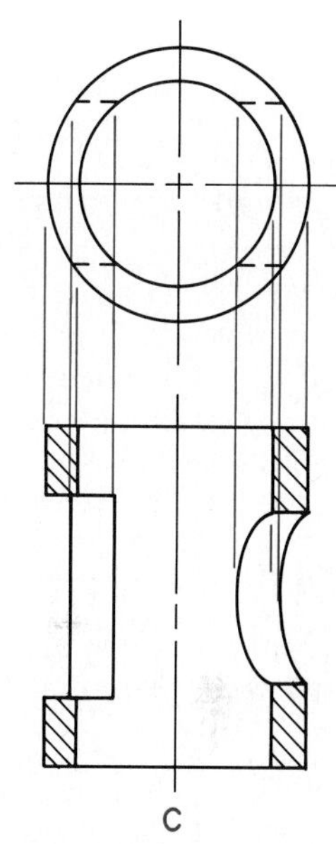
C
LARGE INTERSECTIONS ARE DRAWN IN TRUE PROJECTION.

Fig. 11–36. How to handle intersections on section drawings.

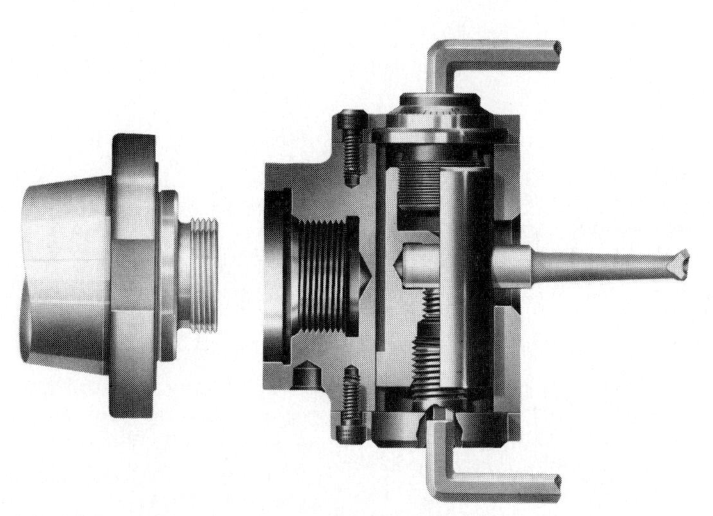

Fig. 11–35. A section through an adjustable boring head with an enclosed boring tool holding an adjusting mechanism. This section was made to appear lifelike by shading. Chapter 20, "Technical Illustration," gives more information on this kind of drawing. (DeVlieg Machine Co.)

Chapter Review

Build Your Vocabulary

You should understand and use the following terms as part of your working vocabulary. Write a brief explanation of what each means.

sectional view
section lining
cutting plane
material symbols
full section
half section
offset section
broken-out section
revolved section
removed section
auxiliary section
phantom section

Study Problems—Directions

The problems that follow will give you the chance to use some of the drafting knowledge that you learned in this chapter. These problems have the needed section indicated. Draw the regular views plus the required section. Do the problems that are assigned to you by your instructor.

Beginning Problems

1. For **Fig. P11-1**, draw the necessary views and indicated sections. Find the sizes by using the scales below the drawing.
2. For **Fig. P11-2**, draw the necessary views and indicated sections. Find the sizes by using the scales below the drawing.

Full Sections

3. **Fig. P11-3** shows a laser shell. Select and draw the views to describe the object. Use a full section to reveal interior details. Select your own scale.
4. **Fig. P11-4** shows a special purpose nut. Draw the front and side views and a full section. Draw full size.
5. For the handwheel in **Fig. P11-5**, draw the views needed to describe the object. Construct a full section. Draw full size.

Half Sections

6. **Fig. P11-6** is an automobile window handle knob. Draw the necessary views and a half section. Draw full size.
7. **Fig. P11-7** is a candle holder. Draw the required views and the half section that best shows the details. Draw to a scale of 2:1.
8. **Fig. P11-8** is a hydraulic hose connection. Draw the view required to describe the object. Use a section to show interior details. Draw full size.
9. **Fig. P11-9** is a porcelain electrical bus support. Draw the required views and a half section. Draw full size.
10. For the steam pipe fitting in **Fig. P11-10**, draw the front and top view. Show interior details with a half section. Draw to a scale of ½″ = 1″.

Offset Section

11. **Fig. P11-11** is a base for a ball float trap. Draw the front and top views. Then draw an offset section. Draw full size.

Broken-out Section

12. For the candlestick in **Fig. P11-12**, draw a front and top view and a broken-out section to show the hidden details. Draw full size.

Revolved Section

13. Draw the front and left side views of the nail set in **Fig. P11-13**. Then draw the indicated revolved sections. Draw double size.
14. Draw the front and top views of the cast crank handle in **Fig. P11-14**. Draw revolved sections 1 inch and 3 inches from the right end of the handle. Draw full size.

Removed Section

15. **Fig. P11-15** is a gate valve handle. Draw the necessary views and a removed auxiliary section taken 50 mm from the end of the handle. Draw full size.

Phantom Section

16. **Fig. P11-16** is a jig borer tool holder. Draw the front and side views. Use phantom section lining to show hidden details. Draw full size.

Assembly Section

17. **Fig. P11-17** shows a union for joining steel and plastic pipe. Draw the necessary views to describe the object. Make a section to show the shape of the three parts and how they are assembled. Draw full size.

18. **Fig. P11-18** is a bearing journal. Draw the front and side views. Draw a section showing interior details. Draw full size.

Other Problems

Study the following problems. Draw the regular views needed, and use sections to clarify hidden details or to replace a regular view.

19. **Fig. P11-19** is a flanged gland. Draw the necessary views to describe the gland. Use a section to show interior details. Draw full size.

20. **Fig. P11-20** is a machinist's center head. Draw the necessary views. Construct an auxiliary section as indicated at A-A. Draw a revolved section at B-B. Draw full size.

21. Draw the front and side views of the sanding drum in **Fig. P11-21**. Construct a section through the drum. Show details of the set screw in the shaft. Draw to a scale of 1:1.

22. **Fig. P11-22** is a plastic mounting bracket. Draw the front and side views. Draw an offset section. Draw full size.

23. **Fig. P11-23** is a cap for a quick-opening gate valve. Draw the necessary views and a half section. Draw full size.

24. **Figs. P11-24A** and **B** are the cap and body of a high-pressure steam trap. Draw the needed views of the cap. Draw a section to show interior details. Draw full size. Then draw the cap and body as assembled. Make a section to show interior details of these parts when assembled. Draw full size.

25. **Fig. P11-25** is a bridge scupper. (A scupper is an opening that permits water to drain off a floor or roof.) Draw the views needed to describe the scupper. Make sections showing hidden details on the front and side views. Draw quarter size.

26. Make an assembly drawing of the air line filter in **Fig. P11-26**. Scale full size.

27. **Fig. P11-27** is a hydraulic accumulator. (The accumulator is part of a hydraulic circuit. It fills with fluid when the circuit is under pressure. The fluid enters at the bottom and forces the piston up against the spring. If the fluid pressure drops, the spring pushes the piston down, forcing fluid into the circuit. This action maintains an even pressure on the fluid.) Draw the necessary views to describe the accumulator. Make an assembly drawing to show interior details. Draw full size.

28. **Fig. P11-28** shows a band saw wheel bearing and hub assembly. Draw the needed views of each part, A-D. Section each to clearly show all details. Select your own scale. As an alternate problem, draw the needed views of the band saw wheel assembly as an assembled unit. Section the assembly to show the construction details and the relationship between parts. Select your own scale.

Study Problems

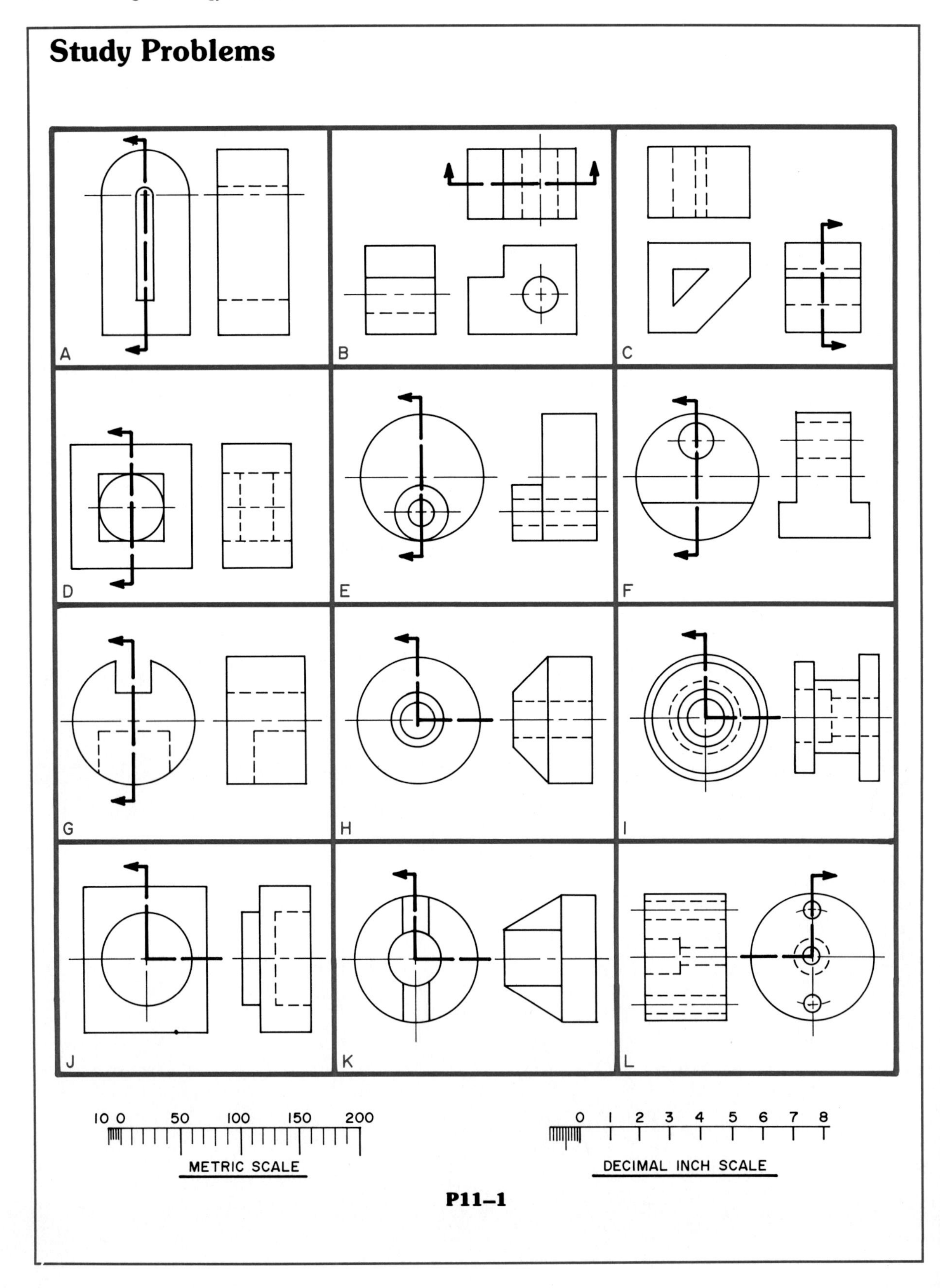

P11–1

Study Problems

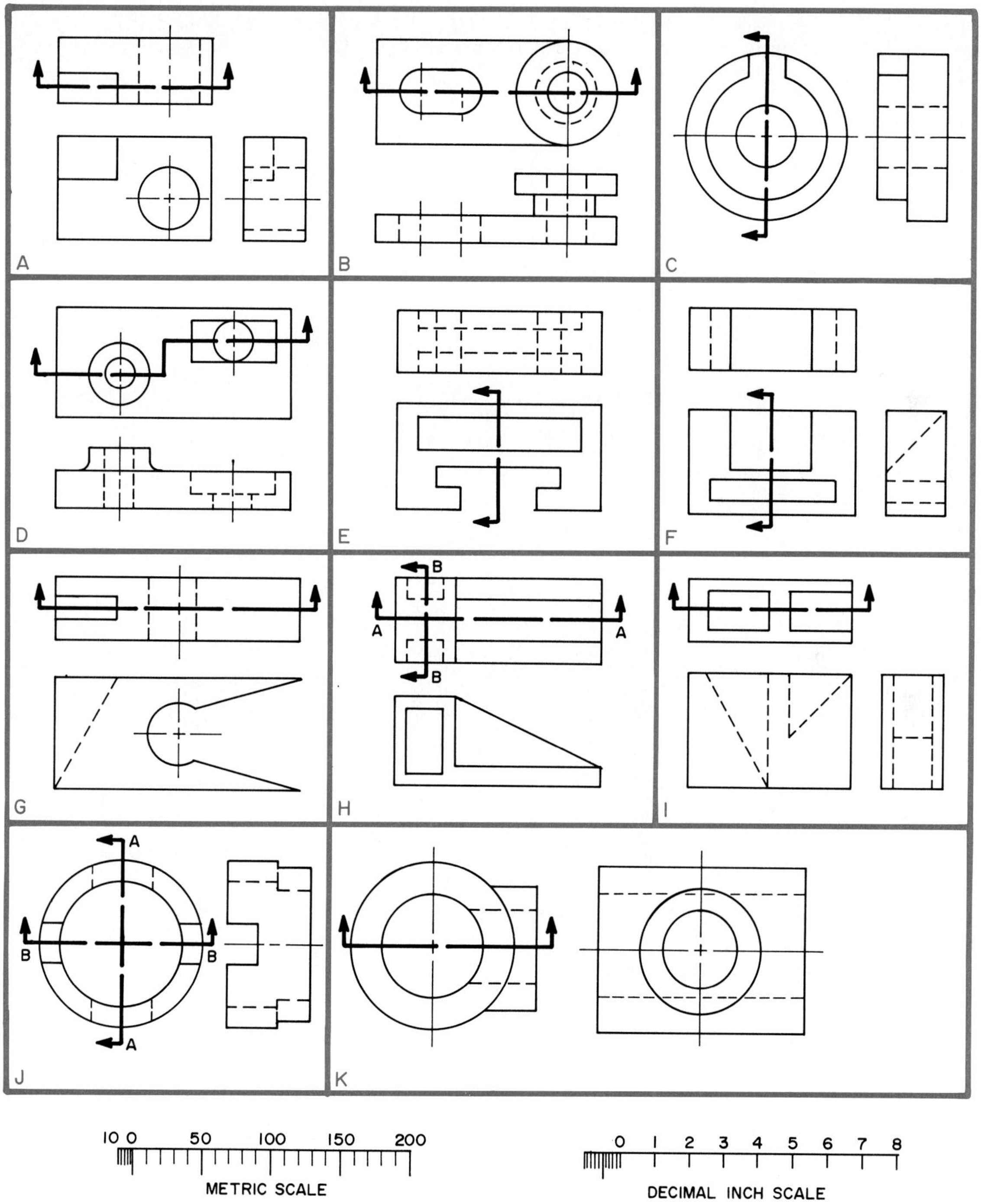

A

B

C

D

E

F

G

H

I

J

K

10 0 50 100 150 200

METRIC SCALE

0 1 2 3 4 5 6 7 8

DECIMAL INCH SCALE

P11–2

Study Problems

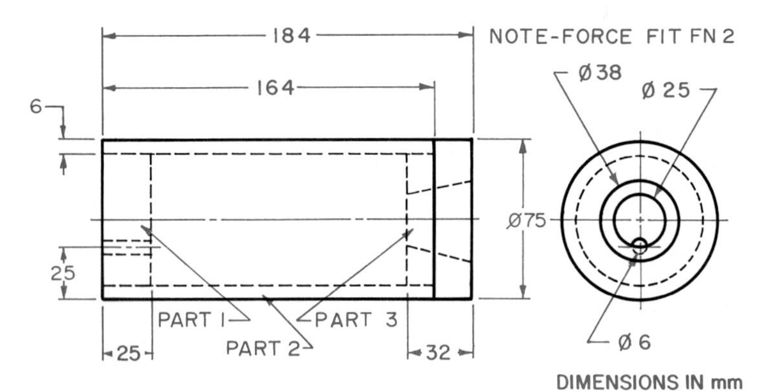

A SHELL FOR A LASER

P11–3

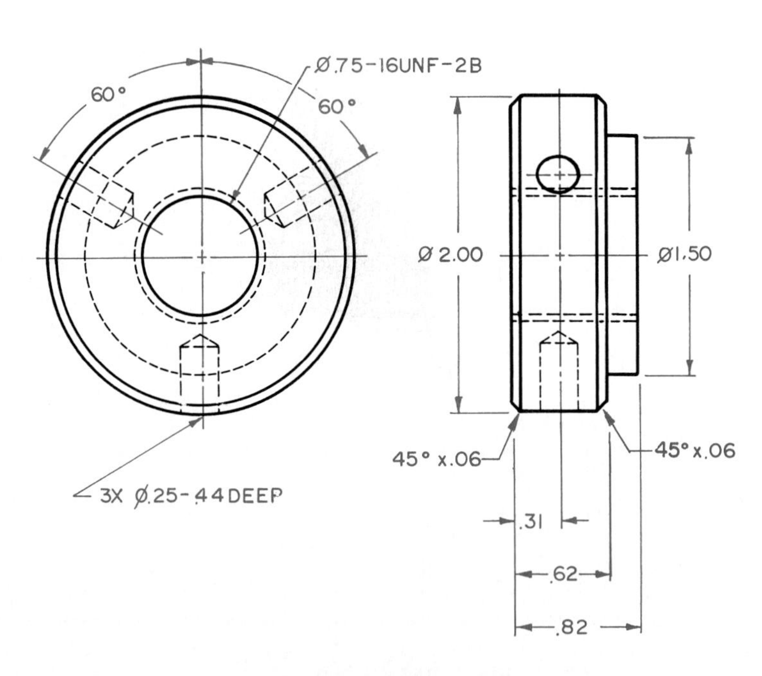

SPECIAL PURPOSE NUT

P11–4

Study Problems

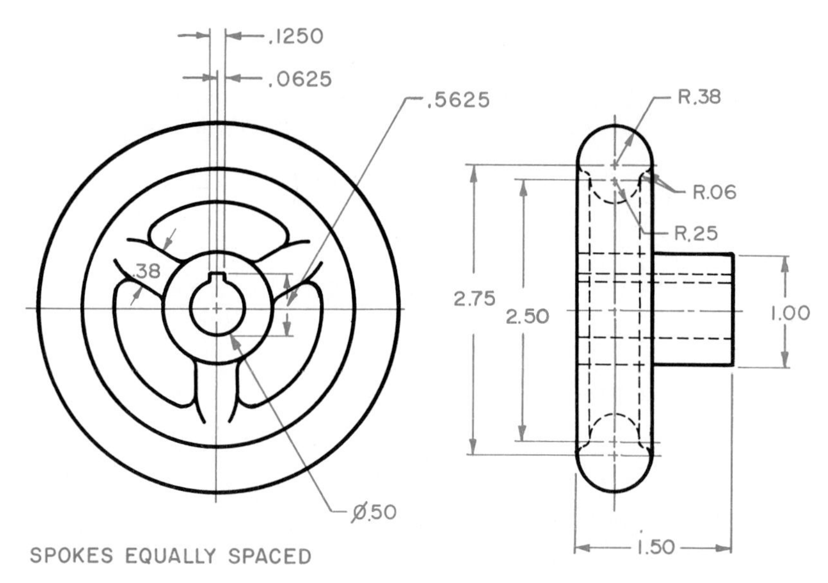

HANDWHEEL
P11–5

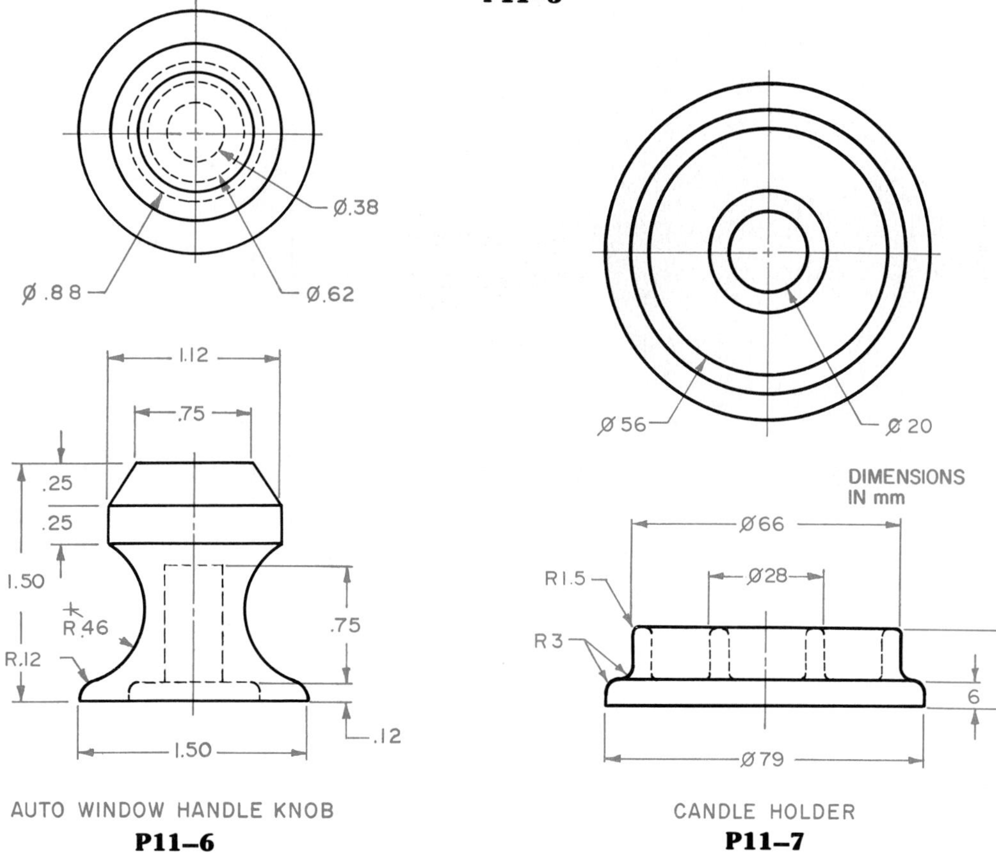

AUTO WINDOW HANDLE KNOB
P11–6

CANDLE HOLDER
P11–7

Study Problems

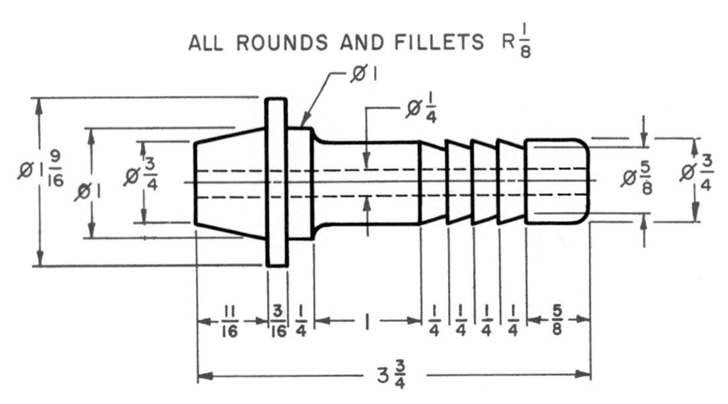

HYDRAULIC HOSE CONNECTION

P11–8

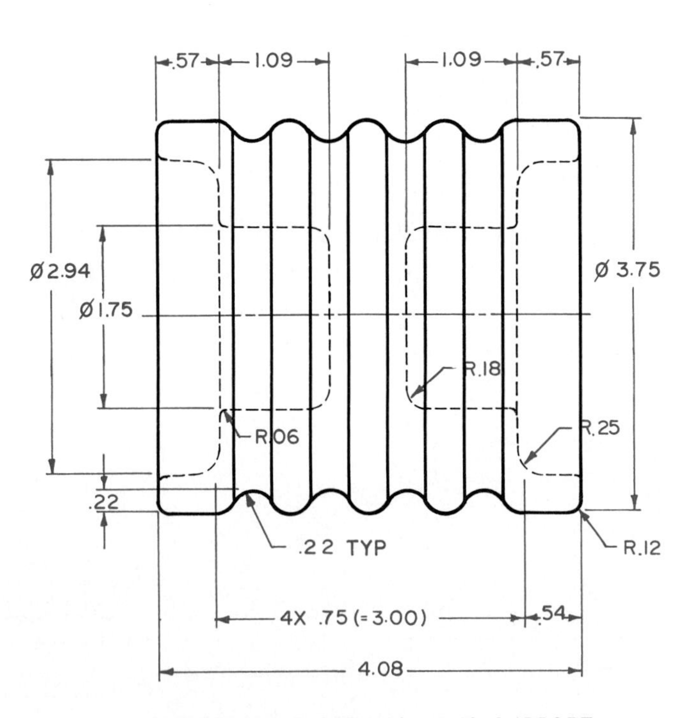

PORCELAIN ELECTRICAL BUS SUPPORT

P11–9

Study Problems

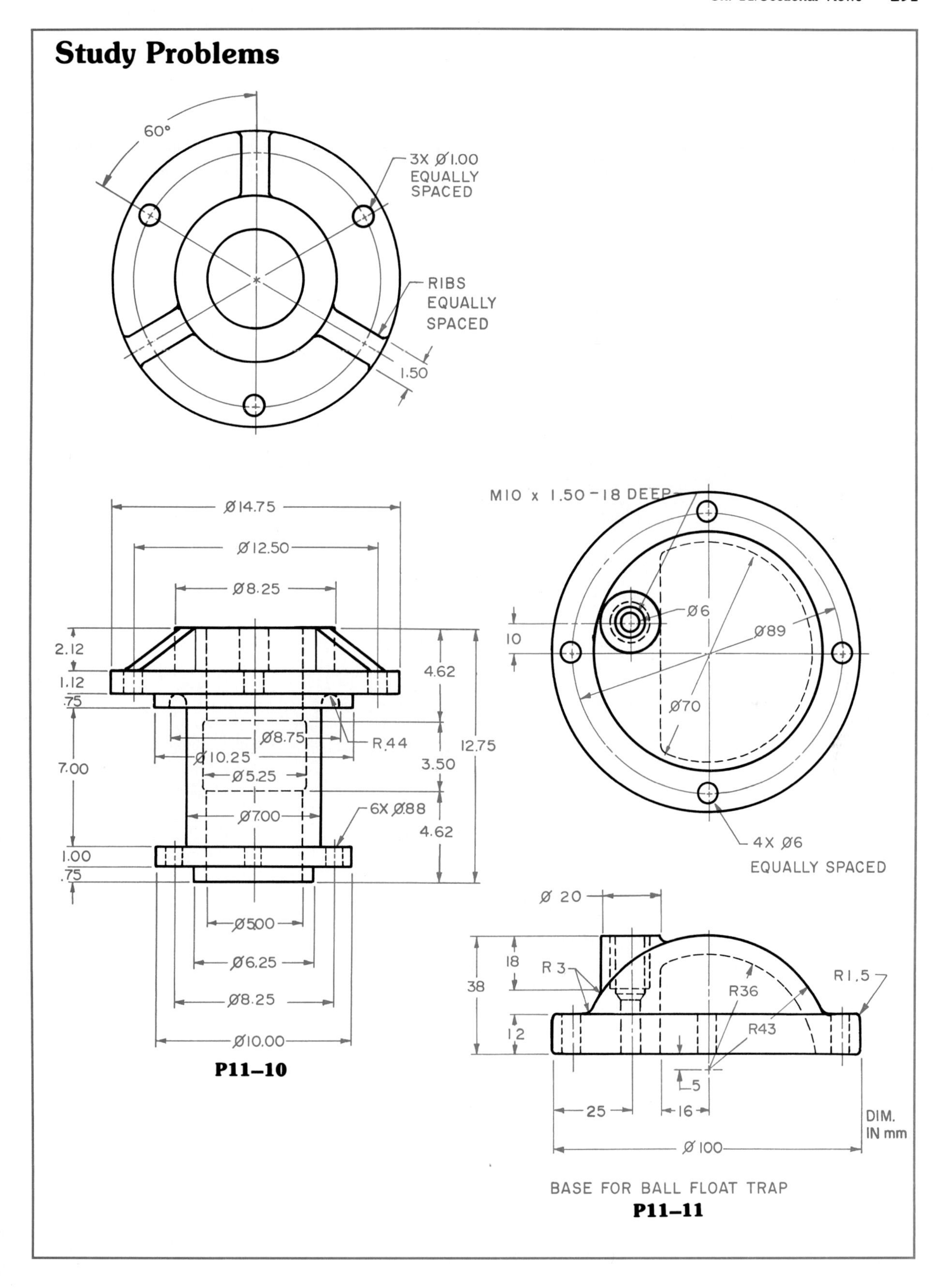

P11–10

BASE FOR BALL FLOAT TRAP

P11–11

Study Problems

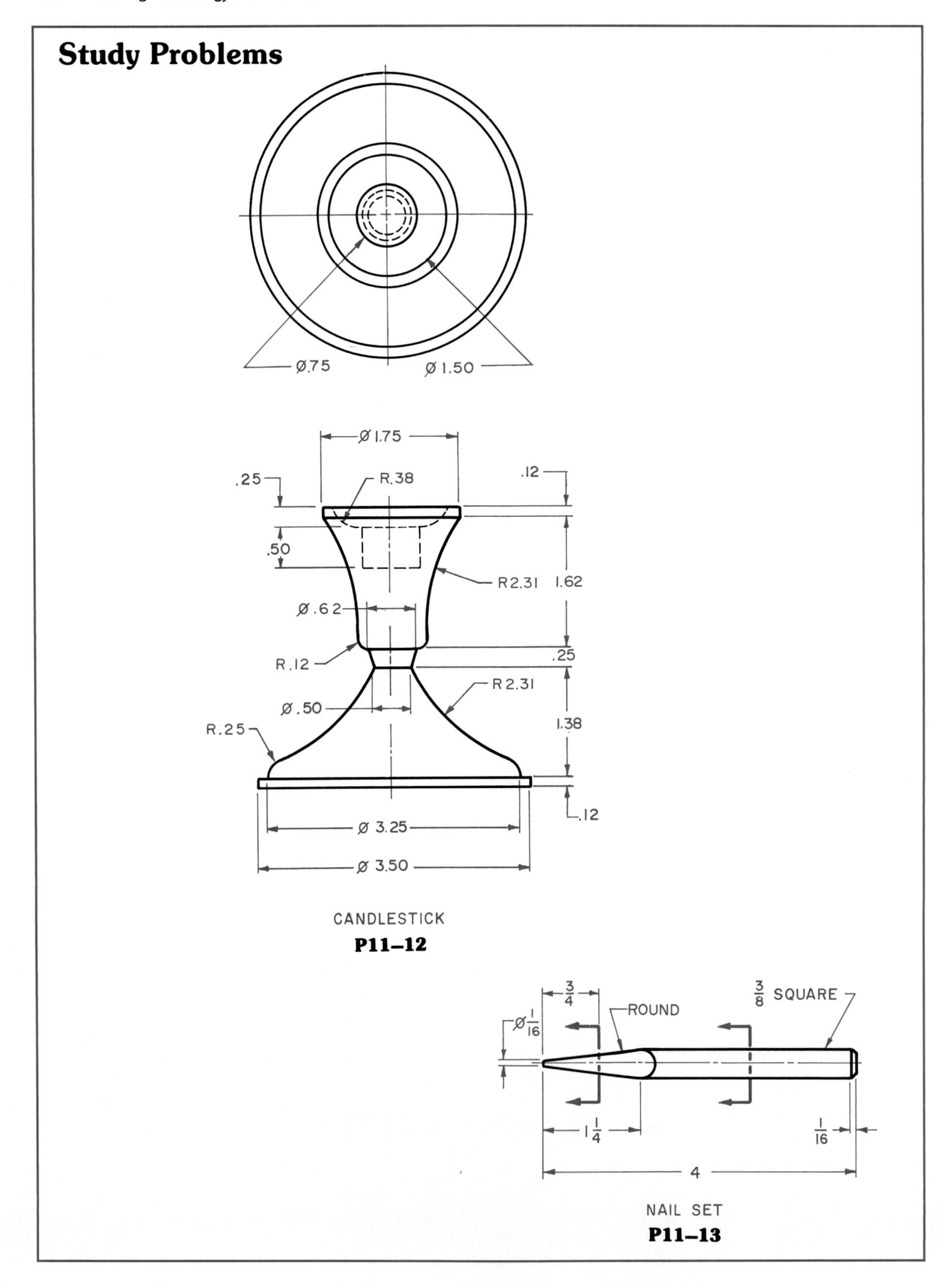

CANDLESTICK

P11–12

NAIL SET

P11–13

Study Problems

ALL FILLETS AND ROUNDS R.10

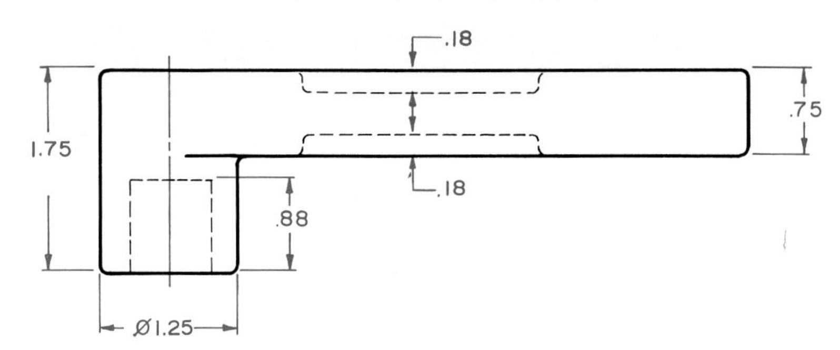

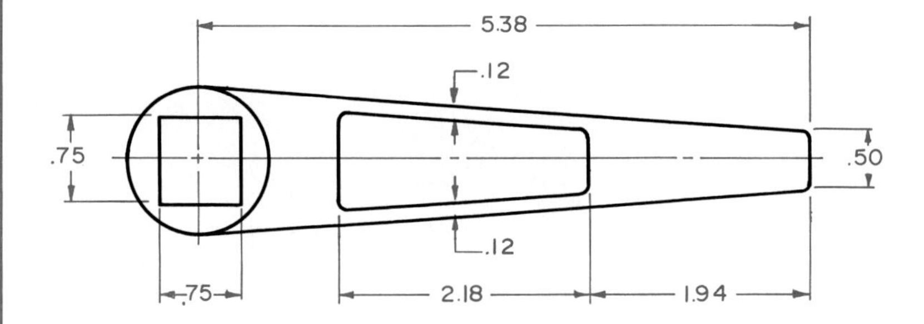

P11–14 CAST CRANK HANDLE

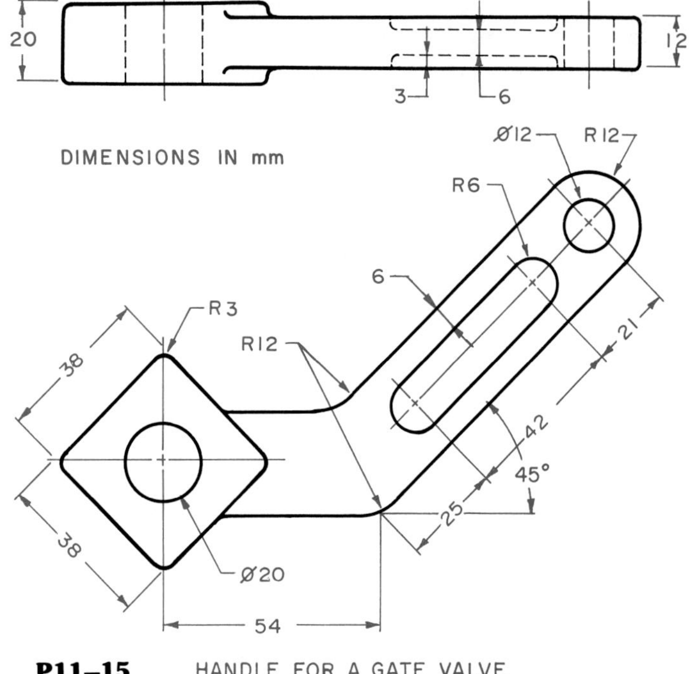

P11–15 HANDLE FOR A GATE VALVE

Study Problems

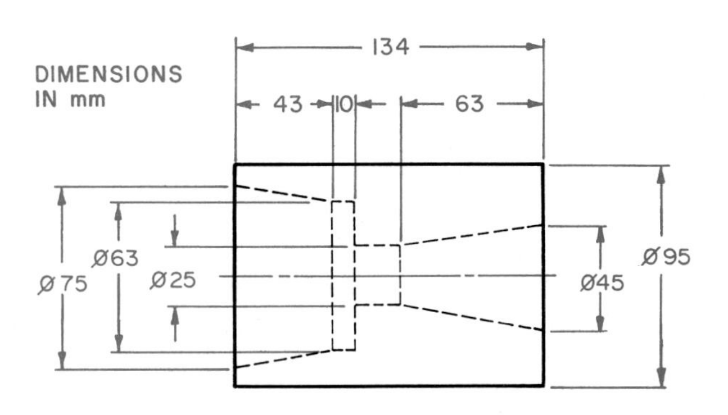

JIG BORER TOOL HOLDER

P11–16

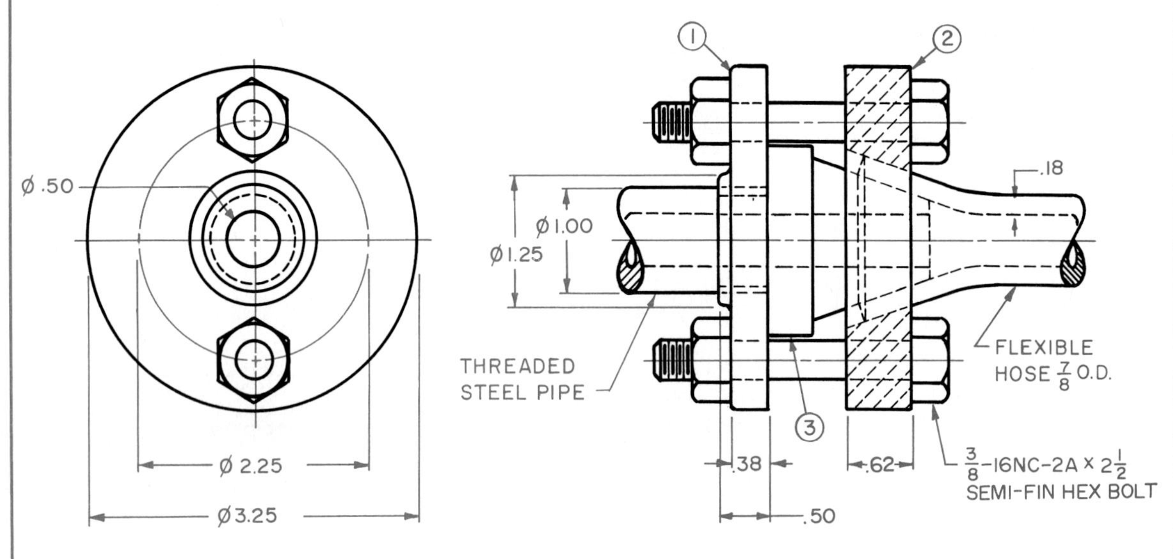

ALL FILLETS AND ROUNDS R.06

NO.	PARTS LIST
1	CAP – CAST IRON
2	ANGLE PRESSURE PLATE – STEEL
3	PRESSURE CONE – PLASTIC

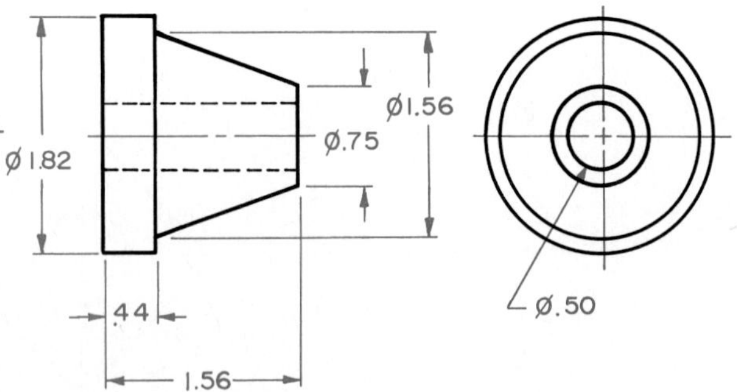

DETAIL OF PRESSURE CONE – PART NO. 3

UNION FOR JOINING STEEL AND PLASTIC PIPE

P11–17

Study Problems

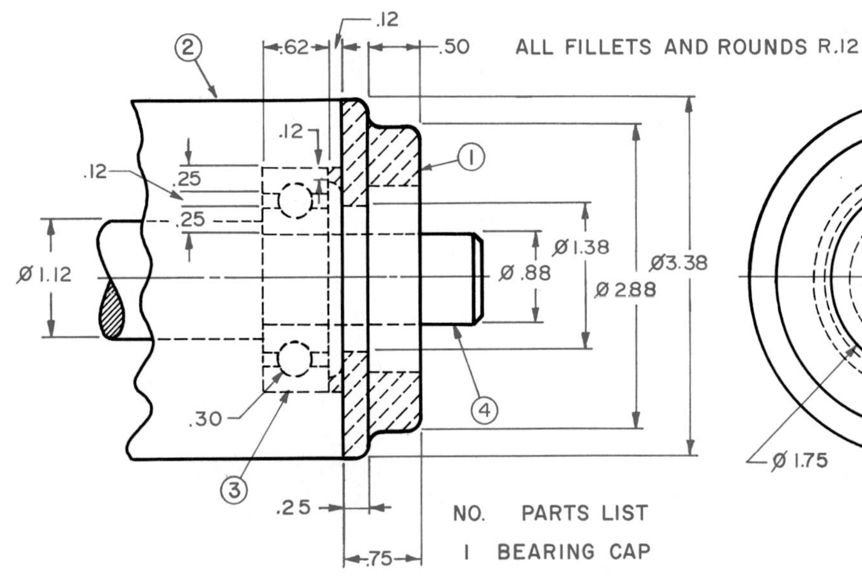

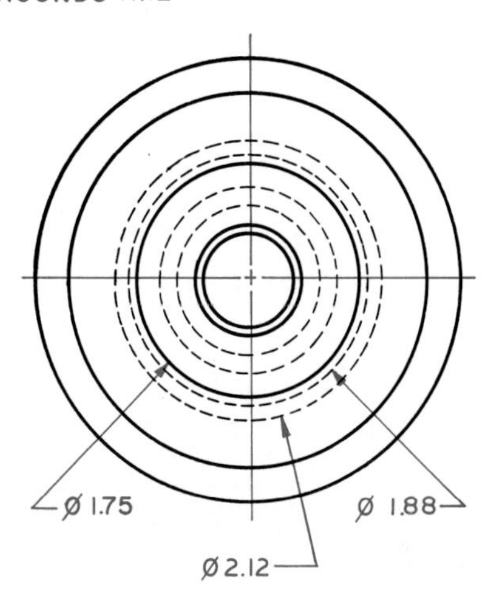

BEARING JOURNAL

P11–18

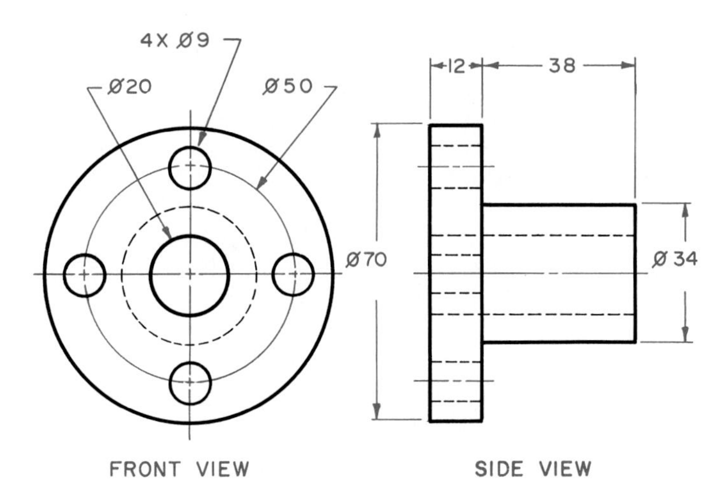

FLANGE GLAND DIMENSIONS IN mm

P11–19

Study Problems

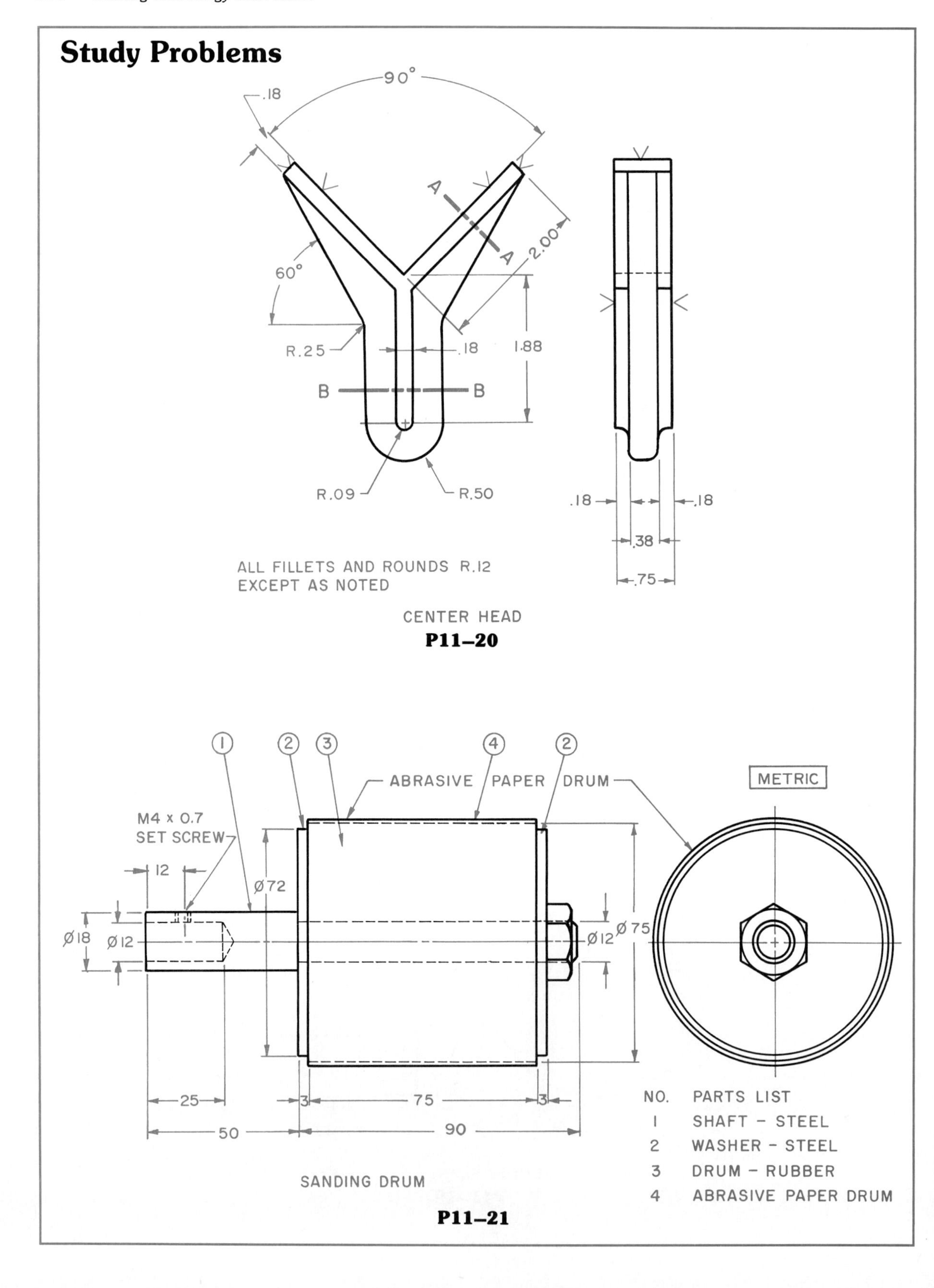

CENTER HEAD

P11–20

SANDING DRUM

P11–21

Study Problems

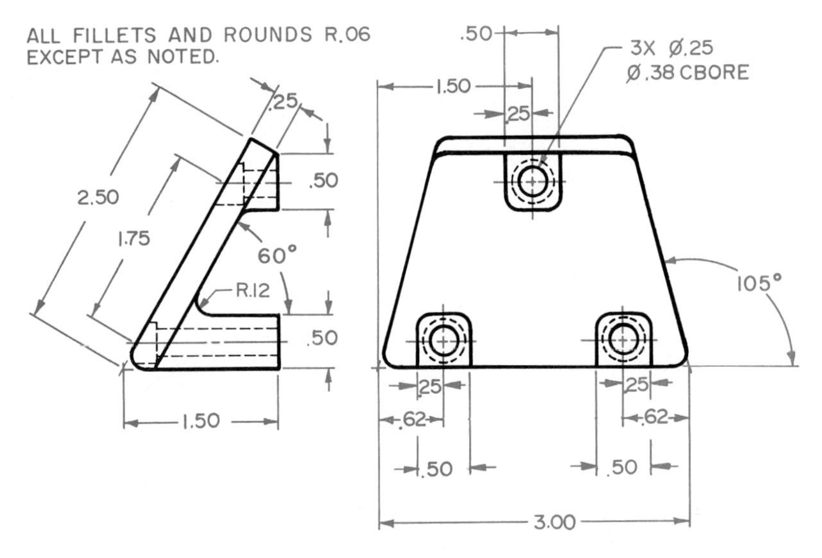

PLASTIC MOUNTING BRACKET

P11–22

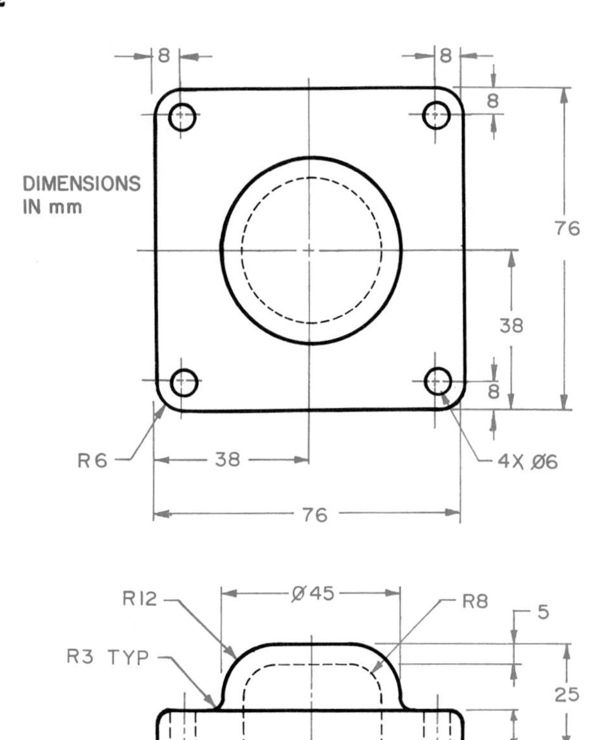

CAP FOR A QUICK-OPENING GATE VALVE

P11–23

Study Problems

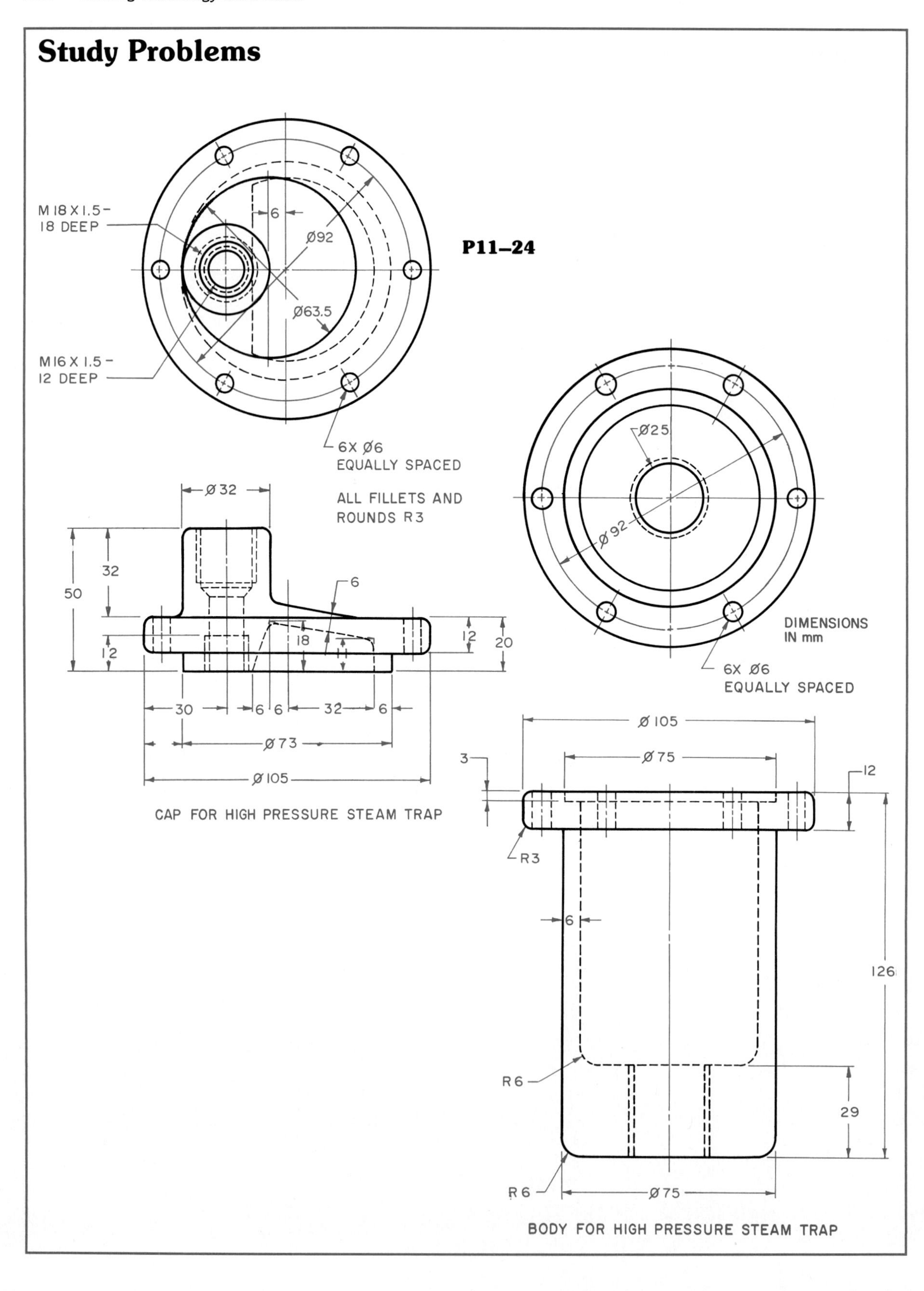

Study Problems

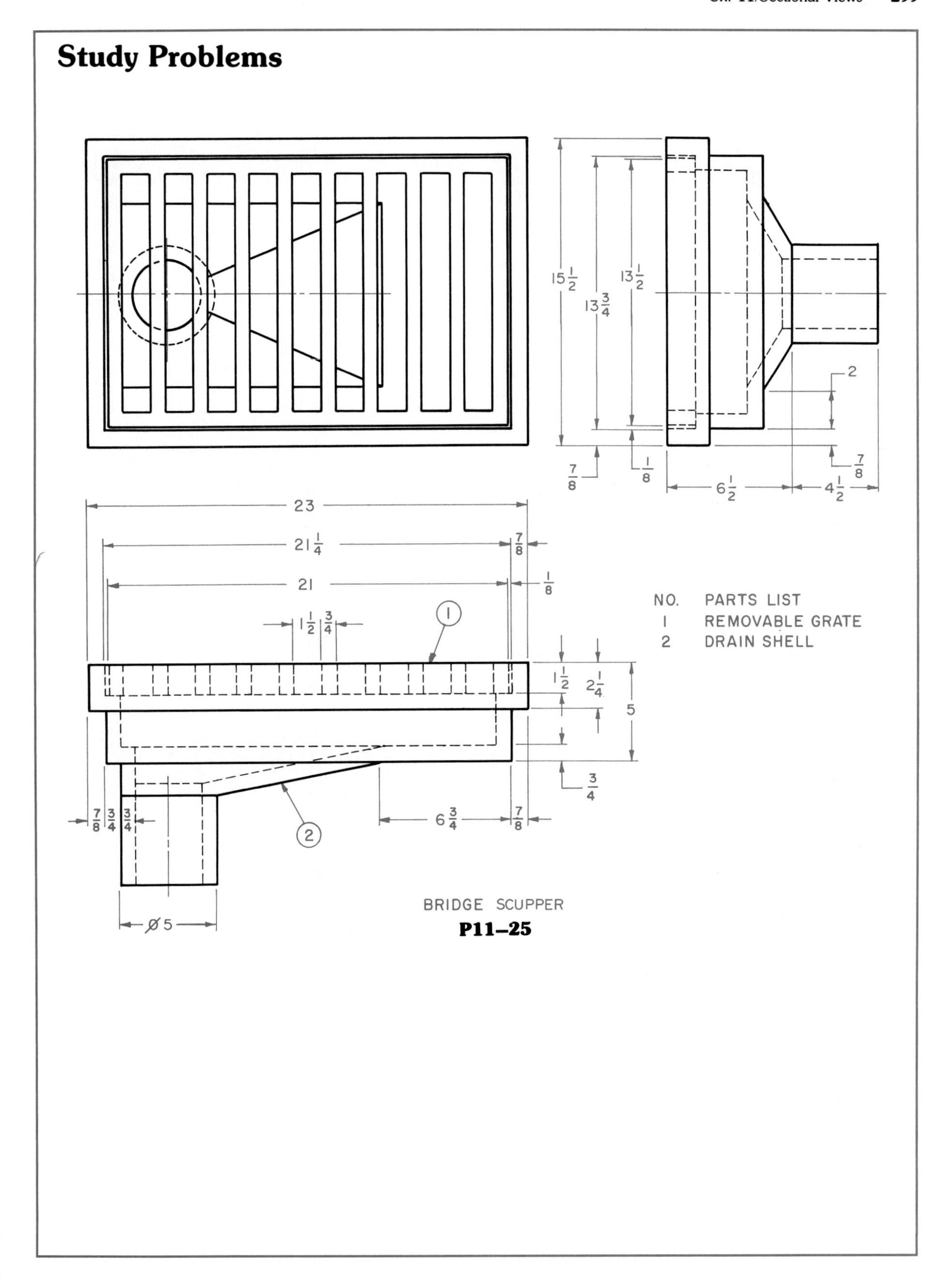

BRIDGE SCUPPER

P11–25

Study Problems

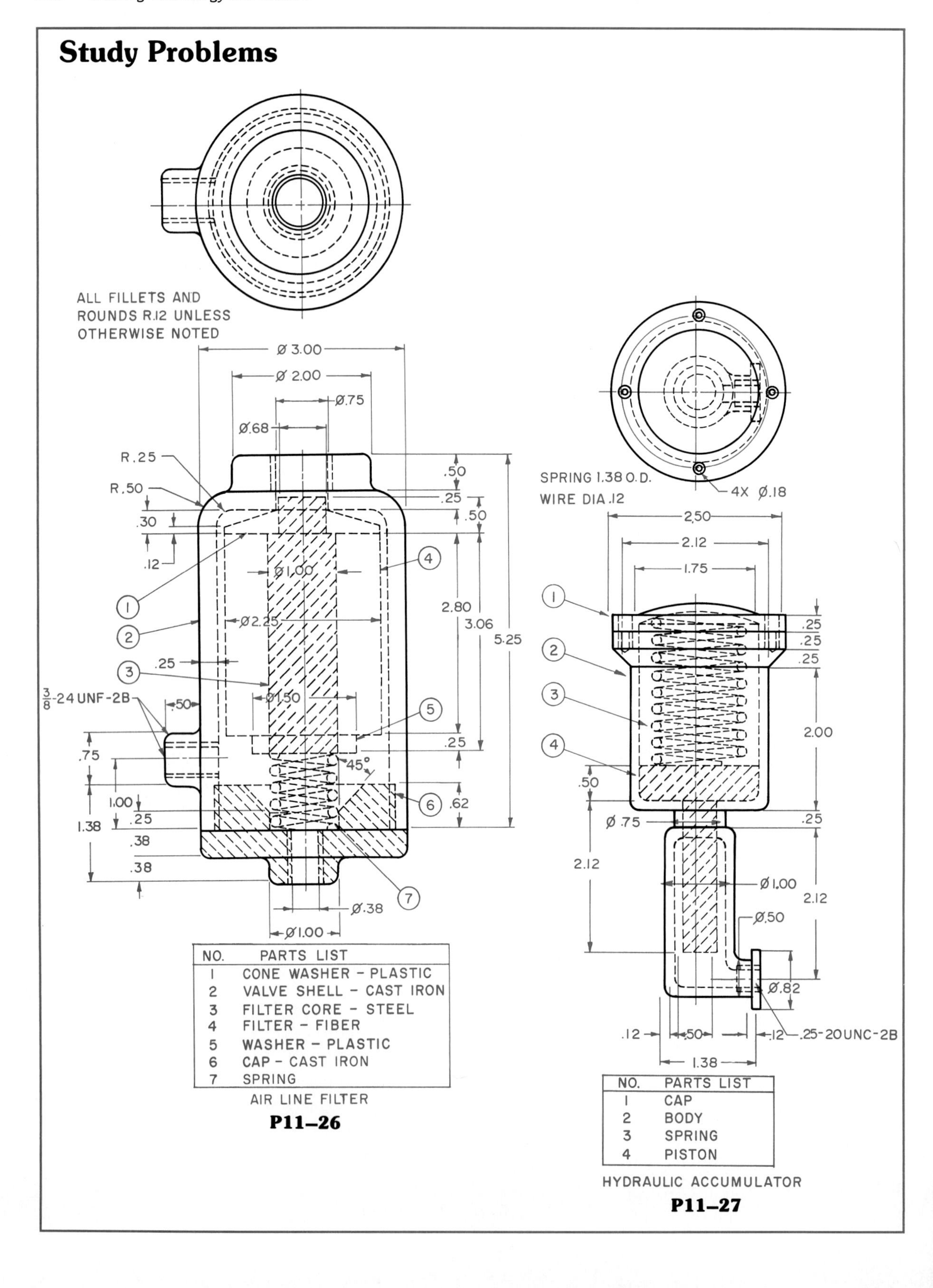

NO.	PARTS LIST
1	CONE WASHER - PLASTIC
2	VALVE SHELL - CAST IRON
3	FILTER CORE - STEEL
4	FILTER - FIBER
5	WASHER - PLASTIC
6	CAP - CAST IRON
7	SPRING

AIR LINE FILTER

P11–26

NO.	PARTS LIST
1	CAP
2	BODY
3	SPRING
4	PISTON

HYDRAULIC ACCUMULATOR

P11–27

Study Problems

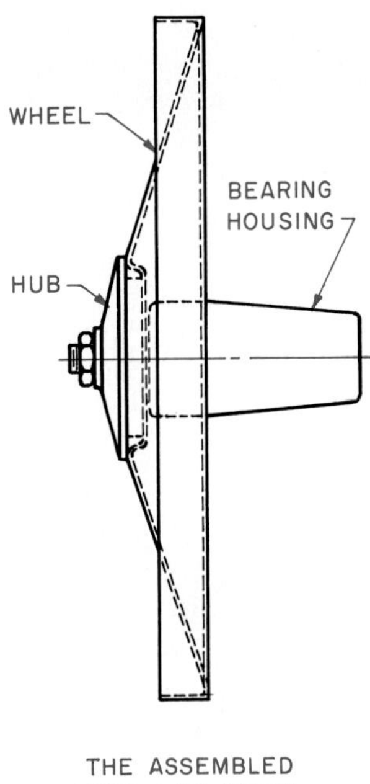

P11–28

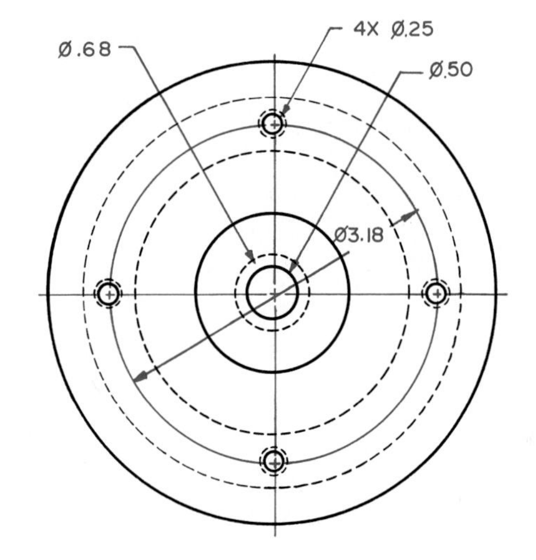

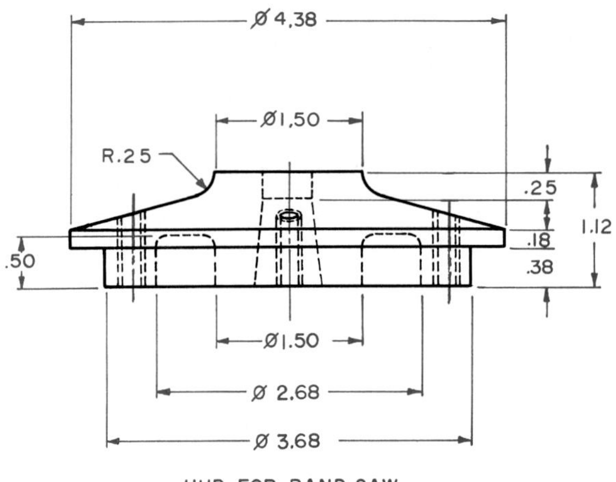

(Continued on next page)

Study Problems

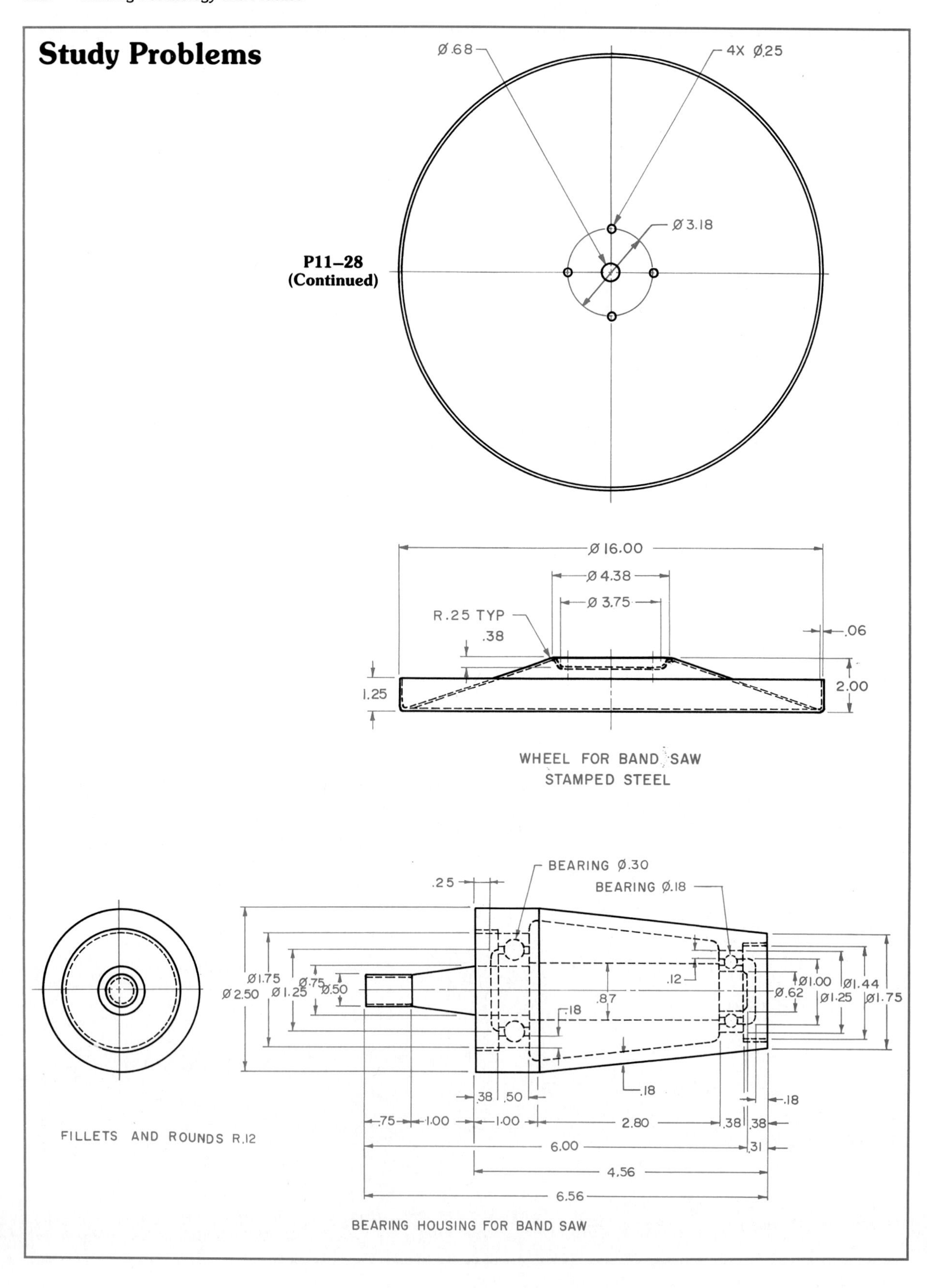

P11–28 (Continued)

CHAPTER 12

Fasteners

After studying this chapter, you should be able to do the following:

- Identify commonly used customary and metric fasteners.
- Write fastener specifications.
- Draw fasteners.

Introduction

As a product is designed, one important decision to be made is the method of fastening, or joining, the parts. Some parts must be fastened so they can be easily separated. Others can be permanently joined. The type of material to be joined influences the type of fastening method to be used. Sometimes the stress upon a joint requires a special method of fastening.

As a drafter, you must know how to draw the various types of fasteners. You must also know how to give the specifications of the fastener on the drawing.

Types of Threads

Threads with customary measurements have been standardized for use in all industries in the United States. They have been approved by the American National Standards Institute (ANSI). Metric threads have been standardized for use in all countries by the International Organization for Standardization (ISO). American National metric thread standards are based on ISO standards.

A thread takes the form of a helix (spiral). It is a ridge formed on the surface of a cylinder or a cone-shaped surface. A thread on a cylinder is called a *straight thread.* A thread on a cone-shaped surface is called a *taper thread.*

Thread Terminology

Since threads are of great importance in the design of products, you should understand their parts. Fig. 12-1.

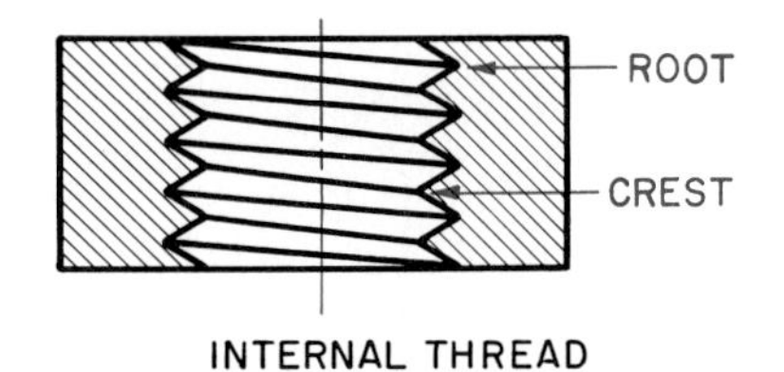

Fig. 12–1. Thread terminology.

An **external thread** is found on the external surface of a cone or cylinder.

An **internal thread** is found on the internal surface of a hollow cylinder or cone.

A *right-hand thread* is one which, when viewed along its axis, winds in a clockwise direction away from you. Fig. 12-2.

A *left-hand thread,* when viewed along its axis, winds in a counterclockwise direction away from you. Fig. 12-2.

The *root of the thread* is the surface at the base of the thread that joins the sides of the thread

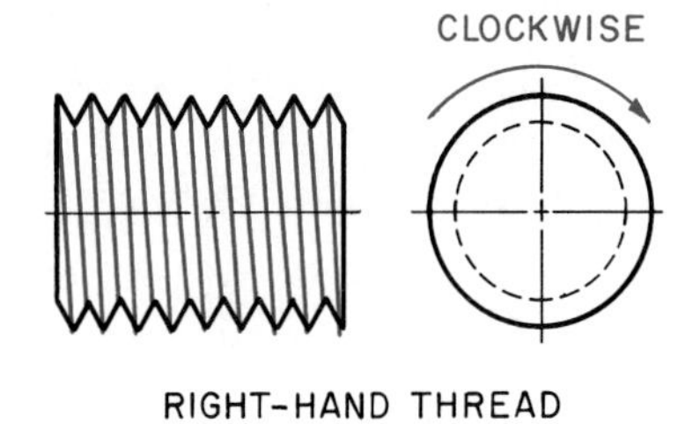

Fig. 12–2. Right-hand and left-hand threads.

form. It is the part of the thread that lies deepest in the material of which the cylinder or cone is made.

The *crest* is a surface that also joins the sides of the thread. It is the part of the thread highest on the material of which the cylinder or cone is made.

Thread *pitch* is the distance, measured parallel to the axis of the thread, between a point on one crest and a similar point on the next crest.

The *lead* is the distance that a point on a threaded shaft moves parallel to the axis in one complete revolution.

Threads per inch is the number of threads in one linear inch along a threaded shaft.

A *single thread* has a lead equal to the pitch.

A *multiple thread* has a lead equal to more than one pitch. For example, a double thread has a lead equal to two pitch distances. Fig. 12-3.

A multiple thread permits a more rapid advance of a part without using a coarser thread form. A double thread will advance twice as far in one revolution as will the same form in a single thread.

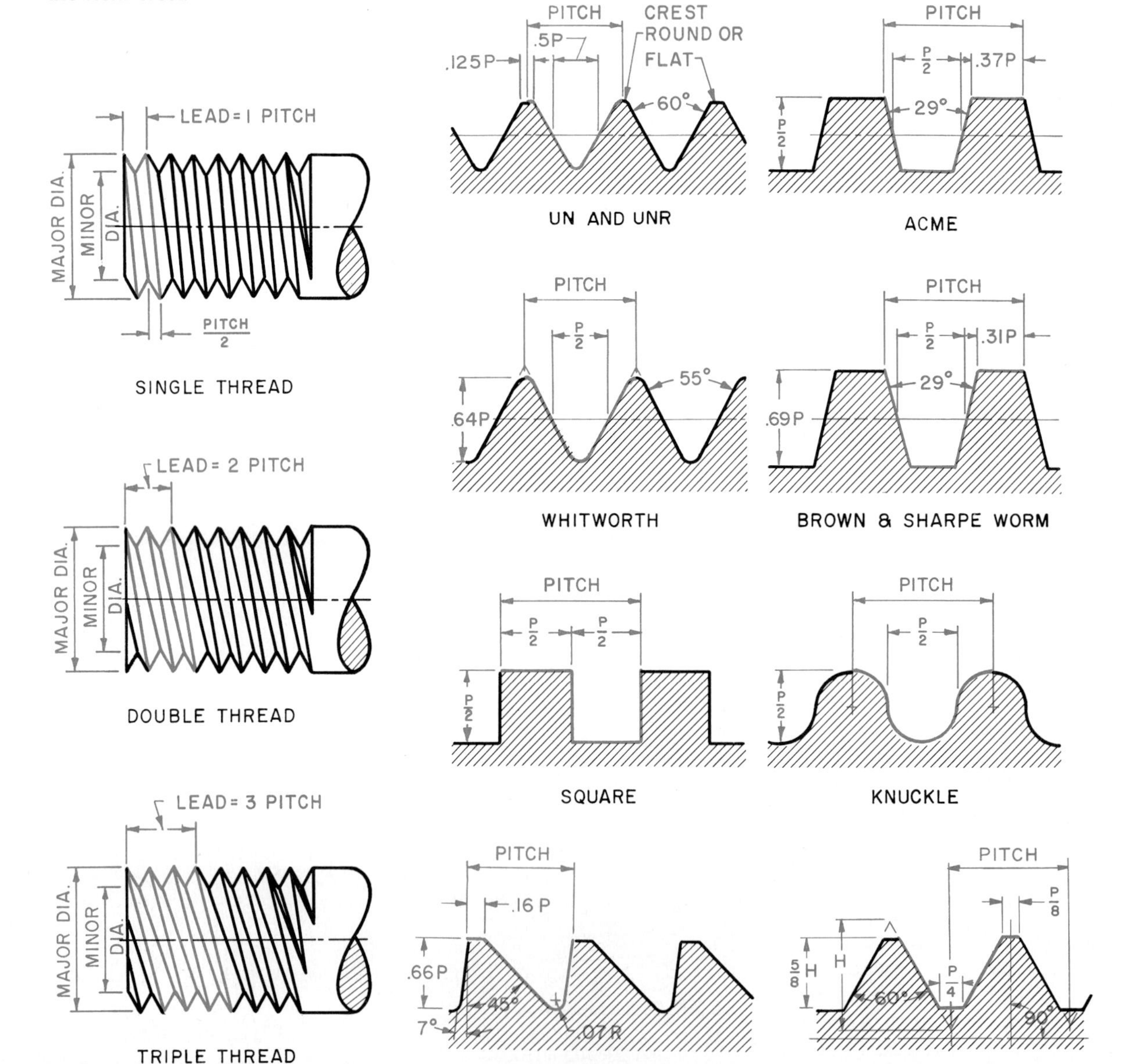

Fig. 12–3. **Single and multiple threads.**

Fig. 12–4. **Common thread forms.**

The *thread form* is the profile of the thread in cross section. Fig. 12-4.

The *depth of thread* is the distance from the crest to the root measured perpendicular to the axis.

The *major diameter* is the largest diameter of the thread.

The *minor diameter* is the smallest diameter of a thread.

The *pitch diameter* is the diameter of an imaginary cylinder that cuts the thread form where the width of the thread and groove are equal.

The *thread angle* is the angle formed between the sides of a thread.

Thread Forms

Threads are used for fastening, adjusting, and transmitting power. Some threads perform these functions better than others. Fig. 12-4 shows the more commonly used thread forms.

The American National Unified (UN and UNR) and ISO metric thread series are used for fasteners and adjustments. Both are being used in the United States. The Whitworth thread serves the same purpose as the American National thread, but is used very little.

The square, Acme, and Brown and Sharpe worm threads are used primarily for transmitting power. The buttress thread transmits power in only one direction.

The knuckle thread is made by casting or by being rolled into thin metal. A common use is on the base of a light bulb.

American National Unified Thread Series

The American National Unified Thread Series is based on customary measurements. The two series are UN and UNR. The thread profiles for both are identical. The UN thread is used for both internal and external threads. The UNR thread is used only for external threads. The UN thread has a flat root contour; however, a rounded contour cleared beyond the 0.25p flat width is optional. The root contour of the external UNR thread has a smooth, continuous nonreversing contour with a radius not less than 0.108p. Data for these are shown in Appendix 2. Following is a description of these screw thread series.

Unified National Coarse (NC or UNC). This is a coarse thread used for all general fasteners. It permits rapid assembly and disassembly.

Unified National Fine (NF or UNF). This is a fine thread. It will not loosen as easily from vibration as the UNC thread and is stronger.

Unified National Extra Fine (NEF or UNEF). This is finer thread than the UNF thread but uses the same thread form as the UNF. It is used where severe vibration is expected or on parts which may be under great stress.

8-Thread Series (8N or 8UN). This series has eight threads per inch regardless of the diameter of the part. It serves much the same purpose as the UNC thread for parts with diameters greater than one inch.

12-Thread Series (12N or 12UN). This series has 12 threads per inch regardless of the diameter of the part. While it is used for parts with a diameter as small as ½ inch, it's generally used for diameters greater than 1½ inches.

16-Thread Series (16N or 16UN). This series has 16 threads per inch regardless of the diameter of the part. It is a fine thread and is used for diameters greater than 1 11/16 inches.

Classes of Fits for Unified Threads

The thread in a nut must be slightly larger than that on the screw to be used with the nut. For some purposes this difference can be great. For other uses, the two threads must fit rather closely, in other words, have a close tolerance. The **class of fit** of a thread is the standard tolerance, or the amount of tolerance and allowance at the pitch diameter.

The Unified National screw thread standards have three classes. External threads have classes 1A, 2A, and 3A. Internal threads have classes 1B, 2B, and 3B.

Classes 1A and 1B have a large allowance and permit a loose fit. They are used where a part must be easily and quickly assembled or disassembled. Classes 2A and 2B are used for general purpose fastening. Most bolts, screws, and nuts are of this class. Classes 3A and 3B have closer tolerances than 2A and 2B. They are used where closer tolerances are needed.

Specifying Unified Threads

Drafters specify a thread on a drawing by giving the diameter, number of threads per inch, letter identification indicating the thread series, class of fit, and whether it is external (A) or internal (B). All threads are considered to be right-handed unless the left-hand designation is given. This is done by lettering LH after the class symbol. Threads are considered single unless specified otherwise. If they are double or triple, letter the word "DOUBLE" or "TRIPLE" at the end of the thread note. You may give the diameter as a common fraction or decimal fraction. Decimal fractions are preferred. Fig. 12-5.

Metric Thread Series

Metric thread sizes are standardized by the International Organization for Standardization (ISO). The United States has adopted some of these as American National Standards. Therefore, metric threads used in the United States are compatible with those in other countries. The metric thread is identified as the M-profile. The diameter-pitch combinations are found in Appendix 3. All of the data are in millimeters. The diameter-pitch combinations shown in bold type in Appendix 3 are those chosen for general use. Those in regular type are used only when unusual requirements occur. Notice that the nominal diameters are divided into three levels of choice. Use the first choice whenever possible.

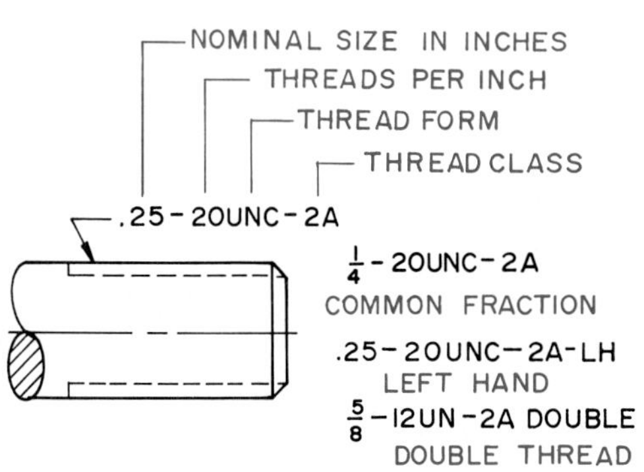

Fig. 12–5. Typical notes specifying unified threads.

Classes of Fits for Metric Threads

Tolerance classes are fine, medium, and coarse. These classes are combinations of tolerance grades, tolerance positions, and length of engagement. In practice, two classes of fits are generally used. The first is for general purpose thread use. It is specified as 6H/6g. The 6H designates the thread as an internal thread and 6g designates an external thread. The other tolerance commonly used is for closer fits. It is classified as 6H/5g6g. The 6H means an internal thread and the 5g6g means an external thread. If the thread symbol on a drawing does not have a tolerance classification, it is assumed to be a general purpose thread. Detailed information on metric thread classifications is available from ANSI.

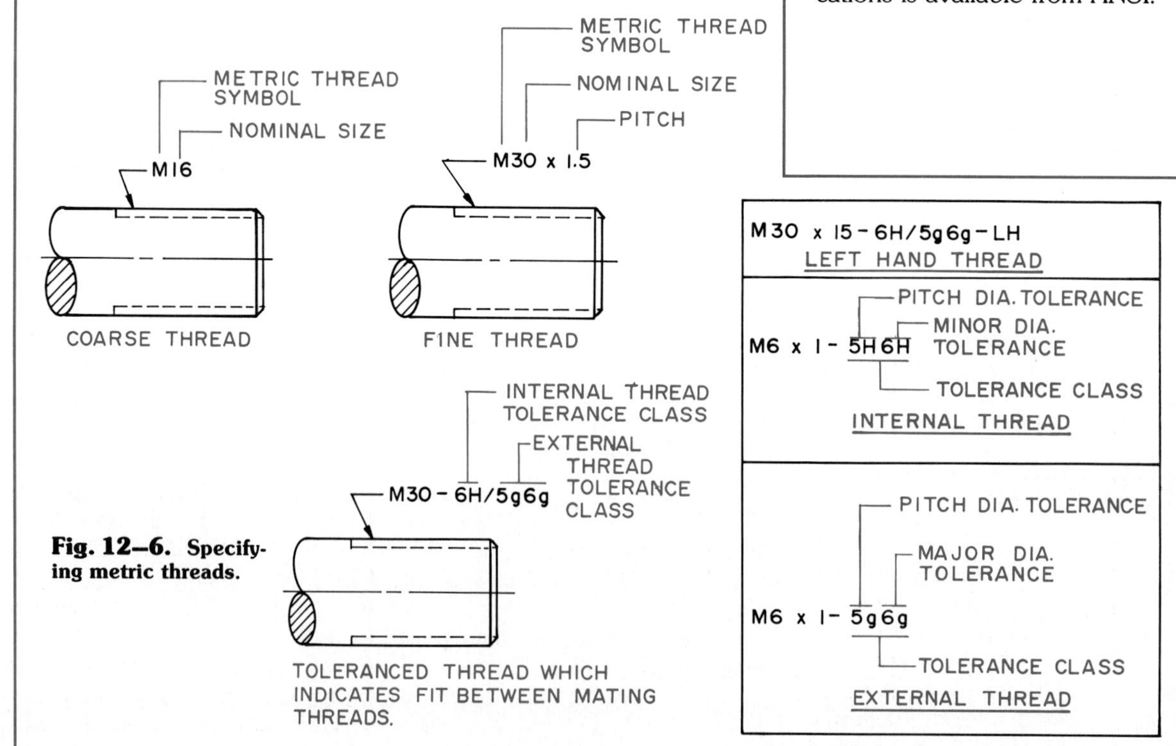

Fig. 12–6. Specifying metric threads.

Specifying Metric Threads

Specify metric threads by giving the nominal size (diameter) and pitch in millimeters. Use the letter M to show the thread as an ISO metric type. When indicating threads, omit the pitch unless the length of the thread is given. Fig. 12-6.

Use an X to separate the diameter and pitch and a dash to separate the pitch from the tolerance class. Follow the tolerance class of a left-hand thread with a dash and the letters LH. If you do not show a designation, the thread is assumed to be a right-hand thread.

Other Thread Forms

The following information deals with several other commonly used thread forms.

Special Threads (NS or UNS)

This thread series is not one of the standard series. Special threads are nonstandard combinations of diameter and pitch.

Square, Acme, and Buttress

Square threads have not been standardized. They are difficult to make and use, and so they are found in few applications.

Acme thread standards are found in Appendix 4. You designate these threads by giving the nominal diameter, the number of threads per inch, the thread form symbol (ACME), and the thread class symbol. Fig. 12-7. The classes of fit for Acme threads are 2G for general purpose threads, 3G and 4G for assemblies that require less backlash and end play, and 5G for a minimum of backlash and end play.

Designate *buttress threads* by giving the nominal diameter, number of threads per inch, thread form symbol (BUTT), and class of fit. Buttress threads are standard when the opposite flank angles are 7 degrees and 45 degrees. Fig. 12-7. See Appendix 5 for buttress thread information.

Drawing Threads

The drafter has three ways of representing threads on a drawing. These are the detailed, schematic, and simplified methods. You must decide which best suits the situation. Your decision will depend upon the purpose and use of the drawing. All three methods can be used on the same drawing.

Detailed Representation

Detailed representation means drawing the threads almost the way they actually appear. Fig. 12-8 shows detailed representation of American National Unified and metric threads. The pitch does not have to be to exact scale. If a part has nine threads to the inch, it can be drawn as eight threads because that number is easier to lay out. You can draw the curved crest and root lines as straight lines. That is, draw the

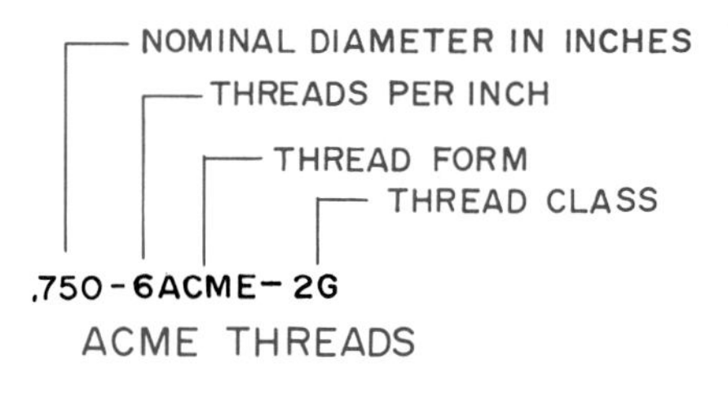

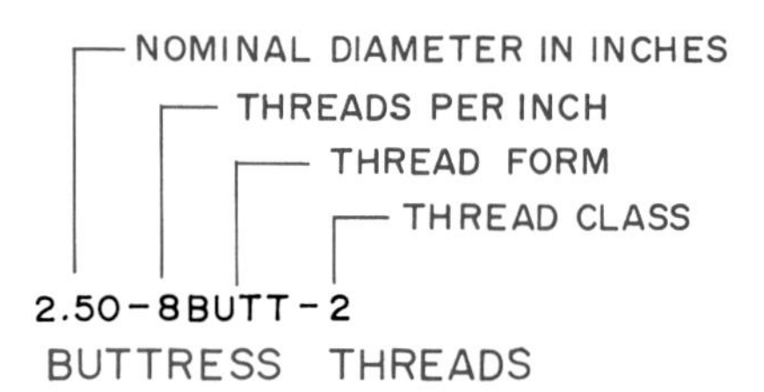

Fig. 12–7. Specifying Acme and buttress threads.

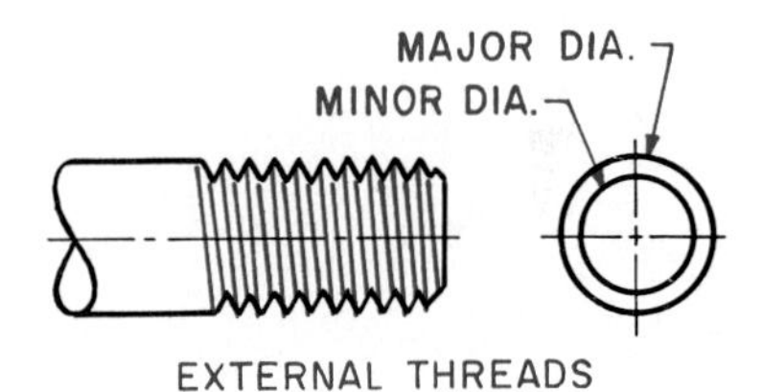

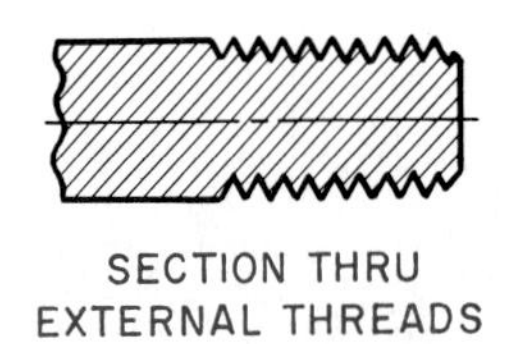

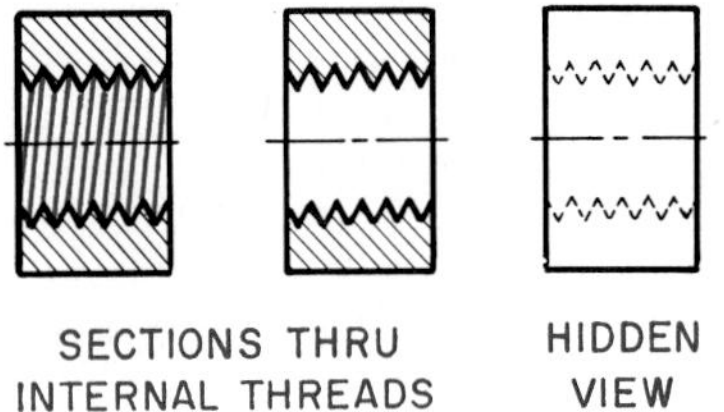

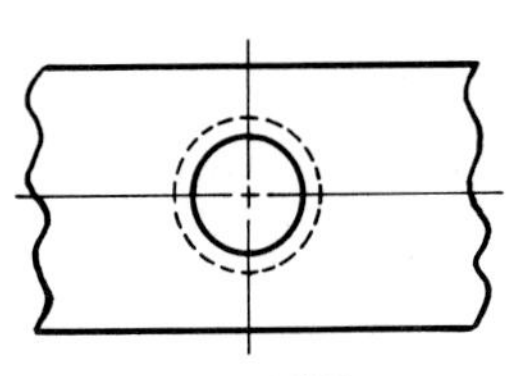

Fig. 12–8. These are detailed representation of American National Unified and metric threads.

crest and root to a sharp "V" shape, even though on the actual thread they are flat or slightly rounded. The steps for drawing detailed National thread representation are shown in Fig. 12-9.

The procedure for drawing a detailed representation of Acme threads is shown in Fig. 12-10. The method of drawing a detailed representation of Square threads is shown in Fig. 12-11.

Schematic Representation

Schematic representation is easier to draw than detailed representation and is still effective. Fig. 12-12. Do not draw the crest and root lines to the exact pitch of the thread. You can space them so they give a pleasing appearance. Usually you space them wider on large diameter shafts.

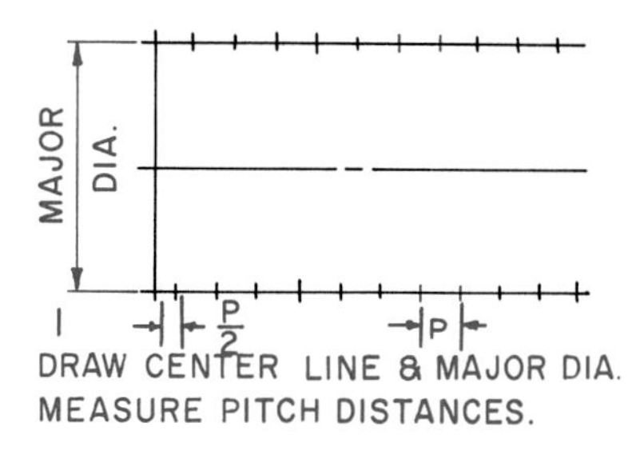

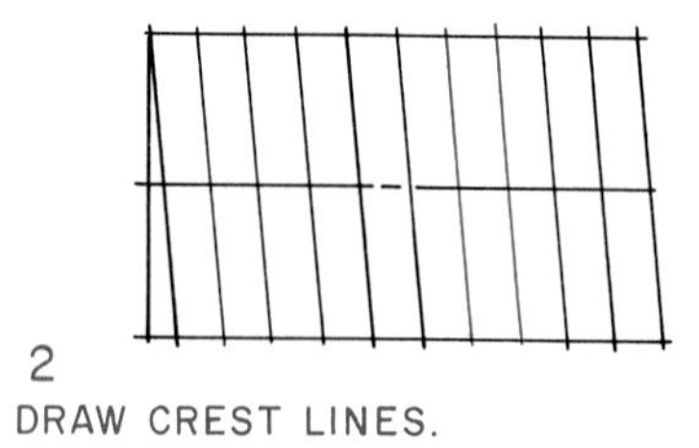

60°
MINOR DIA
3
DRAW SIDES OF THREADS.

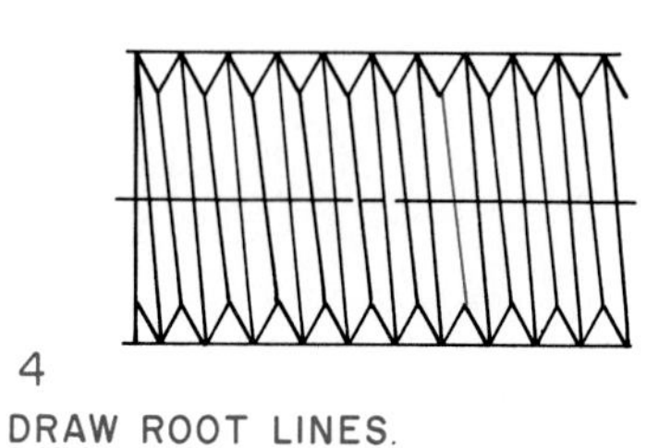

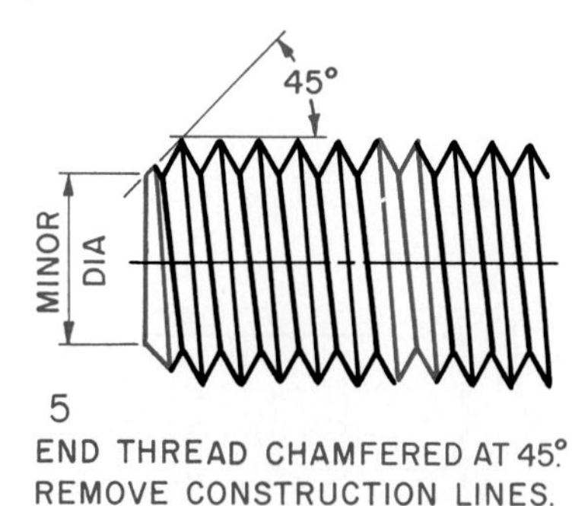

Fig. 12–9. How to draw detailed representation of American National Unified and metric threads.

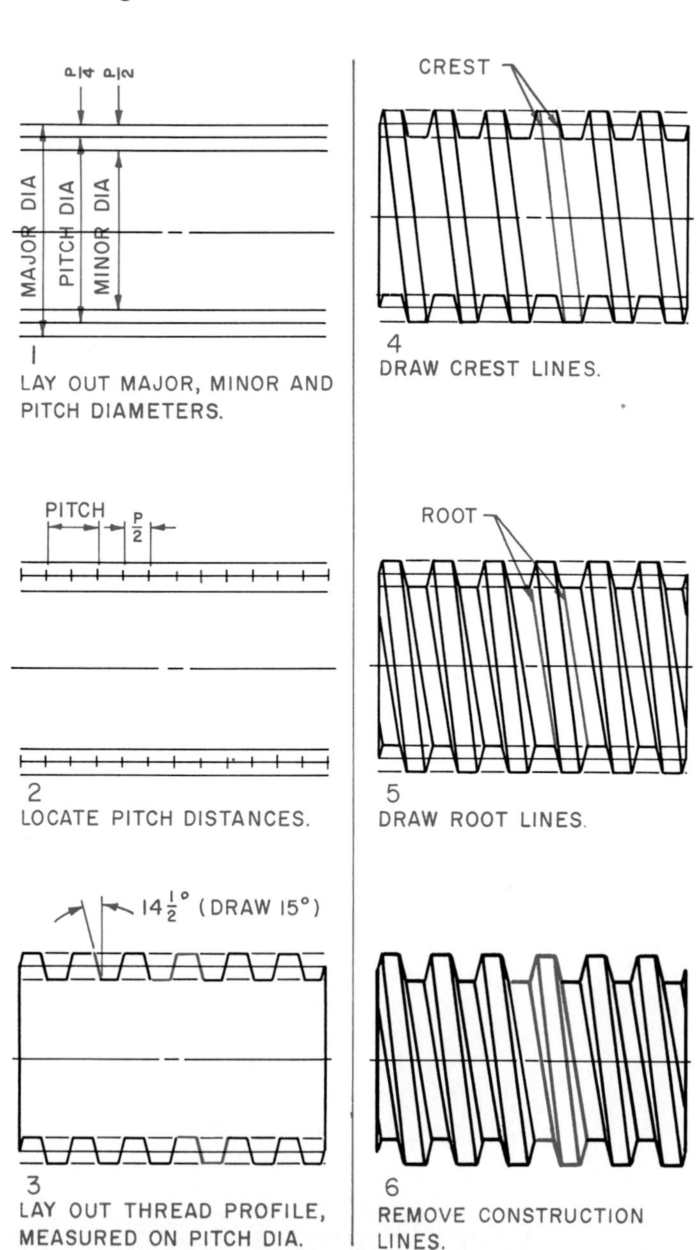

Fig. 12–10. How to draw detailed representation of Acme threads.

It is common practice to draw root lines heavier than crest lines. Root lines are usually drawn perpendicular to the axis of the part threaded. The steps for drawing schematic thread representation are shown in Fig. 12-13.

Simplified Representation

Simplified representation is the easiest and quickest method of indicating threads. It should not be used if the dashed lines might be confused with other hidden lines on the drawing. Fig. 12-14.

Bolts

A **bolt** is a fastening device with a head on one end and threads on the other. It passes through holes in each part to be joined, and a nut is tightened on the threaded end. See Fig. 12-15 for the common types of bolts.

Threaded fastening devices are manufactured with a wide variety of head types. Fig. 12-16.

The flat head is used when the head must be flush with the surface. The oval head, used when the head is exposed to view, improves the appearance of the product. The pan and truss heads are used to cover large clearance holes.

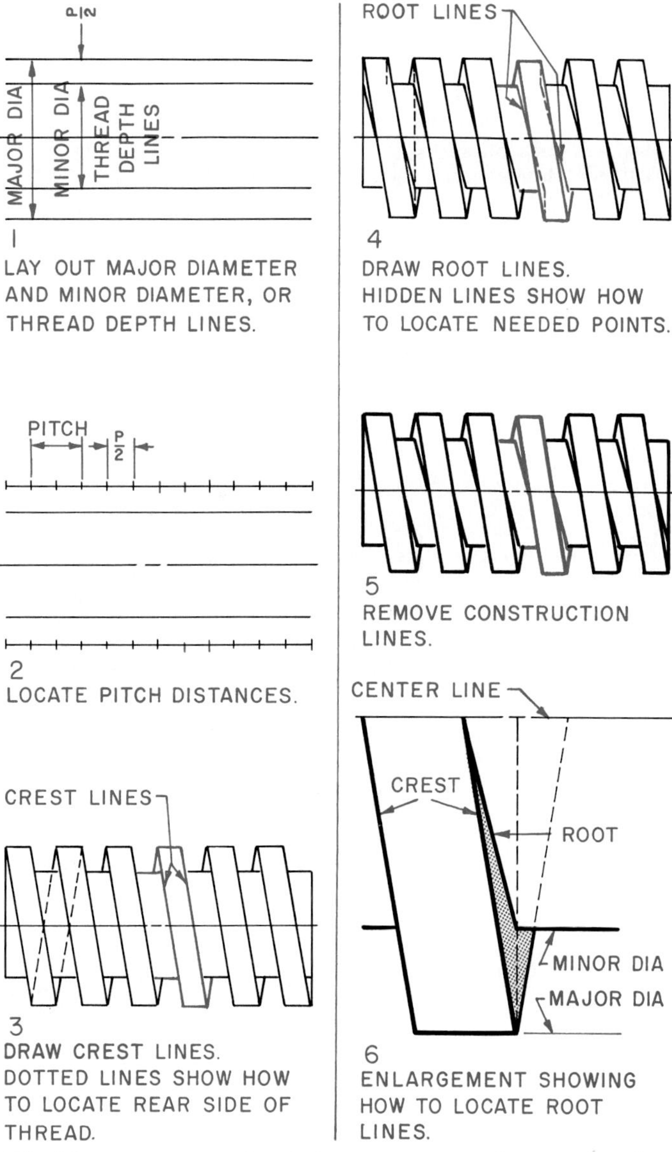

Fig. 12–11. How to draw detailed representation of square threads.

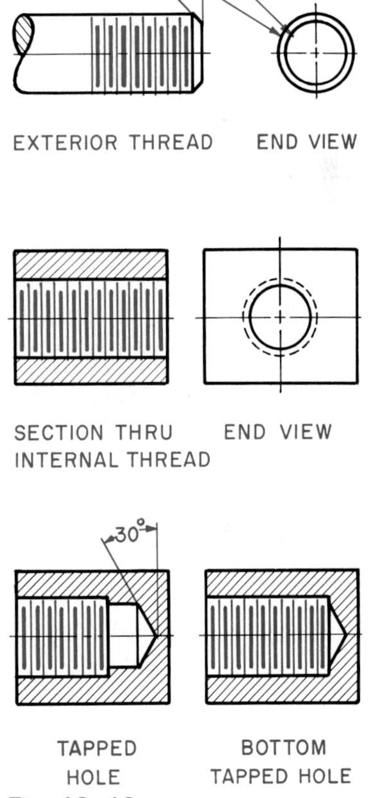

Fig. 12–12. Schematic representation of screw threads.

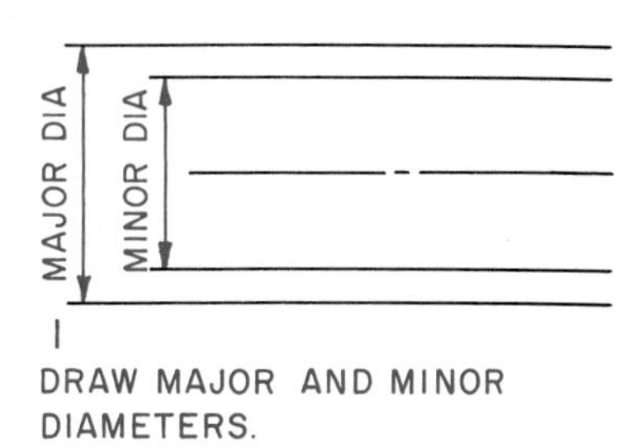

1
DRAW MAJOR AND MINOR DIAMETERS.

45°
MINOR DIA

2
CHAMFER END AT 45° FROM MINOR DIAMETER.

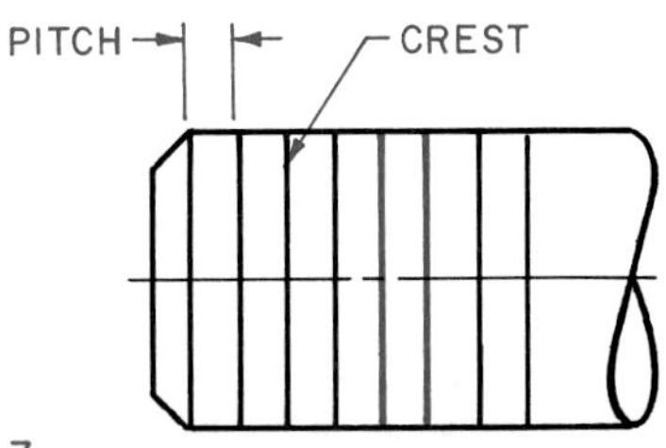

3
MEASURE PITCH DISTANCES AND DRAW CREST LINES.

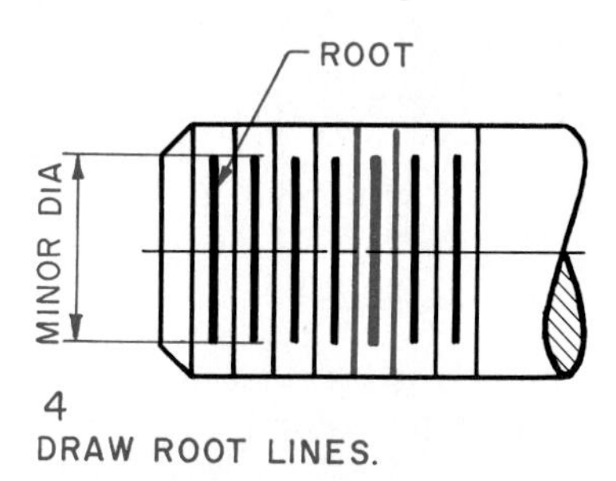

4
DRAW ROOT LINES.

Fig. 12–13. How to draw schematic representation of screw threads.

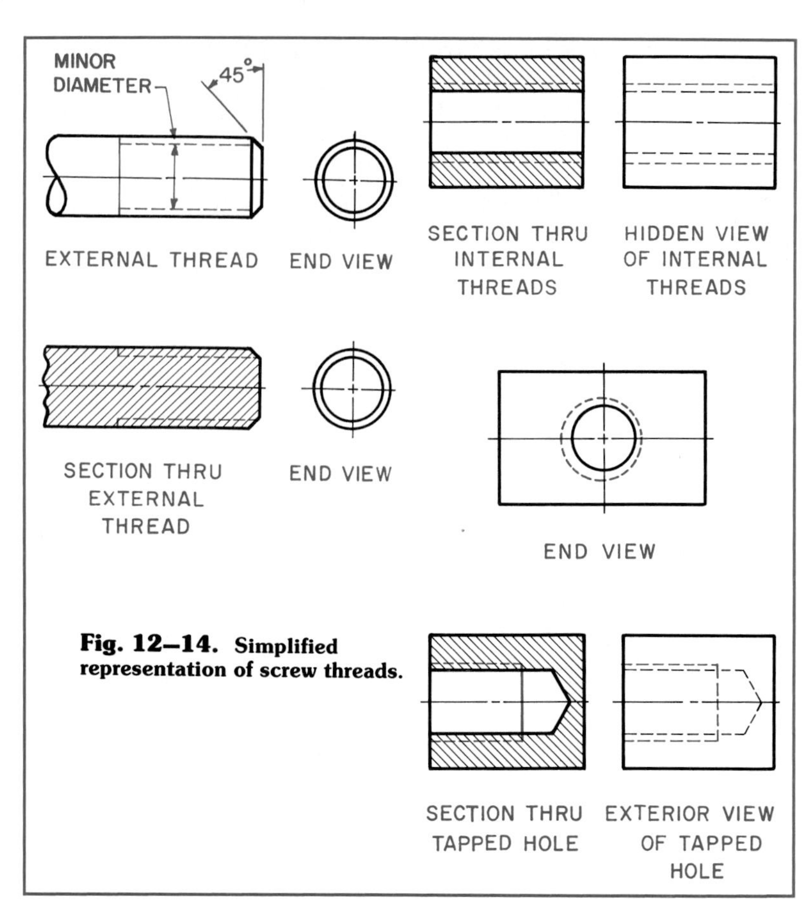

Fig. 12–14. Simplified representation of screw threads.

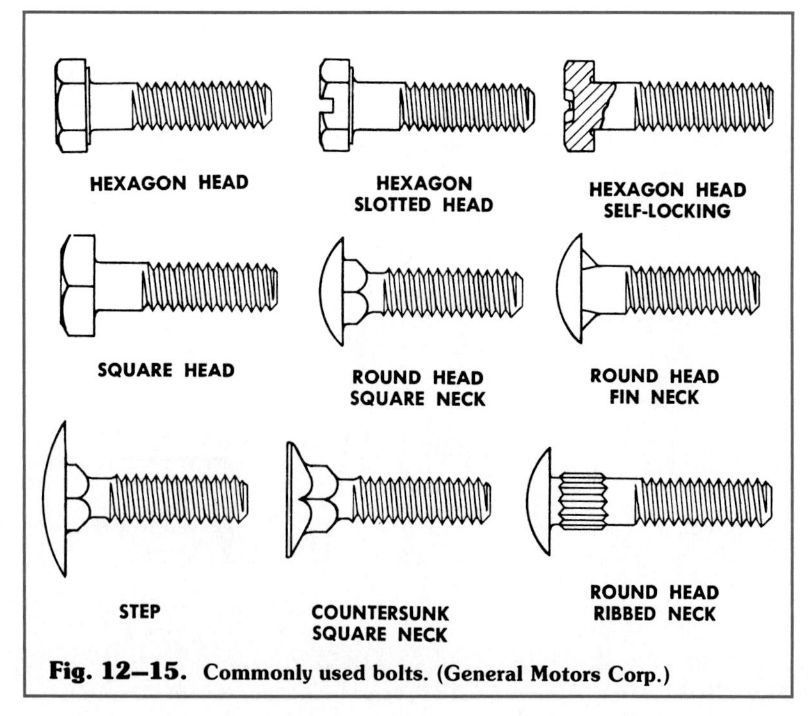

Fig. 12–15. Commonly used bolts. (General Motors Corp.)

Hexagon heads are strong and easy to drive, but they are not pleasing in appearance. If appearance is important, the cross recess head is used. The cross recess head is easier to drive than the slotted head. The clutch recess head is also easy to drive.

The proper type of bolt to use is determined by the job. The need for strength, rapid assembly, and low cost is considered. Bolts have two types of finish for the heads, unfinished and finished. Fig. 12-17. On a *finished* bolt, the bottom of the head (which is the bearing surface) is machined. It is at right angles to the axis of the bolt. It may be chamfered or have a washer face. Washer faces are usually .02 to .05 inches or 0.5 to 1.2 mm thick. The head of an *unfinished* bolt is not machined.

Two types of hexagon bolts, regular and heavy, are available. Data for standard inch and metric hexagon bolts are found in Appendices 6 and 9.

The heavy bolt is used where strength is important. This bolt has a larger head than a regular one and offers a larger bearing surface.

Square head bolts and nuts are available in the regular series only. They are only available unfinished. They are used for less accurate work. Standard square bolt sizes are given in Appendix 6.

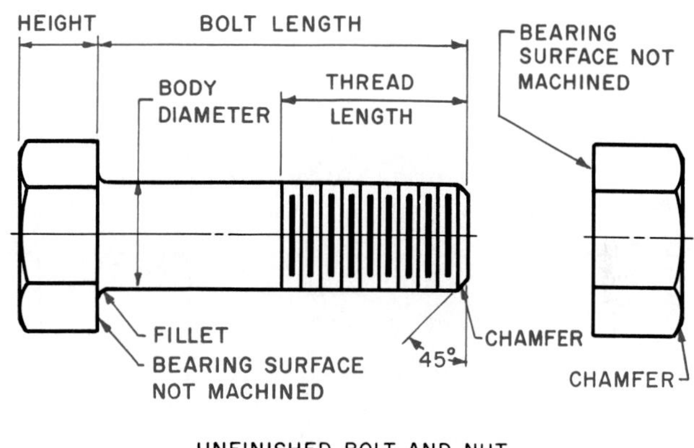

Fig. 12–16. Head types and common driver recess types.

Fig. 12–17. Finished and unfinished hexagon bolts and nuts.

CAD Applications:

Fasteners

Repeatedly drawing fasteners or other details can be a time-consuming process. And the longer it takes the more it costs. Adding details without adding much to the cost of a drawing is one of the real advantages of CAD.

For CAD users, commonly used symbols and details are available in symbol libraries. The libraries are kept in separate computer files. Drafters can create their own symbol libraries, but today a wide variety of ready-made libraries can be purchased. These include:

- residential plumbing fixtures, fittings, and valves
- heating and air conditioning compressors, pumps, and ductwork
- kitchen and bath sinks, faucets, and cabinets
- framing details for footings, foundations, stairs, and roofs
- electrical switches, relays, and connections
- hydraulic line components, gauges, and heaters
- industrial pipe fittings
- welding symbols
- steel structural shapes
- nuts, bolts, screws, and other fasteners

Fastener symbol library by SPOCAD

A commercial symbol library usually consists of computer disks and a printed overlay menu. The software is loaded onto the CAD system. The printed menu, which shows the symbols that are available, is fastened to the digitizer tablet. The menu must then be configured so that the computer will know which places on the overlay correspond to which symbols in the computer memory.

To use the library, the drafter picks a symbol from the overlay with the pointing device and then indicates its position on the drawing. The computer then draws it in the proper spot.

Whether the library is purchased or made by the drafter, symbols in it should meet industry standards. They should also match those commonly used by the design firm. In other respects, the library's usefulness is limited only by the drafter's imagination.

Specifying Bolts

Customary-sized bolts are specified by giving the following data in the order shown:

1. Nominal size (fractional or decimal).
2. Threads per inch.
3. Product length (fractional or two-place decimal).
4. Product name.
5. Material, including specifications when necessary.
6. Protective finish if any.

Following are some examples:
½ - 13 × 1¾ SQUARE BOLT, STEEL, ZINC PLATED
⅝ - 18 × 1.00 HEX BOLT, SAE GRADE 5 STEEL
Metric bolts are specified by giving the following data in the order shown:

1. An "M" to indicate the thread is metric.
2. The nominal diameter in millimeters.
3. The pitch in millimeters.
4. The length in millimeters.
5. Product name.
6. Material and/or protective finish if any.

Following are some examples:
M14 × 2 × 30 HEX BOLT, BRASS
M24 × 3 × 60 HEAVY HEX BOLT, STEEL

Nuts

A common kind of **nut** is a thick, square or hexagonal piece of metal with a hole that is threaded to mate with the threads on a bolt. Nuts are commonly made finished and unfinished in either the regular or heavy series. Both customary and metric sizes are found in Appendices 7, 8, and 10.

Jam nuts are used for locking other nuts in place. These nuts are thinner than regular nuts. They are made in regular, heavy, unfinished, and finished types but are available only in the hexagon shape. Customary sizes are found in Appendix 7.

Specifying Nuts

Customary-sized nuts are specified by giving the following data in the order shown:

1. Nominal size (fraction or decimal).
2. Threads per inch.
3. Product name.
4. Material including specifications when necessary.
5. Protective finish if any.

Following are some examples:
¾ - SQUARE NUT, STEEL, ZINC PLATED
½ - 20 HEX NUT, BRASS
1.000 - 12 HEAVY HEX NUT, SAE GRADE 5 STEEL
Metric nuts are specified by giving the following information in the order shown:

1. An "M" to indicate the thread is metric.
2. The nominal diameter in millimeters.
3. The pitch.
4. The product name.
5. Material specifications.
6. Protective finish if any.

Following are some examples of metric nut specifications:
M14 × 2 HEX NUT, STEEL
M24 × 3 HEAVY HEX NUT, SAE GRADE 5 STEEL

Drawing Bolts and Nuts

When drawing bolts and nuts where accuracy is not important, the sizes of each can be taken from tables in Appendices 6 through 10. If greater accuracy is required, decimal sizes are available in American National Standards Institute and Industrial Fasteners Institute publications.

The steps for drawing hexagon and square bolts are shown in Figs. 12-18 and 12-19. They show how to draw bolts without templates. But use nut and bolt templates if they are available. Computer drafting programs also are available to draw all types of fasteners.

The washer face on finished bolts and nuts is drawn 1/64-inch (0.4 mm) thick.

To draw hexagon head bolts, see Fig. 12-18 and follow these steps:

1. Draw the top view of the bolt head, Fig. 12-18 at A. Draw a circle equal to the dimension across the flats, F.
2. Draw a hexagon tangent to the outside of the circle. Draw the sides on a 60-degree angle.
3. Locate the bearing surface, Fig. 12-18 at B.
4. Measure the head height, H. If it is a semifinished bolt, the washer face is included in the thickness.
5. Project the width of the head and visible corners from the top view.
6. Locate the radius, R, of the curve on the narrow surfaces, Fig. 12-18 at C.
7. Draw the arcs on the small surfaces.
8. Locate the major radius, R, by trial and error. It should meet the small arcs and have a tangent to the top of the bolt.
9. Draw the chamfer on each corner. The chamfer must have a tangent to the arc at an angle of 30 degrees.
10. Remove construction lines and darken the lines, Fig. 12-18 at D.

To draw a square bolt, see Fig. 12-19 and follow these steps:

1. Draw the top view of the bolt head, Fig. 12-19 at A. Begin by drawing a circle equal to the distance across the flats.
2. Draw a square outside and tangent to the circle at a 45-degree angle.
3. Locate the bearing surface, Fig. 12-19 at B.
4. Measure the head height, H.
5. Project the width of the head from the top view.
6. Locate the radius, R, of the curve, Fig. 12-19 at C.

7. Draw the arcs on the bolt head.

8. Draw the chamfer on each corner. The chamfer must be tangent to the arc. It is drawn at an angle of 30 degrees.

9. Remove construction lines and darken the lines, Fig. 12-19 at D.

Studs

A **stud** is a rod that is threaded on each end. It usually passes through a clearance hole in one piece and screws into a tapped hole in a second piece. Fig. 12-20. A nut is screwed on the other end to hold the two pieces together. Customary standards for studs are shown in Appendix 11. Metric stud standards are shown in Appendix 12.

Tapping Screws

Tapping screws are designed to cut their own threads as they are screwed into place. They are

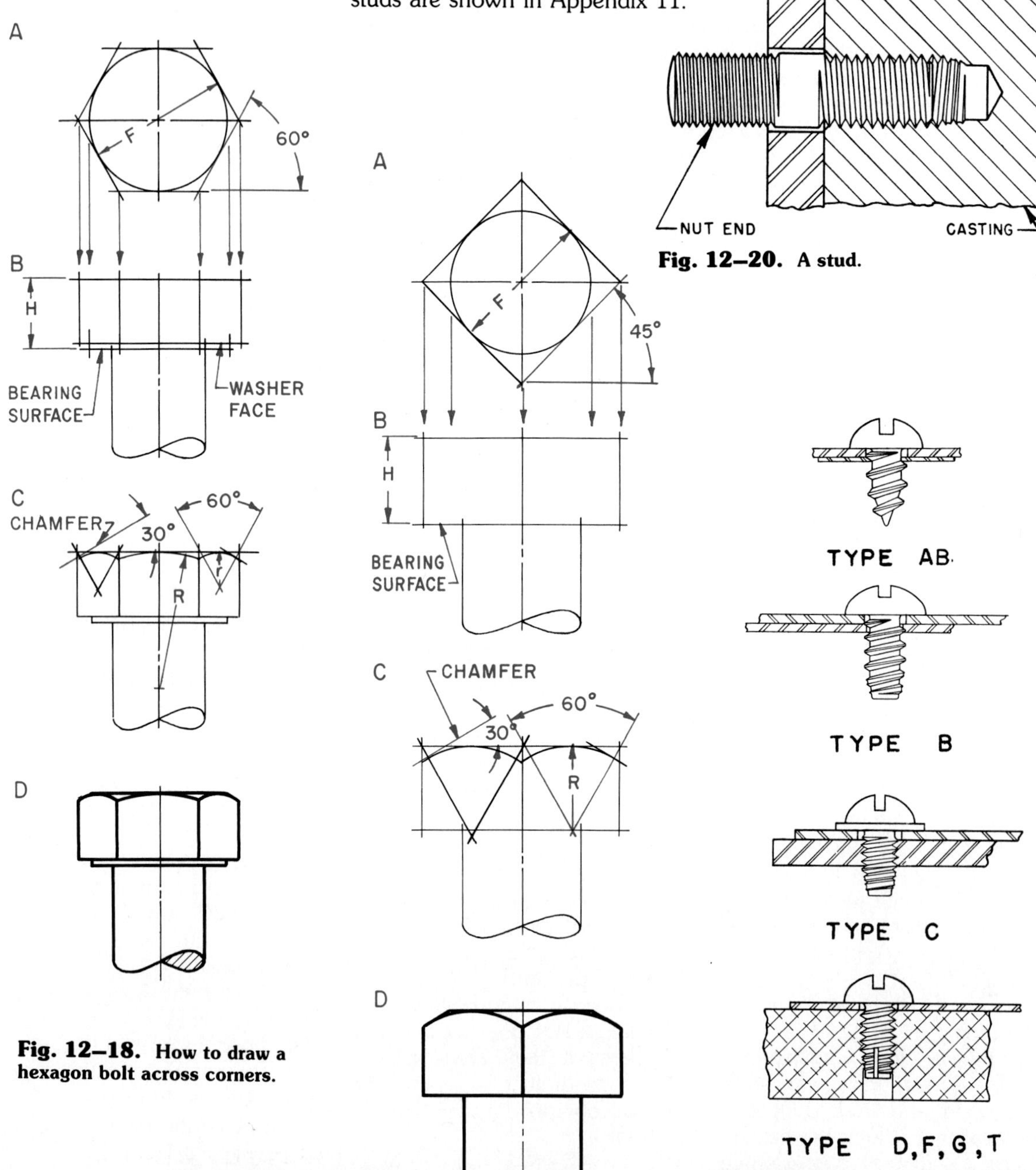

Fig. 12–18. How to draw a hexagon bolt across corners.

Fig. 12–19. How to draw a square bolt across corners.

Fig. 12–20. A stud.

Fig. 12–21. Common tapping screws.

used to fasten both sheet metal and plastic parts. Some are designed to fasten thin materials to soft metal castings.

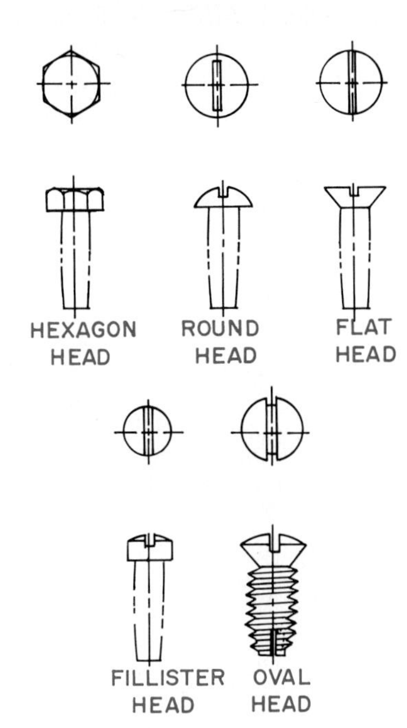

Fig. 12–22. Common heads found on tapping screws.

The most common types of tapping screws are shown in Fig. 12-21. Types AB, B, and C are used to join sheet metals. Types D, F, G, and T are used to join thin material to thick material. They cut threads in thick material. Standard customary and metric tapping screw sizes are found in Appendices 13 and 14.

The most common heads are flat, oval, round, pan, fillister, and hexagon. Both slotted and recessed kinds are available. Fig. 12-22.

Dimension tapping screws by giving length, wire diameter, type, right-hand or left-hand thread, type of head, kind of fastener, and finish. A typical note would read ½ - NO. 4 TYPE B - RH PAN HEAD TAPPING SCREW - NICKEL FINISH.

An example of specifications for a metric tapping screw is 4 × 0.7 × 30, TYPE F, SLOTTED FLAT HEAD TAPPING SCR., NICKEL FINISH

Cap Screws

A **cap screw** passes through a clearance hole in one part and screws into a tapped hole in the other. It does not use a nut. Fig. 12-23.

Common types of heads are round, flat, hexagon, fillister, 12 spline, and socket. Fig. 12-24. Short cap screws are threaded to the head. Long cap screws are threaded only part way to the head. Data for slotted head, hexagon head, and spline and hexagon socket head cap screws with customary dimensions are found in Appendices 15, 16, and 17. Metric cap screw data are found in Appendix 9.

On a drawing you specify cap screws by giving their diameter, number of threads, thread type, class of fit, length, head type, and the type of screw. A typical note would read ⅜-24UNF-2A × ¾ FLAT HD CAP SCR.

Notice in Fig. 12-23 that the slot is drawn on a 45-degree angle. This is common practice for all slotted fasteners.

The metric cap screw data shown in Appendix 9 are common for all head types. Specify metric cap screws as follows: M8 × 1.25 × 30, FLT HD CAP SCR.

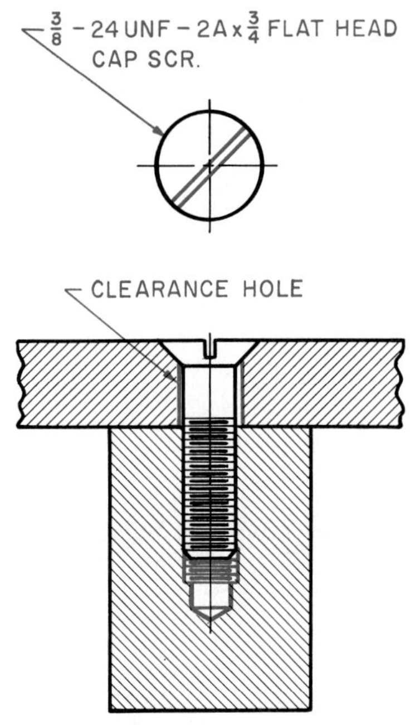

Fig. 12–23. A cap screw with clearance hole.

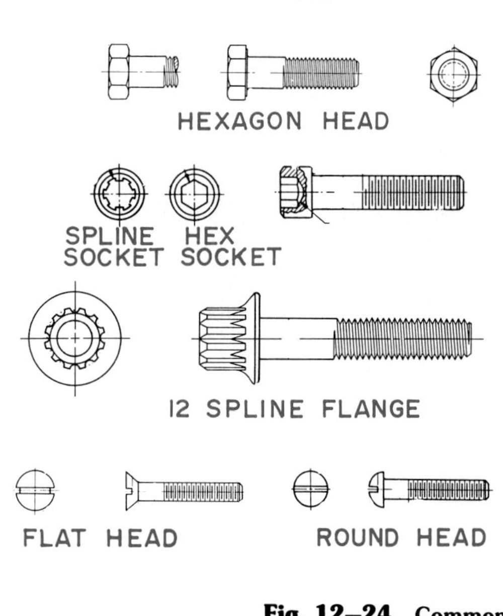

Fig. 12–24. Common heads found on cap screws.

Machine Screws

Machine screws use a nut to fasten parts together. However, they are much like cap screws and can be used in tapped holes. The common head types are flat, oval, round, fillister, pan, and hexagon. Fig. 12-25.

Machine screws are usually used to join thin materials. They have sizes smaller than cap screws. You specify by number those with sizes below ¼ inch in diameter. Metric and customary sizes are shown in Appendix 18.

On a drawing you give the machine screw diameter, number of threads, type of thread, class of fit, length, head type, and the type of fastener. A typical note would read ¼-28UNF-2A × ½ FLAT HEAD MACH SCR.

Metric machine screws are noted on drawings by giving the screw size, thread pitch, length, and type of thread. A typical note would read M4 × 0.7 × 10 FLAT HD MACH SCR.

Set Screws

Set screws are used to keep one part in position against another. One use would be to keep a pulley from slipping on a shaft.

The set screw screws into a threaded hole in the pulley and its point is firmly set against the shaft, which keeps the parts from slipping. Fig. 12-26. The shaft usually has either a flat spot or cone-shaped hole to receive the point of the set screw.

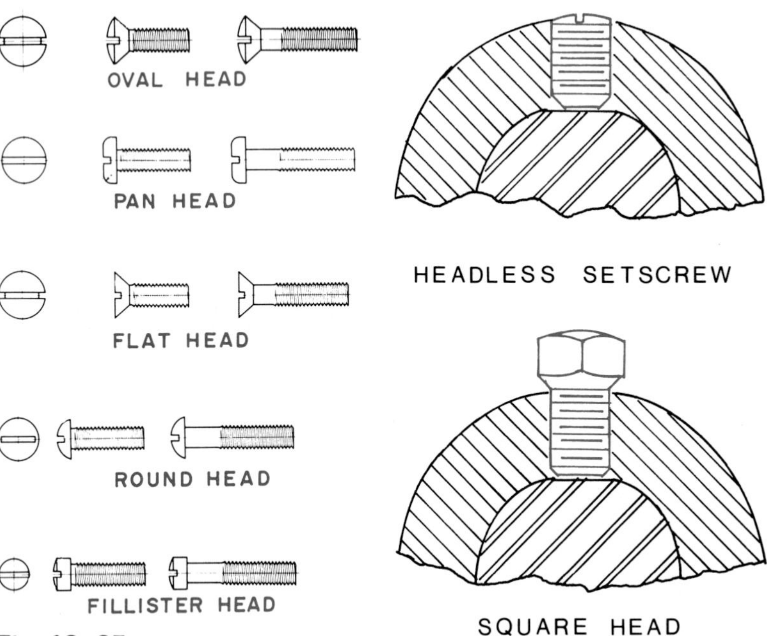

Fig. 12–25. Common heads found on machine screws.

Fig. 12–26. Common set screws.

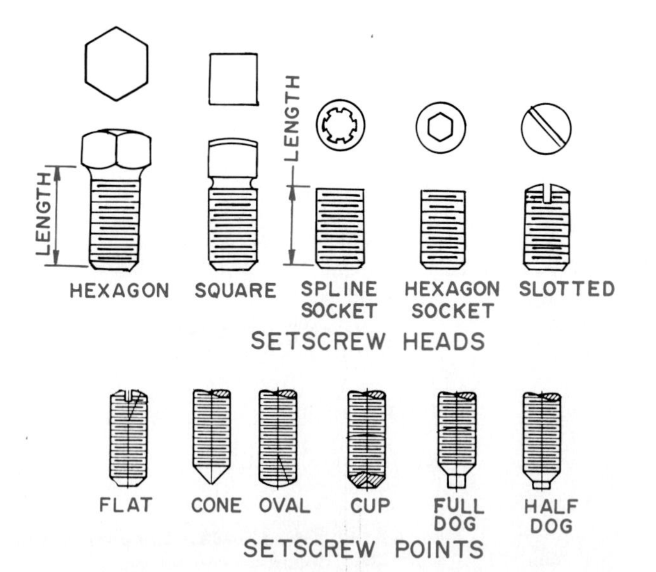

Fig. 12–27. Common types of set screw heads and points.

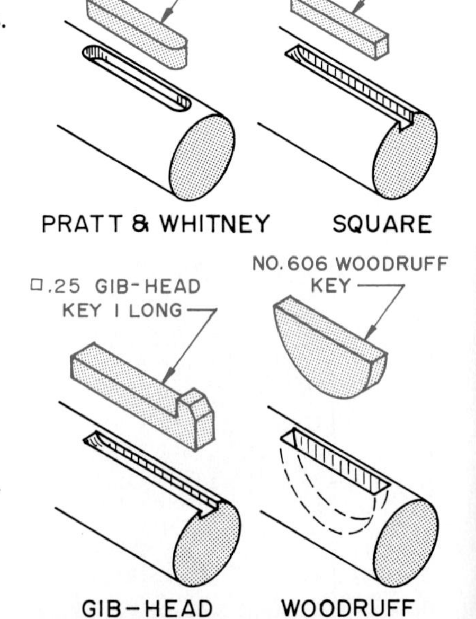

Fig. 12–28. Common types of keys.

The common types of set screws are square head, hexagon socket and spline socket, and slotted headless. Set screws are available with a variety of points—cup, flat, oval, cone, full dog, and half dog. Fig. 12-27.

Headless set screws set flush with the surface of the parts into which they are screwed. These set screws are safer than those with heads. Those with heads can be dangerous when holding rotating parts.

Measure the length of a set screw from under the head to the point.

Specify customary set screws by giving their diameter, number of threads, thread type, class of fit, length, type of head, type of point, and type of screw. A typical note would read 3⁄8-16UNC-2A × 5⁄8 SLOTTED FLAT PT SET SCR. Customary set screw tables are found in Appendices 19, 20, and 21.

Metric set screws have the same diameter-pitch combinations and length increments as metric machine screws. Specify metric set screws by giving the diameter and thread pitch, length, type of head, type of point, and the name of the screw. A typical note would read M8 × 1.25 × 10, SLOTTED FLT PT SET SCR.

Keys and Keyseats

A **key** is a metal part that when placed in a slot on a shaft, known as a keyseat, extends above the shaft and into the hub. It keeps the hub from slipping on the shaft. A **keyseat** is a rectangular groove machined into a shaft or a hub. The common types of keys are square, rectangular, plain taper, Gib head, and Woodruff. Fig. 12-28.

Square keys are preferred for use on shafts 6½ inches in diameter or smaller.

Rectangular keys are used on shafts with a diameter larger than 6½ inches.

Plain taper keys are made in two forms. One tapers the distance the key is in contact with the part. This is the hub length. The other tapers the entire length of the key.

Gib head keys have a large head on one end. They are a form of taper keys. Their heads are used to help remove the keys.

Woodruff keys are almost half round. This kind of key fits into a semicircular slot machined into the shaft. The top of the key fits into a keyseat machined into the part to be held to the shaft. The size of Woodruff key to use for various shaft diameters has not been standardized. A suggested guide is given in Appendix 23.

Sizing Keys

The size of the key used varies with the shaft diameter. Standard customary dimensions and tolerances of different kinds of keys are shown in Appendices 22 through 27. If the shaft is stepped, use the diameter of the shaft at the point where the key is to be placed.

Specifying Keys

You specify customary keys by size and type. In the example 1⁄8 × 3⁄4 SQUARE KEY, the key is 1⁄8-inch square and 3⁄4-inch long. A rectangular key is specified by width × height × length. An example is 1⁄4 × 3⁄16 × 1½ RECTANGULAR KEY. Taper and Gib head keys are specified in the same manner as square and rectangular keys. The taper is a standard 1⁄8 inch in 12 inches and need not be specified. Specify Woodruff keys by number. An example is NO. 606 WOODRUFF KEY. Fig. 12-29.

Metric keys are specified in millimeters by width × height × length. For example, a note might read 10 × 8 × 10 RECTANGULAR KEY. Metric keys are specified in Appendices 26 and 27.

Keyseat depths for customary square, rectangular, plain taper, and Gib head keys are standardized. Data for customary and metric keyseats are found in the appendices for keys.

Pins

Pins are used to hold a part, such as a collar, to a shaft. They are good for light work only. Pins pass through both the collar and shaft. Fig. 12-30.

Common types of pins are the dowel, taper, clevis, straight, and cotter. Fig. 12-31.

Taper pins fit into a tapered hole. They are held in place by friction and are easily removed. The size of pin to use for various shaft diameters is not standardized. The engineer selects the size to use.

Straight pins are not tapered. They are available with chamfered or straight ends and can be removed easily. Straight pins are cheaper to use since a tapered hole is not necessary.

Dowel pins are basically the same diameter their entire length. Their chamfered ends are slightly smaller than the round ends. They are used to hold parts in position or alignment.

Clevis pins are used to hold two parts together. They are held in place with cotter pins.

Cotter pins are used to keep parts from separating. They are placed through a hole drilled in the part. The ends are bent in the same or opposite directions to secure the part in place.

Pin size data are found in Appendices 28 through 31.

Washers

The common types of **washers** are plain, lock, and tooth lock. Fig. 12-32.

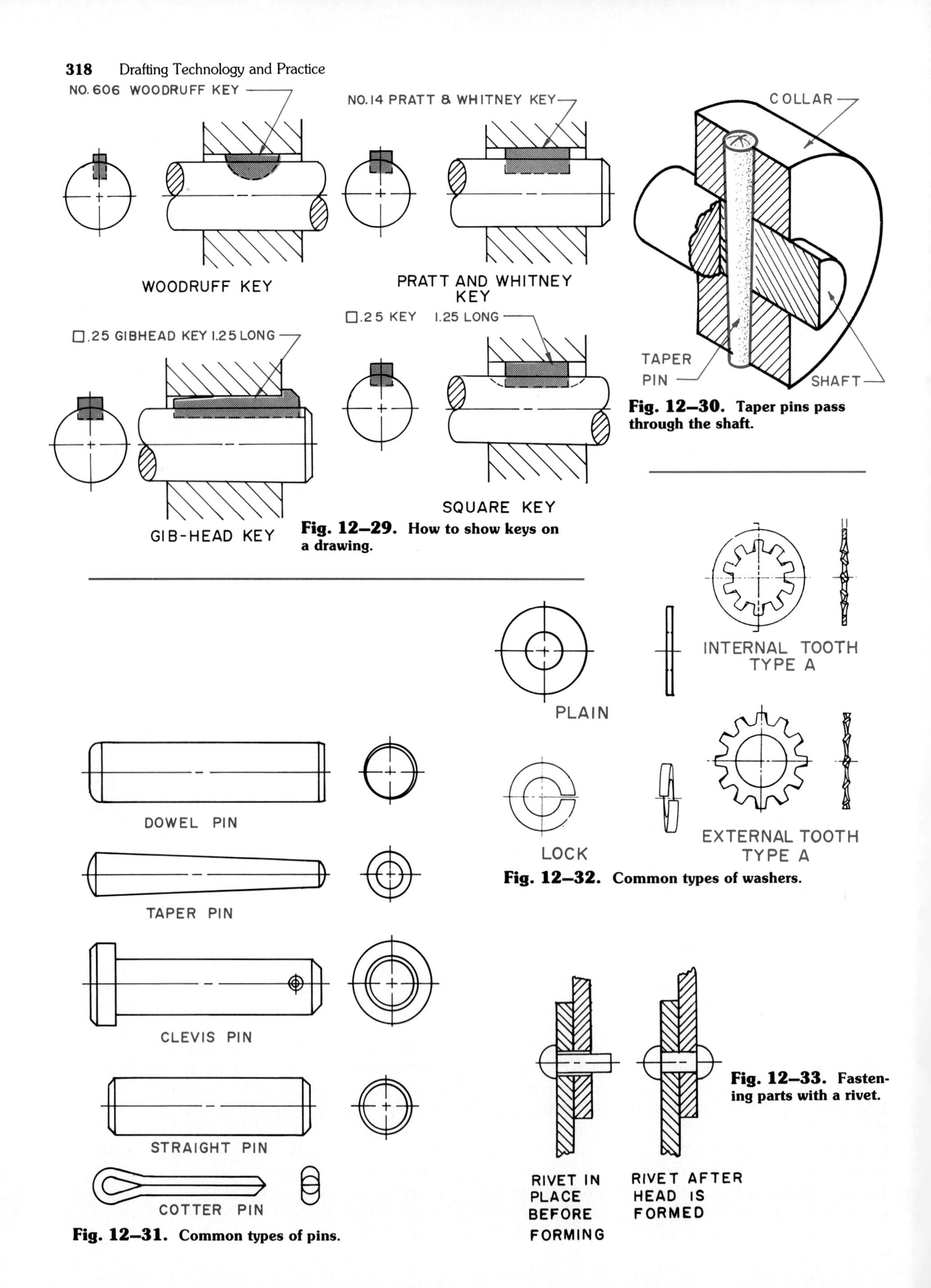

Fig. 12–29. How to show keys on a drawing.

Fig. 12–30. Taper pins pass through the shaft.

Fig. 12–32. Common types of washers.

Fig. 12–31. Common types of pins.

Fig. 12–33. Fastening parts with a rivet.

Plain washers are flat. They are used under the head of a bolt or a nut. They spread the load over a larger area than either the bolt head or nut. Plain washers are available in diameters from 3/16 to 3 inches. On a drawing specify plain washers by giving their inside diameter, outside diameter, and thickness. A typical note would read .406 × .812 × .065 TYPE A PLAIN WASHER.

Lock washers are split and bent in a helical shape. They provide a force due to the spring action that helps hold a part, such as a nut, in place. They also make it easier to unscrew bolted parts. They are available in regular, extra duty, and high-collar series. Their diameters range from 3/16 to 3 inches. Specify lock washers by giving the nominal size (hole diameter) and the series. A typical note would read 3/8″ REGULAR HELICAL SPRING LOCK WASHER.

Tooth lock washers are made in three types. These are external, internal, and internal-external. They are hardened. The hardened teeth are twisted so they can grip the bolt head or nut and keep it from vibrating loose. They are available in diameters from 1/4 to 1 inch. Specify tooth lock washers by giving their nominal size, description, and type. A typical note would read 1/4″, INTERNAL TOOTH, TYPE A.

Tables giving standard washer sizes for the three types are found in the Appendices 32 through 34.

Rivets

Rivets are permanent fasteners. They are used to fasten sheet metal parts and steel plate. They have a head on one end. A rivet is used by placing it through matching holes in the parts to be fastened. The straight end is then formed into another head. Fig. 12-33. Some of the commonly used rivets are shown in Fig. 12-34. Small rivets are made with pan, button, and countersunk heads. These are available with shank diameters from 1/16 inch (1.5 mm) to 7/16 inch (11 mm).

Large rivets are made with button, high button, pan, flat top, countersunk, and cone heads. These are available from 1/2 inch (12 mm) to 1 3/4 inches (43 mm). Rivet tables are found in Appendices 35 and 36.

Showing Rivets on Drawings

Rivets are shown on drawings by standard symbols. Fig. 12-35. They are divided into two types, field installed and shop installed. *Field installed* rivets are fastened in place on the job. The holes are prepared in pieces before they get to the job. *Shop installed* rivets are fastened in place in the plant making the steel structure. Then the assembled pieces are shipped to the job. Some typical riveted joints are shown in Fig. 12-36.

Draw rivets using the symbols shown in Fig. 12-35. Dimension rivets by locating the centers of each rivet hole. Show the size of the rivet in a note. Fig. 12-37. Specify the size of the rivet by giving the shank diameter, length, head type, and material. A customary rivet note would read ∅1/4 × 1/2 FLAT HEAD STEEL RIVET. A metric rivet would be noted ∅6 × 10 PAN HEAD STEEL RIVET.

Wood Screws

Wood screws are used to fasten two pieces of wood together or to secure some other material to wood. The hole in one part should be large enough for the screw to pass freely through it. The hole in the other part should have a diameter approximately the same as the core of the screw. The threads on the screw cut into the sides of the second hole forming their own internal threads. Fig. 12-38.

Screws are easy to install and remove. They can be removed and replaced several times before the threads in the wood are damaged.

Wood screws are made from steel, brass, and aluminum alloy. They have flat, round, and oval heads. The heads may be slotted or recessed. Fig. 12-39. ANSI sizes are given in Appendices 37 and 38.

Screw sizes are noted by giving length, nominal size (diameter of body), type of head, type of screw, and the material. A typical note would read 1 1/2 - NO. 8 FH WOOD SCREW - STEEL.

Springs

Springs are devices used to apply energy. They can either push or pull. Two types of springs are in common use, coil and flat.

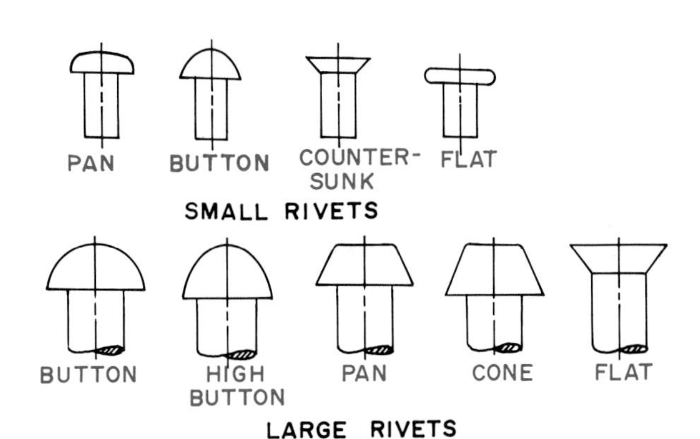

Fig. 12–34. **Common types of small and large rivets.**

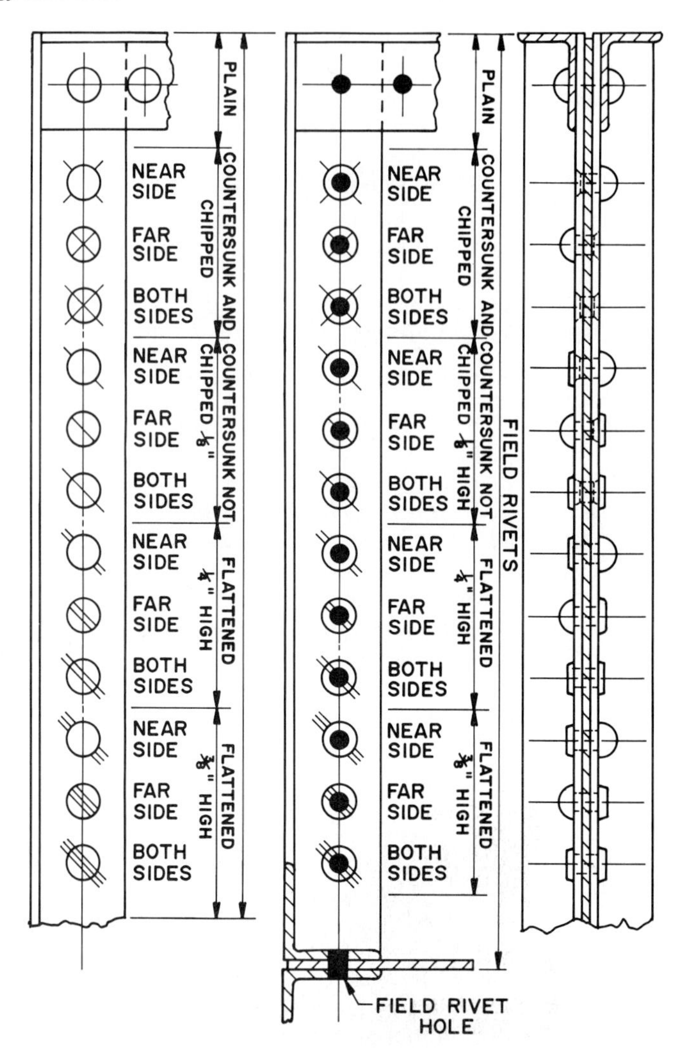

Fig. 12–35. Standard rivet symbols.

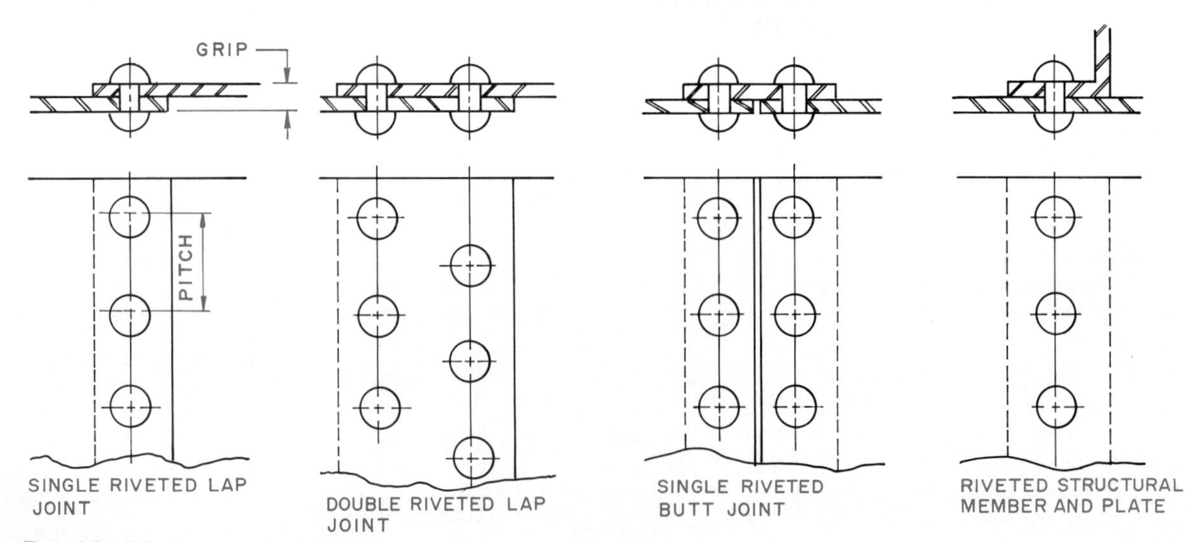

Fig. 12–36. Drawing riveted joints.

Coil springs are formed in the shape of a helix. The common types are compression, extension, and torsion. Fig 12-40. They are made with plain open ends, plain closed ends, ground open ends, and ground closed ends. Fig. 12-41.

Compression springs apply energy (push) when they are squeezed (compressed). In their natural state, the coils are not touching.

Extension coil springs apply energy when they are stretched (pulled). In their natural state, the coils are usually touching. They are commonly made with some type of hook or loop on each end. A screen door spring is an extension spring.

Torsion coil springs apply energy when one end is moved in a circular direction.

Flat springs are made from spring steel. They are commonly used to hold something in place. This kind of spring permits easy removal or movement of the part. A common use is in door catches on cabinets. Such a spring is shown in Fig. 12-42.

Flat springs are designed to serve a special purpose and are not standardized.

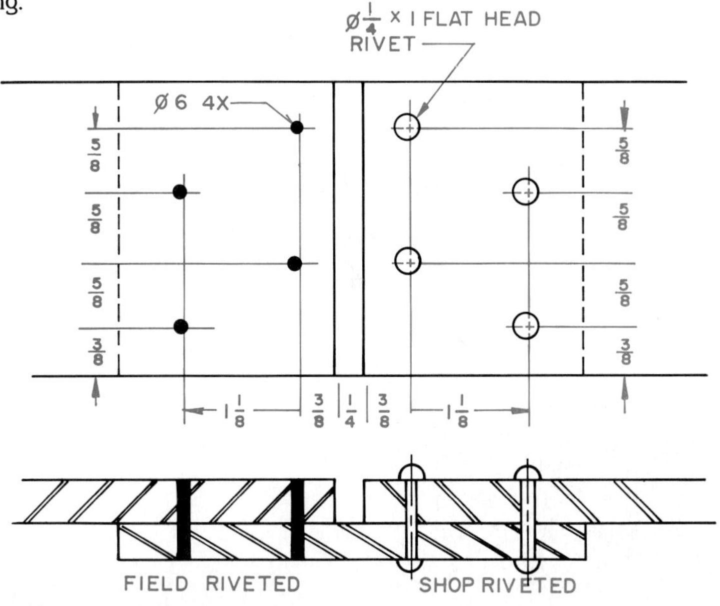

Fig. 12–37. How to draw and dimension shop and field riveted joints.

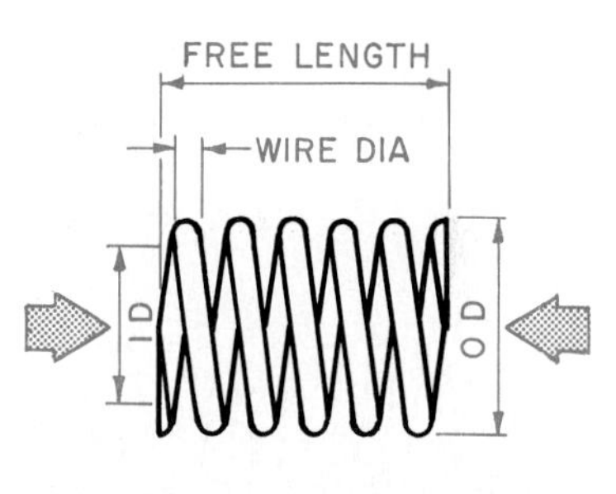

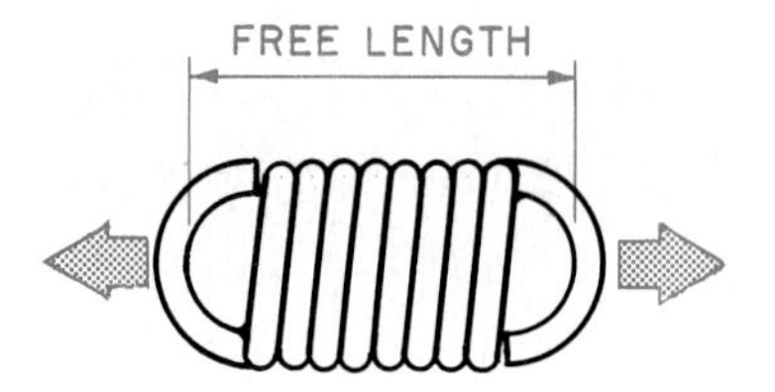

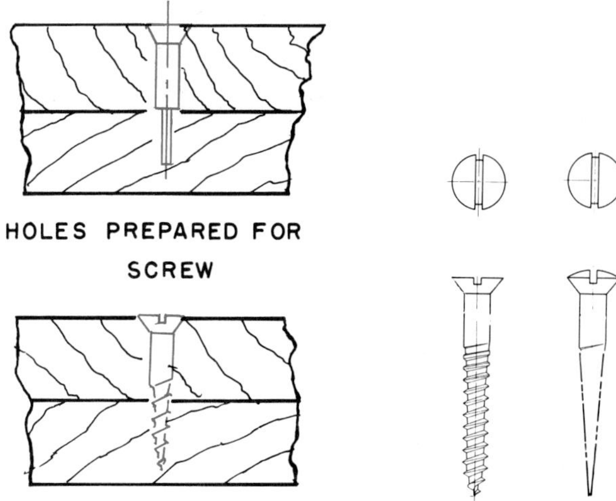

Fig. 12–38. Fastening with wood screws.

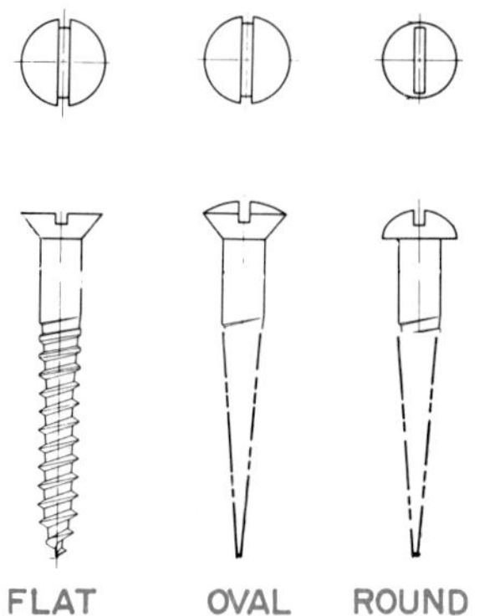

Fig. 12–39. Common types of wood screws.

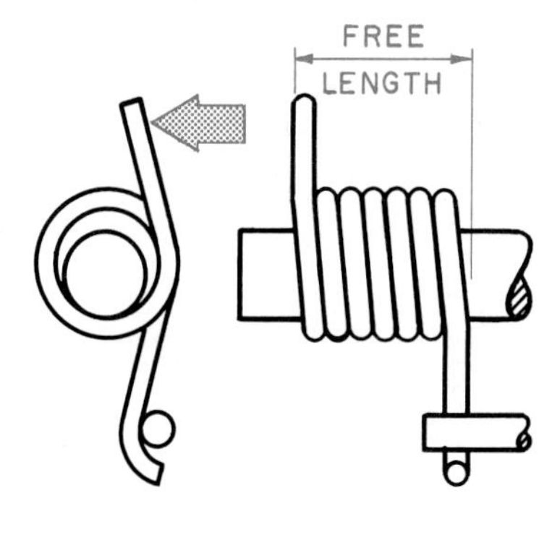

Fig. 12–40. Common types of coil springs.

Spring Terminology

It is important to know the terms used to indicate springs. *Free length* is the length of the spring when it is in its natural, unloaded condition. The free length of a compression spring includes the entire spring. Extension spring free length is measured inside the hooks. Refer to Fig. 12-40. The loaded length is the length under a known load.

Solid length is the length of a compression spring when all the coils are compressed so they touch. *Outside diameter* is the overall diameter of the outside of the coil. The *inside diameter* is measured inside the coil. *Wire size* is the diameter of the wire used to make the spring. A *coil* is one turn of the wire through 360 degrees.

Coil springs are available wound right-hand or left-hand. This is the same as a right-hand or left-hand thread.

How to Draw Coil Springs

You can show coil springs in detailed or schematic representation on drawings. The springs shown in Fig. 12-40 are drawn in detailed form.

Even though the spring wire forms a helix, draw the sides of the coils with straight lines. Doing so is easier and faster than drawing an absolutely accurate spring. To draw a detailed representation of a coil spring, see Fig. 12-43 and follow these steps:

1. Locate the center line of the coil, Fig. 12-43 at A.

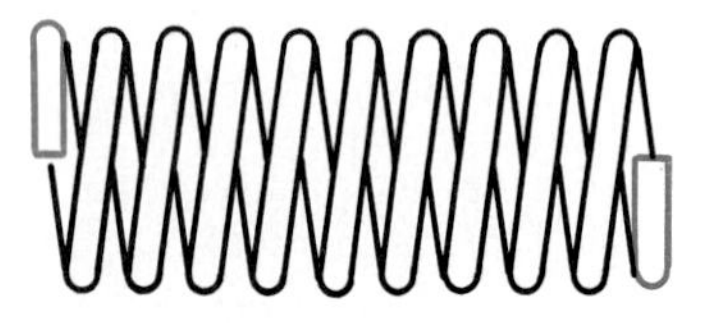

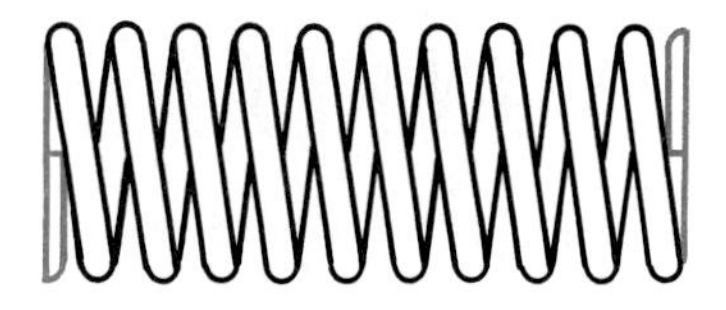

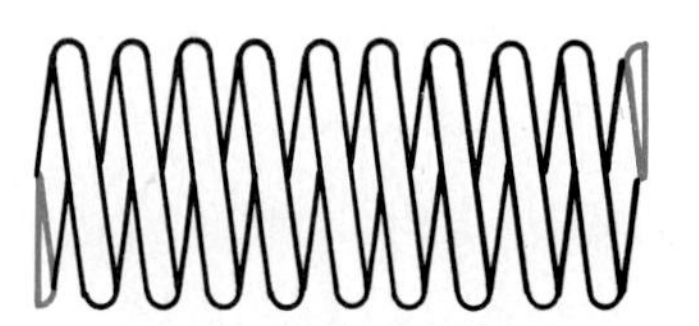

Fig. 12–41. Common ends for coil springs.

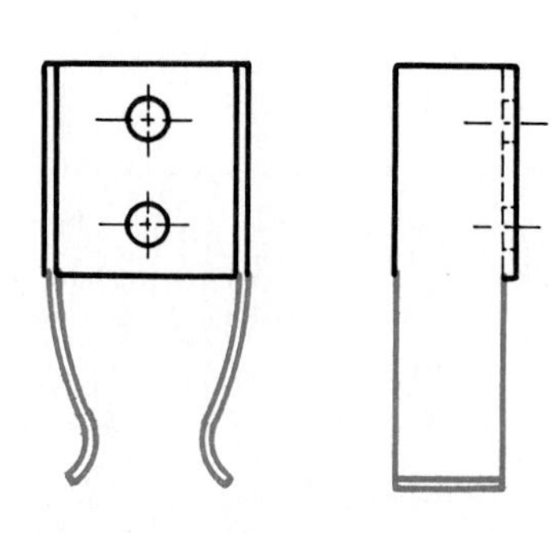

Fig. 12–42. A flat spring.

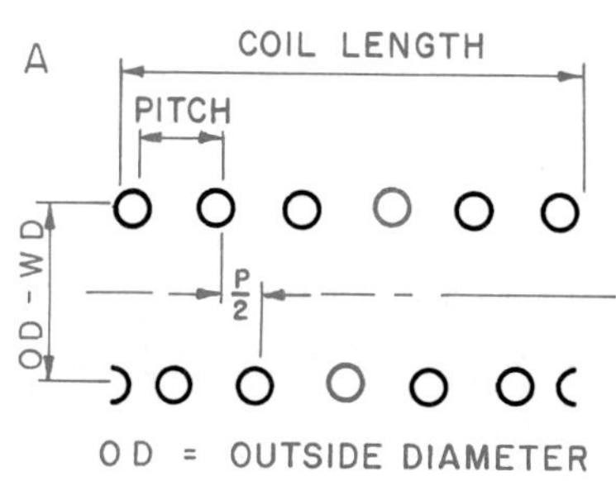

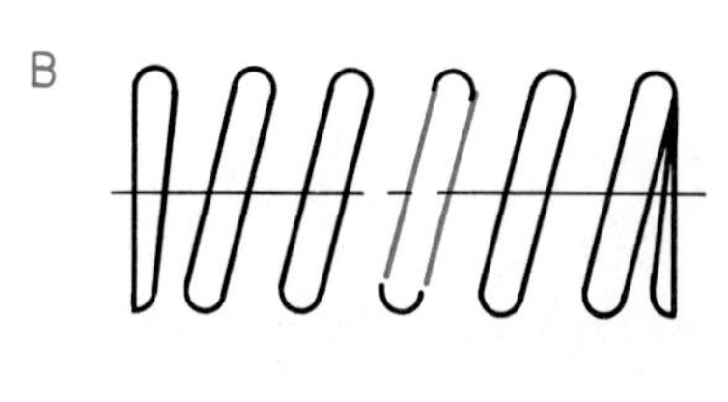

Fig. 12–43. How to draw detailed representation of coil springs.

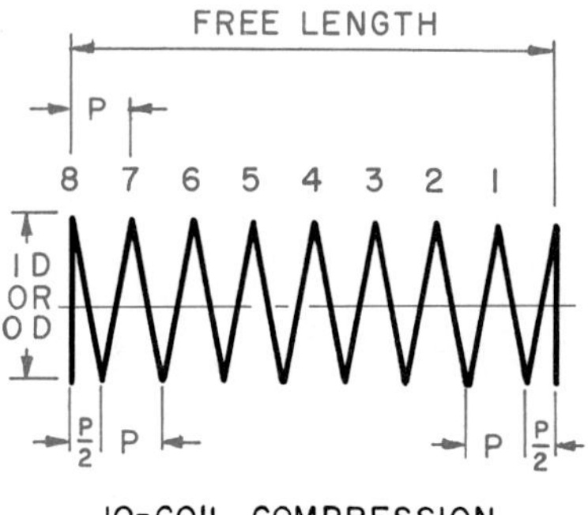

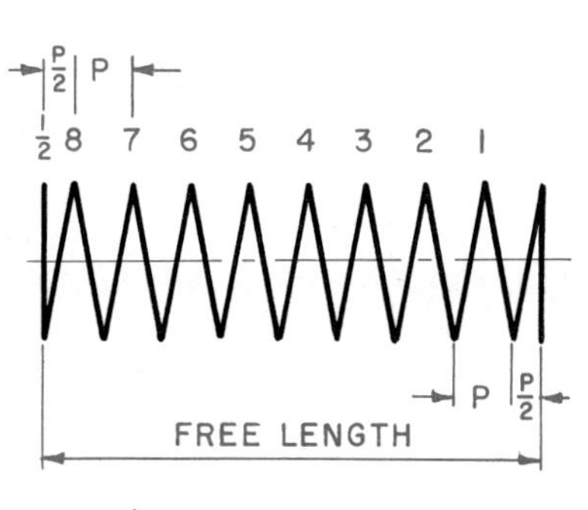

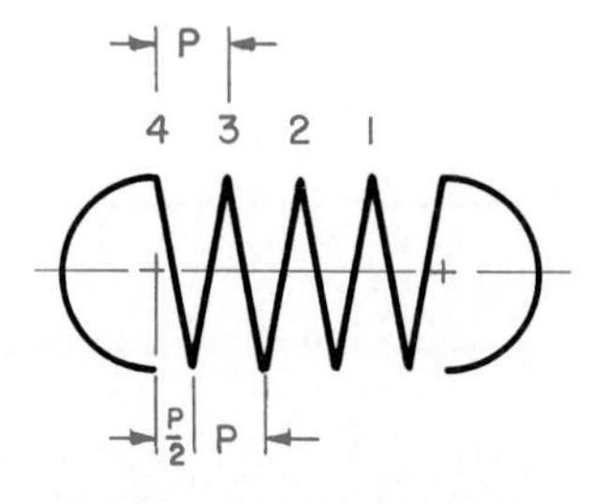

Fig. 12–44. Schematic representation of coil springs.

2. Draw the outside diameter minus the wire diameter (OD-WD).

3. Draw the coil length.

4. Lay out the pitch distance of the coils.

5. Draw light circles for wire diameters.

6. Connect wire diameter circles with straight lines. Fig. 12-43 at B. These are the coils on the front of the spring.

7. Connect the sides of the coils on the rear of the spring. Fig. 12-43 at C.

8. Remove construction lines and darken the lines.

Schematic coil representation is shown in Fig. 12-44. The coils are indicated by straight lines. The distance from one point to the next point represents the pitch, or one coil.

Notice that the ten-coil compression spring is drawn with eight points. Each end of the coil is closed or squared off. The ends do not serve as active coils. Two coils are lost this way. The eight active coils are indicated on the drawing.

In the extension spring, all coils are active. The number of points indicates the number of coils.

Knurling

A **knurled** surface is one with a series of grooves machined into it. One purpose of knurling is to provide a handgrip that can reduce slipping when handling a part. Another use is for fastening two parts together. A knurled shaft held in a smooth hole by a pressed fit has considerable holding power.

Two common knurl patterns are straight and diamond. You draw and dimension them as shown in Fig. 12-45. A simplified method of representing knurls is shown in Fig. 12-46. Here notes are used, and the actual knurl is not drawn.

Information needed for knurling for handgrip purposes is the pitch (distance between grooves), types of knurl, and length of the area to be knurled.

For parts knurled for fastening purposes, you need the tolerance diameter of the shaft before knurling, the pitch, type of knurl, and minimum diameter after knurling.

In the diamond knurl, note that the "96 DP" is the diametrical pitch. This is the ratio of the number of grooves on the circumference (N) to the length of the pitch diameter (D). DP is found by using the formula DP = N/D. Preferred diametrical pitches are 64DP, 96DP, 128DP, and 160DP.

The note P0.6 on the straight knurl indicates the knurling grooves are 0.6 mm apart.

Welding

Welding is one way that metal parts can be fastened together. Steel, aluminum, and magnesium can all be welded.

The two basic welding processes are fusion welding and resistance welding. *Fusion welding* involves melting a metal rod, called welding rod, and combining it with the metal parts to be joined. When the melted rod cools, the parts are permanently joined together. The rod can be melted by electricity or gas. Fig. 12-47.

Resistance welding is done by passing an electric current through the spot to be welded. This is done under pressure. The current heats the metal parts. The pressure plus heat welds them together. This is commonly done on sheet metal parts. Fig. 12-48.

Weld Joints

The commonly used weld joints are shown in Fig. 12-49. These include butt, corner, tee,

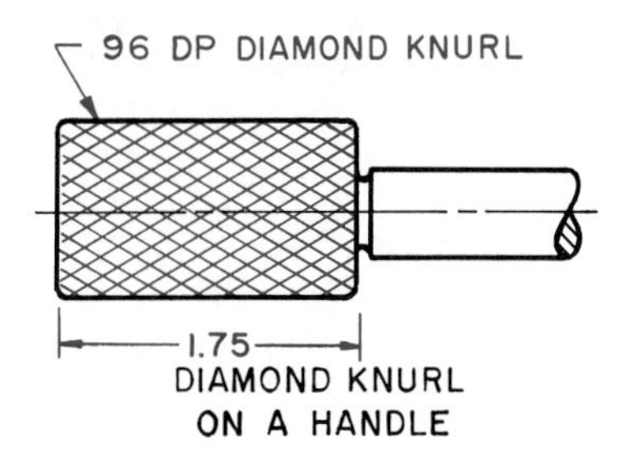

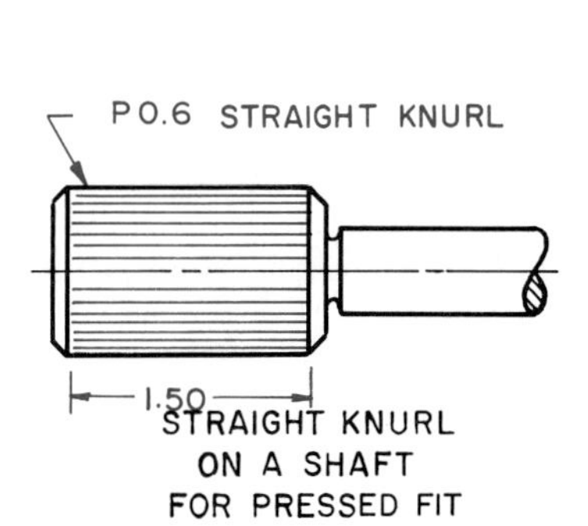

Fig. 12–45. **Common types of knurling.**

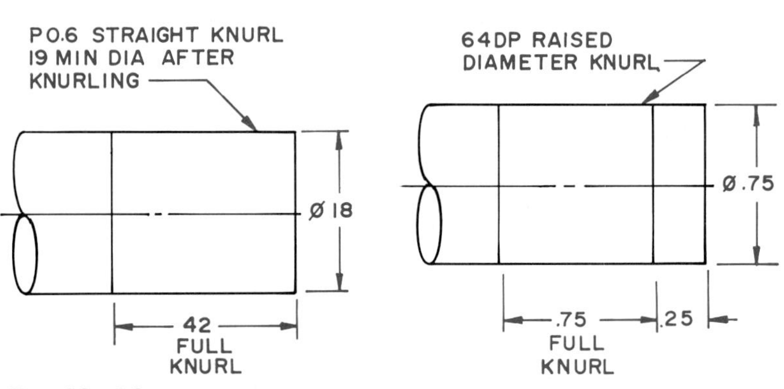

Fig. 12–46. **The actual knurl does not have to be drawn when the knurl is dimensioned.**

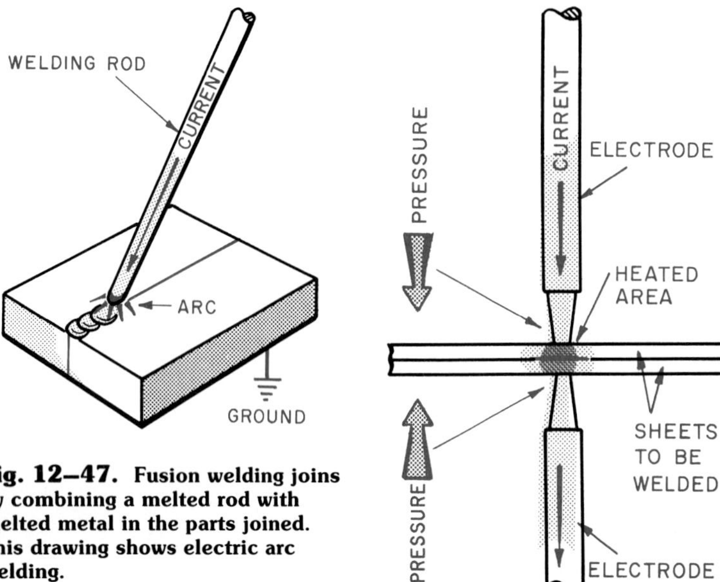

Fig. 12–47. **Fusion welding joins by combining a melted rod with melted metal in the parts joined. This drawing shows electric arc welding.**

Fig. 12–48. **Resistance welding joins by heat and pressure.**

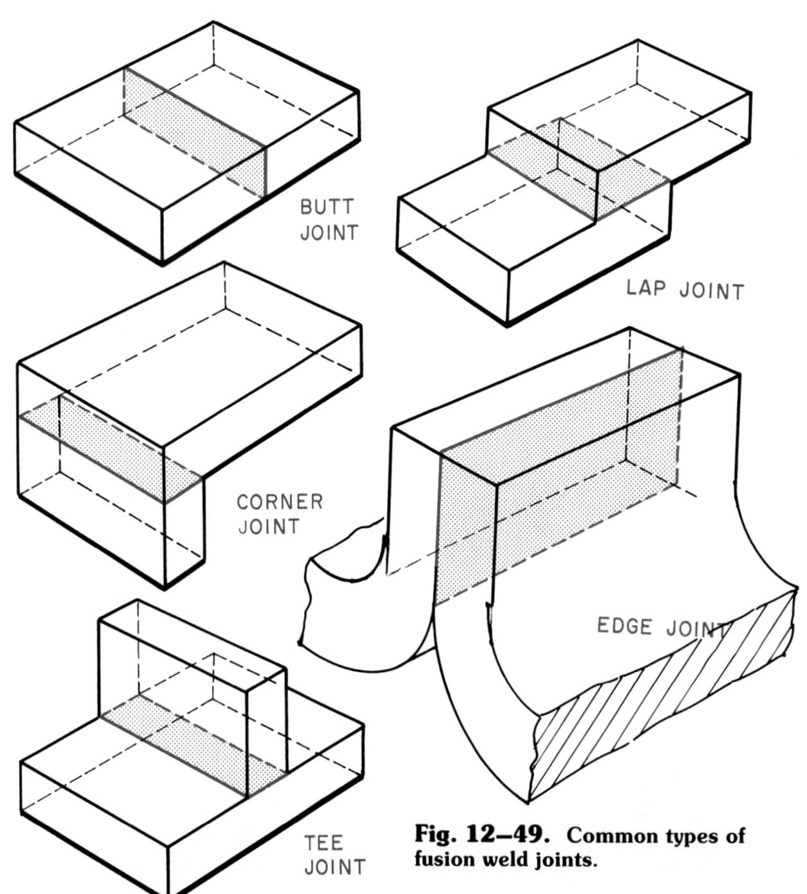

Fig. 12–49. **Common types of fusion weld joints.**

lap, and edge joints. The butt and corner joints can be joined with several types of welds such as the square groove, V-groove, bevel groove, U-groove, and J-groove. The corner joint can also be joined with a fillet weld. Fig. 12-50.

The tee joint can be joined with the bevel groove, J-groove, and fillet welds.

The lap joint can be joined with a fillet, bevel groove, J-groove, slot, spot, projection, or seam weld. The edge joint can be joined with the same welds listed for the lap joint plus the square groove, V-groove, and U-groove.

Welding Symbols

It is important for each weld on a drawing to be fully specified. This is done by using welding symbols. Fig. 12-51. Place a welding symbol on the drawing and connect it to the place where the weld is to occur.

Usually not all of the information shown on the welding symbol in Fig 12-51 is needed. Record whatever parts are needed to clearly specify the weld intended.

Draw the welding symbol as shown in Fig. 12-52. Draw the tail at a 90-degree angle and about ½ inch (12 mm) wide. Place ⅛-inch (3 mm) high lettering about ⅛ inch (3 mm) from the reference line forming the symbol. The symbols indicating the type of weld vary from ⅛ inch (3 mm) to ½ inch (12 mm). Fig. 12-53.

As you study the welding symbol in Fig. 12-51 notice that each bit of information must be located in a special place.

When you place the basic symbol giving the type of weld *above the reference line,* the weld is made on the side opposite the arrowhead. When *below* the reference line, the weld is on the same side as the arrowhead. When on *both sides* of the reference line, the weld is on both sides of the joint.

Always place the basic weld symbols on the reference line with the vertical leg on the left as you view the symbol. Fig. 12-54.

Usually material over ⅛-inch thick requires making a groove for welding. The arrowhead points toward the surface that should have the groove. Fig. 12-55.

Field Weld Symbol. A flag at the end of the reference line means the weld is to be a field weld. This means the welding occurs on the job where the device is assembled rather than in the shop. Fig. 12-56.

Weld All Around Symbol. A circle at the end of the reference line means the joint is welded on all sides. Fig. 12-56.

Symbol Tail. If a weld has a special abbreviation, place the abbreviation in the tail. If no specifications are needed, the tail is not used.

Welding symbols are developed by the American Welding Society (AWS). Complete details are available in their publication *American Welding Society Standard Welding Symbols.* Additional details are found in Appendix 39.

Fusion Welding. The basic symbols indicating the types of welds used in fusion welding are shown in Fig. 12-57 (p. 328). These also include the supplementary symbols that are placed on the welding symbol and give additional information to the welder. Pictorial examples of these welds are shown in Fig. 12-50.

Indicating Weld Sizes. Indicate the size of a weld on the welding symbol. Give the size, strength, groove angle, length, pitch, and number of spot and projection

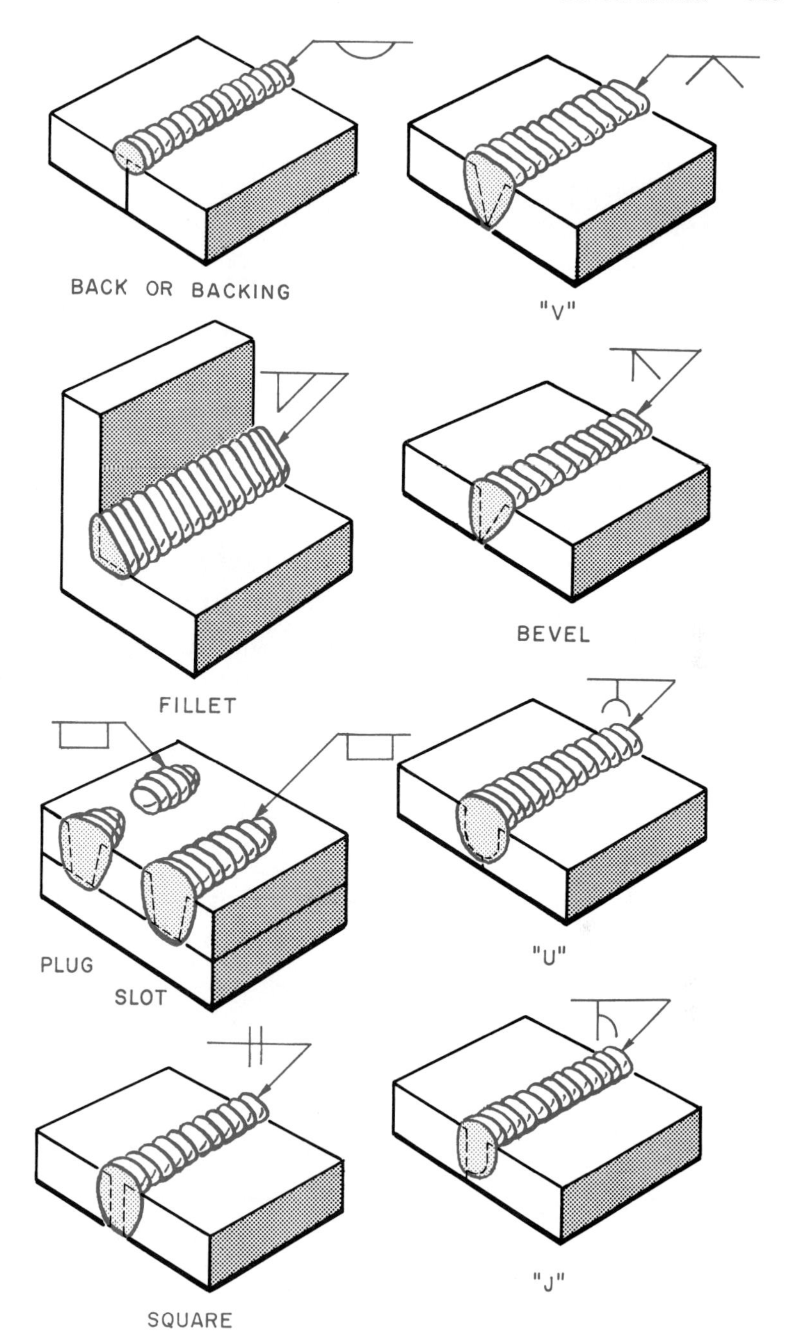

Fig. 12–50. Common arc and gas welds.

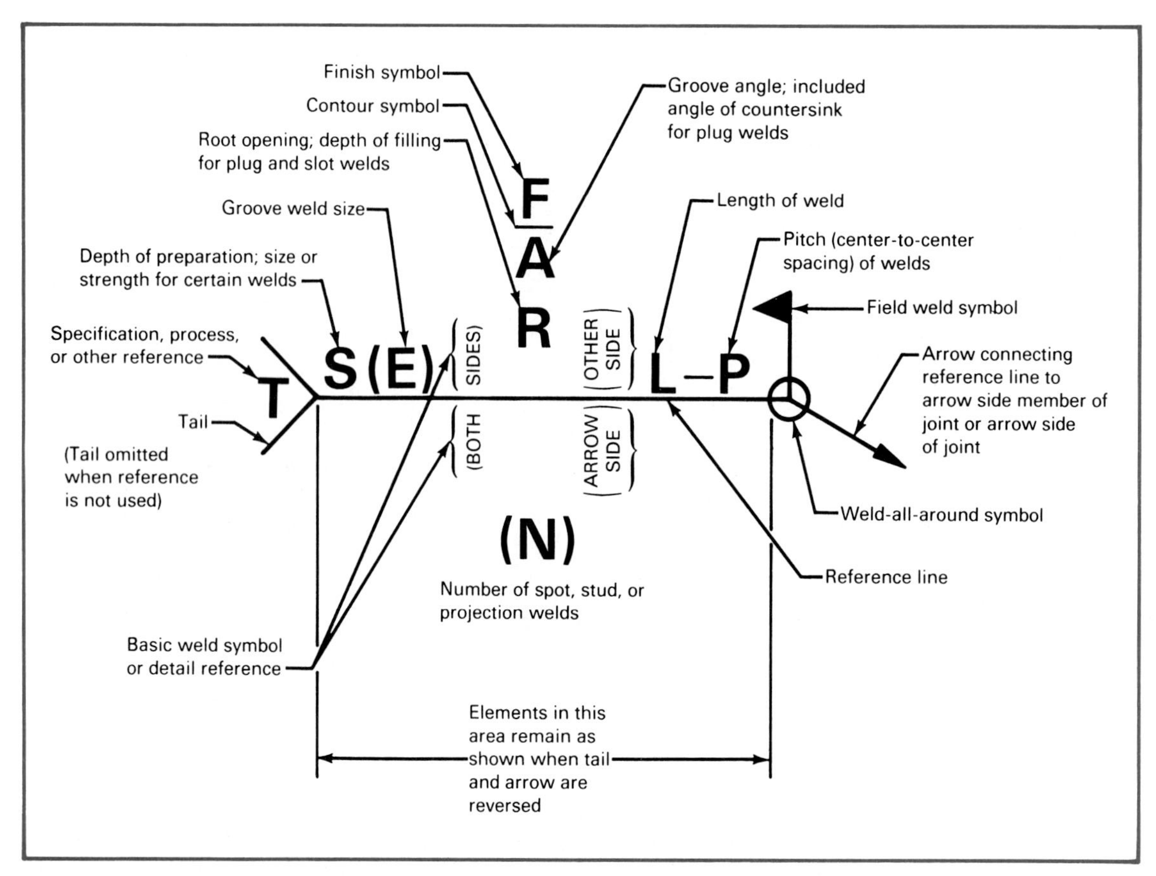

Fig. 12–51. Standard location of elements of a welding symbol. (American Welding Society)

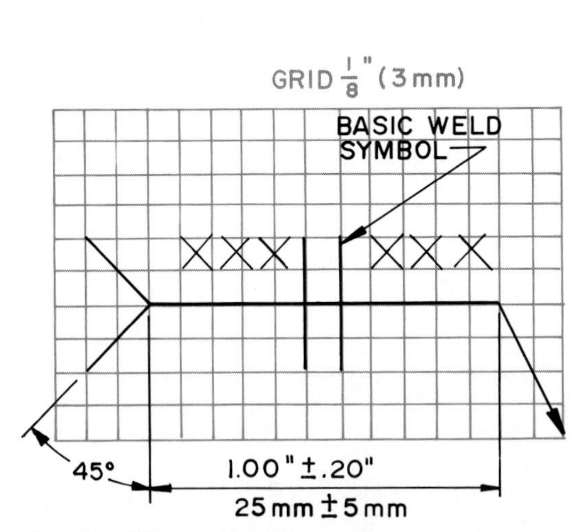

Fig. 12–52. Drawing the welding symbol to size.

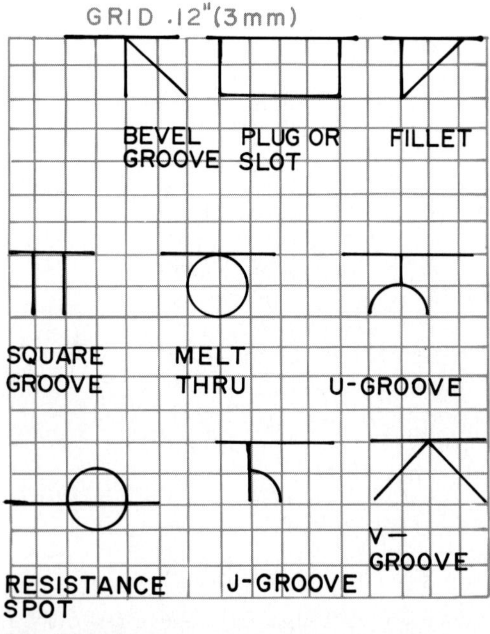

Fig. 12–53. Approximate sizes for drawing welding symbols.

welds as needed. Refer to Fig. 12-51 for the various locations on the welding symbol to place these bits of information.

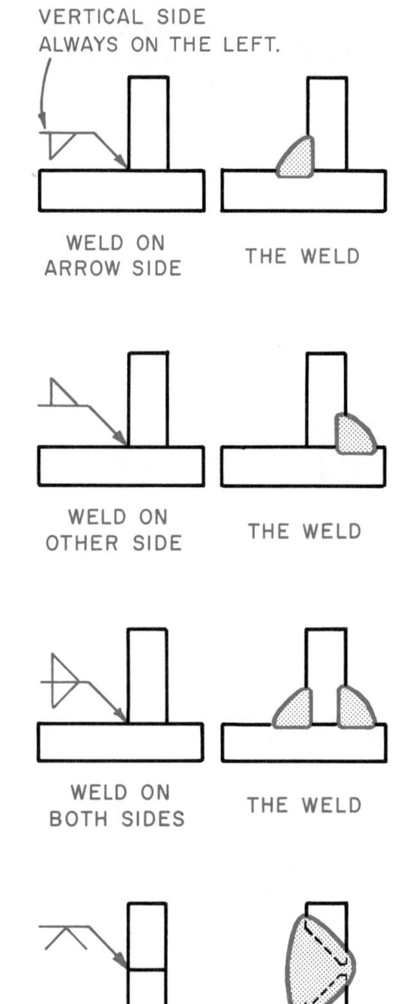

Fig. 12–54. The position of the basic weld symbol indicates whether the weld should be on the same side as the arrowhead or on the opposite side.

Fillet Welds. The location of the size of two fillet welds with both legs the same length is shown in Fig. 12-58 at A.

Fillet welds that are the same size on both sides can be shown as in Fig. 12-58 at B. If the welds are different sizes, they are shown as in Fig. 12-58 at C. If the length of the weld is the same as the length of the part being welded, no length dimension is given.

If the legs of a fillet weld are not the same size, they are dimensioned as shown in Fig. 12-58 at D.

Groove Welds. The proper way to indicate the size of groove welds is shown in Fig. 12-59. The depth of the groove is given to the left of the weld symbol. If the groove is from both sides and is the same size on both sides, only one side need be dimensioned. If the welds differ, both must be dimensioned.

Groove angles are given outside the opening of the weld symbol.

The root opening is given inside the weld symbol. A root opening is the space between the two members being welded. Fig. 12-59.

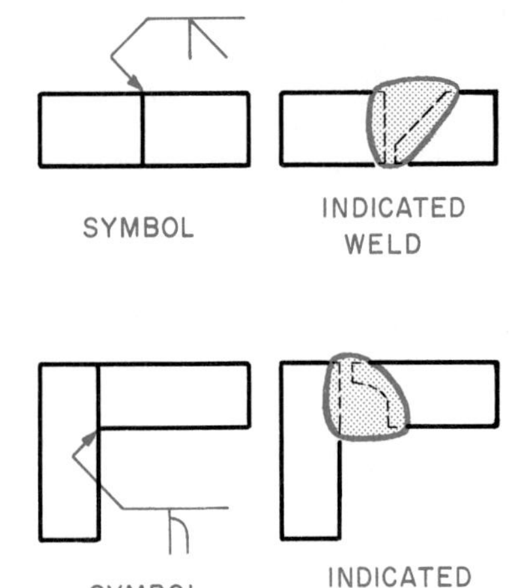

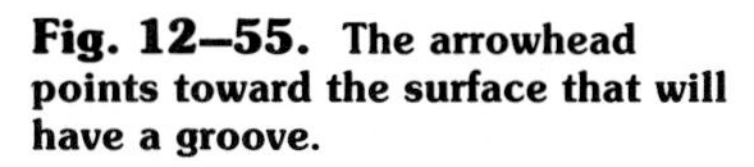

Fig. 12–55. The arrowhead points toward the surface that will have a groove.

If all grooves have the same angle, they need not be dimensioned. A note such as ALL V-GROOVE WELDS 60° GROOVE ANGLE UNLESS OTHERWISE SPECIFIED can be used to give this information.

Resistance Welding. Two types of resistance welding are spot and seam. The AWS resistance symbols are shown in Fig. 12-60 (p. 330).

A spot weld is the fusion of parts in which the area of joining is a small, circular shape. The spot-weld symbol is shown in Fig. 12-61 (p. 330). The size of the spot weld is the diameter. It can be specified in decimal or fractional parts of an inch, or decimal parts of a millimeter. It is always shown to the left of the weld symbol. If it is necessary to show the shear strength in pounds per square inch or kilograms per square meter instead of the weld diameter, this is shown at the left of the weld symbol. The pitch is given in inches or millimeters at the right of the weld symbols. The number of welds (spots) on a joint is indicated above or below the weld symbol. The number is enclosed in parentheses.

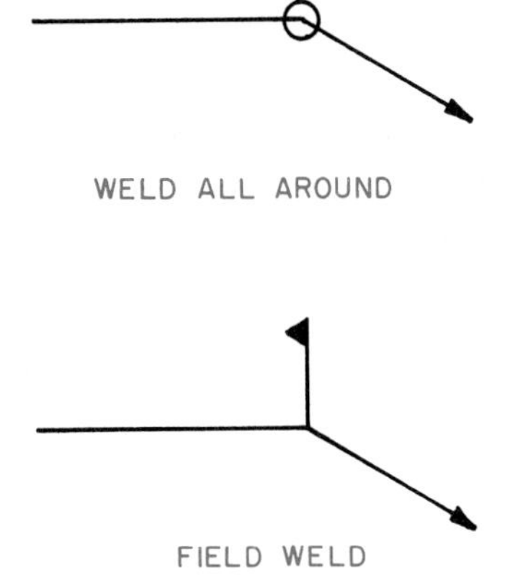

Fig. 12–56. Location of the field weld and weld all around symbols.

When the exposed surface of a resistance spot weld is to be flush, the surface is indicated by adding the flush-contour weld symbol. If the bar is placed below the symbol, the flush surface is on the arrow side. If above the symbol, it is on the other side.

A seam weld is one in which the resistance weld is continuous the entire length. It does not leave unwelded spaces between welds as spot welding does.

The placement of data on the seam welding symbol is shown in Fig. 12-62. The symbols are much like those used for spot welding. The weld symbol is centered on the reference line because there is no arrow side or other side.

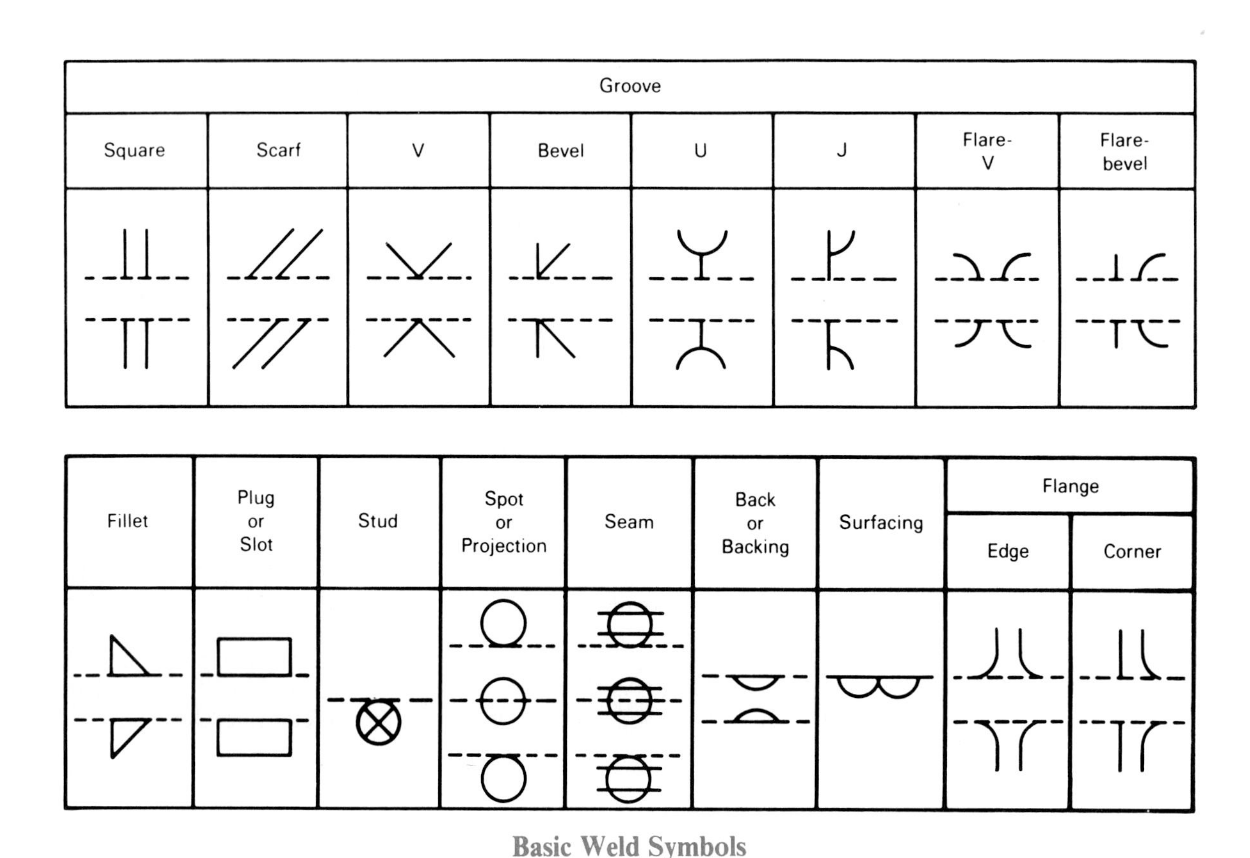

Fig. 12–57. Fusion welding symbols. (AWS A2.4-86, American Welding Society)

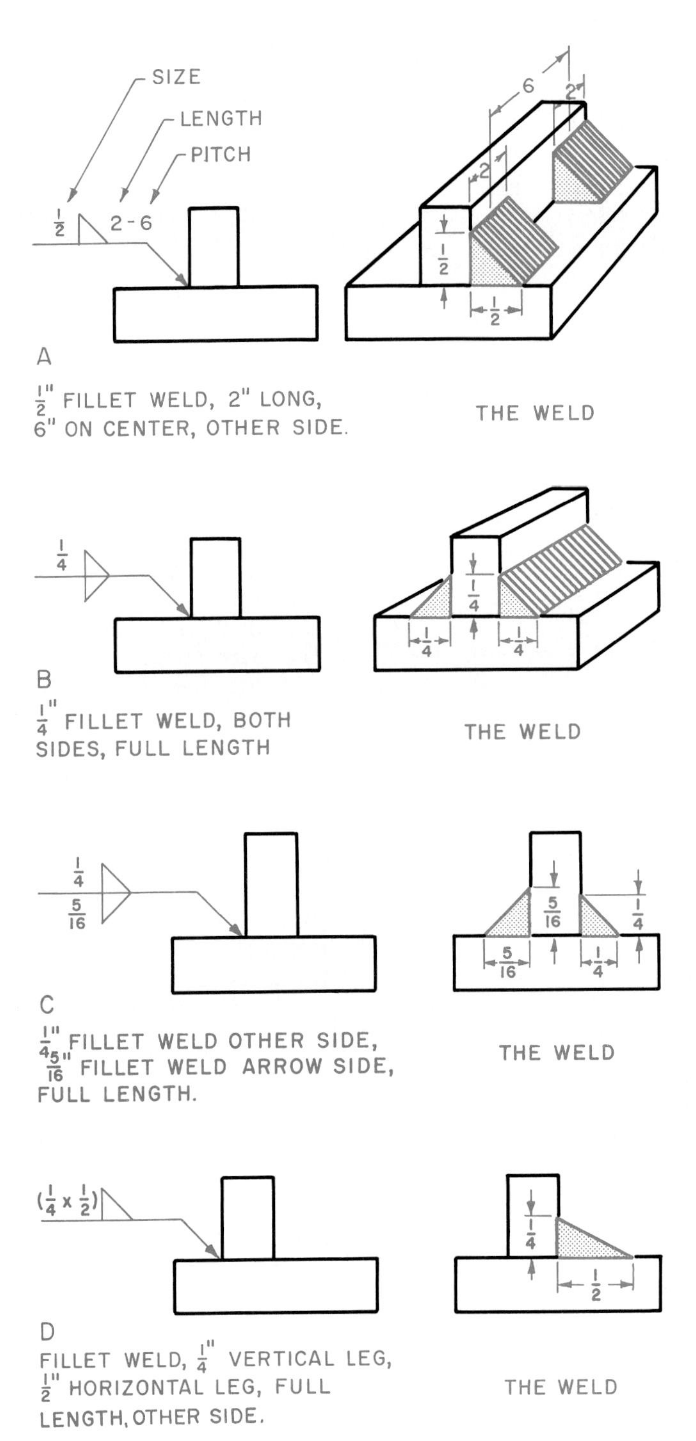

Fig. 12–58. **Dimensioning fillet welds.**

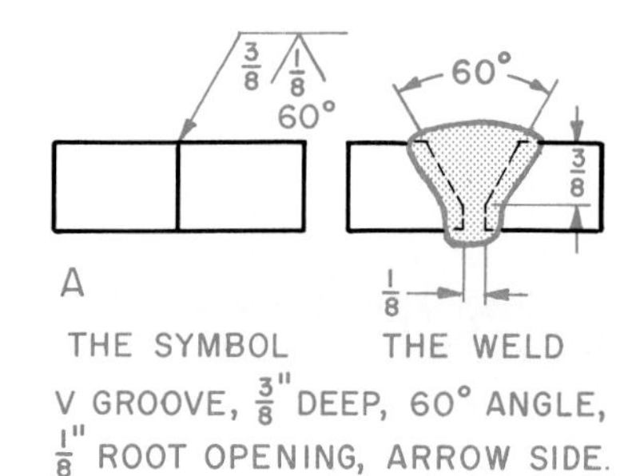

THE SYMBOL THE WELD

V GROOVE, 3/8" DEEP, 60° ANGLE, 1/8" ROOT OPENING, ARROW SIDE.

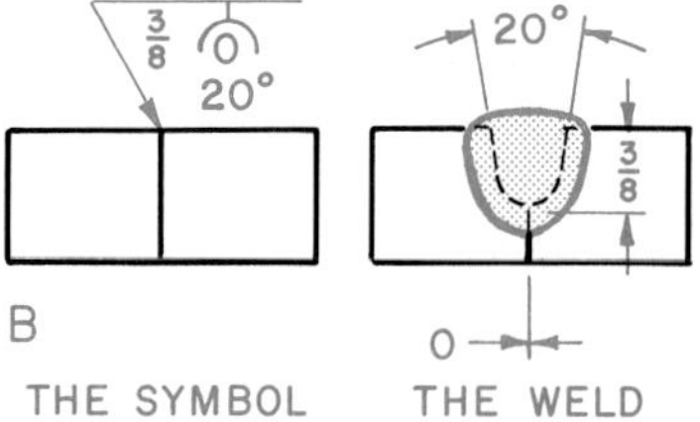

B

0

THE SYMBOL THE WELD

U GROOVE, ARROW SIDE 3/8" DEEP, 20° ANGLE, 0" ROOT OPENING.

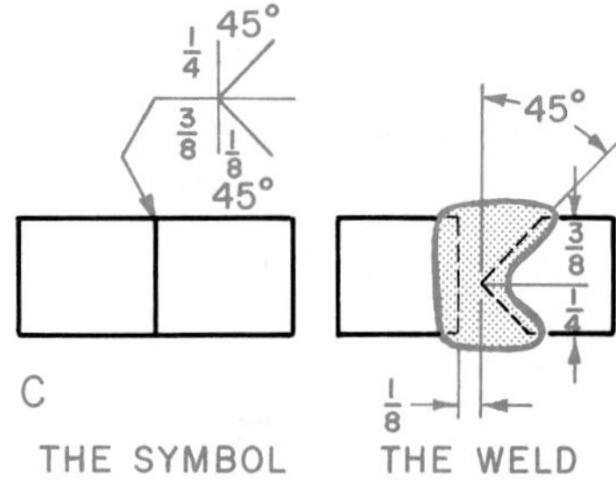

THE SYMBOL THE WELD

BEVEL GROOVE, ARROW SIDE, 3/8" DEEP, 45° ANGLE; OTHER SIDE, 1/4" DEEP, 45° ANGLE, 1/8" ROOT OPENING, ARROW POINTS TO PIECE TO HAVE THE BEVEL.

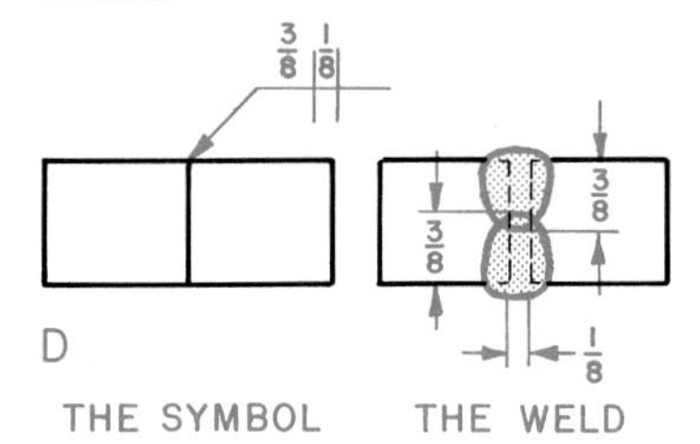

D

THE SYMBOL THE WELD

SQUARE GROOVE, 3/8" PENETRATION FROM EACH SIDE, 1/8" ROOT OPENING.

Fig. 12–59. **Dimensioning groove welds.**

TYPE OF WELD				SUPPLEMENTARY SYMBOLS			
RESISTANCE-SPOT	PROJECTION	RESISTANCE-SEAM	FLASH OR UPSET	WELD ALL AROUND	FIELD WELD	CONTOUR	
						FLUSH	CONVEX
RSW	RPW	RSEW	FW				

Fig. 12–60. Basic resistance weld symbols.

Projection Welds. In projection welding, one member has a dimple stamped into it, forming a pointed projection. This is placed in contact with the member to be joined to it. Electric current is passed through the projection. This plus pressure welds the two parts together. Fig. 12-63.

The projection welding symbol is shown in Fig. 12-64. Notice the location of the data on the symbol. These are much like those for spot welding.

If the projection weld symbol is below the reference line, the dimple is made in the arrow side part. If it is above the reference line, it is on the other side.

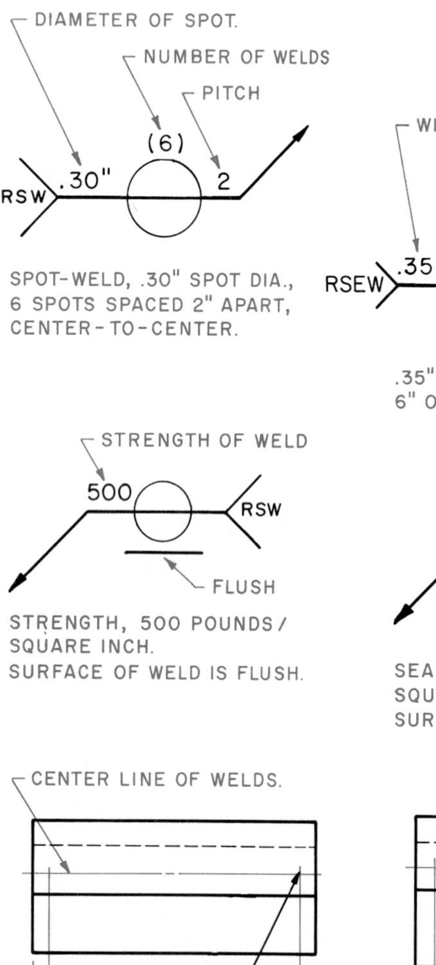

Fig. 12–61 Dimensioning spot welds.

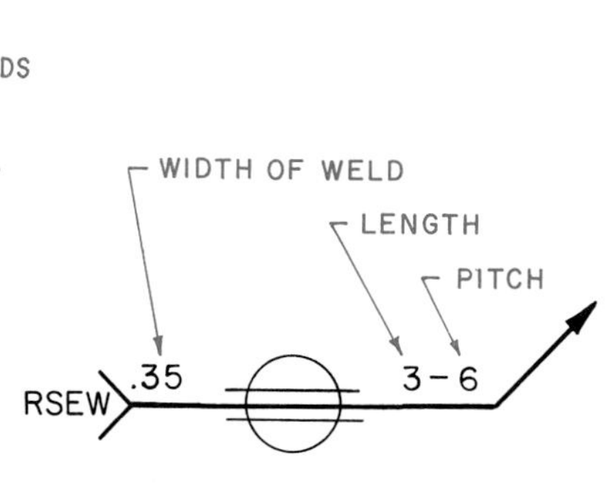

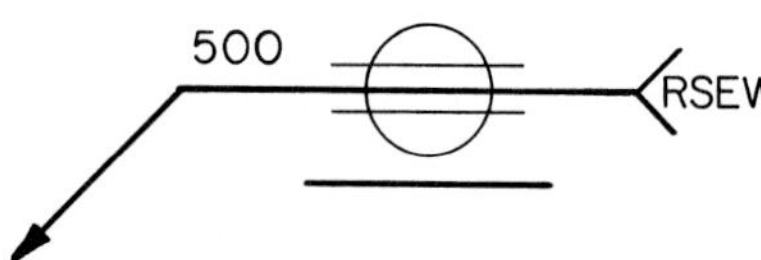

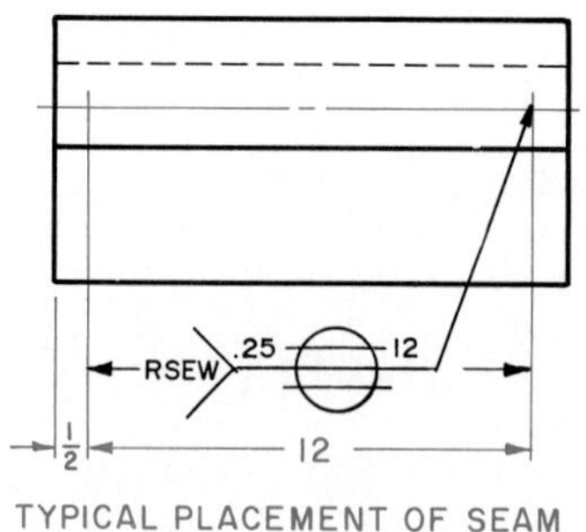

Fig. 12–62. Dimensioning seam welds

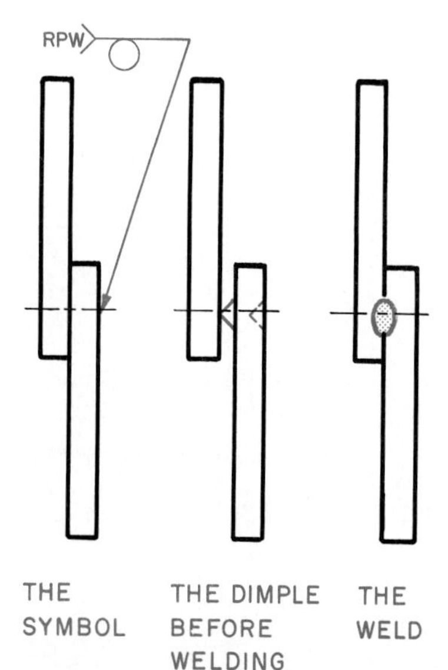

Fig. 12–63. A projection weld.

Flash and Upset Butt Welds. Flash butt welds are made by passing an electric current through two members that are spaced so they do not quite touch. An electric arc is formed between the members, melting their edges. They are then forced together under pressure. Upon cooling, they are welded. Fig. 12-65.

Upset butt welds are made by passing an electric current through two members that are in firm contact with each other. Heat for the weld comes from passing the current through the parts. This heat plus pressure welds the pieces together. Fig. 12-65.

There is no arrow side or other side in flash and upset butt welds. The symbols used are shown in Fig. 12-66. The weld symbol is placed in the center of the reference line. The tail contains information about the process. FW is the flash weld symbol. UW the upset weld symbol. No dimensions are necessary on the welding symbol.

Flash and upset welds can be made flush to the surface of the part. The material forming a bulge at the weld can be removed by machining, rolling, grinding, chipping, or hammering. The first letter of each of these processes is used on the symbol to indicate how the created bulge is to be removed. For example, M means machine the bulge away.

If the surface is to be flush, the flush symbol with the process to be used is given. If the surface is to be convex, the convex symbol and process are given.

DIAMETER
NUMBER OF WELDS.
PITCH
.30
4
(4)
RPW
PROJECTION WELD, .30" DIA., 4 WELDS, 4" ON CENTER, DIMPLE ON ARROW SIDE OF PART.

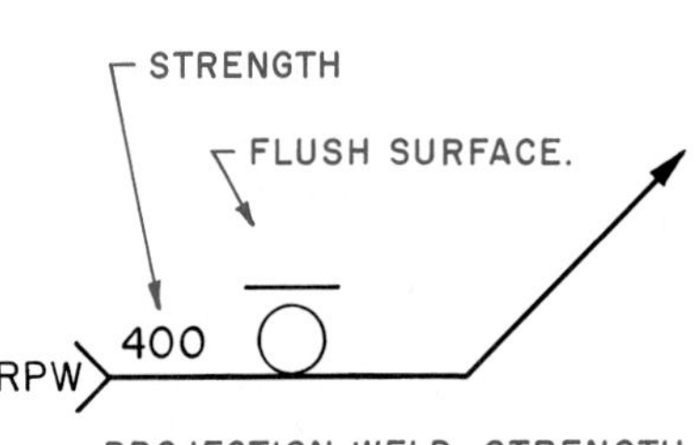

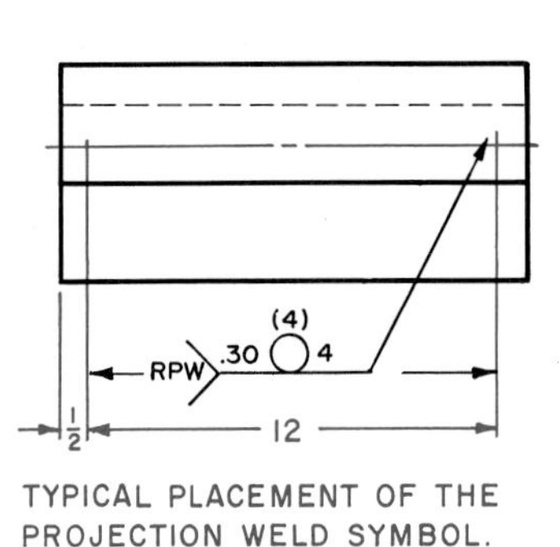

Fig. 12–64. **Projection weld symbols**

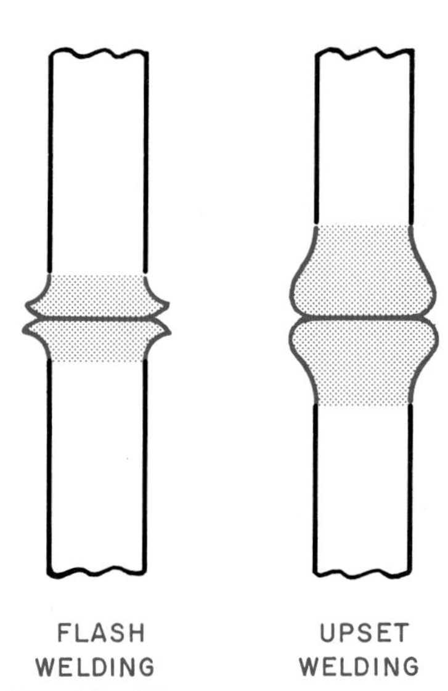

Fig. 12–65. **Bulges caused by flash and upset welding processes.**

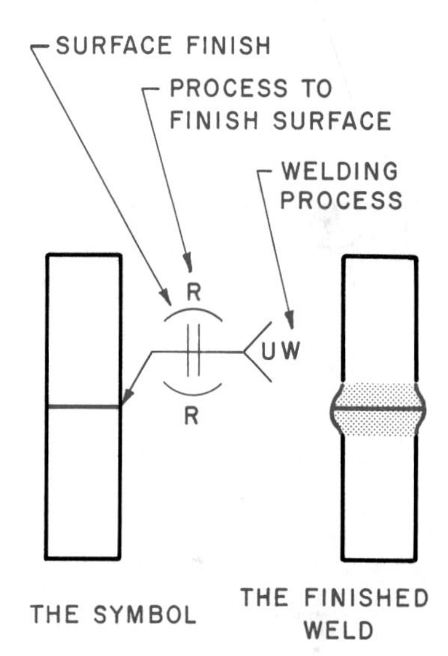

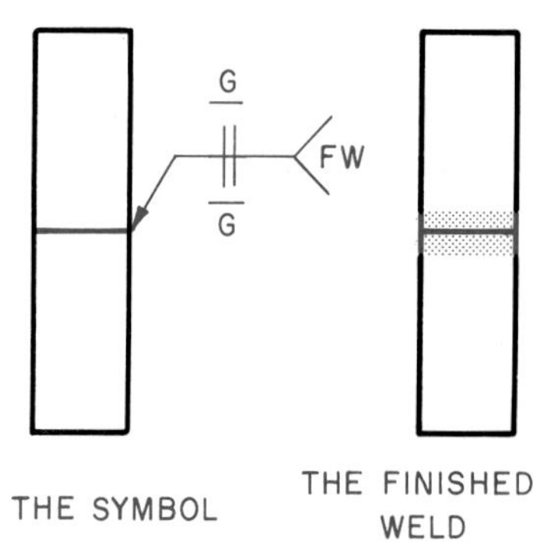

Fig. 12–66. **Typical flash and upset welds and symbols.**

Chapter Review

Build Your Vocabulary

You should understand and use the following terms as part of your working vocabulary. Write a brief explanation of what each means.

external thread
internal thread
class of fit
detailed representation
schematic representation
simplified representation
bolt
nut
stud
tapping screw
cap screw
machine screw
set screw
key
keyseat
pin
washer
rivet
wood screw
spring
knurling

Study Problems—Directions

Make working drawings of the items listed below. Make the joints occurring on these objects as features to be welded together rather than the cast joints as shown. Select an appropriate type of weld and indicate this on the drawing.

Use the following study problem drawings from Chapter 6:

1. **Fig. P6-13.** Angle plate.
2. **Fig. P6-15.** Hydraulic cylinder mounting bracket.
3. **Fig. P6-22.** Miniature Christmas tree stand. Draw to a scale of 1:2
4. **Fig. P6-30.** Angle bracket.
5. **Fig. P6-38.** Universal right-angle iron.
6. **Fig. P6-42.** Yoke end.
7. **Fig. P6-45.** Roadway manhole casing and cover.
8. **Fig. P6-53.** Flanged turret mount.
9. **Fig. P6-55.** Rocker arm. Draw to a scale of 1:4

CHAPTER 13

Drawing for Production: Detail and Assembly Drawings

After studying this chapter, you should be able to do the following:

- Make a detail drawing.
- Make an assembly drawing.

Introduction

When designing a new product, the first drawings made are usually freehand sketches. An engineer may make a large number of sketches before a final design is accepted. Usually the *final freehand sketch* is accompanied by other data, such as load or stress calculations and material specifications. From this information a drafter draws a **preliminary design assembly**. This drawing gives a more accurate picture of the object and aids in working out design details. The design assembly is an instrument drawing. It often is drawn full size. Generally, only major dimensions are given while notes give details about material, finishes, desired clearances, and other needed data.

The drafter uses the accurate preliminary design assembly as the source of information to develop *detail drawings* for each part. When size dimensions are needed, they are taken from the preliminary design assembly. Each part should be drawn and dimensioned so that it will join properly with the other pieces of the device. It is necessary to select the exact sizes of the mating parts and set the tolerances needed so that the assembled device will operate properly. Each part should have sufficient clearance so the device can be assembled and will work. Fig. 13-1.

It is the job of the drafter to draw each part so that it can be manufactured easily. The drafter must understand the shop processes used in manufacturing and how materials are worked.

The detailer must know what standard parts are available. These are parts, such as fasteners, washers, keys, and bearings, that are manufactured in large quantities for general use. A standard part should be used whenever possible. Standard parts are

Fig. 13–1. This wheel loader required many detail and assembly drawings for its production. (Caterpillar Inc.)

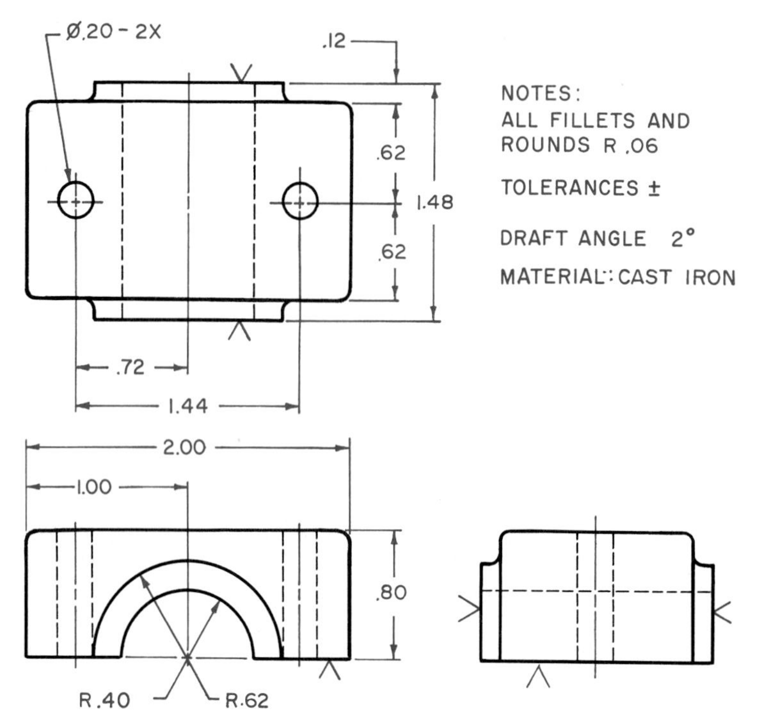

Fig. 13–2. A detail drawing of the bearing cap from the bearing journal in Fig. 13–41. This drawing includes information needed by the patternmaker and the machinist. It is called a casting detail drawing.

Fig. 13–3. This wooden pattern is an exact duplicate of the part to be cast in metal.

usually purchased from another company. They are shown in the parts list for the device but are not detailed. Standard fastening devices are shown in Chapter 12, "Fasteners." Others may be found in the catalogs of the companies that manufacture these items.

Detail Drawings

A **detail drawing** is a complete description of a single part of an object. The drawing may include auxiliary views, sections, or any other descriptive forms necessary. It includes dimensions, tolerances, materials, weight, finish, scale of the drawing, and other information needed to describe the part completely. Fig. 13-2 illustrates a typical detail drawing.

The exact practice to follow when detailing varies with the type of industry and the shop practices used by the company. In general, each detail is placed on a separate sheet. Sometimes a small device with only a few parts has all its detail drawings on one sheet.

In some industries a separate drawing is made for each manufacturing process required to produce a part. This is especially true if the part is large and complicated. For example, if a part is to be cast and then machined, a separate *pattern detail drawing* can be made from which the part will be cast. Then a separate *machining detail drawing* of the finished part will be prepared for use in the machine shop.

Typical types of detail drawings are pattern, machining, casting, forging, welding, and stamping.

Pattern Detail Drawings

A **pattern detail** gives the information needed to make a pattern. A *pattern* is a duplicate of the part to be cast but made from wood, metal, or plastic. Fig. 13-3. Three factors—**draft,**

shrinkage, and **machining**—influence how a pattern detail should be drawn. *Draft* means to give the part a slight taper to help remove the pattern from the sand mold. *Shrinkage* refers to making the pattern slightly larger than the desired size of the finished casting. This extra size is necessary because the metal part shrinks when it cools.

Various metals that are often used in cast parts shrink a standard amount. Examples of these amounts are shown in Fig. 13-4.

Shrink rules for each of these metals have oversize graduations. Fig. 13-5. For example, the one-inch markings on a shrink rule for cast iron castings actually measure 1/96 inch longer than a standard inch. The pattern worker uses this shrink rule when measuring a pattern for a cast iron casting. The pattern will then be enlarged the proper amount to allow for the shrinkage of the metal.

Machining refers to the removal of metal from a surface after the object has been cast.

An adjustment is made in the size of the casting for the material that will be removed in the machining process. The patternmaker makes the pattern larger by adding an allowance. Allowance examples are shown in Fig. 13-6. For example, a part 150 mm long when finished, which is to be made from cast iron and machined, would have an allowance of 2.0 mm. The pattern would have to be made 152 mm long.

In addition, a shrinkage, or casting, tolerance is needed to adjust for shrinkage. Typical shrinkage tolerances are given in Fig. 13-7. For example, a cast iron casting with dimensions up to 200 mm requires a tolerance of ±0.8 mm.

Pattern detail drawings, Fig. 13-8, show any holes to be formed by cores. A **core** is a dry

Fig. 13-4. Standard Shrinkages

Material	Inches per foot	Millimeters per mm
Cast iron, malleable iron	1/8	0.0104
Steel	1/4	0.0208
Brass, copper, aluminum, bronze	3/16	0.0156

Fig. 13–5. A shrink rule used by patternmakers. (L. S. Starrett Co.)

Fig. 13-6. Finish Allowance for Castings—Cast Iron, Aluminum, or Bronze

Pattern Size	Surface Finish Allowance	
	Inches	Millimeters
up to 8″ (200 mm)	0.07	2.0
8″ (200 mm) to 16″ (400 mm)	0.10	2.5
16″ (400 mm) to 24″ (600 mm)	0.14	3.6
24″ (600 mm) to 32″ (800 mm)	0.18	4.6
over 32″ (800 mm)	0.24	6.0

Fig. 13-7. Shrinkage Tolerances for Cast Iron or Aluminum

Casting Size	Tolerances ±	
	Inches	Millimeters
up to 8″ (200 mm)	0.03	0.8
8″ (200 mm) to 16″ (400 mm)	0.05	1.3
16″ (400 mm) to 24″ (600 mm)	0.07	1.8
24″ (600 mm) to 32″ (800 mm)	0.09	2.3
over 32″ (800 mm)	0.12	3.0

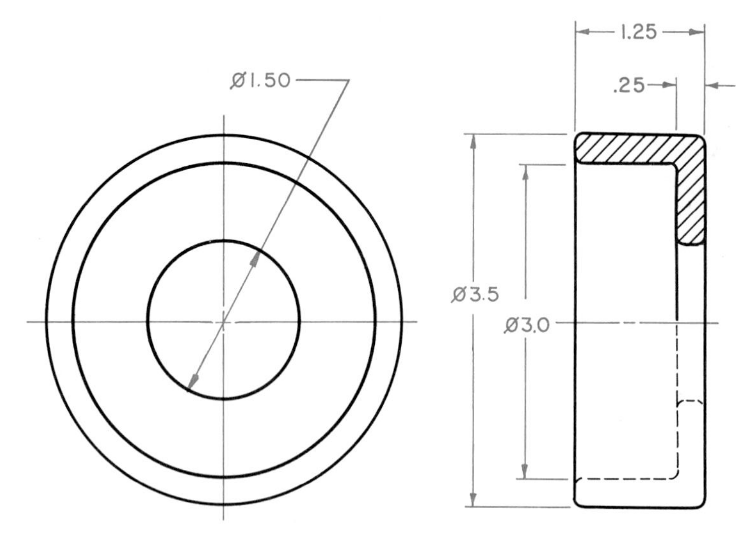

Fig. 13–8. A pattern detail drawing of a bearing cap. It includes allowances for shrinkage and machining. Draft angle is noted. The drawing does not tell where the part is to be machined nor the finished size after machining.

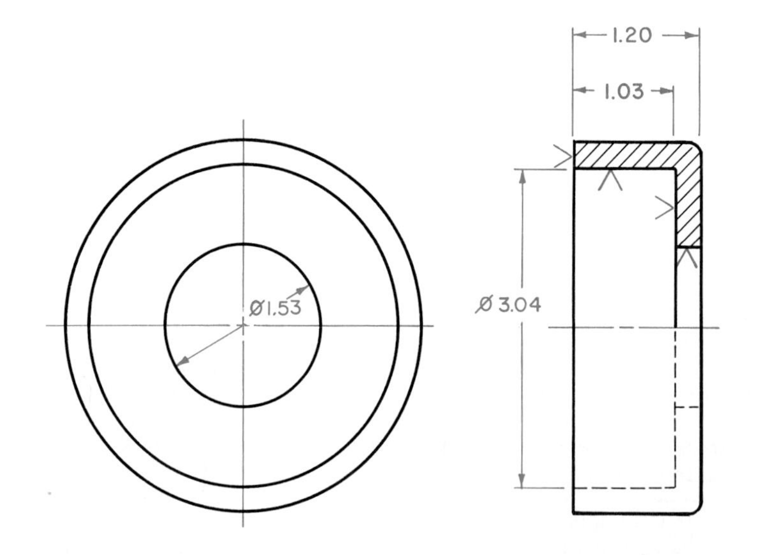

Fig. 13–9. A machining detail drawing of the bearing cap. It gives the information needed by the machinist to produce the finished part. The overall dimensions are not needed since the machinist receives the part after it is cast.

sand form inserted in the mold to form a hole or a cavity on the interior of the casting. Large holes are cored. Small holes are drilled and are not shown on a pattern detail drawing.

The edges of cast parts are rounded because it is difficult to cast sharp corners. The radii of fillets and rounds are indicated on the drawing as are the sizes of any cored parts.

A typical pattern drawing is dimensioned showing the finished size of the casting. The allowances for shrinkage and machining are shown in notes. These allowances are added by the patternmaker as the pattern is built.

Machining Detail Drawings

A **machining drawing** includes only those details necessary for the machine shop to perform the machining operations. This generally includes the finished size, location of holes, finished surface, and tolerances on machined surfaces. Parts of the piece that are left rough (as cast) need no dimensions. Fig. 13-9 shows such a detail.

Notice in Fig. 13-9 how the machining drawing is dimensioned. The dimensions that locate finished surfaces or centers of holes that are machined are related directly in order to maintain needed precision. Since cast holes and surfaces cannot be made to an accuracy greater than ±1/32 (0.8 mm), they must not be used to locate precision dimensions.

Casting Detail Drawings

If a part is not complex, a single drawing can be made that contains all the information the patternmaker and the machinist need. Such a drawing is a combination of pattern and machining detail drawings and is called a **casting detail drawing**. Refer to Fig. 13-2.

Notice that the casting detail drawing contains the overall sizes of the rough casting as well as the precision dimensions needed for machining. The draft angle is noted. The material note CAST IRON tells the patternmaker how much to allow for shrinkage and machining.

Forging Detail Drawings

A **forging detail drawing** shows a part that is to be made by forging. Forging is a process in which the metal is heated and formed to shape with a power hammer. The piece is usually shaped in a die.

A **die** is a metal form with a cavity the shape of the part to be formed. Heated metal is placed between the dies. They are forced closed, forcing the metal into the shape of the cavity. Fig. 13-10.

Several practices used in forgings drawing are different from other types of drawings.

On the plan view, the *draft* is shown by lines drawn as if the corners of the forging were sharp. Fig. 13-11. *Draft* is the slope given to the surfaces to help remove the forging from the die. Standard draft for outside surfaces of all materials is 7 degrees. Inside draft angle for aluminum is 7 degrees and for steel, 10 degrees. Fig. 13-12.

A forging has a parting line. A **parting line** is the line where the two dies meet when forming the part. It is shown on the drawing using the center line symbol. It is labeled PL.

Forging drawings are dimensioned using the datum system. This avoids the accumulation of tolerances. Fig. 13-12. The dimensions should refer to some important part of the object. A center line is most frequently used.

All dimensions parallel to the parting line should refer to the size at the bottom of the die

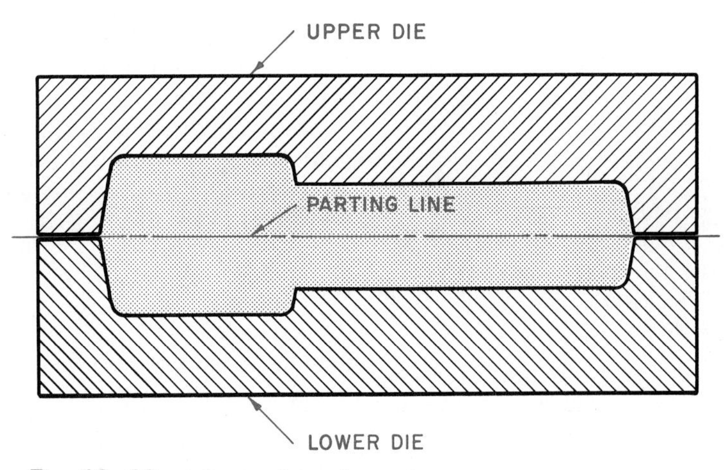

Fig. 13–10. A forging formed in a die.

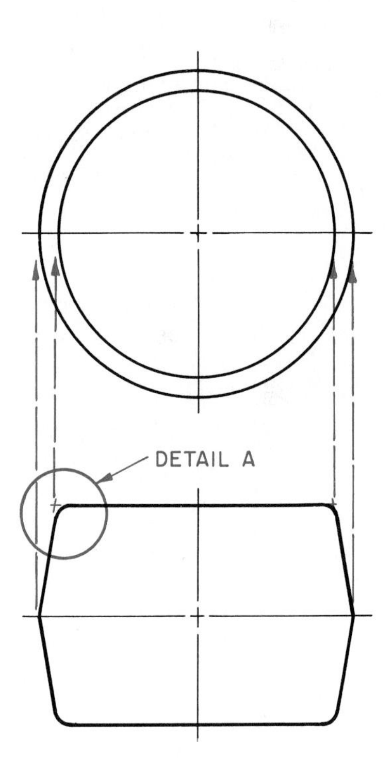

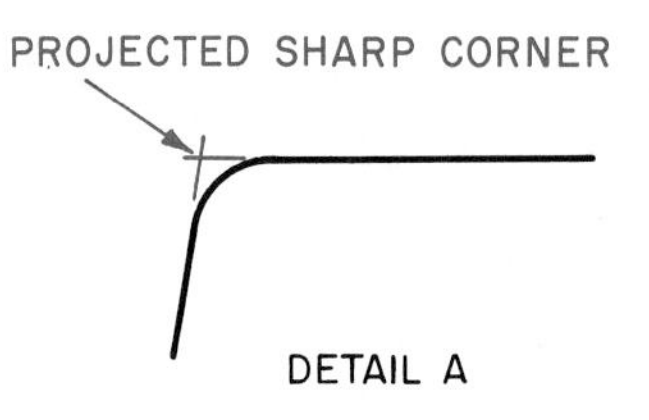

Fig. 13–11. The rounded corners on a forging are projected and drawn as though they are sharp.

impression. This is the narrowest part. The die cavity is widest at the parting line. See the cavity shown in Fig. 13-10.

All dimensions at right angles to the parting line should refer to the parting line. Fig. 13-12.

Forging drawings should be drawn full scale whenever possible. Sectional views are used a great deal.

Tolerances. Forgings are designed with standard tolerances: thickness, shrinkage, die wear, mismatch, and machining allowance.

Thickness tolerances refer to the overall height of the part. This dimension can vary depending upon how closely the dies fit. Fig. 13-13. Standard commercial tolerances and more accurate close tolerances are given in Fig. 13-14. The tolerances for forgings shown in this table and others are those sometimes used in general manufacturing industries. Detailed, specialized standards exist for certain industries, such as the aerospace industry.

Shrinkage tolerances relate to parts that become smaller when the heated metal cools. Standard tolerances are shown in Fig. 13-15. These are figured in a direction parallel to the parting line. Fig. 13-16.

Die wear tolerances relate to the tendency for dies to wear and become larger. Standard tolerances are shown in Fig. 13-15. These are figured in a direction parallel to the parting line. Fig. 13-16.

Mismatch refers to slight size differences caused by the dies not lining up exactly when they close. Fig. 13-17.

Notice in Fig. 13-16 the shrinkage, die wear, and mismatch tolerances.

Machining allowance indicates the material needed to machine the forging to the finished size. Fig. 13-16. This amount is in addition to shrinkage, thickness, and mismatch tolerances. Standard machining allowances are shown in Fig. 13-18.

The length of a forging is the sum of the finished length desired plus machining allowance, shrinkage, die wear, and mismatch tolerance required on each end.

The height of the forging is the sum of the finished height, thickness tolerance, and machining allowance required on each side.

The actual acceptable tolerances are based on the weight of the forging.

Corner and Fillet Radii. The size of the radii used for corners and fillets affects the quality of the forging and the life of the die. Small radii will cause die wear and produce seams or improperly filled sections in the forging. Figs. 13-19 and 13-20 give minimum radii for fillets and corners.

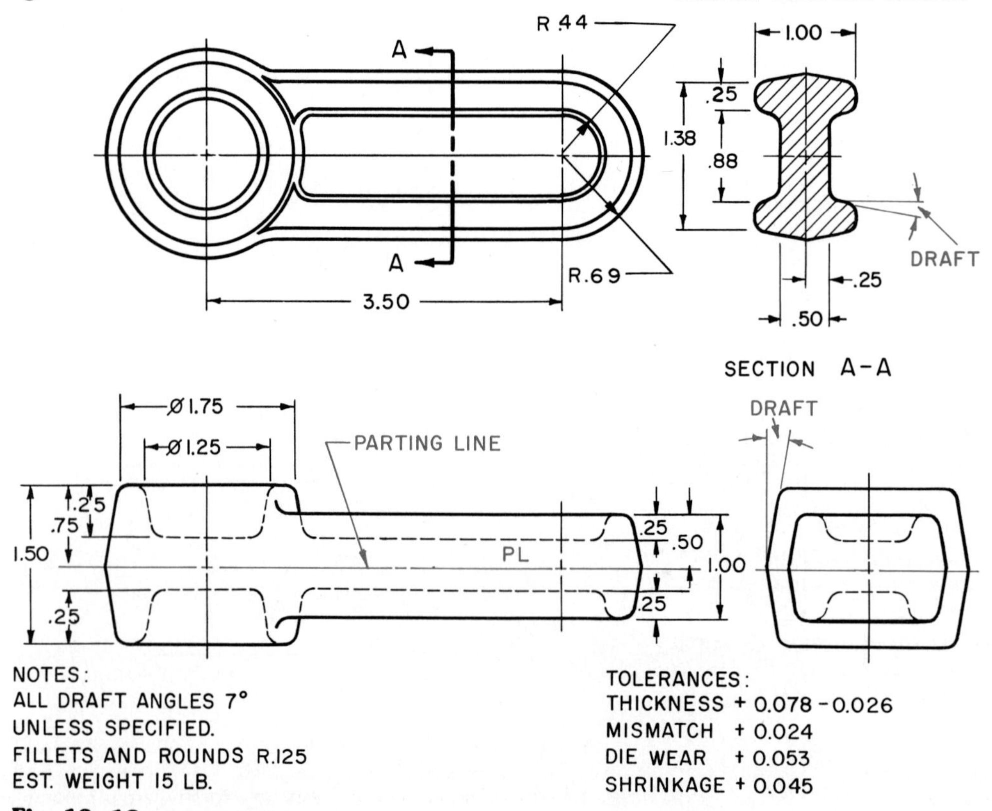

Fig. 13–12. A forging drawing.

Making Forging Drawings. Forging drawings are made in two ways. The most frequently used method of making a forging drawing is to combine the forging and machining information on one drawing. The parts of the rough forging to be removed by machining are indicated by a dashed line. The finished size is indicated by a visible line. The difference between them is the tolerance. Fig. 13-21.

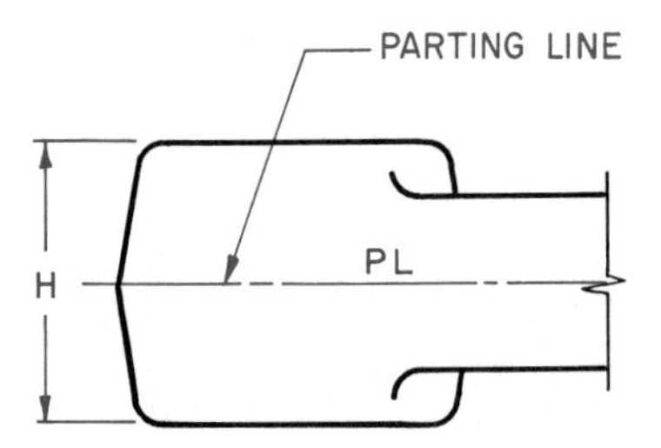

Fig. 13—13. **The thickness tolerance is affected by the closing of the dies.**

Fig. 13-14. Thickness Tolerances in Inches

Net Weight Up to Lb.	Commercial		Close	
	−	+	−	+
0.2	0.008	0.024	0.004	0.012
0.4	0.009	0.027	0.005	0.015
0.6	0.010	0.030	0.005	0.015
0.8	0.011	0.033	0.006	0.018
1	0.012	0.036	0.006	0.018
2	0.015	0.045	0.008	0.024
3	0.017	0.051	0.009	0.027
4	0.018	0.054	0.009	0.027
5	0.019	0.057	0.010	0.030
10	0.022	0.066	0.011	0.033
20	0.026	0.078	0.013	0.039
30	0.030	0.090	0.015	0.045
40	0.034	0.102	0.017	0.051
50	0.038	0.114	0.019	0.057
60	0.042	0.126	0.021	0.063
70	0.046	0.138	0.023	0.069
80	0.050	0.150	0.025	0.075
90	0.054	0.162	0.027	0.081
100	0.058	0.174	0.029	0.087

Fig. 13-15. Shrinkage and Die Wear Tolerances for Forgings in Inches

Shrinkage			Die Wear		
Length or Width Up to In.	Commercial + or −	Close + or −	Net Weight Up to − Lb.	Commercial + or −	Close + or −
1	0.003	0.002	1	0.032	0.016
2	0.006	0.003	3	0.035	0.018
3	0.009	0.005	5	0.038	0.019
4	0.012	0.006	7	0.041	0.021
5	0.015	0.008	9	0.044	0.022
6	0.018	0.009	11	0.047	0.024
Each Additional In. Add	0.003	0.0015	Each Additional 2 Lb. Add	0.003	0.0015

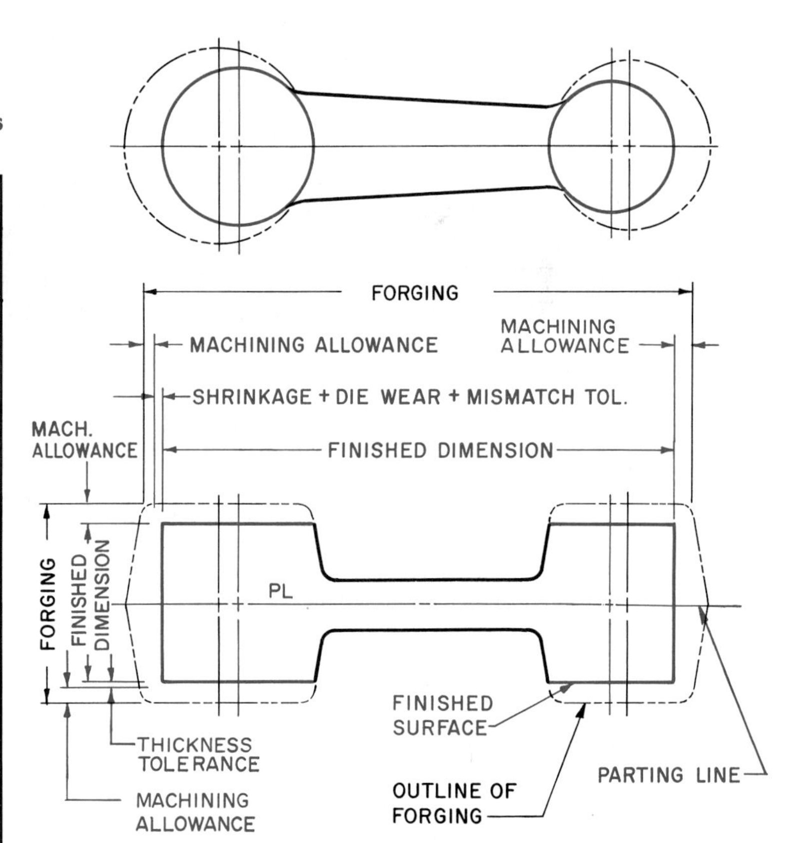

Fig. 13—16 **The length of forged parts includes the finished dimension plus machining allowance, shrinkage, die wear allowance, and mismatch tolerance. The thickness of forged parts includes the finished dimension plus machining allowance and thickness tolerance.**

Mismatch Tolerances in Inches

Net Weight Up to Lb.	Commercial	Close
1	0.015	0.010
7	0.018	0.012
13	0.021	0.014
19	0.024	0.016
Each Additional 6 Lb. Add	0.003	0.002

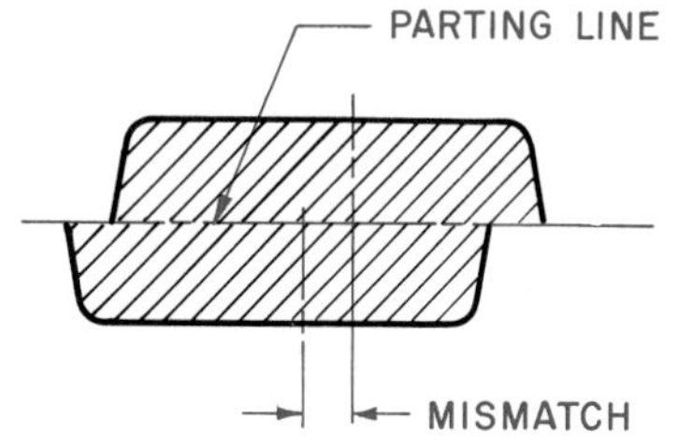

Fig. 13–17. Mismatch is caused by the dies not lining up when they close to form a forging.

The dimensions on the drawing give the size of the finished piece. The tolerances for machining, shrinkage, die wear, thickness, and mismatch are not included. These are added by the person making the die. Tolerances are shown with a note. See Figs. 13-12 and 13-21. Dimensions, which must be accurate, are given in decimal inches or millimeters.

A second way to detail forgings is to make separate drawings of the rough forging and the forging after machining. Fig. 13-22. This is done if the part is so complex that the outline of the forging cannot be shown clearly on one drawing. These drawings are placed on the same sheet if possible. The forging drawing is to the left of the machining drawing.

Fig. 13-18. Machining Allowances for Forgings (in Inches)

	Under 1 Lb.	1 to 10 Lb.	11 to 40 Lb.	41 to 100 Lb.	101 to 200 Lb.
Compact Parts Gears, Discs	1/32	3/64	1/16	3/32	1/8
Thin Extended Parts	1/16	1/16	3/32	1/8	5/32
Long Parts Shafts	1/16	1/16	3/32	1/8	3/16

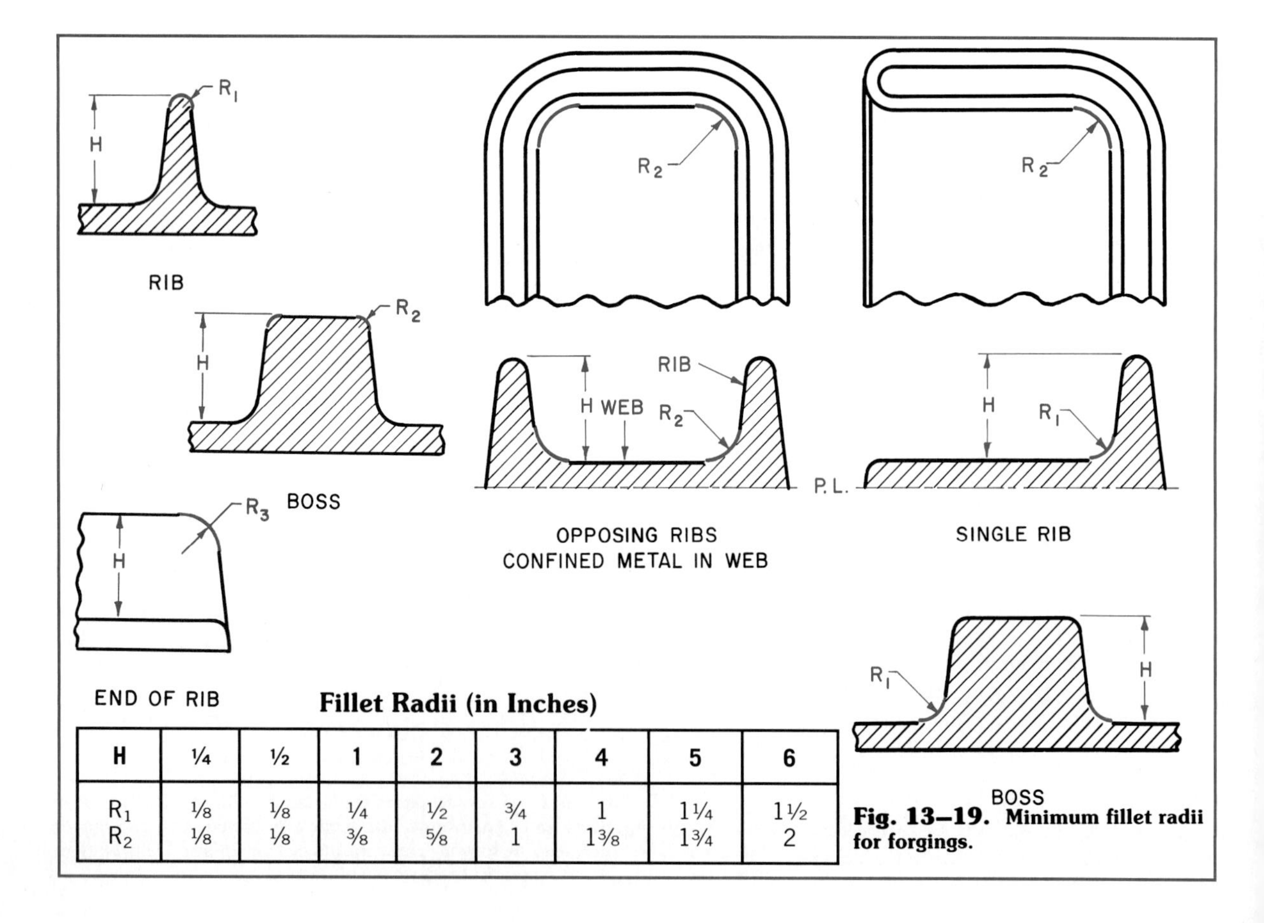

Fillet Radii (in Inches)

H	1/4	1/2	1	2	3	4	5	6
R_1	1/8	1/8	1/4	1/2	3/4	1	1 1/4	1 1/2
R_2	1/8	1/8	3/8	5/8	1	1 3/8	1 3/4	2

Fig. 13–19. Minimum fillet radii for forgings.

On the forging drawing, always indicate some surface of the finished part by phantom lines. This should be related to a forged surface with a dimension.

Forging drawings usually specify draft angles, parting lines, fillets and corner radii, tolerances, allowance for machining, material specifications, heat treatment, and a part identification number. Usually all draft angles are the same and can be specified with a note such as "All draft angles 7 degrees unless otherwise specified."

Fig. 13-20. Minimum Corner Radii for Forgings (in Inches)

H	1/4	1/2	1	2	3	4	5	6	7
R_1	1/16	1/16	1/8	3/16	1/4	5/16	3/8	7/16	1/2
R_2	1/16	1/16	1/8	1/4	5/16	7/16	1/2	5/8	11/16
R_3	3/16	3/16	3/8	1/2	3/4	1	1 1/8	1 1/4	1 1/2

Welding Detail Drawings

A **welding detail** is a drawing of a single part that is made of several pieces of metal joined by welding into a single unit. Each piece of the unit is fully dimensioned so it can be easily made. The parts are drawn assembled as they will appear after welding.

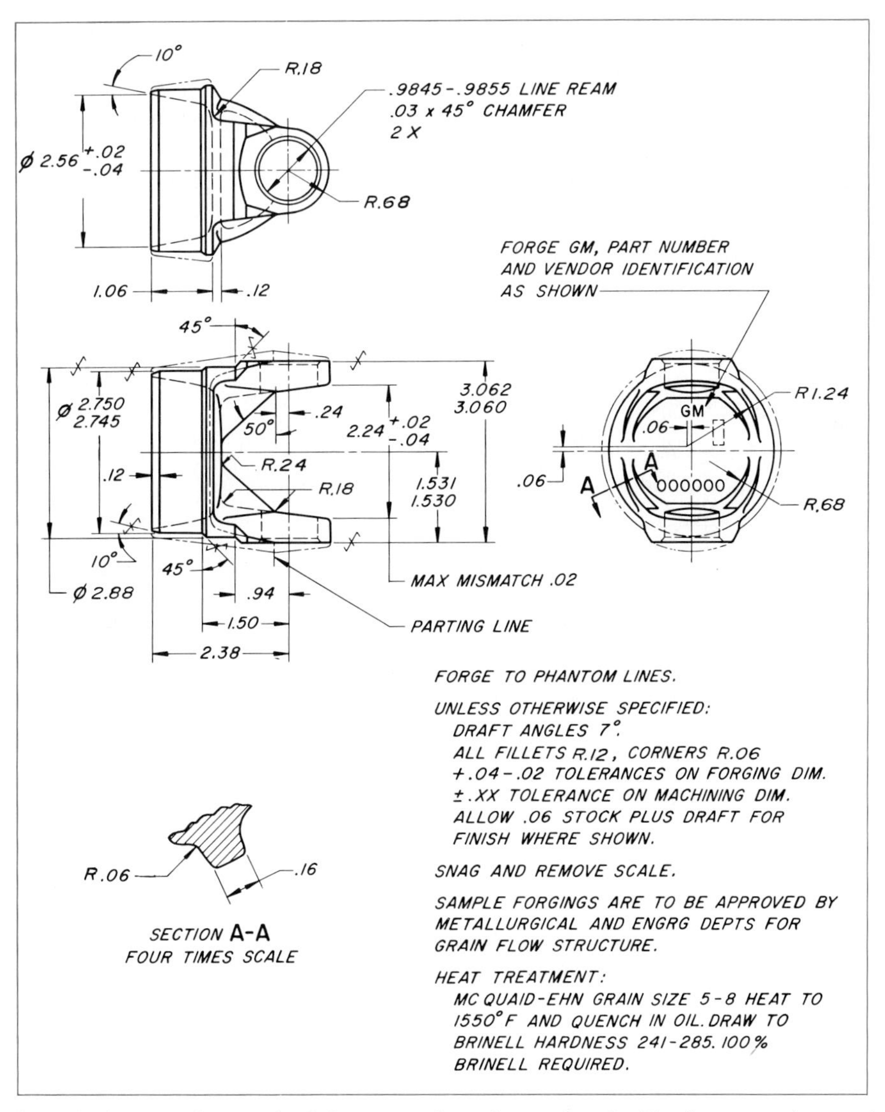

Fig. 13–21. A single drawing showing both forging and machining details. The forging outline is shown by phantom lines, the finished machined surfaces by solid lines. (General Motors Corp.)

No fillets are drawn on a welding drawing since the weld tends to serve the same purpose as a fillet on a cast piece. The pieces are drawn as they appear before welding. Fig. 13-23.

All joints between individual pieces making up the unit are drawn even though they will be covered by a weld on the finished piece. Each joint to be welded must have the proper symbol even though the actual welds are not drawn.

Each individual part of the unit is numbered. The specifications for each numbered part are given in a parts list.

The proper use of welding symbols is explained in Chapter 12, "Fasteners."

Stamping Detail Drawings

Parts produced by **stamping** are made by pressing sheet metal between dies under pressure. This makes the metal bend and in some cases stretch to fit the shape of the die. This process can involve simply forming a precut piece to the desired shape. It also includes shearing or cutting the metal to shape (blanking) and punching holes in it. A typical stamped product is shown in Fig. 13-24.

Some processes involved in making stampings are bending and forming, drawing, coining, blanking, punching, and trimming.

Bending and forming are the processes used to bend, flange, fold, and twist the metal to the desired shape. Usually the thickness of the metal is unchanged.

Drawing metal is a process in which the metal is stretched (drawn) over a form. The form has been made to the shape of the finished part. This usually causes the metal to become thinner in the section drawn. The rest of the part remains the same thickness.

Coining is the process by which metal is caused to flow under great pressure. The metal in the area coined becomes thinner. The metal from the thinned section flows to the areas not coined and they become thicker.

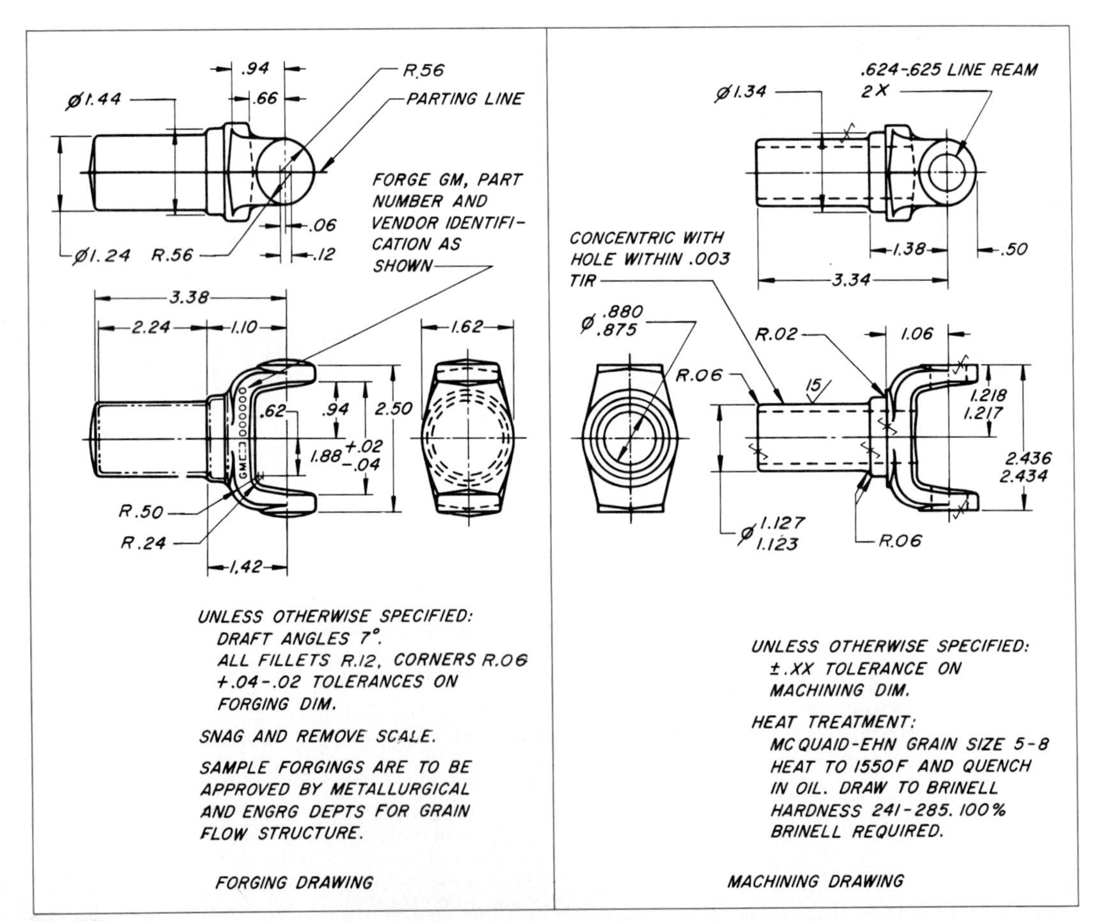

Fig. 13–22. A forging drawing and a machining drawing. Notice the forging drawing has a finished machined surface on the round shaft indicated with phantom lines. It is related to the center line with dimensions. (General Motors Corp.)

CAD Applications:

Assembly Drawings

Any product of industry that has more than one part usually will need an assembly drawing. This drawing illustrates the finished product in its assembled condition. Drawing an assembly on a CAD system allows for the use of many commands well suited to this type of drawing, such as "move," "chamfer," and "hatch." Here is an example of a simple assembly drawing being drawn on a CAD system. The assembly consists of a bronze bushing and a steel pin.

The bushing is drawn using the line command. Then the hatch command is used to crosshatch the wall thickness of the bushing.

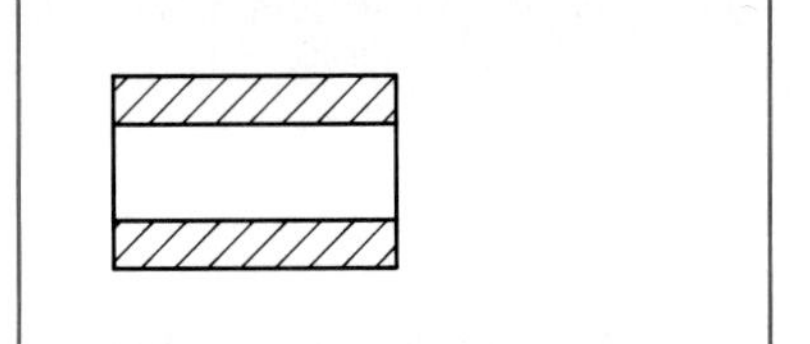

The steel pin is drawn next using the line command. The end is angled with the chamfer command.

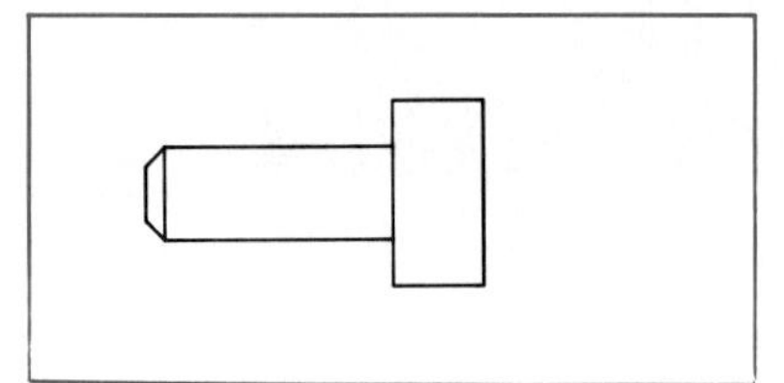

Now we can use a very valuable command: move. This will be used to position the pin into the bushing as illustrated in the next drawing.

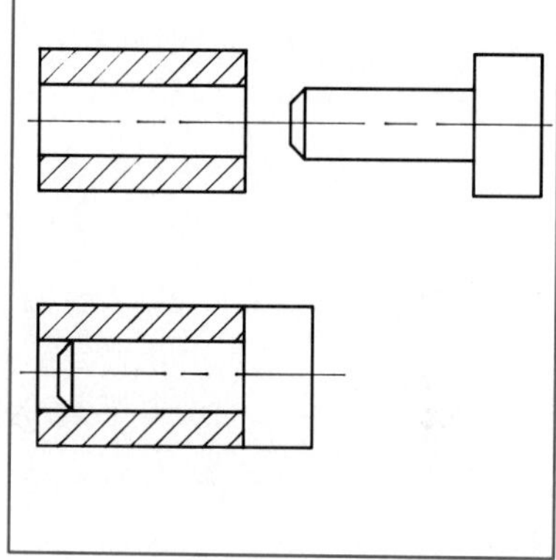

This is just another example of the capabilities of using CAD to create drawings. Remember, though, that not all commands are found on all systems. The chamfer command, for example, will not be found on many CAD systems.

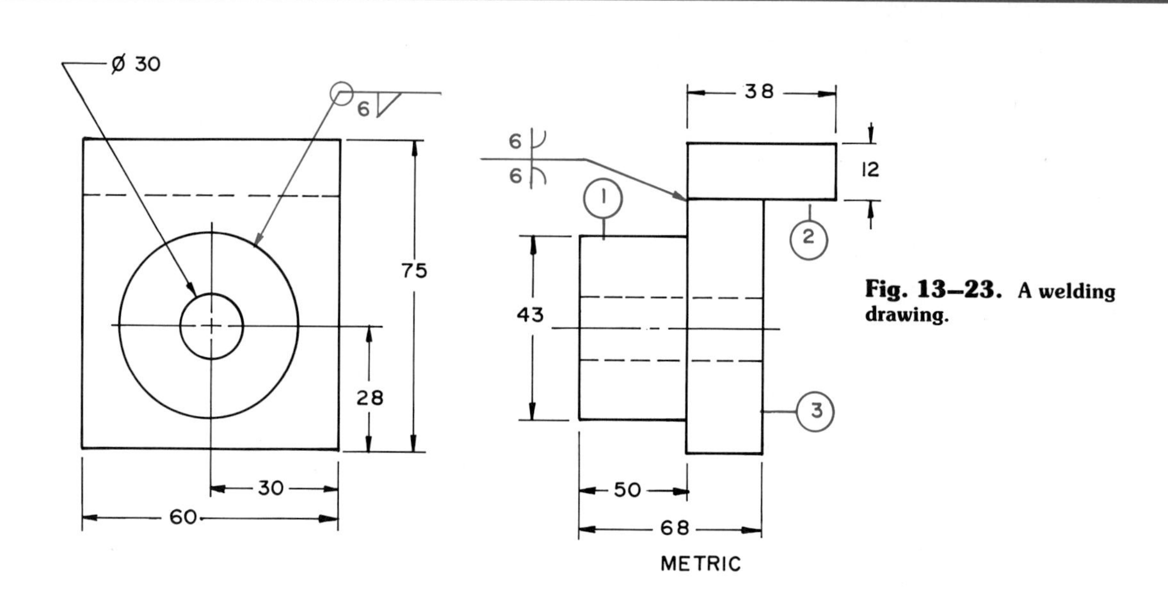

Fig. 13–23. A welding drawing.

Blanking is the process of cutting flat stock to the size and shape of the finished piece.

Punching is a method of internal blanking. It cuts a hole or opening in the piece.

Trimming refers to operations such as cutting off excess material using dies.

Metal Gages. The thickness of metal sheets in inches is indicated by a series of numbers called *gages*. Each gage number stands for a particular thickness. Fig. 13-25 gives the standard inch gage sizes. Metric sheet thicknesses are shown in Fig. 13-26.

The thickness of flat sheet stock under 1/4 inch is usually specified on the drawing in decimals. A larger thickness can be given in fractions. The thickness and type of steel are specified as follows: STEEL, 125-SAE 1010 CR. COMM. QUAL. BRIGHT FINISH. This gives the type of material, steel, decimal thickness of .125, and type of steel.

Forming Holes. Holes in stampings are usually punched, extruded, or pierced. Fig. 13-27. *Punched holes* are made by a punch that shears a clean hole to the finished size. *Extruded holes* are punched. The punch forming the first opening is smaller than the final hole. The punch is enlarged beyond the tip. The enlarged portion is forced into the punched hole forming a flange.

Fig. 13–24. **This hot-air register was developed by drawing sheet metal patterns and produced by the stamping process.**

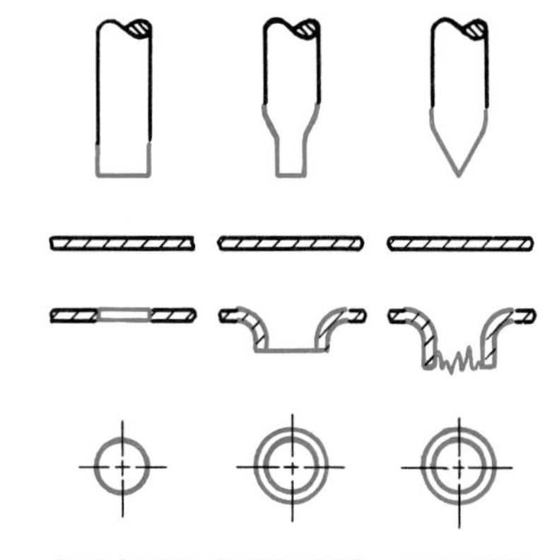

Fig. 13–27. **Methods for forming holes in stamped parts.**

Fig. 13-25. Gages for Sheet and Plate Iron and Steel (Inches)

Gage Number	Approximate Thickness
10	.1406
11	.125
12	.1094
13	.0938
14	.0781
15	.0703
16	.0625
17	.0563
18	.05
19	.0438
20	.0375
21	.0344
22	.0313
23	.0281
24	.025
25	.0219
26	.0188
27	.0172
28	.0156
29	.0141
30	.0125

Fig. 13-26. Preferred Thicknesses for All Flat Metal Products (Millimeters)

Preferred Thickness	Second Preference
0.050	
0.060	
0.080	
0.10	
0.12	0.14
0.16	0.18
0.20	0.22
0.25	0.28
0.30	0.35
0.40	0.45
0.50	0.55
0.60	0.65
	0.70
0.80	0.90
1.0	1.1
1.2	1.4
1.6	1.8
2.0	2.2
2.5	2.8
3.0	3.2
3.5	3.8
4.0	

MINIMUM RADIUS TWICE METAL THICKNESS FOR HIGHLY STRESSED PARTS.

SHARP VERTEX PERMITTED FOR LOW STRESSED PARTS.

Fig. 13–28. **Two types of notches used on stampings.**

Pierced holes are made by a punch with a sharp point. It pierces the metal and forms a flange, leaving a torn edge.

Notches. Two types of notches are used on stampings. One is a sharp V and the other has a rounded vertex. Fig. 13-28. The sharp V is used only on parts with little stress, because the V notch will tend to crack if placed under stress. Highly stressed parts must have a rounded vertex. The radius of the round should be at least twice the thickness of the metal. It should be larger if the design permits.

Bend Allowances. When a stamping is formed, the blank must be made longer than the finished piece to allow for the material consumed in making the bend. This extra allowance is called **bend allowance**. Following are procedures for figuring bend allowance.

Straight-Bend Flat Pattern

When a flat pattern for a sheet metal part with straight bends is developed, the bend allowance depends upon the angle of the bend. In Fig. 13-29 the details for a 120-degree bend are shown.

The size of the flat pattern used to make this part is the sum of each side from the end to the tangent point (distances Y and Z, Fig. 13-29) plus the amount of material needed to form the bend.

Locating the Tangent Point Dimension. In order to find the size of a flat pattern, it is necessary to know the distance from the mold line to the tangent point. This is distance X, Fig. 13-29. The *mold line* is the point at which the sides of the part would meet if extended.

The distance from the mold line to the tangent point is found using the formula (T + R) times the tangent of one-half angle A. T is the thickness of the metal. R is the radius of the curve. Angle A is the angle of bend.

The following example is based on Fig. 13-29.

$X = (T + R) \tan \frac{1}{2}$ angle A

$X = (.0810 + .1875) \tan 60°$

$X = .2685 \times 1.7320$

$X = .4650$

Distances Y and Z are found by subtracting distance X from the overall length, distance W.

Figuring Bend Allowance

Bend allowance is found by using a bend-allowance chart. Fig. 13-30. This chart gives the amount of material needed to bend a sheet metal part through an angle of one degree. The left column gives the radius of the bend. Across the top is the thickness of the metal.

The following problems will show how the actual size of a pattern is found using bend allowance.

Bend Allowance for 90-Degree Bends. Fig. 13-31 shows a sheet metal part .1250 inch thick, formed with a radius of .2500 inch through a 90-degree straight bend. The flat sides are each 1.5000 inches long to the tangent point. These plus the length of the metal in the curved corner give the total width of the flat pattern. The curved corner runs from tangent point A to tangent point B.

To find the amount of material needed to form the curved section, look at the bend-allowance chart shown in Fig. 13-30. The bend allowance for metal .1250-inch thick with a radius of .2500 inch is .00533 inches per one degree bend. Since the bend is through 90 degrees, the length of the curved section is 90 times .00533 inches or .4797 inches.

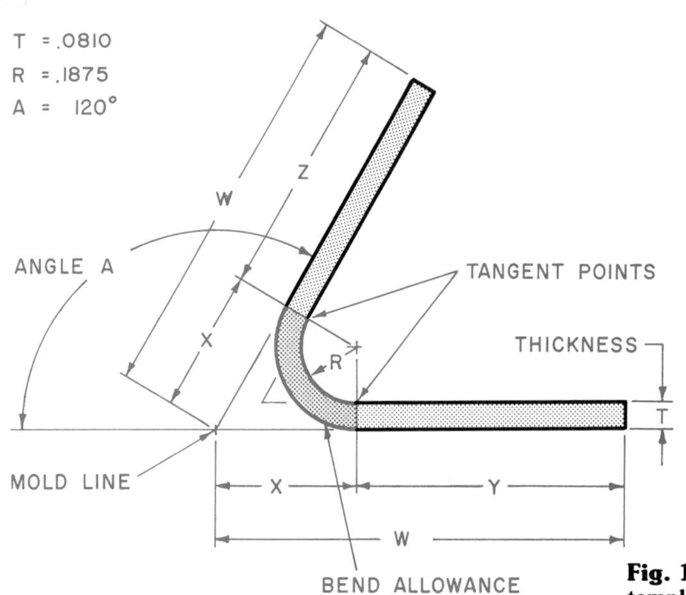

Fig. 13–29. How to find the size of a template with a straight bend. This example is a 120-degree bend.

The total width of the pattern is 1.5000 inches + .4797 inches + 1.5000 inches or 3.4797. See the flat pattern shown in Fig. 13-31.

Bend Allowance for Open-Bevel Bends. An *open-bevel bend* is one forming an obtuse angle. Fig. 13-32. Bend allowance and total size of the pattern are found in the same way as the 90-degree bend. In Fig. 13-30, the bend allowance is .00243 inches per degree of bend. The amount of material needed is .00243 inches times 60 or .1494 inches. The total width of the pattern is 1.00000 + .1494 + 1.5000 or 2.6494 inches.

Bend Allowance for Closed-Bevel Bends. A *closed-bevel bend* is one forming an acute angle. Fig. 13-33. The bend allowance in this problem is .0039. The amount of material needed is .0039 inches times 120 or .4680 inches. The total width of the pattern is 2.2500 + .4680 + 1.5000 or 4.2180 inches.

Fig. 13-30. Bend Allowance for 1 Degree of Bend in Nonferrous Metal

Radius	Thickness of Material											
	.016	*.020*	*.025*	*.032*	*.040*	*.051*	*.064*	*.072*	*.081*	*.091*	*.102*	*.125*
.0625	.00121	.00125	.00129	.00135	.00140	.00145	.00159	.00165				
.1250	.00230	.00234	.00238	.00243	.00249	.00258	.00268	.00274	.00281	.00289	.00297	.00315
.1875	.00339	.00342	.00347	.00352	.00358	.00367	.00377	.00383	.00390	.00398	.00406	.00424
.2500	.00448	.00451	.00456	.00461	.00467	.00476	.00486	.00492	.00499	.00507	.00515	.00533
.3125	.00557	.00560	.00564	.00570	.00576	.00584	.00595	.00601	.00608	.00616	.00624	.00642
.3750	.00666	.00669	.00673	.00679	.00685	.00693	.00704	.00710	.00717	.00725	.00733	.00751
.4375	.00775	.00778	.00782	.00787	.00794	.00802	.00812	.00819	.00826	.00834	.00842	.00860
.5000	.00884	.00887	.00891	.00896	.00903	.00911	.00921	.00928	.00935	.00943	.00951	.00969
.5625	.00993	.00996	.01000	.01005	.01012	.01020	.01030	.01037	.01043	.01051	.01058	.01078
.6250	.01102	.01105	.01109	.01114	.01121	.01129	.01139	.01146	.01152	.01160	.01170	.01187
.6875	.01211	.01214	.01218	.01223	.01230	.01238	.01248	.01254	.01261	.01269	.01276	.01296
.7500	.01320	.01323	.01327	.01332	.01338	.01347	.01357	.01363	.01370	.01378	.01386	.01405
.8125	.01429	.01432	.01436	.01441	.01447	.01456	.01466	.01472	.01479	.01487	.01494	.01514
.8750	.01538	.01541	.01545	.01550	.01556	.01565	.01575	.01581	.01588	.01596	.01604	.01623
.9375	.01646	.01650	.01654	.01659	.01665	.01674	.01684	.01690	.01697	.01705	.01712	.01732
1	.01755	.01759	.01763	.01768	.01774	.01783	0.1793	.01799	.01806	.01814	.01823	.01841

(Cessna Aircraft Co.)

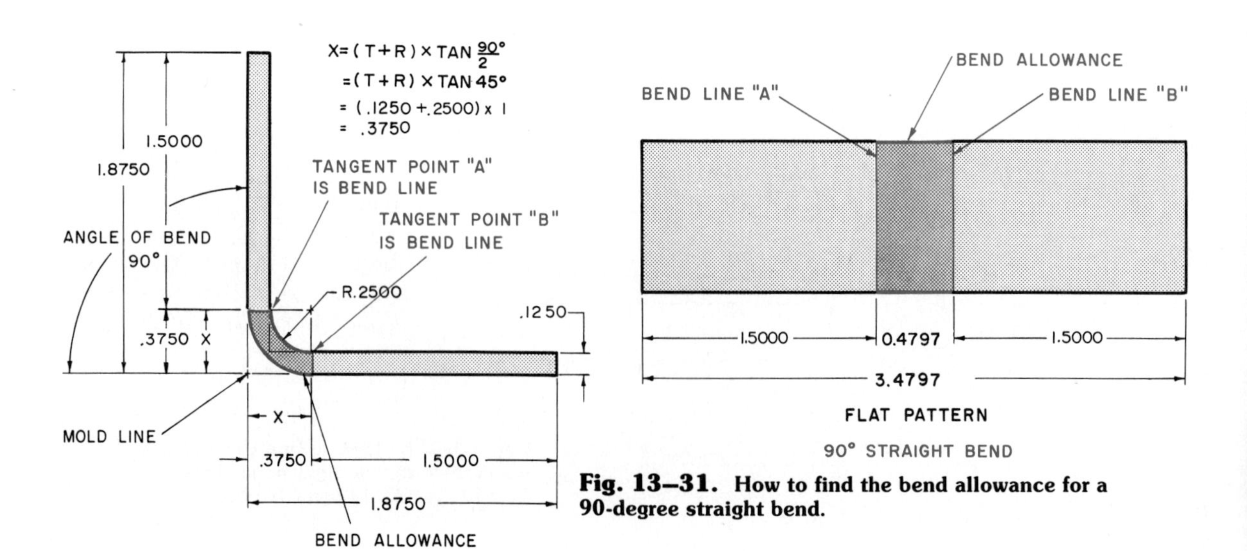

Fig. 13–31. How to find the bend allowance for a 90-degree straight bend.

Bend Radii. The radii to be used when forming sheet metal parts must be carefully chosen. Tests have shown the most practical radii for the type of material being formed. The drafter must always refer to a table of bend radii. Fig. 13-34. The largest permissible radius should be used whenever possible because it is easiest to form and will reduce the chance of the metal becoming cracked.

Corners

Fig. 13-35 shows common ways to form corners on sheet metal products.

Stamping Drawings

The views drawn of a stamping are much the same as any working drawing. Attention should be given to describing clearly the part in its finished form. Usually the shape of the flat piece of metal to be formed is not a part of the drawing. Fig. 13-36 (p. 350).

Dimensioning. Stamping drawings are dimensioned much the same as working drawings. Usually they are dimensioned entirely on one side of the piece. Whenever possible, dimensions should be given to intersections or tangent points.

Hole locations and other critical dimensions must be located from each other rather than from an outside edge. Fig. 13-36.

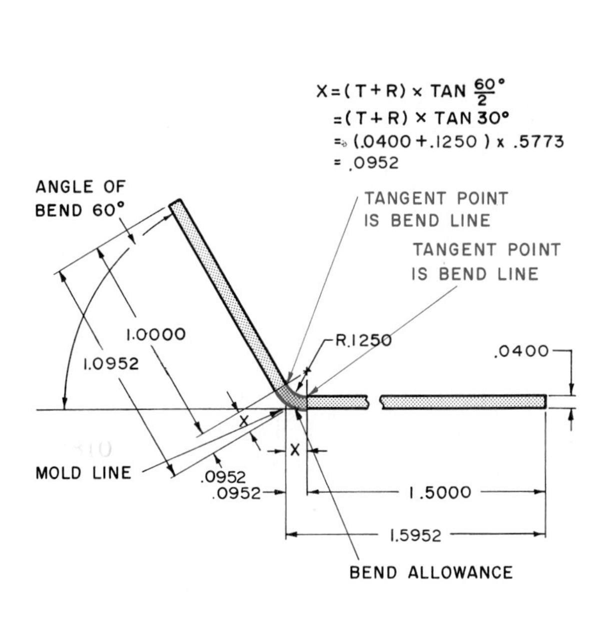

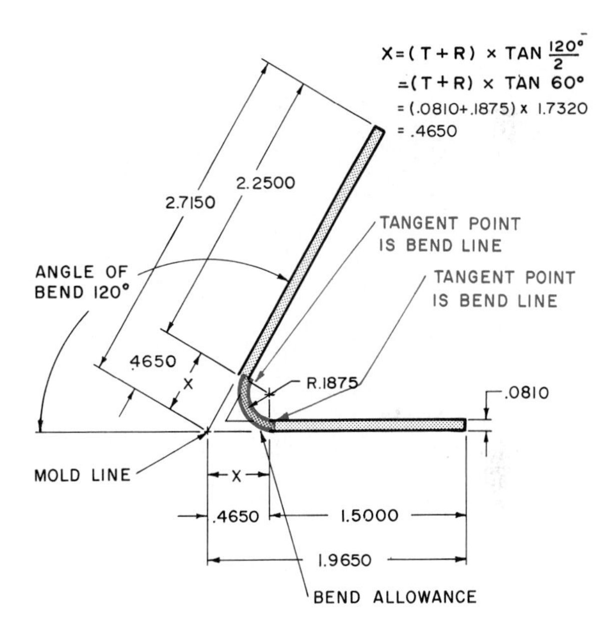

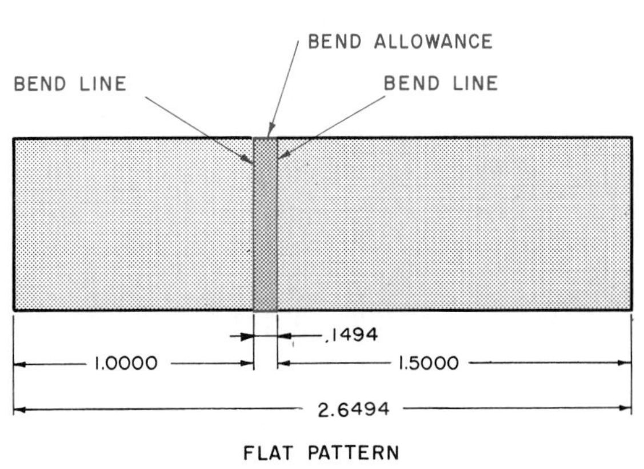

Fig. 13–32. **How to find bend allowance for an open-bevel bend.**

BEND ALLOWANCE
BEND LINE
BEND LINE
2.2500
.4680
1.5000
4.2180
FLAT PATTERN
CLOSED-BEVEL BEND

Fig. 13–33. **How to find bend allowance for a closed-bevel bend.**

Fig. 13-34. Standard and Minimum Bend Radii for Sheet Stock

Gage / Material	.016		.020		.025		.032		.040		.050		.063		.080		.090		.100		.125		.160		.190		.200	
	S	M	S	M	S	M	S	M	S	M	S	M	S	M	S	M	S	M	S	M	S	M	S	M	S	M	S	M
ALUMINUM 2024-0,1100½ H 6061 SW or ST	.06	.03	.06	.03	.06	.06	.06	.06	.09	.06	.12	.06	.12	.09	.19	.12	.19	.12	.22	.16	.25	.19	.38	.25	.50	.34	.50	.34
ALUMINUM 1100-0, 3003-0 6151-0, 6061-0	.06	.03	.06	.03	.06	.03	.06	.06	.06	.06	.06	.06	.09	.06	.12	.09	.12	.09	.16	.12	.16	.12	.19	.16	.25	.19	.28	.19
ALUMINUM 2024-T3 or T4	.06	.06	.06	.06	.09	.06	.12	.09	.16	.09	.19	.12	.22	.16	.34	.25	.38	.28	.44	.34	.53	.44	.75	.66	1.00	.84		1.25
ALUMINUM 7075-0 or SW	.03		.03		.06		.06		.06		.09		.12		.19		.19		.22		.28		.44			.75	.56	
ALUMINUM 7075-T6	.09		.12		.12		.16		.19		.25		.16		.38		.50				.90		1.2				1.3	
STEEL											.09	.06	.12	.09	.16	.12			.25	.16			.31	.25	.38	.28	.44	.34
MAGNESIUM																												
FS-1 SOFT HOT	.06		.06		.06		.06		.09		.09		.16		.19		.19		.25		.25		.38				.38	
FS-1 SOFT COLD	.09		.09		.16		.19		.22		.25		.31		.45		.45		.50		.62		.75				1.00	

S = Standard radius.
M = Minimum radius.
All bend radii apply to inside radius.

(Cessna Aircraft Co.)

Assembly Drawings

It is often useful or necessary to show an object as it appears when all of its parts are assembled. This is called an **assembly drawing.** Assembly drawings are used to check detail drawings, to describe how a machine functions, to show how to install a machine, to give maintenance instructions, to show general design factors for sales purposes, to show subassemblies, or to simplify the assembly process during production. They can be in the form of an orthographic drawing or a pictorial drawing.

Planning an Orthographic Assembly Drawing

An assembly drawing shows all the parts of a device. It is much more complex than a detail drawing. Drafters must understand how the device is supposed to operate and how it is to be assembled. They must decide which views are needed to show it best when it is assembled.

The usual procedure is to make a freehand sketch of the assembled device before making the finished drawing. This will make it possible to try several approaches. Since detail drawings will be made, the assembly drawing does not need to include complete details for each part.

Fastening devices, shafts, and bearings are handled the same as discussed in Chapter 11, "Sectional Views." Dimensions are usually omitted though overall dimensions are sometimes used.

Selection of views depends upon the device. External views, sectional views, auxiliary views, and partial views can be used as required. The selection of the views follows the same principles as those used to make a multiview drawing of a single part. The views must fully describe the entire device rather than one part.

As you draw the views, it is best to block-in the major features of the device in all views and add less important details next. Do not detail each piece one at a time. Draw all preliminary construction lightly and apply finished line work after all details are recorded. This procedure is the same as that for detail drawings.

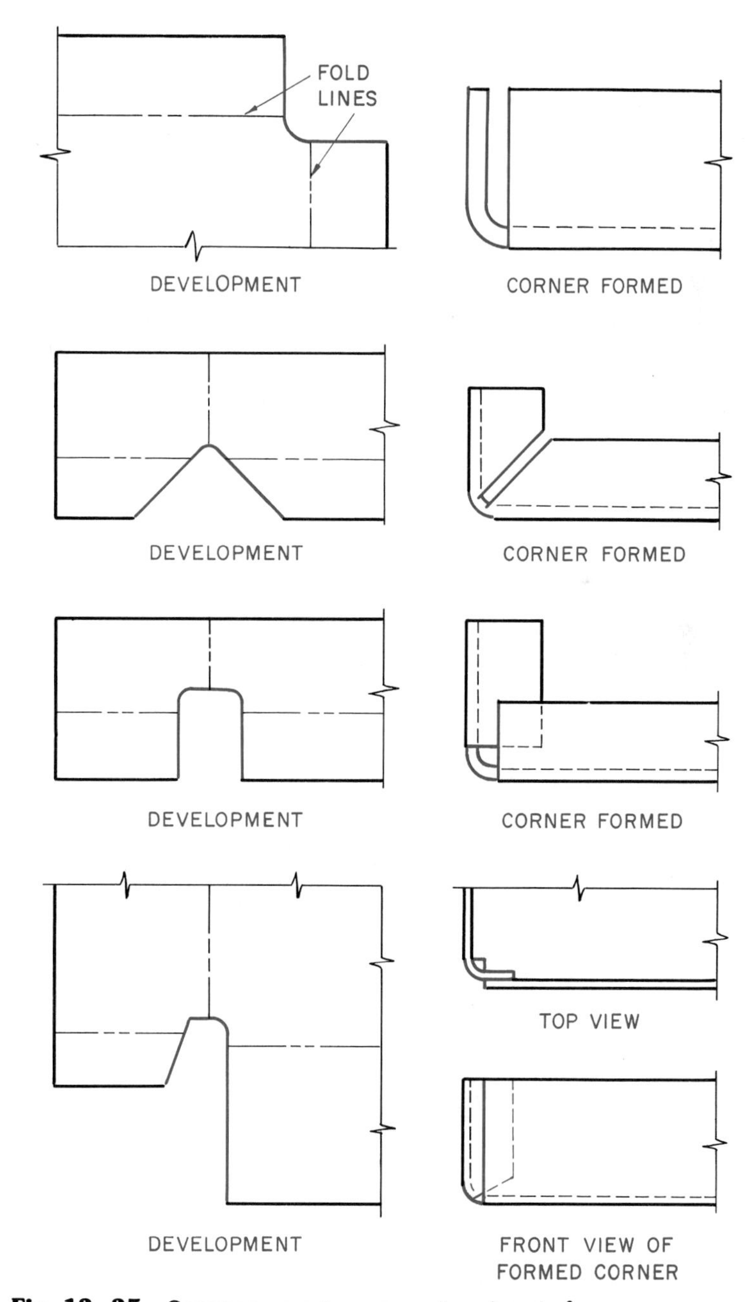

Fig. 13–35. Common ways to prepare stampings to form corners.

Steps in Drawing an Orthographic Assembly Drawing

Lay out assembly drawings in much the same way as detail drawings. Since the assembly is often used to check the correctness of the detail drawings from which it is made, accuracy is important. Lay out all views together. When one item is located on one view, draw it in all views before going on to the next part of the layout.

A good plan to use is as follows:

1. Lay out the principal center lines and major outline of the assembled device. Fig. 13-37 at A.
2. Block in the outlines of the important parts of the device. Use light construction lines and be sure to maintain accuracy. Fig. 13-37 at B.
3. Locate and draw minor details. These are items such as fasteners, holes, ribs, and keys. Fig. 13-37 at C.

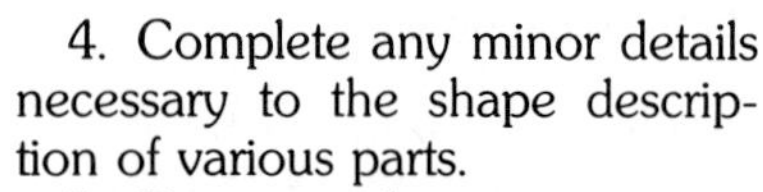

4. Complete any minor details necessary to the shape description of various parts.
5. Remove the construction lines and darken the lines of the drawing. Fig. 13-37 at D.
6. Add section lines to the surfaces cut by the cutting plane. Fig. 13-37 at D.
7. Letter part identification numbers and notes. Fig. 13-37 at D.
8. Prepare the parts list. Fig. 13-38.

Part Identification

Use numbers to identify various parts of an assembly. Usually you assign numbers to parts in the order in which they are assembled. Another system gives the largest piece the smallest number; numbers get larger as parts get smaller.

The usual practice is to record the numbers inside a 3/8 to 1/2 inch (9 to 12 mm) diameter circle on the drawing. Connect the circle to the part with a leader that may or may not have an arrowhead. If possible, arrange the circles in vertical and horizontal rows off the view. They can appear on all sides of the view. The leaders can be at any convenient angle, but don't draw them vertical, horizontal, so that they cross each other, or obscure a detail on the drawing. Draw the leaders so that if they were extended they would cross the center of the numbered circle.

Include a parts list with the assembly drawing. The items you put on the list will vary with company policy, but you usually include the part number, name of the part, the number of parts required for one machine, and the material used to make the part. If the parts list is long, place it on a separate sheet. Fig. 13-38.

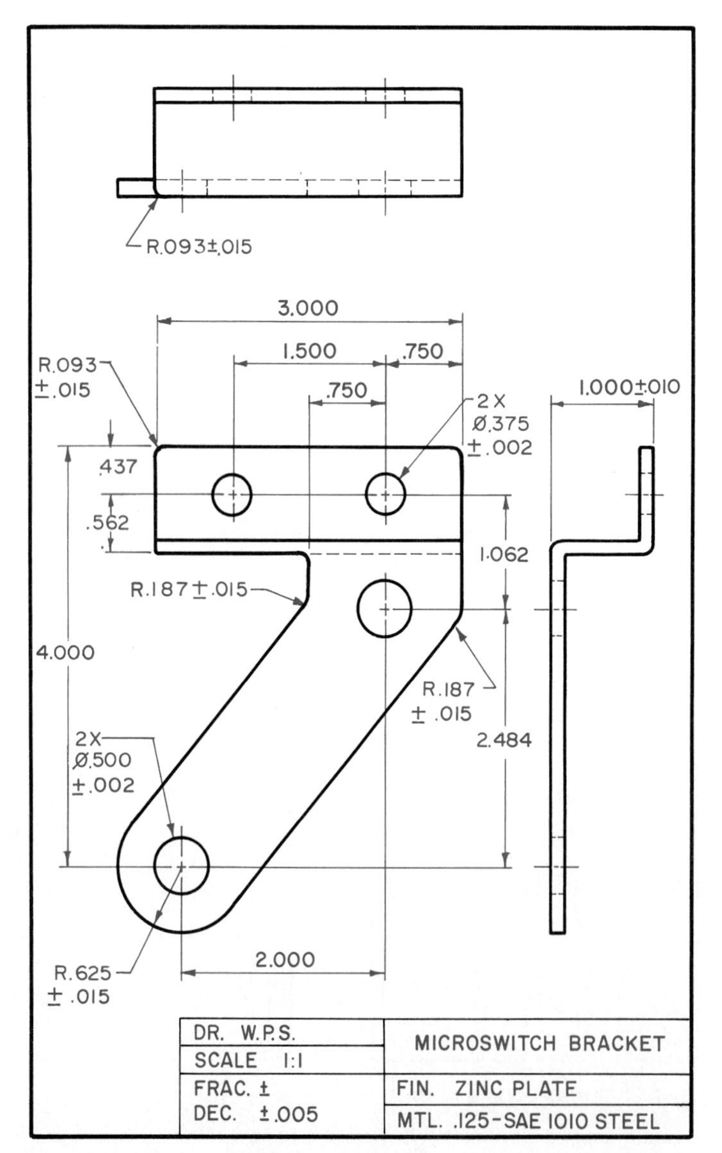

Fig. 13–36. A stamping detail drawing.

Section Lining

Small cylindrical parts, such as shafts, bolts, and pins, are not usually sectioned in assembly drawings even though the cutting plane passes through them. Other items usually not section lined are gear teeth, keys, nuts, and bearings.

The proper section lining symbol for the material from which each part is made is commonly used in assembly drawings. This helps separate the parts and makes reading the drawing easier.

Vary the slope of the section lining on adjacent parts to help clarify the drawing.

If two adjacent parts are spaced 1/32 inch (1 mm) or less apart, draw them with a single line—a shaft in a bushing is an example. If spaced wider than this, draw two lines with a clear space between.

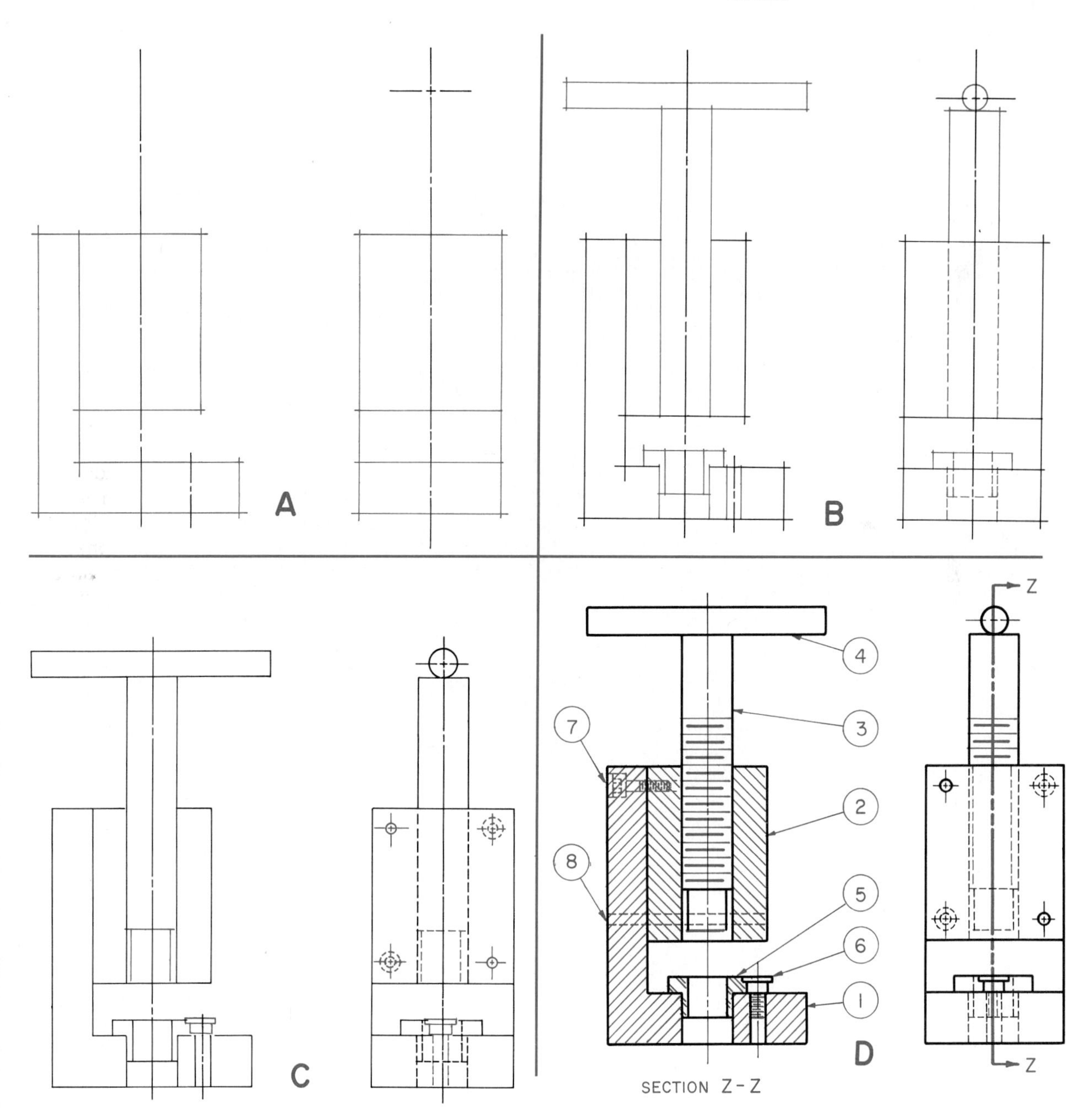

Fig. 13–37. **How to make an assembly drawing. This device is a special tool designed to pierce holes in sheet metal. (Ford Motor Co., Kansas City Assembly Plant)**

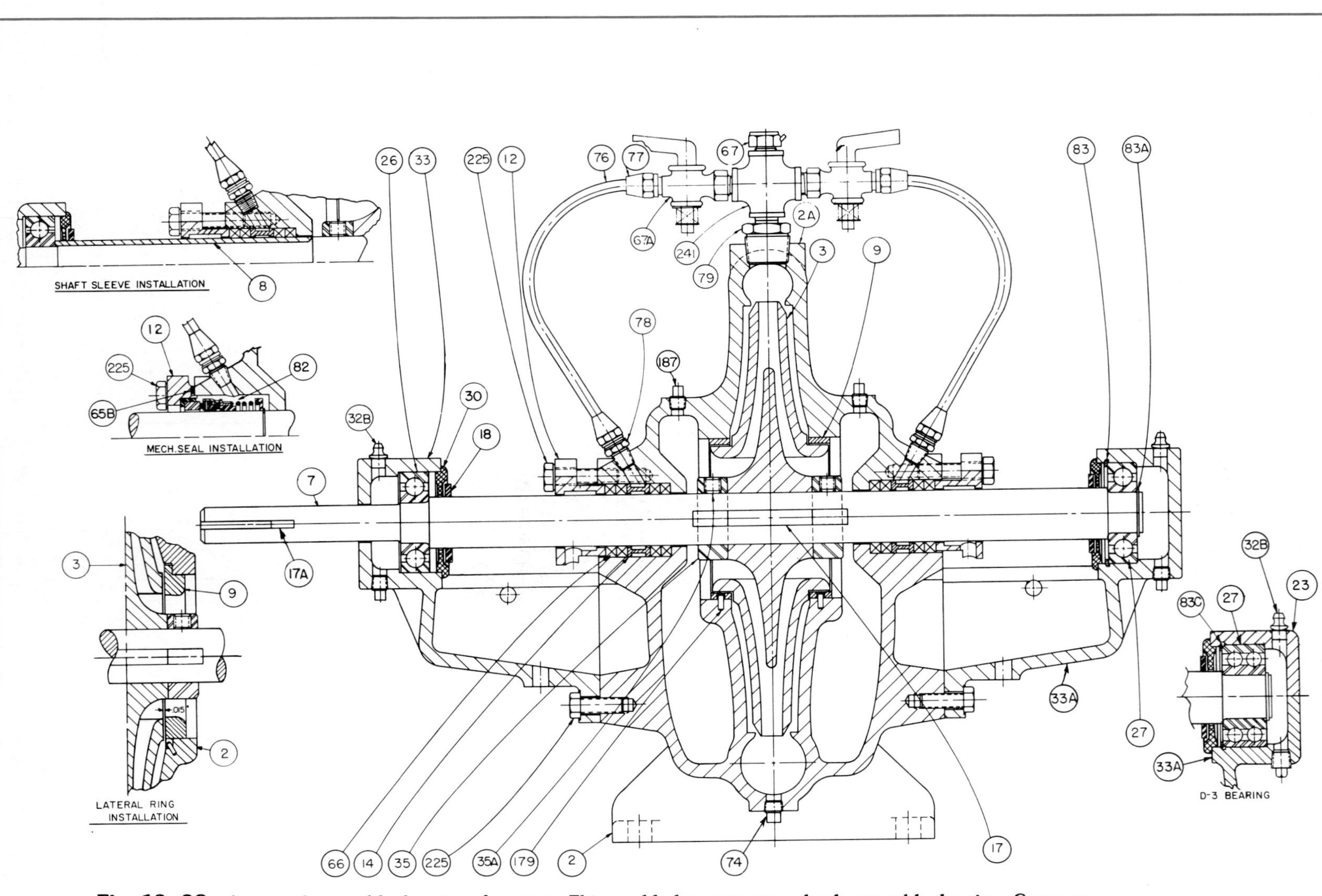

Fig. 13–38. A general assembly drawing of a pump. This could also serve as a check assembly drawing. Compare this assembly with the photo of the pump. Can you see how the assembly was drawn as a section through the pump? (Buffalo Forge Company)

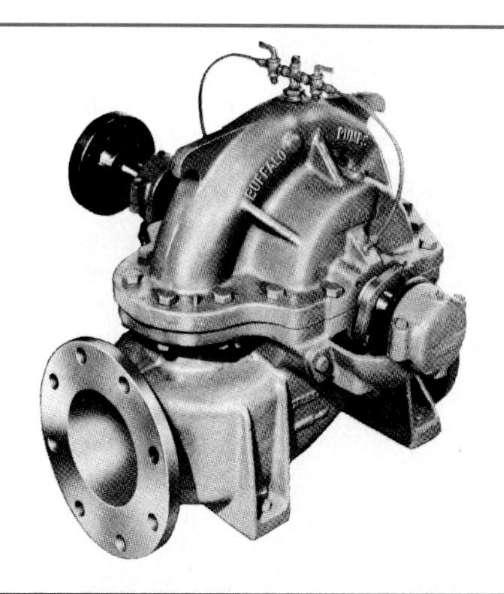

Parts List			
Part Number	**Name**	**Material**	**No. Required**
2	Housing, Lower	Cast Iron	1
2A	Housing, Upper	Cast Iron	1
3	Impeller	Cast Iron	1
7	Shaft	Steel	1
8	Sleeve	Bronze	1
9	Bushing	Bronze	1
12	Flange	Cast Iron	1
14	Seal Spacer	Steel	1
17	Key	Steel	1
17A	Key	Steel	1
18	Bearing Ring	Steel	1
23	Bearing Housing, Upper	Cast Iron	1
26	Bearing		1
27	Bearing		1
30	Bearing Seal	Nylon	1
32B	Grease Fitting	Steel	2
33	Bearing Housing, Upper	Cast Iron	1
33A	Bearing Housing, Lower	Cast Iron	2
35	Impeller Collar	Steel	2
35A	Set Screw	Steel	2
65B	Seal	Nylon	2
66	Seal	Nylon	8
67	Lock Nut	Bronze	1
67A	Valve	Bronze	2
74	Plug	Bronze	1
76	Tubing	Copper	2
77	Fitting	Copper	4
78	Fitting	Copper	2
79	Fitting	Copper	1
82	Spring	Bronze	1
83	Seal	Nylon	1
83A	Ring	Bronze	1
83C	Lock Ring	Steel	1
179	Seal Lock	Steel	2
187	Plug	Bronze	2
225	Bolt	Steel	6
241	T-Fitting	Bronze	1

Indicate the sections on exterior views with a cutting-plane line in the same way you draw detail sections. Such a section drawing is identified with the title "Section" and the identifying letters, such as A-A.

Checking Design

After an engineering staff has designed a machine and drafters have made detail drawings of each part, the detail drawings need to be checked for completeness and accuracy before the object is produced in quantity. This is frequently done by *inspection.* A checker goes over each drawing and checks it for completeness and mating fits and determines if the machine can be made by normal production methods. Inspection requires many calculations checking dimensions and tolerances to see if mating parts will fit and function properly.

Sometimes either a scale model or full-size prototype is built to check the design. This helps the designers to look for problems and improve the product before it gets to the final production stage. Fig. 13-39. Since making a prototype involves constructing dies, patterns, and special tools, this means of checking is often too expensive for a product. Checking a drawing is less expensive and catches many errors before great expense is incurred. However, checking a drawing will not check the performance of the machine. Sometimes performance can be evaluated by computer, and features of the design can also be checked by the computer. Fig. 13-40.

Check Assembly Drawings. Another frequently used method of checking involves making a *check assembly drawing* of the machine. Such a drawing shows the assembled machine with parts in their proper position. This

Fig. 13–39. **Prototypes are built and tested under actual working conditions. (Caterpillar Inc.)**

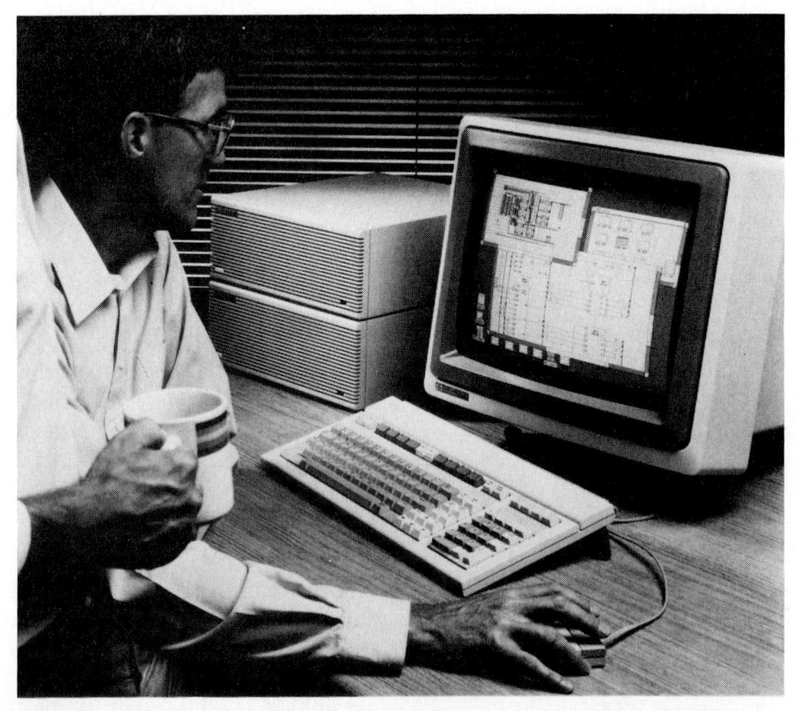

Fig. 13–40. **Product design and verification can sometimes be performed on the computer. (Hewlett-Packard Company)**

method provides a check on dimensions and assembly of the machine and reduces the calculations necessary to complete a check. Inspection is required for a complete check with this type of drawing since tolerances, fit, and manufacturing processes are not evaluated. It is an inexpensive and effective way to check the details. Fig. 13-38.

In making a check assembly drawing, accuracy is of extreme importance since the dimensions of the detail drawings need to be verified. Since the objective of this assembly is to check the details, it need not be as complete in detail as a general assembly drawing.

General Assembly Drawings

The general assembly drawing primarily serves to indicate how the various parts of a machine fit together. It usually includes only the outlines of parts and indicates the movements intended. It is useful in the assembly phase of the production of a machine. The check assembly drawing can often serve as a general assembly. Fig. 13-38.

Additional examples of general assembly drawings are given in Chapter 11, "Sectional Views."

Subassembly Drawings. If a machine is too large to be shown in a single general assembly drawing, it is usually broken down into smaller functioning units called *subassemblies.* They serve the same purpose as the general assembly. An assembly drawing of a water pump is a subassembly of an automobile engine.

Working Drawing Assembly. If an assembled object is not too complex, you can make a single drawing to serve as both detail drawing and assembly drawing. Fig. 13-41. Show the necessary orthographic views along with notes, dimensions, and sectioned portions if needed. If one or two pieces cannot be clearly shown and dimensioned, you may draw them on the sheet as separate details.

Pictorial Assembly Drawings. There are two types of pictorial assembly drawings commonly used. One is the cutaway drawing, as shown in Fig. 13-42. The object is drawn with a section removed to show interior details. It is shaded to appear much like a photograph. This type of work is done by a technical illustrator and is commonly called production illustration.

A second means of presenting a pictorial assembly is with an exploded drawing, as shown in Fig. 13-43. Each part is drawn pictorially and arranged in the order in which it should be assembled. The parts are usually identified by a number and are occasionally shaded. Such a drawing frequently accompanies a machine to assist in assembly and to identify parts for reordering when repairs are needed. Exploded drawings also are used for display purposes.

Additional drawings, examples, and problems on pictorial assemblies are given in Chapter 20, "Technical Illustration."

Installation Assembly Drawings. An installation assembly drawing usually includes only the outline of the machine and the overall dimensions necessary to show the amount of space needed to install it properly. Almost all details are omitted. Frequently, a machine will have a large number of dimensions or be manufactured in several sizes so that the dimensioning becomes complex. In this case, letters can be substituted for the dimensions and the actual dimensions recorded in table form on the drawing, as shown in Fig. 13-44.

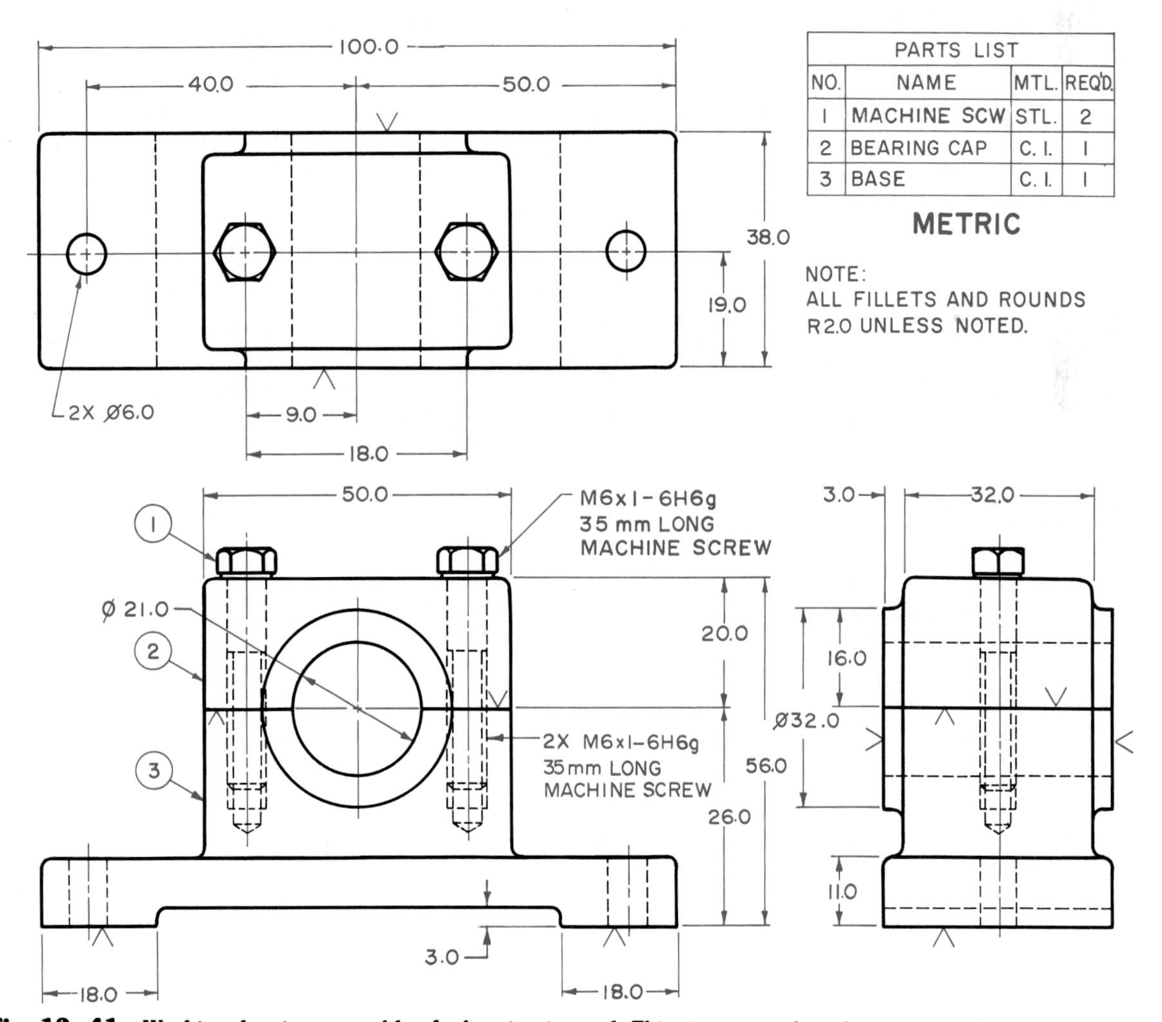

Fig. 13–41. Working drawing assembly of a bearing journal. This gives complete dimensioned details of each part and shows the journal assembled.

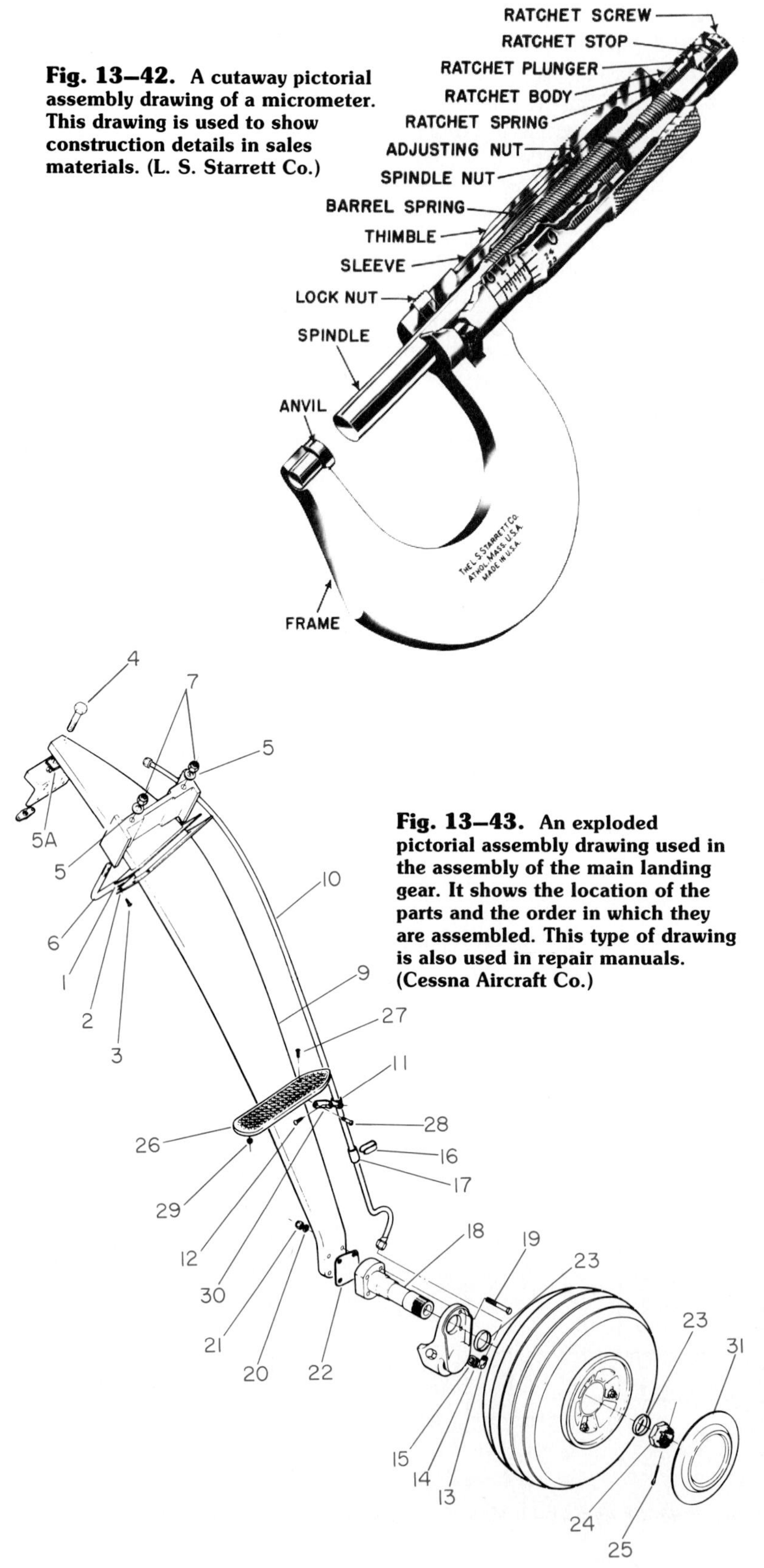

Fig. 13—42. **A cutaway pictorial assembly drawing of a micrometer. This drawing is used to show construction details in sales materials. (L. S. Starrett Co.)**

Fig. 13—43. **An exploded pictorial assembly drawing used in the assembly of the main landing gear. It shows the location of the parts and the order in which they are assembled. This type of drawing is also used in repair manuals. (Cessna Aircraft Co.)**

Maintenance Assembly Drawings. A maintenance assembly can be a general assembly or pictorial assembly. Fig. 13-45. This kind of drawing is used to give lubrication, servicing, and operating information about a machine. Sometimes parts are identified by the manufacturer's part number used for ordering replacement parts.

Manufacturers supply manuals useful to those who must service their products. Assembly drawings are important to show clearly the operation of these units. Fig. 13-46 is an assembly drawing of an automobile engine. It shows the oil circulation system. The drawing is kept as simple as possible. It only shows what is needed to serve its purpose.

Additional examples of maintenance assembly drawings are given in Chapter 20, "Technical Illustration."

Catalog Assembly Drawings. Catalog assembly drawings are made for inclusion in sales literature and parts catalogs. Their purpose is to explain the design and function of the object. Some are similar to general assemblies and have parts, such as shafts, shaded to present a lifelike resemblance. Fig. 13-47.

Pictorial assemblies are also used in catalogs. They are especially useful in parts catalogs. They usually show each part in an exploded pictorial view with the information for identifying each part by name and part number. Fig. 13-43.

Some companies use pictorial drawings showing the exterior of the object or a subassembly in their parts catalogs. Fig. 13-48. These can have parts in exploded positions as well as in final assembled positions. These drawings include a parts list giving the manufacturer's part numbers and the names of the parts.

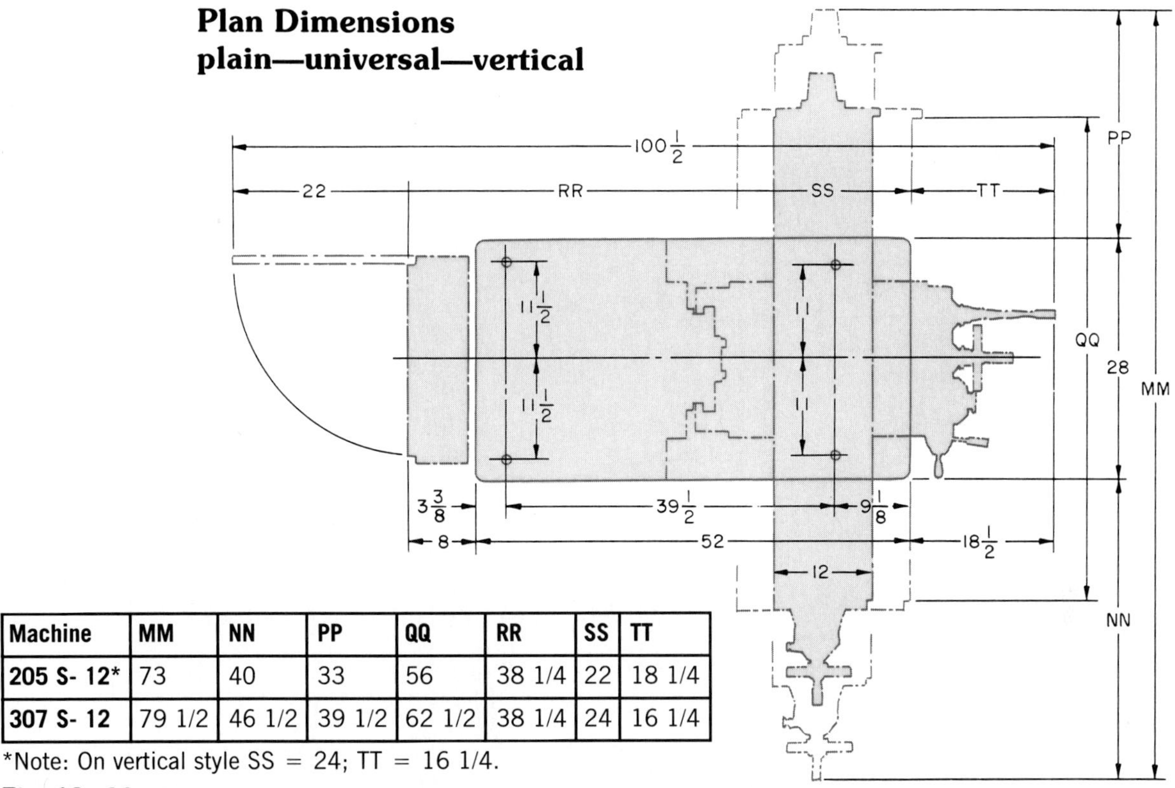

Machine	MM	NN	PP	QQ	RR	SS	TT
205 S- 12*	73	40	33	56	38 1/4	22	18 1/4
307 S- 12	79 1/2	46 1/2	39 1/2	62 1/2	38 1/4	24	16 1/4

*Note: On vertical style SS = 24; TT = 16 1/4.

Fig. 13–44. **This is a plan view of a milling machine showing the sizes for two different models manufactured. Notice that phantom lines are used to indicate the larger machine. (Kearney and Trecker)**

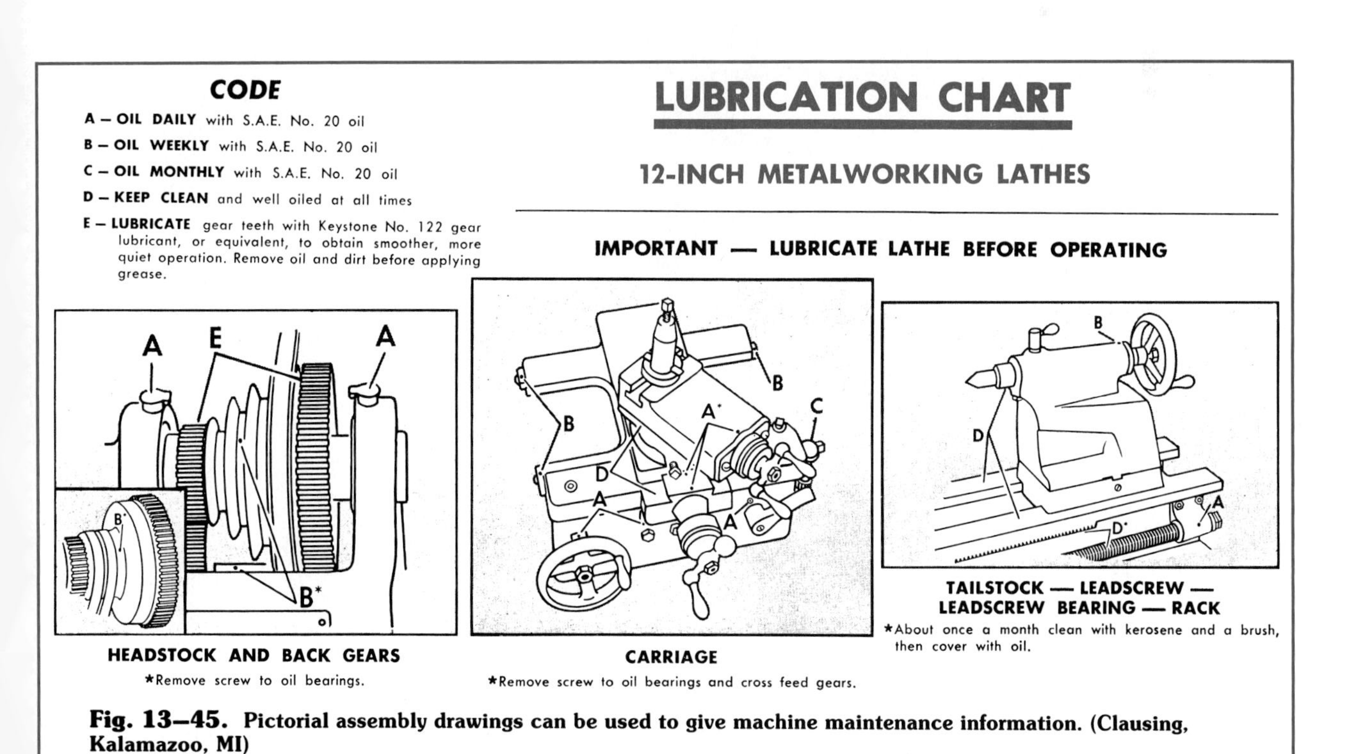

Fig. 13–45. **Pictorial assembly drawings can be used to give machine maintenance information. (Clausing, Kalamazoo, MI)**

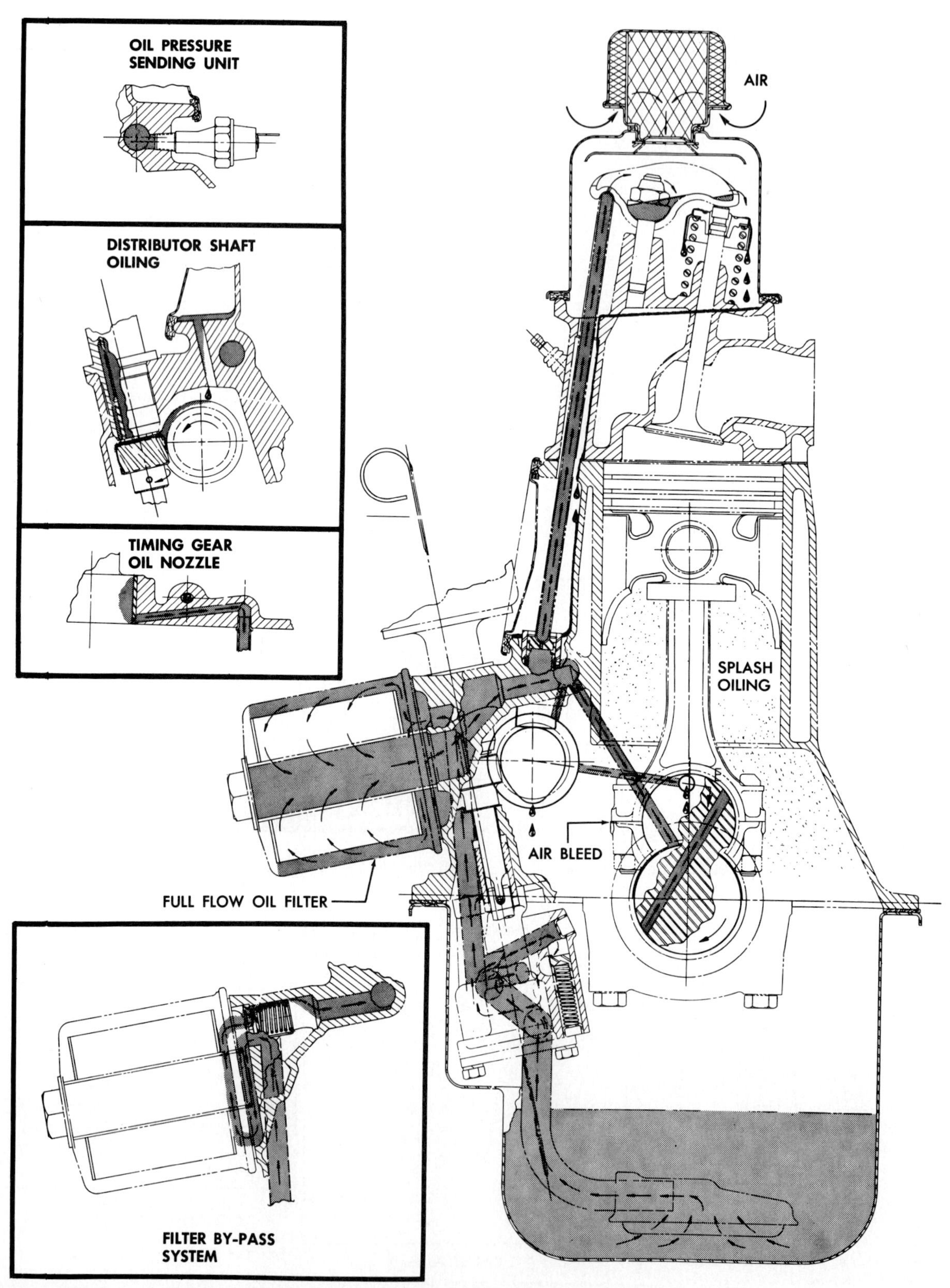

Fig. 13–46. An assembly drawing in section through an automobile engine showing the oil circulation system, and partial assembly drawings showing details not clear on the main assembly. These drawings are used in repair manuals prepared for auto mechanics. (General Motors Corp.)

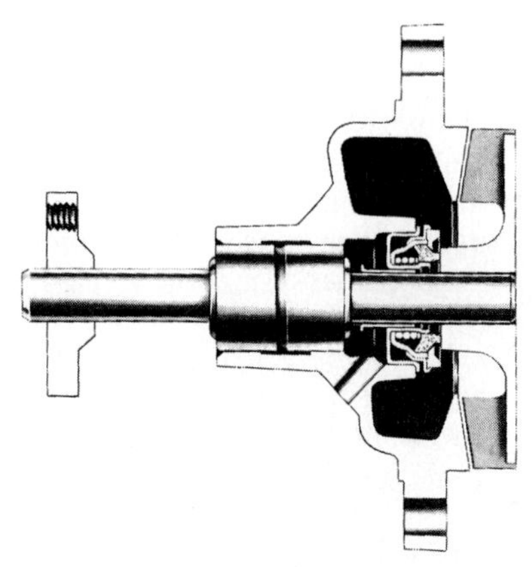

Fig. 13–47. A shaded assembly of an automobile water pump. It is drawn as a full section. This type of assembly drawing is useful in sales literature to show important features. It is also useful in technical manuals for auto repair. (Chevrolet Motor Division, General Motors Corp.)

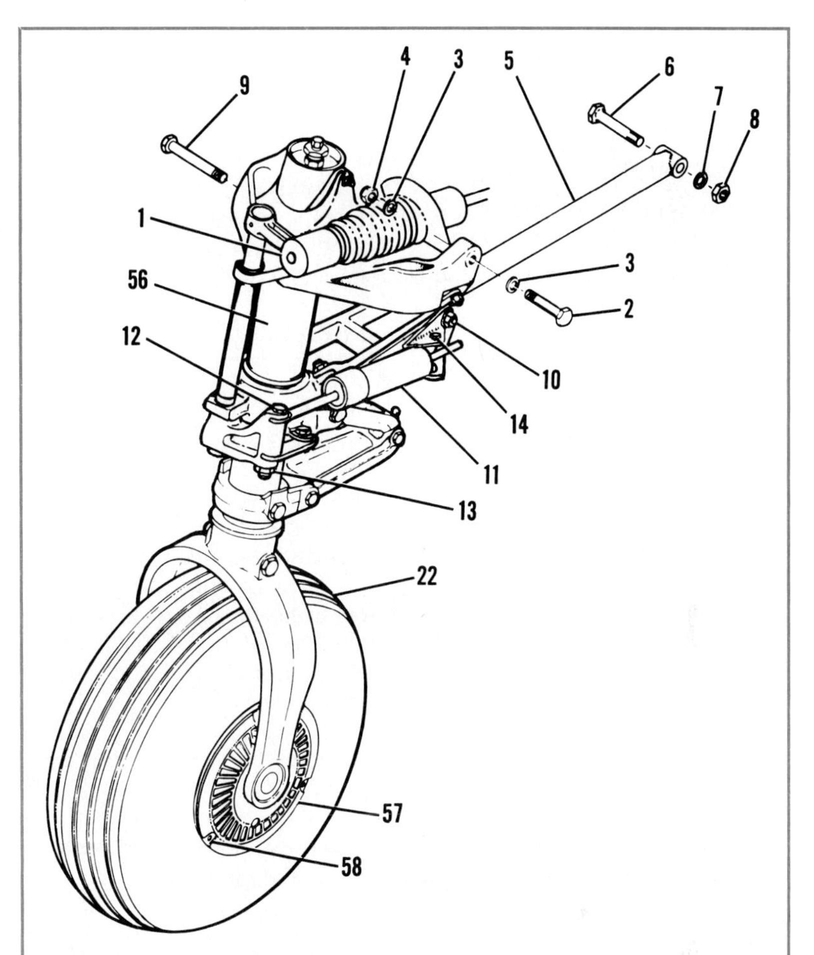

Data for Parts List

Fig. No.	Part No.	Description
1	1260630-1	Nose Gear Steering Assembly
2	AN176-21A	Bolt
3	AN960-616L	Washer
4	NAS679A6	Nut
5	1243600-1	Link-Drag
6	NAS464P5A42	Bolt
7	Ang60-516L	Washer
8	NAS679A5	Nut
9	AN6-32A	Bolt
10	NAS679A6	Nut
11	0743624-2	Shimmy Dampner Attaching Parts
12	NAS464P3A32C	Bolt
13	NAS679A3	Nut
14	AN4H5A	Bolt
22	1241156-12	Wheel Assembly, Nose
56	1243610-2-6	Nose Gear Shock Strut Assembly
57	0743627-1	Cap, Wheel
58	AN936A8	Washer

Fig. 13–48. A pictorial assembly of an aircraft nose gear used in a parts catalog. (Cessna Aircraft Co.)

Chapter Review

Build Your Vocabulary

You should understand and use the following terms as part of your working vocabulary. Write a brief explanation of what each means.

preliminary design assembly
detail drawing
pattern detail
draft
shrinkage
machining
core
machining drawing
casting detail drawing
forging detail drawing
die
parting line
shrinkage tolerance
die wear tolerance
mismatch
welding detail
stamping
drawing metal
bend allowance
assembly drawing

Sharpen Your Math Skills

1. A cast-iron casting is 12″ long. Cast-iron objects must be cast 1⁄96″ per inch longer than finished size to allow for shrinkage. How long should the pattern be made?
2. If a casting 100 millimeters in length has an allowable tolerance of ± 0.8 millimeter, what are its largest and smallest acceptable sizes?
3. A 6″-diameter hole is to be cast in a metal part. Its tolerance is ± .030″. What are the largest and smallest acceptable diameters?
4. A forging is to be 4″ high when finished. Its thickness tolerance is .030″, and the machining allowance on the top and bottom is .060″. What is the total height of the forging?
5. A sheet metal part .05″ is to be bent through a 90° angle. The bend allowance is .12″. The finished piece should measure 3″ long. What is the length of the part before it is bent?

Study Problems—Directions

Unless otherwise noted, make the following drawings full size.

Pattern Detail Drawings

1. Make a pattern detail drawing of the cap of the hoist ring, **Fig. P13-1**. The material is bronze.
2. Make a pattern detail drawing of the base of the optical mount, **Fig. P13-2**. The material is aluminum.
3. Make a pattern detail drawing of the base of the spring stop in **Fig. P13-3**. The material is cast iron.

Casting Detail Drawings

4. Make a casting detail drawing for the cap of the hoist ring, **Fig. P13-1**. The material is steel.
5. Draw a casting detail drawing for the arm of the control handle, **Fig. P13-4**. The material is cast iron.
6. Draw a casting detail drawing for the base of the spring stop, **Fig. P13-3**. The material is cast iron.
7. Draw a casting detail for the hook of the fixture clamp, **Fig. P13-5**. The material is cast iron.

Machining Detail Drawings

8. Draw a machining detail drawing for the arm of the control handle, **Fig. P13-4**. All surfaces are to be finished.
9. Draw machining detail drawings for the four parts of the spray gun nozzle, **Fig. P13-6**. Use a scale of 2:1. All surfaces are to be finished.
10. Draw a machining detail drawing of the optical cup, **Fig. P13-2**. All surfaces are to be finished.
11. Draw a machining detail drawing of the yoke and the handle of the cam clamp, **Fig. P13-7**. All surfaces are to be finished.
12. Draw machining detail drawings for the upper and lower plates of the adjustable angle, **Fig. P13-8**. All surfaces are to be finished.
13. Draw a machining detail of the guide block on the wood shaper hold-down fixture, **Fig. P13-9**. All surfaces except the one with the boss are to be finished.

Forging

14. Draw a forging detail of the pin of the cabinet hinge, **Fig. P13-10**. Scale = 2:1.

15. Draw a forging drawing and a separate machining drawing of the yoke of the foot pedal assembly, **Fig. P13-11**. Be certain to figure tolerances needed. The yoke weighs 0.4 pound.

16. Draw a forging detail of the shaft of the T-handle socket wrench, **Fig. P13-12**. Be certain to figure tolerances needed. The shaft weighs three pounds.

17. Draw a forging detail drawing of the lathe power control link, **Fig. P13-13**. Show machining details on the same drawing. The inside of the bearing hub must be machined on both sides to receive a bearing with a 1⅛-inch outside diameter. A ⅜-inch diameter hole is to be drilled in the center of the bearing hub. All fillets and rounds should be held to the minimum permitted for forgings. The link weighs one-half pound. Be certain to add necessary tolerances to dimensions on the drawing.

18. Draw a forging detail of the foot brake arm, **Fig. P13-14**. Show machining details on the same drawing. All fillets and rounds should be held to the minimum permitted for forgings. The arm weighs three-fourths of a pound. Be certain to add all necessary tolerances to dimensions on the drawing.

Stamping Detail Drawings

19. Draw a stamping detail of the hinge leaves in **Fig. P13-10**.

20. Draw a stamping detail drawing for the brace of the foot pedal assembly, **Fig. P13-11**.

21. Draw a stamping detail of the blade of the bevel square, **Fig. P13-15**.

22. Draw a stamping detail drawing of the seat belt anchor, **Fig. P13-16**.

23. The unit shown in **Fig. P13-17** is formed by stretching the metal during the stamping operation. Make a stamping detail drawing of this unit.

24. The rearview mirror bracket, **Fig. P13-18**, has a stamped rib to give it stiffness. Make a stamping detail drawing of the bracket.

25. Draw a stamping detail drawing of the motor bracket, **Fig. P13-19**.

26. Draw a stamping detail drawing of the flasher bracket, **Fig. P13-20**. Design the corners so they have enough clearance to be formed easily.

Welding Detail Drawings

27. Draw a welding detail drawing showing the union of the pedal and arm in the foot pedal assembly, **Fig. P13-11**.

28. Draw a welding detail drawing of the auto assembly drill fixture, **Fig. P13-21**.

29. Draw a welding detail drawing of the auto assembly fixture, **Fig. P13-22**.

30. An exploded view of a polishing head to be assembled by welding is shown in **Fig. P13-23**. Make a complete set of working drawings. Indicate the welds with the proper symbols.

Check Assembly Problem

31. Details of a spray gun nozzle are given in **Fig. P13-6**. Using the dimensions given, make a check assembly drawing. Make any corrections necessary to make the parts assemble properly.

General Assembly Problems

32. Make a general assembly of the control handle shown in **Fig. P13-4**. Include a parts list.

33. Make a general assembly of the spring stop shown in **Fig. P13-3**. Include a parts list.

34. Draw a general assembly of the fixture clamp shown in **Fig. P13-5**. Include a parts list.

35. Draw a general assembly of the optical cup and optical cup mount with the metal cover over the unit. Detail drawings are shown in **Fig. P13-2**. The optical cup screws into the cup mount. Include a parts list.

36. In **Fig. P13-7** is shown an exploded pictorial view of a cam clamp. Draw this in assembled condition. Use sections where necessary to clarify details. Include a parts list.

Additional general assembly problems are given in Chapter 11, "Sectional Views."

Working Drawing Assembly Problems

37. Make a working drawing assembly of the hoist ring, **Fig. P13-1**.

38. An adjustable angle used in machine shops is shown in **Fig. P13-8**. Draw a working drawing assembly of this tool.

39. Make a working drawing assembly of the salt shaker, **Fig. P13-24**.

40. Make a working drawing assembly of the hinge shown in **Fig. P13-10**. Draw it double size.

Pictorial Assembly Problems

For additional information on pictorial assembly drawings, see Chapter 20, "Technical Illustration."

41. Make a cutaway pictorial assembly drawing of the grease cup, **Fig. P13-25**.

42. Make a pictorial section of the navy pipe union, **Fig. P13-26**. Remove a quarter section to show assembly details.

43. Make an exploded assembly drawing of the tapered nose clamp, **Fig. P13-27**.

Installation Assembly Problem

44. Visit your industrial education shop, select a machine, and make a simple installation assembly drawing of it, showing its overall size.

Maintenance Assembly Problem

45. Select a machine and make maintenance assembly drawings, showing where it should be lubricated. This could be a machine in your industrial education shop.

Catalog Assembly

46. Prepare a catalog assembly drawing of the spring stop, **Fig. P13-3**. Use the principles of shading as explained in Chapter 20, "Technical Illustration." Draw the spring stop as a full section, and draw it double size.

Other Assembly Problems

47. Make an assembly drawing that would be most useful in a shop manual to show a machine operator how to assemble the wood shaper hold-down fixture, **Fig. P13-9**.

48. Make an assembly drawing that would be most useful to a factory worker assembling the bevel square shown in **Fig. P13-15**.

Design Problems

Following are some problems requiring an original solution. There is more than one correct solution. Possible solutions are limited only by the imagination and ingenuity of the designer.

As you consider your solution, remember that simplicity, ease of manufacture, and minimal cost are important factors. Use standard parts, such as bolts, springs, and keys, whenever possible.

The complete solution will include freehand design sketches and assembly, detail, and pictorial drawings as needed to completely describe the solution.

Design Problem 1. **Fig. P13-28** shows panels that are to serve as office partitions. Design a means of connecting these panels so that they can be fastened with only a screwdriver or pliers. The fastener must permit the panels to be taken apart rapidly without damage. This permits easy rearrangement of office space. The fastener should be as inconspicuous as possible.

Design a leg that will hold the panels 12 inches off the floor. The legs should be easily removable.

Design Problem 2. Design a hanger to secure the arm to the machine base, as shown in **Fig. P13-29**. The arm must be able to swing through an arc of 180 degrees and be removed from the machine base without the use of tools.

Design Problem 3. **Fig. P13-30** shows a base for a machine. Shavings from the machine fill the base and have to be cleaned out frequently. Design a sheet metal door and the fastening devices to hold it in place. The door should be completely removable from the base without the use of tools. This design should allow for rapid replacement of the door.

Design Problem 4. Design a fitting that will hold bar A to bar B, as shown in **Fig. P13-31**. The fitting should permit bar B to slide up and down bar A. It should permit bar B to be tightened firmly to bar A anyplace along its length.

Bar B should be able to slide horizontally and lock in place.

Bar B should meet bar A at a 90-degree angle.

Design Problem 5. Design a device to convert the hand grinder, **Fig. P13-32**, into a bench grinder. The holding device should permit the operator to have both hands free to hold the work to be ground. The center of the grinding wheel should be 3½ inches (88 mm) above the table when the grinder is in a horizontal position. The device should permit the grinder to swing above and below the horizontal at least 30 degrees and be locked in any position.

Design Problem 6. Design a fastening device which will enable the steel wall panels in **Fig. P13-33** to be joined quickly, yet permit disassembly so the panels can be moved or replaced. The fastening device should be easily secured in place without the need for tools.

Design Problem 7. The rod shown in **Fig. P13-34** moves in a vertical direction. Design a device or alter the rod or casting so a machine operator can easily raise or lower the rod but still lock it tightly in place. The device should provide positive stops in ½ inch (12 mm) increments. This enables the operator to lock the bar in ½ inch (12 mm) increments without measuring.

Design Problem 8. Design a gland to hold the bearing shown in **Fig. P13-35** in place. A *gland* is a metal part that holds a bearing in place. The gland and the bearing should have a force fit of FN 1. The machine casting can have threaded holes added.

Design Problem 9. Design a pipe hanger that can be adjusted to carry pipes from 2 to 3 inches (50 to 75 mm) in diameter. The part of the hanger that is to be connected to the overhead must be adjustable so the pipe can be carried from 4 to 6 inches (100 to 150 mm) below the overhead. It should be designed so that it can be bolted to wood or metal overhead members. **Fig. P13-36.**

Study Problems

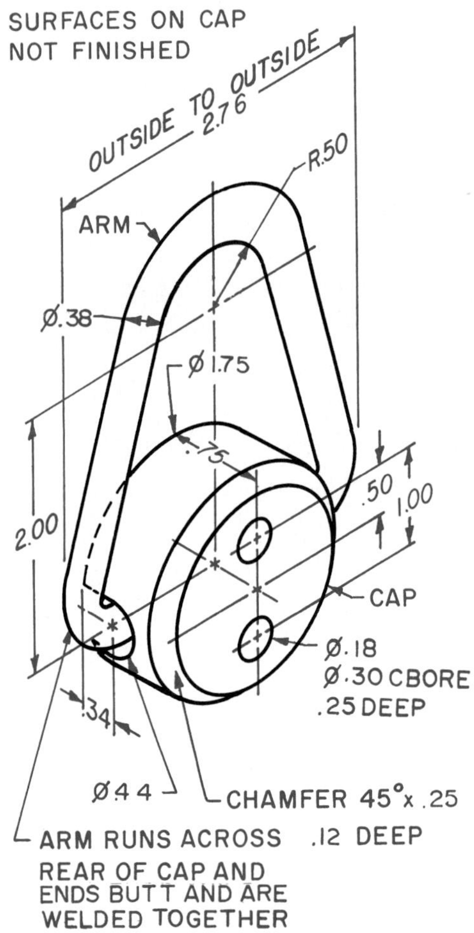

SURFACES ON CAP
NOT FINISHED
OUTSIDE TO OUTSIDE
2.76
R.50
ARM
Ø.38
Ø1.75
.75
2.00
.50
1.00
CAP
Ø.18
Ø.30 CBORE
.25 DEEP
.34
Ø.44
CHAMFER 45° x .25
.12 DEEP
ARM RUNS ACROSS
REAR OF CAP AND
ENDS BUTT AND ARE
WELDED TOGETHER

P13—1. Hoist ring. (Carr Lane Manufacturing Co.)

Study Problems

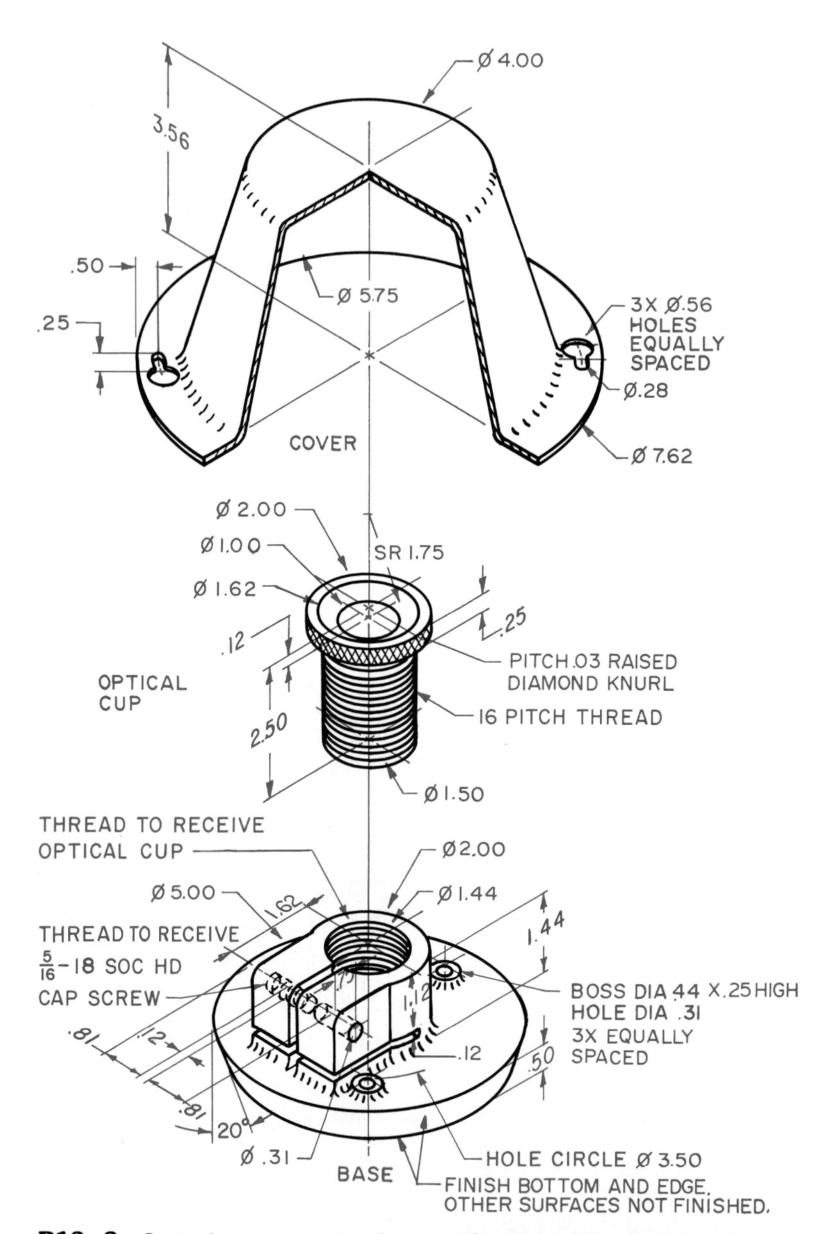

P13–2. **Optical cup, mount, and cover. (Carr Lane Manufacturing Co.)**

Study Problems

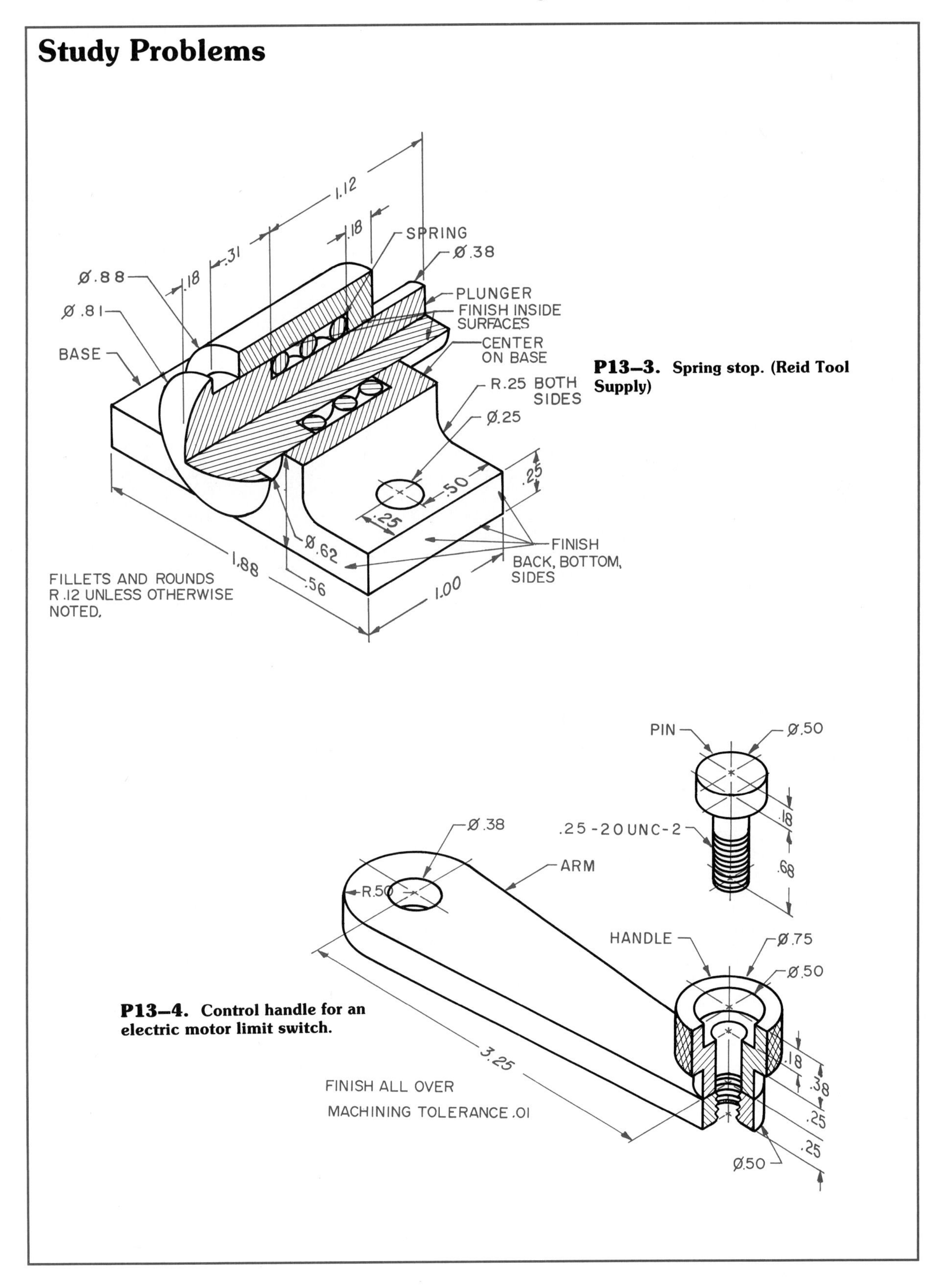

P13–3. Spring stop. (Reid Tool Supply)

P13–4. Control handle for an electric motor limit switch.

Study Problems

P13–5. **Fixture clamp. (Carr Lane Manufacturing Co.)**

P13–6. **Spray gun nozzle. (Delavan Manufacturing Co.)**

Study Problems

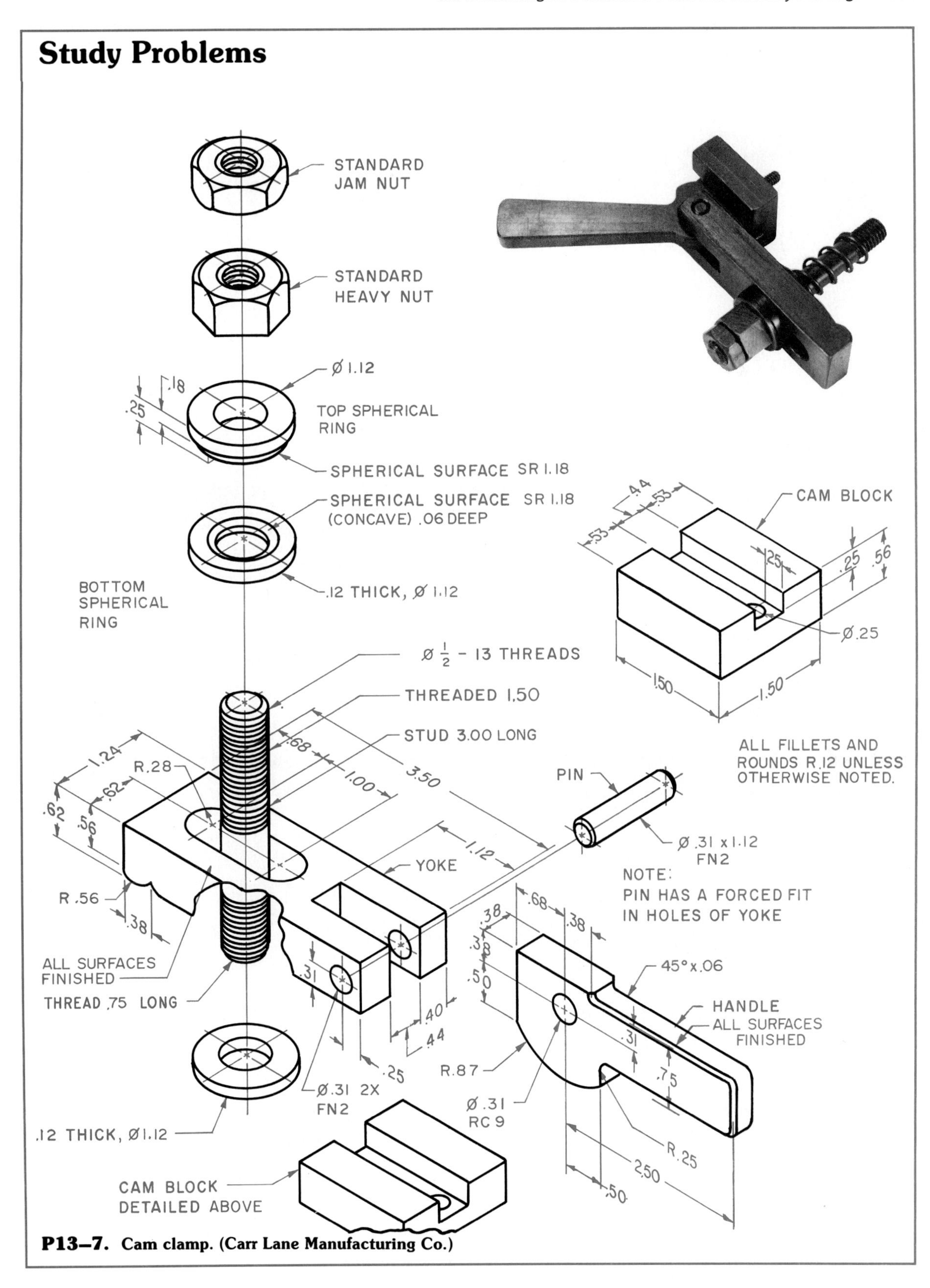

P13–7. Cam clamp. (Carr Lane Manufacturing Co.)

Study Problems

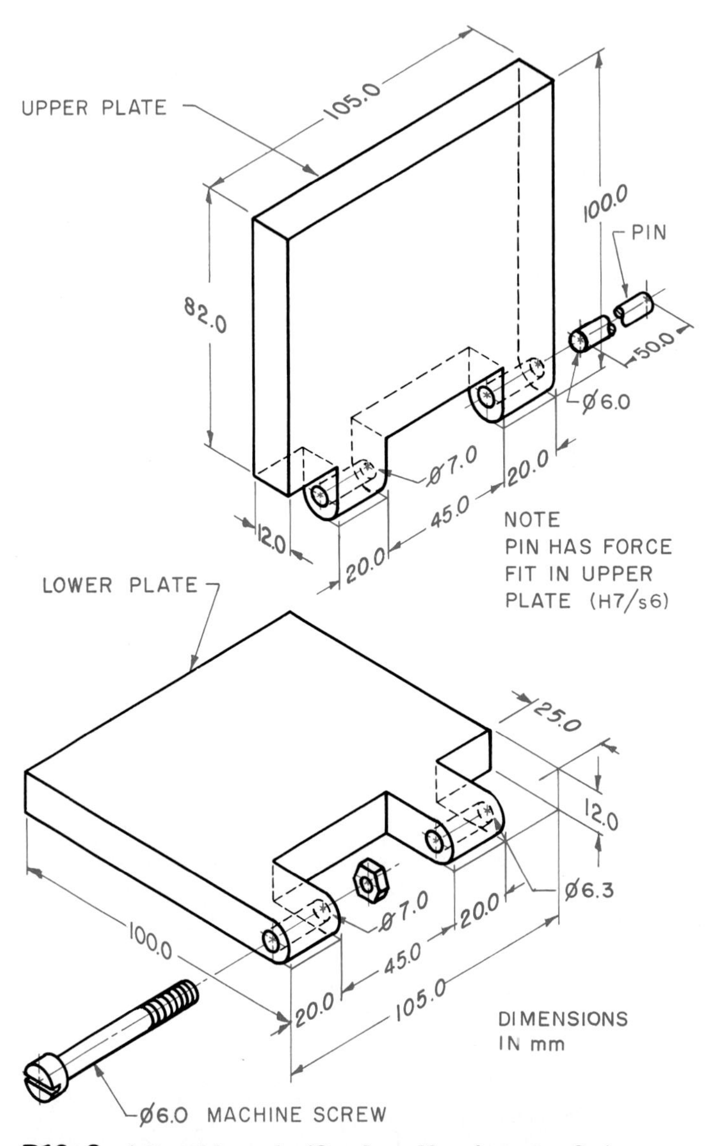

P13–8. Adjustable angle. (Carr Lane Manufacturing Co.)

Study Problems

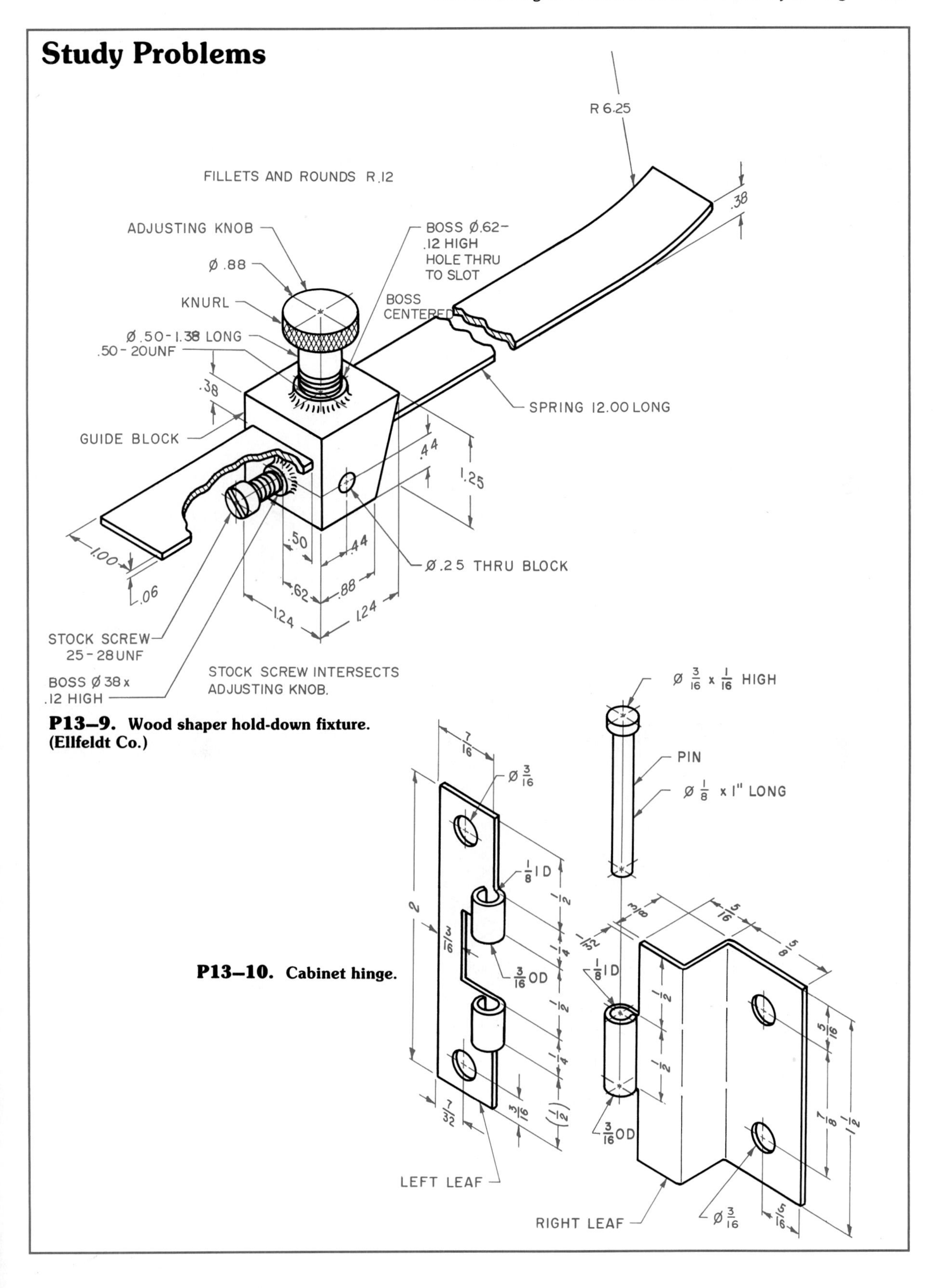

P13–9. **Wood shaper hold-down fixture. (Ellfeldt Co.)**

P13–10. **Cabinet hinge.**

Study Problems

P13–11. Foot pedal assembly. (Ford Motor Co., Kansas City Assembly Plant)

P13–12. T-handle socket wrench. (Billings and Spencer)

Study Problems

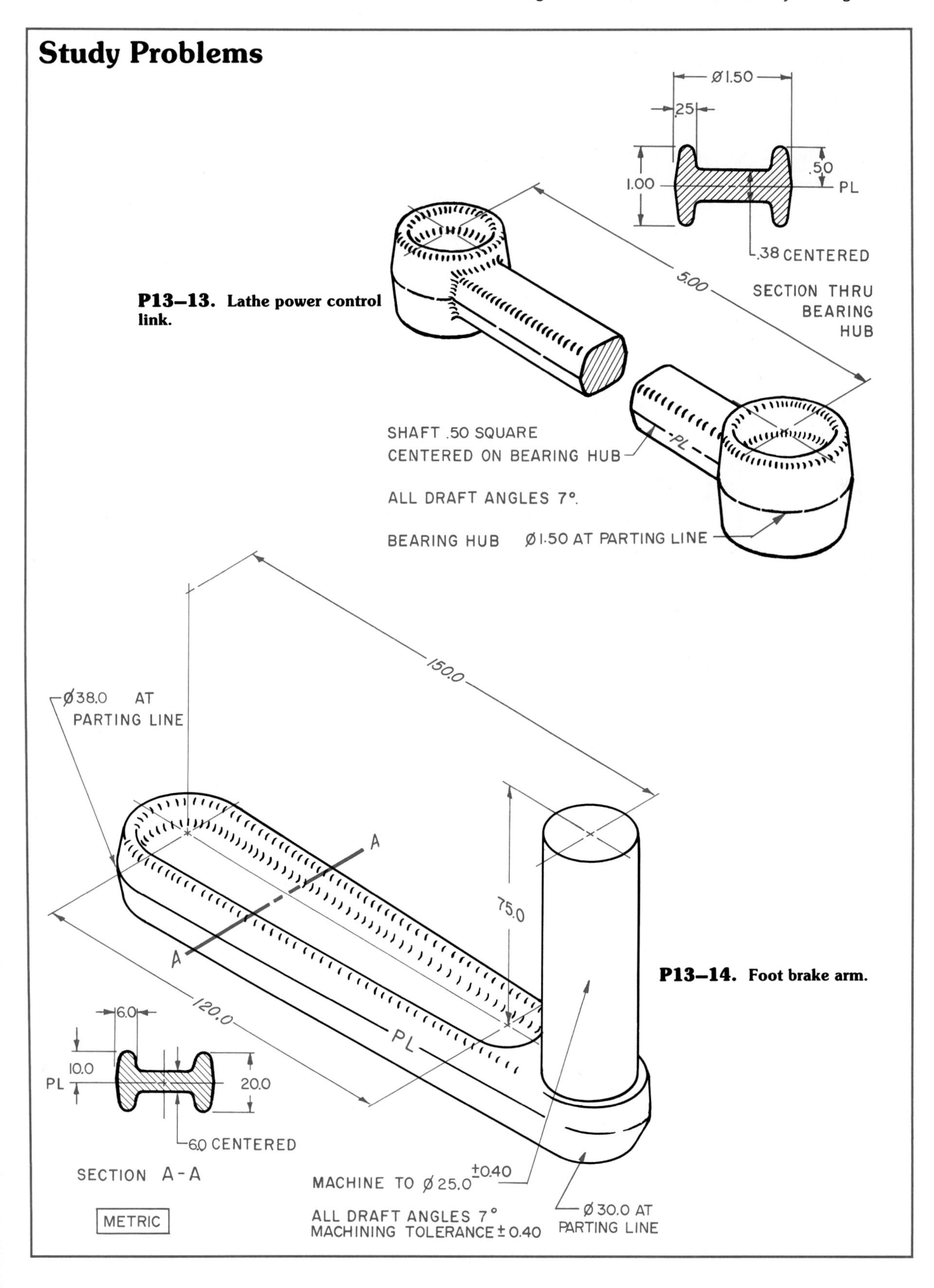

P13–13. Lathe power control link.

P13–14. Foot brake arm.

Study Problems

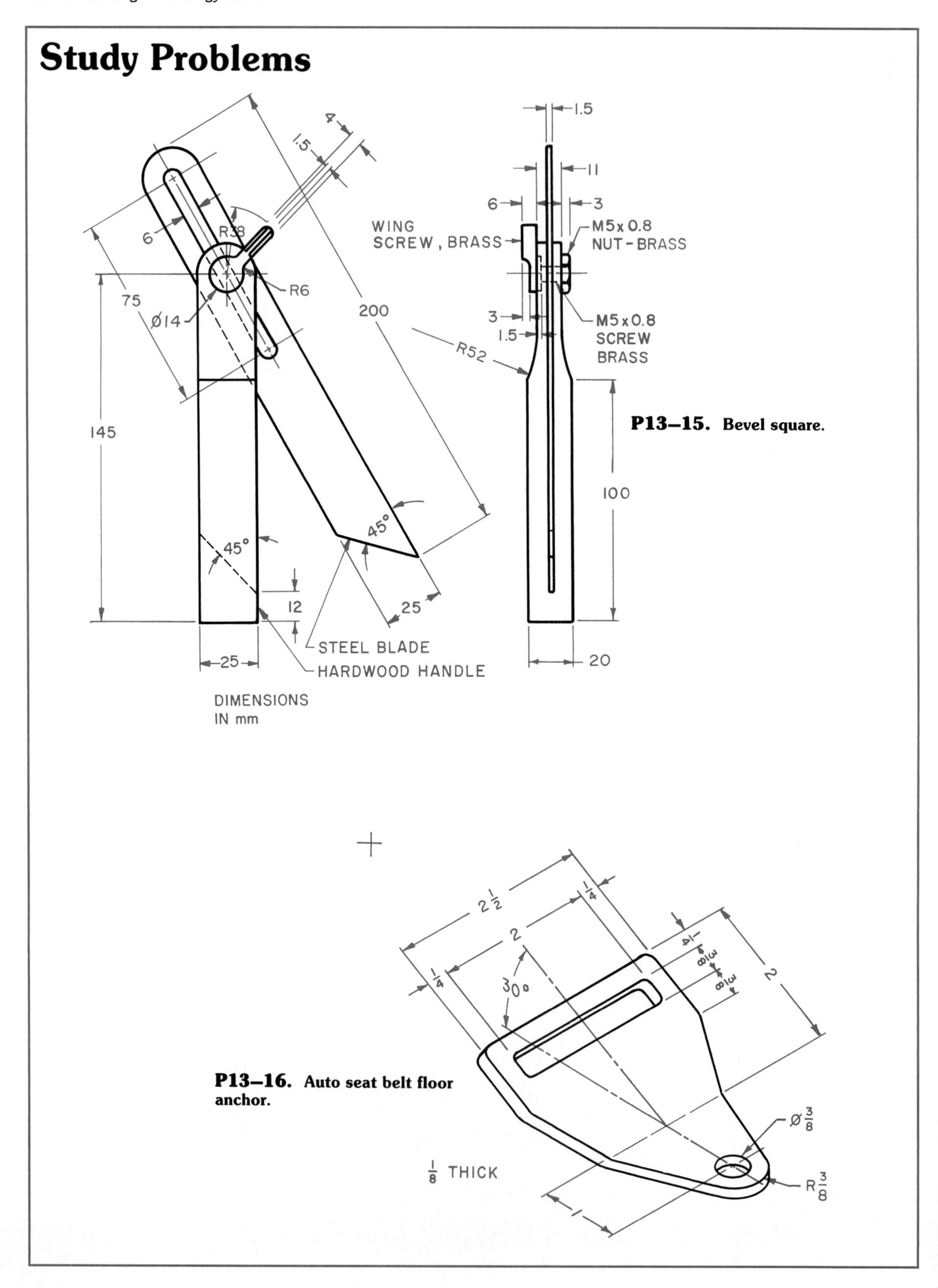

P13–15. Bevel square.

P13–16. Auto seat belt floor anchor.

Study Problems

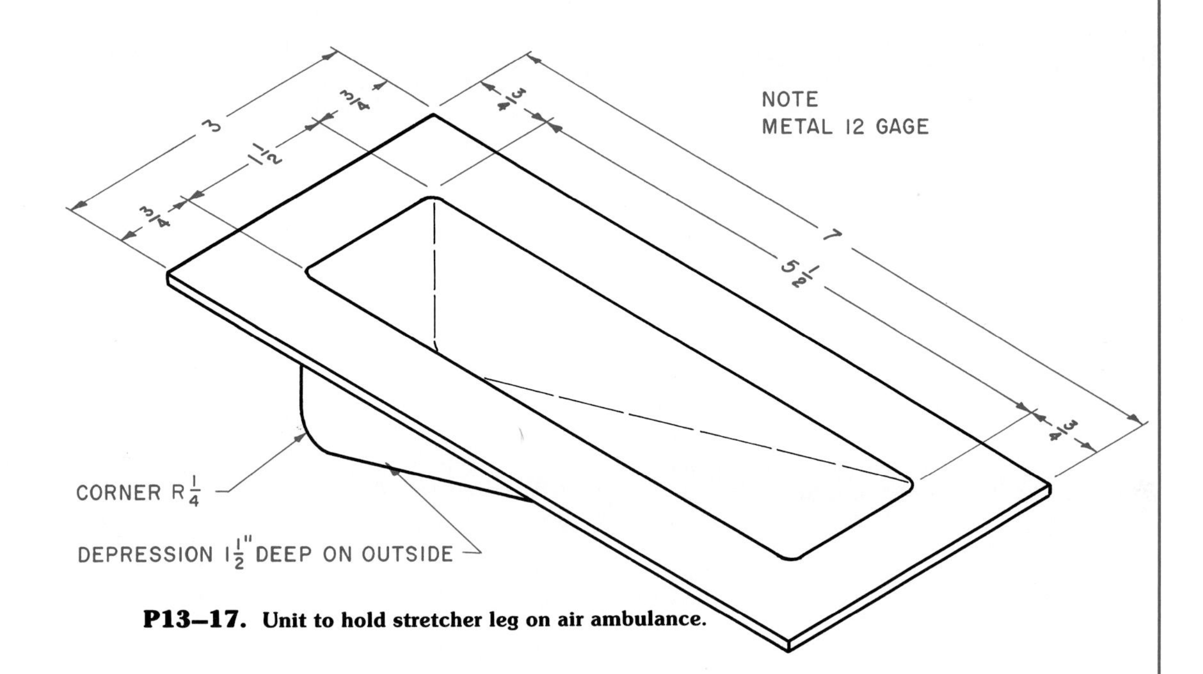

P13–17. Unit to hold stretcher leg on air ambulance.

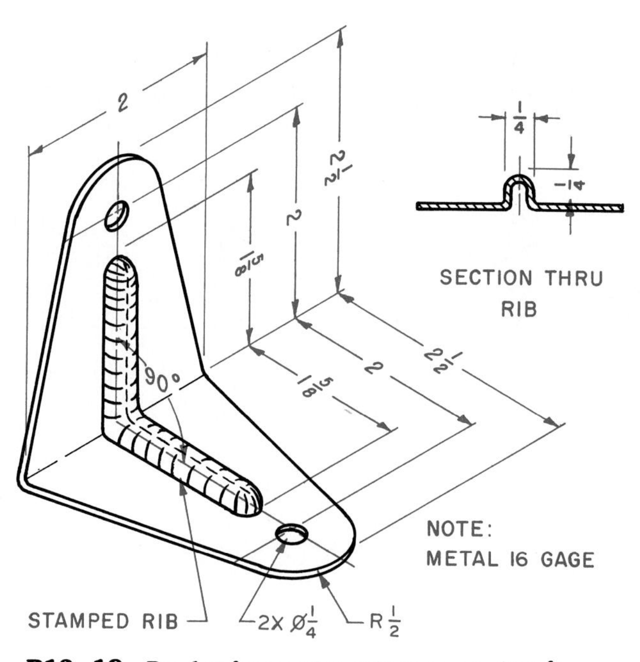

P13–18. Bracket for rearview mirror on an aircraft.

Study Problems

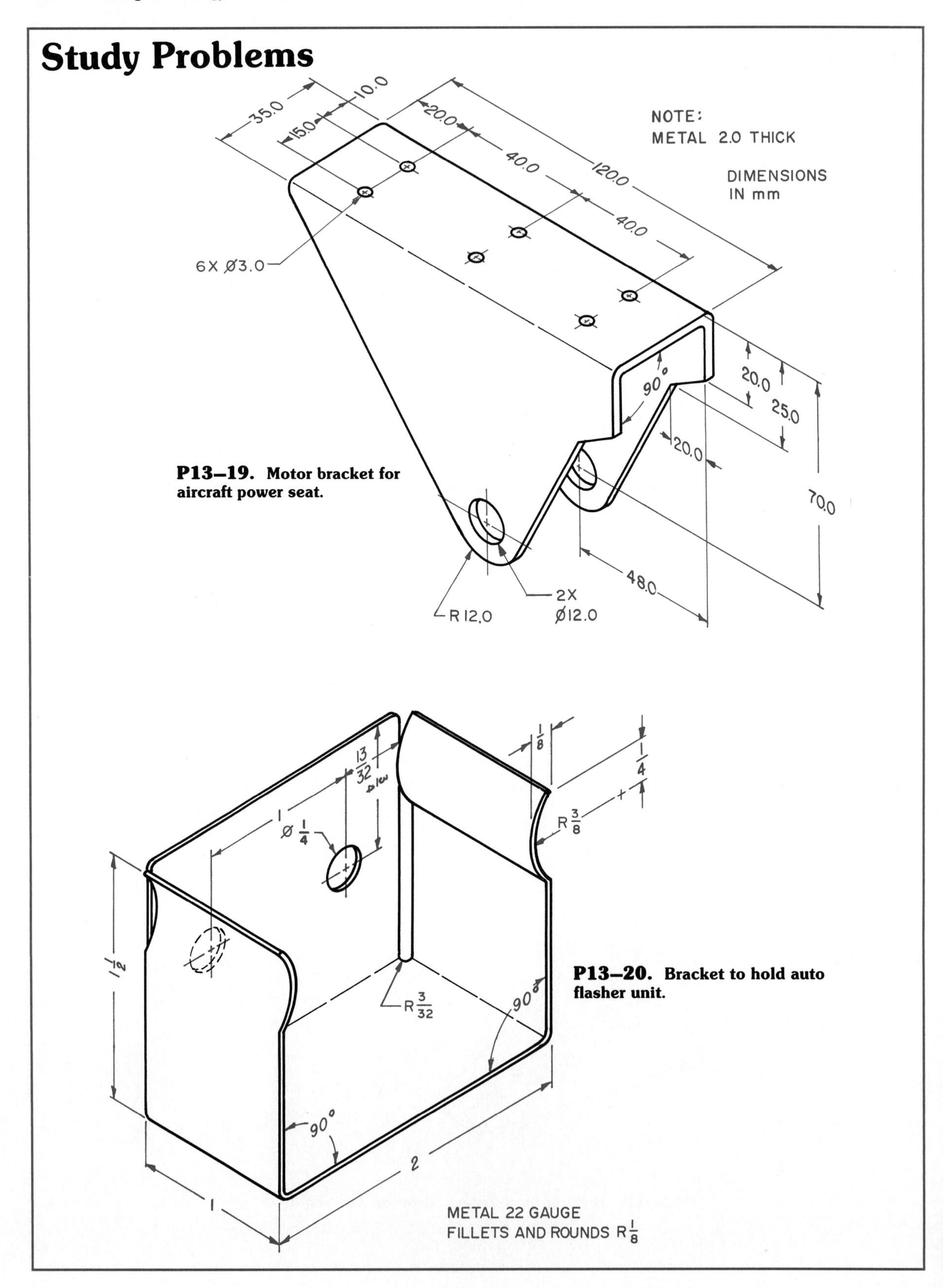

P13–19. Motor bracket for aircraft power seat.

P13–20. Bracket to hold auto flasher unit.

Study Problems

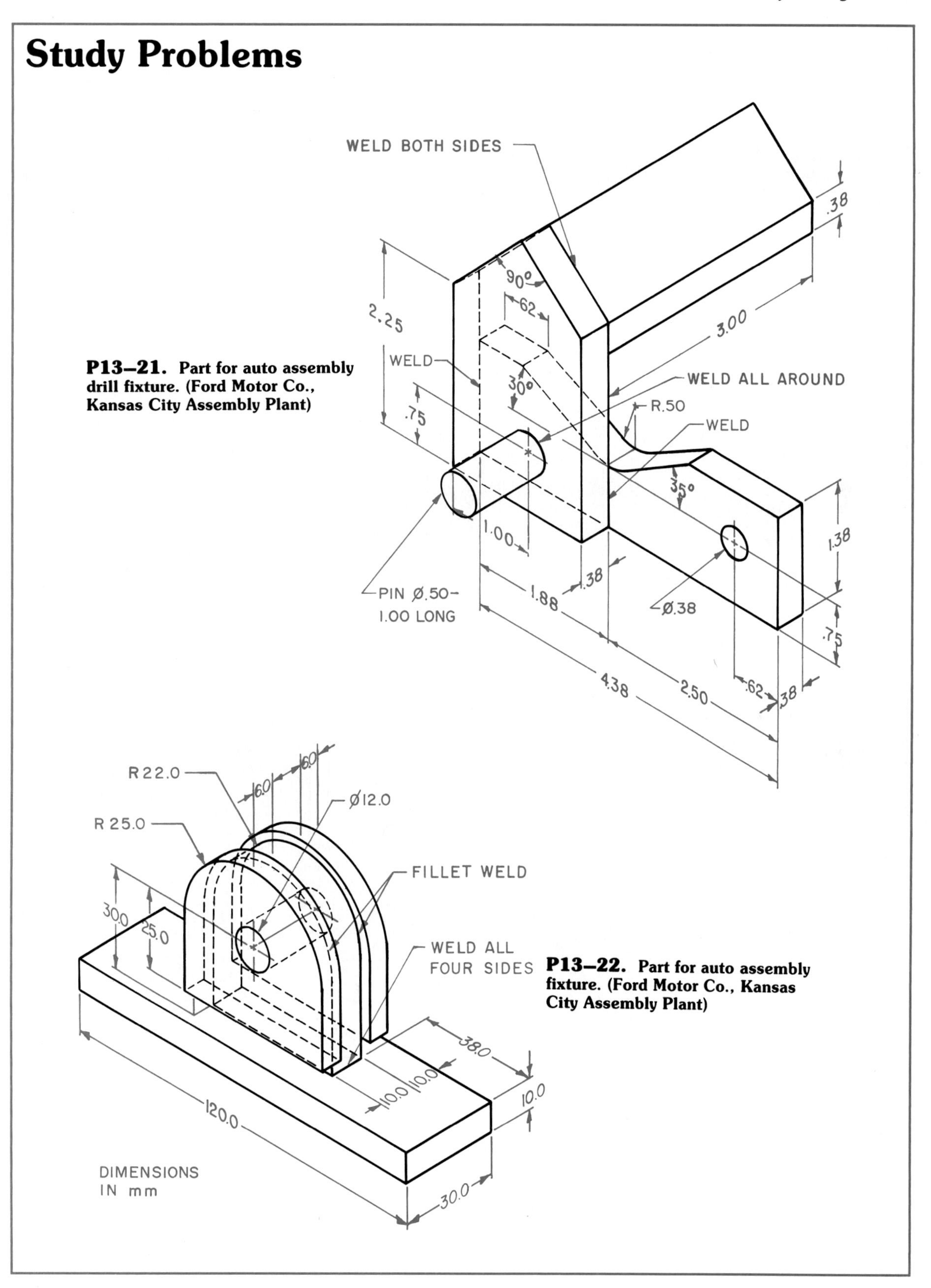

P13–21. Part for auto assembly drill fixture. (Ford Motor Co., Kansas City Assembly Plant)

P13–22. Part for auto assembly fixture. (Ford Motor Co., Kansas City Assembly Plant)

Study Problems

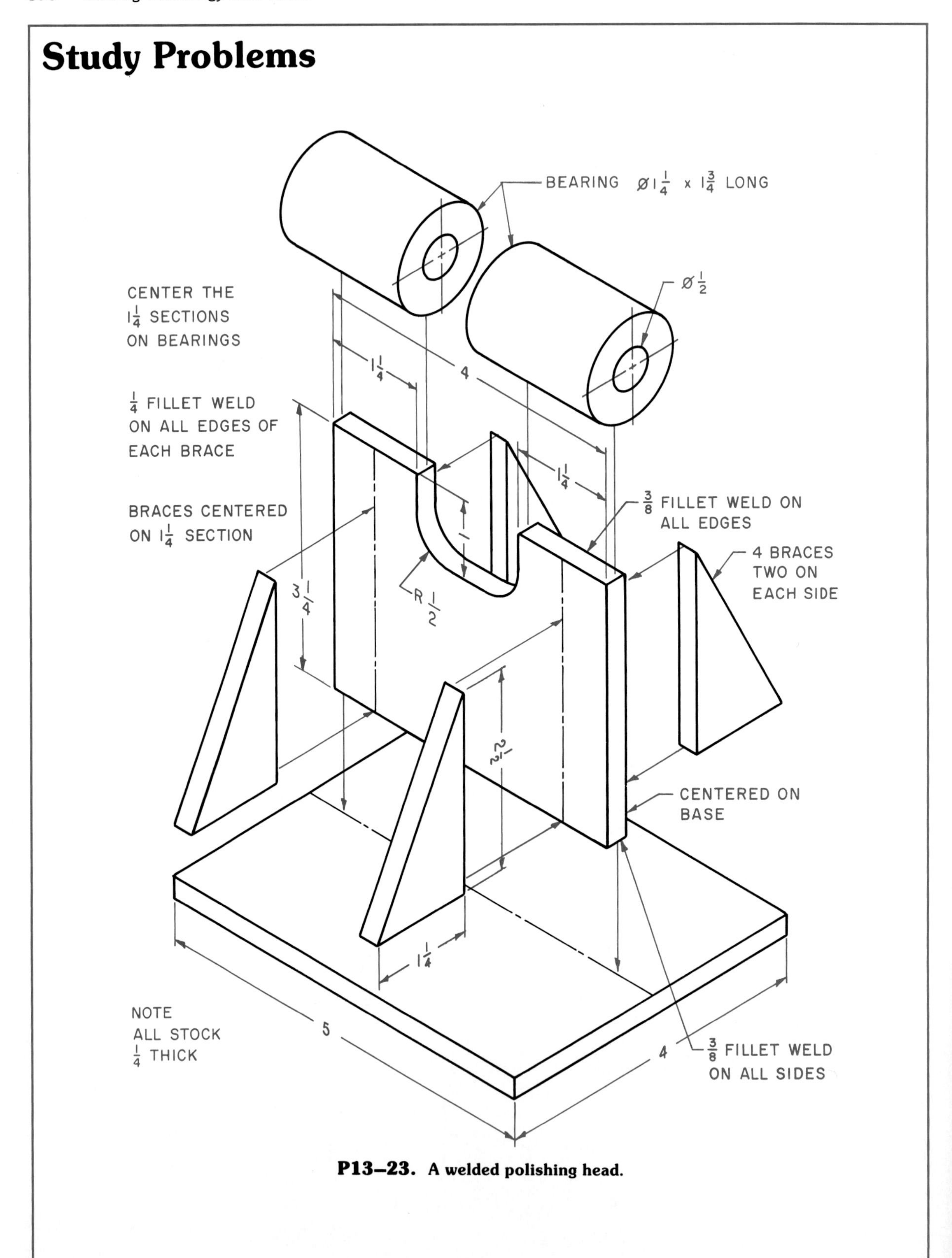

P13–23. A welded polishing head.

Study Problems

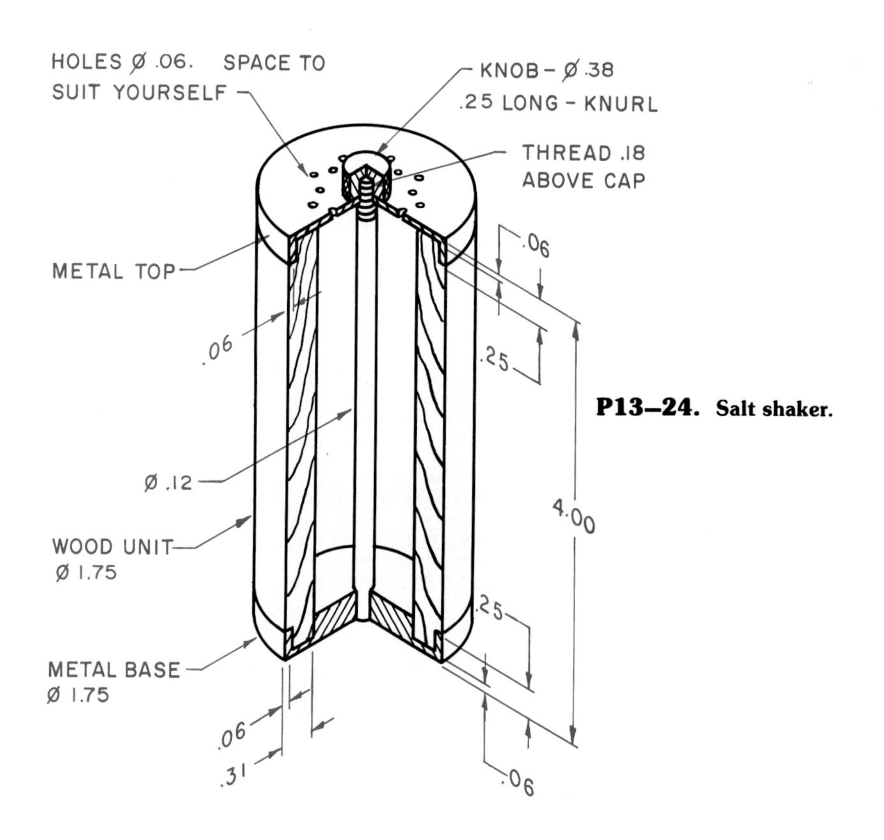

P13–24. Salt shaker.

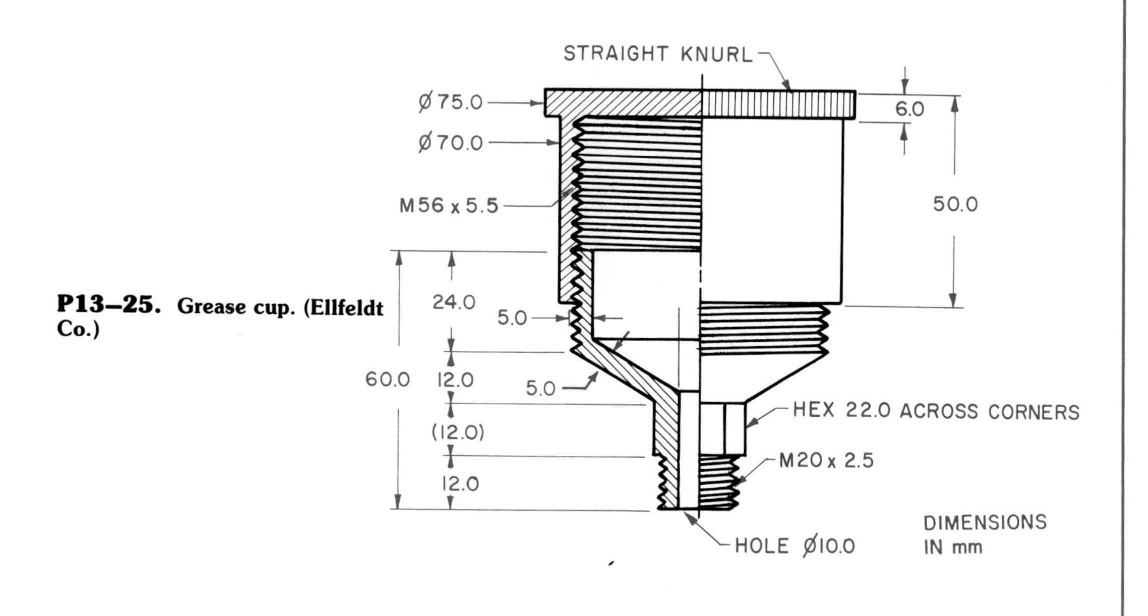

P13–25. Grease cup. (Ellfeldt Co.)

Study Problems

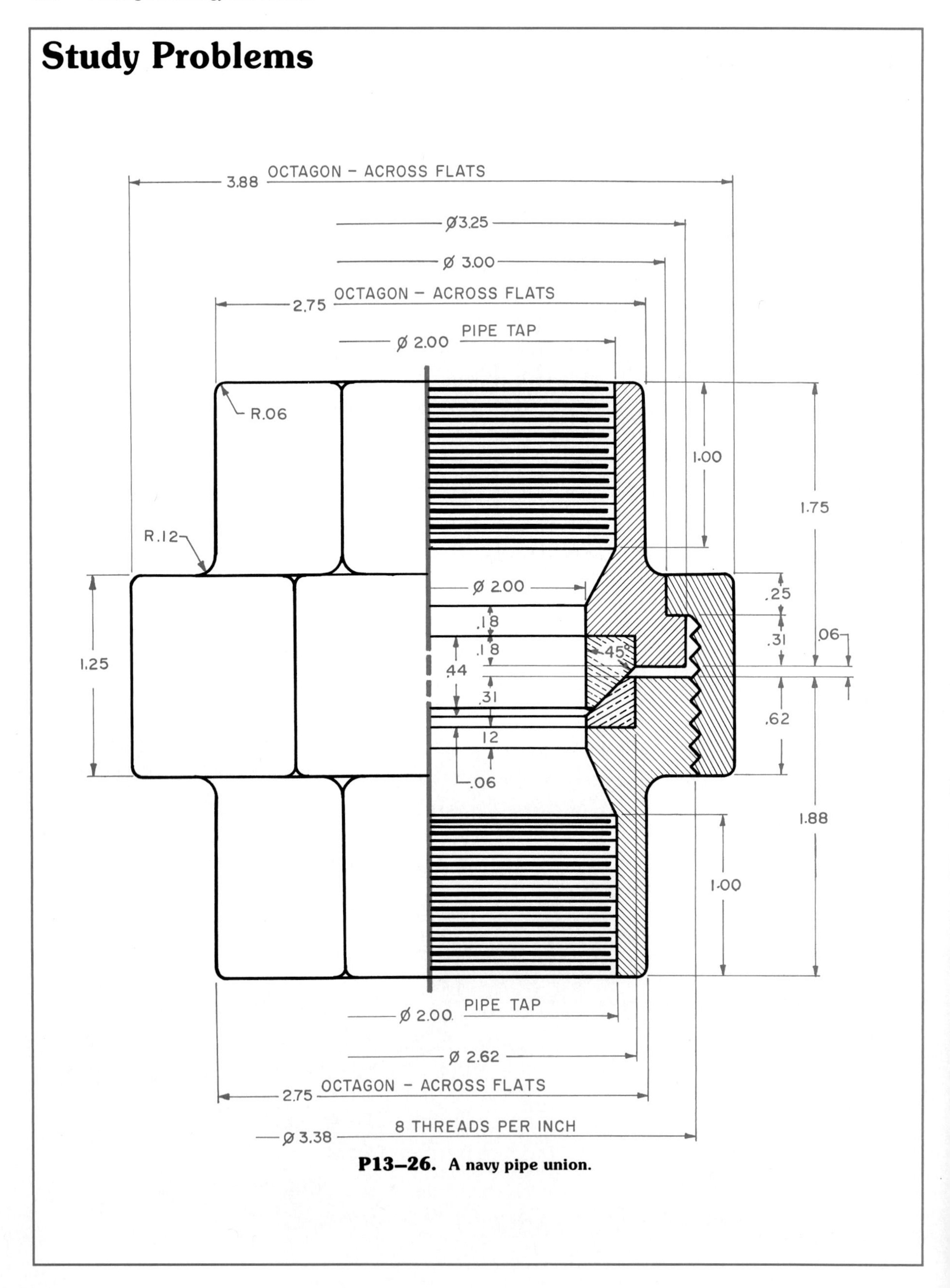

P13–26. **A navy pipe union.**

Study Problems

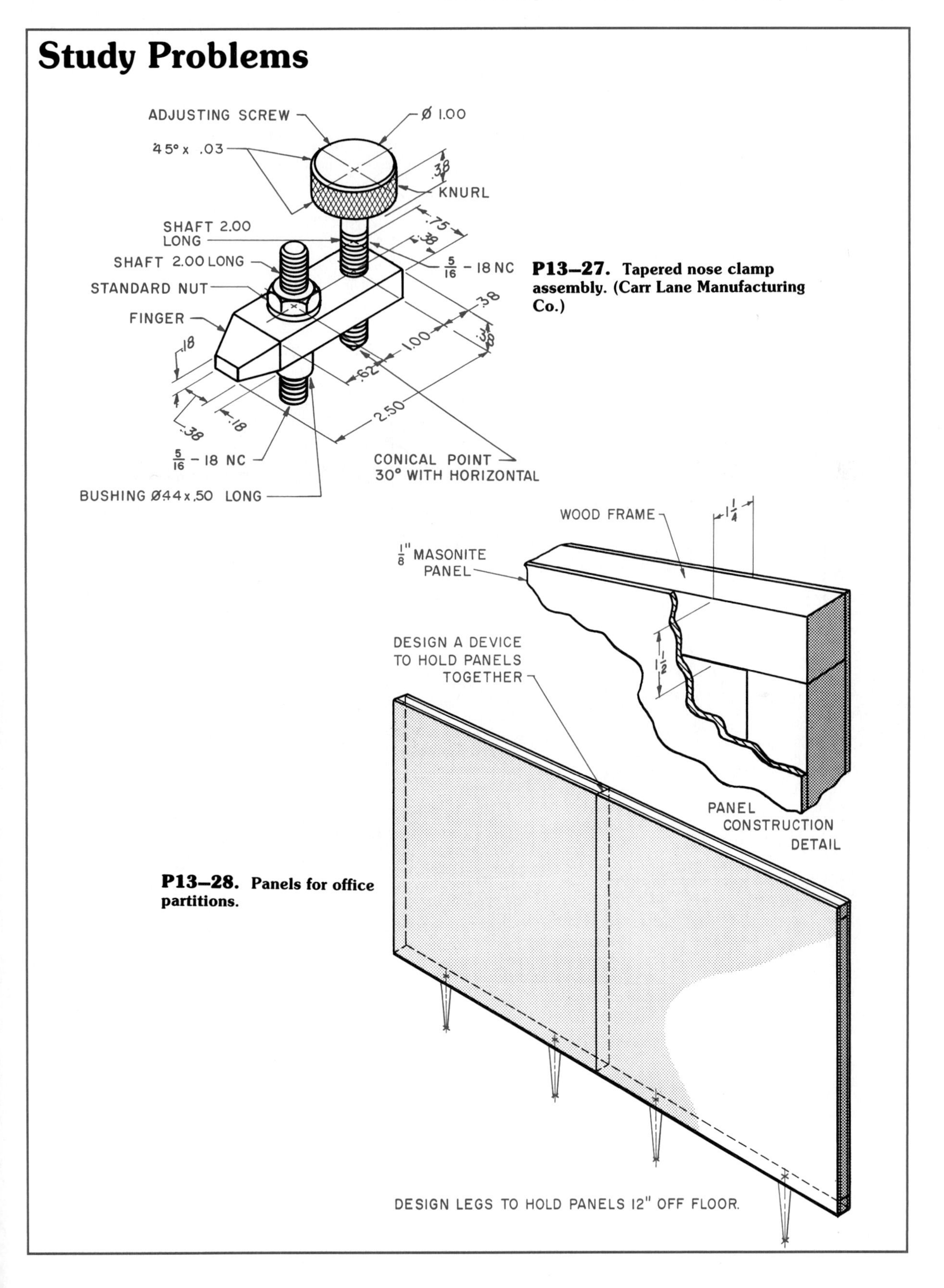

P13–27. **Tapered nose clamp assembly. (Carr Lane Manufacturing Co.)**

P13–28. **Panels for office partitions.**

Study Problems

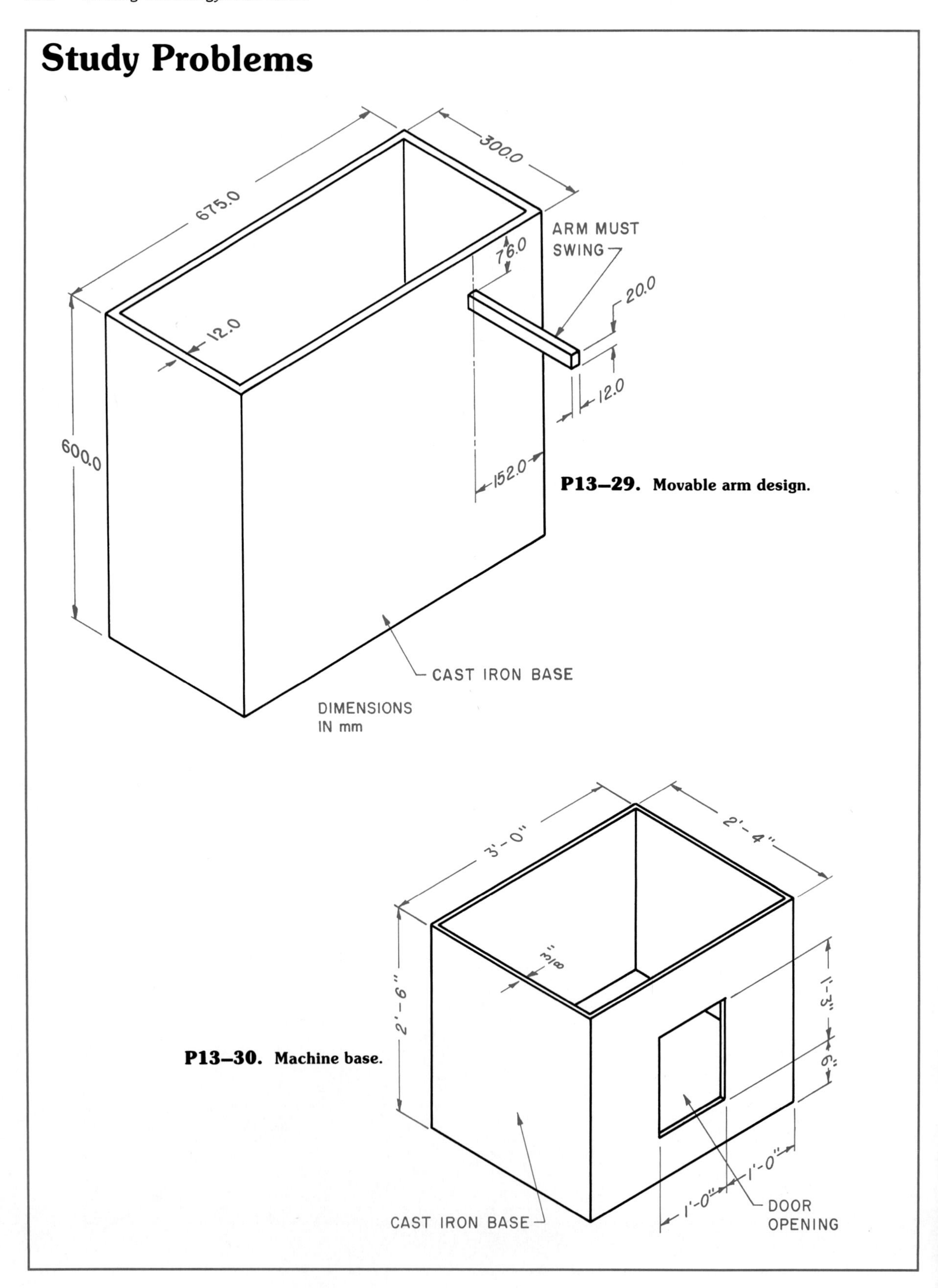

P13–29. Movable arm design.

P13–30. Machine base.

Study Problems

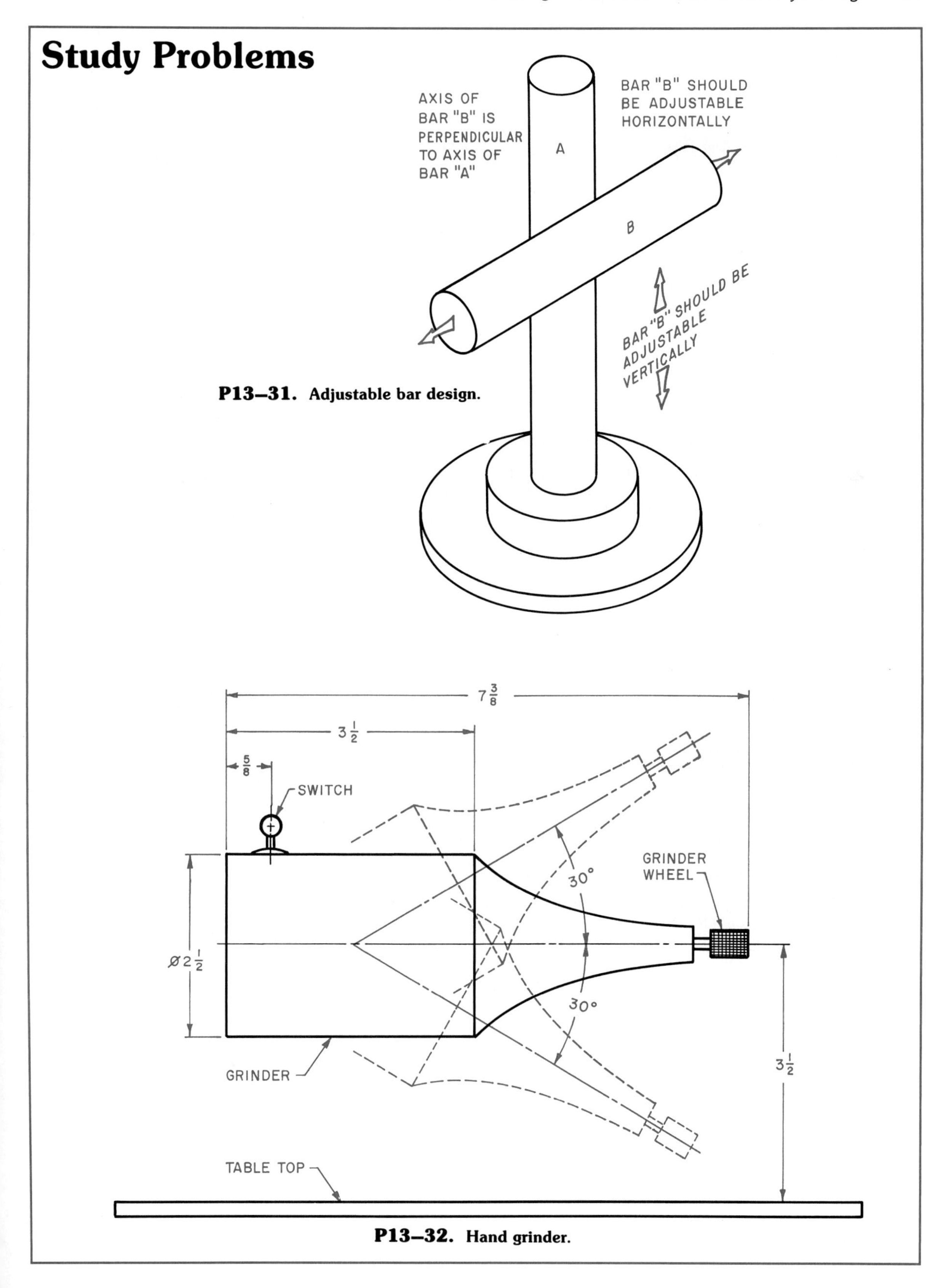

P13–31. Adjustable bar design.

P13–32. Hand grinder.

Study Problems

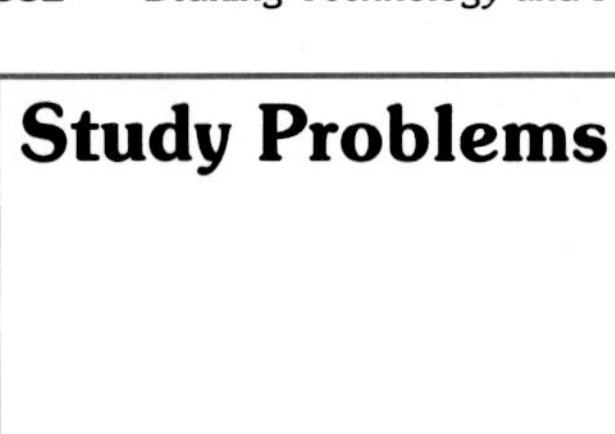

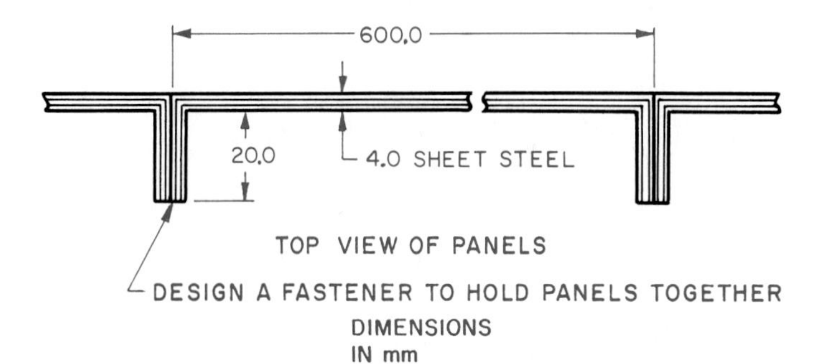

P13–33. **Steel wall panels.**

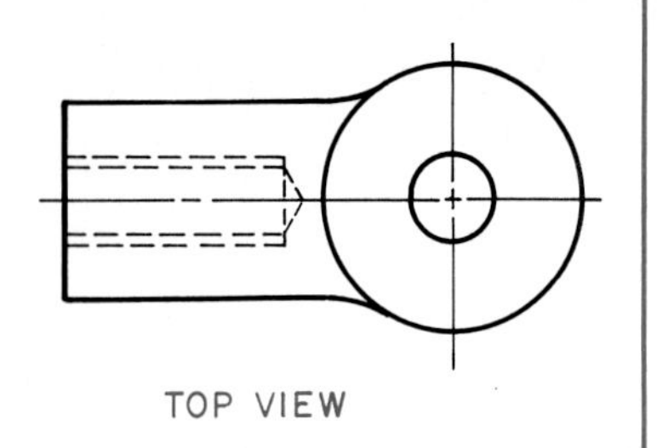

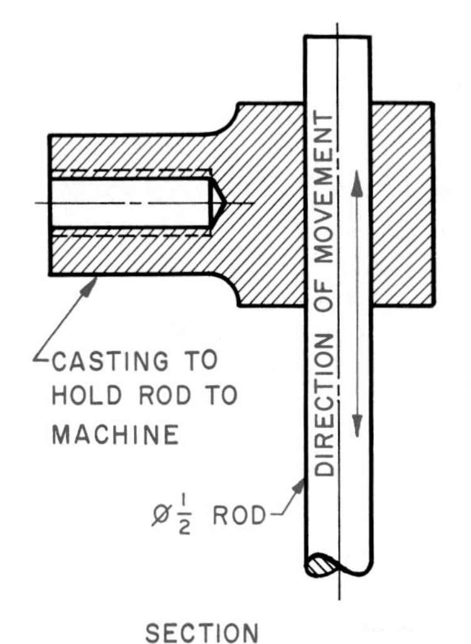

P13–34. **Vertical bar stop lock.**

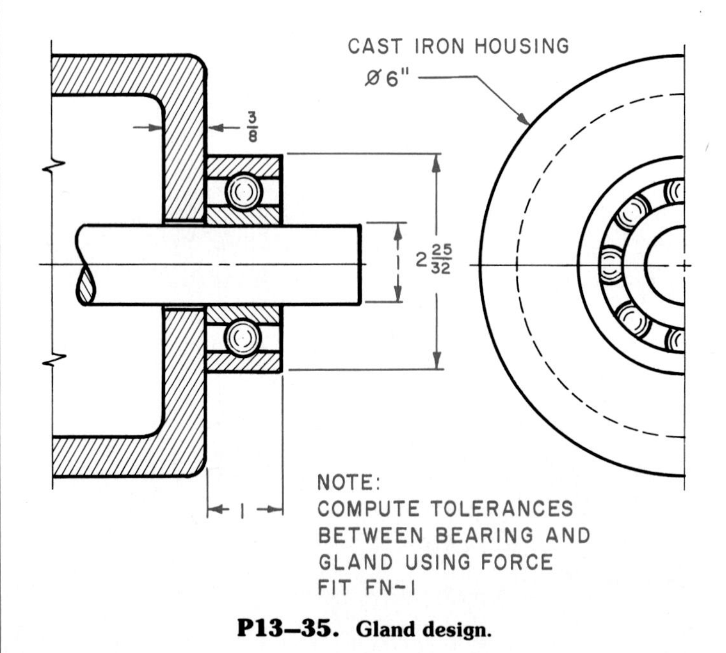

P13–35. **Gland design.**

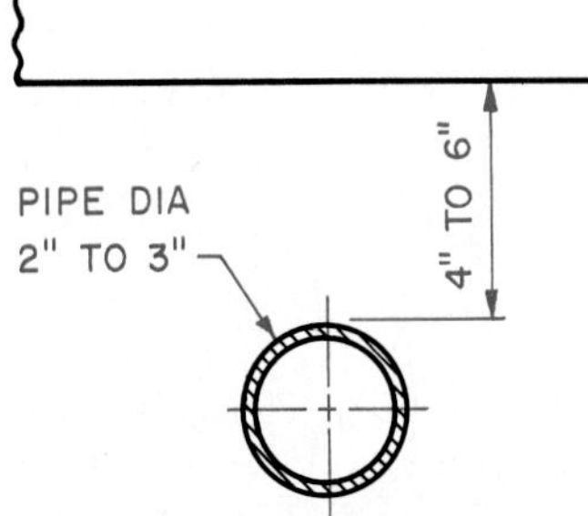

P13–36. **Pipe hanger.**

CHAPTER 14
Pictorial Drawing

After studying this chapter, you should be able to do the following:

- Make isometric, oblique, and perspective drawings.
- Dimension pictorial drawings.

Introduction

Seldom a day passes when you do not see some form of pictorial drawing. Books, newspapers, and magazines are filled with pictorial illustrations. Industry also uses many types of pictorials to present information.

Orthographic projection, as you know, shows only one face of an object at a time. Because each view can show only two dimensions, there is no suggestion of depth. A pictorial drawing, on the other hand, has the advantage of showing three faces in a single view. In other words, it is a three-dimensional drawing. Fig. 14-1.

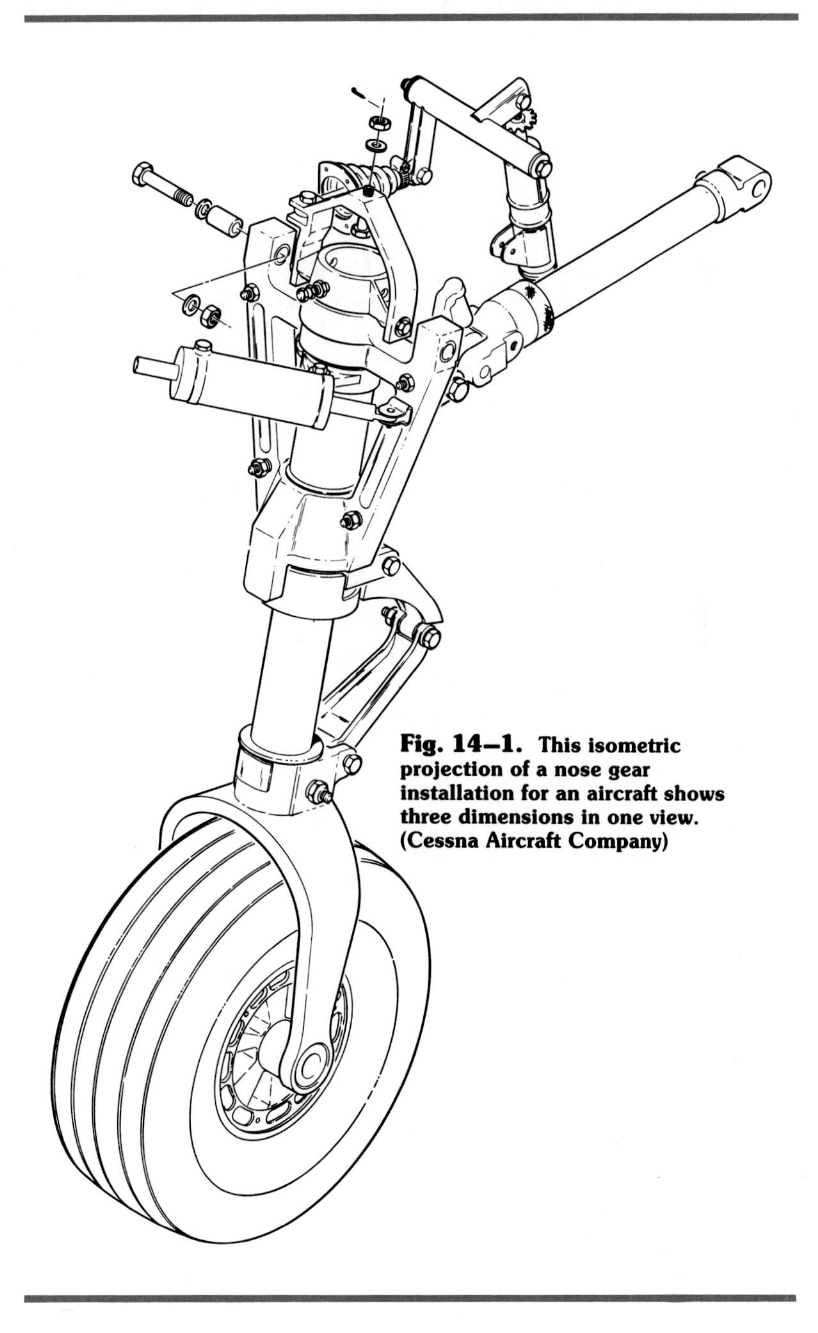

Fig. 14–1. This isometric projection of a nose gear installation for an aircraft shows three dimensions in one view. (Cessna Aircraft Company)

Many people do not understand orthographic drawings. For those who have not had this type of training, pictorial drawings give the needed information. For example, many items people purchase come disassembled and require some form of instruction so they may be correctly assembled. Pictorial drawings help to explain the assembly process.

Types of Pictorial Drawings

Pictorial drawings can be divided into three main types: (1) axonometric, which is divided into isometric, dimetric, and trimetric; (2) oblique, which is divided into cabinet and cavalier; and (3) perspective. See Fig. 14-2. **Axonometric drawings** have the object inclined to the plane of projection. **Oblique drawings** have the face of the object parallel with the plane of projection. **Perspective drawings** allow the object to appear to converge toward some point in the distance.

Isometric Pictorials

The term isometric means *equal measurement*. This term is used to describe this type of pictorial because the drawing has equal angles. There are two types of isometric pictorials: isometric projections and isometric drawings.

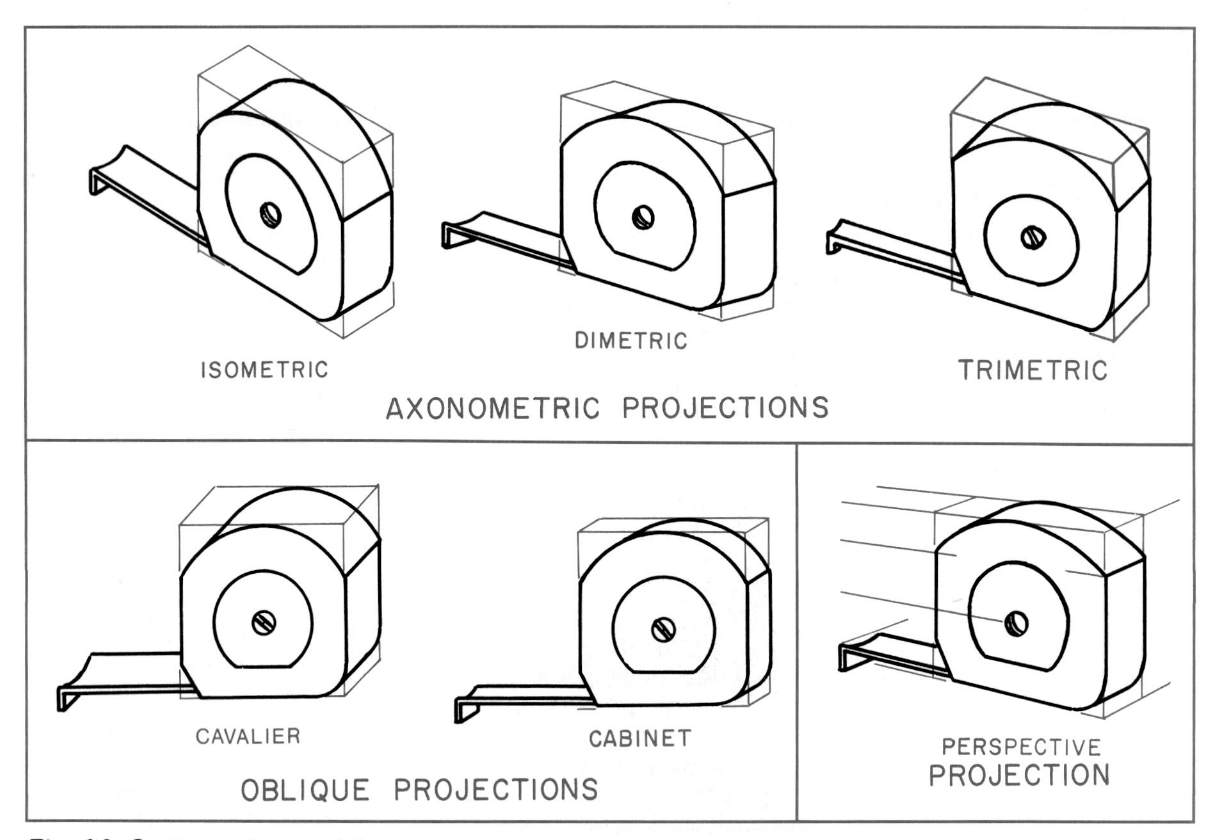

Fig. 14–2. Types of pictorial drawings.

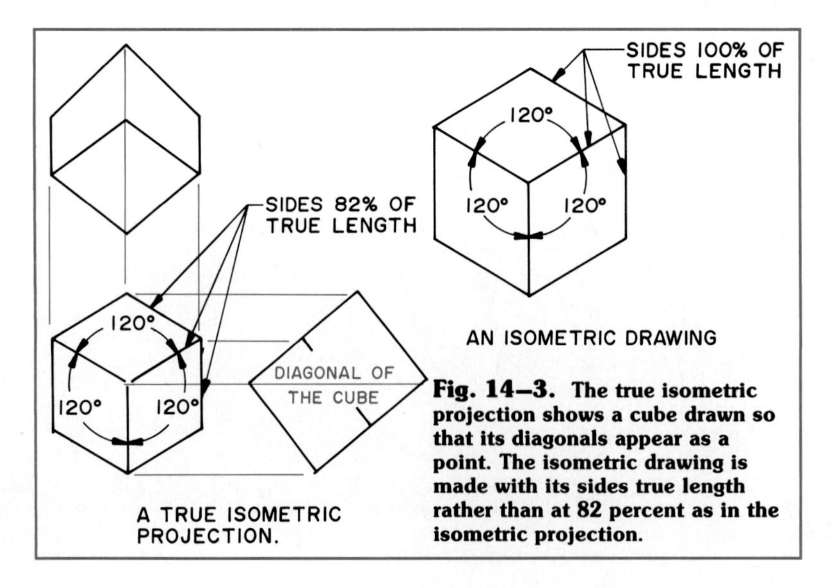

Fig. 14–3. The true isometric projection shows a cube drawn so that its diagonals appear as a point. The isometric drawing is made with its sides true length rather than at 82 percent as in the isometric projection.

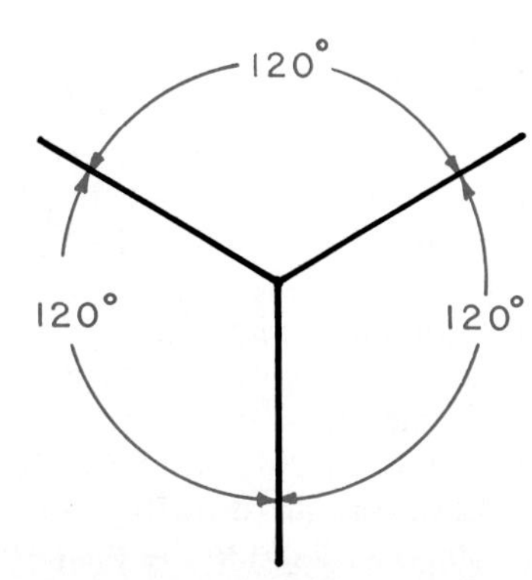

Fig. 14–4. Isometric axes are uniformly spaced at 120 degrees.

Isometric projection is a type of axonometric projection having parallel projectors that are perpendicular to the picture plane. With this projection, the diagonals of a cube appear as a point. The three axes are spaced 120 degrees apart, and the sides are foreshortened to 82 percent of their true length. Fig. 14-3.

Isometric drawing is the same as isometric projection except the sides are drawn true length. This makes it appear like the isometric projection. It is not a true axonometric projection but produces an approximate view. Fig. 14-3.

Making an Isometric Drawing

All isometric drawings are based on three axes spaced 120 degrees apart. Fig. 14-4. The axes may be placed in any position as long as they are 120 degrees apart. Fig. 14-5.

To make an isometric drawing, choose the position of the axes that will best show the object or the side that is most important.

The following procedure may be used in making an isometric drawing. Fig. 14-6.

1. Lay out the isometric axes.
2. Lay out the overall dimensions, blocking in the size of the object.
3. Draw the isometric box enclosing the object.
4. Locate other features by measuring along isometric lines.

Fig. 14–5. The isometric axes can be placed in many positions.

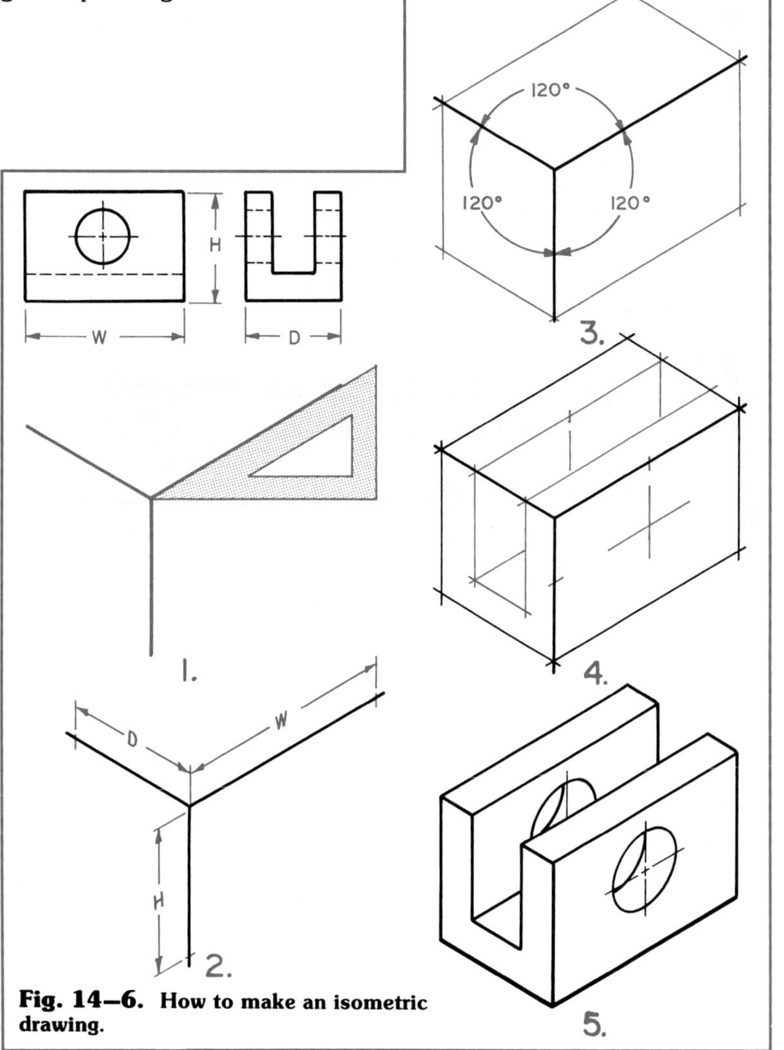

Fig. 14–6. How to make an isometric drawing.

Isometric lines are those drawn parallel with the isometric axes.

5. Remove layout lines and darken the remaining lines to complete the drawing. Use standard line weights like those used on all engineering drawings.

Hidden Lines. Hidden lines are usually not used in any form of pictorial drawing. To show all edges which are not visible would make the drawing difficult to read. In some cases, however, hidden lines may be used to show a feature of the object which is not visible.

Nonisometric Lines. Lines that are drawn parallel with the isometric axes are isometric lines. Lines that are inclined or oblique to the isometric axes are **nonisometric lines**. They cannot be measured directly on the drawing.

To draw a nonisometric line, you must first locate both its ends on isometric lines. Then draw the nonisometric line between the two points. The best method of drawing nonisometric lines is to use the "box method" of construction. Fig. 14-7. Construct an isometric box using the largest dimensions of the object. Then locate the edges of the various surfaces from the edges of the box.

For example, Fig. 14-7 shows two orthographic views of a base for an automotive safety stand. Construct a box using the three largest dimensions. Fig. 14-7 at A. Locate the top of the stand by using dimension B and the offset dimension C. See Fig. 14-7 at B and C. Then draw the nonisometric corners between the top and the bottom of the box.

Angles in Isometric. Occasionally you will find it necessary to lay out an angle in isometric. Angles in isometric, like nonisometric lines, do not show their true size. They cannot be measured in degrees with an ordinary protractor. Special isometric protractors are available for laying out angles in isometric.

If an isometric protractor is not available, you can lay out the angle by the coordinate method. This method uses a series of coordinates parallel to the isometric axes. Step off the size of the angle on the coordinates. Fig. 14-8 shows a partial view of a drill press vise. First, draw the angle in an orthographic view. Make this drawing the same scale as the isometric drawing. Add a third side to the angle, called the *side opposite*, thus forming a right triangle. Use it to step off the correct sizes on the edges of the isometric box.

Isometric Circles and Arcs. Circles in isometric will appear as ellipses. They can be drawn on the three surfaces of an isometric drawing. Fig. 14-9.

Arcs will appear as a part of an ellipse. Rather than plot ellipses by actual projection or coordinates, you may use an "approximate method." This method of drawing ellipses produces the **four-center approximate ellipse**. As the name indicates, the ellipse you draw with it is not accurate but is only approximate. However, the approximate ellipse is good enough for most purposes.

To draw a four-center ellipse, see Fig. 14-10. First, draw an isometric square. Make the length of the side of the square equal to the diameter of the circle. Then bisect each of the four sides of the isometric square. Connect each bisector to the vertex of the opposite large angle. Use the intersections of the bisectors as center points of the large and small arcs. With the compass point at B, draw arc CD. Using the same radius, draw arc EF. Now draw arc DF from point H, and in a similar manner, draw arc CE from point G.

The sides of an isometric square, regardless of its position, may be easily bisected by using a 30-60 degree triangle on a horizontal straightedge. Fig. 14-11. This method of finding the midpoint of each side can save considerable time.

To draw several isometric circles of the same diameter in parallel planes is not as difficult as it may seem. A cylinder of uniform diameter is an example of this problem. Fig. 14-12. Locate the four centers for the approximate ellipse on one end of the cylinder. These are located in the usual manner by first drawing the isometric square. The four centers for the ellipse on the other end are found by moving the centers parallel to the proper axis. The distance the centers are moved is equal to the length of the cylinder.

Arcs in isometric are drawn with the same ease as the four-center approximate ellipse. It is not necessary to draw the entire isometric square since an arc is only part of a circle. The same principle is used in drawing an arc as in drawing the circle. Fig. 14-13. From the corner where the arc is to appear, lay off the radius of the arc on both edges. From each of these points, draw perpendiculars. The intersection of the perpendiculars forms the center of the arc in isometric. Note that the center for the arc is not the center used for dimensioning. Fig. 14-14 illustrates this point.

Irregular Curves in Isometric. Irregular curves in isometric are drawn by plotting a series of points, Fig. 14-15. First, draw an orthographic view of the object to the same scale as the isometric. Next, draw a series of parallel construction lines on the orthographic view. These construction

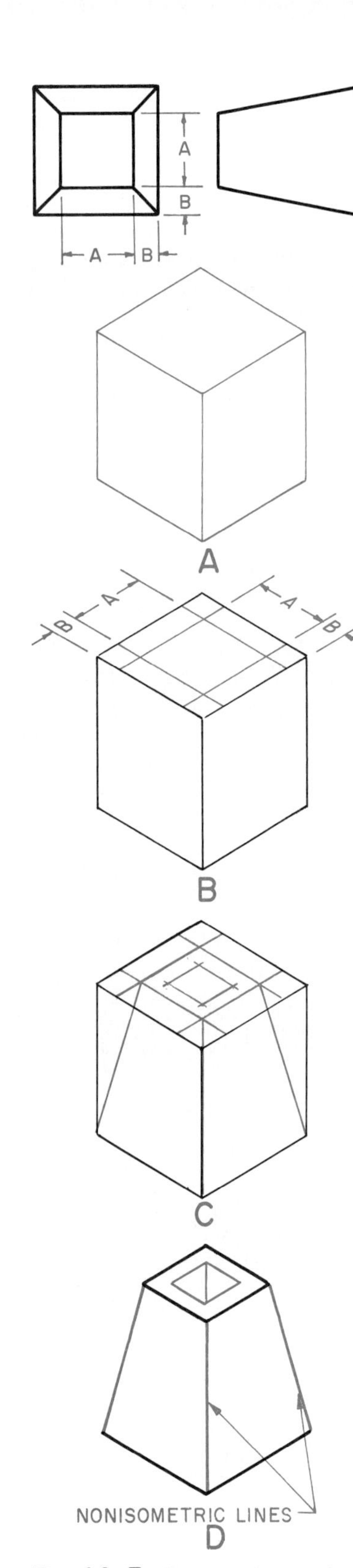

Fig. 14–7. Draw nonisometric lines by locating their corners on isometric lines.

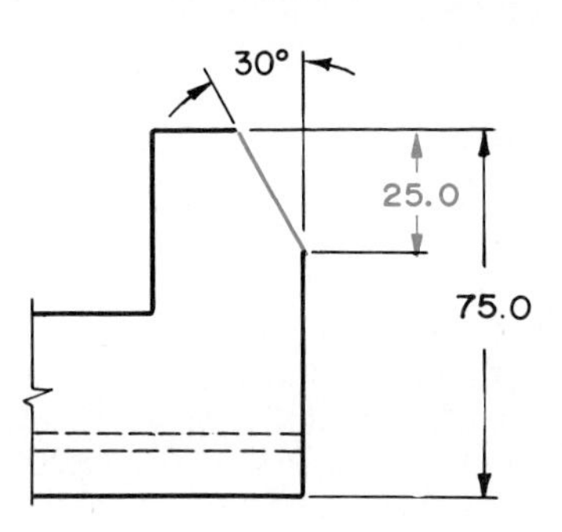

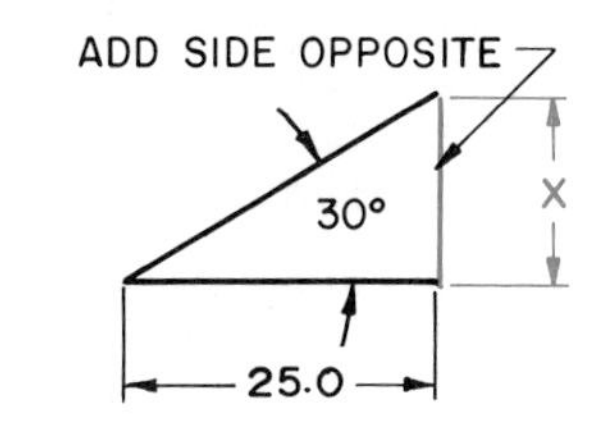

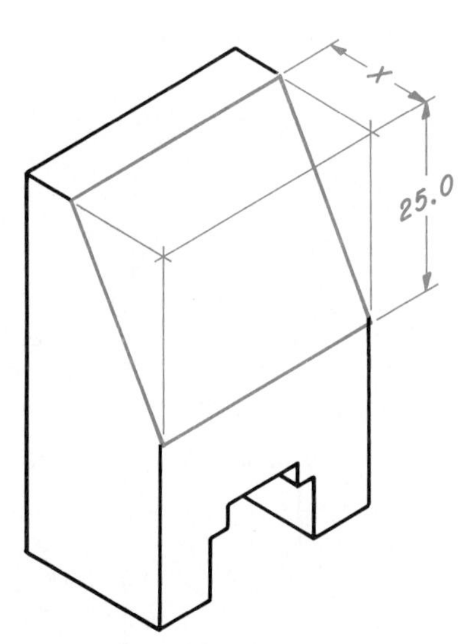

Fig. 14–8. The coordinate method for laying out an angle in isometric.

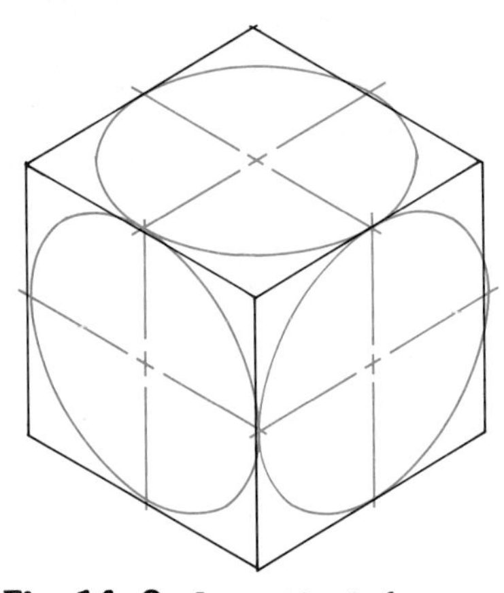

Fig. 14–9. Isometric circles appear as ellipses on the three surfaces of an isometric drawing.

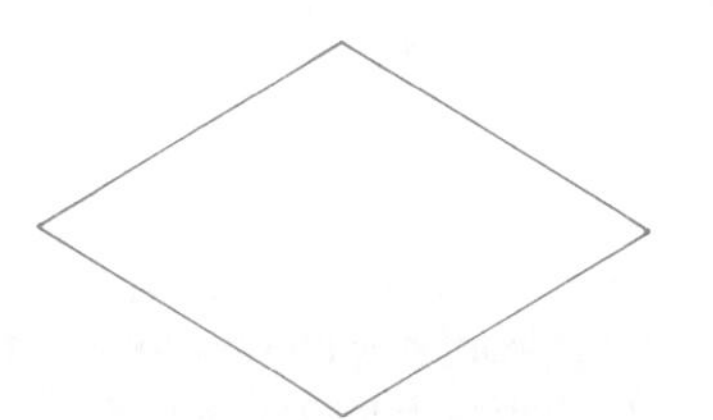

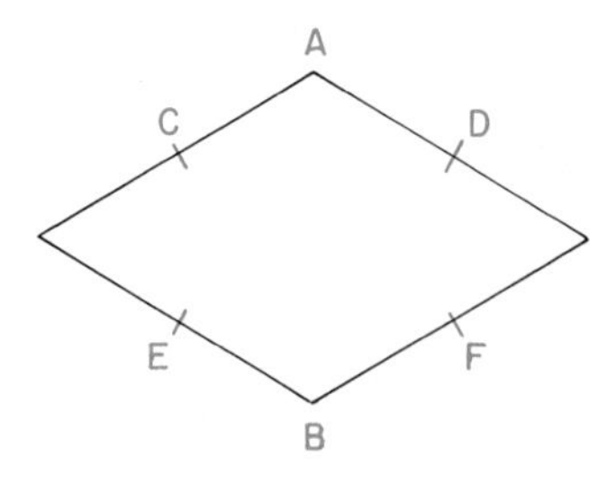

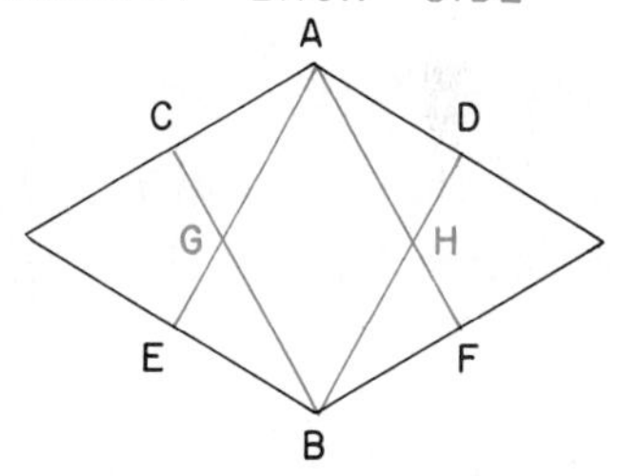

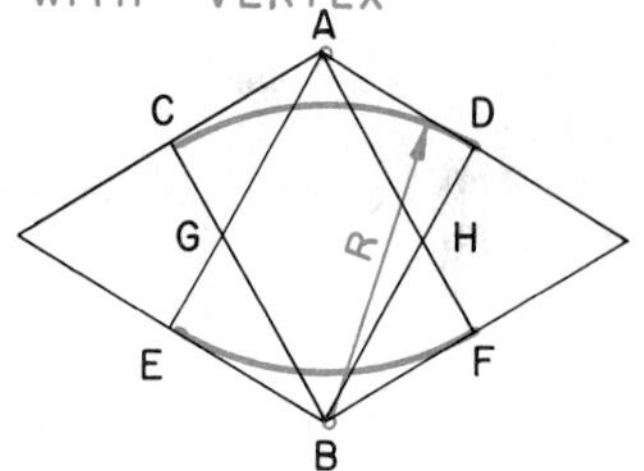

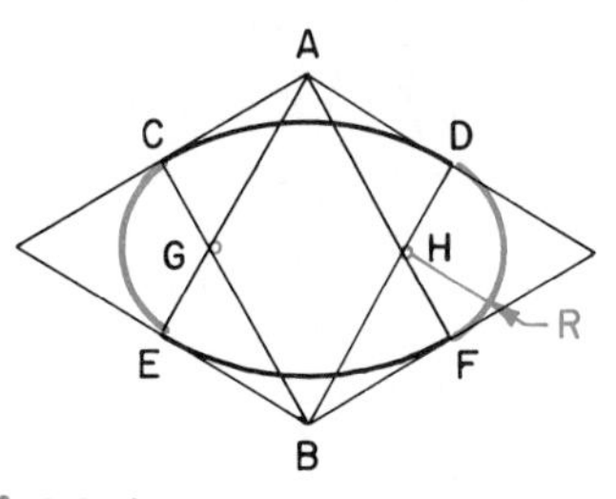

Fig. 14–10. How to draw a four-center approximate ellipse.

lines should be spaced equally. Where the curve intersects each construction line, a coordinate point is formed. Fig. 14-15 at A. Now draw the construction lines in the correct isometric plane. Use the same spacing between the construction lines in the isometric as in the orthographic view. Points on the orthographic view are called *coordinates*. Transfer the coordinates to the isometric view with dividers.

The back face is drawn in a way similar to the method used for parallel isometric circles. Draw a series of parallel construction lines for the depth. Step off the thickness of the object along these construction lines. Now draw the curve by connecting the points. Use an irregular curve.

Centering an Isometric Drawing. There are several methods that you can use to center an isometric drawing on a sheet. One method frequently used is the box method. Fig. 14-16.

On a piece of scrap paper, draw the isometric block that encloses the entire object. Then draw an *enclosing box* around the isometric block. Equalize the distance from the edges of the enclosing box to the borders of the paper upon which the isometric drawing is to be made. Mark the lower corner. The enclosing box is the *reference point* to use for the isometric drawing.

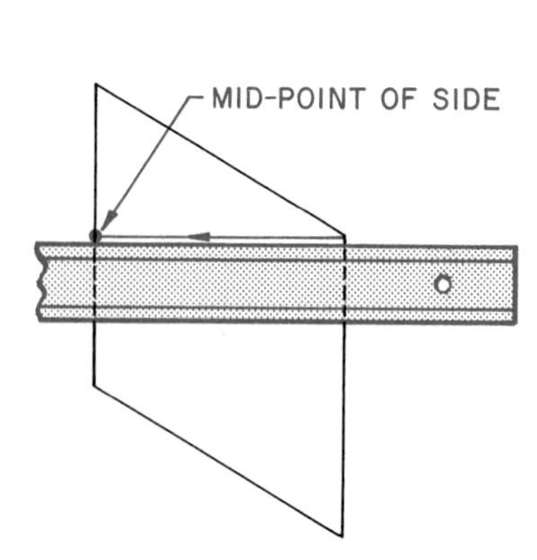

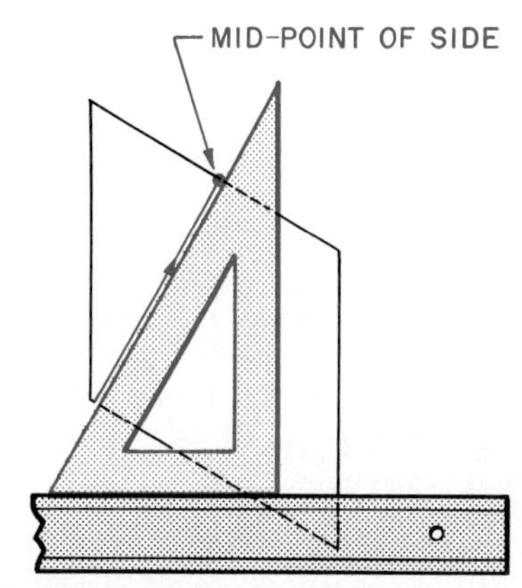

Fig. 14–11. **The isometric square for a four-center ellipse can be bisected easily with a T-square and a 30–60 degree triangle.**

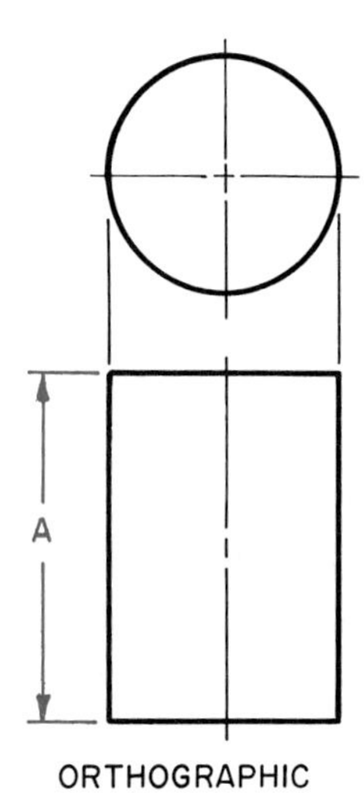

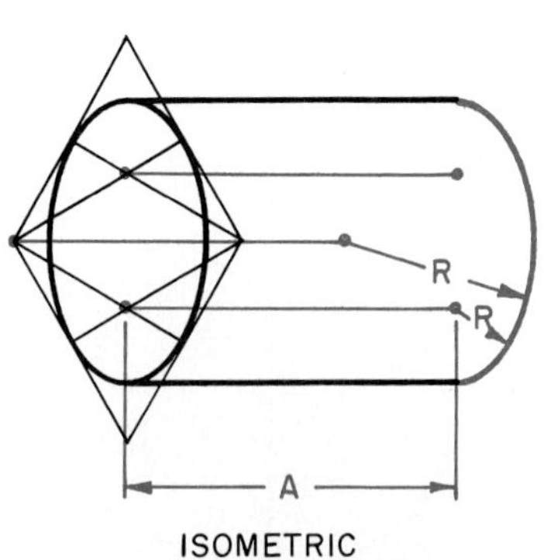

Fig. 14–12. **Ellipses of the same diameter can be drawn by moving the centers of the first ellipse the length of the cylinder.**

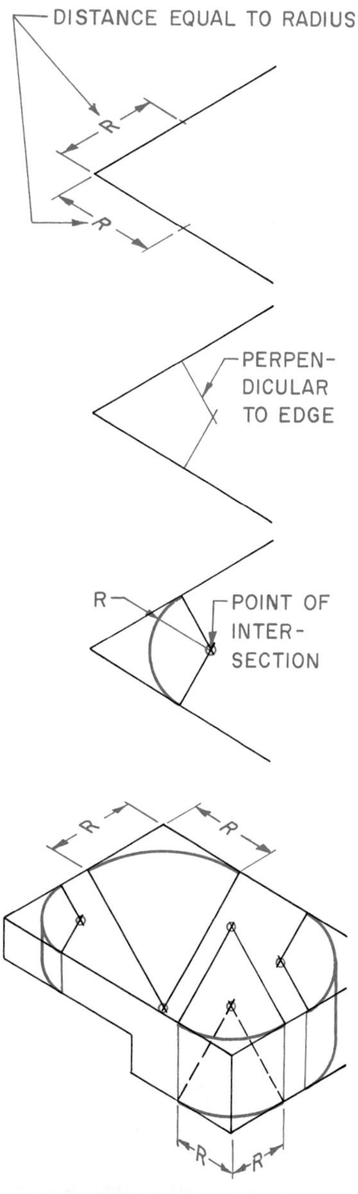

Fig. 14–13. **How to lay out arcs in isometric.**

Isometric Drawing Aids. Templates are available which simplify the drawing of isometric circles. One type of isometric template is shown in Fig. 14-17. All isometric circles are based on the circle's *actual projection*. The circle is placed at an angle of 35°-16′ to the plane of projection. The diameter of the isometric circle is measured along the isometric axis. A regular ellipse is measured along its major axis. Fig. 14-18. Do not confuse the regular ellipse and isometric circle templates.

Isometric grid paper is available to speed the making of an isometric drawing. Fig. 14-19. The grids are printed on both tracing paper and opaque paper.

Isometric grids printed on vellum are usually printed with nonreproducible ink. Any grid printed with nonreproducible ink will not show when reproduced. Only the pencil or ink drawing will print. Tracing paper pads are available with an opaque underlay grid sheet. The underlay sheet is placed beneath the tracing paper. It can be used over and over again.

Fig. 14–14. The center for dimensioning an isometric arc follows the isometric axes.

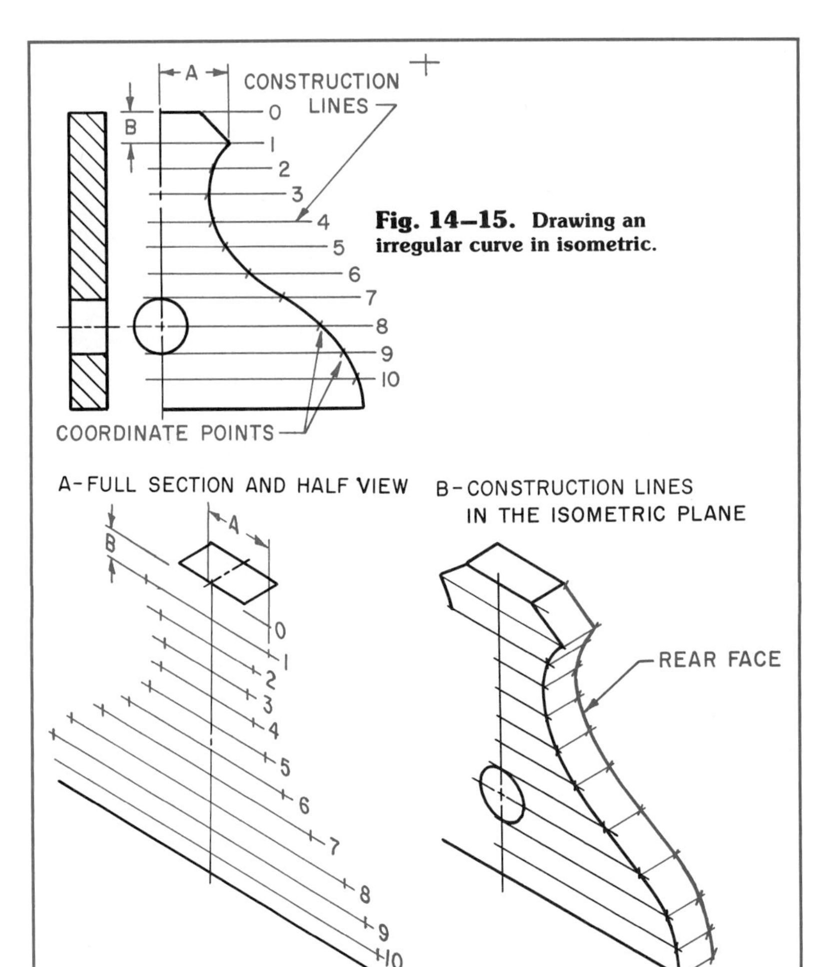

Fig. 14–15. Drawing an irregular curve in isometric.

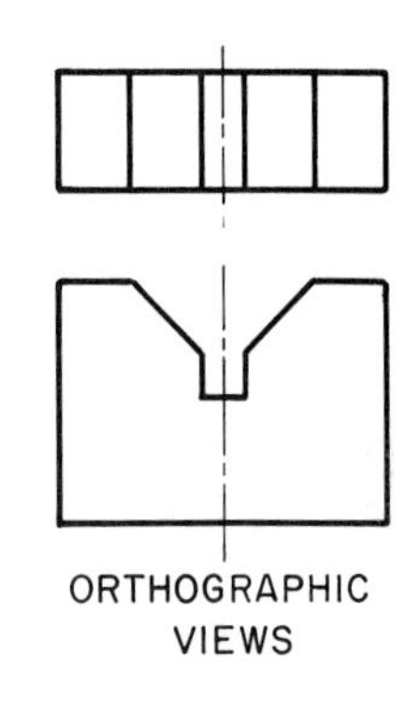

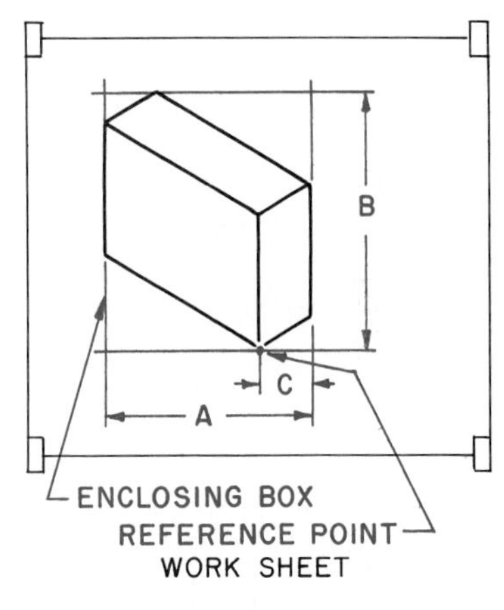

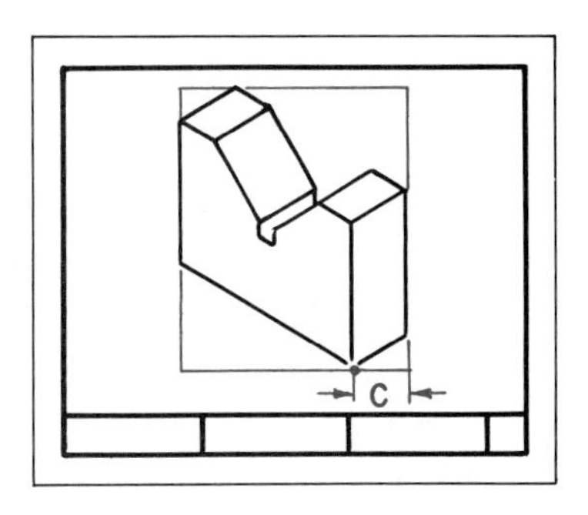

Fig. 14–16. The box method is used to center an isometric drawing on the paper.

Isometric Sections. Isometric sections are used to show interior detail. To try to show interior construction by means of hidden lines would lead to confusion. To avoid confusion, any of the standard sections may be used. See Chapter 11, "Sectional Views," for more information. In some instances the use of a section will eliminate an extra pictorial view.

In making an isometric section, first lay out the entire object. Fig. 14-20. Next, locate the cutting plane. Then block in the section. Now erase the unneeded construction lines. Draw any minor details, such as holes or fillets. Add the section lines and darken all of the other lines.

Pass the cutting plane used in an isometric *full section* through the center of the object. The cutting plane is always parallel to one of the isometric axes. Features that are shown by the section are completed just as any other portion of the piece. Draw section lining at any angle, but be sure it is not parallel to any of the main edges of the cut surface. Section lines placed on an angle of 60 degrees will work in most instances.

In an isometric *half section*, place the cutting planes parallel to *both* receding axes. The cutting planes will intersect at the center line. Fig. 14-21.

Make *broken-out sections* by placing the cutting plane in *any* desired position. If possible, place the cutting plane parallel to an isometric axis. When you make broken-out sections of shafts, holes, or similar features, pass the cutting plane through their axes. Fig. 14-22 is an example of this type of sectioning. Note the treatment of the broken area in the section not cut by the plane.

Intersections. Accurately draw the actual line of intersection between two joining pieces. Fig. 14-23.

Fillets and Rounds. Show fillets and rounds as highlights. Draw them with line shading or curved edge shading. Fig. 14-24 (p. 393).

Threads. Draw threads as a series of isometric circles evenly spaced. Draw them along the center line of the threaded part. The spacing between threads does not have to be the same as the actual number of threads per inch. Shading improves the appearance but is not required.

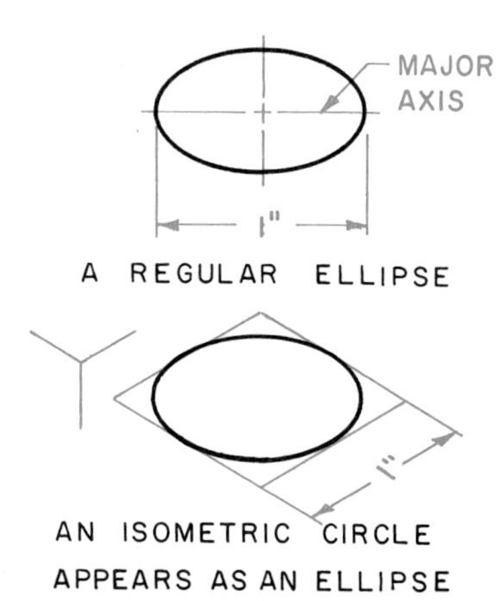

Fig. 14–18. The diameter of an isometric circle is measured along the isometric axis.

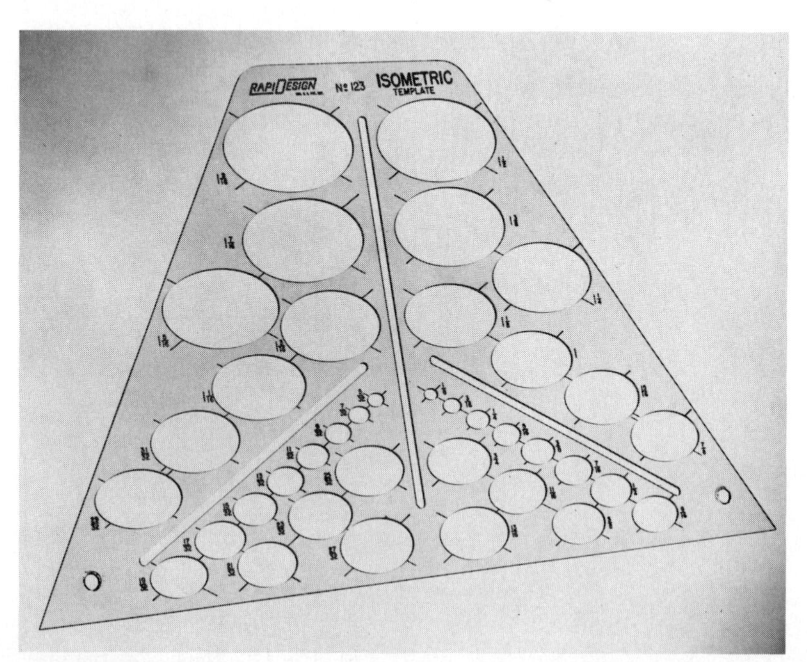

Fig. 14–17. An isometric ellipse template can simplify the drawing of ellipses. (Rapidesign)

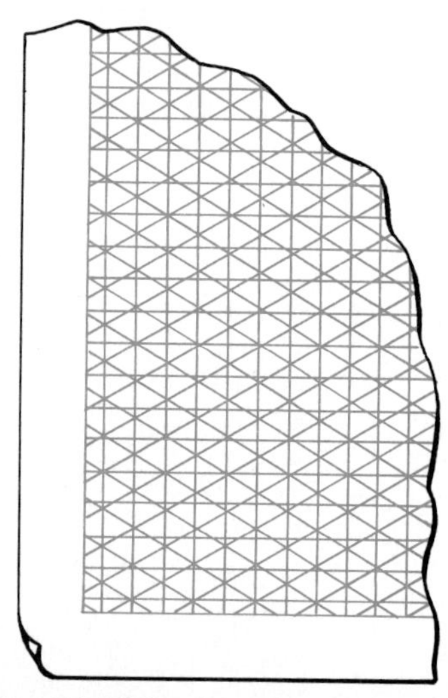

Fig. 14–19. Isometric grid paper simplifies making isometric drawings and sketches.

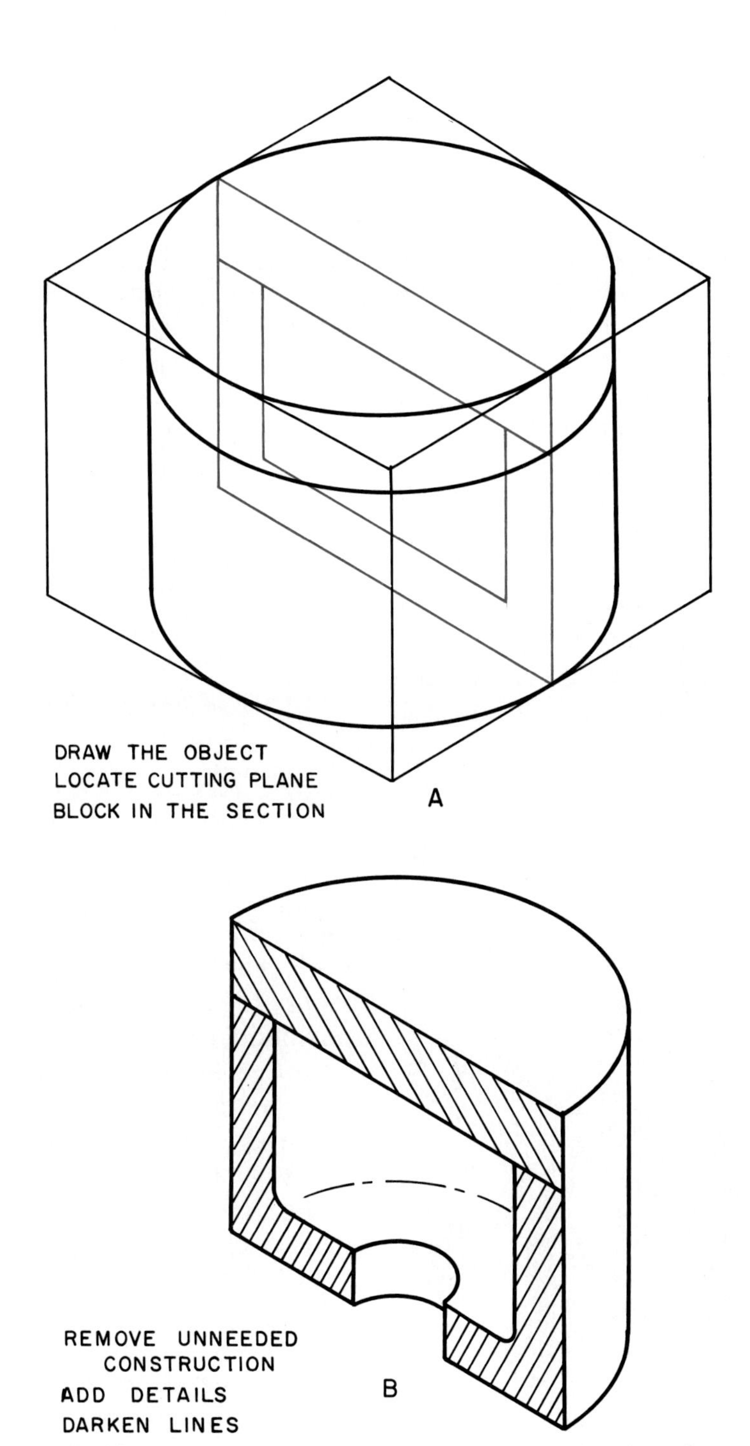
DRAW THE OBJECT
LOCATE CUTTING PLANE
BLOCK IN THE SECTION
A
REMOVE UNNEEDED CONSTRUCTION
ADD DETAILS
DARKEN LINES
B

Fig. 14–20. When making an isometric section, draw the object first; then remove the part not needed.

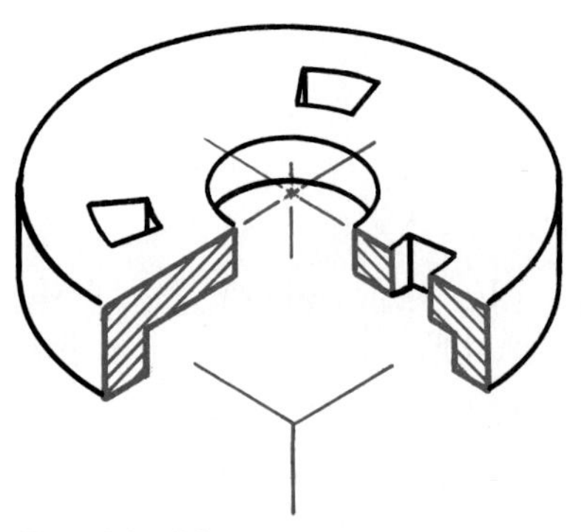

Fig. 14–21. In an isometric half section, the cutting planes follow the isometric axes.

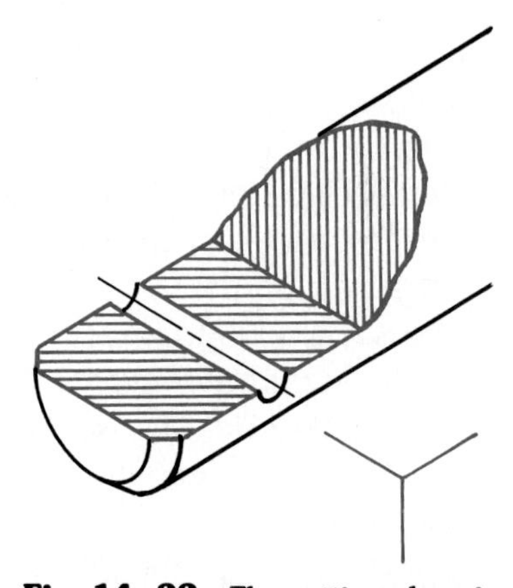

Fig. 14–22. The cutting plane in an isometric broken-out section will follow one of the isometric axes.

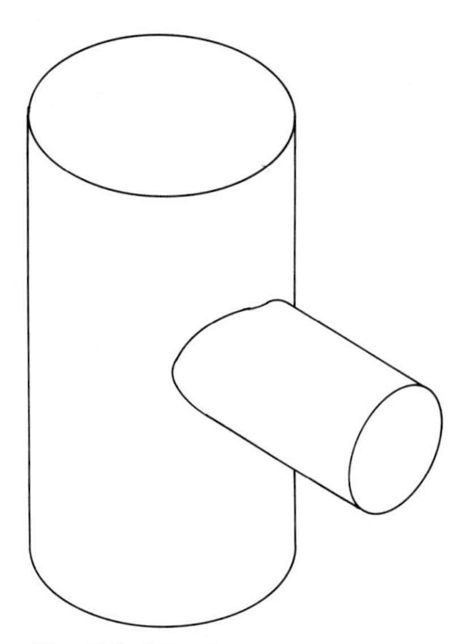

Fig. 14–23. Intersections are shown by drawing the actual line of intersection.

CAD Applications:

Drawing in 3-D

In manual drafting, the quality of pictorial drawings depends on the drafter's abilities. In CAD, it depends not only on the drafter's skills but also on the CAD system's capabilities.

Any CAD system can be used to produce axonometric, oblique, and perspective drawings, but it may not truly be a 3-D system. A true 3-D system allows the user to move the object (model) along the X, Y, and Z axes. It also allows the user to change viewpoints and thereby "move around" the object while it remains in place.

There are three types of 3-D models. The most common is the *wireframe*. A wireframe model consists of lines that represent the object's outline and surfaces. It appears as though the object were made of bent wires.

VersaCAD Corp.

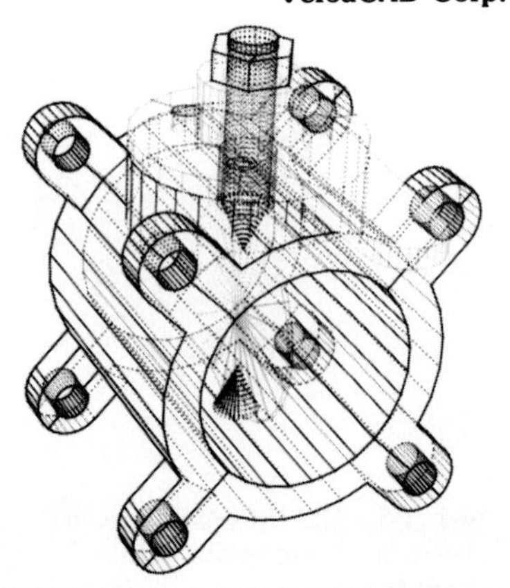

Because a wireframe model shows all lines as visible lines, it can be confusing to look at. A "hidden line removal" command can remove the lines which are below or behind a surface. This produces a more realistic drawing.

If a skin were placed over a wireframe model, the result would be a *surface* model. With a surface model, the drafter can add shading and colors to achieve the appearance of a solid object. However, the model is not mathematically a solid.

VersaCad Corp.

In true *solid* modeling, the object drawn behaves as a solid, real-life object. This not only provides the most realistic appearance but also allows the model to be used for analysis. For example, the model contains information about physical properties such as weight, volume, and density. It can be used in simulations to check how it would fit with other parts or how well it would withstand stresses.

Solid model courtesy of SPOCAD. Created by Douglas Yaeger; photographed by Alan Getz.

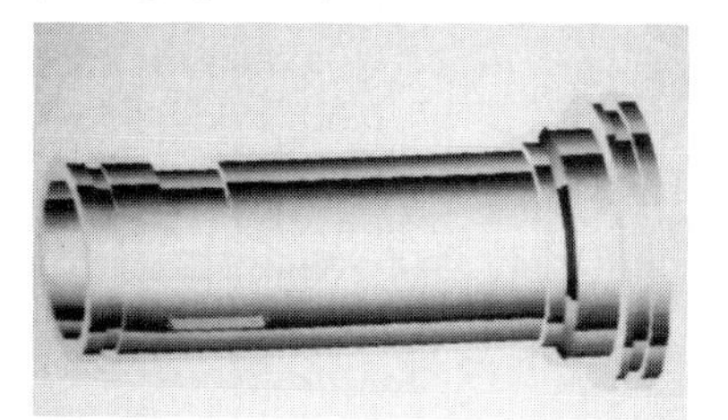

Because true 3-D *is* three-dimensional, a 3-D model cannot be plotted on paper. Ideally, it should be machined or sculpted as a three-dimensional object. In order to produce an image on paper, the file must be converted to a 2-D pictorial.

The 3-D systems are quite a change from traditional drawing, in which everything is done in two dimensions. Users of 3-D systems must learn to think in three dimensions, and that's not always easy. But 3-D systems are a powerful tool for design, engineering, and research and development.

Isometric Dimensioning

Follow the same basic rules for dimensioning an isometric drawing as for multiview drawing. Study Chapter 7, "Dimensioning." All dimensions must be clear and complete. Give sufficient information to describe the piece. There are two systems of dimensioning an isometric drawing. These are the unidirectional and aligned systems.

When using the *undirectional* system, all dimensions are lettered vertically. Fig. 14-25. The unidirectional system is faster and easier to use than the aligned method. Letter notes so that they lie in or parallel to the frontal plane using vertical lettering on horizontal lines. Whenever possible letter off the view.

When using the aligned system, the lettering lies in one of the three principal planes. You letter the dimensions so that they are in the same plane as the surface being dimensioned. Fig. 14-26. Letter notes so that they, too, lie on or parallel with the principal planes of the object. Letter finish symbols in the same planes as notes. Position them on a short mark on the surface to which they apply.

Some basic rules that will help in dimensioning are as follows:

1. Draw dimension lines and extension lines parallel to the isometric axes.
2. Always make sloped lines and guidelines for dimensions parallel to their dimension and extension lines.
3. Arrowheads must lie in the same plane as the extension line and dimension value.
4. Place dimensions on visible features whenever possible.

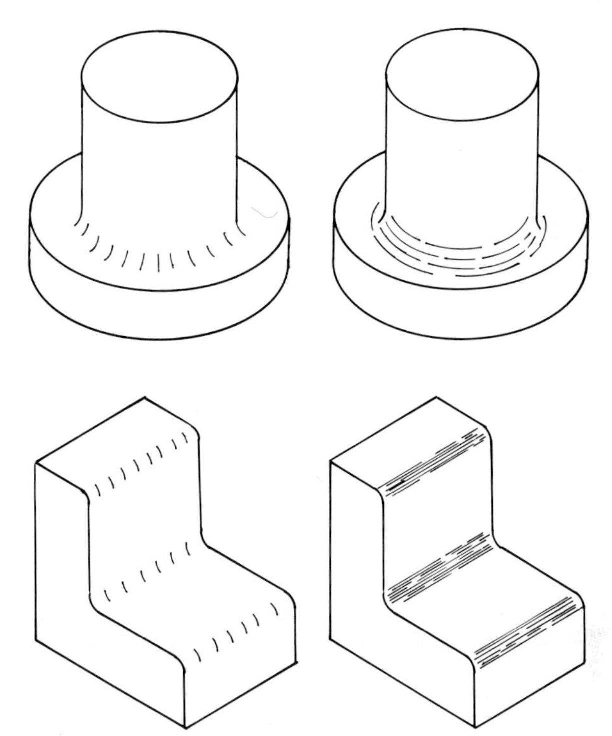

Fig. 14–24. Fillets and rounds can be represented by elliptical arcs or lines that run parallel to them.

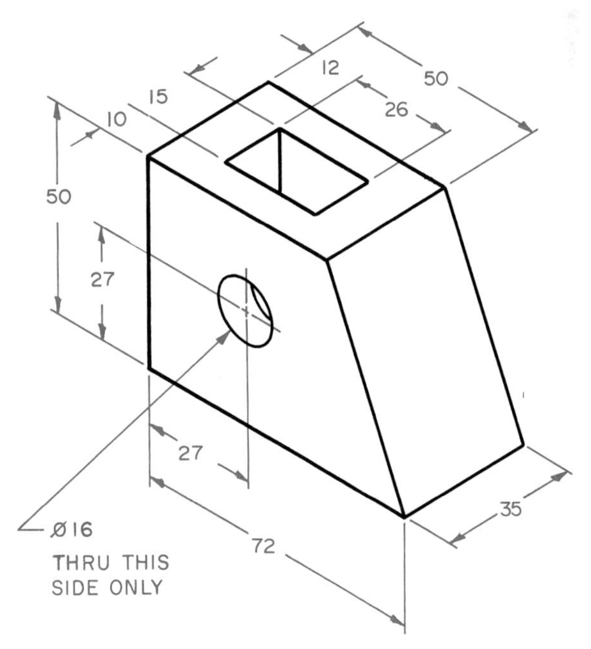

Fig. 14–25. Dimensions using the unidirectional system are lettered vertically.

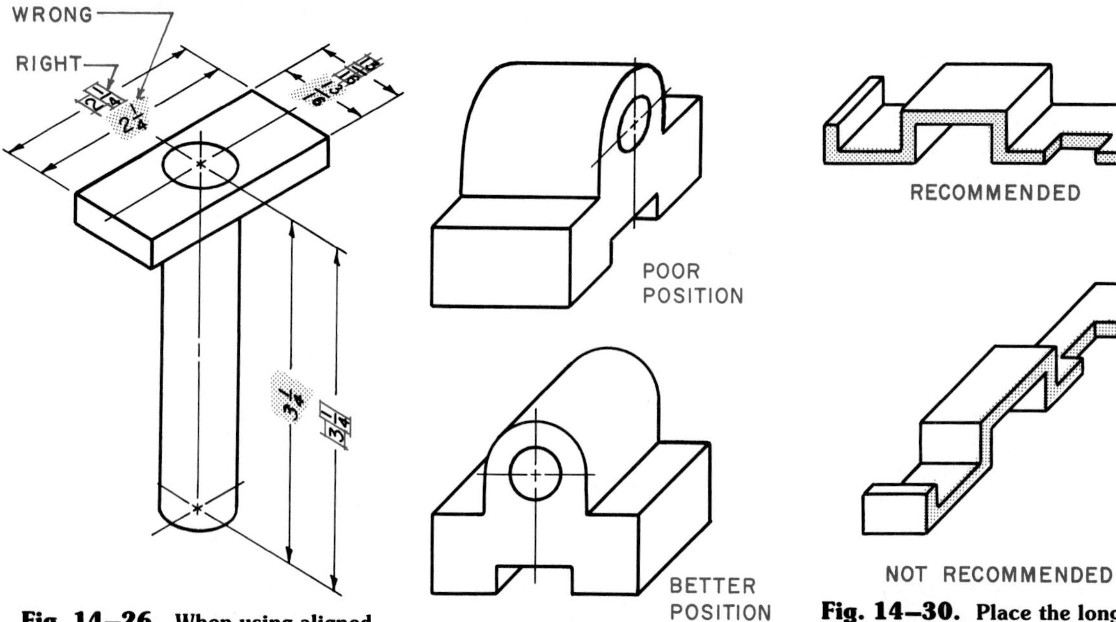

Fig. 14–26. When using aligned dimensioning, place dimensions in the same plane as the surface being dimensioned.

Fig. 14–29. When possible, place circular features parallel to the plane of projection.

Fig. 14–30. Place the long dimension of a piece parallel to the plane of projection.

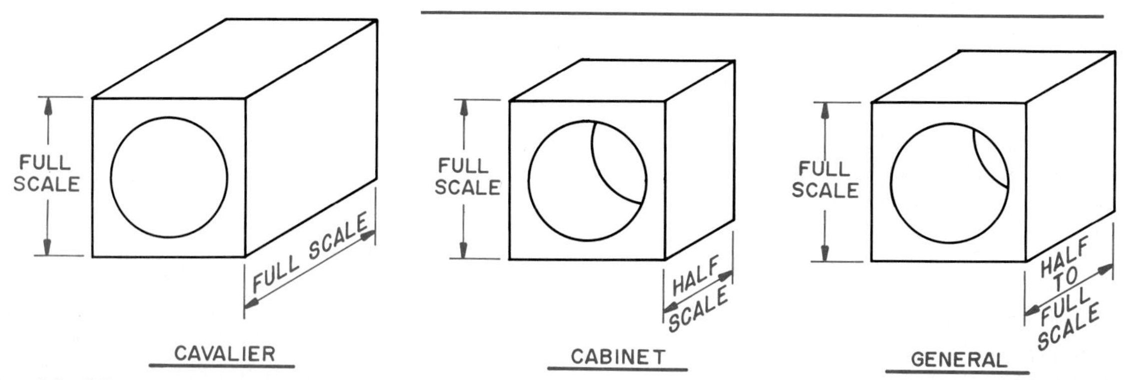

Fig. 14–27. The basic types of oblique drawings.

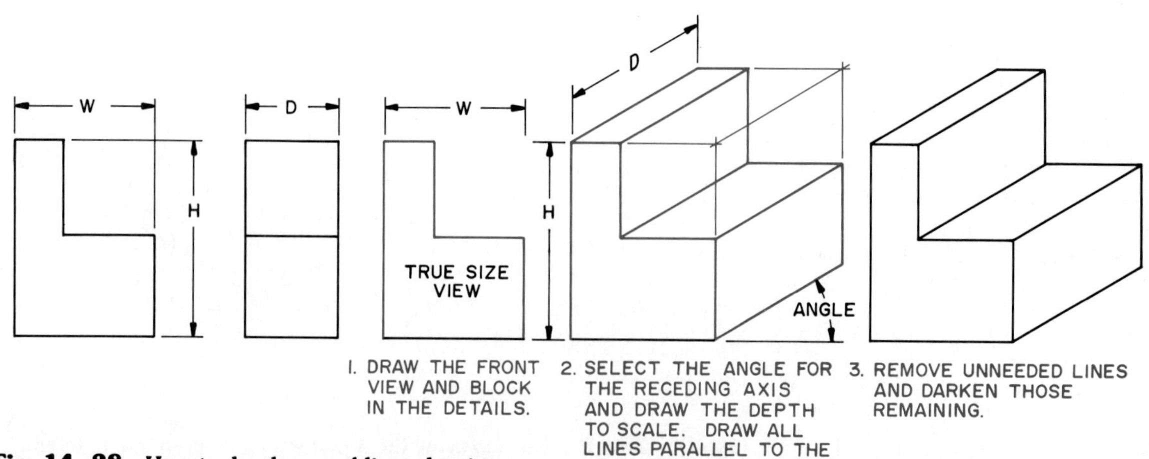

Fig. 14–28. How to develop an oblique drawing.

Oblique Drawings

Perhaps the easiest type of pictorial drawing to make is the oblique. In oblique drawing, one of the faces of the object is parallel to the plane of projection. This face of the object appears its true size and shape.

Oblique drawing uses three axes just like isometric drawing. Two of the oblique axes are placed at right angles to each other. The third axis (receding) is placed at any convenient angle to the horizontal. All receding edges are drawn parallel to this axis.

There are three classifications of oblique: cavalier, cabinet, and general. Each type of oblique drawing has its principal face parallel to the plane of projection. Fig. 14-27.

A **cavalier drawing** is one in which the receding axis makes any angle with the horizontal. Usually this angle is 30 or 45 degrees. The same scale is used on *all* axes. See the cube in Fig. 14-27. Measure the front face and the depth of the cube. The depth of the cube appears to be greater than the width and height. Actually they are equal. Some of this distortion can be eliminated by reducing the angle of the receding axis.

In **cabinet drawing**, the scale of the receding axis is reduced by one-half. Refer to Fig. 14-27. The angle the receding axis makes with the horizontal can be any number of degrees. By shortening the depth of the receding axis, a more natural appearance is obtained. Compare the cabinet drawing with the cavalier. The cabinet drawing gets its name from its popular use in the early days of the furniture industry. Many cabinet drawings of furniture were prepared by crafters and designers.

The *general oblique* drawing is one in which the receding axis scale varies from one-half to full size. The scale is reduced until the object appears most natural. The angle between the receding axis and the horizontal usually is drawn between 30 and 60 degrees.

Making an Oblique Drawing

A good way to make an oblique drawing is by first drawing an oblique box using the overall dimensions of the object. Select the face that is best for the true-size front view. The steps to lay out an oblique drawing are shown in Fig. 14-28.

Positioning the Object

The general shape of the object will have some effect on its position. The most descriptive features of the object should be placed parallel to the plane of projection. This is very important when drawing a piece having irregular curves or circles. Placing the circular features on the front view will not only eliminate distortion but will make the shapes easier to draw since all circles may be drawn with a compass or template. Fig. 14-29. The largest face of the object is usually placed parallel to the plane of projection. This is true also of long objects. When a long object is oriented with its greatest dimension along the receding axis, too much distortion will result. Fig. 14-30.

Angle of the Receding Axis

The angle of the receding axis may vary. For convenience a 30-, 45-, or 60-degree angle is used. The angle chosen should be one in which the details of the receding face are clear. If a great degree of detail appears on the top of the object, the angle of the receding axis would be increased. If this angle were small, 20 degrees for example, the detail could become too compressed. If the detail is too "flat," it may be difficult to read as well as draw. Fig. 14-31. Drafters must decide, based on their own judgment, which angle is best for the object.

Circles and Arcs on Oblique Drawings

Circles and arcs that do not face the plane of projection will appear as ellipses. The four-center approximate ellipse is used for drawing ellipses in oblique. The principle used in drawing this ellipse is the same as that used in isometric. Most ellipses drawn along the receding axis will appear distorted. This is due to the angle of the axis with the horizontal. The four-center approximate ellipse can *only* be used with *cavalier* drawings. Any other type of oblique drawing requires that the ellipse be plotted. The four-center ellipse cannot be used with either cabinet or general oblique drawings.

To draw a four-center approximate ellipse, first draw a square (parallelogram) on the receding face. Fig. 14-32. The sides of the square must be equal to the diameter of the circle. Since this circle is lying in a receding plane, two of the sides of the square are parallel to the receding axis. Next, bisect all four sides of the square. To find the midpoint of a side, use dividers or a scale. From these midpoints, draw perpendiculars to the sides of the square. The centers for the arcs are formed where the perpendiculars intersect. Now, draw arcs tangent to the sides of the square. If the angle of the receding axis is less than 30 degrees, the perpendiculars will intersect *inside* the square. The perpendiculars will intersect *outside* the square if the angle of the receding axis is greater than 30 degrees.

To draw an ellipse in a *cabinet* or *general* oblique drawing, use the *offset method.* An easy method of plotting an ellipse along the receding axis is shown in Fig. 14-33. As with any offset method of plotting, the circle must be drawn its true size and shape. Enclose the circle in a square. Now, draw two diagonals through the square. Where these diagonals intersect the circle, draw a horizontal line to the sides of the square. Fig. 14-33 at A. The horizontal and diagonal lines will be used to locate the coordinates of the circle on the receding plane.

Draw the square on the receding plane. Be sure to reduce the depth of the square according to the rest of the drawing. Draw two diagonals through the square. Fig. 14-33 at B. At the point of intersection, draw two center lines, one vertical and the other parallel to the receding oblique axis. Fig. 14-33 at C. Now, transfer distance A from the orthographic to the pictorial view. Step off this distance above and below the center lines. Draw lines parallel to the receding axis that intersect the diagonals at the transferred points. The intersections thus formed will serve as coordinates in plotting the ellipse. Using eight points to plot the ellipse is sufficient for most circles. Large circles that appear on the receding plane will require more coordinates. Connect the coordinates with an irregular curve.

Arcs in cavalier may be drawn by the same method used in isometric. Refer to Fig. 14-32. Set off the actual radius along the edges. From these points erect perpendiculars to form the center for the arc. Any other method of oblique drawing requires that arcs be set off with coordinates.

Dimensioning an Oblique Drawing

Dimension any oblique drawing in the same manner as an isometric drawing. Either unidirectional or aligned dimensioning may be used. Fig. 14-34.

Angles on Oblique Drawings

Those angles that appear on the front face, or on a face parallel to the front, of an object are laid out in their actual size. Any angle visible in the oblique plane must be drawn first in an orthographic view. Lay out the offset dimensions of the angle on the oblique plane. For an angle to be drawn in cabinet or general oblique, one of the offset dimensions must be shortened. The amount of decrease is the same as the scale of the oblique axis. Fig. 14-35.

Dimetric Projection

A dimetric projection, as stated before, is a form of axonometric projection. When making a dimetric drawing, the object is turned so that two of the axes make the same angle with the plane of projection while the third is at a different angle. Edges which are parallel to the first two axes are drawn to the same scale.

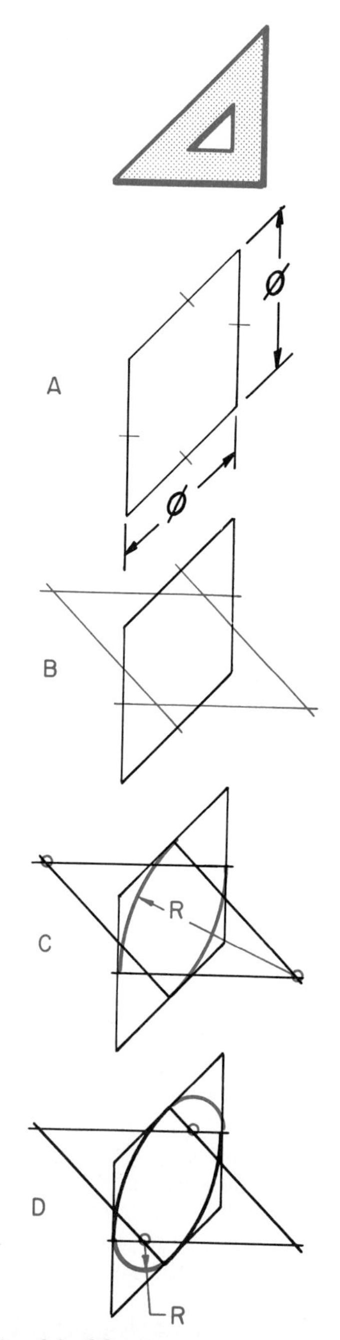

Fig. 14–32. Constructing a four-center approximate ellipse for a cavalier drawing.

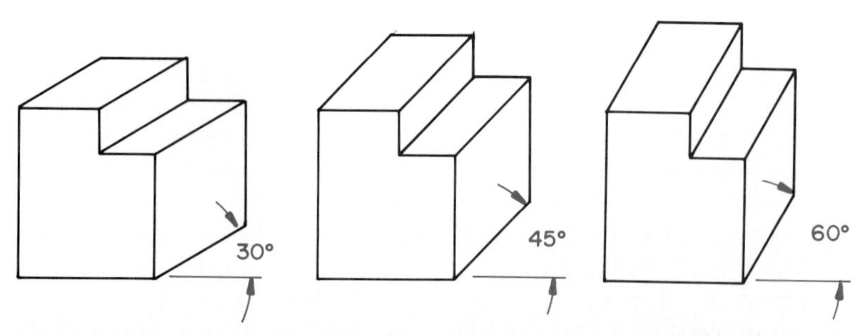

Fig. 14–31. These cavalier oblique drawings were made using standard 30-, 45-, and 60-degree angles on the receding axis. Notice how the top and right side views change in appearance.

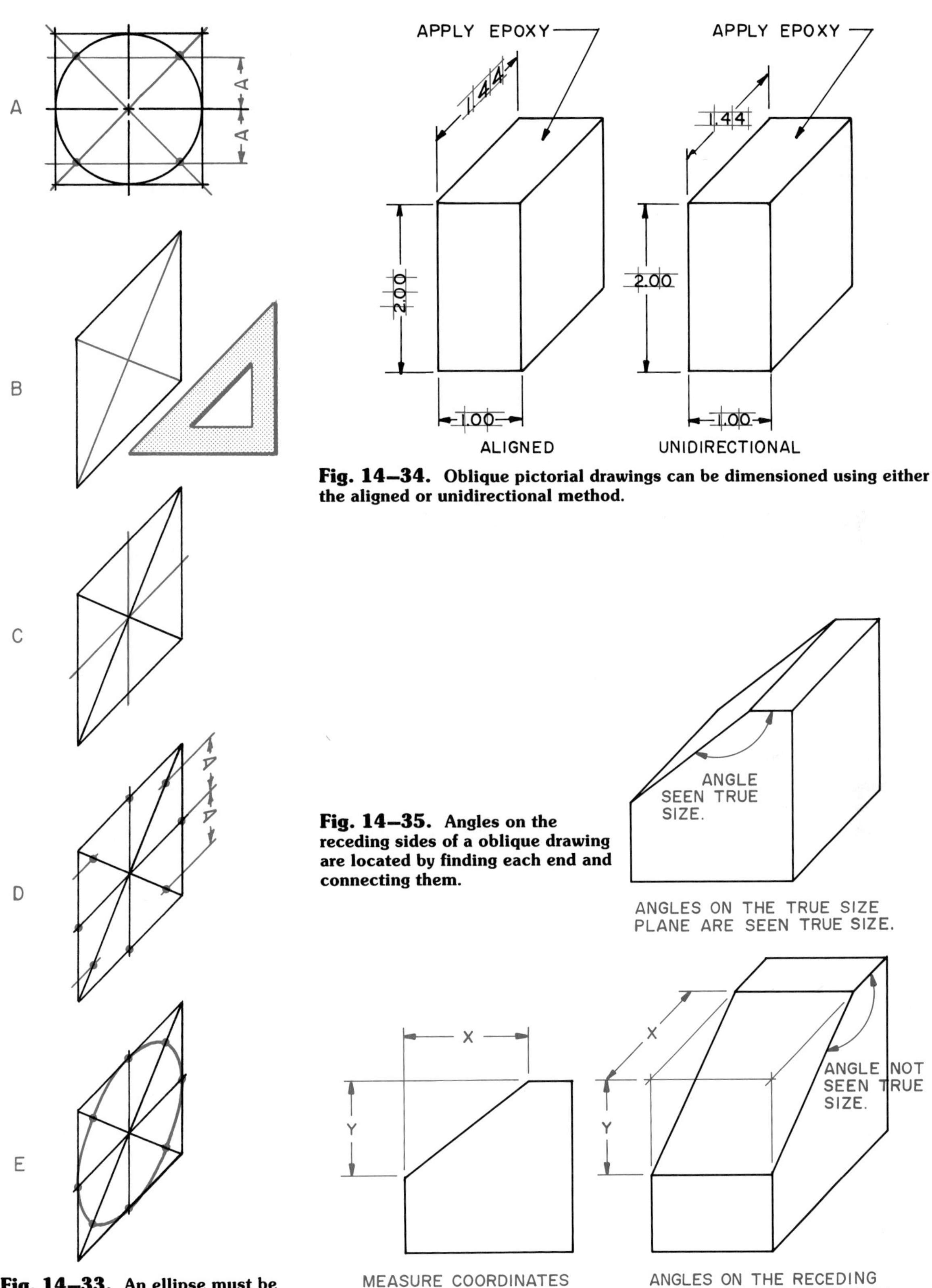

Fig. 14–34. Oblique pictorial drawings can be dimensioned using either the aligned or unidirectional method.

Fig. 14–35. Angles on the receding sides of a oblique drawing are located by finding each end and connecting them.

Fig. 14–33. An ellipse must be plotted by the offset method when used on a cabinet or general oblique drawing.

The edge parallel to the third axis is drawn to a different scale. Since two different scales are used, less distortion is apparent. A dimetric drawing is laid out the same as the isometric drawing. A dimetric drawing is shown in Fig. 14-2.

In dimetric drawings, the angles formed by the receding axes and their scales are many and varied. Fig. 14-36 shows some of the more generally used positions of axes and scales. These scales and angles are approximate. If an adjustable triangle is not available, lay out the angles with a protractor. Use two triangles to draw lines parallel to the axes. Each of these axes may be reversed or inverted to show the piece to its best advantage.

Dimetric templates are available to aid in the construction of dimetric drawings. These templates have not only the proper elliptical openings but the appropriate scales as well. The templates shown are based on axes placed at 39.23 and 11.53 degrees. Scales are full size and .623 to 1. Grid paper is available to simplify sketching in dimetric. Fig. 14-37.

Trimetric Projection

A trimetric projection is a form of axonometric projection. When making a trimetric drawing, the object is turned so that each of the axes makes a different angle with the plane of projection. Each of the axes not only has a different angle but a different scale of reduction. Fig. 14-38. Plastic trimetric angle and scale guides are available for drawing in a variety of positions.

The advantage of trimetric drawings is that they have less distortion than isometric or dimetric drawings. However, they are more difficult to make, especially if circular elements are present.

Perspective Drawing

A perspective drawing is a pictorial drawing that shows the object as the eye sees it when looking from a particular point. A perspective drawing is much like a photograph. It makes the object appear lifelike.

When you view an object, such as a building, the parts farthest from your eye appear smaller than the closer parts. Fig. 14-39. If the edges of the building could be extended, they would appear to meet on the horizon. When you are driving down a highway, the parallel sides of the road appear to come together in the distance on the horizon. This principle of size diminishing with distance is the basis of perspective drawing.

This chapter presents the conventional methods for laying out several types of perspective drawings. In industry these methods are seldom used because they are

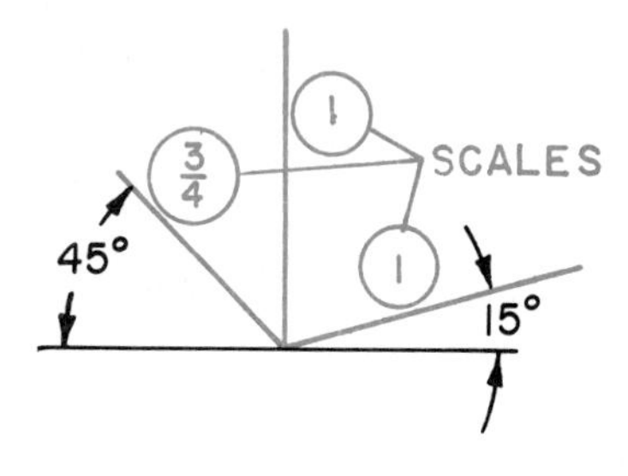

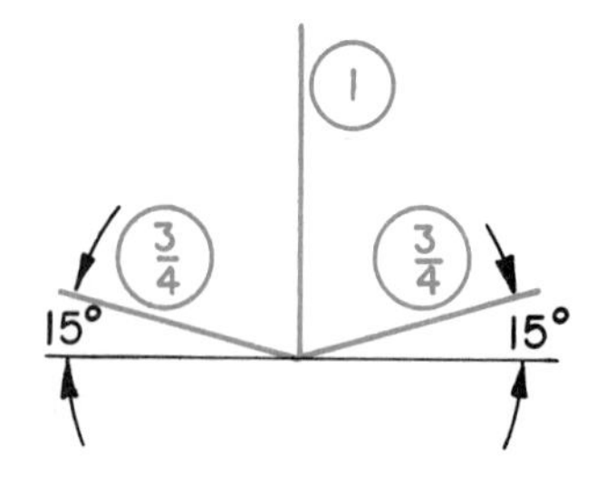

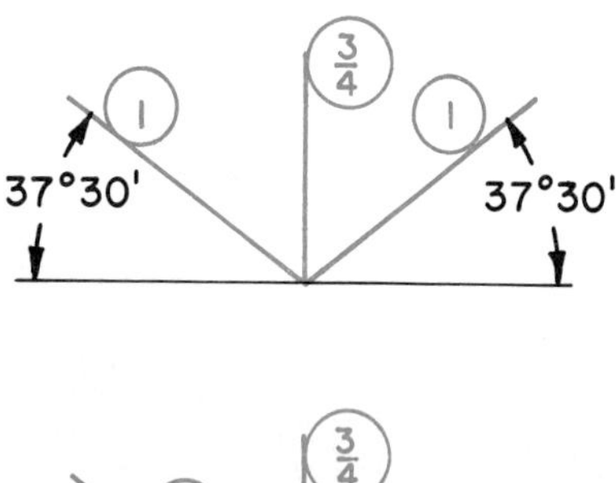

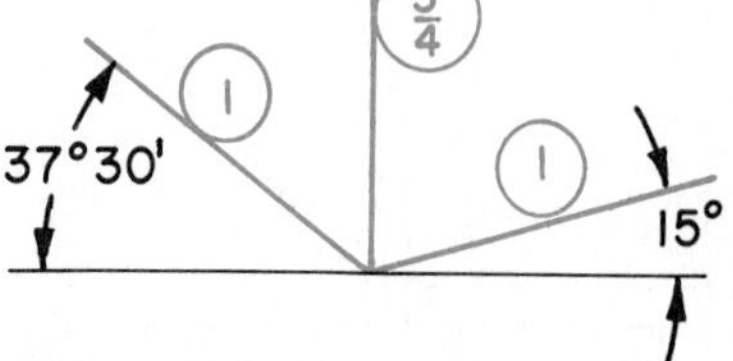

Fig. 14–36. These scales and axes are used most freqently in dimetric drawing.

Fig. 14–37. Dimetric grid paper with lines at 15 and 45 degrees is ideal for sketching.

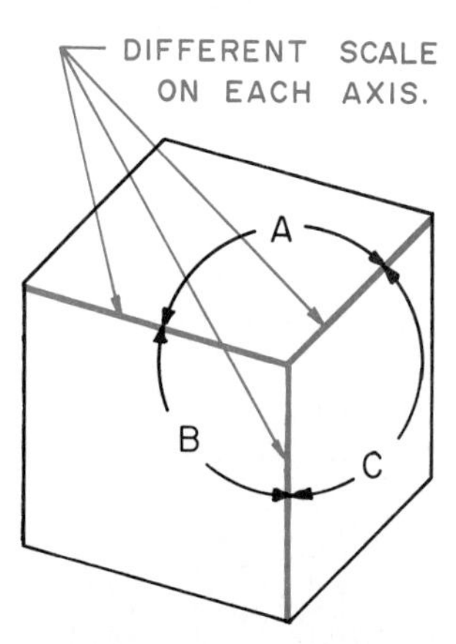

Fig. 14–38. An example of trimetric axes.

difficult and time-consuming. Industry uses mechanical aids or CAD systems to speed the making of perspective drawings. These aids will be described later. It is important to study conventional practices so that you understand the theory of perspective drawing. Knowing the theory makes it possible for you to make proper use of perspective drawing aids.

Understanding Perspective

The forms of pictorial drawing that have been discussed are based on the principle that all projectors are parallel. In perspective drawing, the projectors are not parallel. They originate at a single point. This is called the **station point**, noted on drawings as SP. The station point is the eye of the observer. From his or her eye radiate visual rays to the visible corners of the object. Fig. 14-40.

Fig. 14–39. Objects appear to become smaller the farther they are away from the eye. This is the principle used in perspective drawing. (Whitesitt Hall, School of Technology, Pittsburg State University, Pittsburg, Kansas)

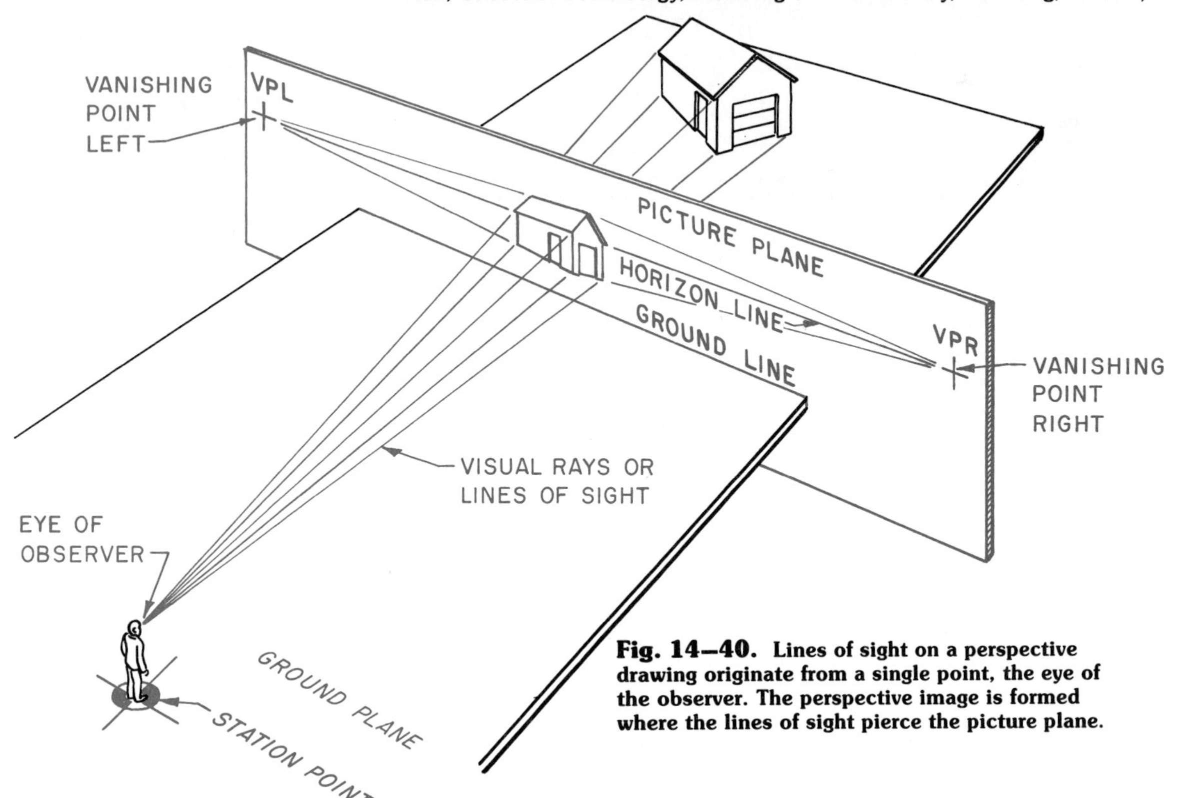

Fig. 14–40. Lines of sight on a perspective drawing originate from a single point, the eye of the observer. The perspective image is formed where the lines of sight pierce the picture plane.

In all forms of perspective drawing, the observer views the object through a plane of projection. This plane of projection is called a **picture plane**, noted on drawings as PP. It is on the picture plane that the object is drawn. This is much like standing in front of a window and drawing on the window what you see through it. The window would act as the picture plane.

The object usually rests on the ground plane. The ground plane intersects the picture plane at the **ground line**. It is noted on drawings as GL.

The **horizon line**, noted as HL, is the line in the distance on which the visual rays meet. It is on eye level above the ground. It can be located above, below, or through the object. Fig. 14-41.

The **vanishing points** are located on the horizon line where the lines of sight meet. These are the points at which the sides of the object would appear to meet if they were extended to the horizon. They are often noted on drawings as simply VP, or as VPL, for vanishing point left, and VPR, for vanishing point right.

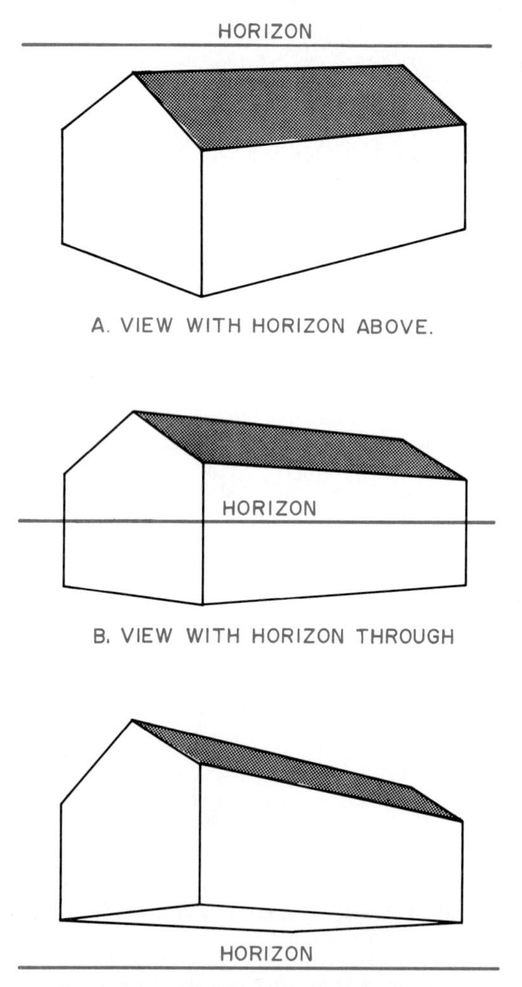

A. VIEW WITH HORIZON ABOVE.

B. VIEW WITH HORIZON THROUGH

C. VIEW WITH HORIZON BELOW.

Fig. 14–41. An object can be placed above, on, or below the horizon.

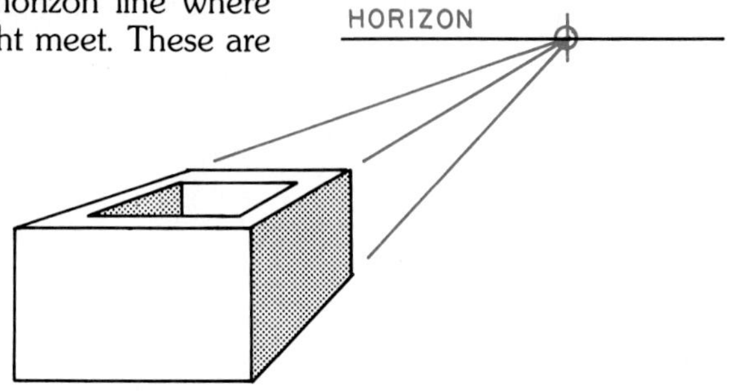

A. PARALLEL OR ONE-POINT PERSPECTIVE

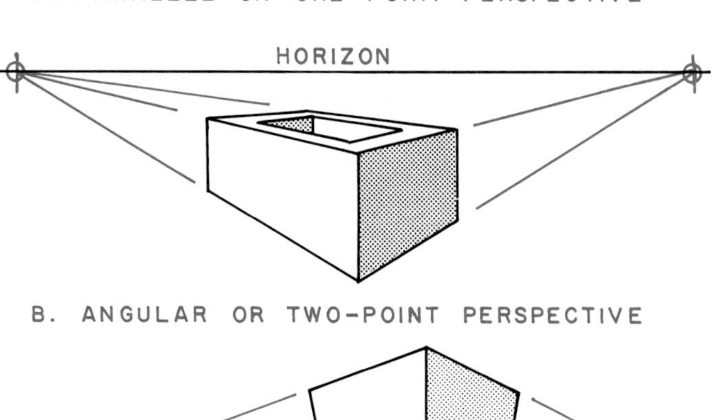

B. ANGULAR OR TWO-POINT PERSPECTIVE

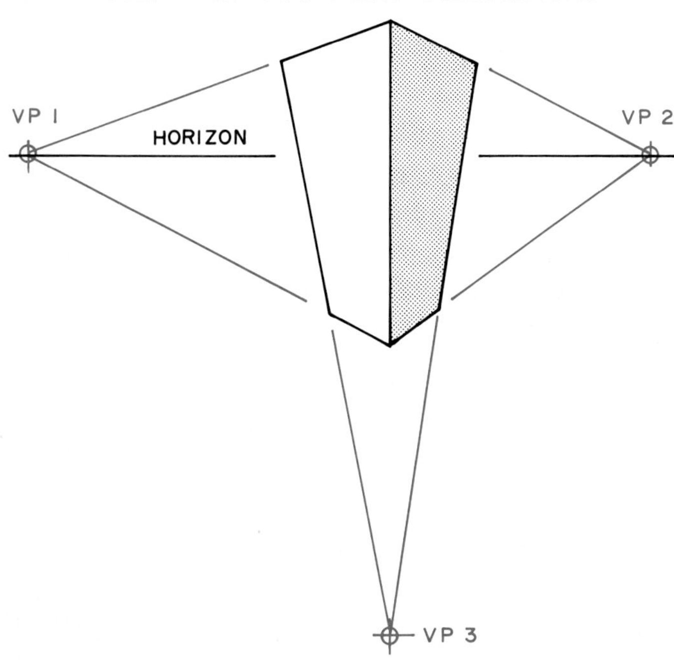

C. OBLIQUE OR THREE-POINT PERSPECTIVE

Fig. 14–42. Each type of perspective has its own advantages.

Kinds of Perspectives

Three types of perspectives are parallel, angular, and oblique (also called three-point.) Fig. 14-42. The **parallel perspective** has two principal axes parallel with the picture plane and the third perpendicular to the plane. The **angular perspective** has one axis, usually the vertical axis, parallel with the picture plane and two axes inclined to it. The *oblique* or *three-point perspective* has all three axes oblique to the picture plane. Each type has its own advantages and disadvantages for a particular situation.

Parallel Perspective. One-point, or *parallel perspective*, gains its name from the fact that one of the faces of the object is parallel to the picture plane. Fig. 14-42 at A.

Since one of the faces is parallel to the picture plane, only one vanishing point can exist. To prove this, hold an object with straight edges so that your line of sight is perpendicular to its front face and slightly below eye level. Notice that the edges that are parallel to your line of sight tend to converge at a single point. Some parallel perspectives have the appearance of being distorted. This distortion is usually caused by poor placement of the station point.

To draw a parallel perspective, as shown in Fig. 14-43, follow these steps:

1. Draw the edge view of the picture plane, PP.

2. Draw the top view of the object with the front face touching the PP. In this position the front face of the object will be its true size and shape. It can be drawn on a sheet of scrap paper since it is not needed after the perspective is finished.

3. Locate the station point, SP, in front and to the side of the object. The SP should be located at least *twice* the length of the object from the PP. The closer the SP is to the PP, the more distortion will occur. In Fig. 14-43 the SP is placed more than three times the length of the object from the PP.

4. Draw in the ground line, GL. It should be located where you want the bottom edge of the perspective on the drawing.

5. Lay out the front view of the object. It is projected from the top view to the ground line. Measure and draw the height using the same scale as for the plan view. (In the plan view, the object appears as seen from above.)

6. Draw the horizon line, HL, above the GL. Locate it to give the type of view wanted. See Fig. 14-41 for examples.

7. Locate the vanishing point, VP. The VP is always located on the HL by projecting the SP perpendicular to the HL.

8. Draw visual rays from the SP to the corners of the object in the top view. The points where they cross the picture plane become the back edge of the object in the perspective view. Project these corners into the perspective area.

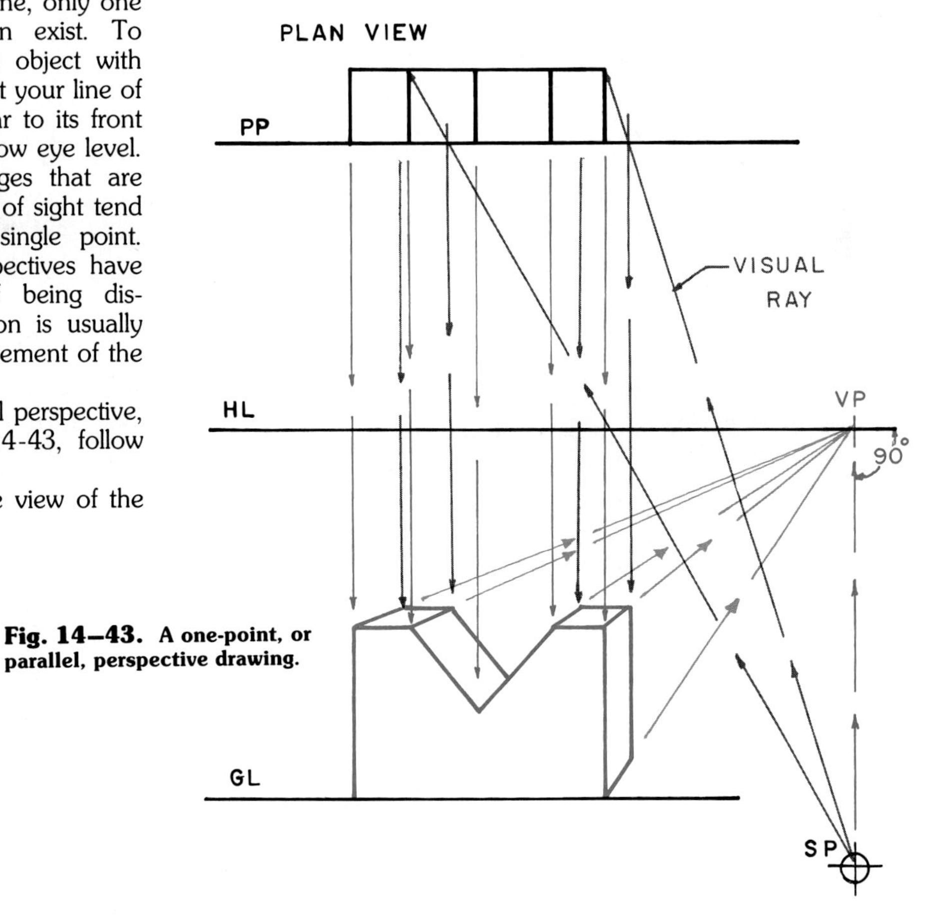

Fig. 14–43. A one-point, or parallel, perspective drawing.

Fig. 14–44. A parallel perspective that is centered on the front door. (Home Planners, Inc.)

Fig. 14–45. A two-point perspective drawing is commonly used in the field of architecture. (Home Planners, Inc.)

9. Project the corners from the front view to the vanishing point. The point of intersection between the line drawn to the vanishing point and the visual ray for the back corner is the location of that corner on the perspective. Repeat this for every corner.

10. Connect the corners to complete the view.

A typical architectural use of parallel perspective is found in Fig. 14-44.

Angular Perspective. Two-point or *angular perspective* gets its name from the fact that the object is placed at an angle to the picture plane. Fig. 14-45. The most frequently used angle is 30 degrees. Other angles commonly used are 15 and 45 degrees. Fig. 14-46. Angular perspective is one of the most frequently used forms. It is readily adapted to almost any object or situation.

One of the most critical features in angular perspective is the location of the station point. The station point should be located far enough away from the object to prevent distortion. Usually this should be two or three times the width of the object. The closer the station point is to the object, the greater the distortion. The angle formed by the visual rays touching the outside edges of the object should not be any greater than 30 degrees. This angle is called the *angle of clear vision.* Fig. 14-47.

Fig. 14-48 shows an angular perspective. To draw this type of perspective, follow these steps:

1. Draw the edge view of the picture plane, PP. Draw the top view of the object so that one corner, AB, touches the PP. This corner will be the only *true length edge* on the perspective.

2. Locate the station point, SP, so the entire object is included in the angle of clear vision. Many times the SP is in line with the corner touching the PP. The SP can be located to the right or left of this point, depending upon which face is to be emphasized.

3. Locate the vanishing point left, VPL, and vanishing point right, VPR, on the PP. To do this, draw lines from the SP to the PP parallel to the sides of the object in the plan view. This locates the vanishing points on the PP. They are marked SPC and SPD.

4. Draw the ground line, GL.

5. Locate the horizon line, HL, at the desired distance above, on, or below the GL. The normal observer's eye is between 5 feet 2 inches and 5 feet 6 inches above the GL. See Fig. 14-41 for examples.

6. Locate the VPL and VPR in the elevation view by dropping points SPC and SPD from the PP to the HL. By dropping verticals from these points to the HL, you are now working in the elevation view of the perspective.

7. Project corner AB from the plan view to the ground line in the elevation view. Measure its true length on line AB.

8. Draw visual rays from the SP to the main features in the top view. See points E and F.

9. Where the visual rays have pierced the PP, drop perpendicular projectors into the perspective area.

10. Draw the basic box shape of the object. Draw lines from points A′ and B′ to the VPL and VPR. Where the visual rays to E and F intersect the PP, drop projectors to the perspective area. The points where they intersect the sides of the box locate the other corners. Corners E and F are now in perspective. Continue locating the remaining corners in the same manner to complete the perspective.

11. Special problems arise when locating surfaces that are not on the edges of the box. For example, in Fig. 14-49, surface Y must be located. To do this, measure true length distance G on corner AB. AB is true length. The true length of measurement G can be measured here and projected to the proper vanishing point. The vertical edges are found by projecting them from the picture plane.

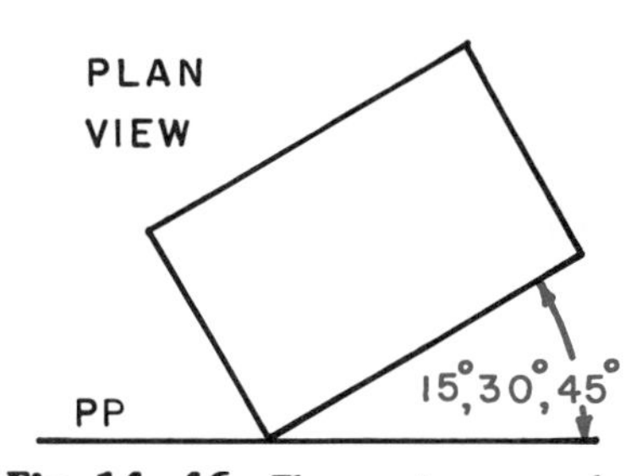

Fig. 14—46. The most commonly used angles on two-point, or angular, perspective drawing.

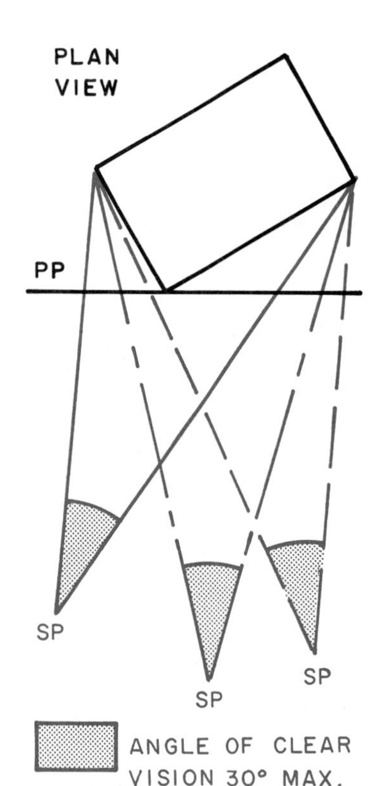

Fig. 14—47. The location of the station point is governed by the face of the object to be emphasized and the angle of clear vision.

Circles and Arcs in Perspective. The arc or circle in perspective is seen as an ellipse. Ellipses vary greatly in appearance from those which resemble a circle to those looking like a cigar. Regardless of the ellipse's shape in perspective, it is more readily constructed in a square (trapezoid). By placing the ellipse in a square, it is easier to visualize its shape in perspective.

To draw the shape of a circle inclined to the picture plane, see Fig. 14-50 and follow these steps:

1. Draw the elevation view of the circle at the side of the perspective area. Divide the circle into any number of equal divisions. A convenient number that is sufficient to draw the ellipse is 12, which means the points are located 30 degrees apart. Number each of these divisions.

2. Draw an auxiliary view of the circle in the plan view behind the picture plane, PP. Divide this circle into the same number of equal parts. Number each of these points identically with those

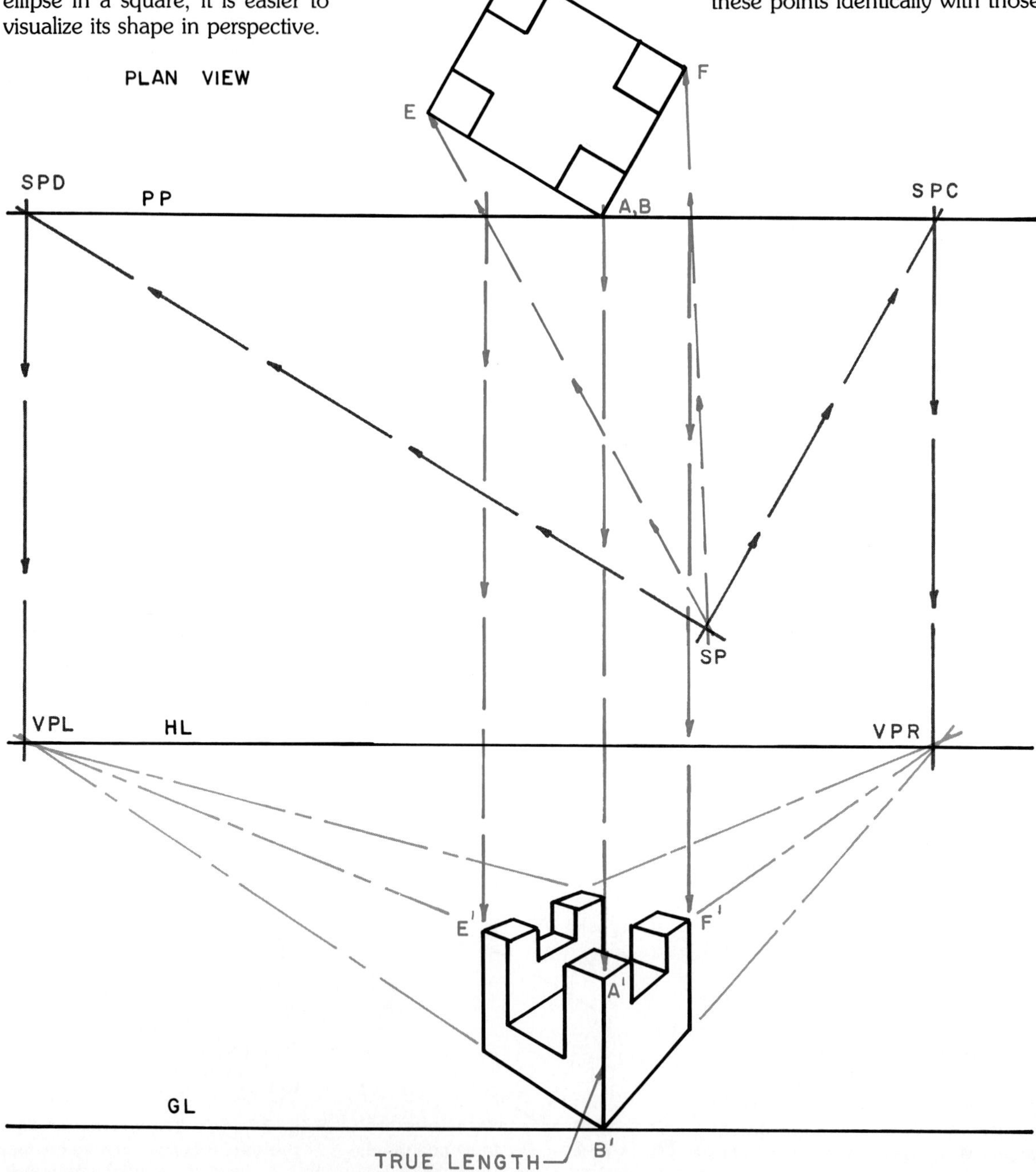

Fig. 14—48. An angular, or two-point, perspective.

in the elevation. Project each point and number to the plan view of the circle.

3. Draw visual rays from the station point, SP, to the numbered points in the plan view.

4. Drop verticals where the visual rays have intersected the PP.

5. Draw the square in perspective by projecting across from the elevation view.

6. Draw horizontals from the points on the circle in the elevation view to the edge of the square. Carry each of these points from the edge of the square to the vanishing point, VP.

7. The coordinate points are located by the intersection of the verticals dropped from the PP and the vanishing lines drawn to the VP.

8. Connect the points with an irregular curve.

One of the least confusing methods used to draw a circle lying on the horizontal plane is by the "ground line" method. In this method the ground line, GL, is temporarily moved from its original position to the height of the circle. Fig. 14-51 shows how to plot a circle in two-point horizontal perspective. Follow these steps:

1. Enclose the circle in a square as seen in the plan view.

2. Divide the circle into equal parts.

3. Extend the sides of the square to the PP points E, F, G, and J. Where the extended sides of the square have touched the PP, drop verticals to the GL. From each of these points, carry lines to the vanishing points right and left, VPR and VPL. The square is thus formed in perspective.

4. Carry each of the points on the circle that are parallel to the sides of the square to the PP. Drop verticals from the PP to the GL. From these points of intersection on the GL, extend lines to the vanishing points.

5. Where the corresponding lines to the vanishing points intersect, the coordinates for the ellipse are formed.

6. Draw a smooth curve through each of the points.

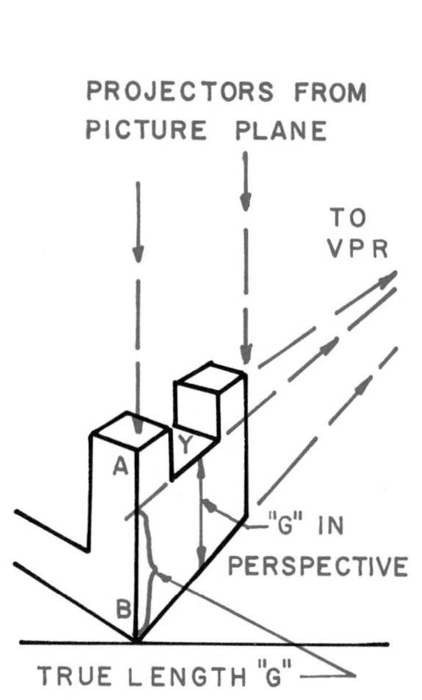

Fig. 14–49. **Vertical distances are found by measuring their true length on the true length corner. The distance is then projected toward the proper vanishing point.**

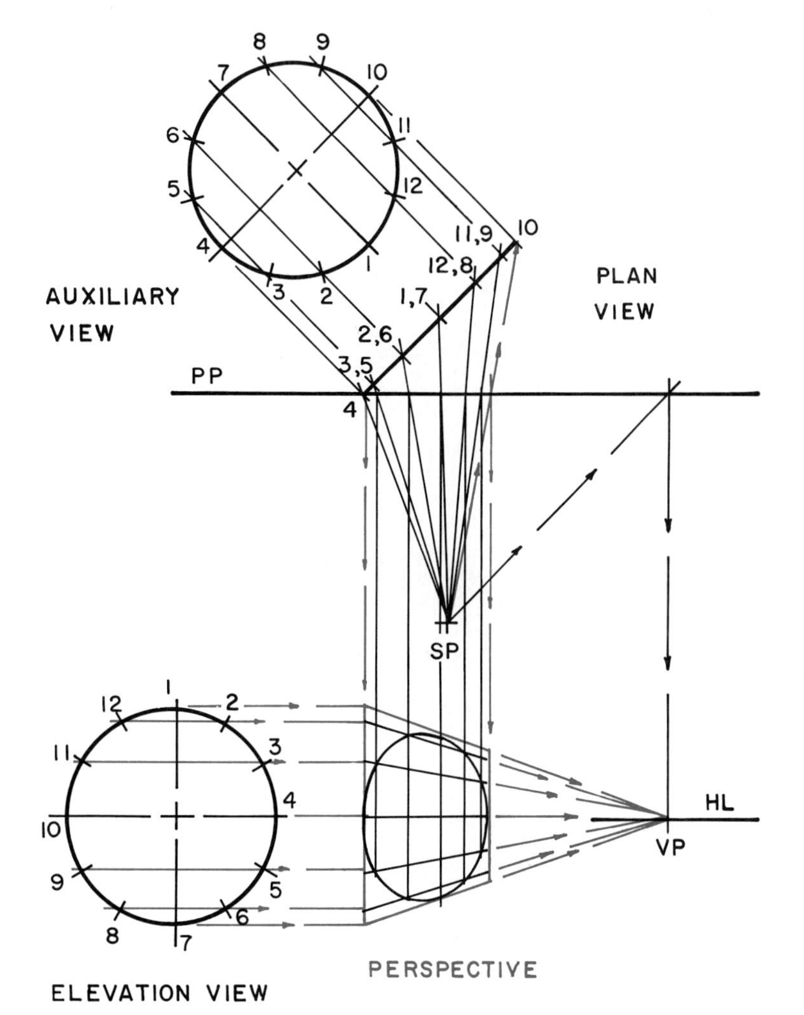

Fig. 14–50. **How to draw a vertical circle in perspective.**

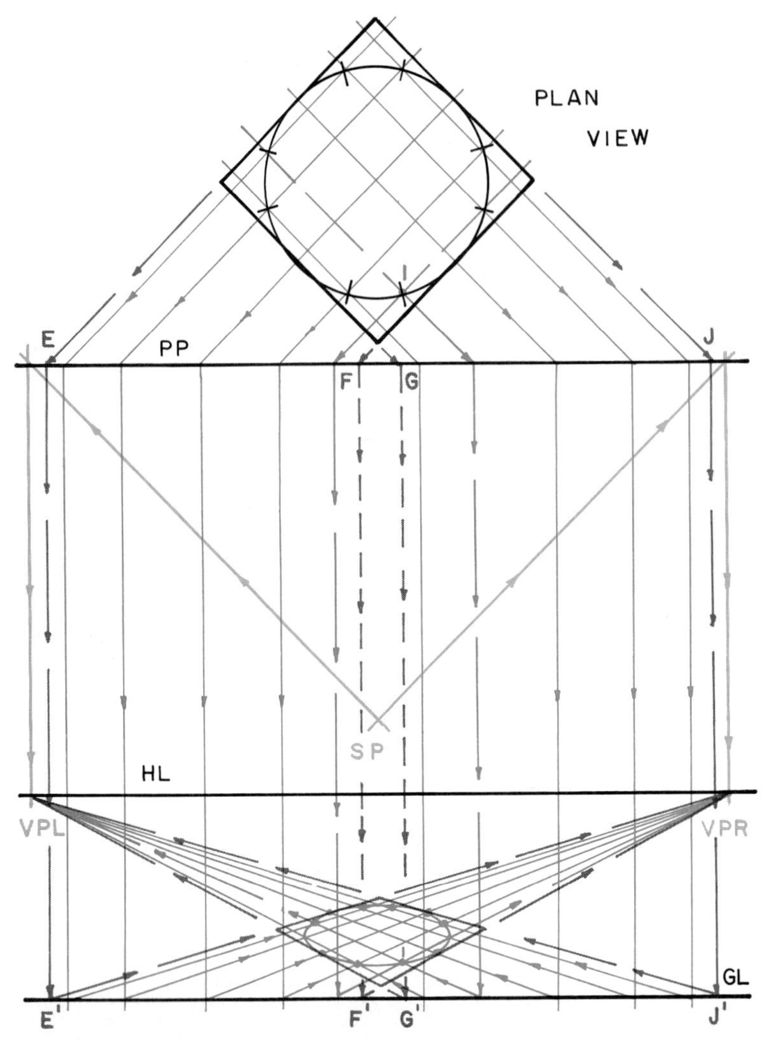

Fig. 14–51. **How to draw a horizontal circle in perspective using the ground line method.**

Perspective Drawing Aids. True perspective foreshortens length, width, and height. One- and two-point perspectives do not shorten the height. Foreshortened height is especially valuable in making architectural perspective drawings of large buildings. The buildings drawn this way are more lifelike and appear as they would to the eye.

There are a number of drafting devices available to use to draw true perspectives. These establish three vanishing points. One of these devices is shown in Fig. 14-52. This perspective drawing board contains a series of graduated scales and eight vanishing points. The extreme left vanishing point is represented by a concave curve. There are five vanishing points on the horizon line. There are two additional vanishing points below the horizon line.

This type of drafting device greatly speeds perspective drawing. It also increases the accuracy of the finished drawing. A true perspective is shown in Fig. 14-53.

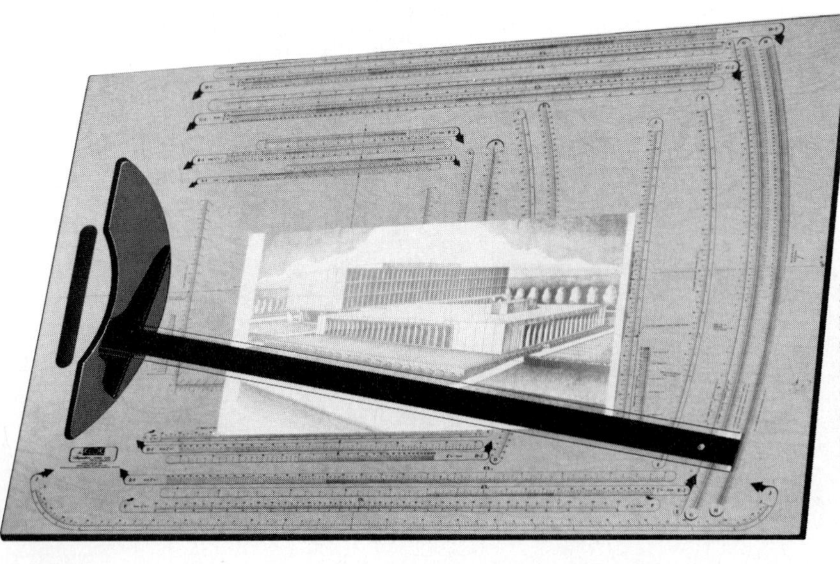

Fig. 14–52. **The Klok Perspector perspective drawing board. The scales permit drawing one-, two-, and three-point perspectives. The drawings are in true perspective, eliminating the distortion found in graphic methods. (The Utley Company)**

Another perspective drawing system is shown in Fig. 14-54. On it the drafter can produce multiview drawings, one-, two-, and three-point perspectives, and isometric-type drawings. Fig. 14-55.

A printed perspective grid is another aid commonly used to produce pictorial drawings. A variety of views can be drawn producing parallel and angular perspectives.

One set of printed grids used for perspective drawings is shown in Fig. 14-56. They are printed on heavy cardboard. The grid is taped to the drawing board. A sheet of vellum or plastic drafting film is taped over it. The drafter uses the grid as a guide when making the drawing. Fig. 14-57.

Isometric-dimetric grids are also available. They are used in the same manner described for perspective grids.

Pictorial drawings are also made by computers. They are developed on the screen, and then hard copies are produced by a printer.

Other Pictorial Techniques. Chapter 20, "Technical Illustration," gives additional details on pictorial drawing. After you master the fundamentals of this chapter, you are ready to try drawing some technical illustrations.

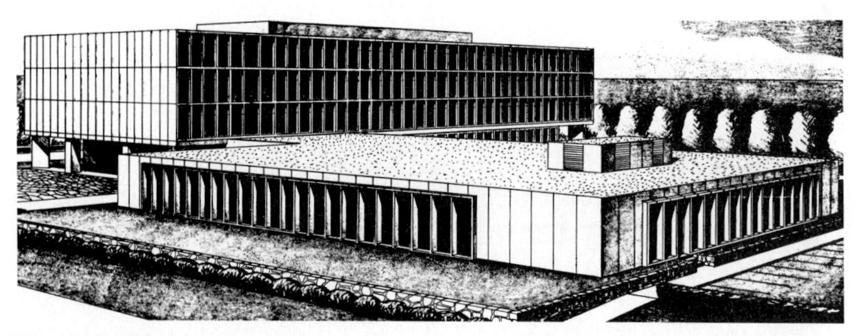

Fig. 14–53. A true perspective drawn on the Klok perspective board. (The Utley Company)

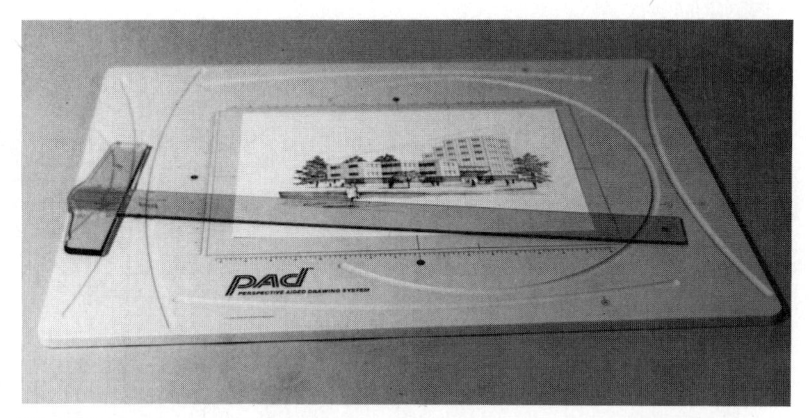

Fig. 14–54. This is the PAD(rmr) drawing system. (Perspective Aided Design International Corp.)

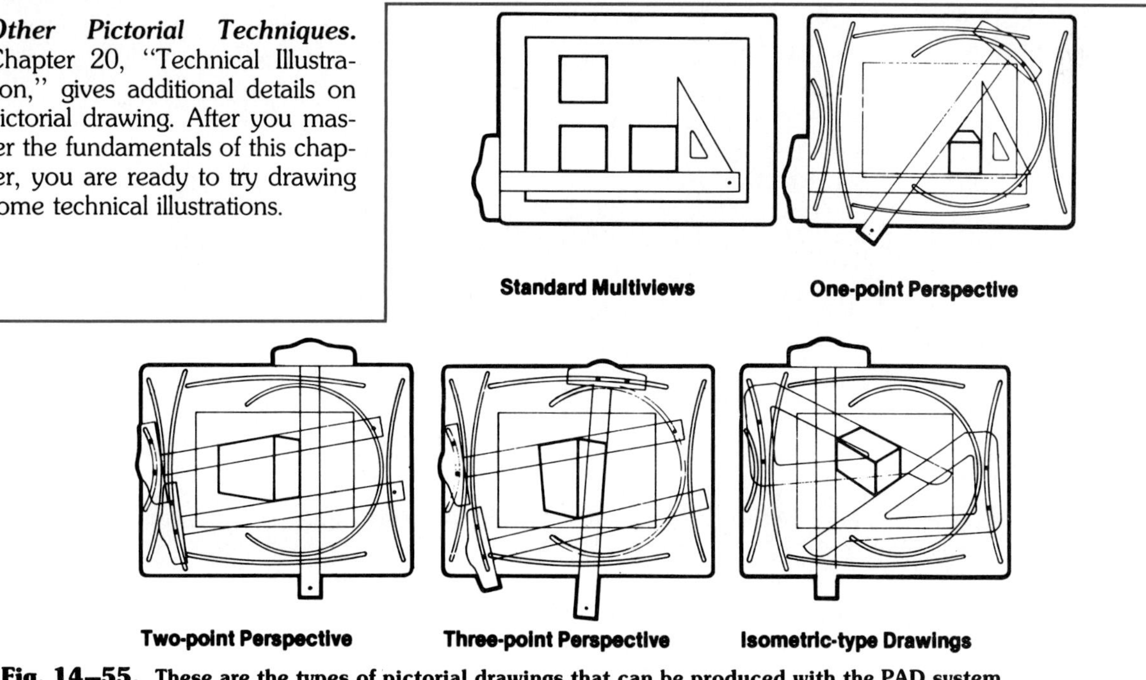

Fig. 14–55. These are the types of pictorial drawings that can be produced with the PAD system. (Perspective Aided Design International Corp.)

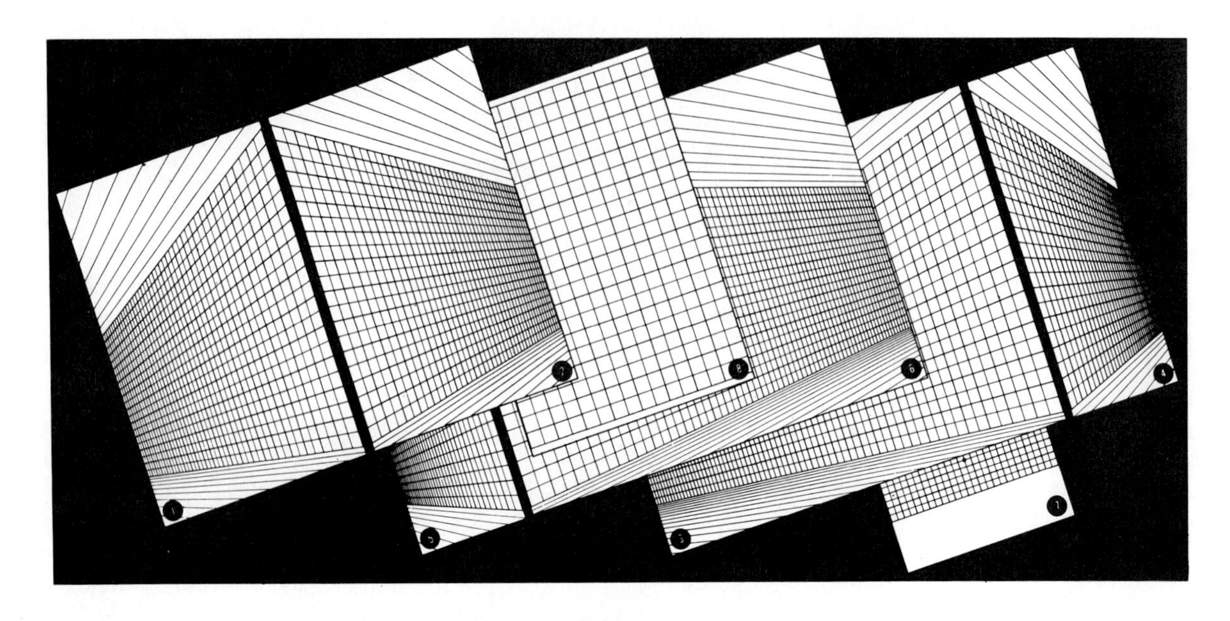

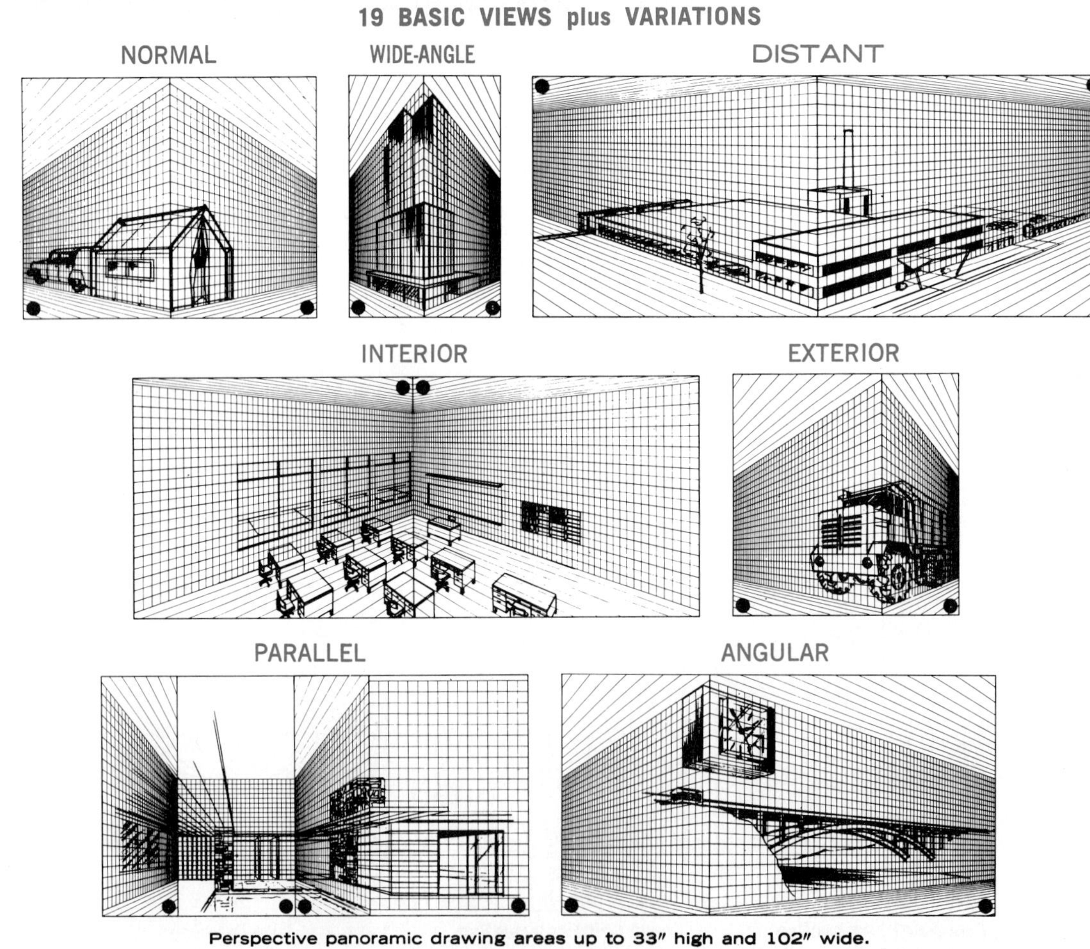

Fig. 14–56. **A coordinated set of eight underlay panels for perspective drawing. (GraphiCraft)**

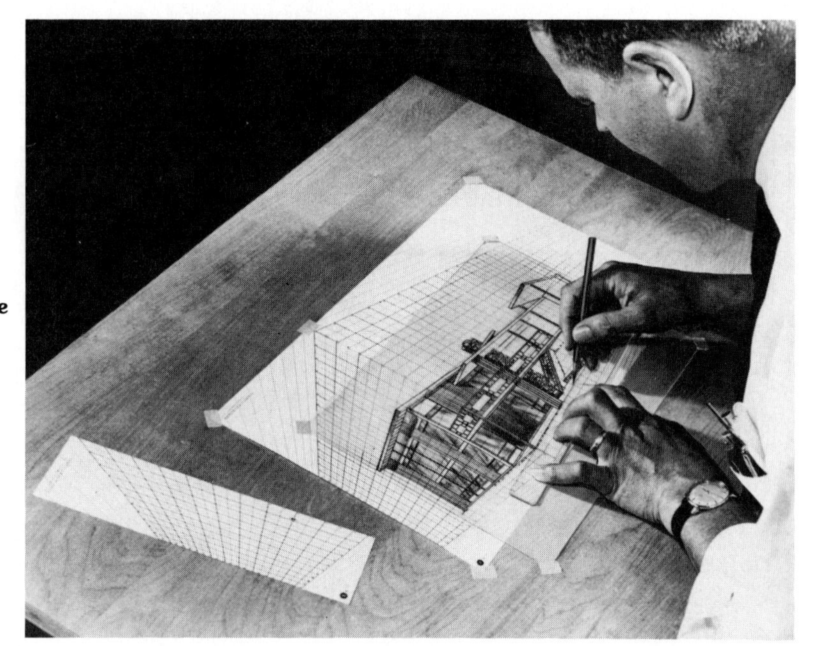

Fig. 14–57. Making a perspective drawing on vellum using one of the perspective panels. (GraphiCraft)

Chapter Review

Build Your Vocabulary

You should understand and use the following terms as part of your working vocabulary. Write a brief explanation of what each means.

axonometric drawings
oblique drawings
perspective drawings
isometric drawings
isometric lines
nonisometric lines
four-center approximate ellipse
cavalier drawings
cabinet drawings
station point
picture plane
ground line
horizon line
vanishing points
parallel perspective
angular perspective

Study Problems—Directions

1. On a sheet of vellum or plastic drafting film, lay out grids for isometric and oblique drawings. Use these when sketching pictorials of Figs. P14-1 through P14-3.

2. Make isometric or oblique drawings of the products in **Figs. P14-1, P14-2,** and **P14-3** as assigned by your instructor.

3. Make an isometric drawing of the try square, **Fig. P14-4.**

4. Make an isometric drawing of the carbide cutting tool inserts, **Fig. P14-5.** Draw twice actual size.

5. Make an isometric drawing of the diamond point cutting tool, **Fig. P14-6.**

6. Make an isometric drawing of the stop block, **Fig. P14-7.**

7. Make an isometric drawing of the machine tool cutter grinding wheel, **Fig. P14-8.**

8. Make an oblique drawing of the grinding wheel, **Fig. P14-8.** Compare it with the isometric drawing you drew for Problem 7. Which was easier to draw? Which looks more natural?

9. Make an isometric drawing of the jaw for an inside micrometer, **Fig. P14-9.** Notice that it has a long irregular curve.

10. Make an isometric drawing of the crane trolley bracket, **Fig. P14-10.**

11. Make an oblique drawing of the crane trolley bracket, **Fig. P14-10.**

12. Make a cabinet oblique drawing of the cabinet shown in **Fig. P14-11.**

13. Make an oblique drawing of the packing fitting, **Fig. P14-12.** Draw it with the axis on 45 degrees. Then draw it with the axis on 30 degrees. Which appears more natural?

14. Study the objects shown in **Figs. P14-13** through **P14-18.** Select the method of pictorial drawing that would best show each. Draw them with the method chosen. Compare your drawing with those of other students who chose a different method or different angle for the receding axis. Was your decision best?

15. Make a perspective drawing of the cabinet shown in **Fig. P14-11.** Draw the plan view to the scale 1″ = 1′0″. Locate the station point ten feet from the picture plane. Position the station point so both sides of the cabinet are clearly visible. Place the horizon 6′0″ above the ground line.

16. Make a perspective drawing of the summer cottage in **Fig. P14-19.** Draw the plan view to the scale ¼″ = 1′0″. Position the station point and horizon at the place you think gives the best perspective.

17. Make a dimetric drawing of the A-frame cottage, **Fig. P14-20.** Select any of the recommended axes.

18. Make pictorial drawings of the study problems found at the end of Chapters 6, 7, and 11. Draw in isometric, oblique, dimetric, trimetric, or perspective as assigned by your instructor.

19. Bring objects to school such as an automobile water pump or a telephone. Make pictorial drawings of them. Use the type of drawing that will show the object the best.

20. Make an enlarged copy of the mid-deck orbiter panel configuration in **Fig. P14-21.** Draw it three or four times as large as printed. Use your dividers to find the needed lengths. (NASA)

Study Problems

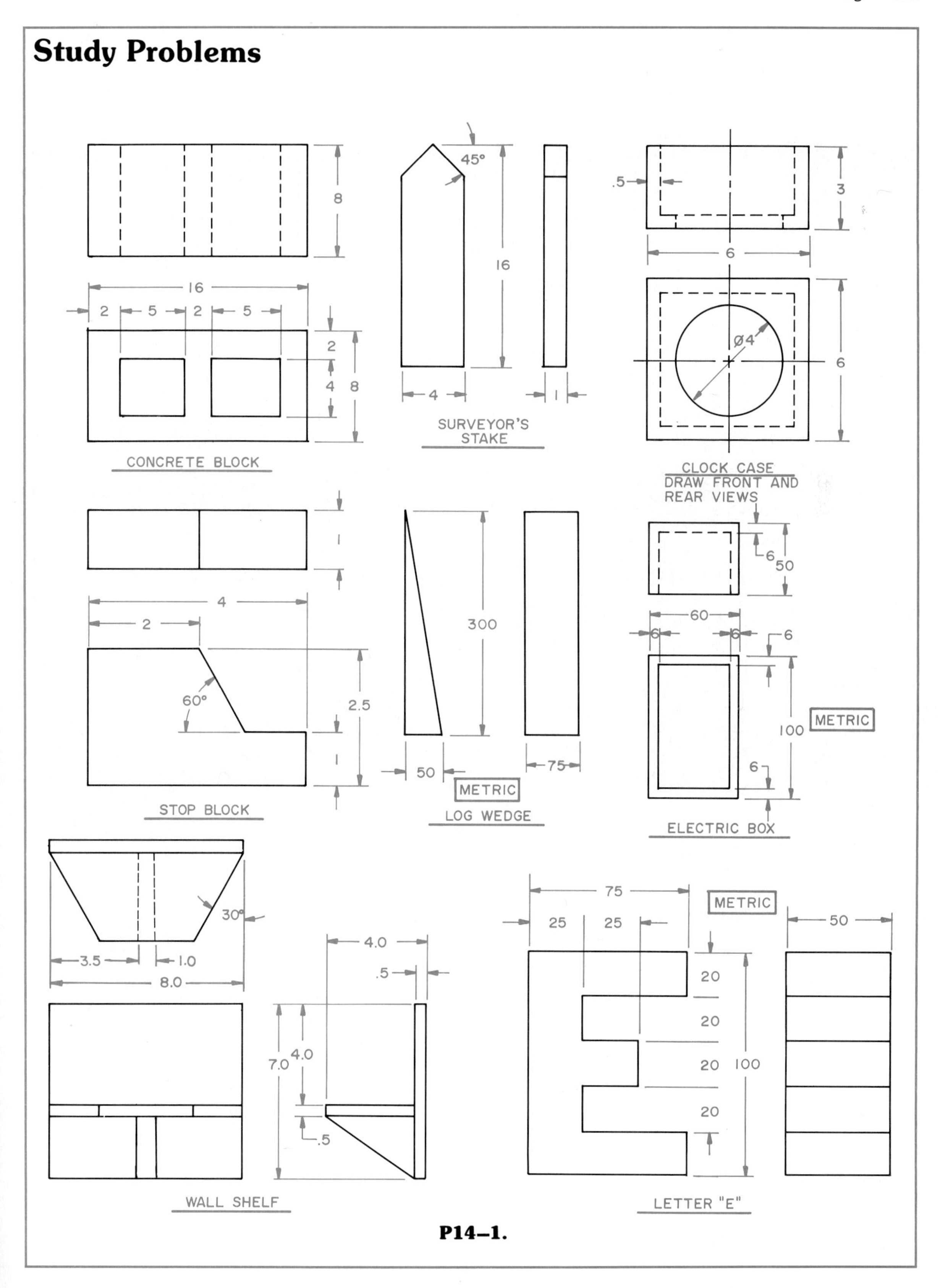

P14–1.

Study Problems

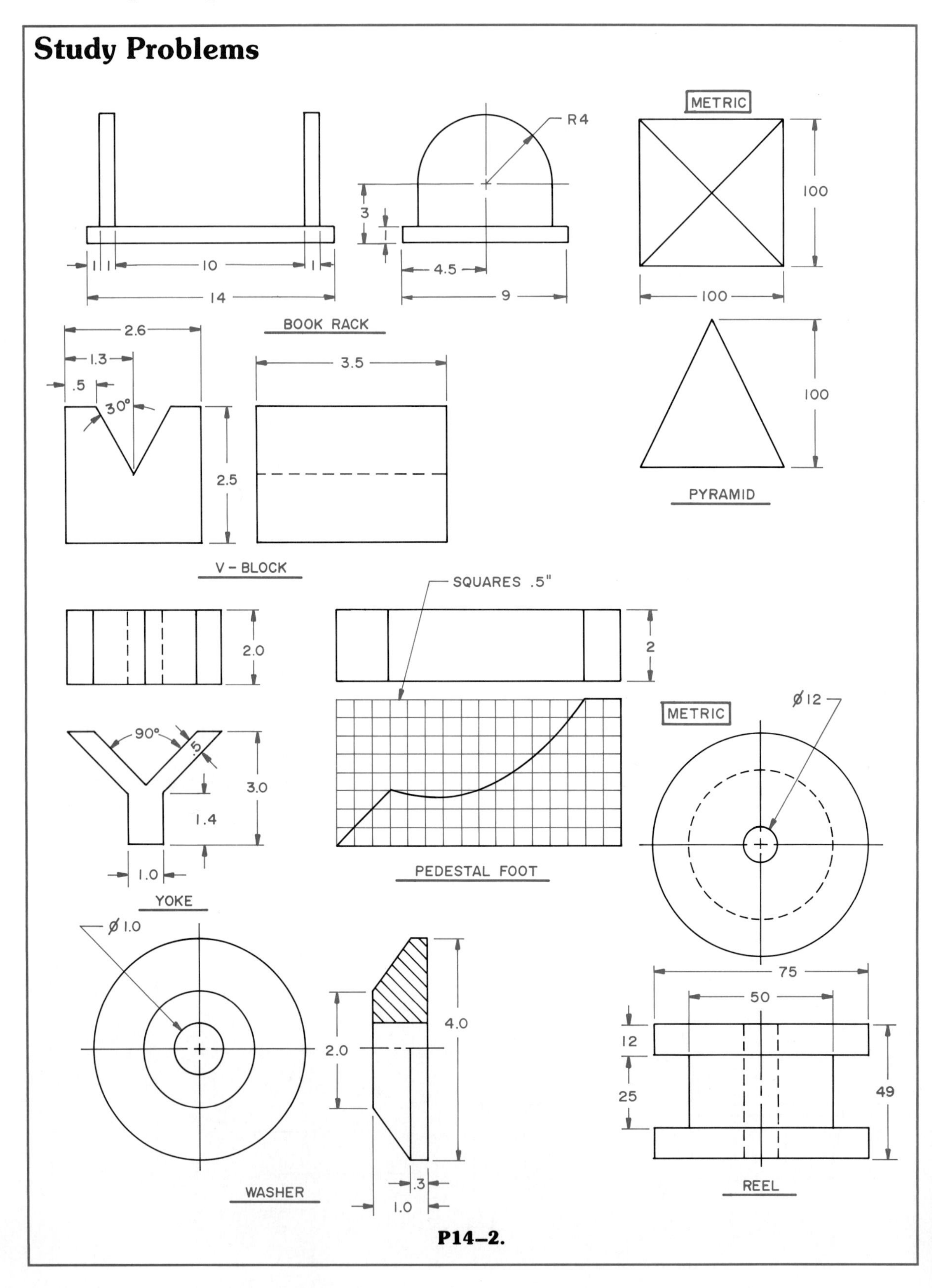

P14–2.

Study Problems

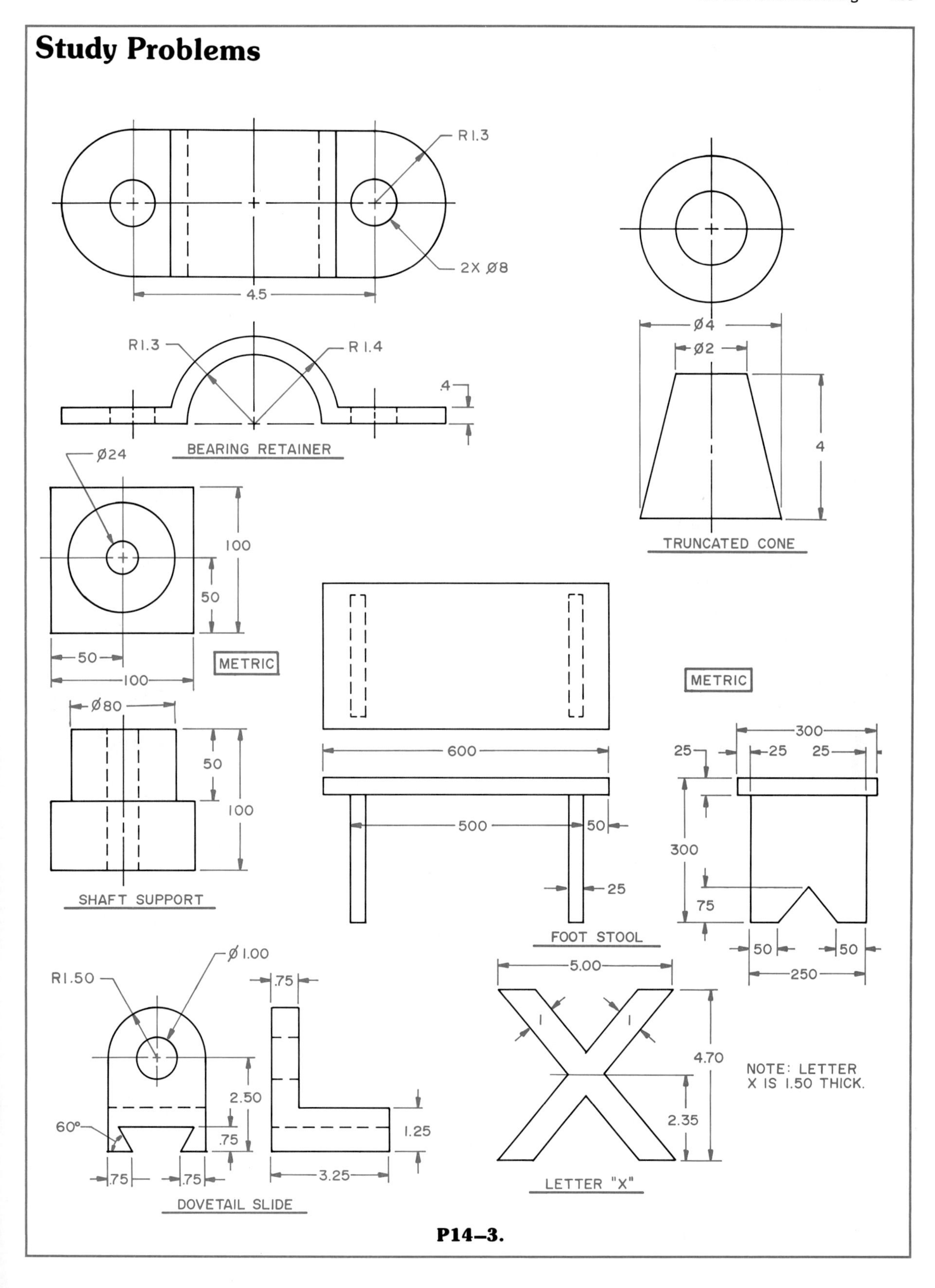

P14–3.

Study Problems

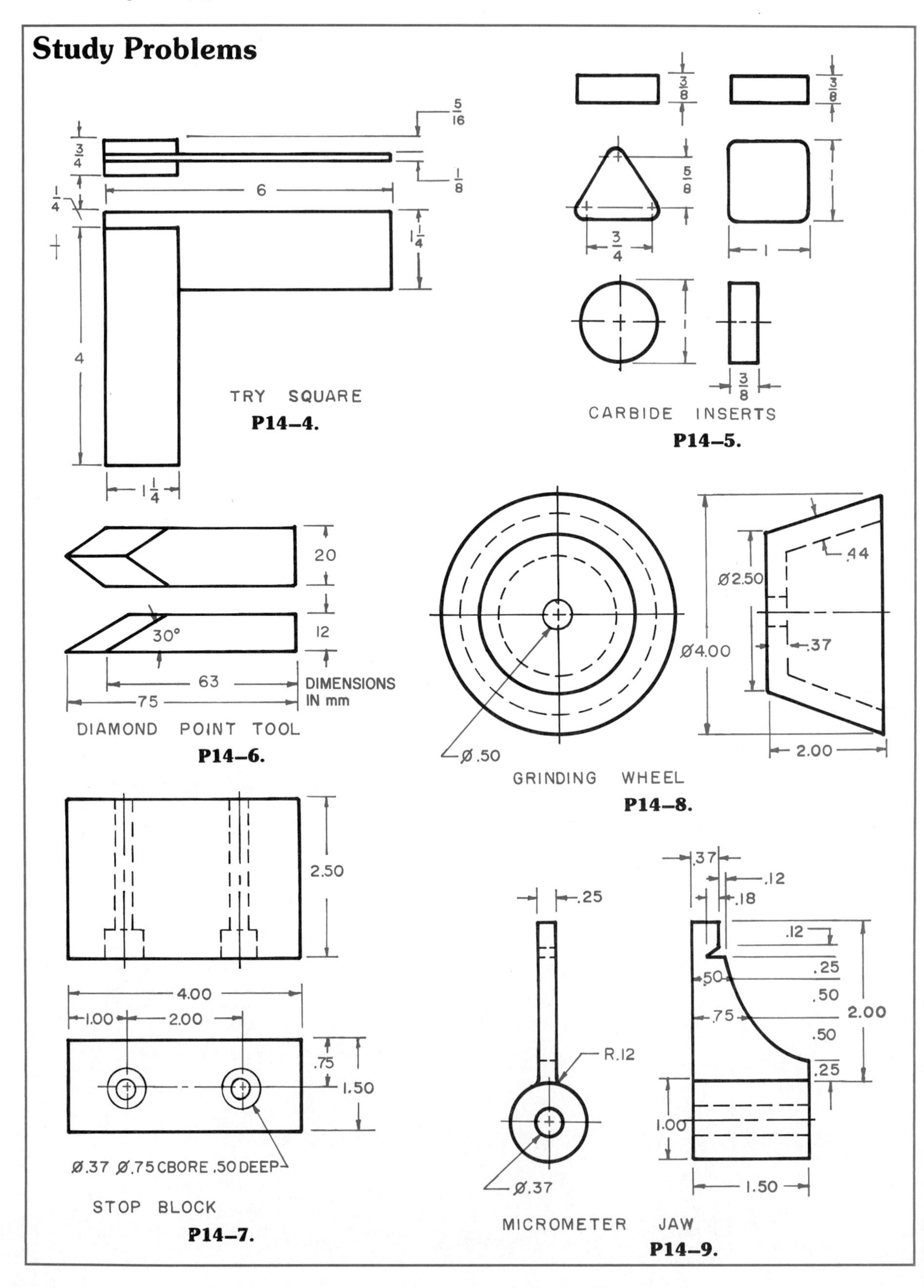

Study Problems

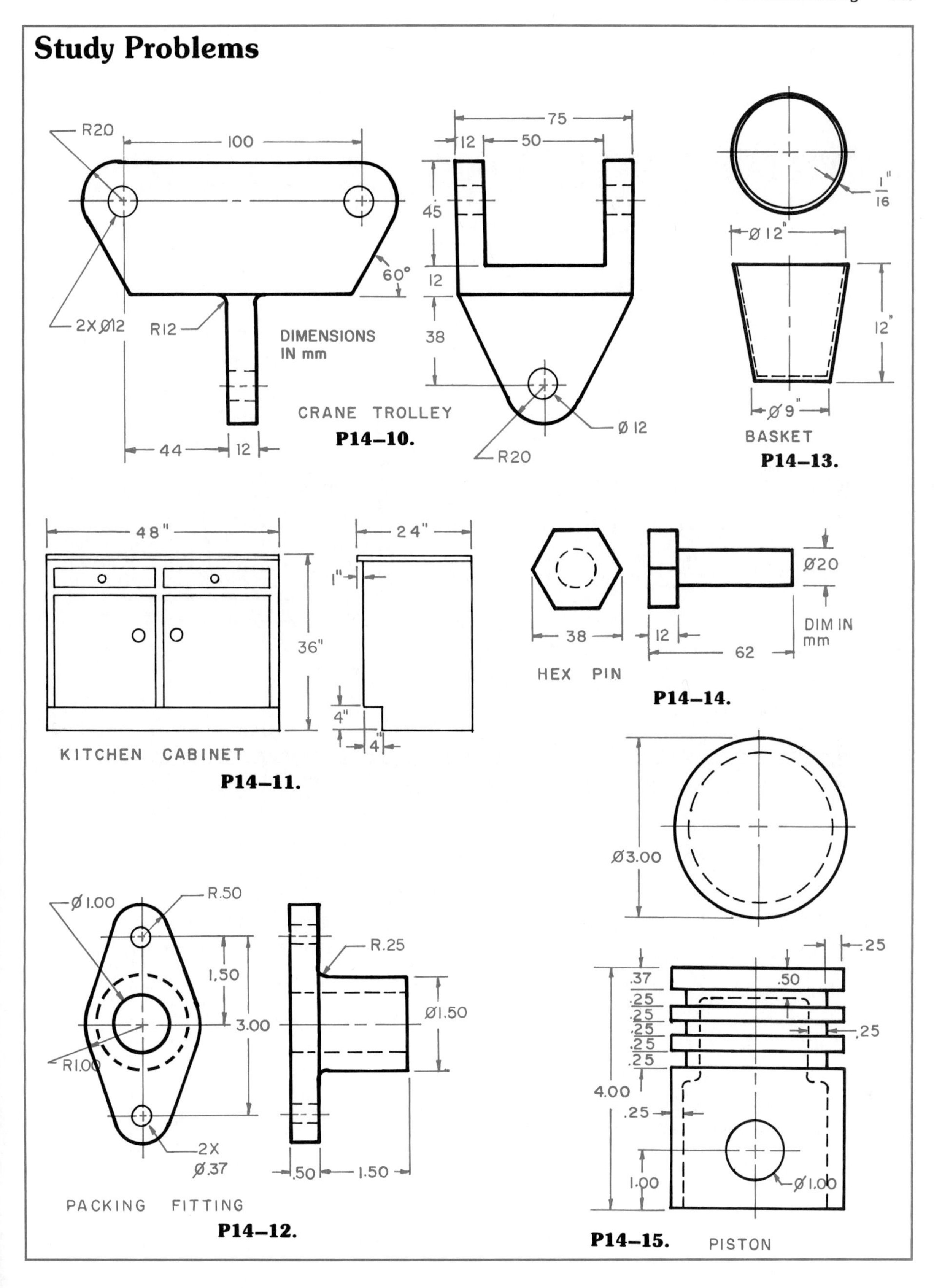

P14–10.

P14–13.

P14–11.

P14–14.

P14–12.

P14–15.

Study Problems

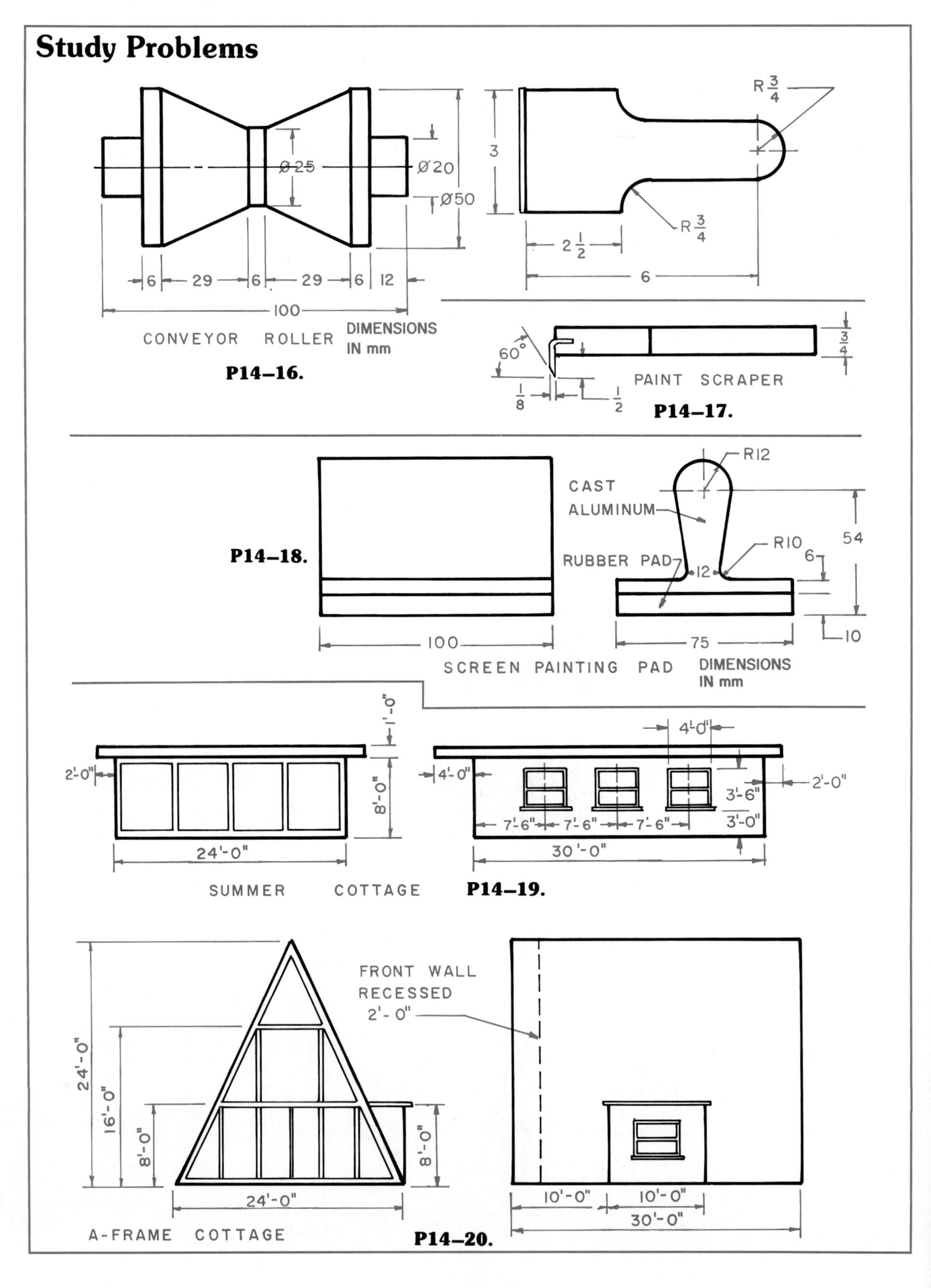

Study Problems

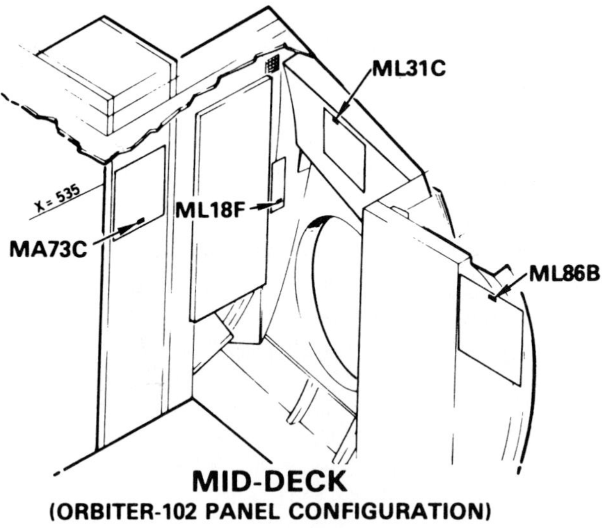

MID-DECK
(ORBITER-102 PANEL CONFIGURATION)

P14–21.

CHAPTER 15

Drawing for Numerical Control

After studying this chapter, you should be able to do the following:

- Explain the principles of two-axis and three-axis numerical control systems.
- Explain the difference between N/C and CNC.
- Prepare drawings for use with point-to-point and continuous-path N/C machines.

Introduction to Numerical Control

Numerical control (N/C) is a means of operating machines following a preprogrammed series of coded instructions. The operations that the machine must perform are determined first. This information is put into a coded form on tape that the numerical control unit can read and understand. The machine accepts and responds to commands from a control unit.

Numerically controlled machines enable industry to increase production while reducing labor costs. They produce parts faster than manually operated machines. The parts are more accurately made, and quality is more consistent. N/C machines are excellent for rapidly producing duplicate parts. Since duplicate parts can be made so easily, the inventory of these parts can be kept low.

Changing the design of a part is much faster and easier using numerical control because the changes are made in the program rather than to the dies and jigs needed for manual production.

Numerical control is being applied to an ever increasing variety of machines. Originally, its major application was to operate machine tools. It is now being used for operations such as bending, forming, stamping, flame cutting, riveting, welding, and inspection.

Programming

The key to successful use of N/C is the preparation of the program. The programmer must understand machining operations as well as be able to read engineering drawings. A knowledge of computer programming is also necessary.

The drafter preparing the drawings must understand the principles of N/C so that the drawings he or she prepares can be used by the programmer with little or no additional calculations.

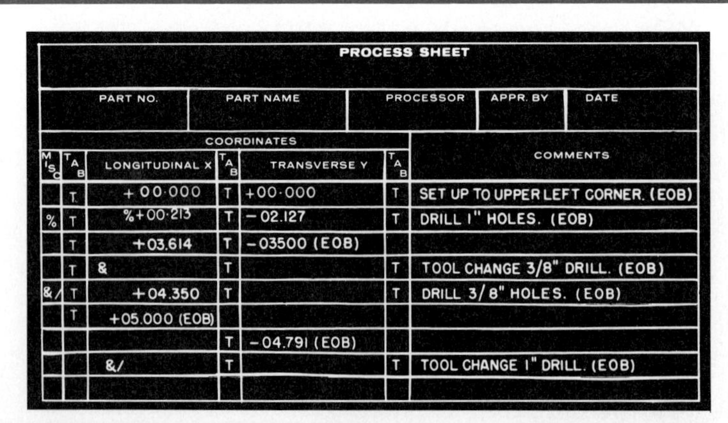

PROCESS SHEET

PART NO.	PART NAME	PROCESSOR	APPR. BY	DATE

MISC	TAB	LONGITUDINAL X	TAB	TRANSVERSE Y	TAB	COMMENTS
	T.	+ 00·000	T	+00·000	T	SET UP TO UPPER LEFT CORNER. (EOB)
%	T	%+00·213	T	− 02.127	T	DRILL 1" HOLES. (EOB)
	T	+03.614	T	−03500 (EOB)		
	T	&	T		T	TOOL CHANGE 3/8" DRILL. (EOB)
&/	T	+04.350	T		T	DRILL 3/8" HOLES. (EOB)
	T	+05.000 (EOB)				
			T	−04.791 (EOB)		
		&/	T		T	TOOL CHANGE 1" DRILL. (EOB)

Fig. 15–1. This is a process sheet upon which the programmer records instructions for the tape punch operator. (Pratt and Whitney Co., Inc.)

How Does N/C Work?

The programmer studies the engineering drawings and writes the program using a standard computer language for N/C such as APT (Automatic Programming Tool). The programmer decides the best sequence in which to perform the operations. This information is recorded on a process sheet. Fig. 15-1. The programmer also establishes the setup point. This is the point from which the N/C machine operator locates the center line of the spindle on the machine in relation to the piece to be machined.

While programs can be manually prepared on tape using a tape typewriter, computer-aided programming greatly reduces the cost of preparing tapes and produces more accurate tapes. Manual tape preparation involves typing the manuscript on the tape typewriter. Fig. 15-2. Electronic circuits allow the typewriter to punch the correct codes in the tape. If the tape is prepared by a computer, the actual punching of the tape is done with computer tape-punching equipment. Fig. 15-3.

The program information is generally punched on paper or plastic tape. The information on the tape conforms to Electronic Industries Association standards. Fig. 15-4. This is one of two tape codes in use. Each hole punched in the tape represents the binary digit one, while a blank space represents the digit zero. Some machines that perform the machining are operated directly from the computer without the use of tape.

Computer Numerical Control

Machines controlled by computer, known as **computer numerical control** (CNC), are the most widely used N/C machines currently being manufactured. Developments in microelectronics and microcomputers have en-

Fig. 15–2. A tape punch operator copies the process sheet by punching the corresponding keys on the tape punch machine. The numbers and instructions are converted into holes in the tape.

Fig. 15–3. A computer tape preparation machine for computer numerical control tapes. (ABR Corporation)

TRACK NUMBER

8	7	6	5	4		3	2	1	
		o			•				0
					•			o	1
					•		o		2
			o		•		o	o	3
					•	o			4
			o	o	•			o	5
			o	o	•		o		6
					•	o	o	o	7
				o	•				8
			o	o	•			o	9
	o	o			•			o	a
	o	o			•		o		b
	o	o	o		•		o	o	c
	o	o			•	o			d
	o	o	o		•	o		o	e
	o	o	o		•	o	o		f
	o	o			•	o	o	o	g
	o	o		o	•				h
	o	o	o	o	•			o	i
	o		o		•			o	j
	o		o		•		o		k
	o				•		o	o	l
	o		o		•	o			m
	o				•	o		o	n
	o				•	o	o		o
	o		o		•	o	o	o	p
	o		o	o	•				q
	o			o	•			o	r
		o	o		•		o		s
		o			•		o	o	t
		o	o		•	o			u
		o			•	o		o	v
		o			•	o	o		w
		o	o		•	o	o	o	x
		o	o	o	•				y
		o		o	•			o	z

o HOLE IN TAPE

• TAPE FEED HOLE

Fig. 15–4. A partial listing of EIA punch tape codes used by N/C controls.

abled the computer to be used as the control unit on numerically controlled machinery. The computer takes the place of the tape reader used on manually programmed N/C machines. Instead of the machine reading punched tape, the program is executed from the machine's computer. CNC machines are easier to program than N/C machines. CNC machines are programmed using an on-board computer keyboard. They also have a tape reader or an electrical connection that permits a program written elsewhere to be transferred to the CNC machine.

N/C Control Systems

There are two types of control systems in use on N/C machines: point-to-point and continuous-path. **Point-to-point machines** move only in straight lines. They are useful for operations such as drilling, reaming, and boring. They can be used to make straight milling cuts that are parallel with the machine axis. If a point-to-point machine is used to cut an arc or angle, it must be programmed to make a series of small, straight cuts. Fig. 15-5.

The **Continuous-path machines** can move their drive motors at varying rates of speed and can cut arcs and angles giving a smooth continuous surface. Fig. 15-6. Most CNC machines are continuous-path.

The Cartesian Coordinate System

The basis for movement of numerically controlled tools is the **Cartesian coordinate system**. The two-axis system is based at the intersection of two mutually perpendicular axes, called X and Y, forming four quadrants. Fig. 15-7. Where the two axes meet is the **zero point**. The horizontal

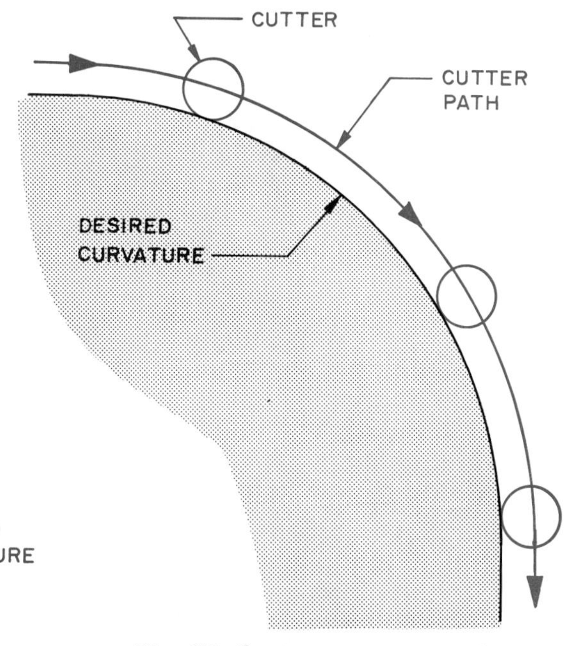

Fig. 15–6. **Cutting a curved surface using a continuous path N/C machine.**

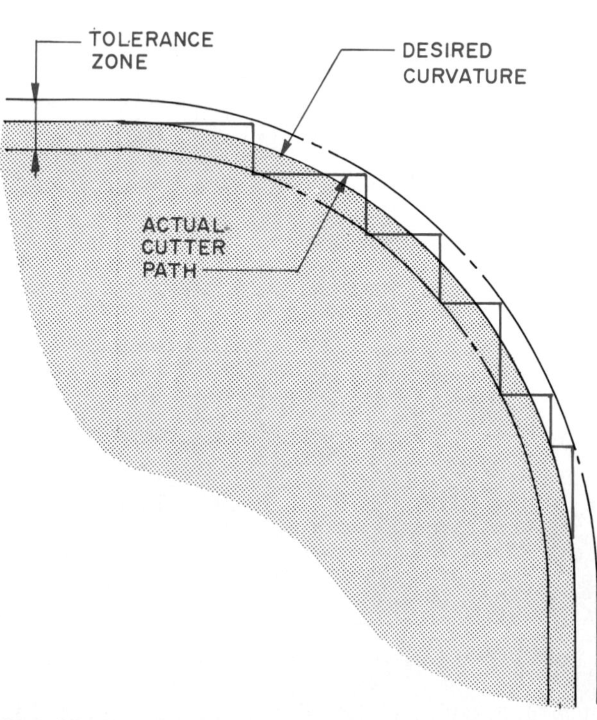

Fig. 15–5. **Cutting a curved surface using a point-to-point machine.**

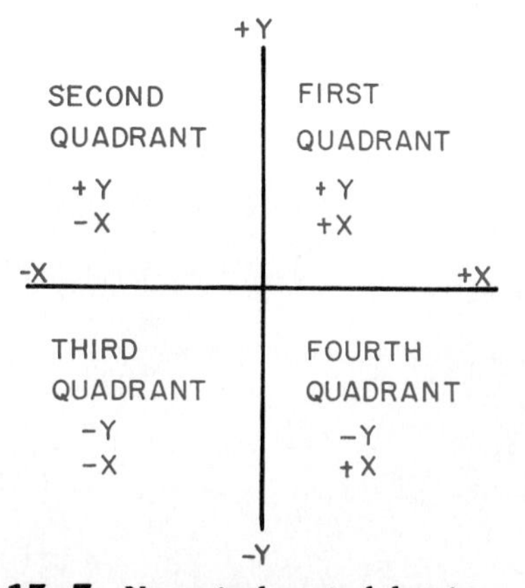

Fig. 15–7. **Numerical control drawings using a coordinate dimensioning system can be placed in any one quadrant or in a combination of the four quadrants.**

axis is the X- axis, and the vertical axis is the Y- axis. X-values to the right of the zero point are positive. Those to the left are negative. Y-values above the zero point are positive, and those below are negative.

Any position can be described by stating its distance from the zero point along the X- and Y- axes. For example, the coordinates 3,4 refer to a position in the *first quadrant*. The first number is always the X- value and the *second* the Y-value. Therefore 3,4 means a + 3X and a + 4Y. This locates the position in the first quadrant. Find the following locations in Fig. 15-8:
3,4; −5,4; −6,−4; 6,−3.

When designing for numerically controlled parts, it is simpler if all work is in the first quadrant because all signs are positive. However, any of the four quadrants can be used.

Numerically controlled machines that can locate positions in only two directions are called two-axis machines. Fig. 15-9. The cutter can be moved in the X and Y directions. The function of numerical control is to move the table or tool of the machine to specified positions and perform an operation there, such as drilling a hole. When the table moves, the work is clamped to the table and the tool is stationary. The N/C control positions the table under the drill so the first hole is drilled according to its coordinates. The table then moves in the X and Y directions to position the tool to drill the next hole. Fig. 15-10.

When the tool moves, the part is clamped to the table. The N/C control locates the first hole by its coordinates, and the hole is drilled. Then the tool moves along the X- and Y- axes to the next position and repeats the operation. Fig. 15-11.

Setting Up the Part

All coordinates are taken from the zero point (where the X- and Y- axes intersect). Some N/C machines have the zero point built into their controls. The machine automatically finds the zero point. It is from this point that the part to be machined is located on the table. Fig. 15-12. This particular setup is in the first quadrant.

When the part is secured to the table, it will not normally be located at the zero point. It will be positioned to make way for the devices that secure it to the table. Because the part cannot be clamped so that one point on its edge is at the zero point, a **setup point** must be located. This point is usually the intersection of two finished surfaces, the center of a hole already in the part or some other important feature. In Fig. 15-12, the setup point was located 75 mm from the zero point on both the X- and Y-axes. In order to accurately perform the machining operations, the setup point must be accurately located.

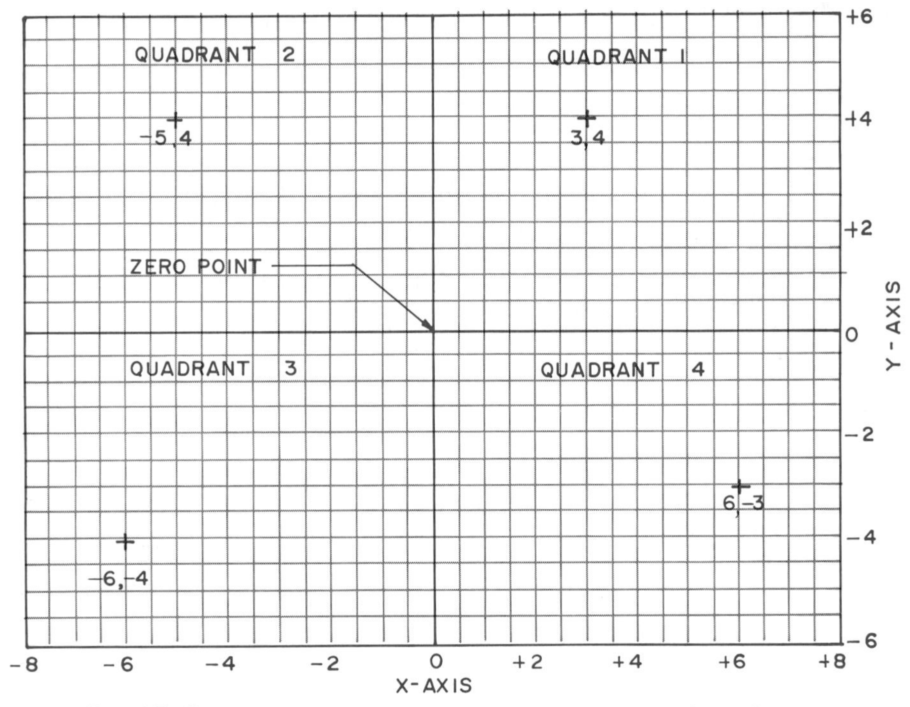

Fig. 15–8. These points are located with two-dimensional coordinates.

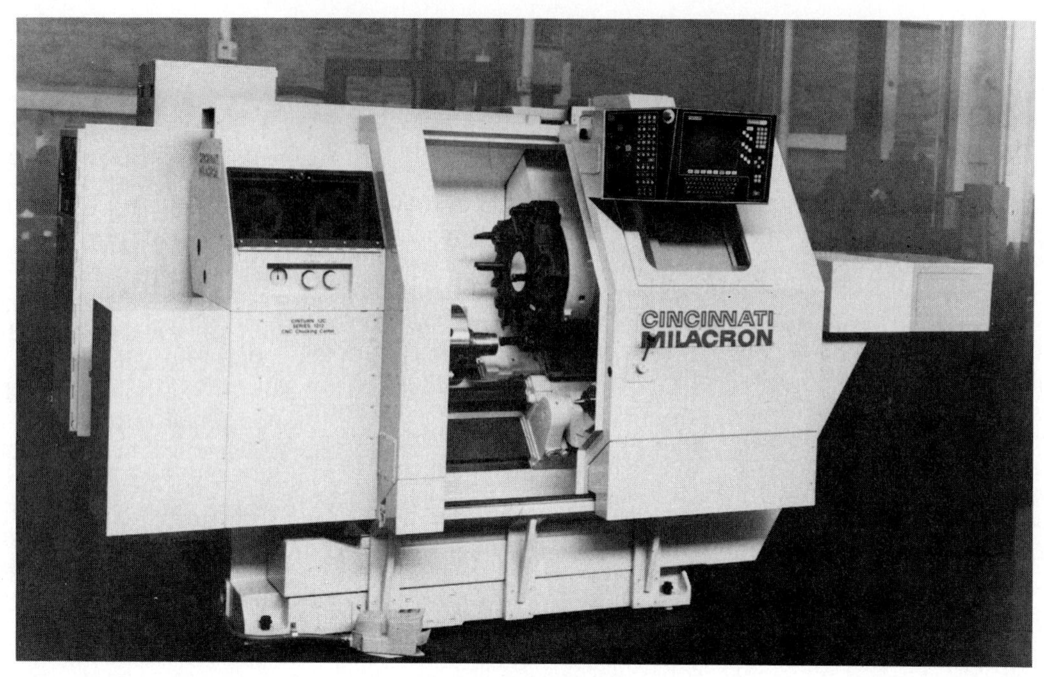
CINCINNATI
MILACRON

Fig. 15–9. A two-axis CNC turning center. (Cincinnati Milacron)

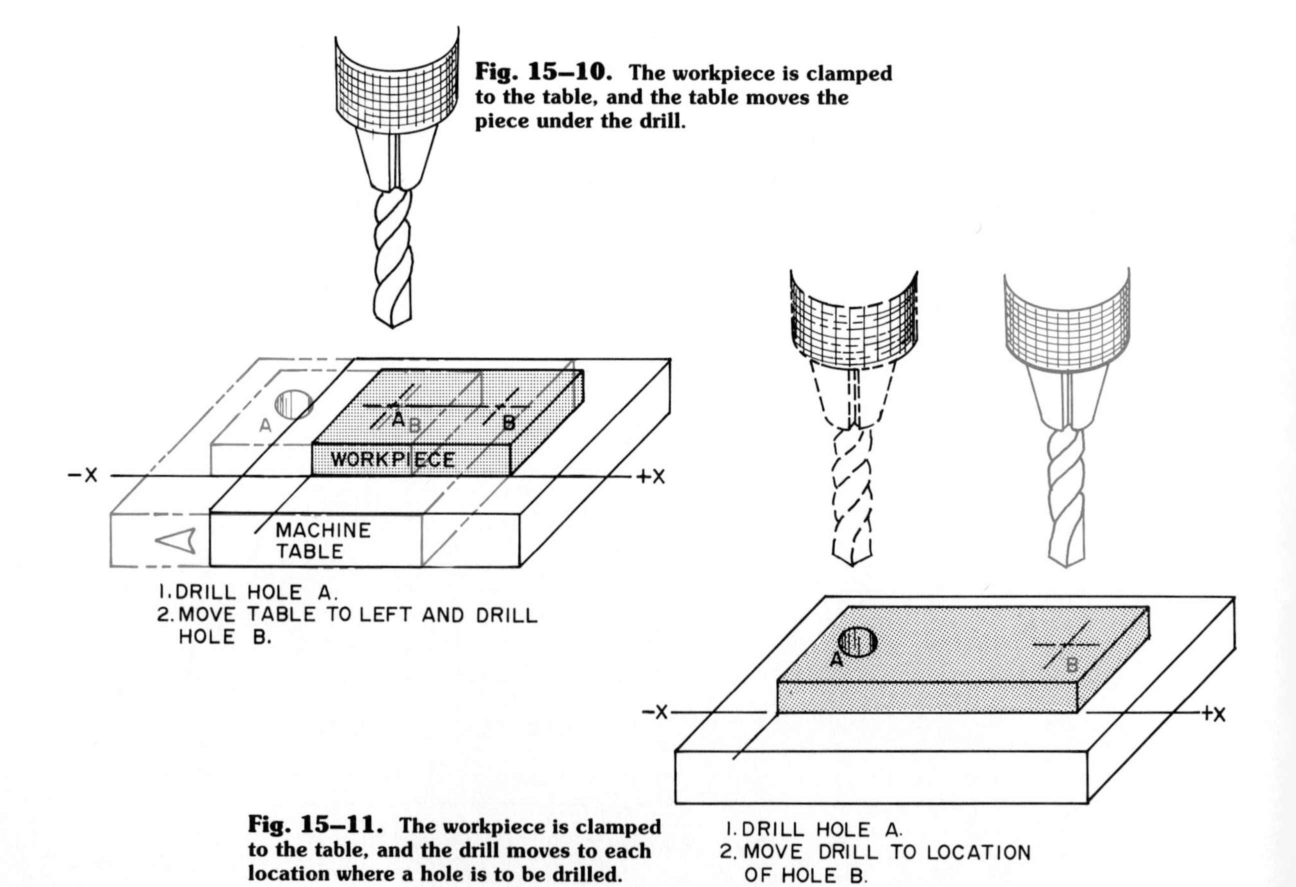
A
B
A
B
WORKPIECE
−X
+X
MACHINE
TABLE
1. DRILL HOLE A.
2. MOVE TABLE TO LEFT AND DRILL
HOLE B.
A
B
−X
+X
1. DRILL HOLE A.
2. MOVE DRILL TO LOCATION
OF HOLE B.

Fig. 15–10. The workpiece is clamped to the table, and the table moves the piece under the drill.

Fig. 15–11. The workpiece is clamped to the table, and the drill moves to each location where a hole is to be drilled.

Positioning Systems

There are two ways N/C machines position themselves in connection with their coordinate systems. These are absolute positioning and incremental positioning.

Absolute positioning requires that all machine locations are taken from one fixed zero point. In Fig. 15-13, the location of the holes is taken from point X0/Y0. The first hole has coordinates of X1.000,Y1.000 while the second hole has coordinates of X1.000,Y2.000. As the machine locates the second hole, it refers back to X0,Y0.

Incremental positioning requires that the X0/Y0 point move with the machine cutter spindle. Fig. 15-14. The first hole is located from the lower left corner, which is point X0/Y0. The second hole is located from the center of the first hole. Therefore, the center of the first hole becomes X0/Y0 for the second hole.

Dimensioning Practices

The two types of dimensioning practices used with N/C machines are datum (or coordinate) and delta (or point-to-point).

Datum, or coordinate, dimensioning is used when absolute positioning is required. It has all dimensions placed so that they relate to the single fixed zero point. In Fig. 15- 15, all dimensions were taken from the lower left corner designated X0/Y0. The actual location of the zero point would depend upon the design considerations of the part. The center of a hole could be used as the zero point if accurate locations from it were necessary to the proper functioning of the part. Dimensions in both the X and Y directions are located from the zero point. The coordinates of each hole, therefore, relate back to the zero point.

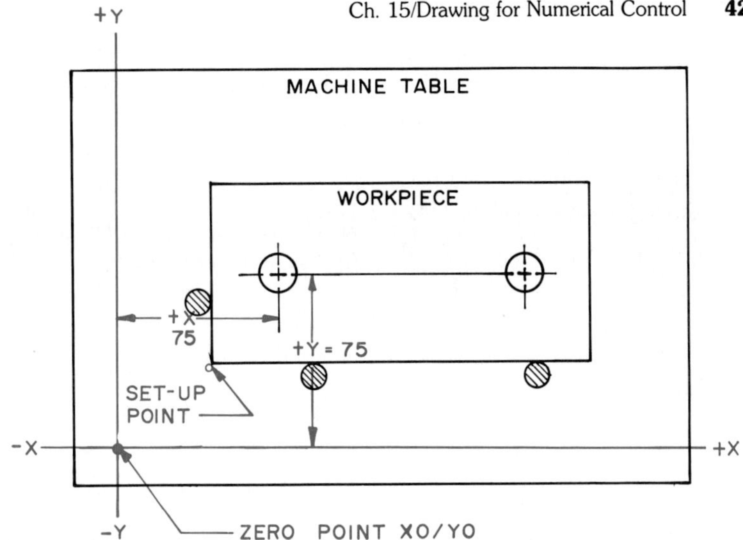

Fig. 15–12. This part is located in the first quadrant, so all X- and Y-values are positive. The setup point is located some distance from the zero point.

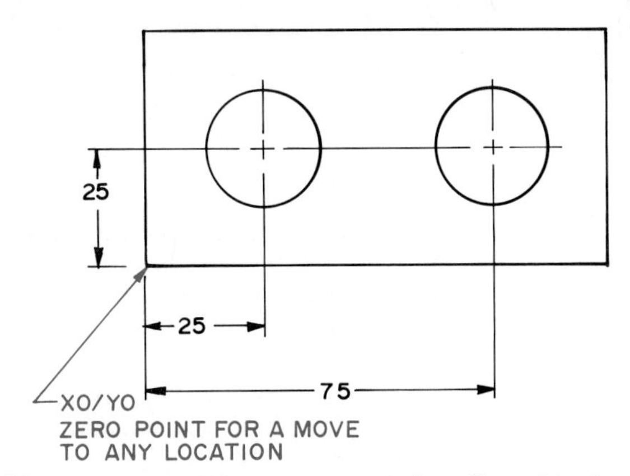

Fig. 15–13. Absolute positioning requires that all machine locations are taken from one fixed zero point.

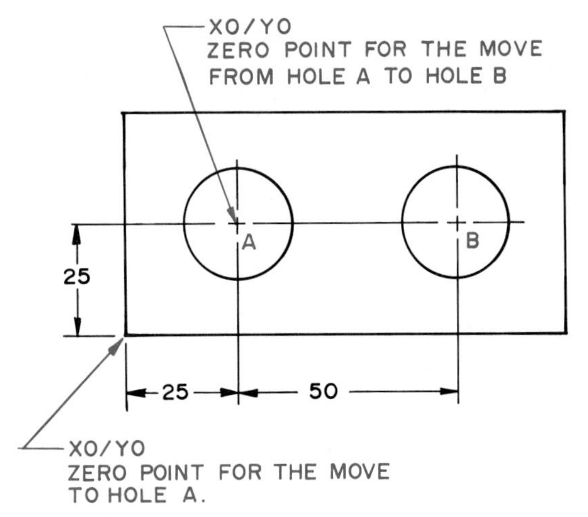

Fig. 15–14. Incremental positioning requires that the XO/YO point move to the next location as the machine spindle moves.

Delta, or point-to-point, dimensioning is used when incremental positioning is desired. Each dimension is located from the previous one. This method of dimensioning is sometimes called *chain dimensioning*. In Fig. 15-16 the coordinates of the hole A are X18,Y18. The coordinates of the hole B use the center of the first hole as the zero point, so they are X32,Y32. The plus and minus signs do not refer to quadrants as in datum dimensioning but to the direction the tool moves. A positive X means move the tool to the right. A negative X means move to the left. A positive Y means move the tool up, while a negative Y moves the tool down. Notice the coordinates for hole C are X37,Y−25.

Three-Axis Control Systems

A three-axis numerically controlled machine will operate in three directions. Fig. 15-17. The machine table and carriage move horizontally in the X- and Y- directions. The tool spindle moves up and down vertically in the Z-direction. Fig. 15-18. The center line of the vertical spindle is used as the Z-axis. It is perpendicular to the X and Y axes. Fig. 15-19. The X and Y axes are indicated as plus and minus as discussed for two-axis machines. The Z axis is positive when it moves away from the workpiece and negative when it moves toward the workpiece.

Points on three-axis machines are located in space from the three axes. In Fig. 15-19, point A has coordinates of 3,4,5 (X,Y,Z). What are the coordinates of point B in this figure?

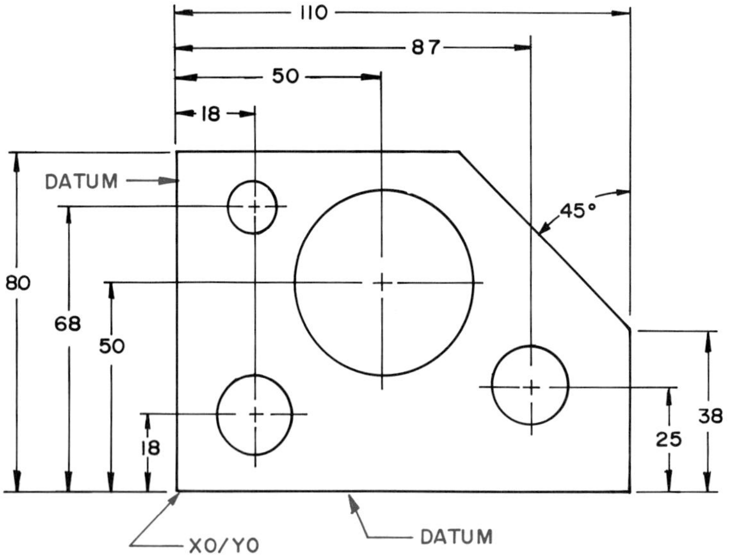

Fig. 15–15. **Datum, or coordinate, dimensioning.**

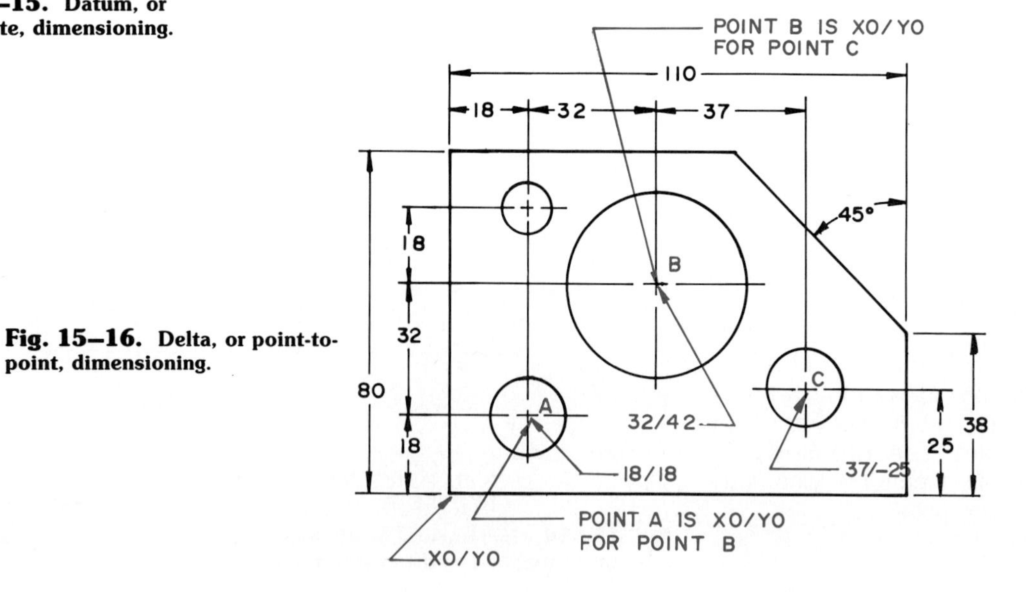

Fig. 15–16. **Delta, or point-to-point, dimensioning.**

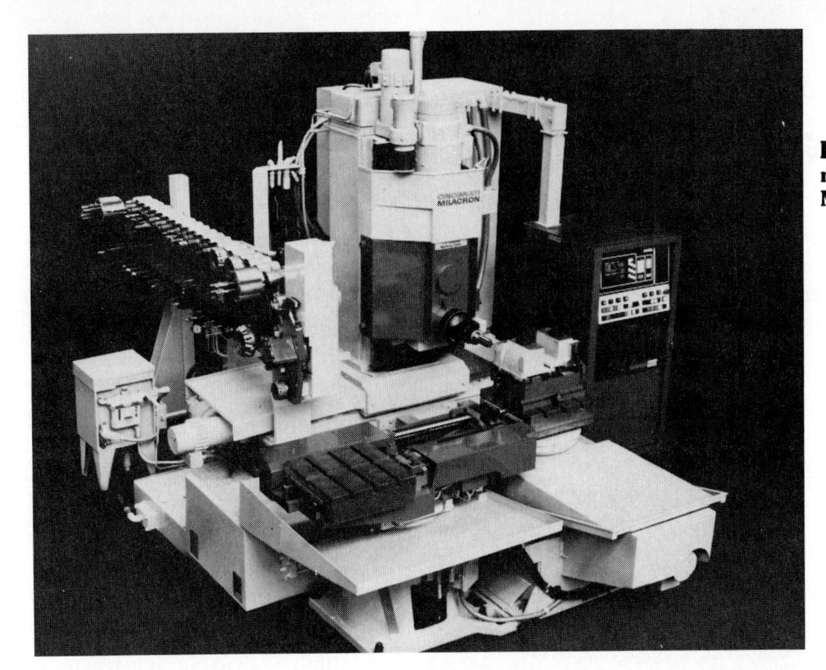

Fig. 15–17. A three-axis CNC machining center. (Cincinnati Milacron)

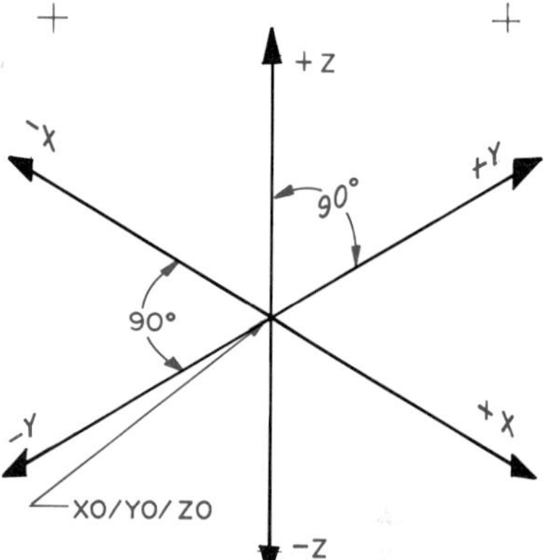

Fig. 15–18. A three-axis coordinate system.

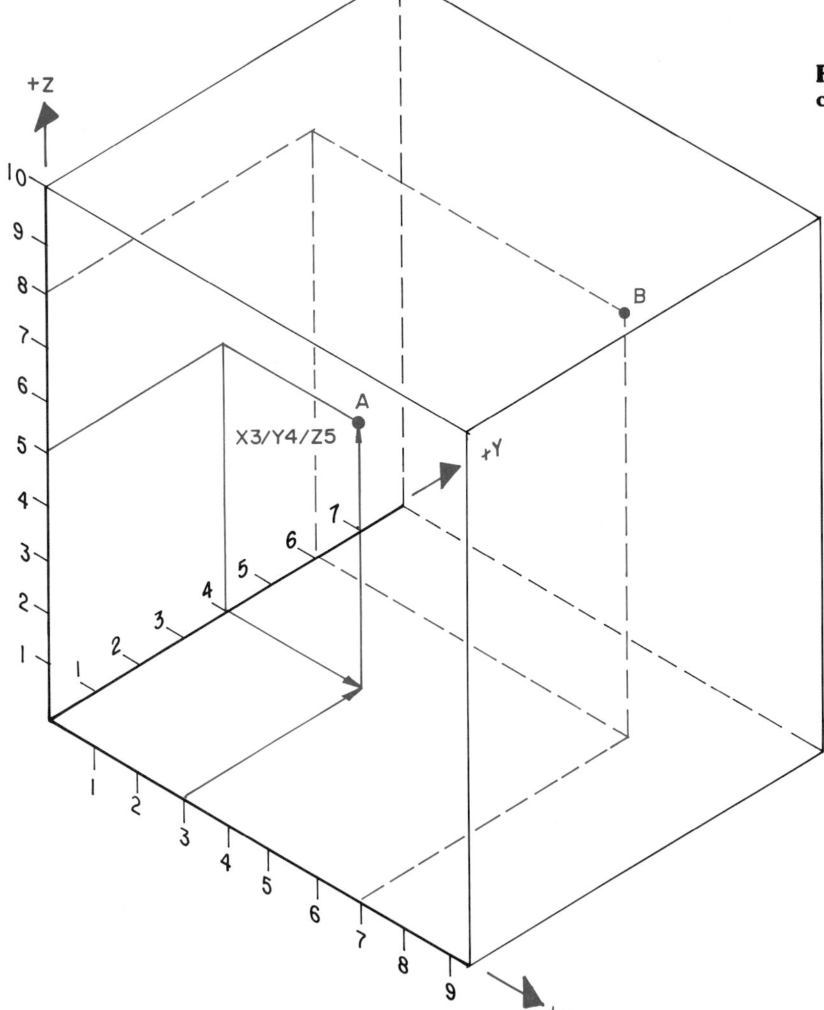

Fig. 15–19 Points on three-axis N/C machines are located in space by the X-, Y-, and Z-axes.

CAD Applications:
Linking CAD with CAM

Imagine designing a part on a CAD system and then sending the drawing electronically to a machine which produces the part. That's the goal of CAD/CAM: to combine computer-aided drafting and design with computer-aided manufacturing. In CAD/CAM the drawing is created on a CAD system and loaded into a CAM system, and the part is machined by a CNC tool.

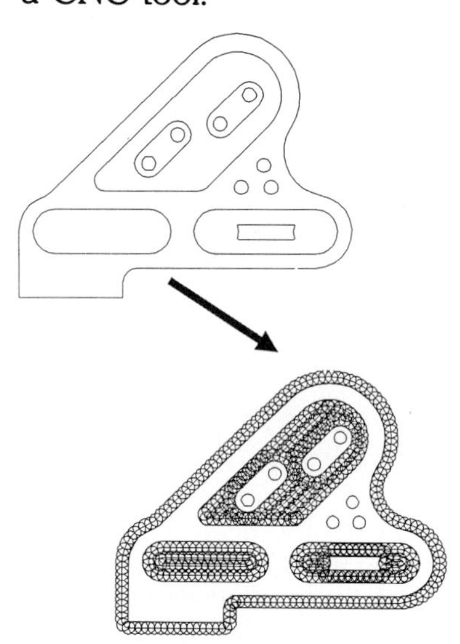

NC Programmer by NC Microproducts, Inc.

Creation of the drawing is similar to other CAD, except the requirements of the CAM system must be kept in mind. For example, it may be necessary to put the non-machined elements of a drawing, such as the dimensions, on a different layer from the actual part outline. The following is a CAD drawing of an aircraft hinge.

SmartCAM by Point Control Co.

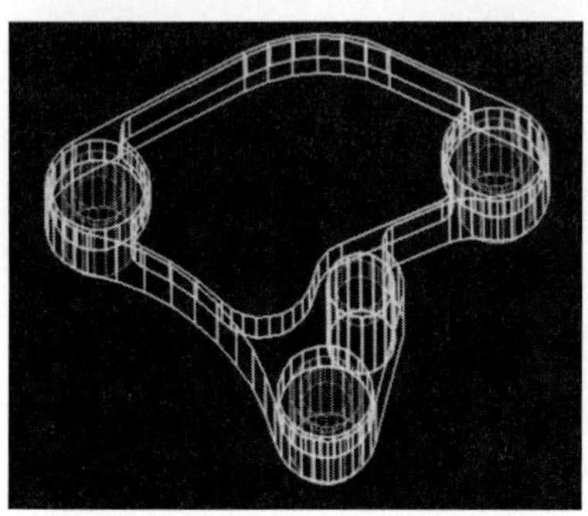

After the part is drafted, the drawing data are sent to the CAM system. The CAM system takes the part geometry created by the CAD system and converts it into instructions for use by the CNC machine. (See page 419.) The picture below shows the aircraft hinge after the CAM system has created the tool path information. This information is needed by the CNC machine to direct the cutting of the part.

SmartCAM by Point Control Co.

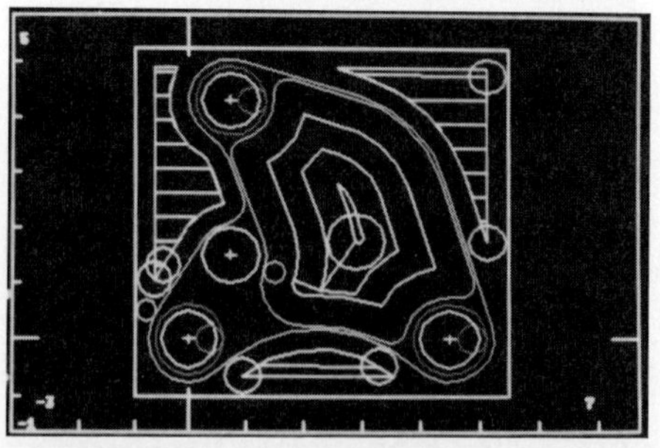

This process is not automatic. It requires a human operator to make decisions about tool size, offsets, and so on. The NC code is produced by the postprocessing software. Since each CNC machine requires somewhat different commands, each one might have its own postprocessor. An alternative is to have a single postprocessor that can be programmed to work with the various machine tool controllers.

Once the code has been generated, it can be sent to the machine tool. The stock is loaded into the tool and the part is machined.

CAD/CAM that uses microcomputers is still fairly new, but rapid progress is being made, both in improving the technology and in lowering the price. The time will come when even small shops can afford a CAD/CAM system.

Chapter Review

Build Your Vocabulary

You should understand and use the following terms as part of your working vocabulary. Write a brief explanation of what each means.

numerical control
computer numerical control
point-to-point machines
continuous-path machines
Cartesian coordinate system
zero point
setup point
absolute positioning
incremental positioning

Study Problems—Directions

The problems that follow will give you the chance to use some of the drafting knowledge that you learned in this chapter. Do the problems that are assigned to you by your instructor.

1. Prepare a chart listing the X- and Y-coordinates and quadrants for the points in **Fig. P15-1**. Make your chart similar to the sample shown in Fig. P15-1.

2. Make two drawings of the product in **Fig. P15-2**. Dimension one using the datum system and the other using the delta system. Use only those dimensions necessary to locate the holes. Use two-place decimals for metric and three-place decimals for inch dimensions. Then prepare a chart giving the X- and Y-coordinates and the quadrant for each hole. Measure the drawing to get the sizes. Make your drawing twice as large as shown.

3. Make a two-view drawing of the product in **Fig. P15-3**. Use datum dimensioning to locate the holes. Then prepare a chart giving their X, Y, Z coordinates. Measure the drawing to get the sizes. Make your drawing twice as large as shown.

Study Problems

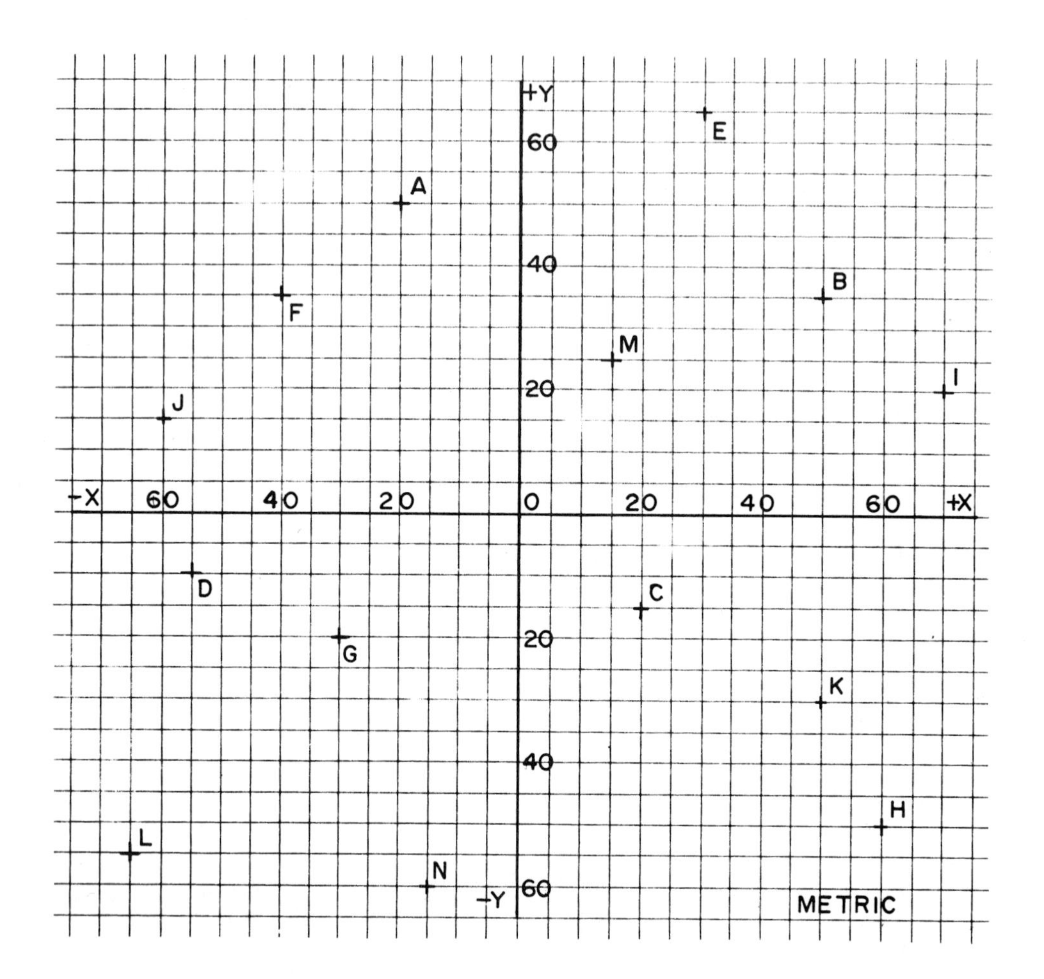

COORDINATE CHART

POINT	X-AXIS	Y-AXIS	QUADRANT
A	-20	50	2
B			
C			

P15–1.

Study Problems

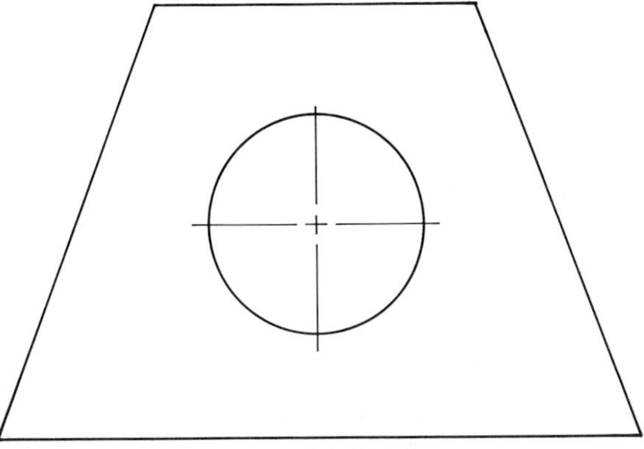

P15–2.

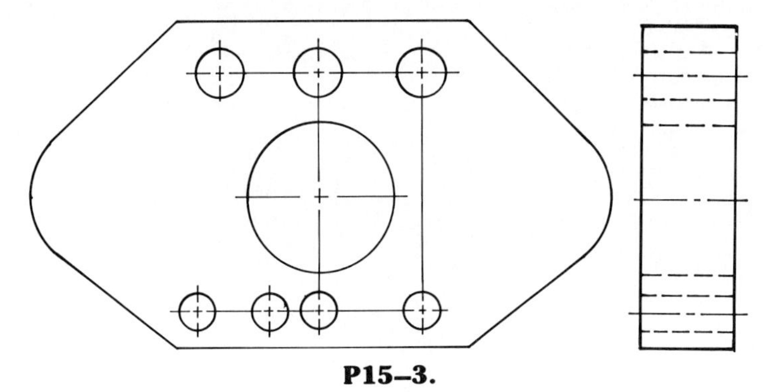

P15–3.

CHAPTER 16

Reproduction of Drawings

After studying this chapter, you should be able to do the following:

- Understand the basic methods of reproducing drawings.
- Use the various types of intermediates.
- Use photo drafting and pin-bar drafting techniques.
- Understand the use of microfilm and microfiche.
- Store drawings properly.

Introduction

Most drawings are made on vellum, tracing cloth, or polyester drafting film. The translucency of these materials makes it possible to produce exact copies. The copies are called *prints*. After a print is made, the original drawing can be stored. This protects it from damage. There are several ways copies can be made. The most common kinds of copies are blueprints and whiteprints.

Blueprints

A **blueprint** has white lines on a blue background. Fig. 16-1. It is made on a very strong paper that will withstand hard use. The blue color fades very little.

Blueprint paper has a chemical coating on one surface. This coating is made from a solution of potassium dichromate and ferrocyanide of potassium. It dries to a light yellow. When ultraviolet light hits these chemicals, the coating turns blue.

A blueprint is made by placing the original drawing next to a piece of undeveloped blueprint paper. Fig. 16-2. Together they are exposed to an ultraviolet light. This causes a chemical reaction to take place wherever the light hits the blue paper. The lines on the drawing protect parts of the blueprint paper so the light does not touch the chemical coating on those parts. After the paper has been exposed to the light for the proper length of time, the blueprint paper is separated from the drawing.

Next, the blueprint paper is developed by washing it in water. The areas protected by the lines wash away. The white paper shows through. The areas not protected are blue. The blueprint

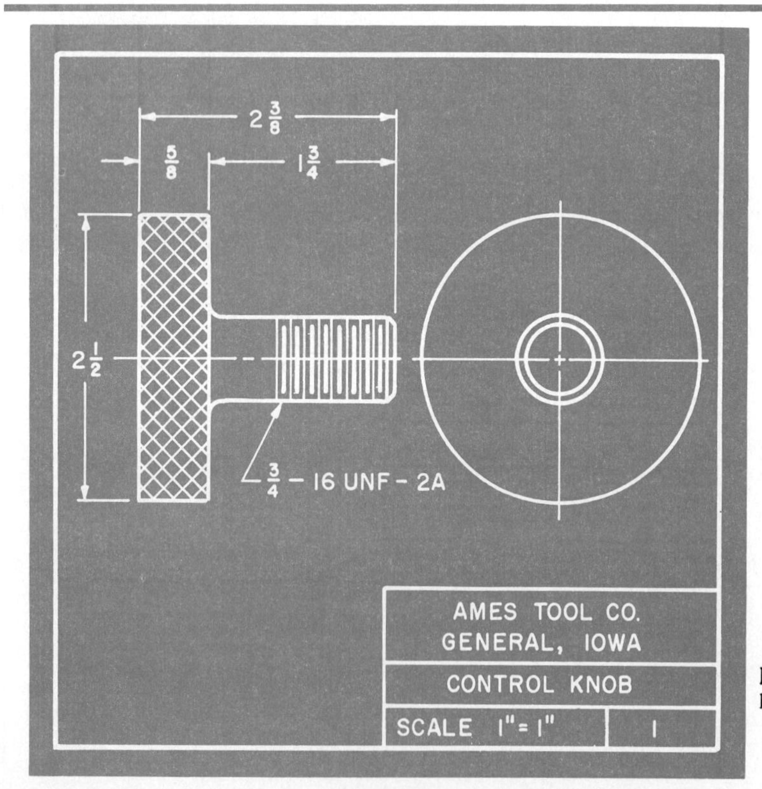

Fig. 16–1. A blueprint has white lines and a blue background.

is then placed in a solution of potassium bichromate. This fixes the print so that it is permanent; that is, no other chemical reaction will take place. It is then washed again in water and dried.

Blueprint machines used in industry do all of these operations automatically. Fig. 16-3 shows a machine with a large roll of blueprint paper. It is protected from light. As a drawing is fed into the machine, the blueprint paper unwinds to travel with it. After exposure to light, the drawing is sent to the tray above the table. Then the exposed blueprint paper goes through all the steps just described to make a final blueprint.

Whiteprints

Whiteprints have colored lines on a white background. The lines are usually blue, black, or maroon. Fig. 16-4.

The whiteprint paper has an azo dye coating. After the paper is exposed to light along with the original drawing, the print is developed by either a dry process or a moist process. Both are called **diazo processes**.

The original drawing must be on translucent material such as vellum or drafting film. During exposure, the light must pass through the drawing to the diazo paper.

To make a print, place the diazo paper with the dye side up. Place the original drawing on top of it with the pencil side up. Feed the two sheets together into the whiteprinter where they are exposed to ultraviolet light. The light that passes through the drawing destroys the dye. The pencil lines block the light, saving the dye below them.

The original and the copy are separated, and the copy is passed through the developing section of the whiteprinter. In a dry process the developer is heated ammonia vapor. When the vapor strikes the dye on the copy, it develops it and produces colored lines. Fig. 16-5.

In a moist process the copy is sent through a developer using a liquid chemical. The copy is fed over fine-grooved rollers containing liquid developer. This develops the diazo dye, producing a print with colored lines on a white background.

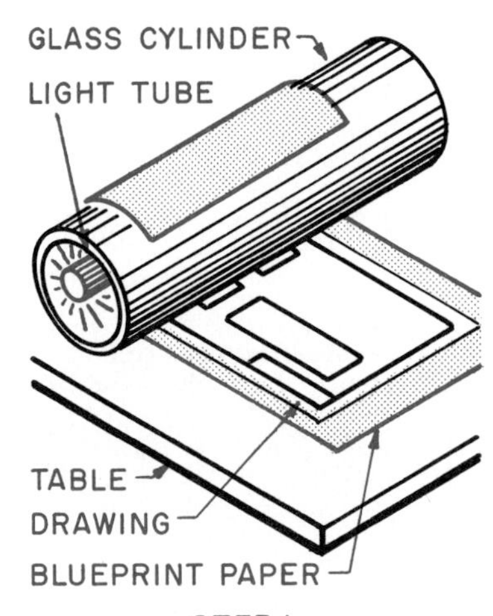

STEP I
THE BLUEPRINT PAPER IS EXPOSED WITH THE DRAWING TO A LIGHT.

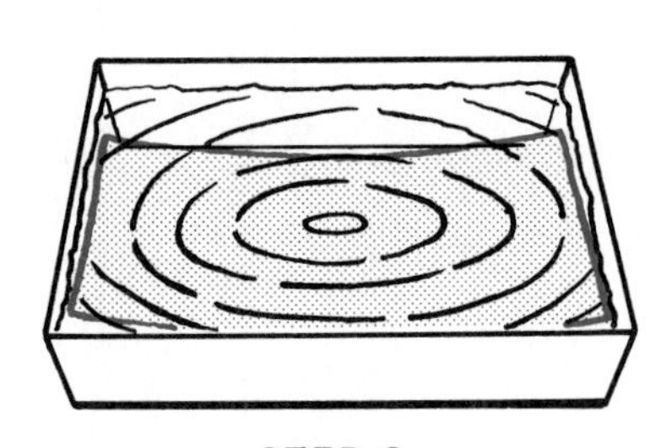

STEP 2
THE BLUEPRINT IS WASHED IN WATER.

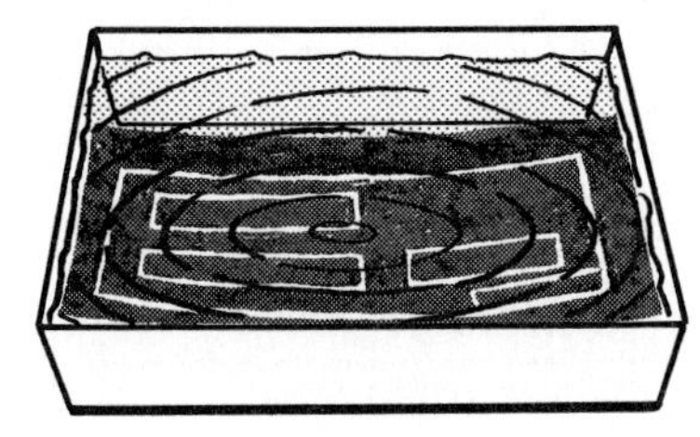

STEP 3
THE BLUEPRINT IS BATHED IN A POTASSIUM BICHROMATE SOLUTION.

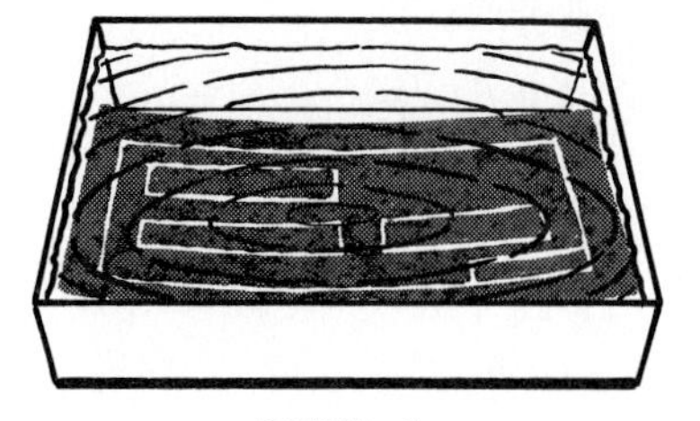

STEP 4
THE BLUEPRINT IS RINSED IN WATER.

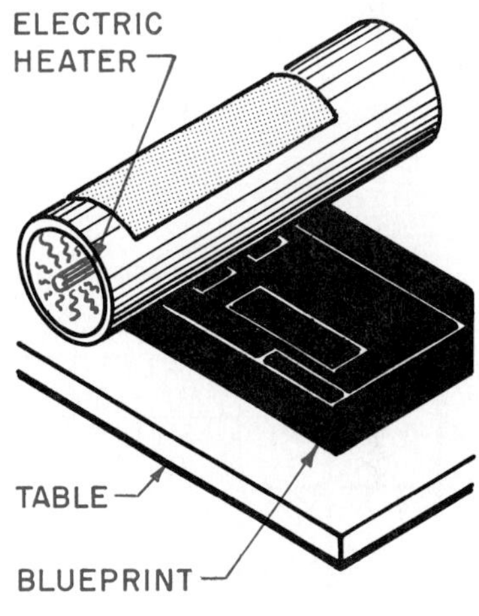

STEP 5
THE BLUEPRINT IS DRIED.

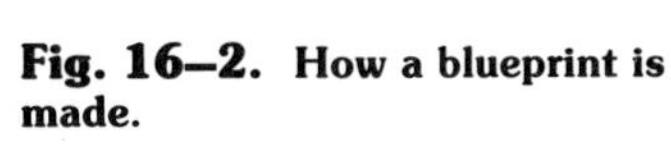

Fig. 16–2. How a blueprint is made.

Fig. 16–3. **This is an industrial blueprint machine. It exposes and develops the blueprint. (Revolute Blueprint Machine, Charles Bruning Co.)**

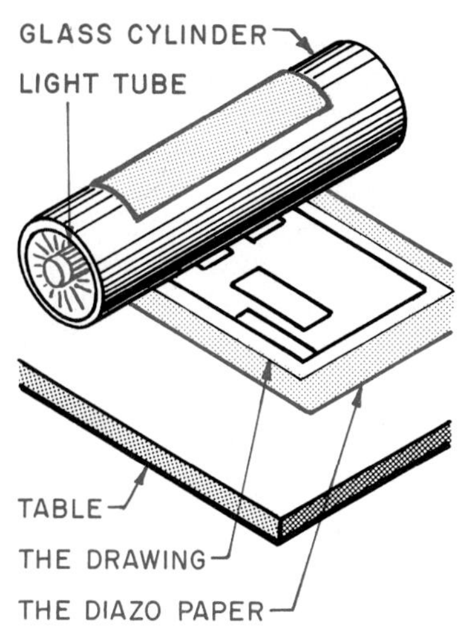

Fig. 16–5. **Steps for printing a dry diazo whiteprint.**

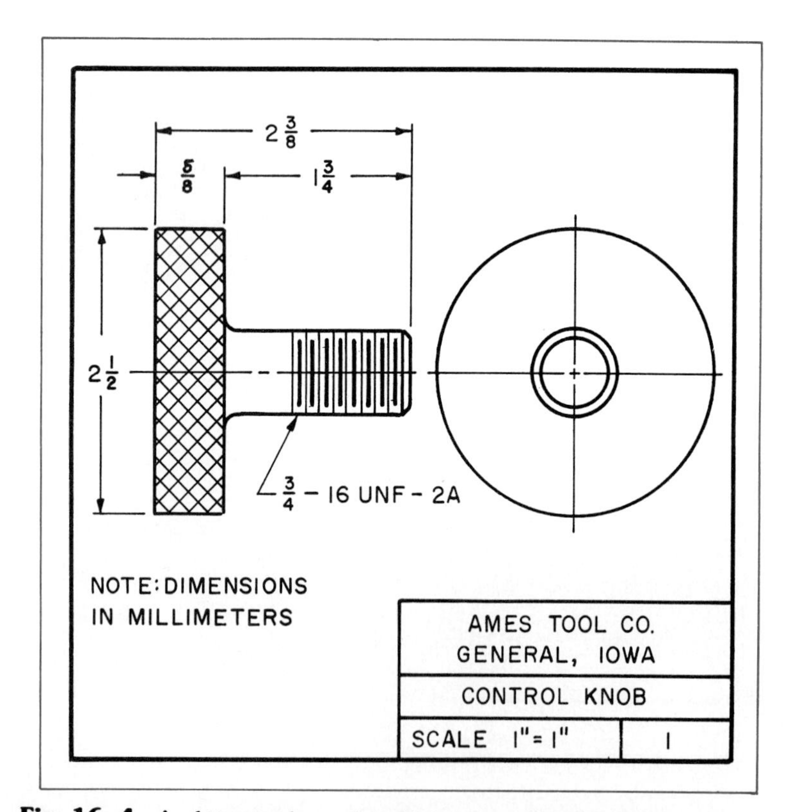

Fig. 16–4. **A whiteprint has colored lines on a white background.**

Also available is a polyester film upon which an original drawing can be reproduced using a standard ammonia whiteprint machine. Fig. 16-6. The image is produced on a matte- finish (dull, rough) surface. This surface can be drawn on with ink or pencil so a copy of a drawing can be made and changes made on the copy. This way, the original is not disturbed. Lines reproduced on the copy can be removed with a special chemical remover so that the drawing can be revised.

Electrostatic Reproduction

Electrostatic reproduction of engineering drawings produces black lines on a white paper. This process is called **xerography**. It is the same process used in office copying machines. Fig. 16-7. This is a very rapid process that produces a dry copy without chemical fumes.

The xerographic machine has a selenium-coated plate that is given a positive electrical charge. The drawing is placed on a glass surface and is projected through a lens onto this plate. Here it is retained as a positive electrical

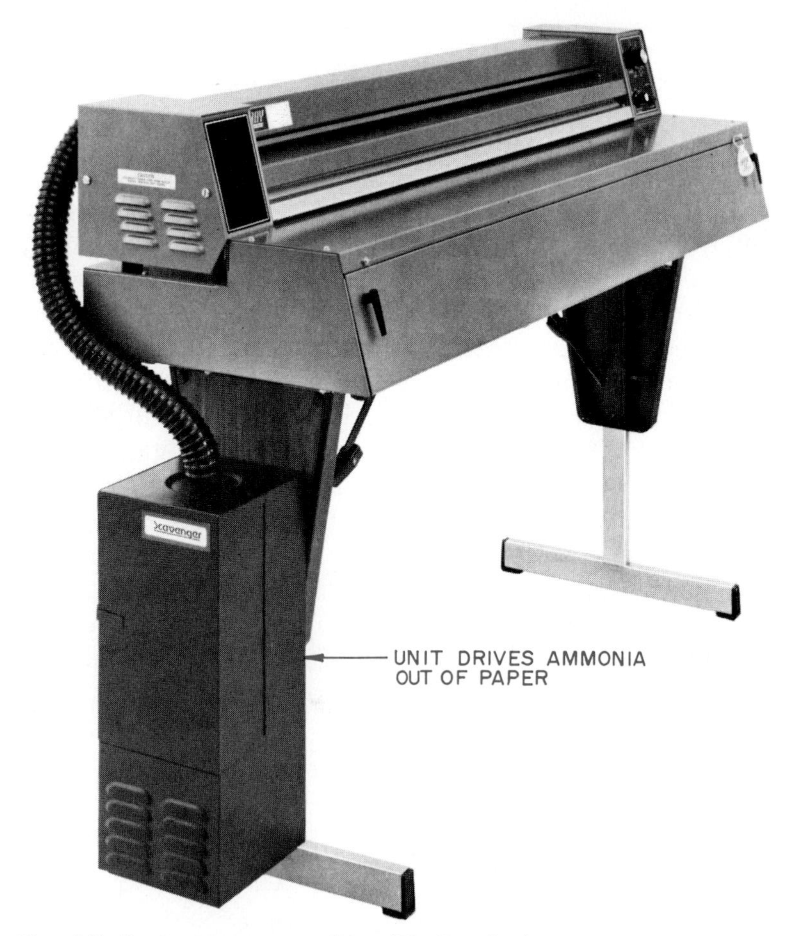

Fig. 16–6. A whiteprint machine. (Blu-Ray, Inc.)

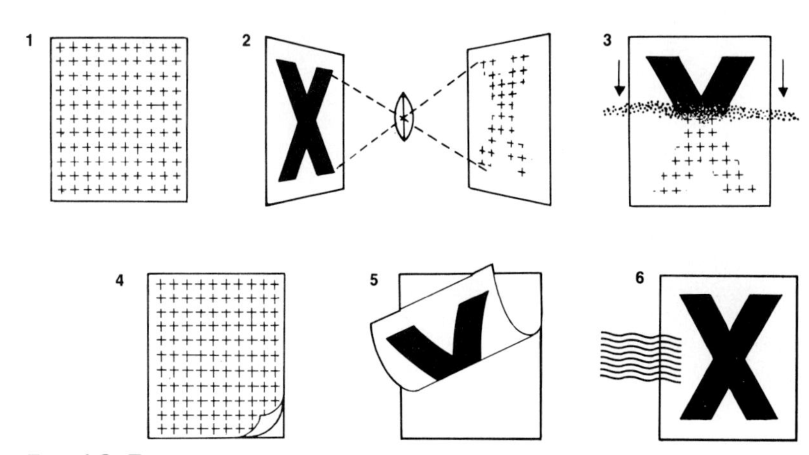

Fig. 16–7. How xerography works. (Xerox Corporation)

Basic Xerography
(1) A photoconductive surface is given a positive electrical charge (+).
(2) The image of a document is exposed on the surface. This causes the charge to drain away from the surface in all but the image area, which remains unexposed and charged.
(3) Negatively charged powder is cascaded over the surface. It electrostatically adheres to the positively charged image area making a visible image.
(4) A piece of plain paper is placed over the surface and given a positive charge.
(5) The negatively charged powder image on the surface is electrostatically attracted to the positively charged paper.
(6) The powder image is fused to the paper by heat.

After the photoconductive surface is cleaned, the process can be repeated.

CAD Applications:

Reproducing CAD Drawings

Today, the ways in which CAD drawings may be reproduced are growing in number. Now drawings may be output in color, on film, and on videotape. They can also be combined with text to produce books and magazines.

Use of color has become quite common. Pen plotters, which are popular for CAD line drawings, use multiple pens to draw in color. Each pen is filled with a different color of ink.

The electrostatic plotters, which are like office copying machines, can also produce color drawings. A drum is electrostatically charged with the image to produce the different colors. When paper passes over the drum, the color clings to it.

Thermal transfer devices sandwich sheets of color with the paper. The "sandwich" is then passed in front of a thermal head that melts the color into the paper.

Ink jet printers spray fine jets of different colored inks onto paper.

Architects and engineers might use film or videotape images of CAD drawings to show to clients. Film recorders are used to produce 35-mm slides and color prints of graphic images. The recorder makes the image into a slide transparency or film negative. The slides are then shown with a projector. The negatives are developed and enlarged into prints. Videotape, made with videotape recorders, requires no processing.

A combination that has recently become popular is CAD joined with desktop publishing. For example, companies might use CAD drawings in technical newsletters, manuals, or company reports. Desktop publishing is done with microcomputers and several programs. Special word processing software allows the user to create text in high-quality typefaces. With other software the typeset copy can be laid out in page form. Illustrations created with CAD are then added. The finished product is usually output on a laser printer.

Laser printer from AST Research, Inc.

charge. The charge on the rest of the plate is drained away by the exposure. Then a negatively charged powder (toner) is distributed lightly over the selenium-coated plate. Since it has a negative charge, it is attracted to the positive image on the plate and sticks to it. The paper is given a positive charge and is placed over the plate. The powder image is attracted to the paper and is fused to it by heat.

Xerographic units are used to produce copies that can be used in scissors drafting to produce revised drawings. With one of these units, a drafter can produce a drawing with a reverse image or a drawing with unwanted areas removed. He or she can reformat a drawing, or create a drawing from composite overlays and produce enlarged or reduced copies of existing drawings. Fig. 16-8.

The drafting process can be made more efficient by using xerographic printers that can enlarge and reduce drawings. These printers can also handle copy developed by pasting details onto an original or by deleting details and redrawing them. Fig. 16-9. A few of the possibilities are shown in Fig. 16-10.

Engineering drawings and reports can be transmitted over long- distance telephone lines, coaxial cable, or microwave transmission facilities. This enables drawings to be sent from one plant to another very quickly.

One system uses ordinary telephone lines to transmit and receive with a facsimile ("fax") terminal. The terminal prints on ordinary plain paper. This terminal, shown in Fig. 16-11, can be programmed to send one transmission to a remote terminal in one place, such as a parts plant in Kansas, and send another to another place, such as a customer's company in California. It can also call in documents from other terminals.

Fig. 16–8. This xerographic copier produces copies of engineering drawings up to 36 inches (914 mm) wide and any manageable length. It prints on paper, vellum, or drafting film. (Xerox Corporation)

Fig. 16–9. This printer prints up to 24 inches (609 mm) wide and any usable length. It will make prints of any size between 45 and 141 percent of the original size. (Xerox Corporation)

A. Scissors-and-Paste Drafting

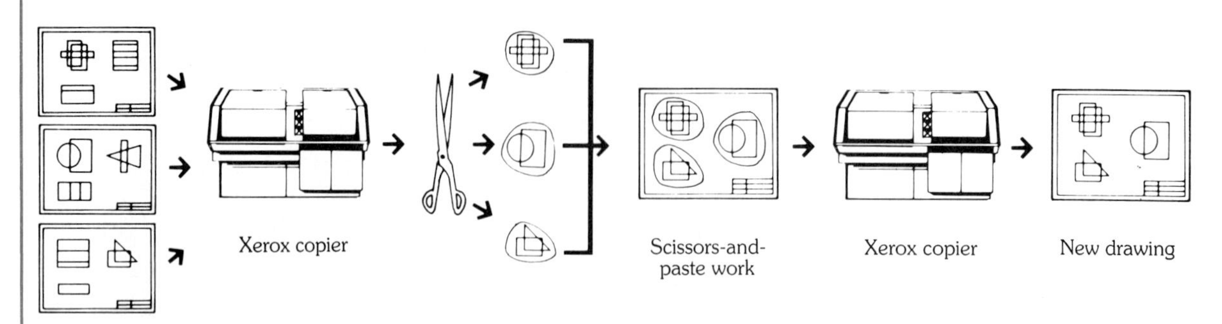

Create a new drawing by merging information and views from existing drawings.

B. Changes/Revisions on Engineering Documents

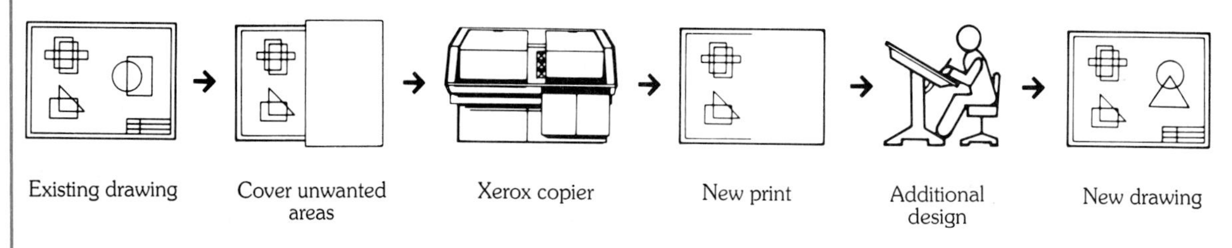

Create new drawings, delete large areas by using masking techniques.

C. Overlay Techniques

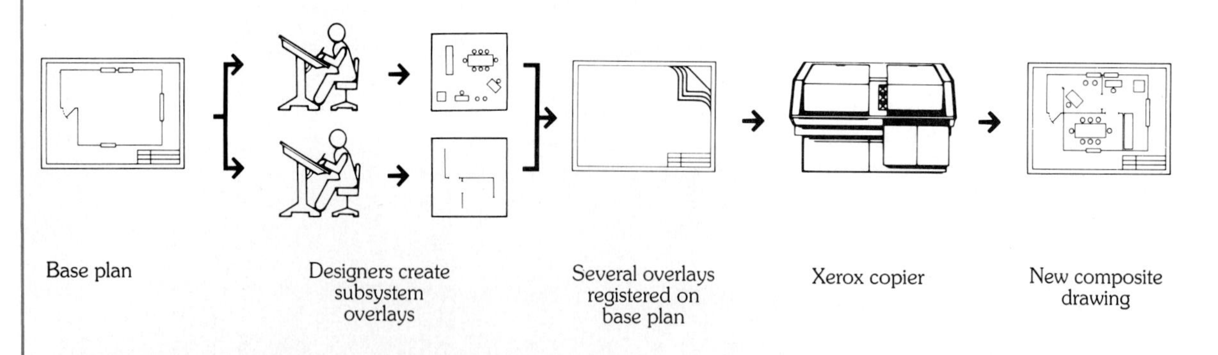

Use pin-bar drafting to create a finished print from composite overlays.

Fig. 16–10. These are some of the aids to drafting that are available using xerographic technology. (Xerox Corporation)

D. Reduction Techniques

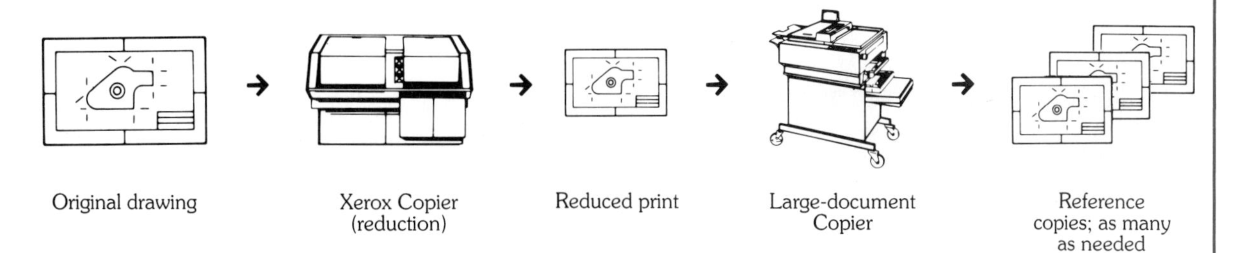

Create many convenient-size copies for low-cost handling and distribution.

E. Enlargement Techniques

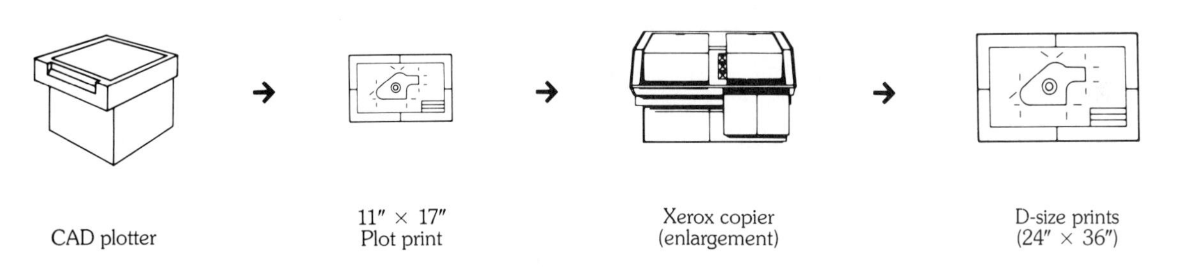

Increase CAD productivity and avoid "plotter-bound" CAD units by plotting smaller prints and then enlarging them to required scale on the copier.

Fig. 16–11. This facsimile terminal uses ordinary paper and can transmit or receive documents over regular telephone lines at the rate of one page every 18 seconds. It employs thermal transfer imaging technology.

Fig. 16–12. A copy camera used to transfer a drawing to microfilm. (ITEK Business Products)

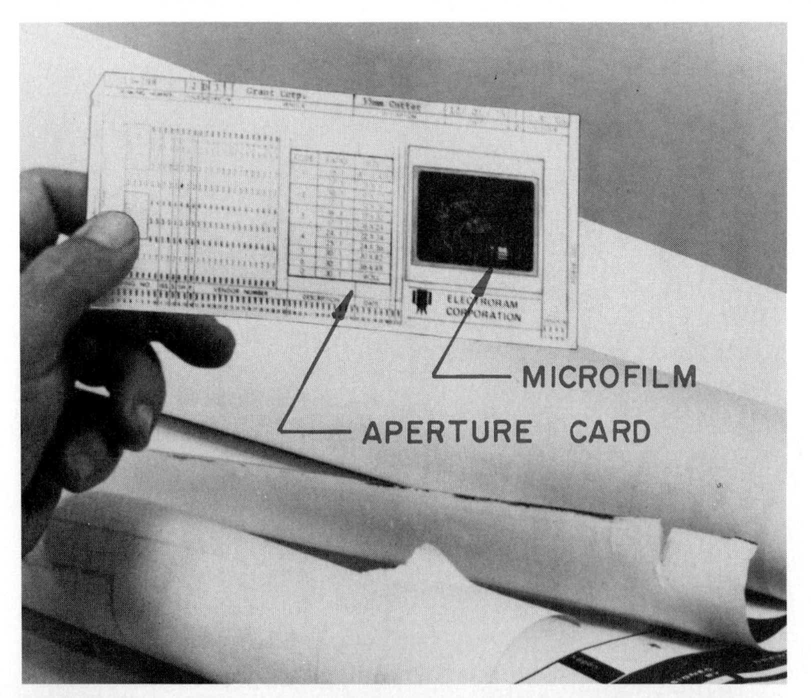

Fig. 16–13. This is a standard 35 mm microfilm in an aperture card. (Microseal Corporation)

Thermographic Reproduction

The thermographic process uses heat instead of light to reproduce a drawing. This process uses the principle that a dark surface (the lines on a drawing) absorbs more heat than a light surface (the open parts of a drawing). The lines and lettering on the original must be metallic or carbon. The thermographic paper and the drawing are fed together into the machine. The lines are heated by infrared light which produces the image on the surface of the thermographic paper. This is a dry process requiring no developer or toner.

Photographic Reproduction

High-quality reproductions are made using photographic film. The drawings are photographed using copy cameras. The film is then developed, and the images are made permanent with developing and fixing solutions. The copy camera can enlarge or reduce the drawing. The film has a matte surface so that it is possible to draw on it with pencil or ink. Fig. 16-12.

Office Copy Machines

Office copy machines can be used to reproduce original drawings on opaque and translucent paper. General purpose bond paper is used. General purpose bond is a specially prepared paper that is visually opaque but is constructed to allow transmission of ultraviolet light. Drawings on this paper appear opaque but can be used to produce copies using a diazo process.

Microfilm

A **microfilm** is a small photographic negative containing a drawing or other information. Fig. 16-13. The negative is produced by photographing the drawing with a microfilm camera and developing the exposed film in a film processor. The negative is mounted in an aperture card. This is a computer card with a window into which the negative is mounted. It is about 3¼ × 7⅜ inches (83 × 187 mm). The aperture card is punched with data used for automatic filing and retrieving.

Microfilm is used to store engineering drawings and data because a large amount of material can be stored in a small space. Also, microfilm is used to preserve valuable originals because after the microfilm is made the original is stored in a safe place. Needed copies are made from the microfilm, and the original is left untouched.

If the engineer wishes to view the drawing, it is placed in a microfilm reader. The drawing is projected greatly enlarged on the screen. A reader-printer will also print a full-size copy of the drawing. Fig. 16-14.

When drawings are to be microfilmed, special attention must be given to drawing lines uniformly dense and making lettering the proper size. Because all lines must be the standard width and density, they are often inked. Lettering must be large enough to permit it to be greatly reduced and enlarged with no loss of clarity. Lettering on A-, B- and C-size sheets should be .125 inch (3 mm) high and on D and E sheets, .150 inch (4 mm) high.

Microfiche

A **microfiche** is a microfilm negative that will hold ninety-eight 8½ × 11 inch (216 × 279.5 mm) pages. Fig. 16-15. The microfiche measures about 4 × 6 inches (100 × 150 mm). Microfiche is used to store drawings, specifications, and other data. Thousands of pages can be stored in a small storage cabinet.

Fig. 16–14. This microfilm viewer—printer displays the drawing greatly enlarged on the screen. It also makes prints up to 18 (×) 24 inches (457 (×) 609 mm) on plain paper or vellum. It uses 35 mm microfilm in either an aperture card or roll format. (Xerox Corporation)

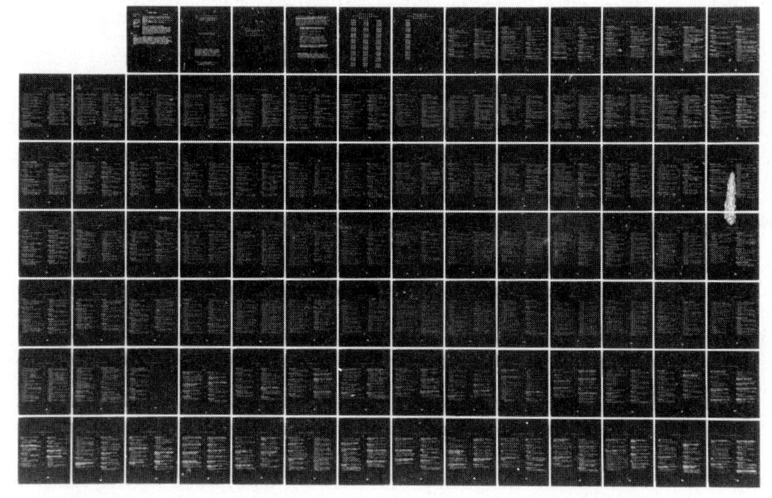

Fig. 16–15. This microfiche holds 98 pages of typewritten material. (Wenrick Manufacturing Co.)

A microfiche can be read by placing it in a reader-printer. Each page can be enlarged to full size, and a copy can then be printed.

Intermediate Drafting Techniques

An intermediate is a reproduction of an original drawing that can take the place of the original drawing. Thus, the original can be stored and prints run using the intermediate. Doing so preserves the original. Most often, an intermediate is used to change or correct an original without changing or destroying that drawing. The changes are made on the intermediate without having to redraw the original drawing.

There are many ways to make and use intermediates. Following are some of those in use:

- An intermediate can be produced by running a copy of an original on polyester film using standard ammonia whiteprint equipment. The surface has a matte finish that can be drawn on with pencil or ink. You can remove the unwanted parts of the printed image with a special fluid and draw changes.
- You could also run (make on a whiteprint machine) an intermediate. Then block out the areas to be removed or redrawn. Cover the area to be blocked out with heavy paper. Finally, run another intermediate and draw the corrections and changes in the blocked-out areas.
- Scissors drafting involves running an intermediate and cutting out the areas to be corrected. Then run a new intermediate, and draw the changes and corrections in the blank area.
- Polyester appliqué sheets that have images printed on them by an ordinary office copy machine are used to add items to drawings. The original art of the item to be added is placed on the copier, and the polyester sheets are fed through just like bond paper. The image is peeled off a paper backing and is stuck onto the drawing in its proper place. Fig. 16-16. This quick copy method is especially useful when adding features that are used frequently on drawings. There is no need to redraw them each time. Just stick a copy on the intermediate.
- Pasteup techniques involve running an intermediate and cutting out the elements you wish to save. These are taped in position on a new sheet, and additions are drawn and notes lettered. Fig. 16-17. Then another intermediate is made from the pasteup.
- The vandyke print (often called a brownprint) is a negative copy of an original drawing. The lines on the drawing appear white against a dark brown background. It is made in the same manner as a blueprint. The vandyke print is then used instead of the original drawing to make blueprints. It produces a print that is the reverse of a blueprint. The lines are blue and the background is white. These are called *blueline prints*.
- Diazo intermediates are available as sepia (brown) or blackline prints. They are used to reproduce other prints. Since these intermediates fade with time they should not be used to record permanent design changes. They are an excellent material for scissors drafting.

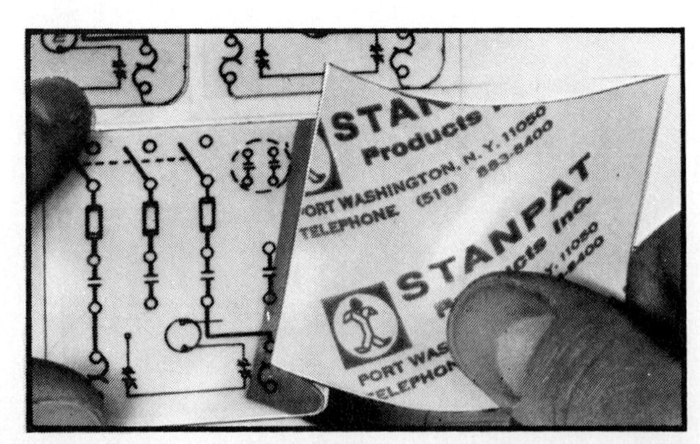

Fig. 16–16. After the image is produced on the polyester applique sheet, it is peeled from the paper backing and pressed in place on the drawing. The sheet has an adhesive on the back. (Stanpat Products, Inc.)

Pinbar Drafting

Pinbar drafting involves the use of overlays to build up the completed drawing. The drafting time can be greatly shortened because several drafters can work on different parts of a project at the same time. The pinbar holds the drawing sheets in register (makes them align correctly).

For example, the floor plan of a building is drawn. Then several duplicate intermediates are made and each given to a different drafter. Each drafter places a sheet of clear drafting film over his or her intermediate floor plan. One drafter proceeds to draw the electrical system, another the heating and air conditioning system and another the plumbing system. These three drawings are placed over the original floor plan to make a complete drawing. This "sandwich" of several overlays is used to produce copies of the completed floor plan. Fig. 16-18.

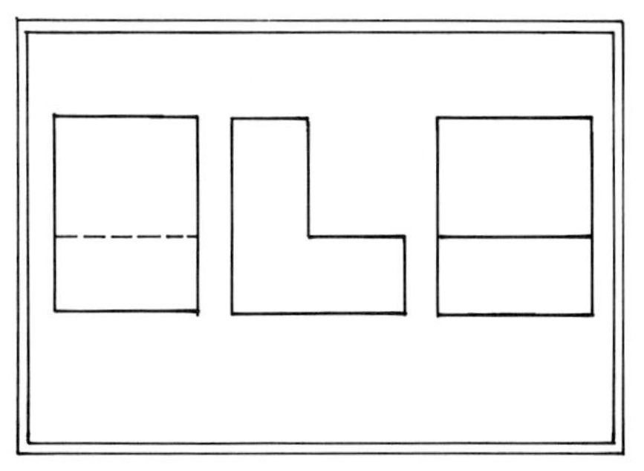

A. THE ORIGINAL DRAWING TO BE REVISED.

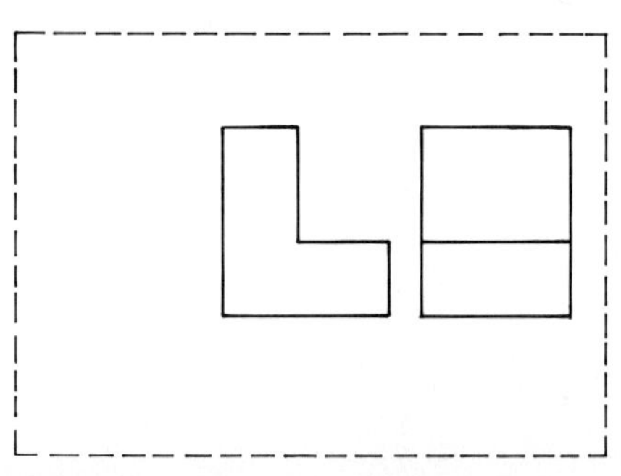

B. MAKE A CLEAR FILM REPRODUCTION OF THE ORIGINAL. CUT OUT THE PARTS TO BE USED IN THE REVISED DRAWING.

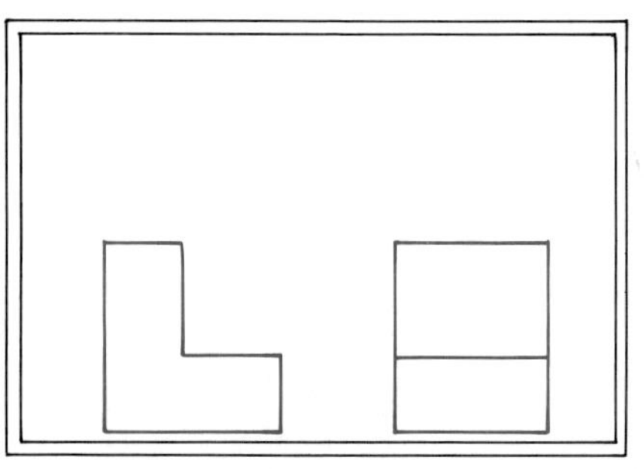

C. TAPE THE PARTS IN POSITION ON A NEW FORM.

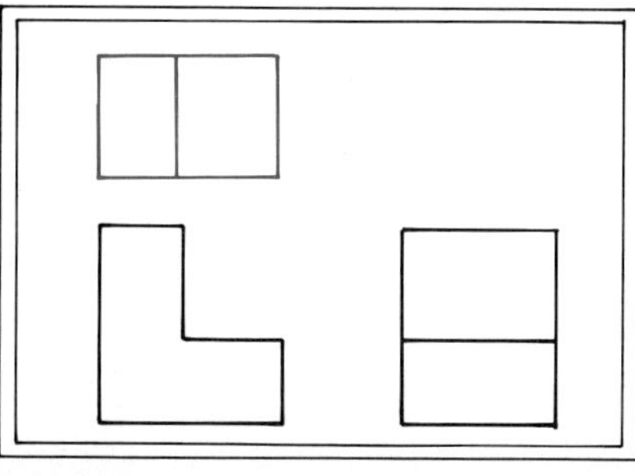

D. DRAW ADDITIONAL VIEWS TO COMPLETE THE NEW DRAWING.

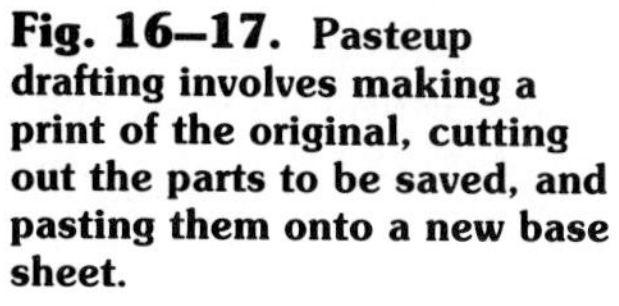

Fig. 16–17. Pasteup drafting involves making a print of the original, cutting out the parts to be saved, and pasting them onto a new base sheet.

Fig. 16–18. Overlay drafting is done on clear sheets placed over the base drawing. The base and the overlay sheets are kept in register by a pinbar.

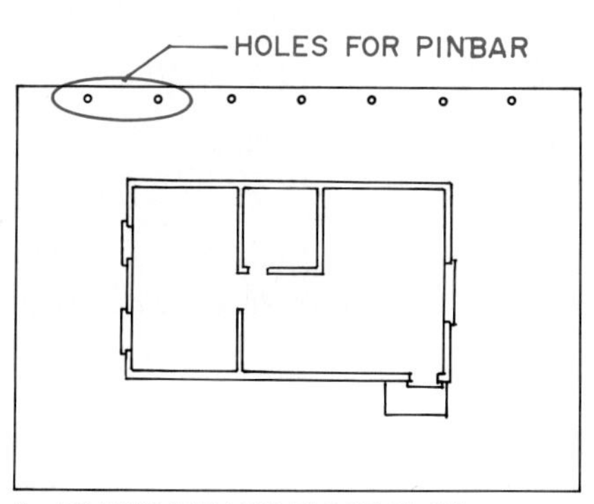

A. DRAW THE BASIC FLOOR PLAN.

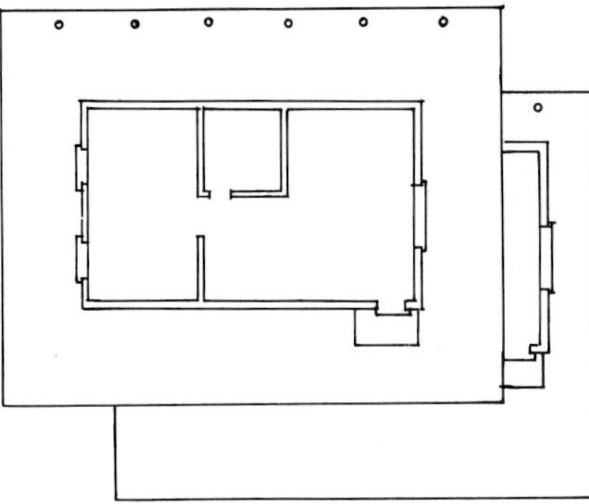

B. PRODUCE DUPLICATE COPIES OF THE FLOOR PLAN.

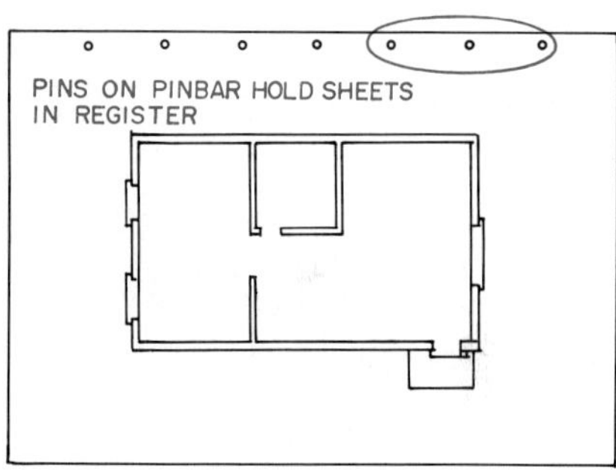

C. PLACE A COPY ON A PINBAR. PLACE A BLANK MATTE OVERLAY SHEET ON TOP. DRAW DESIRED DETAILS ON THIS OVERLAY. IN THIS EXAMPLE THE EXTERIOR DIMENSIONS AND ELECTRICAL PLAN ARE DRAWN ON SEPARATE OVERLAY SHEETS.

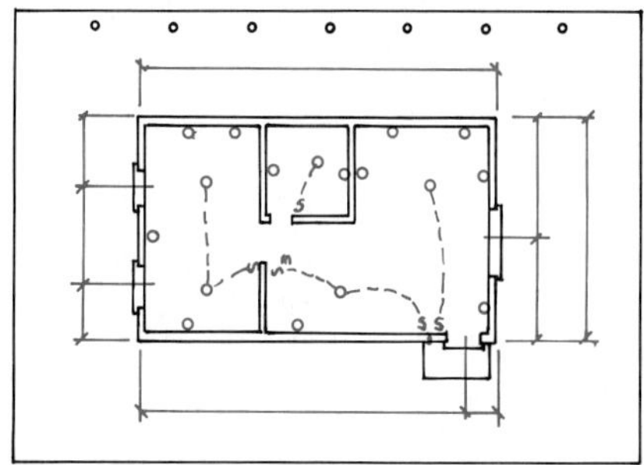

D. THIS SHOWS THE BASE SHEET WITH THE TWO OVERLAYS. NOW MAKE REDUCED SIZE NEGATIVES OF THE ASSEMBLY. THE NEGATIVE PRODUCED IS USED TO MAKE ENLARGED COPIES OF THE DRAWING WITH THE DETAILS ON IT.

Photo Drafting

Photo drafting involves taking a glossy photograph of the object to be described. Make a halftone print of the glossy and tape it in position on a drawing sheet. Photograph it to produce a negative. Make a positive print of the negative on matte film. Add notes or draw other views to produce a finished drawing. Fig. 16-19.

Filing Drawings

Original drawings are very costly to produce and must be preserved in a flat condition. The best methods of storage are using large drawers or hanging in a drawing file. Fig. 16-20. Originals stored rolled tend to be difficult to flatten. Rolled drawings cannot be used for microfilming.

1. MAKE A HALFTONE PRINT OF THE PHOTOGRAPH.

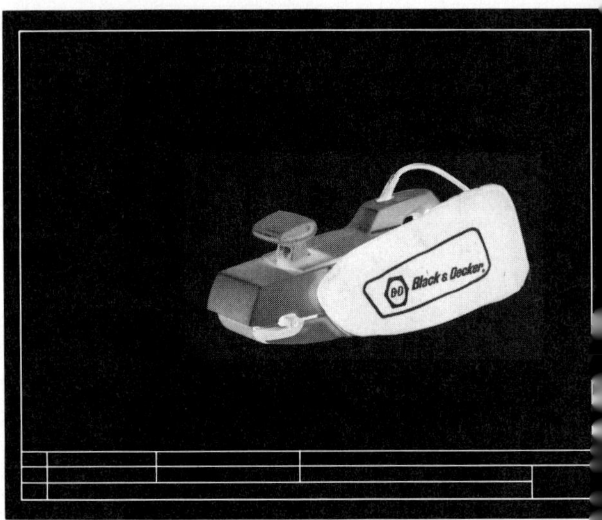

2. TAPE THE HALFTONE PRINT TO THE DRAWING SHEET. PHOTOGRAPH IT PRODUCE A NEGATIVE.

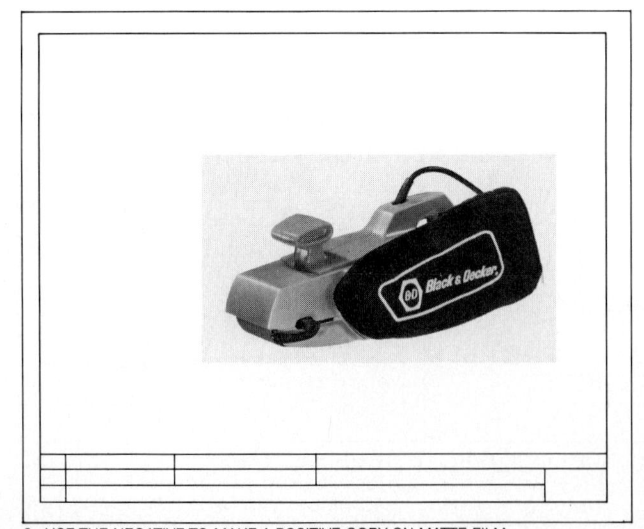

3. USE THE NEGATIVE TO MAKE A POSITIVE COPY ON MATTE FILM.

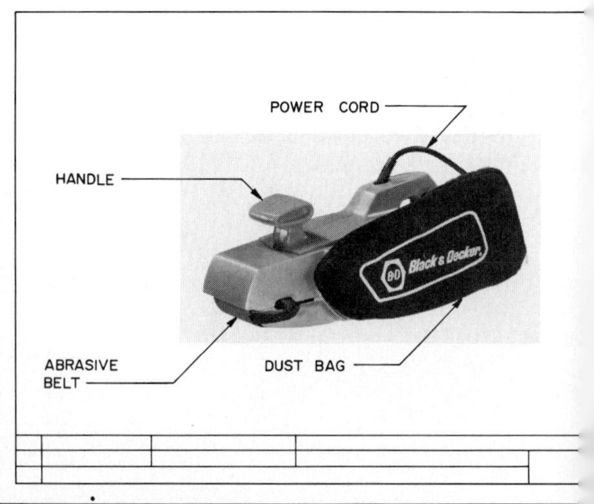

4. ADD NOTES OR DRAWINGS TO PRODUCE THE FINISHED ITEM.

Fig. 16–19. Photo drafting involves the production of positive and negative copies.

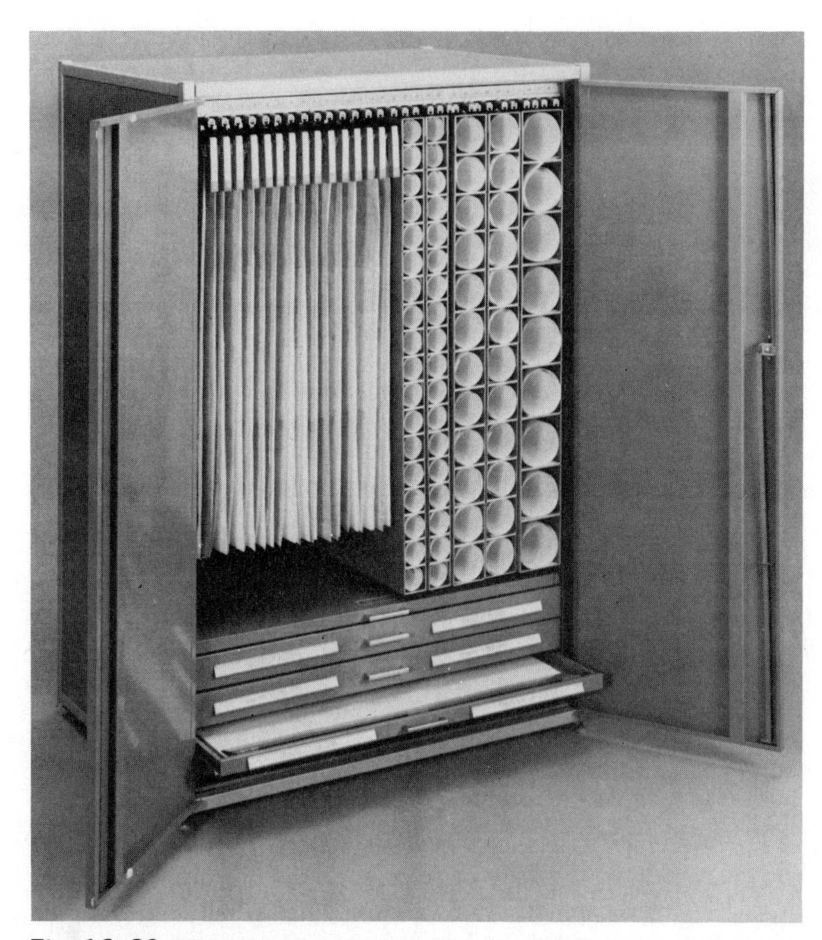

Fig. 16–20. **This drawing storage unit permits drawings to be kept flat in drawers, hung straight, or rolled. (Plan Hold Corporation)**

Chapter Review

Build Your Vocabulary

You should understand and use the following terms as part of your working vocabulary. Write a brief explanation of what each means.

blueprint
whiteprint
diazo process
xerography
microfilm
microfiche
pinbar drafting

CHAPTER 17

Developments and Intersections

After studying this chapter, you should be able to do the following:

- Make parallel-line developments.
- Make radial-line developments.
- Make developments using triangulation.
- Draw intersections between solids.

Surfaces

A surface is the exterior of an object. The outside of a sphere is a surface. The faces of a cube are surfaces. There are several types of surfaces—plane, single-curved, double-curved, and warped. See Fig. 17-1.

In a **plane surface**, if any two points are connected with a straight line, the line lies in the surface. The top of a table is a plane surface. A plane surface may have three or more sides.

A **single-curved surface** is one that can be unrolled to form a plane. A cylinder is an example.

A **double-curved surface** is formed by a curved line. It has no straight line elements. The surface of a sphere is an example. These surfaces cannot be developed exactly, but approximate developments can be made.

A **warped surface** is one that is neither plane nor curved. It, too, cannot be exactly developed, but approximate development can be made. The hoods on most automobiles are warped surfaces.

Surface Development

Surface development is the preparation of patterns or templates which, when properly formed, generate the surface desired. Examples of surface development of four basic geometric shapes are shown in Fig. 17-2.

Surface developments are used in many different industries to form many materials. Patterns are used to mark sheet metal for forming into pipes. Cardboard boxes are formed from developments. Clothing is made from patterns.

The drawing of patterns therefore includes all surfaces of the object to be formed. Patterns also include parts used to hold the formed object together. Fig. 17-3 at A shows a box before it is formed. Notice the tabs used to glue the box together. Fig. 17-4 shows the common joints used to

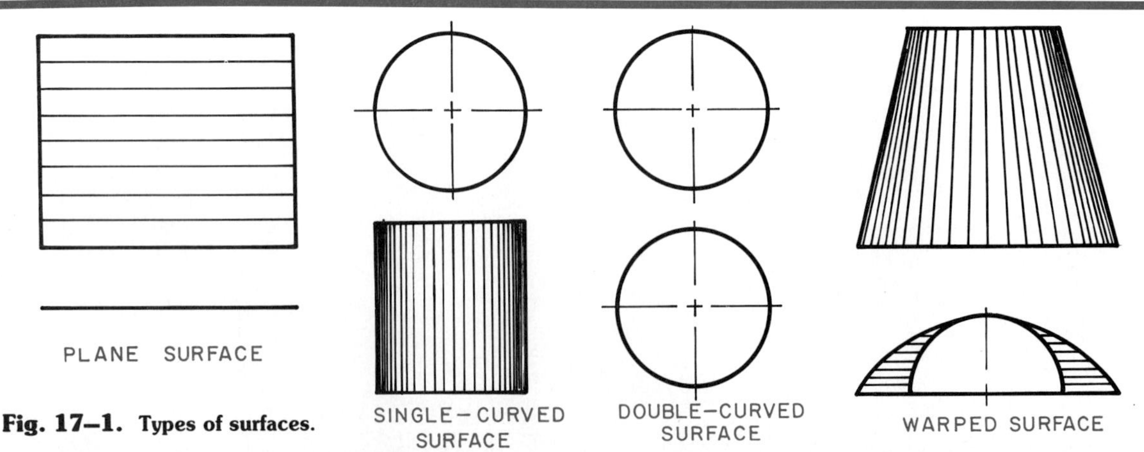

Fig. 17-1. **Types of surfaces.**

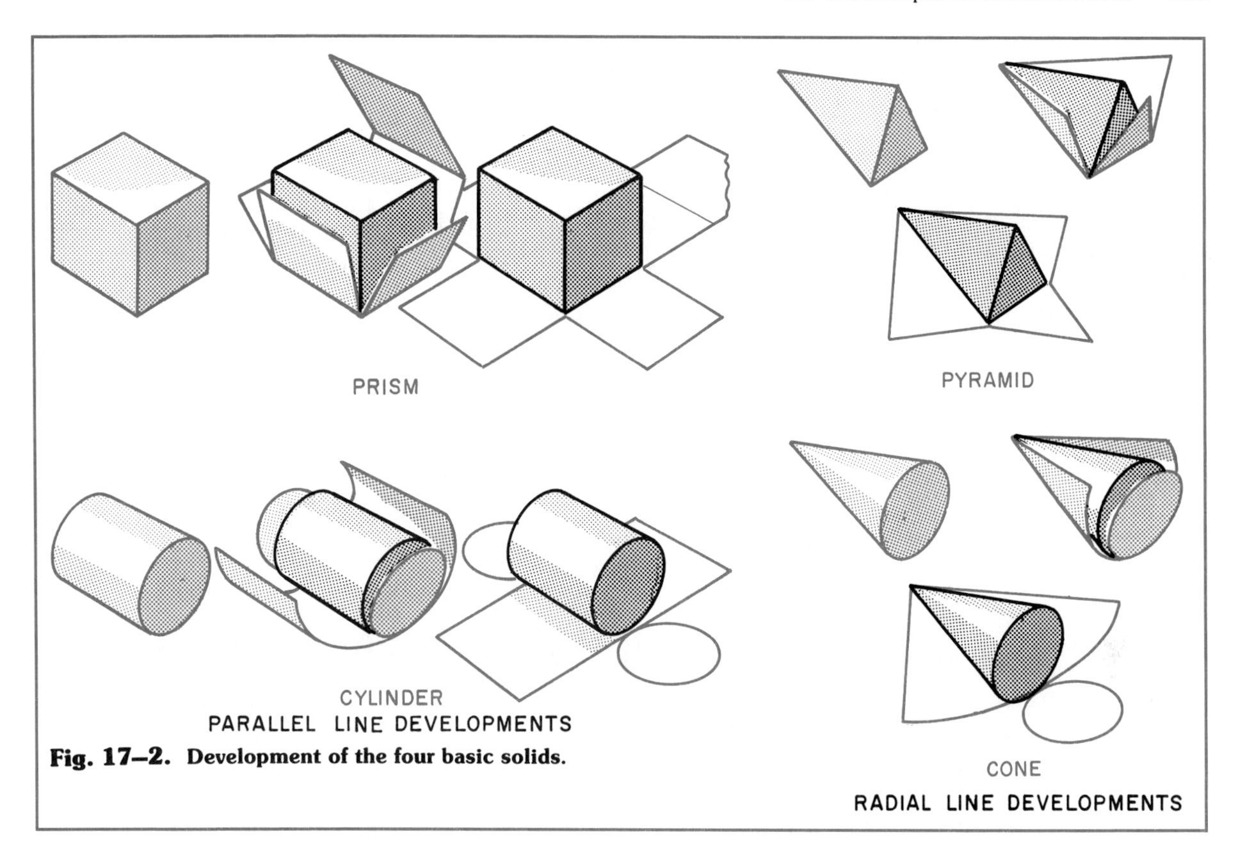

Fig. 17–2. Development of the four basic solids.

Fig. 17–3. **(A) A box formed from a pattern before it is folded into its finished shape. (B) The finished box.**

connect sheet metal surfaces. The method of fastening will vary with the material used in the finished product.

Fig. 17-5 shows a pattern for a cube. Patterns are built around a long line, called a *stretchout line*.

Folds are shown with thin, solid lines. They are sometimes marked with an X to help identify them. The faces of the pattern upon which the fold lines are drawn are the inside faces. When the pattern is folded, these are inside the object.

Before drawing a pattern, decide where the seams will be located. This will depend on the object and the desired appearance. Fig. 17-6 shows a pattern for a mailbox. The vertical seam was placed at the back edge so it would be somewhat hidden to improve the appearance of the product. Notice the metal tabs for joining the corners. The corners of tabs are usually cut on a 30-, 45-, or 60-degree angle. The angle used is the one that makes it easiest to clear the corner for a fold. The tabs on the top edges are used to avoid having sharp metal edges in a place where you would put your hand.

Types of Developments

Most objects fall into one of three types of development. These are parallel-line, radial-line, and triangulation.

Parallel-line development involves preparing patterns for objects having parallel elements. Squares and cylinders are examples. The stretchout line is straight and perpendicular to the fold lines.

Radial-line development involves preparing patterns for objects in which the stretchout line is not a continuous straight line and the lines representing the elements radiate from a single point. Cones and pyramids are examples.

Triangulation is a means of developing a surface that cannot be developed exactly. It provides an approximate development of the surface. Triangulation involves breaking the surface into a series of triangles. Objects made of a combination of curved and plane surfaces are developed by triangulation.

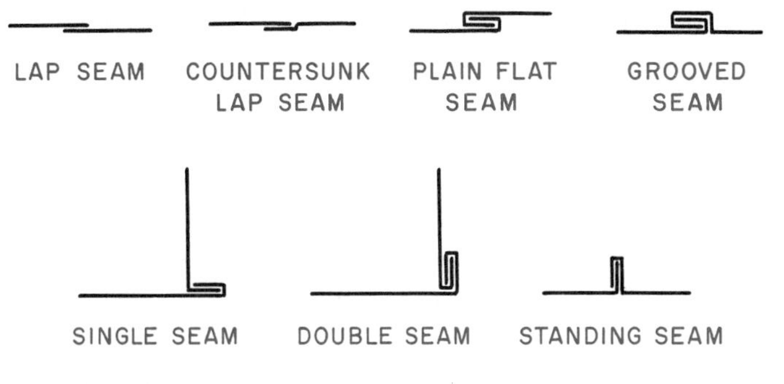

Fig. 17–4. Sheet metal joints.

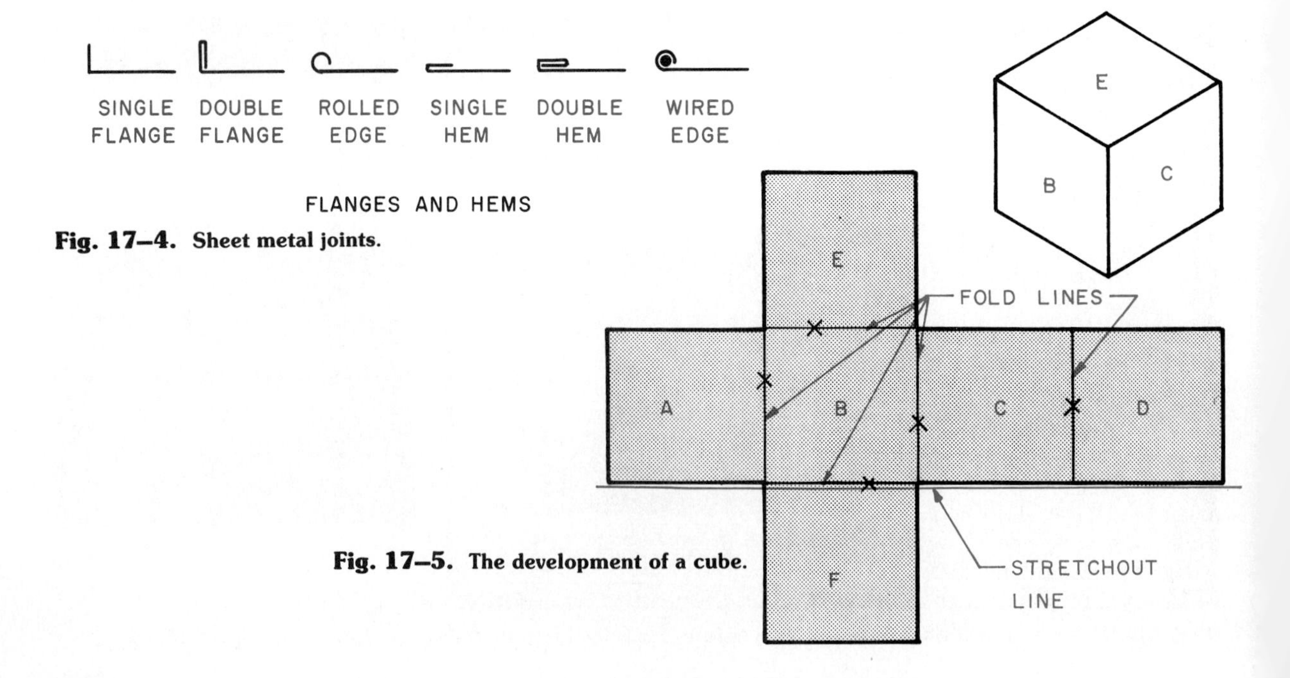

Fig. 17–5. The development of a cube.

Parallel-Line Development

All prisms are developed into rectangular-shaped patterns. Because of this, you use the parallel-line method to develop them.

Development of a Right Prism. A *right prism* has its base perpendicular to its axis. A *truncated prism* has one of its surfaces cut on an angle. The development of a truncated right prism is shown in both Figs. 17-6 and 17-7.

To develop the mailbox shown in Fig. 17-6:

1. Draw the stretchout line. Its length is the sum of the width of the four sides, which is 14 inches. Mark the distance of each side on the stretchout line. These are labeled points 1, 2, 3, 4, and 1.
2. At each of these points, draw vertical fold lines. Measure the length of each corner on these lines. These are points 5, 6, 7, 8, and 5.
3. Connect these points to complete the pattern of the faces.
4. Draw needed tabs.

Fig. 17-7 also shows the development of a truncated right prism. This object, however, has a closed top. The top is an inclined surface. The procedure for drawing the development is the same as explained for the mailbox. The only difference is the inclined surface. This surface is not true size in the top view. On the development, all lines must be true length. One way to find the true size is to make an auxiliary view. Note that the true length of the inclined surface appears on the front view. The true depth is shown on the top view. Measurements can be made in these views.

Development of an Oblique Prism. An *oblique prism* has an axis at the base at an angle other than 90 degrees. Fig. 17-8.

Use the same method to develop the oblique prism as for right prisms. One difference is that the oblique prism development does not unfold in a straight line, so the stretchout line is placed through the center of the prism.

To develop an oblique prism, follow these steps:

1. Draw the top and front views.
2. Draw an auxiliary section. This is perpendicular to the sides of the prism. The auxiliary section gives the true width of the sides.
3. Project the stretchout line from the location of the auxiliary section. In Fig. 17-8 the stretchout line is projected from the front view.

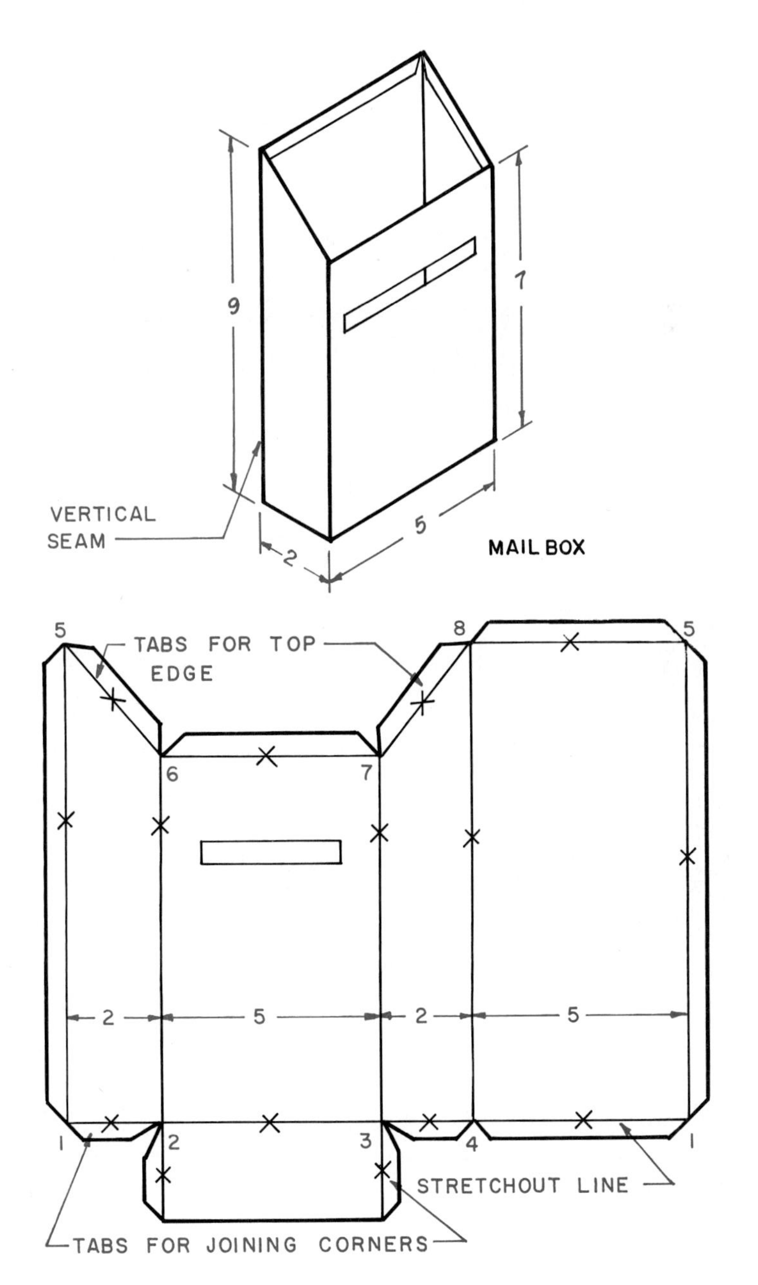

Fig. 17–6. A pattern drawing. This mailbox is a truncated right prism.

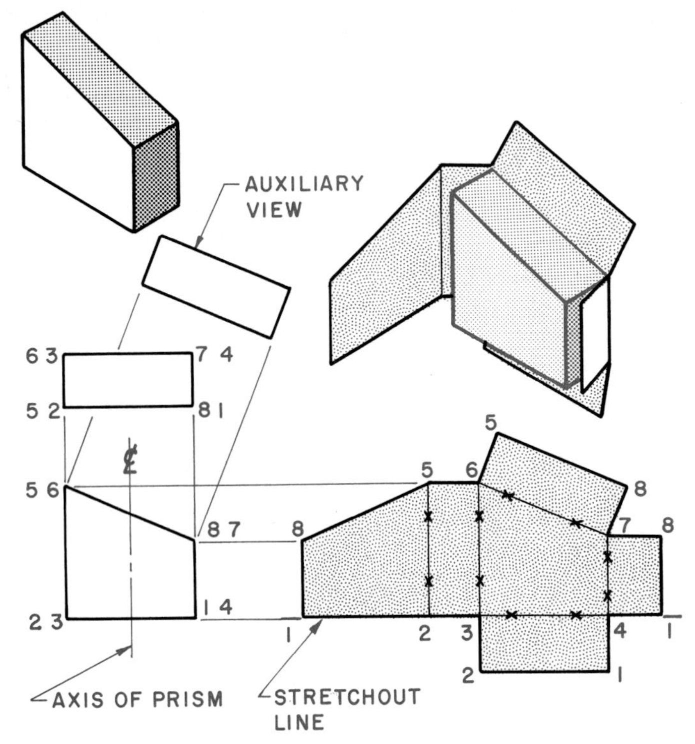

Fig. 17–7. **Development of a truncated right prism.**

4. Measure the true widths of the faces along the stretchout line. These are shown by points 1, 2, 3, 4, and 1.

5. Draw lines through these points, perpendicular to the stretchout line, to form the corners of the prism.

6. Project the end points of the corners to the development. This locates the top and bottom ends of the corners.

7. Connect the corners to finish the development of the oblique prism.

Developing Cylinders

Cylinders may be imagined as prisms. Fig. 17-9. The number of sides of the imaginary prism is infinite. This means there is no limit to the number of sides. The more sides the prism has, the more it resembles a cylinder. Each of the sides of the prism forms an edge. These edges are called elements. An *element* is an imaginary straight line on the surface of a cylinder. All elements on a cylinder are parallel to the axis. Any cylinder whose bases are perpendicular to its axis can be developed from a rectangular pattern. All cylinders are developed by the parallel-line method.

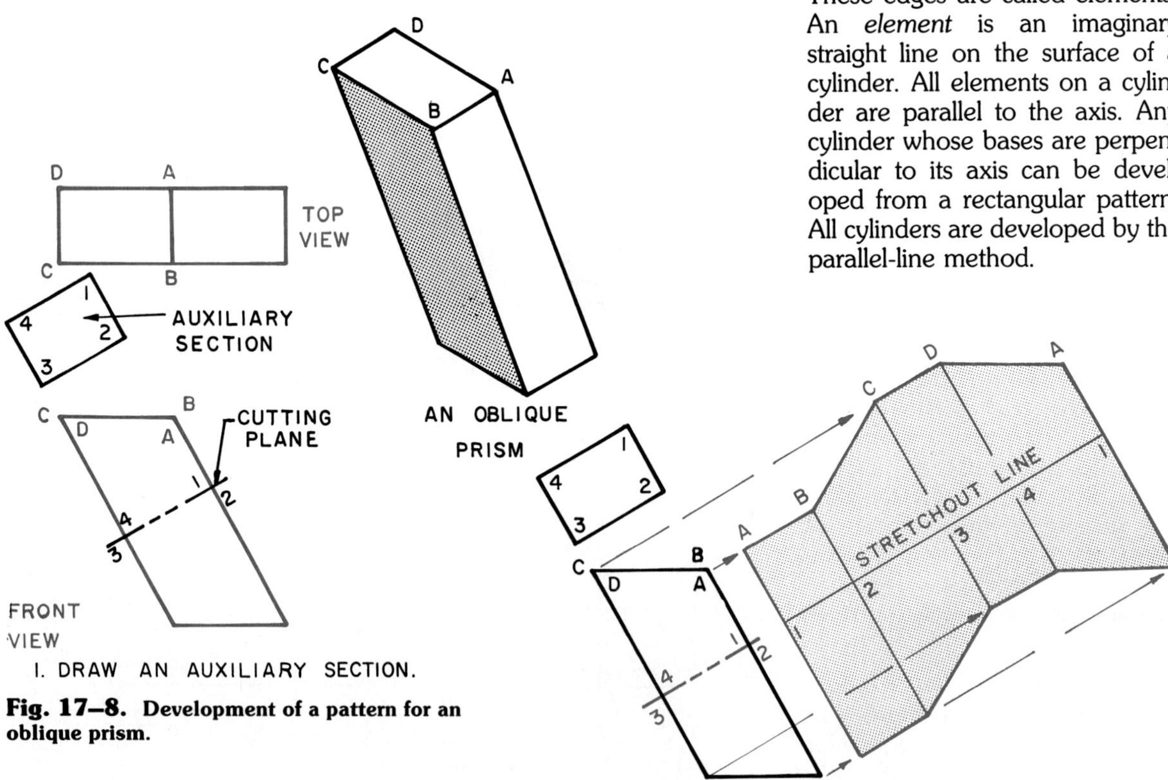

Fig. 17–8. **Development of a pattern for an oblique prism.**

2. DRAW THE STRETCHOUT LINE.
3. LOCATE THE TRUE WIDTHS.
4. LOCATE THE ENDS OF THE CORNERS.

The length of the stretchout line of a cylinder is equal to its circumference. The *circumference* is the distance around the cylinder. The stretchout line is always perpendicular to the axis of the cylinder. If the cylinder is oblique, you must make a right-section view to obtain the circumference. The circumference can be found by multiplying the diameter by 3.14.

Development of a Right Cylinder. To develop a right cylinder, you need two views. Fig. 17-10. One of these views must show the diameter. The other must show the height.

To make an approximate development of a right cylinder, see Fig. 17-10 and follow these steps:

1. Divide the circular view into any number of equal parts. The more divisions, the more accurate the pattern.
2. Number each of the divisions (1 to 12 in the example) on the circular view.
3. Draw the stretchout line. On the left end, draw a perpendicular line. Make it the same length as the height of the cylinder.
4. Measure the straight-line distance between two divisions on the circular view.
5. Step off this distance on the stretchout line. Be certain to step off as many divisions as are on the circular view, including a return to the first part. This gives the circumference.
6. On the last division, draw a perpendicular to the stretchout line. Draw the top edge of the cylinder parallel with the stretchout line.

This method for obtaining circumference is approximate. The distances used are the *chordal* distances. A chord is shorter than its arc. A more accurate method of obtaining circumference is using mathematics. To obtain the circumference, multiply the diameter by π (pi). Pi is equal to 3.1416. For example, a cylinder with a diameter of 2 inches has a circumference of 6.28 inches. The length of the stretchout line therefore, would be 6.28 inches.

Development of a Truncated Right Cylinder. A truncated right cylinder is one that is cut at an angle other than 90 degrees to its axis. The end that is cut will result in a curved line in the development. To make the development, two adjacent views are necessary, commonly the top and front views.

To make this development see Fig. 17-11, and follow these steps:

1. Divide the circular view into any number of equal parts.
2. Project these points into the front view. The seam has been placed on the shortest element.
3. Draw the stretchout line perpendicular to the axis of the cylinder.
4. Measure the straight-line distance between two divisions on the circular view.
5. Step off this distance on the stretchout line. Step off as many divisions as are on the circular view. This gives the circumference.
6. Draw perpendicular lines through each point on the stretchout line.
7. Project the points on the inclined edge in the front view to corresponding lines in the stretchout. Connect these points with an irregular curve.

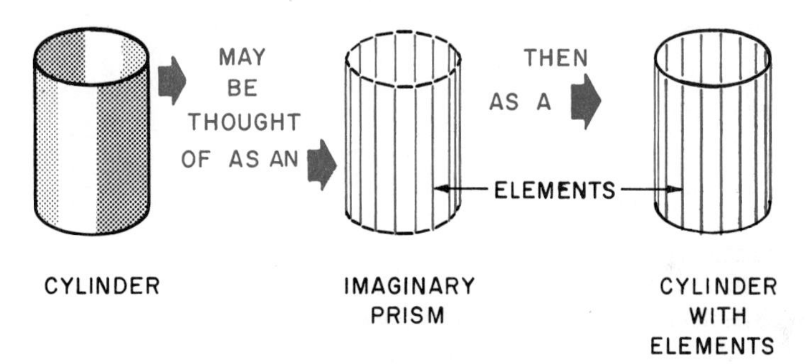

Fig. 17–9. Cylinders may be thought of as imaginary prisms.

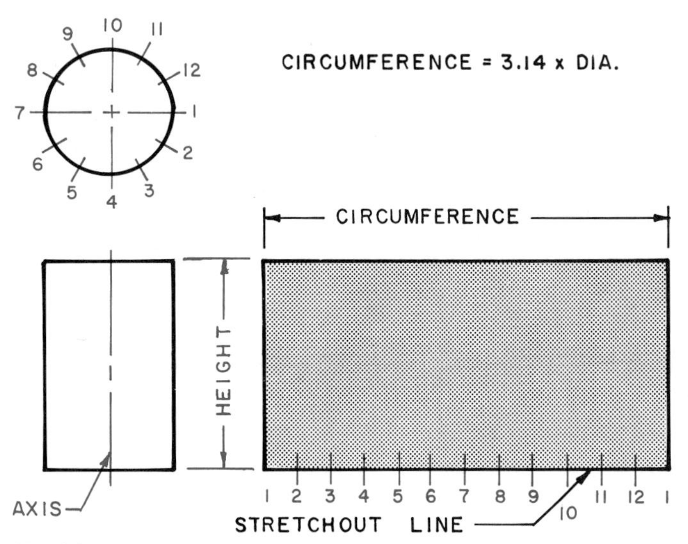

Fig. 17–10. Development of a right cylinder.

Development of Elbows

Elbows are used to change the direction of a pipe. They are made of two, three, four, or more pieces. The number of pieces depends upon the pipe diameter. Elbows are used on round and square pipe. A round two-piece elbow is shown in Fig. 17-11. Three-piece elbows are shown in Fig. 17-12. A round four-piece elbow is shown in Fig. 17-13. To develop an elbow, draw the front view. This shows the number of pieces, the angle of bend, and the true length of the elements of each piece.

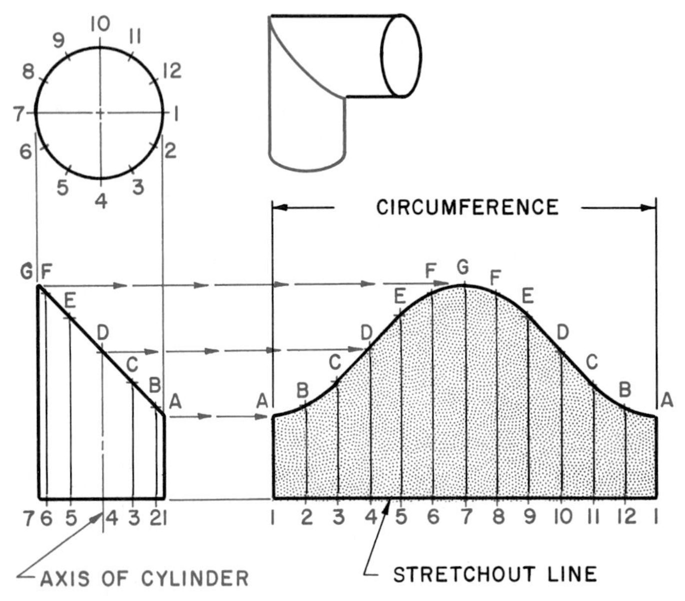

Fig. 17–11. **Development of a truncated right cylinder.**

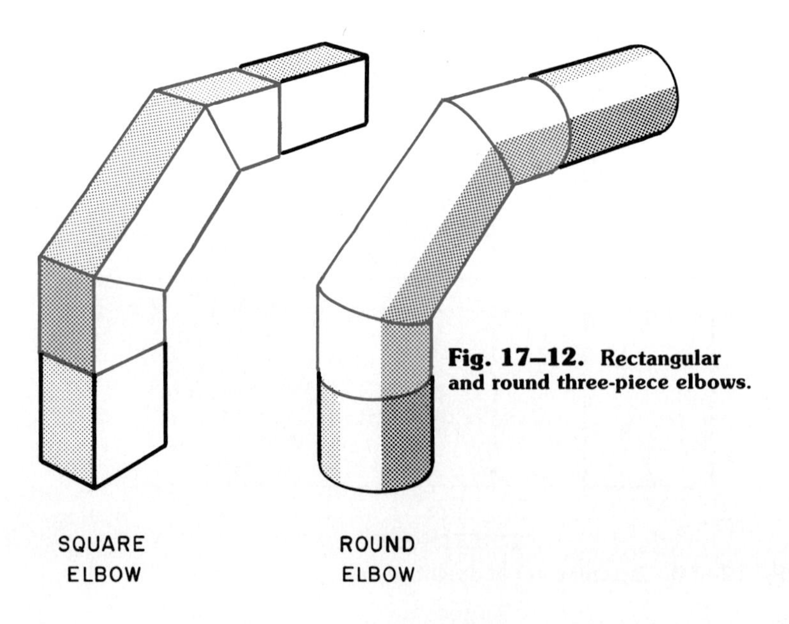

Fig. 17–12. **Rectangular and round three-piece elbows.**

To develop the four-piece round elbow shown in Fig. 17-13, follow these steps:

1. Draw the heel and throat radii. Fig. 17-13 at A.
2. Divide the angle of bend into the proper number of pieces. The middle pieces of the elbow are twice as large as the end pieces. The angles are figured by using the following formula: Number of spaces = (number of pieces × 2) − 2. Since Fig. 17-13 shows a four-piece elbow, the angles are (4 × 2) − 2 = 6. The angle of bend, 90 degrees, is divided by 6. Each angle is 15 degrees.
3. Draw the 15-degree angles on the front view. See Fig. 17-13 at B.
4. Draw the miter lines. These are the places where the pieces of pipe are to be joined. These are lines 1, 3, and 5 in Fig. 17-13 at C. Since the center pieces are twice as large as the ends, they are each 30 degrees.
5. Draw lines tangent to the arcs of each piece. See Fig. 17-13 at C.
6. Draw a half view of the end of the pipe as shown in Fig. 17-13D. Divide it into an equal number of parts.
7. Project these divisions to the front view. Draw them parallel with the surface of each part of the elbow.
8. Lay out the stretchout line. The length is equal to the circumference of the pipe.
9. Lay out the distances between the points on the half-circular view on the stretchout line. Proceed to finish the development using the same technique as explained for a truncated cylinder. Each piece of pipe is a truncated cylinder.

Lay out the pattern as shown in Fig. 17-13. When all four pieces are laid out together, each curve developed serves two pieces of the elbow. This layout also staggers the seams. The patterns A and C, Fig. 17-13, will have seams at the throat. Parts D and B will have seams at the heel.

Radial-Line Development

Some objects are developed by the radial-line method. Radial lines are used to develop forms such as pyramids and cones. Their edges come together at a common point.

Development of a Right Rectangular Pyramid. A right rectangular pyramid is shown in Fig. 17-14. The axis is perpendicular to the base. To draw a development of this figure, follow these steps:

1. Draw the top and front views.

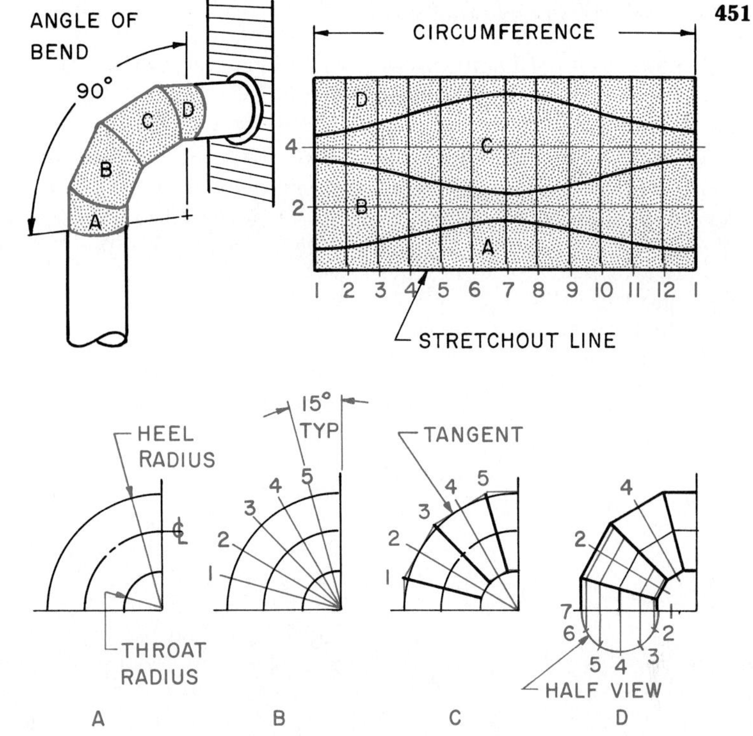

Fig. 17–13. Pattern for a round four-piece elbow.

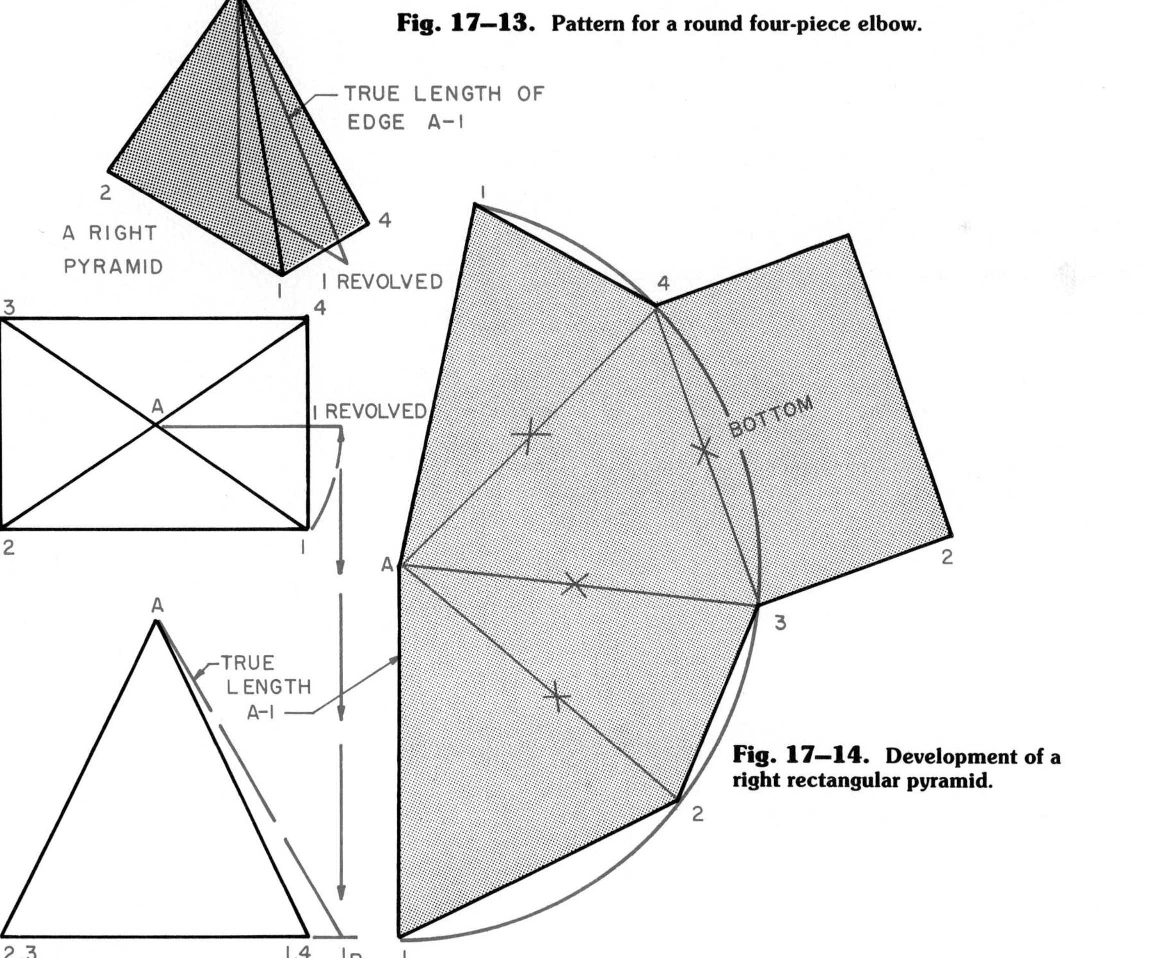

Fig. 17–14. Development of a right rectangular pyramid.

2. Find the true length of one edge by revolution. The steps are shown in Fig. 17-14.

3. Mark point A on the paper. Set a compass with a radius equal to the true length of edge A-1. Swing an arc.

4. Draw edge A-1 from point A to one end of the arc. Step off the four base edges of the pyramid. These are 1-2, 2-3, 3-4, and 4-1. Connect these points with point A to show the fold lines. Connect the base points 1, 2, 3, and 4. This completes the development of the sides.

5. Draw the bottom connected to any of the base edges. In Fig. 17-14, edge 3-4 was used. Draw lines perpendicular to the base edge. Along these lines, measure off the true width of the pyramid. This is line 1-4, found on the top view. Connect the width lines with a line parallel to the base edge to complete the bottom.

Development of a Truncated Right Pyramid. A truncated pyramid is one that is cut on an angle to its axis. This angle is one other than 90 degrees.

Fig. 17-15 shows the steps for making the development. It uses the same procedure as explained for the right pyramid shown in Fig. 17-14. The only difference is that the true length of several edges must be found. In Fig. 17-15, these are OA, OB, and O-1. The true lengths can be found by revolution.

It is best to draw the object as a normal right pyramid. Then locate the true length of each side that is shorter.

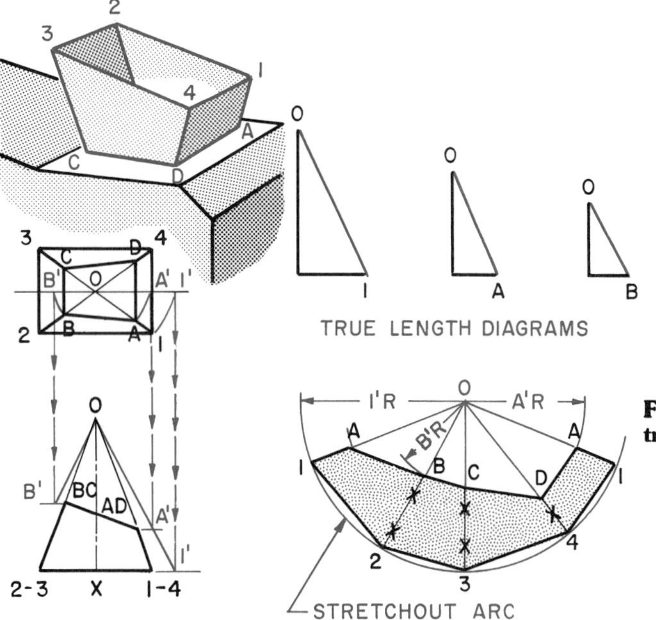

Fig. 17–15. Development of a truncated right pyramid.

Developing Cones

Developing a cone is similar to developing a pyramid. A cone may be thought of as a many-sided pyramid. See Fig. 17-16. Each edge of the imaginary pyramid may be thought of as an element. The development is made by placing the elements next to each other.

Development of a Right Cone. Fig 17-17 shows how to develop a right cone and a frustum of a right cone. The developments are identical except that the frustum of a right cone has the top cut on an angle of 90 degrees to the axis.

To develop a right cone, follow these steps:

1. Draw the top and front views of the cone.

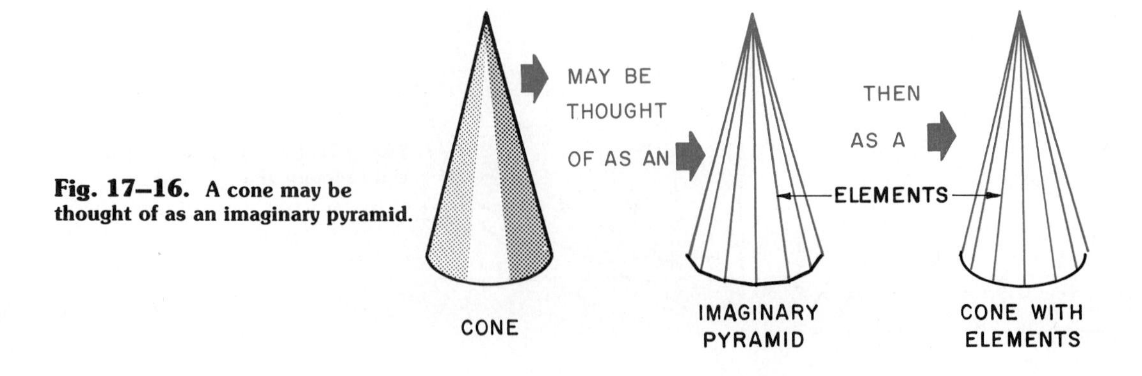

Fig. 17–16. A cone may be thought of as an imaginary pyramid.

2. Divide the circular view into a number of equal parts.

3. Project these points to the base of the cone in the front view.

4. Draw lines from the base to the top of the cone. This is point 0. These lines form the elements of the cone. The only true-length elements are 0-1 and 0-7. They are true length because they are parallel to the frontal plane.

5. Using 0-1 as the radius, draw an arc. This is the stretchout line.

6. Measure the straight line distance (chord) between two of the points located on the top view. For example, use the distance from 1 to 2. Step off this distance along the arc until the arc has the same number of points as are on the top view. This is the circumference of the base of the cone.

7. Connect the ends of the arc to point 0 to complete the development of the right cone.

To draw a frustum of a cone, you follow these same steps. In addition, measure the true-length distance from the top of the cone, point 0, to the flat top surface. This is distance A in Fig. 17-17. Draw an arc on the development with radius 0A to locate the top edge of the frustum.

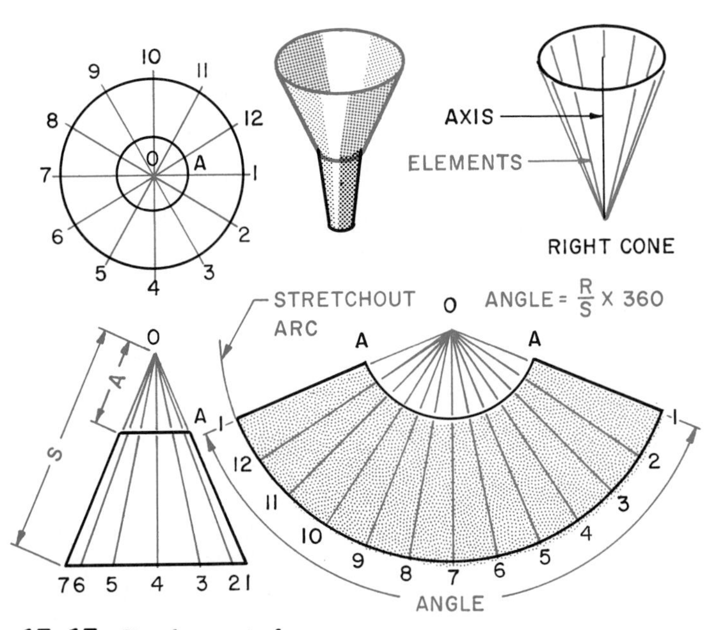

Fig. 17–17. **Development of a right cone and a frustum of a right cone.**

Development of a Truncated Right Cone. A truncated right cone is one that is cut on an angle to its axis. This angle is one other than 90 degrees.

The development is similar to that used for developing a cone. See Fig. 17-18. First develop the object as a cone. Then find the true length of each of the elements from the top, point 0, to the truncated surface. Use the revolution method. Then locate these true lengths on the elements laid out on the development. Connect these points with an irregular curve to form the top edge of the truncated right cone.

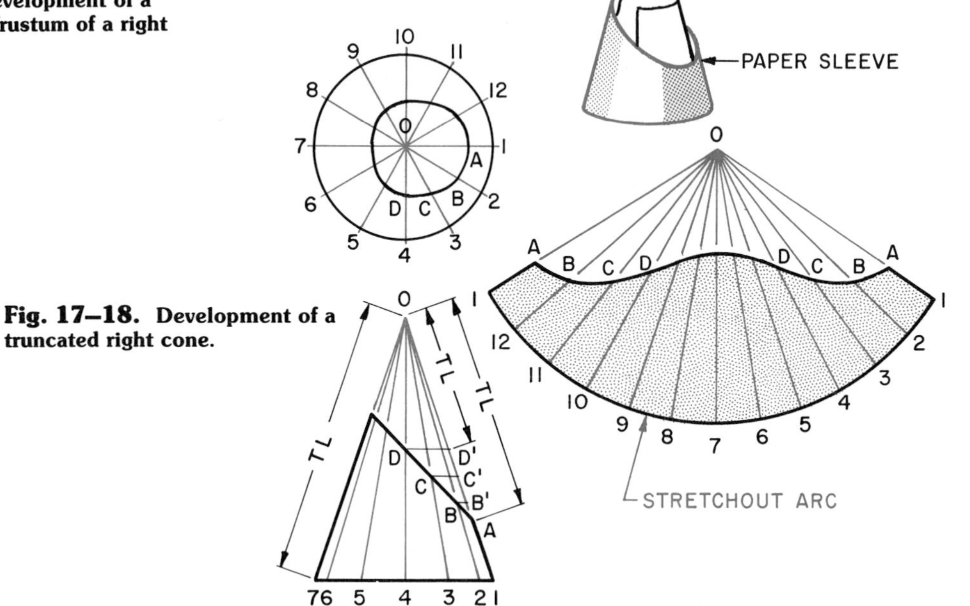

Fig. 17–18. **Development of a truncated right cone.**

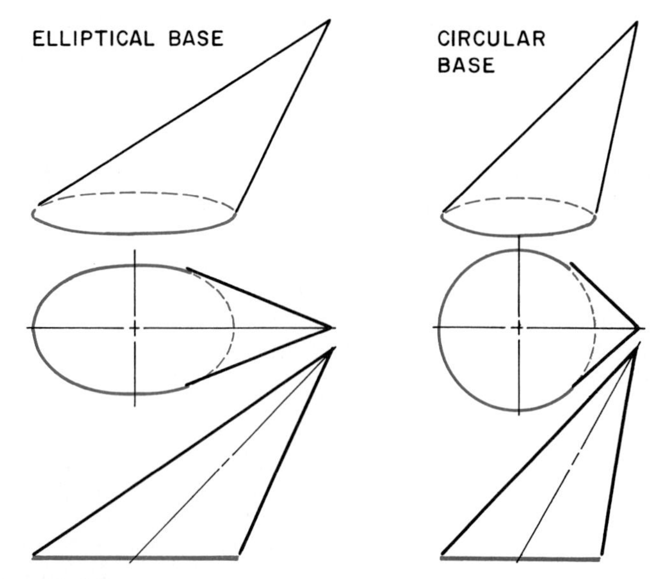

Fig. 17–19. **Oblique cones can have elliptical or circular bases.**

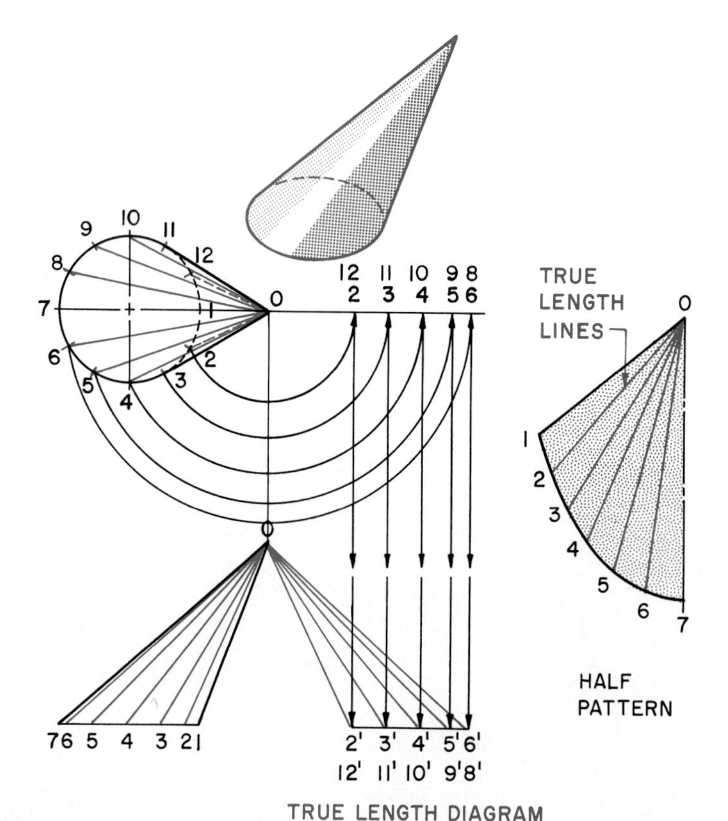

Fig. 17–20. **Development of an oblique cone.**

Triangulation

You can use triangulation to develop surfaces that cannot be developed by the parallel- or radial-line methods. Surfaces developed by triangulation are approximate. With this method, you divide the surface into a series of triangles. The true length of the sides of each triangle must be found. Then the triangles are drawn next to each other on a flat surface. An example of this method is the development of an oblique cone.

Development of an Oblique Cone. Oblique cones can have either a circular or elliptical base. Fig. 17-19. Either type can be developed approximately by triangulation.

To develop an oblique cone, see Fig. 17-20 and follow these steps:

1. Draw the top and front views.
2. Divide the circular view into any number of equal parts.
3. Project these divisions to the base in the front view.
4. Draw the elements from each division on the base to the top of the cone. This divides the surface into a series of triangles.
5. Find the true length of each element. Only 0-7 and 0-1 are true length, as they are drawn in Fig. 17-20. Find the true length of the others by revolution.
6. On the development, lay out the triangles in the order in which they are found on the views of the cone. The first triangle is 0-1-2 in Fig. 17-20. First, lay out true-length side 0-1. Then measure the straight-line distance between 1 and 2 on the top view. Using 1 on the development as a center, swing an arc with this radius. Next, swing an arc with a radius equal to the true length of element 0-2 using 0 as the center. Where the arcs cross is point 2 on the pattern. Repeat these

steps for each element. Connect the points with an irregular curve.

Developing Transition Pieces

Transition pieces are used to connect pipes of different sizes or shapes. Fig. 17-21 shows some examples. Notice that they are made of plane and curved surfaces. Sometimes both are found in one transition piece.

Development of a Rectangular-to-Round Transition Piece. This development is similar to developing an oblique cone. However, it is made of plane triangular surfaces and curved corners. The development is shown in Fig. 17-22. To construct it, follow these steps:

1. Draw the top and front views of the piece.
2. Divide the circular view into a number of equal parts.
3. Project these divisions to the front view.
4. Draw the elements for the curved corners on the front view.
5. Revolve these elements to find their true length.
6. To begin the pattern, lay out one true-length line for one of the plane surfaces. In this example, edge 1′-A in Fig. 17-22 is used. This edge will be the seam. It will appear at the back of the hood, where it is less likely to be seen.
7. Measure the distance from A to B on the top view. Draw it perpendicular to line 1′-A. Connect B with 1′ on the development. Line 1′-B is the true length.
8. Lay out the triangles forming the corner. These are marked B-1′-2′, B-2′-3′, and B-3′-4′.
9. Lay out the next large triangle. This is B-4′-C. The true length of BC is found in the top view. The true length of C-4′ is found by revolution.
10. Continue these steps until all surfaces of the pattern are drawn.
11. Draw a curve through the points forming the circular opening.

Intersections

Intersections are formed when two surfaces meet. The surfaces can be plane, curved, or spherical. The intersection of two plane surfaces is a straight line. The in-

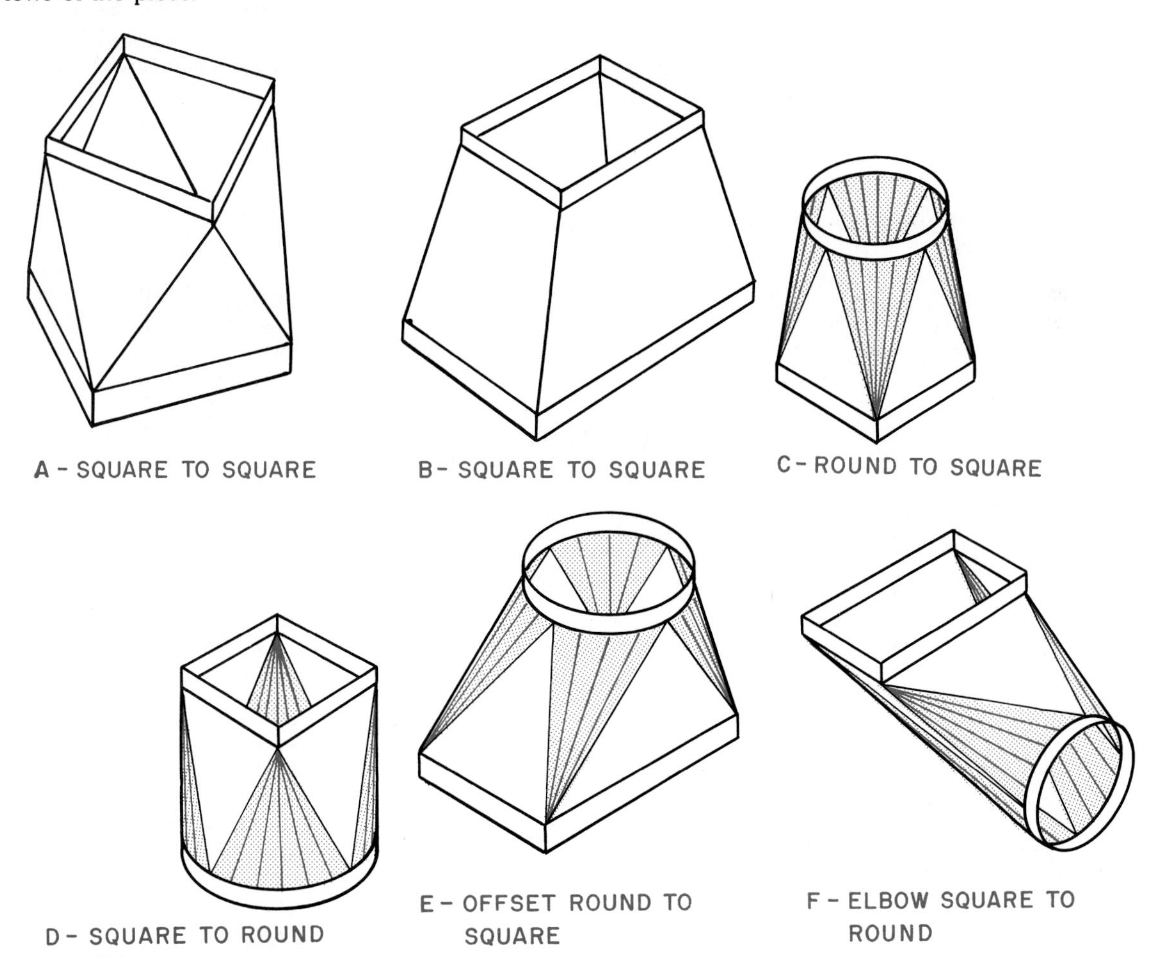

Fig. 17–21. Some typical transition pieces.

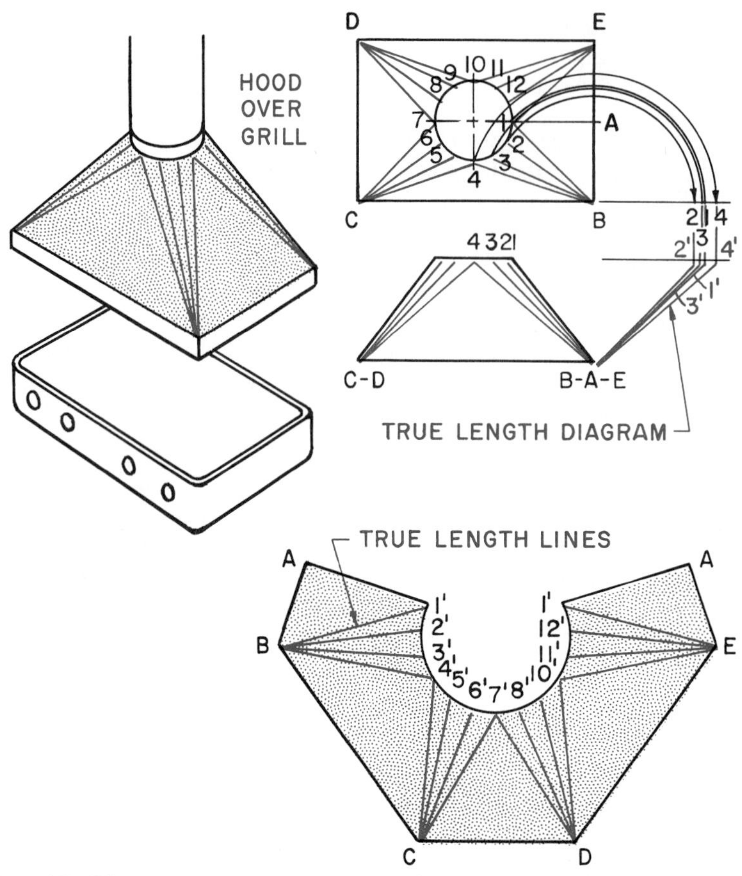

Fig. 17–22. Development of a rectangular-to-round transition piece.

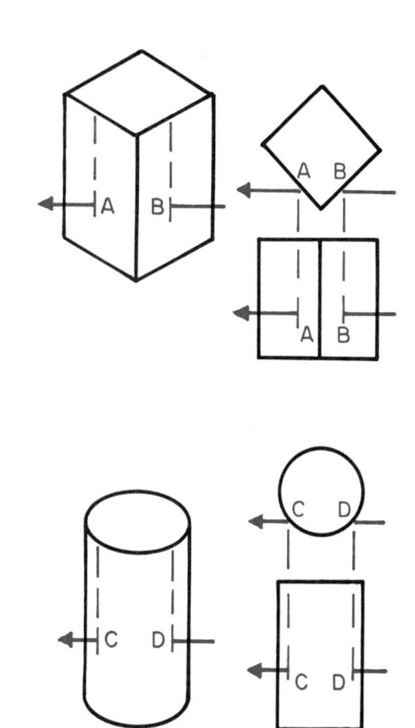

Fig. 17–24. The intersection of a line and a plane or curved surface is a point.

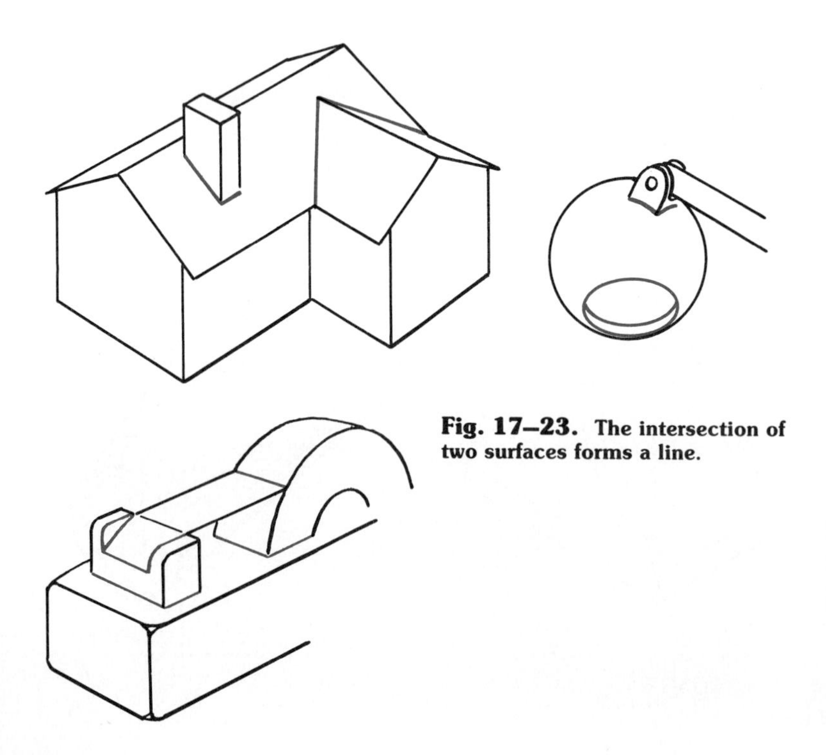

Fig. 17–23. The intersection of two surfaces forms a line.

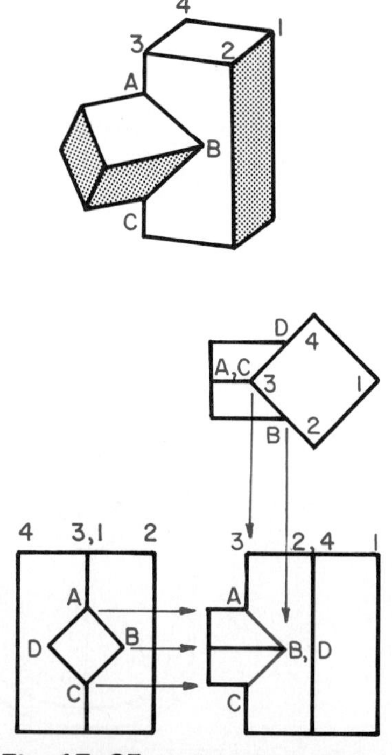

Fig. 17–25. Intersection of two prisms at right angles to each other.

tersection of a plane and a curved or spherical surface is a curved line. The intersection of curved or spherical surfaces is a curved line. See Fig. 17-23.

When a line intersects a plane or curved surface, a point is formed. This is called a **piercing point**. See Fig. 17-24.

Intersecting Prisms

The intersections of the prisms described here are typical of those that drafters often need to find.

Intersection of Two Prisms at Right Angles. The intersection of two prisms can be found by locating the piercing points of the edges of one solid on the surface of the other solid. The piercing point is the point at which the edge touches the surface.

To find the intersection, see Fig. 17-25 and follow these steps:

1. Draw the orthographic views of the two prisms.
2. Project points A, B, C, and D on the side view to the front view.
3. Project points A, B, C, and D on the top view to the front view. Where the projections of corresponding points cross are the piercing points.
4. Connect the piercing points with straight lines.

These same piercing points are used to make a development of the two prisms. A development of the intersecting prisms is shown in Fig. 17-26.

To develop the large prism, follow these steps:

1. Draw the stretchout line. Locate the true-length sizes. True lengths are found in the top view in this example. These are edges 1-2, 2-3, 3-4, 4-1.
2. Measure the true height of the large prism. This is found on the front view.
3. To locate the piercing points (A, B, C, and D), first project them from the front view to the stretchout. Then find the distance between the points. The true distance between the points is found in the top view. Points A and C are located on edge 3. Points B and D are found by measuring their true distance from edge 3 on the top view.
4. Connect the piercing points with straight lines.

The same steps are used to develop the small prism:

1. Draw the stretchout line. True lengths of the small prism's sides are found on the side view for this problem. These are lines 5-6, 6-7, 7-8, and 8-5. Lay out these distances on the stretchout line.
2. The true length of each edge is found on the top and front views. Lay out these distances on the development.
3. Connect the piercing points.

Intersection of Two Prisms Oblique to Each Other. The steps to find this intersection are the same as those for prisms at right angles. See Fig. 17-27 and follow these steps:

1. Draw the orthographic views.
2. Locate the piercing points of the edges of the intersecting planes. These are points A, B, C, and D.
3. Project these to the view on which the intersection is visible. Locate them on the proper edges in this view.
4. Connect the points with straight lines.

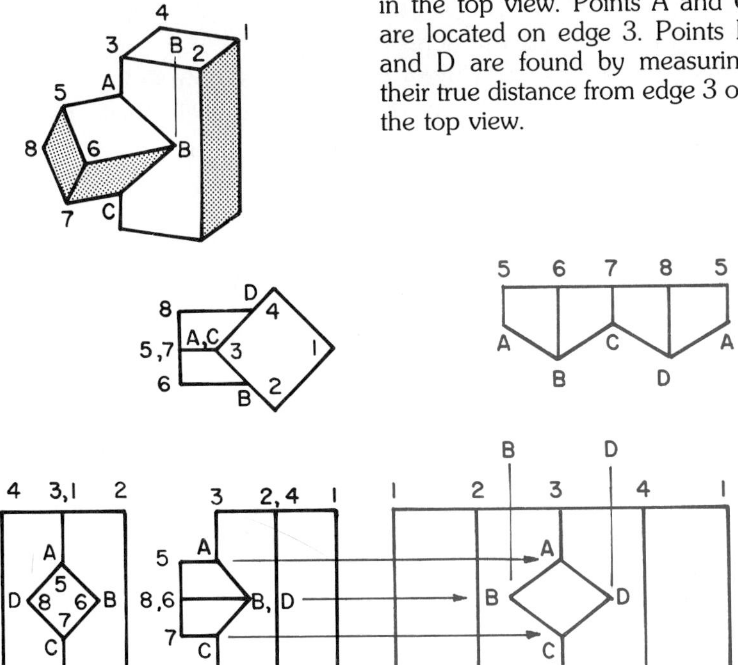

Fig. 17–26. The development of two intersecting prisms.

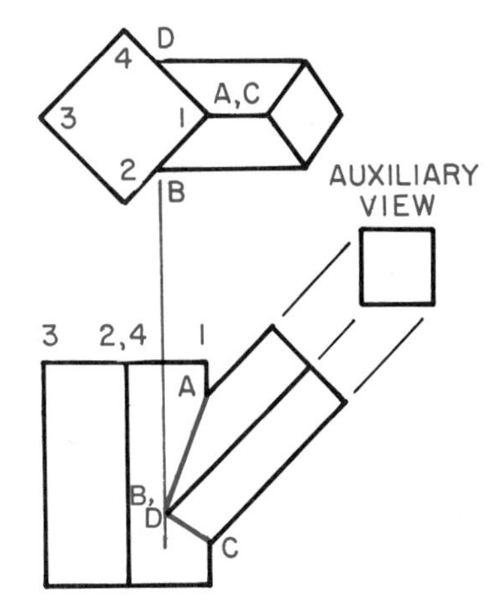

Fig. 17–27. Intersection of two prisms oblique to each other.

Using Cutting Planes with Intersecting Cylinders

Cutting planes can be used to help locate points on a line of intersection. These are imaginary planes like those used in making section drawings. They can be to help find intersections on prisms and cylindrical objects. They are especially useful on cylindrical objects.

Intersection of Two Cylinders at Right Angles. The line of intersection of two cylinders is found by using a series of cutting planes. See Fig. 17-28. To use cutting planes follow these steps:

1. Draw the necessary orthographic views. The circular and side views of both cylinders are needed.

2. Pass several cutting planes through both cylinders. In Fig. 17-28 these are shown on the top view, and they are marked A, B, C, and D.

3. Locate these cutting planes on the circular view of both cylinders. To do this on the front view, a half view of the cylinder is drawn. Since the cylinder is symmetrical, this is all that is needed.

CAD Applications:

Developments

A development is a pattern of a shape that is laid out in a flat plane so that the material can be folded into the desired shape. Sheet metal work is a good example of developments. Let's take a look at the development of a simple cube as drawn on a CAD system. The drawing below shows an isometric view of the cube.

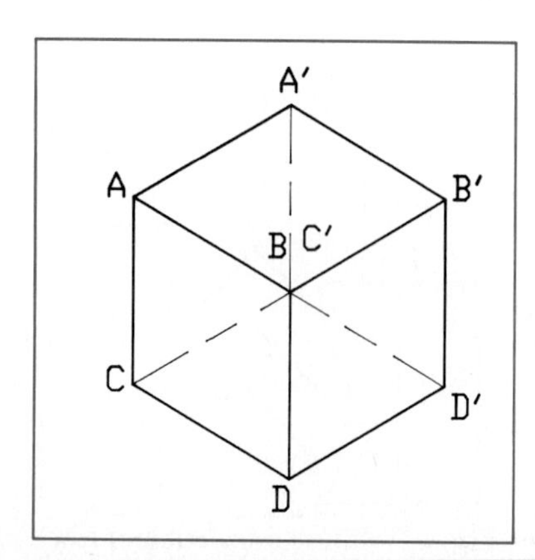

Because a cube has six equal sides, one method of drawing the development would be to use the copy command. This command allows you to duplicate anything already on the screen. First, draw the front view of the cube as shown in the following illustration.

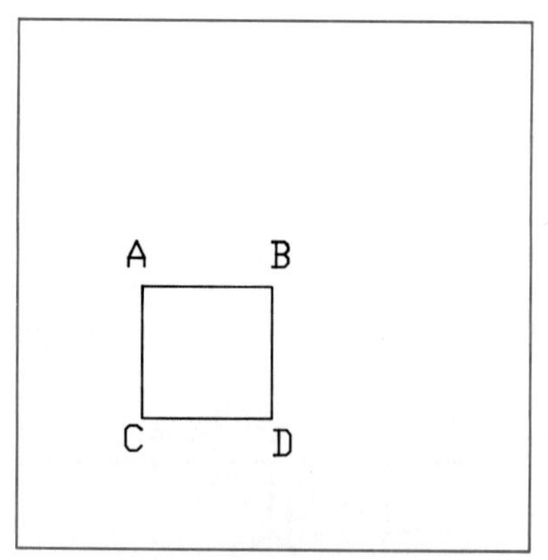

Next, use the copy command. Select three sides of the square to copy and position them in the appropriate places. (If you copy the entire square, you will get a double line where two squares connect.) The final drawing will look like the following.

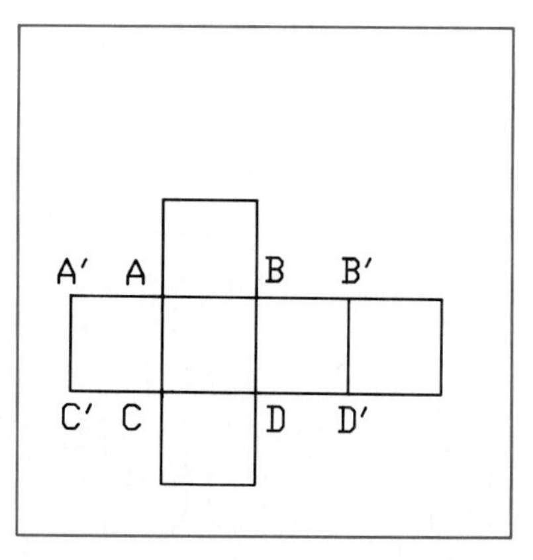

4. Project the points of intersection between the cylinders and the cutting planes to the view on which the intersection will be drawn. In Fig. 17-28 they are shown on the front view. Where these cross are points on the line of intersection. Connect these points in a smooth curve.

Development of Intersecting Cylinders. The points of intersection used to locate the line of intersection are also used when drawing a development. This is shown in Fig. 17-28. Use the following steps:

1. Draw the development of the large cylinder as described in Fig. 17-10.
2. Draw the center line and lay out the lines representing the cutting planes. These are numbered 1 through 7.
3. Project the points of intersection from the front view (located by the cutting planes) to the cutting-plane lines on the development. Connect these points on the development with a smooth curve. This is the shape of the opening into which the small cylinder fits.

Use the procedures shown in Fig. 17-11 to develop the small cylinder.

Intersection of Two Cylinders Oblique to Each Other. The location of the line of intersection of two cylinders oblique to each other is shown in Fig. 17-29. Notice that an auxiliary view of the small cylinder is drawn. The cutting-plane lines from the top view are located on the auxiliary. The projection of points of intersection is the same as explained for Fig. 17-28.

The development of these cylinders is the same as for those explained for Fig. 17-28.

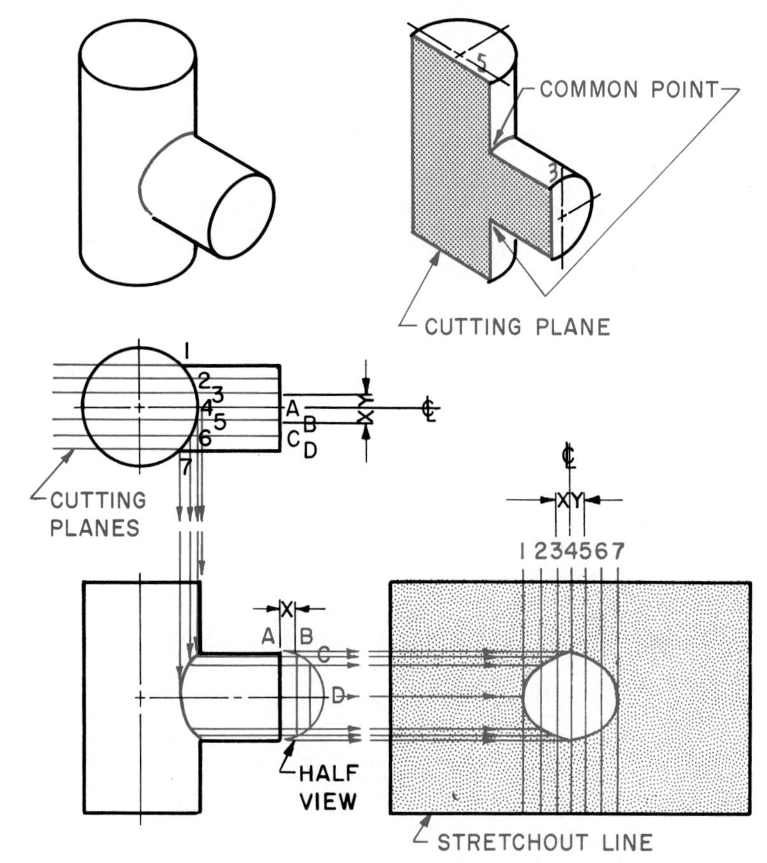

Fig. 17–28. The line of intersection of two cylinders can be found using cutting planes. This figure also shows the development of the intersection.

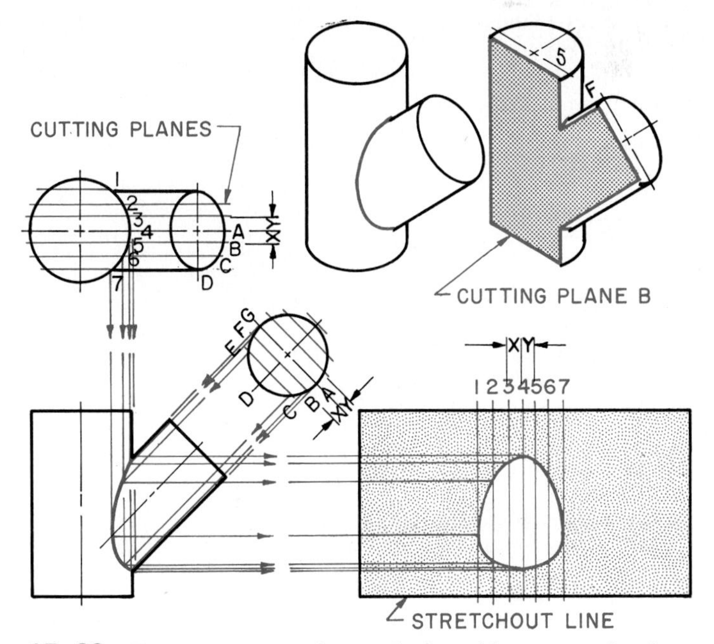

Fig. 17–29. The intersection of two cylinders oblique to each other can be found using cutting planes. This figure also shows the development of the intersection.

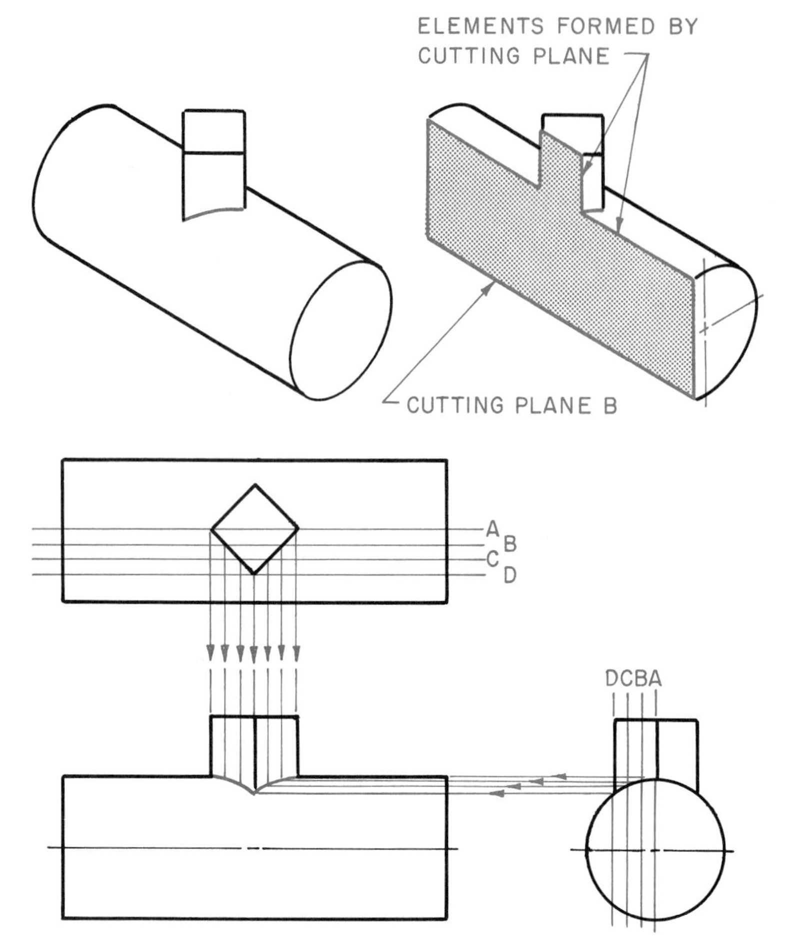

Fig. 17–30. **The intersection of a prism and a cylinder can be found by using cutting planes.**

Intersection of a Prism and Cylinder. The cutting-plane method is used to find the intersection of a prism and cylinder. See Fig. 17-30. Cutting planes A, B, C, and D are used to locate the points of intersection. The steps are the same as explained for Fig. 17-28. Notice that the line of intersection is two curves.

Using Cutting Planes with Cones

There are five ways of cutting cones. These are shown in Fig. 17-31.

If the cutting plane is parallel with the axis of the cone, a curve called a *hyperbola* is formed. If the cutting plane is parallel with the elements of the cone, the curve is a parabola. If the cutting plane passes through the apex of the cone and cuts the base at two points, a triangle is formed. When the cutting plane is perpendicular to the axis, a circle is formed. If the cutting plane passes through the cone at an angle greater than the elements, an ellipse is formed.

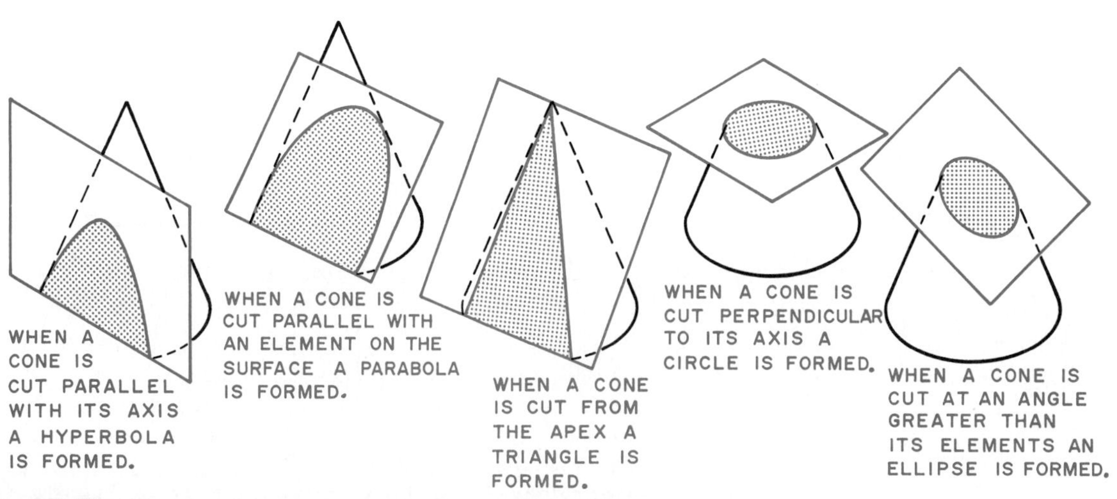

Fig. 17–31. **There are five ways to cut a cone.**

Intersection of a Cone and Cylinder Using Radial Lines. Fig. 17-32 shows how to find the intersection of a cone and cylinder. This method uses radial lines. The cutting planes pass through the apex to the base of the cone. Follow these steps:

1. Draw the orthographic views.
2. Locate the cutting planes on the view in which the cylinder appears as a circle. In Fig. 17-32 this is shown in the side view. The cutting planes are marked 1 through 7. The outside planes, 1 and 7, pass through the point where the center line touches the outside of the cylinder. This is the extreme width. Any number of planes can be used between these two.
3. Locate these cutting planes on the top view.
4. Next, locate the cutting planes on the front view by projecting them from the top view.
5. Now, from the side view project the points where the cutting planes cross the circle to the front view. Where they intersect the same cutting plane are points on the line of intersection.

For example, find cutting plane 6 in the side view. It cuts the circle in two places. These cuts are projected to the front view. Where they cross cutting plane 6 is a point on the line of intersection.

6. Project the points of intersection in the front view to the top view. The same method for locating points is used.
7. Connect the points with a smooth curve.

Intersection of a Cone and a Cylinder Using Cutting Planes Parallel to the Base. Fig. 17-33 shows how to find the intersection of a cone and a cylinder using cutting planes parallel to the base of the cone. As the cone is cut, a circle is formed. Follow these steps:

1. Draw the orthographic views. The top, front, and side views are needed.
2. Draw the cutting planes on the side view. Any number of planes can be used. The more planes used, the more accurate the line of intersection will be.
3. Project the planes to the front and top views. In the top view they will appear as circles.
4. On the side view, measure the distance from the center line to the point where each plane crosses the cylinder. For example, on plane E the points of intersection are marked 1 and 2. They are distance Y from the center line.
5. Project the distance found in the side view to the top view. Where these distances cross the plane locates two points on the line of intersection. To continue with the example, find plane E in the top view, and measure distance Y from the center line. Project this distance until it crosses plane E. This locates points 1 and 2 on the line of intersection. Do this for each plane.

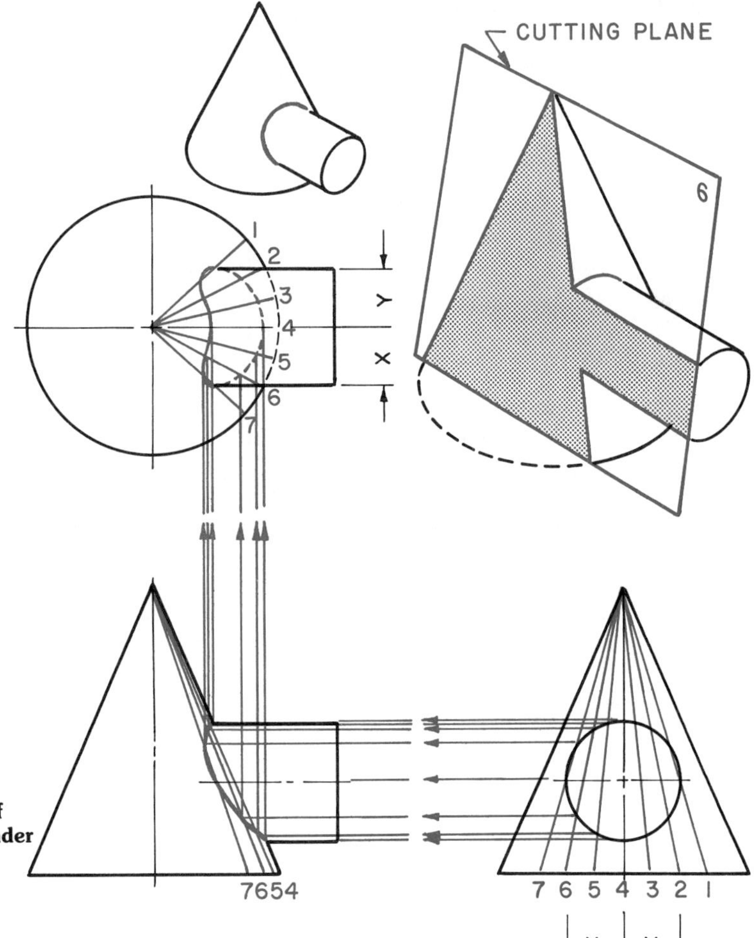

Fig. 17–32. Finding the line of intersection of a cone and a cylinder using radial lines.

6. Project the points located on the top view to the front view. Where they cross their cutting-plane lines are points on the line of intersection.

7. Draw a smooth curve between these points.

Bend Allowance

When sheet metal is bent to form a corner, extra metal is needed. If an accurate pattern is required, the extra metal needed for the corner must be calculated. This is called **bend allowance**. Bend allowance is usually not necessary on very thin metals. On these, the outer surface will stretch while the inner surface will compress enough to form the bend. Thicker metals require a bend allowance.

Fig. 17-34 shows a metal corner. The shaded area shows the bend allowance. Detailed instructions for finding bend allowance are found in Chapter 13, "Drawing For Production: Detail and Assembly Drawings."

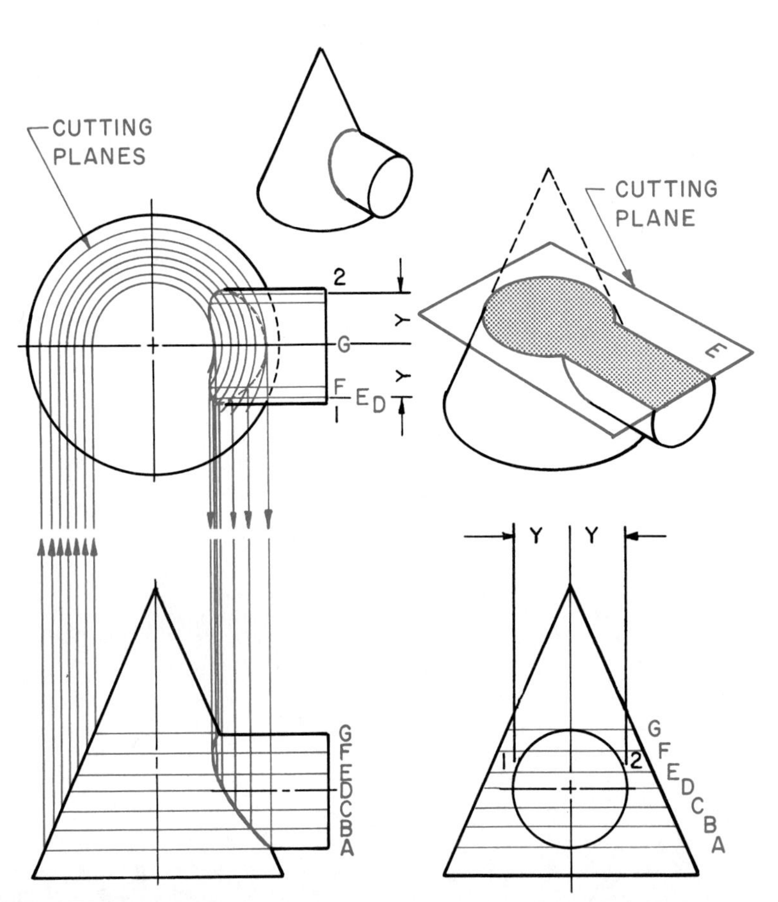

Fig. 17–33. **Intersection of a cylinder and a cone using cutting planes parallel to the base of the cone.**

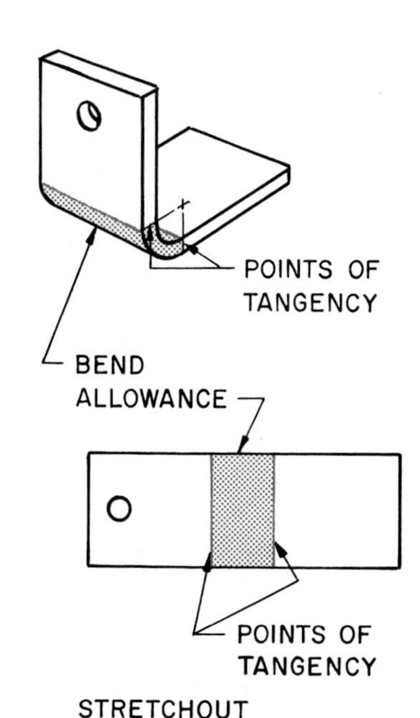

Fig. 17–34. **Bends in thick sheet metal require extra metal when being formed. This is called bend allowance.**

Chapter Review

Build Your Vocabulary

You should understand and use the following terms as part of your working vocabulary. Write a brief explanation of what each means.

plane surface
single-curved surface
double-curved surface
warped surface
surface development
parallel-line development
radial-line development
triangulation
intersection
piercing point
bend allowance

Study Problems—Directions

The problems that follow will give you the chance to use some of the drafting knowledge that you learned in this chapter. Do the problems that are assigned to you by your instructor.

1. Draw developments of problems A through L in **Fig. P17-1**. Draw full size on B-size sheets (11 × 17 or 12 × 18). Draw the top and front views as given and the development for each. Dimension the drawing if directed to do so by your instructor. You may cut out and fold the patterns to produce a model of each object.

2. Draw the given views and the developments of problems A through L in **Fig. P17-2**. Draw full size. Each problem can be placed on a B-size sheet.

3. Draw two views and the development of problems A through D in **Fig. P17-3**. Each problem can be solved on a B-size sheet.

4. Draw two views and the line of intersection for problems A through H in **Fig. P17-4**. Develop the pattern for the large cylindrical or square form showing the line of intersection. These problems can be solved on B-size sheets.

5. Calculate the bend allowance for the parts in **Fig. P17-5**. Draw a pattern showing the allowance. Refer to Chapter 13 for bend allowance tables and formulas.

6. Design a cardboard box to hold the bottle shown in **Fig. P17-6**. Make a full-size pattern. Remember to include tabs to hold the box together. Cut out the pattern and fold it into a finished box.

7. Develop the pattern for a rural mailbox. See **Fig. P17-7**. Design the back so that it can be spot-welded in place. Design the front so that it can be opened with a hinge.

8. Draw patterns for the furnace and piping, **Fig. P17-8**. The heel radius is 1″–1⅛″, and the throat radius is 1⅛″. Cut out the patterns and assemble the installation. Draw to the scale 1″ = 6″.

9. Develop the pattern for the flowerpot, **Fig. P17-9**. Put tabs on the sides. They fold in. Draw the bottom. The bottom is fastened to the tabs.

10. **Fig. P17-10** shows a sheet metal flashing unit. It is used to keep rainwater from leaking into a house where the sewer vent pipe comes through the roof. Make patterns for both pieces.

11. Develop patterns for the three parts of the ventilator head, **Fig. P17-11**. It is used to keep rain from entering the ventilator pipe.

12. **Fig. P17-12** shows an overhead projector. Develop the pattern for the base and brace. Draw the head and lens cylinder together to show the line of intersection. Develop the pattern for the head. Show the opening for the lens cylinder.

13. Develop the pattern for the vacuum cleaner nozzle, **Fig. P17-13**.

14. Develop the pattern for the metal base of the lectern shown in **Fig. P17-14**.

15. Draw the top and front views of the pedestal base, **Fig. P17-15**. Then draw the line of intersection between the column and the base. Next, draw the pattern for the leg.

16. **Fig. P17-16** shows part of a post for scaffolding used by painters. It has two brackets welded to it for horizontal supports. Draw the top and front views of the unit. Find the line of intersection. Then develop the pattern for the bracket.

Study Problems

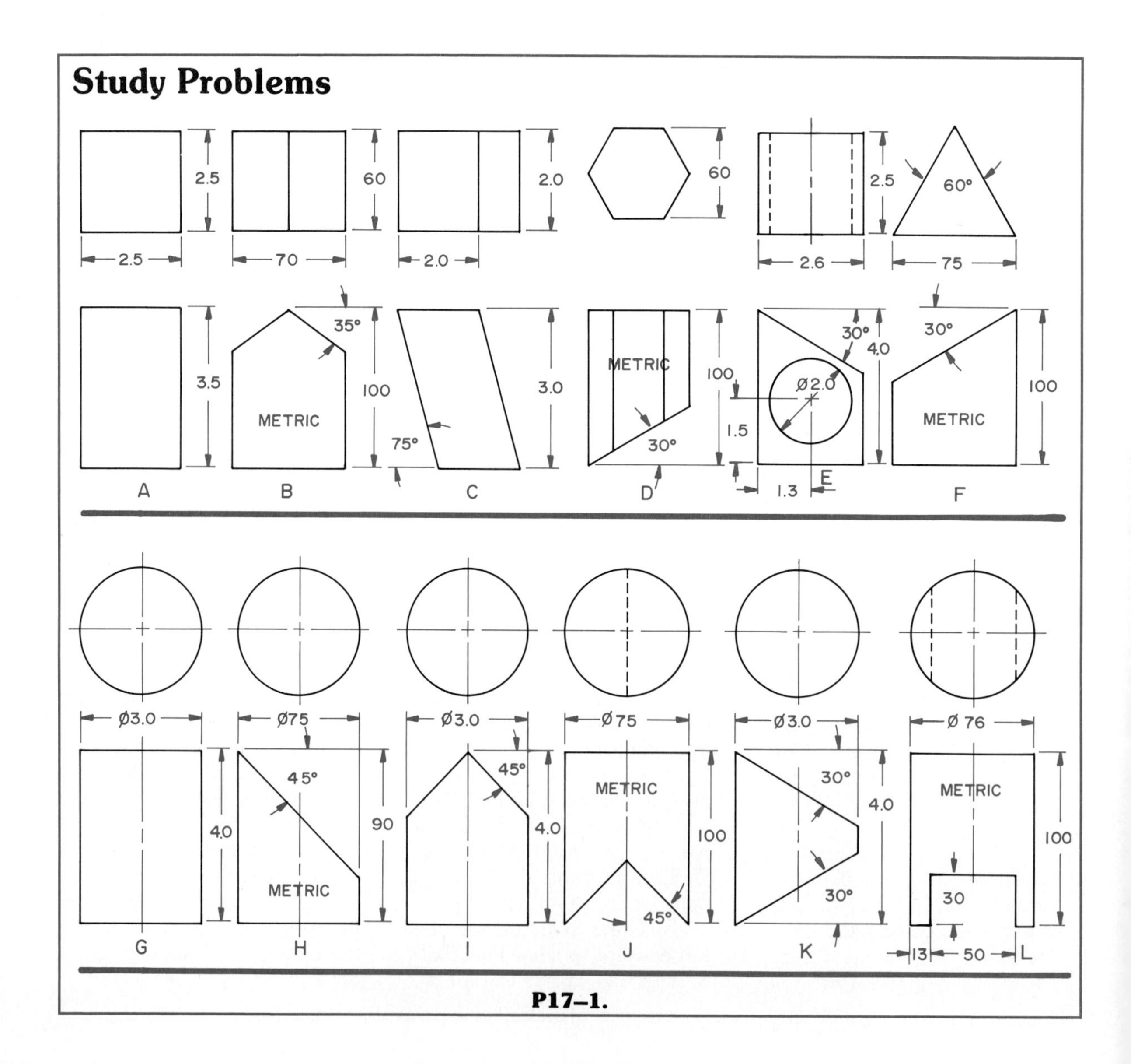

P17–1.

Study Problems

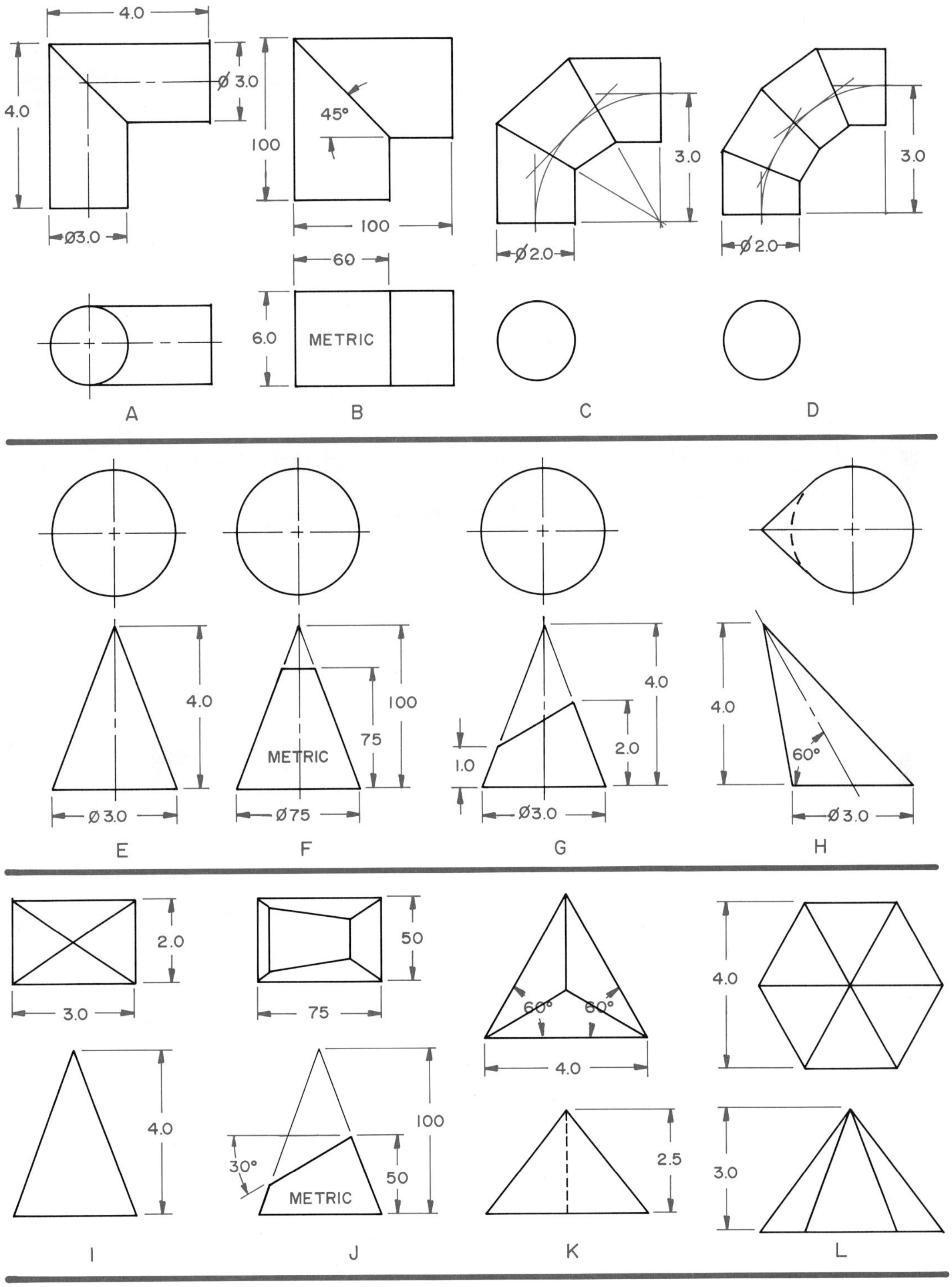

P17–2.

Study Problems

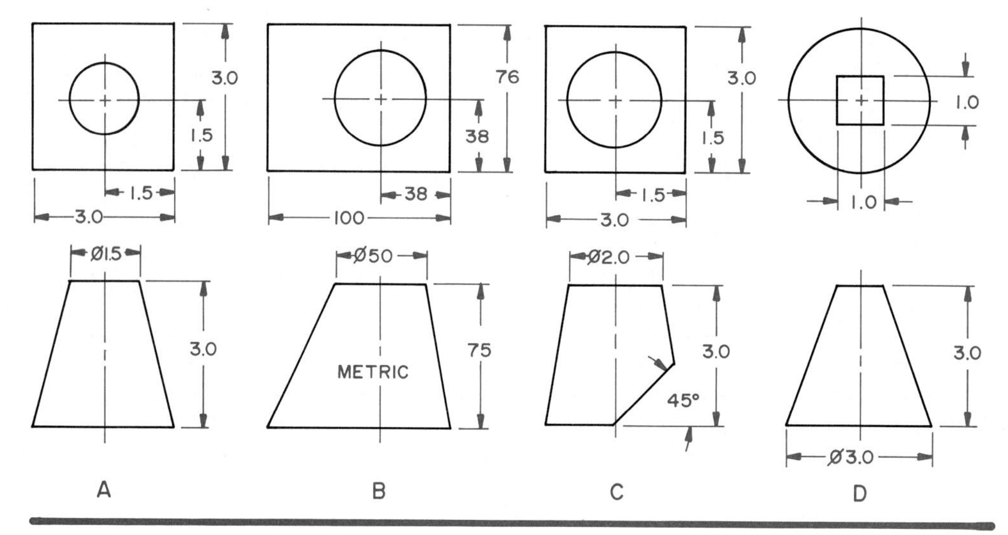

P17–3.

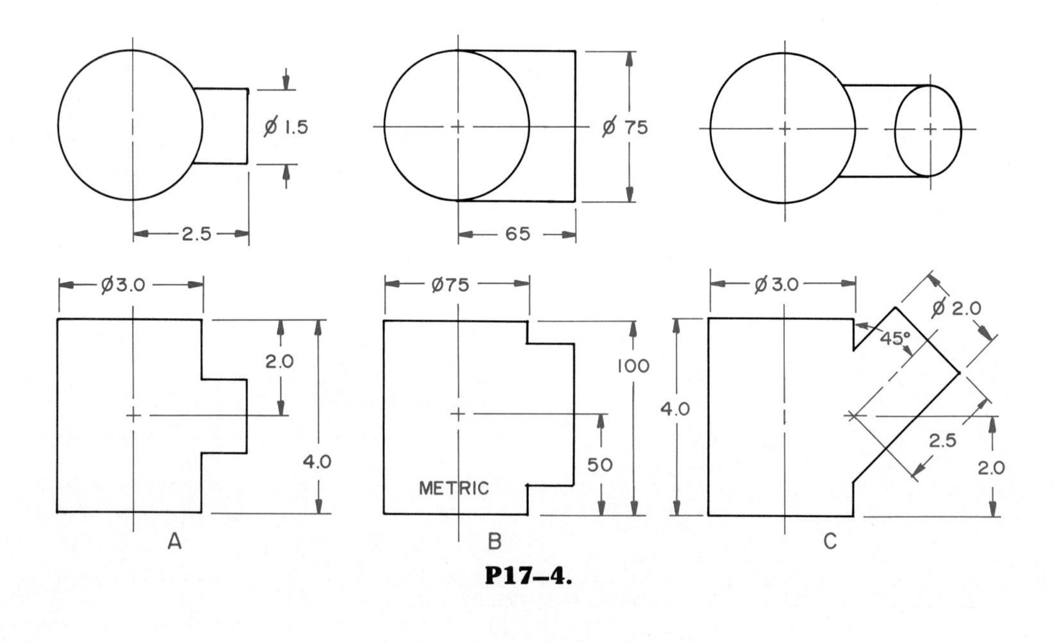

P17–4.

(Continued on next page)

Study Problems

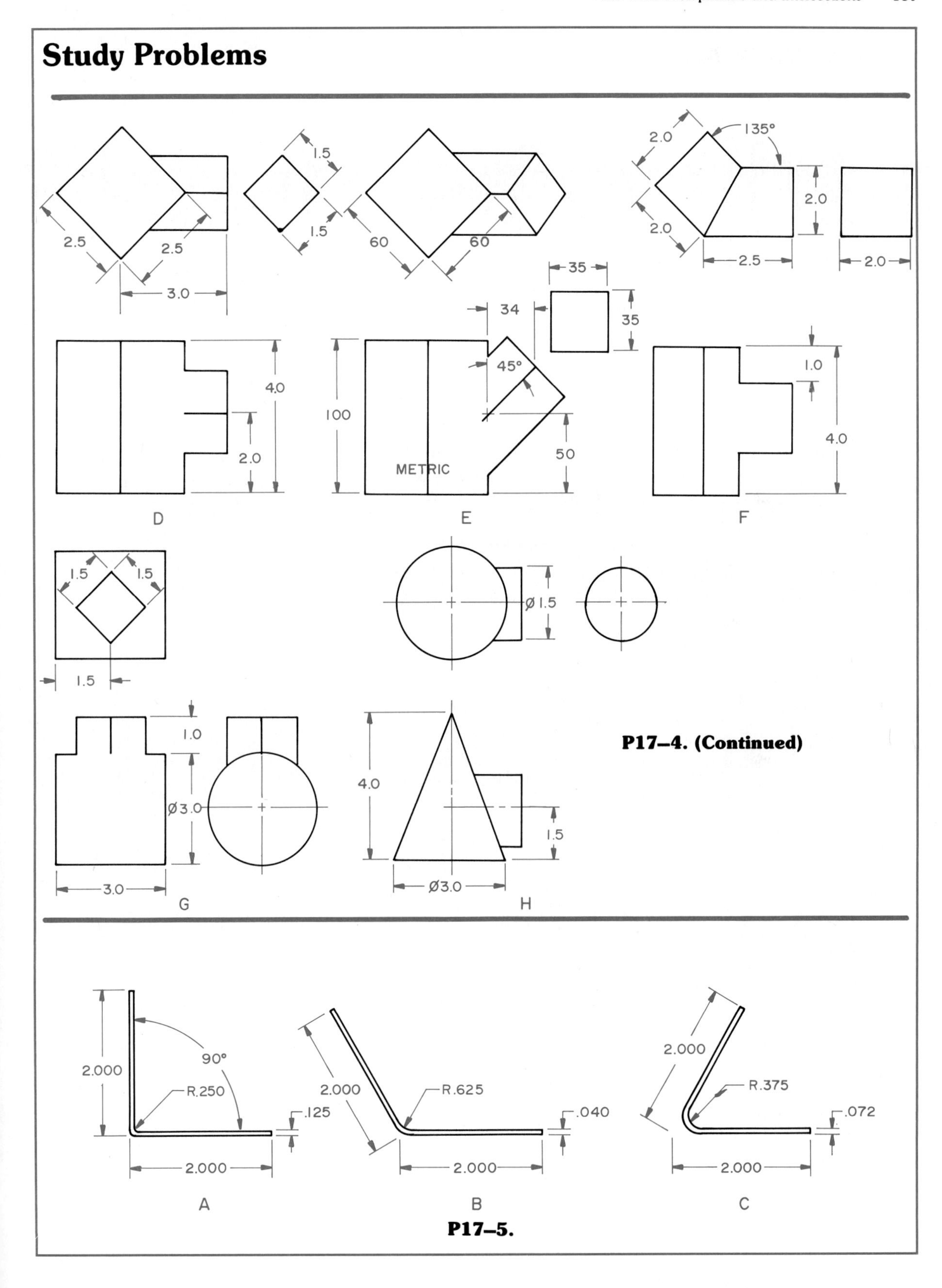

P17–4. (Continued)

P17–5.

Study Problems

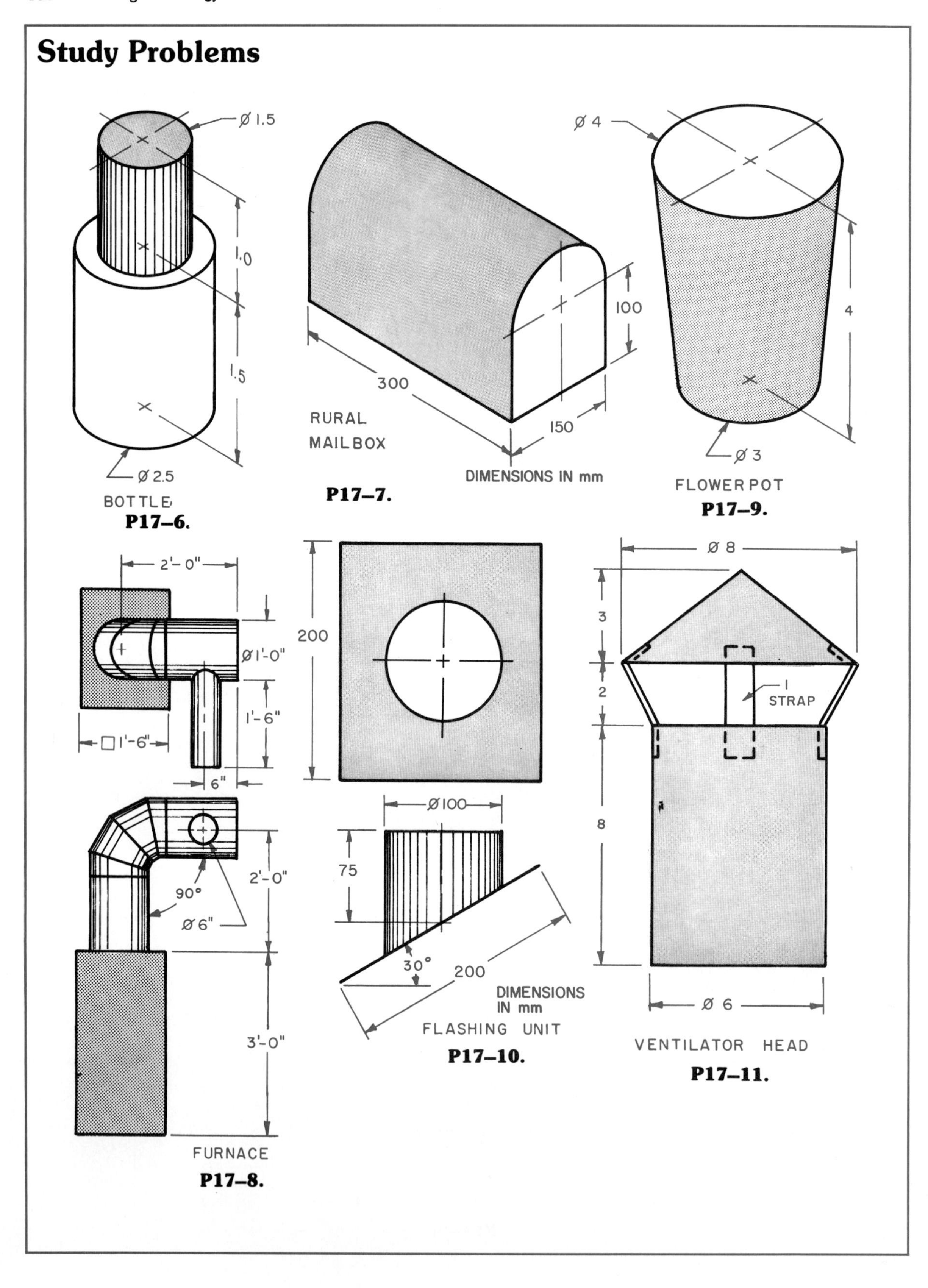

P17–6.

P17–7.

P17–9.

P17–8.

P17–10.

P17–11.

Study Problems

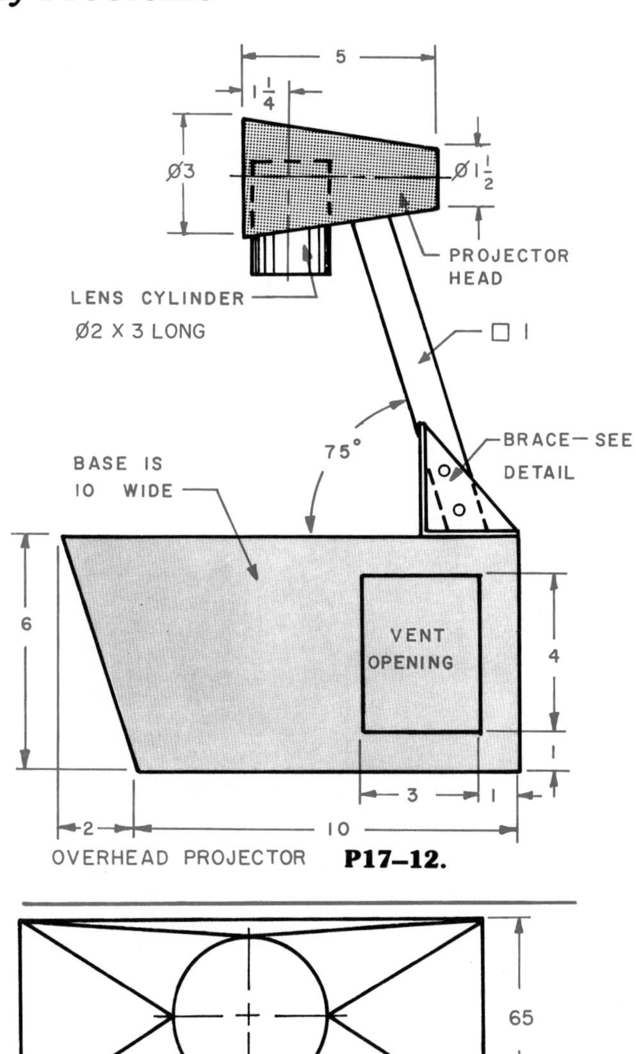

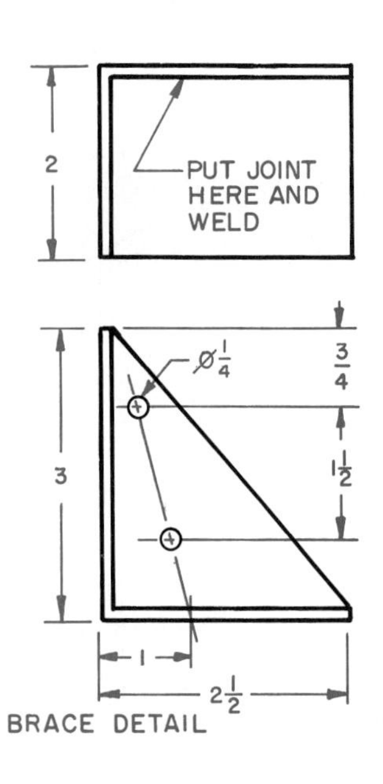

P17–12.

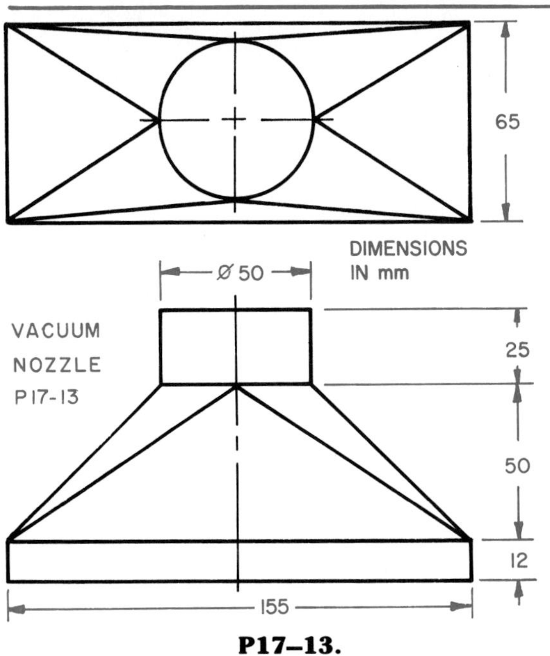

P17–13.

Study Problems

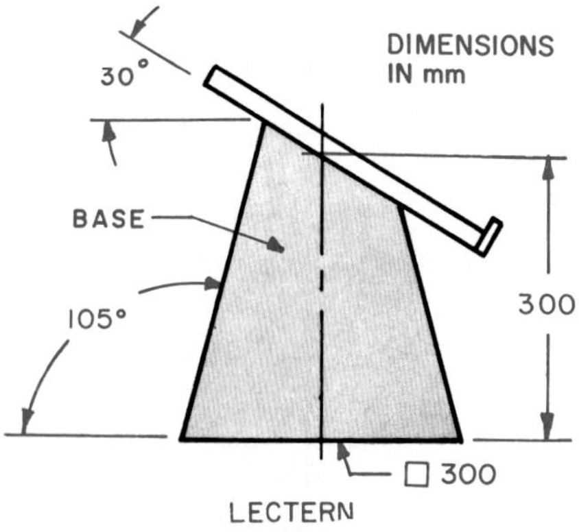

30°
DIMENSIONS
IN mm
BASE
105°
300
□ 300
LECTERN

P17–14.

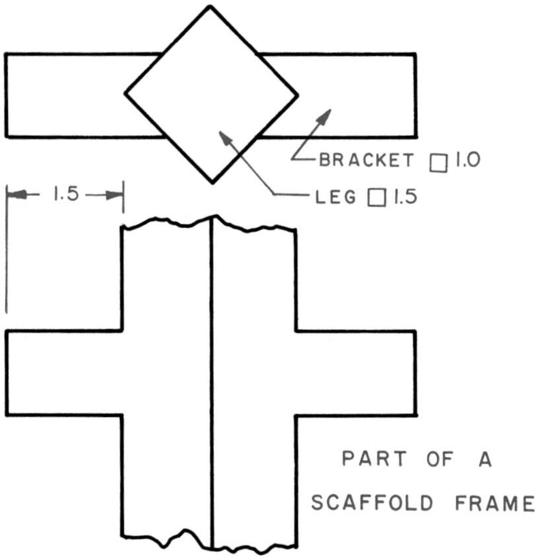

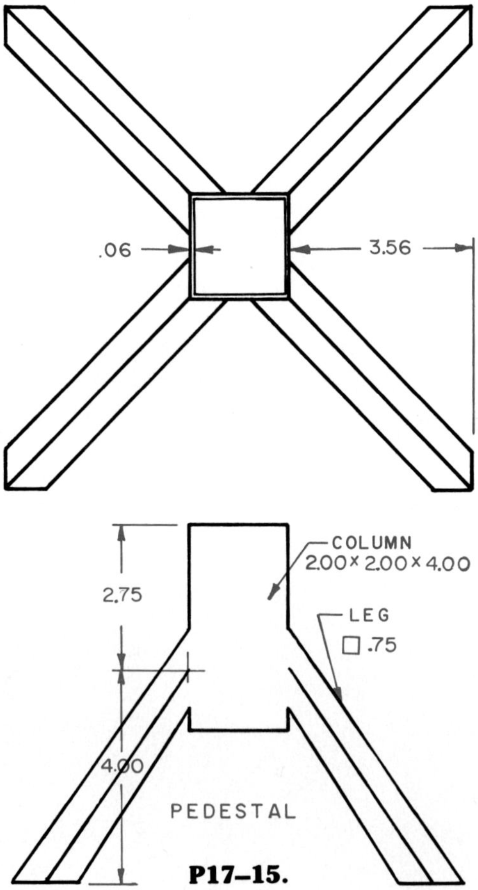

.06
3.56
2.75
COLUMN
2.00 × 2.00 × 4.00
LEG
□ .75
4.00
PEDESTAL

P17–15.

BRACKET □ 1.0
LEG □ 1.5
1.5
PART OF A
SCAFFOLD FRAME

P17–16.

CHAPTER 18
Gears and Cams

After studying this chapter, you should be able to do the following:
- Draw gears of various shapes.
- Understand cam motion.
- Draw cams.

Gears

Gears are wheels with teeth. Two or more gears are used together to transfer rotary motion from one moving shaft to another. They also can be used to change the direction of rotation. By using gears of different sizes, the speed of rotation can be made slower or faster.

Fig. 18-1 shows two gears of the same size that have the same number of teeth. When one gear has power applied to its shaft so that it is rotated, it is called the **drive gear**. The second gear turns because its teeth mesh with the first, and it is called the **driven gear**. When the drive gear makes one revolution, the driven gear does also. Notice that the driven gear rotates in the opposite direction from the drive gear.

Fig. 18-2 shows two gears of different sizes. The small gear has half the number of teeth found on the larger gear. When the smaller drive gear, called a **pinion**, makes one revolution, the larger gear, the **spur gear**, turns one-half of a revolution. This slows down the rotation of the larger gear. This gear combination is a 2:1 gear ratio.

Study the series of gears shown in Fig. 18-3. In which direction does gear D rotate?

The design of gears is a difficult engineering problem. The engineer must decide on the best shape for the gear teeth for the application. He or she must con-

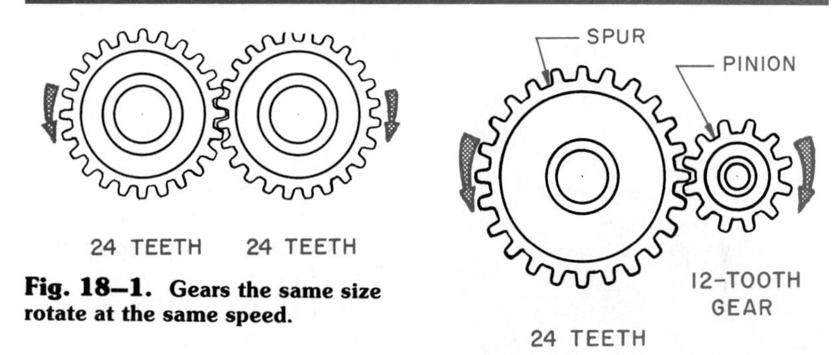

Fig. 18–1. **Gears the same size rotate at the same speed.**

Fig. 18–2. **The speed of rotation is governed by the size of the gear.**

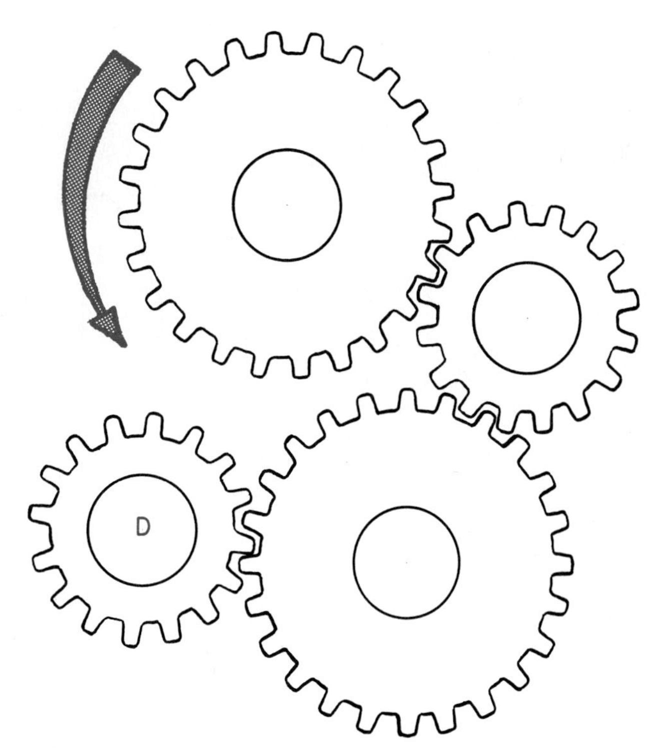

Fig. 18–3. **Can you tell in which direction gear D rotates?**

sider the strength required to operate the mechanism. The problem of tooth wear must be considered. Since gears are complex to manufacture, the designer must consider how they are to be made and the degree of accuracy needed. This chapter will show the details necessary to draw several commonly used gears.

Fig. 18–4. Spur gears. (Boston Gear Division, North American Rockwell)

Gear Nomenclature

Spur gears have teeth that are perpendicular to the face of the gear. Fig. 18-4. The helical gear is much like the spur gear except that the teeth are on an angle to the face of the gear.

The following terms are used to describe gear teeth (Figs. 18-5, 18-6, and 18-7):

Addendum. The distance from the pitch circle to the top of the teeth.

Base diameter. The diameter of the circle from which the involute tooth profiles are derived.

Chordal addendum. The distance from the top of the tooth to the chordal thickness.

Chordal thickness. The thickness of a tooth measured along a chord of the pitch circle.

Circular pitch. The distance along the pitch circle from a point on one tooth to the same point on the next tooth.

Dedendum. The distance from the pitch circle to the bottom of the teeth.

Diametral pitch. The ratio of the number of teeth to the pitch diameter in inches. For example, a four-pitch gear has four teeth per inch of pitch diameter.

Face width. The width of the tooth measured perpendicularly from one face to the other.

Form diameter. The diameter at which the involute tooth profile originates.

Helix angle. The angle between the pitch helix and the axis of the gear.

Pitch circle. An imaginary circle running through the point on the teeth at which the teeth from mating gears are tangent. The size of the gear is indicated by the pitch circle. For example, a six-inch gear has a pitch circle of six inches. Most of the gear dimensions are taken from the pitch circle.

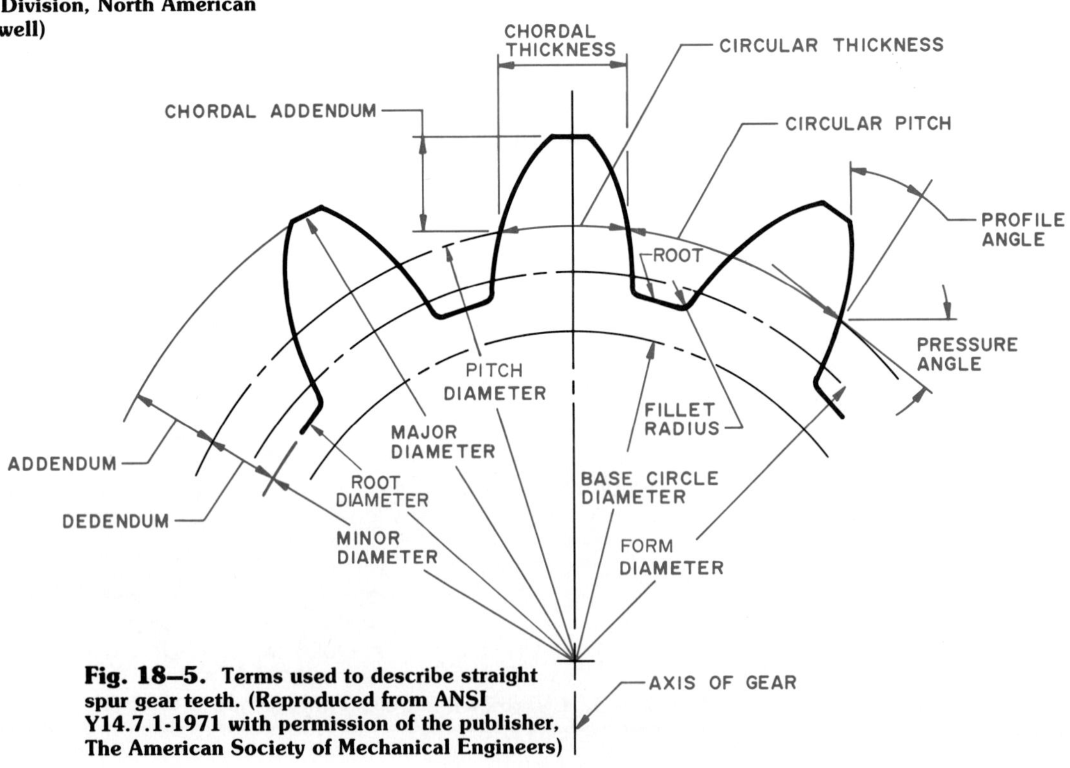

Fig. 18–5. Terms used to describe straight spur gear teeth. (Reproduced from ANSI Y14.7.1-1971 with permission of the publisher, The American Society of Mechanical Engineers)

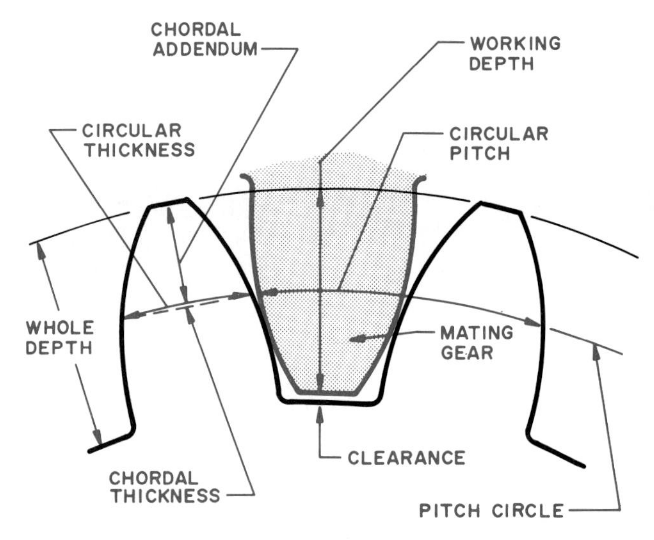

Fig. 18–6. Gear tooth terminology related to the mating teeth of spur gears. (Reproduced from ANSI Y14.7.1-1971 with permission of the publisher, The American Society of Mechanical Engineers)

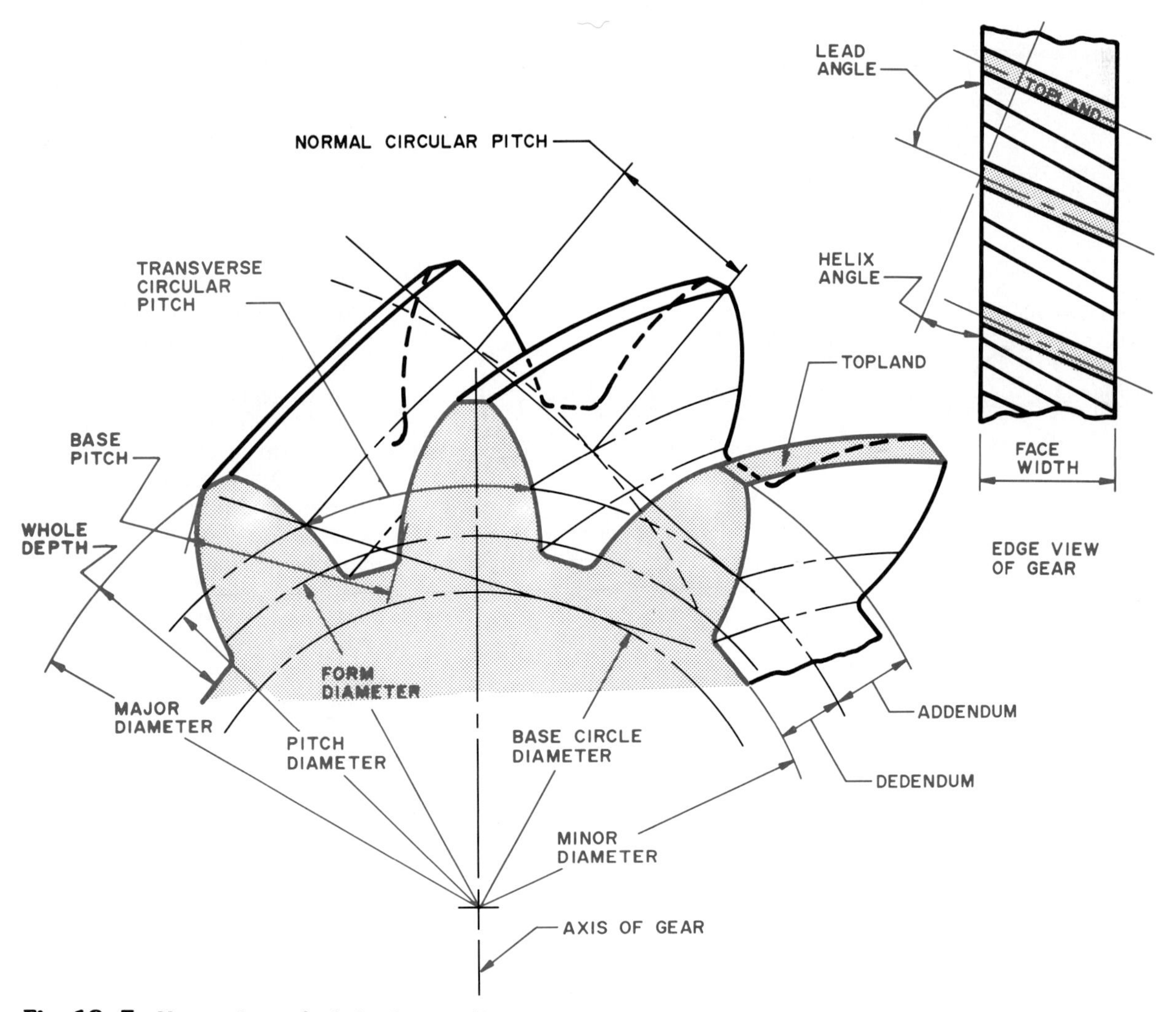

Fig. 18–7. Nomenclature for helical gears. (Reproduced from ANSI Y14.7.1-1971 with permission of the publisher, The American Society of Mechanical Engineers)

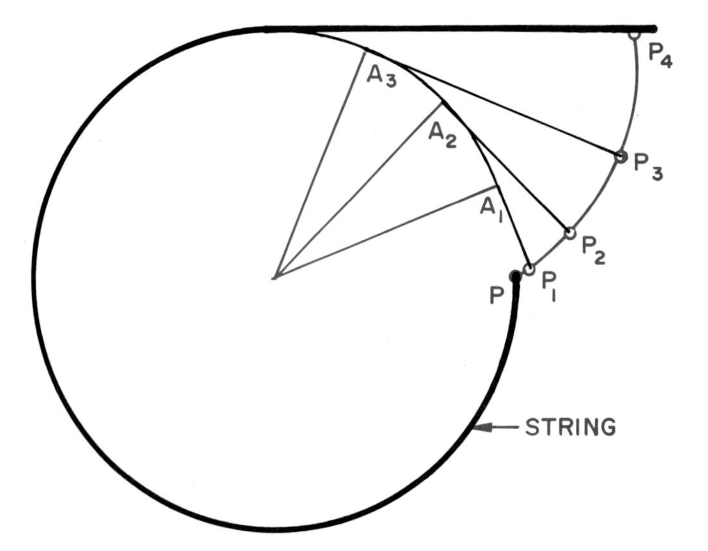

Fig. 18–8. **The involute is formed by point P on the string as the string is unwound. The string is tangent to the cylinder at A1, A2, and A3. These are instantaneous radii of the curvature of the involute.**

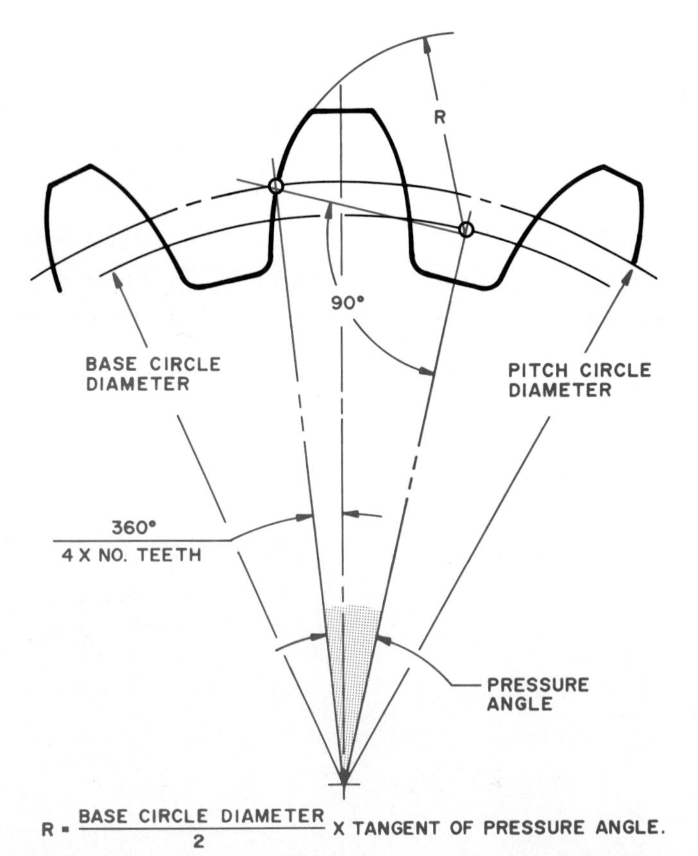

Fig. 18–9. **Use this procedure to draw an approximate involute when detailing gear teeth.**

Pitch diameter. The diameter of the pitch circle.

Pressure angle. The angle used to determine the tooth shape. It is the angle at which pressure from the tooth of one gear is passed to the tooth of another gear. Two angles are in general use. These are 14½ and 20 degrees.

Lead. The axial advance of a helix during one complete turn.

Major diameter. On external gears this is the outside diameter. On internal gears it is the root diameter.

Minor diameter. On external gears it is the root diameter. On internal gears it is the inside diameter.

Root. The area between the teeth at their bottom.

Root circle. A circle passing through the bottom of the teeth.

Root diameter. The diameter formed by the root circle.

Gear Tooth Form

The profile of spur and helical gear teeth directly influences their performance. When properly designed a gear should transmit motion smoothly with a minimum of vibration. The most common tooth form in use is the involute profile. An **involute** is a curve formed by a point on a line as it is unwound (as if it were on a string) from the circumference of the form circle. Fig. 18-8.

The form of the tooth depends upon the *pressure angle.* This is generally 14½ or 20 degrees. The pressure angle determines the size of the form circle. The involute is generated from the form circle. One way to draw an approximate involute when detailing gear teeth is shown in Fig. 18-9.

General Practices for Making Gear Drawings

Materials. Drawings should specify the material of which a gear is to be made.

Heat treatment. When required, heat treatment should be specified and the hardness range shown. Reference can be made to the process standard or a specification.

Marking. Include all required markings. These can include part number, serial number, inspection symbol, or any other identifying number.

Manufacturing process. Whenever possible, this is left up to the company making the gear.

Quality of finish. The method of finishing can be specified, such as ground, lapped, or shaved.

Type of drawing. The type of drawing used depends upon the engineering of the gear and its use. Examples are shown on the following pages.

When drawing external spur and helical gears, use the view parallel with the axis. Do not use the front view unless it is needed to show details or a relationship of the gear teeth to other features. When you draw a front view of external gears, draw the major diameter, pitch diameter, and minor diameter with a line made of one long and two short dashes. Fig. 18-10. Internal gears are drawn as shown in Fig. 18-11.

Use a section when it is necessary to show internal features. Fig. 18-12.

If it is necessary to draw gear teeth, draw them as shown in Fig. 18-13. Usually only two teeth need to be drawn to illustrate their design. They may also be combined with the front view if it has been drawn.

Gears having more than one set of teeth are drawn as shown in Fig. 18-14. If there is no relationship between the sets of gear teeth, indicate this with a note. If there is a relationship, state the angle in decimal degrees.

An accurate way to gage tooth thickness is to measure with micrometers over or between cylindrical gage pins. The pins are placed in diametrically opposite tooth spaces. Fig. 18-15. If this is to be done, the pin measurement figures are placed on the drawing.

Spur and Helical Gear Drawings

A design drawing of an external spur gear is shown in Fig. 18-16. The design data shown are for a format A gear. Gear drawings are classified in four formats, A, B, C, and D. These relate to the degree of specification. Format A is the lowest quality level and the coarsest pitch quality level. A format A design drawing for an internal spur gear is shown in Fig. 18-17.

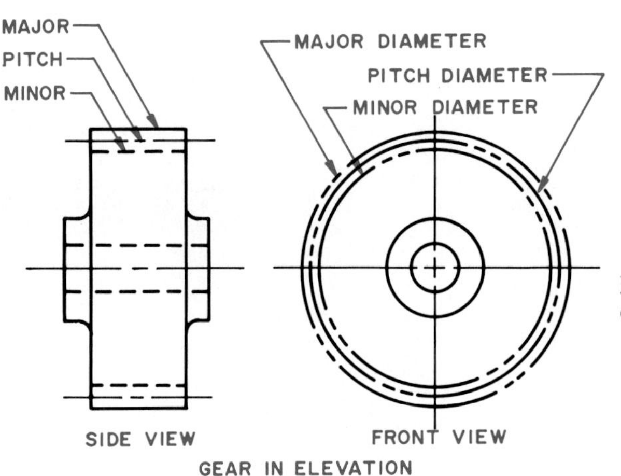

Fig. 18–10. How to draw an external gear in elevation.

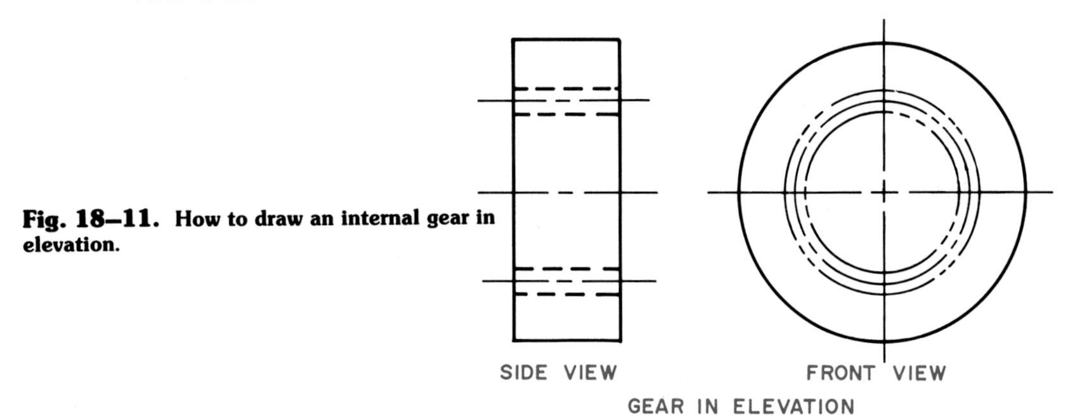

Fig. 18–11. How to draw an internal gear in elevation.

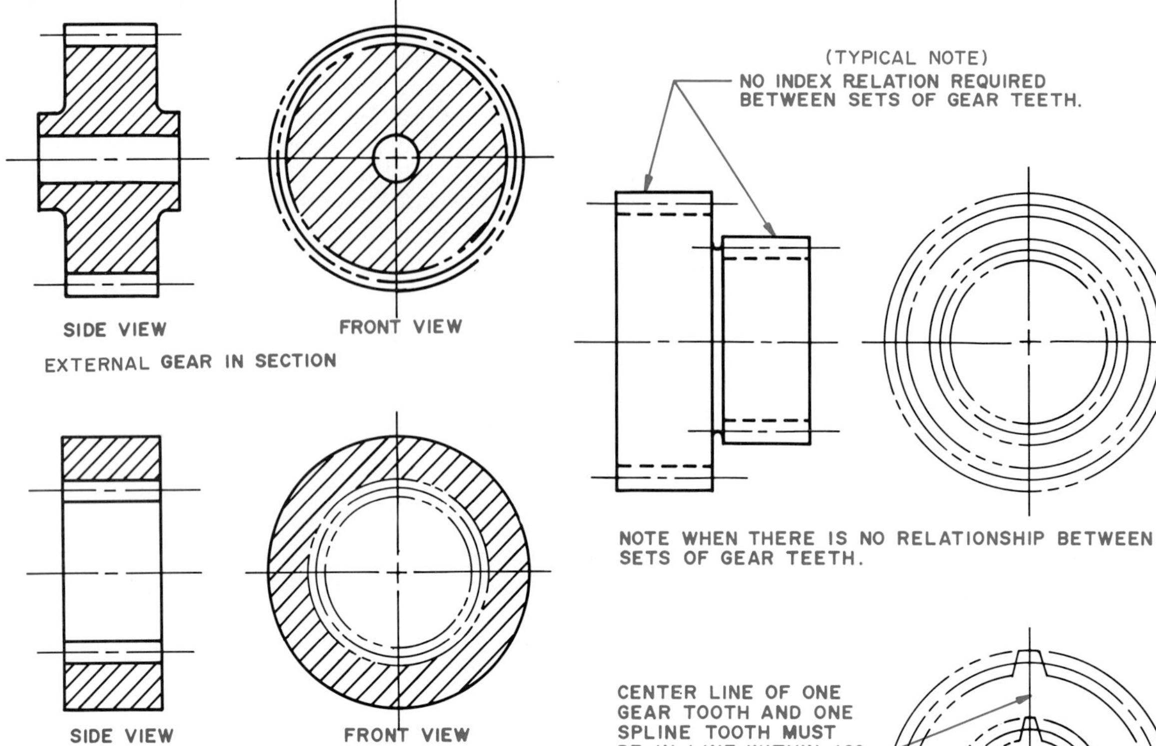

Fig. 18–12. How to draw external and internal gears in section.

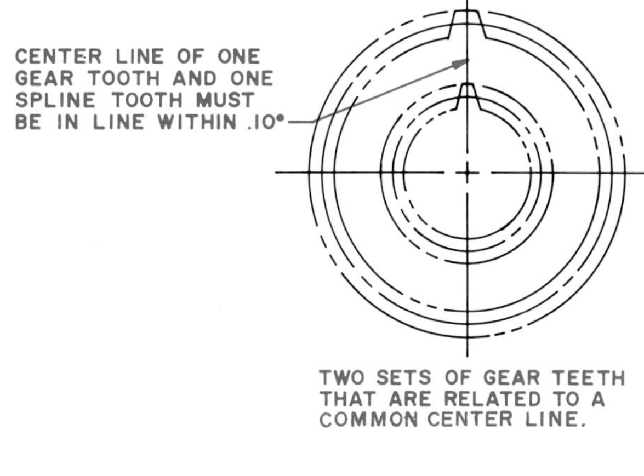

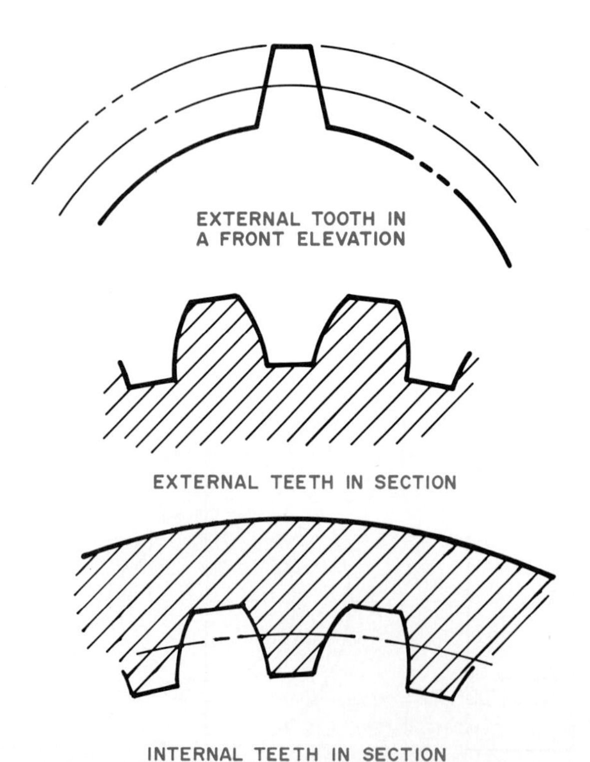

Fig. 18–13. How to represent individual gear teeth.

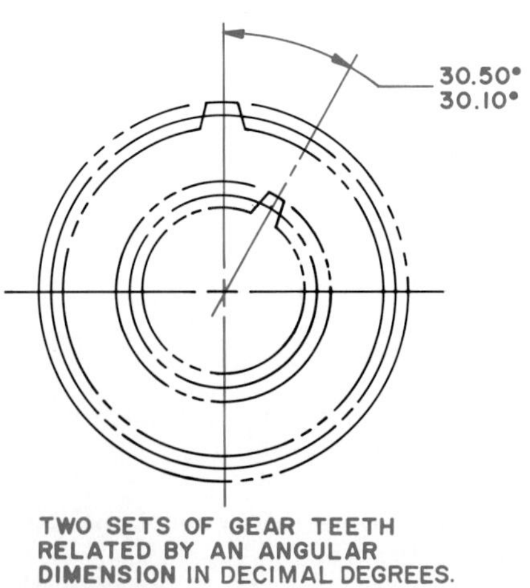

Fig. 18–14. How to indicate the relationship between teeth on gears that have more than one set of teeth. (Reproduced from ANSI Y14.7.1-1971 with permission of the publisher, The American Society of Mechanical Engineers)

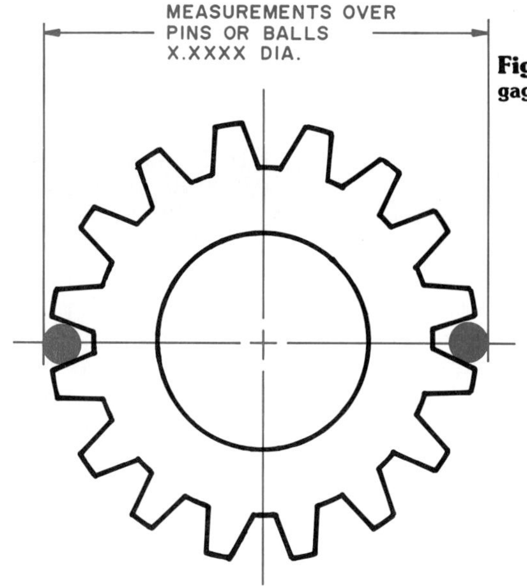

Fig. 18–15. **The use of pins or balls to accurately gage tooth thickness.**

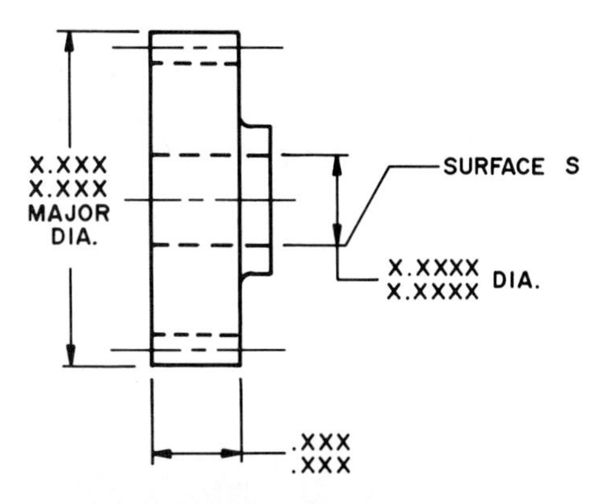

EXTERIOR INVOLUTE SPUR GEAR DATA (1)

__(2)__ ROOT
DIAMETRAL PITCH XX (3)
OR
MODULE X (4)
PRESSURE ANGLE XX.XX°
PITCH DIA. X.XXXXXXXX REF.
BASE DIA. X.XXXXXXXX REF.
FORM DIA. X.XXX
MINOR DIA. X.XXX – X.XXX
ACTL CIRC TOOTH THK AT PITCH DIA. .XXXX–.XXXX
MEAS OVER TWO .XXXXX DIA. PINS X.XXXX–
X.XXXX REF.
MAX RUNOUT .XXXX FIR.
ALL TOOTH ELEMENT SPECIFICATIONS ARE FROM DATUM ESTABLISHED BY THE AXIS OF SURFACE S.

1. DATA SHOWN ARE FOR INCH GEAR DRAWINGS. METRIC GEAR DRAWINGS HAVE ONE LESS DECIMAL PLACE TO THE RIGHT OF THE DECIMAL POINT.
2. SPECIFY WHETHER FULL FILLET, OPTIONAL OR FLAT.
3. USED ON INCH DRAWING ONLY.
4. USED ON METRIC DRAWINGS INSTEAD OF DIAMETRAL PITCH.

Fig. 18–16. **A design drawing for an external spur gear. (Reproduced from ANSI Y14.7.1-1971 with permission of the publisher, The American Society of Mechanical Engineers)**

X.XXX
X.XXX
MINOR
DIA.
SURFACE S
X.XXXX
X.XXXX DIA
.XXX
.XXX

INTERNAL INVOLUTE SPUR GEAR DATA (1)

__(2)__ ROOT
NUMBER OF TEETH XX
DIAMETRAL PITCH XX (3)
OR
MODULE X (4)
PRESSURE ANGLE XX.XX°
PITCH DIA. X.XXXXXXXX REF.
BASE DIA. X.XXXXXXXX REF.
FORM DIA. X.XXX
MAJOR DIA. X.XXX – X.XXX
ACTL CIRC TOOTH THK AT PITCH DIA. .XXXX–.XXXX
MEAS BETWEEN TWO .XXXXX DIA. PINS X.XXXX–
X.XXXX REF.
MAX RUNOUT .XXXX FIR.
ALL TOOTH ELEMENT SPECIFICATIONS ARE FROM DATUM ESTABLISHED BY THE AXIS OF SURFACE S.

1. DATA SHOWN ARE FOR INCH GEAR DRAWINGS. METRIC GEAR DRAWINGS HAVE ONE LESS DECIMAL PLACE TO THE RIGHT OF THE DECIMAL POINT.
2. SPECIFY WHETHER FULL FILLET, OPTIONAL OR FLAT.
3. USED ON INCH DRAWINGS ONLY.
4. USED ON METRIC DRAWINGS INSTEAD OF DIAMETRAL PITCH.

Fig. 18–17. **A design drawing of an internal spur gear. (Reproduced from ANSI Y14.7.1-1971 with permission of the publisher, The American Society of Mechanical Engineers)**

Design drawings for format A external and internal helical gear drawings are found in Figs. 18-18 and 18-19.

A drawing of an external involute double-helical gear is shown in Fig. 18-20.

Rack Drawings

A **rack** is a straight piece of material with teeth cut on one side. Spur and helical racks are used. The teeth are the same as those on spur and helical gears. Fig. 18-21. Racks are usually rectangular. The rack changes rotating motion to reciprocating motion. The terms used to identify the parts of rack teeth are shown in Fig. 18-22. These are the same as used for gears.

A spur rack drawing is shown in Fig. 18-23. The circular pitch of the circular pinion is a straight line called the pitch line. Since a rack is usually very long and all the teeth are identical, draw it with a broken-out section to shorten it. You need to draw only one or two teeth at each end for dimensioning purposes.

Bevel Gears

Bevel gears are gears with teeth cut on an angle to the gear face in order to transfer power between shafts whose axes intersect. Fig. 18-24. The axes may intersect at any angle. The most common intersection, however, is 90 degrees. When bevel gears are the same size and meet at a

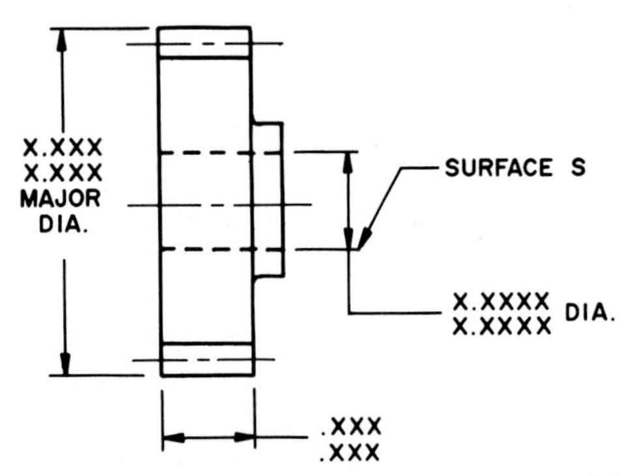

EXTERNAL INVOLUTE HELICAL GEAR DATA (1)

(2) ROOT
DIAMETRAL PITCH – NORM XX (3)
OR
MODULE X (4)
PRESSURE ANGLE – NORM XX.XX°
HELIX ANGLE X.XXXXXXXX° REF.
HAND OF HELIX X
LEAD XX.XXXX
PITCH DIA. X.XXXXXXXX REF.
BASE DIA. X.XXXXXXXX REF.
FORM DIA. X.XXX
MINOR DIA. X.XXX – X.XXX
ACTL CIRC TOOTH THK – NORM AT PITCH DIA. .XXXX – .XXXX
MEAS OVER TWO .XXXXX DIA. BALLS X.XXXX – X.XXXX
MAX RUNOUT .XXXX FIR.
ALL TOOTH ELEMENT SPECIFICATIONS ARE FROM DATUM ESTABLISHED BY THE AXIS OF SURFACE S.

1. DATA SHOWN ARE FOR INCH GEAR DRAWINGS. METRIC GEAR DRAWINGS HAVE ONE LESS DECIMAL PLACE TO THE RIGHT OF THE DECIMAL POINT.
2. SPECIFY WHETHER FULL FILLET, OPTIONAL OR FLAT.
3. USED ON INCH DRAWINGS ONLY.
4. USED ON METRIC DRAWINGS INSTEAD OF DIAMETRAL PITCH.

Fig. 18–18. A design drawing of an external helical gear. (Reproduced from ANSI Y14.7.1-1971 with permission of the publisher, The American Society of Mechanical Engineers)

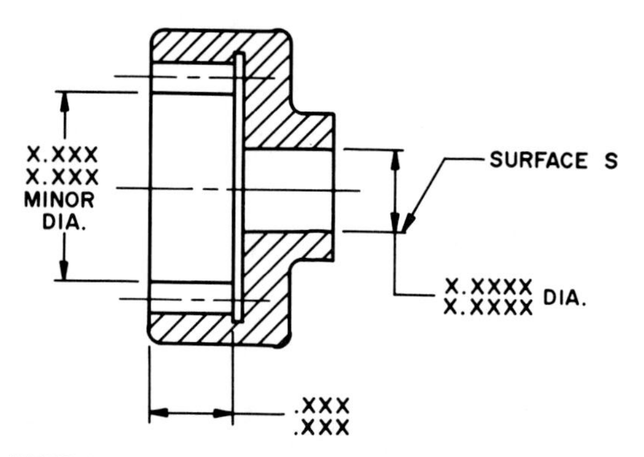

INTERNAL INVOLUTE HELICAL GEAR DATA. (1)

(2) ROOT
NUMBER OF TEETH XX
DIAMETRAL PITCH – NORM XX (3)
OR
MODULE X (4)
PRESSURE ANGLE – NORM XX.XX°
HELIX ANGLE X.XXXXXXXX REF.
HAND OF HELIX X
LEAD XX.XXXX
PITCH DIA. X.XXXXXXXX REF.
BASE DIA. X.XXXXXXXX REF.
FORM DIA. X.XXX
MAJOR DIA X.XXX – X.XXX
ACTL CIRC TOOTH THK – NORM AT PITCH DIA. .XXXX – .XXXX
MEAS BETWEEN TWO .XXXX DIA. BALLS X.XXXX - X.XXXX REF.
MAX RUNOUT .XXXX FIR.
ALL TOOTH ELEMENT SPECIFICATIONS ARE FROM DATUM ESTABLISHED BY THE AXIS OF SURFACE S.

1. DATA SHOWN ARE FOR INCH GEAR DRAWINGS. METRIC GEAR DRAWINGS HAVE ONE LESS DECIMAL PLACE TO THE RIGHT OF THE DECIMAL POINT.
2. SPECIFY WHETHER FULL FILLET, OPTIONAL OR FLAT.
3. USED ON INCH DRAWINGS ONLY.
4. USED ON METRIC DRAWINGS INSTEAD OF DIAMETRAL PITCH.

Fig. 18–19. A design drawing of an internal helical gear. (Reproduced from ANSI Y14.7.1-1971 with permission of the publisher, The American Society of Mechanical Engineers)

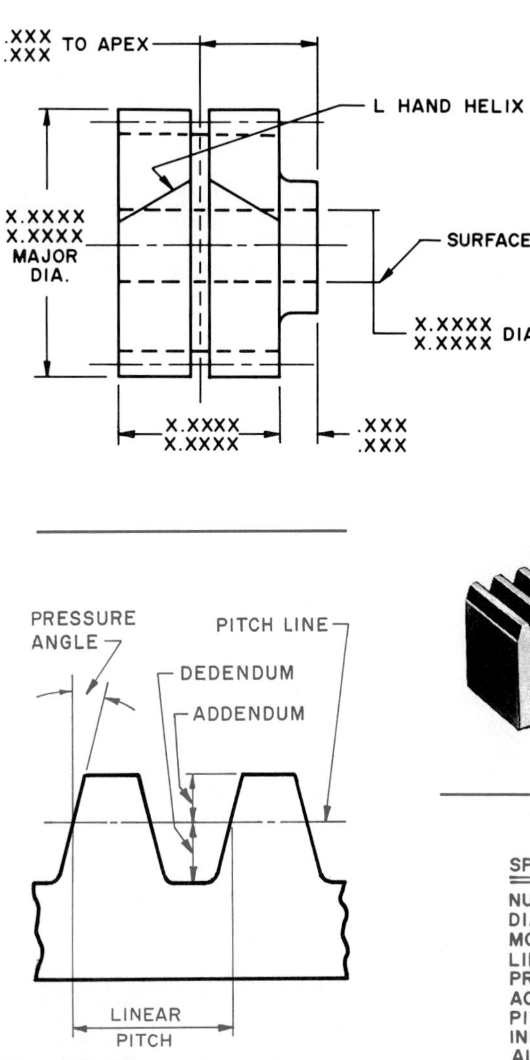

Fig. 18–20. A design drawing of an external involute double-helical gear. (Reproduced from ANSI Y14.7.1-1971 with permission of the publisher, The American Society of Mechanical Engineers)

Fig. 18–22. Terms used to identify the parts of rack teeth.

PINION

RACK

Fig. 18–21. A rack with a spur-gear pinion. (Boston Gear Division, North American Rockwell)

SPUR RACK DATA (1)

NUMBER OF TEETH XX
DIAMETRAL PITCH XX.XXXXXXX REF. (2)
MODULE X (3)
LINEAR PITCH .XXXX
PRESSURE ANGLE XX°
ACTUAL TOOTH THICKNESS AT PITCH LINE .XXXX-.XXXX
PITCH TOL .XXXX
INDEX TOL .XXXX
ALL TOOTH ELEMENT SPECIFICATIONS ARE FROM THE SPECIFIED DATUM.

1. DATA SHOWN ARE FOR INCH GEAR DRAWINGS. METRIC GEAR DRAWINGS HAVE ONE LESS DECIMAL PLACE TO THE RIGHT OF THE DECIMAL POINT.
2. USE ON INCH DRAWINGS ONLY.
3. USE ON METRIC DRAWINGS INSTEAD OF DIAMETRAL PITCH.

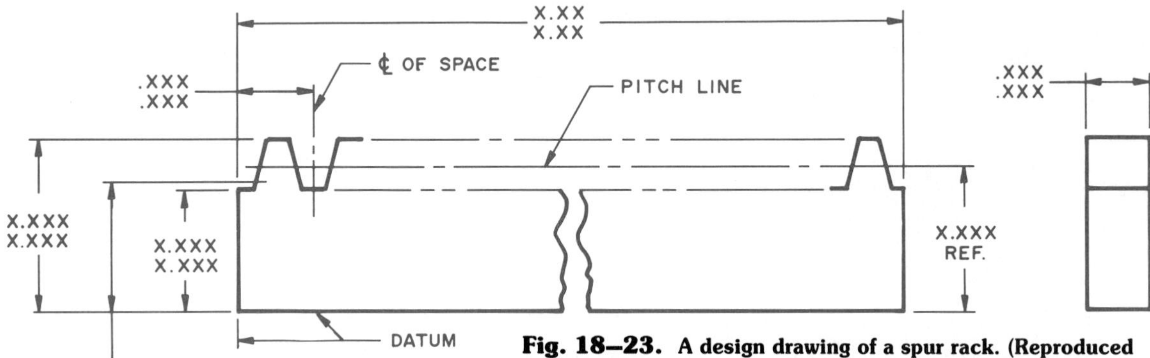

Fig. 18–23. A design drawing of a spur rack. (Reproduced from ANSI Y14.7.1-1971 with permission of the publisher, The American Society of Mechanical Engineers)

right angle, they are called *miter gears*. When a small bevel gear meets a large bevel gear, the small gear is called a *pinion*. Bevel gears are designed in pairs. They use the same involute tooth form as spur gears except the tooth is tapered.

Bevel gear terminology is shown in Fig. 18-25. Most of the terms used for spur gears apply to bevel gears. The general dimensions shown on a bevel gear drawing are diameter, pitch angle, cone distance, and face width. These are shown in the view that shows the gear's axis.

Fig. 18-26 shows a drawing with data for the gear teeth. If the entire gear blank is dimensioned, use standard dimensioning practices. Fig. 18-27 shows a drawing of mating bevel gears. Each gear is dimensioned in the same way shown in Fig. 18-26.

Worms and Worm Gears

Worms and worm gears are used to transmit power between shafts at right angles that do not intersect. A **worm** is a form of a screw. It has a thread the same shape as a rack tooth.

A **worm gear** looks much like a spur gear. However, it has a different tooth form. See Fig. 18-28. The tooth is formed to fit the curvature of the screw thread on the worm. Fig. 18-29 lists the terms used to describe the parts of a worm and worm gear. The worm is described in the same manner as screw threads. The lead angle of the worm must be specified. The lead angle is the angle between the thread helix and the plane of rotation. The pitch diameter of the worm is the mean of the working depth. Fig. 18-30 gives the information needed on a worm drawing.

A worm gear drawing is shown in Fig. 18-31. The data shown are needed to describe the teeth. The other parts of the gear blank are dimensioned following regular dimensioning practices.

Cams

A **cam** is an irregularly shaped disc used to change rotary motion into reciprocating motion. The cam is fastened to a revolving

Fig. 18–24. Bevel gears. (Boston Gear Division, North American Rockwell)

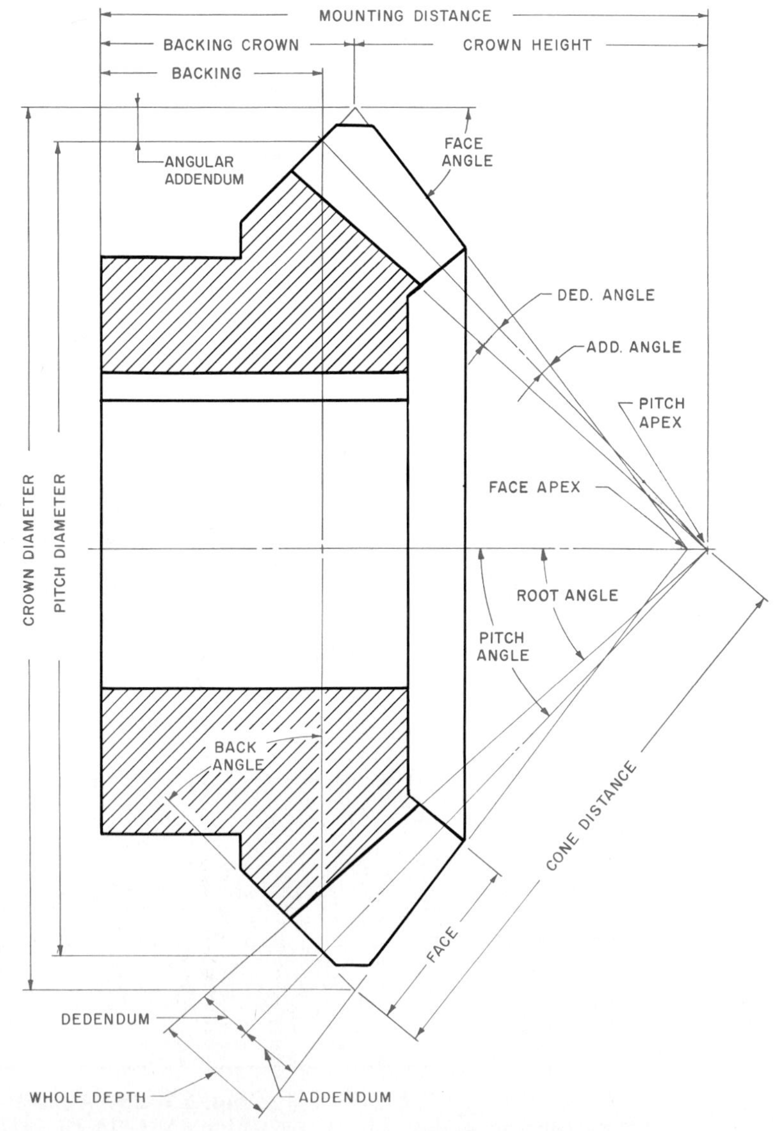

Fig. 18–25. Terms used to identify the parts of a bevel gear.

shaft. A device called a **follower** touches the curved surface of the cam. As the cam rotates, the follower moves along the curved surface of the cam. This causes the follower to move. Fig. 18-32.

There are two general types of cams: disc and cylindrical. Figs. 18-33 and 18-34. With the disc cam, the follower moves in a plane perpendicular to the axis of the shaft. The follower moves up and down as the cam rotates. Fig. 18-32. The cylindrical cam moves the follower back and forth in a plane parallel with the axis of the shaft. Fig. 18-35 p. 484.

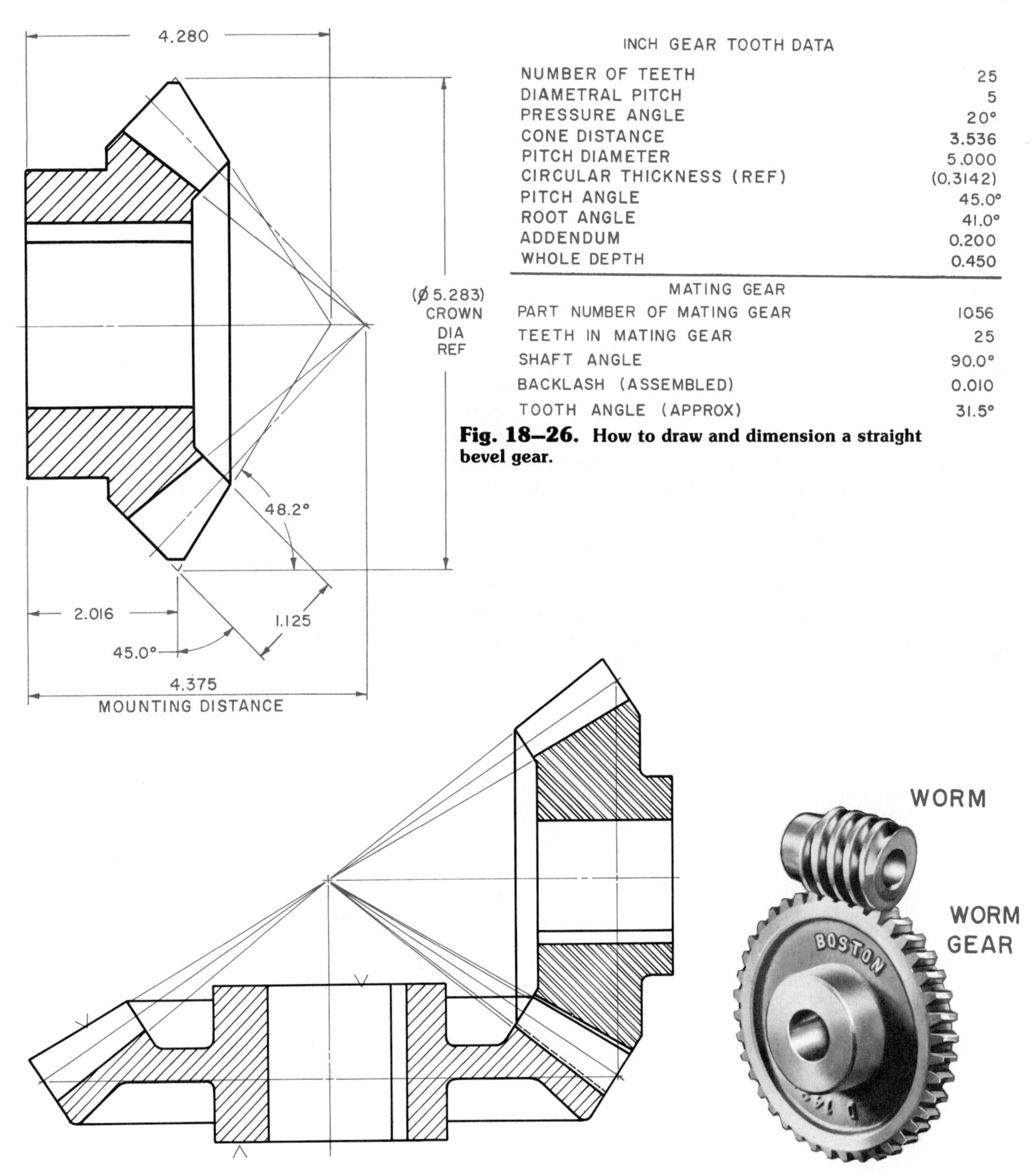

INCH GEAR TOOTH DATA	
NUMBER OF TEETH	25
DIAMETRAL PITCH	5
PRESSURE ANGLE	20°
CONE DISTANCE	3.536
PITCH DIAMETER	5.000
CIRCULAR THICKNESS (REF)	(0.3142)
PITCH ANGLE	45.0°
ROOT ANGLE	41.0°
ADDENDUM	0.200
WHOLE DEPTH	0.450
MATING GEAR	
PART NUMBER OF MATING GEAR	1056
TEETH IN MATING GEAR	25
SHAFT ANGLE	90.0°
BACKLASH (ASSEMBLED)	0.010
TOOTH ANGLE (APPROX)	31.5°

Fig. 18–26. How to draw and dimension a straight bevel gear.

Fig. 18–27. Mating bevel gears are sometimes drawn together. They can be dimensioned using standard dimensioning practices shown in this chapter.

Fig. 18–28. A worm and worm gear. (Boston Gear Division, North American Rockwell)

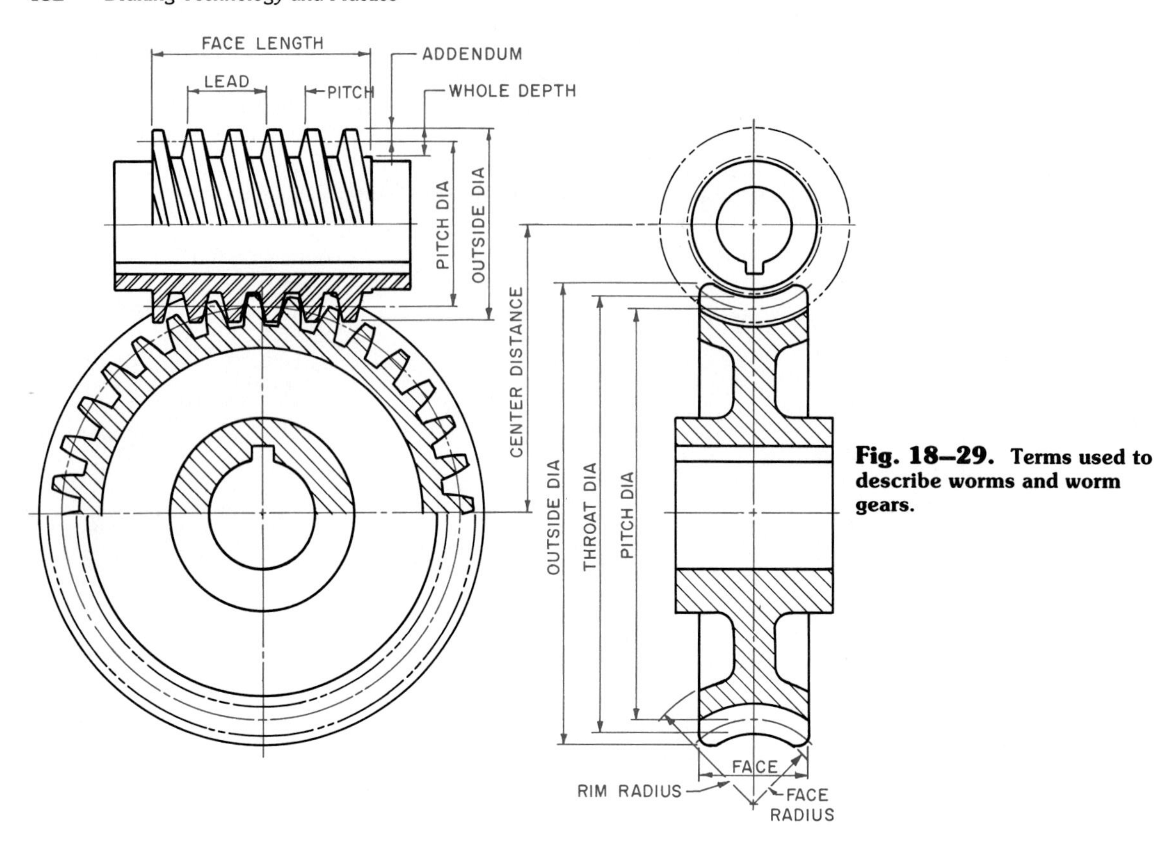

Fig. 18–29. **Terms used to describe worms and worm gears.**

Fig. 18–30. **A worm drawing.**

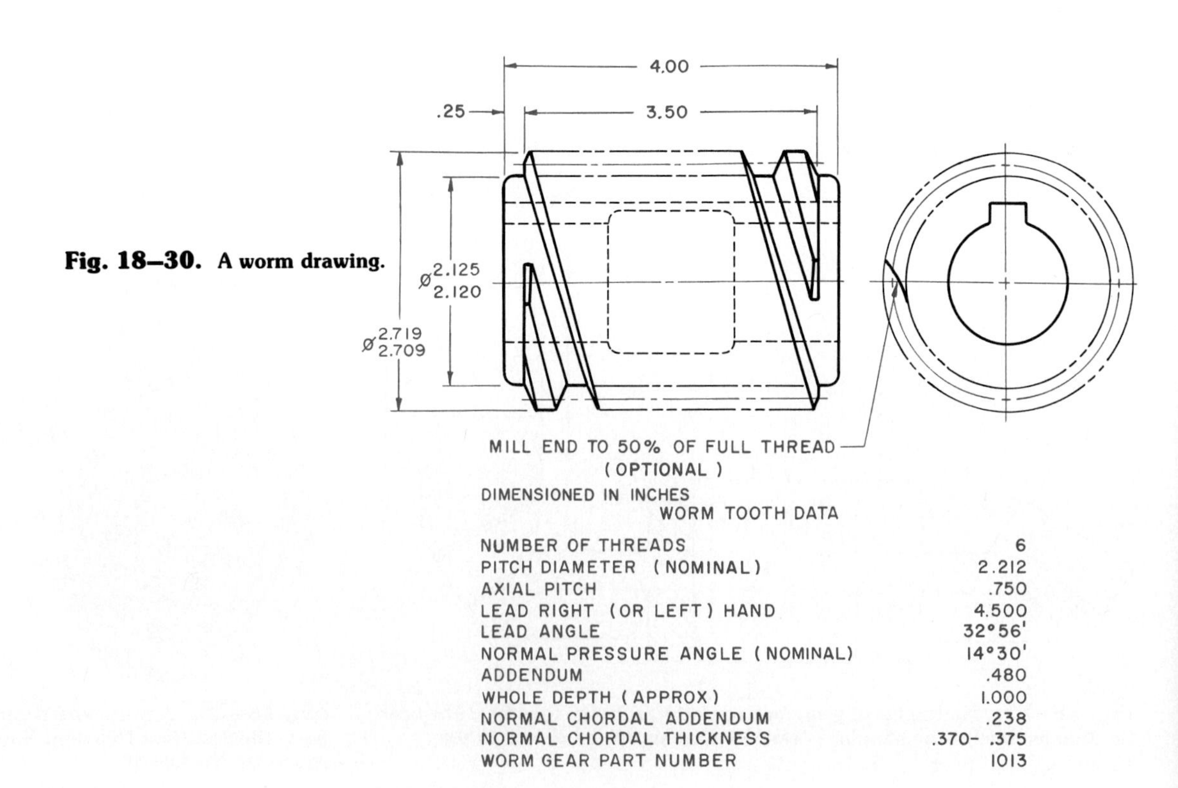

WORM TOOTH DATA

NUMBER OF THREADS	6
PITCH DIAMETER (NOMINAL)	2.212
AXIAL PITCH	.750
LEAD RIGHT (OR LEFT) HAND	4.500
LEAD ANGLE	32°56'
NORMAL PRESSURE ANGLE (NOMINAL)	14°30'
ADDENDUM	.480
WHOLE DEPTH (APPROX)	1.000
NORMAL CHORDAL ADDENDUM	.238
NORMAL CHORDAL THICKNESS	.370- .375
WORM GEAR PART NUMBER	1013

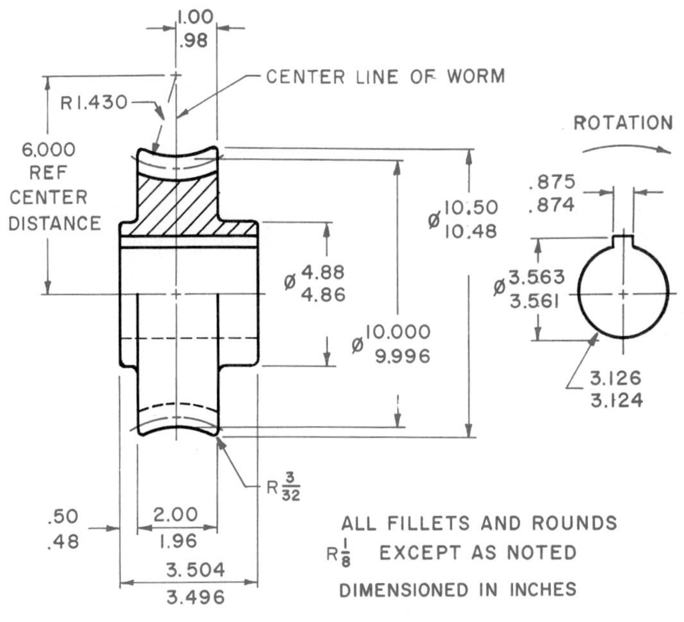

WORM-GEAR TOOTH DATA	
NUMBER OF TEETH	41
PITCH DIAMETER	9.000
ADDENDUM	4.80
WHOLE DEPTH (APPROX)	1.000
WORM PART NUMBER	2103
BACKLASH ASSEMBLED	0.010-0.015
HOB NUMBER	82

WORM DATA (REFERENCE)	
NUMBER OF THREADS	6
AXIAL PITCH	.750
LEAD RIGHT (OR LEFT) HAND	4.500
PITCH DIAMETER (NOMINAL)	2.212
LEAD ANGLE	32° 56'
NORMAL PRESSURE ANGLE (NOMINAL)	14° 30'

Fig. 18–31. A worm gear drawing.

Fig. 18–33. Disc cams.

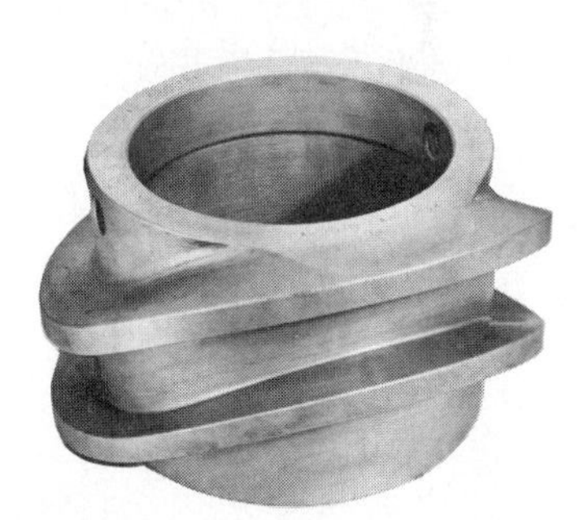

Fig. 18–34. A cylindrical cam.

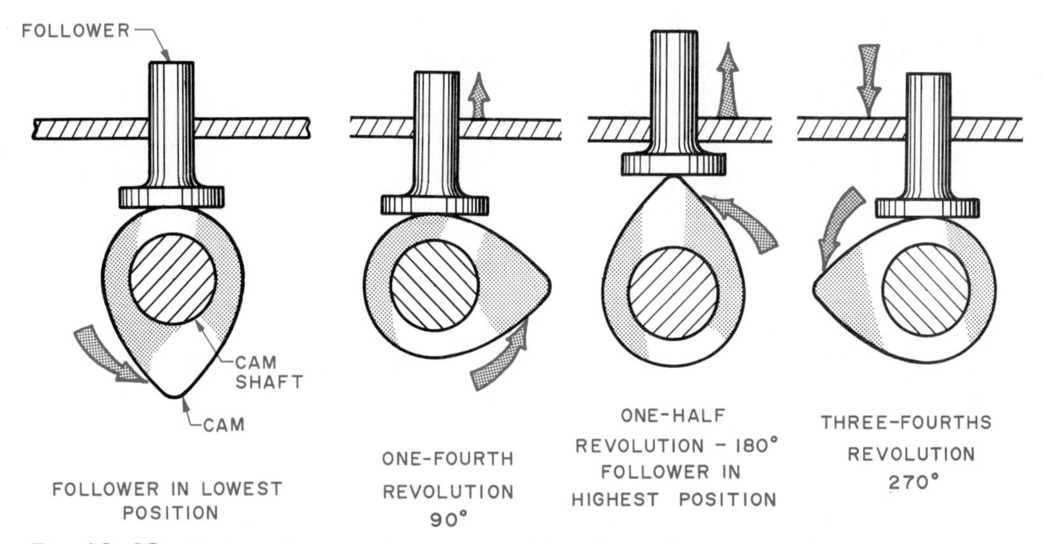

Fig. 18–32. The cam changes rotary movement to reciprocating movement.

There are several common types of cam followers. Fig. 18-36. The type used depends upon design features needed. The speed of rotation and stresses are factors to be considered when selecting a cam and follower.

Cam Motions

Cams produce three kinds of motion. These are uniform, harmonic, and uniformly accelerated and decelerated. The layout of cam motions is shown on a diagram called a displacement diagram.

Displacement Diagrams. A **displacement diagram** is a curve that represents the motion of the follower through successive units of time of one cam rotation. Fig. 18-37. The *length* of the diagram represents one revolution of the cam. Although any convenient length can be used, the length of the diagram is usually found by calculating the circumference of a circle that has a radius equal to the distance from the center of the cam shaft to the highest point on the cam rise. Fig. 18-38. This length equals 360 degrees. It is divided into segments that are often spaced at 10 degrees.

The *height* of the diagram represents the maximum rise of the cam follower. It is drawn to scale. This is called *follower displacement.* This height can be divided into time intervals or angle of cam rotation. The *time interval* is the time it takes the cam to move the follower to that height. The angle of cam rotation is the height the follower moves when the cam rotates a certain number of degrees.

Uniform Motion. **Uniform motion** means the follower rises and falls at a constant speed. It is a straight-line motion. This motion is diagrammed in Fig. 18-39. In this example, the follower is to rise uniformly 2 inches through 180 degrees. Connect the beginning point A with the ending point B with a straight line. It is common practice to draw an arc at each end of a uniform motion diagram. This eliminates an abrupt motion when the follower starts and stops at the beginning and ending of the interval. The curve has a radius of one-third of the rise. Fig. 18-39. The straight line is drawn tangent to these arcs.

Harmonic Motion. A cam having **harmonic motion** is one that lifts the follower gradually from its starting position. The speed of rise increases to a point halfway in the full rise. The speed then decreases until it reaches its full rise. Such a motion relieves the shock of starting and stopping as the follower moves through a rise and fall.

A displacement diagram for harmonic motion is shown in Fig. 18-40. To draw the diagram, follow these steps:

1. Draw the displacement diagram to length.
2. Draw the rise line.

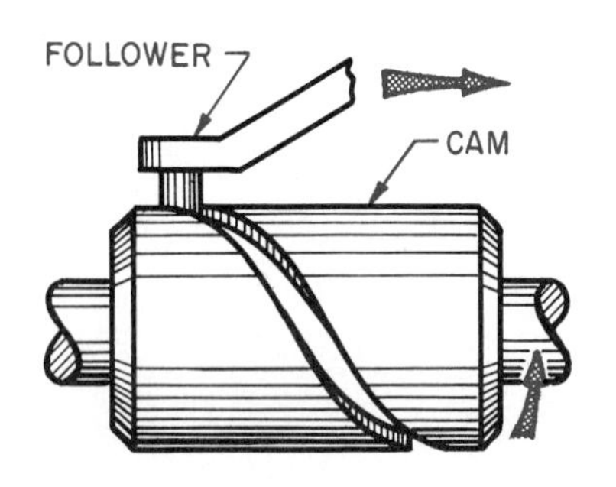

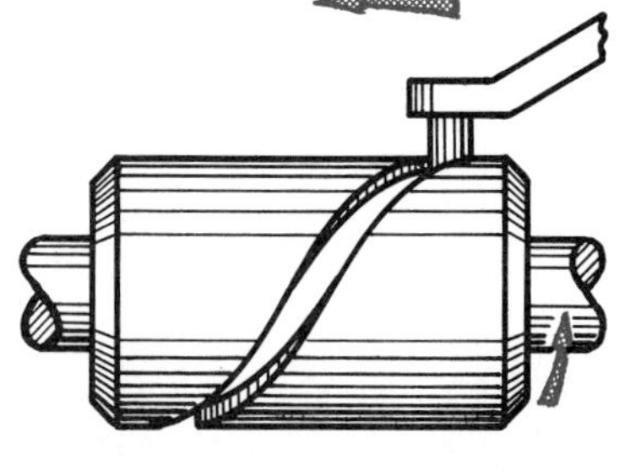

Fig. 18–35. Cylindrical cams produce a movement parallel to the axis of the shaft.

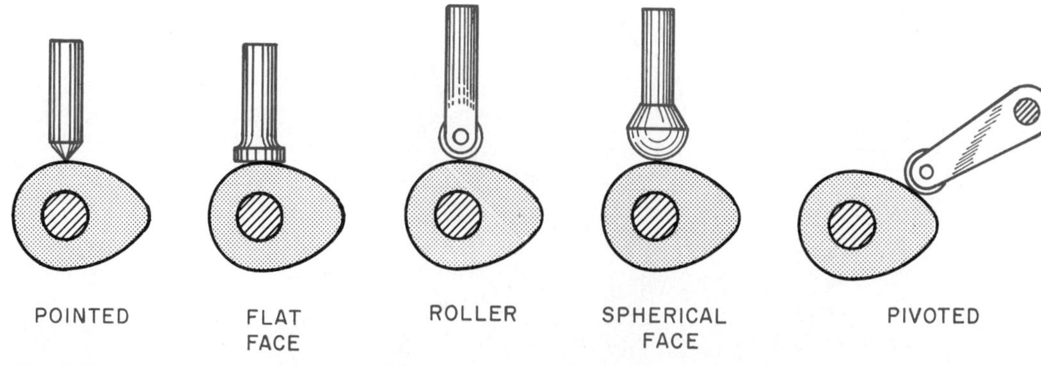

Fig. 18–36. Common types of followers.

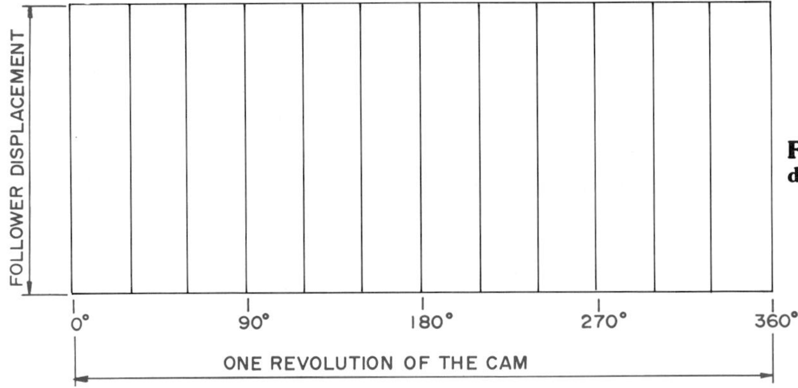

Fig. 18–37. The form for a displacement diagram.

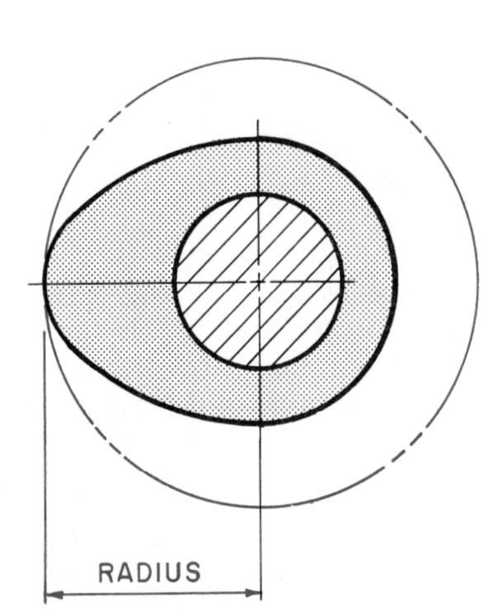

Fig. 18–38. One revolution of the cam can be used as the length of the displacement diagram.

Fig. 18–39. A uniform motion displacement diagram.

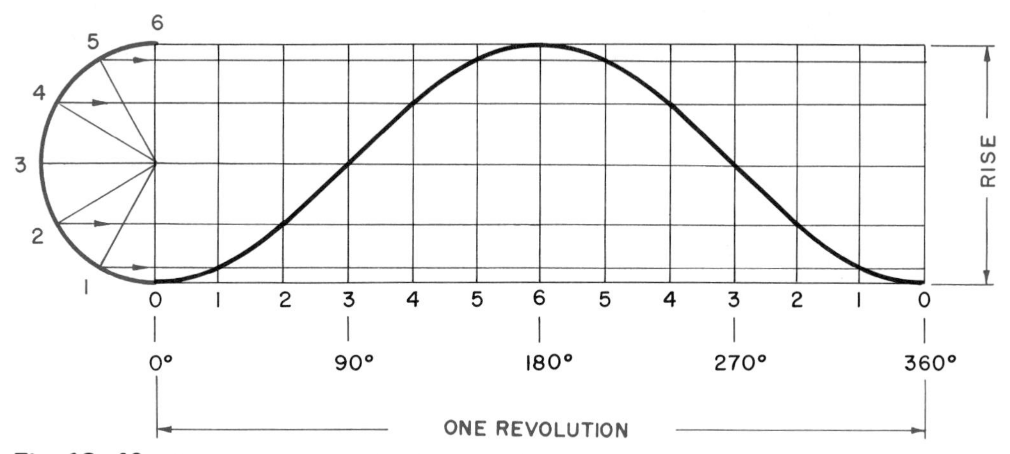

Fig. 18–40. A harmonic motion displacement diagram.

3. Construct a semicircle on the rise line. The radius equals half the rise.
4. Divide the semicircle into a number of equal parts. Number the points.
5. Divide the part of the diagram's length to be used for the harmonic rise into the same number of parts. Number the points. In Fig. 18-40 this is 180 degrees.
6. Project the points on the semicircle across the diagram. Where each intersects the line with the same number is a point on the harmonic curve.
7. Connect the points in a smooth curve.

Uniform Accelerated and Decelerated Motion. **Uniform accelerated and decelerated motion** is similar to harmonic motion. The difference is that uniform acceleration and deceleration occur at a constant speed. Harmonic motion increases and decreases at a constantly changing speed. Uniform accelerated and decelerated motion is smoother than harmonic motion.

A displacement diagram for uniform accelerated and decelerated motion is shown in Fig. 18-41. To draw this diagram, follow these steps:

1. Draw the length and rise of the displacement diagram.
2. Divide the part of the length to be used for uniformly accelerated and decelerated motion into a number of equal parts.
3. Now divide half the rise into distances proportional to the square of the distances on half the length. Square each point (1, 2, and 3) on half the length to get proportional parts (1, 4, and 9) for marking the rise. Draw a line at any angle from point A. Mark off as many equal distances as the largest squared number on the rise. Draw a line from the largest number (9) to the center of the rise. Draw lines parallel to this first line drawn to the rise from the points representing the remaining squared numbers (4 and 1). Project the points marked on the rise across the diagram parallel to the length line. Where the numbered line and the line of its square cross are points on the curve from A through B.
4. To continue the curve to its apex, measure off points 1, 4, and 9 on the rise from D to F but in reverse order, as shown by the dotted lines. The other side of the curve is the reverse of the first side.
5. Connect the points.

Dwell. **Dwell** means a period of time during which the follower does not move. This is shown on a displacement diagram as a horizontal line. Fig. 18-42.

Displacement Diagrams with a Combination of Motions. Cams can contain a combination of the motions just discussed. Each motion is diagrammed as explained in Figs. 18-39 through 18-41. An example of a combination of motions is shown in Fig. 18-42. Here the follower rises 1 inch in 90 degrees using harmonic motion. The follower dwells for 30 degrees. It then rises 1 inch in uniform motion as the cam turns 90 degrees. At the top of the rise it dwells for 30 degrees. It falls us-

CAD Applications:

Parametrics

The *parameters* of an object are the properties whose values determine its characteristics. For example, height, width, thickness, and type of material are some of the parameters of a door. Every door has these parameters, but their values can differ greatly. Thus a door for a kitchen cabinet is not the same as a storm door or a closet door.

With parametrics, it is possible to create many variations of a basic design by specifying different values for the parameters. A parametric program for gear drawings, for example, would prompt the user to enter values for the number of teeth, pitch diameter, root diameter, etc. Once all the values had been entered, the computer would generate the gear drawing.

Parametric programs work with CAD programs. There are numerous applications for parametric programs, especially where many similar products are being designed. Check your library for publications such as *CADENCE, CADalyst,* or *MicroCAD News* for articles on parametric design.

ing uniform accelerated and decelerated motion through 120 degrees.

Drawing a Disc Cam Profile with an Offset Follower

Follow these steps to draw the cam profile shown in Fig. 18-43.

1. Draw the displacement diagram to the same scale as the cam drawing.
2. Locate the center of the shaft on which the cam will operate.
3. Draw the base circle from the center of the shaft. Its radius is equal to the distance from the center of the shaft to the center of the follower wheel at its lowest position. Where it crosses the center line of the follower is the lowest position. This is point 0 on the displacement diagram.
4. Draw the offset circle using the center of the shaft. The radius is equal to the offset.
5. Draw the center line of the follower.
6. Divide the offset circle into the same equal divisions used on the displacement diagram.
7. Number the divisions on the offset circle, beginning with zero, in a direction opposite to the direction of the cam rotation. The 0 point is where the center line of the follower is tangent with the offset circle. Number each division the same as on the displacement diagram.
8. Draw lines through each point on the offset circle tangent to the circle. Number these the same as the points on the circle.
9. Lay out the rise distances for each point on the center line of the follower. These begin at the center of the follower wheel. These distances are taken from the displacement diagram. Number each point like those on the displacement diagram. These positions are the rise of the follower as the cam rotates.
10. Set a compass with a radius equal to the distance from the center of the cam shaft to point 1 on the center line of the follower. Swing an arc until this crosses tangent line 1. Repeat this

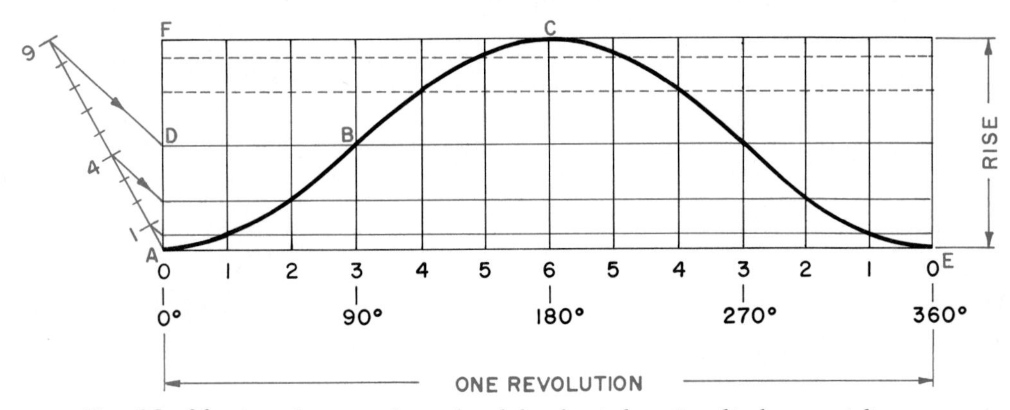

Fig. 18—41. A uniform accelerated and decelerated motion displacement diagram.

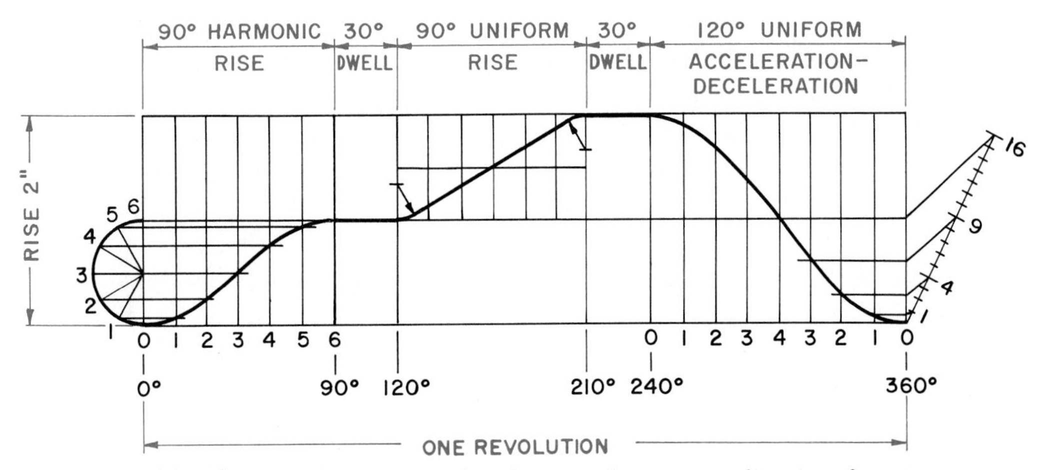

Fig. 18—42. A displacement diagram for a cam having a combination of motions.

for each point on the follower center line. This locates the center of the follower wheel in various positions around the cam.

11. Connect these points with a smooth curve. This forms the *pitch curve*.

12. Set a compass to the radius of the follower wheel. Using the pitch curve as a center, swing a series of arcs. These arcs are tangent to the working face of the cam. The more arcs used, the more accurate the cam profile drawn.

13. Draw a smooth curve tangent to the cam wheel arcs. This forms the finished cam profile.

14. Draw the hub, keyseat, and follower.

Drawing a Disc Cam Profile with a Flat Face Follower

To draw the cam profile shown in Fig. 18-44, follow these steps:

1. Draw the displacement diagram. The drawing shown in Fig. 18-44 uses the displacement diagram shown in Fig. 18-43.

2. Locate the center of the cam shaft.

3. Draw the base circle. Its radius is equal to the distance from

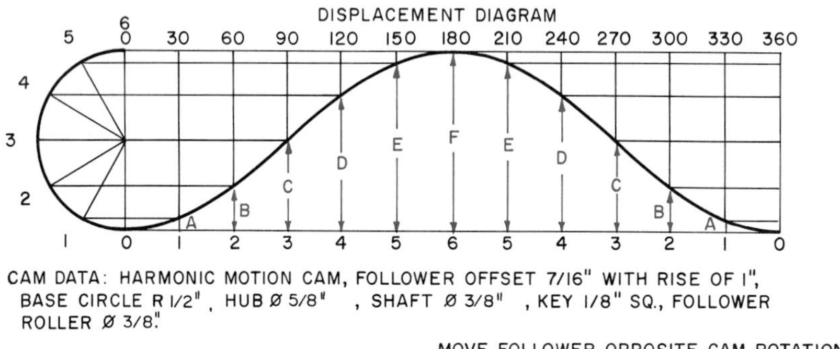

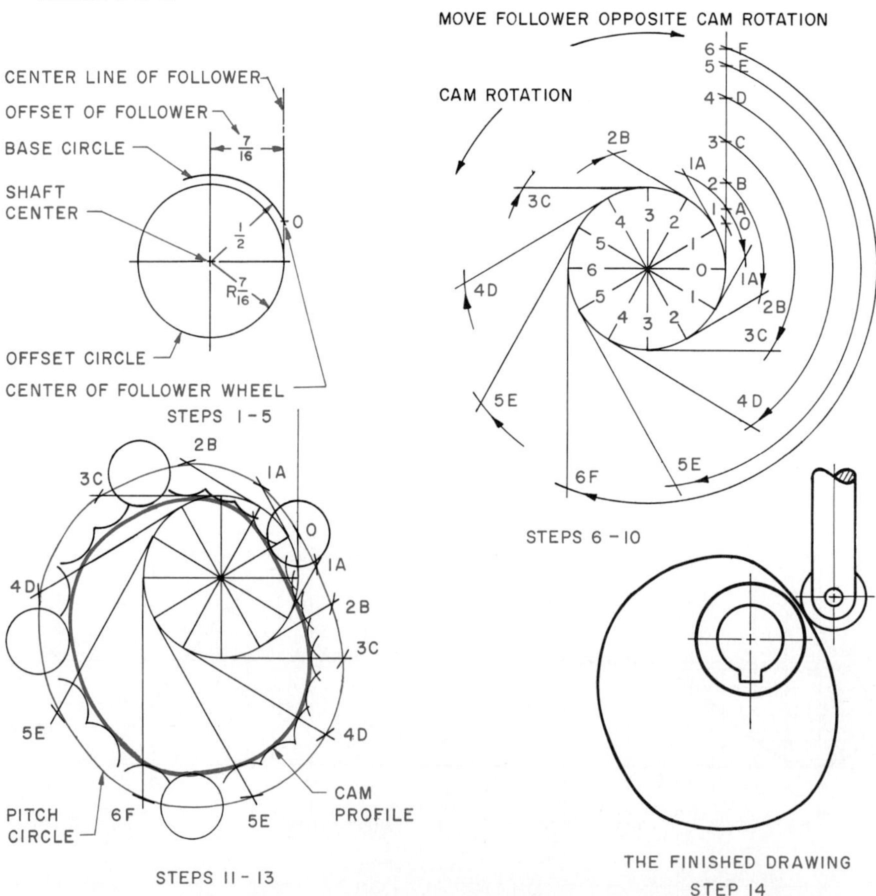

Fig. 18–43. **How to draw a disc cam profile with an offset follower.**

the center of the cam shaft to the face of the follower in its lowest position. This point is where the base circle crosses the center line of the follower.

4. Lay out the follower rise distances on the center line of the follower. These distances are found on the displacement diagram. Identify them in the same way and in the same order as they appear on the displacement diagram. Here they are lettered A through F.

5. From the center of the cam shaft, draw lines at the same degrees as used on the displacement diagram and label them. These lines are the center lines of the followers at each of these locations.

6. Using the center of the cam shaft as center point, swing arcs from each rise distance until they cross the degree line on which they were measured on the displacement diagram.

7. At each point of intersection just found, draw a line perpendicular to the degree line. These intersections represent the position of the follower at these points.

8. Draw the cam profile by constructing a smooth curve that is tangent to each of the perpendicular lines just drawn.

9. Draw the shaft, hub, keyseat, and follower to complete the cam.

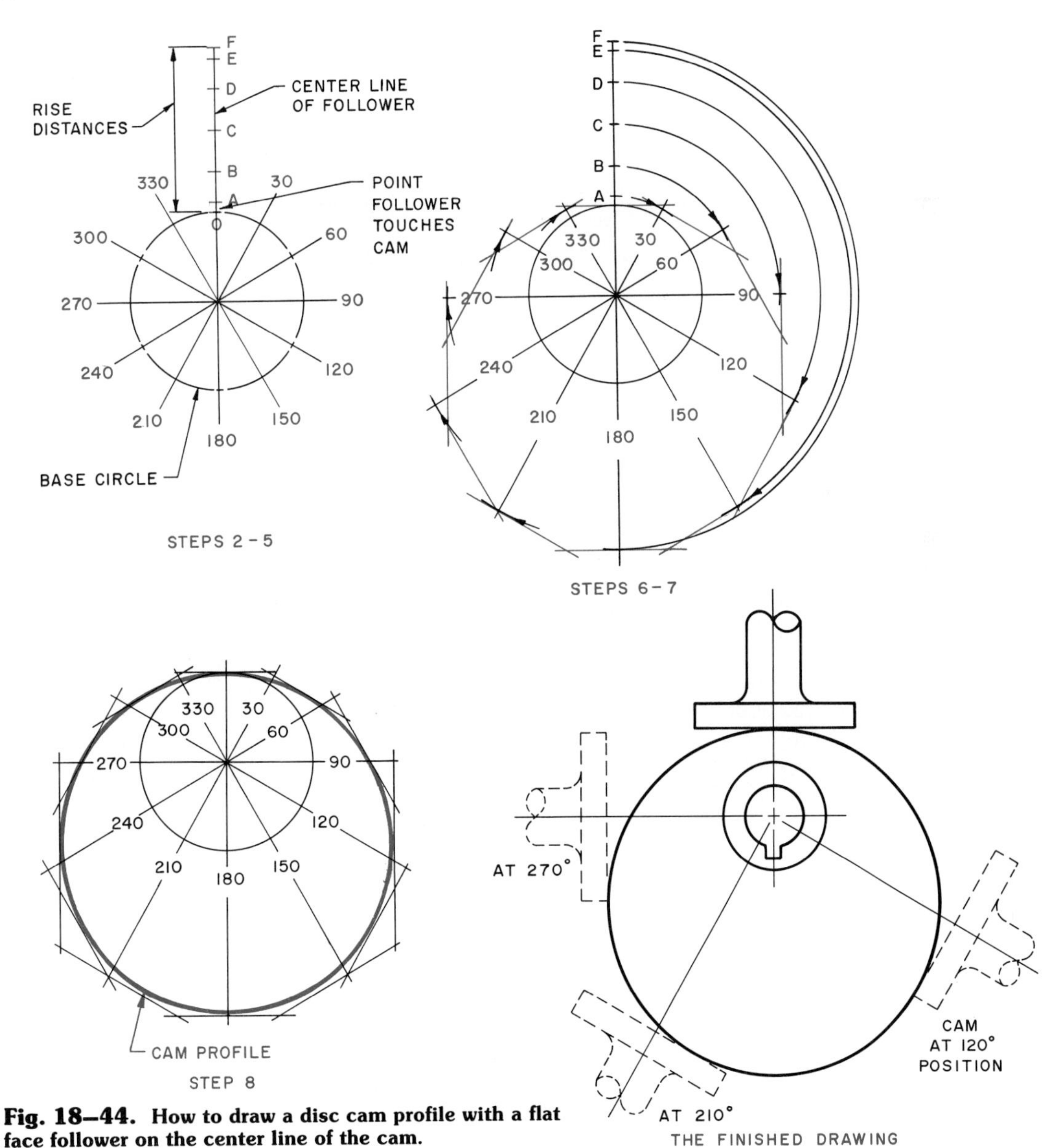

Fig. 18–44. How to draw a disc cam profile with a flat face follower on the center line of the cam.

Chapter Review

Build Your Vocabulary

You should understand and use the following terms as part of your working vocabulary. Write a brief explanation of what each means.

gears
drive gear
driven gear
pinion
spur gear
involute
rack
bevel gears
worm
worm gear
cam
follower
displacement diagram
uniform motion
harmonic motion
uniform accelerated and decelerated motion
dwell

Sharpen Your Math Skills

1. What is the gear ratio of a pinion with 25 teeth and a spur with 100 teeth?
2. A gear has a pitch diameter of 3.0″ and a major diameter of 4.0″. What is the addendum of the gear teeth?
3. The working circle of a cam is 360°. On a drawing it is divided into 10° segments. How many segments would this produce?
4. The circumference of a circle is 3.14 times the diameter. You want to draw a displacement diagram for a cam. The distance from the center of the cam shaft to the highest point on the cam shaft is 3″. Although any convenient length can be used, what is the length most often used to make this drawing?
5. What is the angle of cam rotation when a cam has gone through ½ of a revolution? What is the angle of cam rotation when the cam has gone through ⅝ of a revolution?
6. If a cam follower rises using harmonic motion for 125°, dwells for 30°, and falls using uniform accelerated and decelerated motion to its starting point, how many degrees are used to return to this starting point?

Study Problems—Directions

The problems that follow will give you the chance to use some of the drafting knowledge that you learned in this chapter. Do the problems that are assigned to you by your instructor.

1. Make drawings of the spur gears described in **Fig. P18-1** using the recommended procedures as shown in Figs. 18-16 and 18-17. Draw the front of the gears if assigned by your instructor including two teeth for the sake of clarity. On the front view draw and dimension a standard key seat. Review Chapter 12, "Fasteners," for size and technique. Draw gears 1, 2, 3, and 4 full size. Make gear 5 half size.
2. Draw the front view of the two related gears in **Fig. P18-2**. The center line of the large gear is 12° ± .05° to the left of the center line of the small gear. Make your drawing full size.
3. Make a fully dimensioned, full-size drawing of a bevel gear using the data in **Fig. P18-3**. Develop any dimensions that may be missing by measuring your scale drawing. Use a stock square key in the hub.
4. Make a full-size drawing of the worm described in **Fig. P18-4**. It uses a stock square key.
5. Study the data for the worm gear, **Fig. P18-5**. Make a full-size, dimension drawing. A stock square key is used.
6. Data for a rack are shown in **Fig. P18-6**. Make a full-size, dimensioned drawing of the rack.
7. Draw a displacement diagram for a cam having uniform motion with arcs. Use a circle with a 7-inch circumference. The rise is 2½ inches.
8. Draw a displacement diagram for a cam having harmonic motion. Lay out the displacement diagram length based on a 9-inch circle and a rise of 2 inches.

9. Draw a displacement diagram for a cam having uniform accelerated and decelerated motion. Calculate the length of the diagram using a 6-inch circle and a 3-inch total rise of the follower.

10. Draw the displacement diagram and the cam profile for a harmonic motion cam with an offset roller follower (dimensions in inches): follower offset, 7/8; rise, 1½; base circle radius, 1; hub, ¾ dia; shaft, ½ dia; stock square key; follower roller, 3/16 dia.

11. Draw the displacement diagram and the cam profile for a uniform motion cam with a flat face follower on the center line of the cam (dimensions in inches): rise, 2; base circle radius, 1; hub, ¾ dia; shaft, ½ dia; stock square key.

12. Draw the displacement diagram and the cam profile for a cam with a flat face follower on the center line (dimensions in inches): up 1 inch in 90° harmonic rise; dwell, 60°; up 1 inch in 90° harmonic rise; dwell, 30°; down 2 inches in 90° uniform accelerated and decelerated motion; base circle radius, 1¼; hub, 1 dia; shaft, ¾ dia; stock square key.

Study Problems

Fig. P18-1. Spur Gear Data[1]

Problem Number	1	2	3	4	5
Number of Teeth	12	12	36	24	66
Diametral Pitch	8	4	12	24	6
Face Width	1.500	3.500	1.000	.250	2.000
Addendum	.115	.235	.080	.042	.178
Dedendum	.125	.250	.092	.047	.188
Pitch Circle Diameter	1.500	3.000	3.000	1.000	11.000
Base Diameter	1.165	2.656	2.800	.825	10.620
Form Diameter	1.30	2.850	3.000	.970	10.910
Pressure Angle	20°	20°	14½°	14½°	20°
Maximum Runout	.0050	.0150	.0100	.0005	.0010
Style	Plain	Plain	Web	Plain With Shaft .3800	Spoke
Hole	.7500	1.620	.6250	—	1.2500
Hub Diameter	—	2.250	1.500	—	3.500
Projection	—	1.00	.62	—	1.50

[1]Major diameter equals pitch circle diameter plus the addendum. Minor diameter equals pitch circle diameter minus the dedendum.

Study Problems

Fig. P18-2.

	Major Diameter	Minor Diameter	Pitch Circle Diameter	Pressure Angle	Tooth Width at Pitch Circle
Gear 1	6.0	5.800	5.915	14½°	.2500
Gear 2	4.0	3.800	3.915	14½°	.2500

Fig. P18-3. Bevel Gear Data

Pitch Dia.	3.30
Outside Dia.	3.56
Back Angle	56°
Dedendum Angle	5°–30′
Addendum Angle	4°
Pitch Angle	51°–30′
Face	.62
Mounting Distance	2.12
Crown to Back	.94

Root Angle	47°
Face Angle of Blank	59°
Back Angle Distance	1.00
Hub Projection	.50
Hub Diameter	1.38
Hole Diameter	.87
Hole Length	1.00
Fillets and Rounds	R .38
Backing	.81

Fig. P18-4. Worm Data

Diametral Pitch	8
Hub Projection	.625 each end
Hub Dia.	1.188
Length of Threads (Face)	1.750
Hole	.750
Pitch Dia.	1.500
Lead	.3927 right
Pressure Angle	14½°
Lead Angle	4°–46′
Addendum	.125
Whole Depth	.240

Study Problems

Fig. P18-5. Worm Gear Data

No. of Teeth	20
Diametral Pitch	8
Pitch Dia.	2.50
Outside Dia.	2.94
Throat Dia.	2.68
Style	Plain
Hole Dia.	.38
Face	.75
Hub Dia.	1.00
Hub Projection	.75
Rim Radius	1.37
Face Radius	0.75
Addendum	.09
Whole Depth	.18
Web Width	.38
Fillets & Rounds	.12

Fig. P18-6. Rack Data

No. of Teeth	12
Face Width	.145–.150
Back to Top of Tooth	1.495–1.500
Linear Pitch (Circular Pitch)	.6900
Addendum	.180
Dedendum	.290
Pressure Angle	14½°
Actual Tooth Thickness at Pitch Line	.3100–.3125
Diametral Pitch	1.75
Pitch Tolerance	.0010
Index Tolerance	.0010
Back to Form Line	1.130
End of Rack to Center Line of First Space	.656

CHAPTER 19
Vector Analysis

After studying this chapter, you should be able to do the following:

- Understand a vector.
- Be able to use vectors to analyze stresses on a structure.

Introduction

When the structural system for a product, such as a bridge or a roof truss, has been designed it must be analyzed to determine the stresses that it will have to withstand. This analysis can be made using computer techniques, Fig. 19-1, or found graphically using vector analysis.

A structural system, Fig. 19-2, is subjected to forces such as tension, compression, shear, and moment. When a member is in **tension**, it has forces pulling on it from both ends. When a member is under **compression**, the forces are pushing against it from both ends. Fig. 19-3. **Shear** results when there are forces producing an opposite but parallel sliding motion of the planes of the object. Shear is a cutting, or tearing, force. Fig. 19-4. **Moment of force** is the rotation produced on a body when a force is applied. It is a bending or turning force. Fig. 19-5.

These forces can be analyzed graphically using vector analysis. Vectors can also represent other quantities such as velocity, distance, and electrical properties.

Fig. 19–1. Vector analysis can be performed using a computer. (Hewlett-Packard Company)

Vectors

A **vector** is a line that is given direction and represents a force, velocity, or some other measure of magnitude (quantity). The direction is measured in degrees which could be written as a decimal, such as 35.5°, or in minutes and seconds, such as 40°15′30″. The magnitude of the force is represented by drawing the length of the vector to scale. For example, the vector in Fig. 19-6 has a direction of 30° and a magnitude of 500 lb. The arrow on the end of the vector indicates the sense of the force. *Sense* is the direction in which the force is being applied.

Fig. 19–2. Designing this tower included analyzing the forces that will act on each of its parts when it is standing in place. (McDermott Incorporated)

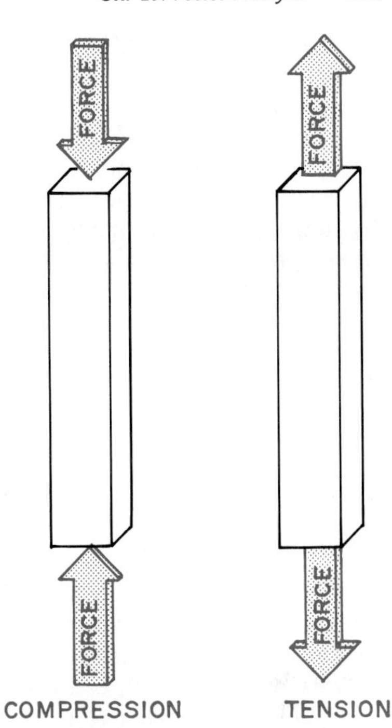

Fig. 19–3. Structural parts may be in compression or tension.

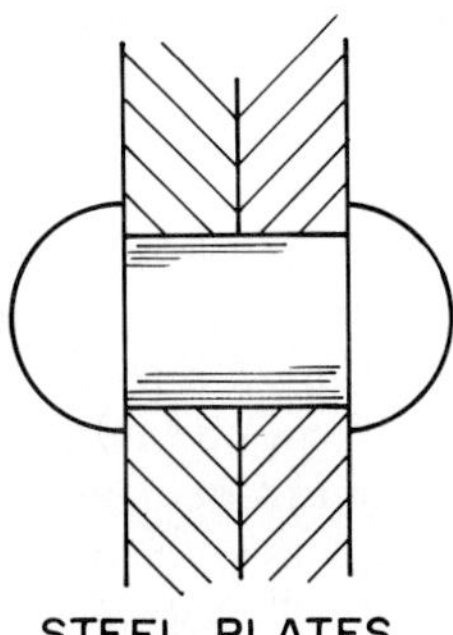

Fig. 19–4. Shear is generated by forces working in opposite directions.

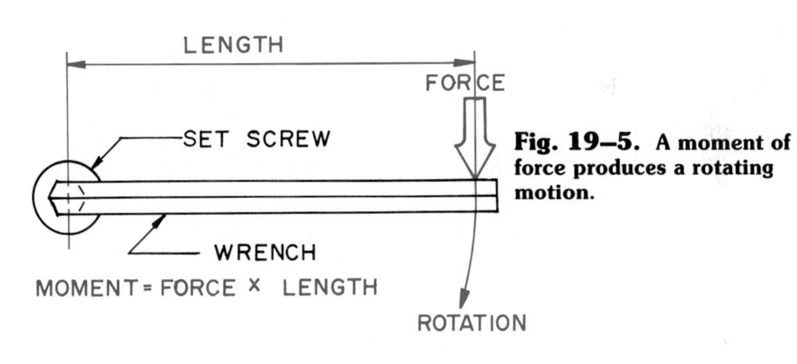

Fig. 19–5. A moment of force produces a rotating motion.

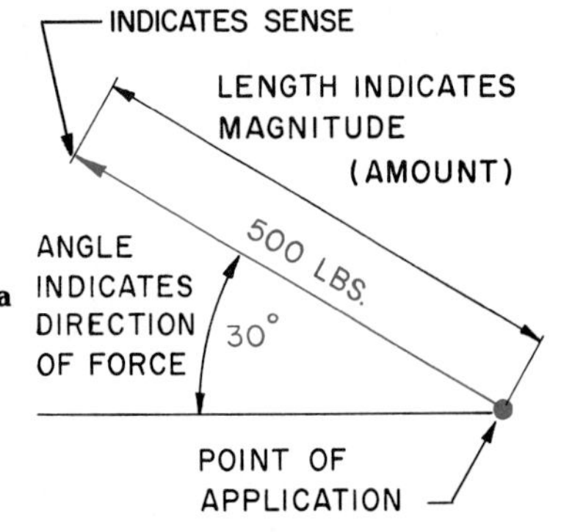

Fig. 19–6. A vector represents a force. It shows the direction and magnitude of the force.

Vectors are used in engineering in an area called *mechanics*. Mechanics is that branch of engineering that deals with the state of rest or motion of bodies under the action of one or more forces. Mechanics is used to study technical problems, such as the strength of roof structures, trusses, and beams; engine performance; and fluid flow. These problems can be solved by mathematical or graphical means. This chapter explains the use of vectors to develop graphical solutions.

Definition of Terms

Before any discussion of vectors can take place, it is necessary to understand certain terms used.

A **force** is something that tends to produce a pushing or pulling motion on an object. It is necessary to know the direction and magnitude of the force.

The U.S. customary unit of force is pounds. For example, in Fig. 19-6 a point has a force of 500 pounds pulling on it. The SI metric unit of force is the *newton* (N). A newton is the unit of force required to accelerate a mass of one kilogram one meter per second. In Fig. 19-7 the vector has a magnitude of 2250 newtons and a direction of 30 degrees with the horizontal.

Velocity is the speed at which an object is moving. The U.S. customary unit for velocity is miles per hour. The SI metric unit of velocity is meters per second (m/s). To convert miles per hour to meters per second, multiply miles per hour by 0.447.

The **sense** of a force is shown by the arrow on the line indicating magnitude. If the arrow points away from the point of application, it indicates tension. If the arrow points toward the point of application, it indicates compression.

Direction of a force is the line in which the force acts. It is shown graphically in degrees from the horizontal. Fig. 19-7.

The **magnitude** is the amount of force. It is expressed in some standard unit. Examples are pounds, newtons, miles per hour, and meters per second.

A force has a **point of application**. This is the point at which the force is applied. It is either pulling or pushing at this point. Fig. 19-7.

Coplanar forces are those acting in a single plane. This could include two or more forces. Fig. 19-8.

Concurrent forces are those that all act upon a single point. They may or may not be in the same plane. Those shown in Fig. 19-8 at A are concurrent and coplanar.

Noncoplanar forces are those that are not all acting in the same plane. In Fig. 19-8 at B the forces are concurrent and noncoplanar.

Nonconcurrent forces are those which do not intersect at a common point. Fig. 19-9.

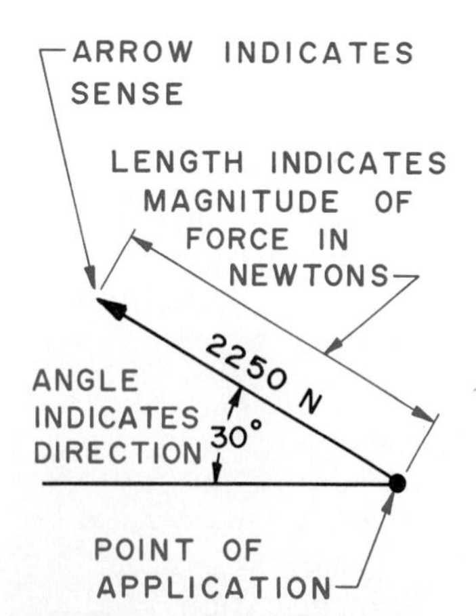

Fig. 19–7. This vector shows the magnitude of the force in newtons.

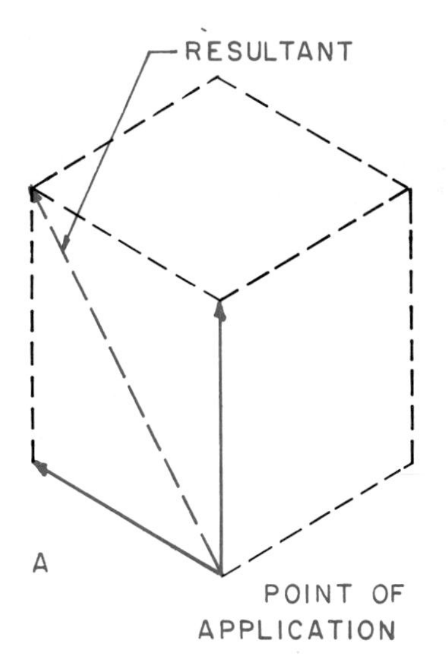

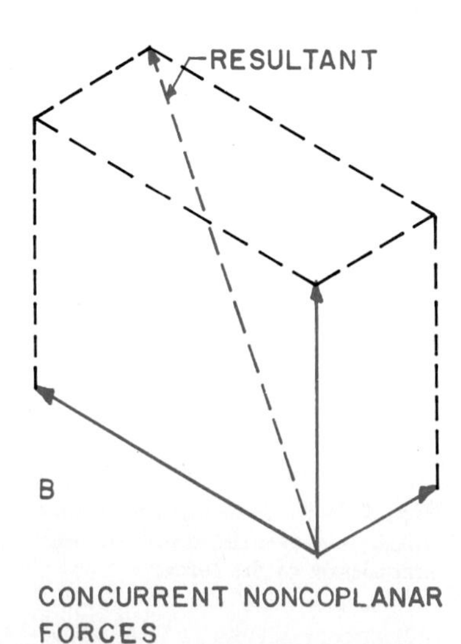

Fig. 19–8. Coplanar forces lie on one plane. Noncoplanar forces lie in different planes.

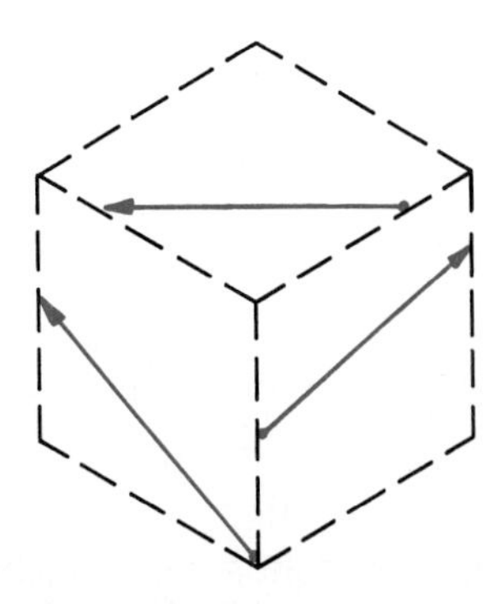

Fig. 19–9. These forces do not intersect at a common point.

The **resultant** is a single vector that can replace the other vectors. Examples are shown in Fig. 19-8.

The **equilibrant** is an equal and opposite balancing force.

Equilibrium is the state in which the object remains at rest at the same time that all forces are acting upon it.

Drawing Vector Diagrams

Simple vector diagrams are shown in Figs. 19-10 through 19-14. The point of application, PA, is located in each. The forces acting upon it must be known. In these problems two forces exist. The forces are represented by vectors. They are drawn from the point of application in the direction, magnitude, and sense in which they are acting. The magnitude of the force is indicated by the length of the vector. The direction of the force is indicated by the angle, and the sense is shown by the arrow on the end of the vector. The direction of the vector is given in degrees. The magnitude of the force is drawn to scale.

When you use vectors, be sure that the scale you select is large enough to permit the solution to be measured accurately. Accuracy in drawing is essential. Use a sharp-pointed pencil.

Addition of Two Concurrent Coplanar Vectors

The following is a simple example of the addition of coplanar forces. Fig. 19-10. Suppose two men find it necessary to move a filing cabinet. One man pushes with a force of 120 pounds. The other man pushes with a force of 90 pounds. The total force being used to move the cabinet is 210 pounds. Both forces were applied in the same direction. Since this was true, both forces were added.

During a picnic two groups of people decide to have a "tug of war." The total force exerted by Team B is 1,000 pounds. Team A put forth a total force of 1,500 pounds. Fig. 19-11 shows the graphical addition of these two forces. Since these two forces are applied in opposite directions, the resultant is 500 pounds. In both examples, the resultants are a simple addition of the quantities.

If the forces are not in the same line, they still may be added. Fig. 19-12 shows two vectors representing two forces acting on point PA. To find the force and direction of pull on PA, add the vectors graphically. To do this, draw lines parallel with the vectors through the ends of the vectors. These lines form a parallelogram. Then draw a diagonal from the point of application. The diagonal is the resultant force. The magnitude of the resultant is found by measuring the length of the diagonal. The resultant is a force that, if applied in the direction shown, would replace the other two forces.

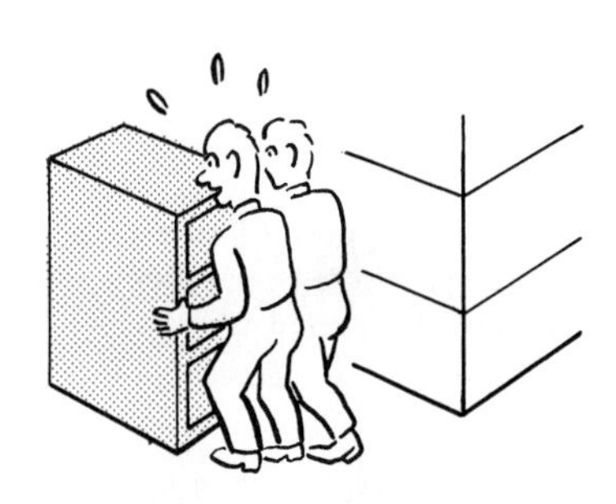

SCALE 1" = 50 LBS

SCALE 1" = 50 LBS
MAGNITUDE
2.4"
1.8"
120 LBS
90 LBS
4.2"
RESULTANT
POINT OF APPLICATION

Fig. 19–10. Forces may be added graphically to determine the total force.

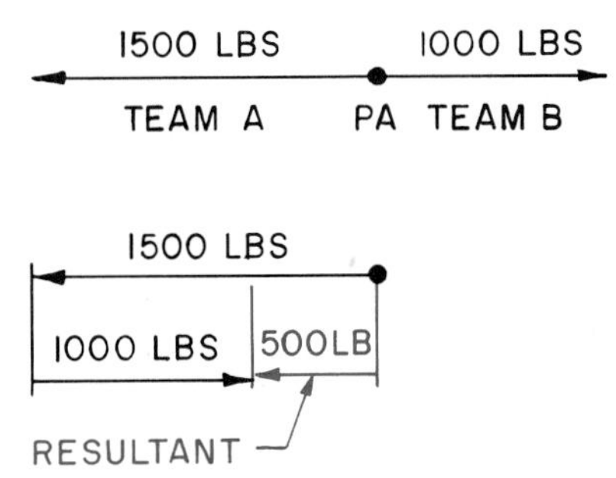

Fig. 19–11. Forces acting in opposite directions can be added graphically.

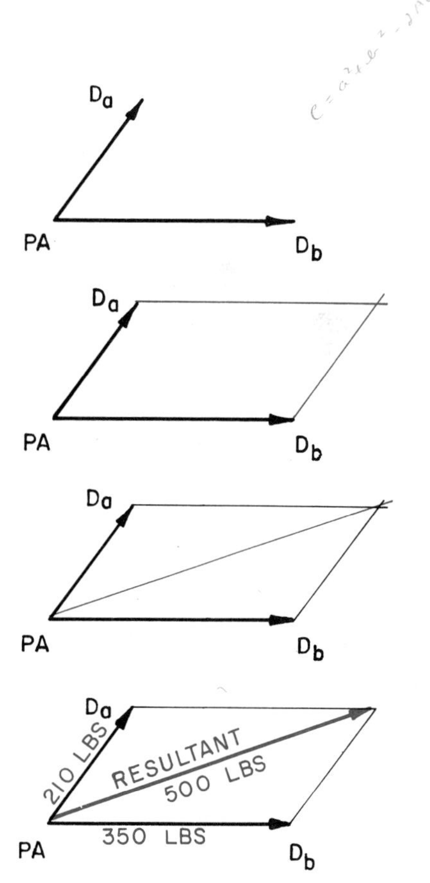

Fig. 19–12. A simple concurrent coplanar vector problem may be solved by closing the parallelogram and drawing a diagonal.

CAD Applications:

Ergonomics

If you have worked with microcomputers, you may be aware of some of the problems that can result when equipment doesn't take human factors into account. Eye strain, muscle aches, and fatigue are common complaints among computer users. Finding solutions to these problems involves ergonomics.

Ergonomics, or biotechnology, uses biological and engineering data to make interaction between humans and machines more efficient, comfortable, and safe.

Some makers of computer hardware have redesigned their equipment to make it more comfortable to use. Examples are monitors that tilt and swivel, keyboards that can be set at an angle, and mice that are shaped to fit the human hand.

Well-designed equipment, however, is only part of the answer. The work area itself must be designed for comfort and ease of use. The following describes some of the key elements in an ergonomically designed work area.

- *Work surface.* There should be plenty of room for the computer, keyboard, monitor, mouse or digitizer, printer/plotter, and other hardware as well as reference materials. Ideally, the keyboard should be somewhat lower than regular desk height. The monitor should be adjustable to various angles. Mice and other input devices should be within easy reach.
- *Chair.* The seat height should be adjustable so that the user's feet touch the floor and his or her thighs are parallel with the floor. The backrest should also be adjustable and should provide good support for the lower back.
- *Lighting.* The overhead lights and large windows of traditional drafting rooms produce glare on computer monitors. For CAD, it's better to have less overall lighting and instead to use smaller lights at the individual work areas. These should be adjustable to focus the light where it is needed.

No matter how advanced the equipment, humans are still needed to operate a CAD system. If the system is ergonomically designed, workers will be healthier, happier, and more productive.

Workspot™ workstation from Integrated Workstations, Inc.

A triangle can be used to save time in solving concurrent coplanar problems. A diagonal (resultant) of a parallelogram cuts it into two equal triangles. Fig. 19-13. The opposite sides of a parallelogram are equal. Since this is the case, a triangle can be used to solve the problem. Note that the sides of the triangle are the same length and direction as in one half of the parallelogram. To plot forces in a triangle, they are placed tip to tail. In this example, the resultant is drawn from PA with its tip touching the tip of vector D_a.

The vectors are moved keeping the same magnitude and direction until they form two sides of a triangle.

Another example of this type of graphical solution is shown in Fig. 19-14. A plane is flying due west at an air speed of 200 mph. The wind is blowing 75 mph from a southeasterly direction (315 degrees from north). Considering these two factors, what will be the air speed and direction of the plane?

By plotting the direction and magnitude (speed) of the plane and the wind, a parallelogram can be drawn. A diagonal from the PA becomes the resultant. In the illustration, the resultant and speed would be 260 mph at 282 degrees or N 78° W.

Addition of Three or More Concurrent Coplanar Vectors

More than two vectors may be added in a manner similar to that shown in Fig. 19-12. Three, four, or more vectors may be added as long as they are in the same plane. One method that is used develops a series of parallelograms.

Parallelogram Method. The parallelogram method is shown in Fig. 19-15. This figure shows how to find the resultant of three vectors in one plane.

First, find the resultant of two of the vectors. In Fig. 19-15, Step 2, vectors D1 and D2 are used. Their resultant is 180 pounds. Then this resultant is made into a parallelogram with vector D3. See Step 3 in Fig. 19-15. The new resultant is 270 pounds. This is the resultant for the three vectors, D1, D2, and D3.

If more than three vectors are involved, the same system is used. It requires the use of additional parallelograms.

Polygon Method. A second method of adding vectors is by the polygon method. This is only used for three or more vectors. A *polygon*, as you recall, is a geometric figure bounded by straight

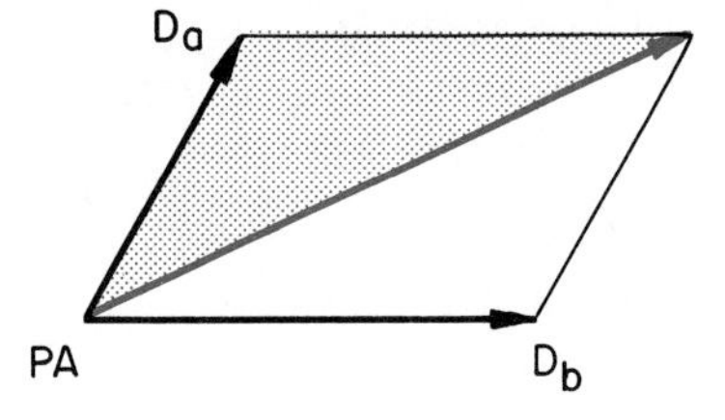

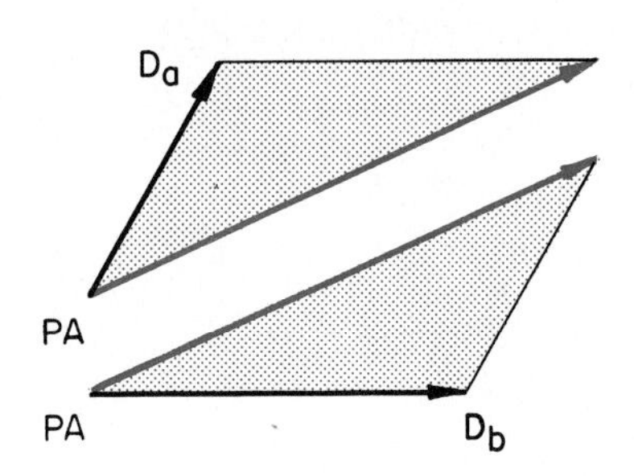

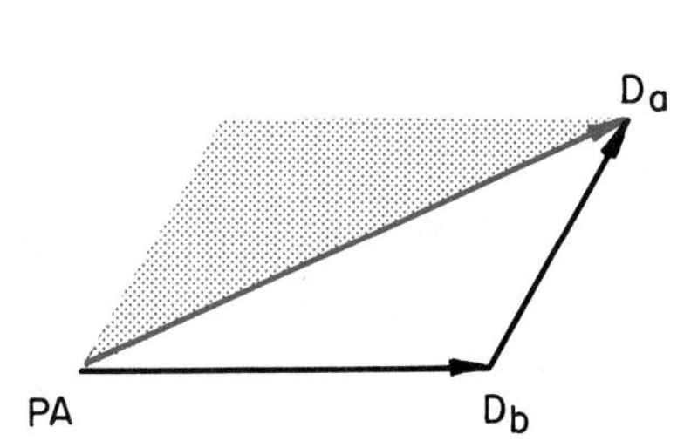

Fig. 19–13. A force triangle can be used to solve simple concurrent coplanar vector problems.

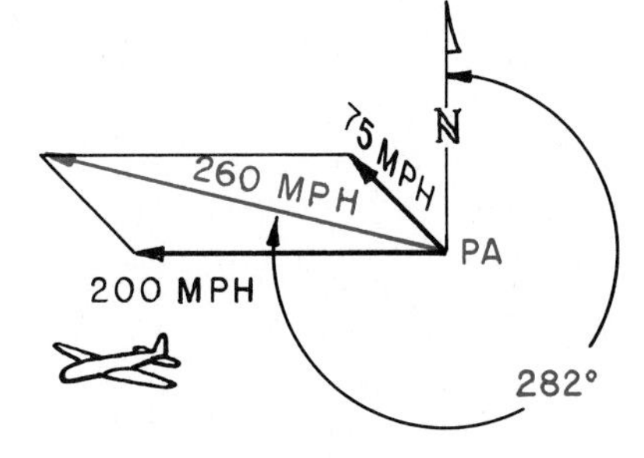

Fig. 19–14. Vectors can be used to solve speed and direction problems.

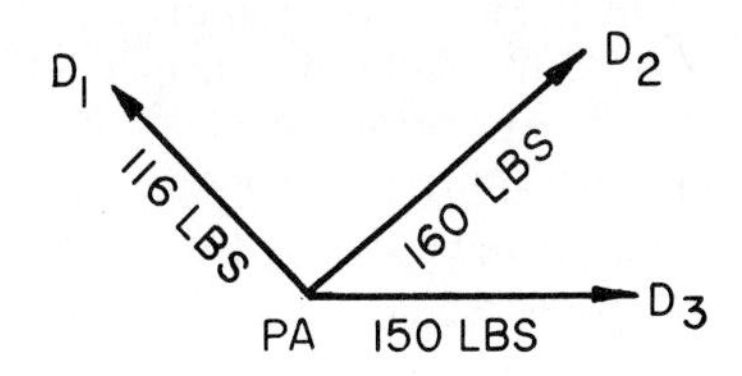

STEP 1

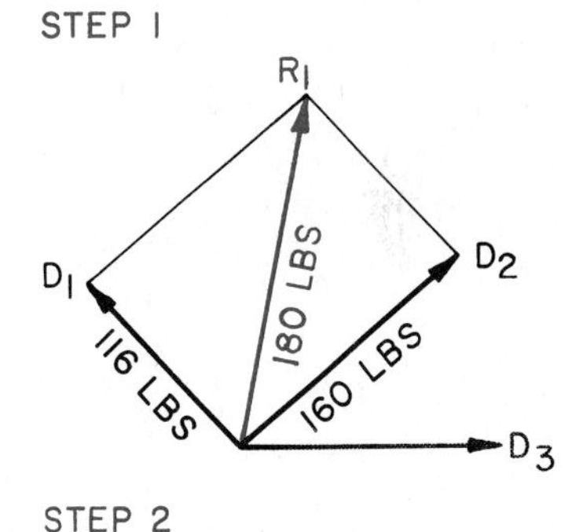

STEP 2

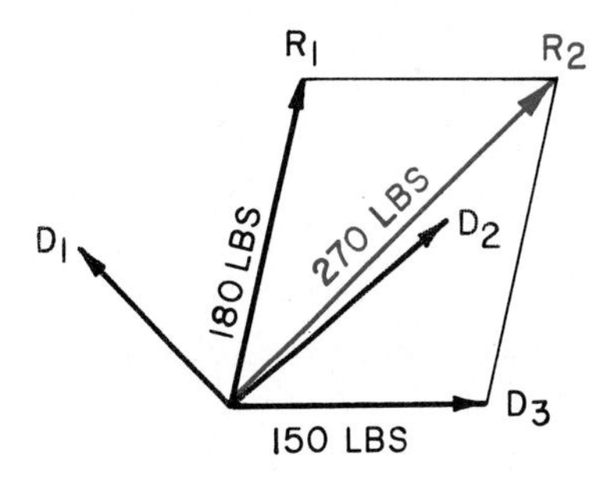

STEP 3

Fig. 19–15. Two or more concurrent coplanar forces can be added by drawing parallelograms.

lines. The polygon method is similar to the triangle method of addition. This method is shown in Fig. 19-16 using four vectors. The resultant can be found by moving the vectors so that the tip of one touches the tail of the next. See Fig. 19-16 at B. As the vectors are moved and connected, keep each in its original direction. In addition, you must accurately keep the magnitude of each. The resultant is found by connecting the tail of one, PA, to the tip of the last vector, D4. This resultant is the magnitude and direction of the four vectors on point PA.

Components

To this point you have learned how separate vectors can be added to yield a resultant vector. Each of the separate vectors is a component of the resultant. Now you will learn the opposite: that is, how to separate a force into its components.

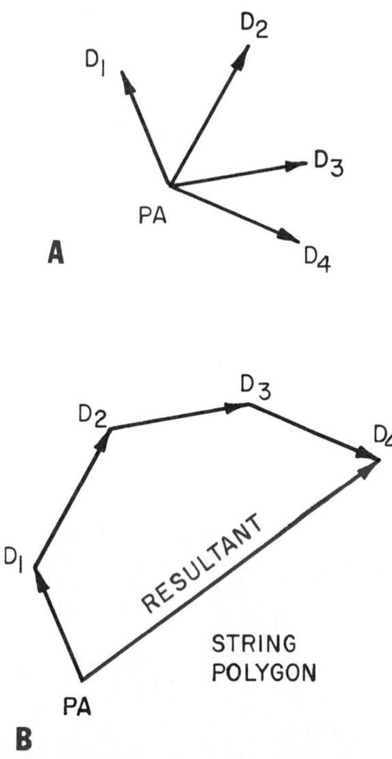

Fig. 19–16. Three or more concurrent coplanar forces can be added by using a string polygon.

Figure 19-17 shows a vector with a magnitude of 400 pounds. It is desired to split this into two components having the directions shown. The directions of the desired vectors must be known.

You can use the parallelogram method. The components form two sides of the parallelogram. Make sure that they are drawn in their specified directions. The known vector (400 pounds) forms the diagonal of the parallelogram. Draw it to scale. Completing the parallelogram will mark the component vectors to length. Measure them, using the same scale used to draw the known vector. This gives the magnitude of each component vector.

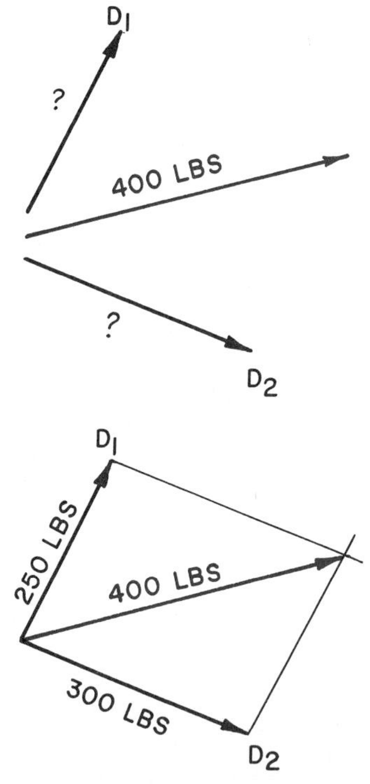

Fig. 19–17. A force can be resolved into its components.

500 LBS 500 LBS
P

Fig. 19–18. These forces are in equilibrium.

Equilibrium

Equilibrium, described earlier, is a condition of balance between opposing forces. Fig. 19-18 shows two equal forces pulling in opposite directions. Since they are equal, they are balanced. Neither force changes the position of point P. These forces are in equilibrium.

Fig. 19-19 shows a space diagram with three vectors drawn to scale. Is this diagram in equilibrium? The answer can be found

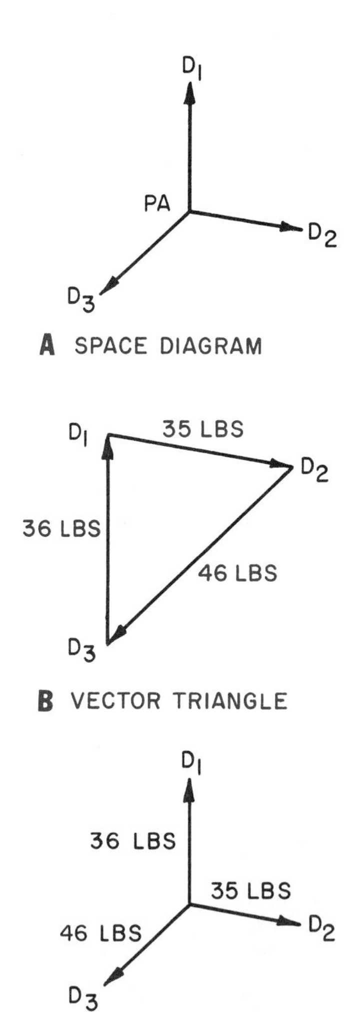

Fig. 19–19. How to determine whether forces are in equilibrium.

by constructing a vector triangle, as shown in Fig. 19-19 at B. Each side of the triangle is drawn with the original magnitude and direction of the vectors. An attempt is made to form a triangle. If a triangle can be formed, the diagram is in equilibrium.

An equilibrant force is one that will balance one or more forces. An equilibrant force acts in the opposite direction from a resultant force.

Fig. 19-20 shows a 300-pound weight supported by two cables. You need to find the forces in the two cables that will support the weight. These forces may be resolved graphically. To do this, draw a vector triangle. Begin by laying off the 300-pound force in its correct magnitude and direction. Next, lay off the two cables in their respective directions. When the triangle closes, scale the distances. These are the forces that are necessary to hold the weight in equilibrium.

Another example of the same type of problem is shown in Fig. 19-21. In this situation, a sign hangs from a support above a building entrance. To determine the correct size cable, you need to know the forces acting on the cable. To resolve the forces, first draw the vertical component in its correct magnitude. Now draw the support and the cable in their proper direction. These are the unknown forces in the diagram. The force polygon will now close since the forces are in equilibrium. Scaling the lines will indicate the forces acting on the cable and support.

The same principles may be used to find the force required to move a wheel over an obstruction. This problem may be solved just as the previous two have been. Fig. 19- 22. A force is applied level with the center of the wheel. The wheel has a 40-inch diameter and weighs 150 pounds. The forces acting on the wheel are unknown. How much force will be required to make the wheel rise over the block?

First, draw the known force. This is a vertical force of 150 pounds. Next, draw the other two forces parallel to the forces they represent. The horizontal member of the force triangle is the amount needed to move the wheel over the block. Suppose the horizontal force were raised above the center. Would the amount be smaller or larger?

Concurrent Noncoplanar Forces

Any series of forces that have the same point of application and are not acting in the same plane are called *concurrent noncoplanar* forces. Fig. 19-23 shows three forces acting from a single point but not in the same plane. In this example, you want to find the resultant of forces OA, OB, and OC. One of the easiest methods to find the needed resultant is to use a series of parallelograms.

Parallelogram Method. When using the parallelogram method, first draw the top and front views of the vectors. Then find the resultant (R_1) of two of the forces (OA and OB) in both views. Fig. 19-23 at *A* and *B*. Next combine R_1 with force OC giving the resultant R_2. Fig. 19-23 at *C* and *D*.

D2
D3
D1
300 LBS
D3
250 LBS
300 LBS
D2
150 LBS
D1

A SPACE DIAGRAM **B** VECTOR TRIANGLE

Fig. 19–20. **A vector triangle is used to find the forces necessary to hold the 300-pound force in equilibrium.**

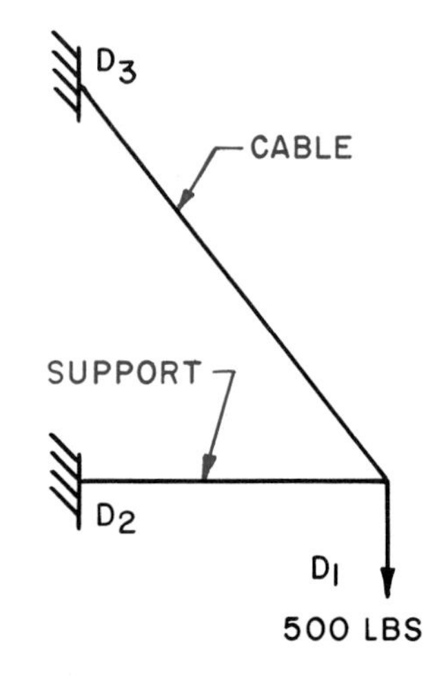

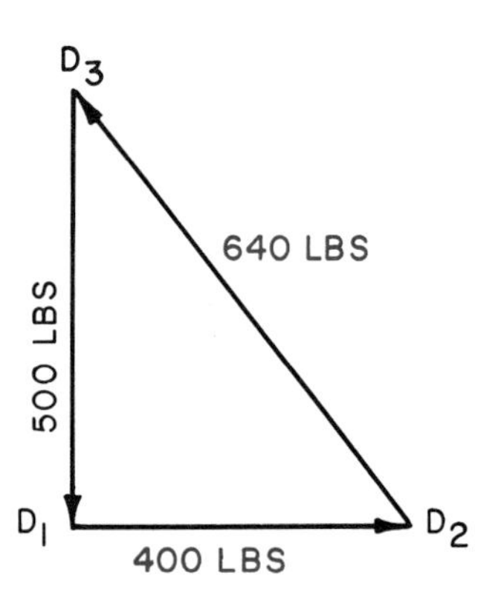

B VECTOR TRIANGLE

Fig. 19–21. **The forces needed in the cable and supports to hold a sign are found by drawing a vector triangle.**

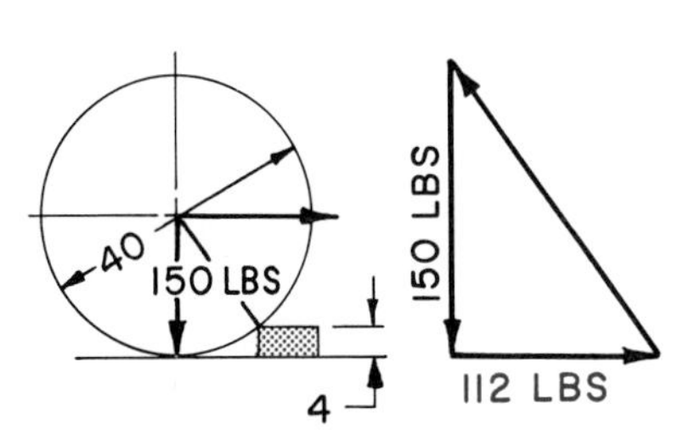

Fig. 19–22. **How much force is required to roll the wheel over the 4-inch block?**

To find the magnitude of the final resultant, revolve R_2 parallel to a plane of projection. In the example, R_2 is revolved parallel to the vertical plane. The true length is then scaled in the front view.

Parallelepiped Method. An alternate method of finding the resultant is by using a parallelepiped. A **parallelepiped** is a prism with a square or rectangular base. Fig. 19-24. The resultant of three concurrent noncoplanar forces is the body diagonal of a parallelepiped. Fig. 19-25 shows a parallelepiped with a body diagonal representing a resultant. This problem uses the same three forces shown in Fig. 19-23. From the terminal points of each force, construct the sides of a parallelepiped. For example, draw edge BX parallel to OA in both views. Next, draw side AX parallel to OB in both views. Then draw edge CY parallel to OB. Finally, draw edges YZ and XZ. When the figure is completed, the body diagonal, OR, is the resultant. To find its magnitude, revolve it parallel to one of the planes of projection. In this case, R is revolved parallel to the horizontal plane.

In some instances you may need to resolve a force into its components. If the direction and sense of the components are known, their magnitude can be determined. Then a parallelepiped may be used to determine the components.

Fig. 19-26 shows three unknown forces and their resultant. It shows their top and front views. In order to find the components, begin by drawing the opposite end of the parallelepiped. Fig. 19-26 at B shows that the parallelepiped is begun at the end of the resultant. In both views, draw two sides of the base, RE and RD, parallel to OB and OA respectively. The third step is to determine where the edge, OC, pierces the base of the parallelepiped.

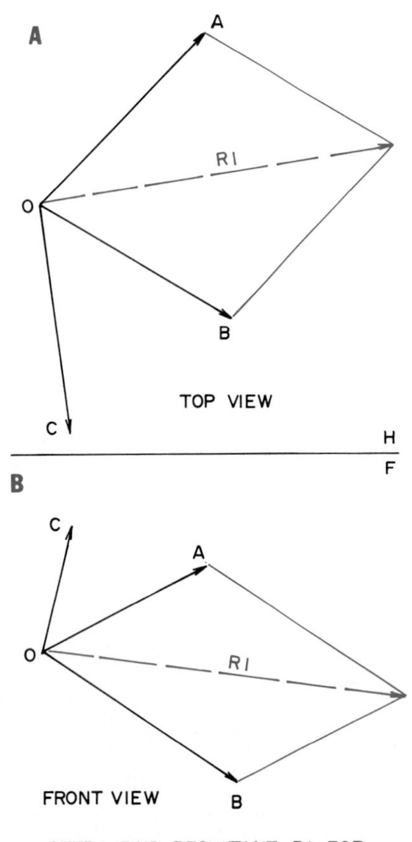

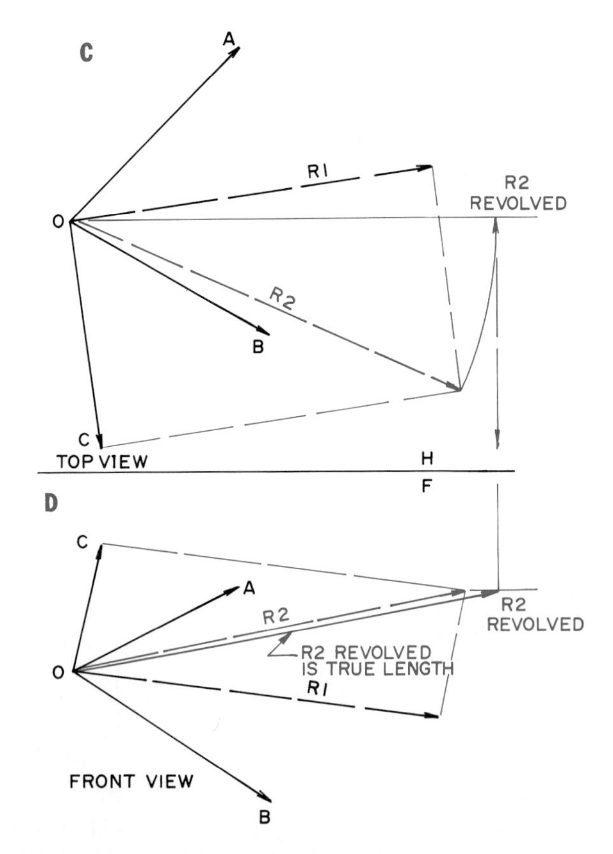

Fig. 19–23. The steps used to determine a force by the parallelogram method.

Part C shows the application of determining where a line pierces a plane. Where line OC pierces the base is the length of one of the edges of the parallelepiped. Next, the base can be completed from point C_p shown in Fig. 19-26 at D. The magnitudes of forces OA, OB, and OC can be scaled at this time; however, the usual steps are to complete the parallelepiped (Part E) and determine the visibility (Part F). To obtain their magnitudes, each force, OA, OB, and OC, must be revolved parallel to a plane of projection.

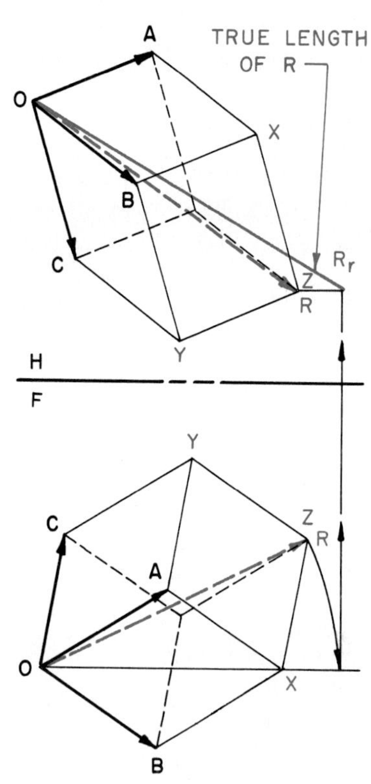

Fig. 19–25. The resultant of three concurrent noncoplanar forces may be found by using a parallelepiped.

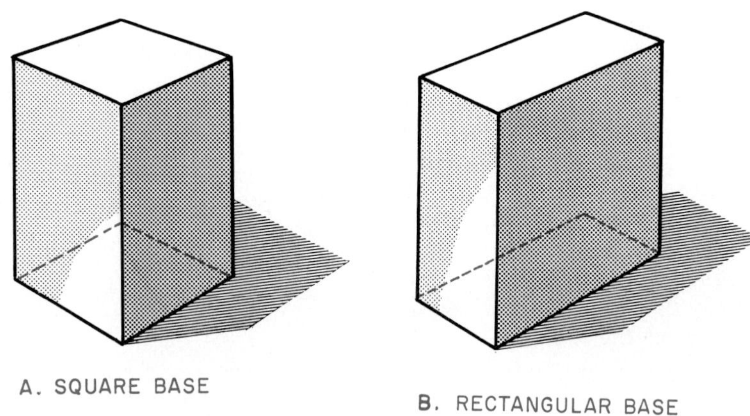

Fig. 19–24. A parallelepiped is a prism based on a parallelogram.

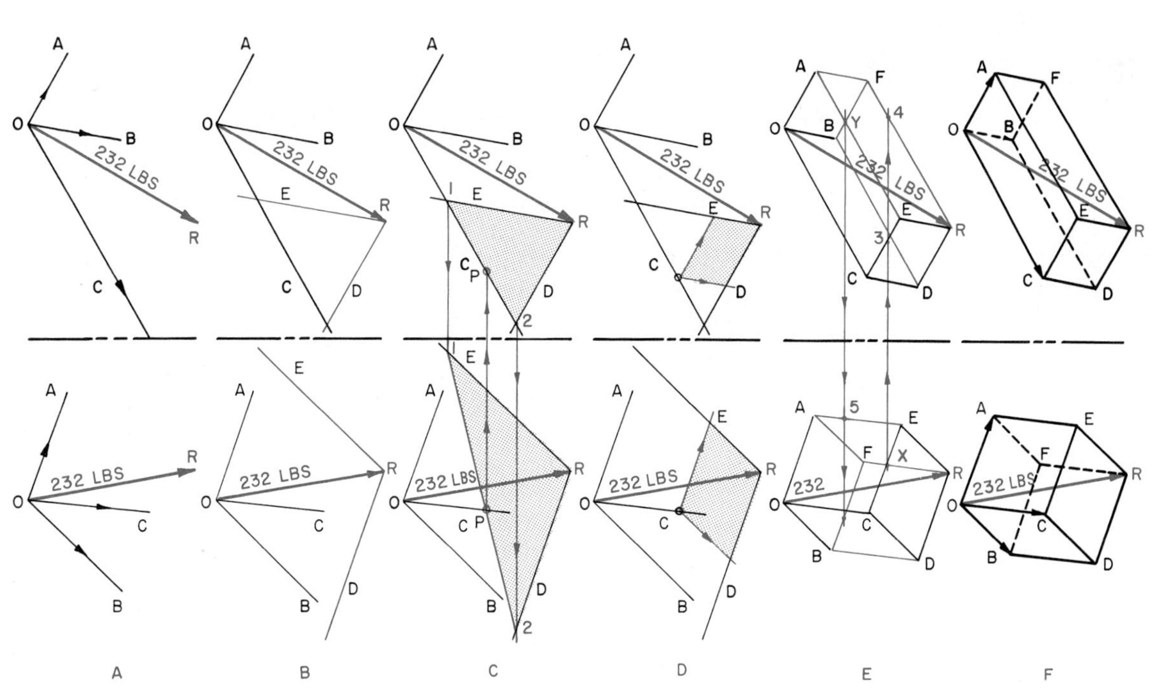

Fig. 19–26. The components of a resultant can be found by drawing a parallelepiped. The resultant is the body diagonal.

Chapter Review

Build Your Vocabulary

You should understand and use the following terms as part of your working vocabulary. Write a brief explanation of what each means.

tension
compression
shear
moment of force
vector
force
velocity
sense
direction
magnitude
point of application
coplanar
concurrent
noncoplanar
nonconcurrent
resultant
equilibrant
equilibrium
parallelepiped

Sharpen Your Math Skills

1. If one vector is on an angle of 35°15′ and another at 72°12′, what size is the angle between them?

2. If a vector has a magnitude of 500 pounds, and ¾″ represents 100 pounds, how long would you draw a line to represent the vector?

3. If a horse pulls on a rope with a force of 2150 pounds, and a tractor pulls on the other end with a force of 3520 pounds, which one wins the tug-of-war? How many extra pounds of force does the victor have?

4. If a vector has a magnitude of 3200 newtons, and 100 millimeters represent 200 newtons, how long would you draw this vector?

5. Two people push a crate. One pushes with a force of 200 pounds and the other 150 pounds. What is the total force used to move the crate?

Study Problems—Directions

The problems that follow will give you the chance to use some of the knowledge of forces that you learned in this chapter. Do the problems that are assigned to you by your instructor. Work carefully and accurately.

1. Two cars are tied together. One pulls with a force of 5,000 pounds. The other pulls with a force of 3,500 pounds. Draw a diagram of these forces. If each car is trying to pull the other under full power, which car will pull the other?

2. Find the resultant of the concurrent coplanar vectors shown in **Fig. P19-1**. Use the parallelogram method.

3. Find the resultant of the concurrent coplanar vectors shown in **Fig. P19-2**. Use the triangle method.

4. Find the resultant of the three concurrent coplanar vectors shown in **Fig. P19-3**. First, solve it with the parallelogram method. Then use the string polygon method. How close are the results?

5. Find the actual speed and direction of flight of the airplane shown in **Fig. P19-4**.

6. Find the coplanar components of the force shown in **Fig. P19-5**.

7. Check the vector diagram, **Fig. P19-6**, to see if it is in equilibrium.

8. Find the force on cables AB and AC, **Fig. P19-7**.

9. Find the true magnitude of the resultant of the three concurrent noncoplanar forces, A, B, and C, in **Fig. P19-8**. The lengths dimensioned on the drawing are the lengths to use to lay out the problem. These vectors do not appear true length on the drawing. Measure the force of the resultant, using the scale 1″ = 200 pounds.

10. Find the weight on each support holding the sign shown in **Fig. P19-9.**

11. Find the resultant of the concurrent noncoplanar forces in **Fig. P19-10**. The lengths shown are to be used to lay out the problem. They are not the true lengths of the components.

12. Find the size of the two forces needed in **Fig. P19-11** to place the known force in equilibrium.

Study Problems

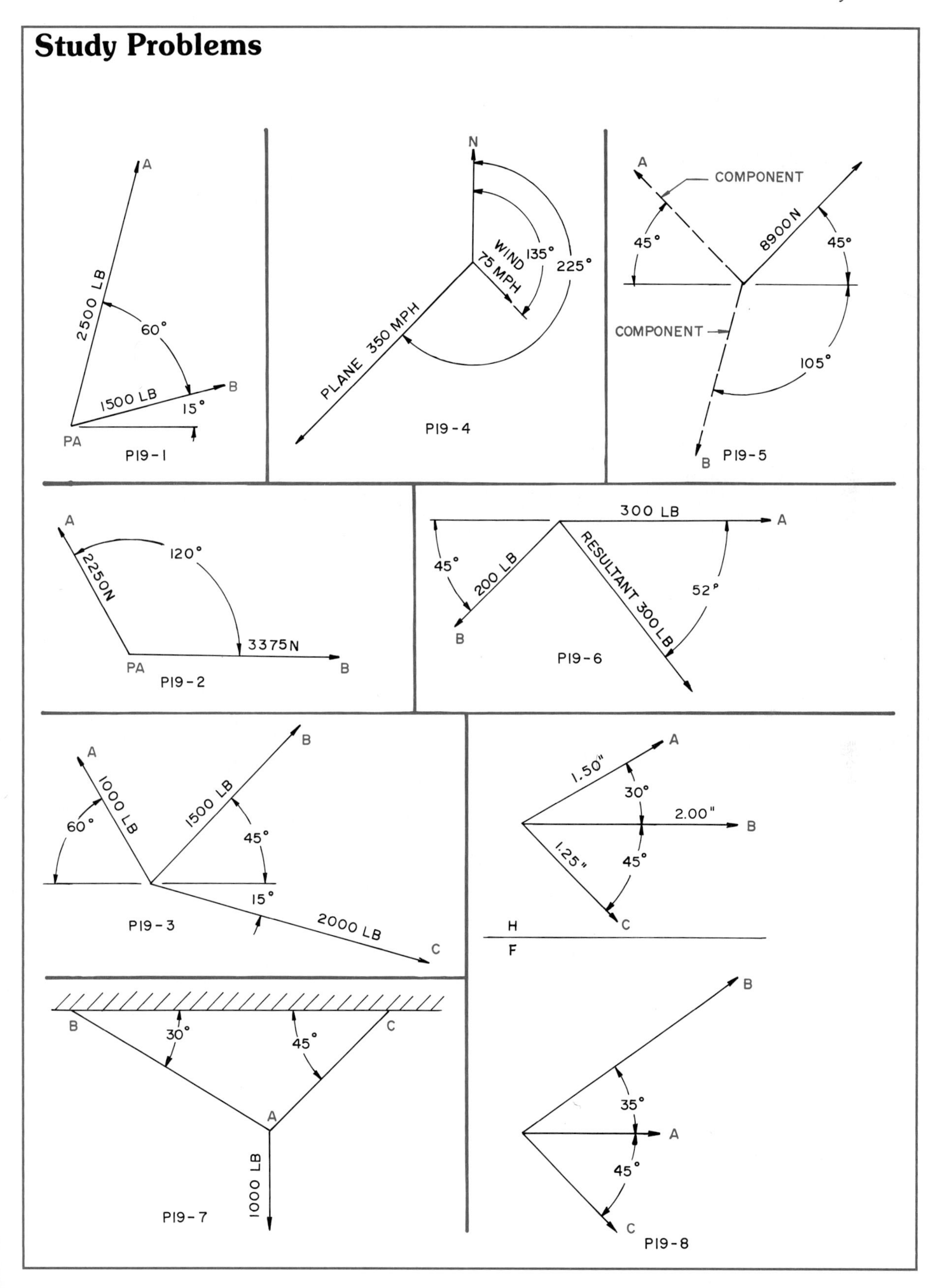

Study Problems

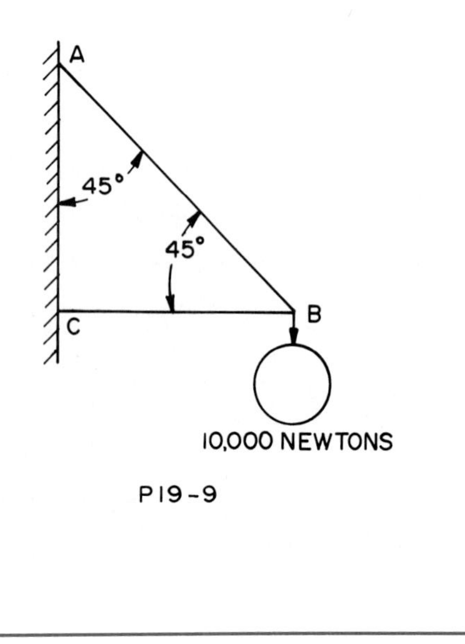

P19-9

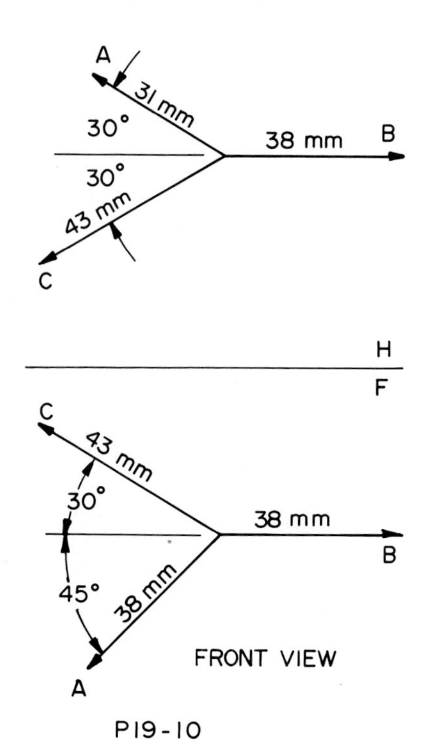

P19-10

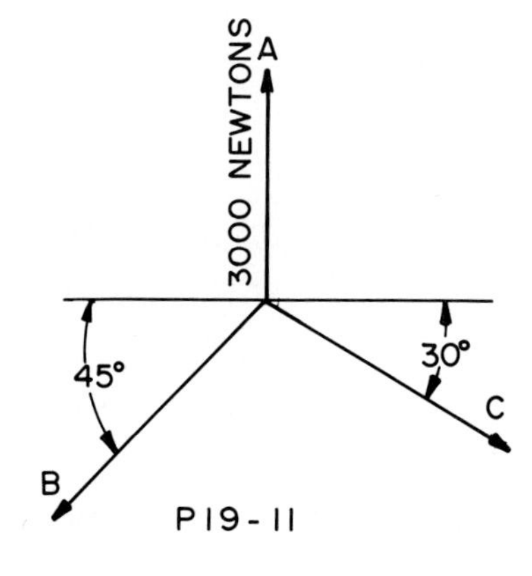

P19-11

CHAPTER 20

Technical Illustration

After studying this chapter, you should be able to do the following:

- Prepare technical illustrations to meet the requirements for a variety of situations.
- Do simple work with an airbrush.
- Apply several types of shading to illustrations.

Introduction

Technical illustrations are pictorial drawings made from orthographic drawings, mockups, models, or photographs. Their purpose is to present technical information in the form of easily understood pictures. For example, they may show how a device is to be assembled. While some may contain a few dimensions, the individual parts are not dimensioned. This information is found on the working drawings.

The common types of pictorial drawings used in technical illustration are isometric, dimetric, trimetric, oblique, and perspective. Study Chapter 14, "Pictorial Drawing," for details about making each of these. Once these pictorial drawing fundamentals are mastered, you can begin to make technical illustrations.

Uses and Types of Illustrations

Two kinds of technical illustrations are common, engineering and publication. **Engineering illustrations** are made for use in manufacturing, assembly, and installation. These drawings are accurately and carefully drawn. They show as simply as possible the information needed. Shading and other ways of making them attractive are not used. Workers having little ability to read working drawings can use engineering illustrations. A simple illustration is shown in Fig. 20-1A.

Publication illustrations are used in catalogs, maintenance and repair manuals, and sales and advertising copy. They are sometimes shaded to present a more lifelike appearance. They do not require the accuracy of engineering illustrations. The parts must be kept in the proper proportion and relationship, but the illustrator frequently estimates their size and shape. It is important to produce a clear and attractive drawing. Fig. 20-1B.

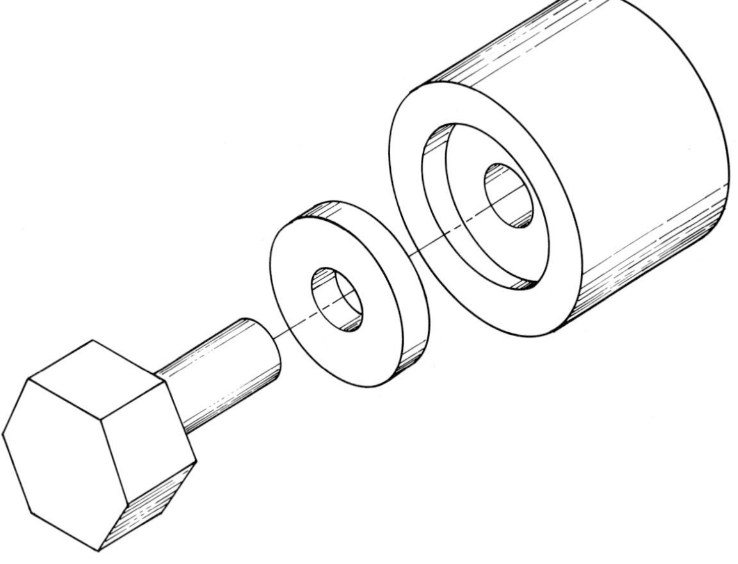

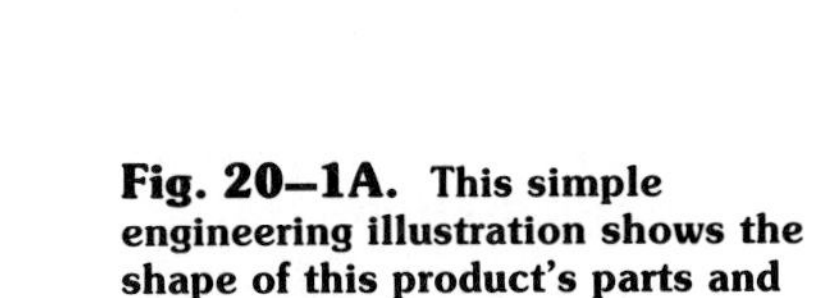

Fig. 20–1A. This simple engineering illustration shows the shape of this product's parts and their relationship to one another.

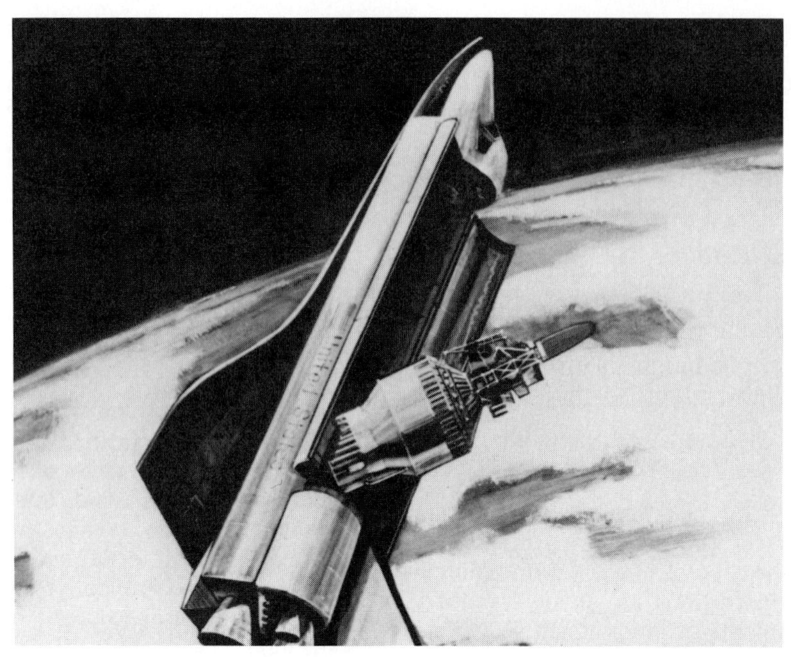

Fig. 20–1B. This technical illustration shows a Centaur launch vehicle, adapted for use from the Space Shuttle orbiter, launching a probe. (NASA)

Fig. 20–2. This artist is making a full-color illustration. He is working from a photograph and a model. This work requires a great deal of skill and artistic talent. (Harlan Krug, Northrop Norair, Hawthorne, CA, and *Industrial Art Methods Magazine*)

Skills Required of the Illustrator

Illustrators must be able to read working drawings. They must be able to use isometric, dimetric, trimetric, and perspective drawing techniques. Illustrators must be able to use an airbrush and know how to work with color. Skill in drafting and freehand sketching as well as a knowledge of art is needed. In Fig. 20-2 an illustrator is shown at work.

Pictorial illustrations can also be made using a computer. The illustrator must be familiar with the procedures for producing illustrations using the computer.

Procedure for Making a Technical Illustration

You should make and follow a definite plan when drawing a technical illustration. While experienced illustrators tend to develop their own procedures, the following plan has proven useful.

1. Understand the purpose of the illustration. Is it to show assembly, installation, maintenance, or repair information?
2. Study available material on the device to be illustrated. This could be a working drawing or the device itself.
3. Make several freehand sketches until a good layout has been developed.
4. Draw an accurate pencil illustration using drafting instruments.
5. Make the final tracing of the illustration.
6. Add lettering and notes.
7. Mount the illustration on cardboard if desired.

Determining the Purpose of the Illustration

It is important to know if the illustration is to be used for assembly, advertising, or some other purpose. Knowing its purpose will help you decide how much time should be spent on accuracy, shading, color, and other factors. If it is made for use under a military contract, Department of Defense specifications must be followed. If it is to be reproduced by printing, the size for printing must be decided. Since drawings are reduced photographically for printing, the amount of reduction must be known. This is usually 2 to 1 or 1½ to 1. A 2 to 1 reduction means the drawing will be reduced one-half (50%) the size it is drawn. For example, a 4 by 6 inch illustration will be printed 2 by 3 inches. A caution here is to realize that when there is a 50% reduction the lines are half as wide as they were drawn.

Studying Available Material

Usually illustrations are made from working drawings. Study these so that you understand each part clearly as well as the relationship between parts. Sometimes the illustration is made from the device itself. It can be disassembled for study.

Making Freehand Sketches

Try several freehand sketches until the object is in a position in which it can most clearly serve its purpose. Usually the object is placed so that it appears in its normal position. Accuracy is not vital in this step but parts should be in proportion to one another. Decide if an isometric, perspective, or other pictorial is best. Figs. 20-3 and 20-4. Make certain no parts have been omitted.

Making the Technical Illustration

After making a number of freehand sketches, select the one that most clearly shows the details of the object. Fig. 20-5 at A. Then, using instruments, block in the overall mass of the various parts. Locate the center lines as needed. These should be very light construction lines. Fig. 20-5 at B. The use of templates speeds up this blocking-in process. Next, add the details to each part such as holes, fillets, rounds, and threads. Fig. 20-5 at C. After the details are located, make the finished drawing. If the layout was done lightly, the excess layout lines can be erased and the remaining lines darkened. Remember to make them wide enough to allow for photographic reduction if the illustration is to be printed. Fig. 20-5 at D. Some illustrators prefer to make the layout with a blue nonreproducible pencil. Such lines will not photograph; therefore, it is not necessary to erase them.

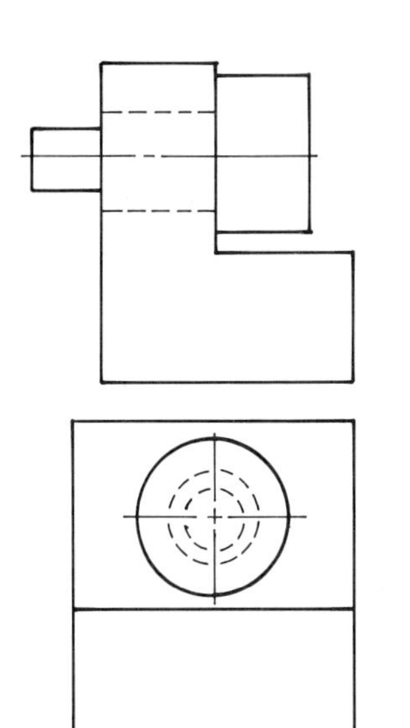

Fig. 20–3. **This two-view drawing might be hard for someone without drafting knowledge to understand.**

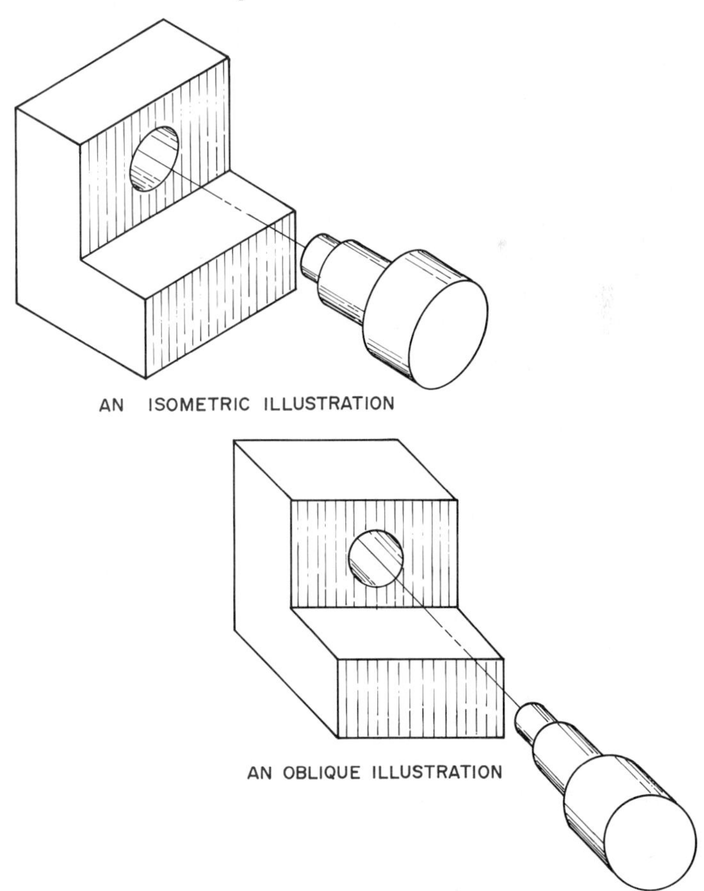

Fig. 20–4. **Here are two technical illustrations of the object in Fig. 20–3. Which of these do you think shows better the object's appearance and the way the parts fit together?**

Other illustrators prefer to place a sheet of vellum or plastic drafting film over the layout and trace the finished illustration in pencil or ink. Plastic drafting film is especially good for inking.

If an object has duplicate parts, draw one and use it to trace the others. This method is fast and insures that all the parts will be the same size.

Since many objects have rounded or cylindrical parts, the task of drawing them pictorially is time-consuming. Use templates to speed up drafting time. If templates are not available, pictorially draw a variety of sizes of circles and trace them where needed.

If the illustration is to be reproduced by the blueprint or diazo process, you can darken the object lines on the instrument drawing in pencil, remove construction lines, and use that drawing for the final copy. If the illustration is to be used for photographic reproduction, you should ink the lines. While the instrument drawing can be inked, it often is rather dirty and so should be traced in ink on a separate sheet.

Making the Final Tracing

You can trace the instrument illustration in ink on vellum, tracing cloth, drafting film, or two-ply Strathmore paper. Before tracing, make certain the instrument illustration is correct. Make any changes before inking.

If you ink on a transparent material, such as vellum, the vellum sheet can be placed over the instrument illustration and traced. If you use an opaque material, such as Strathmore paper, the illustration will have to be redrawn or traced on with carbon paper.

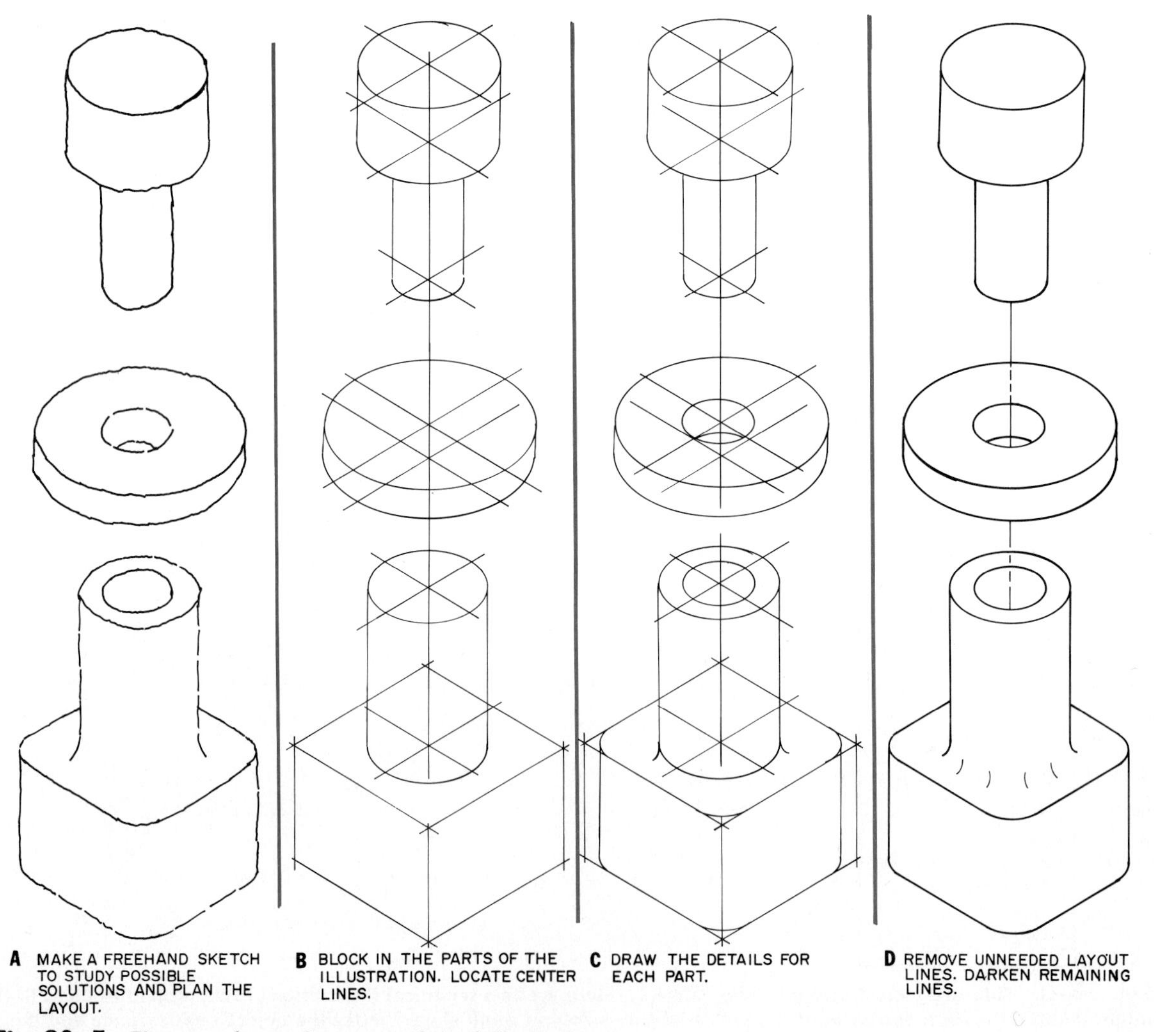

A MAKE A FREEHAND SKETCH TO STUDY POSSIBLE SOLUTIONS AND PLAN THE LAYOUT.
B BLOCK IN THE PARTS OF THE ILLUSTRATION. LOCATE CENTER LINES.
C DRAW THE DETAILS FOR EACH PART.
D REMOVE UNNEEDED LAYOUT LINES. DARKEN REMAINING LINES.

Fig. 20–5. The steps for producing a technical illustration.

Several companies supply transparent sheets with standard fasteners, such as bolts and nuts, printed on them. These are cut from the sheet and pressed into place on the drawing. Their adhesive back makes them stick in place. Fig. 20-6. This speeds up the drafting time and produces high-quality, clear copy.

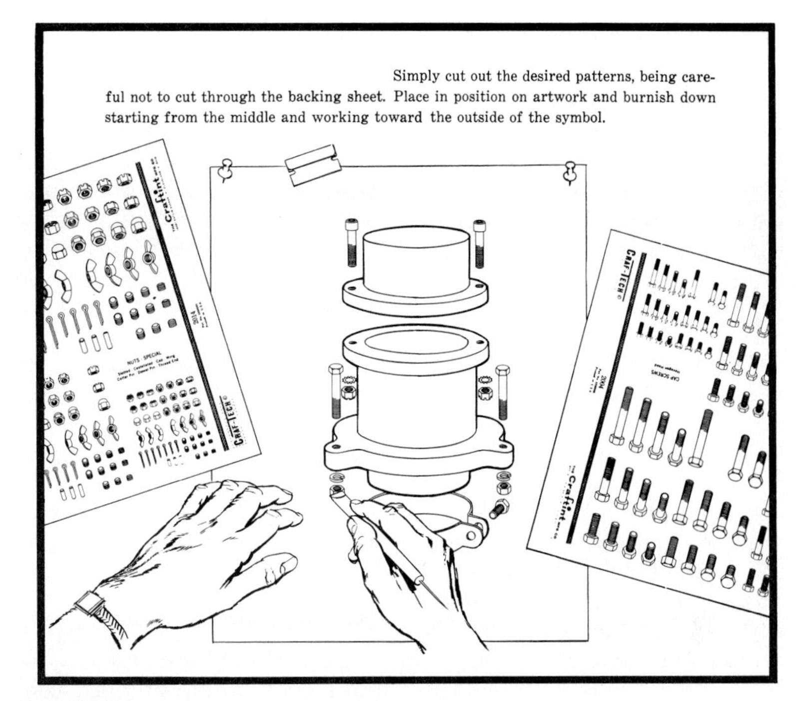

Fig. 20–6. Drawings of fasteners are available on adhesive-backed acetate sheets.

Adding Lettering and Notes

The parts of an object in a technical illustration sometimes require identification. This can be done by numbering each part or lettering the name of the part on the drawing. The numbers may be enclosed in a circle. If some details about the parts must be given, the parts are usually numbered and a parts list is used to give the details. Refer to Fig. 20-32. The placement of the identifying numbers influences the appearance of the drawing and the ease with which it is read. Connect the identifying name or number to the part with a leader. The leader can be straight or angled.

Identifying numbers may be hand-lettered, machine-lettered, or placed on the drawing by cutting them from sheets having the numbers printed on a transparent, adhesive-backed material. The method for using these letters is shown in Fig. 20-7. Other ways for mechanically lettering illustrations are shown in Chapter 3, "Tools and Techniques of Drafting."

If the name of the part is to be lettered on the drawing, place it so as to enhance the appearance. The lettering is usually placed horizontally and tied to the part with a leader.

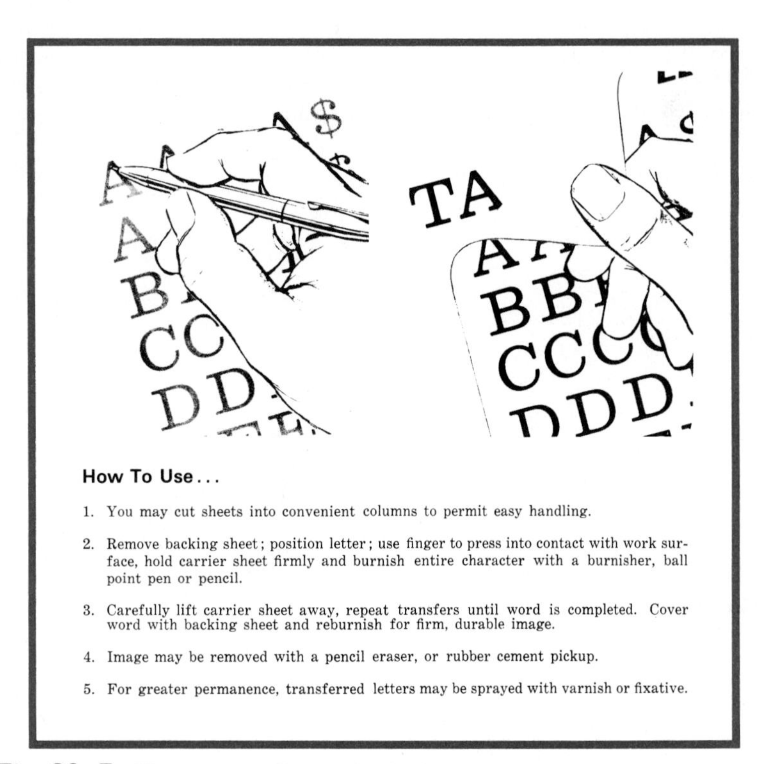

Fig. 20–7. How to use adhesive-backed lettering.

Mounting the Illustration

If the illustration is to be displayed, it can be mounted on a heavy cardboard, called *mountboard*. It can be framed with a mat in the same manner as a painting. Frequently it is covered with a transparent material such as acetate.

Shading Techniques

Shading is a technique of varying light intensity on surfaces by lines or tones. It is used frequently in technical illustrations.

Engineering illustrations are usually not shaded. The only time shading is used is when it makes the drawing easier to understand. Shading takes time and therefore costs a company more money. Publication illustrations usually have some simple form of shading to improve their appearance. Shading provides the appearance of depth.

Study the shading techniques used on the illustrations in this chapter. You will notice some differences in technique, depending upon the skills and desire of the illustrator. All variations are satisfactory if they contribute to the primary purpose of shading, a clear and pleasing appearance.

Source of Light

The direction of the source of light that is assumed to be shining on the object must be decided. While the direction is commonly taken on a 45-degree angle from the upper left corner of the drawing board, any direction can be used. The angle chosen should be the one which will best show the planes of the object.

The sides directly facing the light source will have no shading. Surfaces with small angles receive less light and therefore have some shade. Those surfaces directly opposite the light source are in deep shadow and have the darkest shading.

Exterior Line Shading

The simplest form of line shading is done by drawing the exterior lines of the object considerably heavier than those within the object. The interior lines are frequently broken. Fig. 20-8. It is common practice to break the heavy object line if it is going to intersect another line from a different plane. This helps to show the separation of the planes in the illustration.

Solid Line Shading

Another method of line shading is shown in Fig. 20-9. With **solid line shading**, a shadow effect is created by drawing a wide, solid, black band parallel to the edges of the object farthest from the light source. One or two single lines are drawn beyond the black band. They are spaced by eye.

Surface Line Shading

Still another line shadow technique is shown in Fig. 20-10. The technique of **surface line shading** is to draw lines in the dark areas closer than the lines in light areas. Also, the lines in the areas with the darkest shadows are drawn thicker than those in the lighter areas. The lines are spaced by eye.

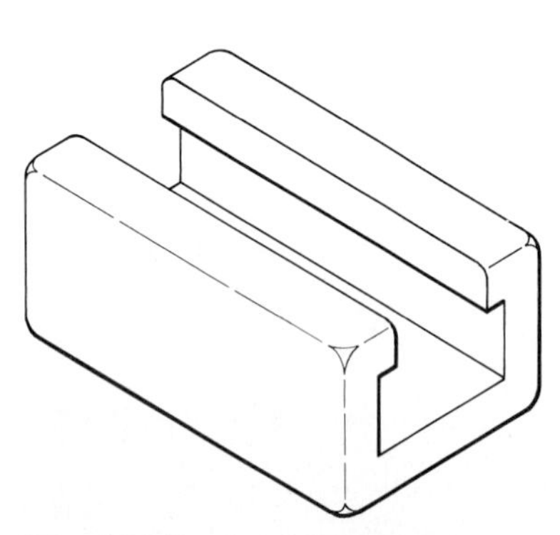

Fig. 20–8. This illustration shows the simplest form of shading. It uses a heavy line to indicate a cast shadow, a medium-width line to indicate visible edges, and a light, broken line to indicate a rounded surface. The light comes from the upper left.

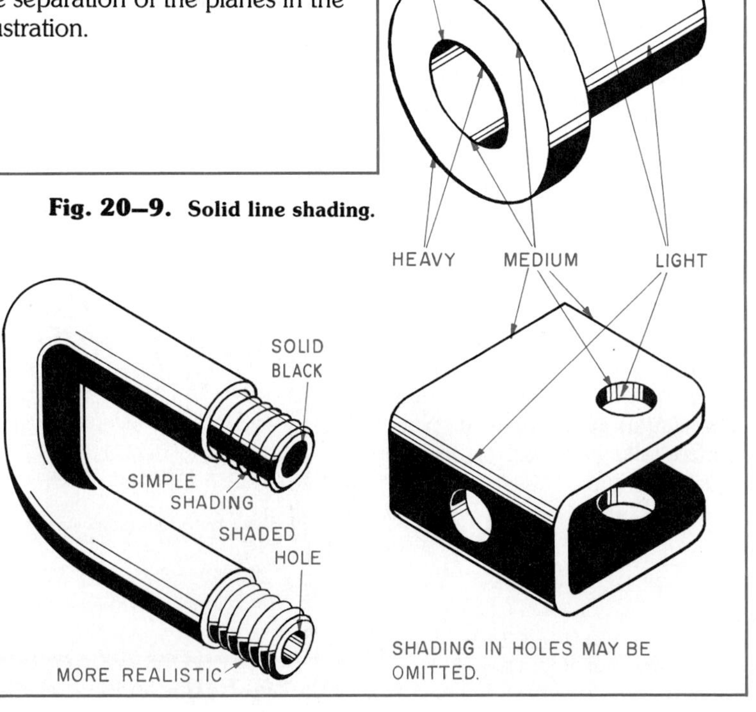

Fig. 20–9. Solid line shading.

The lines should usually parallel the edge of the surface that is to receive the shadow. In other words, vertical surfaces should have vertical shadow lines. Illustrators usually shade lightly one-fifth of the surface nearest the light source, leave two-fifths clear, and shade heavily the two-fifths farthest from the light.

Stippling

An effective shadow technique is achieved by **stippling**. The surface to be shaded is covered with dots spaced at random. The darker areas have the dots closer together. Lighter areas have the dots farther apart. The dots are made freehand. The sharpness of the pencil point or size of the point on the inking device is important. A fine point is used on light areas and a larger point on darker areas. This method takes a lot of time but produces an attractive drawing. Fig. 20-11.

Smudge Shading

If a grayed shading is desired, you can use **smudge shading**. Smudge shading can be used on one surface or a complete part to make it stand out. Fig. 20-12.

An easy way to do smudge shading is to sand some carbon from the lead of a pencil. Put this on a cloth and rub it over the surface to be shaded. Artist's pastel sticks also can be used.

Colored leads can be used to call attention to parts of an object. An example would be to color the hydraulic lines on an illustration of an automobile chassis.

Adhesive-Backed Shading Products

Several companies sell shading patterns of dots, lines, stippling, and other patterns. Fig. 20-13. They are available in black and colors. The patterns are printed on a transparent sheet having an adhesive back. One type is placed over the area to be shaded and rubbed with a smooth object. The pattern sticks to the drawing and pulls loose from the transparent sheet. Fig. 20-14. Another type must be cut to the shape of the surface. It is then pressed to stick on the drawing. Fig. 20-15. Both types are fast and effective.

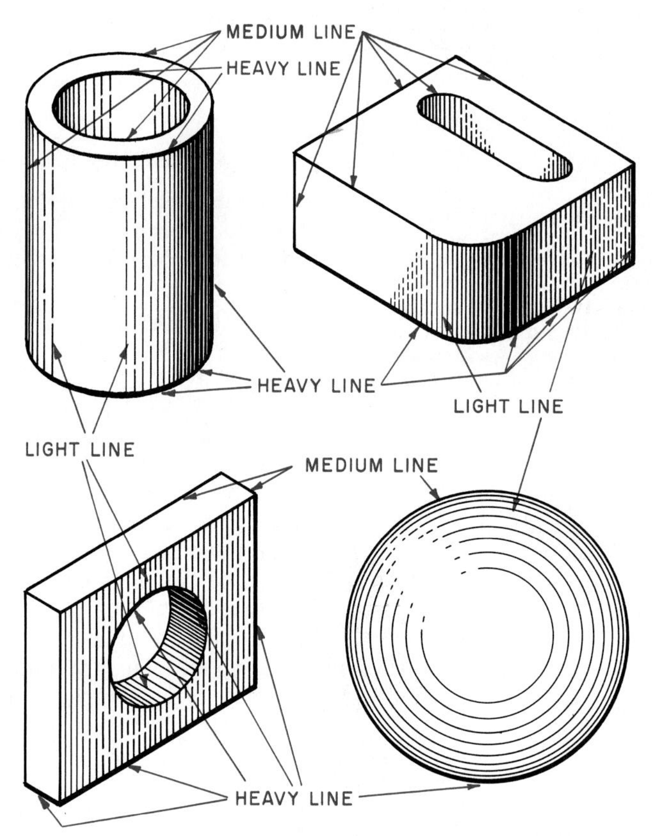

Fig. 20–10. Surface line shading is done by varying both the line width and space between lines. The light is coming from the upper left.

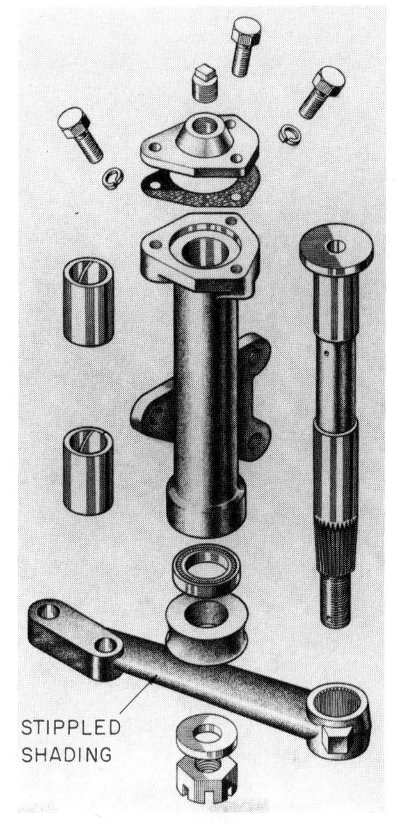

Fig. 20–11. A technical illustration made with stippled shading. Stippled shading is done by varying the size and spacing of dots. (George A. Jovellas, Cambridge, MA)

Fig. 20–12. **An example of smudge shading.**

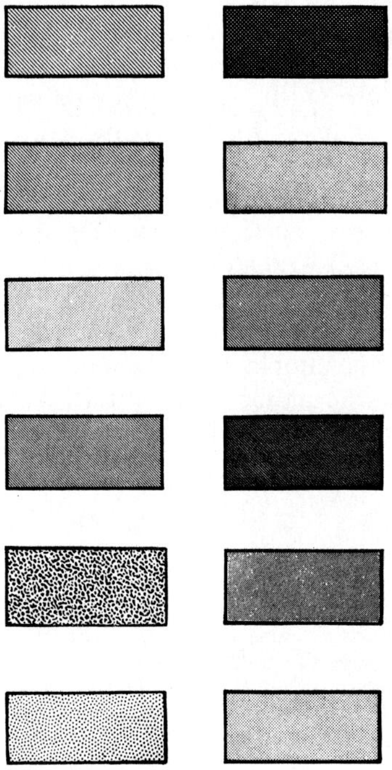

Fig. 20–13. **A few of the patterns available with adhesive-backed shading products.**

Fig. 20–14. **Place the sheet of shading over the area to be shaded. Rub the area with a burnisher. The dots in the burnished area stick to the drawing. (Instantex by Letraset Limited)**

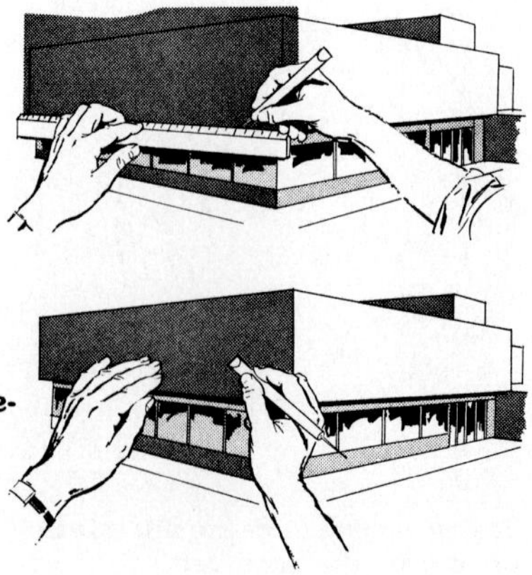

Fig. 20–15. **Applying adhesive-backed shading.**

Applying the Adhesive-Backed Shading. To use adhesive-backed shading, follow these steps:

1. Make sure that the area to be shaded is free from pencil marks and eraser crumbs.
2. Remove a piece of the adhesive-backed shading material from its protective backing sheet.
3. Place the shading material adhesive-side down over the area to be covered.
4. Flatten the shading material and press it lightly so that it sticks lightly.
5. Apply light pressure starting from the lower left corner working upward and to the right. Make certain the shading material is stretched flat.
6. Cut off the excess shading material that runs beyond the edges of the area to be shaded.
7. Peel away the unwanted pieces of shading material.
8. Burnish the entire area that has been shaded so that the material is smooth and perfectly adhered.

Airbrush Shading Techniques

The **airbrush** is a painting tool used to retouch photographs and shade technical illustrations. Fig. 20-16. Retouching involves altering a photograph by spraying special paints on parts of it. Dark areas can be lightened. Parts of the object in the photograph can be darkened to give them increased emphasis. Shadows can be removed. Fig. 20-17.

Original artwork can be produced with the airbrush. Fig. 20-18. Line drawings can have surfaces shaded. This gives them a more lifelike appearance. Cutaway views can be drawn and shaded. For example, a portion of an object can be removed to show inside details. Fig. 20-19.

Good photography eliminates the need for much retouching. If retouching is necessary, glossy prints give the best results. It is best if the photographs are larger than the final size wanted. After they are retouched they are reduced. This improves the quality of the final product. The reduction causes small imperfections or fuzziness to disappear. The lines become more solid.

The basic steps for airbrushing include the following:

1. Prepare the frisket or mask.
2. Select the paint and load the airbrush.
3. Spray the surface.
4. Clean the airbrush.

Preparing the Frisket or Mask. A **frisket** is a sheet of frisket paper that is placed over the drawing or photograph. The area to be airbrushed is cut away. The remaining frisket paper protects the photograph or illustration from the paint spray.

There are two types of frisket paper, prepared and unprepared. Prepared frisket paper is transparent and has an adhesive back. The back is covered with a sheet of glassine or wax paper. This sheet is peeled away exposing the adhesive back. This sticks the frisket to the photograph and holds it there during the airbrushing step.

Unprepared frisket paper has no adhesive on the back. The back is prepared by the artist as follows:

1. Brush two thin coats of rubber cement at right angles to each other on one side of the frisket paper. Allow the cement to dry about one minute. Fig. 20-20.
2. Place the frisket sheet cement side down over the material to be airbrushed. Fig. 20-21. Smooth it out so that all air pockets are removed. Use a straight edge and work from the center to the edges of the sheet. Fig. 20-22.

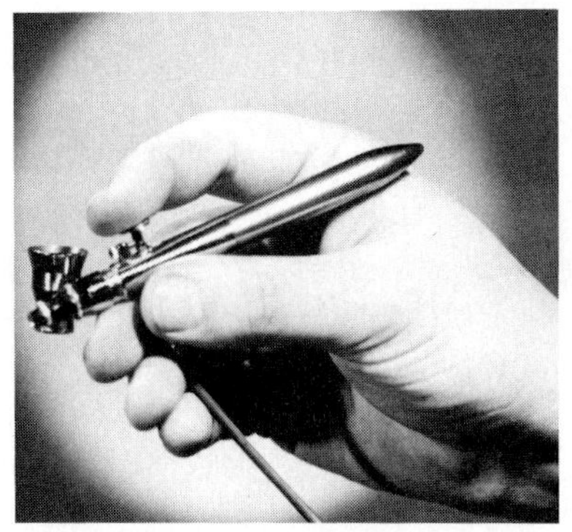

Fig. 20–16. An airbrush is used to shade illustrations. The reservoir on the end holds the paint. Compressed air is used to spray the paint.

Fig. 20–17. This photograph has been retouched with an airbrush. The lower edge in front was darkened. The reflective areas on the round part were enhanced. The holes were shaded so that they are clearly seen.

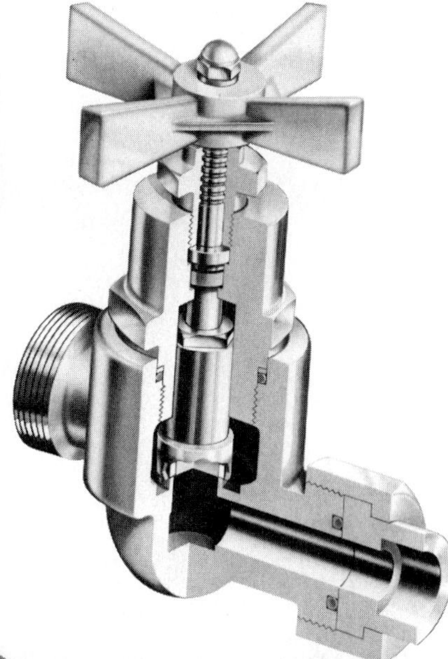

Fig. 20–19. This rendering for sales literature was shaded with an airbrush. (Combination Pump Valve Company)

Fig. 20–18. This award-winning illustration was developed in color using an airbrush, (R. K. Lebedeff, Lockheed Missiles & Space Co.)

Fig. 20–20. Unprepared frisket paper is given two thin coats of rubber cement.

Fig. 20–23. Cut away the area to be sprayed. Do not cut into the artwork below the frisket paper.

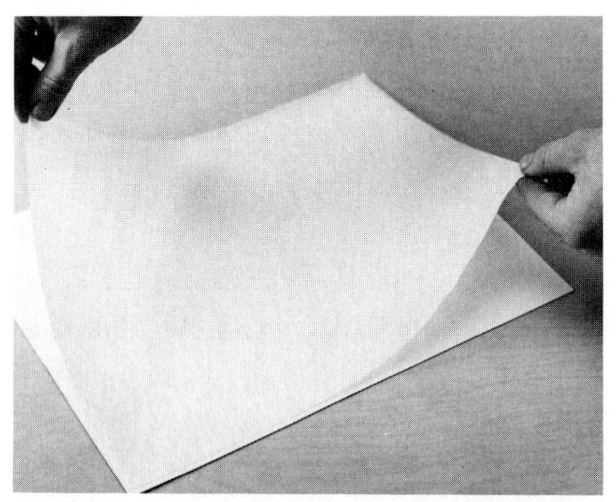

Fig. 20–21. After the rubber cement has dried, place the frisket paper over the material to be sprayed.

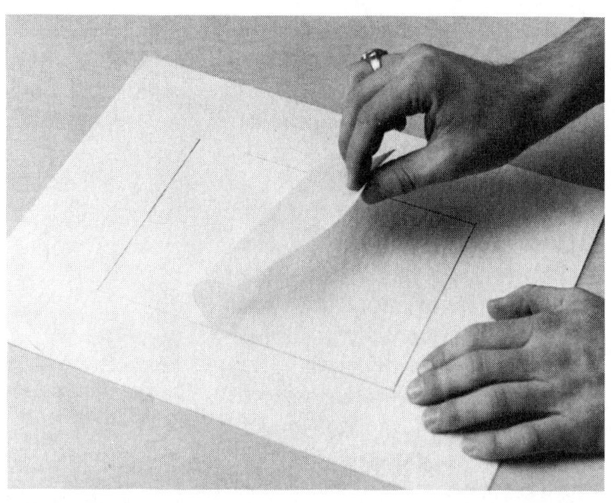

Fig. 20–24. After cutting the frisket paper, peel it off to expose the portion to be sprayed.

Fig. 20–22. Smooth the frisket paper from the center, working toward the edges.

Fig. 20–25. Remove bits of rubber cement that remain on the surface to be sprayed.

3. Now cut out the area to be exposed for painting. Use a sharp frisket knife. A straight edge will help when cutting straight lines. Cut through the frisket paper but *do not* cut into the photograph. Fig. 20-23.

4. Peel away the frisket paper in the areas to be painted. Fig. 20-24.

5. Remove any of the adhesive remaining on the photograph. A piece of drafting tape with some of the stickiness removed is useful for this purpose. A rubber cement pickup can also be used. Fig. 20-25.

6. Now spray the exposed area. Fig. 20-26. (Detailed instructions will be given later in this chapter.) After the paint has dried, remove the remaining frisket paper and any adhesive left on the surface.

Many jobs will require several friskets. For example, if a second color is to be sprayed over part of the first color, a frisket must be made for that area.

Remove the frisket paper as soon as possible. If it is left on too long, it may stick so tightly that the photograph or illustration will be damaged when it is removed.

Masks serve the same purpose as friskets. They are made from sheets of clear acetate. They are different from friskets since they are not held in place with an adhesive. They are held in place with one hand while you operate the airbrush with the other.

Selecting the Paint. Finely ground, water soluble pigment, alcohol colors, or aniline dyes can be used in an airbrush. To be safe, use colors marked for airbrush use on the label. They are sold in jars, tubes, and cakes. The ready mixed liquid paints are the easiest to use.

Mix the paint in a clean container. Use fresh pigment. Do not reuse pigment that has dried. Keep the lids on jars and tubes. The paint should be slightly thicker than milk.

Spraying the Surface. The airbrush is a small paint spray gun. Fig. 20-27. It uses compressed air. The paint is placed in the paint cup. When the trigger is pressed, a needle valve allows the compressed air to rush through. This pulls a small quantity of paint with it. The paint leaves the jet in a fine spray. The more the trigger is pressed, the heavier the spray.

Follow these steps to operate an airbrush:

1. Attach the air hose to the compressed air supply. Blow a stream of air through the hose. This will clear the hose of dust that could clog the airbrush.

2. Attach the hose to the airbrush. Set the pressure for the job to be done. The compressed air supply should be controlled so that the pressure can be regulated from 5 to 40 pounds. Stippled and graduated effects require about five pounds pressure. Medium consistency paints require about 25 pounds pressure. Heavy consistency paints require 40 pounds pressure.

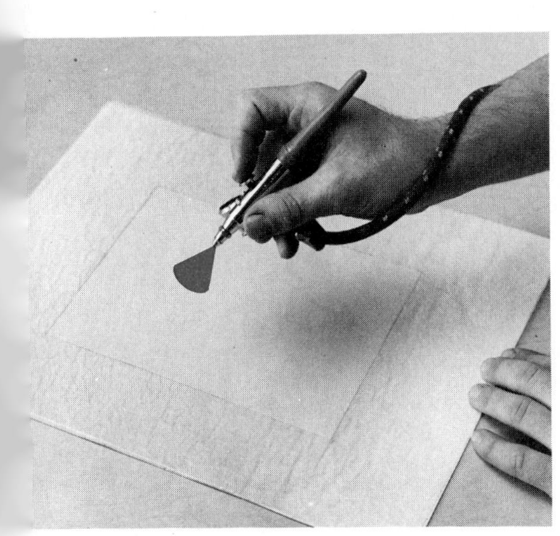

Fig. 20–26. **Spray the exposed area.**

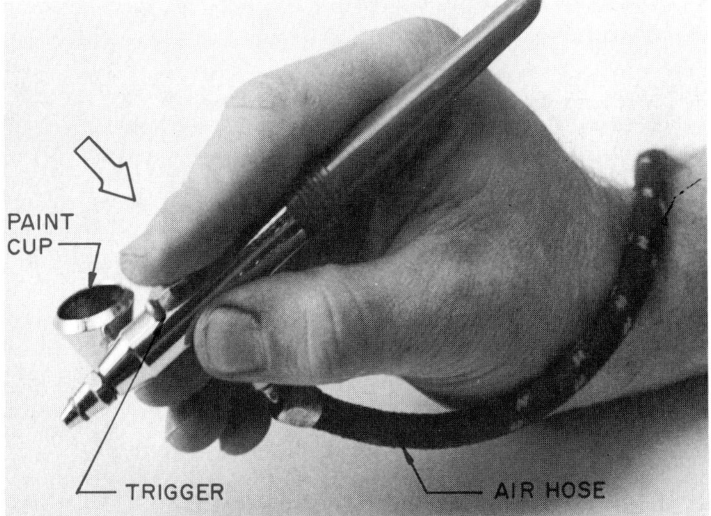

Fig. 20–27. **Hold the airbrush like a pen or pencil. Wrap the air hose around your wrist. Push down on the trigger to release the air.**

CAD Applications:

Presentation Graphics

As you've learned in this chapter, technical illustrations are important for communicating design ideas. When designers want to present their ideas to management or to clients, they prepare presentation drawings that show the design as realistically as possible.

Today a number of packages are available that help designers turn plain 3-D drawings into realistically shaded and colored renderings. The packages are usually some combination of hardware and software.

Some of them start with drawings produced on CAD software. Once the drawings are loaded into the presentation graphics program, the designer selects a viewpoint, distance from the object, direction of the light source, colors, and so on to create a realistic image.

Other packages can "capture" images from video sources, such as a video camera or television set. Once captured, the images can be edited in various ways. They can also be combined with other software.

Here's an example of how this technology might be used. An architectural firm is asked to design an apartment building. The building is drawn using a CAD program. Then a rendering is produced using a presentation graphics package. The building site, a hilly, wooded area, has been photographed with a video camera and the image captured on the computer. An old barn located on the site is removed from the picture. The image of the apartment building is combined with the image of the site.

The interior of the building is also designed using CAD and the presentation graphics package. When the clients arrive, they will "walk through" the building, viewing the hallways and dwelling units on the computer screen.

As you can see, this technology is not only a powerful design tool but an impressive sales tool as well. Clients can see the building's interior and exterior, the landscaping, even the furniture in the model apartment—all before any of it actually exists.

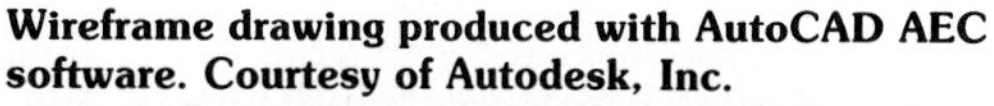

Wireframe drawing produced with AutoCAD AEC software. Courtesy of Autodesk, Inc.

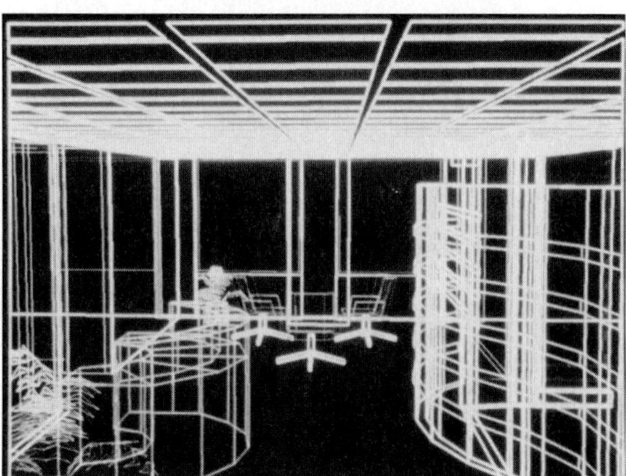

Shaded image produced with AutoSHADE software. Courtesy of Autodesk, Inc.

3. Attach the cup to the airbrush. Fill it about two-thirds full of paint. Fig. 20-28.

4. Hold the airbrush like a fountain pen while resting your forefinger on the trigger. Refer to Fig. 20-27. Loop the air hose over your arm. This helps prevent jarring the airbrush and gives you better control.

5. The air and paint are released by the trigger. When the trigger is in the forward position, the airbrush is off. When it is pushed down, only compressed air is released. When the trigger is held down and pulled back both air and paint are released. Fig. 20-29. The amount of paint released depends upon how far back the trigger is pulled. The paint leaves the airbrush in a cone pattern.

6. It is necessary to regulate the size of spray and volume of air. This adjustment varies with the brand of airbrush used. On one type the tip is rotated to increase or decrease the spray. On another the spray is controlled by the amount the trigger is pulled back.

7. To begin spraying hold the airbrush about 8 inches from the paper. Fig. 20-30. Point it about 1½ inches to the left of the area to be sprayed. Press down on the trigger to release the air. Pull back slowly on the trigger to release a little paint. Move the airbrush parallel with the paper from left to right. Keep your wrist stiff and move your entire arm. The movement should be smooth and continuous. Do not hesitate. Do not stop. If you hesitate or stop, excess paint will be sprayed on that area.

8. As you complete the stroke to the left with the airbrush, slowly push the trigger forward. This cuts off the flow of paint.

9. Now start the next stroke back across the paper from right to left. Follow the procedure just described. Move the airbrush down so the second stroke overlaps the first. Repeat these overlapping strokes until the area is covered. It requires many strokes. If a darker tone is needed go back over the area with additional strokes.

10. Remove the frisket.

Cleaning the Airbrush. Pour the leftover paint into a storage bottle. Wipe the cup clean with a cloth moistened with the proper solvent for the type of paint being used. Use water for water colors and alcohol for alcohol colors and aniline dyes.

Hold the airbrush at a downward angle. Set the paint and air adjustments at their maximum settings. Place a few drops of solvent in the paint cup and run air through it. This will dissolve the paint and blow the airbrush clean.

Solid Shading

In airbrush painting, you build up tones gradually. If a dark area is wanted go over it several times. Allow the paint to dry occasionally. Never get the surface so wet that the paint begins to run.

Fig. 20–28. Fill the paint cup about two-thirds.

TRIGGER FORWARD IN OFF POSITION.

AIR
TRIGGER DOWN IN AIR ONLY POSITION.

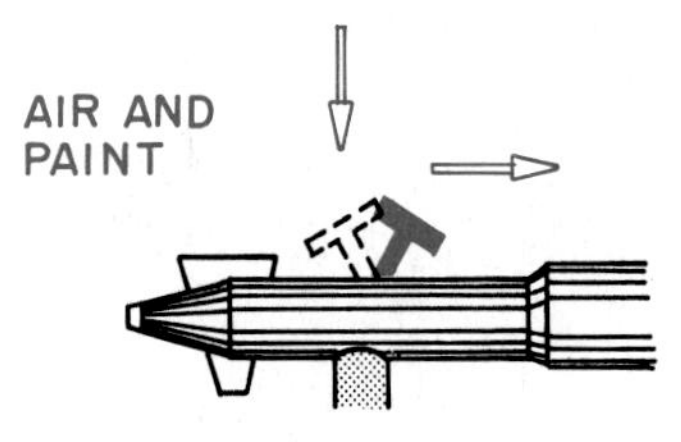
AIR AND PAINT
TRIGGER DOWN AND BACK, RELEASING AIR AND PAINT.

Fig. 20–29. Airbrush trigger positions.

Fig. 20–30. Illustrators at work. Notice how they hold the airbrush. (Meisel Photochrome Corp.)

Graduated Shading

If the shade on one side of an area is to be darker than the other, airbrush the area lightly. Then start a second time. Begin with the area that is to be darkest. Spray a shorter stroke than the first one. Repeat shortening your stroke until the desired shading is built up.

Dot Patterns

To make a small dot pattern, hold the airbrush close to the work. Decrease the spray setting and use a medium air volume. Quickly press and release the trigger. This gives a quick spray with small dots. For large dots, increase the spray setting and use a medium air volume. Hold the airbrush farther away from the work. Give a quick press and release to the trigger.

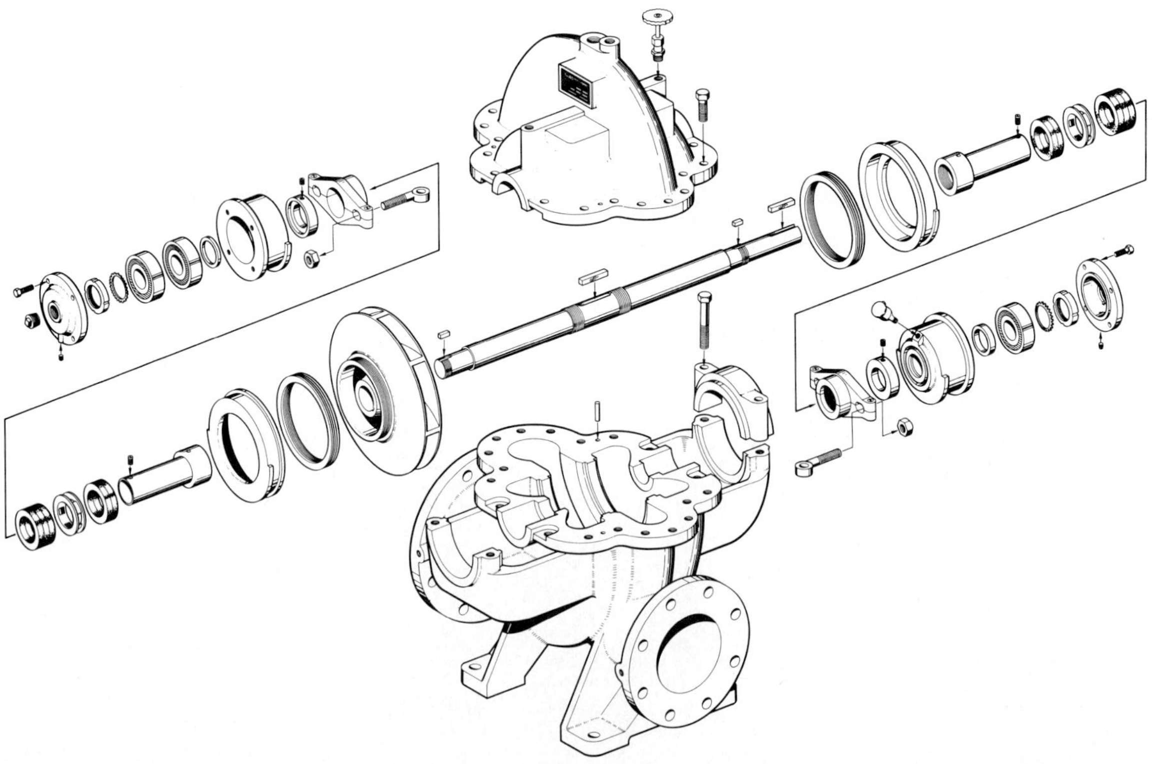

Fig. 20–31. An exploded view. Notice the line shading used to give a more realistic appearance. A center line is used to clarify the sequence of parts. (A. Chandronhait, Raytheon Co.)

Examples of Technical Illustrations

There are many ways an illustrator can show technical information pictorially. After a careful study of the object and the purpose of the illustration, the illustrator decides how to draw it. The most commonly used drawings are exploded, assembly, cutaway, or some combination of these.

Exploded Drawings

Exploded drawings show each part of an assembly drawn separately in a position that shows how they fit together. Center lines sometimes are used to show the order in which parts are assembled. If the order is clear, center lines may be omitted. Fig. 20-31 shows an exploded view.

An effort should be made to keep each part from overlapping with any other part. This is not always possible. Sometimes each part is identified with a number. A parts list gives details about the part.

Exploded views may be shaded if desired. They can be drawn as isometric, dimetric, trimetric, or perspective views.

Cutaway Illustrations

If interior details of an object in its assembled position must be shown, use a **cutaway illustration**. This illustration resembles a sectional view like those found on working drawings but is in pictorial form. The surfaces cut are usually shaded with slanting lines. Fig. 20-32.

Exterior Assembly Drawings

Some illustrations picture the exterior of a product. To be most effective, they are usually shaded. Fig. 20-33 illustrates such a drawing. Notice how the ends of the flexible hoses are shaded to aid in reading the drawing. Examine it carefully and note how the use of different line widths adds to the appearance. The thicker lines give a shadow effect.

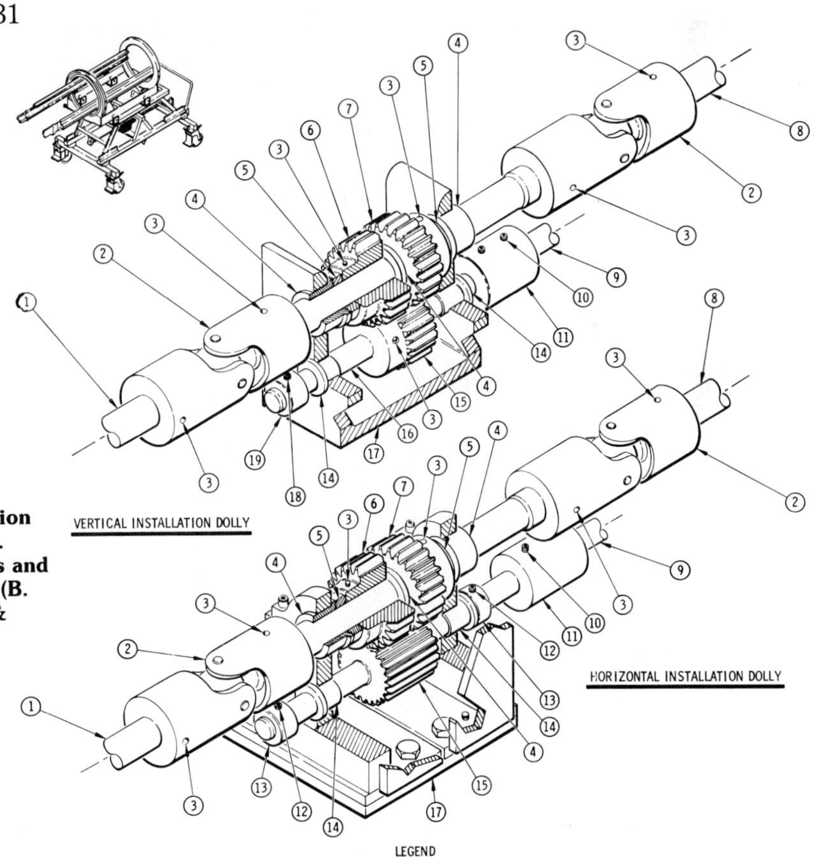

1. SHAFT 1914381 (HORIZONTAL DOLLY) 1914382 (VERTICAL DOLLY)
2. UNIVERSAL JOINT PS-1-2471-2
3. PIN NAS561P6-26
4. BEARING MS35772-84
5. SLEEVE 1914384
6. ELEVATION GEAR 1915767
7. PITCH GEAR 1915768
8. SHAFT 1914383 (HORIZONTAL DOLLY) 1914385 (VERTICAL DOLLY)
9. SHAFT 1914388
10. SETSCREW REF PS-1-2304-12
11. COUPLING PS-1-2304-12
12. SETSCREW REF PS-1-3267-11
13. COLLAR PS-1-3267-11
14. BEARING MS35772-75
15. MAIN GEAR 1914380 (HORIZONTAL DOLLY) PS-2-3078-1 (VERTICAL DOLLY)
16. SHAFT 1955903
17. SUPPORT 1914377 (HORIZONTAL DOLLY) 1955902 (VERTICAL DOLLY)
18. SETSCREW REF PS-2-1180-11
19. COLLAR PS-2-1180-11
20. BOLT NAS608-3-10
21. PIN NAS561P6-20
22. NUT MS21042L6
23. BOLT AN6-13
24. WASHER NAS1099-6

Fig. 20–32. A pictorial section through an assembled product. Notice the use of part numbers and a legend to identify each part. (B. Machado, Lockheed Missiles & Space Co.)

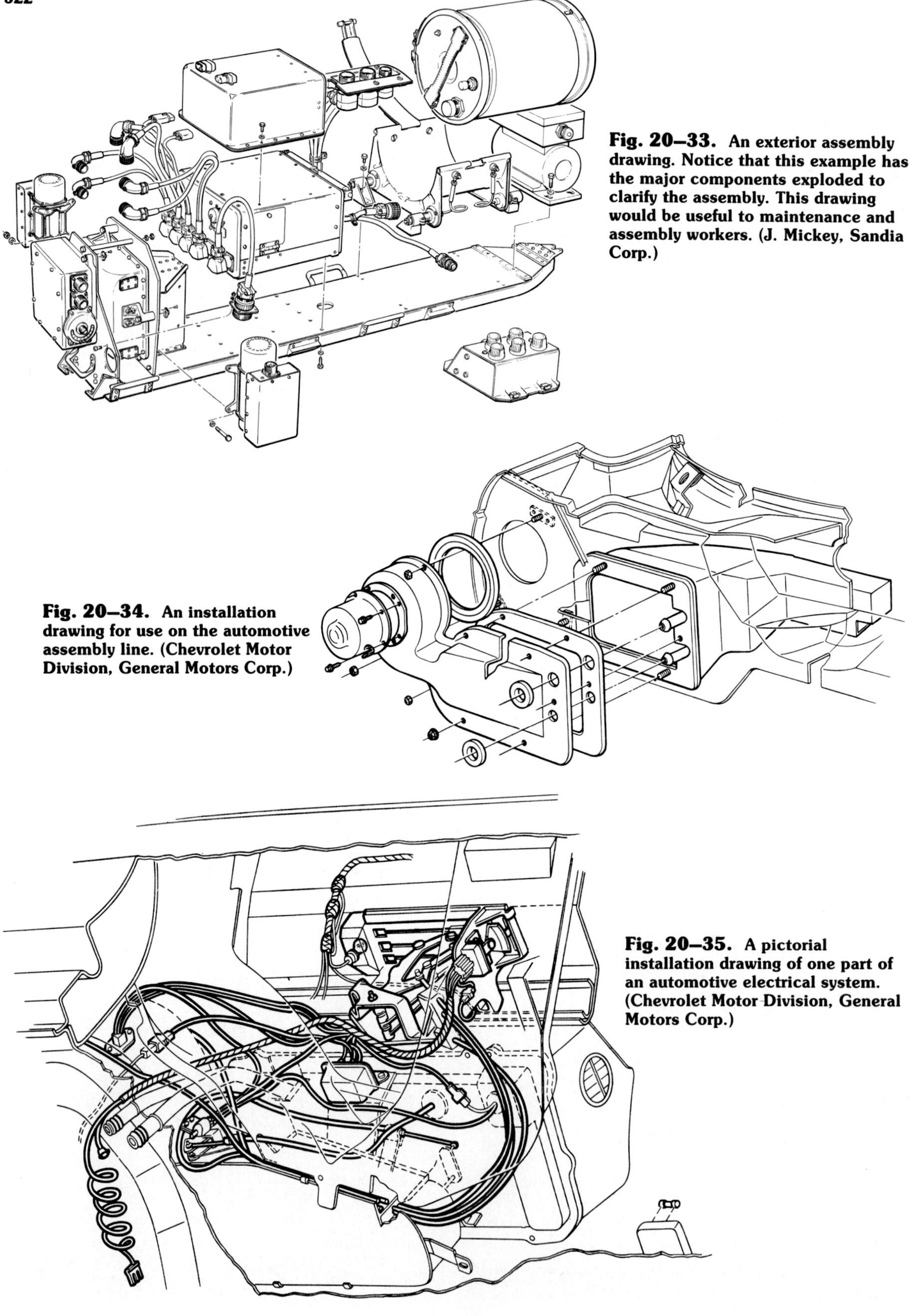

Fig. 20–33. An exterior assembly drawing. Notice that this example has the major components exploded to clarify the assembly. This drawing would be useful to maintenance and assembly workers. (J. Mickey, Sandia Corp.)

Fig. 20–34. An installation drawing for use on the automotive assembly line. (Chevrolet Motor Division, General Motors Corp.)

Fig. 20–35. A pictorial installation drawing of one part of an automotive electrical system. (Chevrolet Motor Division, General Motors Corp.)

Uses for Technical Illustrations

An illustration for an *installation manual* is shown in Fig. 20-34. Its purpose is to help workers understand how a device is assembled. This drawing, when put into the hands of automobile assembly plant workers, shows them how to fasten the unit together. The workers do not need to know how to read a working drawing. Notice the effective use of center lines.

Fig. 20-35 shows an installation drawing for electrical wiring in an automobile. Notice how clearly it shows the worker where the wires are to run and how the connections are to be made. Frequently the wires are colored to match the actual color code on the wiring.

An *illustrated parts breakdown* is shown in Fig. 20-36. It is for use in a parts catalog. Notice that each part is named and the part number given on the drawing. Notice, too, the use of line shading.

A pictorial illustration for use in a *maintenance manual* is shown in Fig. 20-37. See how it uses pictorial drawings to indicate each lubrication point. It has notes telling how to lubricate the parts and what lubricant to use.

Full-Color Illustrations

Sometimes it is helpful to present an idea or concept by making pictorial illustrations in full color. Such drawings present the idea in a most realistic form. They are useful in feasibility studies because they present the total idea before the actual device is much beyond the initial engineering stage. Usually they are rendered in watercolor, pastels, or oil colors. An airbrush is useful for this work. A knowledge of painting, color, and perspective is essential. The illustration shown in Fig. 20-18 was a color illustration.

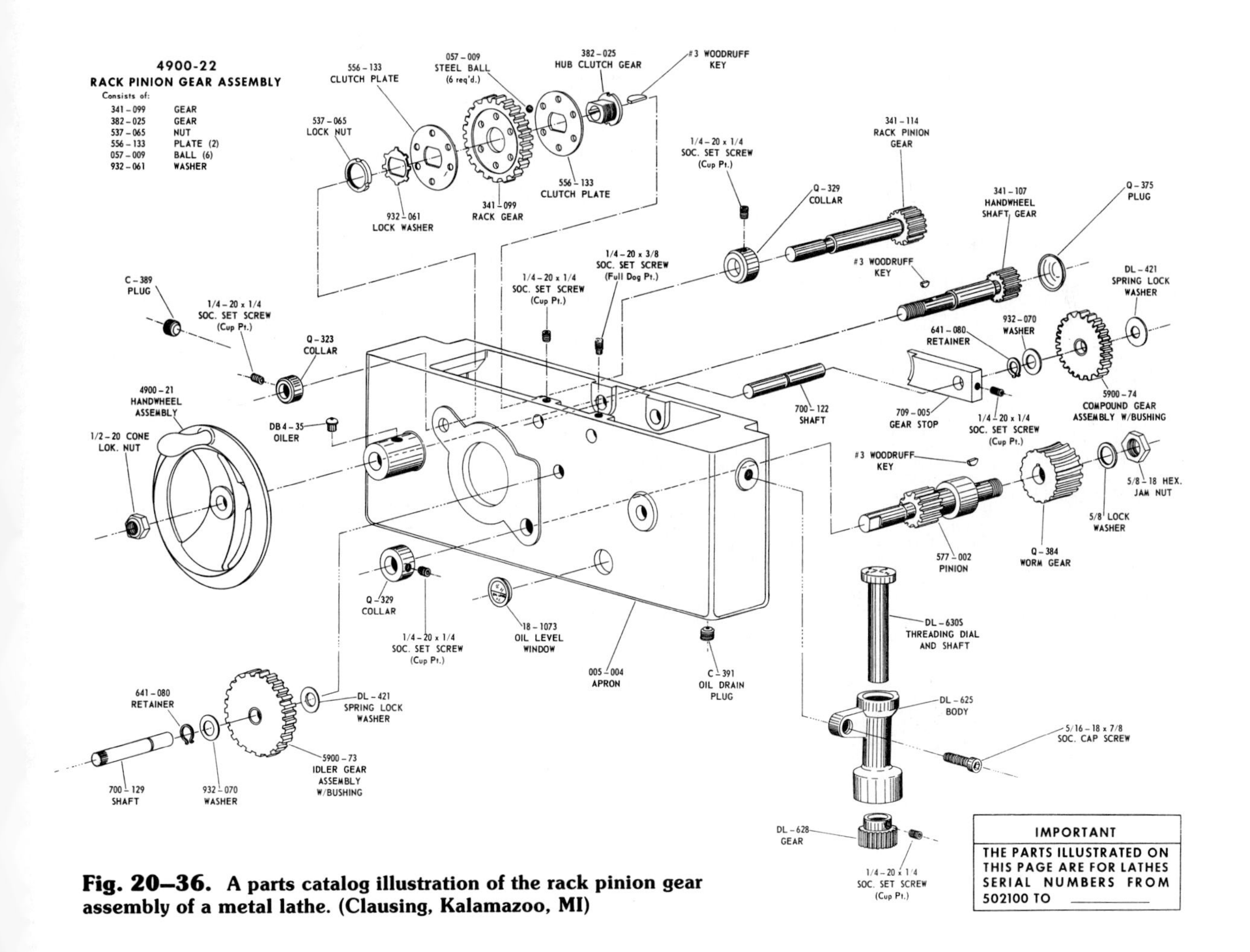

Fig. 20–36. **A parts catalog illustration of the rack pinion gear assembly of a metal lathe. (Clausing, Kalamazoo, MI)**

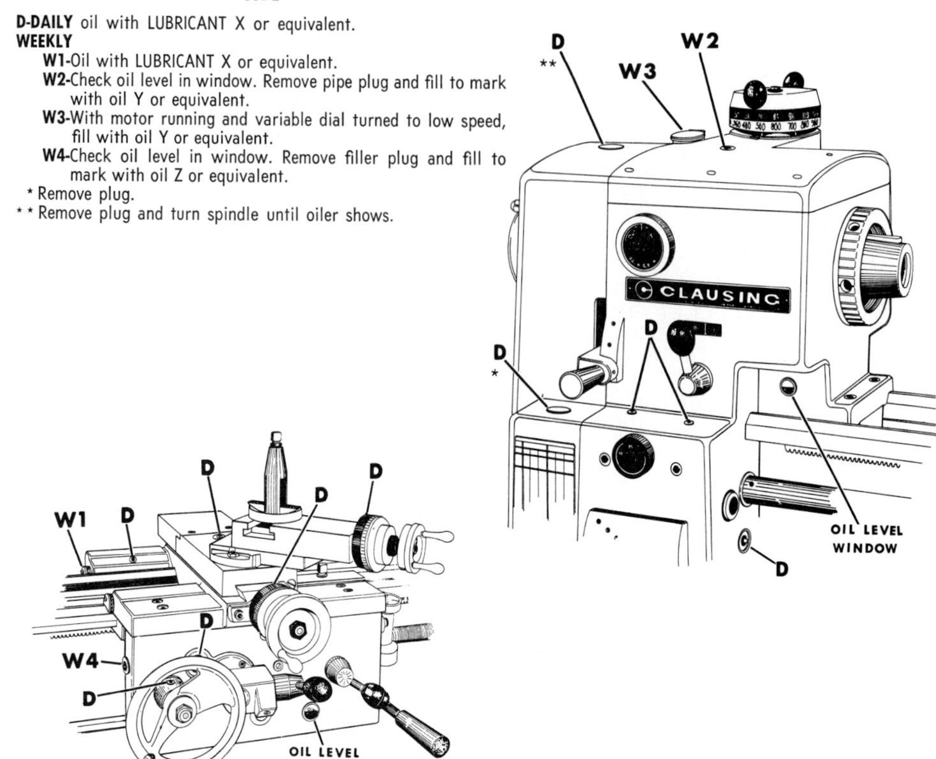

Fig. 20–37. Part of a maintenance manual illustration giving lubrication directions. (Clausing, Kalamazoo, MI)

Chapter Review

Build Your Vocabulary

You should understand and use the following terms as part of your working vocabulary. Write a brief explanation of what each means.

technical illustrations
engineering illustrations
publication illustrations
shading
solid line shading
surface line shading
stippling
smudge shading
airbrush
frisket
exploded drawings
cutaway illustration

Study Problems—Directions

The problems that follow will give you the chance to use some of the drafting knowledge that you learned in this chapter. Do the problems that are assigned to you by your instructor.

1. Draw a simple exploded view of the conveyor hanging bolt, **Fig. P20-1.** Use the exterior line shading technique. Draw it full size.
2. Draw a pictorial illustration of the swivel pad screw, **Fig. P20-2,** that would be suitable for use in a sales catalog. Be certain to show clearly the details of the swivel tip. Draw it twice actual size.
3. Draw an exploded view of the machine tool leveling pad, **Fig. P20-3.** Use any type of shading you desire. Draw it twice actual size.
4. Draw an exploded view of the back gear shaft assembly, **Fig. P20-4.** Shade it using the solid line shading technique. Select a scale that is large enough to show clearly all the parts.
5. Draw an exterior assembly of the control handle for a metal lathe, **Fig. P20-5.** Shade it using the stippling technique. Draw it using a two to one proportion.
6. Draw a pictorial illustration of the steam pipe hanger, **Fig. P20-6,** to show a worker how it is assembled. Select a scale large enough to show it clearly.
7. Draw a pictorial assembly of the electrical wiring pressure connector, **Fig. P20-7.** Pick the dimensions off the drawing with dividers. Draw it using a two to one proportion so that it will fit a 6-inch wide column in a sales catalog.
8. Draw an exploded illustration of the electrical wiring pressure connector, **Fig. P20-7**. Select any scale.
9. Draw a full-size pictorial illustration of the conveyor carrier, **Fig. P20-8**, to show how it is to be assembled.
10. Draw an assembled pictorial view of the conveyor carrier, **Fig. P20-8**, with the necessary section to show hidden details. Shade your illustration using any method desired. Draw 2 times full size.
11. Make a parts catalog illustration of the adjustable clamp, **Fig. P20-9**. Identify each part by a number and prepare a parts list. Pick the dimensions off the drawing with dividers. Draw it twice as large as the printed size.
12. Draw a pictorial section of the tumbler knob assembly, **Fig. P20-10**. Shade it so that it could serve as an illustration for a sales catalog. Draw full size.
13. Draw a parts catalog illustration of the tumbler knob assembly, **Fig. P20-10**. Use shading if it will help clarify the drawing. Identify each part on the drawing. Draw full size.
14. Draw a pictorial section of the water connection for a rock drilling machine, **Fig. P20-11**. Shade it so that it can be used in a sales catalog. Draw to any convenient size.
15. Make a pictorial assembly of the water connection for a rock drilling machine, showing its exterior appearance. Shade it using the stippling technique.
16. Draw pictorially the hose stem of the water connection for a rock drilling machine. Shade it using the solid line technique.
17. Draw a parts illustration of the auto radio antenna shown in **Fig. P20-12**. Select names for each part. Identify each part using a parts list. Draw it half size.
18. Make a pictorial illustration showing the exterior features of the penlight, **Fig. P20-13**. Draw it twice actual size. Use the smudge shading technique.
19. Draw an exploded view of the penlight. Select a scale that will clearly show each part.
20. Bring to your classroom objects that you can take apart easily, such as a ball-point pen, a vise, a door lock, or water faucet. Prepare a technical illustration to serve a particular purpose such as for a catalog or for an assembly worker. Determine dimensions by measuring the parts after they have been disassembled.
21. Draw a pictorial cutaway illustration of the solenoid valve in **Fig. P20-14.** It offers an upstream and a downstream seal that compensate for the differential pressure forces that cause jamming in single-seal valves. Draw your illustration twice as large as printed here. Determine sizes by measuring the printed illustration.
22. Make an isometric section drawing of the turbopump inlet fluid injection unit shown in **Fig. P20-15.** The material is aluminum, so use the proper section lining. Label the drawing to identify the parts and to show the basic fluid flow. Shade if instructed to do so by your teacher. Draw twice as large as printed here. Get the sizes by measuring the printed drawing.

Study Problems

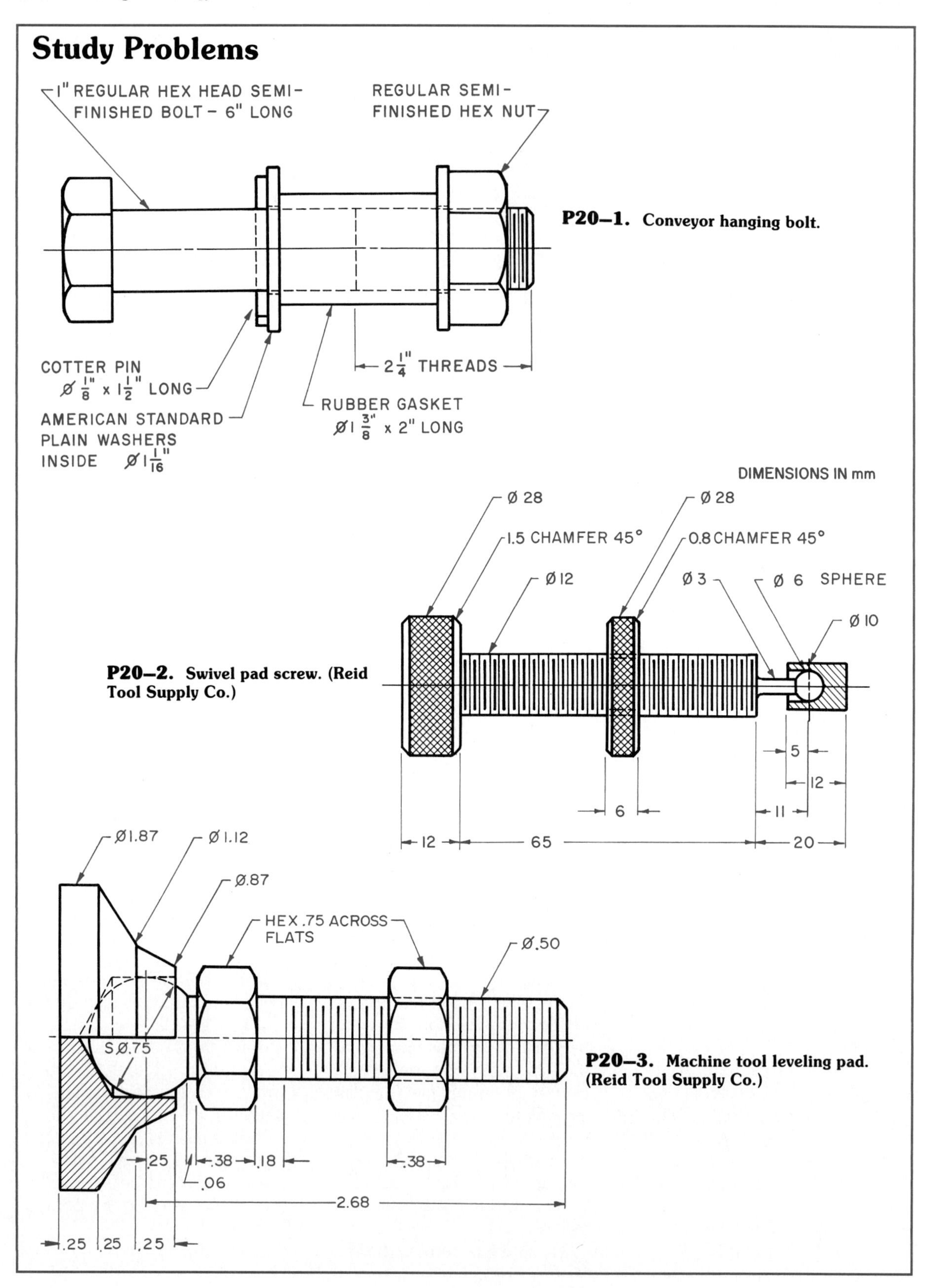

P20–1. Conveyor hanging bolt.

P20–2. Swivel pad screw. (Reid Tool Supply Co.)

P20–3. Machine tool leveling pad. (Reid Tool Supply Co.)

Study Problems

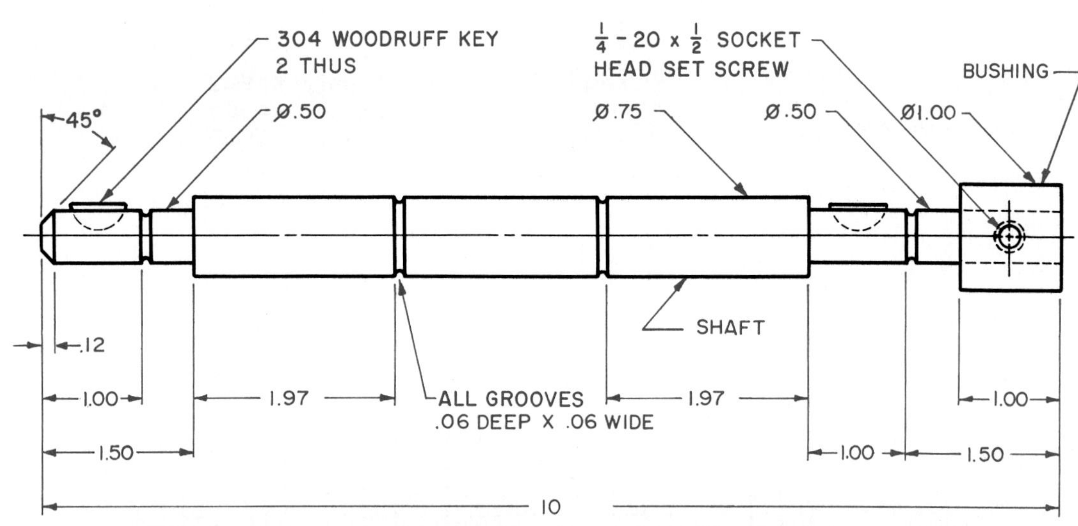

P20–4. Back gear shaft assembly for a metal lathe. (Clausing, Kalamazoo, MI)

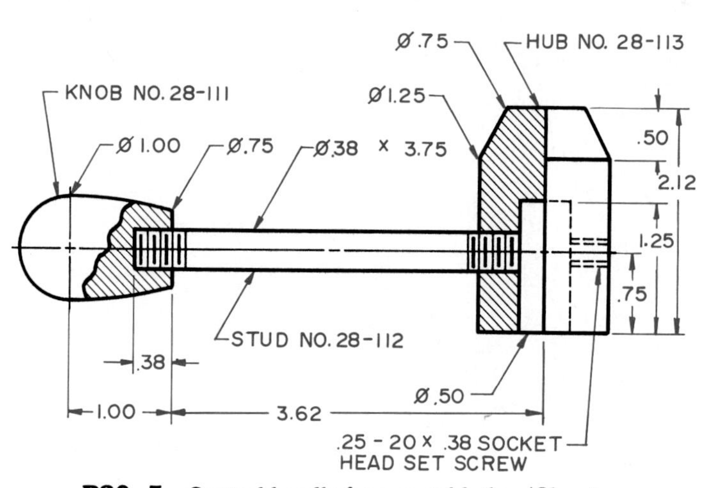

P20–5. Control handle for a metal lathe. (Clausing, Kalamazoo, MI)

Study Problems

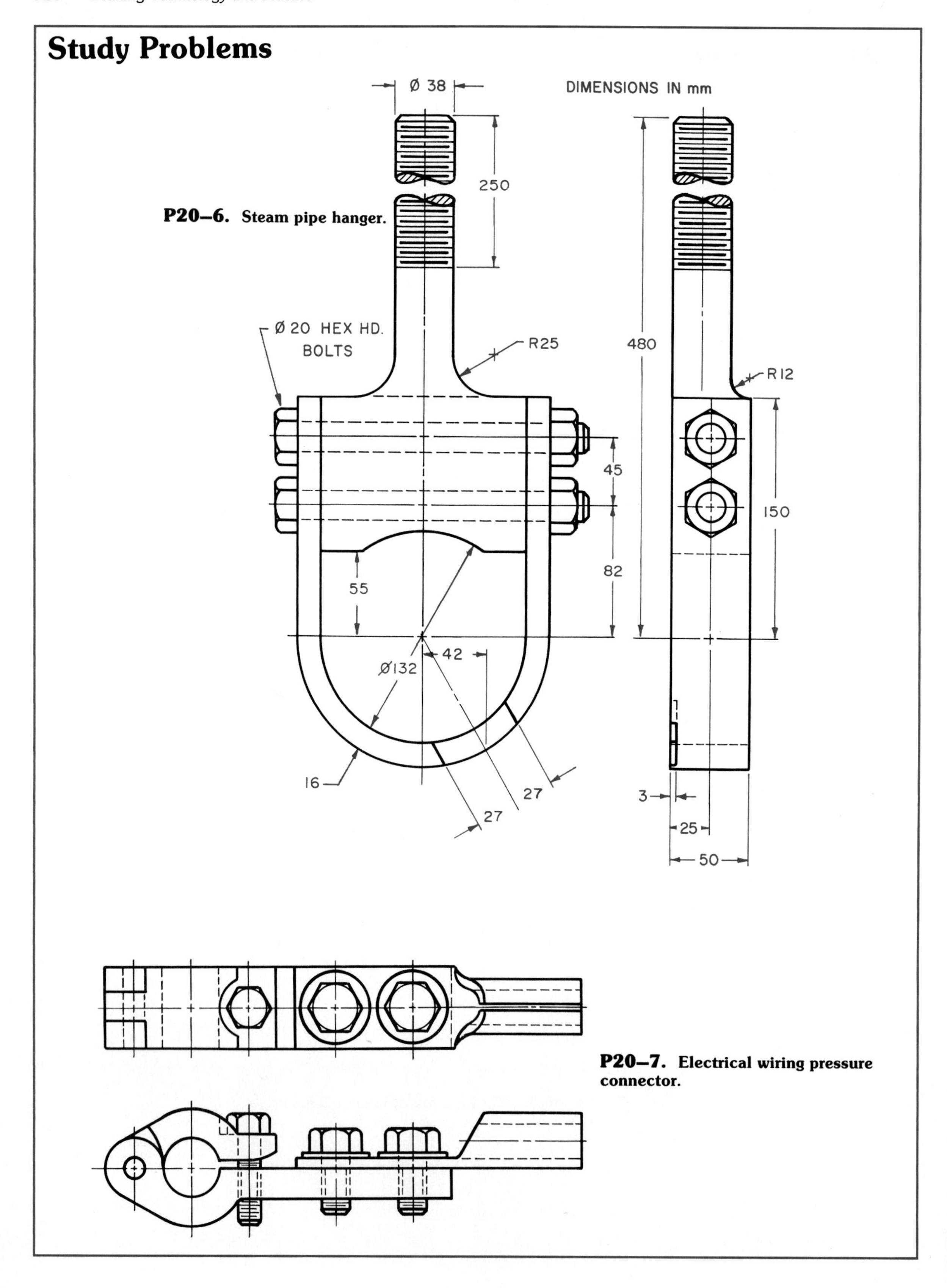

P20–6. Steam pipe hanger.

P20–7. Electrical wiring pressure connector.

Study Problems

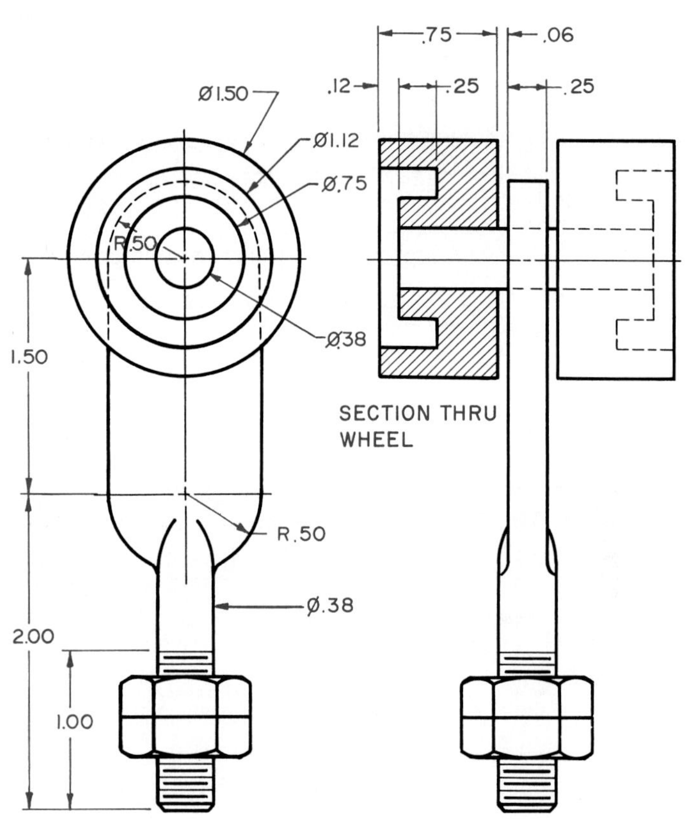

P20–8. Conveyor carrier.

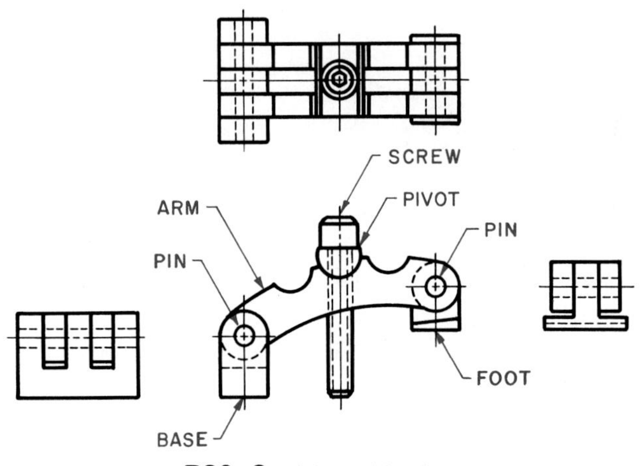

P20–9. Adjustable clamp.

Study Problems

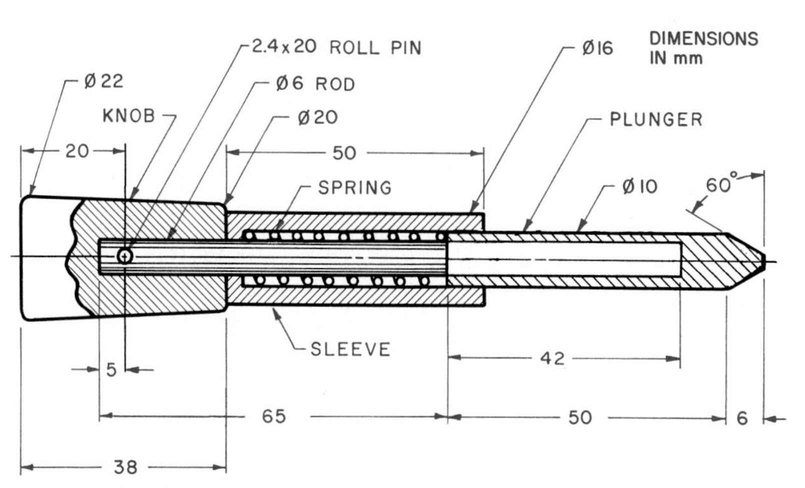

P20–10. Tumbler knob assembly on a metal lathe.

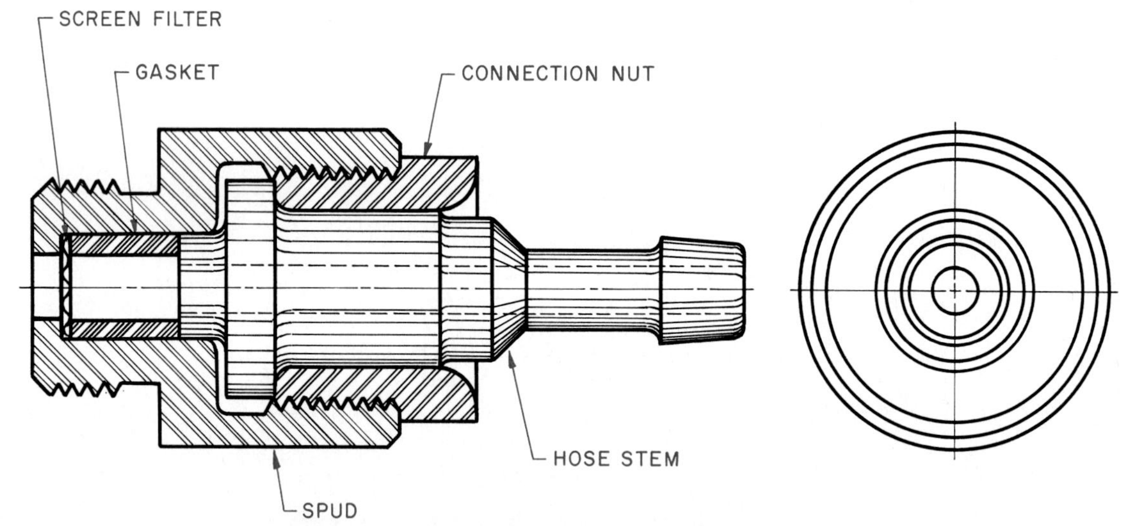

P20–11. Water connection for a rock drilling machine.

Study Problems

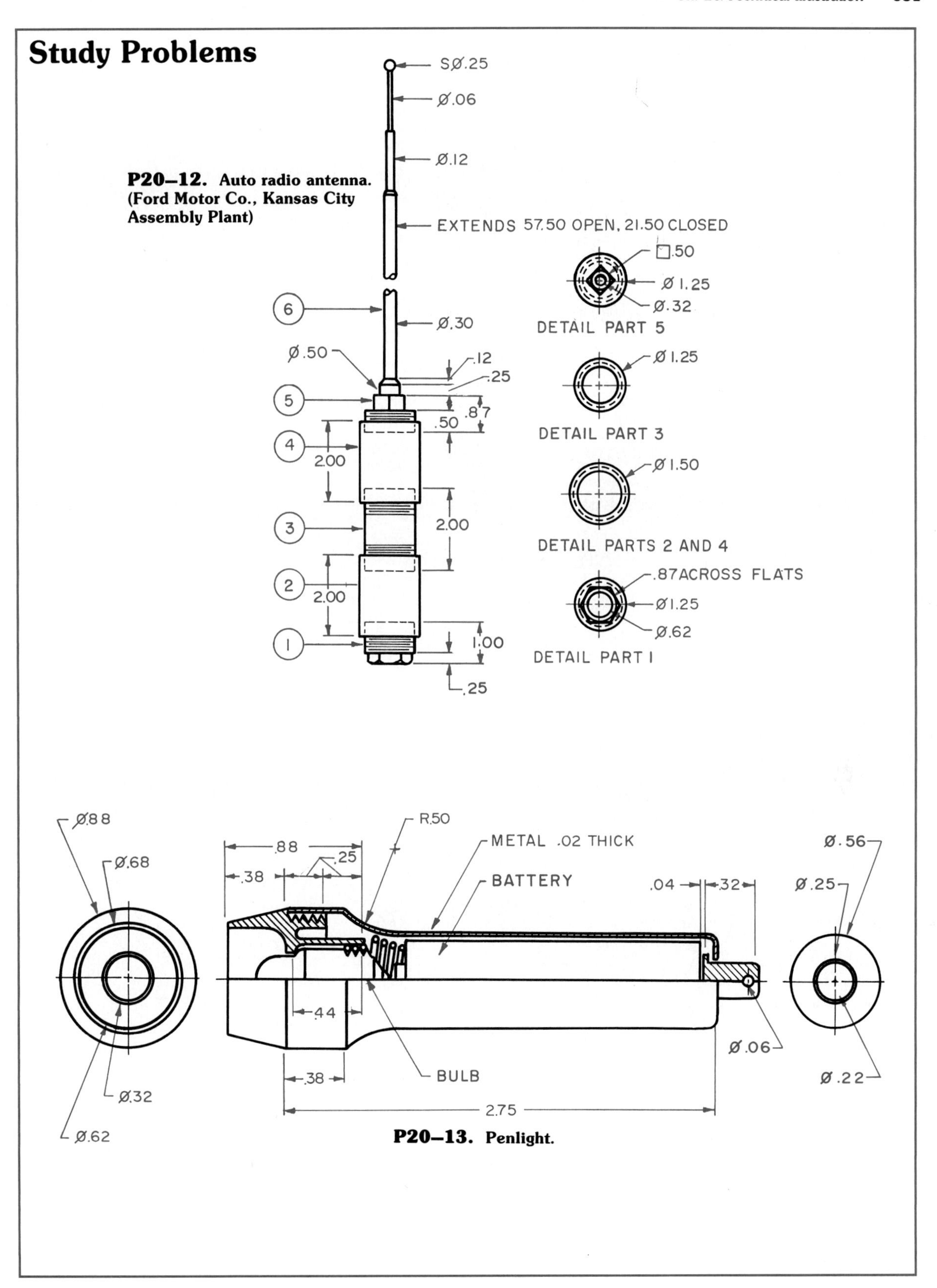

P20–12. Auto radio antenna. (Ford Motor Co., Kansas City Assembly Plant)

P20–13. Penlight.

Study Problems

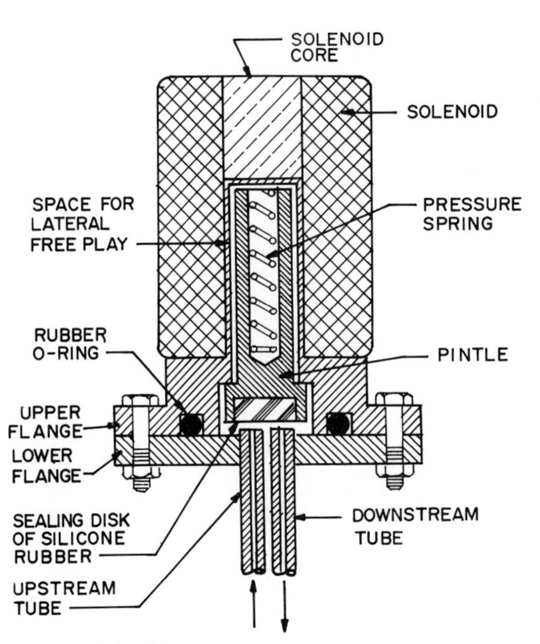

P20–14. Solenoid valve. (NASA)

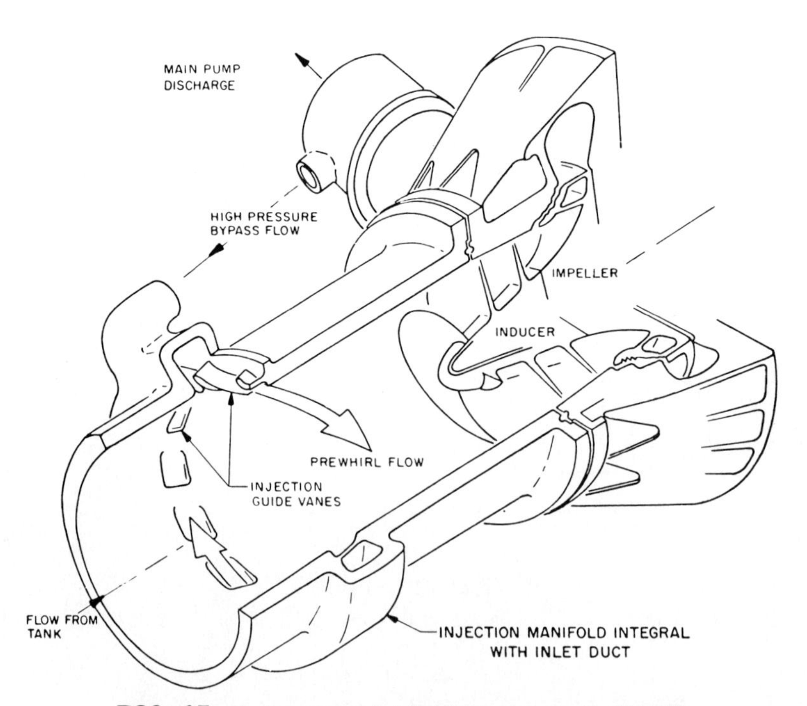

P20–15. Turbopump inlet fluid injection unit. (NASA)

CHAPTER 21
Piping Drawing

After studying this chapter, you should be able to do the following:

- Read different types of piping drawings.
- Draw single-line orthographic piping drawings.
- Draw single-line isometric piping drawings.

Introduction

Piping drawings are the master plans for various types of piping projects. They show the location of equipment, the details of pipe runs, pipe connections, and the necessary dimensions for field construction. Included on piping drawings are such items as pipe, vessels, valves, pumps, motors, and various pipe fittings required for any one project. These drawings must have sufficient detail to clearly indicate pipe clearances, connections, and supports while they provide complete descriptive information for clear instructions to the fabricator.

Piping Design

Designing a piping installation is the work of mechanical engineers and design drafters. Engineers have the responsibility of planning the basic design based on what the installation is to produce. The design of a large plant or refinery involves solving various problems to determine efficient layouts and the proper specifications, processes, flow sheets, pipe selections, pressure vessels, exchangers, storage tanks, instrumentation, mechanical equipment, heating, air conditioning, plumbing, compressors, and pumps. This work requires preparation of various types of drawings and specifications, and, in some cases, building a scale model of the plant. Fig. 21-1 shows a large refinery installation that required all phases of piping design and drafting.

Piping drafters convert the engineering design into plans, elevations, and detail drawings that show all the piping and its supporting equipment.

A piping drafter should be familiar with the operating principles of the process in each piping project. This knowledge is necessary since many of the piping details and connections are determined by drafters.

Fig. 21–1. An example of the complex piping in a refinery. (Phillips Petroleum Co.)

Working with piping problems requires a basic understanding of piping nomenclature and knowledge of the sizes of piping fittings for various working pressures. This work requires reference to standard piping data books and manufacturers' catalogs.

A piping drafter must be skilled in *mechanical drawing* and *lettering*. The drafter will become involved with a wide range of equipment and machinery which require detailing. Since piping problems usually are related to a building, a knowledge of *architectural drafting* is also important. The piping drafter uses conventional drafting practice found in architectural, piping, and mechanical drafting. In addition, mathematics is used for selecting pipe sizes and valves, fluid flow, and stress evaluation.

Kinds of Pipe

There are many different kinds of pipe. Things to consider when selecting pipe are temperature, pressure, and chemical reactions between the pipe material and the gas or liquid to be run through it. Materials that are commonly used to make pipe are copper, steel, cast iron, wrought iron, brass, lead, and plastics.

Copper pipe will withstand corrosion. It can carry oil, gas, steam, and sewage. It is good for carrying drinking water since minerals are less likely to stick to it.

Steel pipe will handle high temperatures and pressures. It is one of the most commonly used types. It is sometimes galvanized (coated with zinc) to prevent rust. When galvanized, steel pipe is used for carrying drinking water.

Cast iron pipe is used underground for water, gas, sewer, and low pressure steam lines. It resists corrosion, but strain and shock will break it.

Wrought iron pipe will withstand corrosion. It is used for the same applications as steel. It is more expensive than steel pipe and is not as strong.

Brass pipe is used for handling corrosive liquids. It will handle high pressures and can be used underground. Brass pipe is more expensive than steel or wrought iron.

Lead pipe is used to handle liquids containing acids. It is flexible and will withstand vibrations and expansion.

Plastic pipe is used to carry drinking water and is resistant to acids and chemicals. It is flexible, but some plastics will not withstand high pressure or heat. Plastic pipe is not used for oil. It can be used underground.

Pipe Sizes

The most commonly used pipes for above-ground installations are wrought iron, steel, and copper.

The sizes of pipes under 14 inches are specified by their nominal inside diameter. For example, a 3-inch pipe has a hole that is 3 inches in diameter. The outside diameter varies with the wall thickness.

Wrought iron and steel pipes are commonly made in three weights. These are standard, extra strong, and double extra strong. The stronger pipe has thicker walls.

Copper and red brass pipes are available in standard and extra strong weights.

Fittings

Pipe **fittings** are used to join lengths of pipe, change the direction of a piping run, or connect pipes of different diameters. Fittings are made from many different materials. Usually, they should be made of the same material as the pipe used. Some fittings, however, permit the connection of pipe of different materials. For example, a fitting is available that permits connecting copper to steel pipe.

A wide variety of pipe fittings are made. This chapter, however, includes only a few selected types. Consult manufacturers' catalogs for a more complete list. The following types of fittings are commonly used: malleable iron threaded fittings, steel butt welded fittings, cast iron flanged fittings and cast brass solder-joint fittings. Sizes for some of these are found in the appendices.

Valves

A piping system directs and controls the flow of liquids and gases. The material flowing through the pipes may be very cold or very hot, under low or high pressure. **Valves** are used to control the flow of the liquid or gas flowing through a pipe.

The basic functions of valves are to start, stop, and regulate flow, prevent back flow, and regulate and relieve pressure.

There are several basic valve designs: gate, globe, angle, ball, cock, butterfly, check, and relief. Fig. 21-2.

A *gate valve* is a free-flow valve. When the gate valve is open, the liquid flows easily. It is not designed to regulate the rate of flow.

A *globe valve* permits regulation of flow and pressure by adjusting to various sizes an opening between the stem and seat.

Angle valves serve the same purpose as globe valves. They are used when the piping system design requires the flow to leave the valve at a 90-degree angle.

A *ball valve* is a quick opening, on-and-off valve that permits a complete, unrestricted flow and is ideal for heavy consistency liquids.

A *cock valve* is used when it is necessary to control liquids through closed conduits under extreme conditions.

A *butterfly valve* derives its name from the flat disk flapper that closes against a resilient liner.

It is an efficient and economical valve for most liquid and gas services.

Check valves permit flow in one direction. For example, a liquid can flow through the valve in one direction but not be forced back through in the other direction.

A *relief valve* is a safety device. It remains closed until the pressure in a system rises to an unsafe level. It then is forced open by the pressure. Most are held closed by a spring.

Process Flow Diagrams

A **process flow diagram** is the first drawing made when planning the piping for a factory or plant. It shows the basic items of major equipment and their relation to each other. The major flow lines are indicated as connecting these pieces of equipment. They help describe how the process operates.

Of great importance in process flow diagrams is simplicity. The symbols used preserve the general physical appearance of the equipment yet require a minimum of drafting effort.

Process flow diagrams are not drawn to scale. The symbols are drawn so they are in keeping with the overall size of the drawing.

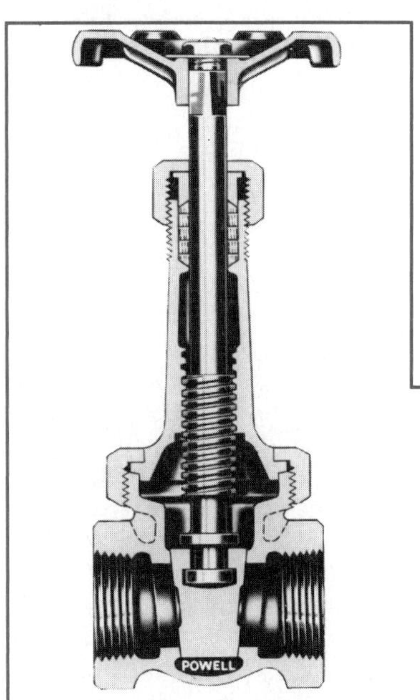

SCREWED-END GATE VALVE (BRONZE)

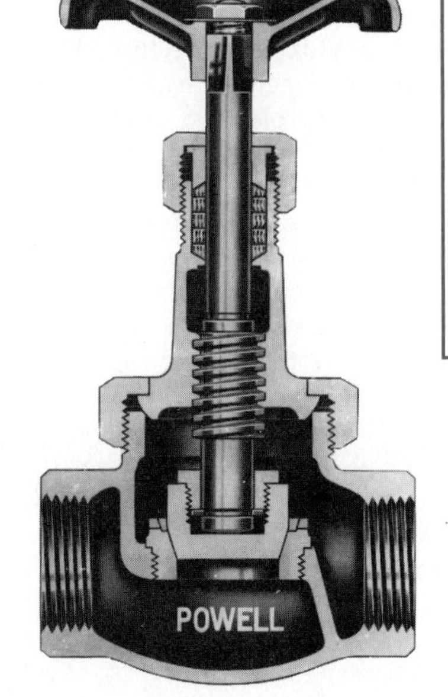

SCREWED-END GLOBE VALVE

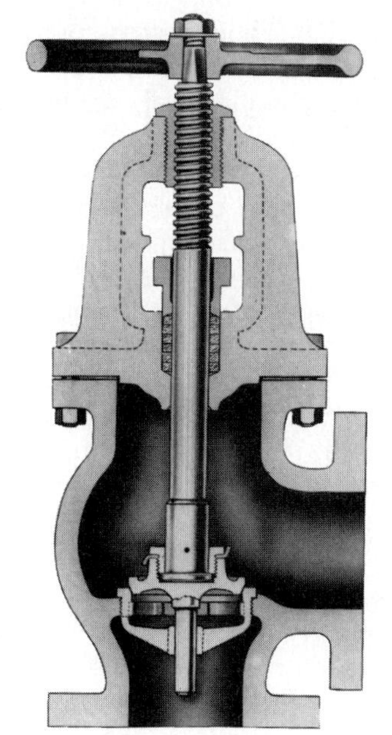

FLANGED-END ANGLE VALVE

SCREWED-END HORIZONTAL LIFT CHECK VALVE.

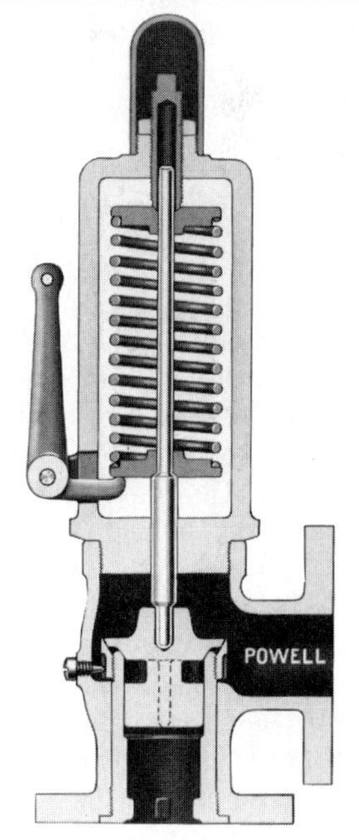

FLANGED-END RELIEF VALVE.

Fig. 21–2. Commonly used valves. (William Powell Co.)

The symbols are arranged on the drawing in a logical sequence of flow. They start where the material to be processed enters the system and follow the process through to the completion of the main product. The diagram does not represent the actual location of the parts of the plant when it is built. Instead the diagram represents the process. Fig. 21-3.

Equipment outlines are drawn with a heavy line. Piping is drawn with a thin line. Usually instrumentation or electrical symbols are not placed on process flow diagrams. The direction of flow is shown by arrowheads. Crossover lines should be kept to a minimum.

Process quantities such as pressure, temperature, and flow are recorded inside symbols given a special shape. For example, a circle, the symbol for pressure, with "100 psi." lettered inside means a pressure of 100 pounds per square inch.

Pipe Drawings

The three methods used for designing and detailing piping systems are the elevation, plan, and section; isometric; and model.

Elevation, Plan, and Section Method

The elevation, plan, and section method uses the same principles of orthographic projection used for detailing objects in machine drawing. This method shows the equipment, pipe runs, pipe support locations and any other features of the system. Small details or complex areas are drawn to a large scale as supporting detail drawings. Since some systems are either large or complex, you may need to use sections frequently to clarify a drawing. Vertical dimensions are given by indicating their elevation. The elevation is their height above sea level. Horizontal distances are shown on the plan view and are often given in feet or meters from a baseline on the drawing. A simple elevation, plan and section drawing is shown in Fig. 21-4.

Developed Views

Another plan used to clarify orthographic drawings that are confusing is to imagine the entire system flattened out on a plane. The vertical parts can be revolved to a horizontal plane, or horizontal parts can be revolved to a vertical plane. All fittings appear turned sideways, and all pipe is true length. Fig. 21-5 shows a developed drawing of the system shown in Fig. 21-4.

Some piping systems do not lend themselves well to orthographic projection. Often pipe lines and related equipment are hidden behind other features and the relationship is not clear. For this reason an *isometric drawing* is often made of the individual pipe lines to show the system pictorially in its entirety.

Isometric Method

The isometric method is used to show pictorially a piping system that is neither large nor complex. For example, a system for a typical residence or a small commercial building such as a motel or restaurant can be easily shown. The isometric method is also used to detail separate systems from large, complex installations that have been drawn using the elevation, plan, and section method. Isometric drawings can show clearances between pipelines and between pipelines and vessels. Because these drawings are so clear, you can dimension them easily as well as make a list of the materials for the system. A simple isometric drawing of the system from Fig. 21-4 is shown in Fig. 21-6.

Model Method

The model method requires that the system be designed using isometric or orthographic drawings. These plus the architectural, electrical, and mechanical drawings are used to develop a scale model of the complete system.

The model maker develops the large construction items such as walls, steel framing, columns, and hangers and bases. Then a piping designer works with the model maker to decide how to run pipe and locate the equipment and vessels related to it. Often the model is built in the drafting area of the company.

The model of the equipment used on the piping system model is shaped to represent the item but not built to include any detail.

As the pipe designer works on the model it is possible to check for clearances or places where there is interference between objects that were not noticed on the drawings. Building the model as the system is being designed helps a great deal to prepare drawings that are free from problems. The model can include not only the piping systems but also other requirements for the job such as the electrical, heating, air conditioning, structural, and other architectural requirements.

Models are usually built to the customary scales of 3/8" = 1', or 1/2" = 1'. Metric sized models use scales such as 1:25 and 1:50.

A model is very good for checking a final solution. The checker can review the drawings and the model to see that they accurately represent the flow diagrams, equipment placement and other drawings developed for the job.

One drawback of a model is that it is not easily reproduced. Photographs may be adequate to show the piping system, but many times the model is moved to the construction site. This helps

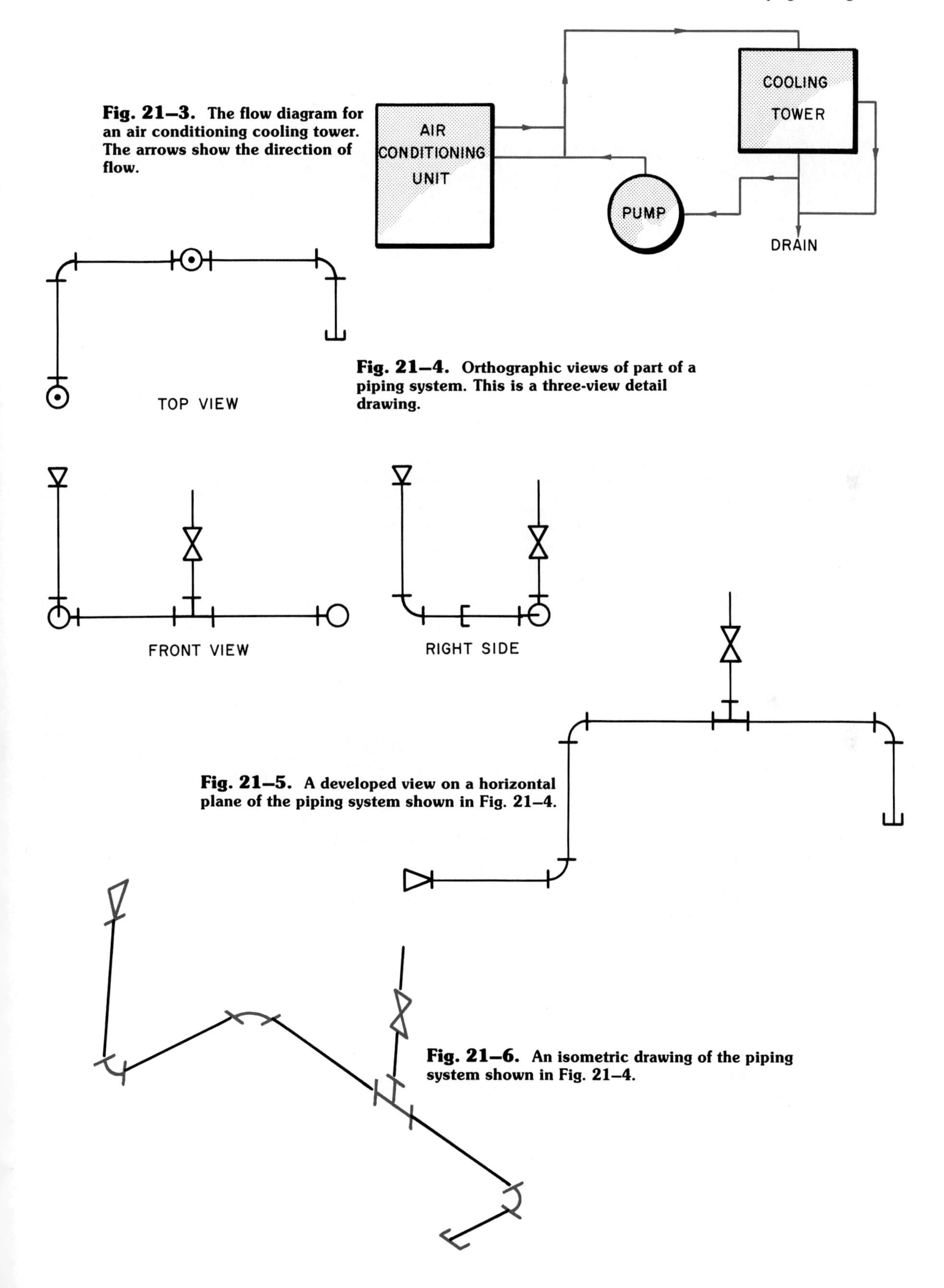

Fig. 21–3. The flow diagram for an air conditioning cooling tower. The arrows show the direction of flow.

Fig. 21–4. Orthographic views of part of a piping system. This is a three-view detail drawing.

Fig. 21–5. A developed view on a horizontal plane of the piping system shown in Fig. 21–4.

Fig. 21–6. An isometric drawing of the piping system shown in Fig. 21–4.

Fig. 21—7. This is a model of a chemical plant. (Monsanto Company)

Fig. 21—8. Computer drafting programs have been developed for piping drawings. (Computervision)

to check the work on the job to see that it progresses according to plan. Some companies keep the model after the plant is operational and use it to train operators and maintenance personnel. Fig. 21-7.

Computers in Pipe Drawing

Piping design can be done with the aid of a computer. The computer can produce the drawings and prepare material lists. The piping designer will usually make sketches of the design and put these into the computer. The computer outputs drawings using a plotter. Fig. 21-8.

Symbols

Piping involves a large number of fittings, valves, and related items that are very time-consuming to draw. These are represented on the drawing by symbols. Making piping drawings can be speeded up by using templates of standard plumbing symbols. Fig. 21-9.

The components of piping systems fall into one of the following types: flanged, screwed, butt welded, socket welded, or soldered. Fig. 21-10. Symbols representing each type, Fig. 21-11, are as follows:

- *flanged valves* and fittings have two short lines perpendicular to the pipe;
- *screw fittings* have one perpendicular line perpendicular to the pipe;
- *welded fittings* are shown with an x or a large black dot;
- *soldered fittings* are shown with a circle at the point where the fitting and pipe meet.

The symbols shown in this book are only a small sampling of those available. Detailed symbol standards are available from The American National Standards Institute. Some industries have had to develop special symbols for items not in the national standards.

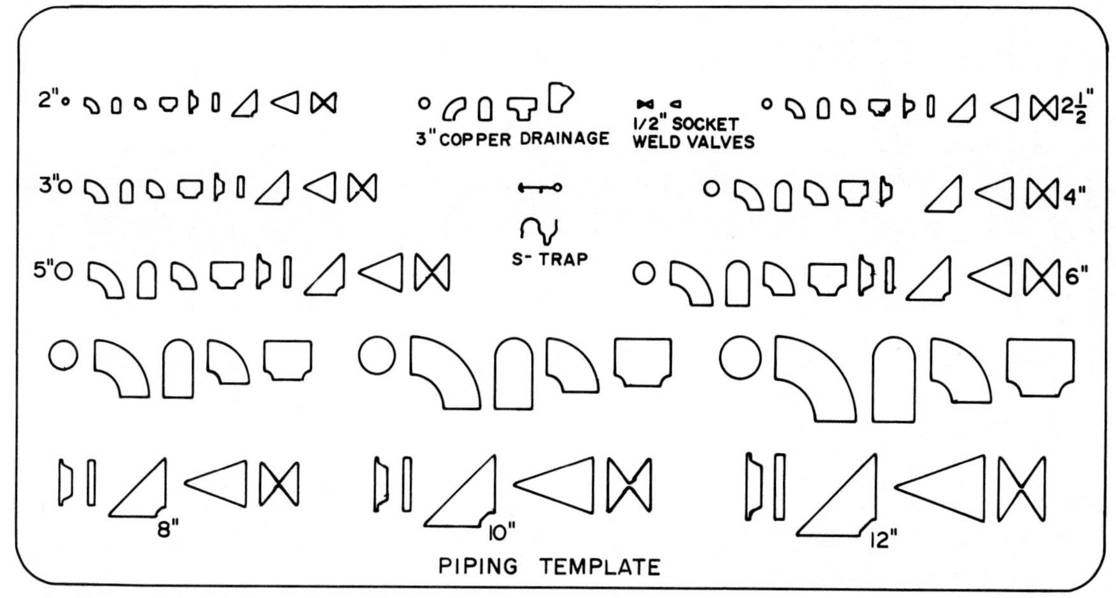

Fig. 21–9. A piping symbol template.

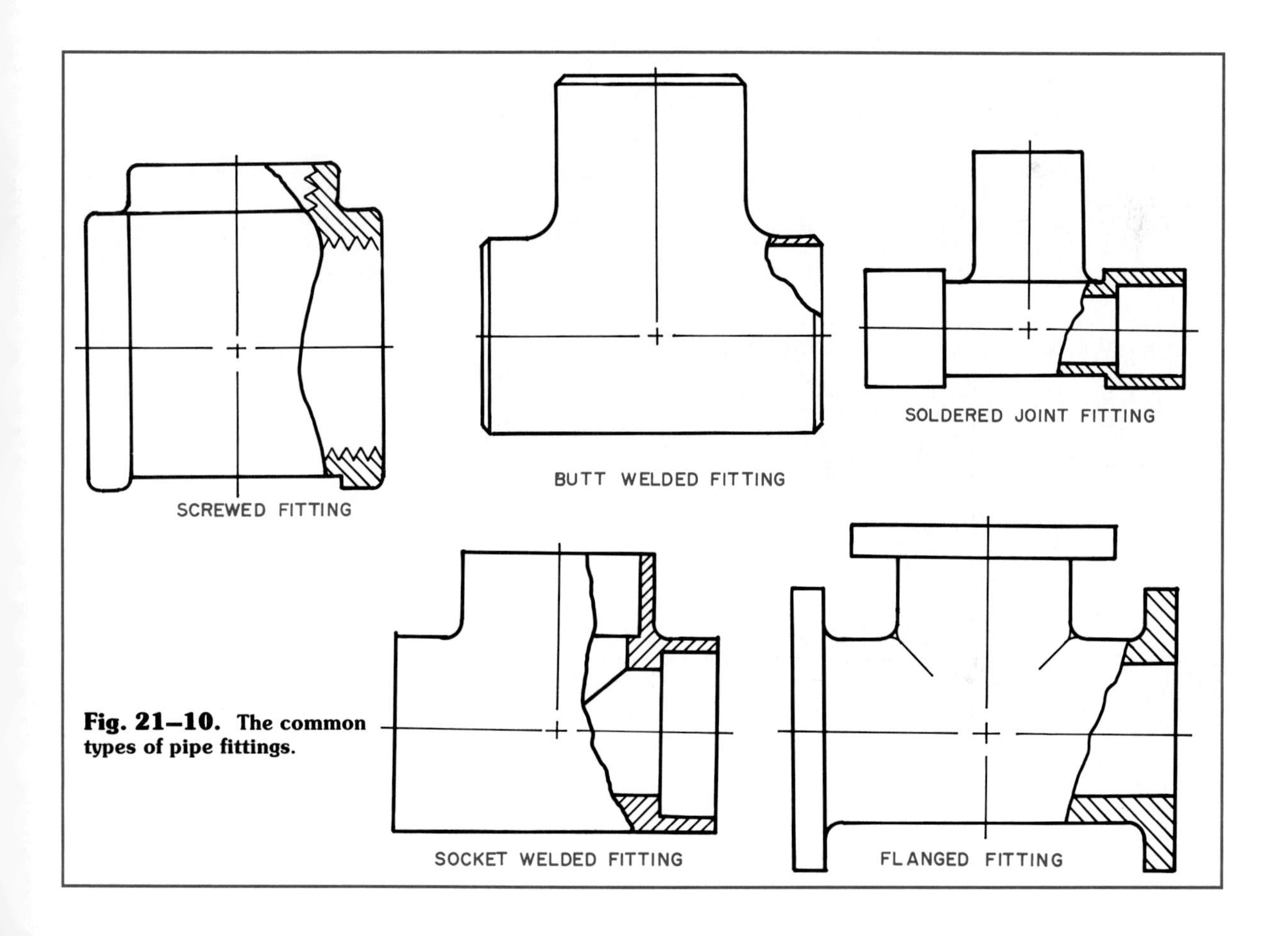

Fig. 21–10. The common types of pipe fittings.

CAD Applications:

Piping Drawing

Piping drawing is a specialized field that presents many opportunities for the use of CAD. Piping drawings are generally diagrammatic. They are drawn orthographically or pictorially. Standard symbols are used to indicate various fittings, connectors, valves, etc. The use of standard symbols makes the use of a shapes or block command very helpful in developing diagram layouts of piping.

If you have access to a CAD system, you may want to try the following exercise. It is a simple piping layout using standard symbols. First, using the line command, draw the following symbols.

GATE VALVE

TEE

LATERAL

ELBOW, 45 DEGREES

ELBOW, 90 DEGREES

Store each of these in a symbol library. Then create the following diagram by using the line command and by inserting the symbols in their appropriate locations.

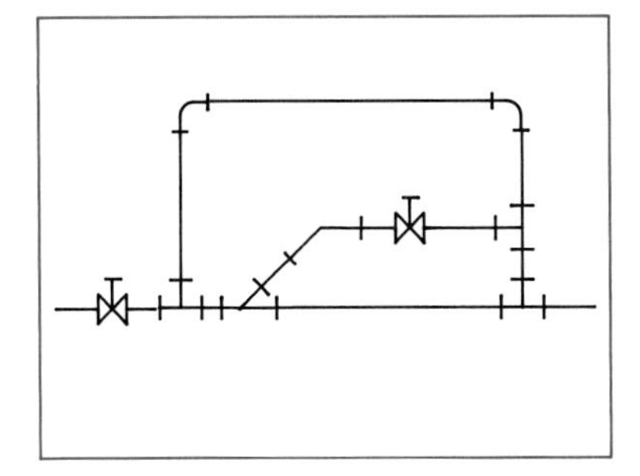

This is an example of a piping drawing using the computer's ability to store and retrieve shapes easily and quickly. A complete shapes library can be constructed. With such a library, complex piping diagrams can be drawn with a minimum amount of time spent on duplicating the various components.

Commercial software specifically designed for piping drawing is also available. It is used in addition to a regular CAD program. The piping program offers such features as parametric design of piping components, interference checks, and automatic generation of isometric drawings from orthographic drawings. Companies that do a lot of piping drawing are finding that such programs greatly increase productivity.

Line symbols are used to identify specialty and utility lines. These are shown in Fig. 21-12. These line symbols are drawn with a medium-width line.

When drawing symbols a medium-width line is also used. They must be drawn carefully so that they are absolutely clear. Many symbols have only small differences; therefore, each must be carefully and fully drawn.

Symbols are laid out on their center lines which stand for the center of the pipes. When making a drawing the center lines are drawn first. The symbols are located at the ends of each section of pipe.

There are three basic types of symbols in use. These are single-line symbols, double-line symbols and pictorial symbols. Fig. 21-13.

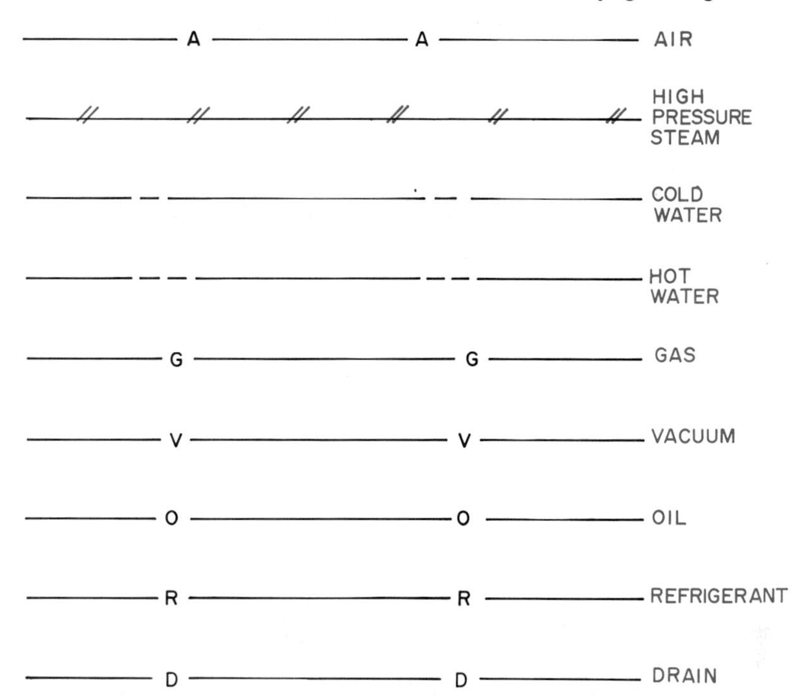

Fig. 21–12. Specialty and utility line symbols are drawn with medium-thick lines. These are just a few of the symbols used for piping.

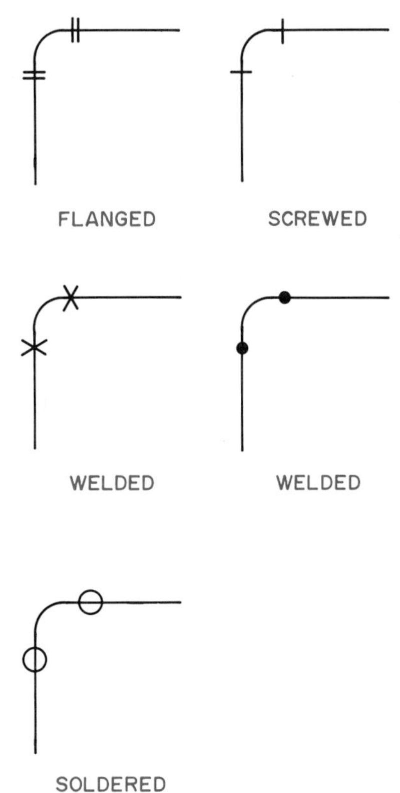

Fig. 21–11. Symbols used to identify the different types of piping systems.

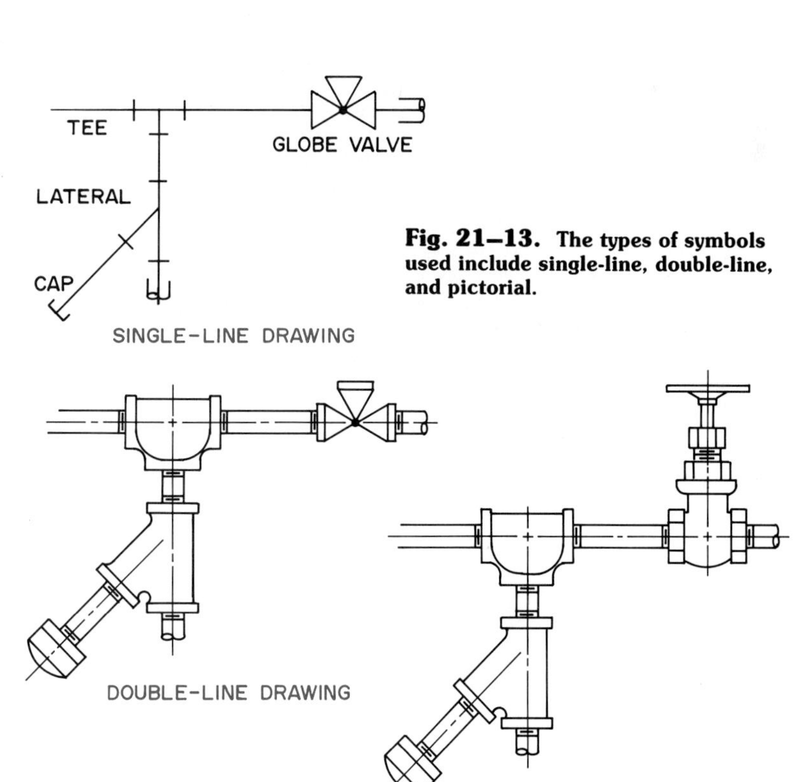

Fig. 21–13. The types of symbols used include single-line, double-line, and pictorial.

	FLANGED	SCREWED	WELDED USES X OR ●	SOLDERED
FITTINGS				
90° ELBOW				
45° ELBOW				
ELBOW TURNED DOWN				
ELBOW TURNED UP				
ELBOW, REDUCING				
CROSS, STRAIGHT				
CROSS, REDUCING				
TEE, STRAIGHT				
TEE, OUTLET DOWN				
TEE, OUTLET UP				
TEE, REDUCING				
LATERAL				
CAP				

TYPE	FLANGED	SCREWED	WELDED USES X OR ●	SOLDERED
REDUCERS				
CONCENTRIC				
ECCENTRIC				
JOINTS				
CONNECTING PIPE			X OR ●	
EXPANSION				
UNION				
VALVES				
CHECK VALVE, STRAIGHTWAY				
GATE				
GLOBE				
SAFETY				

Fig. 21–14. Some of the symbols used on single-line piping drawings.

Single-line symbols are used for most piping drawings. They are the easiest and fastest to draw. When drawing single-line symbols, make an attempt to draw them so that their sizes show a relationship to their actual sizes. They are not drawn to scale. If they were drawn to scale they would be so small they could not be seen. Therefore, *symbols are drawn larger than their actual size and are not in scale with the drawing.* Almost all large orthographic drawings use single-line symbols.

Examples of commonly used single-line symbols are shown in Fig. 21-14.

Double-line symbols are used for drawings where pipe clearances must be checked. The outlines of the pipe and fittings are drawn to scale around a center line. Pipelines that are 12 inches (305 mm) in diameter or larger are almost always drawn using double-line symbols. Make double-line symbol drawings using the largest scale possible. Draw 1½ inch (38 mm) or smaller diameter pipe with single lines. Use single-line symbols on pipe from 1⅝ inches (41 mm) to 10 inches (305 mm) but show the diameter to scale at each end. Fig. 21-15. Symbols for two-line valve drawings are found in Appendix 53.

Pictorial symbols are rarely used. The most frequent use for them is as illustrations in books and other publications so that those not familiar with line symbols can understand the drawing. Fig. 21-16.

Scales

Piping drawings are so large that they must be drawn to scale the same as architectural drawings. The most commonly used scale is ⅜" = 1'—0" (metric 1:25) though scales as small as ⅛" = 1'—0" (metric 1:100) and ¼" = 1'—0" (metric 1:50) are used. Each job will use the scale selected on all the drawings. The scale is lettered near the detail or in the title block if all drawings on the sheet are to the same scale.

Making Elevation, Plan, and Section Drawings

Elevation, plan, and section drawings use the principles of orthographic projection exactly like mechanical drawing. The piping system can be described using top, front, side and section views. In some cases auxiliary views are used for clarity.

Before starting to draw, prepare a sketch of the piping system. Locate the symbols on the center line. This layout will help you decide how large the final drawing will be and what sheet size to use.

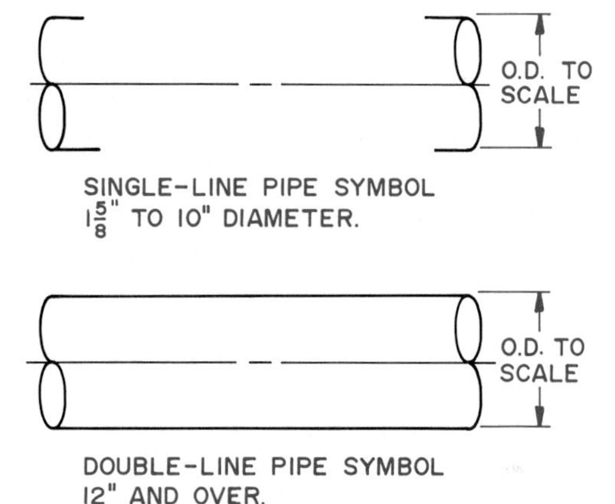

Fig. 21–15. These symbols are used to identify different sizes of pipe.

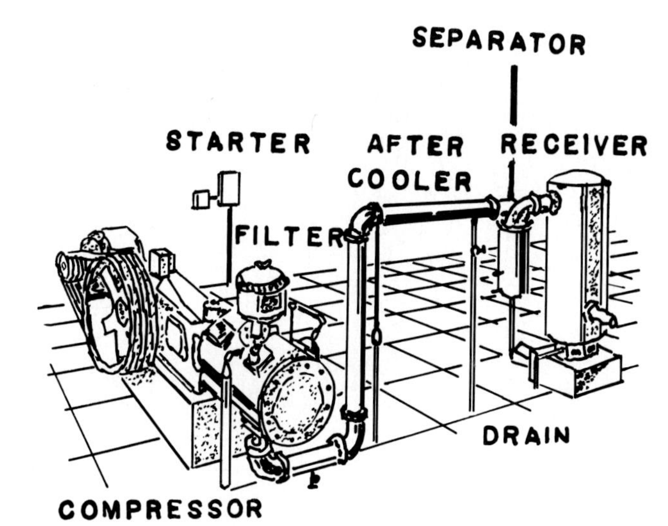

Fig. 21–16. This is a pictorial drawing used to show a typical installation. (Compressed Air and Gas Institute)

First, draw the center lines of the pipes. Then, locate the related equipment. Finally, put the symbols on the center lines. The lengths of pipe sections are not always drawn to scale but are dimensioned accurately. When sizing fittings and pipe it is necessary to refer to manufacturers' catalogs. Selected sizes are shown in Appendices 49 through 53.

A typical orthographic single-line piping drawing with top, front, and right side views is shown in Fig. 21-17. The drawing is oriented to the north with an arrow. All vertical equipment and pipes are located by measuring to their center lines. Horizontal equipment and pipes are located by referring their center lines to the base of the installation. This is usually a concrete slab, pier or footing. Sometimes measurements are taken from the center line of the lowest section of pipe.

Draw equipment, such as a compressor, tower, or a conveyor, with medium-width, dark, solid lines.

Single-line pipes are drawn with heavy lines. This line also represents the center line of the pipe. All structural steel and architectural features are drawn with medium-width, dark, solid lines. All important equipment, such as concrete footings or supports, is drawn to scale. Be certain to show the clearances between pipelines and equipment. When designing exterior pipelines have a 12-foot minimum clearance between them and the ground. Inside buildings pipelines should have a minimum clearance of 7'—6" over main roads and 15'—0" over secondary roads.

Dimensioning Plan, Elevation, and Section Drawings

Dimensioning practices are much like those used for mechanical drawings. Different piping drawings require different dimensioning procedures. Following are the major features used for dimensioning plan, elevation, and section drawings:

1. Place dimensions above a continuous dimension line.
2. Place dimensions so that they can be read from the bottom and right side of the drawing.
3. Do not duplicate dimensions.
4. Place long dimensions outside shorter dimensions.
5. Equipment can be located from the center lines of vessels or structural columns.
6. Use feet and inch marks unless the drawing is metric; then use millimeters. Indicate the metric units with a note; do not place the millimeter abbreviation after each dimension.
7. Show the horizontal dimensions on the plan view.
8. Show vertical dimensions on the elevation views or vertical sections.
9. Show the dimensions between piping and related surfaces such as structural features or equipment.
10. Place dimensions in views where the true size is shown whenever possible.
11. Indicate vertical heights by giving their elevation above the base. Fig. 21-18.

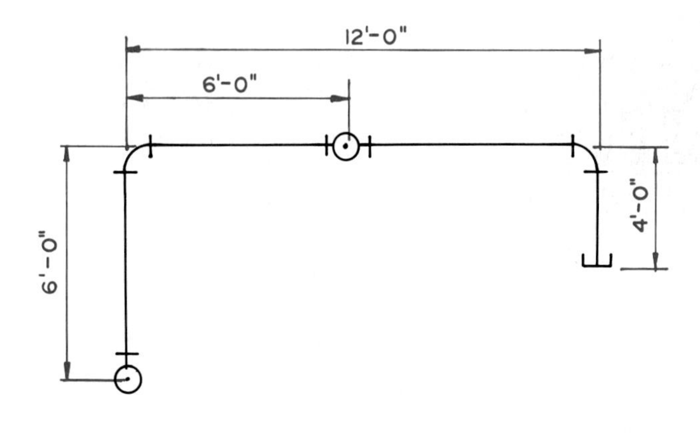

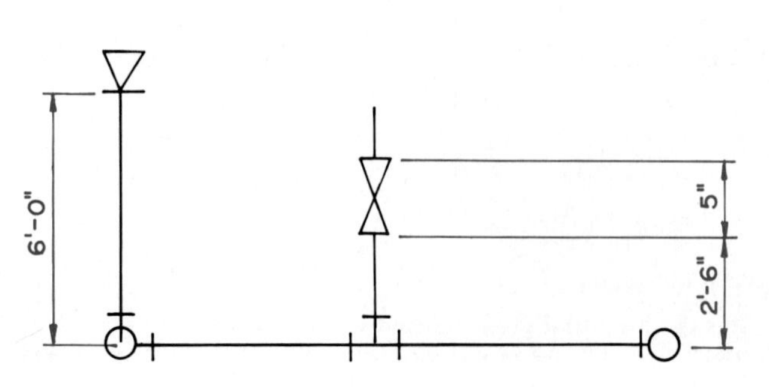

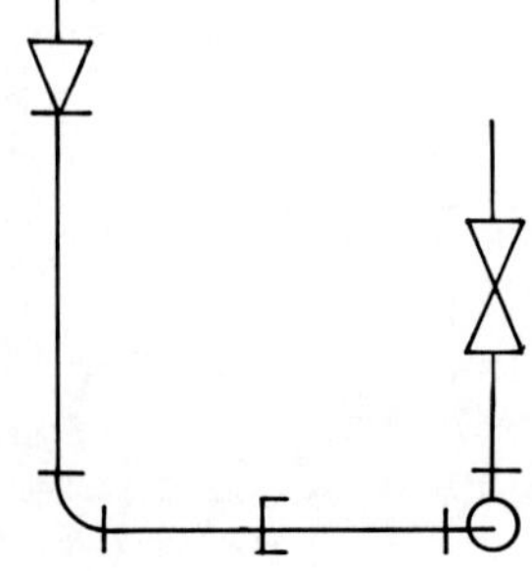

Fig. 21–17. **Orthographic views of part of a piping system. This is a three-view drawing.**

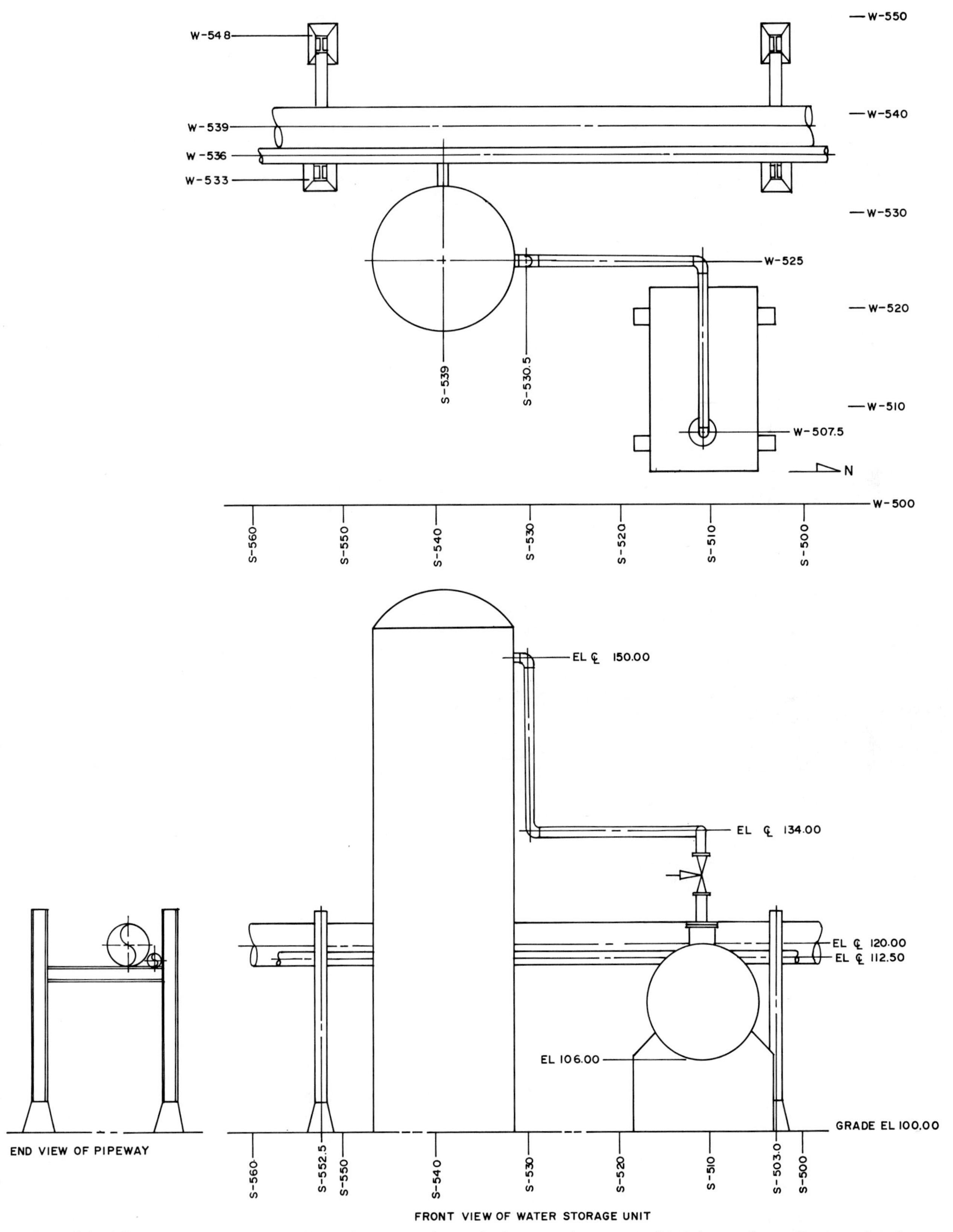

Fig. 21–18. This piping drawing uses elevations above the base to show vertical heights and coordinates related to the north to show horizontal distances.

TAG NO.	QUAN.	SIZE	DESCRIPTION
1	2	1"	TEE
2	2	1"	90° ELBOW
3	1	1"	GATE VALVE

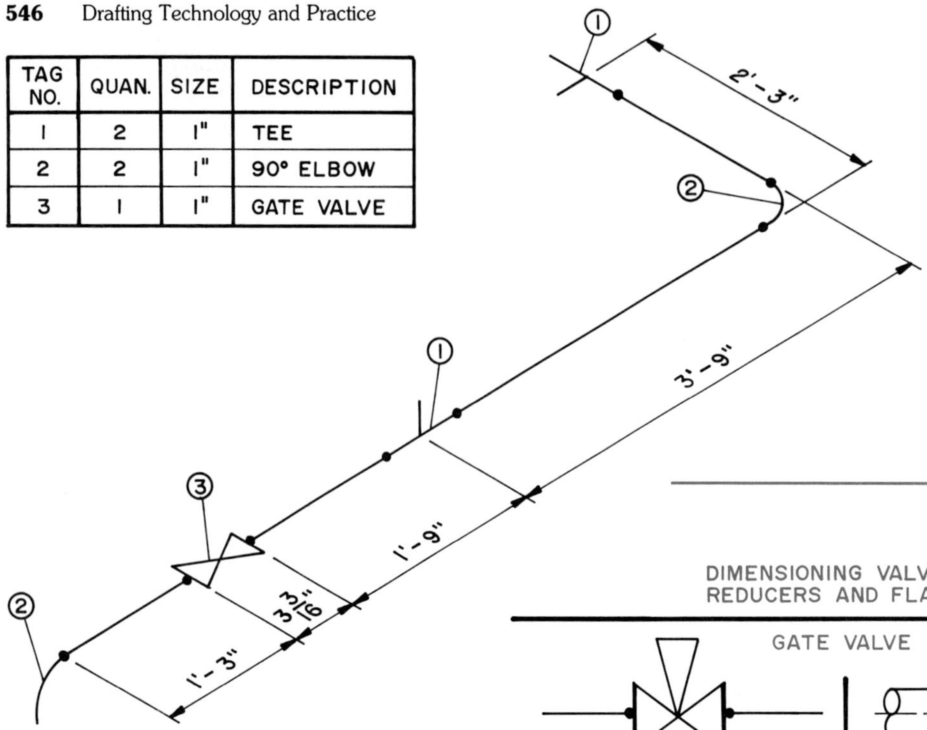

Fig. 21–19. This is a single-line isometric fabrication drawing. This unit will be assembled in the shop and transported to the job site.

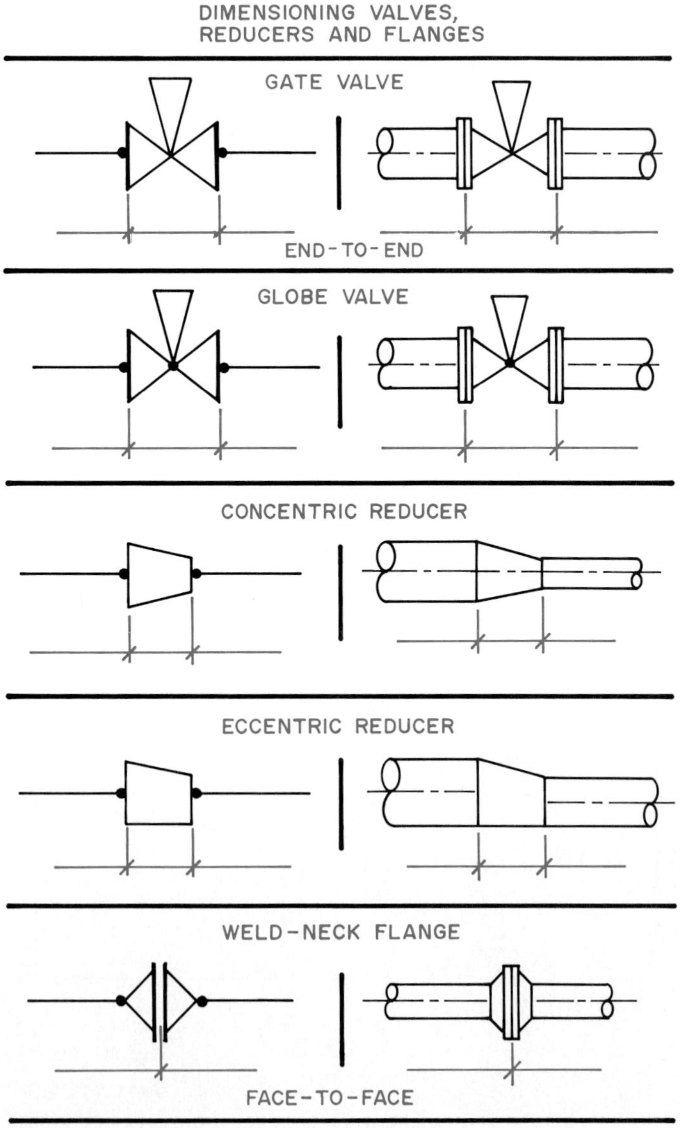

Fig. 21–20. Dimensioning practices for valves, flanges, and reducers.

12. Use coordinates to establish dimensions on the plan view. Fig. 21-18.

13. Sometimes it is necessary to dimension a part of a system using dimension lines in addition to the coordinate system.

14. While the use of arrowheads is preferred, dimension lines may end in a dot or slash.

Making Detail and Fabrication Drawings

Detail drawings are those representing a part of a system that needs to be clarified in greater detail. They are often drawn to a larger scale than the system from which they were taken.

Fabrication drawings show the piping system that is made up in the shop and sent to the job site or assembled on the site. These are often called *spools.* These can be drawn in isometric or orthographic. Double-line drawings are often used but single-line are acceptable. Fig. 21-19.

On detail, fabrication, and isometric drawings, end-to-end and face-to-face dimensions for flanges and fittings are used. Selected examples of how flanged fittings are dimensioned are shown in Fig. 21-20. Face-to-face dimensions run from the face of the outlet port to the face of the inlet port of a valve or fitting. End-to-end is where two flanges butt each other. Locate components and pipes by their centers such as the center of a valve to the center of a pipe. Note in Fig. 21-20 that tic marks are used inside the dimension lines for face-to-face dimensions that include gasket dimensions in the overall size.

Callouts are used to identify special items such as reducing fittings, tees and reducing elbows. All valves and other components are also noted. Fig. 21-21.

Isometric Pipe Drawings

Isometric drawings are widely used in piping drafting. They are prepared either manually or with a computer. When using isometric techniques the elevation, plan, and side views are all shown on one drawing. These drawings can be used for detail and fabrication drawings, especially since they are easy to understand. Fig. 21-22. All of the details mentioned for elevation, plan, and section; detail; and fabrication drawings apply to isometric drawings.

A few of the more commonly used symbols for isometric drawings are shown in Fig. 21-23. The *single- line isometric symbol* is the easiest to use. It is used more than the other types shown. When making isometric drawings the *lines for the flange fittings lie in the plane of the bend of the pipe nearest the valve or fitting.* The *single-line elliptical isometric symbol* is seldom used because it is time-consuming to draw.

The *double-line isometric symbol* allows the drawing to show clearances and interferences between pipes and related features. Fig. 21- 24.

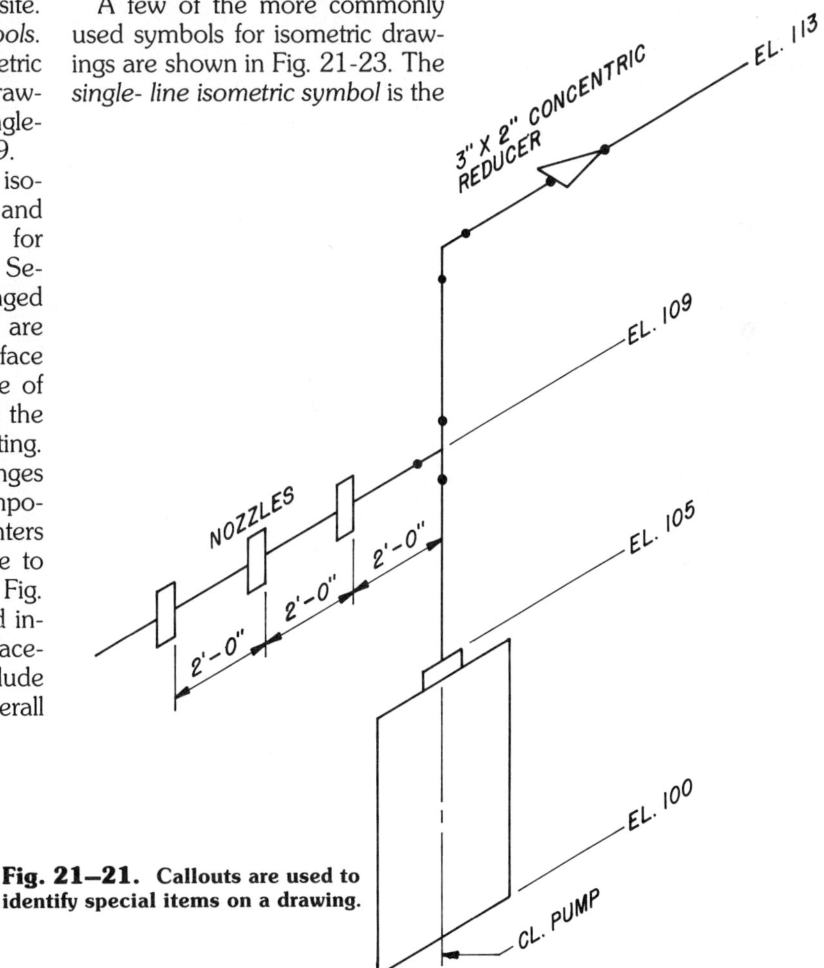

Fig. 21–21. Callouts are used to identify special items on a drawing.

Isometric drawings are usually drawn to a proportion rather than to an accurate scale. Some parts are so large it would be difficult to fit the drawing on a sheet if drawn accurately to scale. Fittings and valves that are very small are drawn large. The principles of isometric drawing found in Chapter 14, "Pictorial Drawing," are used when making these drawings.

Isometric drawings are oriented to the north. Generally one of the horizontal axes to the upper left or right is chosen.

When preparing an isometric drawing, locate large items such as vessels, towers, and pumps first. Then draw the pipelines and components (tees, elbows, etc.), and draw the pipe supports, valves, and fittings last.

Dimensioning Isometric Drawings

Dimension isometric drawings much the same as orthographic drawings. Use the aligned method of dimensioning with the dimension placed above an unbroken dimension line.

Keep dimensions parallel to the pipeline being dimensioned. Most dimensions are taken from the center lines or the end of the pipe to the center of components, such as valves. Face-to-face dimensions between valves are also used.

Locate angles in the vertical or horizontal planes. Fig. 21-25.

Identify components and pipe sizes by a number and a bill of materials.

Slant lettering on an isometric drawing so that it falls in the isometric plane. Use arrows to show direction of flow.

Elbows can be drawn rounded or square. Rounded elbows are more descriptive but square bends are easier to draw.

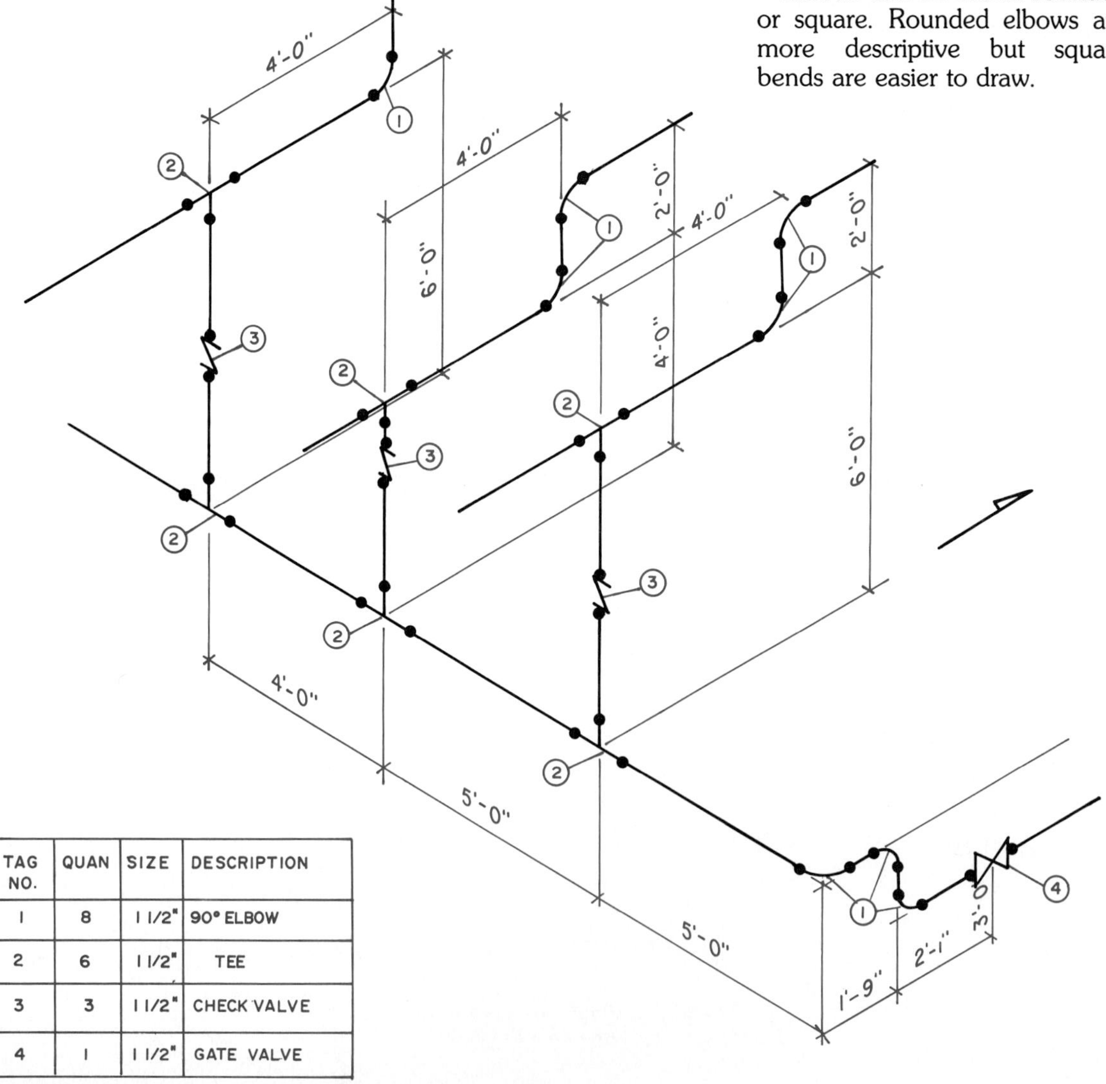

TAG NO.	QUAN	SIZE	DESCRIPTION
1	8	1 1/2"	90° ELBOW
2	6	1 1/2"	TEE
3	3	1 1/2"	CHECK VALVE
4	1	1 1/2"	GATE VALVE

Fig. 21–22. A typical isometric drawing.

	ISOMETRIC	ELLIPSE ISOMETRIC
COUPLING		
STRAIGHT CROSS		
REDUCING CROSS		
45° ELBOW		
90° ELBOW		
STRAIGHT TEE		

	ISOMETRIC	ELLIPSE ISOMETRIC
UNION		
CONCENTRIC REDUCER		
ECCENTRIC REDUCER		
GATE VALVE		
GLOBE VALVE		
SAFETY VALVE		

Fig. 21–23. These are examples of isometric and ellipse isometric symbols.

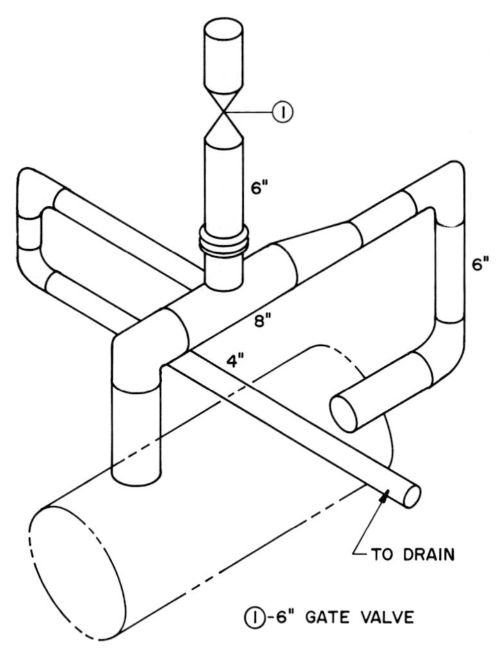

Fig. 21–24. A double-line isometric drawing.

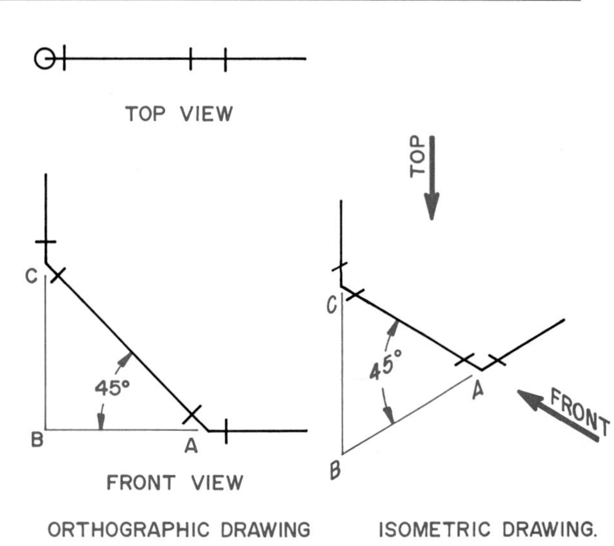

Fig. 21–25. Angles in isometric are located in the isometric planes.

Chapter Review

Build Your Vocabulary

You should understand and use the following terms as part of your working vocabulary. Write a brief explanation of what each means.

fittings
valves
process flow diagram
single-line symbols
double-line symbols
detail drawings
fabrication drawings

Study Problems—Directions

The problems that follow will give you the chance to use some of the drafting knowledge that you learned in this chapter. Do the problems that are assigned to you by your instructor.

Flow Diagram

1. Make a flow diagram for the piping system shown in **Fig. P21-1**.

Single-line Orthographic Drawings

2. On an 8½″ × 11″ sheet draw the following single-line symbols: welded straight cross, flanged lateral, screwed 90° elbow, welded 90° elbow turned down, flanged tee outlet up, soldered union, screwed check valve, welded reducing cross—3″ to 2″, soldered expansion joint, flanged gate valve and a screwed concentric reducer.

3. Make a one-view single-line drawing of the detail shown in **Fig. P21-2**. Select your own scale. The fittings have screw connections.

4. Examine the diagram of the cooling tower pump and condenser connections, **Fig. P21-1**. Draw this as a one-view, single-line diagram. Plan the drawing so the symbols are not crowded. Assume the fittings to be welded. Select your own scale and pipe length.

5. Draw a one-view, single-line drawing of the industrial waste disposal pump and piping, **Fig. P21-3**. Select your own scale and pipe length.

6. Make a single-line orthographic drawing of the piping detail shown in **Fig. P21-4**. Show top, front, and right side views. Make your drawing twice as large as the printed diagram. Position the drawing so the fittings are clearly shown. Select your own scale and spacing between fittings. The fittings are welded.

7. Make an orthographic drawing of the detail of the cooling tower system, **Fig. P21-5**. Make your drawing twice as large as the printed diagram.

Single-line Isometric Drawings

8. Make a single-line isometric drawing of **Fig. P21-6**. Choose your own scale.

9. Make a single-line isometric drawing of **Fig. P21-7**. Draw to the scale ½″ = 1′—0″.

10. Make a single-line isometric drawing of **Fig. P21-8**. Use ellipse isometric symbols. Draw to the scale ¼″ = 1′—0″.

11. Make a single-line isometric drawing of **Fig. P21-9**. Use isometric symbols. Draw three times as large as printed.

Two-line Orthographic Drawings

12. Make a two-line orthographic drawing of the boiler installation, **Fig. P21-7**. Assume the fittings are the flanged type. Select a scale that will clearly show the details.

13. Make a two-line orthographic drawing of **Fig. P21-6**. Draw four times as large as printed.

Two-line Isometric Drawings

14. Make a two-line isometric drawing of **Fig. P21-8**. Use the scale ⅜″ = 1′—0″.

15. Make a two-line isometric drawing of **Fig. P21-9**. Draw it three times as large as the printed diagram.

Study Problems

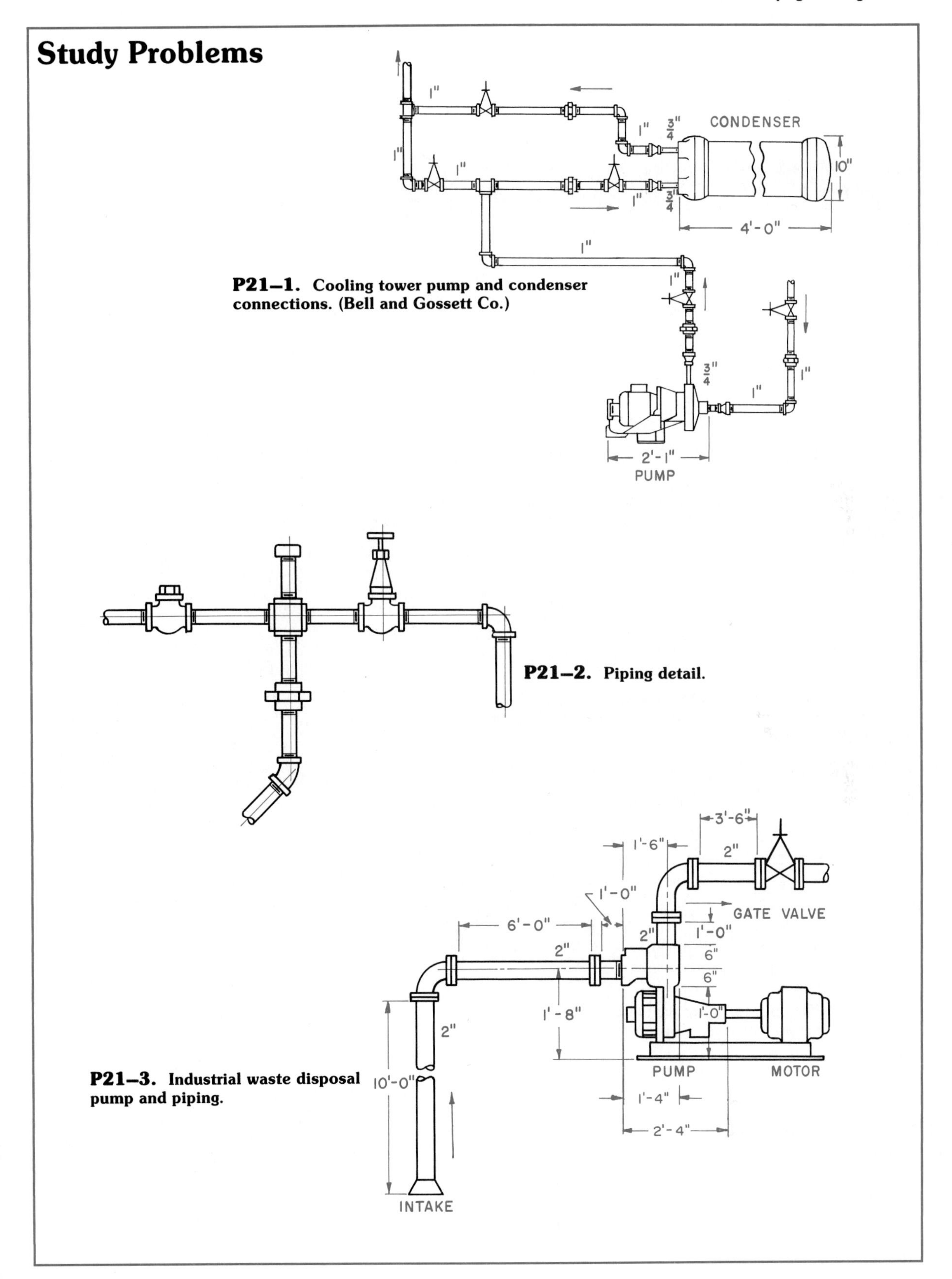

P21–1. Cooling tower pump and condenser connections. (Bell and Gossett Co.)

P21–2. Piping detail.

P21–3. Industrial waste disposal pump and piping.

Study Problems

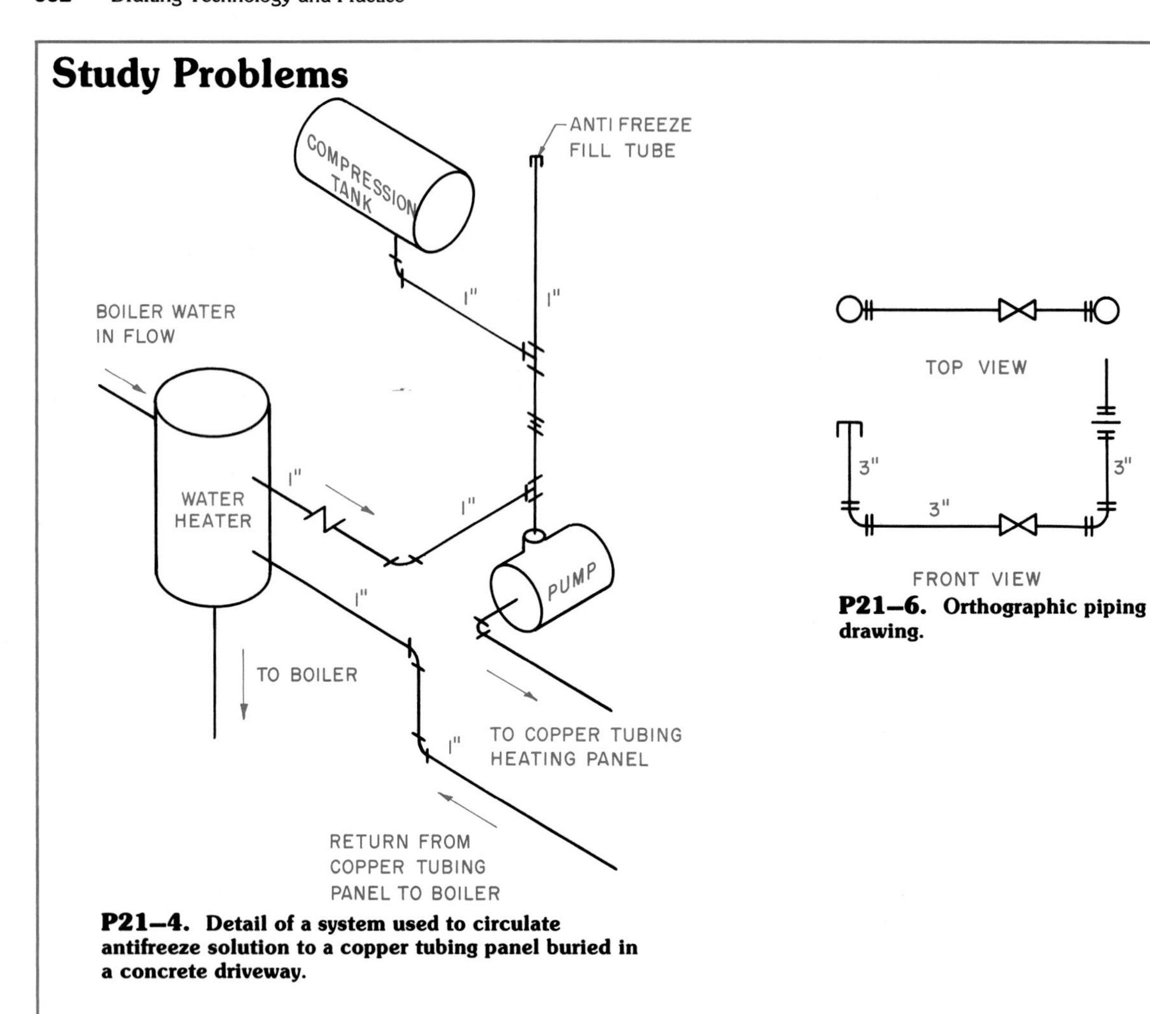

P21–4. **Detail of a system used to circulate antifreeze solution to a copper tubing panel buried in a concrete driveway.**

P21–6. **Orthographic piping drawing.**

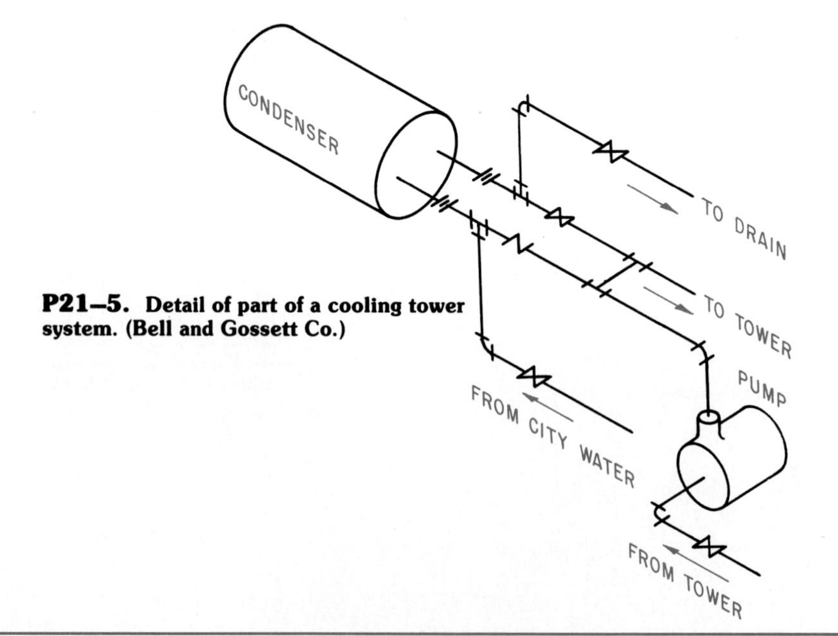

P21–5. **Detail of part of a cooling tower system. (Bell and Gossett Co.)**

Study Problems

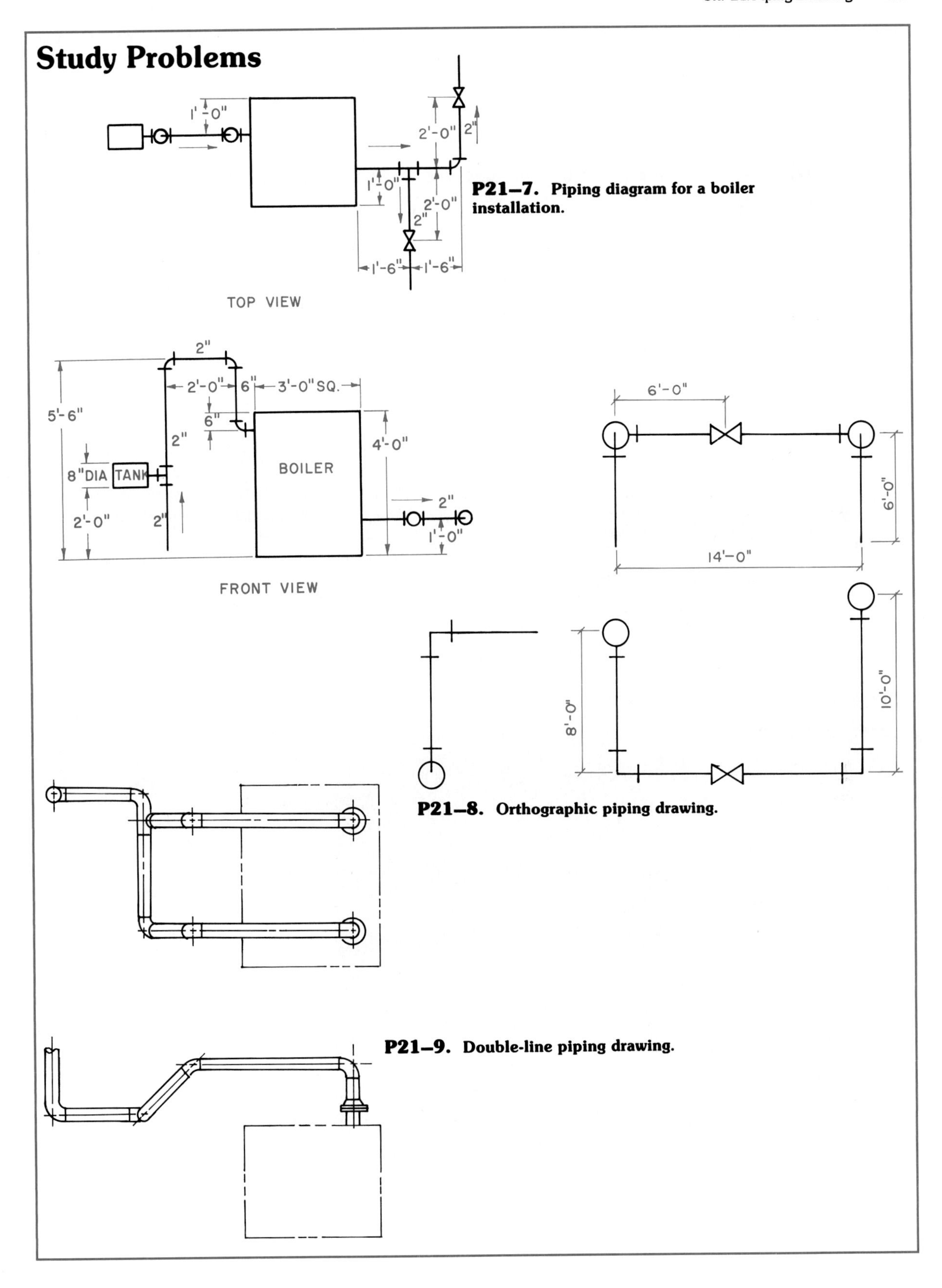

P21–7. Piping diagram for a boiler installation.

P21–8. Orthographic piping drawing.

P21–9. Double-line piping drawing.

CHAPTER 22
Structural Drawings

After studying this chapter, you should be able to do the following:

- Read structural drawings.
- Draw structural drawings using data prepared by a structural designer.

Introduction

Structural drawings show the design of the supporting framework of bridges, towers, and buildings. Fig. 22-1 shows a bridge with a structural steel frame and a concrete deck. Structural drawings give the details needed to make each member. (*Member* refers to any part of a structure.) These drawings also show the fastening methods.

The engineering design is done by engineers. They decide on the placement and sizes of the members. They also decide how the members should be joined.

Structural drafters work from the design information given by the engineer. They prepare structural drawings of the entire framework. They also make working drawings for each member. The structural drafter must understand basic design principles. A knowledge of fastening methods is needed. The structure is no stronger than the fastening devices used.

Structural drafting principles are much the same as those used in other types of drafting. However, the method of presenting information on a structural drawing is different. Some of the methods are unique to this type of drawing.

Structural design and drafting is a very difficult and complex process. Presented in this chapter are some of the easier and more common techniques used. Discussion is limited to selected wood, steel, and poured-in-place concrete drawings.

Fig. 22–1. The Golden Gate Bridge at San Francisco required hundreds of structural drawings. (San Francisco Convention and Visitors Bureau)

Terms and Symbols

Fig. 22-2 shows the steel framework of a building. Before drafters can make the structural drawings, they must understand the terms used. The following sections give some of the more important things to know.

Terms Used in Structural Design

Some structural terms that are commonly used are shown in Figs. 22-3 and 22-4 and described below.

Columns are vertical steel members used to support the roof and floor.

Girders are structural members that run horizontally between columns.

Filler beams are structural members that run horizontally between girders.

Pitch refers to the distance between the center lines of fasteners or holes.

Gage line is a continuous center line passing through holes or fasteners.

Gage distance is the distance the gage line is from the side of the structural member.

Edge distance is the distance the first hole or fastener is from the end of the member.

Slope is an indication of the angle a member makes with the horizontal. It is shown by a triangle with the hypotenuse parallel with the structural member. The longest leg is always 12 inches. The slope symbol example in Fig. 22-4 means that for every 6 inches of rise, the member has 12 inches of run.

Gusset plates are used on trusses to join the structural members.

Steel angles are L-shaped metal structural members.

Working point is the point of intersection of the center lines of holes for rivets and bolts.

Structural Steel Shapes

Structural steel is made in a wide variety of standard sizes and shapes. The most commonly used shapes are shown in Fig. 22-5. These members are made to carry established loads. This information is available from the companies manufacturing the members and from the *Manual of Steel Construction*, American Institute of Steel Construction.

Standard Abbreviations and Symbols

In order to simplify notes on drawings, a system of standard abbreviations has been developed by the American Institute of Steel Construction. This system does not use inch or pound marks. They are understood as the symbol is read. Selected standard abbreviations are shown in Fig. 22-6.

Structural Fasteners

Structural members are fastened with rivets or bolts or by welding.

Fig. 22–2. A building with a structural steel framework. An engineer decides the size and spacing of these members. (The R. C. Mahon Co.)

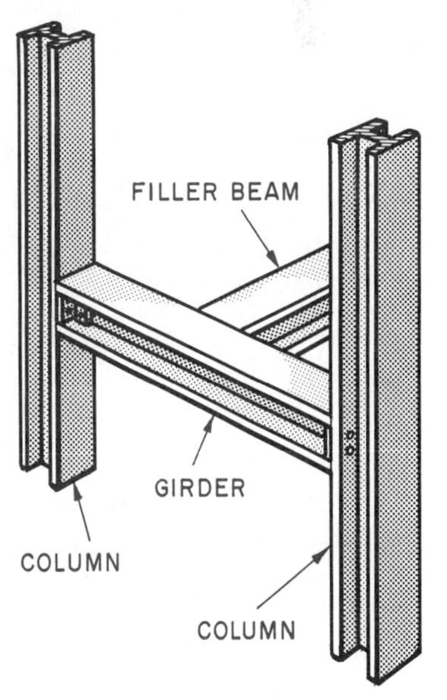

Fig. 22–3. Typical structural members.

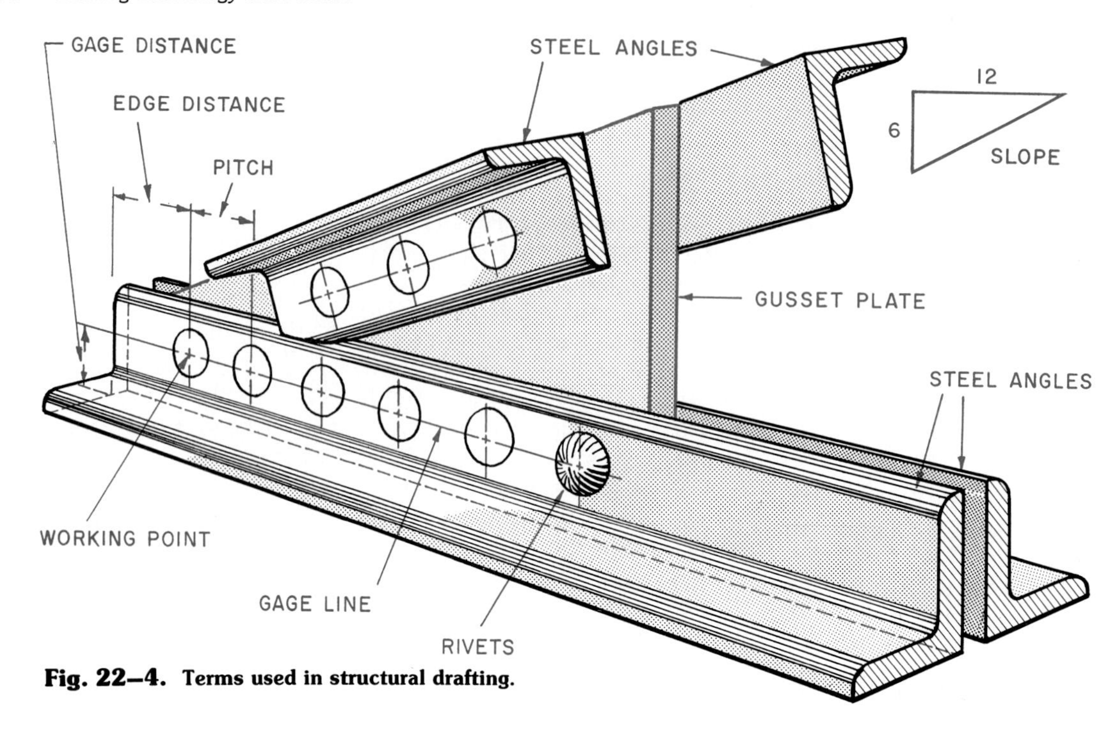

Fig. 22–4. Terms used in structural drafting.

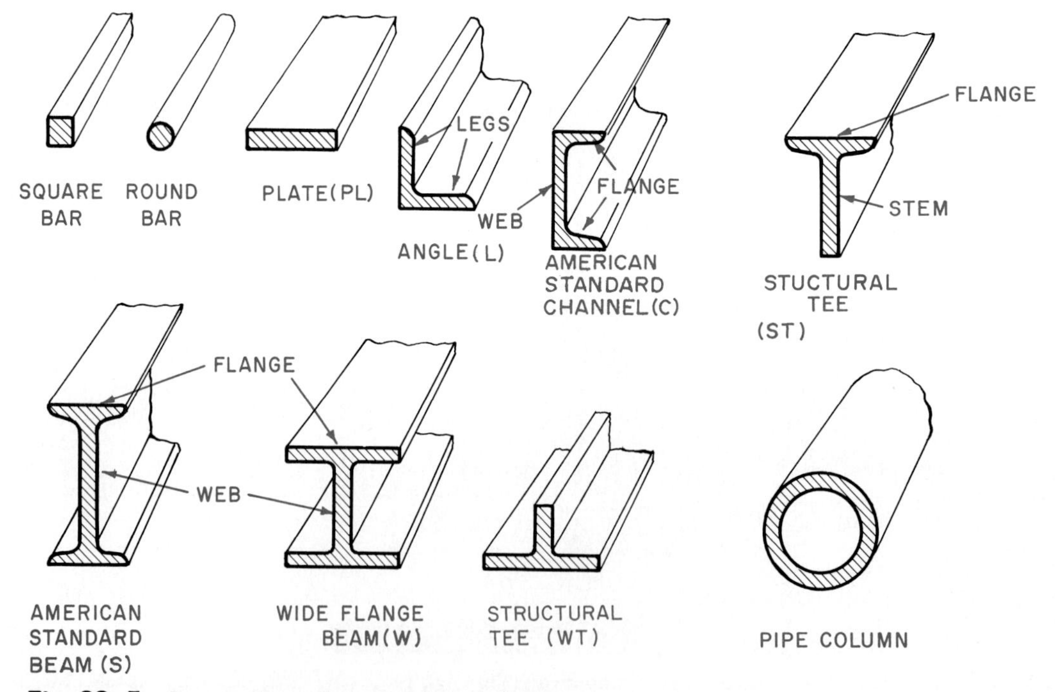

Fig. 22–5. Commonly used structural steel shapes.

Rivets. One method for joining structural steel is riveting. When parts are put together in a shop, the drawing calls for **shop rivets**. Rivets to be installed on the job in the field are called **field rivets**. Shop rivets are shown on the drawing as open circles. Field rivets are drawn solid black. Fig. 22-7.

Rivets are made in a large number of sizes and have a variety of head shapes. The common head shapes are shown in Fig. 22-7. More rivet information can be found in Chapter 12, "Fasteners."

The holes to receive the rivets are made about 1/16 inch larger than the rivet diameter. This permits the rivet to expand as it is driven into place.

A rivet is installed by heating it to a light cherry-red color, inserting it into a hole, and forming a head on the straight end. Fig. 22-8.

High-Strength Steel Bolts. Another method for joining structural steel is using high-tensile steel bolts. Fig. 22-9. Bolting structural members is rapidly replacing riveting for both field and shop assembly. Fig. 22-10.

Standards have been established for bolts suitable for this purpose. Details can be found in *Structural Joints Using ASTM A325 Bolts*, published by the American Institute of Steel Construction.

It is important to get the proper size bolt to carry the load and stand the stress. The bolt must be tightened to the proper tension. The required tension is given in standardized form in the above publication. Hardened washers are placed under the bolt head and nut.

The symbols for drawing bolts are shown in Fig. 22-7.

Spacing Rivets and Bolts. The location of holes for rivets and bolts is controlled by the gage lines and the spacing of the holes on the gage lines. Recall that the *gage line* is the center line that runs parallel to the long edge of the metal member on which the holes for bolts or rivets are placed. See Fig. 22-4.

The location of gage lines has been standardized for most connections. Fig. 22-11 shows the gage line distances for angle connections. These distances vary with the length of the angle's leg. Notice that the gage line is located by measuring from the back of the angle.

The spacing between holes along a gage line of angle connectors is usually in units of 3 inches. This spacing is for drilled and punched holes. See Fig. 22-12 for the spacing of holes on an angle connection. An **angle connection** is used to fasten a girder to a column.

Fig. 22-6. Symbols and Abbreviations for Structural Steel Members

Structural Shape	Symbol	Order of Presenting Data	Sample of Abbreviated Note
Square Bar	⏚	Bar, Size, Symbol, Length	Bar 1 ⏚ 5′-3
Round Bar	Φ	Bar, Size, Symbol, Length	Bar 5/8 Φ 7′-5
Plate	PL	Symbol, Thickness, Weight	PL ½ × 18
Angle, Equal Legs	∠	Symbol, Leg 1, Leg 2, Thickness, Length	∠ 2 × 2 × ¼ × 9′-4
Angle, Unequal Legs	∠	Symbol, Long Leg, Short Leg, Thickness, Length	∠ 4 × 3 × ¼ × 6′-6
Channel	C	Symbol, Depth, Weight, Length	C 6 × 10.5 × 12′-2
American Standard Beam	S	Symbol, Depth, Weight, Length	S 10 × 35.0 × 13′-4
Wide Flange Beam	W	Symbol, Depth, Weight, Length	W 16 × 64 × 20′-7
Structural Tee	ST, WT	Symbol, Depth, Weight	ST 12 × 50
Pipe	NAME OF PIPE	Name, Diameter, Strength	Pipe 3½″ extra strong

(American Institute of Steel Construction)

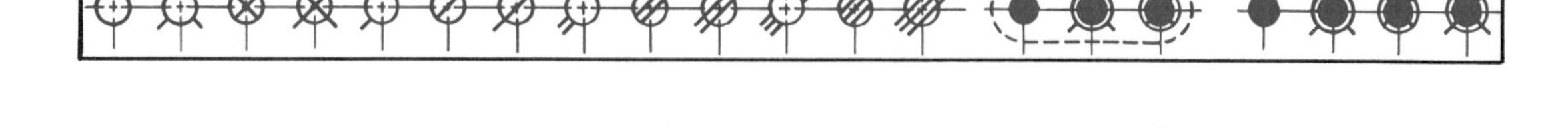

Fig. 22–7. Rivet and bolt symbols. (American Institute of Steel Construction)

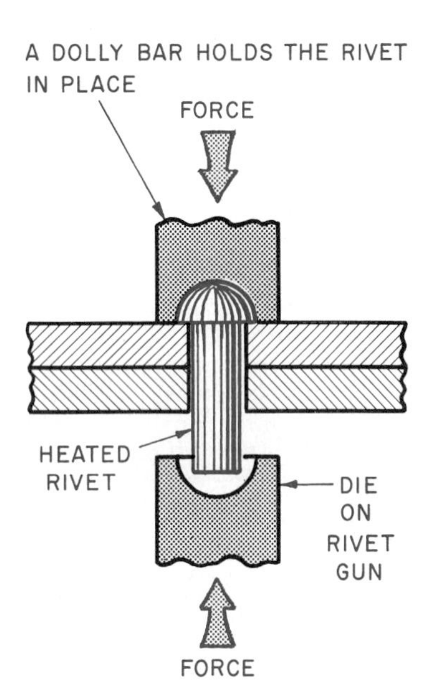

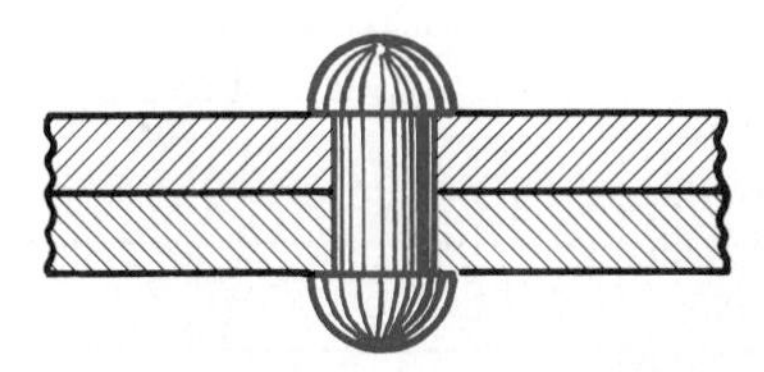

Fig. 22–8. A rivet is installed by heating the rivet, putting it in place, and forming a second head with a die on the rivet gun.

Fig. 22–9. A high-tensile steel bolt used to assemble structural steel framework in buildings. (Russell, Burdsall and Ward Bolt and Nut Co.)

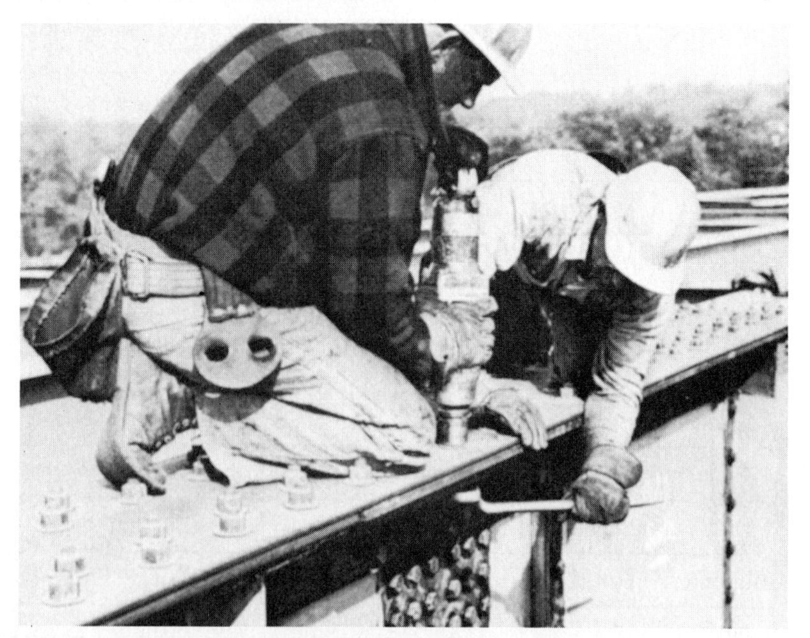

Fig. 22–10. High-tensile steel bolts can be rapidly fastened in place. The operation can be performed by workers with less skill than that required for riveting. (Russell, Burdsall and Ward Bolt and Nut Co.)

Standard spacing of gage lines in W-beams is shown in Fig. 22-13. The gage lines are measured from the center of the flanges as shown in Fig. 22-14. Standard spacing is 3, 3½, and 5½ inches. The spacing of holes along these gage lines is in 3-inch units.

The minimum distance a hole can be from the edge is shown in Fig. 22-15. This distance is measured from the center line of the hole. It varies with the diameter of the rivet. Some riveted steel joint details are shown in Fig. 22-16.

Structural Welding. Welding is another way of joining steel members. The fillet weld is most commonly used. Welding specifications are given in the *Manual of Steel Construction*. A study of welding symbols as used on drawings is found in Chapter 12, "Fasteners." Some welded connections are shown in Fig. 22-16.

Standard Beam and Column Connections

Connections are used to fasten beams to other structural members. Connections have been standardized by the American Institute of Steel Construction. The two common types of beam connections are framed and seated. They are designed to handle the loads and forces that beams must carry. Some of these connections are shown in Fig. 22-16. For a complete description of connections see the *Manual of Steel Construction*.

Drafters must be able to select the proper connections. To do this, they are required to know the strength of the connections. They must know how to use the standards manual.

Scales for Structural Steel Drawings

The scale you select for a structural drawing will depend on whether it is a shop drawing, or a framing or an erection drawing. The scale for a framing drawing or an erection drawing usually ranges from ¼″ = 1′—0 to 1″ = 1′—0″. Shop drawings are usually drawn ¾″ = 1′—0″ to 1″ = 1′—0″. Metric sizes for framing and erection drawings are 1:50 to 1:10. Shop drawings are drawn 1:20 to 1:10.

Dimensioning

Structural steel drawings are dimensioned much the same as other drawings. The following list contains some recommended practices:

- The dimension lines are unbroken. The numerals are placed above the line and near the center.

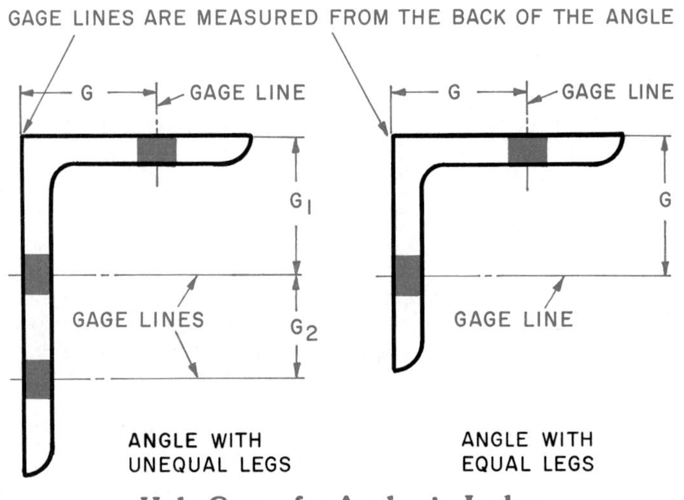

Hole Gages for Angles in Inches

Leg	8	7	6	5	4	3½	3	2½	2	1¾	1½	1⅜	1¼	1
g	4½	4	3½	3	2½	2	1¾	1⅜	1⅛	1	⅞	⅞	¾	⅝
g^1	3	2½	2¼	2										
g^2	3	3	2½	1¾										

Fig. 22—11. Gage lines and hole gages for angles. (American Institute of Steel Construction)

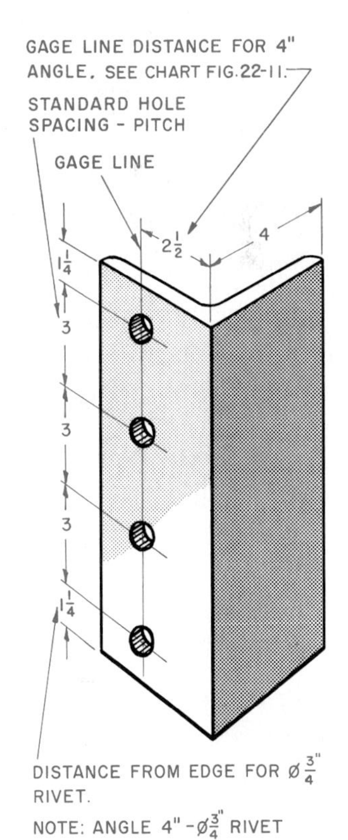

Fig. 22—12. Spacing holes in angle connections.

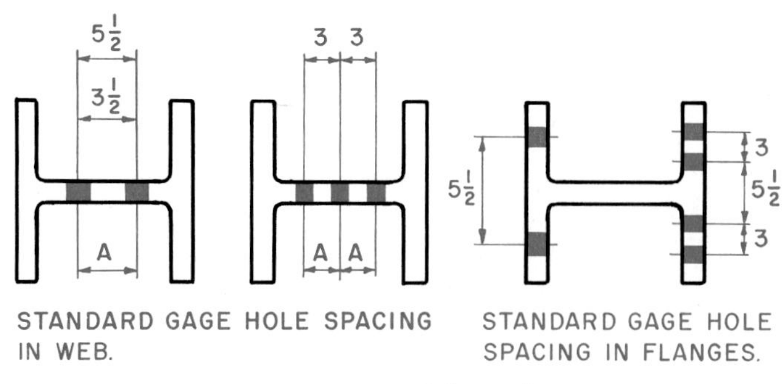

NOTE: MINIMUM DISTANCE FOR "A" IS 3".

STANDARD GAGES FOR DRILLED HOLES

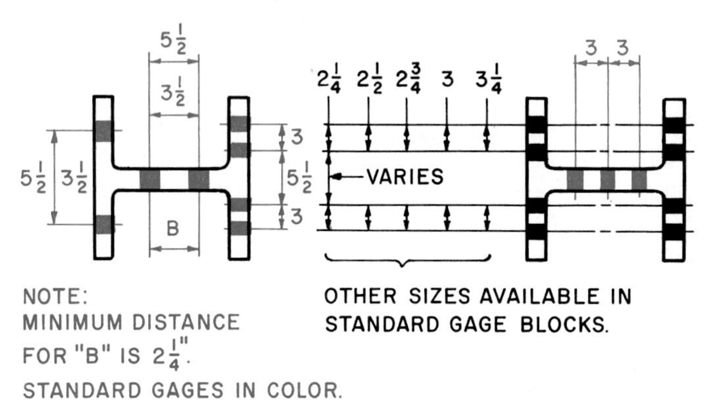

STANDARD GAGES FOR PUNCHED HOLES

Fig. 22–13. Standard spacing of gage lines in W-beams. (American Institute of Steel Construction)

- Small dimensions are placed close to the object outline. Long dimensions are placed outside the small dimensions.
- Dimensions are given to the working points, such as the gage line of a series of holes.
- All dimensions over 1 foot are given in feet and inches except the width dimensions of plates and the depth of structural members. Those are given in inches.
- The foot symbol (') is used on dimensions. Inch marks (") are omitted unless needed for clarity. If the drawing is metric, all dimensions are in millimeters. It is not necessary to put the symbol *mm* after each dimension. A note indicating that the drawing is metric is added to the drawing.
- Fasteners, such as rivets, and holes are dimensioned to their center lines.
- Dimensions are given to the center line of the beams.
- Dimensions are taken on the back of angles and channels.

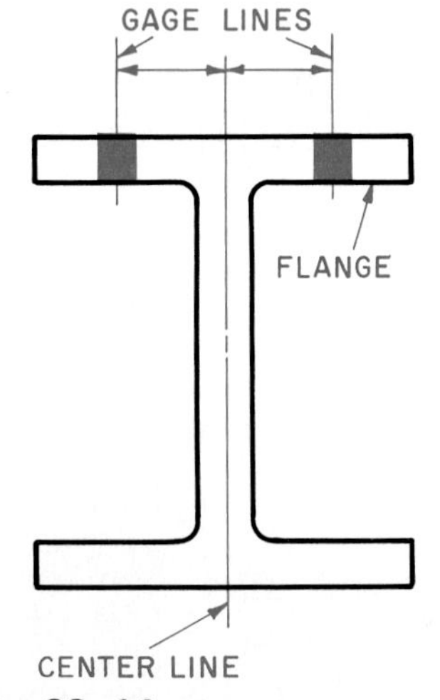

Fig. 22–14. Gage lines are located from the center of the W-beam.

Fig. 22-15. Edge Distances for Rivet Holes

	Minimum Edge Distance (Inches) for Punched Holes		
Rivet Diameter (inches)	In Sheared Edges	In Rolled Edge of Plate	In Rolled Edge of Structural Shapes*
½	1	⅞	¾
⅝	1⅛	1	⅞
¾	1¼	1⅛	1
⅞	1½	1¼	1⅛
1	1¾	1½	1¼
1⅛	2	1¾	1½
1¼	2¼	2	1¾

(American Institute of Steel Construction)

*May be decreased ⅛ inch when holes are near end of beam.

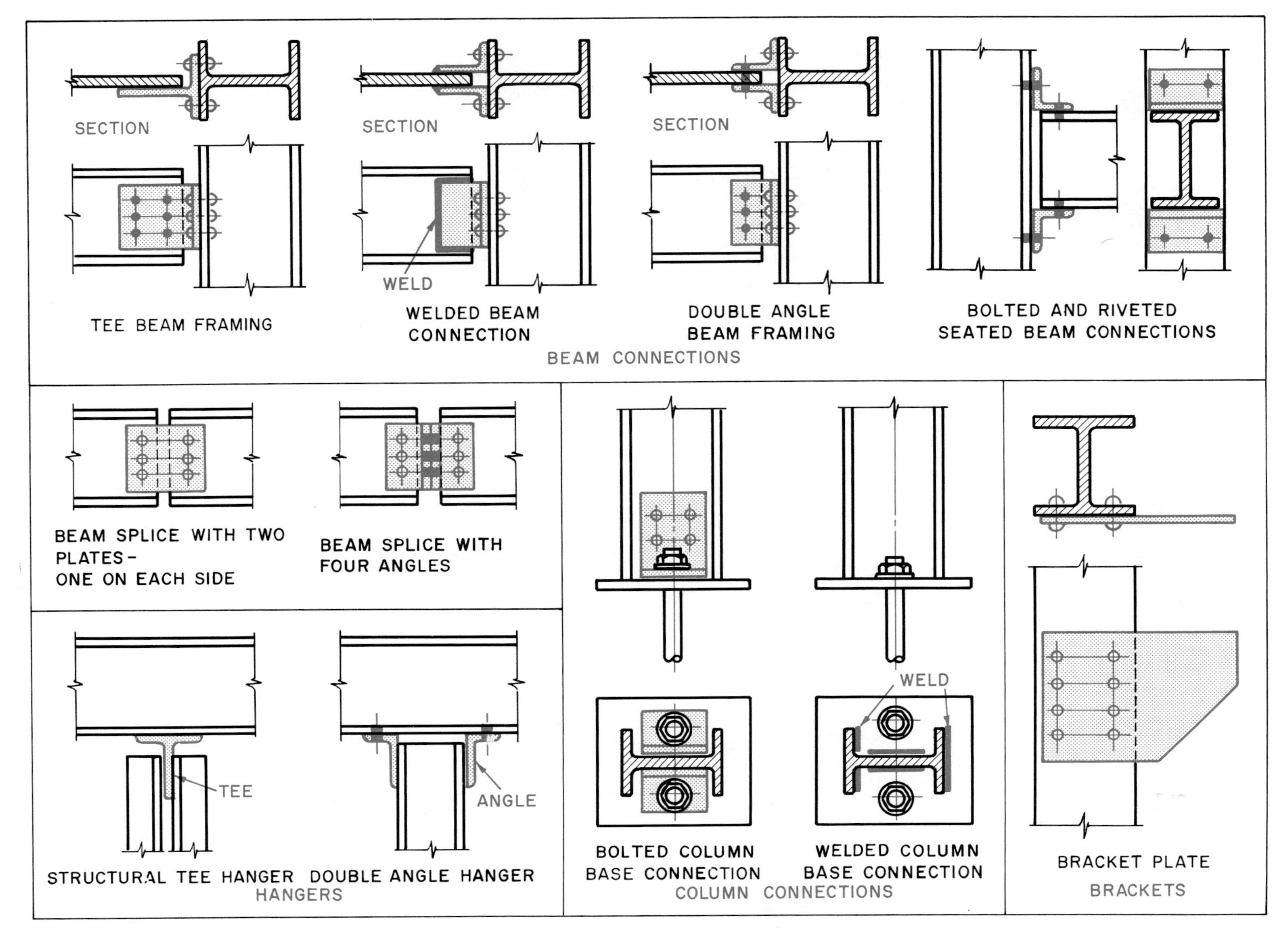

Fig. 22–16. **Selected structural steel framing connections. (American Institute of Steel Construction)**

- Shop drawings should be dimensioned so that those making the members do not have to add or subtract dimensions.
- Details, such as the size of fasteners, hole diameters, and painting instructions, are given as notes.

Structural Steel Drawings

When plans are made for a steel structure, three types of drawings are commonly used. These are steel framing drawings, erection drawings, and shop drawings.

Steel Framing Drawings

The **steel framing drawing** is made by an engineer. It shows the size and location of beams, columns, and other steel members. Examine the roof framing plan, Fig. 22-17. It shows the structural system used for a car haul house. Fig. 22-18.

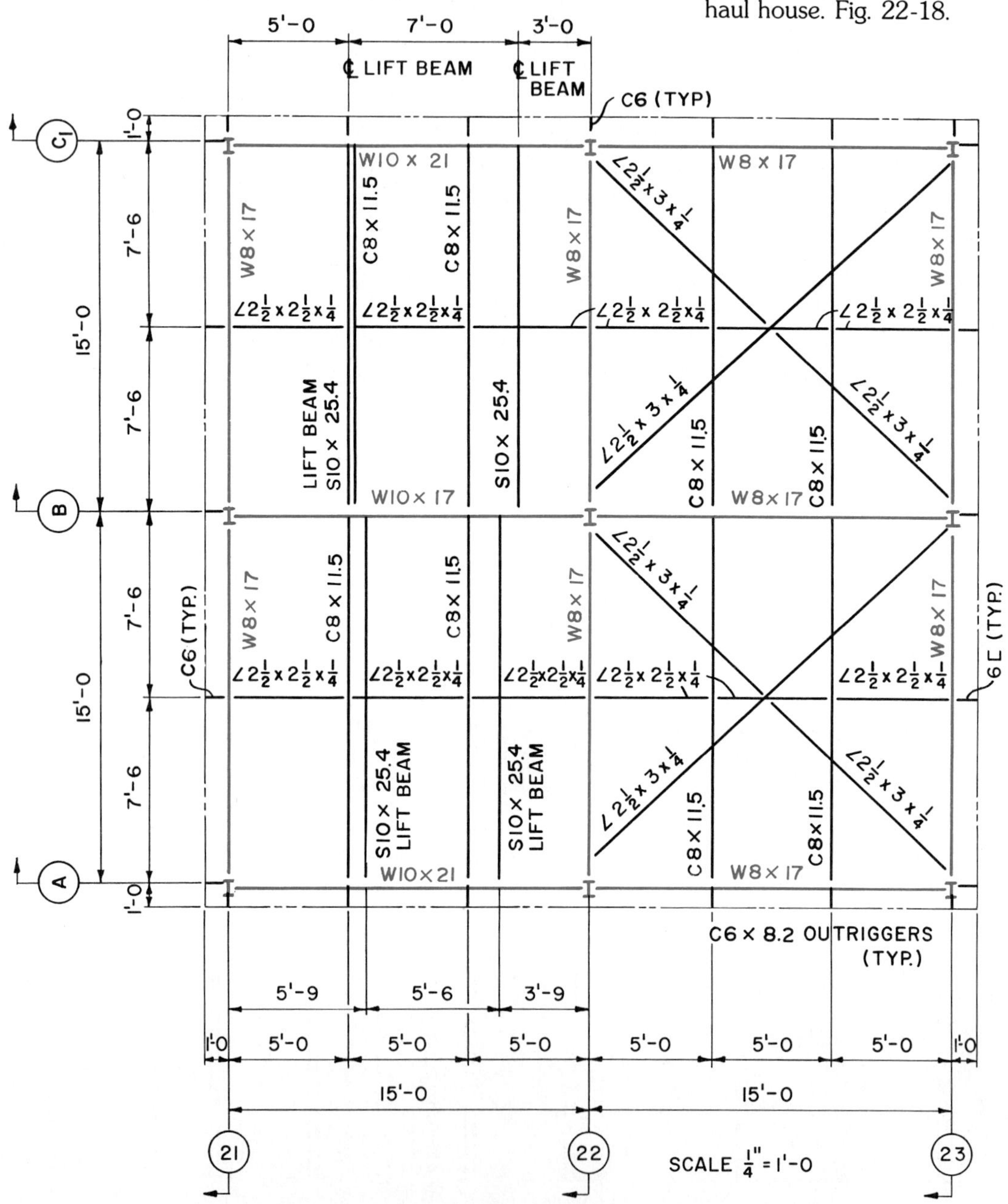

Fig. 22–17. This is a roof framing plan for the car haul house shown in Fig. 22–18. The main structural system consists of W-beam columns and wide flange beams between them. They are printed in color to call attention to them. The rest of the members are angles and channels used to support the roof decking. (McNally-Pittsburgh Mfg. Co.)

Notice that the structural members are indicated by a solid line. The main framework is made from wide flange beams. These carry the roof load. In addition, they support two lift beams. Find these on the drawing. The lift beam is used with a hoist to pick up heavy loads. Notice that the wide flange beams in the area of the lift beams are heavier than those on the other side of the building.

In addition to the beams, channels and angles are used to support the roof. These are indicated by a single line. Their sizes are also given on the drawing. The actual length of the structural members is not shown. This information is found on the shop drawings.

The location of each member is carefully dimensioned. Dimensions are to the center line of the member. The W-beams used as columns are located with overall dimensions.

If a building is several stories high, a separate framing plan is made for each floor plus one for the roof. Framing drawings are made to scale.

Erection Plan

The **erection plan** is an assembly drawing that is drawn to scale. It is used on the job by those putting the framework together. When the steel arrives at the job site, each member has been made to the correct length and numbered. All the holes for riveting or bolting members together have been made. Those erecting the building need to know only where to put each piece as shown on the erection plan.

Examine the erection plan for the car haul house roof shown in Fig. 22-19. Notice that it gives only the dimensions locating the columns. The only other information it gives is where each member is to be placed. Connec-

Fig. 22–18. This is a partial view of a coal processing plant. The small building in the front is a car haul house. Some of the structural steel details for this building are shown in Figs. 22–17, 22–19, 22–20, 22–22, and 22–24. (McNally-Pittsburgh Mfg. Co.)

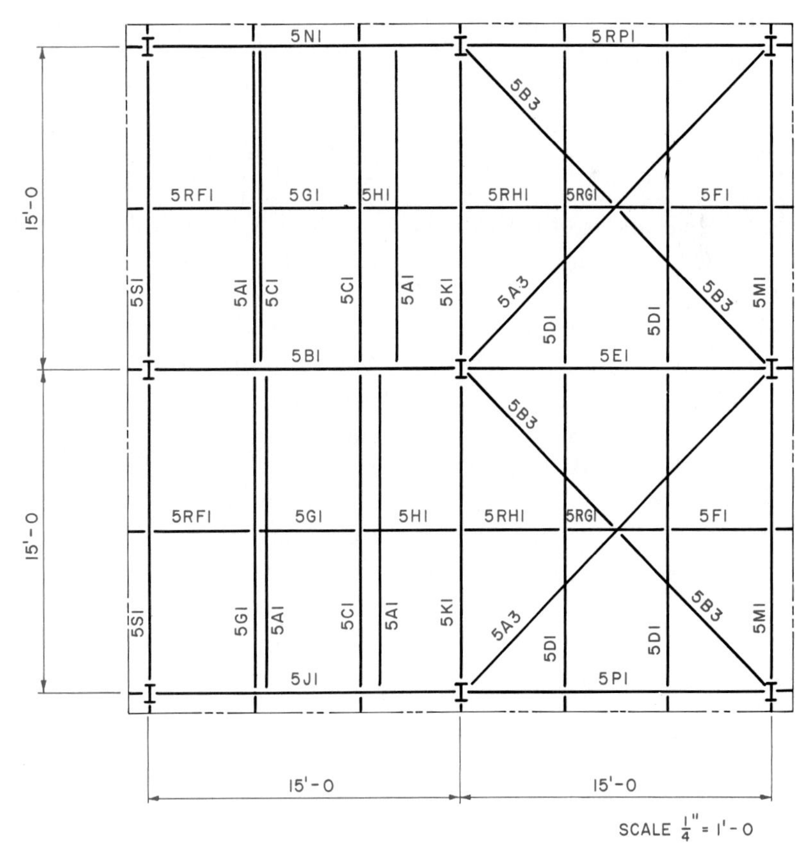

Fig. 22–19. **A roof steel erection plan for the car haul house roof.**

CAD Applications:

Structural Drawings & FEA

In order to build bridges, schools, dams, office buildings, or any other structure, drafters must prepare structural drawings. These drawings show how the supporting framework is to be made and assembled, whether it is of wood, steel, concrete, or some other material.

As with other types of drawings, structural drawings can be made on a CAD system. Features such as semi-automatic dimensioning, the copy command, and symbol libraries are useful for making structural drawings.

Before these drawings are made, however, engineers must design the structure. As explained earlier in this chapter, they decide on the materials, the placement and size of the members, and how they are to be joined.

A valuable aid in structural design is *finite element analysis* (FEA). In FEA, a model of a structure is created and analyzed and the results are interpreted. Finite element analysis helps engineers decide whether the materials and design can withstand the forces to which the finished structure will be subjected.

Programs for FEA are available for microcomputers as well as mainframes. Computerized FEA begins with the creation of a model. The model is created by breaking down the proposed design into finite elements—simple geometric shapes and their functions. After the engineer has created the model, the computer analyzes it by applying a set of equations that determine how the structure will react to heat, static and dynamic loads, and so on. Since this analysis may require solving as many as 50,000 equations at a time, it's easy to see that FEA can be done much faster on a computer than by other methods. The results of the analysis are interpreted to determine the displacements and stresses on the structure. The results are shown in graphic as well as numeric (tabular) form so that the engineer can actually see how the model reacts to loads. An example is shown in this photo.

Elm software by Fujitsu America, Inc.

Computerized FEA saves time because it can do computations quickly. It also reduces the need for building and testing prototypes. This time-saving gives the engineer more freedom to try various designs.

tions and other details are not needed. They appear on the shop drawings.

Some companies combine the framing plan and erection plan on one drawing. Fig. 22-20.

If a building is several stories high, a separate erection plan is made for each floor.

Exterior Walls. Framing and erection plans must be made for the exterior walls of some buildings. These are drawn in the same manner as explained for floor and roof plans. One wall of the car haul house is shown in Fig. 22-20. It has been drawn as a combination framing and erection drawing.

The wall plan also shows the columns. Here, their identification number is shown. The identification number does not appear on the erection plan because the columns are in end view on that drawing.

The other structural members on this elevation are steel channels. The building is covered with metal siding in large sheets. These sheets are fastened to the channels. The steel channels are then fastened to the W-beam columns.

Shop Drawings

After the engineer designs the structure, the drafter details each part. These details are called **shop drawings**. The members are made from these drawings. See Figs. 22-21 through 22-24.

The drafter uses the framing plan to obtain the sizes. The actual length of each member is not shown on the framing plan. The drafter must know how to figure this from the information given.

Shop drawings show the size of the member and the finished length. They show the connections used. The location of all holes is given. They must be dimensioned to the nearest 1/16 inch. The shop drawing includes notes telling how the part is to be made. Instructions on how to assemble the members are also included.

Shop drawings should be drawn to a scale large enough so that they are clear. They must not be crowded. The length of the members is not drawn to scale, but the height is drawn to scale. Since the drawings are completely dimensioned, an accurate scale drawing of the length is not necessary.

More information on making shop drawings is given in the paragraphs on drawing beams and columns.

Numbering Structural Members. Notice that the steel beam 5J1 appears on Figs. 22-19, 22-20, and 22-22. Each part of a structural system is given a number. It is used wherever the piece

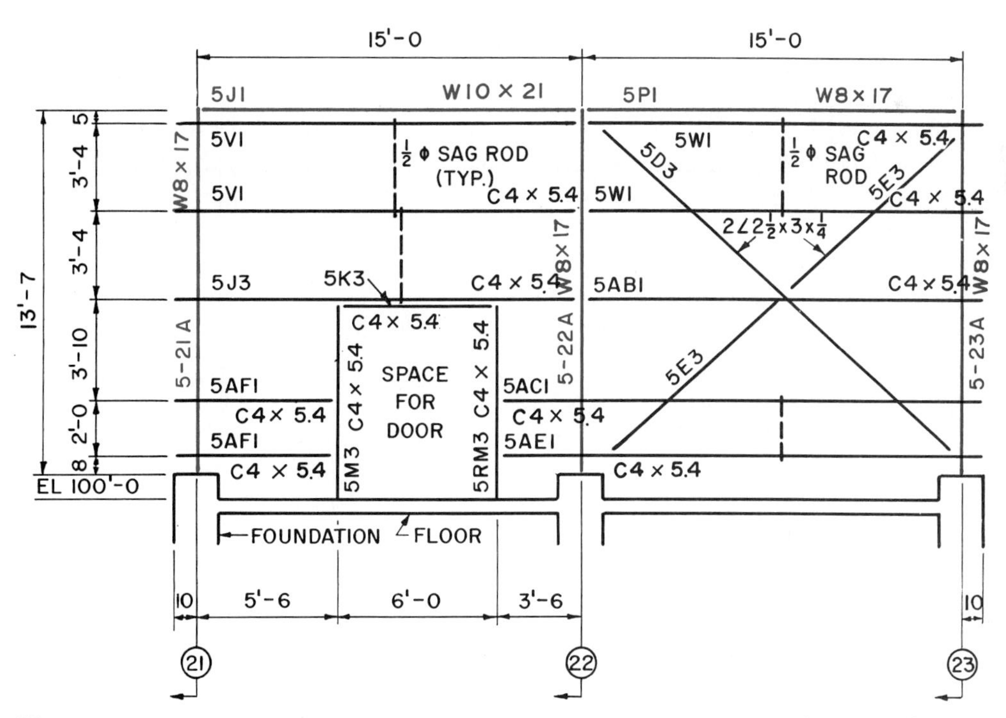

Fig. 22–20. An exterior wall steel framing and erection plan. This is a combination of the steel framing and erection plan of the steel framing and erection plan of the car haul house. It was taken at column line A, shown in Fig. 22–17. Each exterior wall requires a separate drawing. The main structural members are shown in color to call attention to them. (McNally-Pittsburgh Mfg. Co.)

appears on a drawing, whether it is the erection drawing or a shop drawing. The number is painted on the steel member after it is made. These numbers are used by the workers erecting the building. If several members are the same size, they are given the same number.

Drawing Beams. Fig. 22-21 shows a shop drawing of a typical riveted beam. Notice the shop and field rivet symbols. Whenever possible, holes are lined up on common gage lines. It is necessary to locate the gage line for each angle connection. The edge distance must be shown as well as the pitch.

Connections usually extend beyond the beam on each end. This setback is the difference between the overall length of the assem-

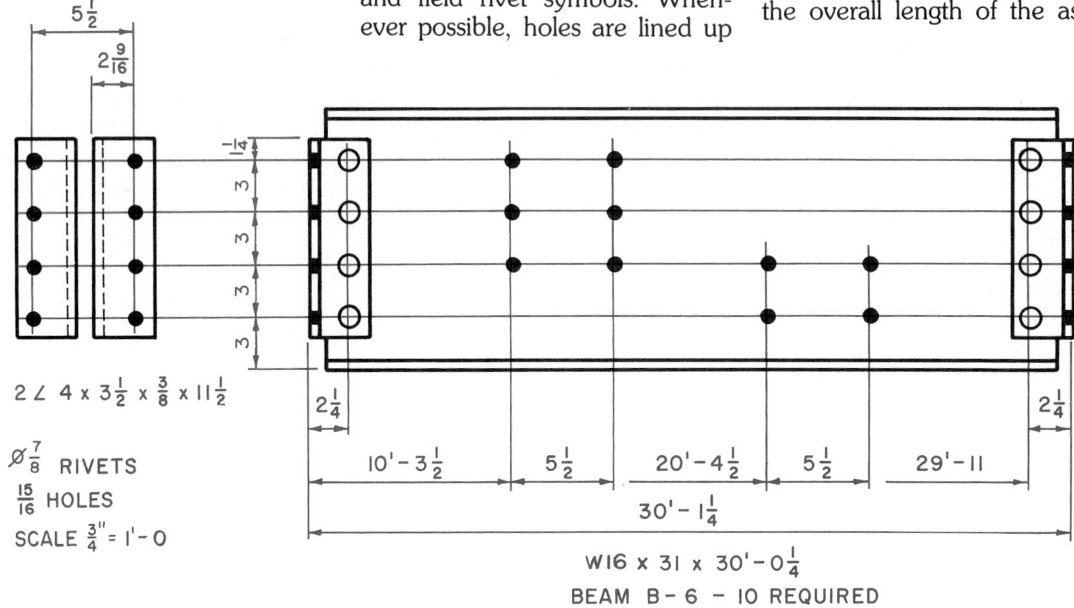

Fig. 22–21. A typical riveted steel beam shop drawing. All parts of this type of drawing are to scale except the length of the beam.

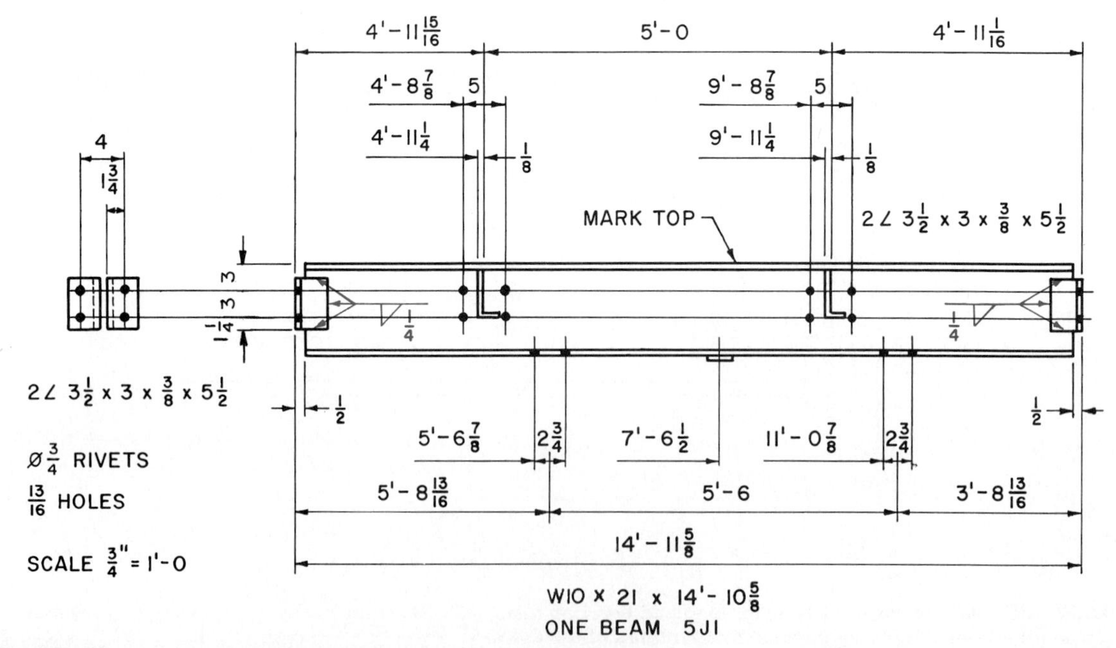

Fig. 22–22. A typical welded steel beam shop drawing. All parts of this type of drawing are to scale except the length of the beam. This is a beam from the car haul house, Fig. 22–17.

bled member and the length of the beam. In Fig. 22-21 the total unit length is 30′—1¼″, and the beam is 30′—0¼″. The setback is ½ inch on each end.

A left-end view is drawn to show the angle connection holes and their dimensions. If the right-end view is the same, it need not be drawn. The beam is not drawn in the end view.

Fig. 22-22 shows a shop drawing for a welded beam. The details are the same as for the riveted beam except for the angle connections. These have welding symbols to tell how they are to be joined. This beam is one from the car haul house, Fig. 22-17. Can you locate it on the plan?

Drawing Columns. Fig. 22-23 shows a shop drawing of a riveted steel column. Usually the column can be described with views of each face plus a view called a *bottom section*. The bottom section is actually a view from the top looking down on the column.

Notice on the column, Fig. 22-23, that the angles and plate are shop-riveted. The angles use buttonhead rivets. The plate uses a rivet with a countersunk head on the bottom and buttonhead on top. The length of the wide flange beam serving as a column is ½ inch shorter than the overall length with the angles in place.

The faces, or columns, are identified by a letter. The surface facing north on the structure is marked on the drawing.

When two faces are the same, only one needs to be drawn. The other can be shown with a center line and a letter. See face C in Fig. 22-23.

Fig. 22-24 shows a shop drawing for a welded steel column. The details are the same as for the riveted beam except for the welded plate and angles. Notice the use of welding symbols. This column is one from the car haul house, Fig. 22-17.

Drawing Steel Trusses. A **truss** is a structural unit made from tees, wide flange beams, or angles with the long legs placed back to back. When a roof is built over a building with a span too great to use girders, trusses can be used economically. They allow large areas of floor space to be free of columns.

The design of trusses is the work of engineers. They must be designed to absorb tension and compression stresses. *Tension* is a force that tries to stretch a member. *Compression* is a force that tries to shorten a member. See Fig. 22-25.

Fig. 22-26 shows the stresses on a Fink truss. Notice that the load on the roof, P, puts some members under tension and other members under compression.

Fig. 22-27 shows common types of steel trusses: the Pratt, Warren, bowstring, Fink, and scissors.

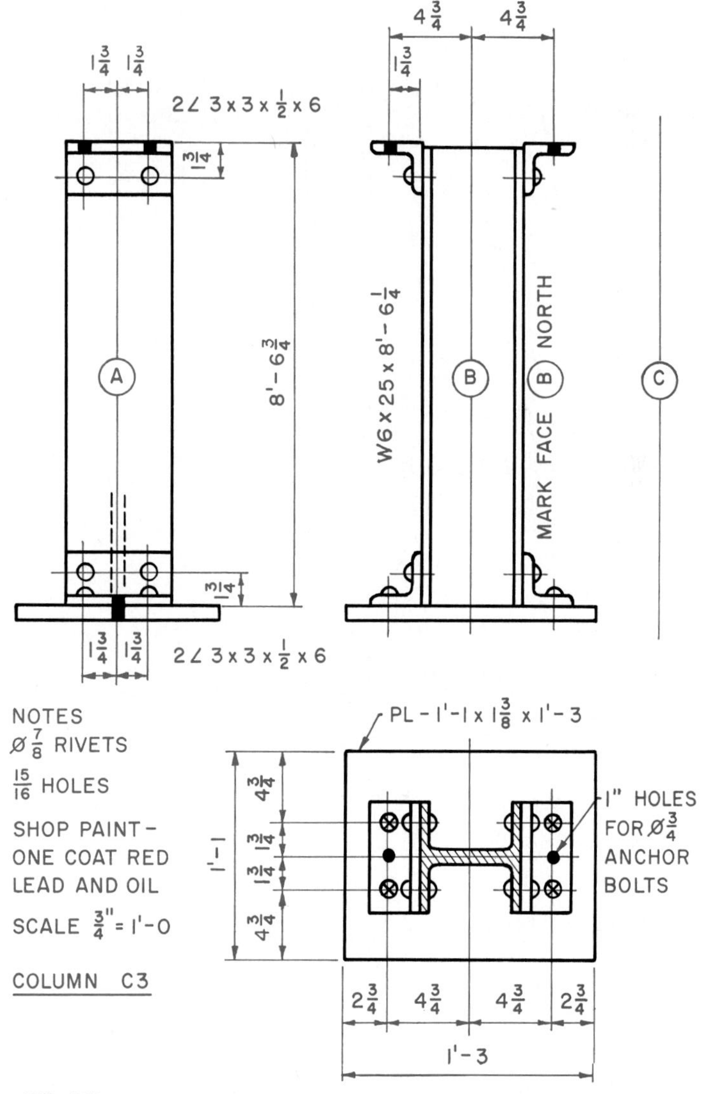

Fig. 22–23. A typical riveted steel column shop drawing.

The Warren and Pratt trusses are used to support floor and roof loads. If a roof has to carry loads other than that of the roof, such as in an overhead crane, a Warren or Pratt truss is used. The Fink truss provides a roof with a good slope. Usually it is not used to carry any other loads except the roof. The bowstring truss can span large distances. It provides a good slope to the roof. The scissors truss is used for high-pitched roofs. A church roof is a good example.

The drafter must know the terms used in describing trusses. These are chord, web member, panel, and span. They are shown in Fig. 22-27. A **chord** is one of the principal structural members braced by web members. **Web members** are the internal braces running between chords. A **panel** is the distance between two web members measured along a chord. **Span** is the distance covered by the trusses with no support from internal columns or walls.

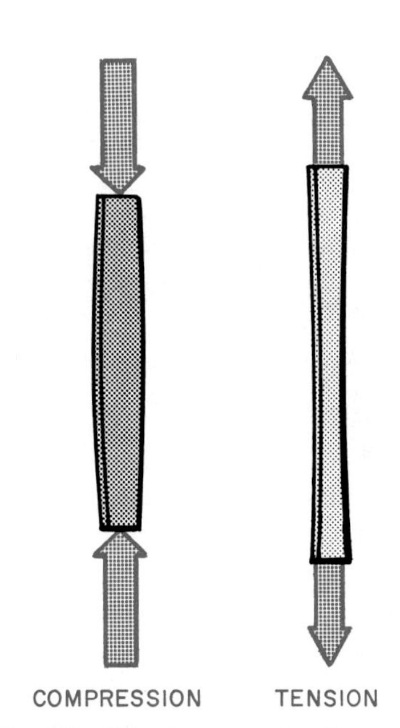

Fig. 22–25. Compression forces tend to shorten a member. Tension forces tend to lengthen a member.

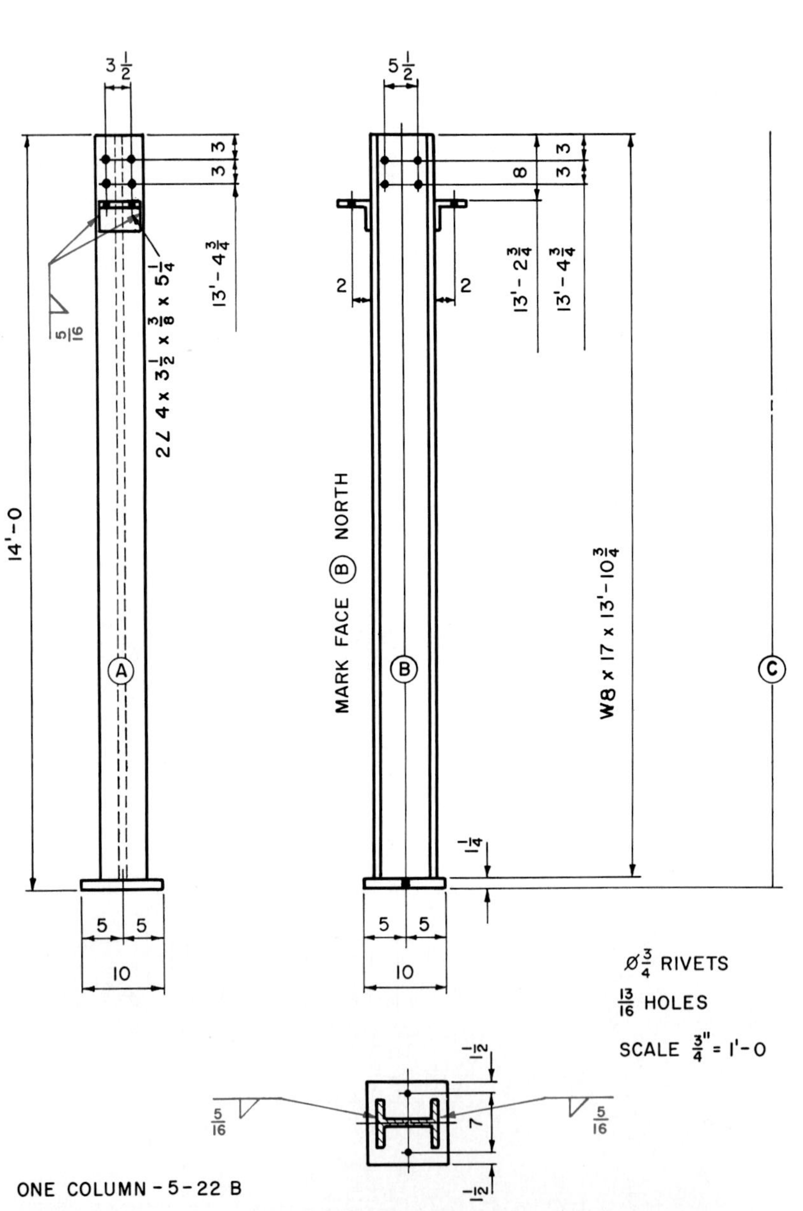

Fig. 22–24. A typical welded steel column shop drawing. This is a column from the car haul house, Fig. 22–17. (McNally-Pittsburgh Mfg. Co.)

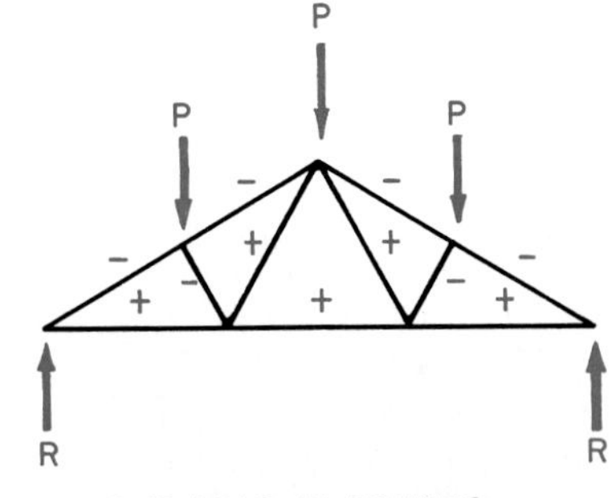

Fig. 22–26. Roof load places members of a truss under tension and compression stresses.

Engineers decide which type of truss is best for the job. They figure the stresses on each member. Then they make a design drawing of the truss. The design drawing usually shows the tension and compression in pounds. Tension is shown by a + and compression by a −. The size of the steel members to be used is shown. See Fig. 22-28. It is from this drawing that the drafter prepares the shop drawing.

A shop drawing of a riveted truss is shown in Fig. 22-29A. Since the individual parts are small and simple, each of them is detailed in the assembled position. Only the left half of the truss is drawn when the other half is the same.

Notice the use of gusset plates to join the members. The drafter locates the rivets according to standard design practice. Fig. 22-29B.

Study Fig. 22-29A. Notice that the location of each rivet is dimensioned. The size of each structural member is shown. This includes the length. Refer to Fig. 22-6 for the standard way to record this information. Each gusset plate size is given. The web members are located by dimensioning to the center line of the rivets. The distance of this center line from the top of the member is dimensioned.

Some truss drawings can become very crowded. See Fig. 22-30. The dimensioning of such a drawing must be carefully planned. The drafter must work carefully so that no details are left off the drawing.

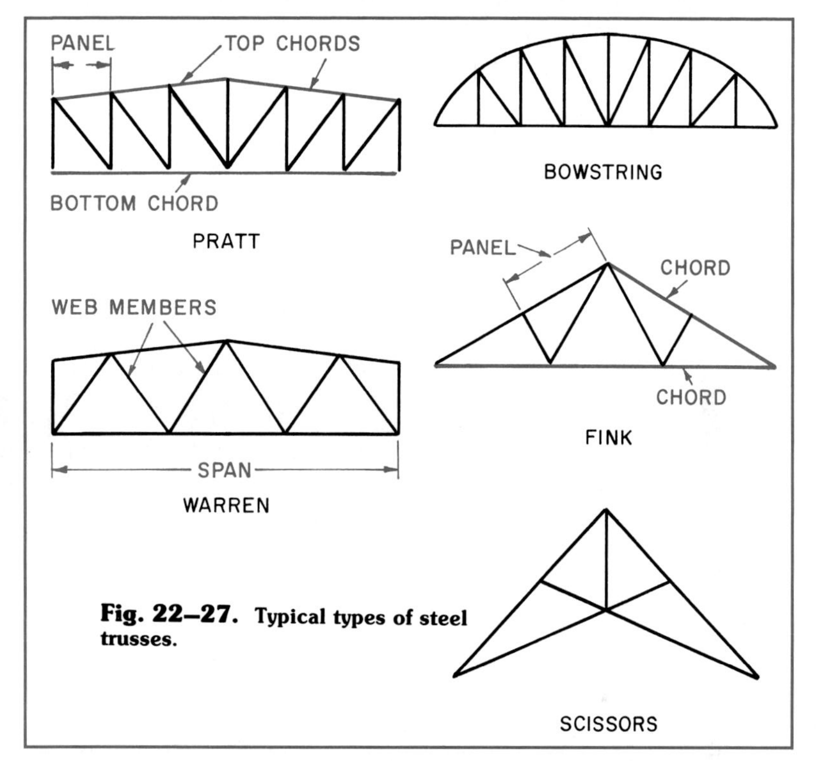

Fig. 22–27. Typical types of steel trusses.

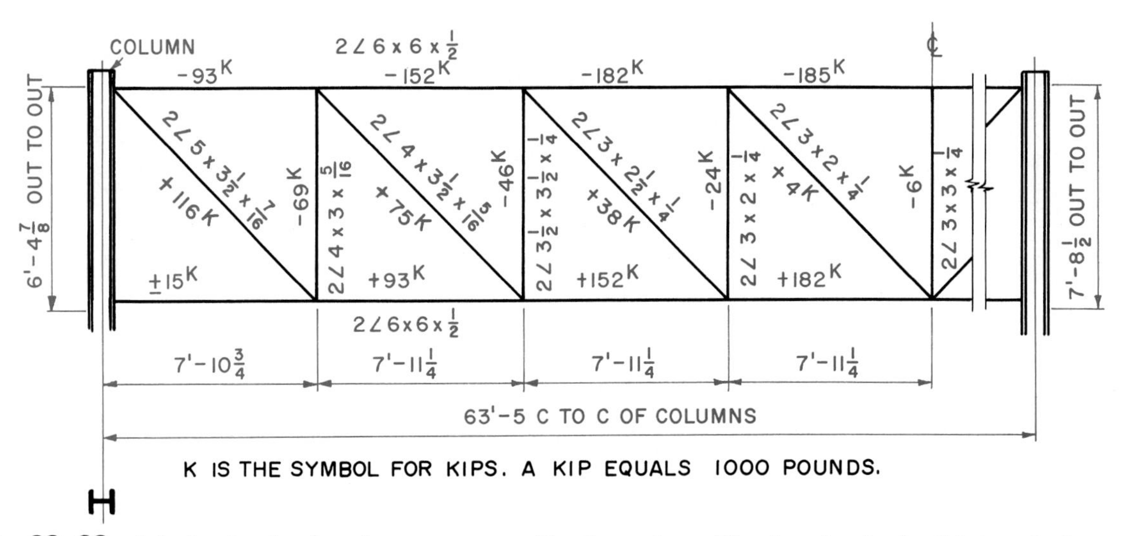

Fig. 22–28. A design drawing for a truss, as prepared by the engineer. The shop drawing for this truss is shown in Fig. 22–30. The term "out-to-out" on the height dimension refers to the outside surface of the top and bottom members. (American Institute of Steel Construction)

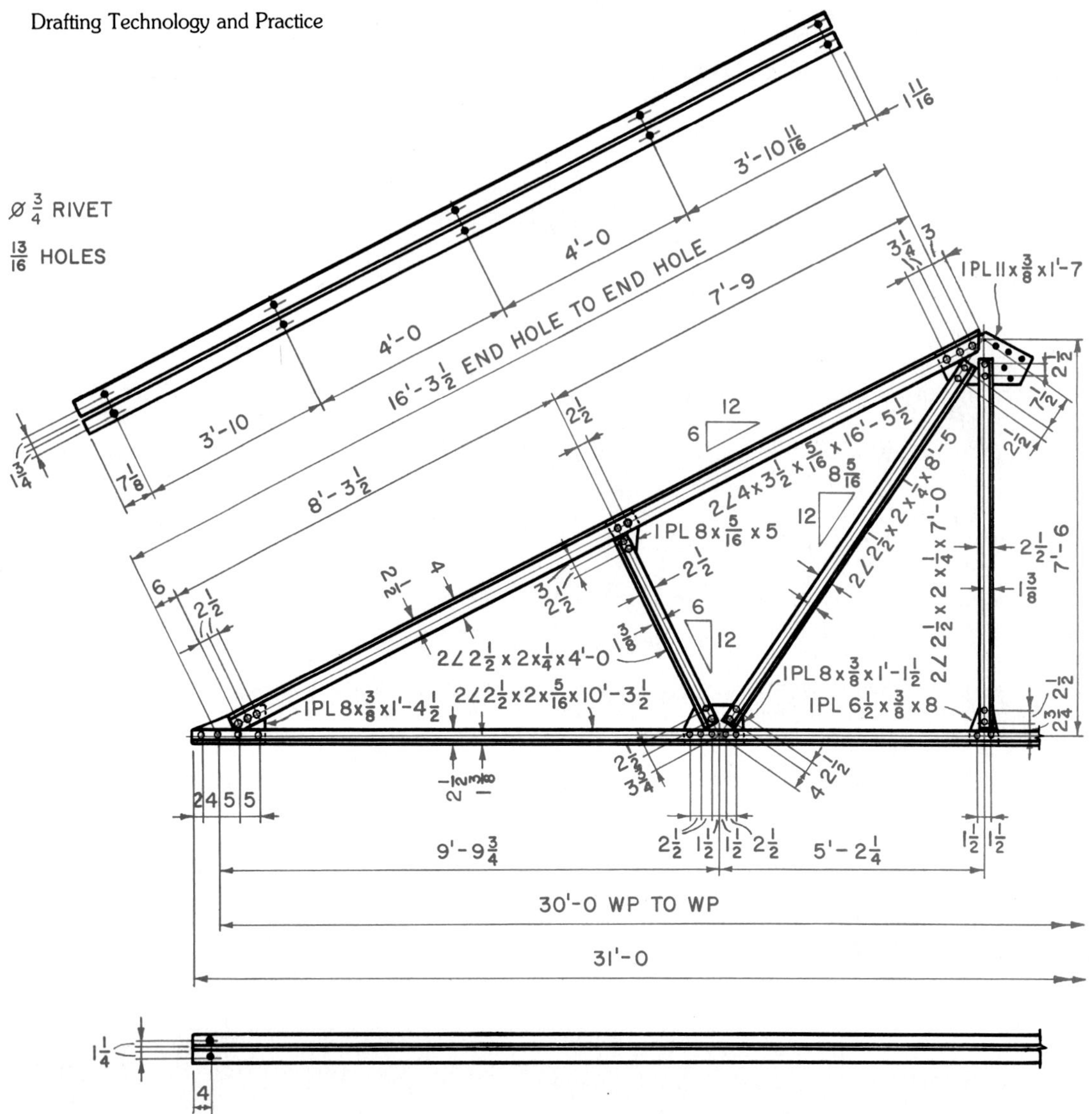

Fig. 22–29A. A simple riveted Fink steel truss that is made of steel angles. One of the gusset plate connections is shown larger in Fig. 22–29B to help show how they are drawn.

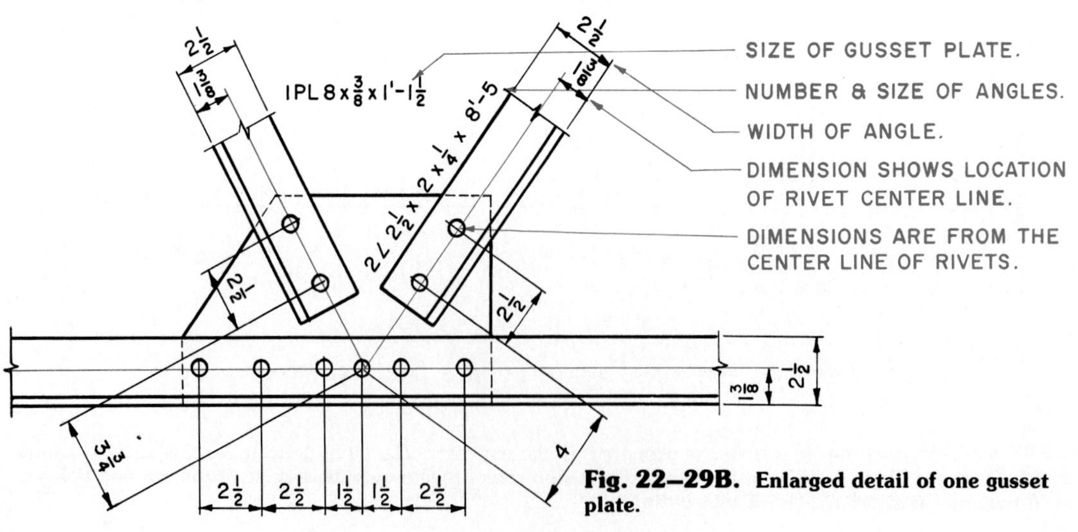

Fig. 22–29B. Enlarged detail of one gusset plate.

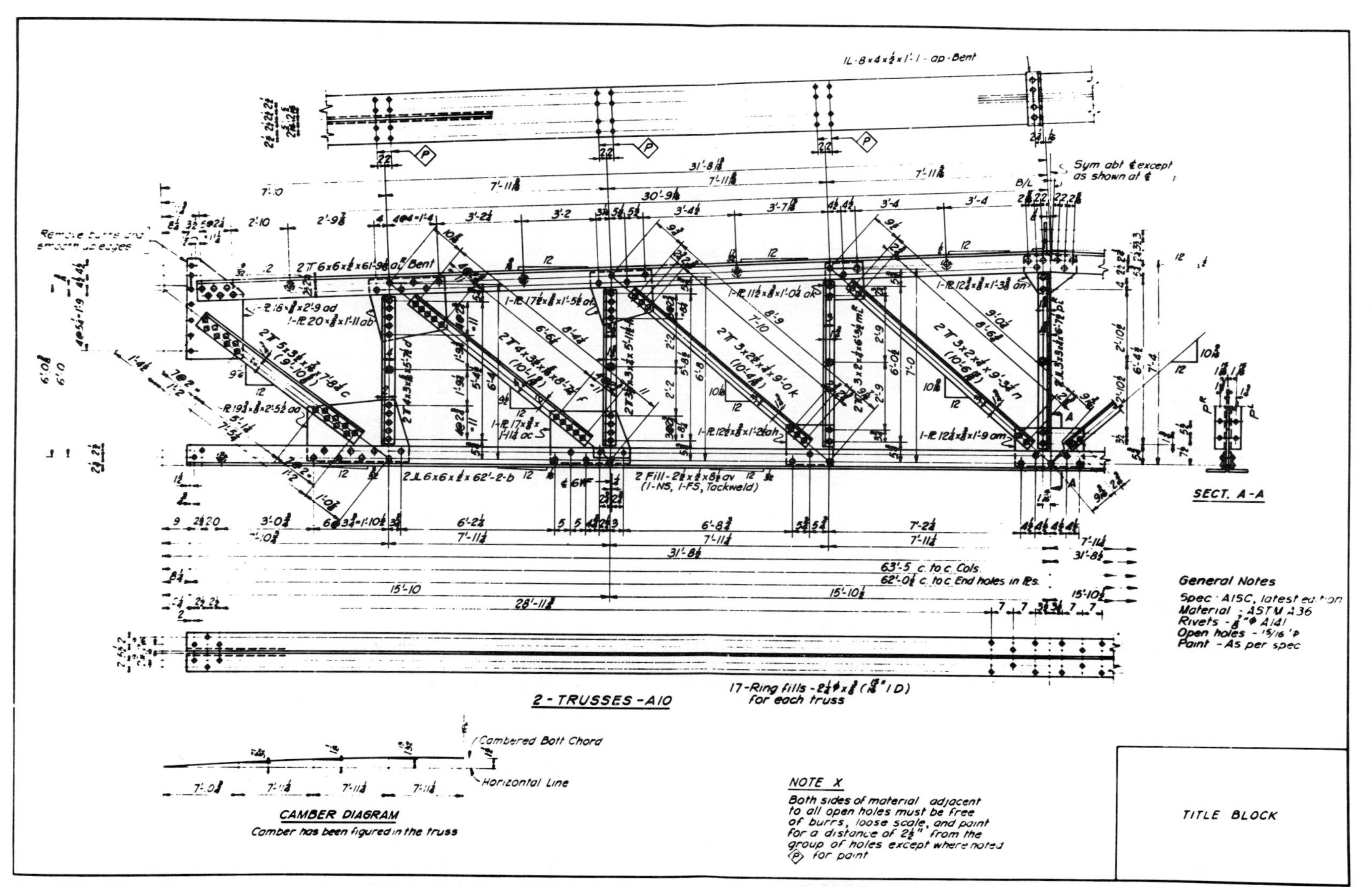

Fig. 22–30. A detail drawing of a Pratt truss. Since such drawings are completely dimensioned assembly drawings, they become crowded. (American Institute of Steel Construction)

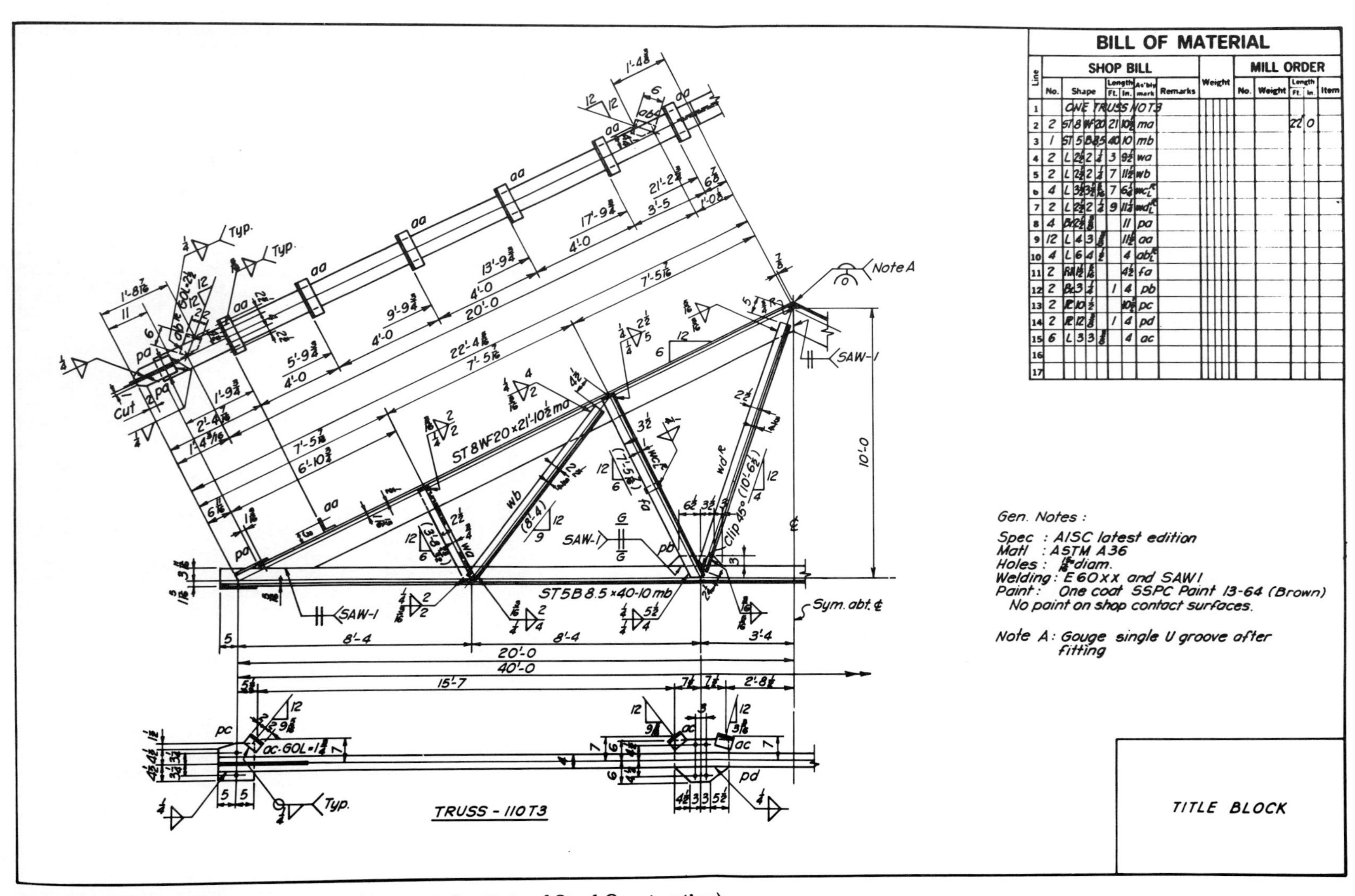

Fig. 22–31. A welded truss drawing. (American Institute of Steel Construction)

Sections are used to clarify construction details and dimensions.

Notice that the dimensions parallel the edges of each part. The slope of members is shown with a triangle drawn on each sloped member. The longest leg of the triangle is always 12 inches because the slope is a ratio of the inches of rise used in 1 foot of run.

Trusses are usually shop assembled. If they are too large to ship, some parts are assembled in the shop. These subassemblies are then put together at the building site. Each individual member of a truss is not usually numbered. Instead, each truss is numbered as a single structural member such as T-1, T-2, and T-3.

All parts of the truss are drawn to the same scale. Detail drawings may be to a larger scale.

Fig. 22-31 shows a detail drawing for a welded truss. It is much the same as the riveted truss. One difference is that gusset plates are not needed.

Notice that the dimensions are measured along a line located just below the top of the angle. Fig. 22-31 shows this dimension as 1 13/16 inches below the top of the inclined chord. It is 1 13/16 inches from the edge of the horizontal chord and is 3/4 inch from the edge of the web members. This distance is found in engineering design manuals, which give distances for the various sizes of angles. The dimensions in a truss drawing refer to these lines. They do not refer to the outside edges of each member.

When designing welded trusses, drafters need to know how many inches of weld are needed at each joint. They have to make certain the angles provide enough edge to give the specified length of weld.

Notice the use of welding symbols in Fig. 22-31. Study the section on welding in Chapter 12, "Fasteners."

On the truss shown in Fig. 22-31 the web members are identified by letters. The size of each member is shown on a bill of materials.

Drawing Rigid Frames. **Rigid frames** are steel girder-like members. When they are assembled, they form the column and roof members. They can span large distances. This provides an area free of columns. See Fig. 22-32.

Rigid frames are made in the shop and shipped to the building site. Here they are bolted together. A large crane lifts them in place. The column is bolted to the concrete footing.

A drawing for a rigid frame is shown in Fig. 22-33. The frame is shown assembled. Detail drawings are used to show things that cannot be seen on this drawing. The details are drawn to a larger scale than on the assembled drawing.

Concrete Structural Drawings

Concrete is made from portland cement, water, sand, and gravel. Its strength varies with the quality of these items. The amount of each item in the mix also influences its strength.

Concrete has a good compressive strength. **Compressive strength** is the ability of a material to withstand forces tending to shorten it. See Fig. 22-25. The tensile strength of concrete is low. **Tensile strength** is the ability of a material to withstand forces tending to lengthen it. To increase the tensile strength of concrete, steel reinforcing bars are cast in the concrete members. The steel resists tension forces. The concrete resists compression forces.

Fig. 22–32. Rigid frames provide a floor area free of columns. (Butler Manufacturing Co.)

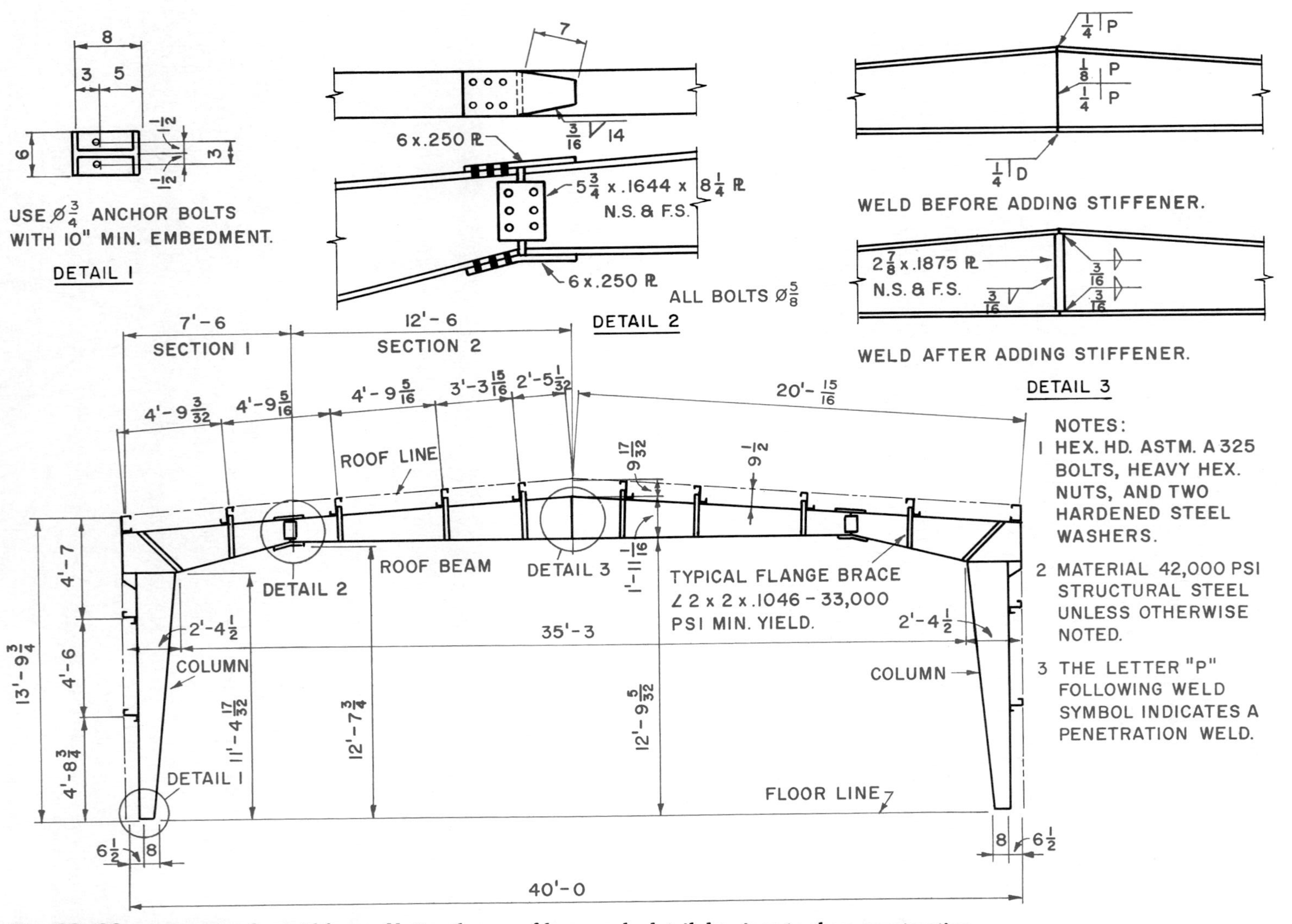

Fig. 22–33. A drawing of a rigid frame. Notice the use of large scale detail drawings to show construction details. (Butler Manufacturing Co.)

Concrete without steel reinforcing is called **plain concrete**. If it has steel reinforcing bars, it is called **reinforced concrete**. Some structural members are cast with the steel reinforcing cables tightly stretched under tension before the concrete is poured. This material is called **prestressed concrete**.

Usually, concrete structural members that are poured in place on the job are reinforced members. Those cast in a factory and shipped to the site as finished members are usually prestressed. Fig. 22-34 shows a steel-framed structure with a precast concrete floor and roof.

Fig. 22–34. Precast concrete floor and roof decking is placed on a bolted steel frame. (Flexicore Co., Inc.)

Terms Used in Concrete Construction

Bars are round steel reinforcing rods placed in a concrete member to strengthen it. They may be smooth or deformed. A deformed bar has ridges on its surface. Bar diameters are specified by a number. The number indicates the diameter in eighths of an inch. Thus, a number 4 bar is 4/8 or 1/2 inch in diameter.

Dowels are round metal bars used to splice together two reinforcing bars in a column. They are usually the same diameter as the bars they join. They are welded in place. Fig. 22-35.

Bar supports are round metal bars used to hold the reinforcing bars the proper distance above the bottom of the form. They prevent the bars from moving when the concrete is poured. There are several standard types in use. Fig. 22-36.

Ties are made from small-diameter metal wire. They are used to hold bars in place in columns. They prevent the bars from moving when the concrete is poured. Fig. 22-37.

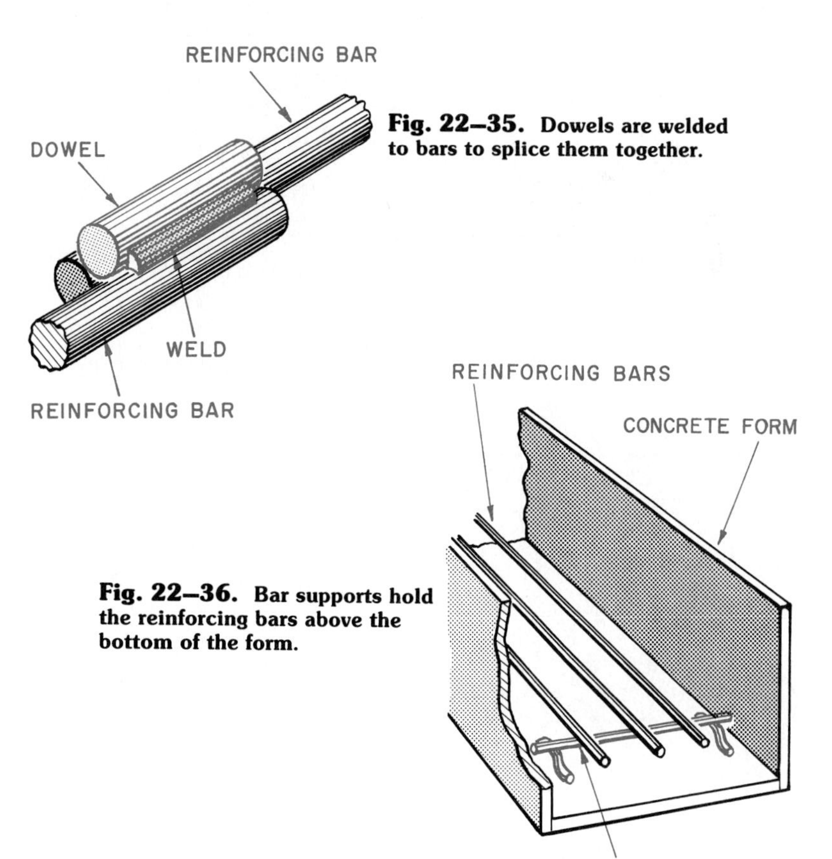

Fig. 22–35. Dowels are welded to bars to splice them together.

Fig. 22–36. Bar supports hold the reinforcing bars above the bottom of the form.

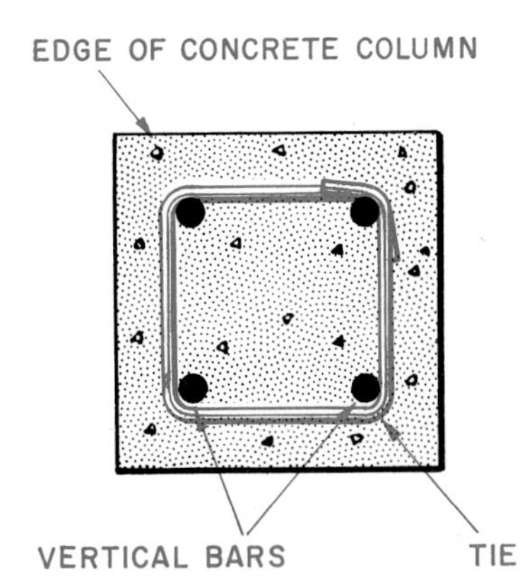

Fig. 22–37. Wire ties are used to hold reinforcing bars in place.

Stirrups are round metal bars usually bent in the shape of a U. They are used to hold bars in the proper position when the concrete is poured. Fig. 22-38.

Mark is a term used for identifying the floor and type of member. See Fig. 22-39 for mark abbreviations. A typical mark would be 2B3. This indicates the second floor, beam number 3.

Beams are the main horizontal structural members. They run from column to column or column to foundation.

Joists are smaller horizontal structural members. They run between beams and provide support for the floor or roof slab.

Bent bar refers to a reinforcing bar that has been bent. Bent bars strengthen the structural members. Fig. 22-40.

Reinforced Concrete Drawings

The design and drawing of reinforced concrete members are complex tasks that require much study. The American Concrete Institute has prepared a book, *A Manual of Standard Practice for Detailing Reinforced Concrete Structures*, that gives standard drafting practices.

The following material covers concrete structural members that are formed and cast on the job. Two types of drawings are commonly used. These are engineering drawings and placing drawings.

Engineering Drawings. Engineering drawings are prepared by the design engineer. They show the general arrangement of the structure, the size of the members, how they are to be reinforced, and notes needed for additional information. Fig. 22-41.

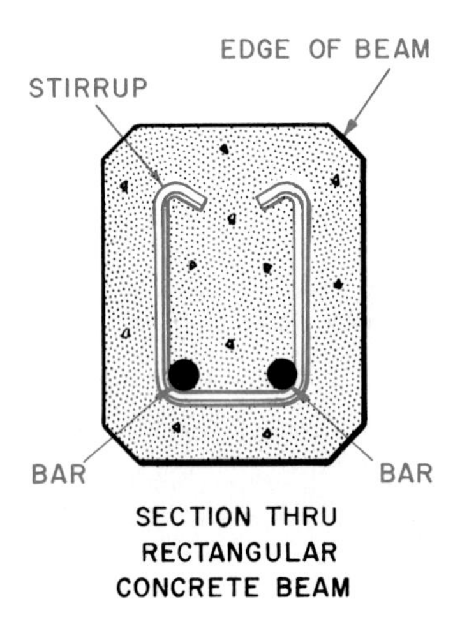

Fig. 22-39. Marks Used in Concrete Construction

Member	Mark
Beam	B
Column	C
Footing	F
Girder	G
Joist	J
Lintel	L
Slab	S
Wall	W

(American Concrete Institute)

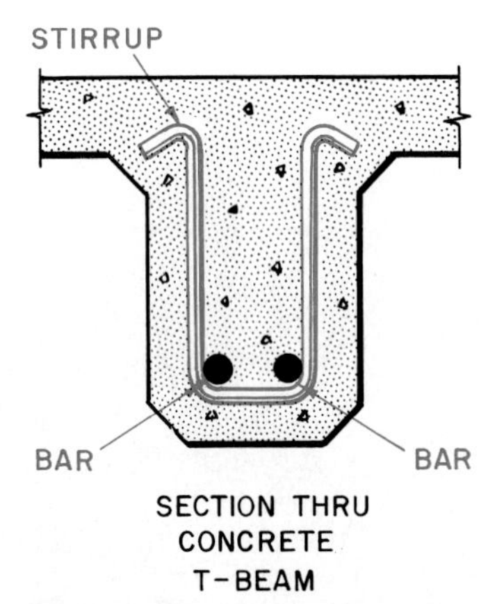

Fig. 22–38. Stirrups are used to locate and hold bars in place.

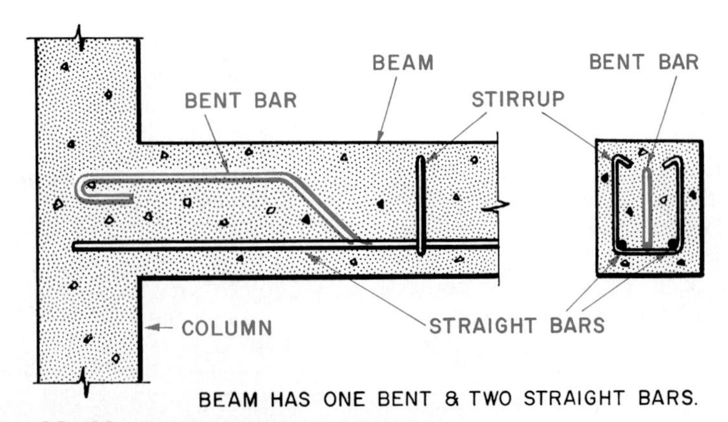

Fig. 22–40. Bent bars strengthen the beam.

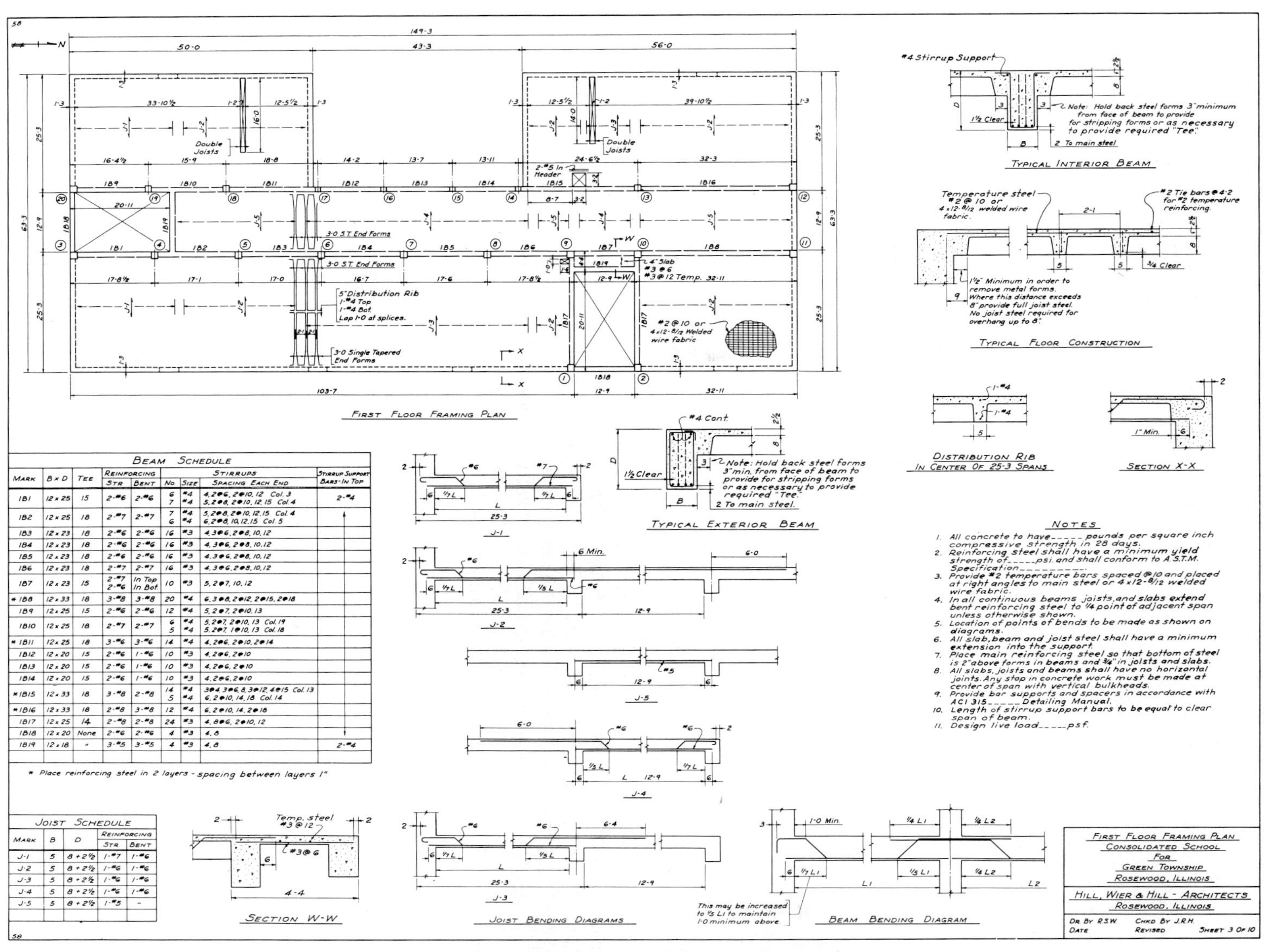

Beam Schedule

Mark	B x D	Tee	Reinforcing Str	Reinforcing Bent	Stirrups No.	Stirrups Size	Stirrups Spacing Each End	Stirrup Support Bars-In Top
1B1	12 x 25	15	2-#6	2-#6	6 7	#4 #4	4, 2@6, 2@10, 12 Col. 3 5, 2@8, 2@10, 12, 15 Col. 4	2-#4
1B2	12 x 25	18	2-#7	2-#7	7 6	#4 #4	5, 2@8, 2@10, 12, 15 Col. 4 6, 2@8, 10, 12, 15 Col. 5	
1B3	12 x 23	18	2-#6	2-#6	16	#3	4, 3@6, 2@8, 10, 12	
1B4	12 x 23	18	2-#6	2-#6	16	#3	4, 3@6, 2@8, 10, 12	
1B5	12 x 23	18	2-#6	2-#6	16	#3	4, 3@6, 2@8, 10, 12	
1B6	12 x 23	18	2-#7	2-#7	16	#3	4, 3@6, 2@8, 10, 12	
1B7	12 x 23	15	2-#7 2-#6	In Top In Bot.	10	#3	5, 2@7, 10, 12	
* 1B8	12 x 33	18	3-#8	3-#8	20	#4	6, 3@8, 2@12, 2@15, 2@18	
1B9	12 x 25	15	2-#6	2-#6	12	#4	5, 2@7, 2@10, 13	
1B10	12 x 25	18	2-#7	2-#7	6 5	#4 #4	5, 2@7, 2@10, 13 Col. 19 5, 2@7, 1@10, 13 Col. 18	
* 1B11	12 x 25	18	3-#6	3-#6	14	#4	4, 2@6, 2@10, 2@14	
1B12	12 x 20	15	2-#6	1-#6	10	#3	4, 2@6, 2@10	
1B13	12 x 20	15	2-#6	1-#6	10	#3	4, 2@6, 2@10	
1B14	12 x 20	15	2-#6	1-#6	10	#3	4, 2@6, 2@10	
* 1B15	12 x 33	18	3-#8	2-#8	14 5	#4 #4	3@4, 3@6, 8, 3@12, 4@15 Col. 13 6, 2@10, 14, 18 Col. 14	
* 1B16	12 x 33	18	2-#8	3-#8	12	#4	6, 2@10, 14, 2@18	
1B17	12 x 25	14	2-#8	2-#8	24	#3	4, 8@6, 2@10, 12	
1B18	12 x 20	None	2-#6	2-#6	4	#3	4, 8	
1B19	12 x 18	"	3-#5	3-#5	4	#3	4, 8	2-#4

* Place reinforcing steel in 2 layers - spacing between layers 1"

Joist Schedule

Mark	B	D	Reinforcing Str	Reinforcing Bent
J-1	5	8 + 2½	1-#7	1-#6
J-2	5	8 + 2½	1-#6	1-#6
J-3	5	8 + 2½	1-#6	1-#6
J-4	5	8 + 2½	1-#6	1-#6
J-5	5	8 + 2½	1-#5	-

Fig. 22–41. An engineering drawing for a concrete joist floor. (American Concrete Institute)

Engineering drawings should be complete in every detail. All design information must be included by either a drawing, a note, or a schedule. Engineers do not usually prepare complete drawings for each member. They do not usually schedule the bending of every bar. They do have to give the drafter this information through notes and diagrams or references to standard manuals.

The engineering drawings include a typical concrete slab, joist, beam, and column detail. These show where the reinforcing steel is put.

The engineer does not work out all the dimensions. He or she does show where the bars are to be bent and to what points they should be extended.

Engineers must show the quality of concrete to be used. The type and grade of reinforcing bars must be shown. Engineers note the live loads the structure will carry and other design data, such as the load-bearing capabilities of the soil upon which the structure is to be built.

Study the engineering drawing, Fig. 22-41, showing the plan of the first floor of a building. The dimensions of the building and column spacing are given. The direction of the joists is shown. The joists are identified by number.

The *beam schedule* shows all the information a drafter needs for making the placing drawing. The following is an explanation of beam 1B1 in the schedule. The mark 1B1 means first floor, beam number 1. B × D gives the width and thickness of the beam. This beam is 12 inches by 25 inches.

The beam uses two No. 6 straight bars. It also uses two No. 6 bars that are to be bent on the end. The beam contains six stirrups made from No. 4 bars. These six are for the end of the beam next to the No. 3 column. The first stirrup is 4 inches from the end of the beam. The next two are spaced 6 inches apart. The next two are spaced 10 inches apart. The last stirrup is 12 inches from the last 10-inch one.

From the end of the beam next to column 4 are seven stirrups made of No. 4 bars. The first stirrup is 5 inches from the end of the beam. The next two are spaced 8 inches apart. The next two are 10 inches apart. The next one is spaced 12 inches, and the last is 15 inches.

The last entry for beam 1B1 means that two No. 4 bars are used to support the stirrups. These directions are detailed on the placing drawing shown in Fig. 22-42. See the detail for beam 1B1. Here the stirrups are drawn in their exact location and dimensioned.

Notice the engineering drawing shows a typical section through a beam. It shows the two straight bars as round, black circles with the bent part shown by dashed lines.

Also shown are the beam- and joist-bending diagrams, which indicate how the engineer wants the bars bent.

Sections are shown through the floor slab. This shows how the floor joins the exterior foundation (Section X-X). It also shows the slab in connection with the floor joists.

Examine the notes shown in Fig. 22-41. They give information that is not drawn.

Placing Drawings. **Placing drawings** are detail drawings that show the floor plan and the details for making the structural members. They also show the size, shape, and placing of the reinforcement bars. Refer to Fig. 22-42. The only dimensions needed are those showing the location of the bars.

The drafter follows very detailed information on the engineering drawing when making placing drawings.

The reinforcement of beams, joists, and girders is shown on a beam schedule. The placing drawing schedule gives the number, mark, and size of the member; number, size, and length of straight bars; number, size, total length, mark, and bending detail of bent bars and stirrups; spacing of stirrups; bar supports; and other placement information that may be needed.

Concrete slab reinforcement is drawn on the plan. A schedule is used to show the bars needed. The schedule is the same as the beam schedule. Sometimes the number of the bars is not shown. This is found on the plan of the slab.

Column reinforcement is shown on a column schedule. This schedule has a section showing ties and bent bars and diagrams showing the arrangement and bending of the ties.

Bar supports are listed on placing drawings. They are usually a part of the beam schedule. The schedule shows their number, size, and length.

Ties are used in columns to prevent bars from buckling. Ties hold the bars in place as the column is cast. Standards have been established giving the maximum spacing of ties. These are available in the American Concrete Institute standards manual.

Examine the placing drawing, Fig. 22-42. Of special importance are the detail drawings showing bar and stirrup spacing. Notice that each beam and joist is completely dimensioned. Drafters get these dimensions from the floor framing plan. They also dimension the bent bars. The information for figuring these dimensions comes from the *beam-bending diagram* on the engineering drawing. The engineer does not calculate these dimensions.

The floor plan of a placing drawing can be traced from the plan on the engineering drawing.

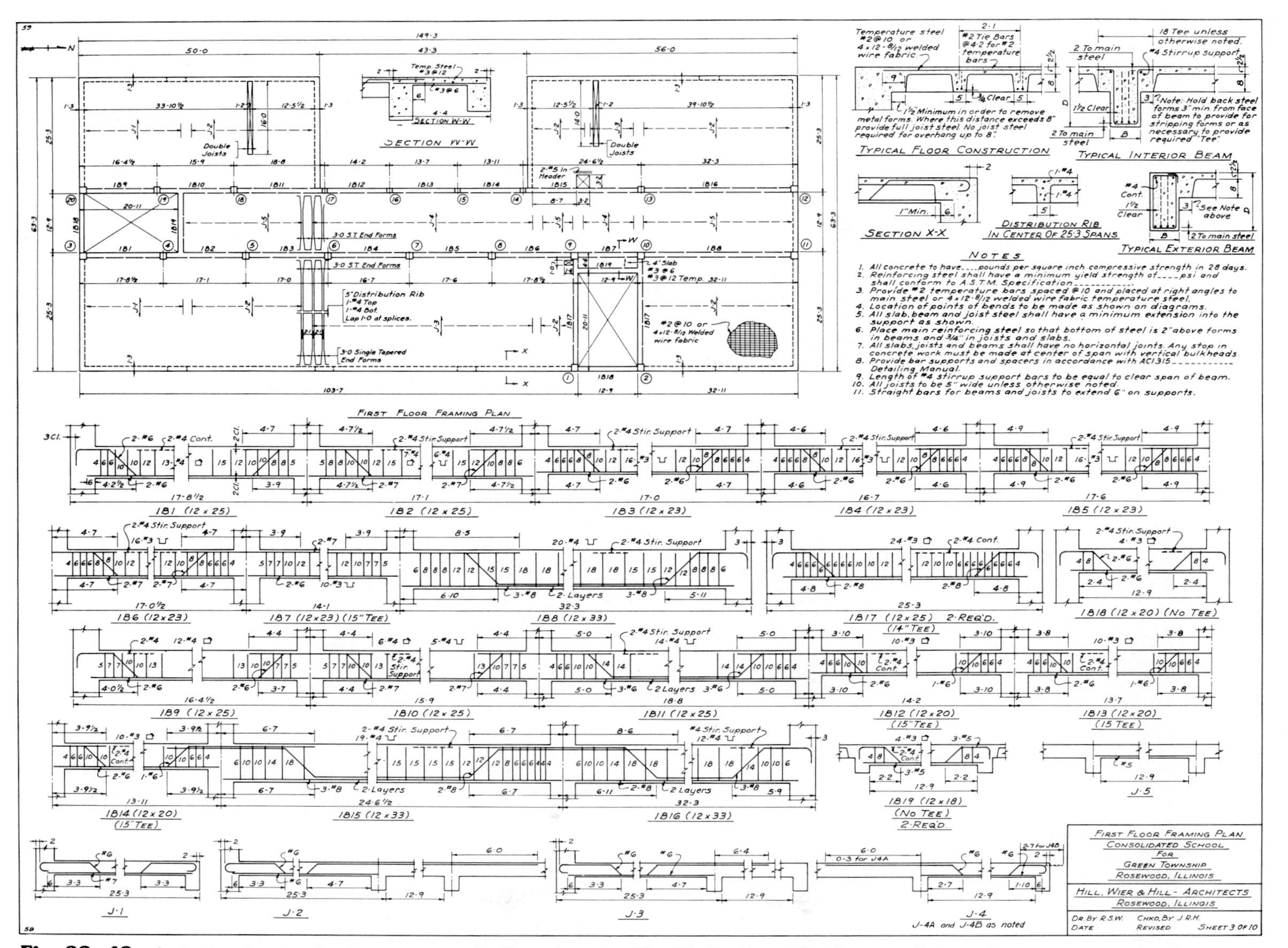

Fig. 22–42. A placing drawing for the concrete joist floor shown in Fig. 22–41. (American Concrete Institute)

Fig. 22–43. Steel forms being set in place. Beams, joists, and a floor slab will be cast as a single unit. (The Ceco Corporation)

Fig. 22–44. The steel forms are removed from below after the concrete has hardened. (The Ceco Corporation)

Scale

All reinforced concrete drawings are drawn to scale. This includes both the engineering drawings and the placing drawings. Remember that structural steel shop drawings need not be to scale in their long direction.

The most commonly used scale is ¼″ = 1′—0″. A larger scale is usually used for sections and other detail drawings.

Dimensions

Dimensions are placed in the same manner as structural steel drawings. They are in feet and inches. The foot mark is used. The inch mark is not used unless needed for clarity.

Casting the Structural Members

Wood or steel forms are used to form the beams, joists, and floor slabs. One way this is done is to pour the columns first. Then forms are placed, and beams, joists, and floor slab are poured together. In this way they form a single unit. This is called **monolithic casting**.

Fig. 22-43 shows steel forms being put into place. The reinforcing bars for the beams and joists are placed in the recesses left between the forms. The bars for the slab are placed on top of the forms. The concrete is poured over the entire structure.

After the concrete has set, the forms are removed from below, Fig. 22-44. Fig. 22-45 shows the underside of the floor with the joists and beams after the forms have been removed. Fig. 22-46 shows a section through the slab.

Wood Structural Drawings

Research in wood technology has produced findings that have made wood more useful for structural members. Engineers can accurately figure tension and compression forces in wood members. They can design members to carry specified loads. Wood members can span wide distances. Two commonly used types are wood trusses and laminated wood members.

Wood Trusses

Two common types of wood trusses are shown in Fig. 22-47. These are the king-post and the Fink.

The king-post truss requires fewer pieces. On large spans it usually requires 2- by 6-inch chords. The Fink truss has more pieces. It can span greater distances with 2- by 4-inch chords. These trusses are used in house construction.

When designing wood trusses, the engineer must decide what type of wood to use. Yellow pine and fir are often used. The grade of the wood must also be known. The various types and grades of wood have different load-carrying qualities. This information has been found by experimentation. It is available in table form. A good source is *Wood Structural Design Data*, published by National Lumber Manufacturers Association, Washington, D.C.

The method for fastening truss joints influences their strength. Two common methods are using plywood and metal fasteners.

Plywood gussets are glued and nailed or stapled. See Fig. 22-48. They are placed on both sides of each joint. The size of the gusset is carefully engineered. It must have the proper amount of gluing and nailing surface. The size of nails or staples must follow a standard. The spacing of the nails is specified. All of these factors must be correct if the truss is to carry the design load.

Notice that the web members are cut square on the end. This simplifies making the truss.

Fig. 22–45. The finished ceiling after the forms are removed and the surface is painted. (The Ceco Corporation)

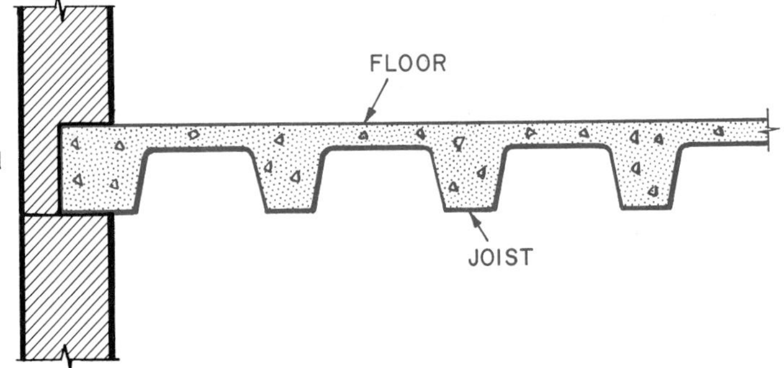

Fig. 22–46. A typical floor and joist section.

Metal gusset plates are shown in Fig. 22-47. They are made of 60-gage sheet steel. Special 8-penny nails, 1½-inches long, are used. Metal gusset plates are made in a wide variety of sizes.

Split-ring connectors are another means of strengthening trusses. Fig. 22-49. They are inserted halfway into each member. When two web members join a chord, a split ring is used for each one.

A drawing of a truss with split-ring connectors is shown in Fig. 22-50.

Drawing Wood Trusses. A drawing for a wood truss is shown in Fig. 22-50. This truss is designed for house construction. The drawing is made of several parts including a design drawing, hardware list, wood specification notes, and the truss production drawing.

Study the *design drawing*. The members are shown with a single line. This is the center line of each member. The load data computed by an engineer is given. Also shown are the dimensions of the chords, web members, and panels. These dimensions are to the center line of the split-ring connectors. The connectors are on the center line of each member.

The allowable stresses the wood must carry are shown in a note. The wood used to make the truss must equal or exceed these specifications.

A hardware list is given. The length of the bolts depends upon the number of members the bolts must hold together.

The *production drawing* shows the members drawn to scale. The thickness and width of each member before the wood is planed are given. Wood is rough-cut to these sizes in a mill. After it is dried, it is planed (smoothed). This makes the finished member smaller. The actual size of a 2″ × 4″ becomes 1½″ × 3½″ after planing. See Fig. 22-51 for standard sizes.

Split-ring connectors are used at critical connections. When two web members join a chord, each has a split ring. The split rings are shown on the drawing as solid circles. The number needed at each joint is given. The number of bolts is also noted. The bolt length is given in the hardware table.

Other connections are made with a 1- by 4-inch wood gusset, glued and nailed in place. The dimensions show the distances between the intersection of the center lines of the members. These points are also the center of the split ring. The dimensions are placed on the design drawing. The thickness and width of each member are placed on the production drawing.

Views of the horizontal and inclined chords are drawn. The horizontal chord is drawn with the truss sectioned just above the chord. Both views are drawn

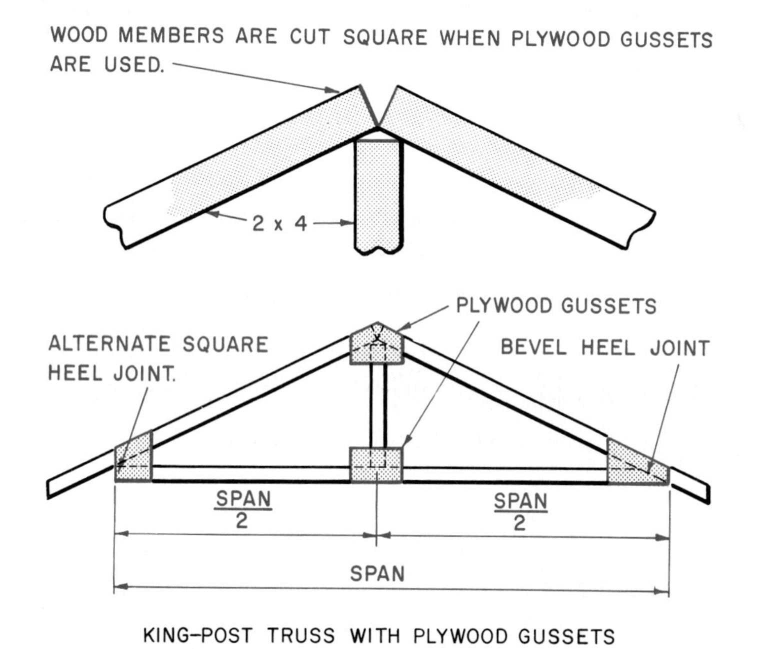

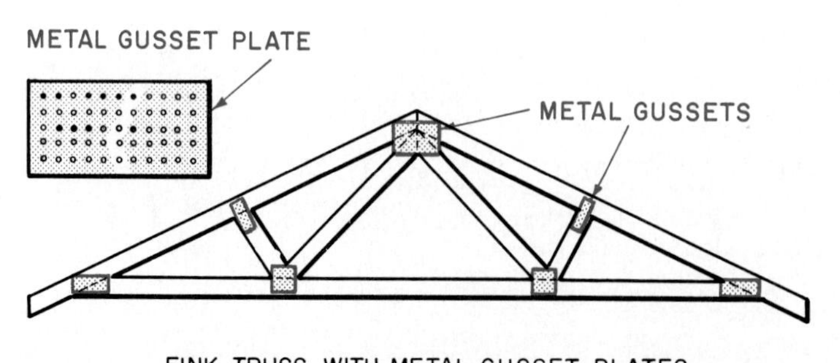

Fig. 22–47. Common wood trusses used in house construction.

looking down on the chord from above. Colored projection lines show how the members are projected to these views.

When members are long, they must be made from two pieces that are joined by splicing. The standard spacing for connectors is shown on the drawing and must be carefully dimensioned.

Laminated Wood Structural Members

Laminated wood structural members are made by gluing together layers of wood. The most commonly used layers are 3/4 inch and 1 5/8 inch. Fig. 22-52. They are about one-third stronger than solid wood members of the same size. The layers can be seasoned better than solid members. Defects, such as knots, can be reduced or eliminated.

Engineers design laminated members. After they decide the loading and type of wood to be used, they compute the size of the member needed. The drafter works with these data.

Fig. 22-53 shows a warehouse under construction. The structural system is made up entirely of laminated wood members. The structural drawings used to design this warehouse are shown in Figs. 22-54 through 22-57.

Fig. 22-54 (p. 586) shows a *roof framing plan* for the warehouse. It shows all the columns, girders, beams, and purlins (horizontal members) that form the structure to support the roof decking. The dimensions have been left off this drawing because they are so complex.

In Fig. 22-55 (p. 587) the complete system of dimensioning is shown. This is only a small corner of the building. The roof framing plan, when drawn to a large scale, would be completely dimensioned in this way.

Notice that each row of columns is indicated by either a letter or number. Each beam, girder, and purlin is identified by a letter.

Fig. 22–48. **Plywood gussets are glued and nailed or stapled on both sides of a truss. (American Plywood Association)**

Fig. 22–49. **Split-ring connectors are recessed into the two members to be joined. (Timber Engineering Co.)**

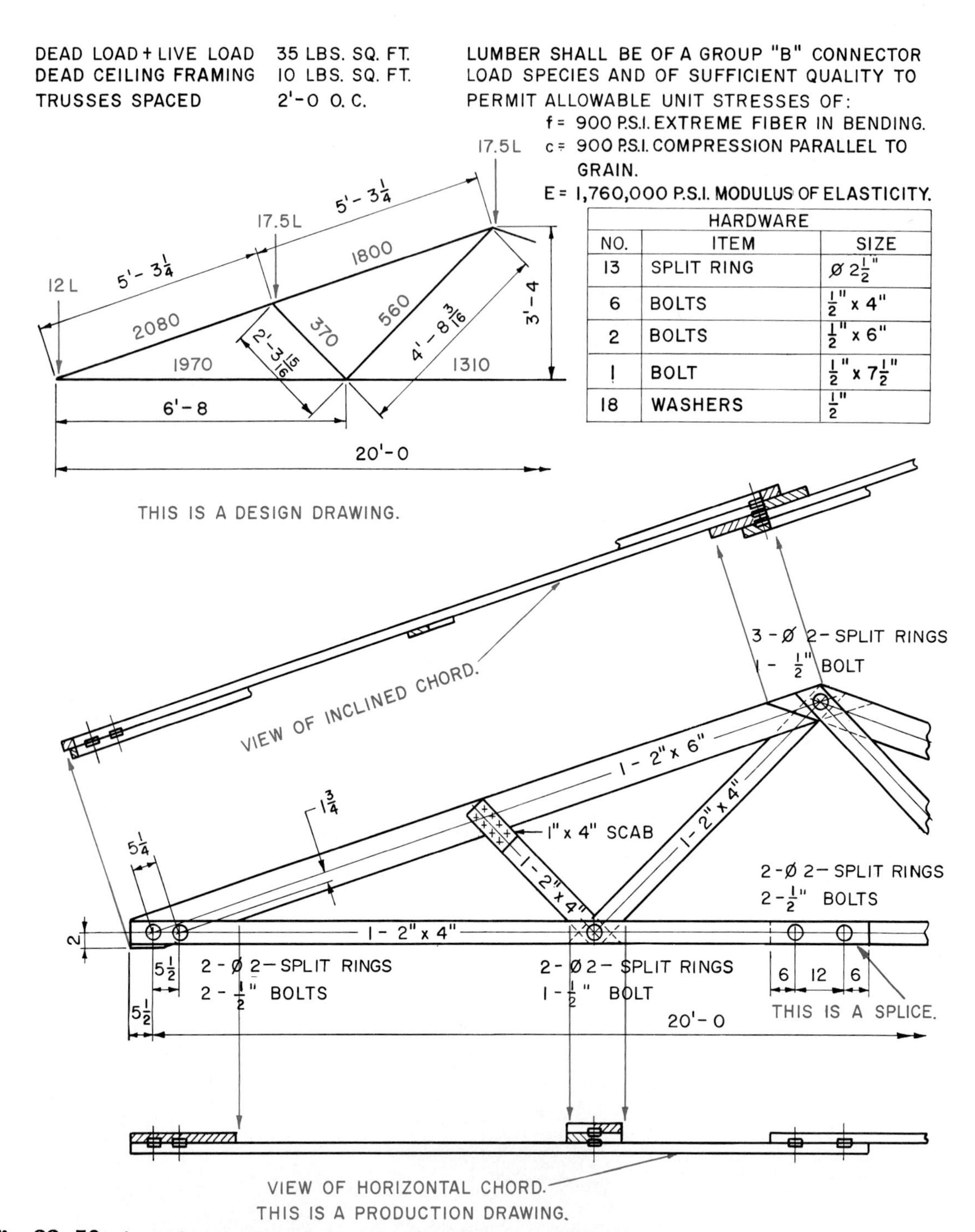

HARDWARE		
NO.	ITEM	SIZE
13	SPLIT RING	Ø 2 1/2"
6	BOLTS	1/2" x 4"
2	BOLTS	1/2" x 6"
1	BOLT	1/2" x 7 1/2"
18	WASHERS	1/2"

Fig. 22–50. **A wood truss with a slope of 4″ in 12″. (Based on a design by timber engineering Co.)**

Fig. 22-51. Nominal and Dressed Lumber Sizes

Nominal Size*	Dressed Size**
2 × 2	1½ × 1½
2 × 4	1½ × 3½
2 × 5	1½ × 4½
2 × 6	1½ × 5½
2 × 8	1½ × 7¼
2 × 10	1½ × 9¼
2 × 12	1½ × 11¼

(National Lumber Manufacturers Association)

*Nominal size is the rough size before the lumber is planed smooth. It is the size shown on the drawing.
**Dressed size is the actual size of the member after it is planed smooth. The members used in the truss are actually this size.

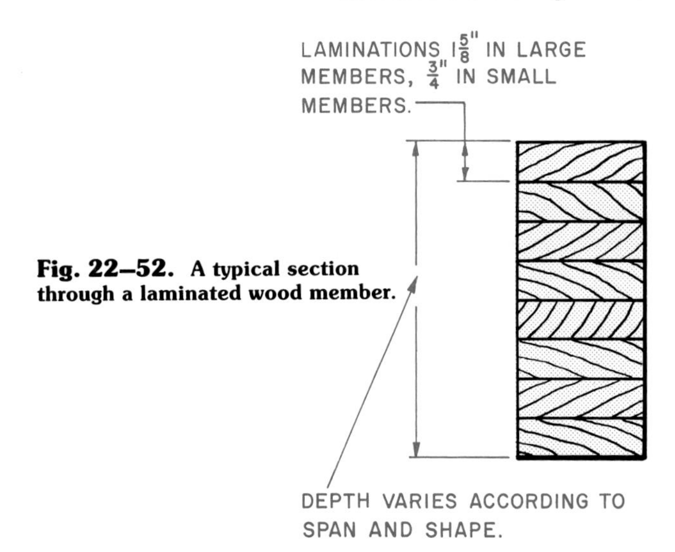

Fig. 22–52. **A typical section through a laminated wood member.**

Fig. 22–53. **A warehouse being framed with laminated wood members. The columns, beams, and purlins are all wood. In this picture the beams have been painted white. The purlins are dark. The second row of columns and beams is shown in section view in Fig. 22–56. Compare the column base set in the corrugated metal drum shown in the photograph with the column base detail shown in Fig. 22–56. (Timber Structures, Inc.)**

The details of each structural member are shown by making sections through the building. One section is shown in Fig. 22-56. The section shown is taken only through the corner detailed in Fig. 22-55. On a large scale drawing, the section would continue through the entire building.

The size of the roof's structural members is shown on the section. This includes the thickness, height, and length. A complete listing of all beams, girders, purlins, and columns is shown by a schedule. Figs. 22-57A and B are partial schedules for the beams and columns shown in Fig. 22-55.

Included on the drawing with the section are details. In Fig. 22-56 two details are shown: how the columns are connected to the beams and how the columns are connected at the base. Other details may be included in a drawing if the engineer decides they are needed.

Another type of laminated wood structural member is the arch. One type is shown in Fig. 22-58. Arches are made in a wide variety of shapes. They can span large distances.

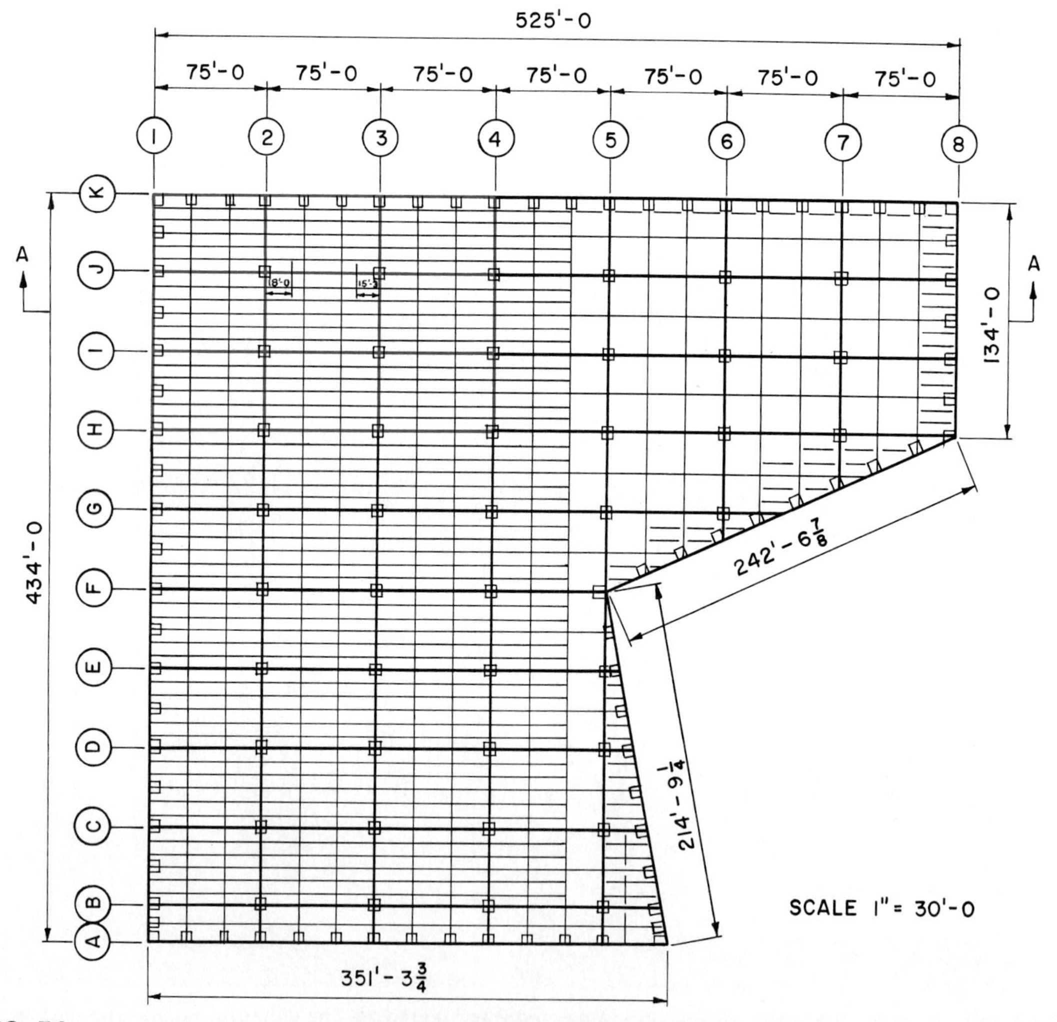

Fig. 22–54. **This is the roof framing plan for the warehouse shown in Fig. 22–53. Dimensions and identification of parts have been left off because they cannot be shown on this reduced drawing. One section in color is drawn in full detail in Fig. 22–55 to show how the total drawing would be dimensioned. (Timber Structures, Inc.)**

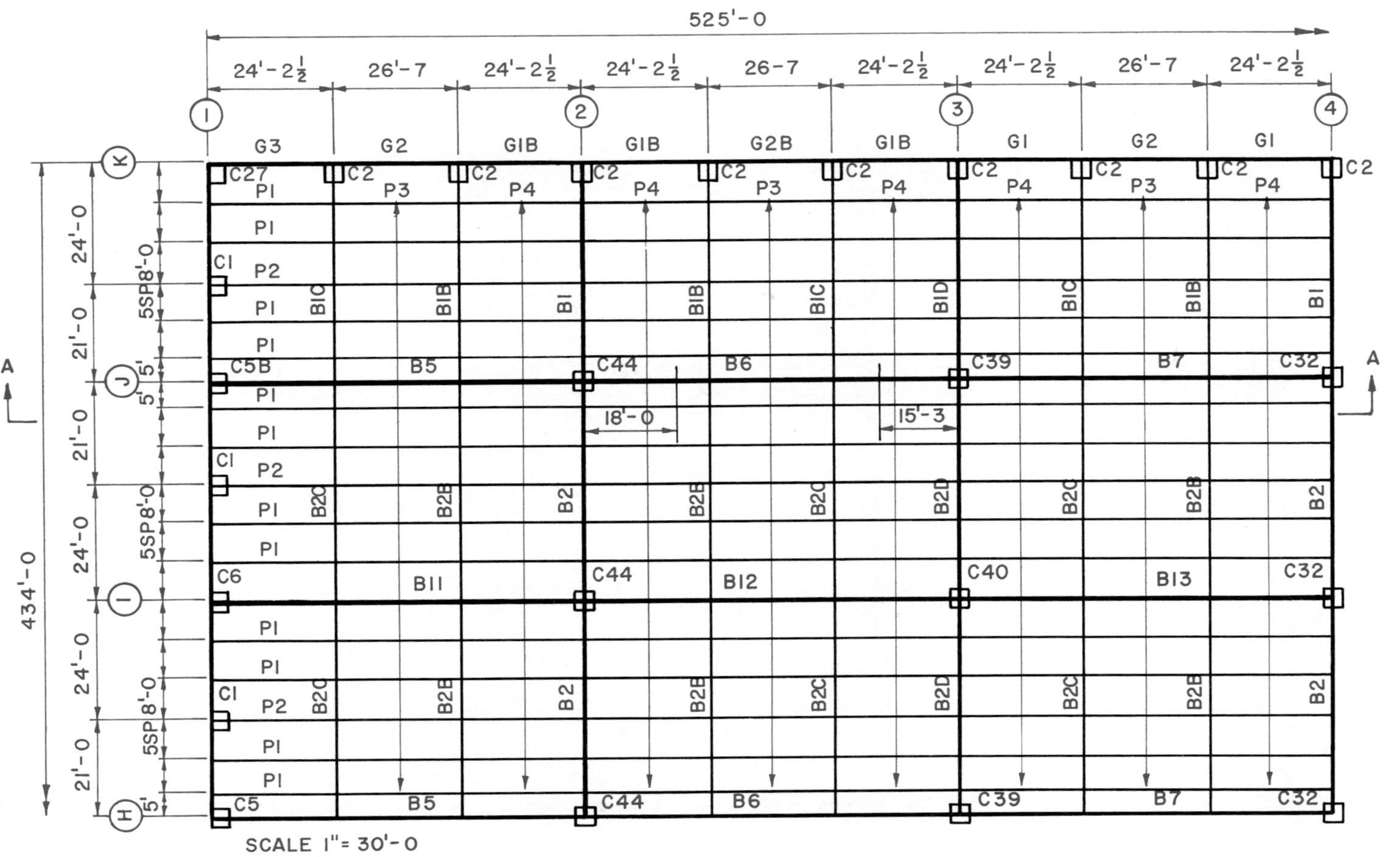

Fig. 22–55. This is a roof framing plan for one corner of the warehouse shown in Figs. 22–53 and 22–54. Columns are marked with a "C," beams with a "B," purlins with a "P," and girders with a "G." (Timber Structures, Inc.)

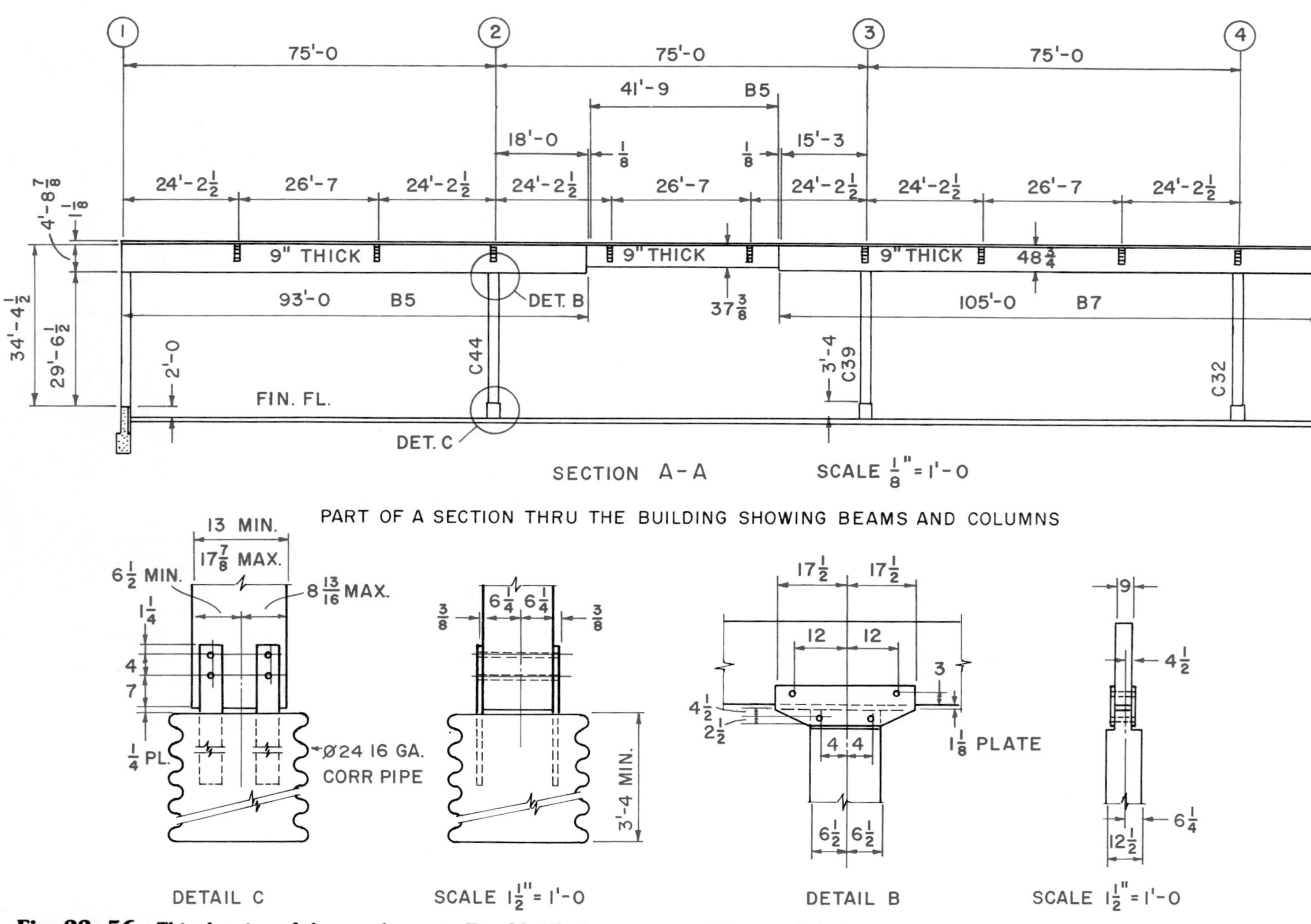

Fig. 22–56. **This drawing of the warehouse in Fig. 22–53 shows a part of Section A-A. It contains column and beam connections.**

Fig. 22-59 is the detail drawing for a three-hinged Tudor arch. Use of this arch in a church is shown in Fig. 22-60.

Each part of the arch must be dimensioned so that it can be laid out and cut to shape. The width at each end is given. The curved member has the radius of the curve given. The thickness is shown with a note on the member.

The connection of the two members is of great importance. The location of each fastener is carefully dimensioned. Notice the sections through the members in Fig. 22-59. They reveal that steel plates are set in grooves. On the vertical member, the plate is covered with wood. On the roof member, the decking covers the steel plate.

Since the arch is symmetrical, only half needs to be drawn. The center line is used for dimensions locating the vertical member.

Fig. 22-57A. Beam Schedule

Beam No.	Size	Length	No. Required
B1	9″ × 30″	45′-0″	3
B2	9″ × 30″	45′-0″	3
B5	9″ × 56 7/8″	93′-0″	3
B6	9″ × 37 3/8″	41′-9″	3
B7	9″ × 48 3/4″	105′-0″	2
B11	9″ × 56 7/8″	75′-0″	5
B13	9″ × 48 3/4″	75′-0″	3

Fig. 22-57B. Column Schedule

Col. No.	Size	Length	No. Reqd.	Top Fab	Bottom Fab
C1	9″ × 17 7/8″	33′-9 3/16″	5	Detail of C1	and P2
C2	9″ × 16 1/4″	32′-8 11/16″	15	Detail of C2	and B1
C5	9″ × 17 7/8″	29′-6 1/2″	3	Typ. of C1	Typ. of C1
C6	9″ × 17 7/8″	30′-8 9/16″	6	" "	" "
C27	9″ × 9 3/4″	35′-0 5/16″	1	Det. T	Corner Detail
C32	12 1/2″ × 14 5/8″	28′-5 13/16″	8	Typ. Int. Col.	Cap and Anchor
C39	12 1/2″ × 14 5/8″	27′-9 3/4″	2	"	"
C44	12 1/2″ × 16 1/4″	28′-9 11/16″	4	"	"

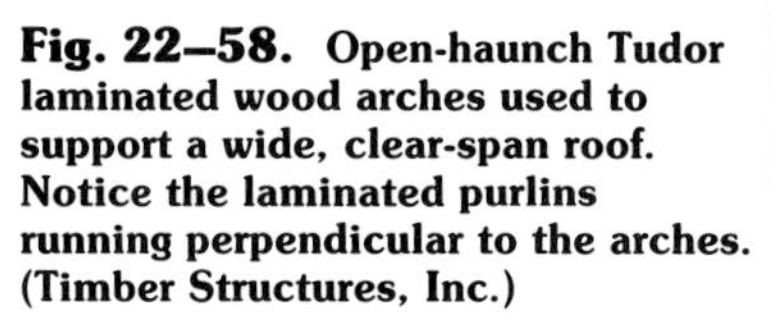

Fig. 22–58. Open-haunch Tudor laminated wood arches used to support a wide, clear-span roof. Notice the laminated purlins running perpendicular to the arches. (Timber Structures, Inc.)

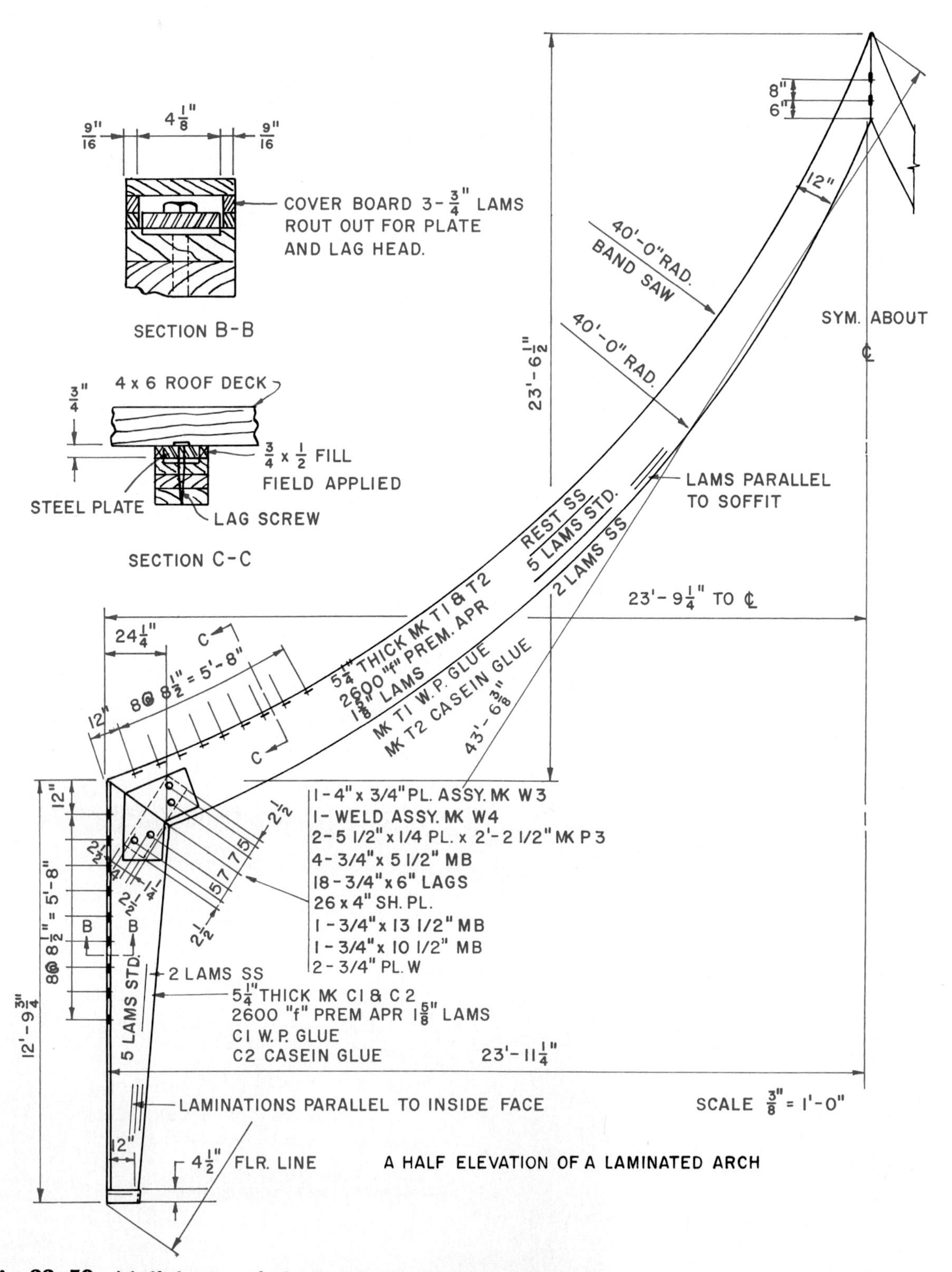

Fig. 22–59. A half elevation of a laminated arch. (Timber Structures, Inc.)

Fig. 22–60. Three-hinged Tudor wood-laminated arches were used to form the roof and walls of this church. The roof decking is several inches thick and spans the distance between the arches without support of purlins. (Timber Structures, Inc.)

Chapter Review

Build Your Vocabulary

You should understand and use the following terms as part of your working vocabulary. Write a brief explanation of what each means.

columns
girders
filler beams
pitch
gage line
gage distance
edge distance
slope
gusset plates
steel angles
working point
shop rivets
field rivets
angle connection
steel framing drawing
erection plan
shop drawings
truss
chord
web members
panel
span
rigid frames
compressive strength
tensile strength
reinforced concrete
prestressed concrete
placing drawings
monolithic casting
laminated wood structural members

Study Problems—Directions

The problems that follow will give you the chance to use some of the drafting knowledge that you learned in this chapter. Do the problems that are assigned to you by your instructor.

1. **Fig. P22-1** shows an angle connector used to join steel beams. Note that it has unequal legs. Make a two-view drawing of this connector. Locate the rivet holes and dimension them. The holes have a 3/4-inch diameter.

2. **Fig. P22-2** shows the connection of a 12-inch and 14-inch W-beam. The 14-inch W-beam has two rivet holes in the bottom flange. These are to have standard spacing. The beams are joined by a 3½ × 3½ × 8-inch angle connector.

   Make a full-size drawing of this detail. Locate all the holes in the 14-inch beam and the angle connector. Use standard spacing. Dimension the hole locations.

   The rivets have a 3/4-inch diameter. The members have rolled edges.

3. Dimension the W-beam shown in **Fig. P22-3**. All rivets have standard spacing. Rivet diameter is 3/4 inch. Members have sheared edges.

4. Draw a combined roof framing plan and erection drawing of the building shown in **Fig. P22-4**.

   All beams are W 8 × 24 (8 by 6½ inches). They are connected to the columns with 3½ × 3½ × ½ × 5 inch angle connectors prepared to receive two 3/4-inch diameter bolts. The angle connectors are welded to the beam. The top of the beam is flush with the top of the column.

   All columns are W 8 × 17 (8 by 5½ inches) 8′—0″ long. The W member is 12′—3½″ long. The base is a 10- by 10-inch plate, 1¼-inches thick. It has two 7/8-inch diameter holes. The base is welded to the column.

   The channels are C 8 × 25 (8 by 2½ inches). They are connected to the beams with angle connectors. The connectors have two 3/4-inch diameter holes for bolting to the beams. They are welded to the channel.

5. Make shop drawings for all the beams in the structure shown in **Fig. P22-4**.

6. Make shop drawings for the columns in the structure shown in **Fig. P22-4**.

7. Draw a completely dimensioned detail drawing of the welded truss shown in Fig. 22-30. Draw to the scale 1″ = 1′—0″ or 1:10.

8. Study the riveted truss shown in Fig. 22-29A. Prepare a parts list giving the size of each member. This list should include gusset plates, structural members, and rivets. Record the number of each part needed to make one complete truss.

9. Study the riveted truss shown in Figs. 22-29A and B. Write the answers to the following questions:

   A. What is the rivet spacing on the web members?
   B. What is the distance between the center lines of the columns to hold this truss?
   C. What is the overall height of the truss at the center line? This is the outside-to-outside dimension.
   D. What is the overall height of the truss at the end that connects to the column? This is the outside-to-outside dimension.

10. Draw a completely dimensioned detail drawing of the riveted Fink truss shown in Figs. 22-29A and B. Draw to the scale 1″ = 1′—0″ or 1:10.

11. Make a dimensioned drawing of the rigid frame shown in **Fig. P22-5**.

12. Study the concrete engineering drawing shown in **Fig. P22-6** and make a placing drawing. Draw to the scale 1/4″ = 1′ —0″ or 1:50. Include a beam and joist schedule. Dimension the location of each stirrup and bend. Identify each bar.

13. Make a production drawing of the wood truss shown in the design drawing, **Fig. P22-7**. Tables are provided giving data for various spans. Select one of the spans and make the dimensioned production drawings for it. This design uses split-ring connectors.

14. Make a detail drawing of the laminated wood arch shown in Fig. 22-58. Draw to the scale 3/8″ = 1′—0″ or ½″ = 1′ —0″, 1:50 or 1:25.

15. Make a roof framing plan for the building shown in **Fig. P22-8**. Draw the section shown. Prepare a column, beam, girder, and purlin schedule.

Study Problems

P22–1. **An angle connector.**

P22–2. **A connection between two W-beams**

P22–3. **A steel wide flange beam.**

Study Problems

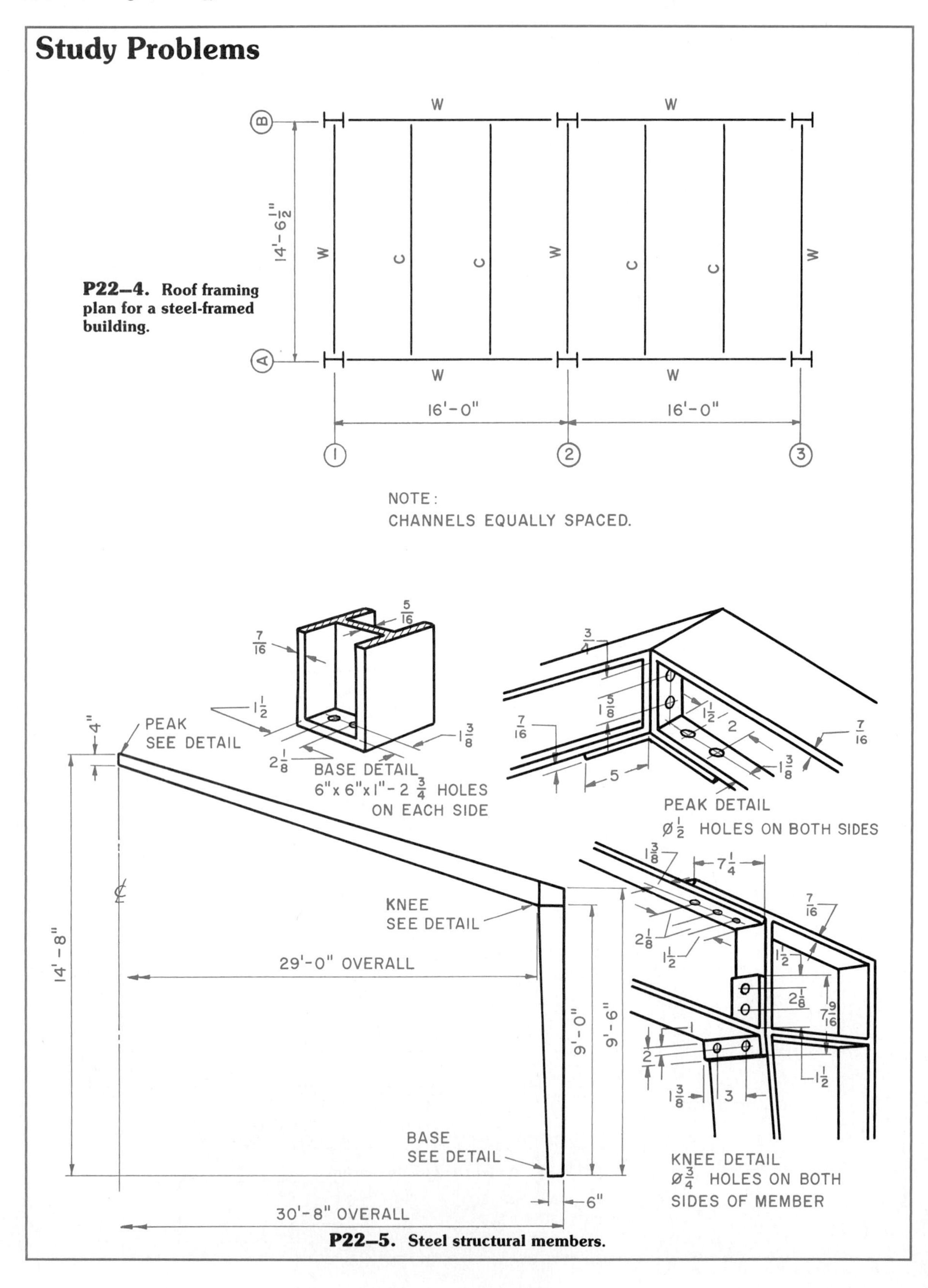

P22–4. **Roof framing plan for a steel-framed building.**

P22–5. **Steel structural members.**

Study Problems

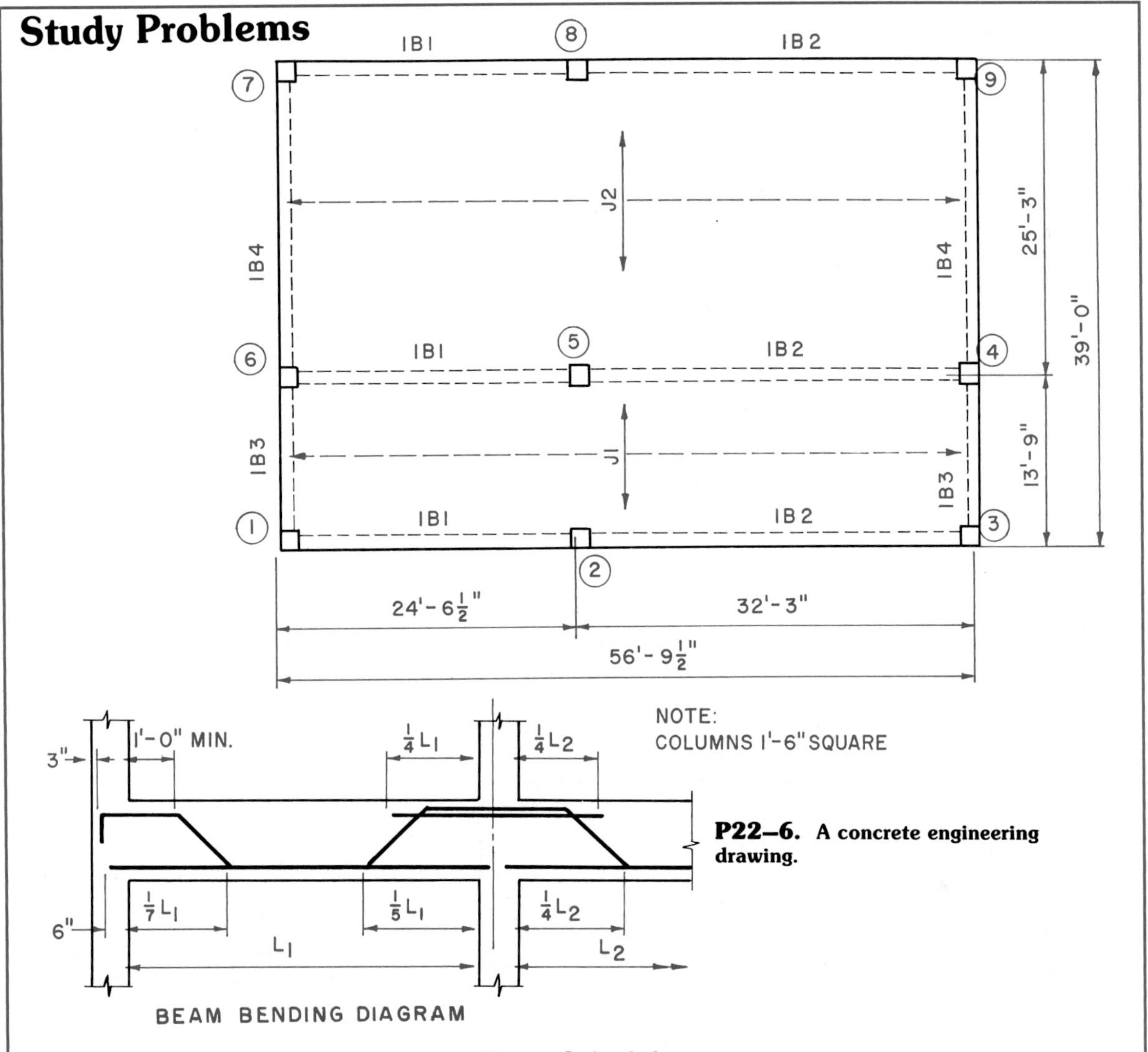

P22–6. A concrete engineering drawing.

Beam Schedule

Mark	B × D	Reinforcing		Stirrups			Stirrup Support Bars in Top
		Str.	Bent	No.	Size	Spacing Each End	
1B1	12 × 33	3-#8	2-#8	14	#4	3@ 4, 3@ 6, 3@ 8, 3@ 12	2-#4
1B2	12 × 33	2-#8	3-#8	12	#4	2@ 10, 2@ 14, 2@ 18	2-#4
1B3	12 × 20	2-#6	2-#6	4	#3	1@ 4, 1@ 8	2-#4
1B4	12 × 33	3-#8	2-#8	14	#4	3@ 4, 3@ 6, 2@ 8, 2@ 12	2-#4

Joist Schedule

Mark	B	D	Reinforcing	
			Str.	Bent
J1	5	8 + 2½	1-#6	1-#6
J2	5	8 + 2½	1-#6	1-#6

Study Problems

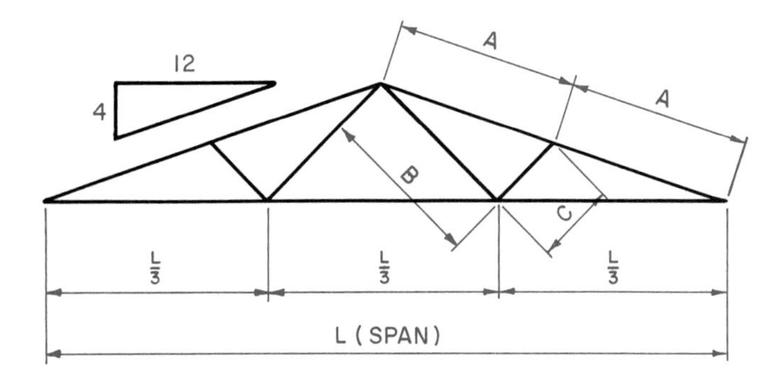

Span	Lumber		
	Inclined Chords	Horizontal Chord	Web
20′-0″ 26′-0″ 32′-0″	2 × 6 2 × 6 2 × 8	2 × 4 2 × 4 2 × 6	2 × 4 2 × 4 2 × 4

Span	Dimensions		
	A	B	C
20′-0″ 26′-0″ 32′-0″	5′-3½″ 6′-10³⁄₁₆″ 8′-5³⁄₁₆″	4′-8³⁄₁₆″ 6′-1³⁄₁₆″ 7′-6³⁄₁₆″	2′-3¹⁵⁄₁₆″ 3′-0⁷⁄₁₆″ 3′-8⅞″

P22–7. **A wood truss design. (Design by Timber Engineering Co.)**

Study Problems

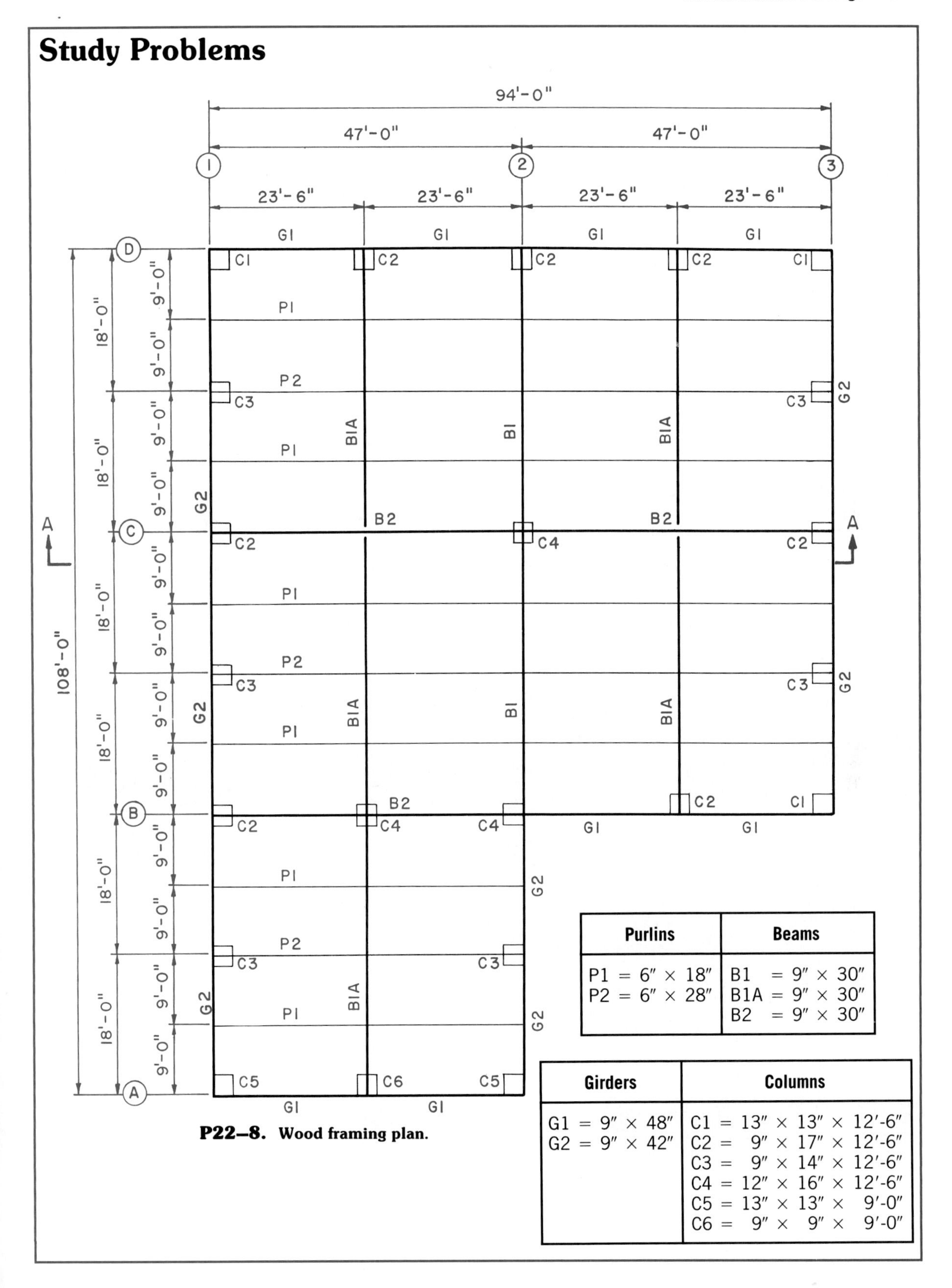

P22–8. **Wood framing plan.**

Purlins	Beams
P1 = 6″ × 18″ P2 = 6″ × 28″	B1 = 9″ × 30″ B1A = 9″ × 30″ B2 = 9″ × 30″

Girders	Columns
G1 = 9″ × 48″ G2 = 9″ × 42″	C1 = 13″ × 13″ × 12′-6″ C2 = 9″ × 17″ × 12′-6″ C3 = 9″ × 14″ × 12′-6″ C4 = 12″ × 16″ × 12′-6″ C5 = 13″ × 13″ × 9′-0″ C6 = 9″ × 9″ × 9′-0″

CHAPTER 23
Electrical and Electronic Diagrams

After studying this chapter, you should be able to do the following:
- Draw basic schematic diagrams.
- Draw printed circuit drawings.
- Critically analyze the layout of electrical and electronic drawings.

Introduction

An important part of the plans for the manufacture of machines, communications equipment, and controlling devices is drawings that show the electrical or the electronic requirements. Electrical and electronic drawings show the wiring diagrams or circuits. These drawings are called **schematic diagrams**. They show the components to be used and how they are to be connected. A **component** is a single unit, such as a transistor.

Electrical and electronic drawings are used extensively in all parts of industry. Notice in Fig. 23-1 the complex wiring system of a jet engine being checked by technicians. The design, production and maintenance of this engine require many large, detailed schematics.

Fig. 23–1. Technicians are using electrical drawings to check this jet engine. (NASA)

The design of electrical and electronic devices is done by electrical and electronic engineers. Their work is recorded as freehand sketches. The drafter makes finished drawings from these sketches and notes.

Fig. 23-2 shows a sketch (at A) and the finished schematic drawing (at B) for an RF impedance transformer.

Drafters are more effective in their work if they understand the fundamentals of electricity and electronics. They should have a basic idea of circuitry. They should understand the purpose served by the components in the system. They must also know the symbols used to show the components. The purpose of this chapter is to show the commonly used symbols and how to draw schematics.

What Is Electrical Energy?

The source of electrical energy is the *atom*. All matter is made up of atoms. A substance made up of only one kind of atom is an *element*. There are over one hundred known elements. Examples of elements are copper, silver, and iron.

Every atom has a core, called a *nucleus*. The nucleus contains positively charged particles called *protons* and neutral particles called *neutrons*. Revolving around the nucleus are negatively charged particles called *electrons*. Fig. 23-3.

An **electric current** is a stream of electrons moving along a conductor, such as a copper wire. Fig. 23-4. The electrical energy is transferred through a conductor by the free movement of electrons from one atom to the next atom. As an electron enters one end of a conductor, it displaces an electron that was already there. The displaced electron moves to another atom, where it in turn displaces another electron, and so on. Thus, for every electron that enters one end of the conductor, another leaves the opposite end.

Electrical Measurements

Electrical currents and power can be measured. The units used to do this are volts, ohms, amperes, and watts. A **volt** is a measure of electrical *pressure*. It is the force that causes electricity to flow along a conductor. This pressure is sometimes called *electromotive force*. The symbol for voltage is V.

An **ohm** is a measure of *resistance* to the flow of electrical current. The symbol for the ohm is Ω.

An **ampere** is a measure of the amount of electrical current flowing. The symbol for the ampere is A.

One ohm of resistance will allow one ampere of current to pass when one volt of pressure is applied. This can be shown by the formula, $\Omega = V/A$. If two of these units are known, the third can be found by entering the two known units into the formula and solving for the unknown unit.

A **watt** is a unit of electrical *power*. Its symbol is W. Electrical power is the rate at which electrical energy is delivered to a device. The power (wattage) is equal to the current (amperes) multiplied by the force (voltage). This is represented by the formula $W = A \times V$.

Horsepower is a measure of how much work is done. One horsepower equals 746 watts.

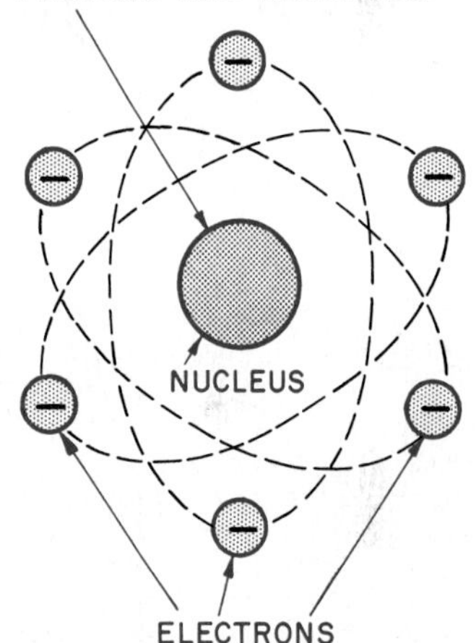

Fig. 23-3. An atom has electrons revolving about a nucleus.

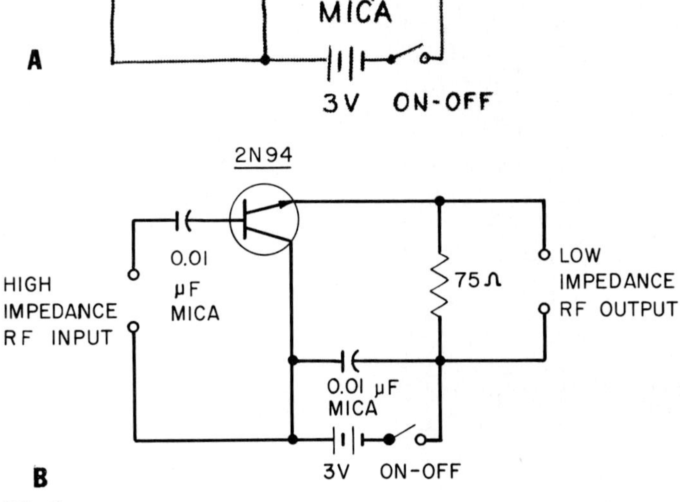

Fig. 23—2. (A) A rough sketch of a proposed circuit of an RF impedance transformer. (B) The finished schematic of the RF impedance transformer. (GTE-Sylvania Electric Products, Inc.)

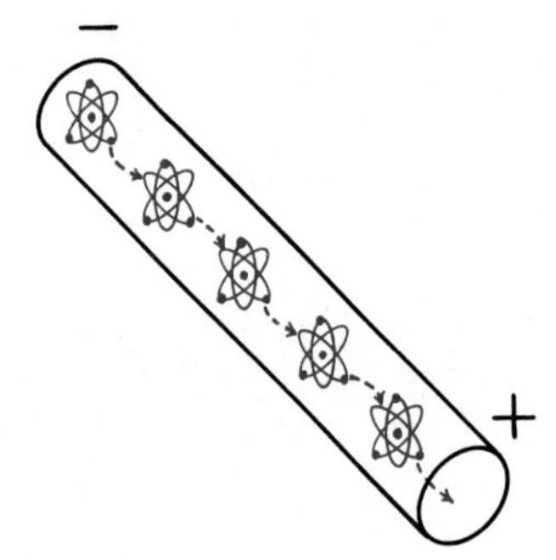

Fig. 23—4. Electric current is the movement of electrons through a conductor.

Fig. 23-5. CAD systems can be used to draw electrical and electronic schematic diagrams. (Hewlett-Packard Company)

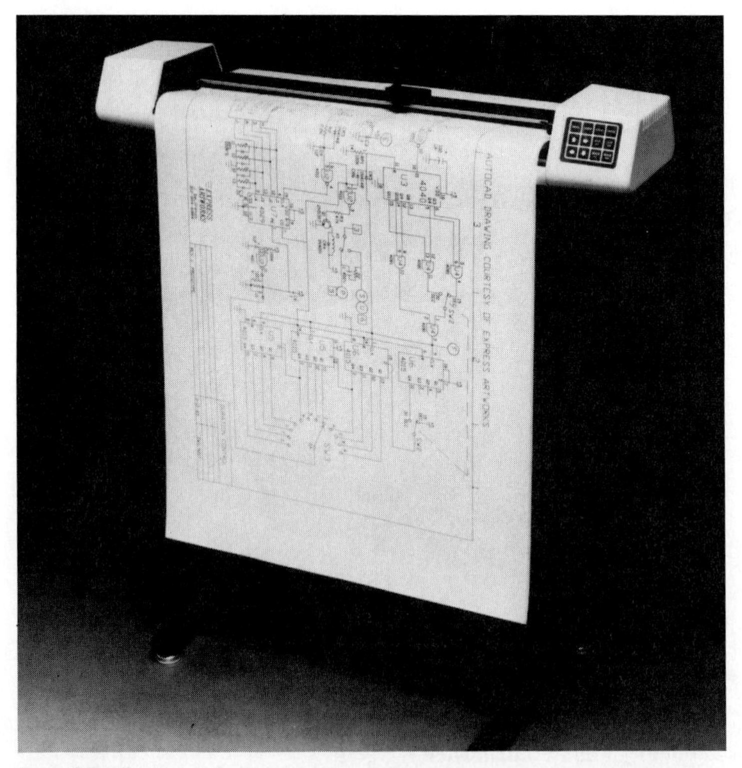

Fig. 23-6. Hard (paper) copies of the schematic are printed by the plotter. (Houston Instrument)

Electrical Conductors

As stated earlier, a *conductor* is a material that permits electrical current to pass through it. Some materials permit an easier flow of electrical power than others. In other words, they offer little resistance to the current flow. The best conductors are silver, copper, and aluminum. Materials such as tungsten, iron, and carbon offer considerable resistance to the flow of electricity and are therefore poor conductors.

Other materials will not conduct electrical energy. They are called *insulators*. Glass and plastics are examples.

Computer Drawings

Electrical and electronic diagrams can be rapidly and accurately drawn using CAD systems. Fig. 23-5. As it is drawn the schematic is displayed on the screen of the monitor. The designer can rearrange, add, or remove components or adjust the schematic's layout in any way necessary using the input devices. The changes or corrections are quickly made, and the finished diagram is constantly displayed.

The schematic diagram can be printed on a plotter. Fig. 23-6. Design changes can be sketched on this paper copy. Then they can be entered on the program stored in the computer. The final design can be permanently stored on a floppy disk. This disk can be used to produce additional hard copies. Other changes may be made to the schematic on the disk at any time.

Block Diagrams

A **block diagram** is one that uses rectangular blocks to represent the major parts of a circuit. The part is identified by a label inside the block. It is quite different from a schematic diagram since no graphic symbols are used to show the individual components that make up the device.

The block diagram in Fig. 23-7 shows the overall circuit plan of a transistor radio receiver. It represents in a simple way the relationships between each major part of the circuit. It does not give details about the design of each major part. The diagram explains that the signal is received through the antenna (represented by a symbol), and moves through the mixer, through first and second intermediate-frequency stages (IF), and on to the speaker. It shows a feedback circuit (AGC) and an oscillator (OSC) connected into the main circuit.

A block could represent a complete unit in an installation, such as a television camera or video tape recorder in a television studio.

Drawing a Block Diagram

The symbol generally used to represent the major units on a block diagram is a rectangular block. Sometimes squares and triangles are used. Some components are identified by graphic symbols rather than blocks. These are units that are located on the end of the drawing, such as a speaker or antenna. Fig. 23-7.

Drafters must decide on the relationship between the blocks. They must decide the best way to present these on a drawing. Usually a number of sketches on graph paper are made before a good solution is found.

The size of the blocks will vary with the lettering that must be placed on them and the space available for the drawing. The blocks are usually made the same size throughout the diagram.

As the plan is developed, the flow of current or the signal should be made to run from left to right. The input is usually at the upper left corner and the output at the lower right corner. The flow of the current or signal from one block to another is represented by a single line connecting the blocks. The direction of the flow is indicated by an arrow. The line weight for flow should be as heavy as that used for the block. The line for flow is usually a rather thick line similar to a visible line on working drawings.

Schematic Diagrams

Electrical circuits may be illustrated by a pictorial wiring diagram or by a schematic diagram. Fig. 23-8 shows a pictorial drawing of a simple circuit for a dome light in an automobile. The battery, door switch, and light are the components. The lines connecting them are leads (wires). The battery supplies the electrical current. When the auto door is opened, the push button closes. The current runs through the push button, through the dome light, and through the frame of the auto back to the battery. This makes a complete circuit.

This same circuit is shown in Fig. 23-9 as a schematic diagram. The schematic diagram uses symbols to represent the electrical parts of the circuit. This greatly simplifies the drawing. The person using the schematic diagram is not interested in what each part looks like but only in the design of the circuit.

Graphic Symbols

An electrical or electronic schematic diagram involves the use of symbols to represent the parts of the circuit. These symbols are a type of technical shorthand enabling a complex circuit to be represented as clearly and briefly as possible. They usually do not represent a picture of how the part really looks. Industry and government agencies have devel-

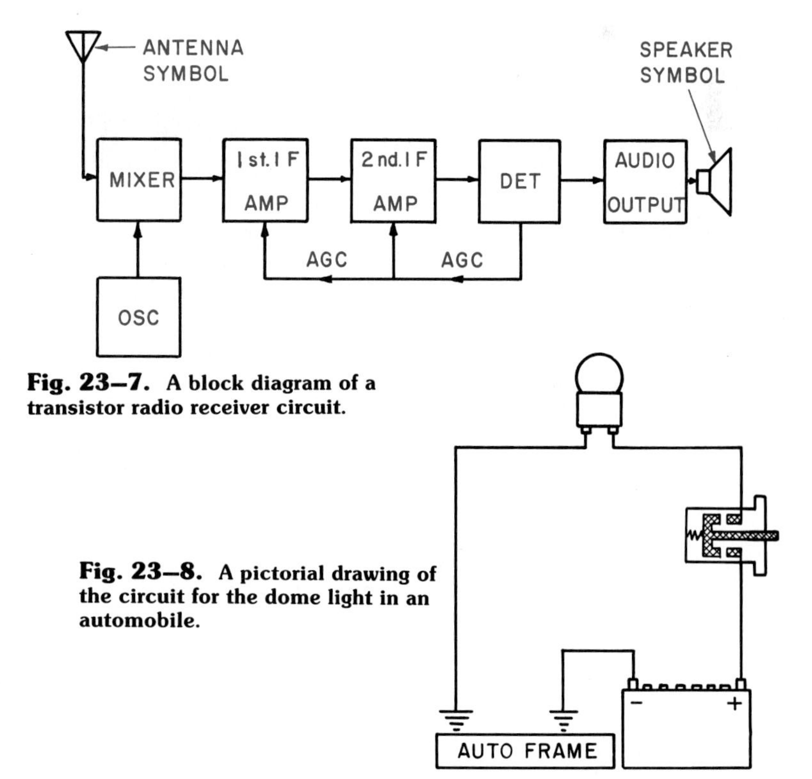

Fig. 23–7. A block diagram of a transistor radio receiver circuit.

Fig. 23–8. A pictorial drawing of the circuit for the dome light in an automobile.

oped standard symbols to be used on drawings. Commonly used sets of symbols are found in ANSI/IEEE 315-1975 and ANSI/IEEE 315A-1986, *Graphic Symbols for Electrical and Electronics Diagrams*.

Graphic symbols are made of several basic elements. The size and location of each element indicate the purpose served by the symbol. Fig. 23-10 shows the basic elements used to form graphic symbols.

Most symbols can be placed on the drawing in any position desired. Studying schematic diagrams is a good way to learn how symbols are placed.

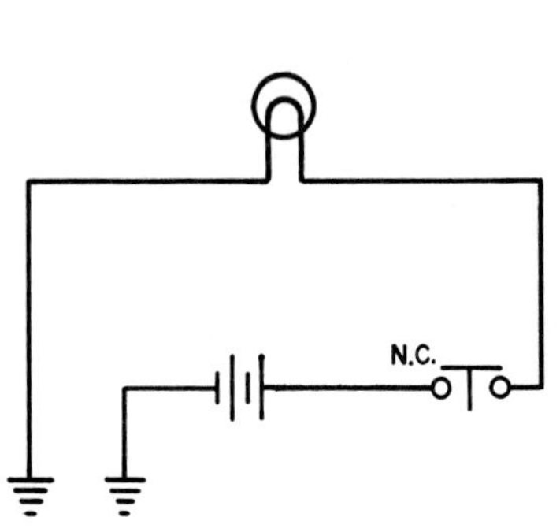

Fig. 23–9. A schematic diagram of the circuit for the dome light in an automobile.

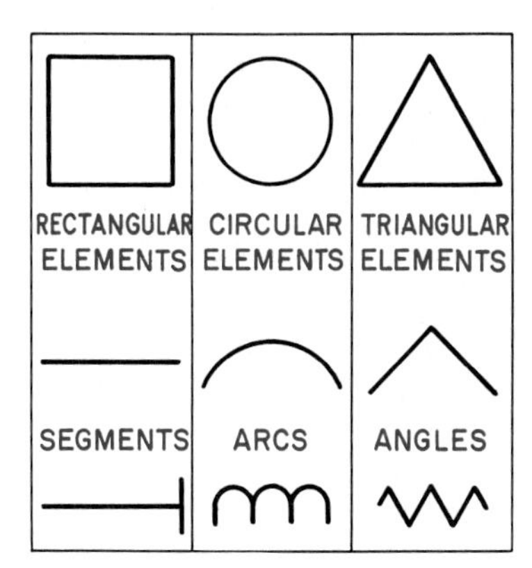

Fig. 23–10. Symbols are based on rectangular, circular, and triangular elements.

Line Width

Generally the weight of lines used to draw symbols is the same as for the leads. Recall that a lead is the line used to show the connections between symbols. If the symbols should stand out, they can be drawn with a heavier line than the leads. The width of the line should be in proportion to the size of the drawing and the lettering to be used.

Symbol Size

Standard practice is to use templates to draw symbols. See Fig. 23-11. This sets a standard size for all symbols on the drawing. Any symbol drawn should be the same size wherever it appears on the drawing. A good size for symbols is shown in Fig. 23-12. These can be drawn a little smaller or larger depending upon the size of the drawing. If templates are not available, copy these symbols on vellum; then slide the vellum under your drawing, and trace the symbols.

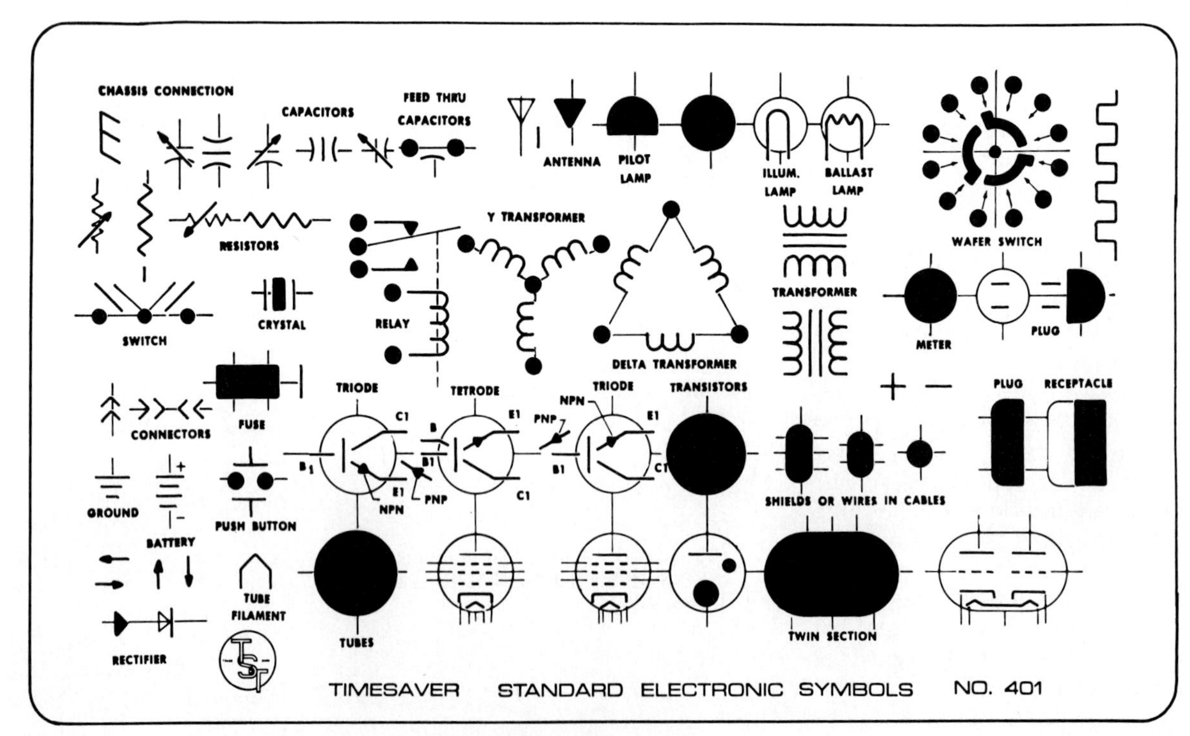

Fig. 23–11. An electronic symbol template. (Alvin and Co., Inc.)

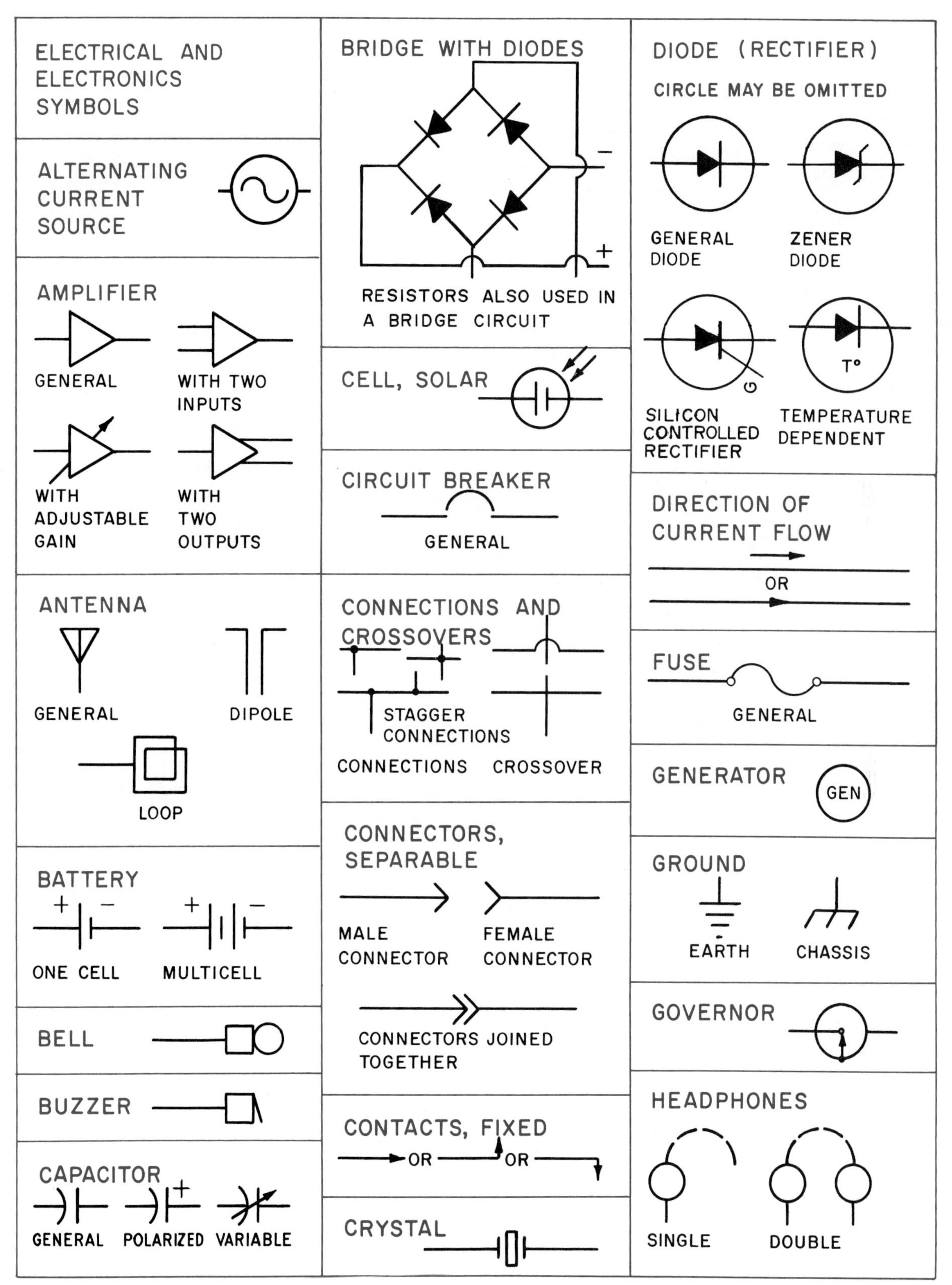

Fig. 23–12. **Symbols most commonly used for electrical and electronic diagrams.**

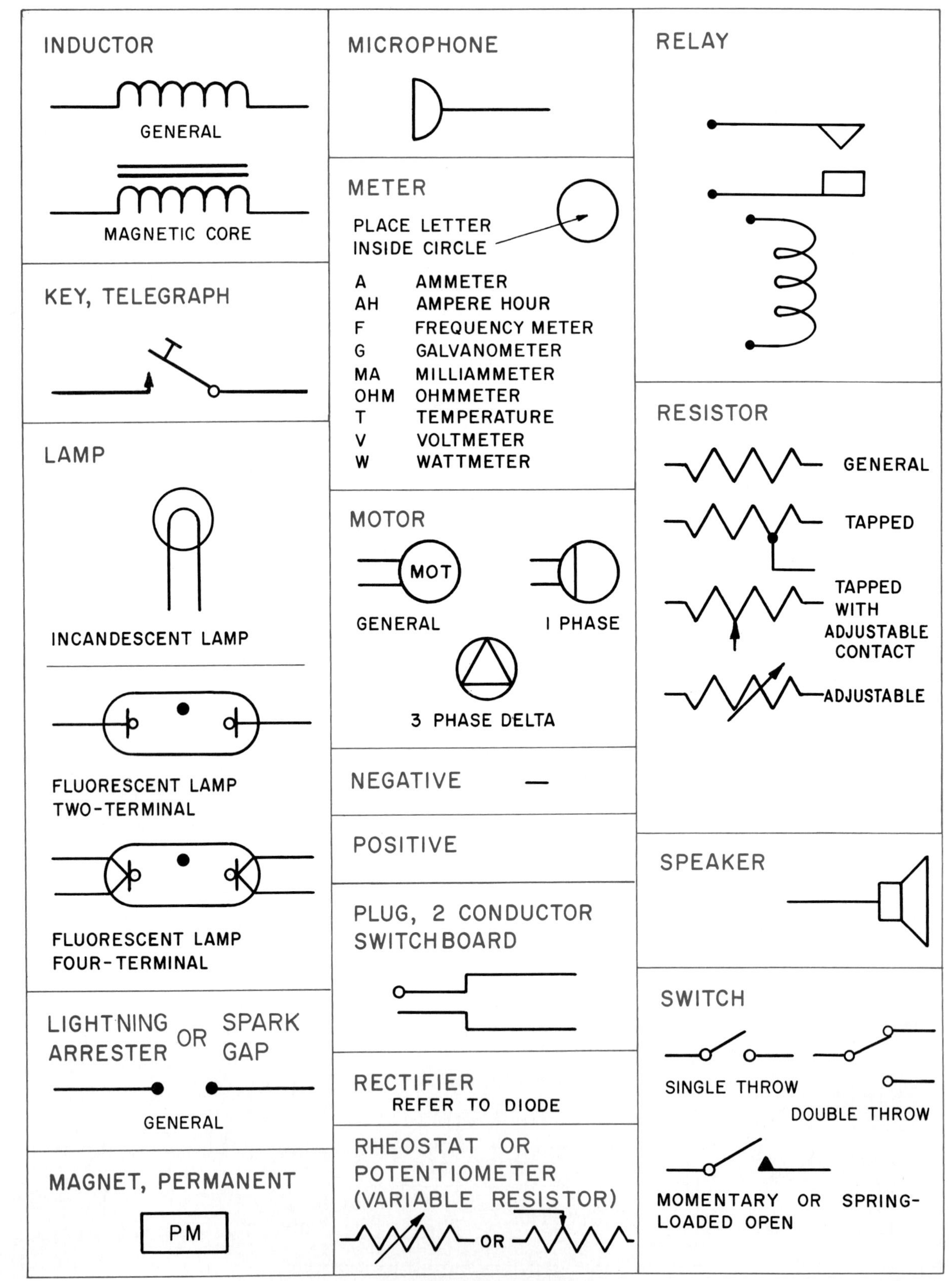

Fig. 23–12. Continued

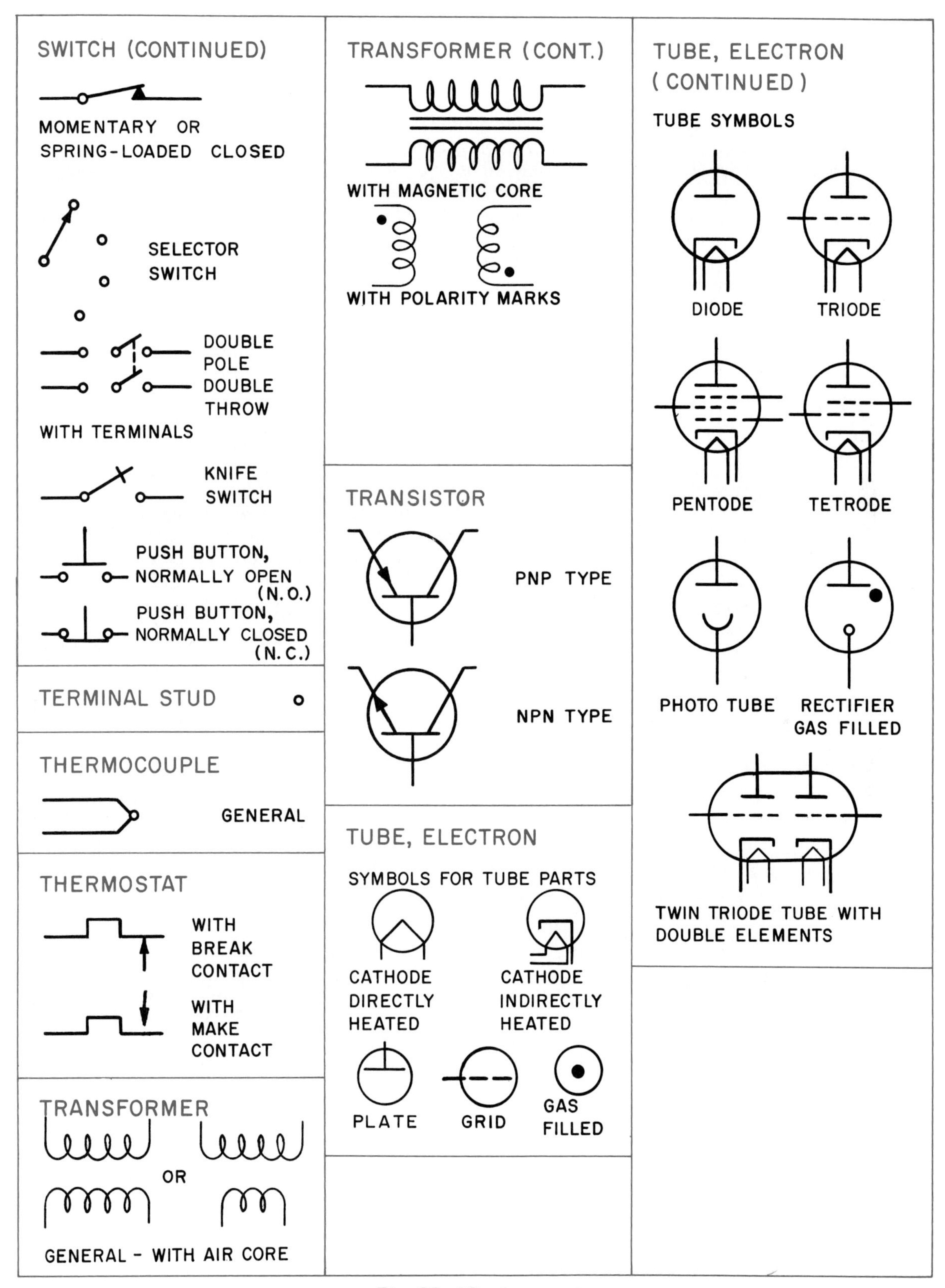

Fig. 23–12. Continued

Sometimes you must letter some type of identification with each symbol. The lettering might show the type of part, size, electrical value, or other necessary data. A standard symbol system and approved abbreviations should be used.

The symbols shown in Fig. 23-12 are the most commonly used for electrical and electronics diagrams. A standards manual gives a more complete listing. Symbols for house wiring can be found in Chapter 25, "Architectural Drawing."

Drawing a Schematic Diagram

The location of the components when a device is manufactured is different from that given on a schematic. The schematic shows the components and how they are connected. When the device is made, the location of components is influenced by such factors as the desired shape and size of the finished device. For example, a radio could be manufactured either long and low or tall and thin. The same schematic could be used for both.

The features desired in a schematic diagram are (1) correct grouping of the symbols; (2) uniform alignment of the symbols, leads, and lettering; (3) leads that are as short as possible to make, (4) crossing of as few leads as possible, (5) uncrowded lettering; and (6) balance in the drawing.

Remember that there is more than one satisfactory way to lay out any schematic diagram.

Errors are difficult to notice on schematic diagrams; therefore, care and accuracy in drawing are important.

Correct Grouping of Symbols

Electronic devices are made of subunits. When these are put together, they make a complete device. The symbols for each unit are grouped together on the drawing. Before starting to make the drawing, the symbols that are to be grouped must be marked. Then you must decide how they are to be placed on the drawing. This is done by studying the first sketches.

The schematic of an audio amplifier is shown in Fig. 23-13. On this figure are seven subunits. They are a first amplifier, driver amplifier, phase inverter, power amplifier, output transformer, speaker output, and power supply.

The symbols for each subunit are kept together on one section of the diagram. The blocks are then related to each other so that the connections between them can be shown. This requires an understanding of electricity and electronics. Since most students do not have this understanding, the schematics shown in this book will be properly grouped. As engineers design circuits, they tend to group these subunits properly. However, a drafter should try to improve the diagram by rearranging it so the final drawing is easy to read.

Alignment of Symbols, Leads, and Lettering

Symbols should be located according to a plan and placed so that there is enough space for lettering the data needed. Two

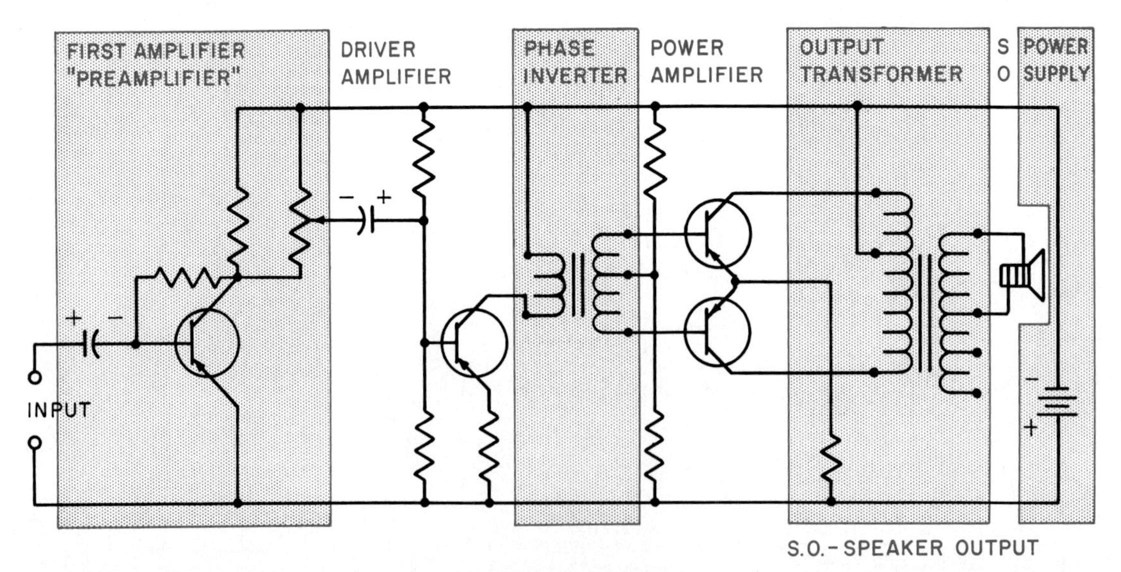

Fig. 23–13. Schematic diagram for an audio amplifier showing symbols grouped in subunits. (GTE-Sylvania Electric Products, Inc.)

plans may be used. The symbols may be grouped in alignment or staggered. Fig. 23-14 shows symbols that are in *alignment.* Notice that the resistors are lined up horizontally. In Fig. 23-15 the resistors are *staggered* so that data can be lettered without crowding.

Schematic diagrams should be kept as simple as possible. One means of doing this is to record any data other than that specifically related to a symbol on a specification chart. The chart is placed on the drawing but not on the diagram itself. If this data is placed on the diagram, it becomes more difficult to read. Notice in Fig. 23-14 that some data are grouped on the side of the diagram.

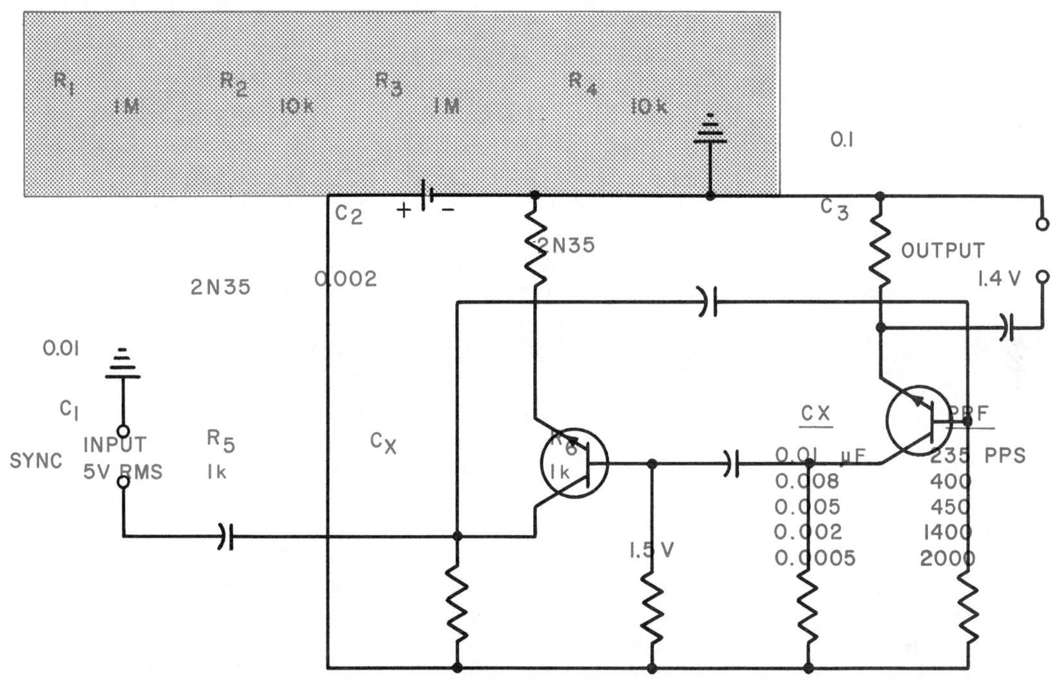

Fig. 23–14. A multivibrator. Notice the alignment of the resistors. (GTE-Sylvania Electric Products, Inc.)

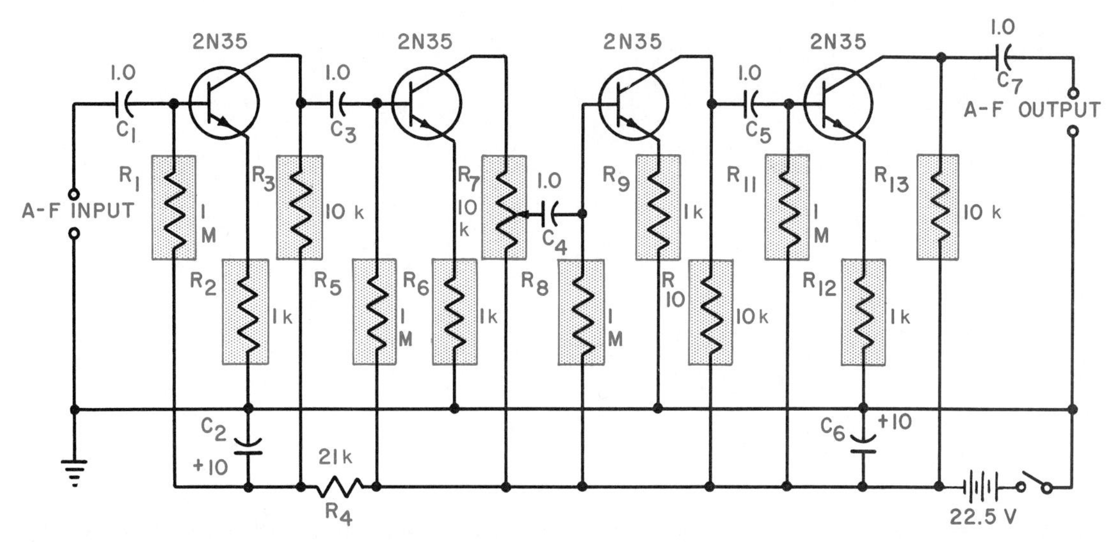

Fig. 23–15. Cascaded RC coupled audio amplifier. Notice how the staggered resistors aid in recording values. (GTE-Sylvania Electric Products, Inc.)

The lettering that is to be on the diagram should be placed in the largest open space that is close to the symbol.

Another good practice is to use a *common lead* instead of drawing a number of parallel leads. Fig. 23-16 shows this simplification. Notice that the four parallel leads have been replaced by one lead.

Sometimes several leads can be drawn to a single point. This point serves as a common terminal for related circuits. This is shown in Fig. 23-17. This practice is not always possible because of design requirements but should be used whenever possible.

Ground leads are usually drawn short. This simplifies the drawing and saves space.

Notice that leads change direction by making a 90-degree corner. This makes the diagram easier to read.

Balancing the Schematic Diagram

The spacing between symbols must be planned so that one side of a drawing is about the same as the other. A typical error is to begin the drawing on one side of a sheet and then crowd the symbols on the other side because not enough space was left. This error is shown in Fig. 23-18.

In order to avoid crowding, examine the sketch of the circuit to be drawn. Look for a pattern or a series of blocks in the design. Make a trial layout on graph paper before starting to make the final drawing. It is best to make several trial plans.

Notice that the layout of the schematic in Fig. 23-18 is in three major blocks. The space allowed for this drawing should be divided into three approximately equal areas. This allows room for the needed data and will result in a balanced appearance.

Another problem occurs when a schematic is stretched over too long an area. While it is not as bad to stretch a drawing as to crowd it, the appearance suffers. The proper spacing to make a pleasing drawing results from experience and through studying properly drawn diagrams.

Abbreviations and Letter Symbols

If a schematic diagram calls for identification of various parts of the circuit, these must be labeled. These identifying terms are shown by abbreviations and letter symbols.

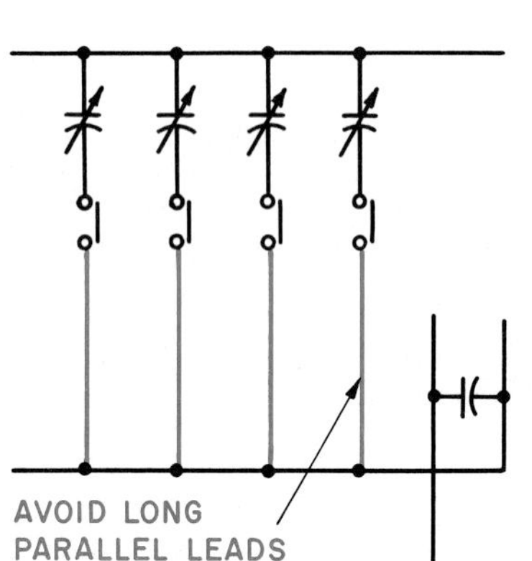

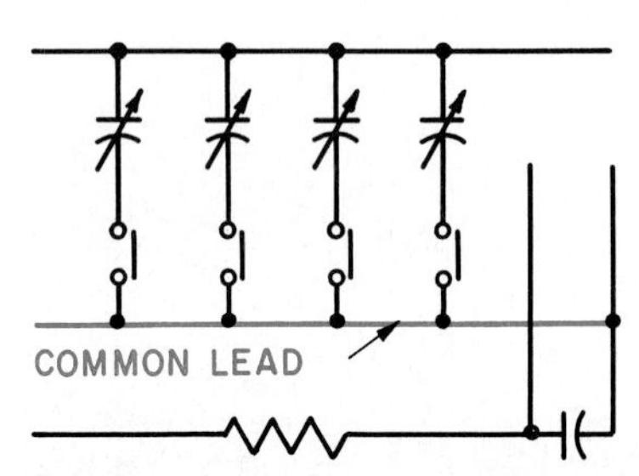

Fig. 23–16. **(A) Long parallel lines add confusion to a schematic diagram. (B) A common lead replaces long parallel leads and simplifies the schematic.**

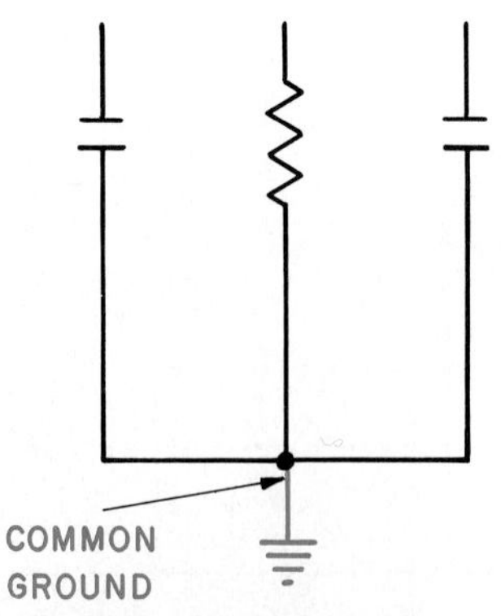

Fig. 23–17. **Use a common ground when design permits.**

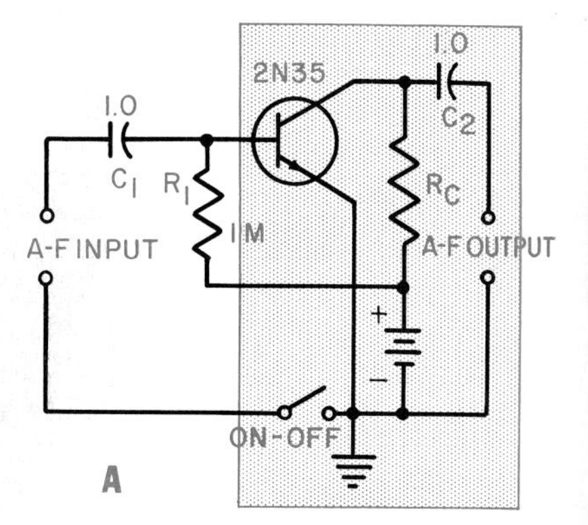

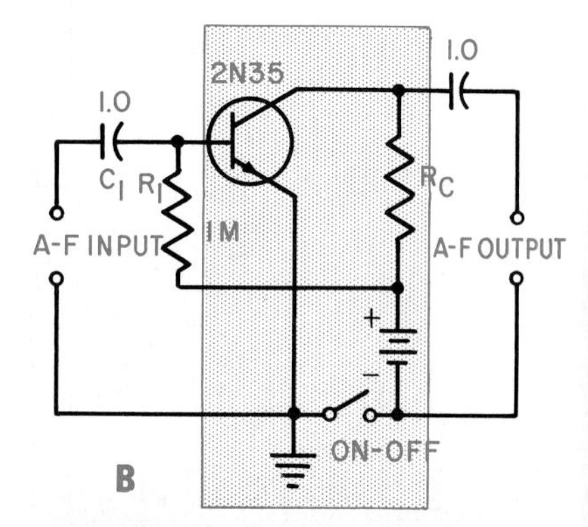

Fig. 23–18. **(A) This schematic is crowded on the right side. (B) The revised schematic. Notice that the crowding is reduced.**

Labeling Electrical Units on Schematic Diagrams

Electrical values are indicated on drawings by using letters to represent those values. The schematics in this chapter show the use of letter symbols. Fig. 23-19 gives the symbols for electrical units.

If you desire to give the actual electrical values of the units on a drawing, use the numerical values instead of the letter symbols. This is of special help to those servicing electrical and electronic units.

Color Codes

The use of a color code is another way of presenting information on schematics.

The wires are color-coded to assist in identifying the circuits in an automobile on the schematic in Fig. 23-38. This information is especially useful to those doing the actual wiring. One standard color code system is used by the Electronic Industries Association. Fig. 23-20. This system enables letter abbreviations or numbers to be used to identify wire colors. This simplifies the drawing.

Printed Circuit Drawings

The invention of the integrated circuit and the development of the printed circuit have enabled the electronics industry to advance rapidly into miniaturization. Without this development our small radios, hearing aids, and other miniature electronic devices could not be built.

A **printed circuit** is one that is reproduced upon a base made of an insulating material with a foil surface. A printed circuit board with components is shown in Figs. 23-21 and 23-22. The printed circuit replaces hand wiring. Circuit characteristics can be closely maintained. They can be mass-produced, which reduces the cost of making the circuits.

Fig. 23-19. Symbols for Electrical Units

Unit	Letter Symbol
Ampere (current)	A
Ohm (resistance)	Ω
Volt (electromotive force)	V
Farad (capacitance)	F
Watt (power)	W

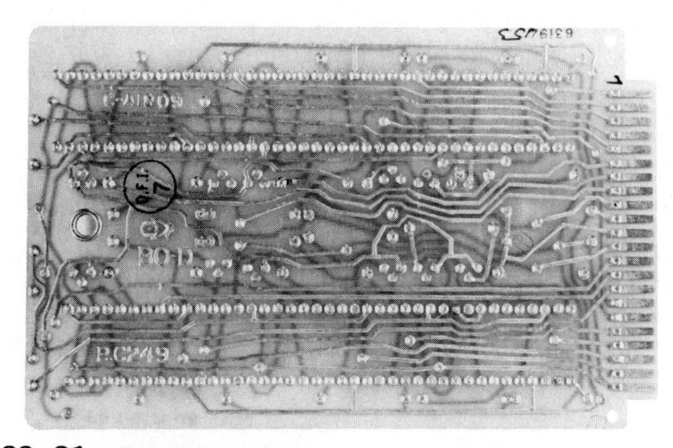

Fig. 23–21. The bottom of a printed circuit board showing the etched board, the foil conductors, and the soldered connections.

Fig. 23-20. Electronic Industries Association Standard Color Code

Color	Abbreviation	Number
Black	Blk	0
Brown	Brn	1
Red	Red	2
Orange	Orn	3
Yellow	Yel	4
Green	Grn	5
Blue	Blu	6
Violet	Vio	7
Gray	Gra	8
White	Wht	9

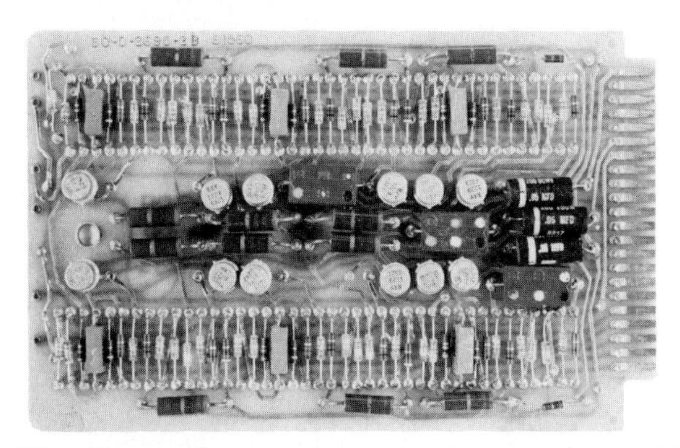

Fig. 23–22. The top of the printed circuit board shown in Fig. 23–21. It shows the components in place.

CAD Applications:

Routing Circuits Automatically

CAD systems have already been a big help in designing printed circuit boards. Components can be moved around and other changes can be easily and quickly made. Commonly used symbols are available in libraries. And some programs can generate terminal blocks and numbering. Now, however, it is also possible to use CAD to design a circuit board automatically.

Special autorouting software is required to design circuit boards. The software needs two kinds of information: (1) component layouts and schematic drawings already in the user's computer files and (2) design rules. Design rules include user-defined specifications such as trace widths and clearances. The computer then automatically lays out the board, drawing all the symbols and connections in the proper places. The best groupings are selected, parallel or common leads are used as needed, and the design is properly balanced. Routing programs seek the highest quality design as well as the one most profitable to manufacture.

Additionally a single design problem can be input and rerouted several different ways using different rules and layouts. Or multiple boards can be routed at the same time. Some routing programs will also operate on data prepared with a text editor. In this case no schematics are required.

All drawings done with a routing program need a certain amount of editing and finishing. The amount needed depends on the speed at which the drawing was done and its quality. Although routing programs are fully automatic, a drafter must still be experienced in working with circuit board layouts and with CAD.

Hewlett-Packard Company

Conductors in Printed Circuits

Many things must be considered when drawing the layout for a printed circuit. The size of the board upon which it will be made must be decided. This is influenced by the purpose of the device and by the size of the components to be fastened to the board.

The drafter must know the basic conductor shapes. These are shown in Fig. 23-23. The conductor is a thin flat metal strip. It serves in place of a wire to carry the electric current.

The actual circuit is difficult to design. The conductors should be kept as *short as possible*. However, they should not cross over each other. Crossovers are used only as a last resort. This is done by installing short jumpers over the etched conductors. Usually these must be insulated.

The *width and thickness* of the conductors must be decided. This allows the conductors to carry the required current. For suggested sizes, see Fig. 23-24.

The *spacing* between conductors is also important. The greater the voltage potential difference, the farther apart the conductors should be. For example, conductors with a potential difference of 0 to 150V should be at least 0.031 inches (0.787 mm) apart. This is generally a safe minimum for circuit wiring. For other suggested spacing, see Fig. 23-25.

Long conductors should be placed near the edge of the circuit board. Pad diameters should be large enough to give at least 0.025 inches (0.64 mm) of conductor material around the mounting hole. The size of the hole varies with the size of the wire to be put through it. A common size for the hole is 0.020 inches (0.51 mm). The most frequently used pad sizes are 0.100 and 0.125 inches (2.54 and 3.18 mm). The pad must be large enough to provide a good soldering base.

If an eyelet is to be placed through the pad, the pad diameter should be 0.025 inches (0.64 mm) larger than the flange of the eyelet.

Conductors should meet pads or join another conductor smoothly by means of a radial fillet or straight fillet. Refer to Fig. 23-23. Sharp corners should always be avoided. Tees and elbows should have rounded corners.

Sometimes brass eyelets are placed in the holes in the pads. The eyelet can also be used to connect conductors on both sides of the circuit board. Fig. 23-26.

Making a Printed Circuit Layout

A regular schematic drawing of a circuit gives information about the components needed and the proper connections. Fig. 23-27 is the schematic for a logic circuit test probe. From this information the needed drawings can be made.

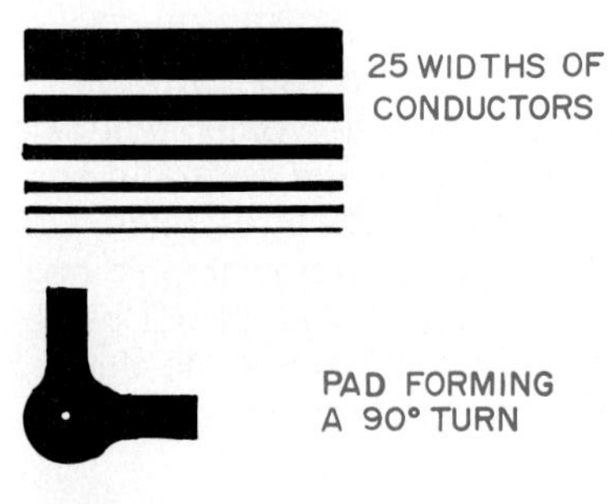

Fig. 23–23. These conductors are available on precut, pressure-sensitive tape.

Fig. 23-24. Width and Thickness of Conductors

Current (amperes)	0.00135 Inch (0.035 mm) Thick Conductor	0.0027 Inch (0.069 mm) Thick Conductor
1.5	0.015″ (0.381 mm) wide	—
2.5	0.031″ (0.787 mm) wide	0.015″ (0.381 mm) wide
3.5	0.062″ (1.575 mm) wide	0.031″ (0.787 mm) wide

Fig. 23-25. Conductor Spacing Standards

Minimum Spacing Between Conductors	Voltage Potential Difference
0.031 inches (0.787 mm)	0–150V
0.062 inches (1.575 mm)	151–300V
0.125 inches (3.175 mm)	301–500V

From Electrical and Electronics Drawing by C. J. Baer, McGraw-Hill Book Co. Used by permission.

The detailed directions for laying out a drawing for a printed circuit are too complex to present in this text. The following description will give an overview of a procedure that uses pressure-sensitive adhesive tapes. There are other ways the pads and conductors can be produced.

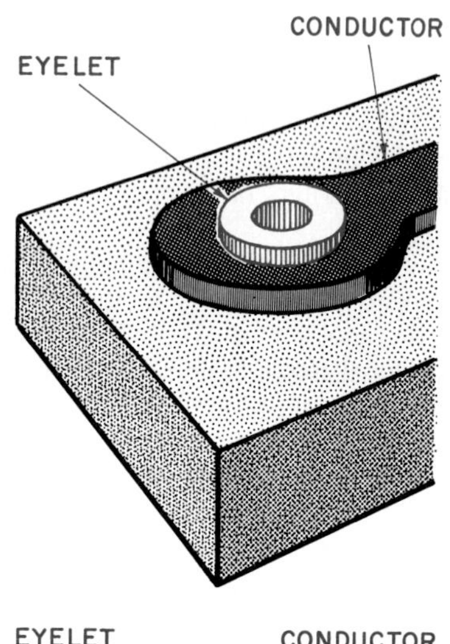

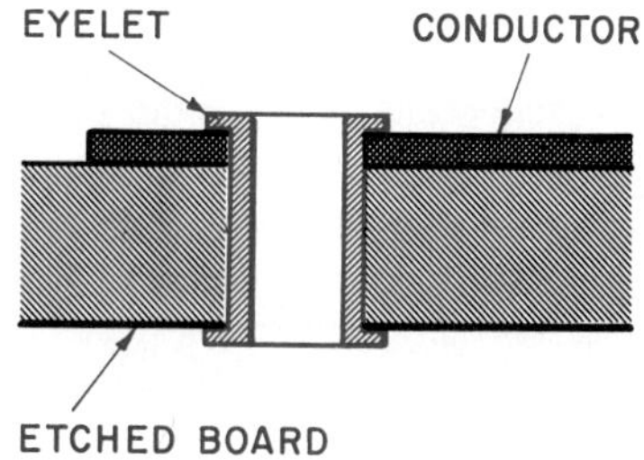

Fig. 23–26. An eyelet in a printed circuit board.

Greatest accuracy is obtained by using a printed grid as a base. This is taped to the drawing board. A sheet of plastic drafting film is placed over this, and the outline of the circuit board and the limits placed on the components are drawn. Fig. 23-28. These drawings are made larger than the desired finished size so that the quality of the lines can be improved by photographic reduction. Scales commonly used are 2:1 and 4:1. A 2:1 scale means that it is drawn 2 times larger than the finished size.

Printed circuit templates are used to draw patterns of each component. These templates are made in various scales such as 2:1 and 4:1. The patterns are cut out and placed on the circuit board drawing. They are rearranged until an acceptable layout is achieved. Fig. 23-29. A knowledge of electronics is necessary to do this. Once a solution has been reached, a drawing is made that accurately locates each component. A finished component and conductor drawing is shown in Fig. 23-30.

Place a sheet of plastic drafting film over the component drawing. First, adhere all the pads and outlets for components. Then place the tape over the lines representing the conductors. Keep the line approximately in the center of the tape. Be certain the tape is pressed firmly to the drafting film. Avoid bulges and irregular places. Fig. 23-31. Then add border delineation marks, grid locators used for alignment, and artwork reduction dimensions. This produces a finished master drawing. Fig. 23-32.

The art is photographed to make the negative for exposing the image to the foil on the back of the printed circuit board. The negative produced is a mirror image of the art. Fig. 23-33. This is done because the artwork is drawn from the component side of the board. The pads and conductors will actually be printed on the back. The photographic negative made of this art will be reversed when exposed to the foil side (back) of the circuit board.

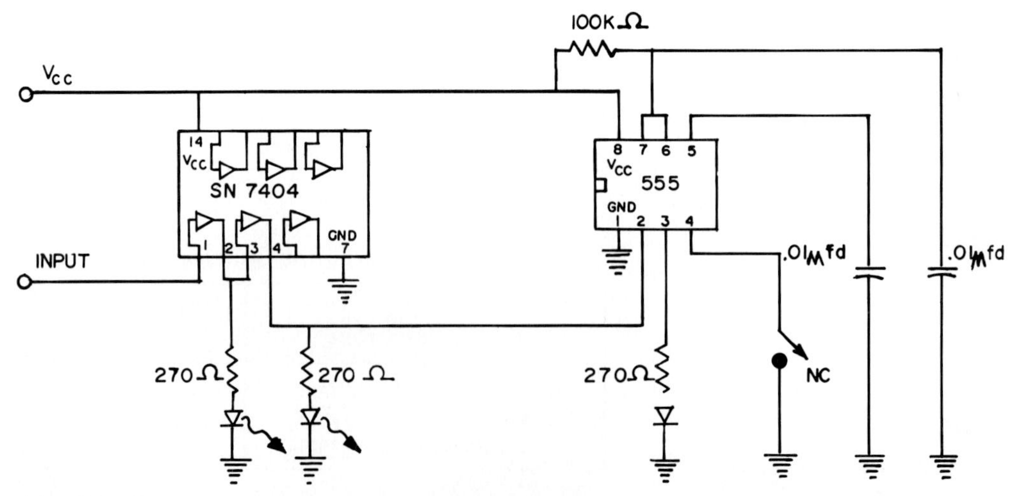

Fig. 23–27. The schematic of a logic circuit test probe.

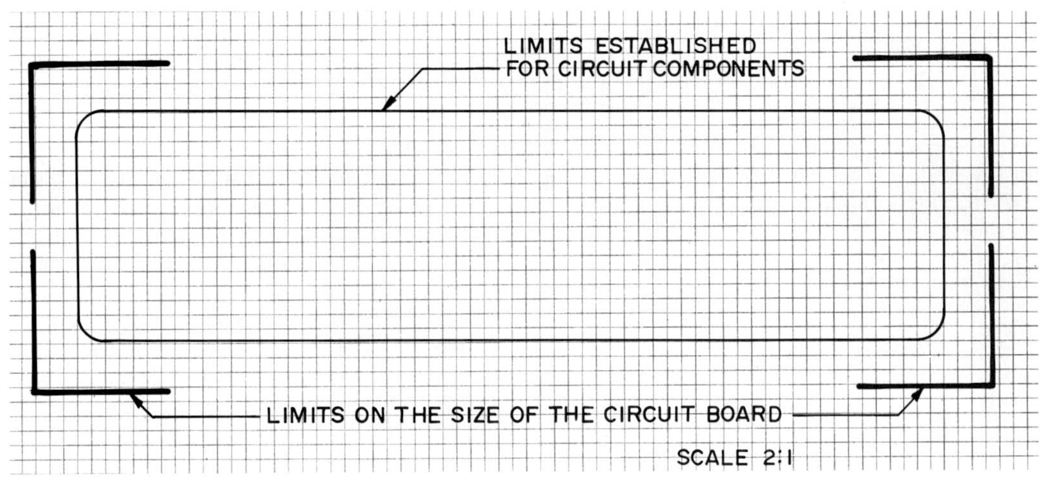

Fig. 23–28. Lay out the limits of the available area for the circuits and components, and set the maximum size for the circuit board.

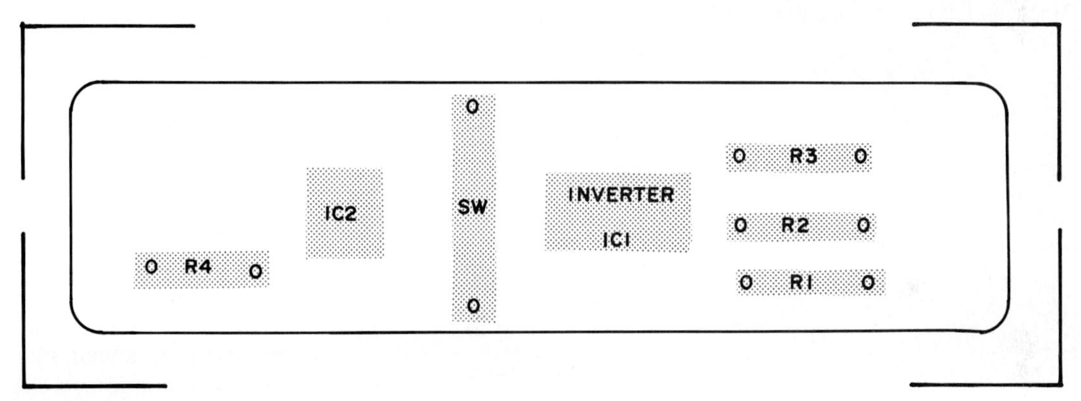

Fig. 23–29. Make a trial layout by cutting out templates of the components and arranging them on the sheet.

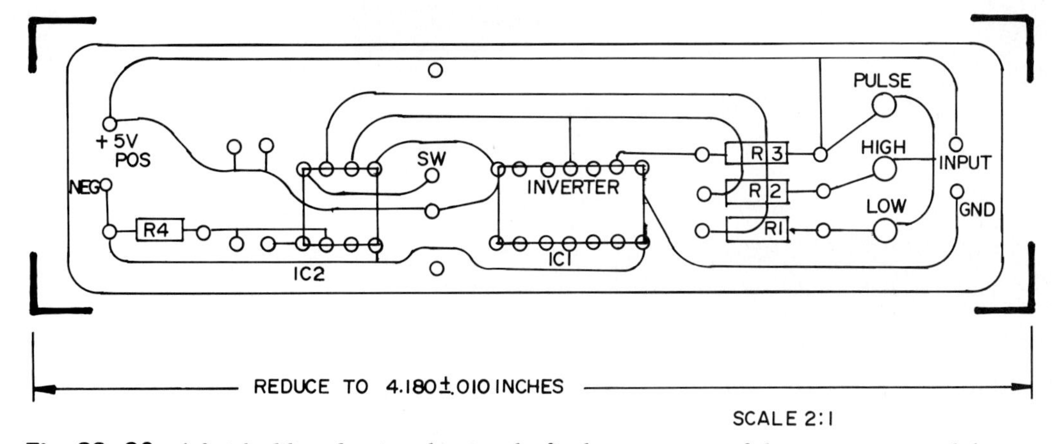

Fig. 23–30. A finished line drawing showing the final arrangement of the components and the connecting conductors. The printed circuit is made from this drawing.

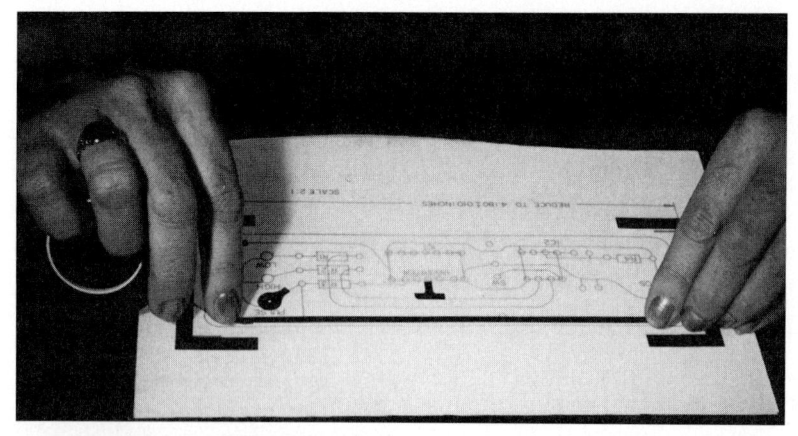

Applying tape to form conductors.

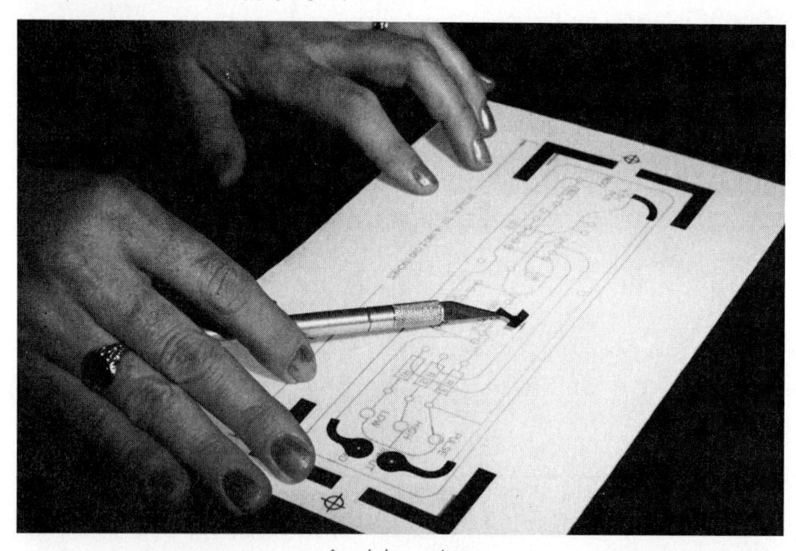

Applying a tee.

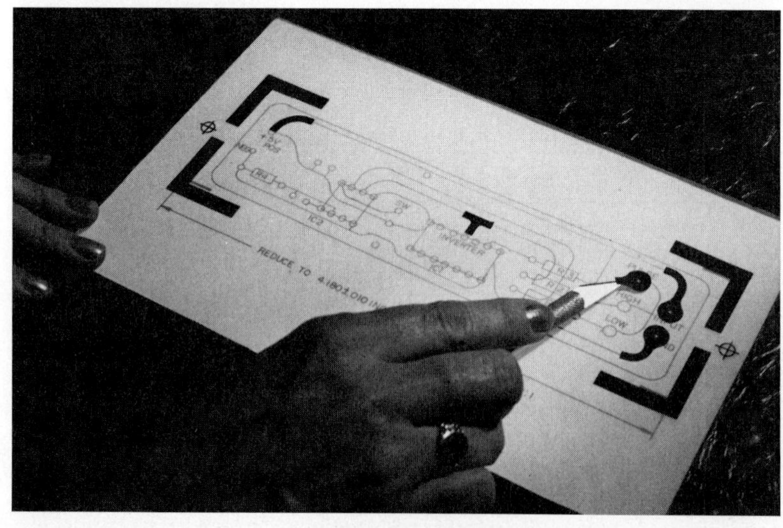

Applying a pad.

Fig. 23–31. Applying the adhesive-backed pads and tape.

A close-up of the foil side of the etched circuit board is seen in Fig. 23-34. The board is shown installed in its plastic case in Fig. 23-35.

Computer-aided drafting systems are excellent for drawing printed circuit master layouts. The basic information such as pad sizes and conductor widths can be stored in memory. The designer can place these parts where needed on the design and review the layout on the screen of the monitor. The design can be easily rearranged. Fig. 23-36. Once the layout is acceptable, it can be printed in ink on the computer plotter.

Manufacturing the Printed Circuit Board

The printed circuit is made on specially laminated paper, phenolic, or lucite. These boards are coated on one side or both sides with a thin layer of conducting material called *foil.* Fig. 23-37 at A. The boards are available in thicknesses from 1/32 to 1/4 inches (0.794 to 6.350 mm). The foil is usually copper, aluminum, copper-clad aluminum, or brass. The standard thicknesses of the foil are 0.00135, 0.0027, and 0.0040 inches (0.035, 0.069, and 0.102 mm).

From the master drawing the pads, elbows, tees, conductors, and other parts are reproduced directly on the foil on the circuit board. This is usually done by offset printing, photoengraving, or silk screen. Fig. 23-37 at B. The circuit is reproduced on the board with an acid-resistant coating. The board is placed in a solution which etches away the foil not covered with the acid-resistant coating. See Fig. 23-37 at C. All areas covered remain, thus forming a conductor. Fig. 23-37 at D.

After the board is etched, the necessary holes are drilled to receive the components. The components are set in place and soldered. The entire board can then be coated with a protective layer to keep it free from dust and moisture.

Electrical Drawings

Some of the common types of electrical drawings are (1) point-to-point, (2) highway, (3) baseline, and (4) single-line and construction drawings.

A **point-to-point diagram** shows the components pictorially as they appear to someone looking at them. They are connected with lines to represent the conductors. Fig. 23-38. This drawing is used to give assembly information to workers. Each conductor runs from a terminal on one component to the connecting terminal on another component. Each conductor is drawn entirely separate from the others.

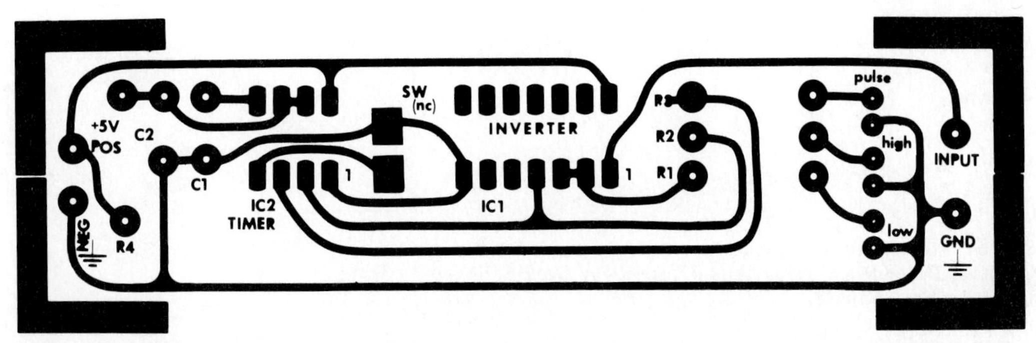

Fig. 23–32. **The finished master drawing of the printed circuit board. (Courtesy Dr. William Studyvin and Gregory A. Talkin, Pittsburg State University, Pittsburg, Kansas)**

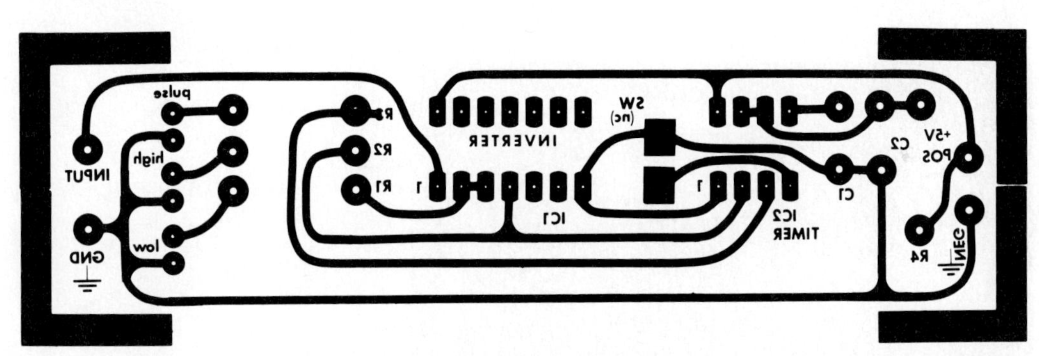

Fig. 23–33. **When the art is exposed to make a negative, it is reversed because it will be reproduced on the back of the board. (Courtesy Dr. William Studyvin and Gregory A. Talkin, Pittsburg State University, Pittsburg, Kansas)**

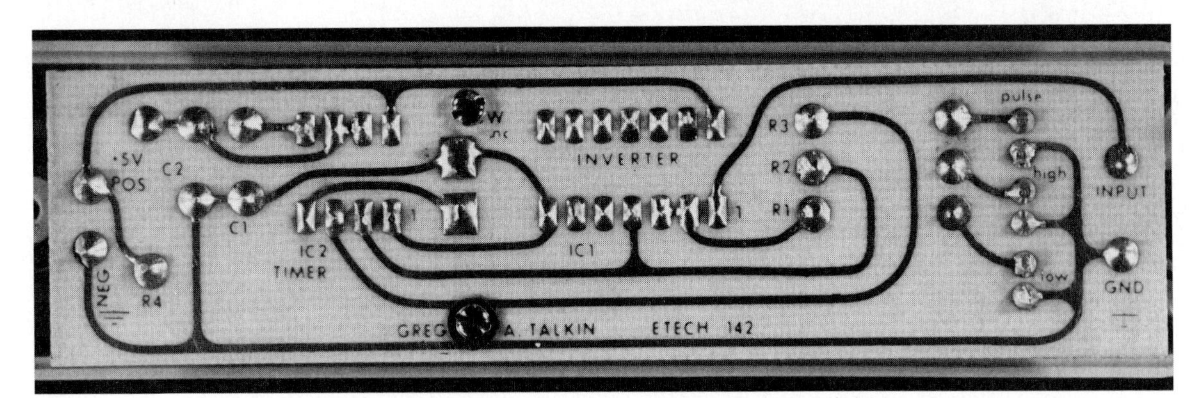

Fig. 23–34. **A close-up view of the finished circuit board. (Courtesy Dr. William Studyvin and Gregory A. Talkin, Pittsburg State University, Pittsburg, Kansas)**

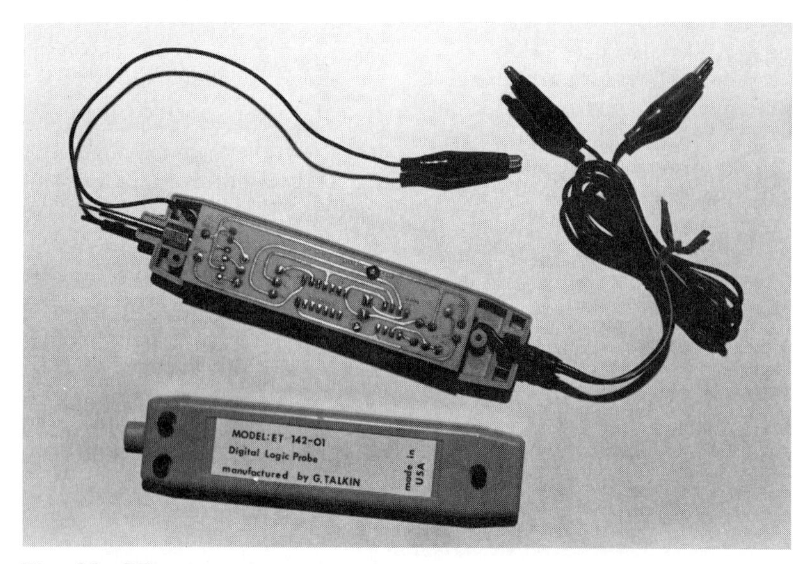

Fig. 23–35. The circuit board installed in the logic circuit test probe case. (Courtesy Dr. William Studyvin and Gregory A. Talkin, Pittsburg State University, Pittsburg, Kansas)

Fig. 23–36. The printed circuit master layout is displayed on the screen. (Hewlett-Packard Company)

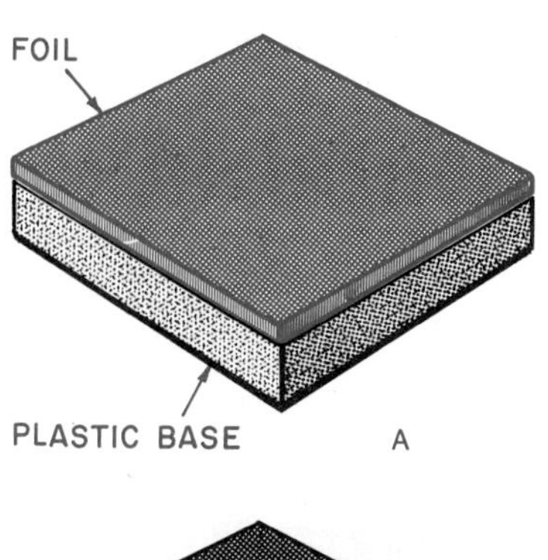

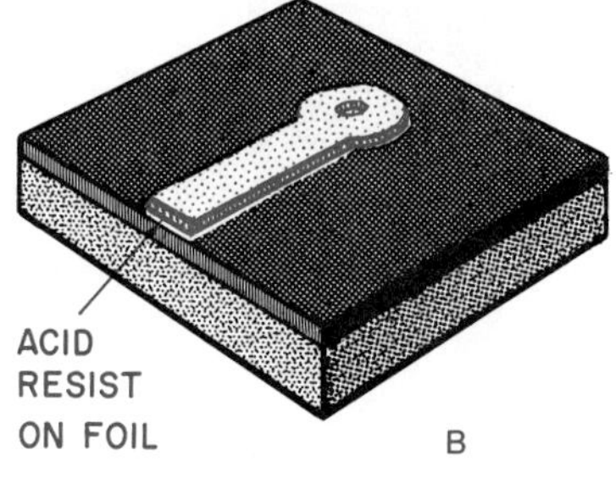

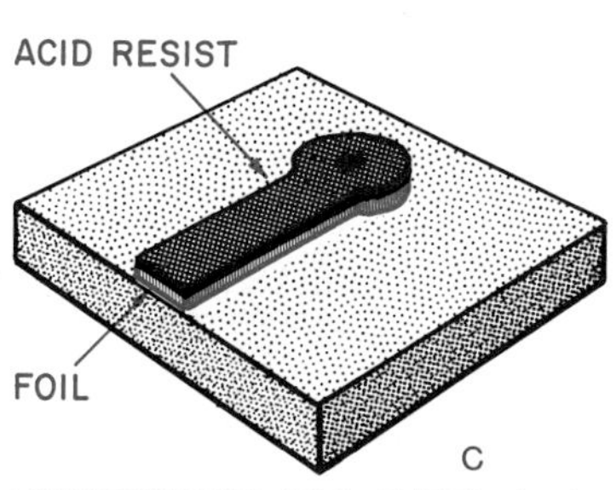

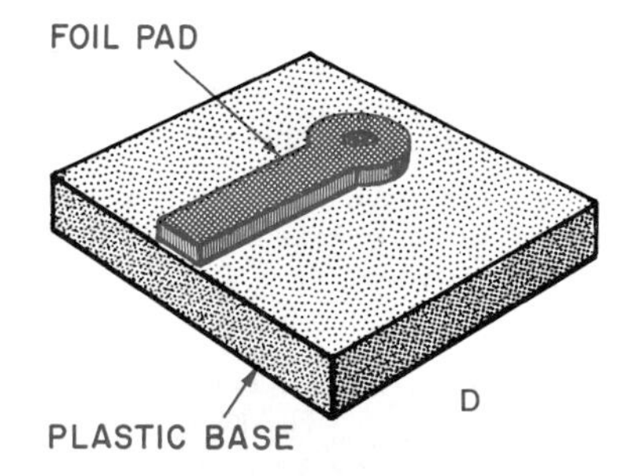

Fig. 23–37. Steps in making a printed circuit board.

The **highway diagram** is much like the point-to-point diagram. The components are drawn pictorially rather than with symbols. The main difference is that on a highway diagram all the conductors leading from one component to another are joined into a single line, called a *highway*. They are separated when they reach the component to which each conductor is connected. In Fig. 23-39 find the conductors 12 Wht-D1, 12 Blk-D2, and 12 Red-D3. Notice how they join to form a highway and are separated when they reach the motor. This lets each conductor be connected to the motor at the proper terminal. It is not necessary to draw three long conductors running parallel to the same component. Highways simplify a drawing. Notice that the conductors are drawn curved whenever they connect to a highway.

Each conductor on a highway drawing must be carefully labeled so that it can be easily followed. Read the paragraphs following the description of baseline drawings for directions on how to read the lead codes.

Baseline drawings are much like highway drawings. They are built around a horizontal or vertical line, called the *baseline*. This is an imaginary line, not an actual cable. It is used to simplify the drawing. Lead lines from the components run to the baseline. They meet and leave it at right angles. Examine the baseline drawing, Fig. 23-40. Find component B. Notice one lead is 12 Wht-D1. This lead leaves component B and connects with the baseline. It leaves the baseline and connects to terminal 1 on component D.

Identification of each component, terminal, and lead is vital to all types of electrical drawings. In Fig. 23-38 the wire size and color are given for each lead. Since the leads all run directly to the terminal on each component, they are not numbered.

Highway and baseline drawings must have complete identification since each lead does not go directly from one component to another.

One system is to label each component with a large letter. Each terminal is numbered. The leads are identified by color and size. For example, in Fig. 23-39 the motor shown has three leads. One of these is 12 Red-B1. This means the lead is wire size 12, red in color, goes to component B, and connects with terminal 1.

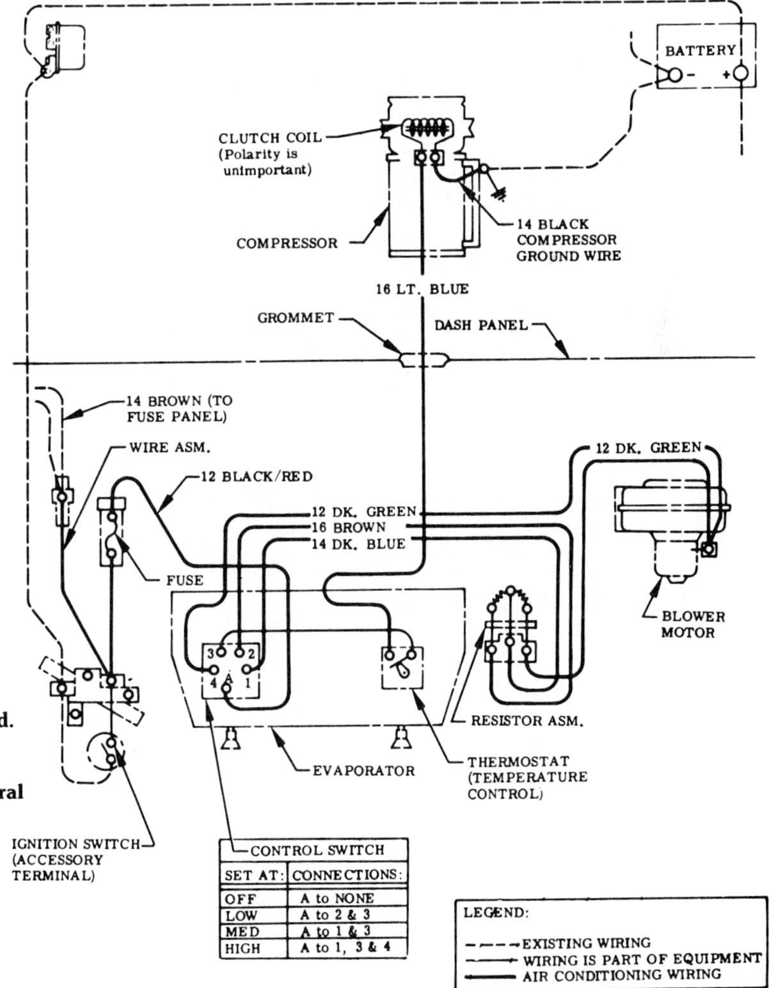

SET AT:	CONNECTIONS:
OFF	A to NONE
LOW	A to 2 & 3
MED	A to 1 & 3
HIGH	A to 1, 3 & 4

Fig. 23–38. Automotive schematic with leads color-coded. This is a form of point-to-point pictorial production drawing. (Chevrolet Motor Division, General Motors, Corp.)

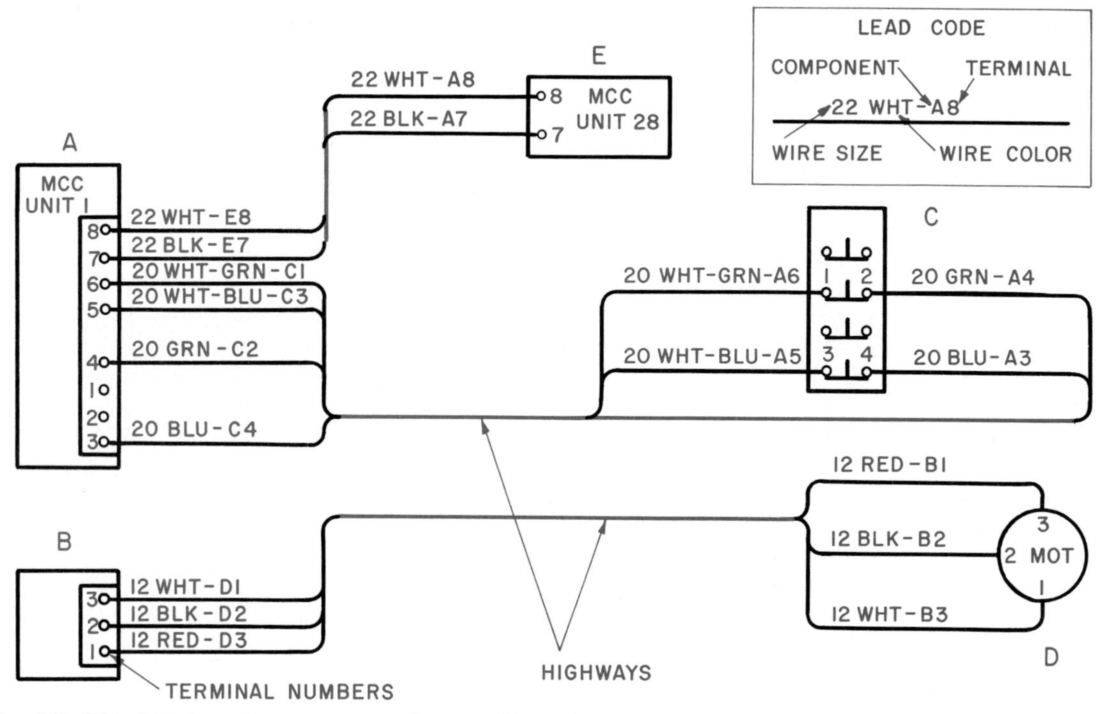

Fig. 23–39. A highway line connection diagram. Notice how several conductors are merged into one highway and are then divided when they reach their component. (National Aeronautics and Space Administration)

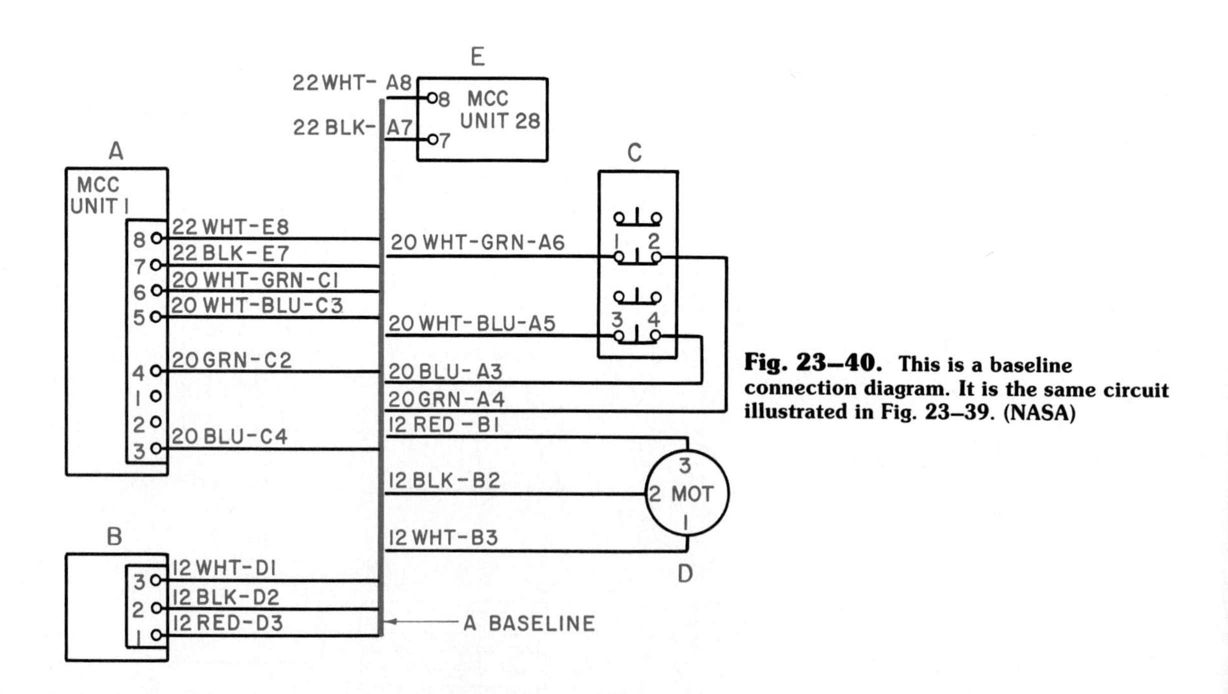

Fig. 23–40. This is a baseline connection diagram. It is the same circuit illustrated in Fig. 23–39. (NASA)

Examine the lead identification at component B, terminal 1. It reads 12 Red-D3. This means it is number 12 wire, red in color, and connects with component D (the motor) at terminal 3.

A **single-line diagram** shows in simple form the major pieces of equipment for a complete project, such as an electrical power substation. It includes information about each major component, such as voltages of potential transformers, ampere rating of current transformers, interrupting capacity and trip ratings of circuit breakers, and motor horsepower ratings. A single-line diagram can include wire and cable information and descriptions of the major components.

Symbols are used to represent the components. The single line shows the order of connection. It can represent several conductors in a cable.

A typical single-line diagram is shown in Fig. 23-41. When drawing line diagrams, the main circuits are usually shown in the most direct path. They are placed in the order in which they occur in the circuit. The single line connecting the symbols runs in either a vertical or horizontal direction. Spacing is the same as that on schematics. Enough background is needed to avoid crowding the symbols. Special attention must be given to allow room for notes. Notes are an important part of a single-line diagram. They contain information about types, ratings, and other data related to the parts of the circuit.

Construction wiring diagrams show electrical systems of buildings. Items shown include lights, outlets, switches, telephones, and special features such as electric dryer outlets. These are shown on the drawing with symbols. See Chapter 25 for symbols and wiring diagrams. The symbols are placed on the drawing wherever an electrical feature is wanted.

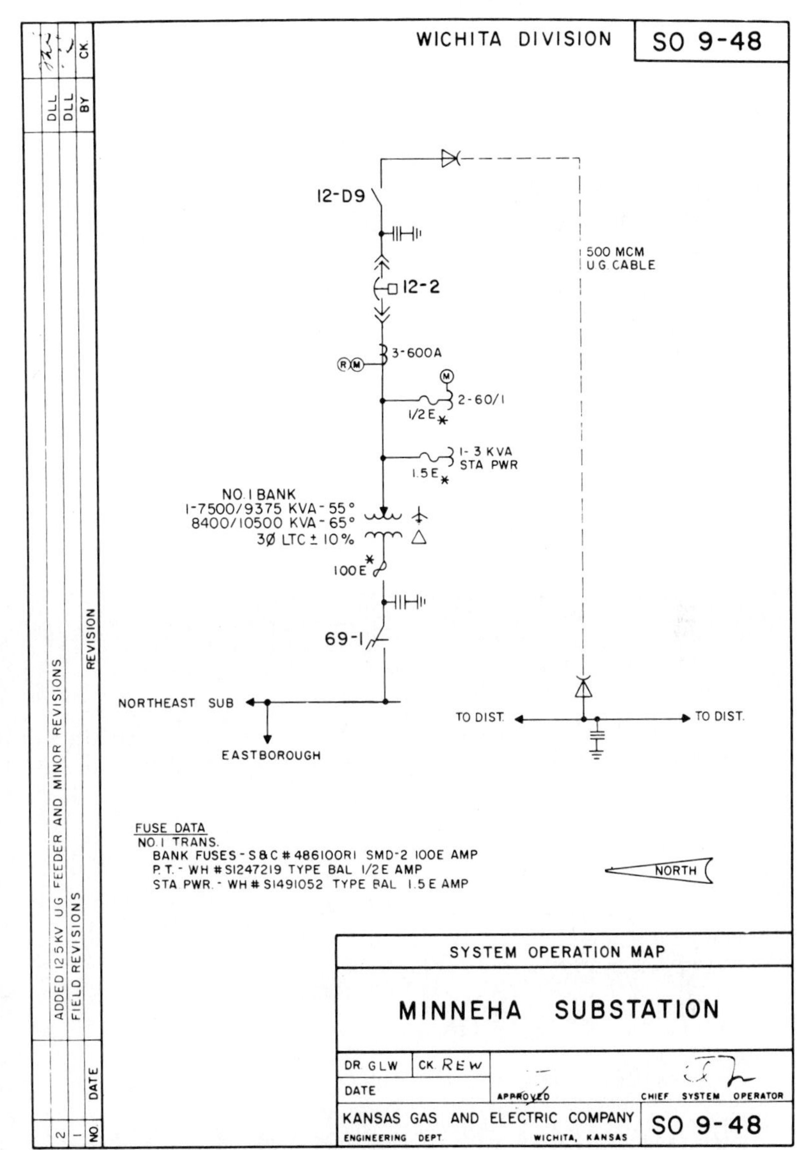

Fig. 23—41. A single-line drawing of a substation. Reading from the bottom left, the units are an arc resisting switch (69—1), lightning arrester and ground, fused disconnect switch, power transformers, two nondisconnecting fuses and current transformers, three current transformers at 600 amperes each, electrically operated air circuit breaker (12-2), lightning arrester and ground, disconnect switches (12-D9), and a cable termination to which an underground cable is connected. (Kansas Gas and Electric Co.)

Light fixtures are connected to switches with dashed lines. The types of fixtures are identified by letters on the drawing and referenced to a table.

Drawings for the Electrical Power Field

Many different types of drawings are used in the electrical power field.

A typical small substation is shown in Fig. 23-42.

When a small substation is in the planning stage, a single-line diagram is one of the first drawings made. Fig. 23-41 shows such a drawing. It gives the overall plan for the substation. Notice that it is located on the building site by a reference to the north.

The details of the substation are made up in the form of very detailed schematic diagrams. Fig. 23-43 shows a portion of such a drawing. This is a control circuit for opening a high voltage circuit breaker in the substation. The symbols used are in most cases like those in the standards.

Many other detail drawings are necessary. Fig. 23-44 shows an elevation of the towers and the power unit in place.

Fig. 23–42. A unit substation in which the transformer and control circuits are contained in a single cabinet. This type of installation is used in rural and industrial areas. (Kansas Gas and Electric Co.)

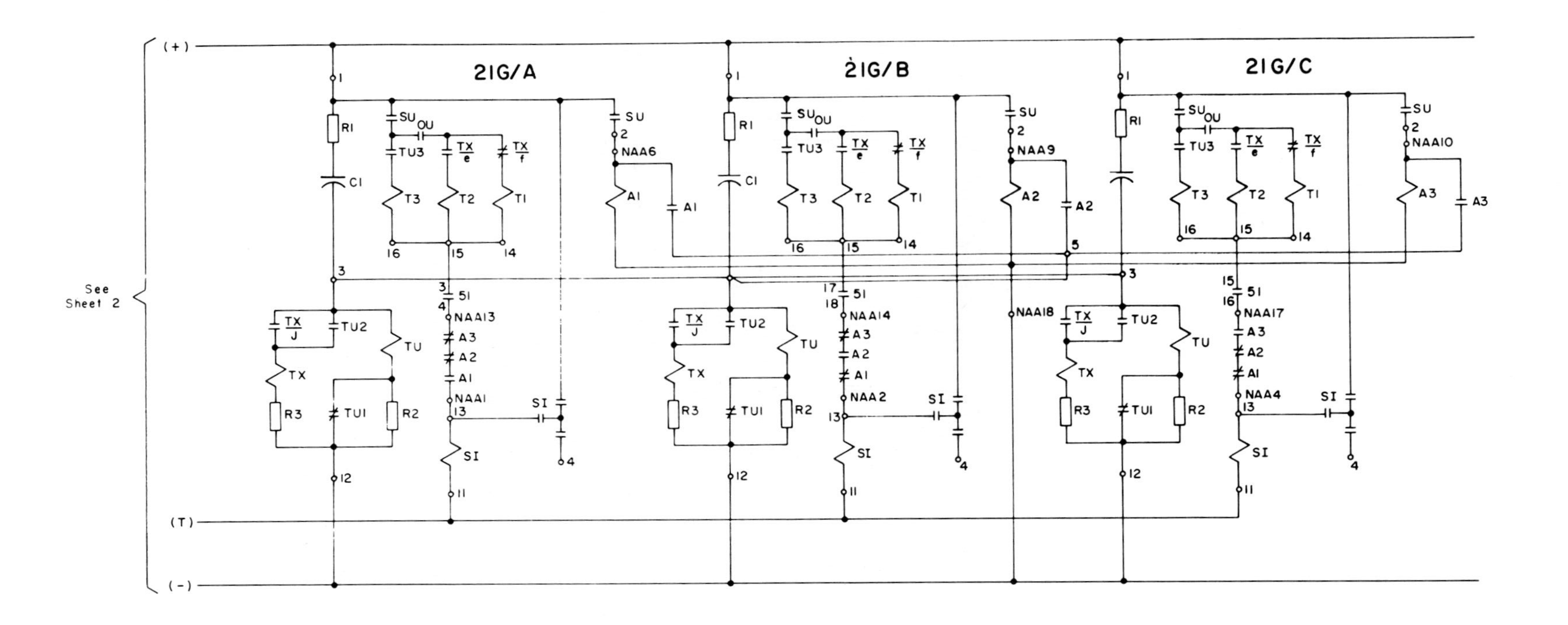

69-6 TRIP CIRCUIT

GCX15 GROUND RELAYS

Fig. 23–43. A schematic of part of a substation. This is a control circuit for a high-voltage circuit breaker. (Kansas Gas and Electric Co.)

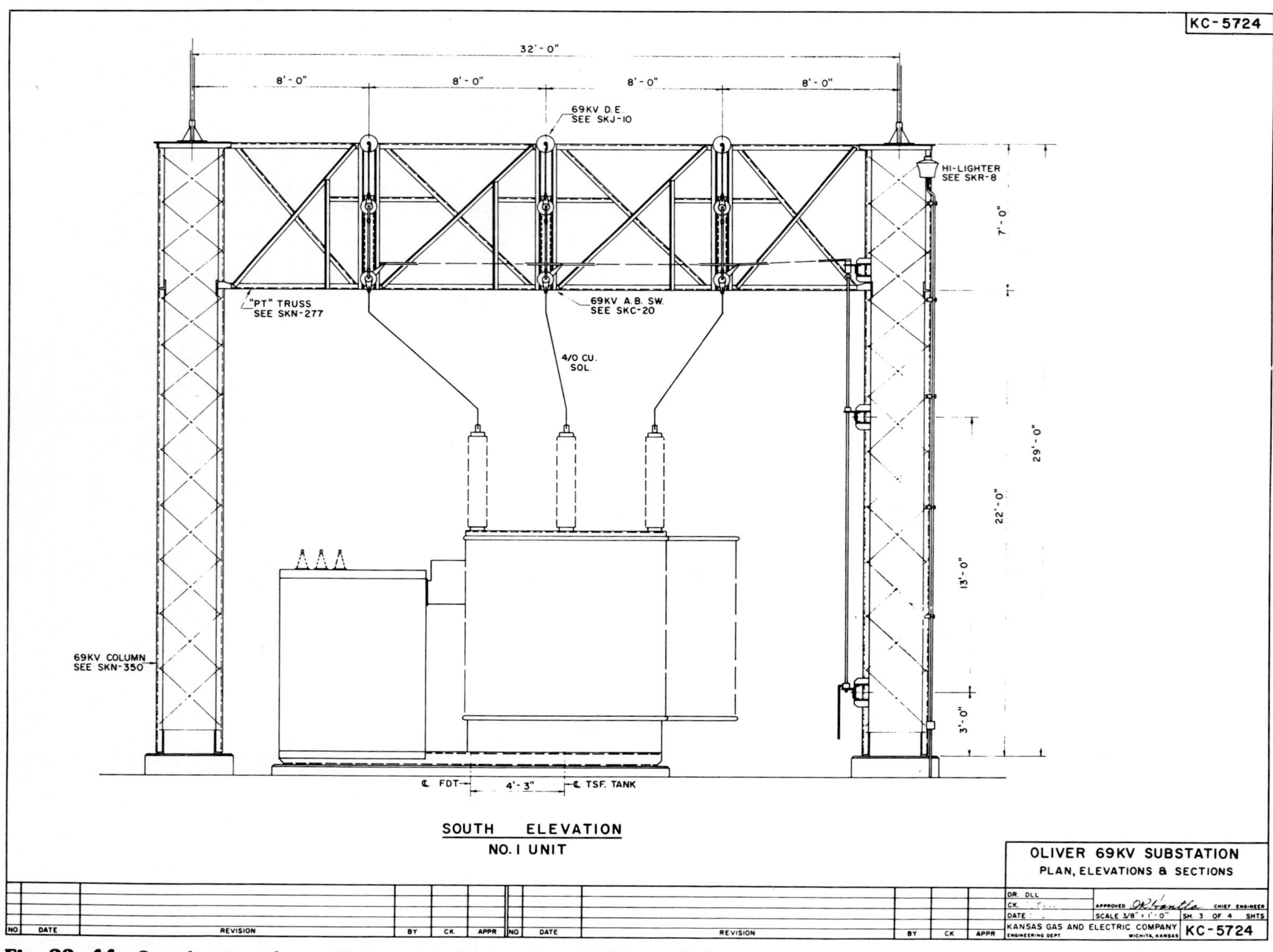

Fig. 23–44. One elevation of a small substation. (Kansas Gas and Electric Co.)

Chapter Review

Build Your Vocabulary

You should understand and use the following terms as part of your working vocabulary. Write a brief explanation of what each means.

schematic diagrams
component
electric current
volt
ohm
ampere
watt
block diagram
printed circuit
point-to-point diagram
highway diagram
baseline drawing
single-line diagram

Sharpen Your Math Skills

1. A watt equals amperage × voltage. If a device requires 11 amps at 120 volts, how many watts are consumed?
2. Amperes equal watts divided by volts. If a device requires 240 volts and consumes 1000 watts, how much amperage is needed? Round your answer to 2 decimal places.
3. Voltage required equals watts divided by amperage. If a device consumes 1800 watts at 15 amperes, what is the required voltage?
4. A horsepower is equal to 746 watts. If the device in Problem 3 is an electric motor, what is the horsepower? Round your answer to 2 decimal places.
5. If a finished printed circuit board is to measure 1.25″ × 1.75″, and you draw it at a scale of 4:1, what size is the drawing?

Study Problems—Directions

The problems that follow will give you the chance to use some of the drafting knowledge that you learned in this chapter. Do the problems that are assigined to you by your instructor.

Block Diagrams

1. Study the connections shown on the pictorial drawing of a part of an auto electrical system, **Fig. P23-1**. Draw a block diagram showing the system.
2. Make a block diagram of the pictorial drawing of the turntable, preamplifier, amplifier, and speaker, **Fig. P23-2**.

Reading Electronic Schematics

Following are some schematics used to illustrate solutions to typical design problems. Study these and answer the questions about each drawing.

3. The receiver, **Fig. P23-3**, uses a single audio transistor and a diode detector, which enables it to be built in a unit smaller than a cigarette package. Study the diagram and answer the following questions:
   a. What is the power source?
   b. What is used to enable a person to hear the signal?
   c. What type of switch is specified?
   d. How many resistors are used?
   e. Does it use tubes or transistors?
   f. What is the tube or transistor specification number?
   g. How many capacitors are used?
   h. What is the diode specification number?
   i. What type of transformer is used?
   j. What is the rating of the variable capacitor?

4. The transmitter, shown in **Fig. P23-4** is an 80-meter crystal oscillator-type CW transmitter. Study the diagram and answer the following questions:
   a. How many transistors are needed to build this transmitter?
   b. What type of transistors are used?
   c. How many resistors are in the circuit?
   d. Where is it grounded?
   e. What type of tuner is used?
   f. What voltage is required for operation?
   g. What device is used to send the signal?
   h. How many capacitors are used?
   i. How many chokes are used?
   j. What type of crystal is used?

Symbol Drawing

Draw the following schematics. Enlarge them so that normal size symbols can be used.

5. Make a freehand sketch of the automatic safety flasher, **Fig. P23-5**. Complete it by inserting the proper symbols.

6. Make an instrument drawing of the schematic of the battery charger, **Fig. P23-6**, and complete it by inserting the proper symbols. Next to each symbol give its value.

7. Make an instrument drawing of the schematic of the S meter, **Fig. P23-7**, and complete it by inserting the proper symbols.

Planning Schematic Diagrams

The schematics in the following problems are correctly designed. They have been purposely drawn in a poor arrangement. There is no need to redesign the plan of circuitry. The assignments do require that the indicated connections and components be drawn to show more clearly the schematic and present a pleasing, balanced drawing.

8. Revise the schematic of the constant current generator, **Fig. P23-8**. The placement of the components on the diagram is poor and should be improved.

9. Revise the schematic of the TV control box, **Fig. P23-9**. The placement of the components on the diagram is poor and should be improved.

10. Revise the schematic for the light amplifier, **Fig. P23-10**. This device increases the light intensity. Use is made of a solar battery to convert light energy to electrical energy. The electrical energy is amplified by the device and used to furnish power for a light bulb.

Draw the proper component symbols and rearrange the diagram so it is pleasing in appearance. Place the components so they are in a better relationship.

11. Revise the schematic of the class A audio power amplifier, **Fig. P23-11**. Rearrange the diagram so it is pleasing in appearance and the components are in a better relationship.

Printed Circuit Drawings

12. Study the schematic for the volter, **Fig. P23-12**. It consists of two resistors (Ø2 mm, length 5 mm), a penlight battery cell (Ø13 mm, length 50 mm), and a battery holder. An assembly drawing gives the size of the circuit board and the location of the components on the board. Design a master layout for a printed circuit for this unit. Draw it at a 4:1 scale.

13. Study the schematic for the code practice oscillator, **Fig. P23-13**. Notice the required connections. A component assembly drawing is given. It is drawn on a grid of 0.10. Make a master layout printed circuit. The proper connections are shown with solid lines. Rearrange it so there are no crossovers. Draw it at a 4:1 scale.

14. Make a master layout printed circuit board drawing from the schematic diagram of the microphone preamplifier, **Fig. P23-14**. The outside dimensions of the finished board are 1¼ by 2 inches.

Industrial Electricity Drawing

15. A small portion of a schematic for a business office machine is shown in **Fig. P23-15**. Simplify by redrawing it as a baseline diagram. Identify each unit and lead, using a standard system.

16. Simplify the partial schematic of an industrial control, **Fig. P23-16**. Make it a highway diagram. Identify each component, terminal, and lead, using a standard system.

17. Draw a block diagram of the emergency-control system shown in **Fig. P23-17**. This system processes the collective-pitch pilot command, lateral-stick positions, angle of attack, and pitch-rate signals to generate a new collective-pitch signal that restores neutral lift and partial control of roll and collective pitch. Choose block sizes and spacing to produce a clear, neat layout. You may draw the flow lines in color if your instructor approves.

18. Draw the block diagram of the control circuit in **Fig. P23-18.** This circuit is for a high-power, brushless DC motor. The control circuit channels the flow of electrical energy among the power supply, motor, and load resistors. Choose block sizes and spacing to produce a clear, neat layout. You may draw the flow arrows with a colored pencil if your instructor approves. (NASA)

Study Problems

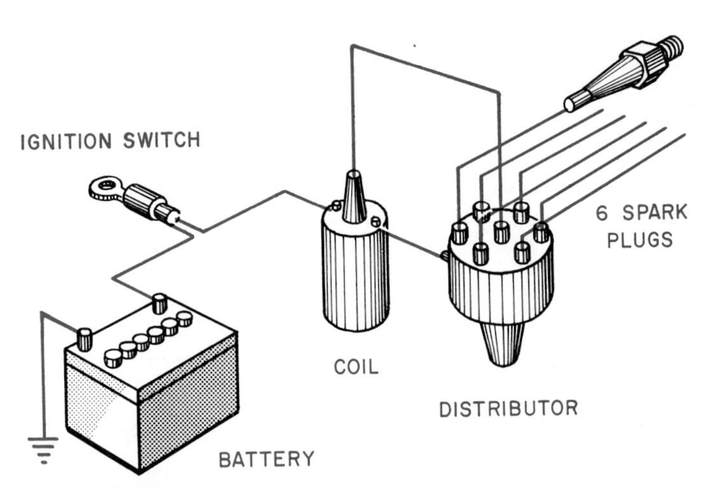

P23–1. Part of an auto electrical system.

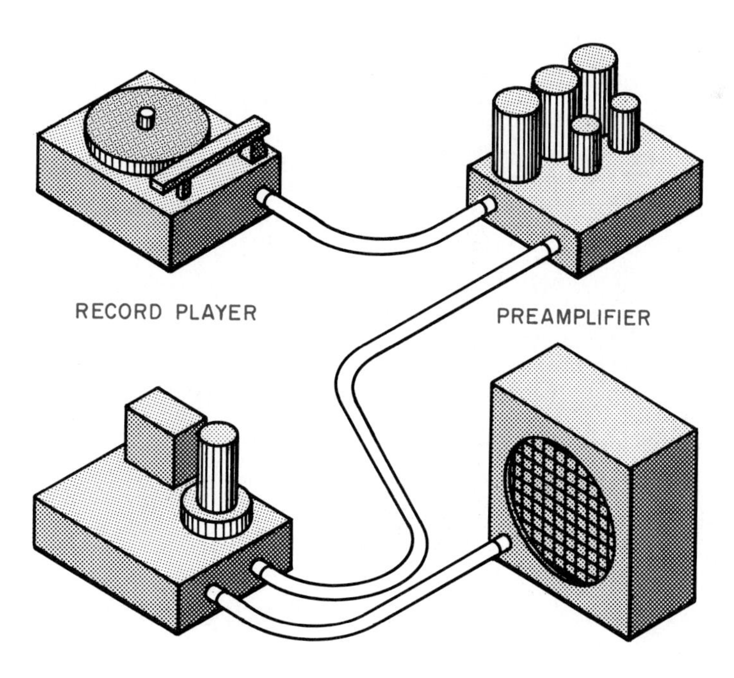

P23–2. Record amplifying system.

Study Problems

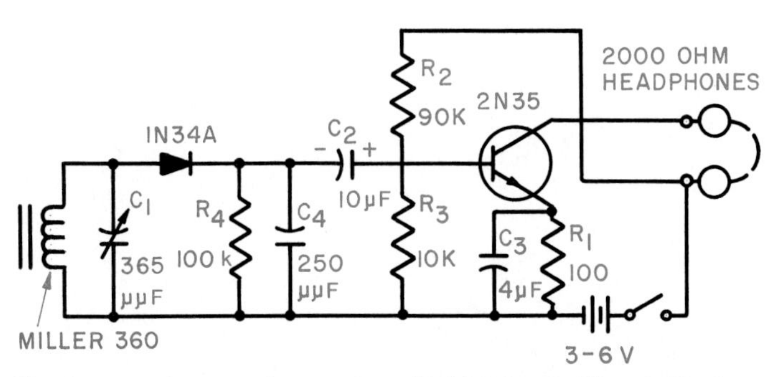

P23–3. **One-transistor pocket receiver. (GTE-Sylvania Electric Products, Inc.)**

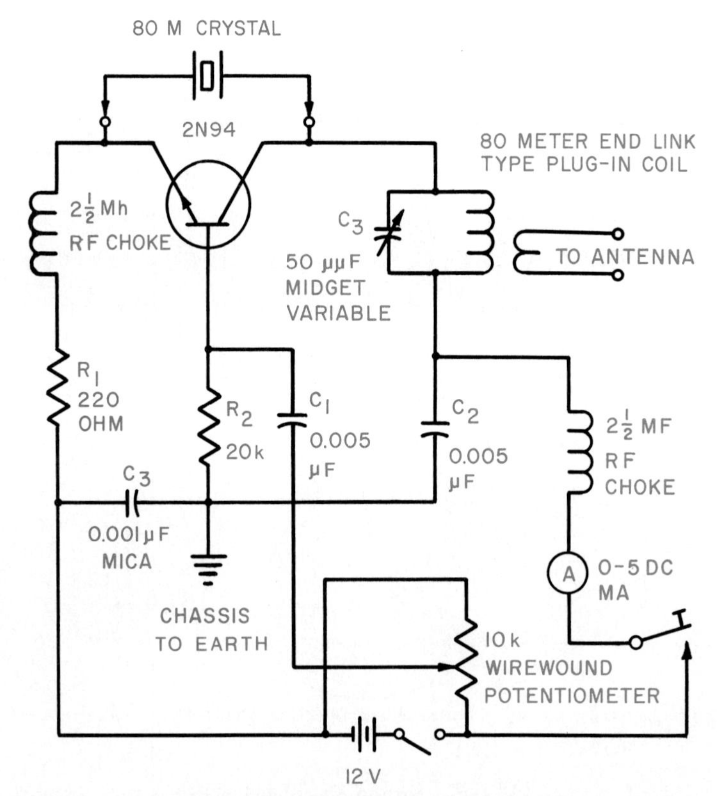

P23–4. **Novice CW transmitter. (GTE-Sylvania Electric Products, Inc.)**

Study Problems

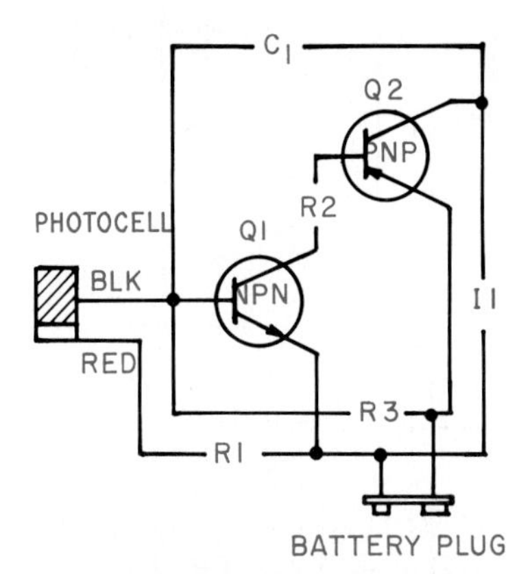

Parts List

C1—30 μf, 10 volt capacitor
I1—Pilot light

Q1—2N229 NPN transistor
Q2—2N187 PNP transistor
R1—1,200 ohm, ½ watt resistor
R2—470 ohm, ½ watt resistor
R3—47,000 ohm, ½ watt resistor

P23–5. Automatic safety flasher. *(Popular Electronics Electronic Experimenter's Handbook)*

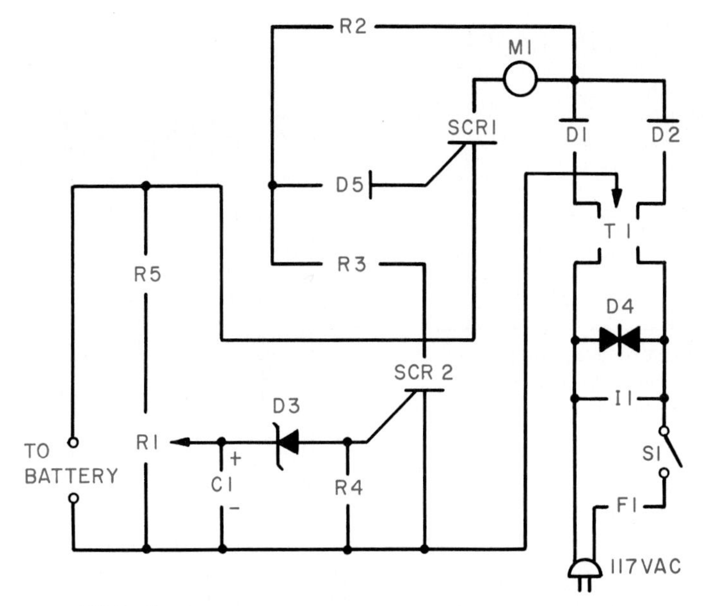

P23–6. Auto battery charger. *(Popular Electronics Electronic Experimenter's Handbook)*

Parts List

C1—100 μf, 25 volt capacitor
D1, D2—15 amp, 50 volt silicon rectifier
D3—8.2 volt, 1 watt zener diode
D4—Transient voltage suppressor
D5—100 volt, 600 MA silicon rectifier
F1—2 amp fuse
I1—120 volt neon indicator light
M1—0–10 amp meter
R1—500 ohm, 2 watt linear scale potentiometer
R2, R3—27 ohm, 3 watt resistor
R4—1,000 ohm, ½ watt resistor
R5—47 ohm, 2 watt resistor
S1—S.P.S.T. toggle switch
SCR1—Silicon controlled rectifier
SCR2—Silicon controlled rectifier
T1—Power transformer, primary 117V, secondary 24V

Study Problems

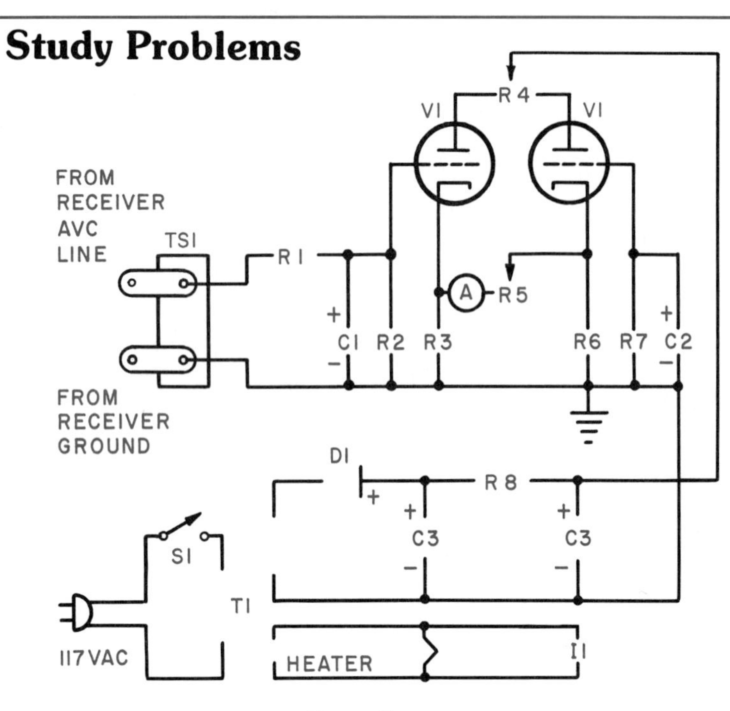

Parts List

C1, C2—0.0.1 μf., 150 volt capacitor
C3—Dual 20 μf, 150 w.v.d.c. capacitor
D1—1N 2069 diode
I1—6.3 volt pilot lamp
M1—2⅜" square illuminated S meter, 1 MA movement
R1—2.2 megohms
R2, R7—10 megohms, ½ watt resistor
R3, R6—3,300 ohms, ½ watt resistor
R4, R5—20,000 ohm potentiometer
R8—1,000 ohm, 5 watt resistor
S1—S.P.S.T. switch
T1—Power transformer: primary 117 volts, A.C., secondaries, 125 volts
TS1—2 lug terminal strip
V1—12 AU 7 tube

P23–7. S meter to measure signal strength of communications receivers. ***(Popular Electronics Electronic Experimenter's Handbook)***

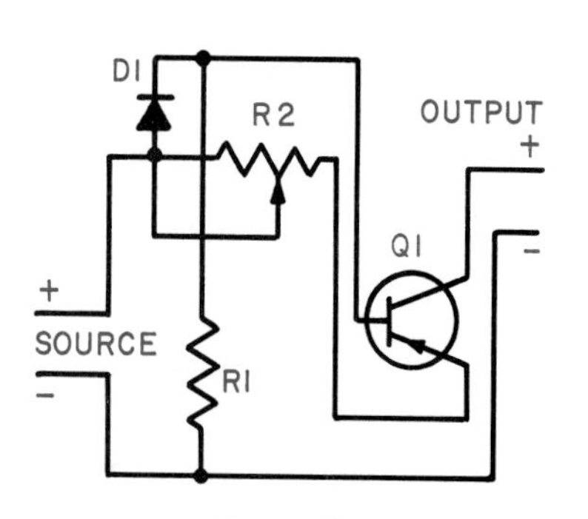

Parts List

D1—1N456 silicon diode
R1—1,000 ohm, ½ watt resistor
R2—1,000 ohm potentiometer
X1—PNP transistor 2N404

P23–8. Constant current generator circuit purposely arranged in a poor layout. (Howard W. Sams and Co., Inc.)

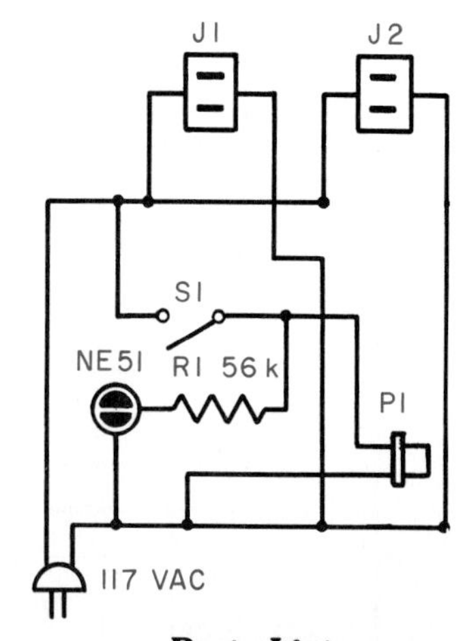

P23–9. Schematic diagram for a television control box used when servicing the set with the back removed. The box provides power to operate the set and two outlets for accessories, such as a soldering iron. The circuit is purposely arranged in a poor layout. ***(Popular Electronics Electronic Experimenter's Handbook)***

Parts List

J1, J2—Accessory sockets, A.C.
S1—Toggle switch
NE51—Indicator lamp
R1—56,000 ohm resistor
P1—Plug to fit connection on rear of television set

Study Problems

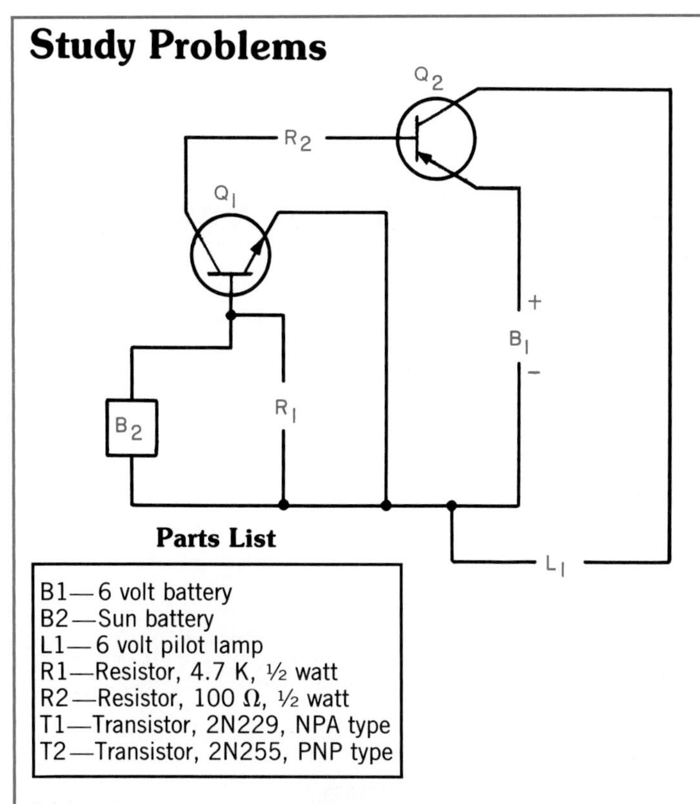

Q2
R2
Q1
+
B1
−
B2
R1
L1
Parts List
B1—6 volt battery
B2—Sun battery
L1—6 volt pilot lamp
R1—Resistor, 4.7 K, ½ watt
R2—Resistor, 100 Ω, ½ watt
T1—Transistor, 2N229, NPA type
T2—Transistor, 2N255, PNP type

P23–10. A light amplifier circuit purposely arranged in a poor layout. (GTE-Sylvania Electric Products, Inc.)

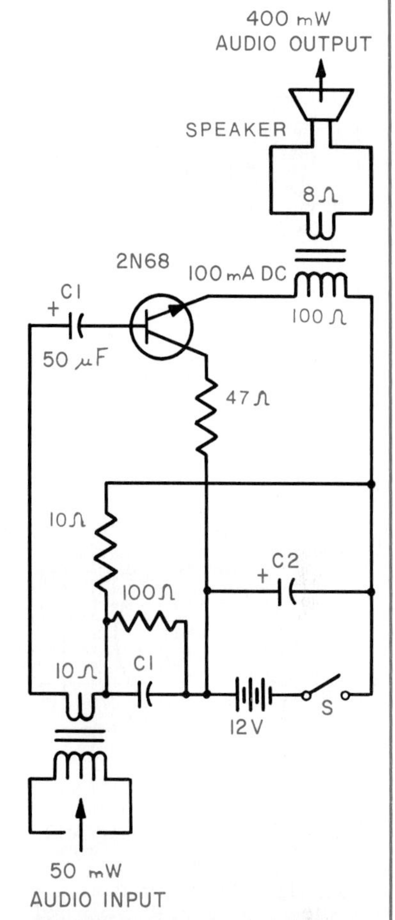

P23–11. Class A audio power amplifier purposely arranged in a poor layout. (GTE-Sylvania Electric Products, Inc.)

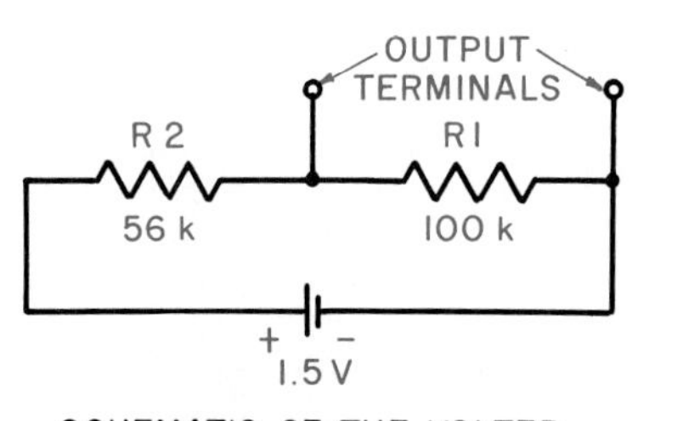

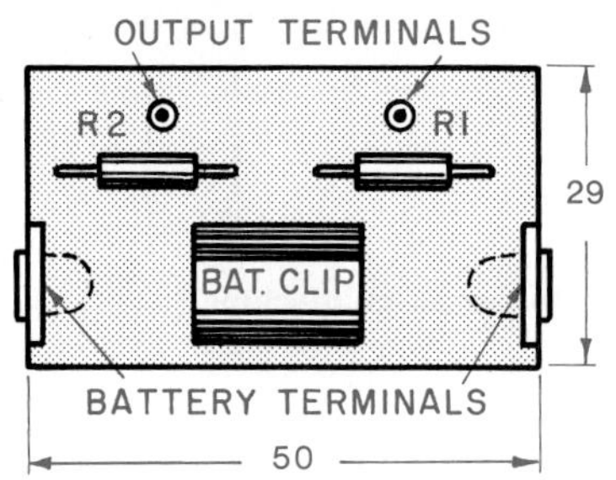

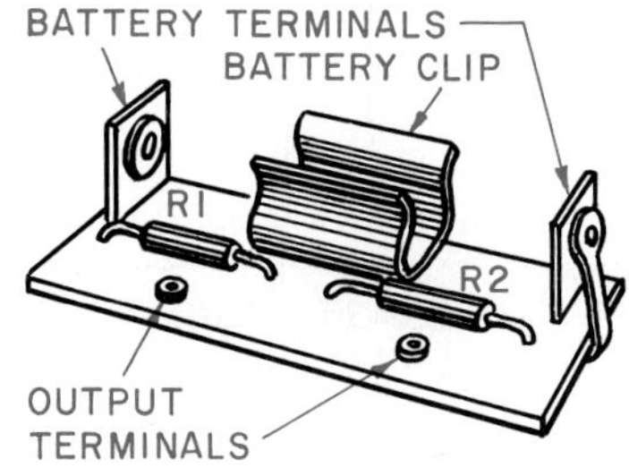

P23–12. Printed circuit problem. (Howard W. Sams and Co., Inc.)

Study Problems

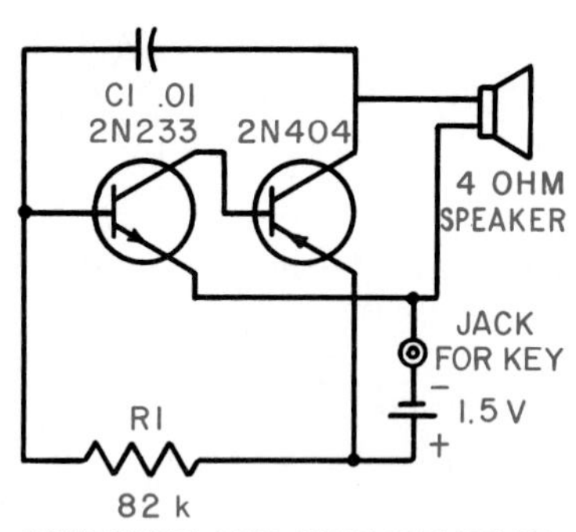

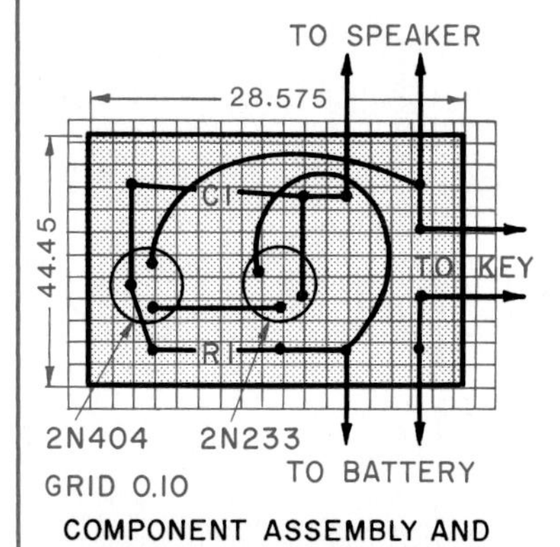

P23–13. A code practice oscillator. (Howard W. Sams and Co., Inc.)

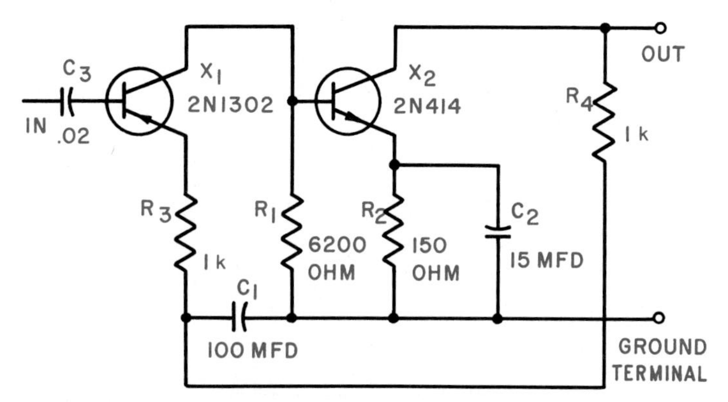

P23–14. Schematic of microphone preamplifier. (Howard W. Sams and Co., Inc.)

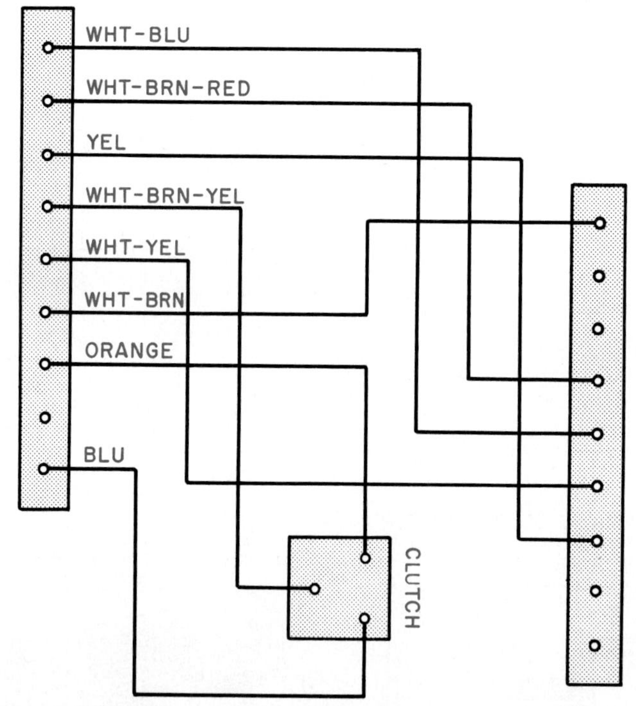

P23–15. Partial schematic for an office machine.

Study Problems

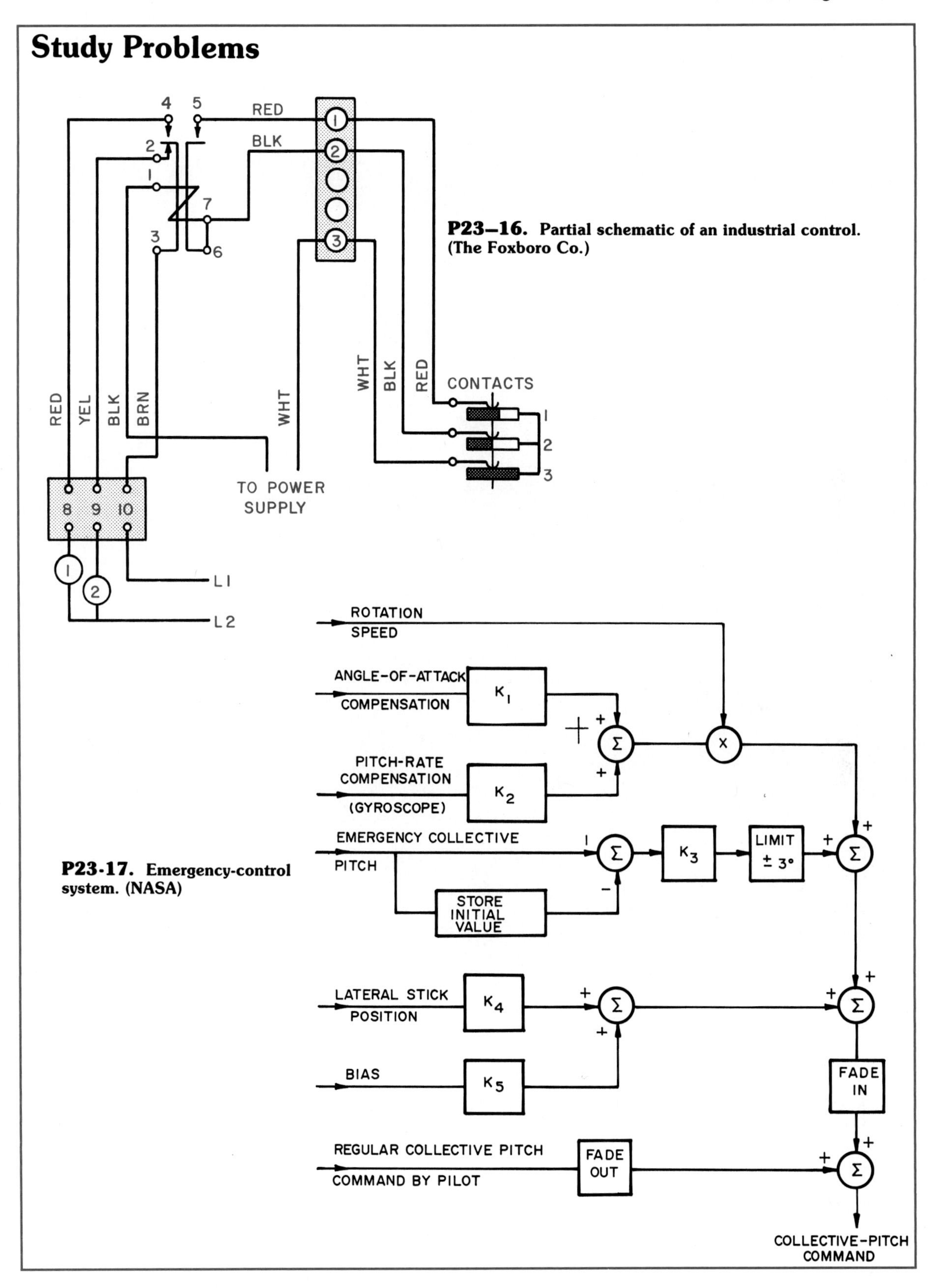

P23–16. Partial schematic of an industrial control. (The Foxboro Co.)

P23-17. Emergency-control system. (NASA)

Study Problems

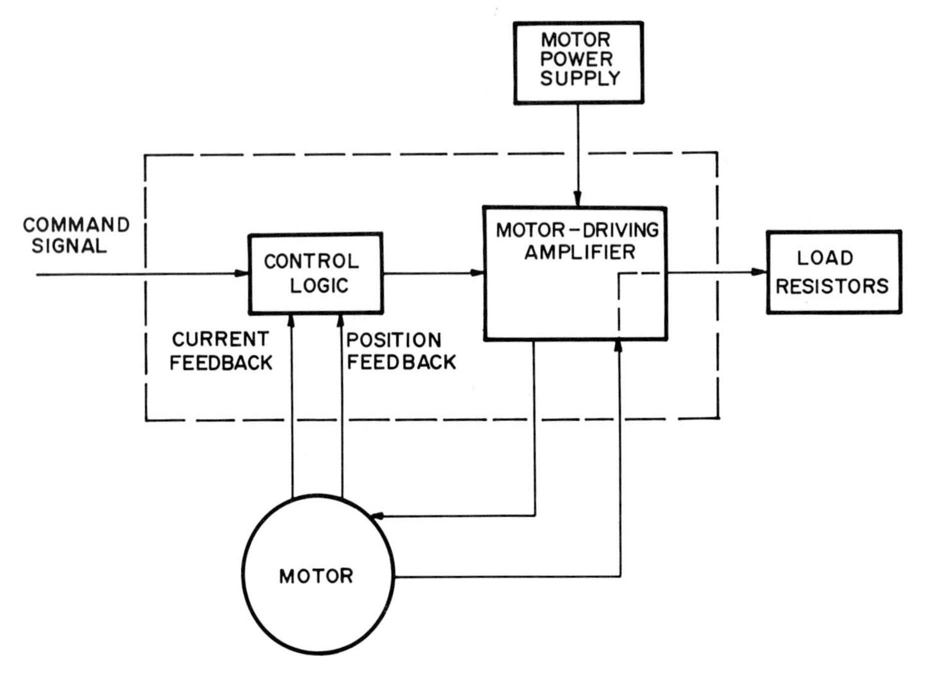

P23-18. **Control circuit. (NASA)**

CHAPTER 24
Mapping

After studying this chapter, you should be able to do the following:

- Recognize various types of maps.
- Be familiar with some of the symbols used on maps.
- Draw some of the simpler types of maps.

Introduction

There are many different kinds of maps, each serving a special purpose. Most maps show a portion of the surface of the earth. Maps are also made of other bodies, such as the moon, and the vastness of space is mapped. The features on maps vary a great deal, depending upon the purpose. A person who designs and draws maps is called a **cartographer**.

Geographers and astronomers make extensive use of maps. Surveyors and builders use land survey maps. Navigation and aeronautical maps have special features to aid navigators and pilots. Our space program has enabled us to map vast areas of space and to improve the accuracy of Earth maps. Fig. 24-1.

This chapter presents some of the more common types of maps, their features, and uses.

Computers in Mapping

Computer drafting equipment is very useful when producing maps of all kinds. It can be used to lay out the straight boundary lines of land surveys as well as plot the irregular curves of contour lines and letter the notes and dimensions. Fig. 24-2.

Scale

Maps are drawn to scale. **Scale** refers to the distance that one unit on the map represents. The unit may be inches or millimeters, depending upon which measurement system you are using to draw the map. The scale of most maps drawn in the United States is in inches. Unless noted otherwise, the scale of the maps in this chapter is in inches.

The scale of a map is usually expressed as a ratio, or fraction. Consider the scale 1:63,360. This means that one inch on the map equals one mile on the ground. The figure, 63,360, is the number of inches in a mile. This ratio, or fraction, is called the *natural scale*.

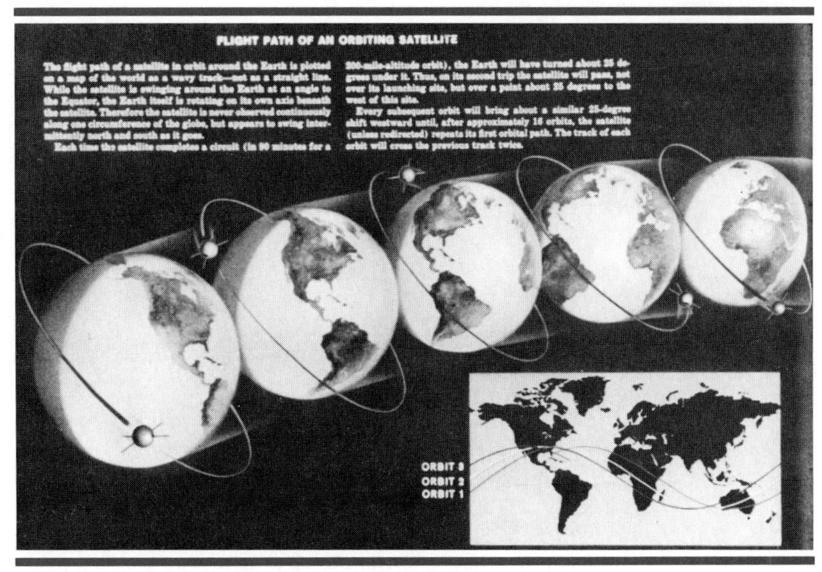

Fig. 24-1. The inset map shows the flight path of an orbiting satellite. (*Great World Atlas*, Reader's Digest Association)

The scale is sometimes shown by indicating the number of miles that are represented by one inch. For example, one inch to 39 miles means one inch on the map *represents* 39 miles. One inch may also represent kilometers (km). One kilometer equals 0.624 mile. One mile equals 1.6093 km.

Metric maps measure linear distances in meters (m) or kilometers. When land area is considered, the square kilometer (km^2) is used.

Most maps also show graphic scale. Fig. 24-3. This is a line drawn to the same scale as the map. It is divided into lengths that equal distances on the map so that straight line distances can be measured easily. Curves, such as those that represent a highway, can be measured by setting dividers to a small distance, such as one-half mile. This small distance can then be stepped along the curve and the number of them added to find the total distance. A special tool called a *cartometer* is used for this purpose also. Fig. 24-4.

Fig. 24–2. **Maps of all kinds can be laid out using computer drafting equipment. (California Computer Products, Inc.)**

Scale Size

Large scale maps are those having scales larger than four miles to the inch (1:250,000). Three miles to the inch is a larger scale than four miles to the inch and shows more detail.

Medium scale maps are those with scales from four to 16 miles to the inch. These scales are 1:250,000 to 1:1,000,000.

Small scale maps are those with more than 16 miles to the inch. Such maps can show only major features, such as rivers and mountains.

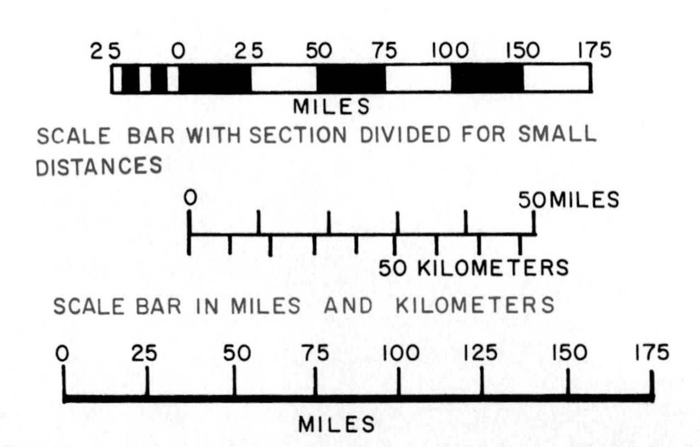

Fig. 24–3. **Typical examples of scale bars. You can mark the scale on a strip of paper and use it to measure distances on a map.**

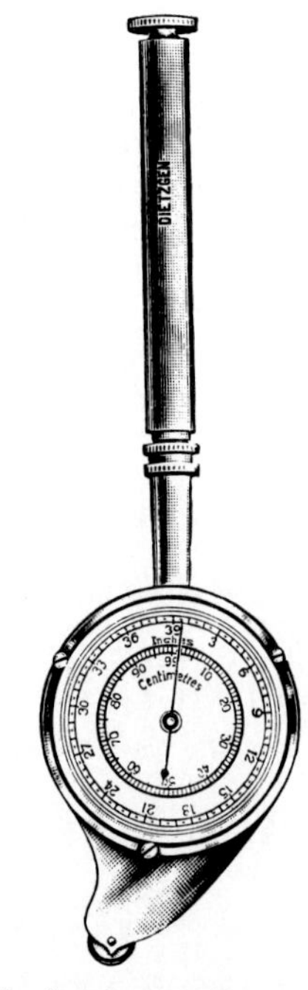

Fig. 24–4. **A cartometer is used to measure mileage on maps. (Eugene Dietzgen Co.)**

When considering the size of the scale to use for your map, remember that as the number of miles per inch gets larger, the scale gets smaller.

Features on Maps

The features found on a map vary with its purpose and scale. Features that are more important than others are emphasized. For example, the features that are most important to pilots are emphasized on an aeronautical map. Looking at the aeronautical map in Fig. 24-32 you can see that emphasis is given to airports and landmarks easily seen from the air.

Features that do not contribute to the usefulness of the map should be omitted. For example, a contour map will not show highways.

The cartographer must decide which features to show and emphasize. Every feature must be located accurately in relation to the other features. Features that are included should not be crowded to the extent that the map is difficult to read.

Features are given emphasis by using heavy lines, drawing large symbols, or using bright colors. Sometimes special lettering is used. Arrows can be drawn pointing to these features.

The details on a map depend upon the scale. Small scale maps show general features. Large scale maps can show more detail. For example, houses are not shown on small scale maps.

Using Symbols

Maps use symbols to represent various features. Small scale maps need more symbols. For example, on a large scale map a city could be shown by drawing its outline and some major features. On a small scale map a city may be represented by a dot.

Small scale maps use simpler symbols than those on large scale maps. For example, on a large scale map, a railroad is shown by parallel lines with short cross lines. On the small scale map, it is a single line with cross lines.

Symbols should have a simple design and be easy to draw. Their meaning should be easily recognized. The symbols for most map features have been standardized.

Symbols may be divided into several types:

- Those representing water features. These are usually shown in blue.
- Those representing constructed features, such as buildings. These usually appear in black or red.
- Those representing relief features, such as mountains. Relief features show the various heights of the land. These features are usually shown in brown.
- Those features that represent vegetation are usually printed in green.
- Special symbols for unusual features, such as a shrine or ruins. These often have a legend to call attention to them. They usually appear in a bright color such as orange.

Water Feature Symbols

Water feature symbols are shown in Fig. 24-5. Usually a cartographer will draw rivers, lakes, or shore lines first. From these, the other features are located. While water feature symbols are usually blue, black also can be used.

The thickness of the lines must vary. For example, the thickness of the lines representing rivers varies. Branches of the river are thinner than the main stream. The main stream gets wider as more branches connect to it.

Some rivers are dry part of the year. The intermittent river symbol shows this. Some rivers have a series of dams that control the flow of river traffic. These are called *canalized* rivers. The location of the dam should be drawn on the river symbol.

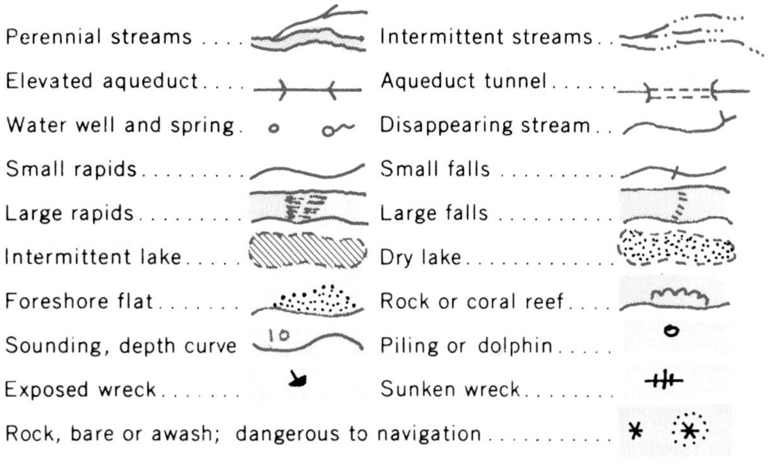

Perennial streams . . . . | Intermittent streams . .
Elevated aqueduct . . . . | Aqueduct tunnel . . . . .
Water well and spring . | Disappearing stream . .
Small rapids . . . . | Small falls . . . .
Large rapids . . . . | Large falls . . . .
Intermittent lake . . . . . | Dry lake . . . .
Foreshore flat . . . . | Rock or coral reef . . . .
Sounding, depth curve 10 | Piling or dolphin . . . . .
Exposed wreck . . . . | Sunken wreck . . . .
Rock, bare or awash; dangerous to navigation . . . .

Fig. 24–5. Water feature symbols.

Buildings (dwelling, place of employment, etc.)	
School, church, and cemetery	Cem
Buildings (barn, warehouse, etc.)	
Power transmission line	
Telephone line, pipeline, etc. (labeled as to type)	
Wells other than water (labeled as to type)	Oil . . . . . Gas
Tanks; oil, water, etc. (labeled as to type)	Water
Located or landmark object; windmill	
Open pit, mine, or quarry; prospect	
Shaft and tunnel entrance	

Fig. 24–6. **Human-created feature symbols.**

Fig. 24–7. **Topographical features such as these are shown on maps by symbols.**

It is not usually possible to show every turn in a river. The general features and major changes in direction should be shown.

Human Creation Feature Symbols

There are a large number of standard symbols in use to show features created by people. Fig. 24-6. These are very simple in design. Whenever possible, they are made to look like a typical feature. For example, a tank is shown by a picture of a cylinder. The actual tank may be somewhat different in shape, but the meaning of this symbol is clear. These symbols can be drawn as viewed from directly above or in pictorial form.

Relief Feature Symbols

Relief feature symbols show contour lines and land forms. Relief maps cannot show every hill and valley, but they do show the general topography. Fig. 24-7. The symbols are drawn viewing the land from about 45 degrees above the surface. Land form symbols are used only on small scale maps. Larger scale maps use the block diagram method of showing relief features. Other examples are shown in the section "Land Forms," later in this chapter.

Vegetation and Surface Feature Symbols

There are a number of standardized vegetation feature symbols. These try, in a very simple way, to represent the actual vegetation. They are drawn in color. If they are drawn in black, they should be used sparingly. If too many symbols are placed on an area, the map becomes crowded and hard to read. The symbol patterns show the vegetation from the side view. Examples are shown in Fig. 24-8.

Road, Railroad, Canal, and Dam Symbols

Road symbols show the number of lanes and the type of surface. They are drawn with black lines. Hard-surfaced roads use red on the symbol. Road and railroad bridge symbols are drawn in black. Dams are shown with a solid black line. Canals are drawn in blue. Examples of these symbols are shown in Fig. 24-9.

Boundary Symbols

Boundary symbols are combinations of long and short dashes or dots. The thickness of these lines varies with the symbol. The type of boundary is indicated by various symbols. For example, a state boundary uses a different symbol than a county boundary. The line thickness also varies. Some symbols are black while others are red. Fig. 24-10.

Special Feature Symbols

In addition to the previously mentioned symbols, some special symbols are used to show features not occurring commonly. These are generally found on maps used for a special purpose. For example, on aeronautical maps, air markers are noted to aid the pilot. (Air markers are ground location signs that can be seen from the air.) Often the cartographer has to develop a symbol to show these special features. They are usually drawn in a bright color. Their meaning is often lettered beside the symbol.

Legends

Many symbols are clear with no additional explanation. Others require a **legend** to explain their meaning. For example, the size of cities is often shown by a circular symbol. The size of the symbol is used to show the size of the city. This is explained on the title portion of the map. Fig. 24-11. The legend is used to explain the use of colored areas. Often the heights of land surfaces are given different shading or different colors. A legend explains what each represents. Fig. 24-12. Legends are developed by the cartographer.

How Much Detail?

Even on large scale general maps, it is not possible to show every house or garage separately. Generally, large scale maps can only show the area of a farm or park. Details, such as park buildings, would be so small they might be just a dot. On medium scale maps, an entire city might be shown as a shaded outline area. Only main roads and railroads can be shown.

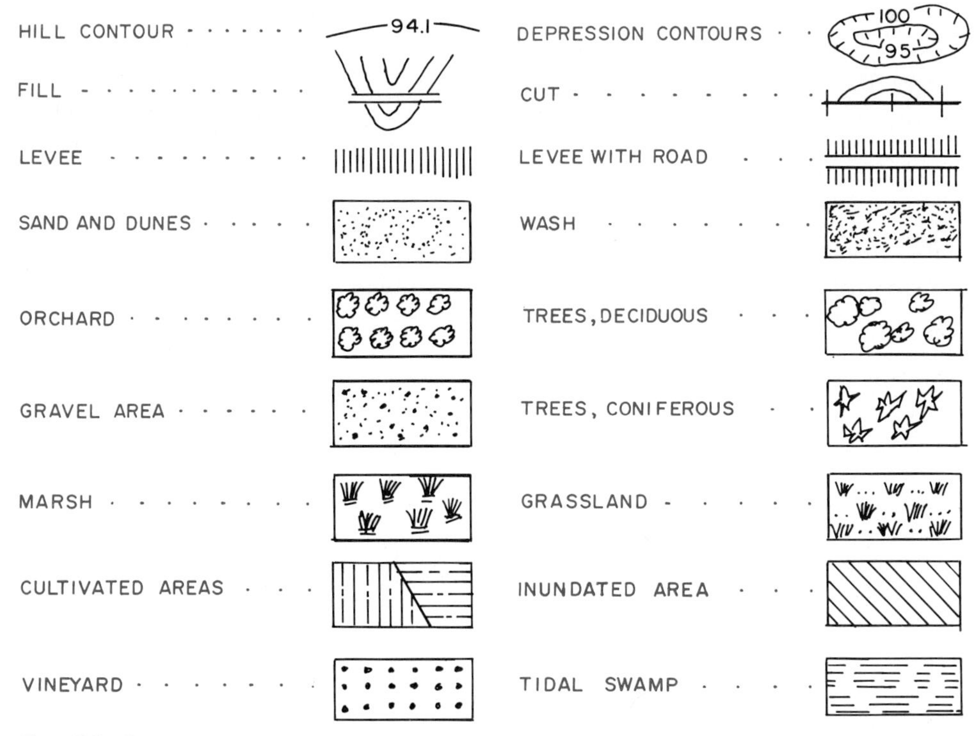

Fig. 24—8. Symbols for vegetation and surface features.

Small scale maps do not even show roads. They can only show the location of cities by a square or circular symbol. It is not possible to show the shape of the area of the city. Fig. 24-13.

Cartographers must use judgment in deciding on details. They decide how to combine features so that the map serves its intended purpose. Some details are omitted entirely.

About the only time each building and all the streets can be shown is when a very small area or several city blocks are drawn to a large scale. This is a common practice when recording the actual land sizes found by a careful survey of an area. These maps are used for legal descriptions of property. City engineers also use these maps.

Color on Maps

Most maps are in color. Proper use of colors greatly helps in the reading of maps. For example, color can be used to emphasize the important features of a map. As stated in the discussion of symbols, water is shown in blue, forests and vegetation in green. Ice is shown in white, deserts in reddish brown.

Cartographers must understand the principles of color and use them to advantage. They must try to avoid too much contrast between large areas of color so that these areas do not compete for the reader's attention. For example, strong, bright greens and browns are not generally used. They are toned down so that too much contrast is not developed. The use of pastel colors on maps is more effective than bright colors.

Cities and roads are usually shown in red or black. They stand out clearly against the usual green and brown land areas. They are small and must be strong so they can be easily seen. Lines used to indicate boundaries, such as those between two states, are usually red or black.

The cartographer will vary the hue of the color in areas that are similar. For example, green may be used for corn and a lighter green for hay. A wheat field might be indicated by light brown or yellow-brown.

Color can be used effectively to separate features, such as the countries of Europe. The color, in this case, has no reference to the actual ground. It is used to show clearly the shape and size of each country. Such maps do not generally use a different color for each country. The colors might be limited to two or three. The intensity is varied for contrast. Fig. 24-14.

Hard surface, heavy duty road, four or more lanes
Hard surface, heavy duty road, two or three lanes
Hard surface, medium duty road, four or more lanes
Hard surface, medium duty road. two or three lanes
Improved light duty road
Unimproved dirt road and trail
Dual highway, dividing strip 25 feet or less
Dual highway, dividing strip exceeding 25 feet
Road under construction

Railroad, single track and multiple track
Railroads in juxtaposition
Narrow gage, single track and multiple track
Railroad in street and carline
Bridge, road and railroad
Drawbridge, road and railroad
Footbridge
Tunnel, road and railroad
Overpass and underpass
Important small masonry or earth dam
Dam with lock
Dam with road
Canal with lock

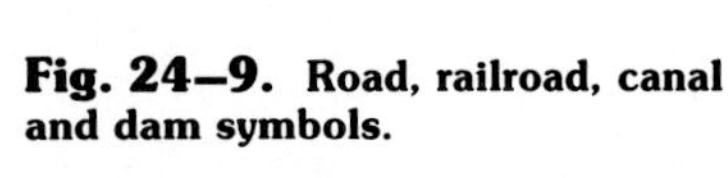

Fig. 24–9. Road, railroad, canal and dam symbols.

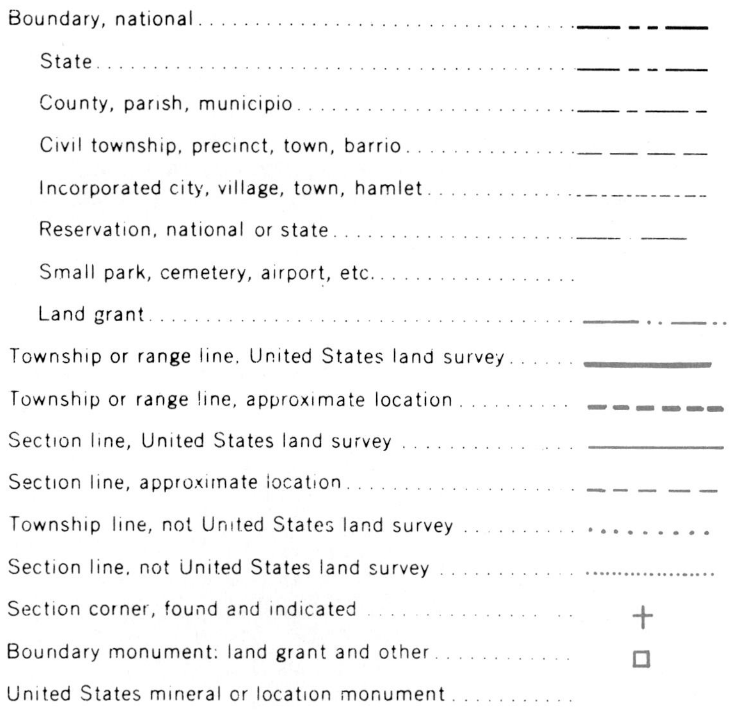

Boundary, national

- State
- County, parish, municipio
- Civil township, precinct, town, barrio
- Incorporated city, village, town, hamlet
- Reservation, national or state
- Small park, cemetery, airport, etc.
- Land grant

Township or range line, United States land survey

Township or range line, approximate location

Section line, United States land survey

Section line, approximate location

Township line, not United States land survey

Section line, not United States land survey

Section corner, found and indicated +

Boundary monument: land grant and other □

United States mineral or location monument

Fig. 24–10. **Boundary symbols.**

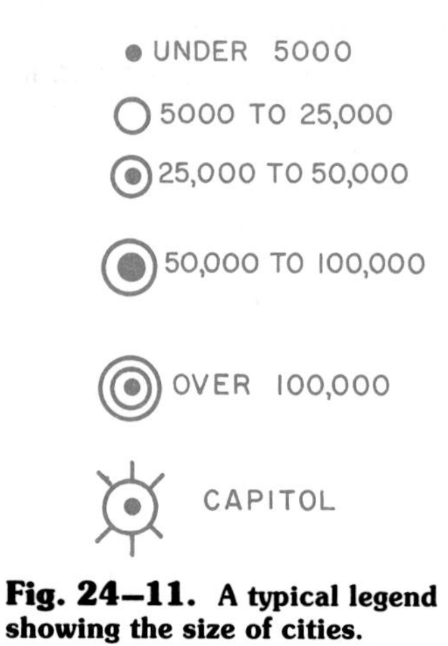

Fig. 24–11. **A typical legend showing the size of cities.**

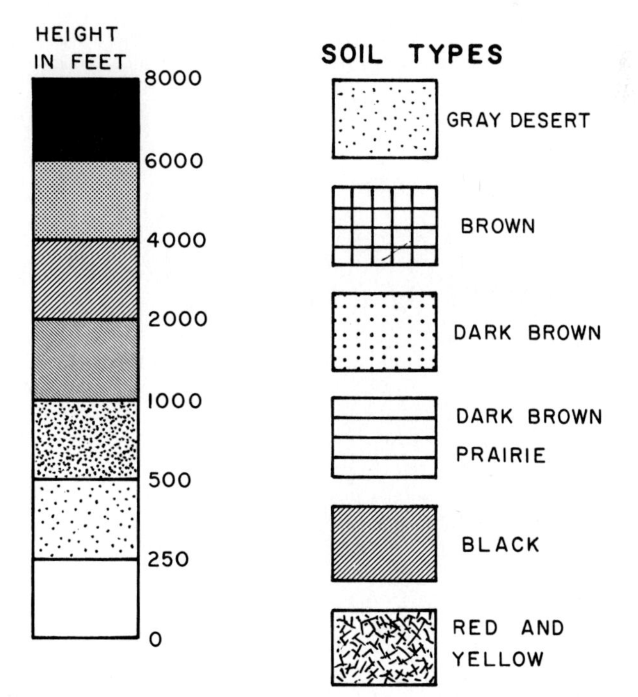

Fig. 24–12. **Legends are used to explain the meaning of the shading patterns.**

Another effective approach is to color a narrow strip along each border. That way the map does not appear like a gaily colored pinwheel. The colors chosen should blend harmoniously and not clash. Notice the gray areas in Fig. 24-15.

One effective medium used to color maps is transparent watercolors. The beginner should practice on scrap paper before trying to color a map. To be effective, much experience is needed.

Generally watercolors are applied starting at the top and sides of the area. Paint toward the center and down to the bottom of the area. Keep the brush well filled with paint. A dry brush does not produce effective results. Be careful not to overlap areas. It is best to let an area dry before painting the one next to it.

Colored pencils can be used also. Pastel sticks are good for shading.

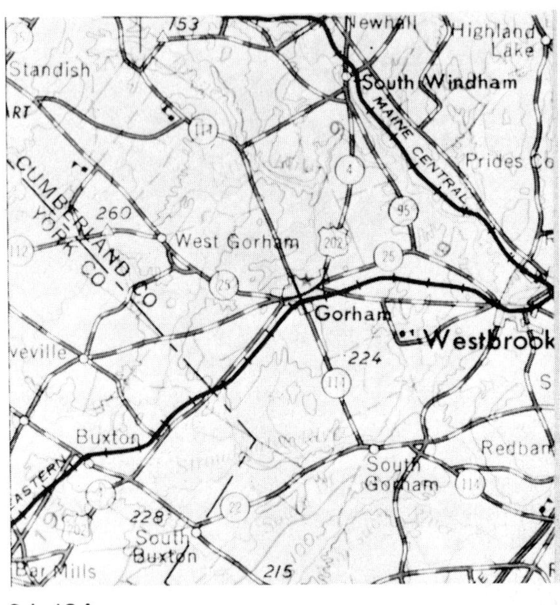

24-13A

1:250,000 scale
1 inch = nearly 4 miles
Area shown = 107 sq. miles

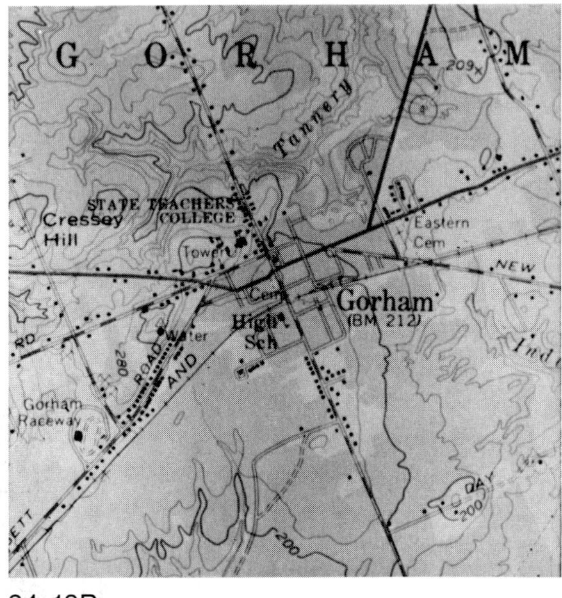

24-13B

1:62,500 scale
1 inch = nearly 1 mile
Area shown = 6¾ sq. miles

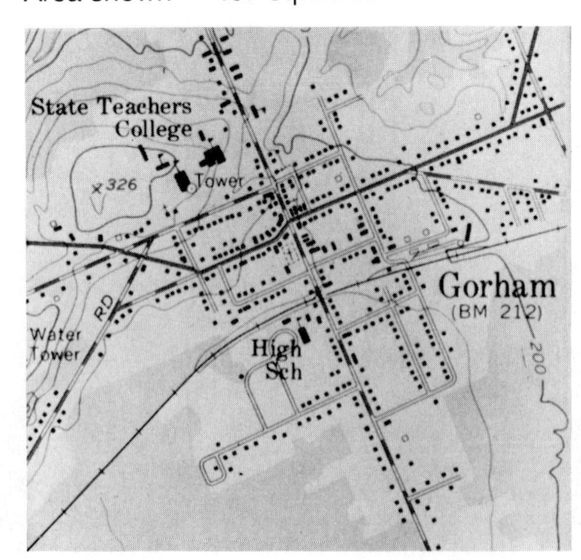

Fig. 24–13A, B, C. The details shown vary with the scale of the map.

24-13C

1:24,000 scale
1 inch = 2000 feet
Area shown = 1 sq. mile

Tools

The tools used by the cartographer are the same as those used by drafters. Usually soft drafting pencils are preferred. The H and HB grades of lead are popular. Various types of inking pens are used. The T-square or parallel rule with drafting triangles form the basic layout equipment. Watercolors and watercolor brushes are necessary.

Paper used for maps should be of best quality. It should have a high rag content. The surface should not be too slick or glossy. Plastic drafting film is also an excellent material upon which to draw maps.

Of considerable use in map making are the various shading materials available. They are printed on transparent sheets and have an adhesive back. Pieces are cut to the shape of the area to be shaded and are pressed into place. See Chapter 20, "Technical Illustration," for more information.

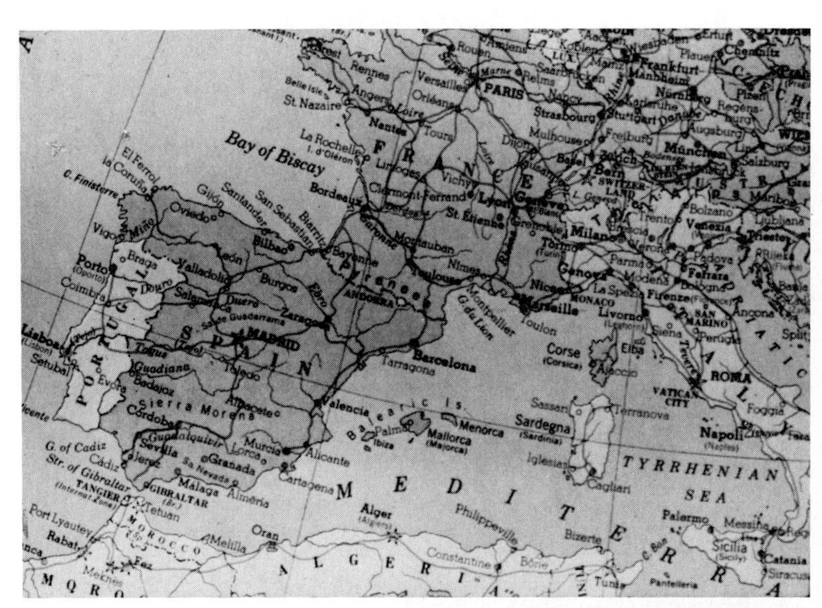

Fig. 24–14. **When printed in color, this map shows a different color over each country to emphasize the shapes and sizes. (Hammond, Inc.)**

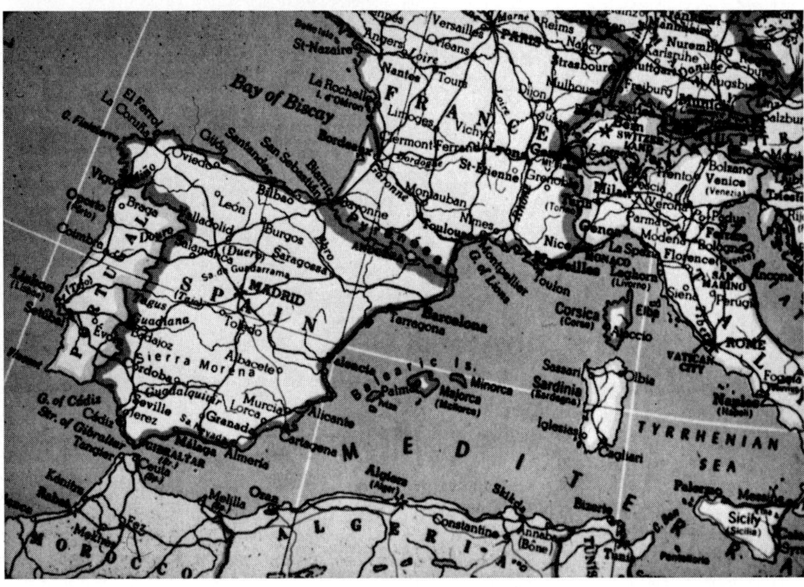

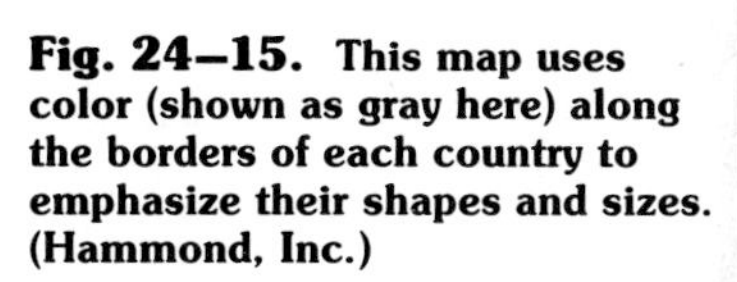

Fig. 24–15. **This map uses color (shown as gray here) along the borders of each country to emphasize their shapes and sizes. (Hammond, Inc.)**

CAD Applications:

Expanding the Territory of Maps

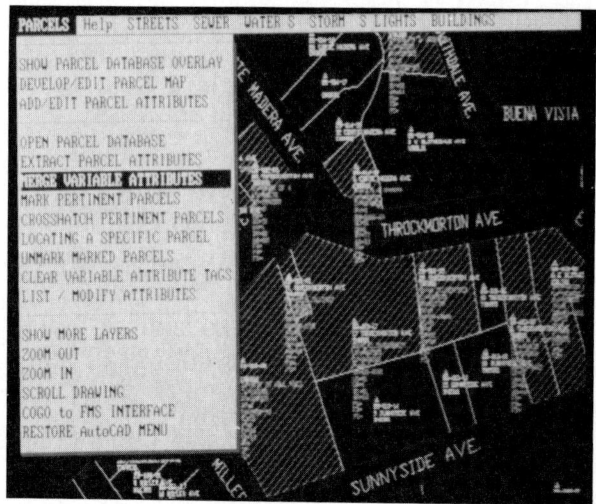

FMS/AC™ facility mapping system by Dennis Klein & Associates

Many local and state governments keep file cabinets full of records that go quickly out of date and are costly to maintain. One practical way to record such information is on a map. But ordinary maps are limited in the information they can show. However, by combining data stored in a computer with a good map, the usefulness of both is greatly increased. Using CAD, variations in different "base" maps can be produced that give specialized information. The information is stored on layers that can be updated easily and printed separately.

For example, the same city street map could be adjusted to show locations of manhole covers, sewers, utility lines, or fire hydrants. Populations of school-age children could be shown on the map to indicate where new schools need to be built or existing ones closed. Information about soil types, underground springs, etc., would be useful when planning new construction. The map could even show the location of individual addresses.

Several software programs are now available that can handle such data ranges. They offer flexible map design using 16 or more colors, several hundred hatch patterns, and options for legends and labels. Other mapping software can measure like a calculator. For example, a lake could be measured and the islands at its center subtracted out. The length of an irregular line such as a road or stream can be accurately told without using a ruler.

By linking maps with computers, data are more manageable and cost is reduced. And as more and more people become acquainted with these techniques, their usefulness will increase.

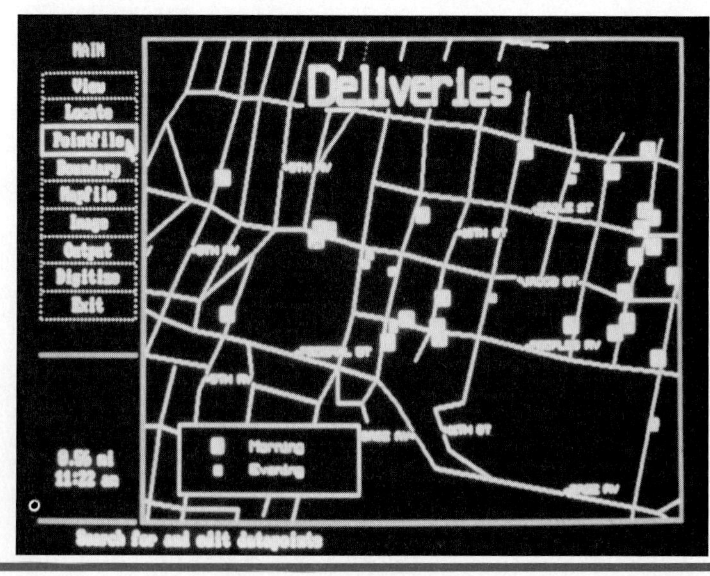

MapInfo by Mapping Information Systems Corp., Troy, NY

Parallels and Meridians

The earth's surface has been divided into a system of coordinates. These are parallels of latitude and meridians of longitude. **Latitude** is an arc distance measured in degrees from the equator. Those parallels north of the equator are called *north latitude*. Those south are called *south latitude*. The equator is zero degrees latitude. Fig. 24-16.

Longitude is the arc distance measured in degrees from the prime meridian. The prime meridian is a specially selected longitude running through Greenwich, England. It is designated as zero degrees longitude. Longitude is measured east and west of the prime meridian. Fig. 24-17. The United States is in west longitude. The Soviet Union is in east longitude.

Latitude and longitude are measured in degrees, minutes, and seconds. For example, New York City is 74°0′ west longitude and 40°45′ north latitude. Fig. 24-18.

Latitude and longitude are essential in mapping. They provide an accurate means for locating features on a map.

Designing Maps

There are many questions to be raised before drawing a map. Cartographers must do considerable study and research. They must make many decisions. Following are some of the things they must know.

- Purpose of the map.
- Scale.
- Best layout to use for the drawing space available.
- Features to be shown.
- Amount of detail needed.
- The kind of map to use.
- Method to be used to reproduce the map.
- Land area to be included.
- Features that need special emphasis.
- Is a legend necessary?
- What lettering will be needed? Is there room for it?
- What is the best way to show features? With standard symbols or relief land form symbols?
- Should it be in color? How many colors? What colors?
- How much contrast is needed in shaded areas?
- What shading medium should be used?
- Where to find material on the area to be drawn.

Making the Layout

Always make a preliminary layout in pencil on tracing paper or graph paper. Use the same scale as planned for the finished drawing. Plan space for the title area. Freehand letter the words to be in the title. Letter the more important words larger than the others. The title usually includes the name of the area, the type of map, the cartographer's name, date, scale, legend, and any other remarks considered necessary. The title is placed in one of the corners. It must be in an area where no important features are located.

Plan for a border. Most maps are enclosed with some type of border. The border can be very simple. Two parallel lines are commonly used. Sometimes part of the map will cut through the border. The border should be broken to permit this. The inside line of the border is the actual edge of the map. It is called the *neat line*.

If a map has an inset, plan where to place it. A typical example is an enlarged map showing the main streets of a major city on a state highway map.

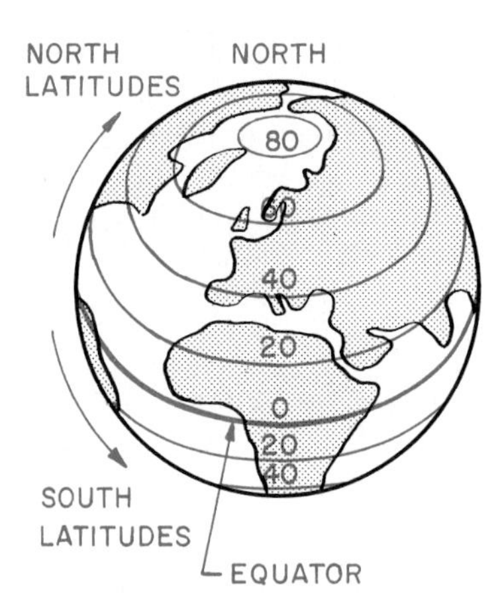

Fig. 24—16. Latitude is measured north and south of the equator.

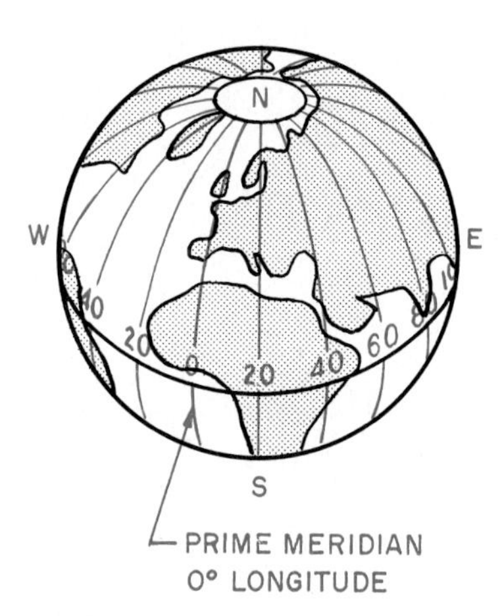

Fig. 24—17. Longitude is measured east and west of the prime meridian.

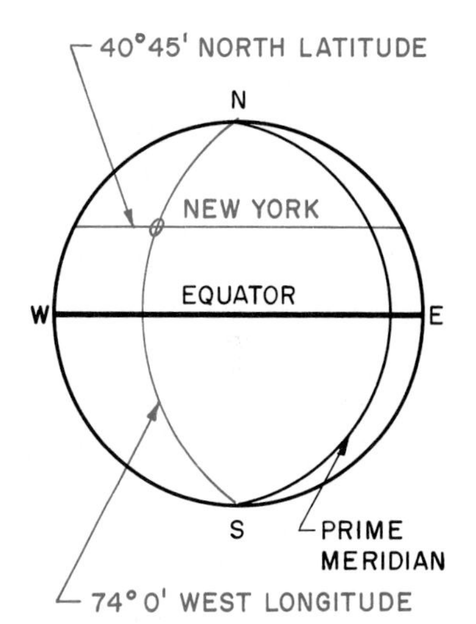

Fig. 24—18. Latitude and longitude are used to locate points on the surface of the earth.

Maps have a symbol indicating north. The top of the map is always north. Plan an area for this symbol. If the parallels and meridians are shown, the north arrow is sometimes omitted.

After the rough layout is made and all details are planned, the finished map can be drawn. Draw it carefully in pencil. Use a sharp, soft pencil. The map is not inked until it is completely drawn in pencil. The following list is a suggested general order for laying out the parts of the map:

1. Draw the neat line. This sets the area of the map.
2. Draw the parallels and meridians.
3. Draw the water features such as rivers, lakes, and shore lines.
4. Draw roads and railroads.
5. Draw hills and mountains.
6. Locate the symbols for cities and other special features.
7. Letter any words needed. Place them in clear, uncluttered areas as near to the symbol or location as possible. These can be lettered lightly in freehand. They will be inked more accurately with lettering templates.
8. Draw guidelines for the lettering in the title area.
9. Now ink the drawing. Some cartographers recommend that inking be done in the reverse order of the pencil drawing. Ink the title, legend, and words first, then the symbols, and on through the list in reverse order. This reverse inking procedure can help prevent mistakes. For example, a parallel or meridian can be broken if it crosses a word or other important feature. A river can be broken if it is necessary to letter a name across it. If a map is to be printed in color, it is not always necessary to break these lines. Those items that might conflict can be printed in a light color so that the darker words or symbols will print clearly over them.
10. After the inking is finished, remove all pencil lines. This must be done carefully so the inked lines are not damaged.
11. If areas are to be shaded, apply shading now. Watercolor areas can now be painted.
12. It is wise to mount the map on heavy cardboard to prevent damage.

Topographic Maps

Topographic maps are general maps of a large area. They show all the important land features. Since they show a large area, they are usually drawn to a large scale. Some are developed at a medium scale.

Topographic maps are in relief. They involve the elevation above sea level, height of an area above adjacent areas, and the steepness of slopes. There are several ways to show relief. Common ways are with contour maps, land form maps, and photographs of scale models.

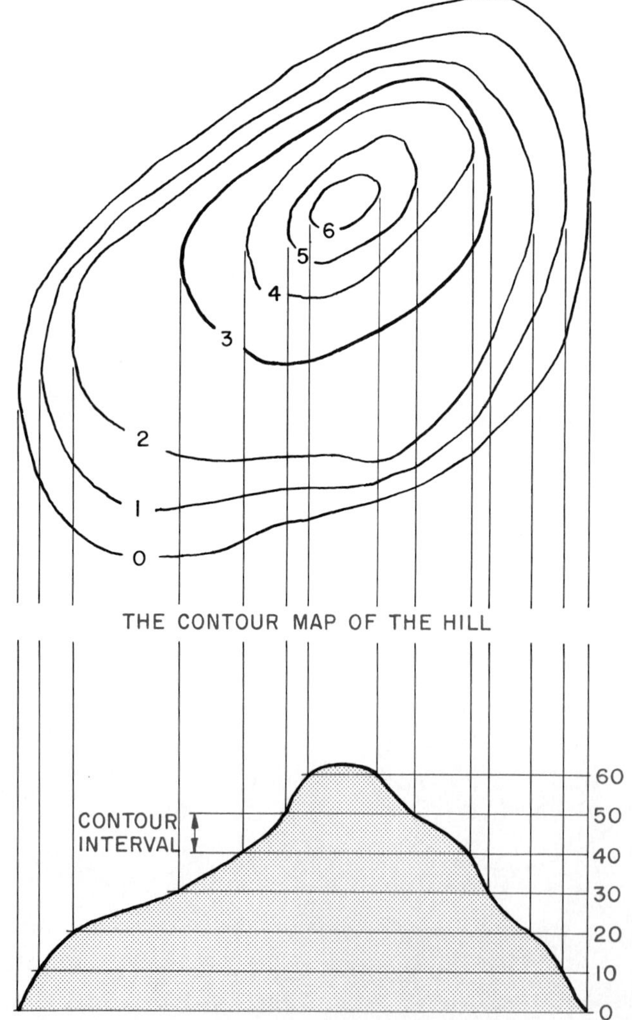

Fig. 24–19. A contour map.

Planographic maps show the same features as a topographic map. The difference is that they are not in relief. Color is used to show relief.

Contour Maps

A contour map is difficult to read unless you understand what each line represents. Fig. 24-19 shows how to read contours. Examine the hill. It is cut in horizontal slices at regularly spaced distances. The shape of the hill at each cut is the **contour** at the height above sea level. This contour map shows the hill from above with each contour line drawn. The height above sea level is lettered on the contour line. The contour line passes through all the points on the hill that are at the same height.

Contour maps are made to show the depth of water as well as height of hills and mountains.

The distance between contour levels varies with the scale of the map. Intervals of 10 to 20 feet are commonly used. Larger intervals are used on high, steep mountains. Smaller intervals are used on rolling hills and flat land.

Contour data are sometimes difficult to obtain. A surveying crew may measure elevations of the most important points and estimate the contours in between. This requires making actual field drawings as they observe the area. Accurate contour maps can be made from aerial photographs using stereovisual equipment. This special equipment enables the photograph to be seen in three dimensions.

Contours should be drawn in pencil. Draw every fourth or fifth contour with a heavy line. Letter the level of the contour on the line. Break the line at the number. Contour lines do not cross each other.

Fig. 24-20 shows a contour map made from a land survey. A hollow or depressed area is shown with a special contour line.

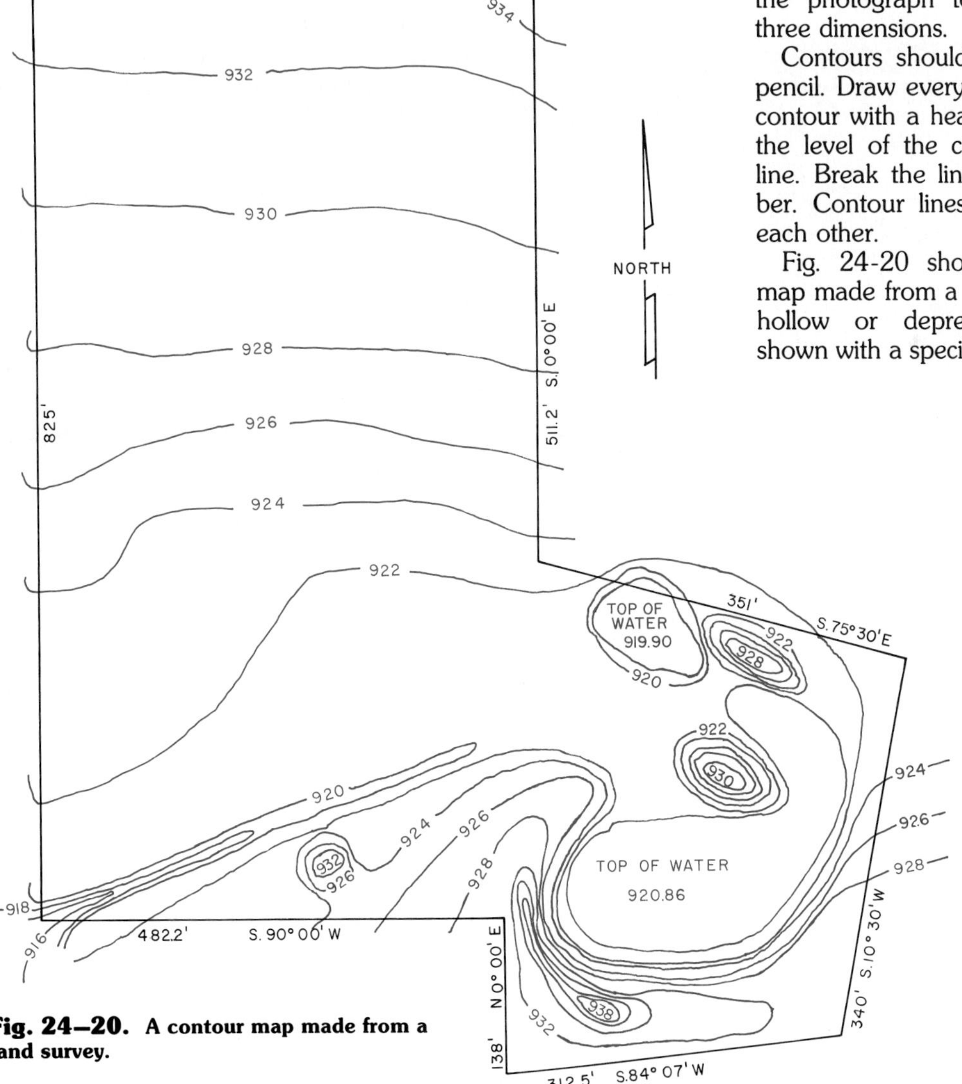

Fig. 24–20. A contour map made from a land survey.

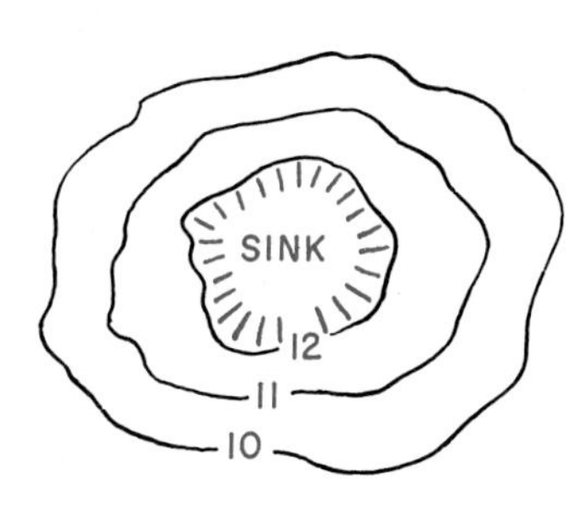

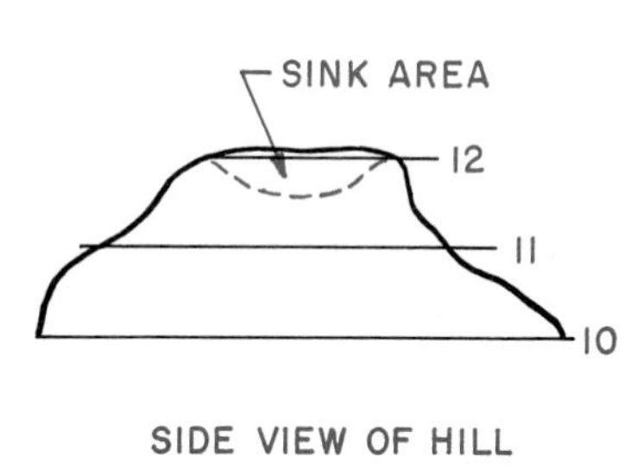

Fig. 24–21. A sink is a depressed contour area.

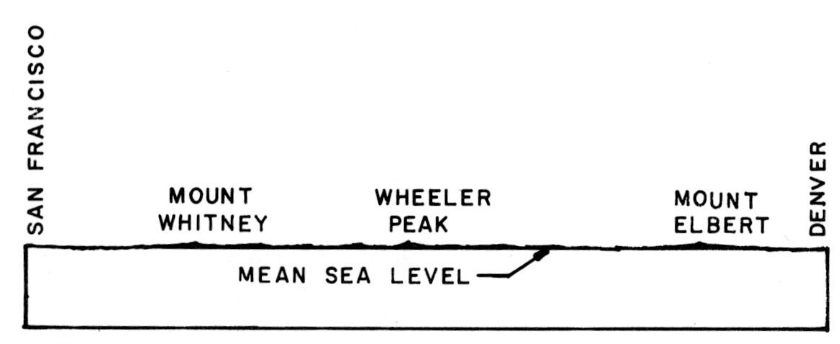

Fig. 24–22. This profile has the same vertical and horizontal scale. It is of little value in showing the land profile.

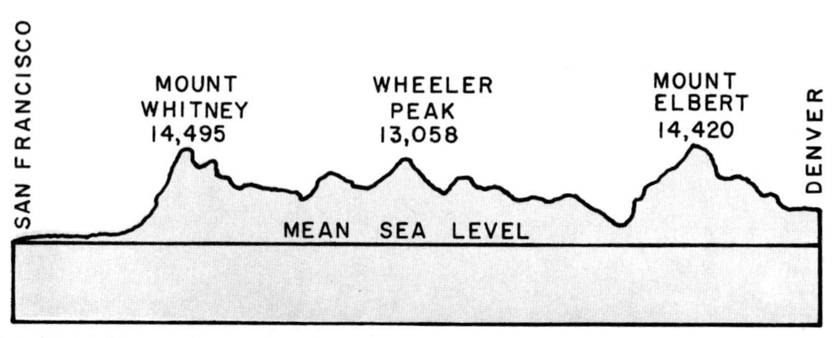

Fig. 24–23. A profile through part of the Rocky Mountains. The vertical scale is 40 times greater than the horizontal scale.

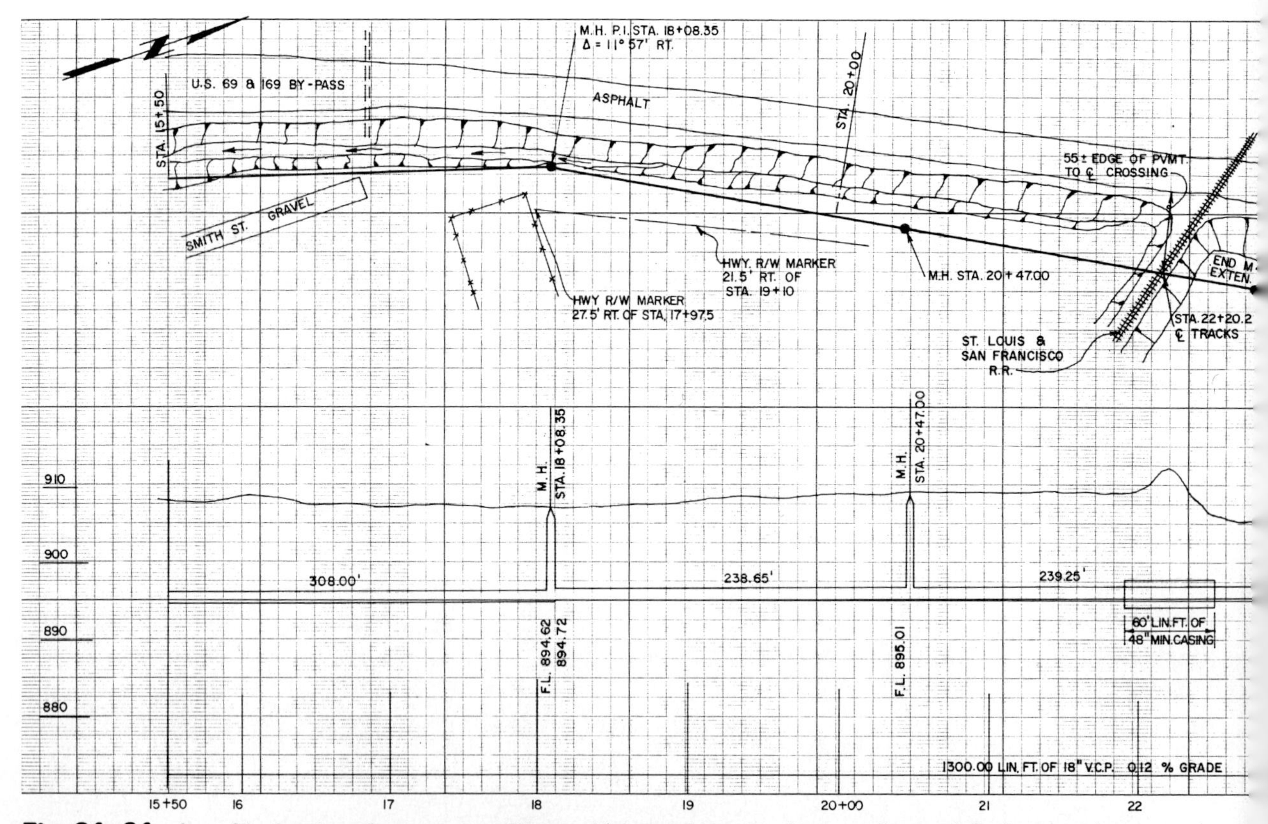

Fig. 24–24. A profile drawing of a sewer installation. A pictorial view is shown at the top left. At top right is a plan view. The bottom drawing is the profile. It was drawn from the field notes of the surveying crew. The abbreviation, "M.H." refers to manholes. Notice the construction of supports to carry the sewer pipe across a creek. Notice the difference in the horizontal and vertical scales. (City Engineering Department, Pittsburg, Kansas)

It has short lines drawn perpendicular to it. Often it is labeled "sink" to call attention to it. Fig. 24-21.

Profiles

Profiles are vertical cuts through a land area. They are much like a section through a machine part. Profiles are often made from contour maps. If a contour map is used, the location of the cut is marked. This is much like making a section on a machine drawing.

A profile that has the vertical scale equal to the horizontal scale is not very useful. With these scales even high hills and mountains appear quite small. Fig. 24-22. To emphasize the height, cartographers use a larger vertical scale. In mountainous areas, less exaggeration is needed than in flatter areas. Fig. 24-23.

Fig. 24-24 is an engineer's drawing showing a profile of a sewer line. It was plotted on graph paper from an engineering survey. The surveyor's field notes were used to plot the profile.

Relief Models

Cartographers build scale relief models. The models clearly show the hills, mountains, valleys, and rivers. They are photographed to make a relief map. To get good photographs, the models are lighted from one side. This creates light and dark areas. The photographs show the relief effect clearly. Fig. 24- 25. Models are often colored. The colors are used to emphasize the different heights. Low areas can be green, plains and hill areas tan to yellow, and mountains a red-brown.

Land Forms

A **land form map** shows basically the same thing as a contour map. It is in pictorial form and easier for most people to understand. No attempt is made to show every hill and valley. It shows the general topography of the area. Fig. 24-26. Land form maps are used for small scale drawings. Larger scale drawings are drawn as block diagrams.

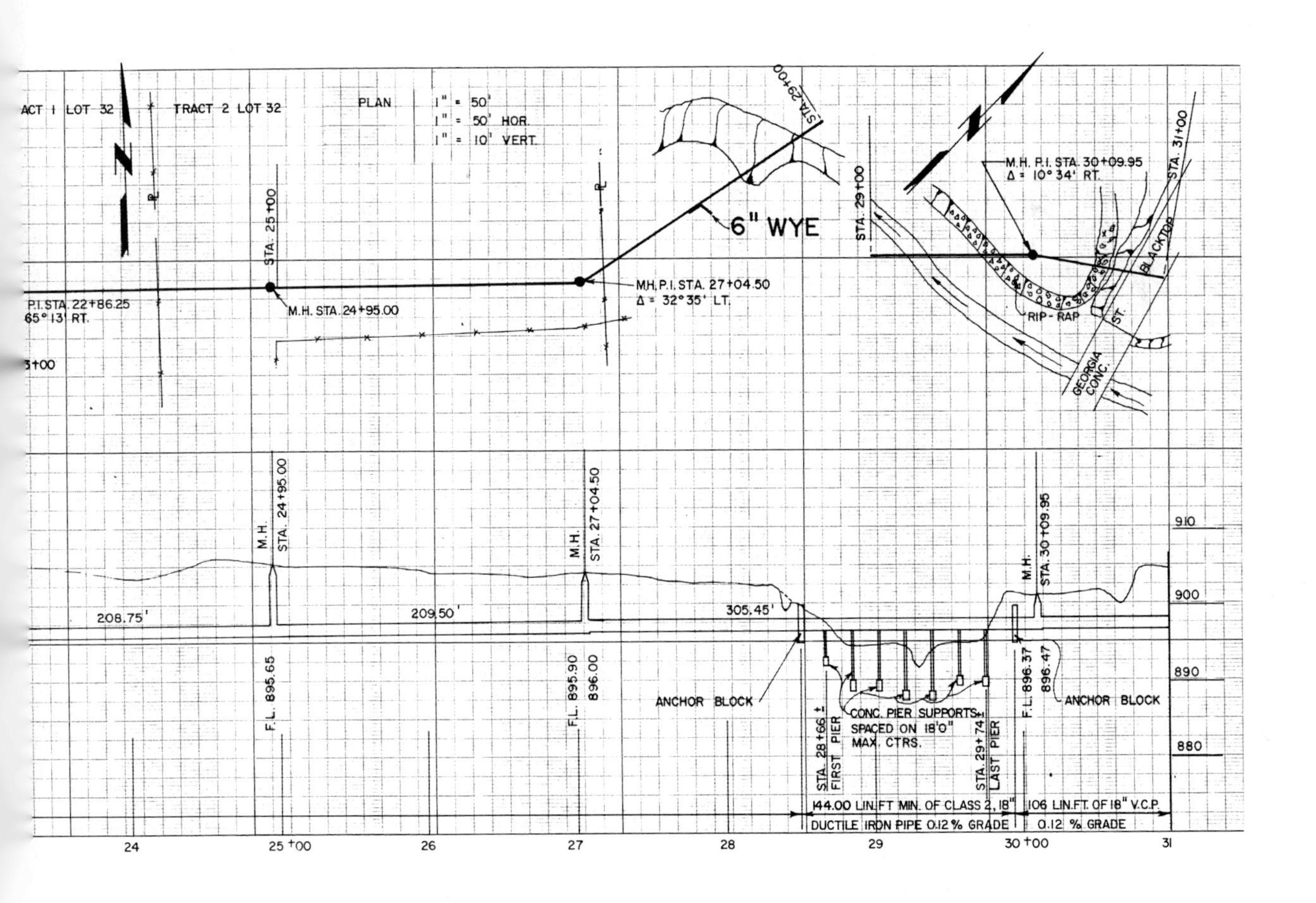

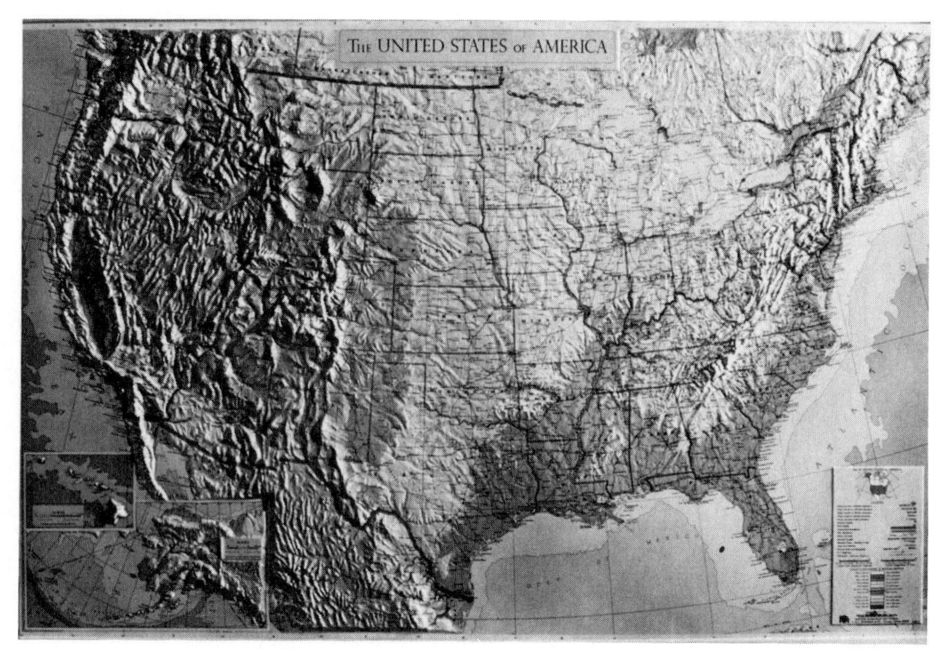

Fig. 24–25. A raised relief map showing the earth's surface as it would appear to an astronaut. The relief on this map was magnified 20 times greater than its actual height. (A. J. Nystrom and Co., Chicago)

*Portion of a physiographic diagram (Cities designated by initials)**

Drawn by A. K. Lobeck. Courtesy of The Geographic Press.

Fig. 24–26. Part of a physiographic diagram using land form symbols. (Reproduced by permission from *Mapping* by David Greenhoon, The University of Chicago Press)

The size of the symbols drawn will vary with the size of the mountain. Mountains are not drawn in proportion to each other. Very high mountains are usually drawn half of their actual height in relationship to surrounding smaller hills and mountains. If true proportion were used, the smaller hills and mountains would be very small. Some land forms would not be seen at all. The height above sea level is sometimes given at different levels. They are usually lettered in black ink.

It takes a great deal of work to develop the information needed for a land form map. The cartographer secures information from many sources. This may involve flying over the area. Photographs can be examined. Topographic maps and other special maps of the area can be examined.

Common land form symbols, Fig. 24-26, are formed by drawing lines in the direction of the slope. This is the direction rain would drain when it falls on the land. These lines are called **hachure lines**. Notice how the hachure lines are carefully spaced and located.

Flat plains can be left blank. If plains are rolling, short horizontal or curved lines can be drawn. Hachure lines are used to show the earth sloping to a river. The rivers are drawn first. Then the hachure lines are adjusted to them.

Plateaus usually have bluffs at their edges. The slopes generally have valleys carved by erosion. Some plateaus are rounded and worn. See the top drawing in Fig. 24-27.

Some mountains are old and worn. While they have high peaks, they tend to be rounded. Mountains formed by glaciers tend to have sharper peaks and ridges. The surfaces are concave. They have deep troughs running to glacial lakes or rivers.

Deserts have a wide variety of land forms. They can have rock plateaus and rounded mountainous areas. For the most part, they have large basins and areas of sand dunes.

A shaded land form drawing, Fig. 24-27, uses shading instead of hachure lines. This drawing was made from the contour drawing shown below it.

Block Diagrams

Block diagrams are another way to show the earth's surface in relief. They show a section of the earth removed. The drawing is viewed from above and at an angle. The more commonly used block diagrams are drawn in isometric, or one-point or two-point perspective. See Chapter 14, "Pictorial Drawing," for details on how to make these kinds of drawings.

The isometric block diagram is the easiest to draw and is reasonably accurate. The information for a block diagram is taken from a contour map. Draw a light grid of squares over it. Fig. 24-28, Step 1. Next draw the contour map in isometric, Step 2. This places all of the contour lines in isometric. Now draw an isometric box and locate each contour height, as shown at Step 3.

Start the finished drawing by laying out the highest contour first. If the drawing is on transparent paper, this can be traced from the isometric drawing made in Step 2. Slide the isometric contour drawing down and copy the contour at each level. Connect the ends of the contour lines. Label each with the height above sea level, Fig. 24-28, Step 4.

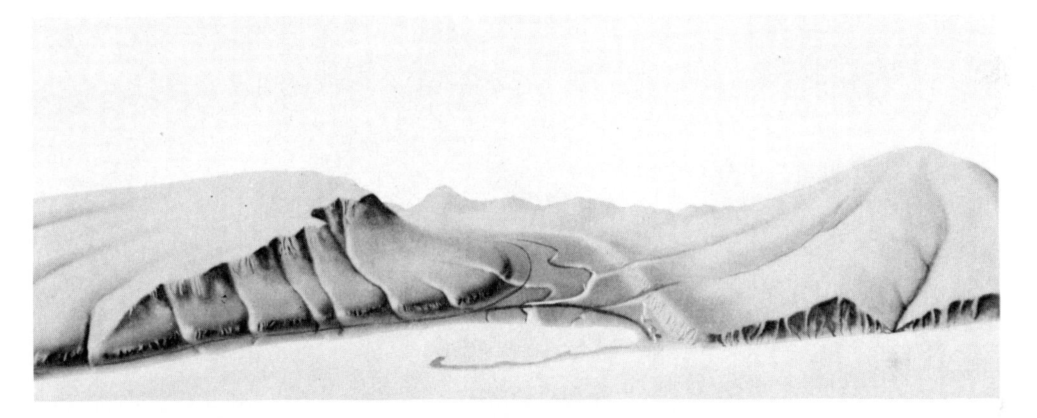

A LAND FORM DRAWING

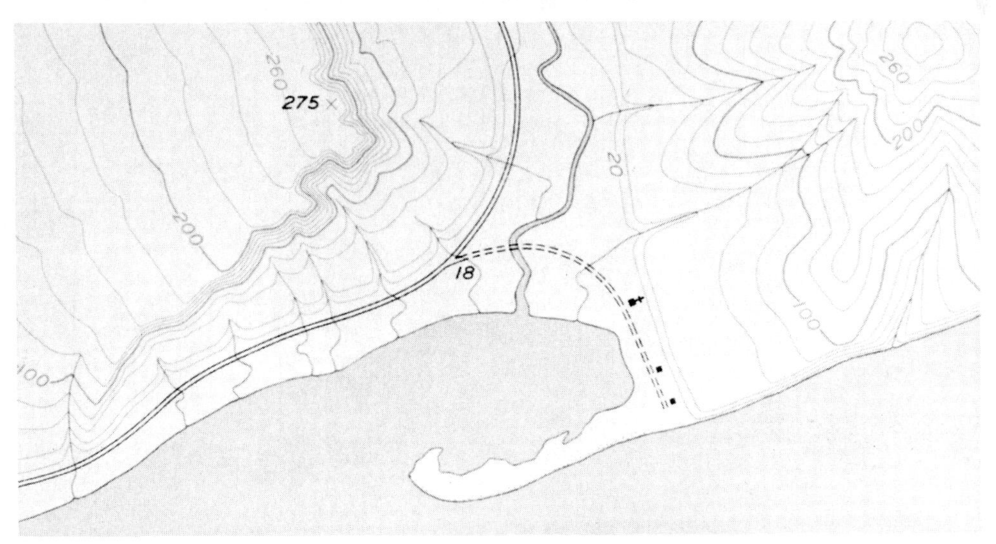

THE CONTOUR DRAWING

Fig. 24—27. A land form drawing using shading to give emphasis to the contour. Can you relate the land form drawing to the contour drawing below? (United States Department of the Interior, Geological Survey)

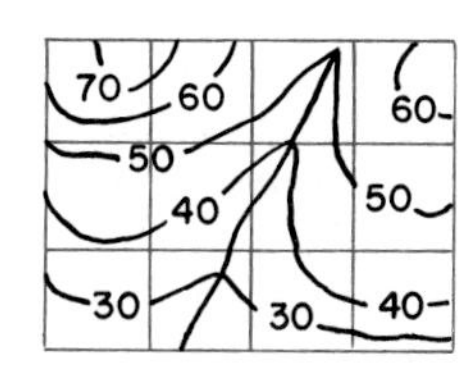

STEP 1
THE CONTOUR DRAWING OF TWO STEEP HILLS WITH A VALLEY.

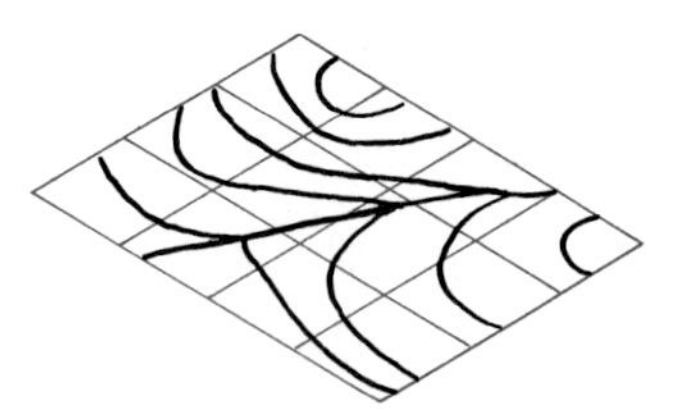

STEP 2
THE ISOMETRIC DRAWING.

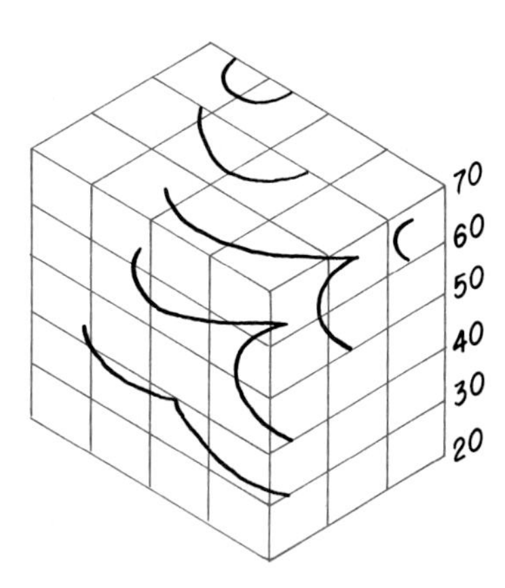

STEP 3
THE ISOMETRIC BLOCK WITH THE CONTOURS LOCATED.

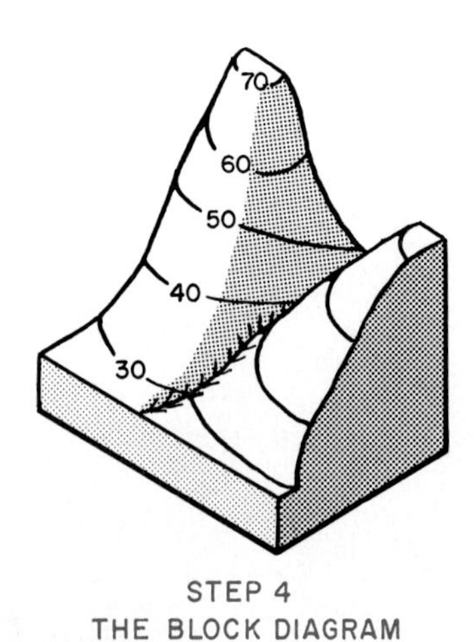

STEP 4
THE BLOCK DIAGRAM

Fig. 24–28. How to draw a block diagram.

Physical Maps

Physical maps show altitudes by tinting various areas different colors. The colors selected should be representative of the type of surface. For example, areas that are suitable for farming can be shown in green. If a map is not in color, the various heights can be shown by using different hatching on each one. (*Hatching* means to shade with fine lines or dots). Fig. 24-29. A legend is placed on the map to show the meaning of each color or hatched symbol.

Nautical and Aeronautical Charts

Nautical maps, known as **charts**, are used by navigators to guide their ships and avoid dangerous conditions. They show such things as accurate shore line contours, features of the coast line, depth indications, navigation lights, tides, currents, shoals and other obstructions, and harbor

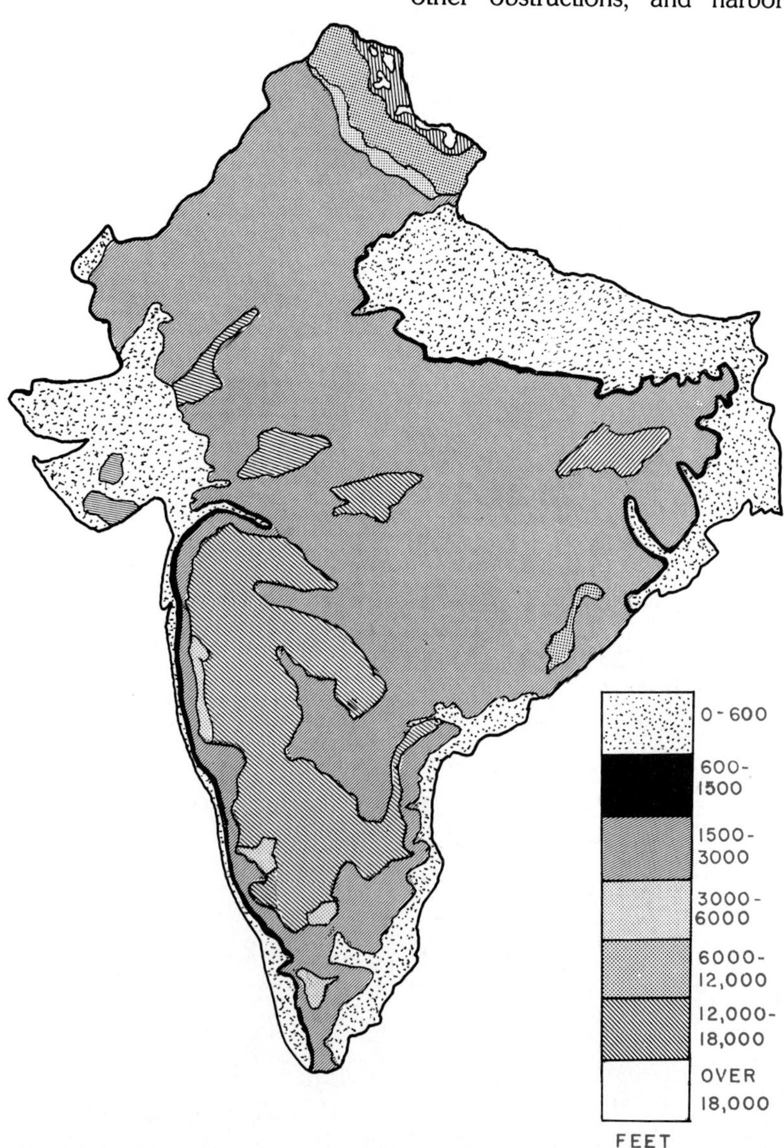

Fig. 24–29. Heights can be shown using different hatching in the contour planes.

details. Fig. 24-30. The scales used vary considerably. A very detailed harbor chart will be on a large scale. One of an entire ocean would be on a small scale. Some of the symbols used are shown in Fig. 24-31.

Aeronautical charts are used by aircraft navigators. They show such things as the altitude of mountains, navigation beacons, airports, visual landmarks such as lakes, location and extent of navigation beacons, high obstructions such as a tower, and the general shape of cities. Generally the scale is small. Fig. 24-32.

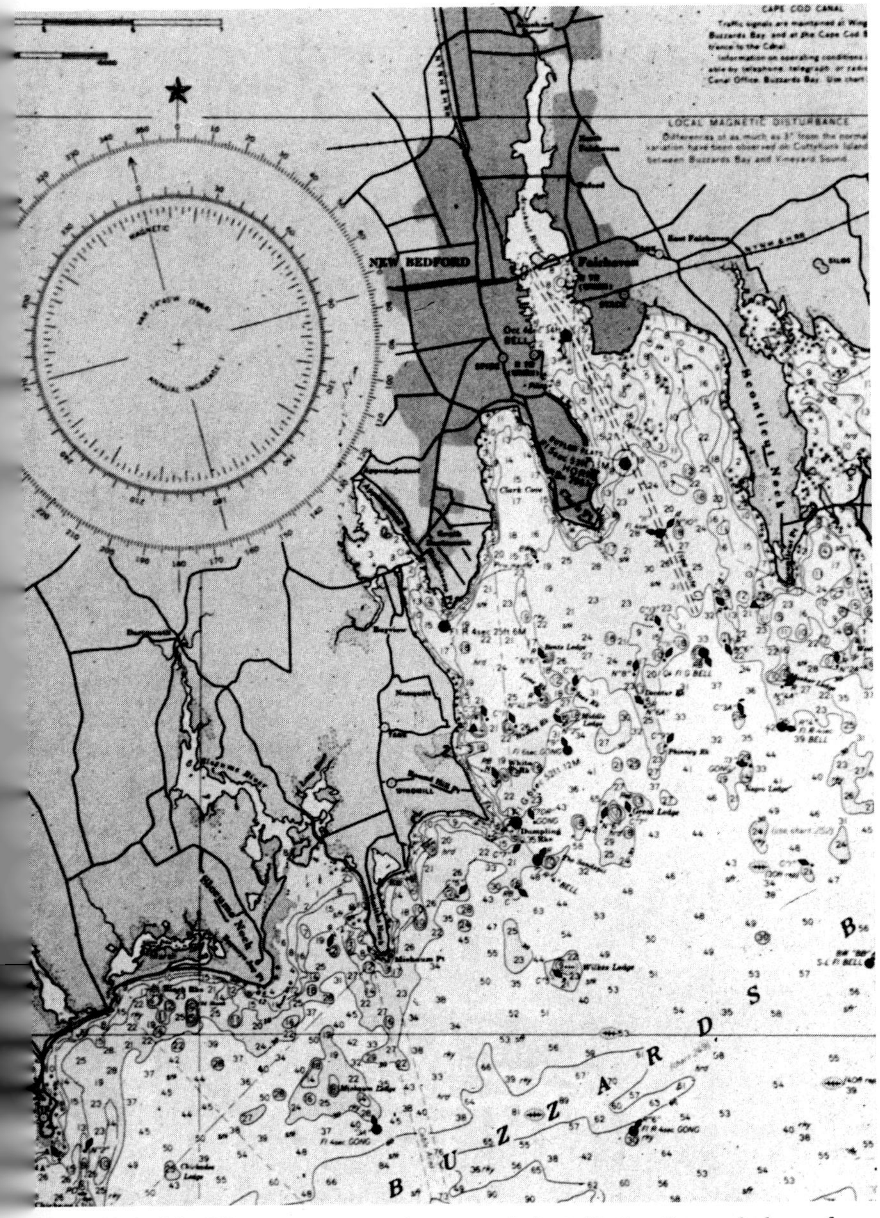

Fig. 24–30. A small part of a nautical chart. Notice the symbols used.

Other Maps

There are many other types of maps in use. There are too many to cover in this brief chapter. Following is a sampling of some of these. Maps are designed to serve hundreds of purposes. As stated before, the cartographer must decide which form to use and what information is needed.

Regional maps show large areas such as a country, continent, or the entire world. They are drawn to a medium or small scale. Detail is very limited.

Weather maps show the temperature, precipitation, wind direction, pressure fronts, snow, and other such data. They are issued daily. Separate maps are developed for surface weather, temperatures and precipitation. A system of standard symbols is used. Fig. 24-33.

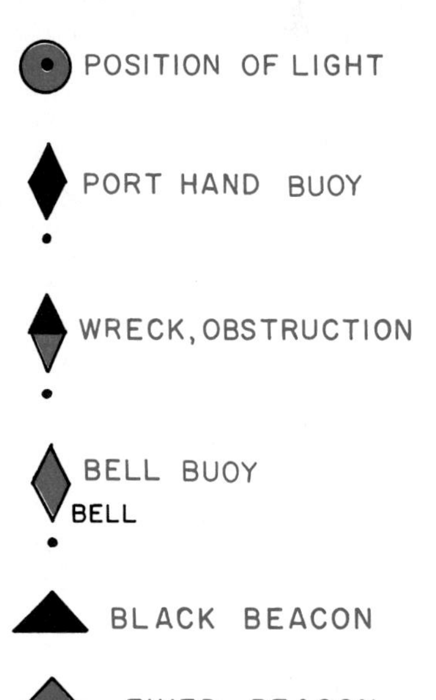

Fig. 24–31. A few of the many symbols used on nautical charts.

Single purpose maps are designed to show the special features for which they are to be used. The special features are emphasized. All unnecessary details are eliminated. For example, a map showing the streets of a city does not need to show the contours of the land. Some examples of special purpose maps are geological, vegetation, statistical, land use, city, transportation, political, science, and land ownership maps.

There are many types of *statistical maps*. They are used to report data such as rainfall, population, and farm production. One example is shown in Fig. 24-34. Color is used extensively on statistical maps.

Maps are in use that show the bottom of the ocean, soil use, erosion, and vegetation. Maps can show the wealth and education of the population of an area.

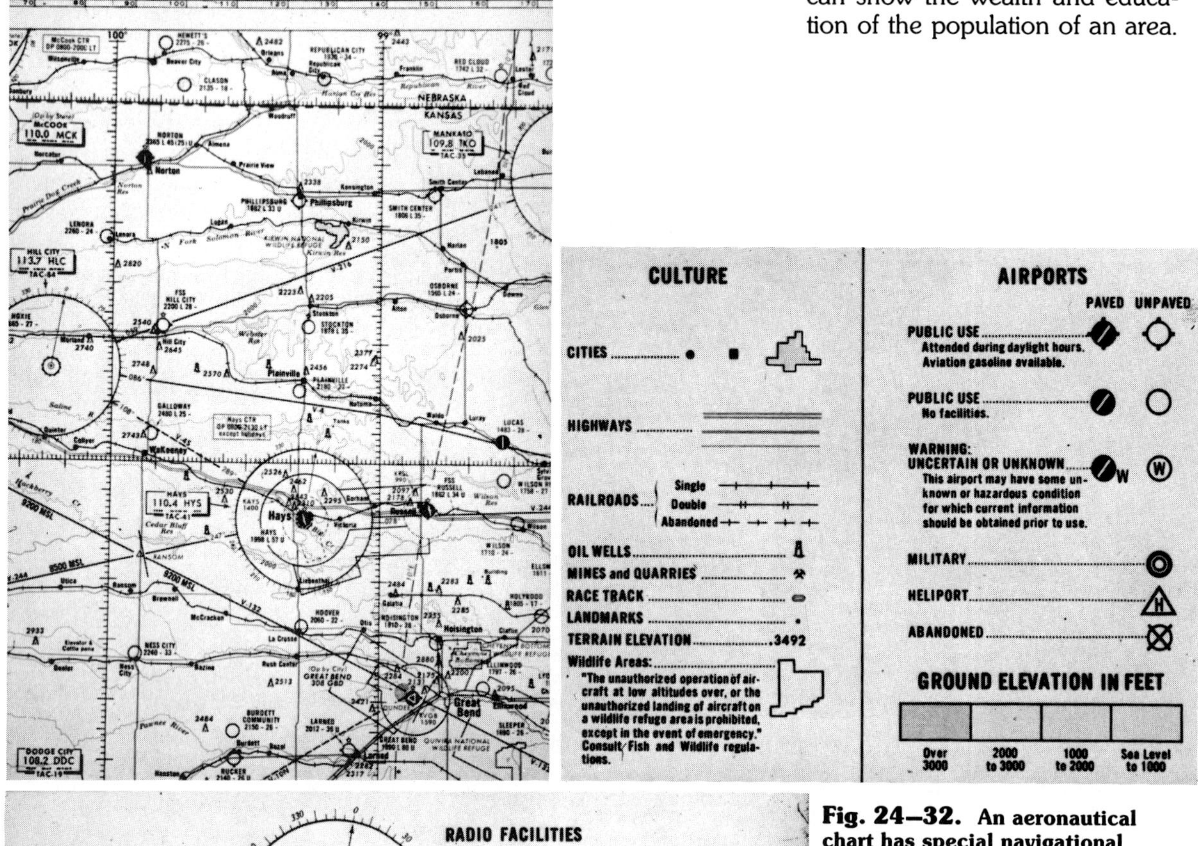

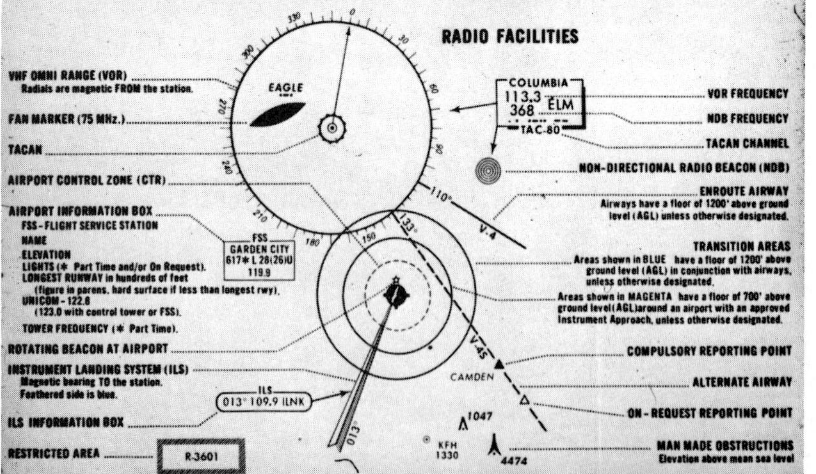

SYMBOLS USED ON AERONAUTICAL CHARTS

Fig. 24–32. An aeronautical chart has special navigational symbols.

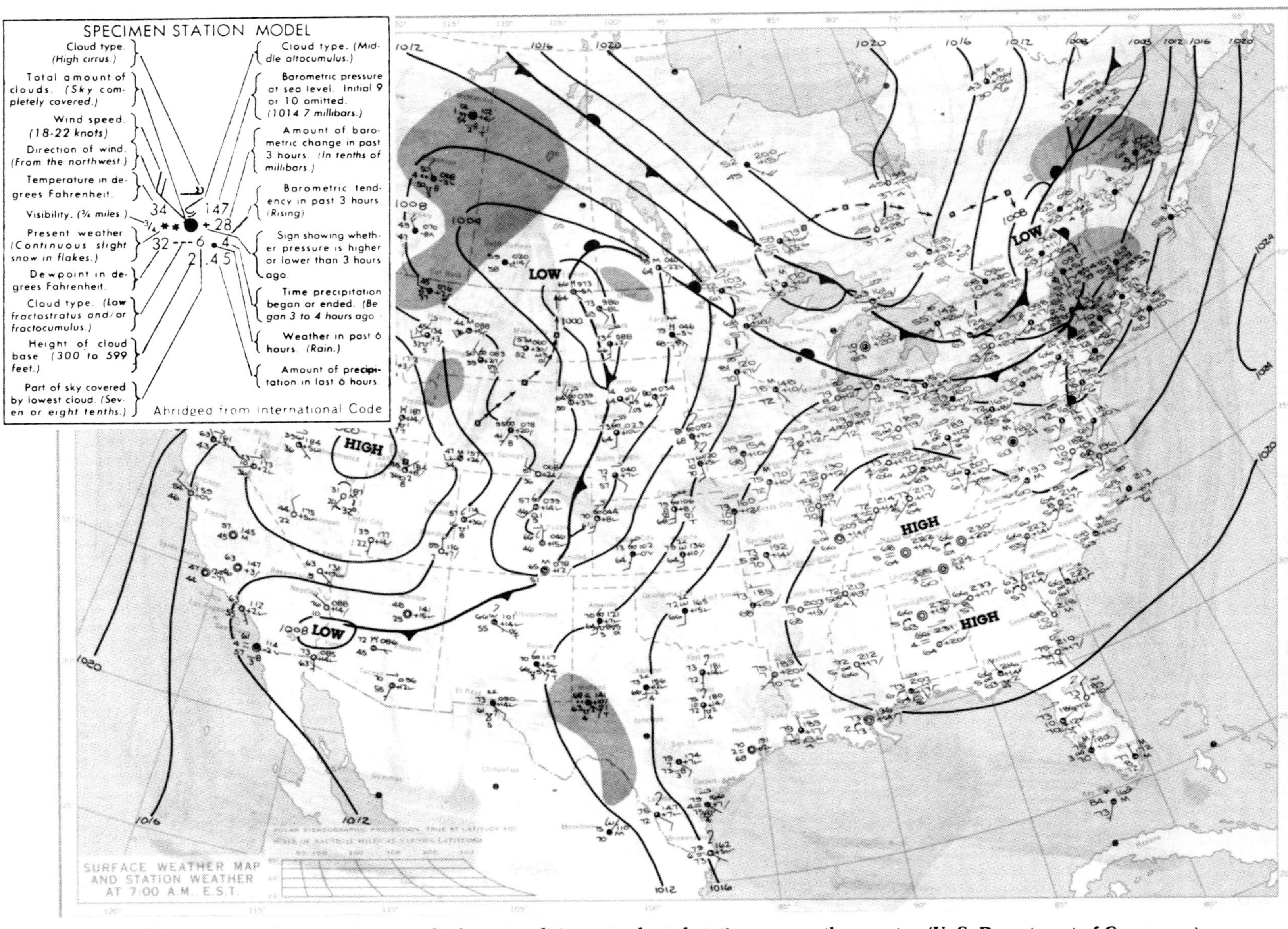

Fig. 24—33. A daily surface weather map. It shows conditions at selected stations across the country. (U. S. Department of Commerce)

The distribution of stars can be mapped. A common map that everyone uses is the auto road map. The number and variety of maps are almost limitless.

Fig. 24-35 shows a map used by the National Aeronautics and Space Administration to show which ground stations are in contact with spacecraft at various locations.

Land Survey

One common type of map that almost everyone uses is the land survey. When a house is sold, the land included in the sale is mapped. Entire cities are mapped. Farms and industrial properties are mapped.

The survey process requires too much explanation to go into much detail in this text. Basically, a **land survey** provides two types of data needed to draw the map. The compass direction of each boundary of the property is needed. The length of each side must be measured.

Land surveys use feet as the unit for measuring distance and acres as the unit for area. When metric measures are used, the hectare (ha) is the unit of area. A hectare equals 2.471 acres and an acre equals .405 hectare.

Compass direction is taken using the north-south axis as the base. The angle formed is called the **bearing angle**. Study the bearing angles in Fig. 24-36 at A. They are located in the four quadrants formed by the N-S and E-W axes. For example, the bearing angle is 30 degrees east of north, in part A of Fig. 24-36. It is read N 30 E. If the 30-degree bearing angle was in each of the other quadrants, it would read as shown at B, C, and D. The bearing angle cannot be over 90 degrees.

Another way the direction of a line can be shown is with azimuth angles. The **azimuth** is an angle measured from the north or south axis in a clockwise direction. The axis is zero. All angles are measured from the north axis in Fig. 24-37. The north axis is generally used for azimuth angles, but the south axis can also be used. Then the south axis is zero and the angles are measured clockwise from it, as shown in Fig. 24-38.

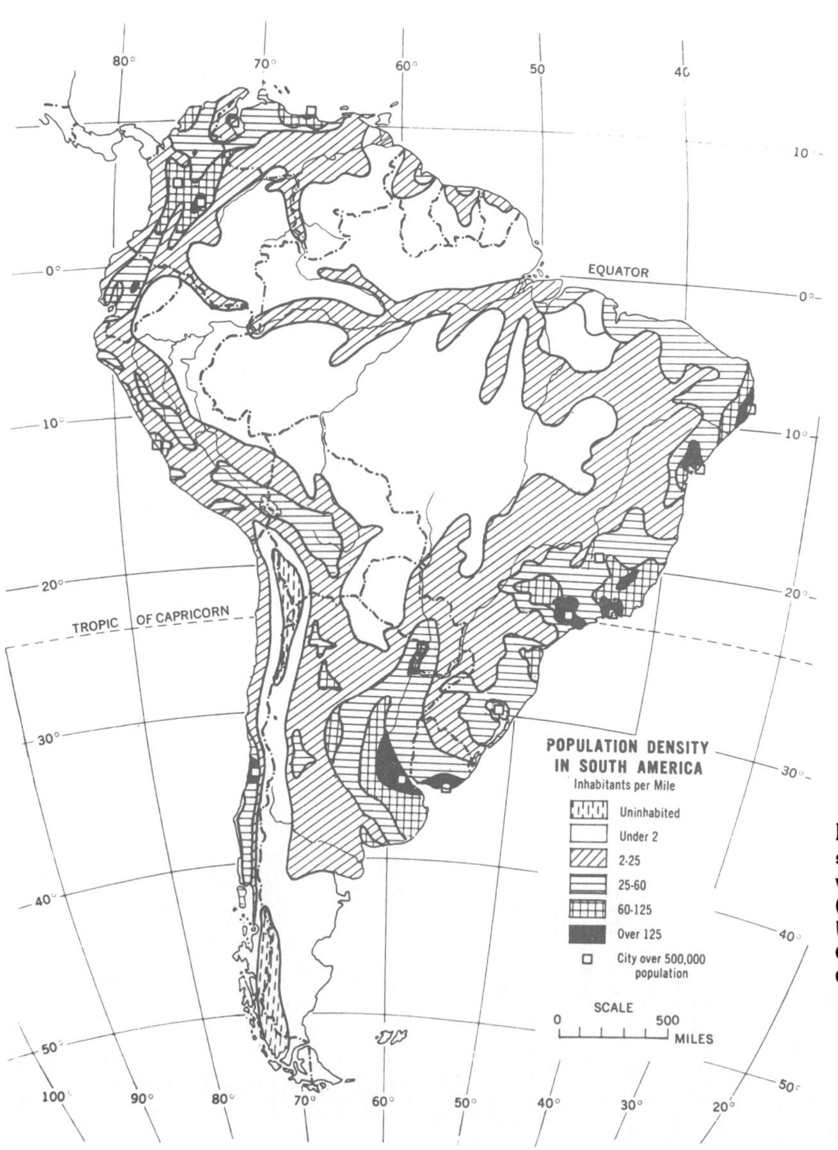

Fig. 24–34. A statistical map. It shows the population density in various regions in South America. (Reproduced with permission from *World Geography*, J. W. Morris and O. W. Freeman, McGraw-Hill Book Company)

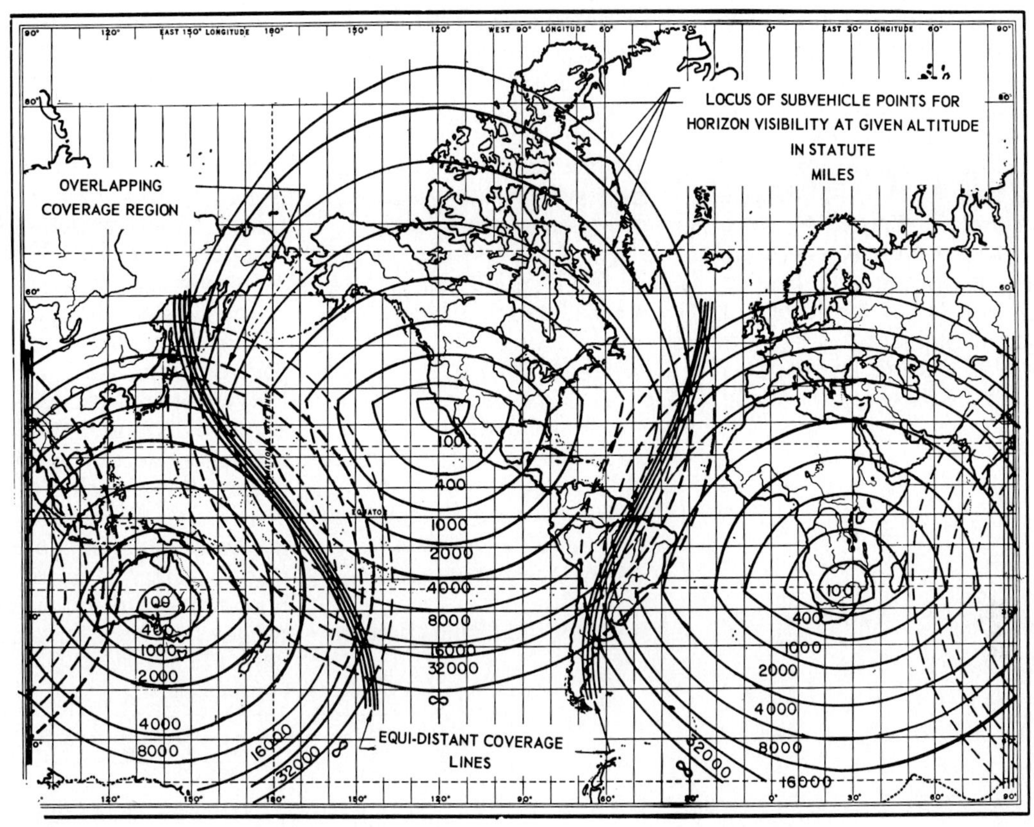

Fig. 24–35. This is a map developed to show the coverage of a spacecraft by three stations as the spacecraft increases its altitude. The three stations are Goldstone, California; Woomera, Australia; and Johannesburg, South Africa. (NASA)

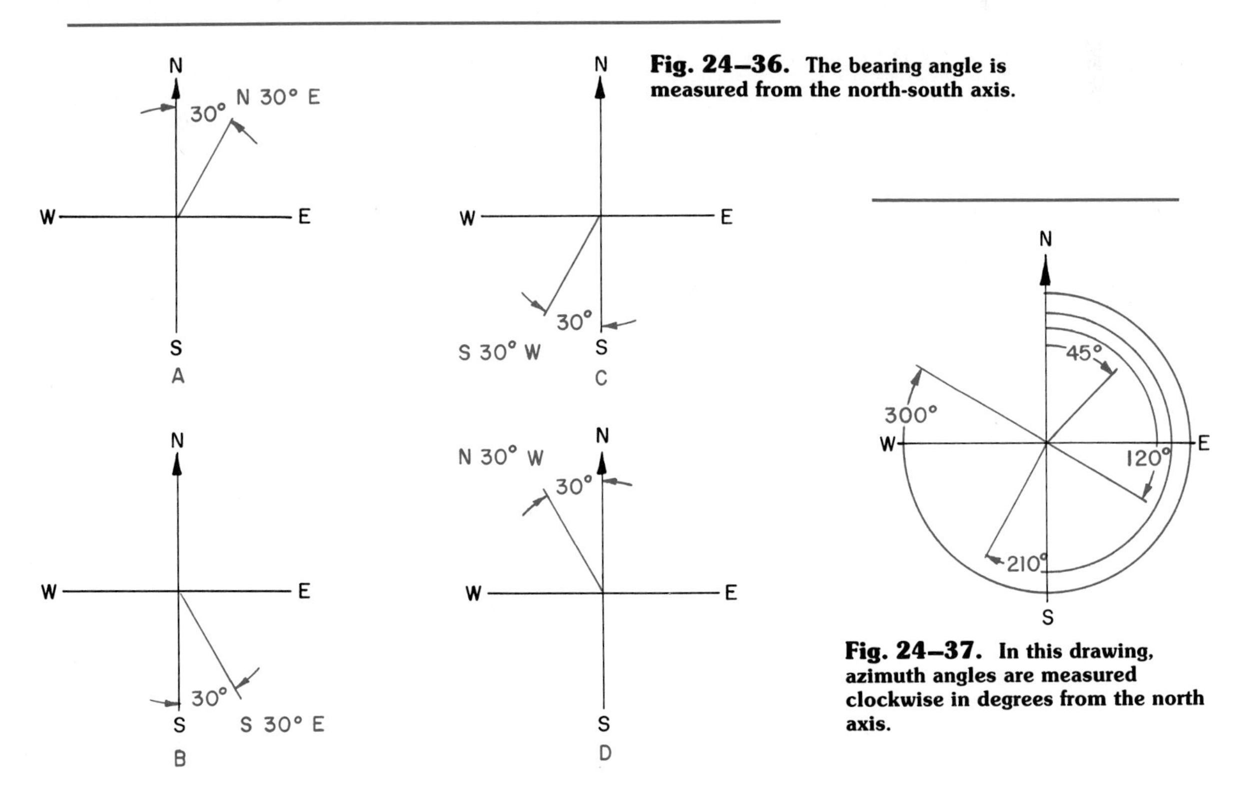

Fig. 24–36. The bearing angle is measured from the north-south axis.

Fig. 24–37. In this drawing, azimuth angles are measured clockwise in degrees from the north axis.

The map using azimuth angles should always tell whether the north or south axis is used.

In beginning a survey, the surveyor must accurately locate one corner. This is usually found by starting from a point that is known and working to a corner of the land to be surveyed. A transit is used to sight the bearing of one side. Then the length of the side is measured. A stake is driven at that point. The transit is moved to this stake and the angle and distance of the next side are measured. These angles and distances are usually obtained from the legal description of the land. This information is part of the deed to the land. Many landowners drive metal stakes at each corner to preserve these locations.

A typical land survey map is shown in Fig. 24-39. It is sometimes called a plat. It should contain the following:

- Direction of each side.
- Length of each side.
- Acreage in the property.
- Location and description of each permanent marker.
- Location of roads, streams, and rights-of-way.
- Division lines within the tract, if any exist.
- Names of adjoining property owners.
- A dedication. This is a word description of the bearings and length of sides of the property.
- A title, scale, date of survey, name of surveyor, official seal of surveyor, descriptive information, such as a lot number, location of property, such as county or city and state, and name of the owner.

The bearing angles and length of sides of a tract of land are recorded in the deed. The deed is filed with the recorder of deeds in a city or county. This makes a permanent record of this information. When such a tract is surveyed and the corners located, the surveyor often does not repeat locating the bearing angles. Sometimes the angles are given between the sides in degrees. Fig. 24-40.

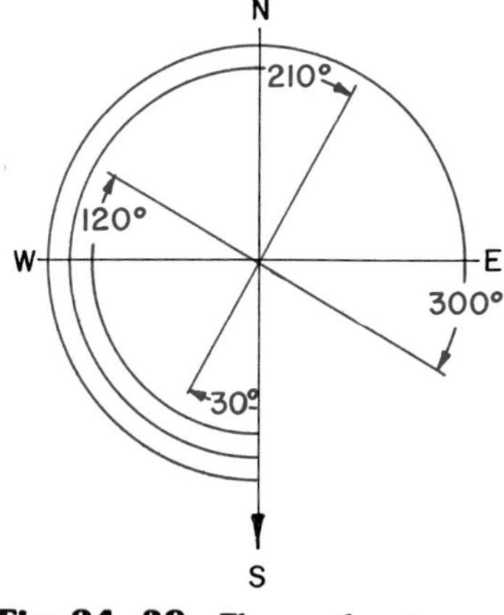

Fig. 24–38. **The south axis can be used to measure azimuth angles.**

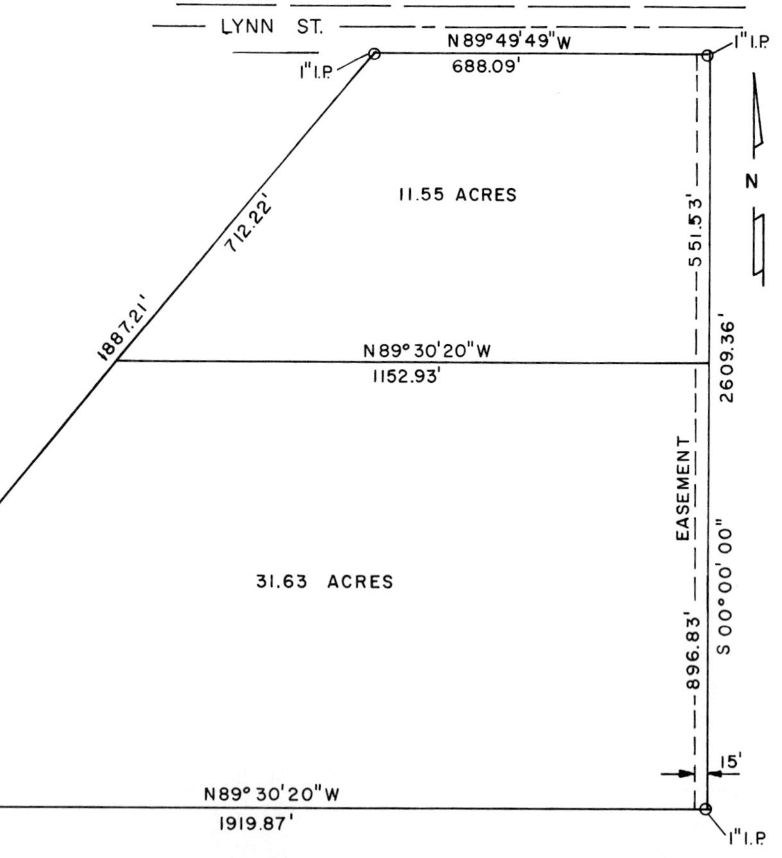

Fig. 24–39. **A land survey of acreage. Bearings are given using a bearing angle. Each corner is located with an iron pipe. This is shown on the drawing as "I.P."**

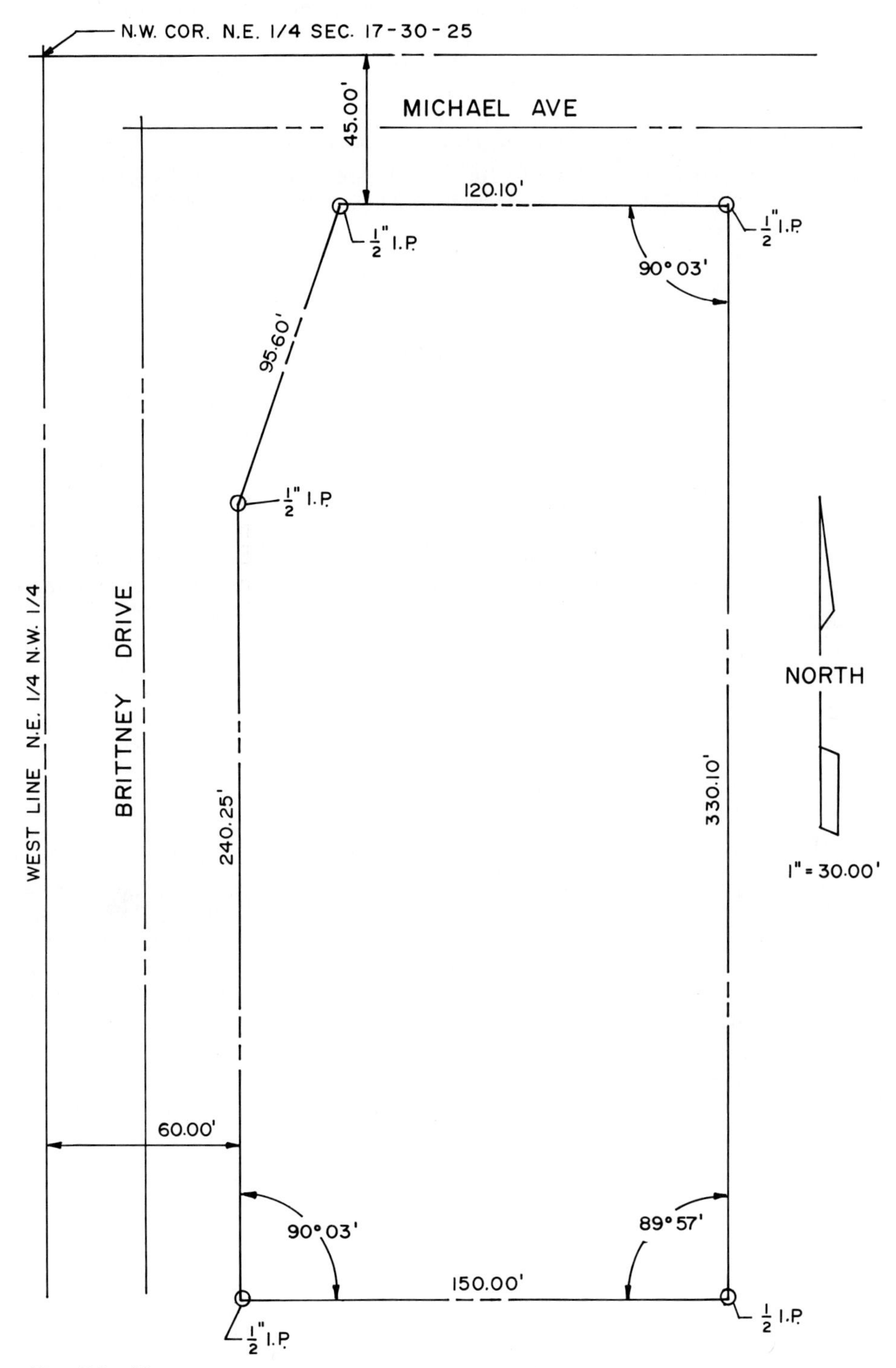

Fig. 24—40. **A land survey showing the direction of the sides using corner angles.**

After a land survey is made, the engineer writes a legal description. Fig. 24-41. It gives the general location of the land area. Then it cites the point of beginning. This is the point from which all measurements were taken. The description gives the direction and lengths of the sides of the tract of land. Try to match the legal description with the land survey shown in Fig. 24-40.

Another type of land survey is a plat of land divided into lots. This is done whenever acreage is divided to start a subdivision. This plat is filed with the city or county recorder. It shows the size and location of each lot. All permanent stakes are noted. These stakes are called monuments. Each lot is numbered. They are referred to by these numbers. For example, Fig. 24-42 shows that a section of land identified as "Lot 8" in Baker's Subdivision was divided into smaller lots. The legal description for lot 8 is in Fig. 24-43. As each of the 17 subdivided areas or individual lots is sold, a legal description will be written for it.

Cities make plats of all lots and acreage. They show streets and all major buildings. These plats are also used to show the location of gas, sewer, and water lines. They are used to plan improvements, such as new or rebuilt streets. Sometimes contours are shown. Two types of city plats are shown in Figs. 24-44 and 24-45.

Aerial Mapping

Aerial photographs are a good source of information for a cartographer. The photographs cover a small area of land surface. A large area is mapped by photographing a series of land areas. The photographs are taken so that they overlap. Fig. 24-46. Photographs taken perpendicular to the earth's surface give a true record of directions and distances. These are taken through a hole in the floor of the aircraft. The aircraft must fly as straight a course as possible. When it doubles back for a parallel run, it must be in a position so that the photographs overlap. Usually the end overlaps are from 50 to 60 percent. Side overlaps are from 15 to 50 percent.

DESCRIPTION

A PART OF THE N.E. 1/4 OF THE N.W. 1/4 OF SECTION 17, TOWNSHIP 30 SOUTH, RANGE 25 EAST, PITTSBURG, CRAWFORD COUNTY, KANSAS, MORE PARTICULARLY DESCRIBED AS FOLLOWS:

BEGINNING AT A POINT 580 FEET SOUTH AND 70 FEET EAST OF THE N.W. CORNER OF SAID N.E. 1/4 OF THE N.W. 1/4, THENCE EAST AND PARALLEL TO THE NORTH LINE OF SAID N.E. 1/4 OF THE N.W. 1/4 227.0 FEET, THENCE NORTH AND PARALLEL TO THE WEST LINE OF SAID N.E. 1/4 OF THE N.W. 1/4 550.0 FEET TO A POINT 30 FEET SOUTH AND 297 FEET EAST OF THE N.W. CORNER OF SAID N.E. 1/4 OF THE N.W. 1/4, THENCE WEST AND PARALLEL TO THE NORTH LINE OF SAID N.E. 1/4 OF THE N.W. 1/4 207.0 FEET, THENCE SOUTHWESTERLY 189.86 FEET TO A POINT 218.8 FEET SOUTH AND 70 FEET EAST OF THE N.W. CORNER OF SAID N.E. 1/4 OF THE N.W. 1/4, THENCE SOUTH AND PARALLEL TO THE WEST LINE OF SAID N.E. 1/4 OF THE N.W. 1/4 361.2 FEET TO THE POINT OF BEGINNING. CONTAINING 2.82 ACRES MORE OR LESS.

CERTIFICATE

THIS IS TO CERTIFY THAT I HAVE SURVEYED AND MARKED THE CORNERS OF THE ABOVE TRACT, AND THE ABOVE IS A TRUE RECORD THEREOF.

Gene E. Russell
GENE E. RUSSELL P.E. 2322

Fig. 24–41. A legal description of a land survey.

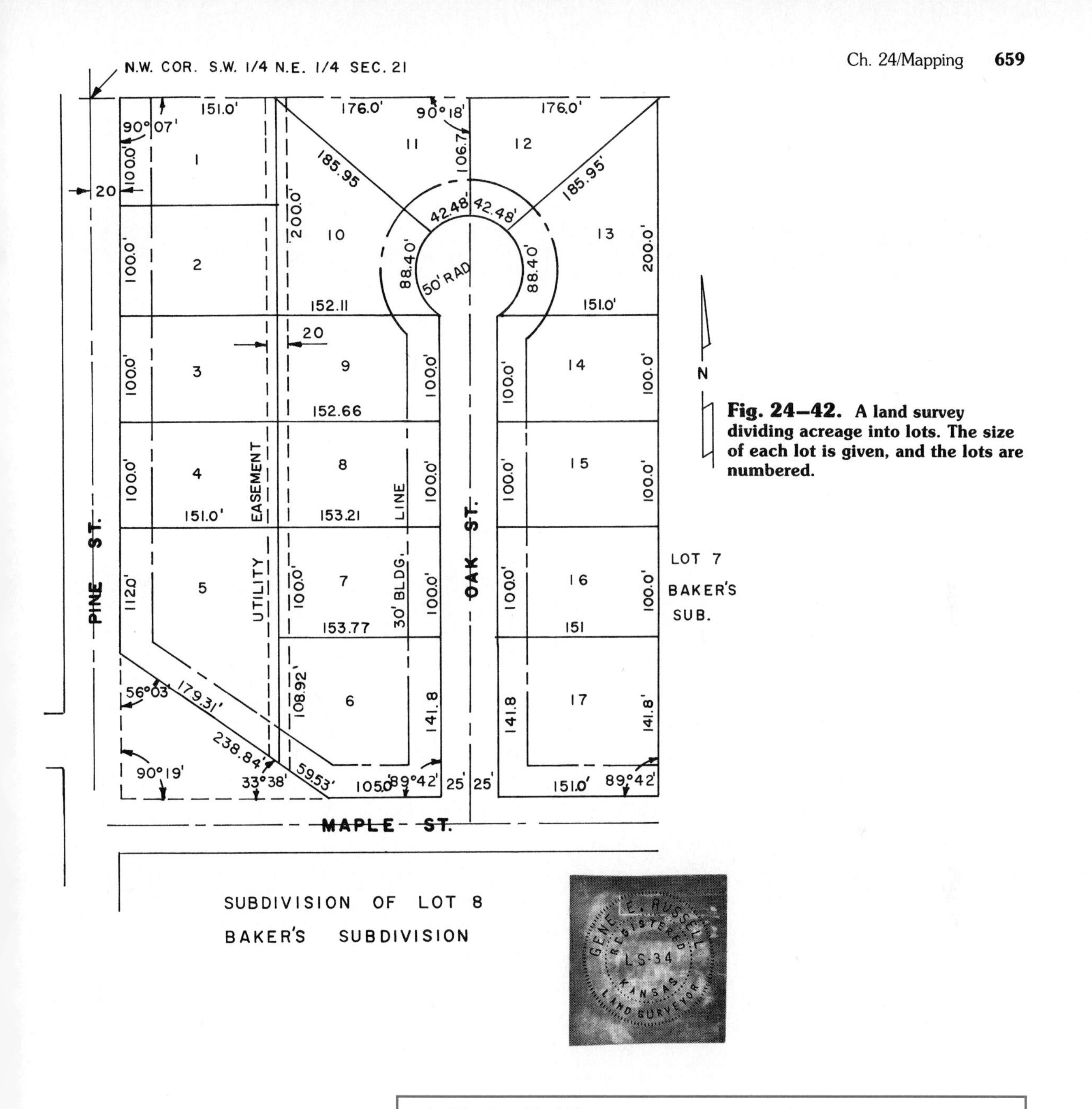

Fig. 24–42. A land survey dividing acreage into lots. The size of each lot is given, and the lots are numbered.

DESCRIPTION

A TRACT OF LAND IN LOT 8 BAKER'S SUBDIVISION, SECTION 21, T.W.P. 30 SOUTH, RANGE 25 EAST, LAWTON COUNTY, MISSOURI, MORE PARTICULARLY DESCRIBED AS FOLLOWS: BEGINNING AT A POINT 20' EAST OF THE NORTHWEST CORNER SOUTHWEST 1/4, NORTHEAST 1/4 OF SAID SECTION 21, THENCE EAST 503' TO THE NORTHEAST CORNER OF SAID LOT 8, THENCE SOUTH 641.8' TO THE SOUTHEAST CORNER OF SAID LOT 8, THENCE WEST 306' ALONG SOUTH LINE OF LOT 8, THENCE NORTHWESTERLY 238.84' TO A POINT ON THE WEST LINE OF SAID LOT 8 130' NORTH OF THE SOUTHWEST CORNER, THENCE NORTHERLY 512' MORE OR LESS TO THE POINT OF BEGINNING 7.14 ACRES MORE OR LESS.

Fig. 24–43. A legal description of a lot in the subdivision.

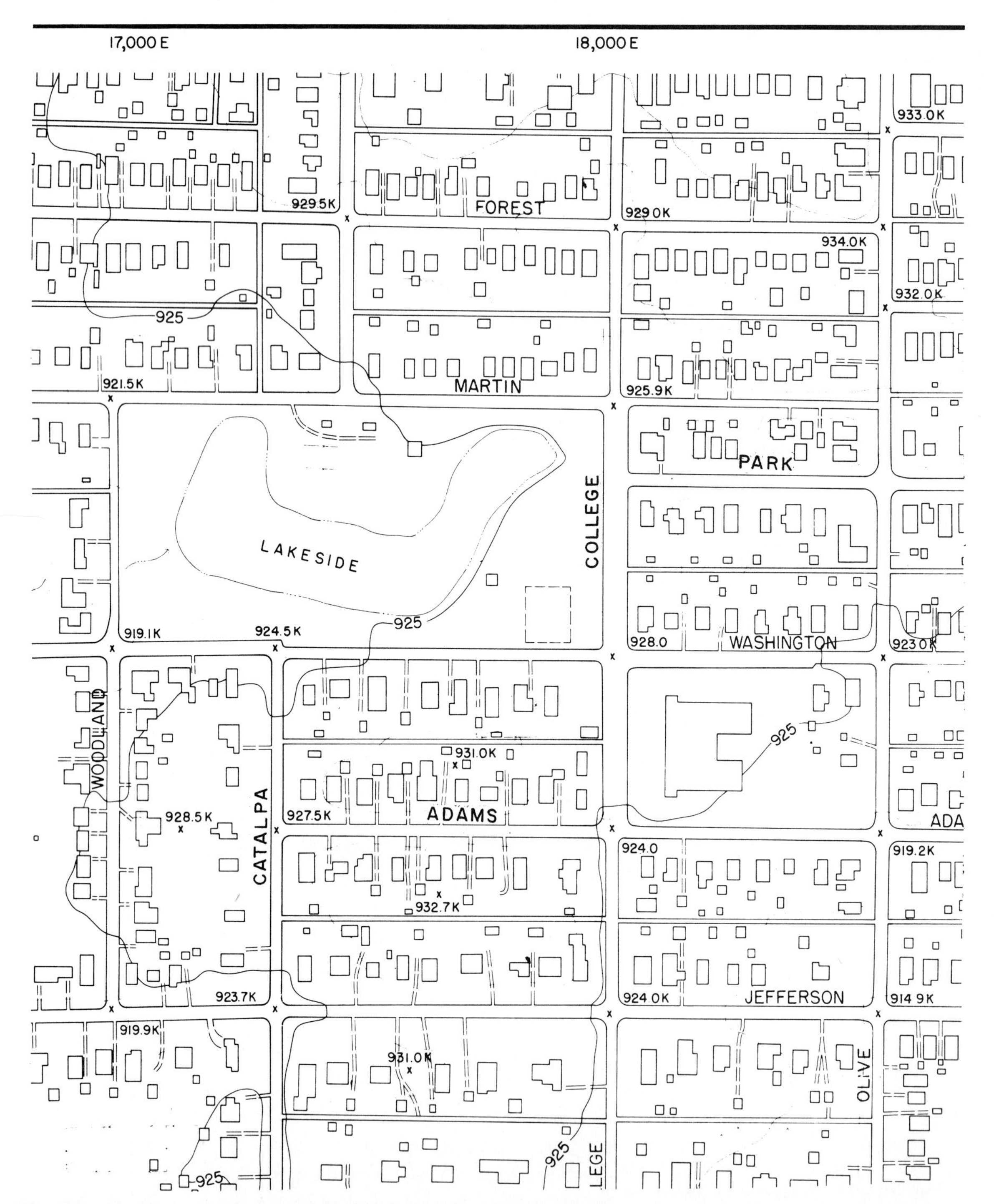

Fig. 24–44. Part of a city map that shows the location of every building. Contour lines are given. This basic drawing is used to show utilities, such as sewer and water lines. Copies of this basic drawing have these recorded on them. (City Engineering Department, Pittsburg, Kansas)

After photographs are taken and developed, the cartographers begin to work. The photos are interpreted. Then they are selected and arranged into one complete photo of the land area. This is called a *photomosaic*. It is rephotographed to produce the final product. Fig. 24-47. From this a line map is developed. Fig. 24-48. Study the photo and see if you can identify the important features. Then study the map. Find the rivers and streams. Notice the villages and roads. One area is wooded. A large area is farm- and grassland.

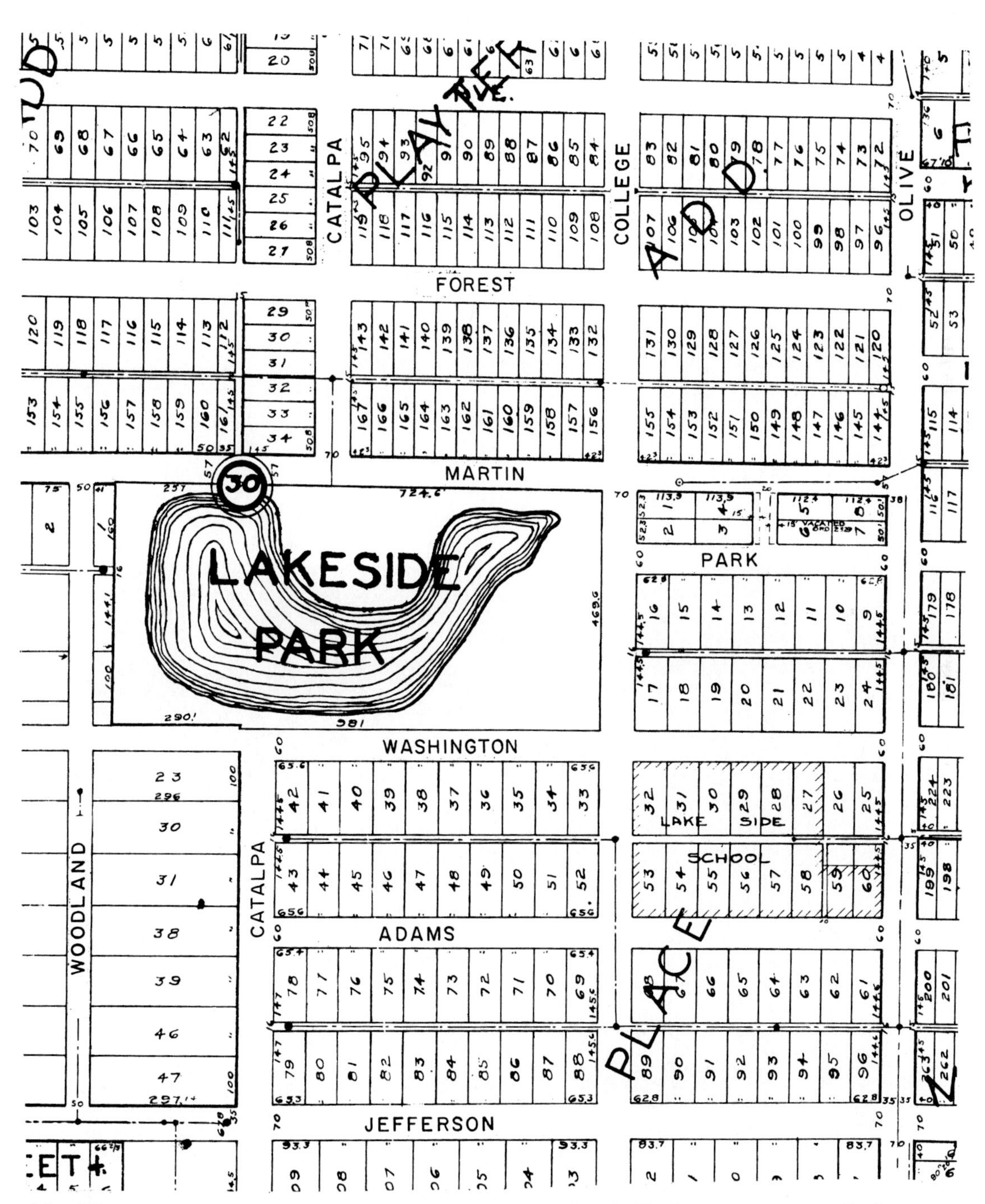

Fig. 24–45. Another form of a city map shows each lot in each block. The lot sizes are given. Each lot is identified by a number. This is a lot plan of the same area shown in Fig. 24–44. (City Engineering Department, Pittsburg, Kansas)

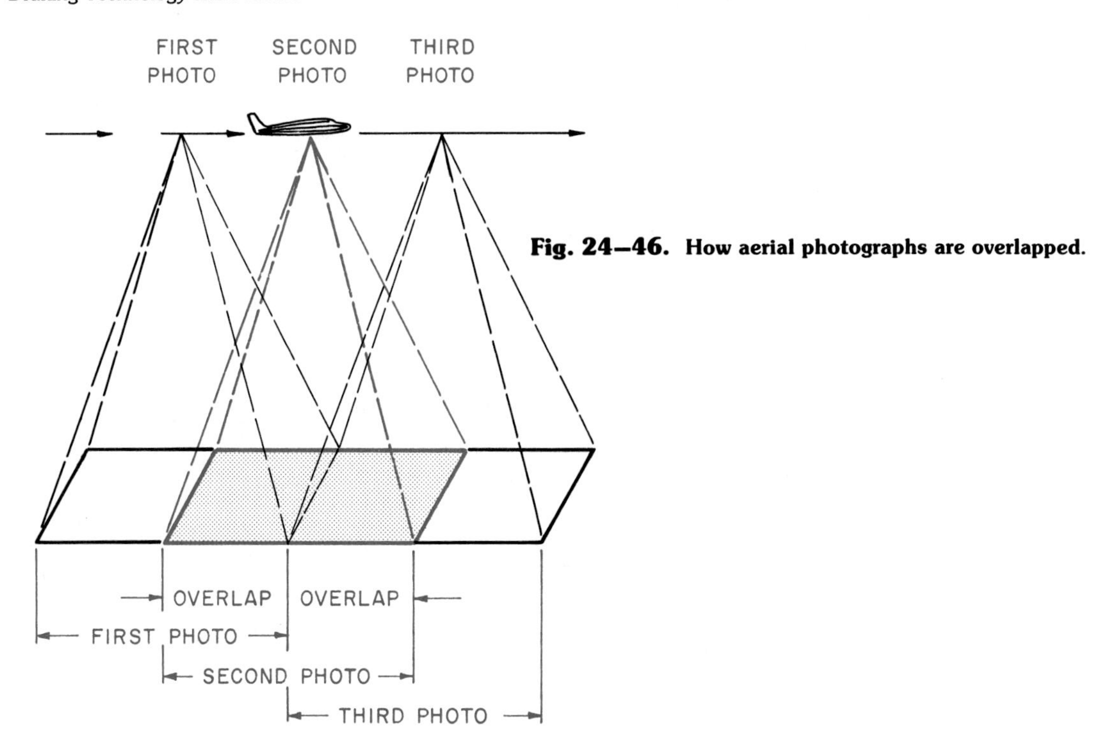

Fig. 24–46. How aerial photographs are overlapped.

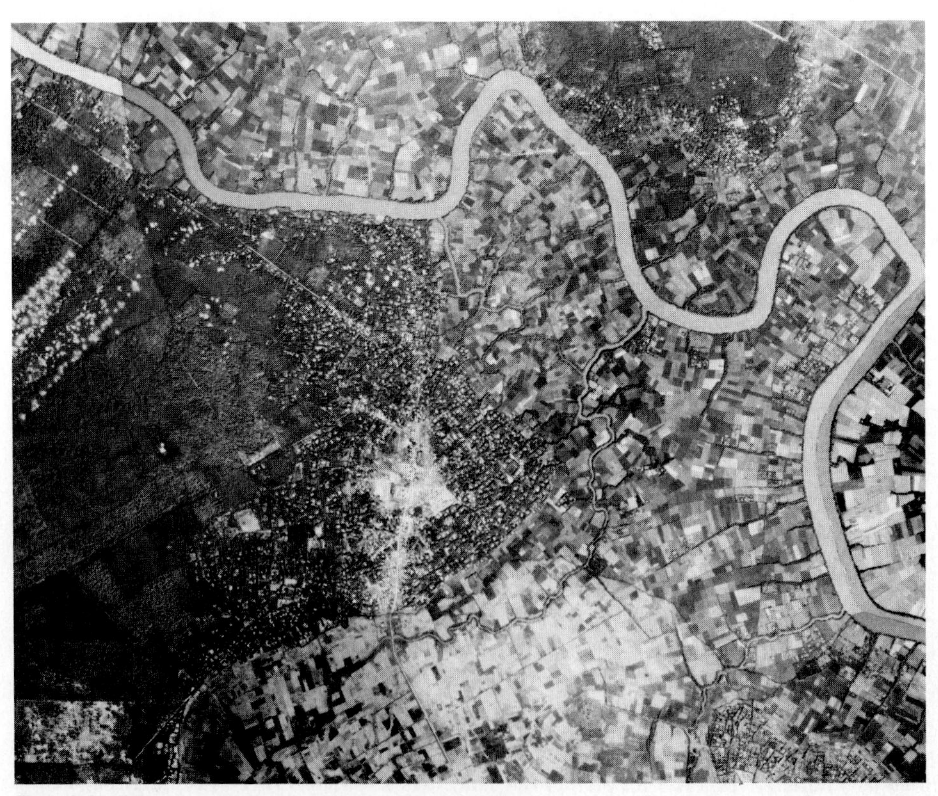

Fig. 24–47. Photomosaic of an area in Vietnam.

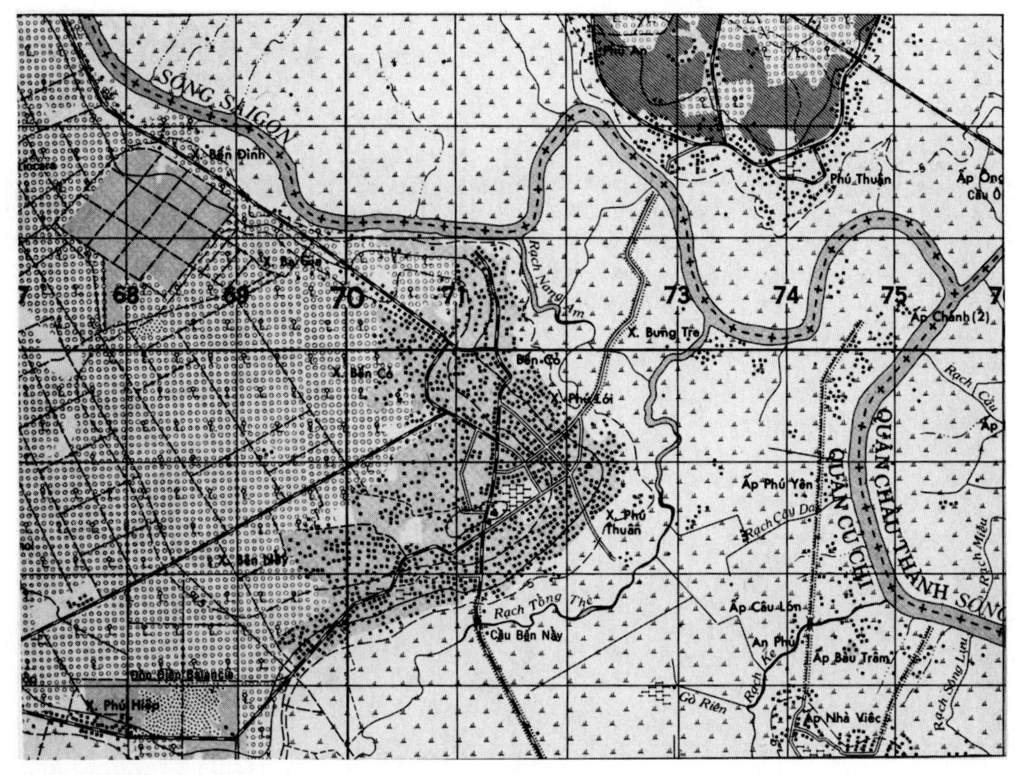

Fig. 24–48. Line map of an area in Vietnam. This photo covers the same area as the photomosaic shown in Fig. 24–47.

Chapter Review

Build Your Vocabulary

You should understand and use the following terms as part of your working vocabulary. Write a brief explanation of what each means.

cartographer
scale
legend
latitude
longitude
topographic maps
contour
profiles
land form map
hachure lines
charts
land survey
bearing angle
azimuth

Sharpen Your Math Skills

1. If you draw a map to the scale 1″ equals 40′, what size drawing is required to draw a map of an area 2500′ × 4000′?

2. If you draw a map to the scale 4 miles per inch, how many miles does a 36″ line represent?

3. A kilometer equals .624 mile. How many kilometers are there in 1325 miles? Round your answer to 1 decimal place.

4. One square kilometer is equal to .386 square mile. How many square kilometers are there in 100 square miles? Round your answer to 1 decimal place.

Study Problems—Directions

The problems that follow will give you the chance to use some of the drafting knowledge that you learned in this chapter. Do the problems that are assigned to you by your instructor.

1. Get a highway map of your state, and then do the following:
   a. List all the information it shows.
   b. Find your county on the map. Draw it to a scale so that the finished map will fit in an 8- by 10-inch drawing area. Locate all cities, roads, rivers, airports, and special features. Record the scale used. Use color to improve the clarity of the map.
2. Find the parallel of latitude and longitude of your town or city.
3. Find the parallel of latitude of the northern and southern borders of your state. Use a world atlas to find which countries in other parts of the world are located within these parallels.
4. Make a map of your state. Locate each county. Find the population for each county. Using color or shading, indicate the population densities. Develop your own legend.

Illustrations P24-1 through P24-6 are line drawings of states and countries. They are generalized. You can draw them by using the grid method. More detailed information can be found in a world atlas.

5. Draw a map of South America, as shown in **Fig. P24-1**. Locate each country. Develop a legend to show the national average food consumption in calories per day. Following are the data to be charted.
Brazil: 2,400 to 2,900
Bolivia: under 2,400
Peru: under 2,400
Ecquador: under 2,400
Colombia: under 2,400
Venezuela: under 2,400
Guyana: under 2,400
Suriname: under 2,400
Paraguay: under 2,400
Chile: 2,400 to 2,900
French Guiana: under 2,400
Argentina: 2,900 and over
Uruguay: 2,900 and over
6. Secure a highway map of your state. Based on this, make a map showing your state's river system. Also show the major highways. Locate any special features such as parks, camping grounds, and historical sites. Locate all cities with over 25,000 population, and locate the state capital. Develop a legend to identify the special features and the size of the cities.
7. Cut the weather map from your daily newspaper. Make a large drawing of this map. Use color to emphasize the important parts.
8. A map of the United States (minus Alaska and Hawaii) is shown in **Fig. P24-2**. Make a map showing any statistical or economic data you desire. This could be the population of each state. It could be the dollar volume of manufacturing. Data can be found in an atlas or encyclopedia. Develop a legend. Use color.
9. A map of Arkansas is shown in **Fig. P24-3**. The heights of various areas are shown. Draw a topographic map using land form symbols. This map is greatly simplified.
10. Find a map of your state in an encyclopedia or geography book. Study it to find the heights of its various parts. Draw a relief map using land form symbols. Use color to give emphasis to the various heights. Draw a color legend in a clear corner.
11. Copy the map of Libya, **Fig. P24-4**. Make a planographic map to show elevations. Carefully select the colors used to represent the heights.
12. Data for a contour map are given in **Fig. P24-5**. Enlarge this map to use most of an 8- by 10-inch drawing area. The contour intervals should be 10 feet. Make a contour map connecting points of elevation so the 10-foot contour intervals are shown.
13. Make a profile drawing using the data for the contour map in Problem 12. The location of the profile cut is shown on the drawing. Select a scale that clearly shows the profile.
14. Make a line map like the one shown in Fig. 24-48. The data for the map are given in **Fig. P24-6**. Use standard symbols to indicate the features. To find distances, use the scale with the line drawing layout.
15. The data in **Fig. P24-7** are from a surveyor's field notes. Make a plot of this acreage. Start the drawing by locating a point 1 inch from the left edge of your paper and 1 inch from the bottom. The bearings and distances are taken from this point. It is Point 1. The angles given are bearing angles. Use the scale 1″ = 100′.

16. Use the data shown in **Fig. P24-8** to draw a plot for a subdivision. Start the drawing in the lower left corner of the paper. The subdivision has the following specifications:

Draw a 50-foot wide street right-of-way, 175 feet from the west boundary and parallel to it. Then, draw another 50-foot wide street 350 feet east from the first street. These streets divide the site into three tracts. Beginning on the west side of the site, divide the first tract into 100-foot wide lots. Divide the center tract into 100-by 175-foot lots, each facing a street. Divide the third tract into 150-foot wide lots running their depth to the east boundary. Add any extra footage left over on the widths to the lot on the north boundary. Number each lot and letter the length of each side on the drawing.

Study Problems

P24–1.

Study Problems

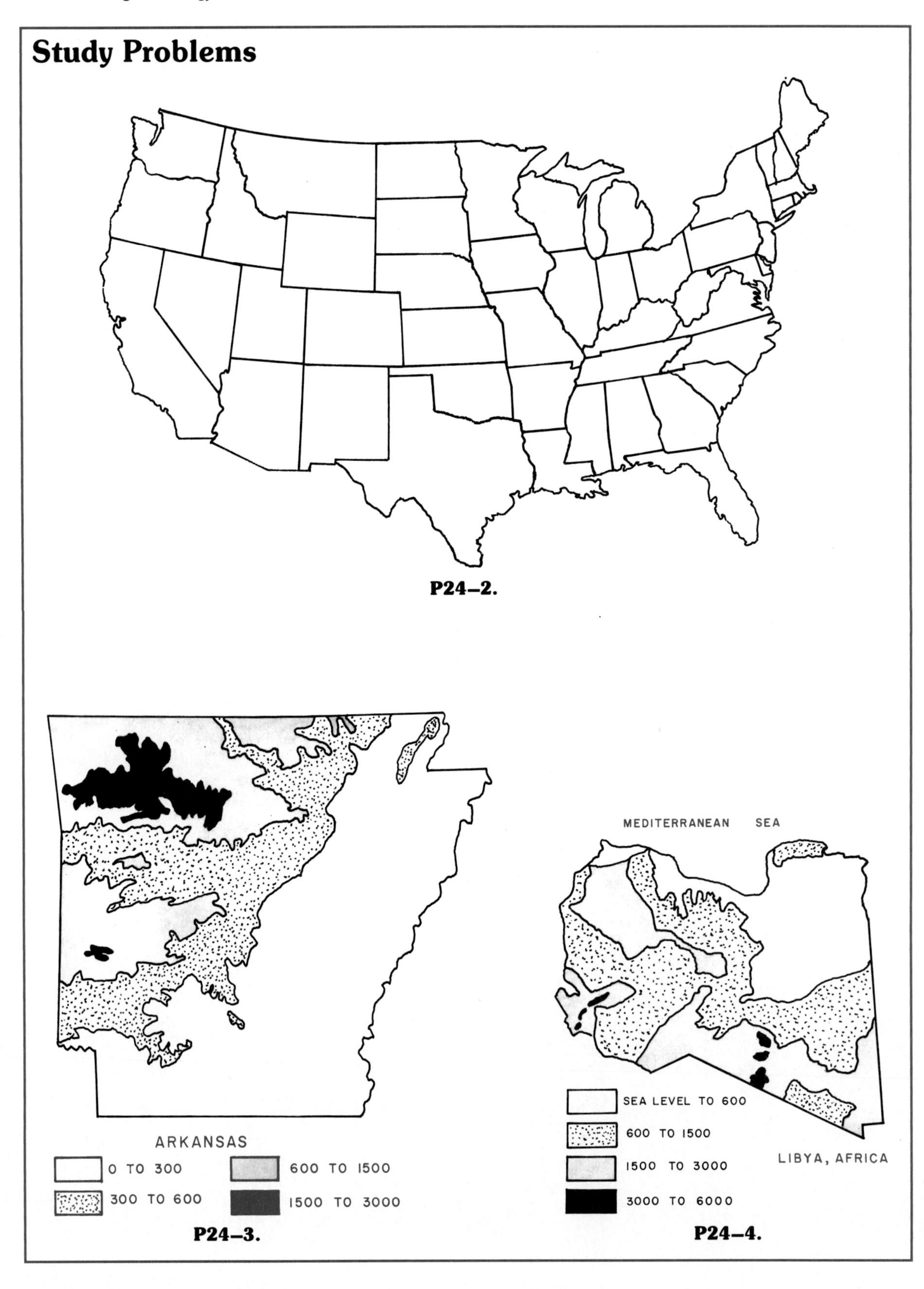

P24–2.

P24–3.

P24–4.

Study Problems

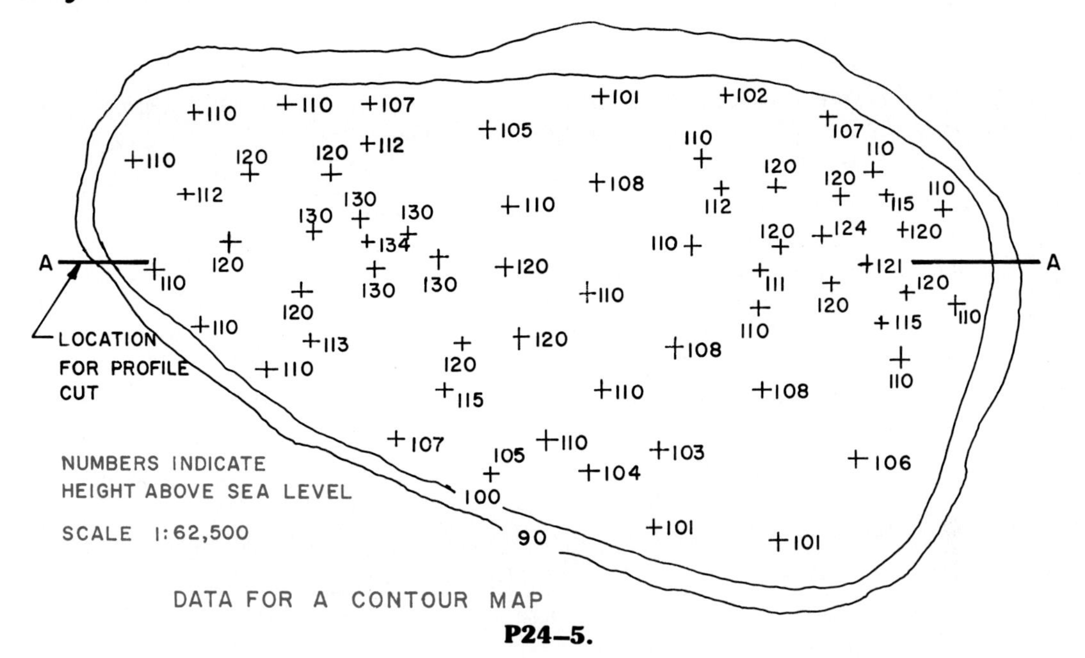

DATA FOR A CONTOUR MAP

P24–5.

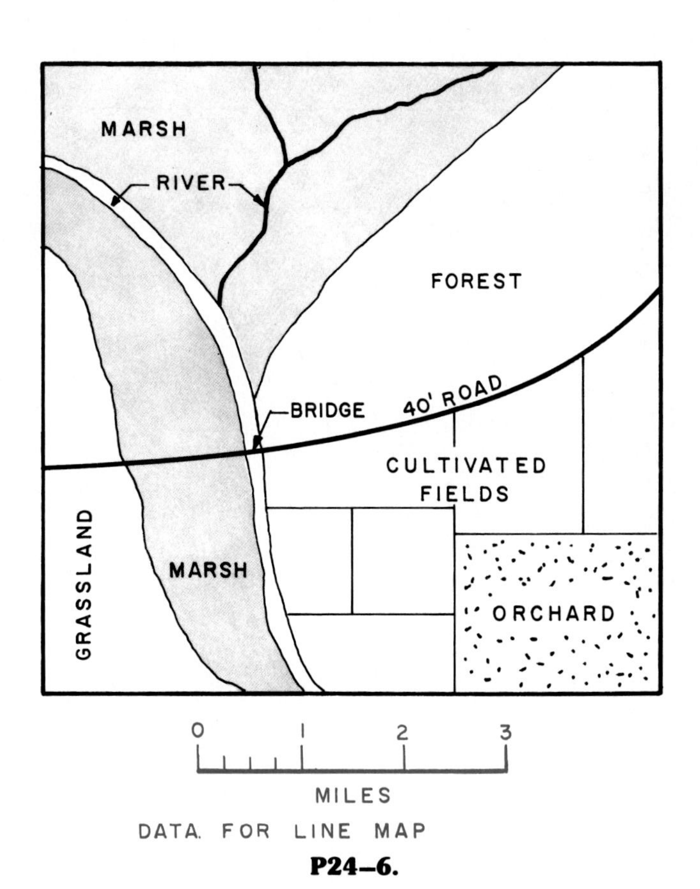

DATA FOR LINE MAP

P24–6.

Study Problems

P24–7.

	Bearing	Distance
Point 1 to 2	North	575′
Point 2 to 3	N 45°– 0′ E	200′
Point 3 to 4	S 70°– 0′ E	625′
Point 4 to 5	S 30°– 0′ E	350′
Point 5 to 6	S 25°– 0′ W	300′
Point 6 to 1	To closure at Point 1	

P24–8.

	Bearing	Distance
Point 1 to 2	N 5°– 0′ E	700′
Point 2 to 3	S 85°– 0′ E	900′
Point 3 to 4	S	531′
Point 4 to 1	To closure at Point 1	

CHAPTER 25
Architectural Drawing

After studying this chapter, you should be able to do the following:
- Design simple floor plans.
- Select and draw structural details for several types of houses.
- Design stairs.
- Draw floor plans, foundation plans, elevations, and details.

Introduction

The design of houses and commercial buildings is the work of an **architect**. Architectural drafters make the actual drawings. They must know how buildings are constructed. They must be able to use the graphic symbols of architecture.

Planning a House

The architect is responsible for designing a house to meet the needs of the client. In order to do this, the architect must determine the following things:
- How much the family can afford to spend.
- The style of house they like.
- The size of the family.
- The ages of the members of the family.
- Recreational and social activities they enjoy.
- The shape and slope of the lot upon which the house is to be built.
- Building codes and deed restrictions.

To plan a house, many decisions must be made. Some of these are:
- The style of the house such as one-story, two-story, or split-level.
- The number and types of rooms.
- The type of foundation: basement, concrete slab floor, or crawl space.
- The type of foundation construction: concrete block or poured concrete.
- The type of wall construction: wood frame or masonry.
- The type of roof and roofing material.
- The type of windows.
- The type of heating and cooling systems.
- The size of the garage.

A house is designed around three areas: food preparation, living, and sleeping. Fig. 25-1. When the architect designs a floor plan, these sections of a house must be planned so they are separate areas.

The food preparation area includes food storage, preparation, cleanup, and serving. All these activities involve the movement of food and dishes and a certain amount of odor and mess. Therefore, these activities should not conflict with the living and sleeping activities.

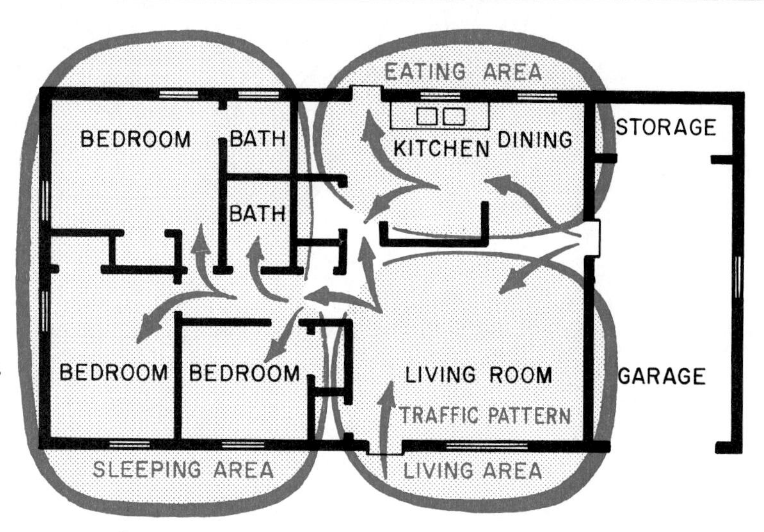

Fig. 25–1. **A house is designed around three areas: sleeping, eating, and living.**

Fig. 25–2. An open floor plan. Notice the family room in the background. The house appears much larger than it is because of the lack of walls that block off each area. (Armstrong)

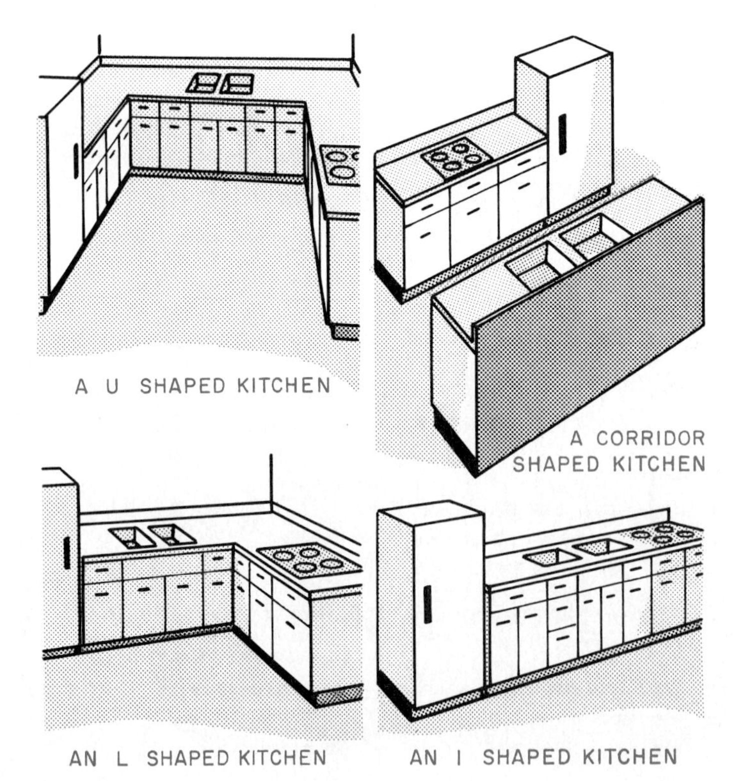

Fig. 25–3. Common kitchen shapes.

The living area also includes noisy activities. The living room, family room, game room, or hobby space are typically found in this area. This noise, while a necessary part of living, would tend to disrupt those desiring to rest in the sleeping area unless living area is planned carefully.

The sleeping area should be located away from the areas full of activity. It should be located on the cool side of the house if possible.

Bathrooms must be located so that they can be reached easily from all areas. It should not be necessary to disrupt activity in any area to reach a bathroom.

The traffic patterns in a house should be planned. **Traffic patterns** are the natural paths that people use to go from one area of the house to another. Those entering the front door should be able to go to any part of the house without going through one of the areas. For example, it should not be necessary to walk through the living room to get to the bedroom area. Groceries should be carried directly into the kitchen. The stairs to the basement should be centrally located and reached easily from any area in the house. Notice the traffic patterns on the plan shown in Fig. 25-1.

Small houses can be made to appear larger by reducing the number of walls. This is called **open planning**. For example, the kitchen and dining area can be one large room. The family room and laundry can be together. The living room and dining area can be combined. Fig. 25-2.

Planning the Kitchen

Kitchens are the center of many activities such as food preparation, dining, laundry, and sewing. Kitchens are usually arranged in "I," "L," "U," and corridor shapes. Fig. 25-3.

Kitchens are planned around three appliances: the refrigerator, range, and sink. Cabinets are built between these units.

The sink is usually placed near the food storage area. It is used to prepare food and clean up afterwards. It should not be too far from the dining area.

The range is placed near the dining area. The hot food is served from the range to this area.

The refrigerator is the center of the food storage area. It should be near the sink. It is best if it is also near the door through which groceries enter the kitchen.

The total walking distance among the three major appliances should total no more than 22 feet. If it is more than this, the kitchen plan should be revised.

Ideally, the kitchen should have at least 15 feet of free counter space. This is in addition to the surfaces taken up by the sink and range. The sink and range should have 2 to 3 feet of counter space on each side. It is helpful to have some counter space beside the refrigerator. Study the kitchens shown in the plans in this chapter.

If laundry facilities are in a kitchen, they should not interfere with the food preparation areas.

A small eating area is often part of the kitchen. It is used for light snacks and breakfasts. Fig. 25-4.

The size of kitchen cabinets is standardized. Cabinet details and sizes are given in Fig. 25-5.

Planning the Dining Area

The dining area can be a separate room near the kitchen, as shown in Fig. 25-6, or part of the kitchen. Sometimes it is combined with a family room or living room.

A major factor in designing the dining area is making it large enough to hold the furniture. Room must be left behind each chair so that a person can be seated easily. Fig. 25-7. The dining area should have a pleasant atmosphere. Large windows are helpful.

Plans should include storage space for linens, china, and silverware in or near the dining area.

Planning the Living Area

The living room is one of the most used rooms in a house. It can serve many purposes. For some families, it is a sitting room, music room, study, and library. It can even serve as an extra bedroom for short periods of time.

Usually this room is placed facing the best outside view. Large areas of windows are popular. However, they reduce the amount of wall space available for furniture. The living room should have a central focal point. This is often a fireplace or a picture window. Furniture is grouped so conversation can be held easily. Persons seated over six feet apart have difficulty carrying on a conversation. Fig. 25-8. It is best if the front entrance does not open directly into the living room. It makes part of the room into a hall and reduces the space available for use.

Do not crowd the furniture. Leave aisles of at least 3′—6″. This helps people move about the room.

Fig. 25—4. A small eating area in a kitchen. The atmosphere is bright and cheerful. (Armstrong Cork Company)

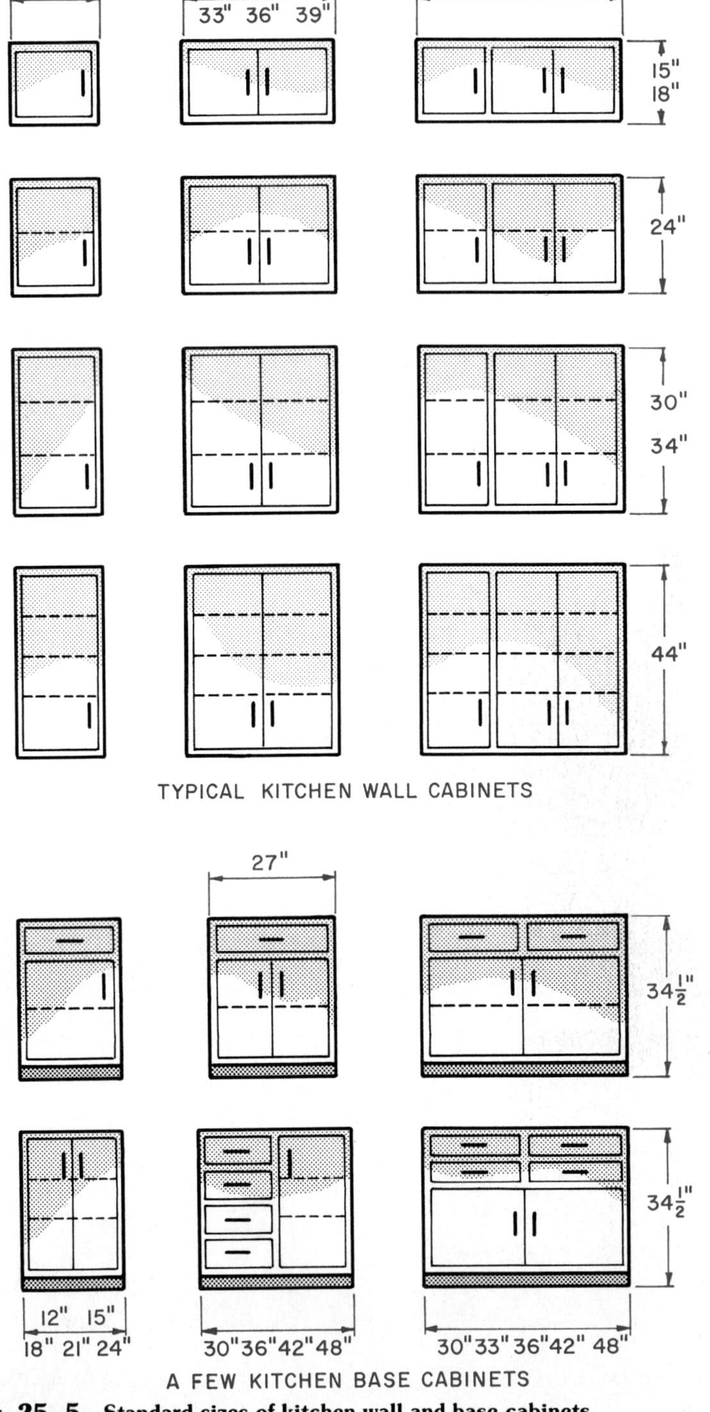

Fig. 25-5. Standard sizes of kitchen wall and base cabinets.

Planning the Family Room

The family room is used for hobbies and relaxation. It is an informal room. It can be used for television viewing, laundry, and games. It can include a dining area. It can open onto a patio or screened porch. Fireplaces are often placed in the family room. Fig. 25-9.

Planning the Bedrooms

Bedrooms must be large enough to hold the furniture planned. Aisles for traffic must be provided. Aisles should be at least 24 inches wide. Space for dressing is needed. Closets are essential. The area in front of closets should be free of furniture. Fig. 25-10 shows minimum spacing for furniture.

Windows are needed for proper ventilation. Some prefer windows placed 4 to 5 feet above the floor. This insures privacy and makes the wall available for placing furniture below the window.

Bedrooms should be placed so they are entered from a hall. It should never be necessary to go through one bedroom to get to another.

It is convenient to have a full or half bath off the master bedroom.

Planning Closets

Closets are essential in every house. They are used in every bedroom. Hall closets are used for linens and other storage. A closet is needed near the front entrance. This gives a place for visitors to hang their coats.

Bedroom closets should be at least 2 feet deep. Fig. 25-11. It is best if 3 to 4 feet of closet space is provided for each person using the bedroom. Sliding or folding doors are generally used.

Each bedroom closet should have a clothes rod. At least one shelf should be above the rod. It is desirable to have a light in each closet.

Fig. 25—6. An attractive dining room. Notice that the table and chairs are not crowded. The atmosphere is pleasant.

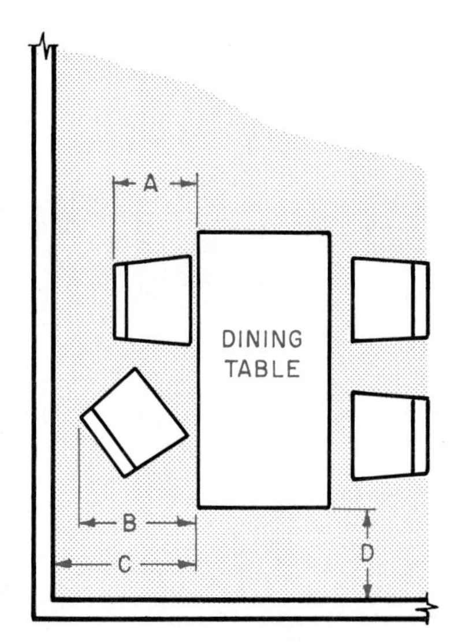

A- PERSON SEATED NEEDS 1'-6" TO 1'-10" FROM CHAIR BACK TO TABLE
B- AVERAGE PERSON NEEDS 2'-6" TO 3'-1" TO RISE FROM TABLE.
C- TABLE SHOULD BE 3'-6" FROM WALL IF AN AISLE IS PLANNED BEHIND THE CHAIRS.
D- AISLE SHOULD BE 2'-0" IF NO ONE IS TO BE SEATED AT END OF TABLE.

Fig. 25—7. Adequate dining space dimensions.

Fig. 25—8. A living room designed for comfort and conversation. The seating is grouped around a circular table. Windows provide a pleasant outside view. (Armstrong Cork Co.)

Fig. 25–9. A family room provides for a variety of activities. It should have durable furnishings and be made of materials that will withstand hard use. (Armstrong Cork Co.)

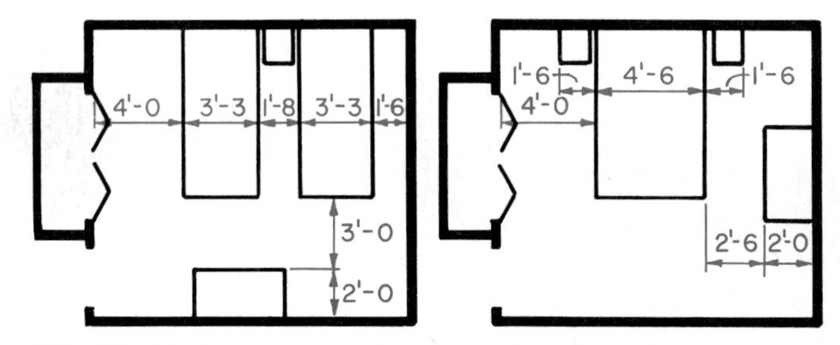

Fig. 25–10. Minimum spacing for twin and double beds.

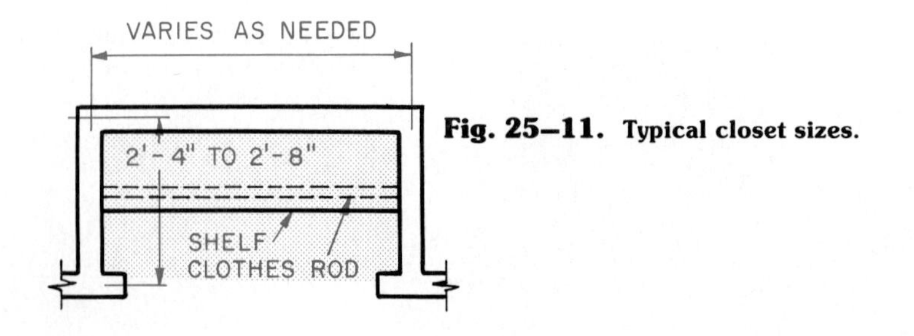

Fig. 25–11. Typical closet sizes.

Linen closets can be as shallow as 1 foot. The length can vary. The average house needs a linen closet at least 2 feet long.

The closet doors should open the entire width. Fig. 25-11. If they do not, space is wasted.

Planning a Bath

Most houses have one and a half baths, and two full baths are common. A four-bedroom house should have two full baths. Houses built on several levels should have a bath or water closet on each level.

Fig. 25-12 gives the most commonly used fixture sizes. When planning baths for a house, select the fixtures first. Then design the bath around the actual sizes of the units. Using a shower in a second bath instead of a tub saves floor space. Fig. 25-13 gives the minimum spacing for bath fixtures. Each fixture should be easily accessible.

Putting two baths side-by-side is economical. Fig. 25-14. Another way to reduce plumbing costs is to back up the bath to the kitchen.

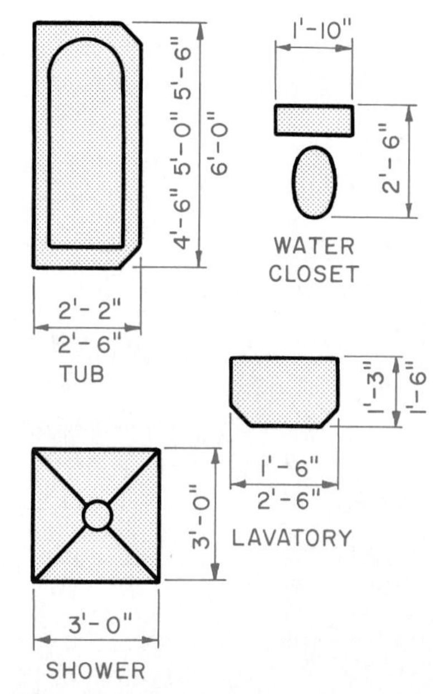

Fig. 25–12. Standard bath fixture sizes.

A bath can be planned so that two people can use it at the same time. In this arrangement, the water closet is placed in a separate room. Fig. 25-15.

The door to the bath should be placed so that the water closet is not seen when the door is open. To avoid drafts, windows are not usually placed over a tub. Such windows are also difficult to reach and open. If a house has one

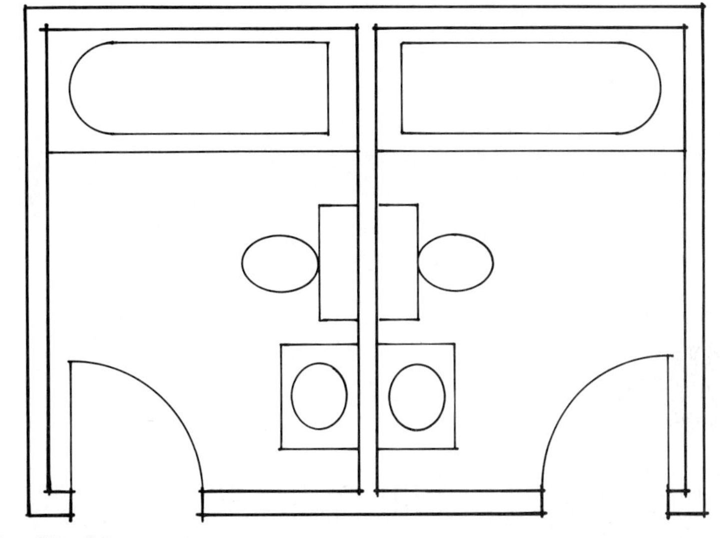

Fig. 25–14. **Two baths can be placed side-by-side to lower plumbing costs.**

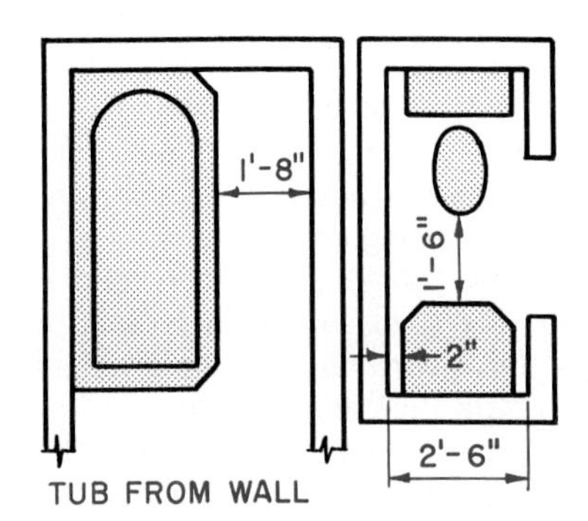

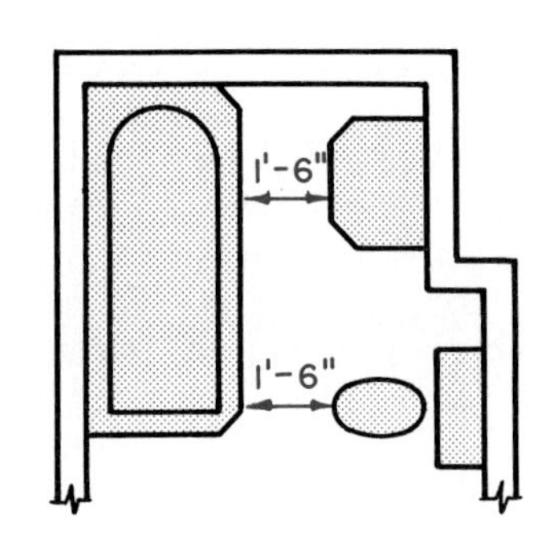

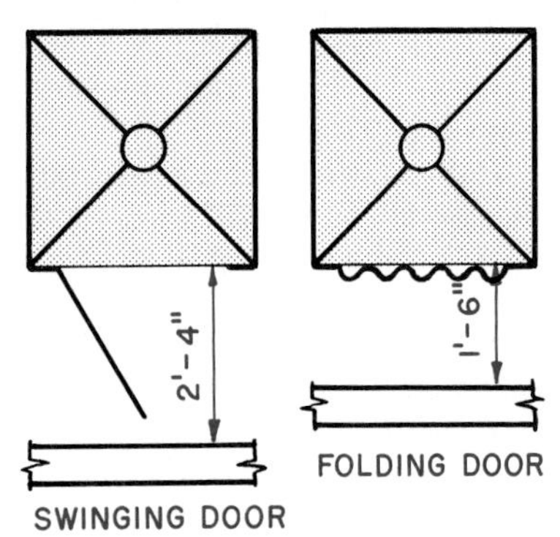

Fig. 25–13. **Minimum spacing for bath fixtures.**

Fig. 25–15. **A bath planned for use by several persons. Two lavatories are available. The water closet is in a room by itself just beyond the tub. (American Olean Tile Co.)**

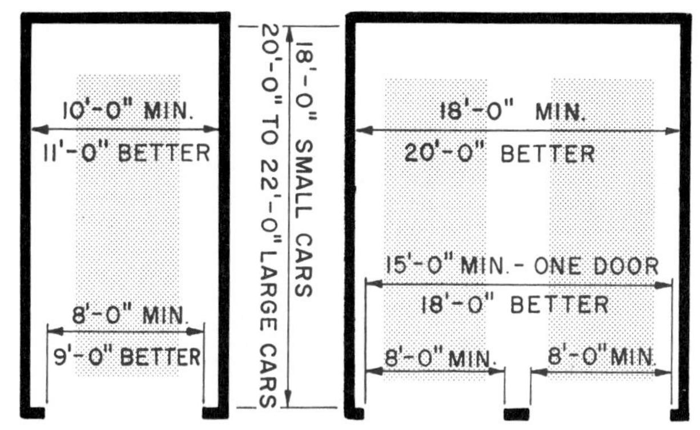

Fig. 25–16. Typical garage sizes.

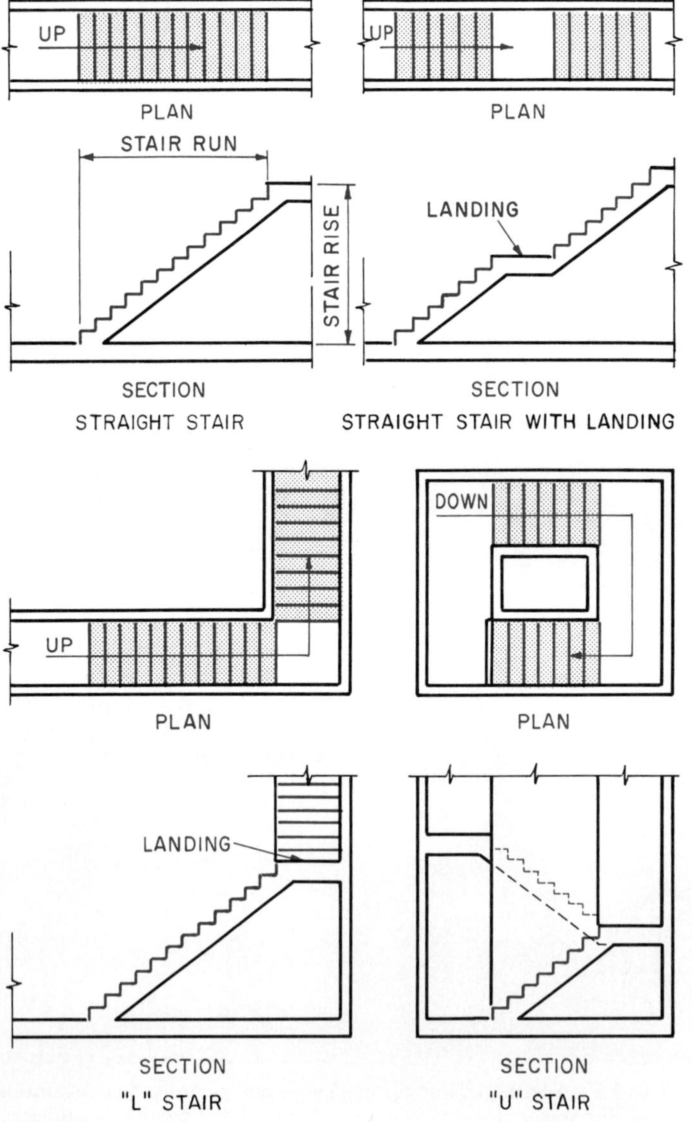

Fig. 25–17. Common stairs.

bath, it should be placed near the bedrooms. It should be easily reached from other parts of the house without going through another room.

Planning Garages and Carports

Carports are used in warm climates. Garages are essential in areas having severe winters. Standard garage sizes are shown in Fig. 25-16. A garage attached to a house should be near the kitchen.

Space should be provided to store garden equipment. This space is in addition to the sizes shown in Fig. 25-16.

The garage should be located on the plan so that it is easy to enter. Sharp turns make it difficult to park the cars.

Two-car garages can have two single doors or one large double door. Usually a door that slides overhead is used. It is easy to open and is out of the way.

Planning Stairs

The three most commonly used plans for stairs are straight, "L," and "U." Fig. 25-17.

When planning a building, floor area must be saved for stairs. The amount of space needed depends on the size and type of stair. The straight stair is the easiest and least expensive to make. It can have a landing. The "L" and "U" stairs require a landing in order to change the direction of the stair.

The main stairway in a house should be at least 2′—8″ wide, clear of the handrail. It should have 6′—8″ clear headroom. Basement stairs should be at least 2′—6″ wide, with 6′—4″ clear headroom.

The parts of a stairway are shown in Fig. 25-18. The size of the tread varies with the height of the riser. Commonly used tread and riser sizes for interior stairs in residential buildings are shown in

Fig. 25-19. Notice that the maximum rise is 8¼″ and the minimum tread is 9″.

To find the number of risers, the total rise of the stair is divided by the height of the riser selected. The total rise is the distance from one floor to another. The number and height of the risers in Fig. 25-20 were determined as follows. The total rise is 106 inches. A trial riser size of 7½ inches was chosen. Divide 106 by 7½ inches and you find you need 14 risers and have a little left over. Each riser must be the same size. To account for the extra distance, divide 106 by 14 risers to find the final riser size which is 7⁹⁄₁₆ inches.

Orientating the House

The comfort and enjoyment of a house can be improved by proper orientation. **Orientation** refers to the placement of the house on the lot and the location of rooms as related to the sun, winds, and view.

In the northern hemisphere, the south wall of the house receives the hot summer sun most of the day. The north wall receives very little sunlight. The west wall receives much sun late in the day.

The living room can face east or south. If it faces south, it will become rather warm in the evening. Generally, the garage, kitchen, or utility room is placed on the north side. The kitchen and dining areas can be placed on the east side. Any rooms facing west will be very hot in the summer. A porch or garage can be placed here to help break the sun's rays.

It is important to consider some means of shielding the house on the hot sides. Sometimes a large roof overhang is used. Trees are also good.

The direction of the prevailing wind must be considered. The summer breeze can be used to cool the bedroom area, if it is on that side of the house. Winter winds can be broken by placing the garage on that side.

Preliminary Sketches

Begin planning the house by making **preliminary sketches**. These are freehand sketches of the floor plan made on graph paper. Paper with ¼-inch squares is convenient to use since each square can represent 1 foot. Sketch the rooms planned for the house. Relate them to each other. Check to see if you have followed the principles of good planning. Change the plan until it is satisfactory.

Make certain the rooms are large enough to hold the furniture. It is helpful to make paper templates to scale of the furniture and appliances. Place these on the drawing and see if the rooms are large enough.

As you develop the floor plan, consider the exterior of the house. Whenever you move one room, you change the outside appearance of the house as well. Make sketches of the exterior of the house. You may find it desirable to change the plan slightly to improve the outside appearance. The location of windows is important to appearance.

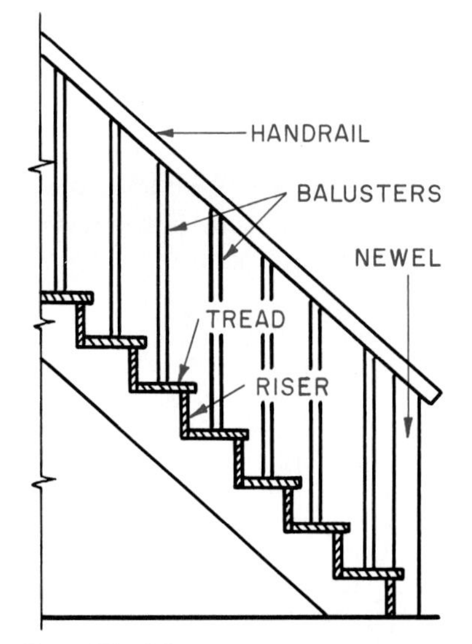

Fig. 25–18. **Parts of a stairway.**

Fig. 25-19. Common Tread and Riser Sizes

Tread Width	Riser Height
9″ min.	8¼″ max.
10″	7½″
11″	7″

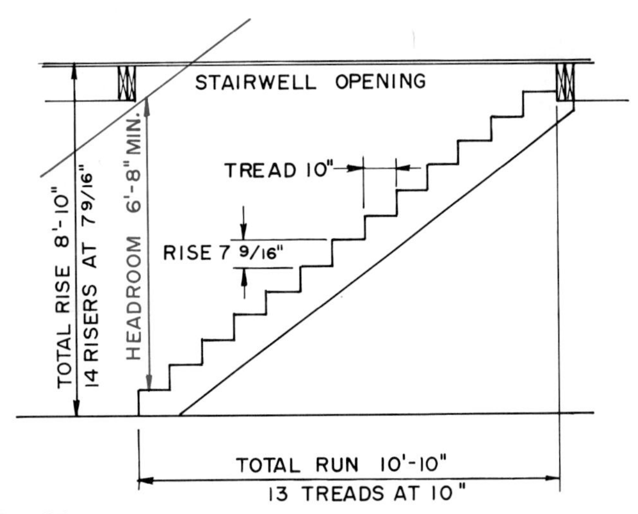

Fig. 25–20. **A typical stair detail showing total rise, run, and headroom.**

The type of roof used greatly affects the appearance of a house. The common roof types are shown in Fig. 25-21.

Working Drawings

When you have developed a workable floor plan and a pleasing exterior, you can make the working drawings. Before these can be drawn, however, you must know the symbols used in architectural drawing. You must also understand the construction details of a house.

Symbols in Architecture

The working drawings for a building are complex. Items such as doors, windows, and electrical fixtures cannot be drawn on the plan in detail. They are represented by symbols so that the drawing is easy to make.

Selected **architectural symbols** are shown in Figs. 25-22 to 25-24. More detailed symbol lists can be found in architectural standards manuals.

Construction Details

As you plan a building, you must consider the construction details. You may design a plan that requires very extensive construction. Knowing construction details enables you to develop the desired building at a minimum cost.

Selected construction details are shown in Fig. 25-25. Three common ways to construct walls are shown: frame, solid masonry, and brick veneer. A frame wall is usually 5 inches thick. A solid masonry wall is usually 8 inches thick. A brick veneer wall is usually 10 inches thick.

Details for concrete slab construction are shown in Fig. 25-25 (p. 682). If the slab is built in a cold climate, the edge of the floor must be insulated. Slabs can be supported by the ground or by piers. If the ground has had much fill, you should use piers. Piers 8 inches square should be spaced 6 feet on center throughout the floor.

Windows

Some of the common types of windows are shown in Fig. 25-22. The type you select depends upon the style of house. For example, awning windows would look strange on a colonial style house.

Windows are the major source of air to ventilate the house. The location of the windows influences where furniture can be placed in a room. Exterior appearance is also affected by window location. Windows should be balanced for a pleasing exterior.

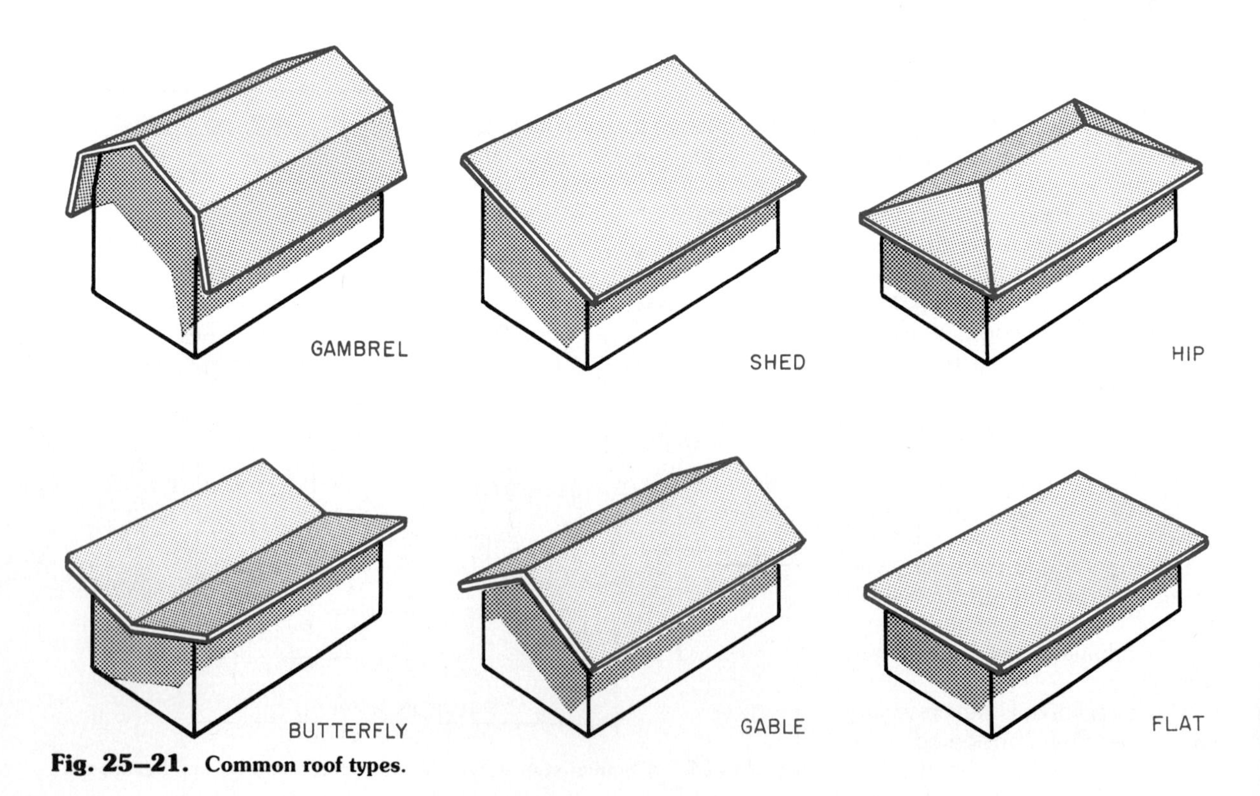

Fig. 25–21. Common roof types.

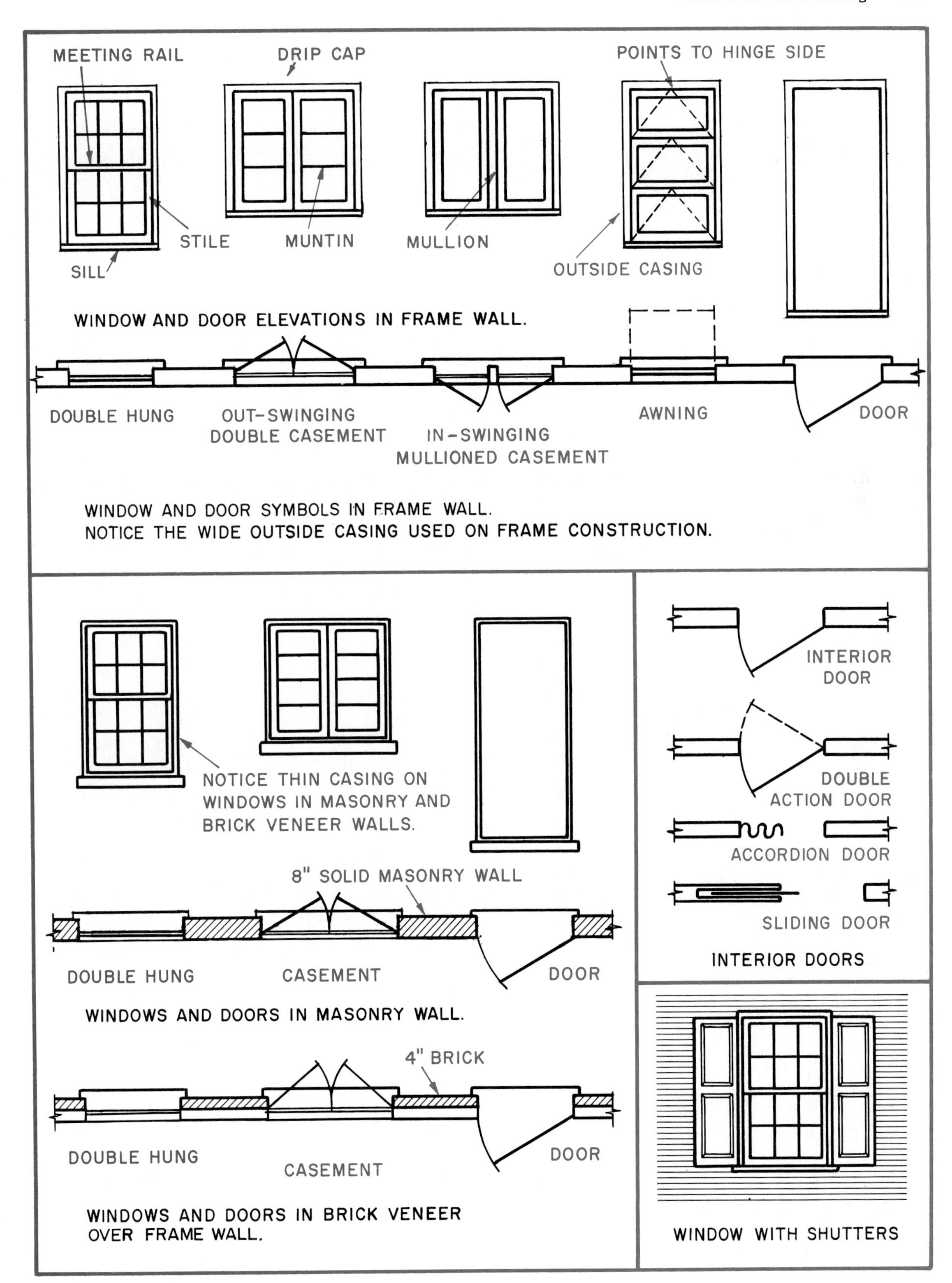

Fig. 25–22. Window and door symbols.

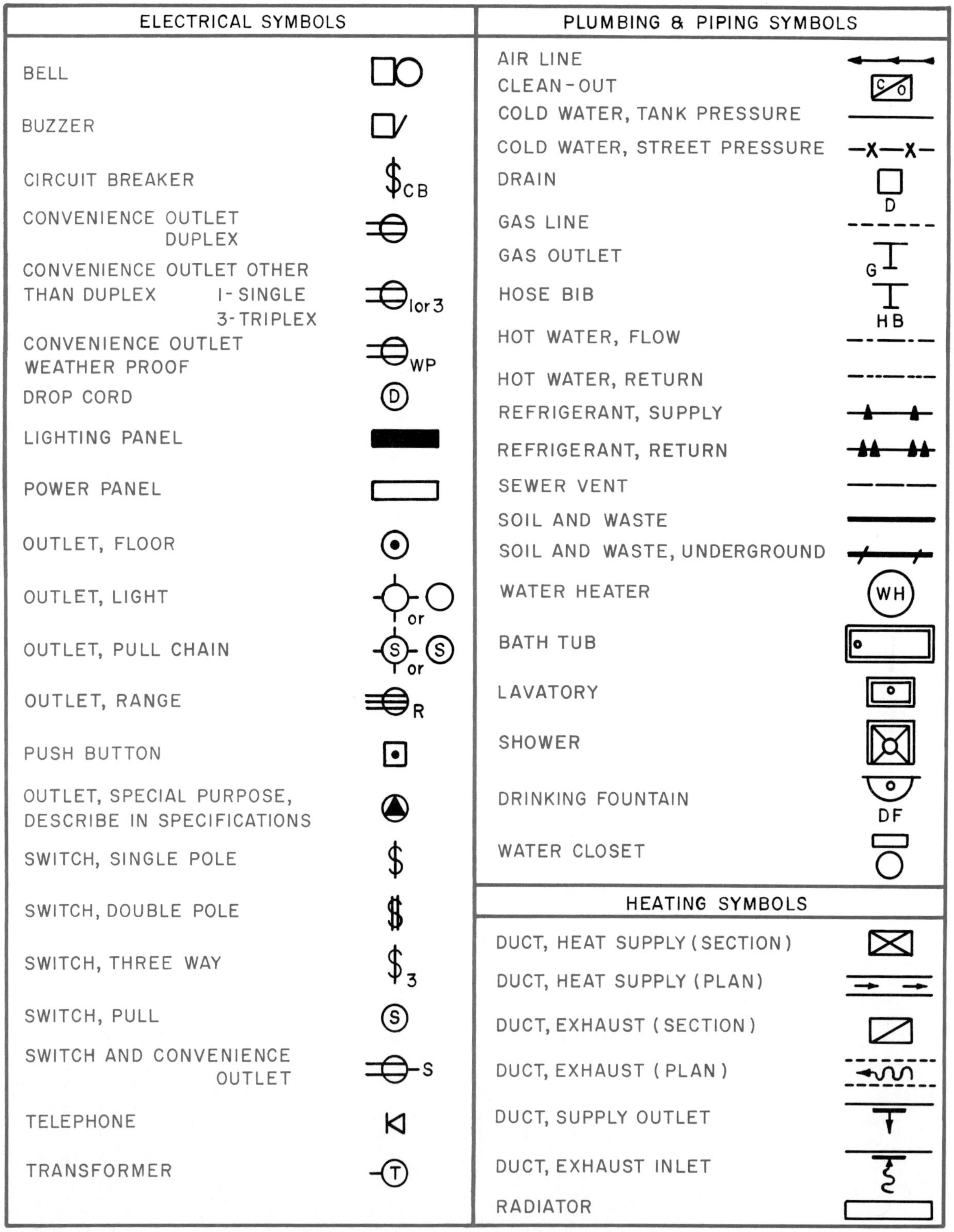

ELECTRICAL SYMBOLS	
BELL	
BUZZER	
CIRCUIT BREAKER	$_{CB}$
CONVENIENCE OUTLET DUPLEX	
CONVENIENCE OUTLET OTHER THAN DUPLEX 1-SINGLE 3-TRIPLEX	1 or 3
CONVENIENCE OUTLET WEATHER PROOF	WP
DROP CORD	(D)
LIGHTING PANEL	
POWER PANEL	
OUTLET, FLOOR	
OUTLET, LIGHT	or
OUTLET, PULL CHAIN	(S) or (S)
OUTLET, RANGE	R
PUSH BUTTON	
OUTLET, SPECIAL PURPOSE, DESCRIBE IN SPECIFICATIONS	
SWITCH, SINGLE POLE	$
SWITCH, DOUBLE POLE	
SWITCH, THREE WAY	$_3
SWITCH, PULL	(S)
SWITCH AND CONVENIENCE OUTLET	S
TELEPHONE	
TRANSFORMER	(T)

PLUMBING & PIPING SYMBOLS	
AIR LINE	
CLEAN-OUT	C O
COLD WATER, TANK PRESSURE	
COLD WATER, STREET PRESSURE	—X—X—
DRAIN	D
GAS LINE	
GAS OUTLET	G
HOSE BIB	HB
HOT WATER, FLOW	
HOT WATER, RETURN	
REFRIGERANT, SUPPLY	
REFRIGERANT, RETURN	
SEWER VENT	
SOIL AND WASTE	
SOIL AND WASTE, UNDERGROUND	
WATER HEATER	WH
BATH TUB	
LAVATORY	
SHOWER	
DRINKING FOUNTAIN	DF
WATER CLOSET	

HEATING SYMBOLS	
DUCT, HEAT SUPPLY (SECTION)	
DUCT, HEAT SUPPLY (PLAN)	
DUCT, EXHAUST (SECTION)	
DUCT, EXHAUST (PLAN)	
DUCT, SUPPLY OUTLET	
DUCT, EXHAUST INLET	
RADIATOR	

Fig. 25–23. Electrical, plumbing and piping, and heating symbols.

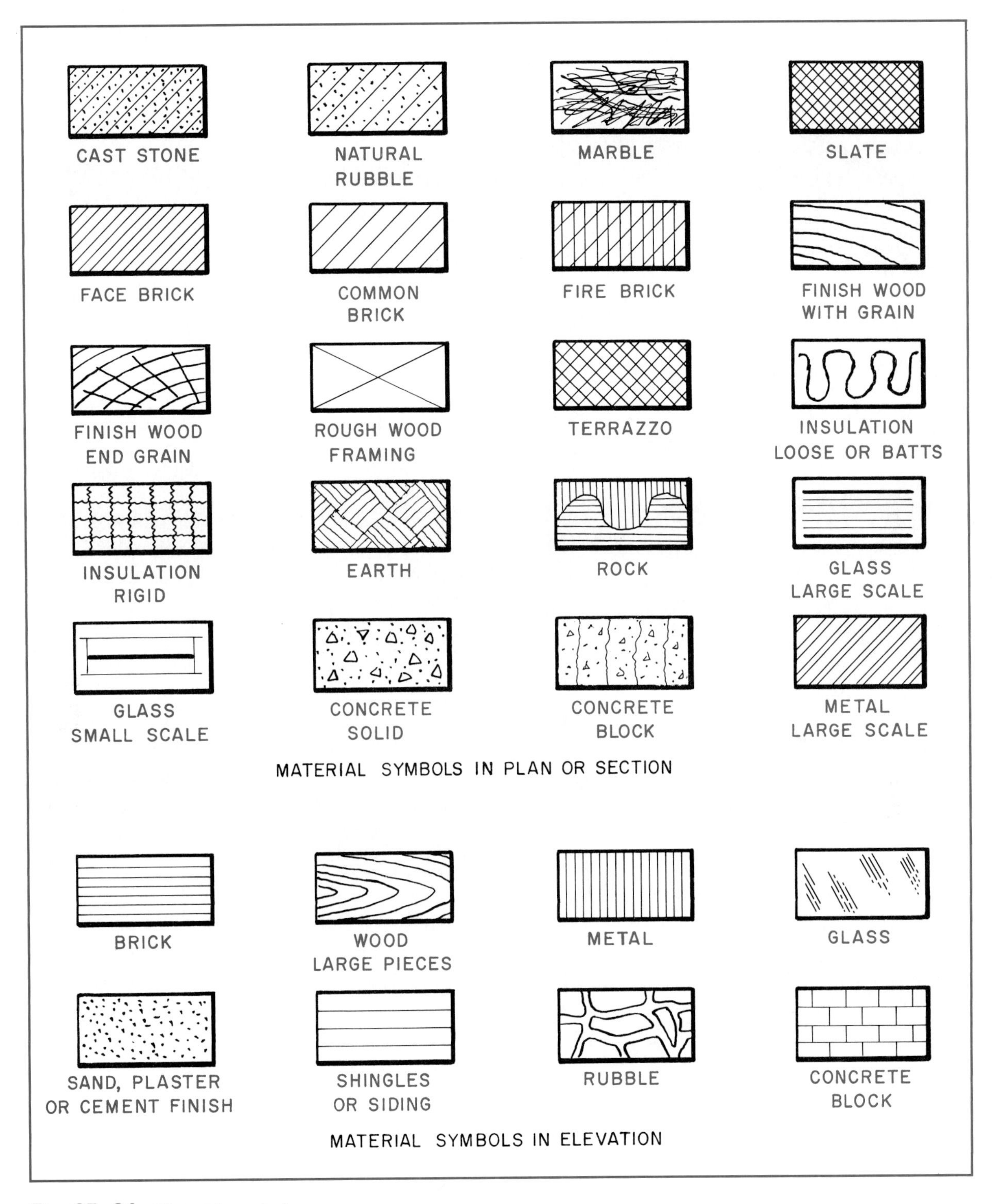

Fig. 25–24. **Material symbols.**

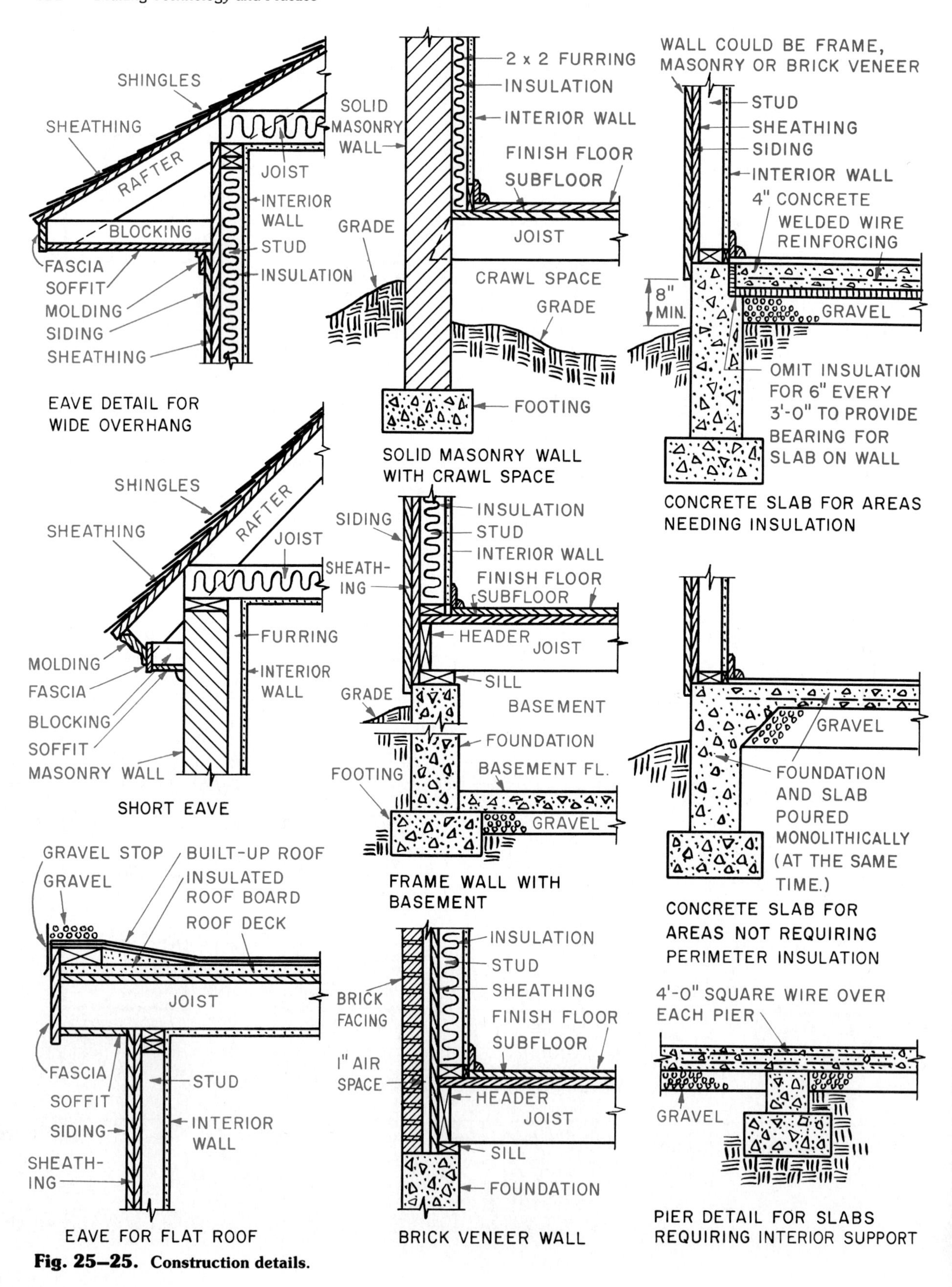

Fig. 25–25. Construction details.

CAD Applications:

Architecture

CAD applications in architecture are numerous and growing. Many CAD programs have features and commands that are especially useful in architectural drafting. For example, there are commands that enable the CAD user to stretch or trim lines or produce double lines. There are also add-on programs specially designed for architectural applications. Some of these are described here.

Several window manufacturers offer symbol libraries of their products. Instead of drawing a window, the architect calls it up from the symbol library. The drawing includes not only the graphic representation of the window but also information such as the model number, size, and so on. This information can be used later to create a window schedule.

Parametric design (see Chapter 18) can be applied to architectural drawings, as can finite element analysis (Chapter 22). Database and spreadsheet programs use the data from CAD programs to generate door and window schedules, bills of materials, and estimates.

Architectural 3-D modeling systems are used to create presentation drawings and to analyze mass relationships. For example, it is possible to determine interference—two parts occupying the same space. It's also possible to look at the computer model from different viewpoints and in different settings. Using a computer model helps architects refine the design before a paper and cardboard model is built, and it's much easier to alter.

The trend is toward integrated systems. These combine design, drafting, analysis, estimating, specifying, rendering—all the functions involved in planning and drawing a building. Used in this way, a CAD system becomes much more than an electronic drafting board. It's a whole new way of designing and creating.

Versa CAD Corporation

The Electrical System

The electrical system is shown on the floor plan of small buildings. Larger buildings require a separate electrical drawing.

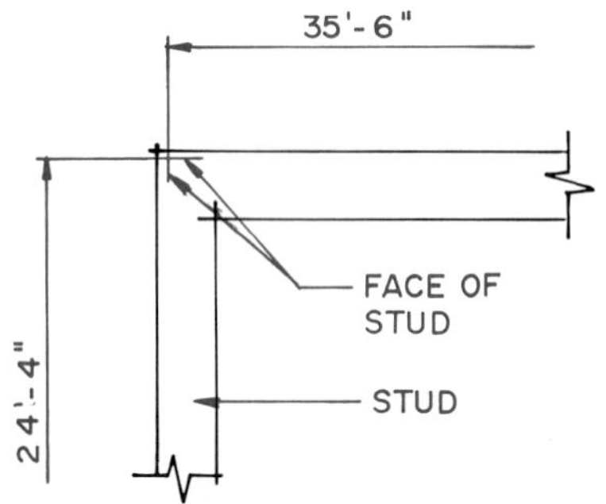
35'-6"
FACE OF STUD
STUD
24'-4"
FRAME EXTERIOR WALLS ARE DIMENSIONED TO THE OUTSIDE FACE OF THE STUD.

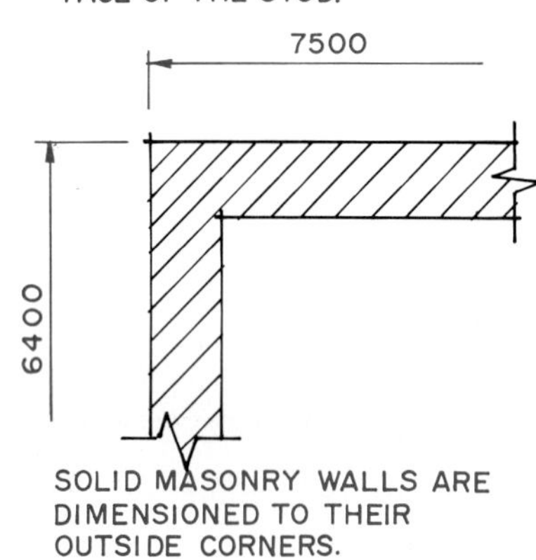
7500
6400
SOLID MASONRY WALLS ARE DIMENSIONED TO THEIR OUTSIDE CORNERS.

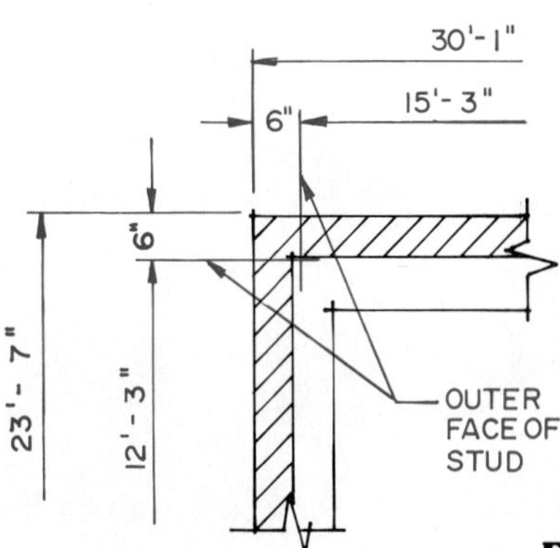
30'-1"
6"
15'-3"
6"
23'-7"
12'-3"
OUTER FACE OF STUD
MASONRY VENEER OVER FRAME CONSTRUCTION. THE MASONRY IS DIMENSIONED TO ITS OUTER FACE. THE FRAME WALL IS DIMENSIONED OVER THE STUD.

Fig. 25–26. Methods used to locate the corners of buildings.

Standard symbols are used to represent lights, outlets, switches, and other electrical units. Fig. 25-23. The floor plan in Fig. 25-31 shows how these are placed on a drawing.

Space convenience outlets in the wall so that no point along the wall is over 6 feet from an outlet. Place light switches on the latch side of doors.

Lettering

Most architectural drafters use capital Gothic letters. These can be vertical or slanted. Lettering is usually 1/8 to 3/16 inch (3 to 5 mm) high. See the lettering section in Chapter 3, "Tools and Techniques of Drafting."

Dimensioning

Using the customary system of measurement, dimensions over 12 inches are shown in feet and inches, such as 10′—6″.

Using metric measure, the floor plan, foundation plan, and details are dimensioned in millimeters. The site plan is dimensioned in meters.

Study the dimensioning on the drawings in this chapter. The dimensions are placed above an unbroken line.

Dimensions are placed on all sides of the foundation and floor plan. The overall length is given on each side. See Fig. 25-26 for the methods used to locate the corners in the different types of construction.

Locate windows, doors, and interior partitions in frame construction by their center lines. In solid masonry construction, locate windows and doors by dimensioning to the sides of the openings and the faces of partitions. Identify each room by name. Show the number of stair treads and risers. Show the size and direction of ceiling joists on the floor plan. Show floor joists on the foundation plan.

Scale

Foundation plans, floor plans, and elevations for small buildings are drawn to the scale 1/4″ = 1′—0″ or, if metric, 1:50. Larger buildings are drawn to the scale 1/8″ = 1′—0″, or 1:100 if metric. Construction details are drawn to the scale 3/4″ = 1′—0″, or 1:20 using metric measure.

Title Blocks

The design of the title block varies from company to company. Fig. 25-27 is typical of those in use.

Types of Working Drawings

After the house has been planned, final working drawings can be made. These include *site plan, foundation plan, floor plan, elevations*, and *details*.

Architectural drawings are made by manual and computer drafting methods. With an architectural program placed in computer memory, the drafter enters commands using a keyboard and a digitizer tablet. Fig. 25-28. The drawing is displayed on the screen of the monitor. Once the drawing is complete it is printed by the plotter. Fig. 25-29.

A set of drawings for a small house is shown in Figs. 25-30 through 25-35.

Floor Plan

The **floor plan** is a horizontal view of the house as seen from above. The house is sectioned through the windows. The floor plan shows:

- Room size and arrangement.
- Location and type of each window and door.

- Exterior and interior wall thickness.
- Stairs.
- Electrical system.
- Kitchen cabinets and fixtures.
- Bath fixtures.
- Fireplace.
- Any special built-in items, such as shelving and counters.

Usually the floor plan is the first drawing made. It provides the information needed for the foundation plan.

Study the floor plan in Fig. 25-31. How many duplex outlets are in the house? What types of windows are used? What is the size of the living room?

Doors and windows are indicated on the floor plan with marks. A mark is a circle with a letter or number inside. Often, letters are used for doors and numbers for windows. All doors or windows that are exactly the same have the same mark. For example, there may be four closets on the floor plan that use the same size and style of door. Each of them would be shown with the same mark. The mark is used to identify the door or window specifications on the door or window schedule.

The door and window schedules contain information needed to describe the doors and windows such as size and type. Scale elevations of the doors are drawn below the door schedule. Fig. 25-32 (p. 688). These schedules are placed on the same sheet as the floor plan.

Follow these steps when drawing the floor plan:

1. Lay out the overall length and width of the house.
2. Lay out the thickness of the outside walls.
3. Locate and draw the interior walls.
4. Locate and draw windows.
5. Draw the stairs, fireplace, and kitchen cabinets.
6. Draw the porches and steps.
7. Indicate the wall materials.
8. Locate the electrical outlets and fixtures.
9. Draw the doors.
10. Dimension the outside of the house.
11. Dimension the inside of the house.
12. Place the code for windows and doors if a schedule is used.
13. Complete the door and window schedule.
14. Label the items that need identification such as the range and refrigerator.
15. Letter any other notes needed.

Foundation Plan

A **foundation plan** is a top view of the foundation of the house. It shows the foundation before the house is built on it. This plan shows the following:

- Size of the footings.
- Size of the foundation wall.
- The location and size of beams and columns.
- Wall material (concrete or concrete block).
- Wall openings (vents, windows, or doors).
- Electrical outlets and lights if it is a basement.

JOHN C. KING AND ASSOCIATES
ENGINEERS AND ARCHITECTS
KANSAS CITY MISSOURI

MEMORIAL STATE HOSPITAL
STATE BUREAU OF HEALTH
JEFFERSON CITY~MISSOURI

REVISIONS	DATE	DRAWN BY	SHEET
NONE		B.S.H.	1
	JOB NUMBER 50 - 13245		

Fig. 25–27. A typical title block.

Fig. 25–28. This is a complete computer drafting station that can be used for any type of drawing, including architectural drawings. (Hewlett-Packard Company)

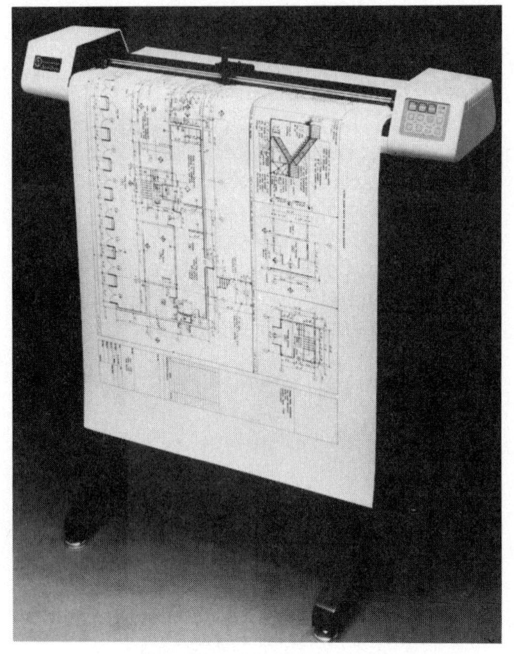

Fig. 25–29. The architectural drawing is printed on a plotter. The drawing coming off this plotter shows a floor plan and a stair detail. (Houston Instrument)

- Stairs, furnace, hot water heater, or any other such items.
- Size and direction of floor joists.
- Dimensions for all these items.

Study the foundation plan shown in Fig. 25-33. Try to identify each part. Notice that the dimensions are to the corners of the foundation wall. The front of the building is drawn facing the bottom of the drawing.

Elevations

Elevations are views of the outside of the house. Normally a view is made of each side. See Fig. 25-34 (pp. 690, 691). They are usually labeled according to the direction they face.

Elevations show the roof line and slope. The windows and doors are drawn. The type of wall material is shown with the proper symbol. Sometimes it is also written on the drawing. The entire surface need not be covered with the symbol. It is drawn in a few places on the large surfaces.

Very few dimensions are given. Usually the finished floor and ceiling lines are dimensioned. Overhangs are sometimes dimensioned.

The finished grade is shown. This is the level to which the soil is to be graded.

Before drawing the elevations, complete the foundation plan, a floor plan, and a sectional drawing through the wall of the building up through the eave. Begin the elevations by first drawing the end elevations. Doing so locates the lines of the roof. Then, draw the front and rear elevations.

To draw an elevation, follow these steps:

1. Lay out the finished floor line.
2. Locate the finished ceiling height.
3. Draw the grade line.
4. Draw the extreme right and left sides of the house.
5. Locate any breaks in the exterior wall such as in an L-shaped house.
6. Block in the roof. The slope must be found on the side elevation. This distance can be transferred to the front elevation.
7. Locate the doors and windows and draw them.
8. Locate and draw chimneys, porches, and other items.
9. Draw the foundation and footing.
10. Indicate materials using the proper symbols.

Fig. 25–30. An architectural rendering of the house shown in Figs. 25–31 through 25–35. (Home Planners, Inc.)

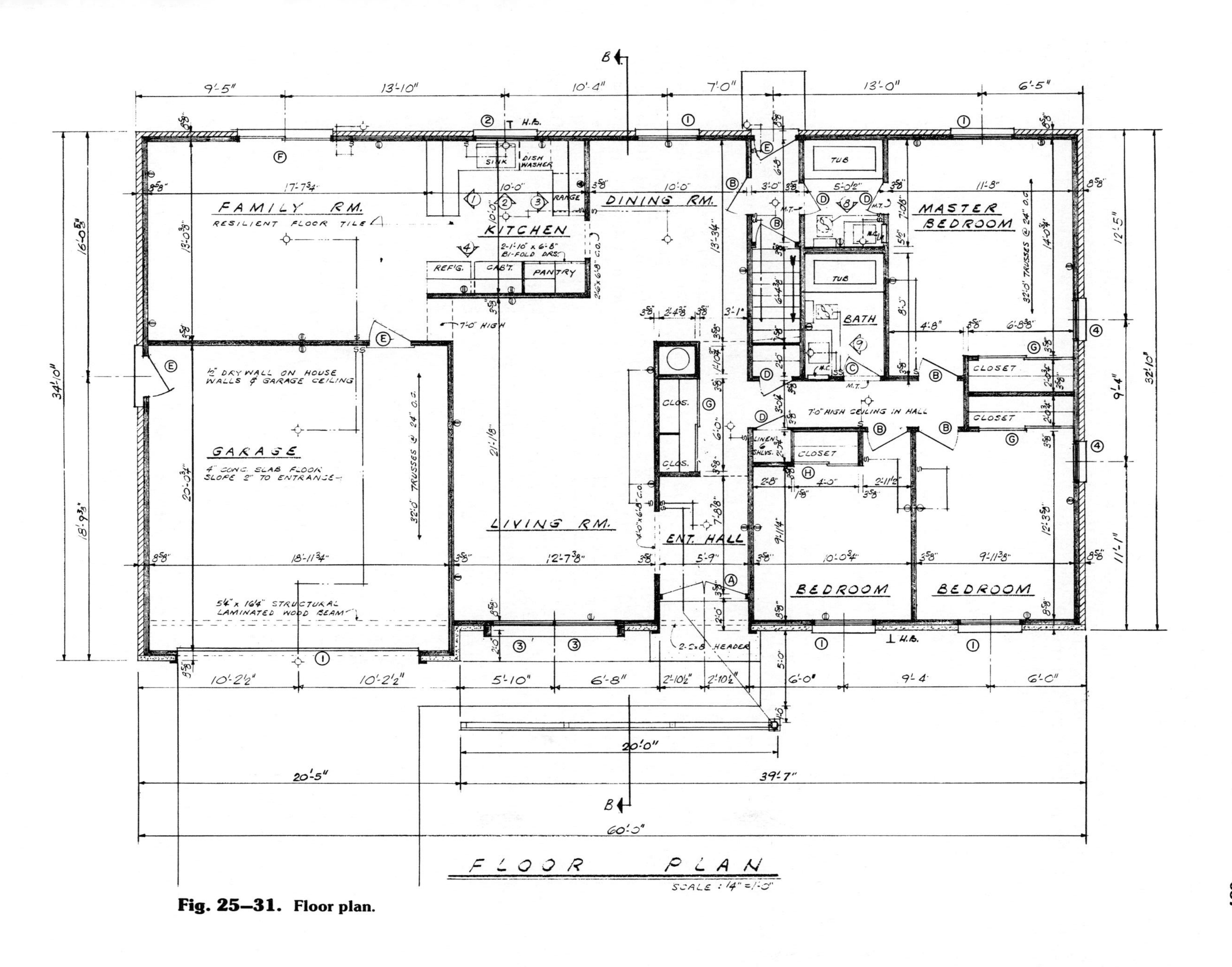

Fig. 25–31. Floor plan.

Details

It is necessary to draw **details** whenever something on the plans is not clear. Usually the following details are needed:

- A section through the wall of the house. This runs from the footing through the eave. It shows the size of the footing, foundation thickness, wall construction, and how the eave is to be made. It is dimensioned. All parts are labeled. See Fig. 25-35 (p. 692).
- Elevations of the kitchen cabinets. This shows how the cabinet doors and drawers are to be arranged. An elevation is made of each wall with cabinets. See Fig. 25-36 (p. 693). These are usually placed on the same sheet as the floor plan.
- Elevations and sections through the fireplace. These are dimensioned. See Fig. 25-37 (p. 694).
- Stair details. See Fig. 25-35.
- Elevations or sections of any other cabinets or unusual construction details.
- Sections to clarify construction. A cross section through the house is shown in Fig. 25-36.

The Site Plan

A **site plan** is a top view of the building site. Fig. 25-38 (p. 695). It shows the location of the house on the site. Also shown are sidewalks, driveways, porches, and carports. The size of the site is dimensioned in decimal feet or meters. The building line is noted and dimensioned. This is the minimum distance a house can be built from the street.

The site plan shows the elevations of the contours of the land. The elevation is the number of feet above the local datum level. The datum level is an assumed basic level above sea level. It is used in measuring heights.

Study the site plan in Fig. 25-38. Notice that the *original* contours are shown with a *dashed* line. The *new* contours are shown with a *solid* line. These are the desired levels after the site is graded. The dimensions and contours for the site plan are established by a surveyor.

WINDOW SCHEDULE

MARK	NO.	UNIT SIZE	TYPE	REMARKS
1	4	3'-4" X 4'-0"	D.H.	DOUBLE GLAZED
2	1	3'-0" X 3'-4"	D.H.	DO[1]
3	2	3'-4" X 4'-6"	D.H.	DO
4	2	2'-0" X 4'-0"	D.H.	DO

1. DO SYMBOL MEANS DITTO.

DOOR SCHEDULE

MARK	NO.	SIZE	TYPE	REMARKS
A	2	2'-8" X 6'-8"	PANEL	PINE
B	5	2'-6" X 6'-8"	FLUSH H.C.	BIRCH
C	1	2'-4" X 6'-8"	FLUSH H.C.	BIRCH
D	4	2'-0" X 6'-8"	FLUSH H.C.	BIRCH
E	3	2'-8" X 6'-8"	FLUSH S.C.	BIRCH
F	1	6'-0" X 6'-8"	SLIDING	ALUMINUM
G	6	3'-0" X 6'-8"	FLUSH H.C.	BIRCH
H	2	2'-0" X 6'-8"	FLUSH H.C.	BIRCH
I	1	7'-0" X 15'-0"	PANEL	PINE

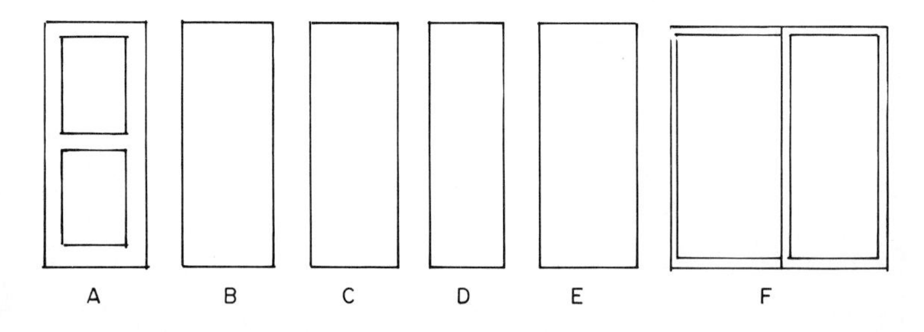

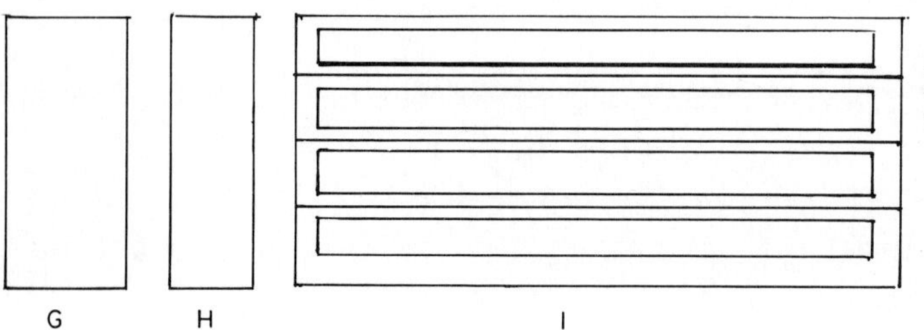

Fig. 25–32. Typical door and window schedules.

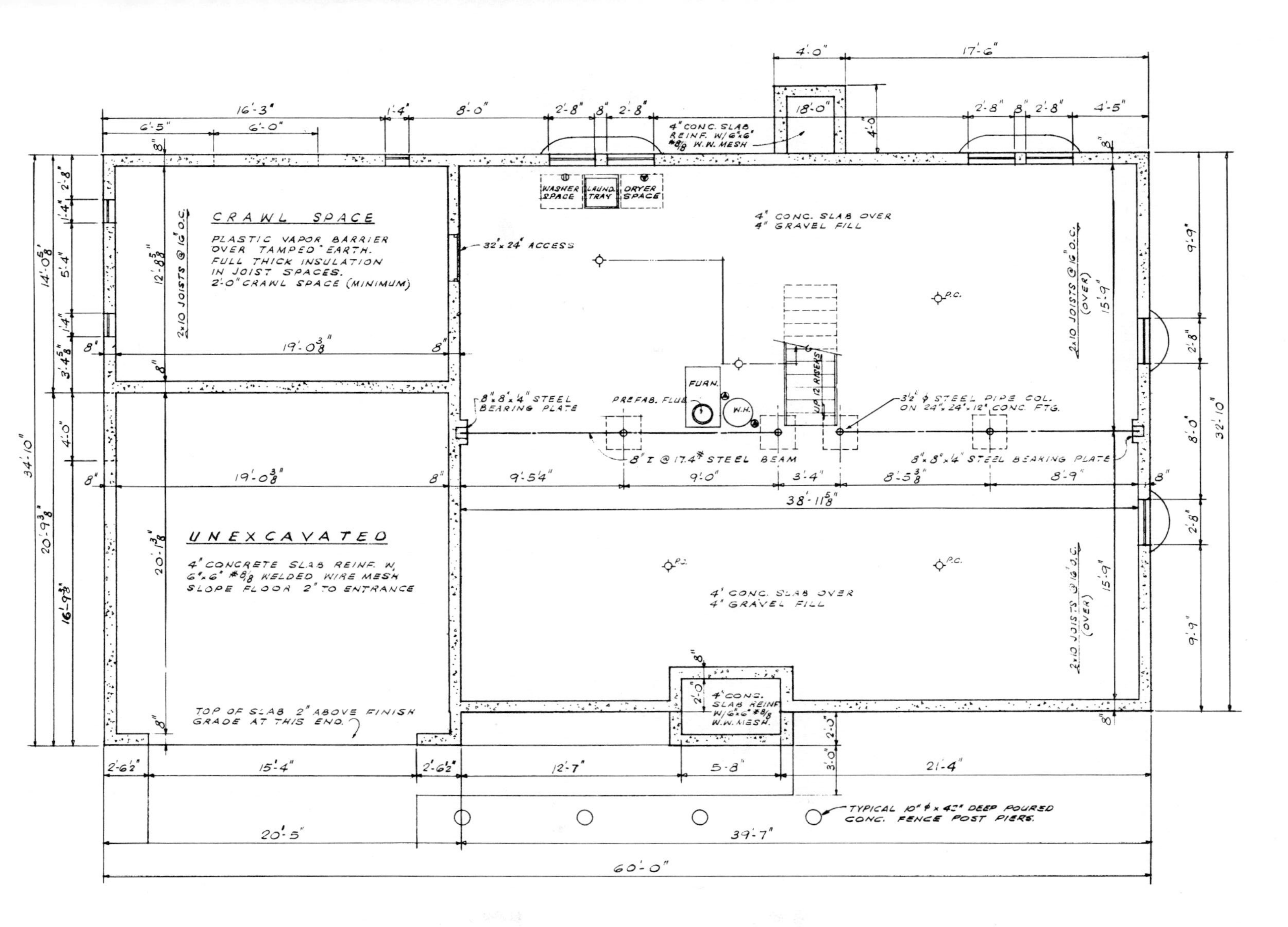

Fig. 25–33. Foundation plan for a basement.

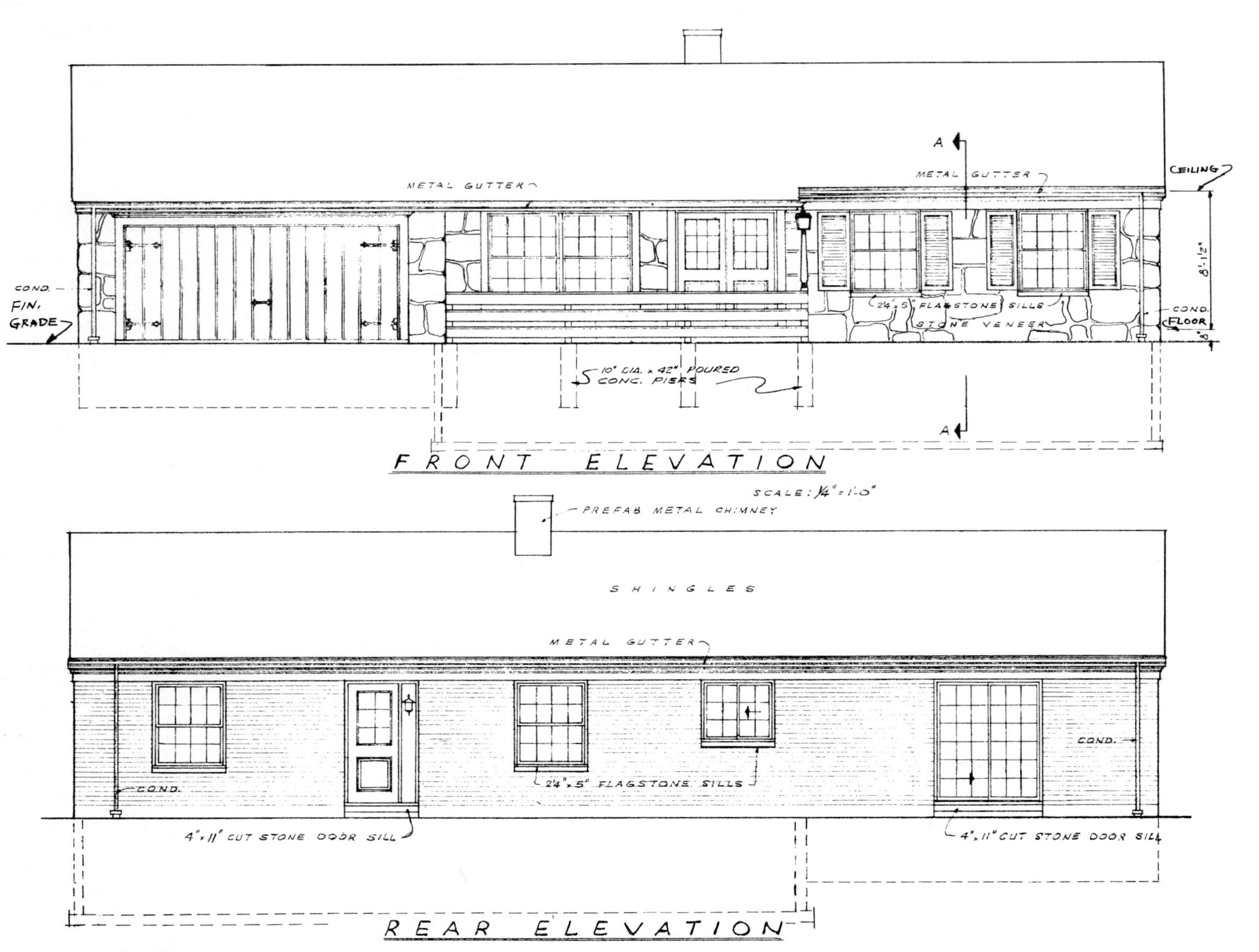

Fig. 25–34. Front and rear elevations.

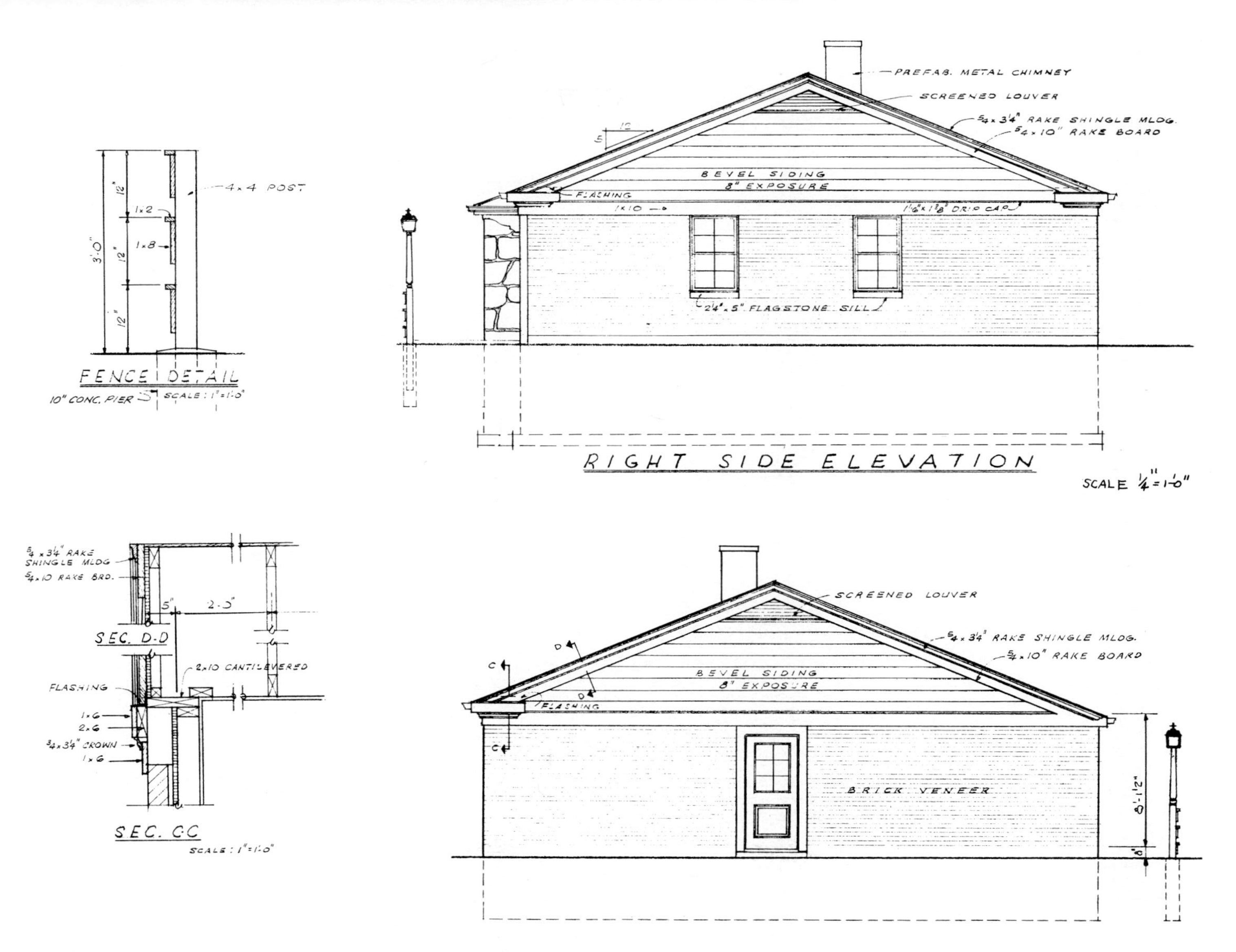

Fig. 25–34 (continued). Left- and right-side elevations.

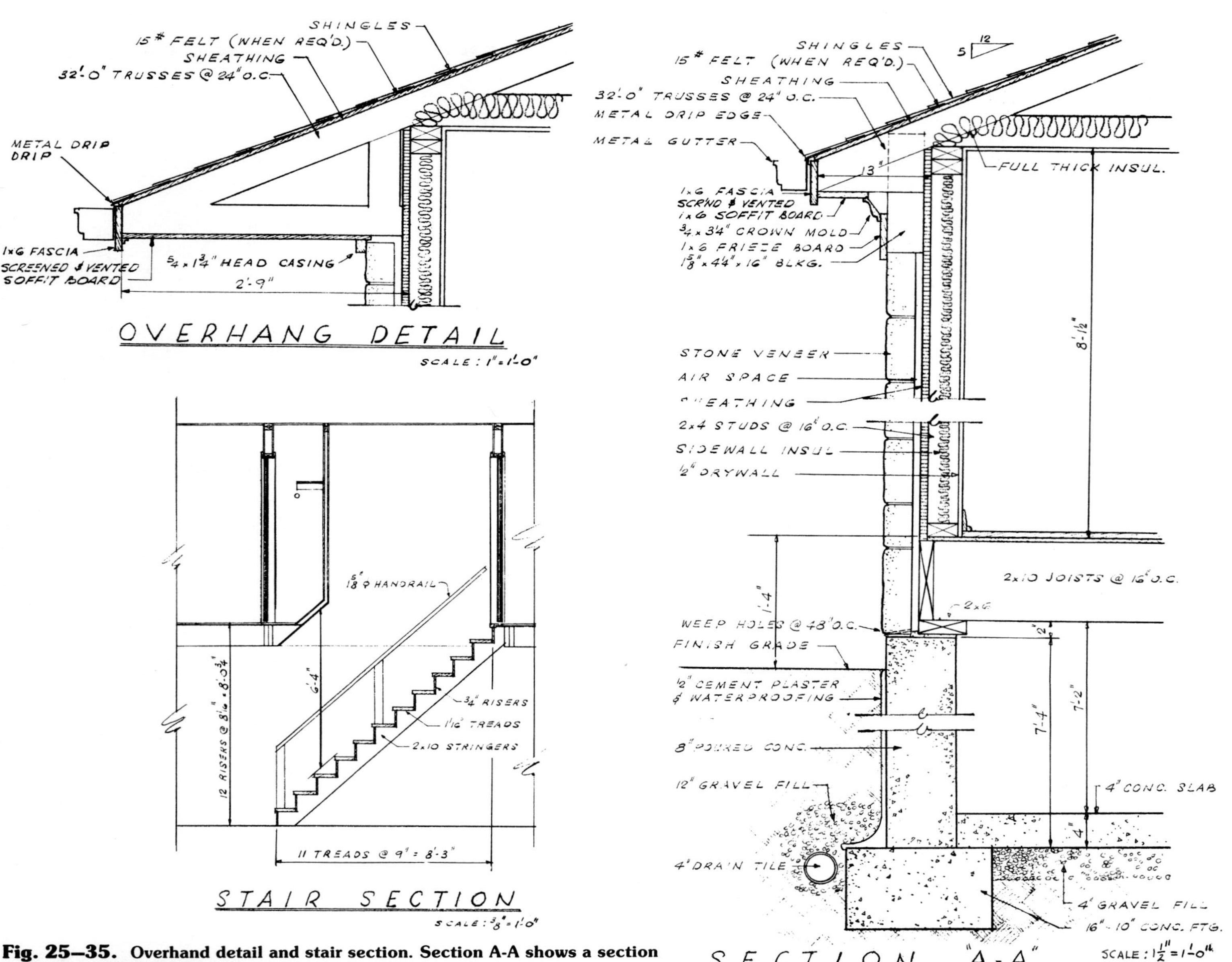

Fig. 25–35. **Overhand detail and stair section. Section A-A shows a section through the wall of the house.**

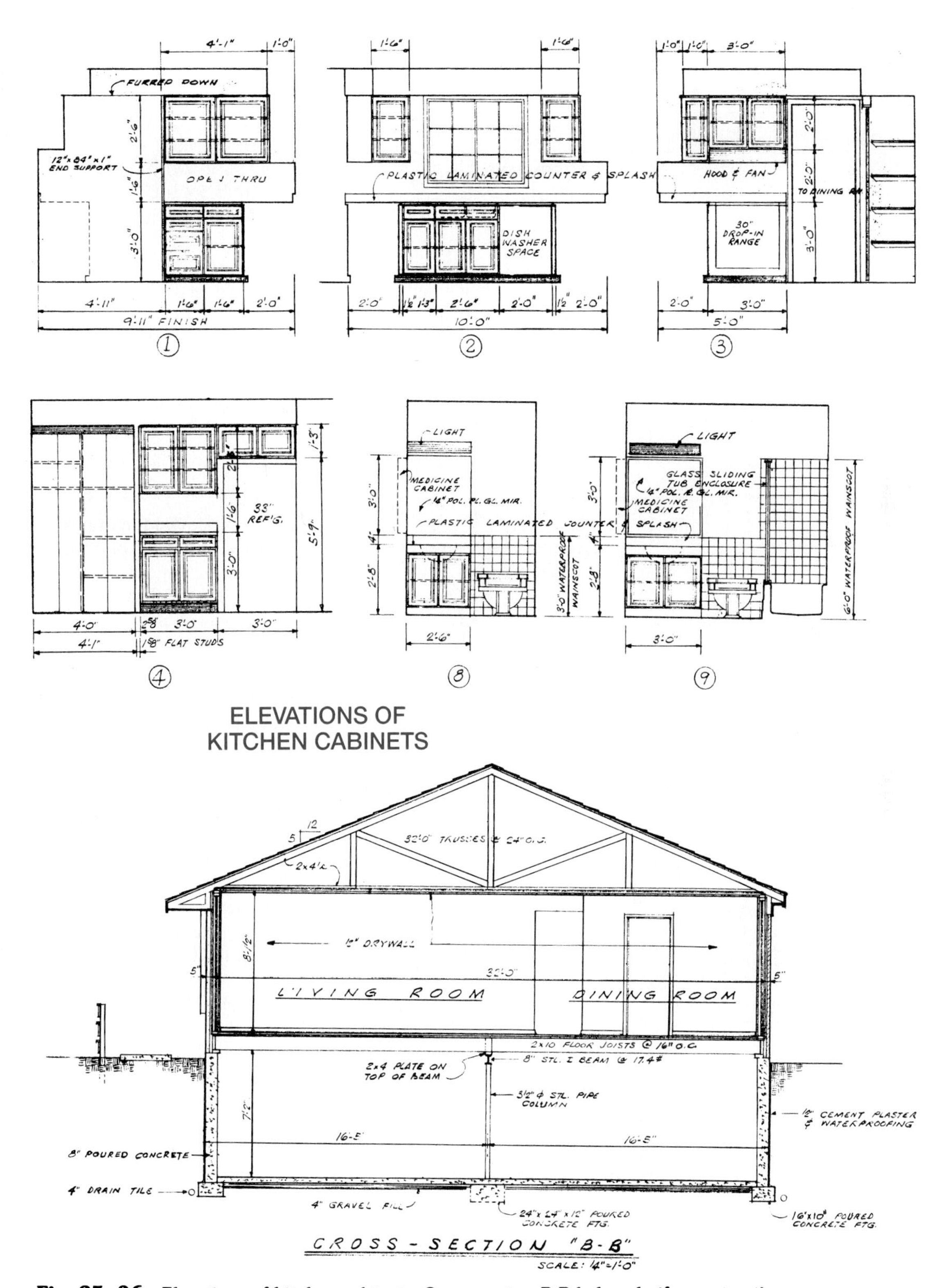

Fig. 25–36. **Elevations of kitchen cabinets. Cross section B-B helps clarify construction.**

A	B	C	D	E	F	G	H	I	J	K	L
24	11	16	12	8	14	18	$8\frac{3}{4}$	24	19	32	10
30	17	16	12	8	14	18	$8\frac{3}{4}$	29	24	38	13
36	23	16	12	12	14	23	$8\frac{3}{4}$	29	27	44	16

DESIGN SIZES FOR FIREPLACES
DONLEY BROTHERS CO.

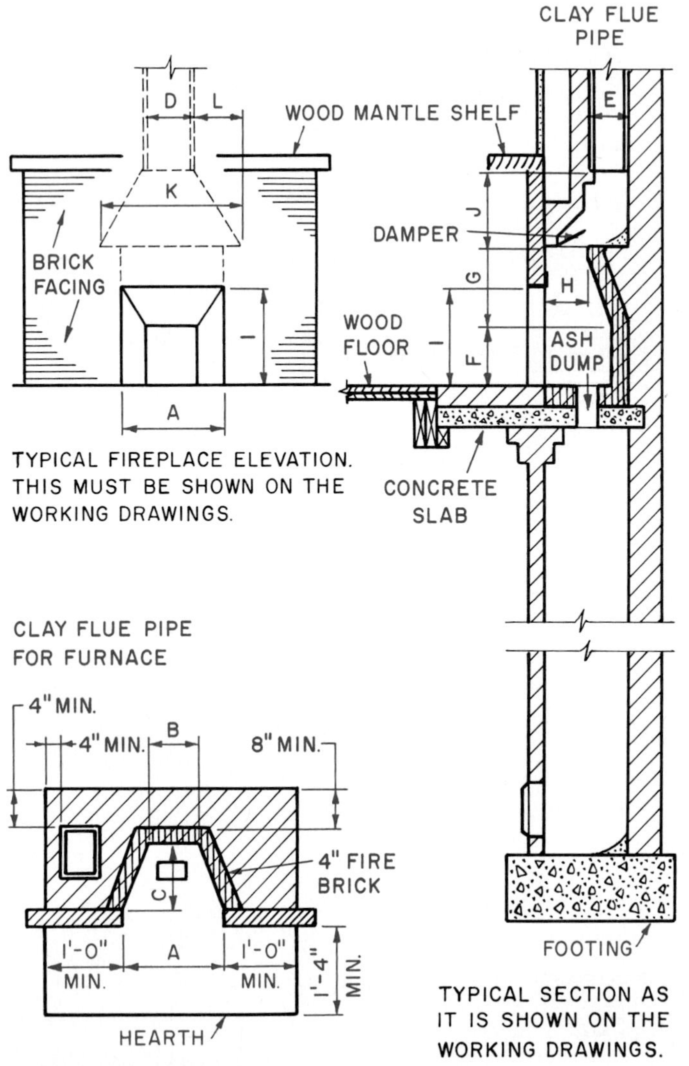

Fig. 25–37. Details needed to show a fireplace on the working drawings.

Display Drawings

Display drawings are made to show the proposed building in its finished form. They are used to sell a client on the plan developed by the architect. Usually they are only made for large commercial buildings.

These drawings show one or more perspective views of the proposed building. They show the floor plan but with little detail. The plans are made attractive by drawing trees and shrubs around the building. Watercolors and ink are used. This gives the building a realistic look. Fig. 25-47 is an example of a display drawing for a house.

Styles of Houses

The following figures show elevations and floor plans for typical houses of various styles. As you study these illustrations, notice how the architect related the various areas on the floor plans. See, too, how the front elevations work well with the floor plans.

One-Story House

Fig. 25-39 shows a one-story house. This style is sometimes called a ranch house. A true ranch, however, usually has a more rambling, irregular floor plan than is shown here. Small city and suburban lots have forced the original ranch house into a tight floor plan.

The house has a brick veneer exterior. The posts on the front are like those used on early Spanish-style homes found in the southwestern United States. This house can be built with or without a basement.

Another one-story house, Fig. 25-40, has very simple lines. Some call this contemporary styling, meaning it is designed for modern living. The floor plan does not follow that of older houses.

One-and-One-Half Story House

This house, Fig. 25-41 (p. 698), is like the one-story house except that the roof is higher. This provides space to build several rooms on the second floor. However, because the ceilings slope, only part of the area on the second floor can be used for living space. Fig. 25-42 (p. 699). Usually the side walls are made 5 feet high.

Light and air are brought to these second floor rooms by dormer windows. These are windows cut through the roof. They have sides and small roofs built around the windows.

This type of house offers a lot of floor space at a low cost. The cost is lower than a full two-story house because the roof also serves as side walls. In addition, this style takes less material to build than a two-story house. The style of house shown in Fig. 25-41 is called Cape Cod.

Two-Story House

The two-story house shown in Fig. 25-43 is of frame construction. The first floor front elevation has a brick veneer. It is designed after the style of colonial American houses.

Notice on the floor plan how the stairs to the basement and second floor are located. This style of house has a balanced front elevation. This fact must be remembered by the architect as the room locations are planned. This type of house can be built with or without a basement.

A contemporary two-story house is shown in Fig. 25-44. Notice the differences between it and the more traditional style shown in Fig. 25-43.

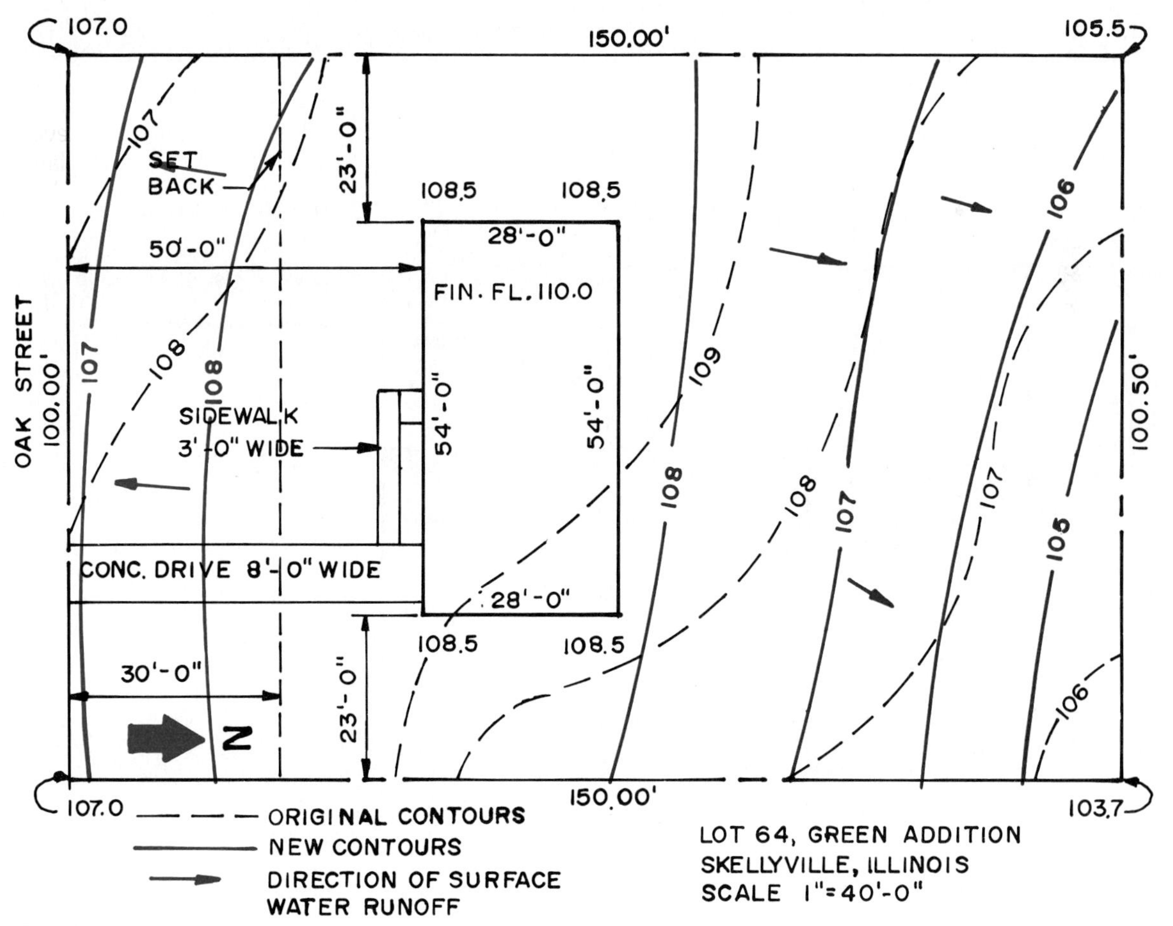

Fig. 25–38. A site plan.

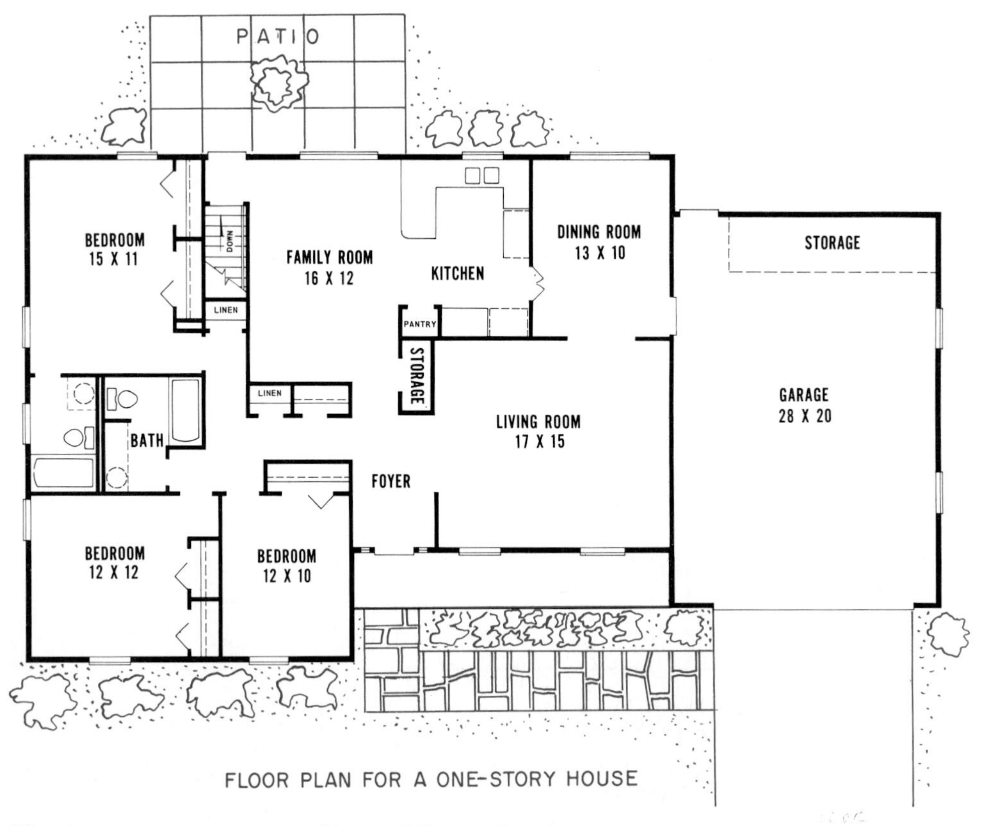

Fig. 25–39. A one-story ranch house. (National Homes Corp.)

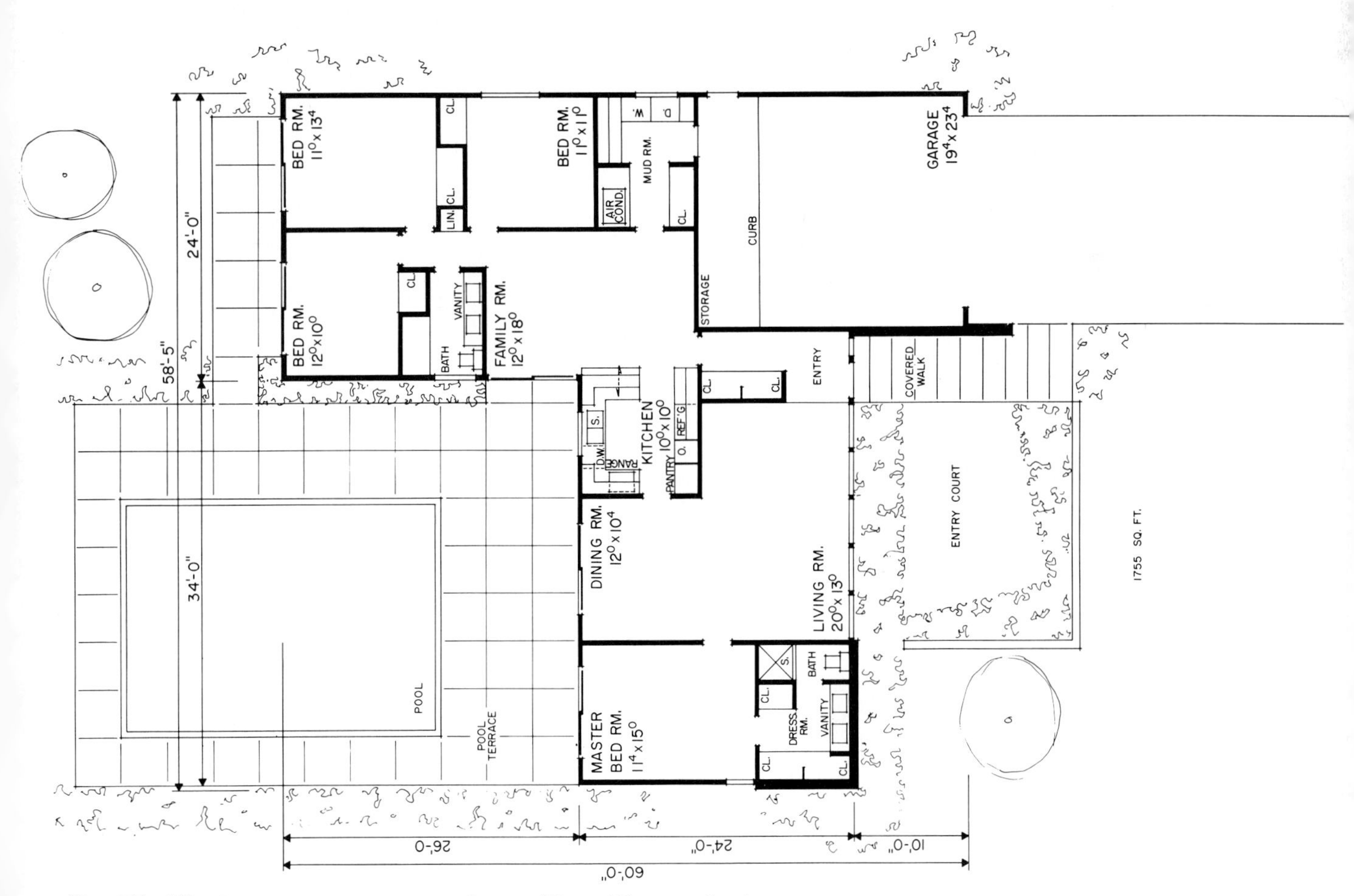

Fig. 25–40. A one-story contemporary house. (Home Planners, Inc.)

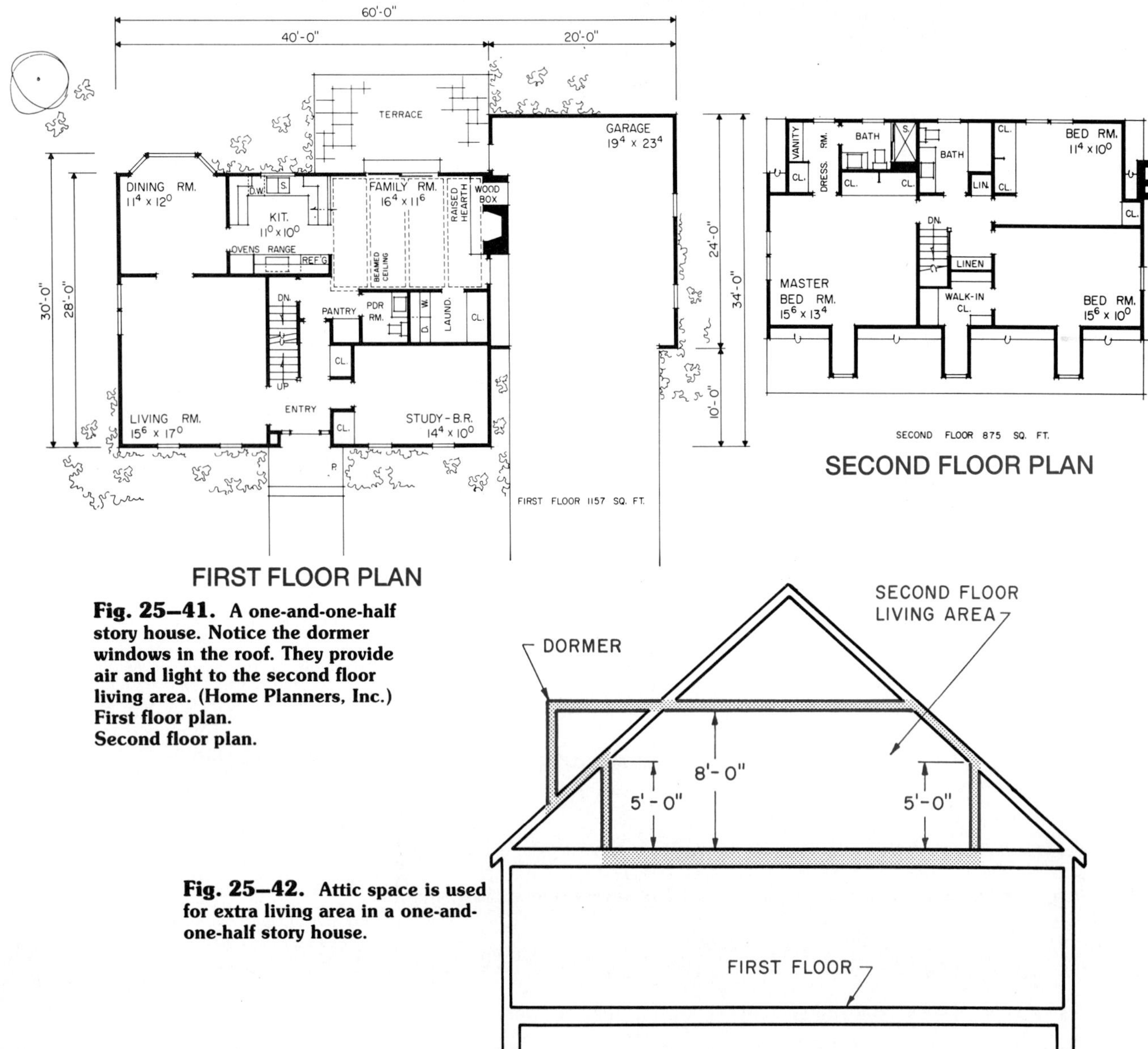

Fig. 25–41. A one-and-one-half story house. Notice the dormer windows in the roof. They provide air and light to the second floor living area. (Home Planners, Inc.) First floor plan. Second floor plan.

Fig. 25–42. Attic space is used for extra living area in a one-and-one-half story house.

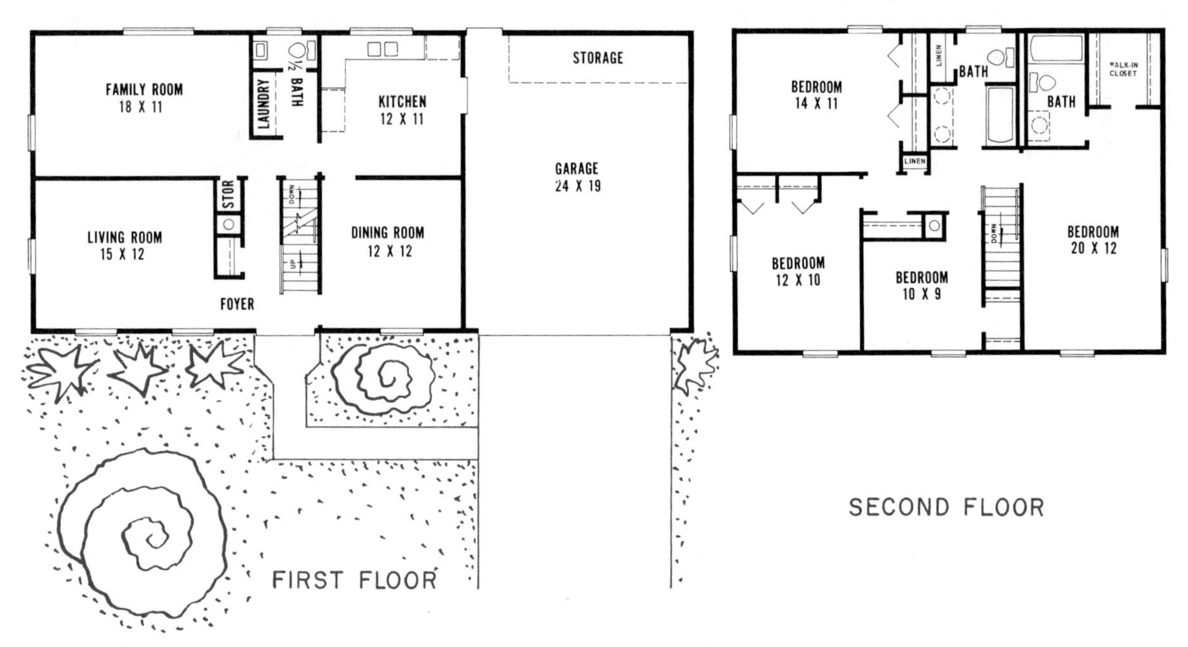

Fig. 25–43. **A traditional two-story house. (National Homes Corp.)**

Fig. 25–44A. A contemporary two-story house. (Home Planners, Inc.)

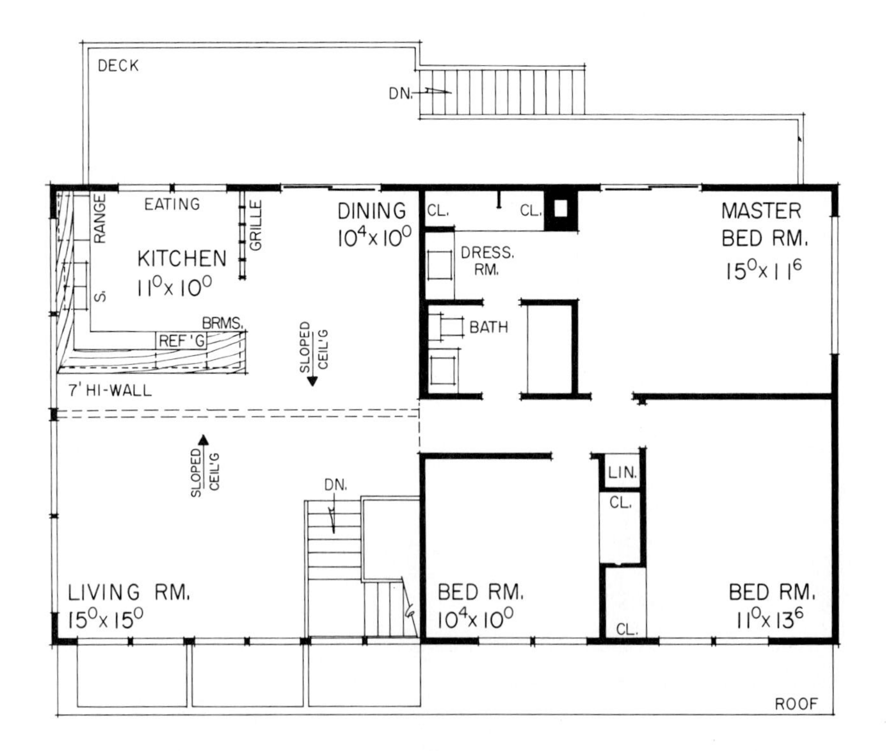

Fig. 25–44B. Upper level floor plan.

House with Split-Foyer Entrance

The house shown in Fig. 25-45 is the same as a one-story house with a basement. The difference between them is that the basement level in this house is raised several feet above grade level. This allows the use of regular windows in this area. The basement thus becomes planned, useful living area.

The main entrance to the house is between the upper and lower levels, about on the level of the ground. To reach the upper level, a half flight of stairs is needed. A half flight of stairs is also needed to reach the lower level. Fig. 25-46.

Split-Level House

Fig. 25-47 shows one type of split-level house. This house has three levels. Notice that the ground around this house slopes. The first level is the garage. It is on the lower ground level. The second level is the living area. The third level is the sleeping area, located above the garage.

In Fig. 25-48 the common levels used on split-level houses are shown. The basement can be omitted. This was done in the house shown in Fig. 25-47.

The front entrance can be on the garage level as it is in Fig. 25-47. However, the front entrance is often located on the living level.

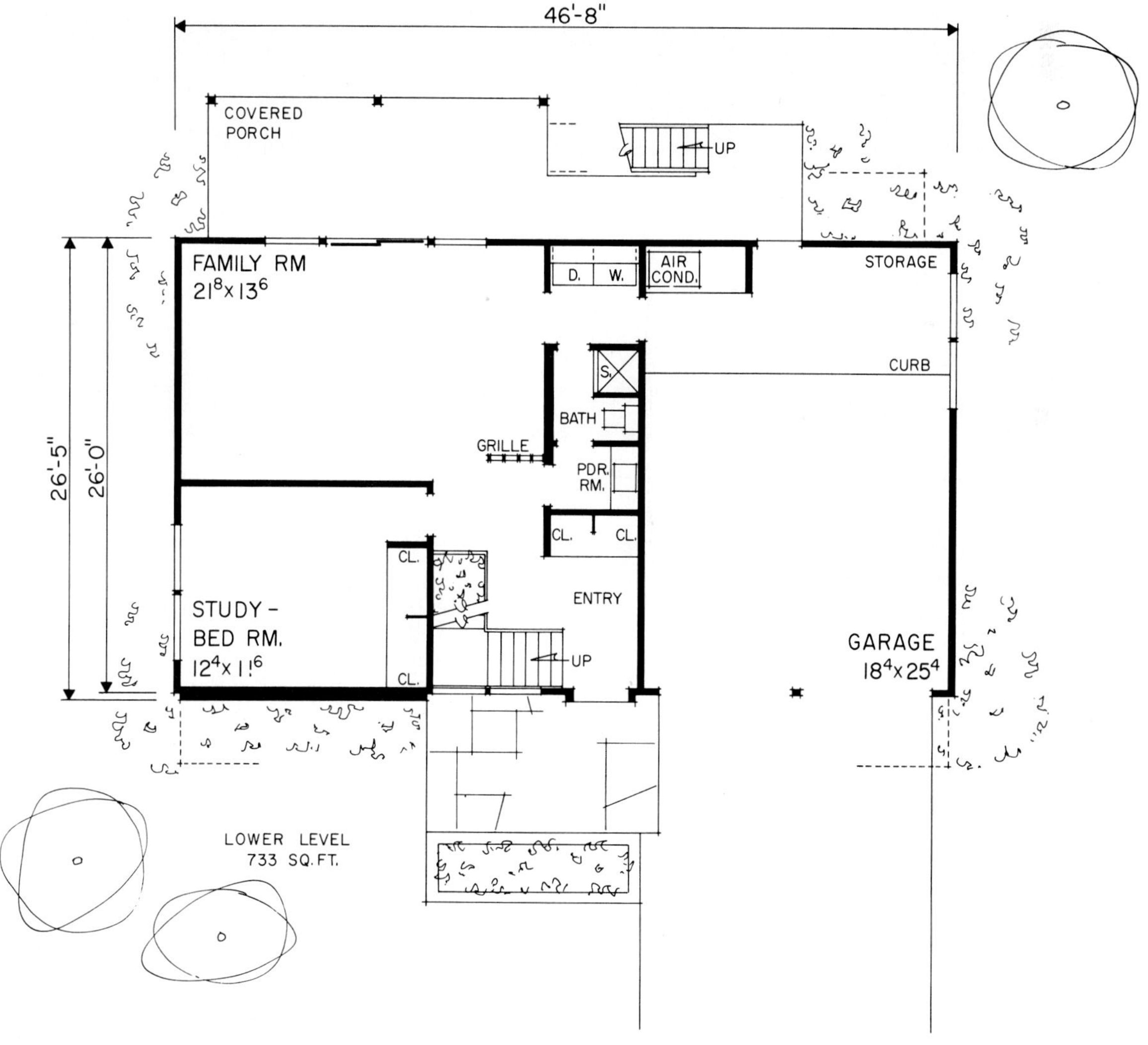

Fig. 25–44C. Lower level floor plan.

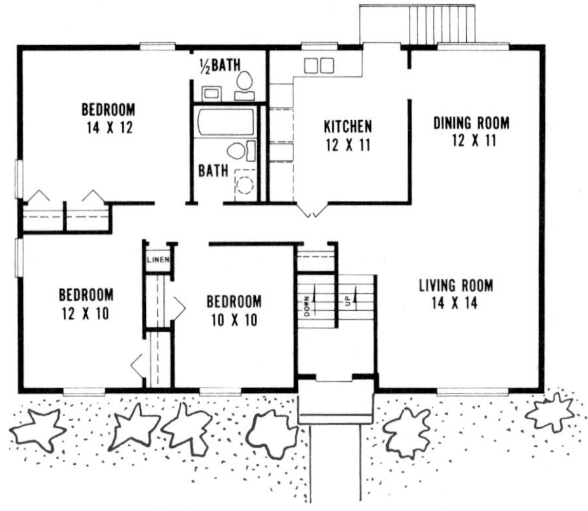

LAUNDRY
½BATH
DOWN
UP
LOWER LEVEL

Fig. 25–45. A house with a split-foyer entrance. (National Homes Corp.)

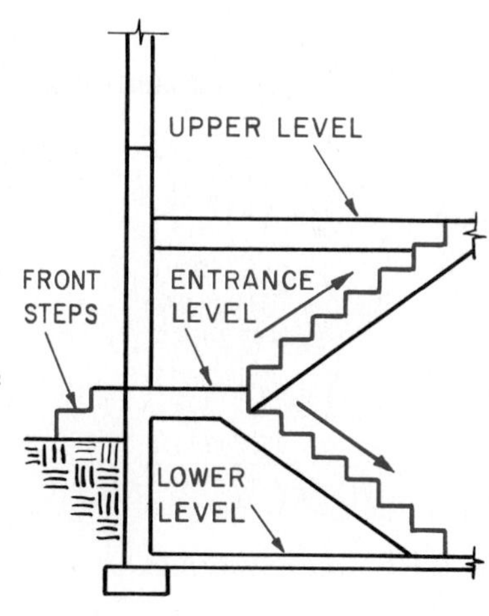

Fig. 25–46. Stairs in a house with a split-foyer entrance.

Fig. 25–47A. **Front elevation of a split-level house with three levels. (Home Planners, Inc.)**

Fig. 25–47B. **Rear elevation of the split-level house.**

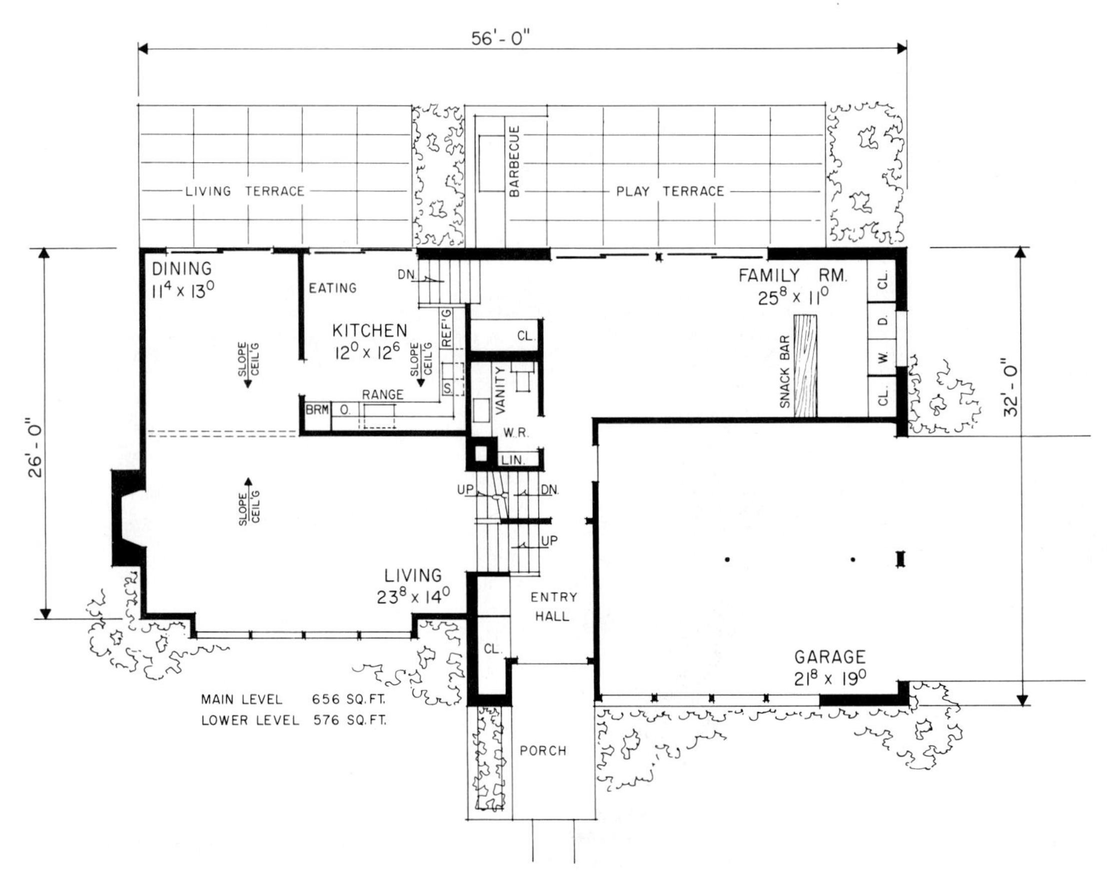

Fig. 25–47C. Floor plan of main and lower levels.

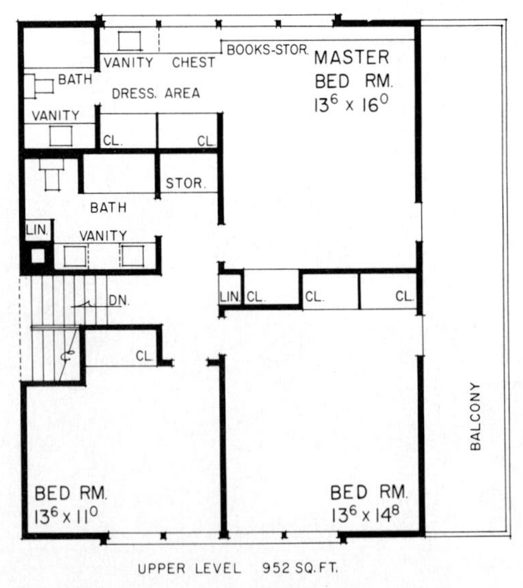

Fig. 25–47D. Floor plan of upper level.

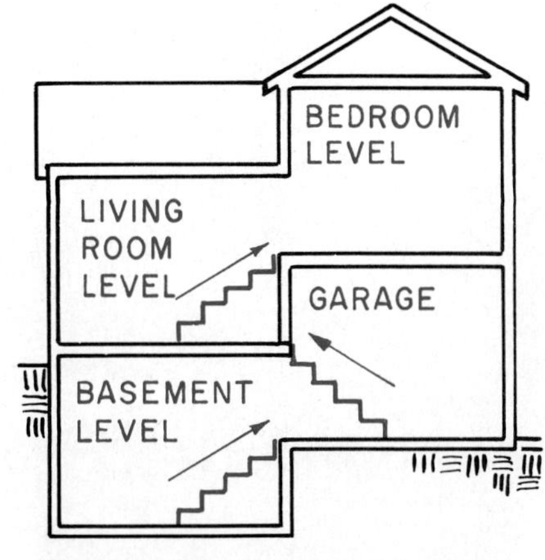

Fig. 25–48. The levels in a split-level house. The basement level can be omitted, resulting in a three-level house.

Chapter Review

Build Your Vocabulary

You should understand and use the following terms as part of your working vocabulary. Write a brief explanation of what each means.

architect
traffic patterns
open planning
orientation
preliminary sketches
architectural symbols
floor plan
foundation plan
elevations
details
site plan
display drawings

Sharpen Your Math Skills

1. A stair has a total rise of 8′–9″. If the stair riser is 7″, how many risers are there in the stair?
2. If the stair in Problem 1 has treads 10″ long, what is the total run of the stair?
3. If you draw a house to the scale ¼″ = 1′–0″, and the house is 63′–0″ long, how long in inches is the drawing?
4. If a house is drawn to the metric scale 1:50 and is 7300 millimeters wide, how long is the line representing the width?
5. One foot equals .305 meters. If a building site measures 150′ × 375′, what is its size in meters?

Study Problems—Directions

The problems that follow will give you the chance to use some of the drafting knowledge that you learned in this chapter. Do the problems that are assigned to you by your instructor.

1. Plan a bathroom that can be used by two people at the same time. Keep the floor area as small as possible. Draw the floor plan for this solution. Dimension the plan, giving the exact location of each fixture.
2. Study the floor plan for the house in Fig. 25-39. On graph paper, sketch two different front elevations for this plan. You may change window size, type, and location if it will not hurt the use of the rooms.
3. Draw the details for the fireplace in the house shown in Fig. 25-41. The opening is to be 4 feet. The hearth is raised 8 inches. Design the front elevation and mantel to suit yourself.
4. Study the kitchen in the house shown in Fig. 25-43. Try to redesign the kitchen so that a small dining area is in front of the windows. Draw the plan and elevations of the cabinets.
5. Design and prepare a full set of drawings for a small lakeside cottage that has the following features:
   - Low cost materials that require little upkeep.
   - Sleeping, eating, and living area for four people.
   - Floor area as small as possible yet large enough to serve four people.
   - A minimum size bathroom.
   - An inexpensive source of heat.
   - Storage for items commonly needed at the lake, such as fishing gear, life preservers, and lawn furniture.
   - Electricity and water service are available.
   - Lot slopes toward the water on a ten-degree angle.
6. Design a carport that can be built next to the cottage in Problem 5. Include a storage wall as part of the structure.
7. Draw a full set of working drawings for the one-story house in Fig. 25-40. The front walls are stone veneer. It is built with a concrete slab floor. The roof is tar and gravel.

CHAPTER 26
Charts

After studying this chapter, you should be able to do the following:

- Convert data into chart form.
- Be able to draw bar, line, and pie charts.

Introduction

Charts are used to describe the relationships among facts and statistics in a manner that is easy to understand. Charts are made of lines, bars, or other descriptive drawings.

The terms *charts*, *graphs*, and *diagrams* are very close in meaning and so they tend to be used interchangeably in industry.

Charts have become increasingly important in all areas of business and industry. The stockbroker keeps a record of stock prices with charts. A medical doctor records the characteristics of a heartbeat with a chart. The engineer uses charts to record test results. Fig. 26-1.

Fig. 26–1. Charts—with and without graphics—are widely used in business and industry. (Chartpak)

Preparing Charts on a Computer

The computer can be used to prepare charts and other types of graphics. Often engineering data are already in the computer. These can be called upon when making charts to show comparisons or reach engineering decisions. Fig. 26-2. The computer can also be used to prepare visuals such as bar charts for the nontechnical viewer. The charts can be printed in a wide range of colors complete with scales and descriptive material lettered.

Plotters that print in color have four or more pens that are filled with different colored ink. The computer signals the plotter to lower a pen, such as the one with red ink, and directs it to draw all the red lines. Fig. 26-3. It then raises the red pen off the paper and lowers the pen with the next color of ink needed. This continues until the chart is finished.

It is important that the person developing the charts on the computer know how to design the different types of charts. Without this basic knowledge the person cannot tell the computer what to do to create the chart needed. Fig. 26-4.

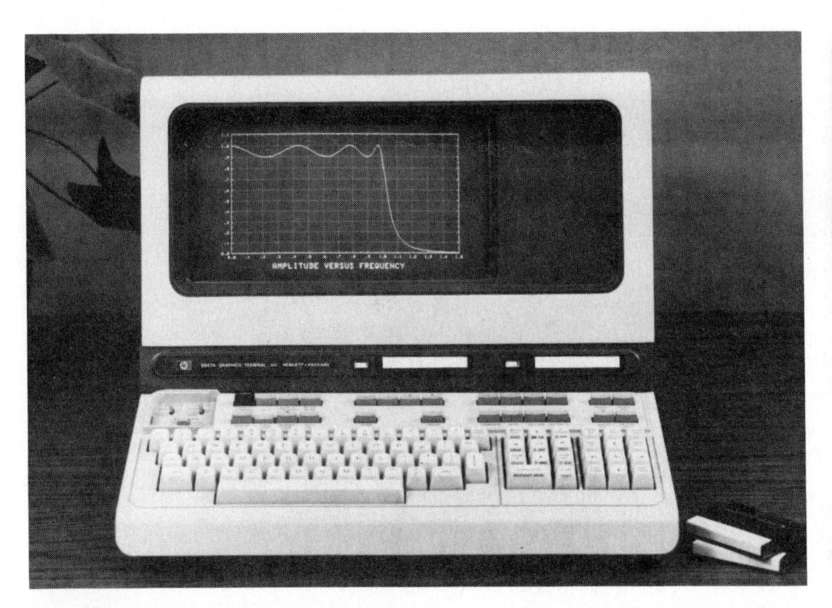

Fig. 26–2. This computer screen shows a line chart developed to present engineering data. (Hewlett-Packard)

Fig. 26–3. The computer directs the plotter to lower a pen and draw a line. (Data Technology, Inc.)

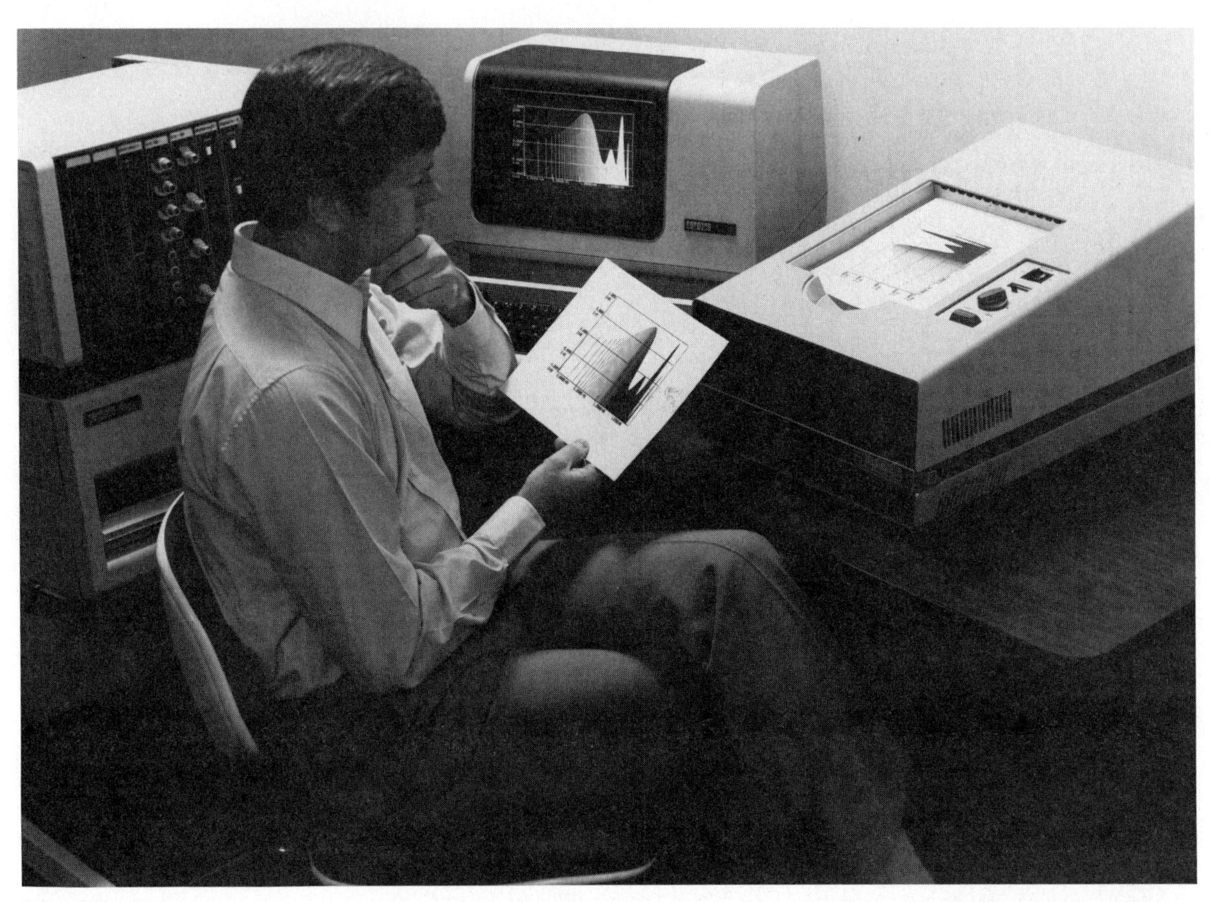

Fig. 26–4. This desktop video hardcopy unit produces attractive, high-resolution copies of all types of graphics. However, designing charts that communicate information well still requires human skills, too. (Tektronix, Inc.)

Bar Charts

Bar charts are used to show comparisons of amounts. The numerical value is shown by the length of the bar. There are several types of bar charts that are commonly used. Bar charts are used to present data for the nontechnical reader.

One-Column Bar Charts

The column of a one-column bar chart equals 100 percent. It is divided into parts to represent the percentages of each item. Each area can be emphasized by using different shading patterns or different colors.

The method used to draw this type of chart is shown in Fig. 26-5. To draw a one-column bar chart, follow these steps:

1. Assemble the data to be recorded.

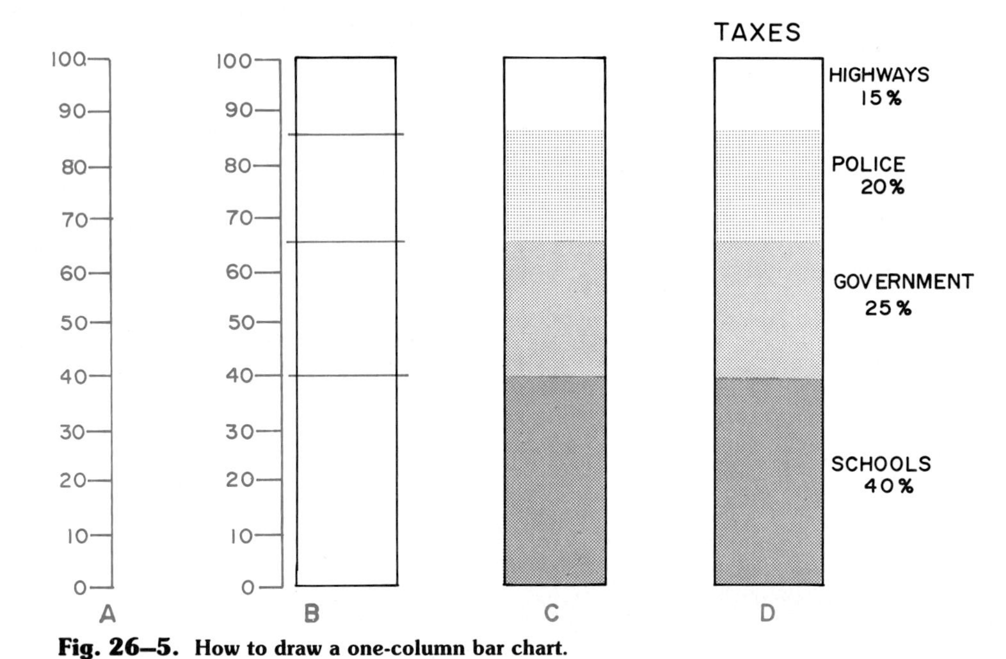

Fig. 26–5. **How to draw a one-column bar chart.**

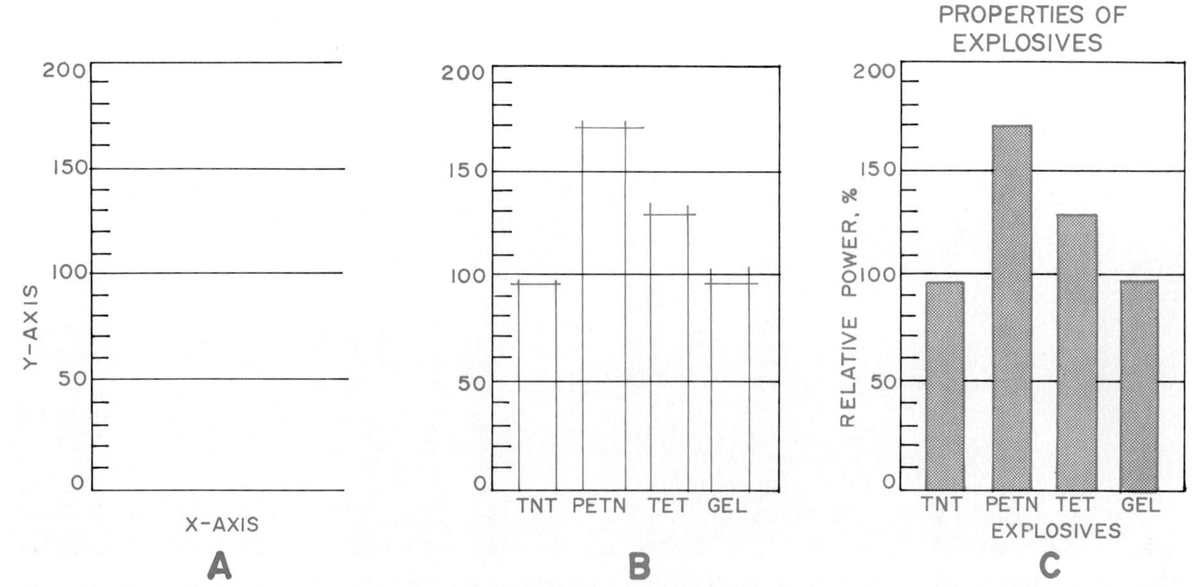

Fig. 26–6. **The steps used to draw a multiple-bar chart.**

2. Decide on the length of the column and divide it using a scale equal to 100 units. Fig. 26-5 at A.

3. Draw the column to the desired width. Fig. 26-5 at B.

4. Locate the percentage of each part and draw a line across the bar to represent that amount. Fig. 26-5 at B.

5. Shade, color, or crosshatch each part. Fig. 26-5 at C.

6. Letter the title and other identifying data. Fig. 26-5 at D.

Multiple-Column Bar Charts

Multiple-column bar charts are used to compare a number of related factors. Each factor is represented by the length of the bar drawn. Its value is shown by the scale on the chart. The steps for drawing a multiple-column bar chart are shown in Fig. 26-6. To draw a multiple-column bar chart, follow these steps:

1. Assemble the data to be shown.

2. Lay out the X- and Y-axes and the scale to be used on each axis. Fig. 26-6 at A.

3. Draw the bars to size to show the required data. Fig. 26-6 at B.

4. Shade or color the bars and add the title and other identifying data. Fig. 26-6 at C.

There are several different types of bar charts in use. The bars can run *vertically*, as in Fig. 26-6, or *horizontally*, as in Fig. 26-7. The *over-and-under bar chart* shows the relationship between two series of numbers. Plus and minus amounts are usually shown. Fig. 26-8. The chart has a zero point with data above and below. A *range-bar chart* shows the high and low figures over a period of time. The difference between the high and low is called the *range*. Fig. 26-9.

Planning Bar Charts

When planning bar charts consider the following factors:

- The bars should be of equal width.
- A scale can be used along the side of the chart to show values.
- Values can be lettered on or next to the bar.
- Bars showing different things should have different surface indications. Color is very useful for this purpose.
- The scale selected should show the differences in values. It can be placed on the horizontal or vertical axis.
- A title should be lettered on each chart. It should be placed above the chart and off the grid. The title should be larger than the lettering on the chart.
- The meaning of the chart should be clear even though it is separated from the text material it is illustrating.
- Sometimes a key is needed to tell what each bar represents. Fig. 26-10 is an example.
- The bars can be drawn pictorially as done in Fig. 26-11.

Line Charts

A **line chart** is an easy way to show change in data. It is built around two axes. The vertical axis is the Y-axis. The horizontal axis is the X-axis. Follow these steps to draw the line chart shown in Fig. 26-12:

1. Lay out the X- and Y-axes.

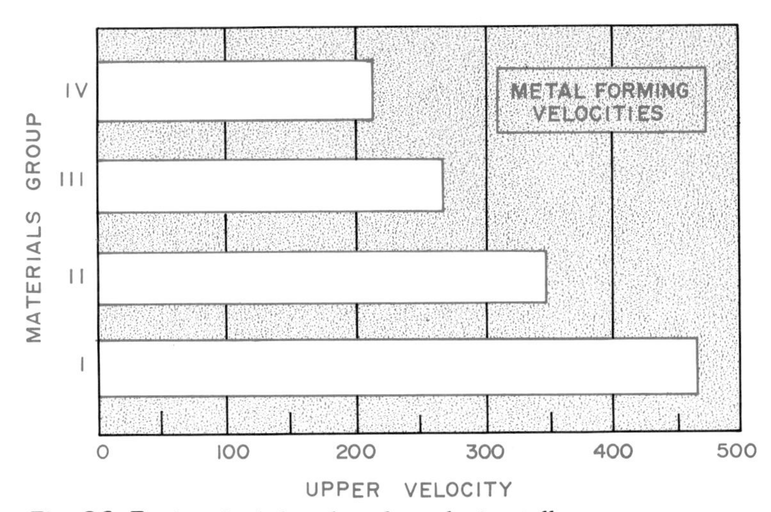

Fig. 26–7. A multiple-bar chart drawn horizontally.

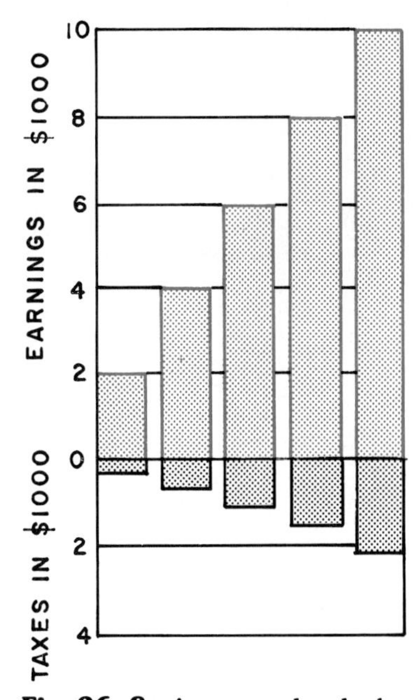

Fig. 26–8. An over-and-under bar chart.

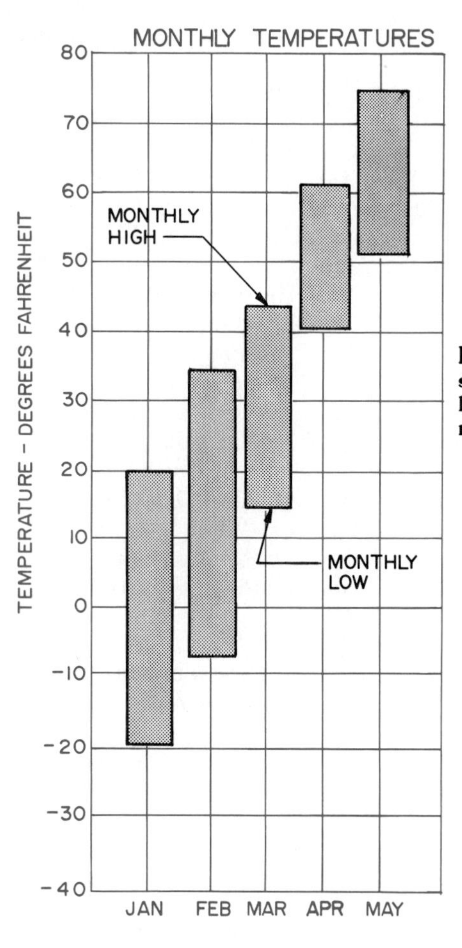

Fig. 26–9. A bar chart showing the range (high to low) of temperatures by month.

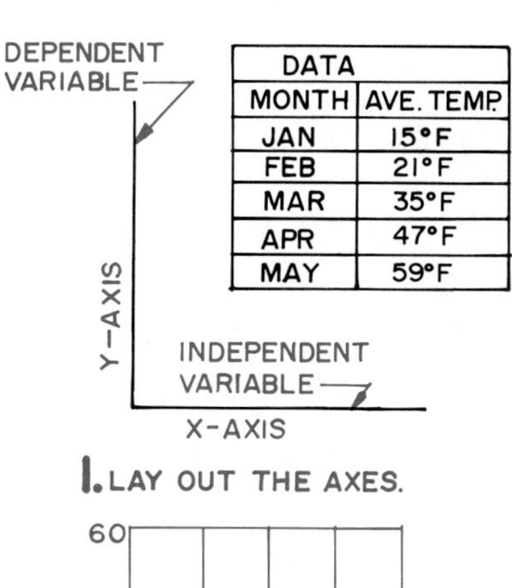

DATA	
MONTH	AVE. TEMP.
JAN	15°F
FEB	21°F
MAR	35°F
APR	47°F
MAY	59°F

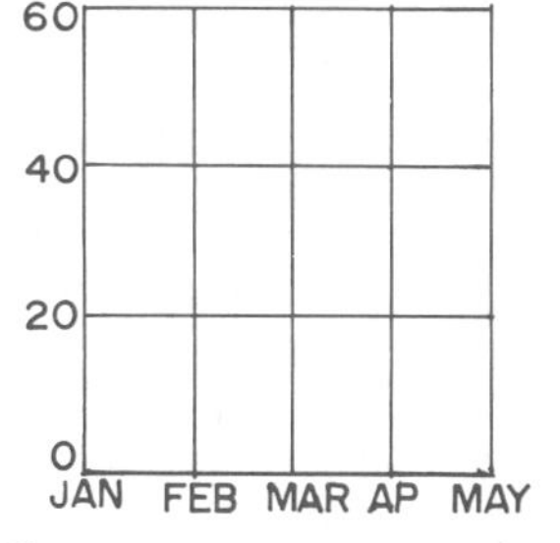

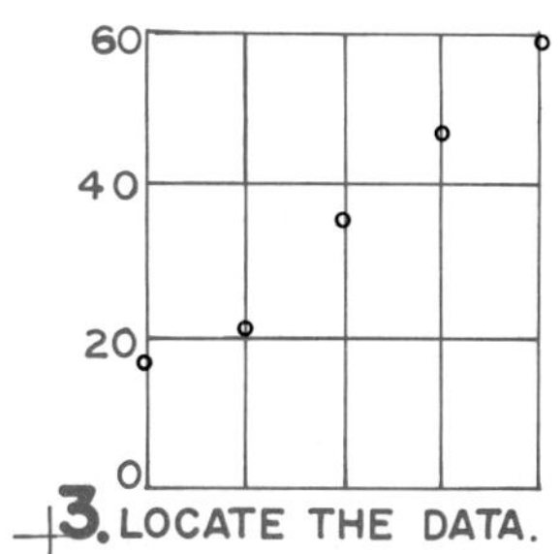

CONSUMPTION OF RAW MATERIALS

MILLIONS OF NET TONS

150

100

50

0

1989 1990

IRON ORE

COKING COAL

Fig. 26–10. Grouped bars can be used to make comparisons. Notice the use of a key to identify what each bar represents.

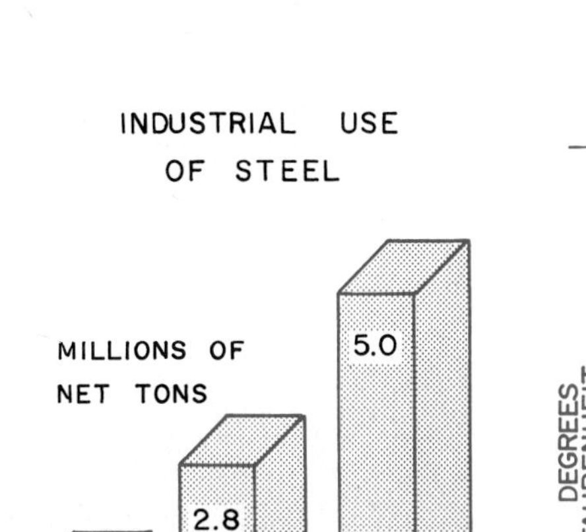

Fig. 26–11. A pictorial bar chart. (American Iron and Steel Institute)

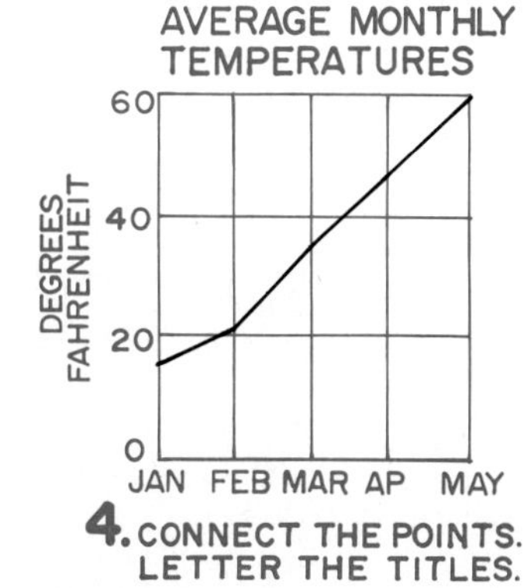

Fig. 26–12. The steps to draw a line chart.

2. Develop the scale and mark it on the axes.

3. Locate the data on the chart.

4. Connect the points and letter the titles.

Variables

Generally the independent variable is placed on the X-axis. The **independent variable** is the information that is controlled. In Fig. 26-12 the months are controlled and are therefore the independent variable. The temperatures vary from month to month and are the dependent variable. **Dependent variables** are data that are not controlled, and they are placed on the Y-axis.

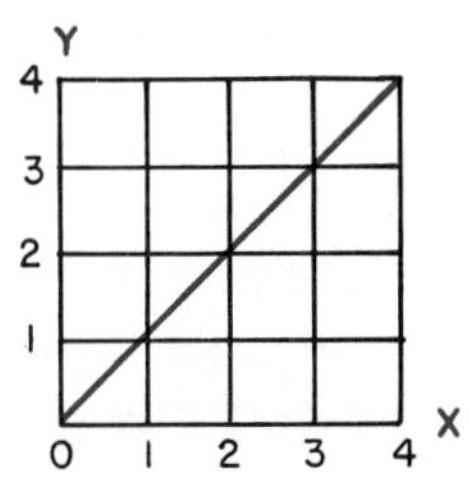

THIS IS THE WAY THE GRAPH SHOULD APPEAR

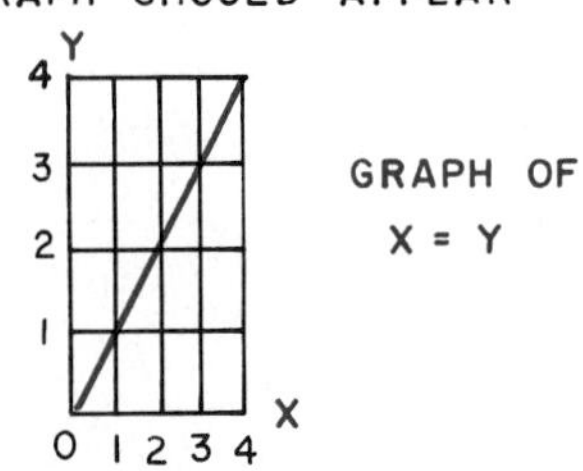

X-AXIS SCALE IS HALF THE Y-AXIS SCALE

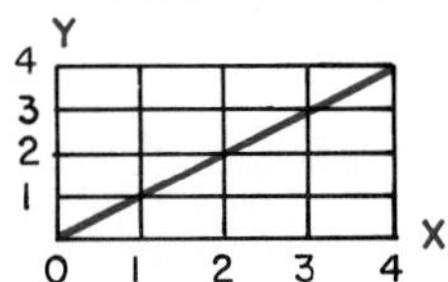

Y-AXIS SCALE IS HALF THE X-AXIS SCALE

Fig. 26–13. The selection of the scale for the grid influences the impression given by the plotted data.

Scale

The selection of the scale is very important. The shape of the grid will influence the shape of the curve. The top chart in Fig. 26-13 shows equal values on each axis. With this condition the line drawn through units that increase by one on each scale is on a 45-degree angle. Notice what happens to the angle of the line as the scales are changed. Angles greater than 45 degrees give the impression of a fast or important rise. Those angles less than 45 degrees indicate a slow or unimportant rise. When preparing your scale, study the significance of the change to be reported. Choose a scale that shows the importance of the change.

Grids for charts involving time are usually rectangular. Proportions of two to three or three to four are commonly used. A grid that is wider than it is high is best for time charts having many points to be plotted. Such a scale is also used when the chart covers a long period of time. Fig. 26-14. A grid that is higher than it is wide is used for time charts that cover short periods or show a rapid change. Fig. 26-15.

Plotted Points

Line charts that are used for general information do not have each point noted. The points are located where the ends of the plotted lines meet. Fig. 26-15. The line used to show the curve is usually drawn thick.

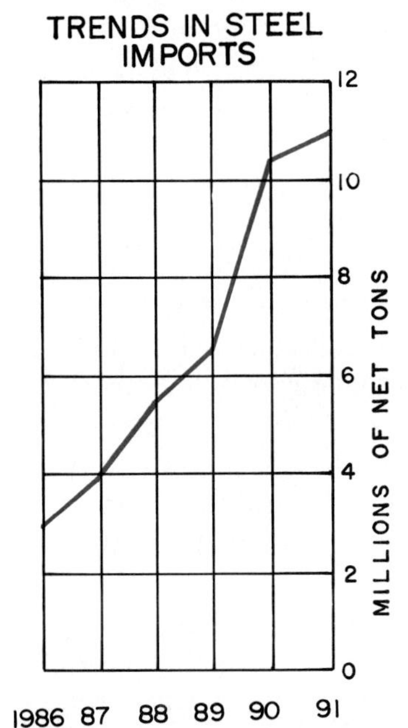

Fig. 26–15. Charts that show a rapid rise use a grid that is higher than it is wide. (American Iron and Steel Institute)

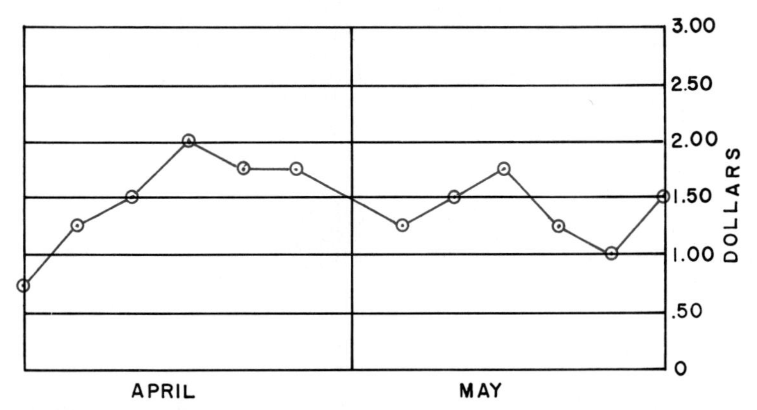

Fig. 26–14. Charts with many points should use a grid that is wider than it is high.

CAD Applications:

Computer Charting

People whose jobs require them to make many charts are finding computer charting software to be a valuable aid. With this software, they can make colorful line, pie, bar, pictorial, or combination charts quickly and easily.

The charting program can accept data imported from a spreadsheet or database program, or the user can start "from scratch." After inputting data, the user selects the format. In seconds, the program draws the chart.

Many charting programs have features that allow the user to be more creative. For example, some can produce exploded pie charts, 3-D effects, or overlapping charts. Many offer a variety of text styles for adding words. There are programs that allow the user to make drawings for a chart or to import drawings from other programs.

Finished charts may be output to a printer, plotter, or film recorder (a device that makes 35-mm slides). Some programs can become part of a desktop publishing system. With such a system, words, charts, drawings, and pictures are combined on the computer and printed out in a page format.

If you have drawn charts, you know it takes time to make them attractive, neat, and easily readable. Considering that charting programs can draw a bar or pie chart in 16 seconds or less, it's easy to see why this software is becoming popular.

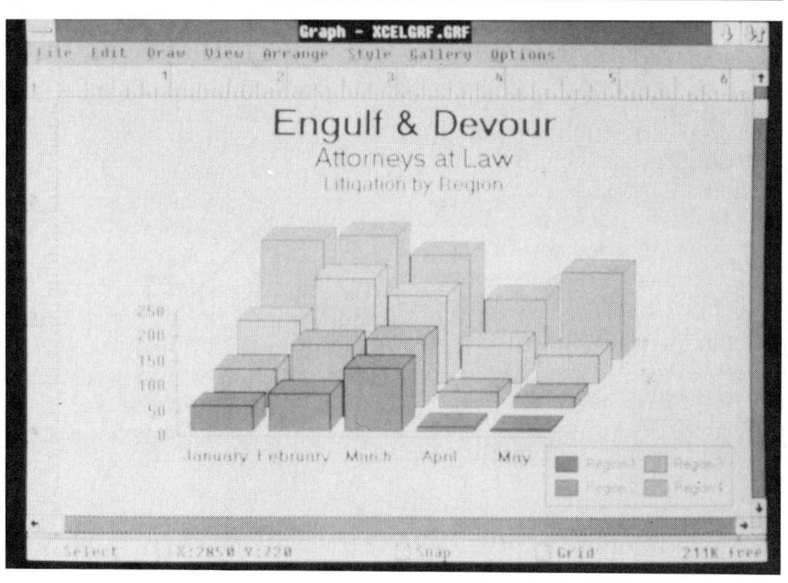

Graph Plus software by MICROGRAFX

Pixie software by Zenographics

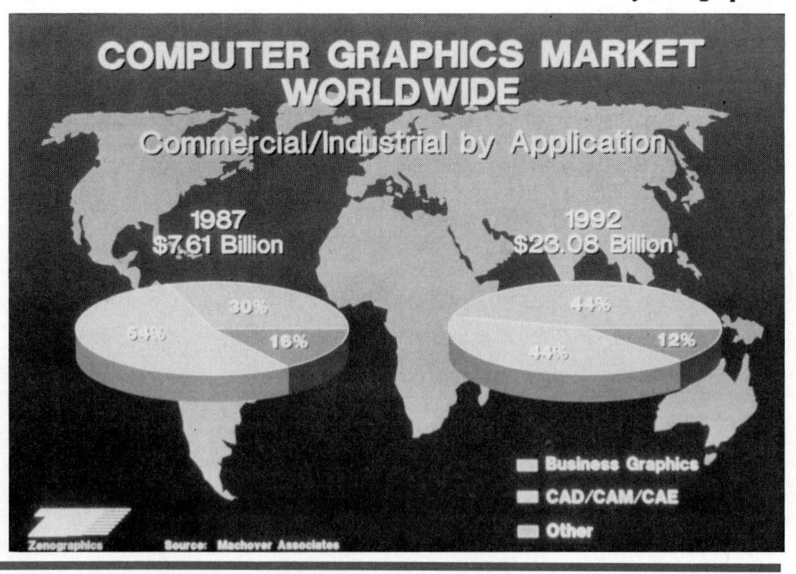

If the value of each point is important, a dot within a small circle is used. The dot accurately locates the value. The line forming the curve touches the circle and stops. Fig. 26-14. Usually this kind of curve is drawn with a thin line.

Stepped-Line Chart

A *stepped-line chart*, Fig. 26-16, is used when the data shown retain a fixed value over a period of time. Fig. 26-16 shows the interest rate began the first quarter at about 10.4 percent. It went up and then down over the next three quarters ending at about 10.3 percent.

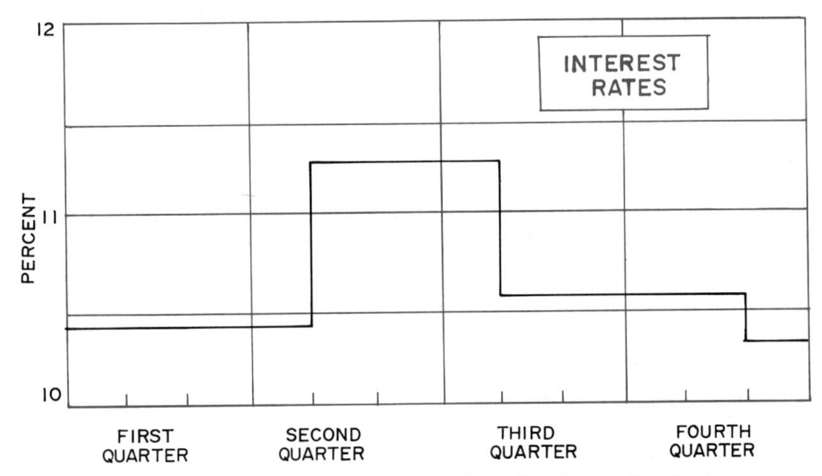

Fig. 26–16. A stepped-line chart shows data that have a fixed value over a period of time.

Smooth-Curve Graph

You should plot graphs that show continuous data with a smooth curve to connect their points. In continuous data we find a mathematical relationship. As one variable changes, the other also changes. For example, Fig. 26-17 shows the change in temperature as amperes are increased. The more amps of electric current, the higher the temperature generated.

In some cases, the points will not fall exactly on a smooth curve. However, the curve is still drawn smoothly even though some points will be missed. Generally the curve is drawn so that some points fall on each side of the line. Fig. 26-18.

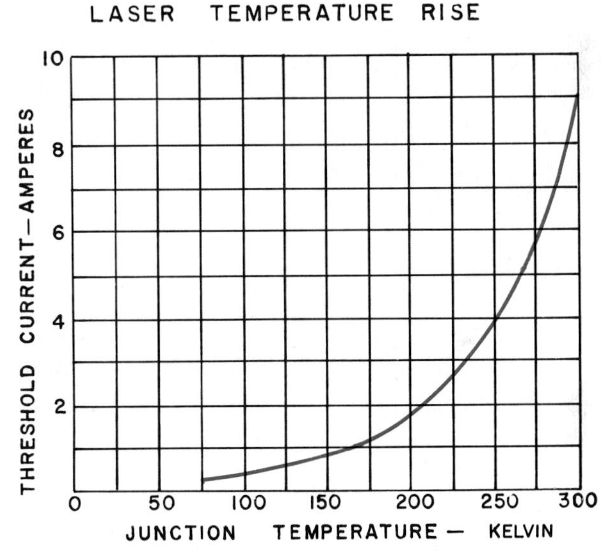

Fig. 26–17. This is a smooth curve graph. Changes in one variable cause changes in the other variable.

Multiple-Line Charts

A line chart can be used to show more than one set of related data, but the lines used to show the data should be different. They can be solid, dotted, dashed, or of different colors. Fig. 26-19.

On multiple-line charts you must explain what each curve represents. This can be done with a key. Fig. 26-20. Another way is to letter a caption near each curve. Fig. 26-21.

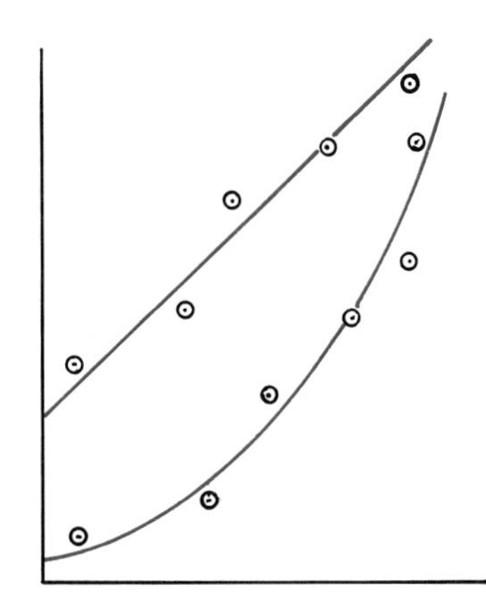

Fig. 26–18. Drawing a smooth curve for continuous data when all points do not fall on the curve.

Surface Charts

A **surface chart** shows values by the height of a shaded surface. This technique can be used to give emphasis to a single-line chart. Fig. 26-22. Surface charts can show the difference between several sets of data. In Fig. 26-23, a comparison of the rainfall in three states over a four-month period is quickly seen. Fig. 26-24 shows a surface chart made with pressure-sensitive shading materials.

Quadrants

Most line charts are drawn in the first quadrant. Some data, especially mathematical data, require the use of several quadrants. Fig. 26-25.

Each quadrant represents a positive and negative value of X and Y. The point of intersection of axes X and Y is zero. Values to the right of the Y-axis are positive. Those to the left are negative. Values above the X-axis are positive. Those below are negative. Each point has an X- and Y-value.

Study the data shown in Fig. 26-26. Notice how each point was plotted using its X- and Y-value. It is essential to pay attention to the positive or negative sign on each value.

Lettering on Charts

Titles and captions on the scales are placed outside the axes. The scale captions are lettered using the aligned system. The titles and captions can be lettered to the left or right of the chart or along its bottom.

The vertical axis caption is usually lettered on the left. The vertical axis scale and captions are read from bottom to top.

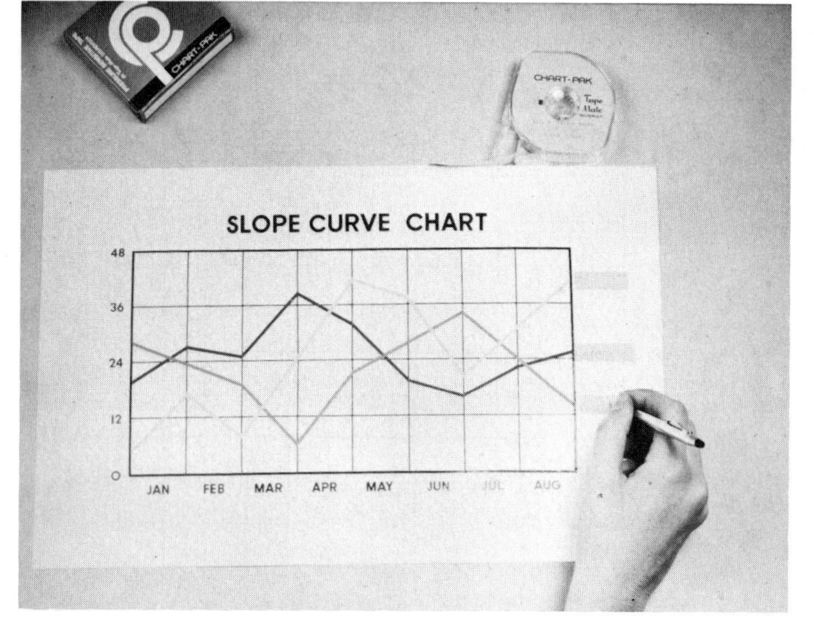

Fig. 26–19. Multiple-line charts show several sets of related data. This chart was made with pressure-sensitive tape. (Chartpak)

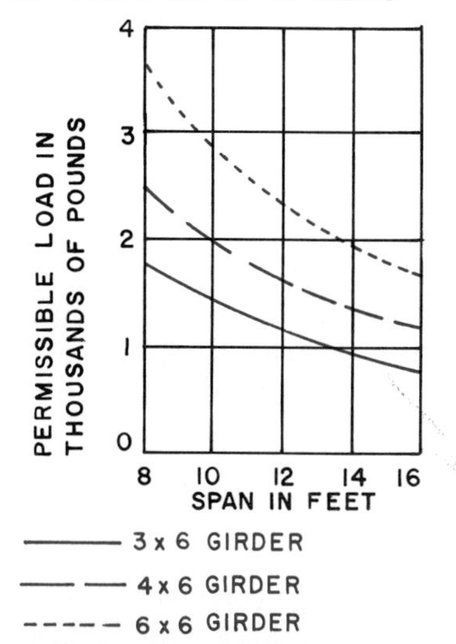

Fig. 26–20. The key identifies what each line represents on a multiple-line chart.

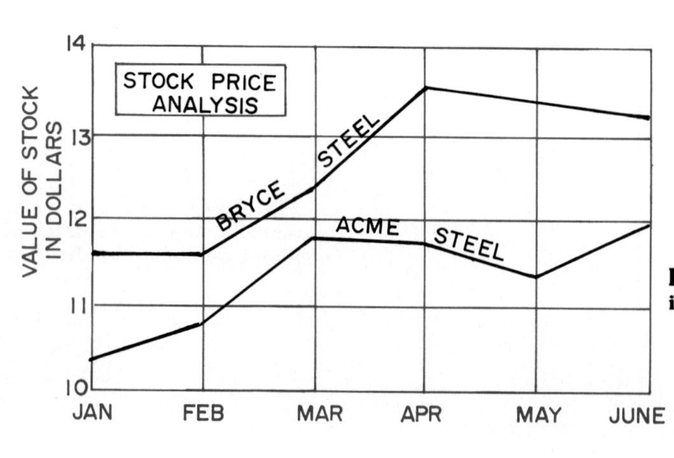

Fig. 26–21. Line chart curves can be identified by lettering along the curve.

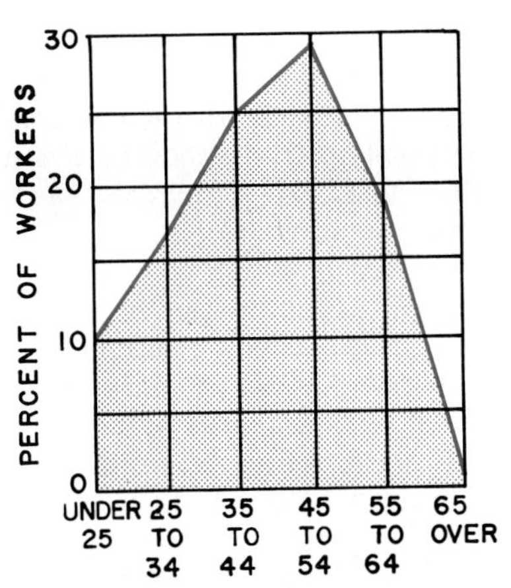

Fig. 26–22. A surface chart gives added emphasis to data. (American Iron and Steel Institute)

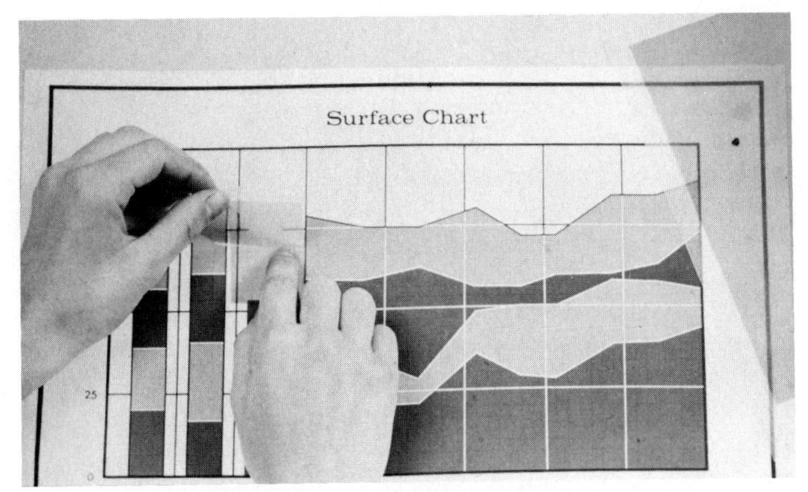

Fig. 26–24. Pressure-sensitive shading materials can be used to make surface charts. (Chartpak)

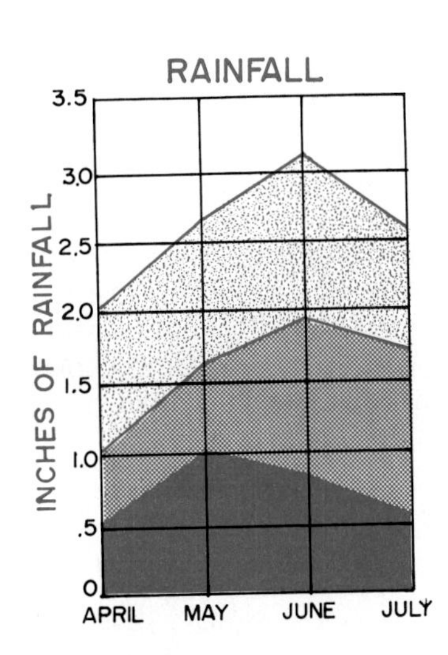

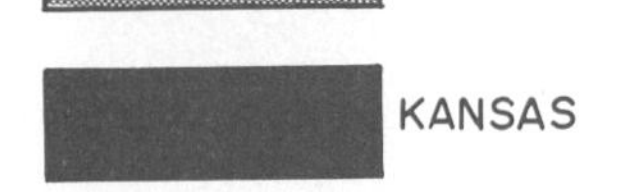

Fig. 26–23. A surface chart can be used to compare data.

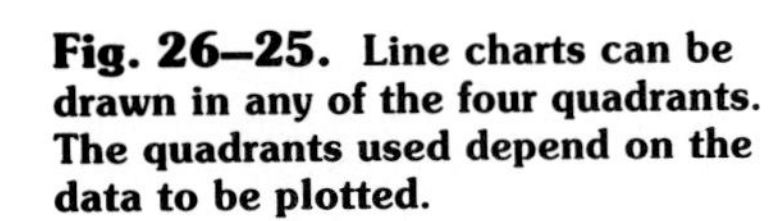

Fig. 26–25. Line charts can be drawn in any of the four quadrants. The quadrants used depend on the data to be plotted.

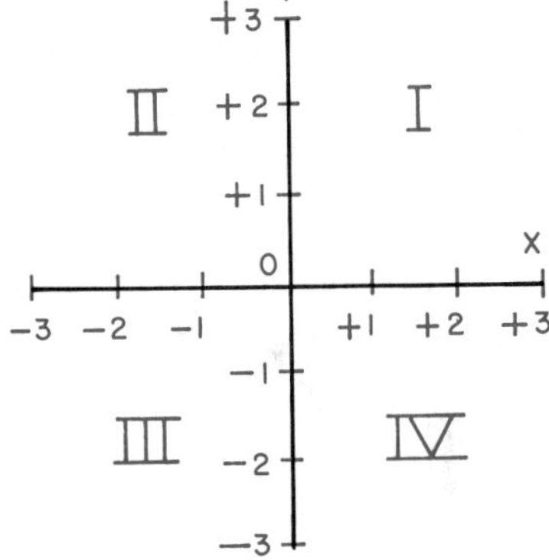

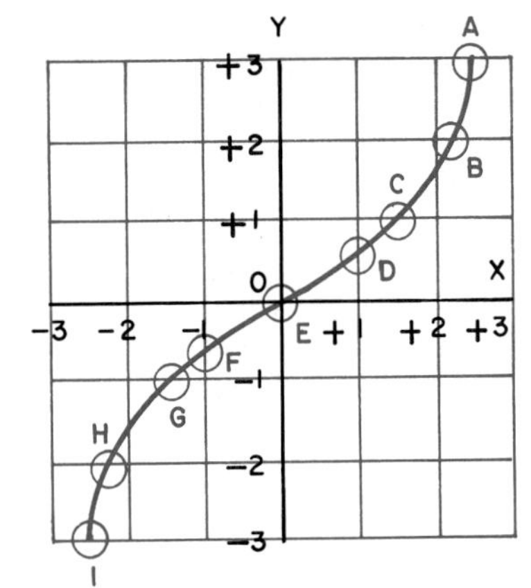

	DATA	
	X	Y
A	2.5	3.0
B	2.25	2.0
C	1.5	1.0
D	1.0	0.6
E	0	0
F	−1.0	−0.6
G	−1.5	−1.0
H	−2.25	−2.0
I	−2.5	−3.0

Fig. 26–26. This line chart shows data having positive and negative values. The first and third quadrants were used.

Fig. 26–27. Transfer lettering can be used on charts and graphs. (Chartpak)

Both the scale and the caption on the horizontal axis are lettered along the bottom. They are also read along the bottom from left to right. The title for the line chart is usually placed above the chart. See Fig. 26-23 for an example of lettering on charts.

In many drafting rooms, all lettering is typed on the chart. If the chart is on vellum, a piece of carbon paper is placed beneath the sheet. This produces a darker image because the carbon leaves an impression on the back of the sheet. Transfer lettering can be used on charts. Fig. 26-27.

Pie Charts

Pie charts are sometimes referred to as *percentage charts*. A **pie chart** shows how a given amount is divided into specific parts or percentages. Fig. 26-28. The entire circle ("pie") represents 100 percent of any specified amount.

Pie charts are most effective when there are no more than eight divisions. If more than eight parts are used in the chart, it become difficult to understand. The circle should be large enough to permit lettering without crowding.

RAW STEEL PRODUCTION IN THE UNITED STATES			
PROCESS	TONS	%	DEGREES
OPEN HEARTH	45,172,200	55.6	200.1
BASIC OXYGEN	26,485,900	32.6	117.4
ELECTRIC FURNACE	9,586,000	11.8	42.5

A

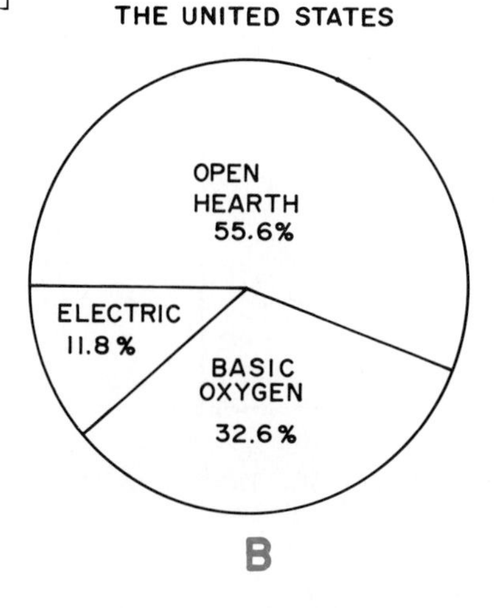

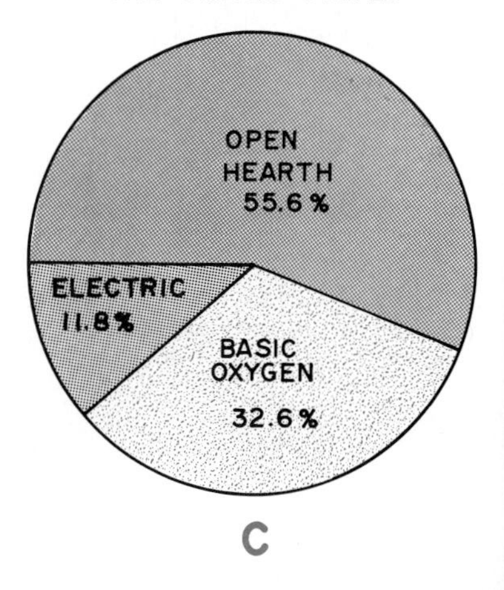

Fig. 26–28. The steps to draw a pie chart.

It is standard practice to place all lettering horizontally on the sector of the circle. If several of the sectors are too small, place the lettering on the outside of the circle. Then connect the lettering to the sector with a leader.

It is a simple matter to convert percentage to degrees of a sector. Since there are 360 degrees in a circle, simply multiply 360 degrees by the percentage. This will give the number of degrees to be used in a sector. For example, find the number of degrees in 35 percent:

First, convert the percent amount to a decimal, 35% = .35.

Second, multiply 360° by the percent, .35 × 360° = 126°.

The areas of the pie chart are usually shaded. Each sector should be shaded in a different manner. The use of color is especially effective.

In some cases, a pie chart may be made in pictorial form. Fig. 26-29. When you draw a pictorial pie chart, you may want to separate the several sectors of the circle. The pictorial form is not effective for charts having more than five or six segments.

How to Draw a Pie Chart

To draw a pie chart, see Fig. 26-28 and follow these steps:

1. Convert the data to be presented to percentages and then to degrees.
2. Draw a circle of the size desired. Lay out the radial lines forming each section.
3. Letter identifying information and titles.
4. Shade or color the sections as desired.

Pictorial Charts

Pictorial charts are sometimes referred to as *pictographs*. These are classed as nonmathematical charts. A **pictorial chart** is a visual form of presenting information to the nontechnical reader. These charts give comparisons by means of pictures. Each picture represents a specific unit or quantity. Fig. 26-30. The pictorial chart is a form of bar chart. The row of representative symbols makes up the bar.

The pictorial chart has a limited degree of accuracy. Usually the amount represented by the symbols is given next to the last symbol. In Fig. 26-30, each symbol shown equals one million vehicles. Always give a key on the chart to show the value of the symbol used.

WORLD STEEL PRODUCTION

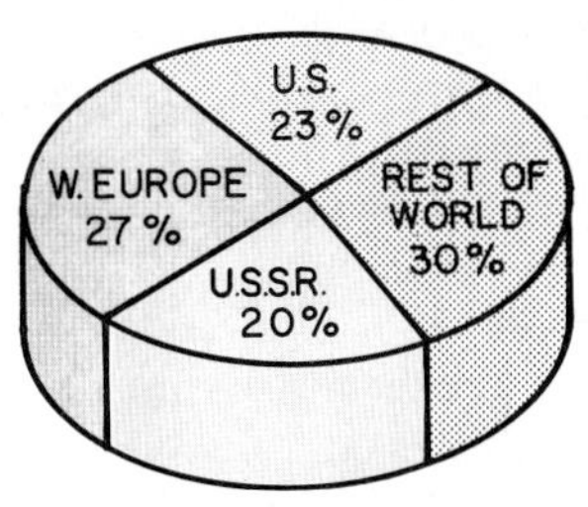

A PICTORIAL PIE CHART

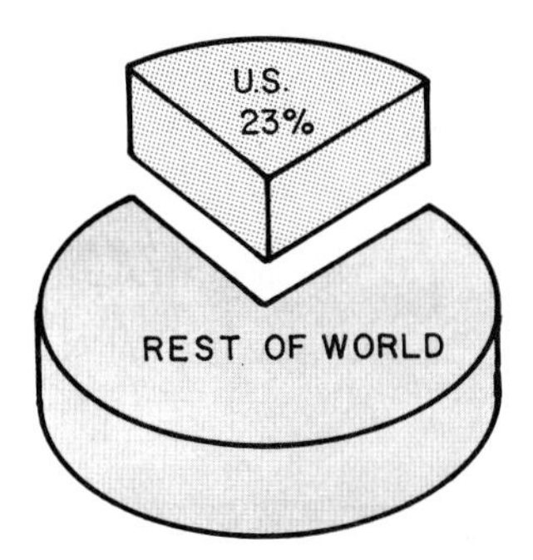

SECTORS CAN BE REMOVED FOR ADDED EMPHASIS

Fig. 26–29. Pictorial pie charts can be drawn in many ways. (American Iron and Steel Institute)

MOTOR VEHICLE OUTPUT

EACH SYMBOL EQUALS ONE MILLION VEHICLES

Fig. 26–30. A pictorial chart. This one is in bar chart form.

Some forms of pictorial charts use symbols of various sizes to represent quantity. The comparison of data is made by observing the difference in the size of the symbols. Fig. 26-31. These charts can be misleading. The reader frequently does not know whether the drafter intended the symbol to be concerned with area or linear size. As an example, a symbol measuring 2 inches by 2 inches would have an area of 4 square inches. If the drafter wanted to show twice the amount, he or she might draw the symbol 4 inches by 4 inches. But then the area shown is 16 square inches or four times the original size, rather than twice the amount. This would be misleading to the reader. Therefore, it is necessary to give the value that each symbol represents.

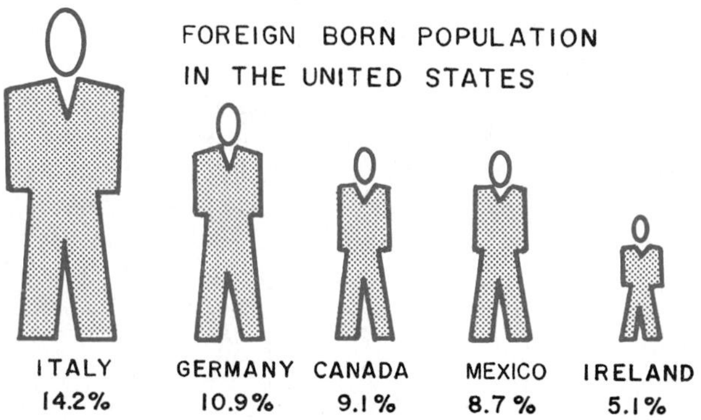

FOREIGN BORN POPULATION
IN THE UNITED STATES

ITALY 14.2%
GERMANY 10.9%
CANADA 9.1%
MEXICO 8.7%
IRELAND 5.1%

Fig. 26–31. A picture chart in which the size of the figure is used to make a comparison of the data.

Flow and Organization Charts

Both the flow chart and organization chart are similar in purpose. They are nonmathematical charts. The advantage of a flow or organizational chart is that it replaces lengthy verbal or written description.

A **flow chart** shows a schematic representation of a process. A simple form of a flow chart is made with a series of rectangles. The name of each step in the process is lettered in the rectangle. Fig. 26-32. More complex forms use pictures that resemble the units involved to describe the process. Fig. 26-33.

One use of a flow chart is to represent the flow, or distribution, of materials as they pass through a manufacturing plant. For example, the processing, production, packaging, and marketing of a product can be easily described with a flow chart.

Organization charts show the line of authority within an organization such as a company. They show the responsibility of the ex-

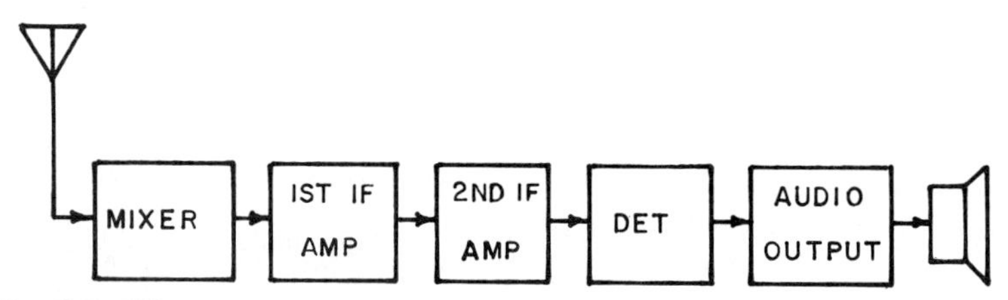

Fig. 26–32. This flow chart shows the major components of a transistor radio receiver.

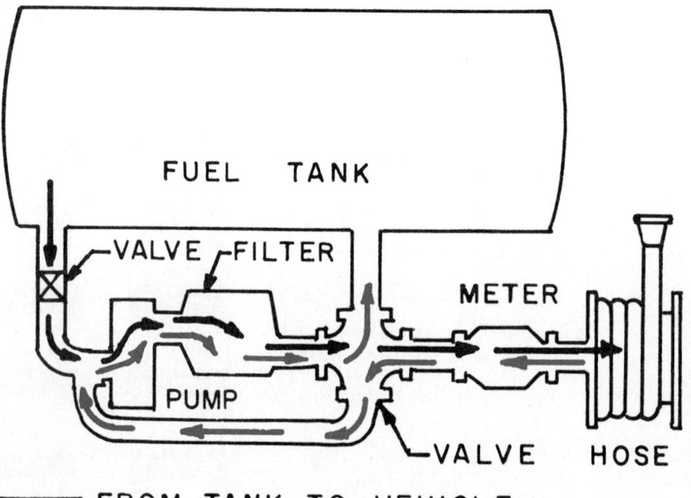

Fig. 26–33. A flow chart shows the movement, or flow, involved in a process. This diagram shows the flow of fuel from a tank truck to an aircraft and the reverse flow to remove fuel from an aircraft. (The Gormann-Rupp Co.)

ecutive to a division and to a divisional head. Interrelated and interlocking activities of an organization can be easily diagrammed.

An organization chart may also be used to show the separate functions of a company. The names of divisions or functions are enclosed in rectangles. For variety, use both rectangles and circles. Fig. 26-34.

Coordinate Paper

The easiest way to lay out a chart is on graph, or coordinate, paper. Some of the types available have rectangular rulings of 4, 5, 6, 8, 10, 12, 16 and 20 spaces to the inch and 1 millimeter squares. They are available on opaque paper and vellum.

Special types of graph paper are available for pie charts. Polar paper is used for plotting curves by polar coordinates. Triangular coordinate paper is used for plotting a curve based on three variables. Logarithmic graph paper is divided on the basis of logarithms. This paper is used when a percentage of change is more important than the quantity of change.

Drafting Aids

In recent years, drawing charts has been greatly simplified by special tools. Pressure-sensitive tapes, shading films, and other components have been developed that reduce the time it takes to make a graph. Adhesive-backed tapes can be used on charts. Tapes range in size from 1/16 to 2 inches in width. They are available in assorted colors, densities, and surfaces.

If a chart is to be used on an overhead projector, transparent tape in various colors can be used for grids, curves, and bars. Glossy tapes are used primarily for direct presentation.

Curves can be "drawn" by using thin tape that comes in rolls. It is held in a special container that makes it easy to lay out a curve. Fig. 26-35.

Pressure-sensitive screening, Fig. 26-36, is also used to make charts. Various dot and line pattern screens are used to identify different parts of a pie or multiple-bar chart. Screens may also be used for bar charts. Most of the charts in this chapter were shaded with this material.

Pressure-sensitive lettering, Fig. 26-37, is sometimes used for identification on all types of charts. This material is applied directly to the chart. It is available in a wide variety of sizes and type faces.

Symbols are available for use in pictorial charts. These symbols are used when figures must be repeated. Some examples of symbols are shown in Fig. 26-38.

When pressure-sensitive material is used, errors may be corrected very easily. The tape or screen can be lifted off with a sharp knife or a single-edge razor blade.

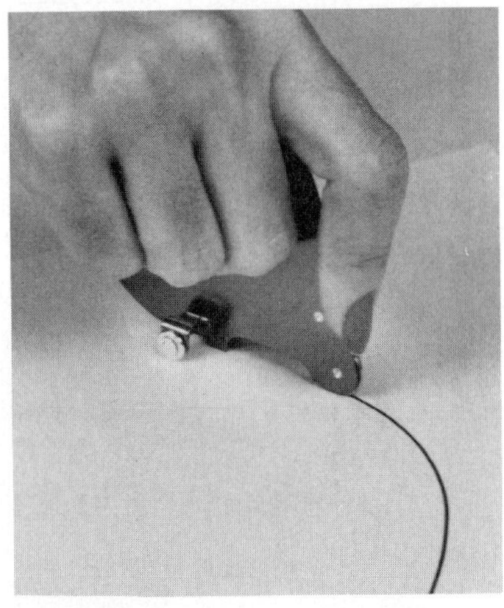

Fig. 26–35. Lines can be drawn on charts using pressure-sensitive tape held in a dispenser. (Chartpak)

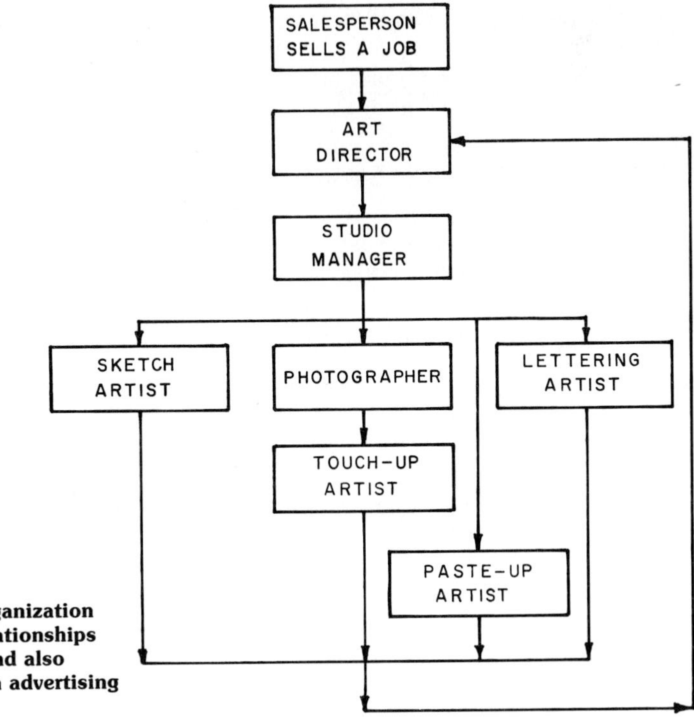

Fig. 26–34. This is both an organization and a flow chart. It shows the relationships among parts of an organization and also shows how a job flows through an advertising agency.

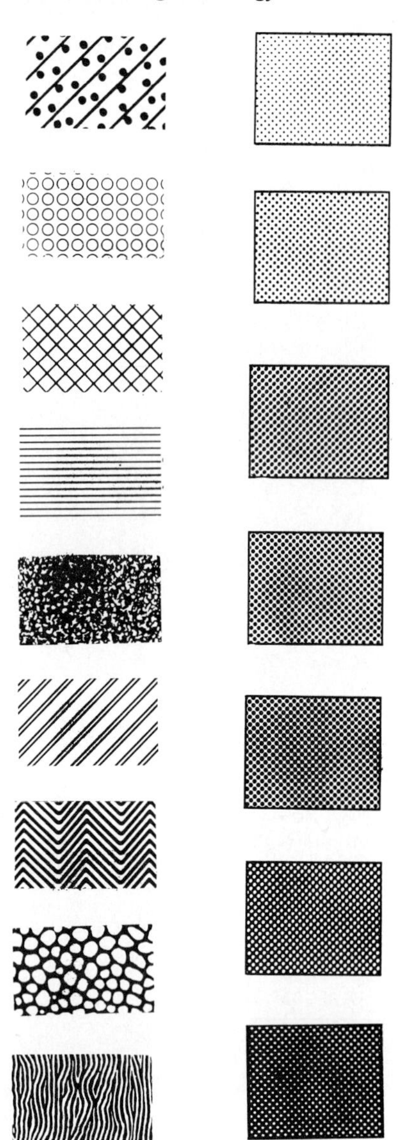

Fig. 26–36. These are a few of the shading patterns available on pressure-sensitive sheets. (Chartpak

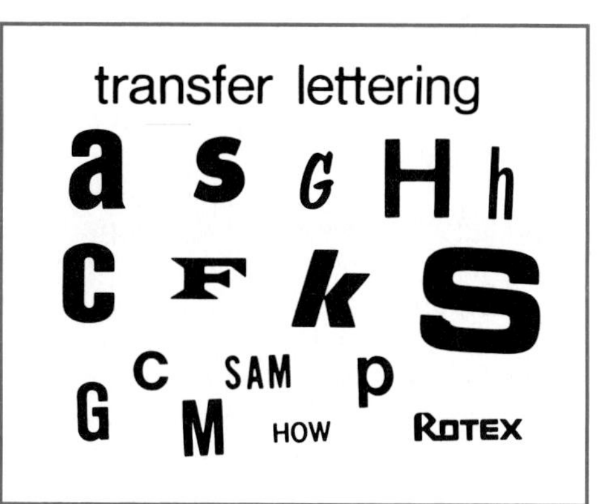

Fig. 26–37. A few of the sizes and styles of pressure-sensitive letters. (Chartpak)

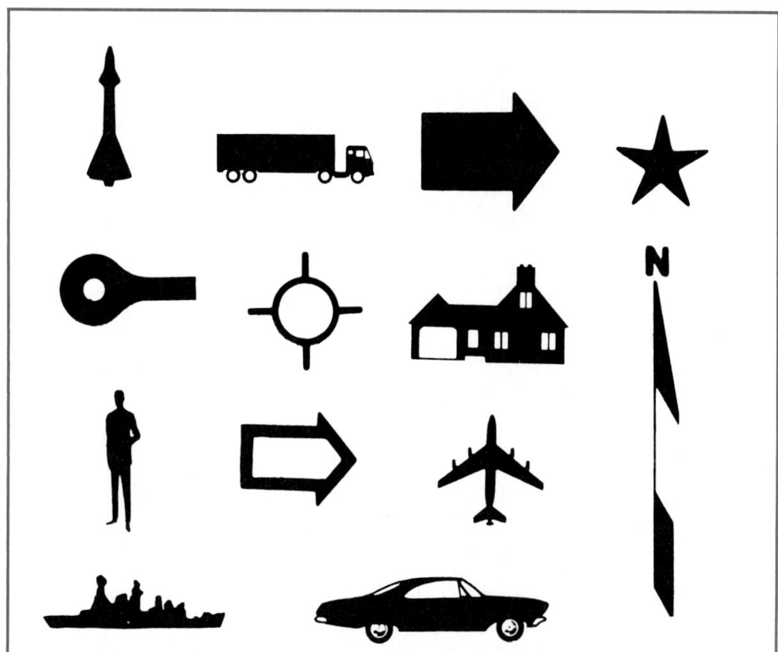

Fig. 26–38. A few of the symbols available on pressure-sensitive sheets. (Chartpak)

Chapter Review

Build Your Vocabulary

You should understand and use the following terms as part of your working vocabulary. Write a brief explanation of what each means.

bar chart
line chart
independent variable
dependent variable
surface chart
pie chart
pictorial chart
flow chart
organization chart

Sharpen Your Math Skills

1. A pie chart shows three energy sources: coal, oil, and gas. If coal provides 28%, and oil provides 47%, how much energy does gas contribute?

2. A single-line bar chart represents 100 percent. If you draw the chart 6″ long, how many inches do you need to represent each energy source in Problem 1?

3. You are to draw the three energy sources in Problem 1 using a three-bar chart. If each bar is drawn 1″ wide, and the bars are spaced 3/8″ apart, how wide is the chart?

Study Problems—Directions

The problems that follow will give you the chance to use some of the drafting knowledge that you learned in this chapter. Do the problems that are assigned to you by your instructor. Note: Statistics given in the following charts are hypothetical.

1. Make a bar chart showing the telephones in service from the data in **Fig. P26-1**.
2. Make a line chart showing the trend of U.S. investments abroad as shown in **Fig. P26-2**.
3. Make a multiple-line chart comparing the tourist arrivals given in **Fig. P26-3**.
4. Make a pie chart or percentage bar chart showing the world production of nickel, **Fig. P26-4**.
5. Make a chart comparing the ten largest nations by area, **Fig. P26-5**.
6. Make a multiple-bar chart showing the monetary reserves of the government of West Germany, **Fig. P26-6**.
7. Make a pictorial chart to show the data in **Fig. P26-7**. Develop symbols for the vehicles reported.
8. Plot a curve using the data shown in **Fig. P26-8**.
9. Make a surface chart to report the data on the number of wells drilled as shown in **Fig. P26-9**.
10. Make an organization chart showing the line of authority in your school system. Your teacher or principal can help trace the line of authority. Begin with the voters who elect the school board.
11. Make a flow chart showing the generation and distribution of electricity to residential areas. The electricity is generated in a generating plant. It then goes to a step-up transformer. This increases the voltage and sends it over high voltage wires to a substation with step-down transformers. At the substation the voltage is lowered for distribution in a local area, such as a city. The current then travels over wires to a small distribution station. This station sends the current over wires to local residential areas. In these areas are small step-down transformers. They lower the voltage further and send it to a home.
12. The tables in **Fig. P26-10** through **P26-18** contain a variety of data to be charted. Select the type of chart you think will best show the data. Draw the chart accurately.
13. Bring to school data you find in magazines and newspapers. Decide the best way to show this data using charting techniques. Draw the chart.

Study Problems

P26-1.

Telephones in Service (per 1000 population)			
Australia	489	West Germany	383
Brazil	72	Japan	479
Canada	693	U.S.S.R.	93
Finland	522	United States	789
France	498		

P26-2.

U.S. Investments Abroad*	
1988	59,491
1989	64,983
1990	71,016
1991	78,090

*Millions of Dollars.

P26-3.

Estimated International Tourist Arrivals				
Country	1988	1989	1990	1991
Africa	1,833,000	1,926,000	2,000,500	2,105,000
Europe	95,975,000	100,322,000	100,501,000	101,397,000
Central & N. America	23,753,000	27,379,000	29,479,000	29,970,000
Middle East	3,138,000	2,510,000	2,102,000	2,000,000

P26-4.

World Production of Nickel*	
Canada	89
U.S.S.R.	170
Rest of World	330
World Total	589

*Thousand Tons.

P26-5.

Ten Largest Nations by Area*	
U.S.S.R.	8,649,489
Canada	3,851,809
China	3,691,502
United States	3,615,211
Brazil	3,286,470
Australia	2,967,877
India	1,232,560
Argentina	1,072,156
Sudan	967,491
Algeria	919,590

*Square Miles.

P26-6.

Monetary Reserves-West Germany*		
	Gold	Foreign Exchange
Jan.	2.5	3.1
Apr.	4.2	2.8
July	3.9	3.3
Oct.	4.4	3.1

*Billions of Marks.

Study Problems

P26-7.

Vehicle Production	
Automobiles	6,500,000
Motorcycles	3,950,000
Off-the-road vehicles	1,250,000

P26-8.

Points on a Curve When X = Y		
Point	X	Y
1	−5	+5
2	−4.0	+3.0
3	−3.0	+1.8
4	−1.5	+ .75
5	0	0
6	+1.5	− .75
7	+3.0	−1.8
8	+4.0	−3.0
9	+5.0	−5.0

P26-9.

Wells Drilled	
Oil	17,025
Gas	9,100
Dry holes	13,700

P26-10.

Voltages Produced by Different Sources	
Source	Voltage Range
Batteries	6–12
Auto Generators	6–12
Train Generators	32–60
Residential Transformers	120–440
City Distribution Transformers	2,300–4,200
Power Transmission Lines	13,200–287,500

P26-11.

Ten Largest Nations by Population	
China	1,075,400,000
India	774,921,000
U.S.S.R.	281,238,000
United States	242,651,000
Indonesia	168,827,000
Brazil	140,696,000
Japan	122,134,000
Bangladesh	104,676,000
Pakistan	100,704,000
Nigeria	98,150,700

Study Problems

P26-12.

U.S. Corporate Bond Prices*		
	1990	1991
January	77.2	85.9
February	77.5	86.4
March	76.9	85.6
April	76.2	85.4
May	75.3	83.4
June	76.6	81.7
July	76.1	81.1
August	78.1	80.3
September	78.4	80.0
October	77.0	78.5
November	75.7	76.8
December	75.0	75.9

*Average Price in Dollars Per $100 Bond.

P26-13.

World Production of Tin*	
Malaysia	52,342
Indonesia	33,800
Bolivia	26,773
Thailand	26,207
Australia	12,615
Brazil	8,300
Rest of World	30,863

*Tons.

P26-14.

Plastics Production*		
	1990 Monthly Average	1991 Monthly Average
United States	475	510
Japan	340	390
West Germany	230	250
U.S.S.R.	90	110
United Kingdom	90	95

*Thousand Metric Tons.

Study Problems

P26-15.

World Petroleum Production*	
North America	13
South America	4
Western Europe	4
Africa	5
Middle East	11
Asia	5
Eastern Europe	12
World Total:	54

*Millions of Barrels Daily.

P26-16.

Strength of Various Steels	
Steel Number	**Tensile Strength PSI**
C1018	69,000
B1112	82,500
B1113	83,500
Ledloy 375	79,000
C1045	103,000
C1095	145,000

P26-17.

Composition of German Silver	
Copper	50%
Nickel	20%
Zinc	30%

P26-18.

Melting Points of Metals	
Metal	**Temperature Degrees Fahrenheit**
Pewter	420
Tin	449
Lead	621
Zinc	787
Aluminum	1218
Bronze	1675
Gold	1945
Cast Iron	2200
Steel	2500

Appendix Contents

Appendix 1. Standard Abbreviations Used on Drawings

A

Allowance	ALLOW
Alloy	ALY
Alternate	ALTN
Alternating Current	AC
Aluminum	AL
Ampere	AMP
Anneal	ANL
Approximate	APPROX
Assemble	ASSEM
Assembly	ASSY
Average	AVG

B

Back to Back	B TO B
Balance	BAL
Ball Bearing	BBRG
Base Line	BL
Bevel	BEV
Both Faces	BF
Both Sides	BS
Both Ways	BW
Bottom	BOT
Bottom Face	BF
Brass	BRS
Break	BRK
British Thermal Units	BTU
Bronze	BRZ
Bushing	BUSHG
Button	BTN

C

Capacity	CAP
Cap Screw	CAP SCR
Case Harden	CH
Casting	CSTG
Cast Iron	CI
Cast Steel	CS
Center	CTR
Center Line	CL
Center to Center	C TO C
Chamfer	CHAM
Chord	CHD
Circle	CIR
Circumference	CRCMF
Clear	CL
Clockwise	CW
Cold Drawn	CD
Cold Drawn Steel	CDS
Cold Rolled	CR
Cold Rolled Steel	CRS
Concentric	CNCTRC
Connect	CONN
Counterbore	CBORE
Counterclockwise	CCW
Countersink	CSK
Cross Section	XSECT

C (con't)

Cubic	CU
Cubic Foot	CU FT
Cubic Inch	CU IN
Cubic Metre	CU m
Cubic Millimeter	CU mm

D

Decimal	DEC
Degree	DEG
Detail	DET
Diagonal	DIAG
Dimension	DIM
Drawing	DWG
Drill	DR

E

Elevation	EL
Estimate	EST
Extra Strong	X STR
Extrude	EXTD

F

Fabricate	FAB
Face to Face	F TO F
Fahrenheit	F
Feed	FD
Feet	FT
Figure	FIG
Fillet	FIL
Fillister	FIL
Finish	FNSH
Finish All Over	FAO
Flange	FLG
Flat	FL
Flat Head	FLH
Foot	FT
Front	FR

G

Gage	GA
Gallon	GAL
Gasket	GSKT
General	GENL
Glass	GL
Grade	GR
Grind	GRD
Groove	GRV

H

Hard	H
Head	HD
Headless	HDLS
Heat	HT
Heat Treat	HT TR
Heavy	HVY
Hexagon	HEX
Horizontal	HORIZ

H (con't)

Horsepower	HP

I

Inch	IN
Inside Diameter	ID
Interior	INTR
Irregular	IRREG

J

Joint	JT
Junction	JCT

K

Key	K
Keyseat	KST

L

Left	L
Left Hand	LH
Length	LG
Length Over All	LOA
Light	LT
Line	L
Locate	LCT
Long	L

M

Machine	MACH
Material	MATL
Maximum	MAX
Metal	MET
Meter	m
Miles	MI
Miles Per Hour	MPH
Millimeter	mm
Minimum	MIN
Minute	MIN

N

Negative	NEG
Neutral	NEUT
Nominal	NOM
Normal	NORM
Not to Scale	NTS
Number	NO

O

Obsolete	OBS
Octagon	OCT
Outside Diameter	OD
Overall	OA

P

Part	PT
Pattern	PATT
Perpendicular	PERP
Pitch	P
Pitch Circle	PC
Pitch Diameter	PD

(Continued on next page)

Appendix 1. Standard Abbreviations Used on Drawings (continued)

P (con't)

Word	Abbreviation
Plate	PL
Point	PT
Pound	LB
Pounds Per Square Inch	PSI

Q

Word	Abbreviation
Quadrant	QUAD
Qualify	QUAL
Quarter	QTR

R

Word	Abbreviation
Radial	RDL
Radius	R
Rectangle	RECT
Reinforce	REINF
Revolution	REV
Revolutions per Minute	RPM
Right Hand	RH
Rivet	RVT
Rough	RGH
Round	RND

S

Word	Abbreviation
Screw	SCR

S (con't)

Word	Abbreviation
Section	SECT
Shaft	SFT
Sheet	SH
Single	SGL
Small	SM
Spring	SPR
Square	SQ
Standard	STD
Steel	STL
Stock	STK
Surface	SURF

T

Word	Abbreviation
Tangent	TAN
Taper	TPR
Template	TEMPL
Thick	THK
Thousand	M
Thread	THD
Through	THRU
Tolerance	TOL
Tooth	T

U

Word	Abbreviation
Unit	U
Universal	UNIV

V

Word	Abbreviation
Valve	V
Versus	VS
Vertical	VERT
Volt	V
Volume	VOL

W

Word	Abbreviation
Washer	WSHR
Watt	W
Weight	WT
Width	WD

X Y Z

Word	Abbreviation
Yard	YD
Year	YR

Abbreviations for use on drawings and in text, ANSI Y1.1—1972 (R1984). With permission of the publisher. *The American Society of Mechanical Engineers.*

Appendix 2. American National Standard Unified Inch Screw Threads

Sizes		Basic Major Diameter	Threads per Inch											
			Series With Graded Pitches			Series With Constant Pitches								
Primary	Secondary		Coarse UNC	Fine UNF	Extra Fine UNEF	4UN	6UN	8UN	12UN	16UN	20UN	28UN	32UN	Sizes
0		0.0600	—	80	—	—	—	—	—	—	—	—	—	0
	1	0.0730	64	72	—	—	—	—	—	—	—	—	—	1
2		0.0860	56	64	—	—	—	—	—	—	—	—	—	2
	3	0.0990	48	56	—	—	—	—	—	—	—	—	—	3
4		0.1120	40	48	—	—	—	—	—	—	—	—	—	4
5		0.1250	40	44	—	—	—	—	—	—	—	—	—	5
6		0.1380	32	40	—	—	—	—	—	—	—	—	UNC	6
8		0.1640	32	36	—	—	—	—	—	—	—	—	UNC	8
10		0.1900	24	32	—	—	—	—	—	—	—	—	UNF	10
	12	0.2160	24	28	32	—	—	—	—	—	—	UNF	UNEF	12
1/4		0.2500	20	28	32	—	—	—	—	—	UNC	UNF	UNEF	1/4
5/16		0.3125	18	24	32	—	—	—	—	—	20	28	UNEF	5/16
3/8		0.3750	16	24	32	—	—	—	—	UNC	20	28	UNEF	3/8
7/16		0.4375	14	20	28	—	—	—	—	16	UNF	UNEF	32	7/16
1/2		0.5000	13	20	28	—	—	—	—	16	UNF	UNEF	32	1/2
9/16		0.5625	12	18	24	—	—	—	UNC	16	20	28	32	9/16
5/8		0.6250	11	18	24	—	—	—	12	16	20	28	32	5/8
	11/16	0.6875	—	—	24	—	—	—	12	16	20	28	32	11/16
3/4		0.7500	10	16	20	—	—	—	12	UNF	UNEF	28	32	3/4
	13/16	0.8125	—	—	20	—	—	—	12	16	UNEF	28	32	13/16
7/8		0.8750	9	14	20	—	—	—	12	16	UNEF	28	32	7/8
	15/16	0.9375	—	—	20	—	—	—	12	16	UNEF	28	32	15/16
1		1.0000	8	12	20	—	—	UNC	UNF	16	UNEF	28	32	1
	1 1/16	1.0625	—	—	18	—	—	8	12	16	20	28	—	1 1/16
1 1/8		1.1250	7	12	18	—	—	8	UNF	16	20	28	—	1 1/8

UN thread form has flat or rounded root contour.
UNR thread form used for external threads. It has only rounded root contour.
UNC, UNCR—unified national coarse.
UNF, UNRF—unified national fine.
UNEF, UNREF—unified national extra fine.
Constant pitch series shown as 4UN or 4URN, 6UN or 6URN, etc.
Table heading shows only UN designation. The UNR thread form may be specified by substituting UNR in place of UN for all external threads.

Extracted from *Unified Inch Screw Threads* (ANSI B1.1–1982) with the permission of the publisher, The American Society of Mechanical Engineers.

Appendix 3. ISO Metric Screw Thread Series

Nominal Diameters			Pitches										
Col. 1 1st Choice	Col. 2 2nd Choice	Col. 3 3rd Choice	Coarse	Fine									
				3	2	1.5	1.25	1	0.75	0.5	0.35	0.25	0.2
1.6			**0.35**	. . .	. . .	. . .	. . .	. . .	. . .	. . .	. . .	. . .	0.2
	1.8		0.35	. . .	. . .	. . .	. . .	. . .	. . .	. . .	. . .	. . .	0.2
2			**0.4**	. . .	. . .	. . .	. . .	. . .	. . .	. . .	. . .	0.25	. . .
	2.2		0.45	. . .	. . .	. . .	. . .	. . .	. . .	. . .	. . .	0.25	. . .
2.5			**0.45**	. . .	. . .	. . .	. . .	. . .	. . .	. . .	0.35	. . .	. . .
3			**0.5**	. . .	. . .	. . .	. . .	. . .	. . .	. . .	0.35	. . .	. . .
	3.5		**0.6**	. . .	. . .	. . .	. . .	. . .	. . .	. . .	0.35	. . .	. . .
4			**0.7**	. . .	. . .	. . .	. . .	. . .	. . .	0.5	. . .	. . .	. . .
	4.5		0.75	. . .	. . .	. . .	. . .	. . .	. . .	0.5	. . .	. . .	. . .
5			**0.8**	. . .	. . .	. . .	. . .	. . .	. . .	0.5	. . .	. . .	. . .
		5.5	. . .	. . .	. . .	. . .	. . .	. . .	. . .	0.5	. . .	. . .	. . .
6			**1**	. . .	. . .	. . .	. . .	. . .	0.75	. . .	. . .	. . .	. . .
		7	1	. . .	. . .	. . .	. . .	. . .	0.75	. . .	. . .	. . .	. . .
8			**1.25**	. . .	. . .	. . .	. . .	**1**	0.75	. . .	. . .	. . .	. . .
		9	1.25	. . .	. . .	. . .	. . .	1	0.75	. . .	. . .	. . .	. . .
10			**1.5**	. . .	. . .	. . .	**1.25**	1	**0.75**	. . .	. . .	. . .	. . .
		11	1.5	. . .	. . .	. . .	. . .	1	0.75	. . .	. . .	. . .	. . .
12			**1.75**	. . .	. . .	**1.5**[4]	**1.25**	**1**	. . .	. . .	. . .	. . .	. . .
	14		**2**	. . .	. . .	**1.5**	1.25[1]	1	. . .	. . .	. . .	. . .	. . .
		15	. . .	. . .	. . .	1.5	. . .	**1**	. . .	. . .	. . .	. . .	. . .
16			**2**	. . .	. . .	**1.5**	. . .	1	. . .	. . .	. . .	. . .	. . .
		17	. . .	. . .	. . .	1.5	. . .	**1**	. . .	. . .	. . .	. . .	. . .
	18		2.5	. . .	2	**1.5**	. . .	1	. . .	. . .	. . .	. . .	. . .
20			**2.5**	. . .	2	**1.5**	. . .	**1**	. . .	. . .	. . .	. . .	. . .
	22		**2.5**[3]	. . .	2	**1.5**	. . .	1	. . .	. . .	. . .	. . .	. . .
24			**3**	. . .	**2**	1.5	. . .	1	. . .	. . .	. . .	. . .	. . .
		25	. . .	. . .	2	**1.5**	. . .	1	. . .	. . .	. . .	. . .	. . .
		26	. . .	. . .	. . .	1.5	. . .	. . .	. . .	. . .	. . .	. . .	. . .
	27		3[3]	. . .	**2**	1.5	. . .	1	. . .	. . .	. . .	. . .	. . .
		28	. . .	. . .	2	1.5	. . .	1	. . .	. . .	. . .	. . .	. . .
30			**3.5**	(3)[5]	**2**	**1.5**	. . .	1	. . .	. . .	. . .	. . .	. . .
		32	. . .	. . .	2	1.5	. . .	. . .	. . .	. . .	. . .	. . .	. . .
	33		3.5	(3)[5]	**2**	1.5	. . .	. . .	. . .	. . .	. . .	. . .	. . .
		35[2]	. . .	. . .	. . .	**1.5**	. . .	. . .	. . .	. . .	. . .	. . .	. . .
36			**4**	3	**2**	1.5	. . .	. . .	. . .	. . .	. . .	. . .	. . .
		38	. . .	. . .	. . .	1.5	. . .	. . .	. . .	. . .	. . .	. . .	. . .
	39		4	3	**2**	1.5	. . .	. . .	. . .	. . .	. . .	. . .	. . .

NOTES:

(1) Only for spark plugs for engines.
(2) Only for nuts for bearings.
(3) Only for high strength structural steel fasteners.
(4) Only for wheel studs and nuts.
(5) Pitches shown in brackets are to be avoided as far as possible.
(6) Diameter-pitch combinations in bold face are those chosen for general use.

Extracted from ANSI/ASME B1.13M–1983 with the permission of the publisher, The American Society of Mechanical Engineers.

(Continued on next page)

Appendix 3. ISO Metric Screw Thread Series (continued)

Nominal Diameters			Pitches					
				Fine				
Col. 1 1st Choice	Col. 2 2nd Choice	Col. 3 3rd Choice	Coarse	6	4	3	2	1.5
		40	. . .	. . .	. . .	3	2	**1.5**
42			**4.5**	. . .	4	3	**2**	1.5
	45		4.5	. . .	4	3	2	**1.5**
48			**5**	. . .	4	3	**2**	1.5
		50	. . .	. . .	. . .	3	2	**1.5**
	52		5	. . .	4	3	2	1.5
		55	. . .	. . .	4	3	2	**1.5**
56			**5.5**	. . .	4	3	**2**	1.5
		58	. . .	. . .	4	3	2	1.5
	60		5.5	. . .	4	3	2	**1.5**
		62	. . .	. . .	4	3	2	1.5
64			**6**	. . .	4	3	**2**	1.5
		65	. . .	. . .	4	3	2	**1.5**
	68		6	. . .	4	3	2	1.5
		70	. . .	6	4	3	2	**1.5**
72			. . .	**6**	4	3	**2**	1.5
		75	. . .	. . .	4	3	2	**1.5**
	76		. . .	6	4	3	2	1.5
		78	. . .	. . .	. . .	. . .	2	. . .
80			. . .	**6**	4	3	**2**	**1.5**
		82	. . .	. . .	. . .	. . .	2	. . .
	85		. . .	6	4	3	**2**	. . .
90			. . .	**6**	4	3	**2**	. . .
	95		. . .	6	4	3	**2**	. . .
100			. . .	**6**	4	3	**2**	. . .
	105		. . .	6	4	3	**2**	. . .
110			. . .	6	4	3	**2**	. . .
	115		. . .	6	4	3	2	. . .
	120		. . .	6	4	3	**2**	. . .
125			. . .	6	4	3	2	. . .
	130		. . .	6	4	3	**2**	. . .
		135	. . .	6	4	3	2	. . .
140			. . .	6	4	3	**2**	. . .
		145	. . .	6	4	3	2	. . .
	150		. . .	6	4	3	**2**	. . .
		155	. . .	6	4	3	. . .	. . .
160			. . .	6	4	**3**	. . .	. . .
		165	. . .	6	4	3	. . .	. . .
	170		. . .	6	4	**3**	. . .	. . .
		175	. . .	6	4	3	. . .	. . .
180			. . .	6	4	**3**	. . .	. . .
		185	. . .	6	4	3	. . .	. . .
	190		. . .	6	4	**3**	. . .	. . .
		195	. . .	6	4	3	. . .	. . .
200			. . .	6	4	**3**	. . .	. . .

Appendix 4. Standard General-Purpose Acme Threads

Size	Threads per Inch	Size	Threads per Inch	Size	Threads per Inch	Size	Threads per Inch
1/4	16	3/4	6	1 1/2	4	3	2
5/16	14	7/8	6	1 3/4	4	3 1/2	2
3/8	10	1	5	2	4	4	2
7/16	12	1 1/8	5	2 1/4	3	4 1/2	2
1/2	10	1 1/4	5	2 1/2	3	5	2
5/8	8	1 3/8	4	2 3/4	3	—	—

Extracted from *Acme Screw Threads* (ANSI B1.5-1977), with the permission of the publisher, The American Society of Mechanical Engineers.

Appendix 5. Buttress Threads

Selected Diameter-Pitch Combinations for 7°/45° Threads	
Major Diameter Range	**Threads Per Inch**
.05-0.75	12, 16, 20
over 0.75-1.0	10, 12, 16
over 1.0-1.5	8, 10, 12
over 1.5-2.5	6, 8, 10
over 2.5-4	5, 6, 8

Appendix 6. Hexagon, Heavy Hexagon, and Square Bolts

Nominal Size or Major Thread Dia.	Regular Hexagon		Heavy Hexagon		Regular Square	
	F Across Flats	H	F Across Flats	H	F Across Flats	H
1/4	7/16	11/64			3/8	11/64
5/16	1/2	7/32			1/2	13/64
3/8	9/16	1/4			9/16	1/4
7/16	5/8	19/64			5/8	19/64
1/2	3/4	11/32	7/8	11/32	3/4	21/64
5/8	15/16	27/64	1 1/16	27/64	15/16	27/64
3/4	1 1/8	1/2	1 1/4	1/2	1 1/8	1/2
7/8	1 5/16	37/64	1 7/16	37/64	1 5/16	19/32
1	1 1/2	43/64	1 5/8	43/64	1 1/2	21/32
1 1/8	1 11/16	3/4	1 13/16	3/4	1 11/16	3/4
1 1/4	1 7/8	27/64	2	27/32	1 7/8	27/32
1 3/8	2 1/16	29/32	2 3/16	29/32	2 1/16	29/32

Hexagon bolts and hexagon cap screws are considered the same. See Appendix 16 for sizes.

Extracted from *Square and Hex Bolts and Screws Inch Series,* (ANSI B18.2.1-1981), with the permission of the publisher, The American Society of Mechanical Engineers.

Appendix 7. Hexagon Nuts[1] and Jam Nuts[2], Inch

Diameter	Regular F	Nut H	Jam Nut H
1/4	7/16	7/32	5/32
5/16	1/2	17/64	3/16
3/8	9/16	21/64	7/32
7/16	11/16	3/8	1/4
1/2	3/4	7/16	5/16
9/16	7/8	31/64	5/16
5/8	15/16	35/64	3/8
3/4	1 1/8	41/64	27/64
7/8	1 5/16	3/4	31/64
1	1 1/2	55/64	35/64
1 1/8	1 11/16	31/32	5/8
1 1/4	1 7/8	1 1/16	3/4
1 3/8	2 1/16	1 11/64	13/16
1 1/2	2 1/4	1 9/32	7/8

Extracted from *Square and Hex Nuts* (ANSI B18.2.2-1972, R1983), with the permission of the publisher, The American Society of Mechanical Engineers.

[1]Refer to drawing to find location of sizes marked F and H.
[2]Jam nut "across the flats" dimension is the same as the regular nut.

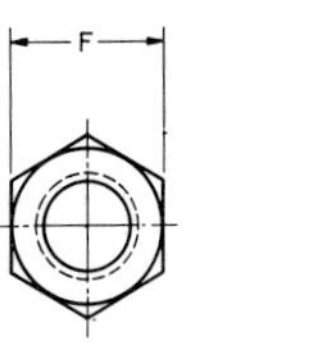

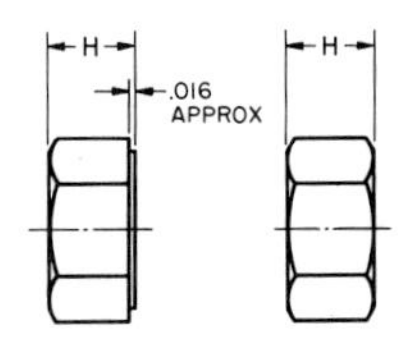

FINISHED HEXAGON NUT

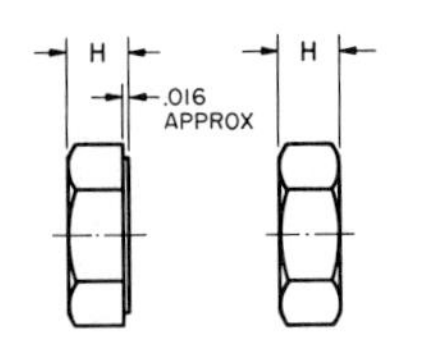

FINISHED HEXAGON JAM NUT

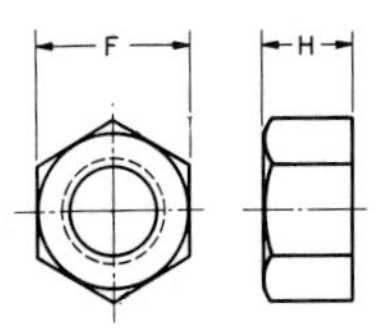

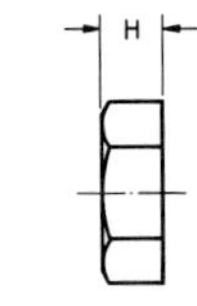

UNFINISHED HEXAGON NUT

UNFINISHED HEXAGON JAM NUT

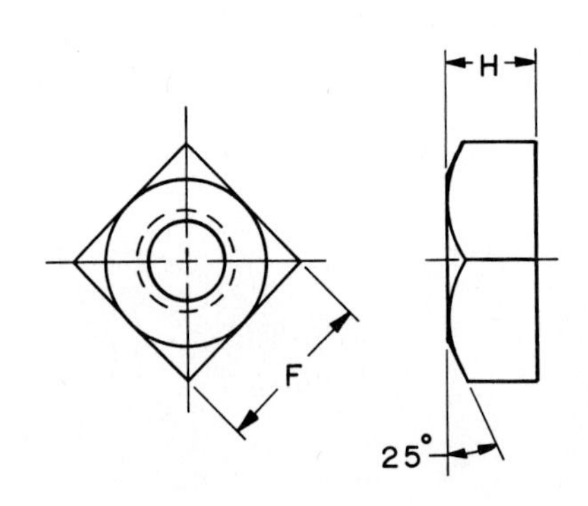

Appendix 8. American Standard Regular and Heavy Square Nuts (Inch Sizes)

Diameter		Regular F	Regular H	Heavy F	Heavy H
1/4	0.2500	7/16	7/32	1/2	1/4
5/16	0.3125	9/16	17/64	9/16	5/16
3/8	0.3750	5/8	21/64	11/16	3/8
7/16	0.4375	3/4	3/8	3/4	7/16
1/2	0.5000	13/16	7/16	7/8	1/2
5/8	0.6250	1	35/64	1 1/16	5/8
3/4	0.7500	1 1/8	21/32	1 1/4	3/4
7/8	0.8750	1 5/16	49/64	1 7/16	7/8
1	1.0000	1 1/2	7/8	1 5/8	1

Extracted from *Square and Hex Nuts* (ANSI B18.2.2-1972, R1983), with the permission of the publisher, The American Society of Mechanical Engineers.

Appendix 9. Metric Hex Cap Screws and Hex Bolts*

Nominal Bolt Size & Thread Pitch	F	H
	Width Across Flats	Head Height
M5 × 0.8	8.00	3.88
M6.0 × 1	10.00	4.70
M8 × 1.25	13.00	5.73
M10 × 1.5	15.00	6.86
M12 × 1.75	18.00	7.99
M14 × 2	21.00	9.32
M16 × 2	24.00	10.56
M20 × 2.5	30.00	13.12
M24 × 3	36.00	15.68

*Dimensions in millimeters

METRIC HEX CAP SCREWS AND HEX BOLTS

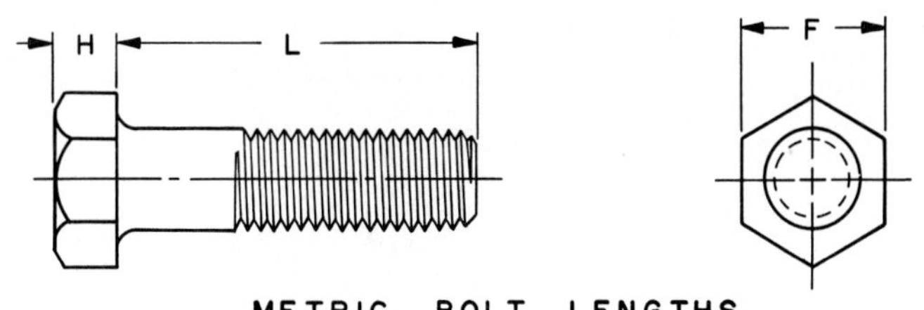

METRIC BOLT LENGTHS

FROM 25 mm TO 90 mm BY 5 mm INCREMENTS
FROM 90 mm TO 190 mm BY 10 mm INCREMENTS
FROM 190 mm TO 290 mm BY 20 mm INCREMENTS

Appendix 10. Metric Hex Nuts*

Nominal Nut Size and Thread Pitch	F	G	O	H	H_1
	Width Across Flats	Width Across Corners	Bearing Face Dia.	Nut Thickness	
				Style 1	Style 2
M5 × 0.8	8.00	9.24	7.0	4.5	5.3
M6.0 × 1	10.00	11.55	8.9	5.6	6.5
M8 × 1.25	13.00	15.01	11.6	6.6	7.8
M10 × 1.5	15.00	17.32	13.6	9.0	10.7
M12 × 1.75	18.00	20.78	16.6	10.7	12.8
M14 × 2	21.00	24.25	19.4	12.5	14.9
M16 × 2	24.00	27.71	22.4	14.5	17.4
M20 × 2.5	30.00	34.64	27.6	18.4	21.2
M24 × 3	36.00	41.57	32.9	22.0	25.4

*All dimensions in millimeters

METRIC HEX NUTS

F H G O O

H 0.4 APPROX O

STYLE 1

H_1 H_1 0.4 APPROX O O O

STYLE 2

Appendix 11. Inch Stud Sizes

E	B	D
Nominal Diameter -Inches-	Threads per Inch	Maximum Thread Length
1/4	20	.3750
5/15	18	.5000
3/8	16	.5625
7/16	14	.6875
1/2	13	.7500
9/16	12	.8750
5/8	11	.9375
3/4	10	1.1250

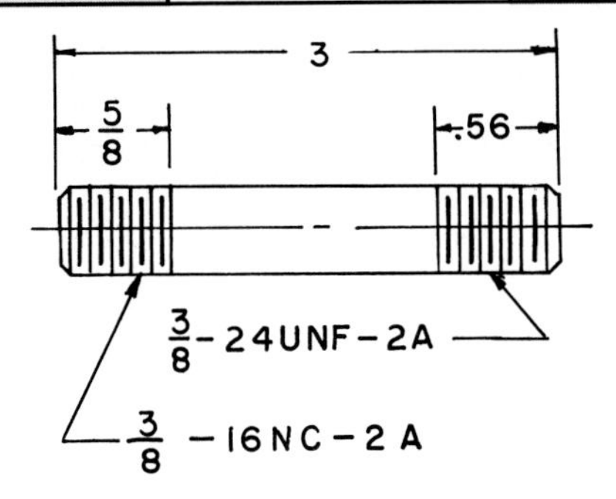

Appendix 12. Metric Stud Sizes*

Nominal Stud Size (E) and Thread Pitch	G	G
	For Studs Up to and Including 125 mm Length	For Studs Over 125 to 200 mm Length
M6.0 × 1	22.6	28.6
M8 × 1.25	27	33
M10 × 1.5	32	38
M12 × 1.75	37	43
M14 × 2	42	48
M16 × 2	46	52
M20 × 2.5	56	62
M24 × 3	—	72
M30 × 3.5	—	86
M36 × 4	—	—

*All dimensions in millimeters

METRIC STUDS

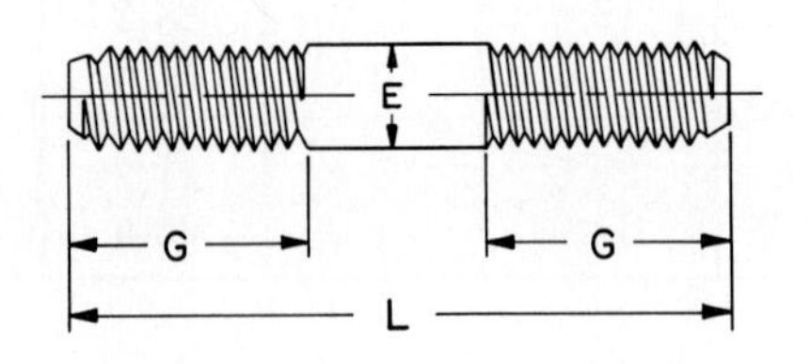

Appendix 13A. American Standard Slotted Head Tapping Screws[2] Inch Sizes

Nominal Size[2] or Basic Screw Diameter	Applicable to Screw Types[1]	Flat Heat		Oval Head		Pan Head	
	Code Symbols	A	H	A	H	A	H
0 0.0600	● ▲	0.119	0.035	0.119	0.035	0.116	0.039
1 0.0730	● ▲	0.146	0.043	0.146	0.043	0.142	0.046
2 0.0860	● ▲ ■	0.172	0.051	0.172	0.051	0.167	0.053
3 0.0990	● ▲ ■	0.199	0.059	0.199	0.059	0.193	0.060
4 0.1120	● ▲ ■	0.225	0.067	0.225	0.067	0.219	0.068
5 0.1250	● ▲ ■	0.252	0.075	0.252	0.075	0.245	0.075
6 0.1380	● ▲ ■	0.279	0.083	0.279	0.083	0.270	0.082
7 0.1510	● ▲	0.305	0.091	0.305	0.091	0.296	0.089
8 0.1640	● ▲ ■	0.332	0.100	0.332	0.100	0.322	0.096
10 0.1900	● ▲ ■	0.385	0.116	0.385	0.116	0.373	0.110
12 0.2160	● ▲ ■	0.438	0.132	0.438	0.132	0.425	0.125
1/4 0.2500	● ▲ ■	0.507	0.153	0.507	0.153	0.492	0.144
5/16 0.3125	● ▲ ■	0.635	0.191	0.635	0.191	0.615	0.178
3/8 0.3750	▲ ■	0.762	0.230	0.762	0.230	0.740	0.212
7/16 0.4375	▲	0.812	0.223	0.812	0.223	—	—
1/2 0.5000	▲	0.875	0.223	0.875	0.223	—	—

[1] Indicates type of thread for screw size. See Appendix 13B for thread dimensions.

- ● Type AB thread forming
- ▲ Type B thread forming
- ■ Type C thread forming and types D, F, G, and T thread cutting

[2] See the standard for other slotted types and sizes and for all recessed slot types.

Extracted from *Screws, Tapping and Metallic Drive, Inch Series* (ANSI B18.6.4–1981), with the permission of the publisher, The American Society of Mechanical Engineers.

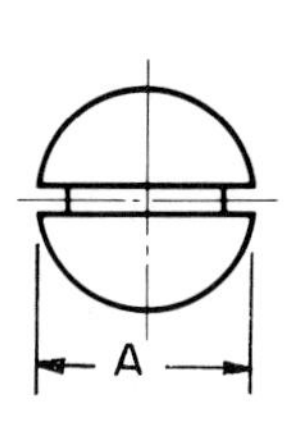

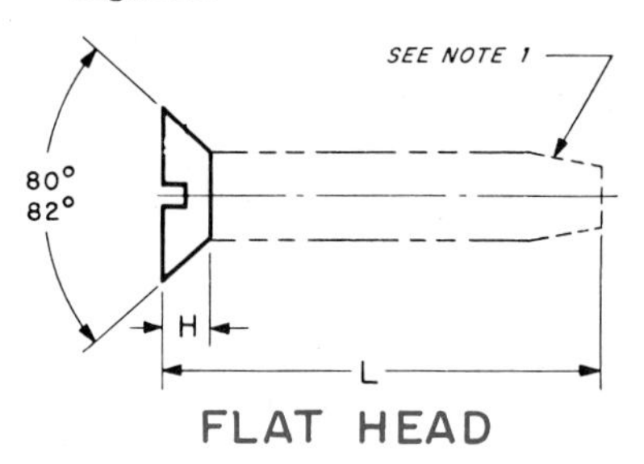

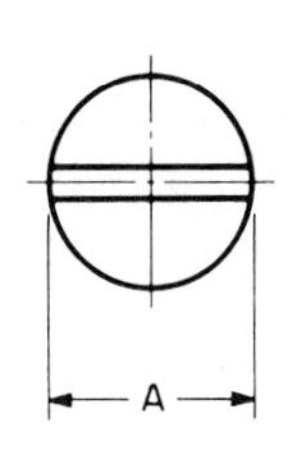

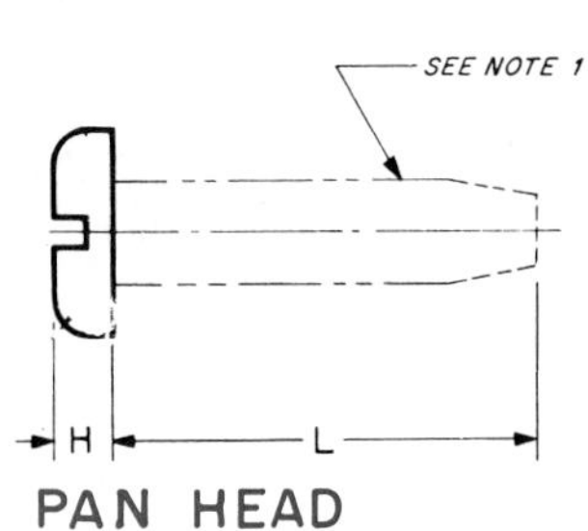

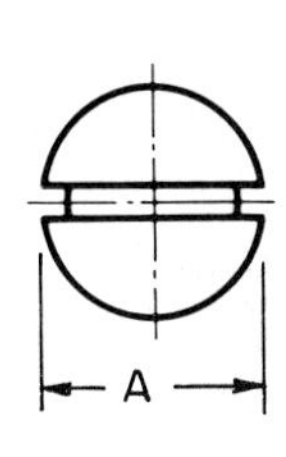

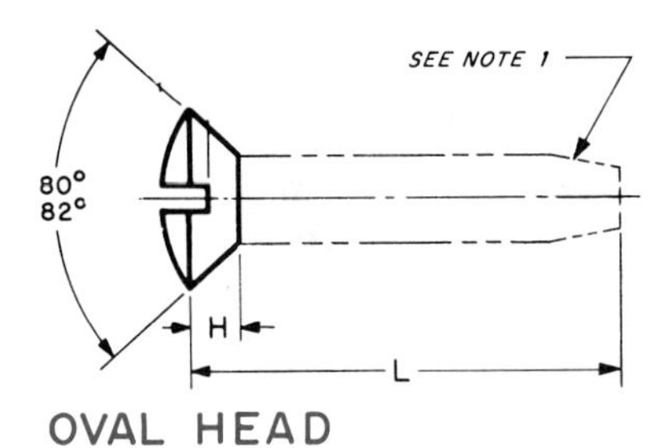

OVAL HEAD

Appendix 13B
Thread Forms for Standard Tapping Screws, Types AB, B, C, D, F, G, and T

Inch Sizes

Type B		Type AB	Type C				Type D,F,G,T
Screw Size	Threads Per Inch	D	D	Screw Size	Threads Per Inch	D	D
0	48	0.060	0.060	2	56	0.0860	0.0860
1	42	0.073	0.075	2	64	0.0860	0.0860
2	32	0.086	0.088	3	48	0.0990	0.0990
3	28	0.099	0.101	3	56	0.0990	0.0990
4	24	0.112	0.114	4	40	0.1120	0.1120
5	20	0.125	0.130	4	48	0.1120	0.1120
6	20	0.138	0.139	5	40	0.1250	0.1250
7	19	0.151	0.154	5	44	0.1250	0.1250
				6	32	0.1380	0.1380
				6	40	0.1380	0.1380
8	18	0.164	0.166	8	32	0.1640	0.1640
10	16	0.190	0.189	8	36	0.1640	0.1640
12	14	0.216	0.215	10	24	0.1900	0.1900
				10	32	0.1900	0.1900
				12	24	0.2160	0.2160
				12	28	0.2160	0.2160
1/4	14	0.250	0.246	1/4	20	0.2500	0.2500
5/16	12	0.3125	0.315	1/4	28	0.2500	0.2500
3/8	12	0.375		5/16	18	0.3125	0.3125
7/16	10	0.4375		5/16	24	0.3125	0.3125
1/2	10	0.500		3/8	16	0.3750	0.3750
				3/8	24	0.3750	0.3750

Extracted from *Screws, Tapping and Metallic Drive, Inch Series* (ANSI B18.6.4-1981), with the permission of the publisher, The American Society of Mechanical Engineers.

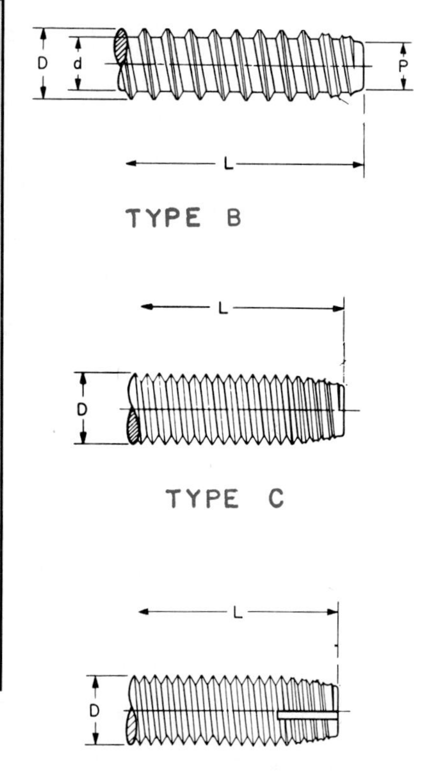

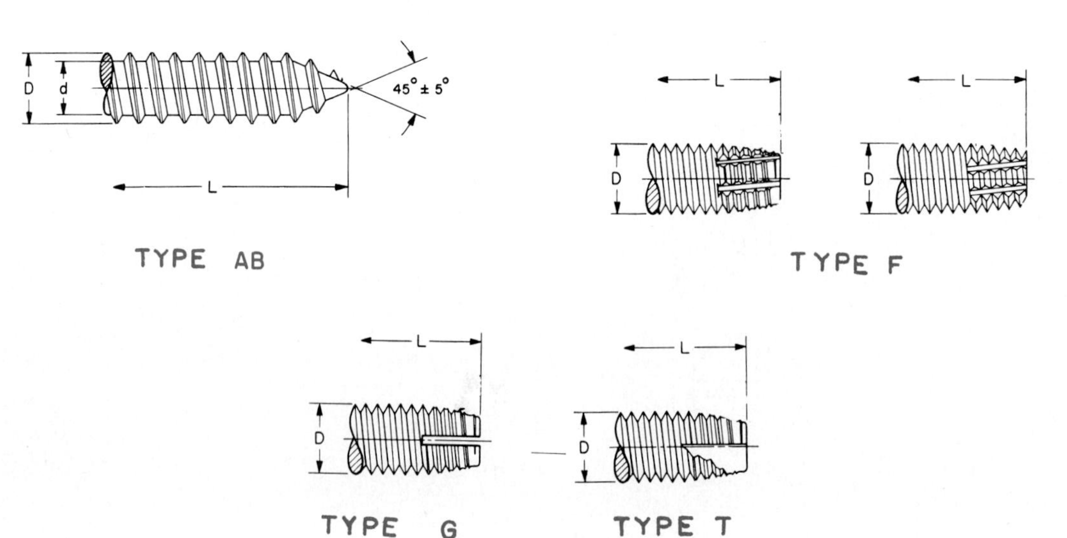

Appendix 14. Metric Tapping Screws, Types D, F, and T*

	D	Type D, F, T — L: Minimum Practical Screw Lengths	
Nom. Screw Size and Thread Pitch	Major Diameter	Pan, Hex and Hex Washer Heads	Flat and Oval CTSK Heads
2 × 0.4	2.00	3	5
2.5 × 0.45	2.50	4	6
3 × 0.5	3.00	5	7
3.5 × 0.6	3.50	6	8
4 × 0.7	4.00	6	9
5 × 0.8	5.00	8	11
6.0 × 1	6.30	10	13
8 × 1.25	8.00	12	16
10 × 1.5	10.00	15	20
12 × 1.75	12.00	18	24

*All dimensions in millimeters. See Appendix 13B for drawings of thread types.

Appendix 15. American Standard Slotted Head Cap Screws, Inch Sizes

	D	Flat Head		Round Head		Fillister Head		
Nominal Size	Diameter of Screw	A	H	A	H	A	H	O
1/4	0.250	0.500	0.140	0.437	0.191	0.375	0.172	0.216
5/16	0.3125	0.625	0.177	0.562	0.245	0.437	0.203	0.253
3/8	0.375	0.750	0.210	0.625	0.273	0.562	0.250	0.314
7/16	0.4375	0.8125	0.210	0.750	0.328	0.625	0.297	0.368
1/2	0.500	0.875	0.210	0.812	0.354	0.750	0.328	0.413
9/16	0.5625	1.000	0.244	0.937	0.409	0.812	0.375	0.467
5/8	0.625	1.125	0.281	1.000	0.437	0.875	0.422	0.521
3/4	0.750	1.375	0.352	1.250	0.546	1.000	0.500	0.612
7/8	0.875	1.625	0.423			1.125	0.594	0.720
1	1.000	1.875	0.494			1.312	0.656	0.803
1 1/8	1.125	2.062	0.529					
1 1/4	1.250	2.312	0.600					
1 3/8	1.375	2.562	0.665					
1 1/2	1.500	2.812	0.742					

Extracted from *Slotted Head Cap Screws, Square Head Set Screws, and Slotted Headless Set Screws* (ANSI B18.6.2–1972 (R1983), with the permission of the publisher, The American Society of Mechanical Engineers.

Appendix 16. American Standard Hexagon Head Cap Screws, Inch Sizes*

Nominal Size or Basic Major Diameter of Thread		F Width Across Flats	H Height
1/4	0.2500	7/16	5/32
5/16	0.3125	1/2	13/64
3/8	0.3750	9/16	15/64
7/16	0.4375	5/8	9/32
1/2	0.5000	3/4	5/16
9/16	0.5625	13/16	23/64
5/8	0.6250	15/16	25/64
3/4	0.7500	1 1/8	15/32
7/8	0.8750	1 5/16	35/64
1	1.0000	1 1/2	39/64
1 1/8	1.1250	1 11/16	11/16
1 1/4	1.2500	1 7/8	25/32
1 3/8	1.3750	2 1/16	27/32
1 1/2	1.5000	2 1/4	15/16

*Metric hexagon head cap screws shown in Appendix 10.

Sizes above dark line are unified dimensionally with British and Canadian standards. Bearing surface always chamfered or has washer face.

Thread Length

Screw lengths up to 6 inches have thread length equal to 2D + 1/4 inch.

Threads may be National coarse, National fine, or 8-Thread series, class 2A fit.

Extracted from *Square and Hex Bolts and Screws* (ANSI B18.2.1–1981), with the permission of the publisher, The American Society of Mechanical Engineers.

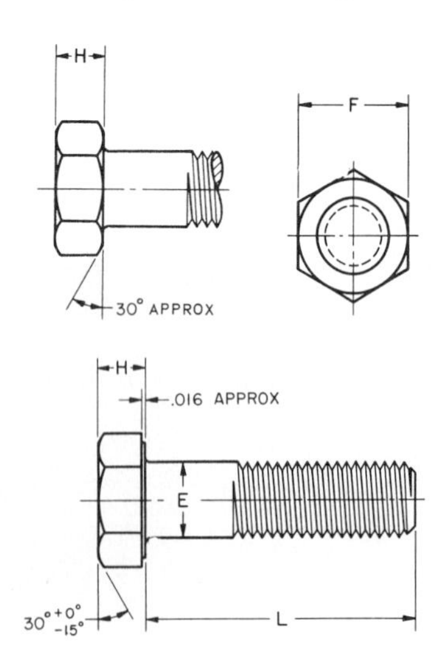

Appendix 17. American Standard Hexagon and Spline Socket Head Cap Screws, Inch Sizes

	D	A	H	M	J	T
Nominal Size	Body Diameter	Head Diameter	Head Height	Spline Socket Size	Hexagon Socket Size	Key Engagement
	Max	Max	Max	Nom	Nom	Min
0	0.0600	0.096	0.060	0.060	0.050	0.025
1	0.0730	0.118	0.073	0.072	1/16	0.031
2	0.0860	0.140	0.086	0.096	5/64	0.038
3	0.0990	0.161	0.099	0.096	5/64	0.044
4	0.1120	0.183	0.112	0.111	3/32	0.051
5	0.1250	0.205	0.125	0.111	3/32	0.057
6	0.1380	0.226	0.138	0.133	7/64	0.064
8	0.1640	0.270	0.164	0.168	9/64	0.077
10	0.1900	0.312	0.190	0.183	5/32	0.090
1/4	0.2500	0.375	0.250	0.216	3/16	0.120
5/16	0.3125	0.469	0.312	0.291	1/4	0.151
3/8	0.3750	0.562	0.375	0.372	5/16	0.182
7/16	0.4375	0.656	0.438	0.454	3/8	0.213
1/2	0.5000	0.750	0.500	0.454	3/8	0.245
5/8	0.6250	0.938	0.625	0.595	1/2	0.307
3/4	0.7500	1.125	0.750	0.620	5/8	0.370
7/8	0.8750	1.312	0.875	0.698	3/4	0.432
1	1.0000	1.500	1.000	0.790	3/4	0.495
1 1/8	1.1250	1.688	1.125	....	7/8	0.557
1 1/4	1.2500	1.875	1.250	....	7/8	0.620
1 3/8	1.3750	2.062	1.375	....	1	0.682
1 1/2	1.5000	2.250	1.500	....	1	0.745

Screw head shall be flat and chamfered. Chamfer E shall be at an angle of 30° ± with surface of the flat.

Screw lengths:

Screw lengths 1/8 to 1 inch available in 1/8 inch increments.
Screw lengths 1 to 3 1/2 inches available in 1/4 inch increments.
Screw lengths 3 1/2 to 6 inches available in 1/2 inch increments.

Thread lengths:

National Coarse threads—thread length, L_T, equal 2D + 1/2 inch.
National Fine Threads—thread length, L_T, equal 1 1/2D + 1/8 inch.
Thread class of fit is 3A.

Extracted from *Socket Cap, Shoulder, and Set Screws* (ANSI B18.3–1983), with permission of the publisher, The American Society of Mechanical Engineers.

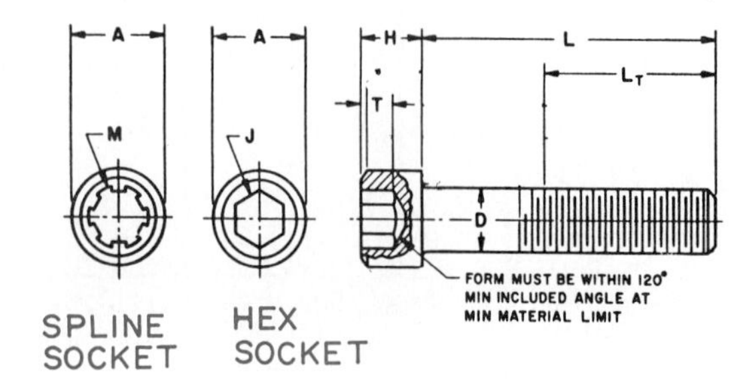

Appendix 18A. American Standard Slotted Head Machine Screws, Inch Sizes

Nominal Size	D	Flat Head		Pan Head		Oval Head			Fillister Head		
	Diameter of Screw	A	H	A	H	A	H	O	A	H	O
0	0.0600	0.119	0.035	0.116	0.039	0.119	0.035	0.056	0.096	0.043	0.055
1	0.0730	0.146	0.043	0.142	0.046	0.146	0.043	0.068	0.118	0.053	0.066
2	0.0860	0.172	0.051	0.167	0.053	0.172	0.051	0.080	0.140	0.062	0.083
3	0.0990	0.199	0.059	0.193	0.060	0.199	0.059	0.092	0.161	0.070	0.095
4	0.1120	0.225	0.067	0.219	0.068	0.225	0.067	0.104	0.183	0.079	0.107
5	0.1250	0.252	0.075	0.245	0.075	0.252	0.075	0.116	0.205	0.088	0.120
6	0.1380	0.279	0.083	0.270	0.082	0.279	0.083	0.128	0.226	0.096	0.132
8	0.1640	0.332	0.100	0.322	0.096	0.332	0.100	0.152	0.270	0.113	0.156
10	0.1900	0.385	0.116	0.373	0.110	0.385	0.116	0.176	0.313	0.130	0.180
12	0.2160	0.438	0.132	0.425	0.125	0.438	0.132	0.200	0.357	0.148	0.205
1/4	0.2500	0.507	0.153	0.492	0.144	0.507	0.153	0.232	0.414	0.170	0.237
5/16	0.3125	0.635	0.191	0.615	0.178	0.635	0.191	0.290	0.518	0.211	0.295
3/8	0.3750	0.762	0.230	0.740	0.212	0.762	0.230	0.347	0.622	0.253	0.355
7/16	0.4375	0.812	0.223	0.863	0.247	0.812	0.223	0.345	0.625	0.265	0.368
1/2	0.5000	0.875	0.223	0.987	0.281	0.875	0.223	0.354	0.750	0.297	0.412
9/16	0.5625	1.000	0.260	1.041	0.315	1.000	0.260	0.410	0.812	0.336	0.466
5/8	0.6250	1.125	0.298	1.172	0.350	1.125	0.298	0.467	0.875	0.375	0.521
3/4	0.7500	1.375	0.372	1.435	0.419	1.375	0.372	0.578	1.000	0.441	0.612

Extracted from *Machine Screws and Machine Screw Nuts* (ANSI B18.6.3–1972, R1983), with the permission of the publisher, The American Society of Mechanical Engineers.

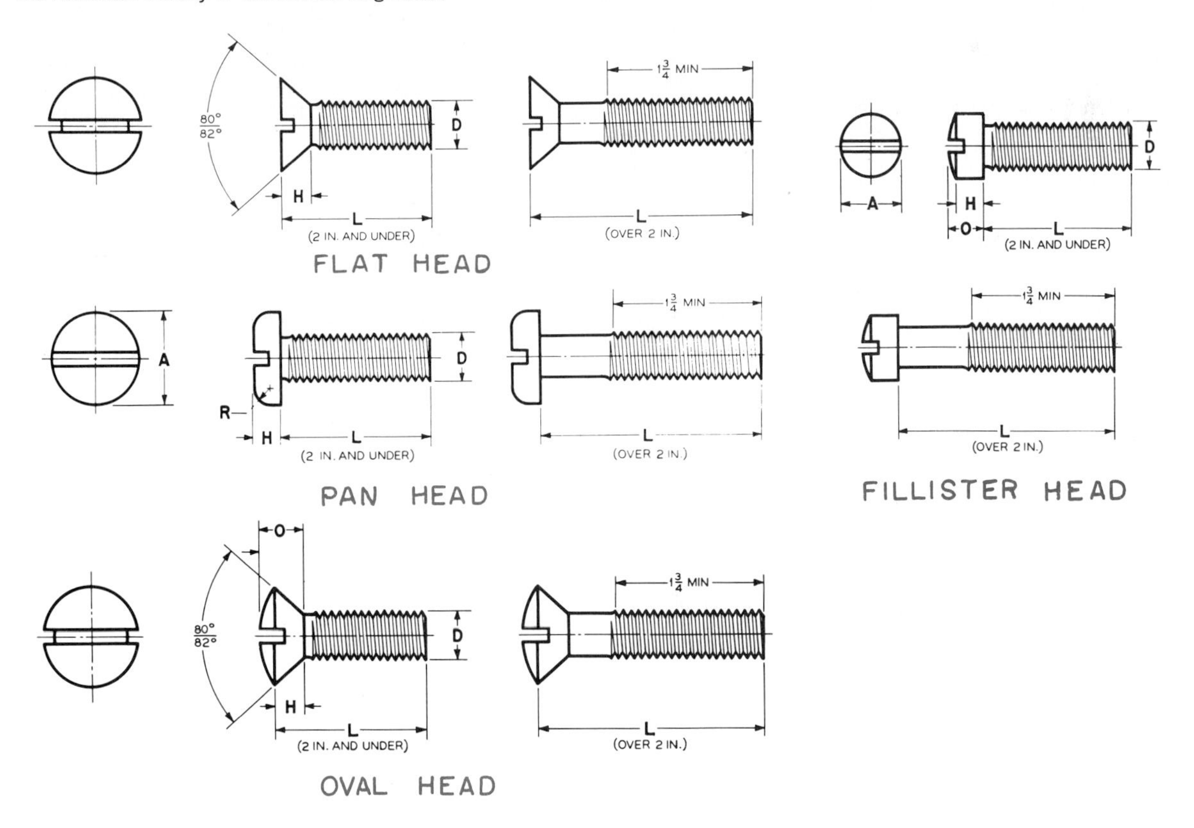

Appendix 18B. Metric Slotted Countersunk Machine Screws*

Nom. Screw Size and Thread Pitch	Flat and Oval Head		Pan Head		Hex Head	
	A	H	A	H	A	H
M2×0.4	3.60	1.20	3.90	1.35	3.20	1.27
M2.5×0.45	4.60	1.50	4.90	1.65	4.00	1.40
M3×0.5	5.50	1.80	5.80	1.90	5.00	1.52
M3.5×0.6	6.44	2.10	6.80	2.25	5.50	2.36
M4×0.7	7.44	2.32	7.80	2.55	7.00	2.79
M5×0.8	9.44	2.85	9.80	3.10	8.00	3.05
M6.0×1	11.87	3.60	12.00	3.90	10.00	4.83
M8×1.25	15.17	4.40	15.60	5.00	13.00	5.84
M10×1.5	18.98	5.35	19.50	6.20	15.00	7.49
M12×1.75	22.88	6.35	23.40	7.50	12.00	9.50

*All dimensions in millimeters.

Appendix 19. American Standard Square Head Set Screws, Inch Sizes

Nominal Size		F	H		G
		Width Across Flats	Nom.	Max	Width Across Corners
#10	0.190	0.188	9/64	0.148	0.247
1/4	0.250	0.250	3/16	0.196	0.331
5/16	0.3125	0.312	15/64	0.245	0.415
3/8	0.3750	0.375	9/32	0.293	0.497
7/16	0.4375	0.438	21/64	0.341	0.581
1/2	0.500	0.500	3.8	0.389	0.665

Square head set screw points are the same types and sizes as shown for slotted headless set screws.

Extracted from *Slotted Head Cap Screws, Square Head Set Screws,* and *Slotted Headless Set Screws* (ANSI B18.6.2–1972, R1983), with the permission of the publisher, The American Society of Mechanical Engineers.

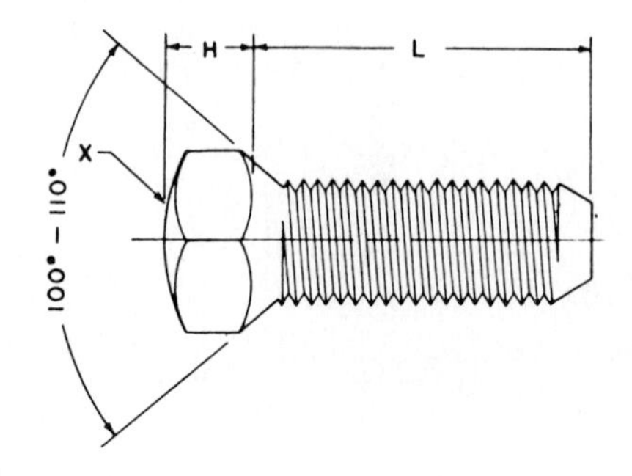

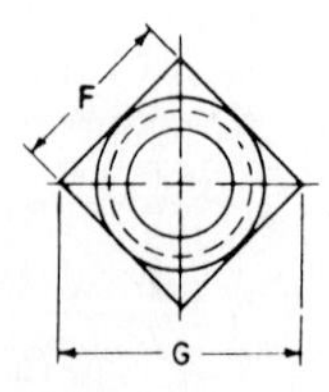

Appendix 20. American Standard Slotted Headless Set Screws, Inch Sizes

D		I	J	T	R	C	P	Q	q
Nominal Size		Radius of Headless Crown	Width of Slot	Depth of Slot	Oval Point Radius	Diameter of Cup and Flat Points	Diameter of Dog Point	Length of Dog Point (see note)	
						Max	Max	Full	Half
5	0.125	0.125	0.026	0.036	0.094	0.067	0.083	0.063	0.033
6	0.138	0.138	0.028	0.040	0.104	0.074	0.092	0.073	0.038
8	0.164	0.164	0.032	0.046	0.123	0.087	0.109	0.083	0.043
10	0.190	0.190	0.035	0.053	0.142	0.102	0.127	0.095	0.050
12	0.216	0.216	0.042	0.061	0.162	0.115	0.144	0.115	0.060
1/4		0.250	0.049	0.068	0.188	0.132	0.156	0.130	0.068
5/16		0.312	0.055	0.083	0.234	0.172	0.203	0.161	0.083
3/8		0.375	0.068	0.099	0.281	0.212	0.250	0.193	0.099

All dimensions given in inches.
Where usable length of thread is less than the nominal diameter, half-dog point shall be used.
When L (length of screw) equals nominal diameter or less, = 118 deg ± 2 deg; when L exceeds nominal diameter, Y = 90 deg ± 2 deg.
Point Angles. W = 80 deg to 90 deg; X = 118 deg ± 5 deg; Z = 100 deg to 110 deg.

Extracted from *Slotted Head Cap Screws, Square Head Set Screws,* and *Slotted Headless Set Screws* (ANSI B18.6.2–1972, R1983), with the permission of the publisher, The American Society of Mechanical Engineers.

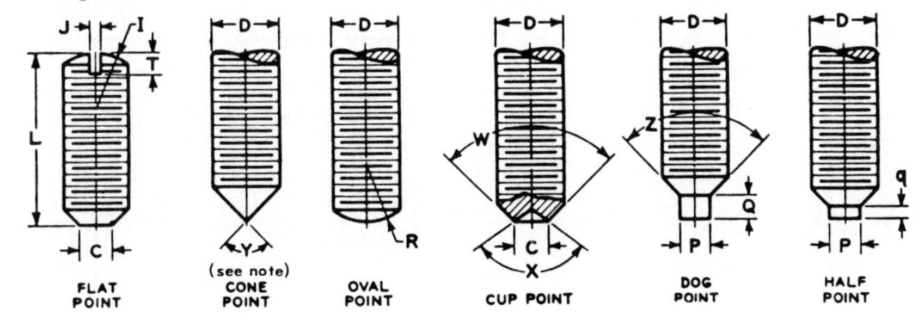

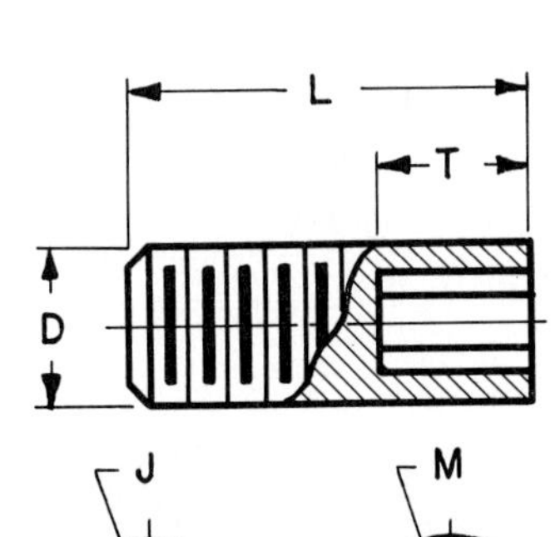

Appendix 21. American Standard Hexagon and Spline Socket Set Screws, Inch Sizes

D		J	M	T
Nominal Size		Hex Sockets Width Across Flats	Spline Socket Major Diameter	Key Engagement
5	0.125	0.0625	0.072	0.080
6	0.1380	0.0625	0.072	0.080
8	0.1640	0.0781	0.096	0.090
10	0.1900	0.0938	0.111	0.100
1/4	0.2500	0.125	0.145	0.125
5/16	0.3125	0.156	0.183	0.156
3/8	0.3750	0.188	0.216	0.188

Extracted from *Standard Socket Cap, Shoulder, and Set Screws* (ANSI B18.3, 1982), with the permission of the publisher, The American Society of Mechanical Engineers.

Appendix 22. American Standard Parallel, Plain Taper, and Gib Head Keys, Inch Sizes

Key			Nominal Key Size		Tolerance	
			Width, *W*			
			Over	To (Incl)	Width, *W*	Height, *H*
Parallel	Square	Bar Stock[1]	—	3/4	+0.000 −0.002	+0.000 −0.002
			3/4	1 1/2	+0.000 −0.003	+0.000 −0.003
			1 1/2	2 1/2	+0.000 −0.004	+0.000 −0.004
	Retangular	Bar Stock[1]	—	3/4	+0.000 −0.003	+0.000 −0.003
			3/4	1 1/2	+0.000 −0.004	+0.000 −0.004
			1 1/2	3	+0.000 −0.005	+0.000 −0.005
Taper	Plain or Gib Head Square or Rectangular		—	1 1/4	+0.001 −0.000	+0.005 −0.000
			1 1/4	3	+0.002 −0.000	+0.005 −0.000

[1]Two types of stock are used for parallel keys. One is a bar stock with a negative tolerance. Another is a key stock with a close plus tolerance.

Extracted from *Standard Keys and Keyseats* (ANSI B17.1–1967, R1973), with permission of the publisher, The American Society of Mechanical Engineers.

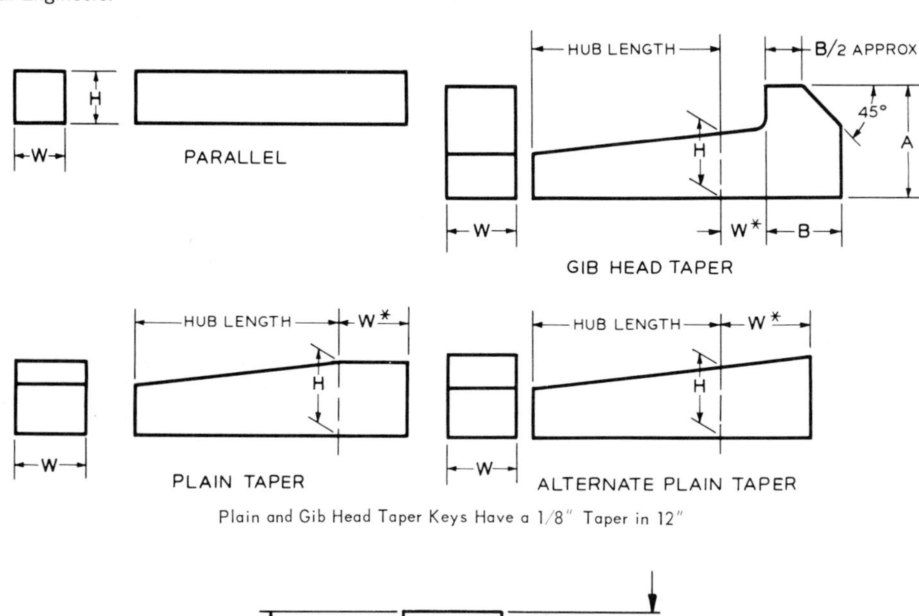

Plain and Gib Head Taper Keys Have a 1/8″ Taper in 12″

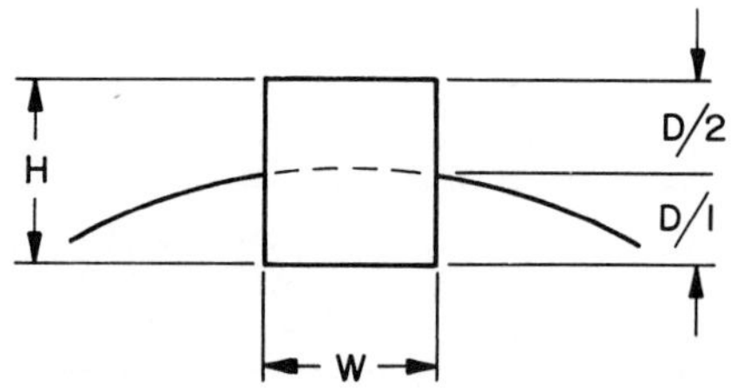

Appendix 23. American Standard Key Sizes in Inches for Selected Shaft Diameters for Parallel, Plain Taper, and Gib Head Keys

Nominal Shaft Diameter		Nominal Key size			Nominal Keyseat Depth	
			Height, *H*		*H*/2	
Over	To (Incl)	Width, *W*	Square	Rectangular	Square	Rectangular
5/16	7/16	3/32	3/32		3/64	
7/16	9/16	1/8	1/8	3/32	1/16	3/64
9/16	7/8	3/16	3/16	1/8	3/32	1/16
7/8	1 1/4	1/4	1/4	3/16	1/8	3/32
1 1/4	1 3/8	5/16	5/16	1/4	5/32	1/8
1 3/8	1 3/4	3/8	3/8	1/4	3/16	1/8
1 3/4	2 1/4	1/2	1/2	3/8	1/4	3/16
2 1/4	2 3/4	5/8	5/8	7/16	5/16	7/32
2 3/4	3 1/4	3/4	3/4	1/2	3/8	1/4
3 1/4	3 3/4	7/8	7/8	5/8	7/16	5/16
3 3/4	4 1/2	1	1	3/4	1/2	3/8
4 1/2	5 1/2	1 1/4	1 1/4	7/8	5/8	7/16
5 1/2	6 1/2	1 1/2	1 1/2	1	3/4	1/2
6 1/2	7 1/2	1 3/4	1 3/4	1 1/2	7/8	3/4
7 1/2	9	2	2	1 1/2	1	3/4
9	11	2 1/2	2 1/2	1 3/4	1 1/4	7/8
11	13	3	3	2	1 1/2	1

Sizes and dimension in the unshaded area are preferred.

Extracted from *Keys and Keyseats* (ANSI B17.1–1967 R1978), with the permission of the publisher, The American Society of Mechanical Engineers.

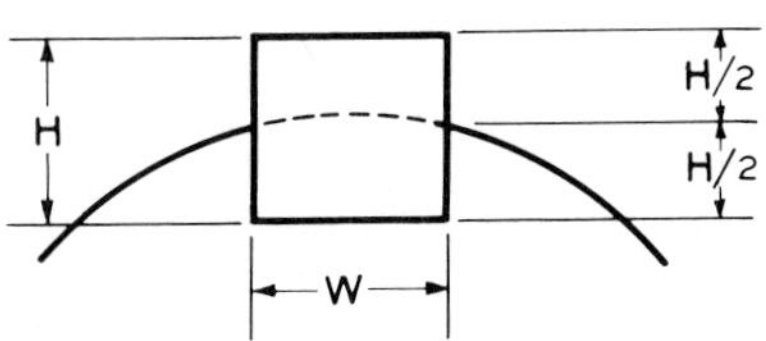

Appendix 24. American Standard Woodruff Keys and Keyseats

Key No.[1]	Nominal sizes			Maximum Sizes				Key No.[1]	Nominal Sizes				Maximum Sizes		
	W × B	E	F	D	A	C	H		W × B	E	F	D	A	C	H
204	1/16 × 1/2	3/64	1/32	.194	.1718	.203	.0422	808	1/4 × 1	1/16	1/8	.428	.3130	.438	.1360
304	3/32 × 1/2	3/64	3/64	.194	.1561	.203	.0579	809	1/4 × 1 1/8	5/64	1/8	.475	.3590	.484	.1360
305	3/32 × 5/8	1/16	3/64	.240	.2031	.250	.0579	810	1/4 × 1 1/4	5/64	1/8	.537	.4220	.547	.1360
404	1/8 × 1/2	3/64	1/16	.194	.1405	.203	.0735	811	1/4 × 1 3/8	3/32	1/8	.584	.4690	.594	.1360
405	1/8 × 5/8	1/16	1/16	.240	.1875	.250	.0735	812	1/4 × 1 1/2	7/64	1/8	.631	.5160	.641	.1360
406	1/8 × 3/4	1/16	1/16	.303	.2505	.313	.0735	1008	5/16 × 1	1/16	5/32	.428	.2818	.438	.1672
505	5/32 × 5/8	1/16	5/64	.240	.1719	.250	.0891	1009	5/16 × 1 1/8	5/64	5/32	.475	.3278	.484	.1672
506	5/32 × 3/4	1/16	5/64	.303	.2349	.313	.0891	1010	5/16 × 1 1/4	5/64	5/32	.537	.3908	.547	.1672
507	5/32 × 7/8	1/16	5/64	.365	.2969	.375	.0891	1011	5/16 × 1 3/8	3/32	5/32	.584	.4378	.594	.1672
606	3/16 × 3/4	1/16	3/32	.303	.2193	.313	.1047	1012	5/16 × 1 1/2	7/64	5/32	.631	.4848	.641	.1672
607	3/16 × 7/8	1/16	3/32	.365	.2813	.375	.1047	1210	3/8 × 1 1/4	5/64	3/16	.537	.3595	.547	.1985
608	3/16 × 1	1/16	3/32	.428	.3443	.438	.1047	1211	3/8 × 1 3/8	3/32	3/16	.584	.4065	.594	.1985
609	3/16 × 1 1/8	5/64	3/32	.475	.3903	.484	.1047	1212	3/8 × 1 1/2	7/64	3/16	.631	.4535	.641	.1985
807	1/4 × 7/8	1/16	1/8	.365	.2500	.375	.1360	.....	....	....	....	.....	......	.....	......

[1]The last two numbers of the key number indicate the diameter (B) in eighths of an inch. The other one or two numbers in front of these indicate the width of the key (A) in thirty-seconds of an inch. For example, key number 608 means the diameter is 8/8 or 1 inch and the thickness is 6/32 or 3/16 inch.

Extracted from *Woodruff Keys and Keyseats,* ANSI B17.2–1967 (R1978), with the permission of the publisher, The American Society of Mechanical Engineers.

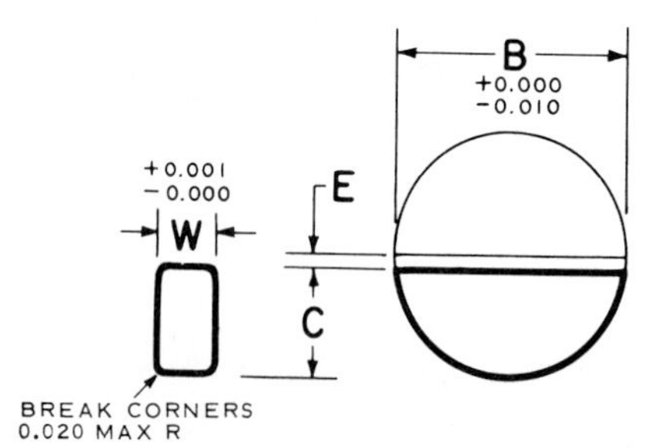

FULL RADIUS TYPE

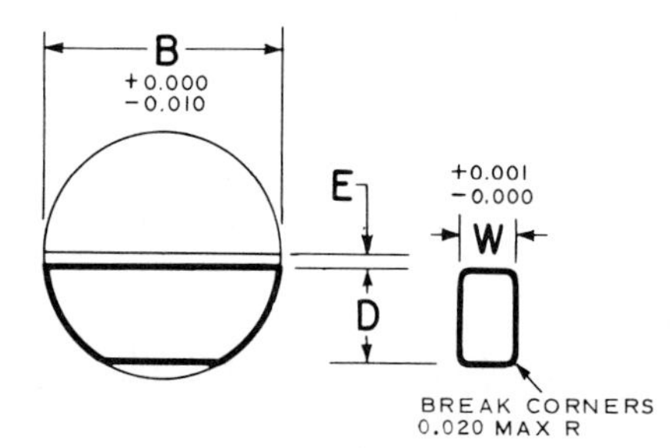

FLAT BOTTOM TYPE

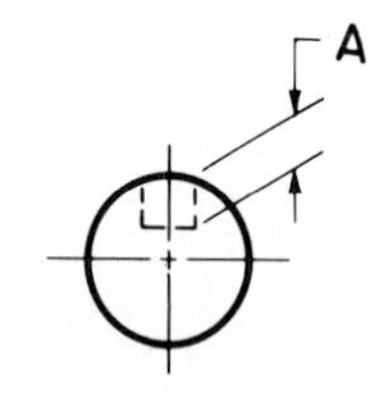

KEYSEAT-SHAFT

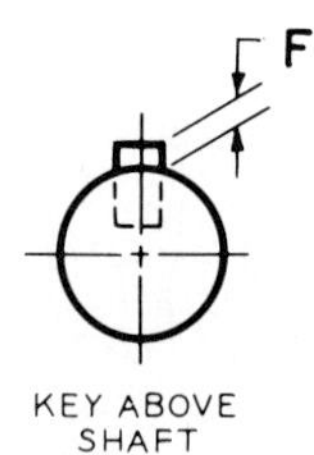

KEY ABOVE SHAFT

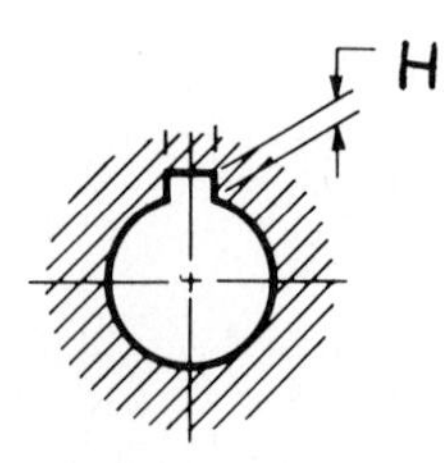

KEYSEAT-HUB

Appendix 25. American Standard Woodruff Key Sizes for Selected Shaft Diameters

Shaft Diameter	5/16 to 3/8	7/16 to 1/2	9/16 to 5/8	11/16 to 3/4	13/16
Key Numbers	204	304 305	404 405	404 405 406	505 506
Shaft Diameter	**7/8 to 15/16**	**1**	**1 1/16 to 1 1/8**	**1 3/16**	**1 1/4 to 1 5/16**
Key Numbers	505 506 507	606 607 608	606 607 608 609	607 608 609	607 608 609 810
Shaft Diameter	**1 3/8 to 1 7/16**	**1 1/2 to 1 5/8**	**1 11/16 to 1 3/4**	**1 13/16 to 2 1/8**	**2 3/16 to 2 1/2**
Key Numbers	608 609 810	808 809 810 812	809 810 812	1011 1012	1211 1212

Selected sizes from General Motors Engineering Standards.

Appendix 26. Metric Pratt and Whitney Key Sizes

KEY NO.	L	W	H	D	KEY NO.	L	W	H	D	KEY NO.	L	W	H	D
1	12	1.6	2.4	1.6	17	28	7	8	5	27	50	6	10	6.5
2	12	2.4	3.6	2.4	18	28	6.5	10	6	28	50	8	12	8
3	12	3.2	5	3.2	C	28	10	12	8	29	50	10	14	10
4	16	2.4	3.6	2.4	19	32	5	7	5	54	55	6	10	6.5
5	16	3.2	5	3.2	20	32	7	8	5	55	55	8	12	8
6	16	4	6	4	21	32	6.5	10	6	56	55	10	14	10
7	20	3.2	5	3.2	D	32	10	12	8	57	55	11	16	12
8	20	4	6	4	E	32	12	14	10	58	65	8	12	8
9	20	5	7	5	22	35	6.5	10	6	59	65	10	14	10
10	22	4	6	4	23	35	10	12	8	60	65	11	16	12
11	22	5	7	5	F	35	12	14	10	61	65	12	20	13
12	22	6	8.4	7	24	38	6.5	10	6	30	75	10	14	10
A	22	6.5	10	6.5	25	38	10	12	8	31	75	11	16	12
13	25	5	7	5	G	38	12	14	10	32	75	12	20	13
14	25	6	8.4	6	51	45	6.5	10	6	33	75	14	22	14
15	25	6.5	10	6.5	52	45	10	12	8	34	75	16	24	16
B	25	8	12	8	53	45	12	14	10					
16	28	5	7	5	26	50	5	7	5					

All dimensions given in millimeters.

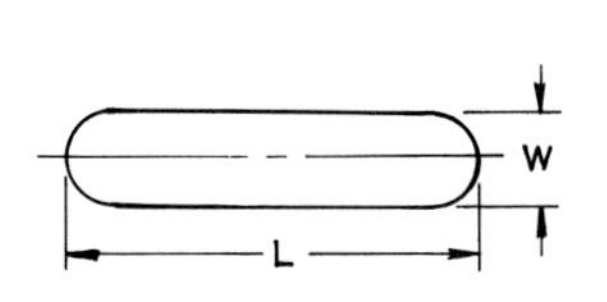

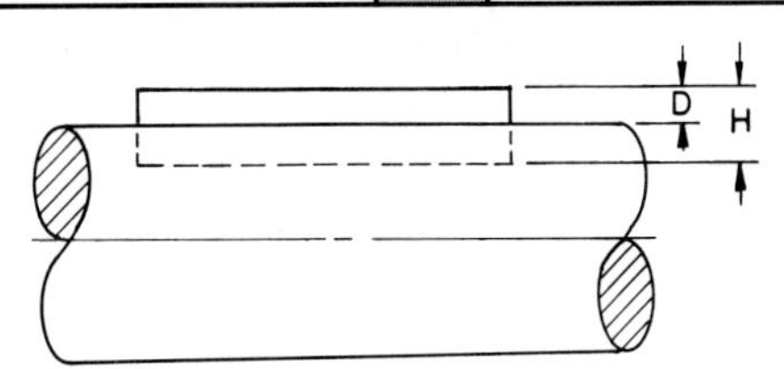

Appendix 27. Metric Square and Rectangular Parallel and Taper Keys for Round Shafts up to 50 Millimeters*

Nominal Shaft Dia.		Nominal Key Size		Nominal Seat Depth		
Over	To Include	Width W	Height H	Shaft D_1	Hub D_2	Hub (Taper) D_2
6	8	2	2	1.2	1.0	0.5
8	10	3	3	1.8	1.4	0.9
10	12	4	4	2.5	1.8	1.2
12	17	5	5	3.0	2.3	1.7
17	22	6	6	3.5	2.8	2.2
22	30	8	7	4.0	3.3	2.4
30	38	10	8	5.0	3.3	2.4
38	44	12	8	5.0	3.3	2.4
44	50	14	9	5.5	3.8	2.9

*In Millimeters

Appendix 28A. American Standard Taper Pins, Inch Sizes

Pin Size Number and Basic Pin Diameter	A				R	
	Major Diameter (Large End)				End Crown Radius	
	Commercial Class		Precision Class			
	Max	Min	Max	Min	Max	Min
7/0 0.0625	0.0638	0.0618	0.0635	0.0625	0.072	0.052
6/0 0.0780	0.0793	0.0773	0.0790	0.0780	0.088	0.068
5/0 0.0940	0.0953	0.0933	0.0950	0.0940	0.104	0.084
4/0 0.1090	0.1103	0.1083	0.1100	0.1090	0.119	0.099
3/0 0.1250	0.1263	0.1243	0.1260	0.1250	0.135	0.115
2/0 0.1410	0.1423	0.1403	0.1420	0.1410	0.151	0.131
0 0.1560	0.1573	0.1553	0.1570	0.1560	0.166	0.146
1 0.1720	0.1733	0.1713	0.1730	0.1720	0.182	0.162
2 0.1930	0.1943	0.1923	0.1940	0.1930	0.203	0.183
3 0.2190	0.2203	0.2183	0.2200	0.2190	0.229	0.209
4 0.2550	0.2513	0.2493	0.2510	0.2500	0.260	0.240
5 0.2890	0.2903	0.2883	0.2900	0.2890	0.299	0.279
6 0.3410	0.3423	0.3403	0.3420	0.3410	0.351	0.331
7 0.4090	0.4103	0.4083	0.4100	0.4090	0.419	0.399
8 0.4920	0.4933	0.4913	0.4930	0.4920	0.502	0.482
9 0.5910	0.5923	0.5903	0.5920	0.5910	0.601	0.581
10 0.7060	0.7073	0.7053	0.7070	0.7060	0.716	0.696

Extracted from *Taper Pins, Dowel Pins, Straight Pins, Grooved Pins, and Spring Pins* (ANSI B18.8.2-1978), with permission of the publisher, The American Society of Mechanical Engineers.

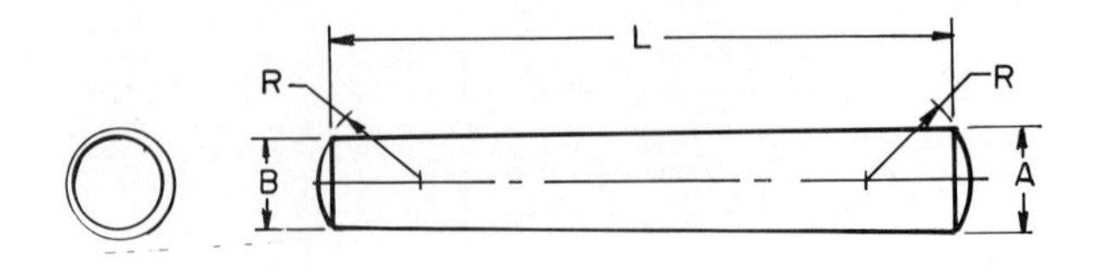

Appendix 28B. Standard Sizes and Lengths of Taper Pins, in Inches

Nominal Length	Pin Size Number																
	7/0	6/0	5/0	4/0	3/0	2/0	0	1	2	3	4	5	6	7	8	9	10
1/4	X	X	X	X	X												
3/8	X	X	X	X	X												
1/2	X	X	X	X	X	X	X										
5/8	X	X	X	X	X	X	X										
3/4	X	X	X	X	X	X	X	X	X	X	X						
7/8	X	X	X	X	X	X	X	X	X	X	X						
1	X	X	X	X	X	X	X	X	X	X	X	X					
1 1/4		X	X	X	X	X	X	X	X	X	X	X	X	X	X	X	
1 1/2		X	X	X	X	X	X	X	X	X	X	X	X	X	X	X	X
1 3/4				X	X	X	X	X	X	X	X	X	X	X	X	X	X
2				X	X	X	X	X	X	X	X	X	X	X	X	X	X

Extracted from *Taper Pins, Dowel Pins, Straight Pins, Grooved Pins, and Spring Pins* (ANSI B18.8.2-1978), with permission of the publisher, The American Society of Mechanical Engineers.

Appendix 29. Suggested Shaft Diameters for Use with Taper Pins, Inch Sizes

Pin No.	7/0	6/0	5/0	4/0	3/0	2/0	0	1	2	3	4	5	6	7	8
Suggested Shaft Dia.		7/32	1/4	5/16	3/8	7/16	1/2	9/16	5/8	3/4	13/16	7/8	1	1 1/4	1 1/2

Appendix 30. American Standard Hardened and Ground Dowel Pins, Inch Sizes

Length (L)	Nominal Diameter								
	1/8	3/16	1/4	5/16	3/8	7/16	1/2	5/8	3/4
	(A) Basic Diameter of Standard Pins ±0.0001								
	0.1252	0.1877	0.2502	0.3127	0.3752	0.4377	0.5002	0.6252	0.7502
	(A) Basic Diameter Oversize Pins ±0.0001								
	0.1260	0.1885	0.2510	0.3135	0.3760	0.4385	0.5010	0.6260	0.7510
1/2	X	X	X	X					
5/8	X	X	X	X					
3/4	X	X	X	X	X		X		
7/8	X	X	X	X	X	X			
1	X	X	X	X	X	X	X		
1 1/4	X	X	X	X	X	X	X	X	
1 1/2	X	X	X	X	X	X	X	X	X
1 3/4	X	X	X	X	X	X	X	X	X
2	X	X	X	X	X	X	X	X	X
2 1/4			X	X	X	X	X	X	X
2 1/2			X	X	X	X	X	X	X
3					X	X	X	X	X
3 1/2							X	X	X
4							X	X	X
4 1/2								X	X
5								X	X
5 1/2									X

All dimensions are given in inches.

These pins are extensively used in the tool and machine industry and a machine reamer of nominal size may be used to produce the holes into which these pins tap or press fit. They must be straight and free from any defects that will affect their serviceability.

Extracted from *Taper Pins, Dowel Pins, Straight Pins, Grooved Pins, and Spring Pins* (ANSI B18.8.2-1978), with the permission of the publisher, The American Society of Mechanical Engineers.

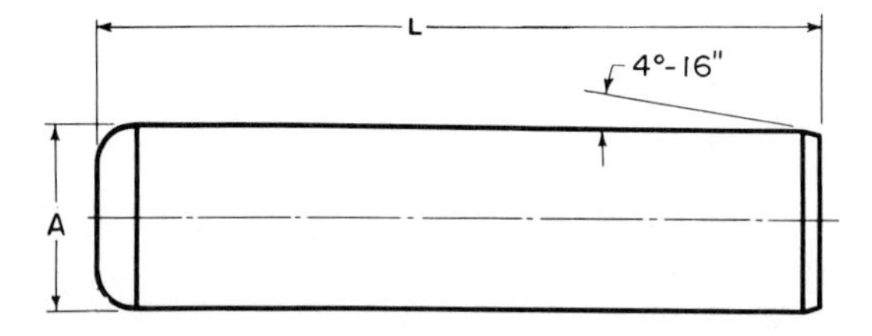

Appendix 31. American Standard Cotter Pins, Inch Sizes

Diameter Nominal A	Outside Eye Diameter B Min.	Hole Sizes Recommended	Diameter Nominal A	Outside Eye Diameter B Min.	Hole Sizes Recommended
0.031	1/16	3/64	0.188	3/8	13/64
0.047	3/32	1/16	0.219	7/16	15/64
0.062	1/8	5/64	0.250	1/2	17/64
0.078	5/32	3/32	0.312	5/8	5/16
0.094	3/16	7/64	0.375	3/4	3/8
0.109	7/32	1/8	0.438	7/8	7/16
0.125	1/4	9/64	0.500	1	1/2
0.141	9/32	5/32	0.625	1 1/4	5/8
0.156	5/16	11/64	0.750	1 1/2	3/4

All dimensions are given in inches.

Extracted from *Clevis Pins and Cotter Pins* (ANSI B18.8.1–1972, R1979), with the permission of the publisher, The American Society of Mechanical Engineers.

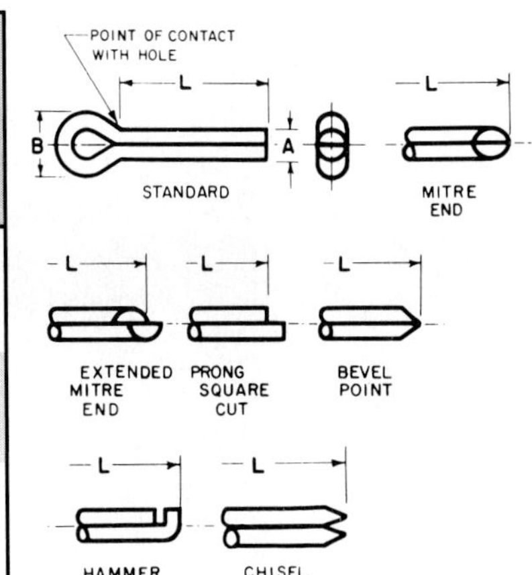

Appendix 32. American Standard Plain Washers, Inch Sizes[1]

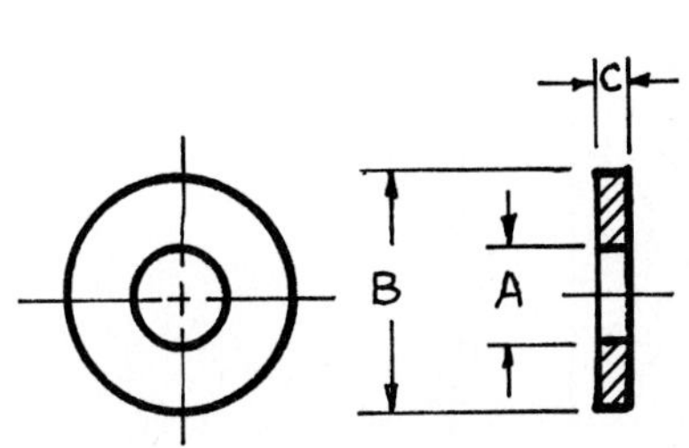

Nominal Washer Size			Inside Dia. A	Outside Dia. B	Nominal Thickness C
1/4	0.250	N	0.281	0.625	0.065
1/4	0.250	W	0.312	0.734	0.065
5/16	0.312	N	0.344	0.688	0.065
5/16	0.312	W	0.375	0.875	0.083
3/8	0.375	N	0.406	0.812	0.065
3/8	0.375	W	0.438	1.000	0.083
7/16	0.438	N	0.469	0.922	0.065
7/16	0.438	W	0.500	1.250	0.083
1/2	0.500	N	0.531	1.062	0.095
1/2	0.500	W	0.562	1.375	0.109
9/16	0.562	N	0.594	1.156	0.095
9/16	0.562	W	0.625	1.469	0.109
5/8	0.625	N	0.656	1.312	0.095
5/8	0.625	W	0.688	1.750	0.134
3/4	0.750	N	0.812	1.469	0.134
3/4	0.750	W	0.812	2.000	0.148
7/8	0.875	N	0.938	1.750	0.134
7/8	0.875	W	0.938	2.250	0.165
1	1.000	N	1.062	2.000	0.134
1	1.000	W	1.062	2.500	0.165
1 1/8	1.125	N	1.250	2.250	0.134
1 1/8	1.125	W	1.250	2.750	0.165
1 1/4	1.250	N	1.375	2.500	0.165
1 1/4	1.250	W	1.375	3.000	0.165
1 3/8	1.375	N	1.500	2.750	0.165
1 3/8	1.375	W	1.500	3.250	0.180
1 1/2	1.500	N	1.625	3.000	0.165

[1] "N" is narrow series. "W" is wide series. Additional sizes are in the standards.

Extracted from *Plain Washers* (ANSI B18.22.1–1965, R1981), with the permission of the publisher, The American Society of Mechanical Engineers.

Appendix 33. American Standard Helical Spring Lock Washers, Inch Sizes

Nominal Washer Size		Inside Diameter A		Outside Diameter B	Washer Section Width W	Washer Section Thickness $\frac{T+t}{2}$
		Min	Max	Max**	Min	Min
No. 2	0.086	0.088	0.094	0.172	0.035	0.020
No. 3	0.099	0.101	0.107	0.195	0.040	0.025
No. 4	0.112	0.115	0.121	0.209	0.040	0.025
No. 5	0.125	0.128	0.134	0.236	0.047	0.031
No. 6	0.138	0.141	0.148	0.250	0.047	0.031
No. 8	0.164	0.168	0.175	0.293	0.055	0.040
No. 10	0.190	0.194	0.202	0.334	0.062	0.047
No. 12	0.216	0.221	0.229	0.377	0.070	0.056
1/4	0.250	0.255	0.263	0.489	0.109	0.062
5/16	0.312	0.318	0.328	0.586	0.125	0.078
3/8	0.375	0.382	0.393	0.683	0.141	0.094
7/16	0.438	0.446	0.459	0.779	0.156	0.109
1/2	0.500	0.509	0.523	0.873	0.171	0.125
9/16	0.562	0.572	0.587	0.971	0.188	0.141
5/8	0.625	0.636	0.653	1.079	0.203	0.156
11/16	0.688	0.700	0.718	1.176	0.219	0.172
3/4	0.750	0.763	0.783	1.271	0.234	0.188
13/16	0.812	0.826	0.847	1.367	0.250	0.203
7/8	0.875	0.890	0.912	1.464	0.266	0.219
15/16	0.938	0.954	0.978	1.560	0.281	0.234
1	1.000	1.017	1.042	1.661	0.297	0.250
1 1/16	1.062	1.080	1.107	1.756	0.312	0.266
1 1/8	1.125	1.144	1.172	1.853	0.328	0.281
1 3/16	1.188	1.208	1.237	1.950	0.344	0.297
1 1/4	1.250	1.271	1.302	2.045	0.359	0.312
1 5/16	1.312	1.334	1.366	2.141	0.375	0.328
1 3/8	1.375	1.398	1.432	2.239	0.391	0.344
1 7/16	1.438	1.462	1.497	2.334	0.406	0.359
1 1/2	1.500	1.525	1.561	2.430	0.422	0.375

Extracted from *Lock Washers* (ANSI B18.21.1–1972), with the permission of the publisher, The American Society of Mechanical Engineers.

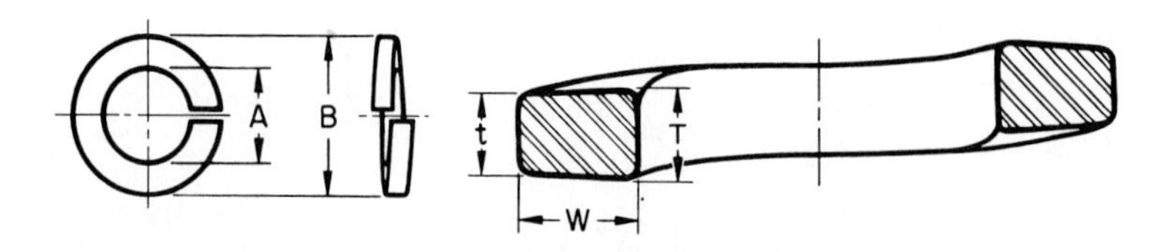

Appendix 34. American Standard Internal Tooth Lock Washers, Inch sizes

Dimensions of Internal Tooth Lock Washers				
Nominal Washer Size		A Inside Diameter	B Outside Diameter	C Thickness
		Max	Max	Max
No. 2	0.086	0.095	0.200	0.015
No. 3	0.099	0.109	0.232	0.019
No. 4	0.112	0.123	0.270	0.019
No. 5	0.125	0.136	0.280	0.021
No. 6	0.138	0.150	0.295	0.021
No. 8	0.164	0.176	0.340	0.023
No. 10	0.190	0.204	0.381	0.025
No. 12	0.216	0.231	0.410	0.025
1/4	0.250	0.267	0.478	0.028
5/16	0.312	0.332	0.610	0.034
3/8	0.375	0.398	0.692	0.040
7/16	0.438	0.464	0.789	0.040
1/2	0.500	0.530	0.900	0.045
9/16	0.562	0.596	0.985	0.045
5/8	0.625	0.663	1.071	0.050
11/16	0.688	0.728	1.166	0.050
3/4	0.750	0.795	1.245	0.055
13/16	0.812	0.861	1.315	0.055
7/8	0.875	0.927	1.410	0.060
1	1.000	1.060	1.637	0.067
1 1/8	1.125	1.192	1.830	0.067
1 1/4	1.250	1.325	1.975	0.067

Extracted from *Standard Lock Washers* (ANSI B18.21.1–1972), with the permission of the publisher, The American Society of Mechanical Engineers.

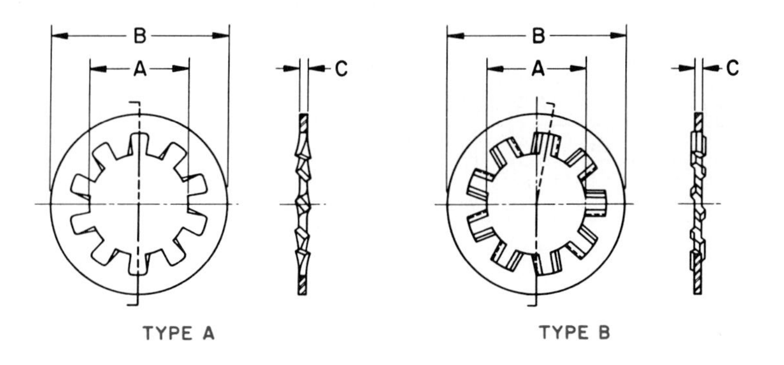

Appendix 35. American Standard Small Rivets, Inch Sizes

Nominal Size or Basic Shank Diameter		Diameter of Shank *D*	Pan Head					Button Head			Countersunk Head		Flat Head	
			A	H	R_1	R_2	R_3	A	H	R	A	H	A	H
1/16	0.062	0.064	0.118	0.040	0.019	0.052	0.217	0.122	0.052	0.055	0.118	0.027	0.140	0.027
3/32	0.094	0.096	0.173	0.060	0.030	0.080	0.326	0.182	0.077	0.084	0.176	0.040	0.200	0.038
1/8	0.125	0.127	0.225	0.078	0.039	0.106	0.429	0.235	0.100	0.111	0.235	0.053	0.260	0.048
5/32	0.156	0.158	0.279	0.096	0.049	0.133	0.535	0.290	0.124	0.138	0.293	0.066	0.323	0.059
3/16	0.188	0.191	0.334	0.114	0.059	0.159	0.641	0.348	0.147	0.166	0.351	0.079	0.387	0.069
7/32	0.219	0.222	0.391	0.133	0.069	0.186	0.754	0.405	0.172	0.195	0.413	0.094	0.453	0.080
1/4	0.250	0.253	0.444	0.151	0.079	0.213	0.858	0.460	0.196	0.221	0.469	0.106	0.515	0.091
9/32	0.281	0.285	0.499	0.170	0.088	0.239	0.963	0.518	0.220	0.249	0.528	0.119	0.579	0.103
5/16	0.313	0.317	0.552	0.187	0.098	0.266	1.070	0.572	0.243	0.276	0.588	0.133	0.641	0.113
11/32	0.344	0.348	0.608	0.206	0.108	0.292	1.176	0.630	0.267	0.304	0.646	0.146	0.705	0.124
3/8	0.375	0.380	0.663	0.225	0.118	0.319	1.286	0.684	0.291	0.332	0.704	0.159	0.769	0.135
13/32	0.406	0.411	0.719	0.243	0.127	0.345	1.392	0.743	0.316	0.358	0.763	0.172	0.834	0.146
7/16	0.438	0.443	0.772	0.261	0.137	0.372	1.500	0.798	0.339	0.387	0.823	0.186	0.896	0.157

Small rivets are available in length increments of 1/32.

Extracted from *Small Solid Rivets* (ANSI B18.1.1–1972, R1977), with permission of the publisher, American Society of Mechanical Engineers.

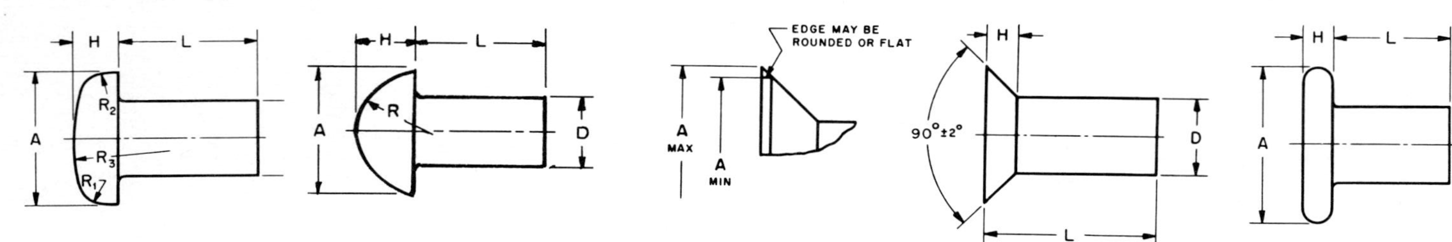

Appendix 36. American Standard Large Rivets, Inch sizes

	Button			High Button				Pan			Cone			Flat	
Nominal Size	A	H	G	A	H	F	G	A	B	H	A	B	H	A	H
½	0.875	0.406	0.443	0.781	0.500	0.656	0.094	0.800	0.500	0.381	0.875	0.469	0.469	0.936	0.260
⅝	1.094	0.500	0.553	0.969	0.594	0.750	0.188	1.000	0.625	0.469	1.094	0.586	0.578	1.194	0.339
¾	1.312	0.593	0.664	1.156	0.688	0.844	0.282	1.200	0.750	0.556	1.312	0.703	0.687	1.421	0.400
⅞	1.531	0.687	0.775	1.344	0.781	0.937	0.375	1.400	0.875	0.643	1.531	0.820	0.797	1.647	0.460
1	1.750	0.781	0.885	1.531	0.875	1.031	0.469	1.600	1.000	0.731	1.750	0.938	0.906	1.873	0.520

Head dimensions are for manufactured head after driving. Large rivets are available in length increments of ⅛ inch.

Extracted from *Large Rivets* (ANSI B18.1.2–1972, R1977), with the permission of the publisher, The American Society of Mechanical Engineers.

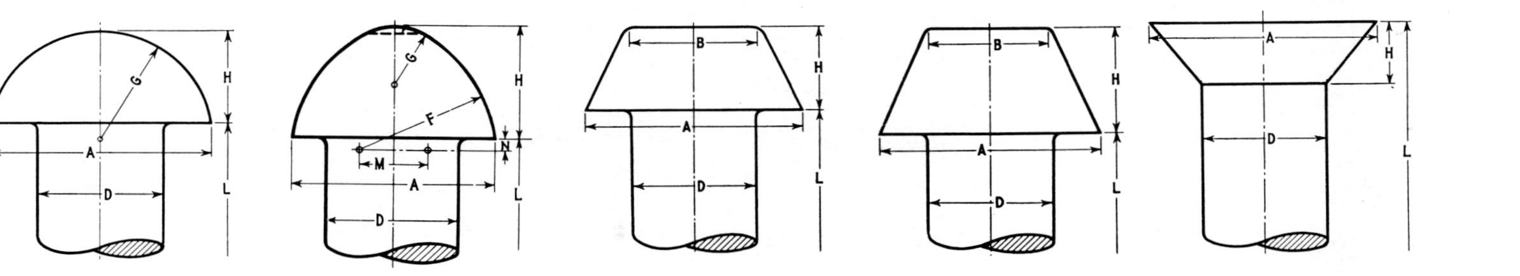

Appendix 37. American Standard Slotted Head Wood Screws, Inch Sizes

Nominal Size	Threads per Inch	Screw Dia. D.	Flat and Oval Head A	Flat and Oval Head H	Flat and Oval Head O	Round Head A	Round Head H
0	32	0.064	0.119	0.035	0.056	0.113	0.053
1	28	0.077	0.146	0.043	0.068	0.138	0.061
2	26	0.090	0.172	0.051	0.080	0.162	0.069
3	24	0.103	0.199	0.059	0.092	0.187	0.078
4	22	0.116	0.225	0.067	0.104	0.211	0.086
5	20	0.129	0.252	0.075	0.116	0.236	0.095
6	18	0.142	0.279	0.083	0.128	0.260	0.103
7	16	0.155	0.305	0.091	0.140	0.285	0.111
8	15	0.168	0.332	0.100	0.152	0.309	0.120
9	14	0.181	0.358	0.108	0.164	0.334	0.128
10	13	0.194	0.385	0.116	0.176	0.359	0.137
12	11	0.220	0.438	0.132	0.200	0.408	0.153
14	10	0.246	0.491	0.148	0.224	0.457	0.170
16	9	0.272	0.544	0.164	0.248	0.506	0.187
18	8	0.298	0.597	0.180	0.272	0.555	0.204
20	8	0.324	0.650	0.196	0.296	0.604	0.220
24	7	0.376	0.756	0.228	0.344	0.702	0.254

Screw lengths: ¼ to 1 inch by ⅛ inch increments.
1 to 3 inches by ¼ inch increments.
3 to 5 inches by ½ inch increments.
Thread lengths equal to ⅔ screw length.

Extracted from *Slotted and Recessed Head Wood Screws* (ANSI B18.6.1–1981), with the permission of the publisher, The American Society of Mechanical Engineers.

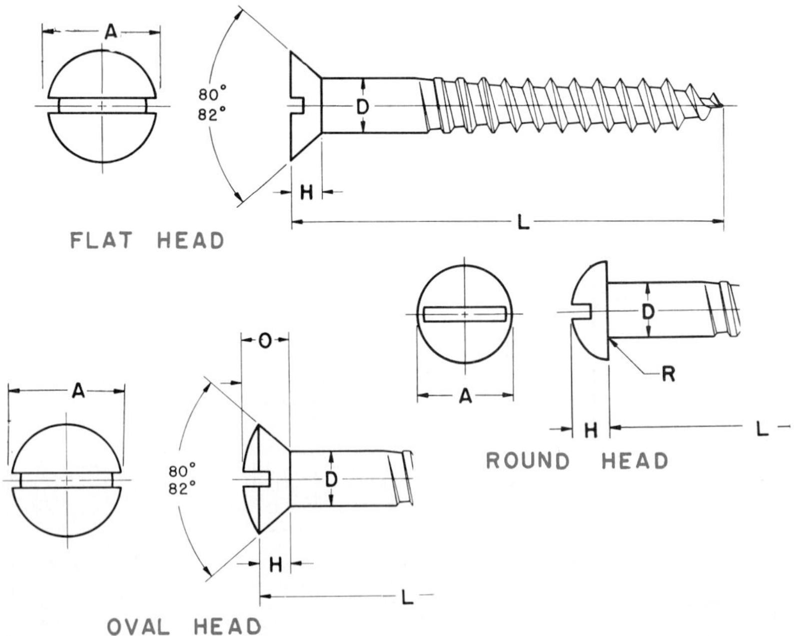

Appendix 38. American Standard Recessed Flat Head Wood Screws, Inch Sizes

Nominal Size	M	T	N
0	0.079	0.039	0.021
1	0.097	0.051	0.023
2	0.114	0.062	0.026
3	0.131	0.073	0.028
4	0.148	0.082	0.031
5	0.165	0.094	0.033
6	0.182	0.105	0.036
7	0.199	0.116	0.038
8	0.216	0.122	0.041
9	0.234	0.133	0.043
10	0.251	0.145	0.046
12	0.286	0.167	0.051
14	0.320	0.182	0.056
16	0.354	0.204	0.061
18	0.388	0.226	0.066
20	0.423	0.249	0.071
24	0.491	0.293	0.081

Head Dimensions are the same as those of slotted head wood screws.

Extracted from *Slotted and Recessed Head Wood Screws* (ANSI B18.6.1–1981), with the permission of the publisher, The American Society of Mechanical Engineers.

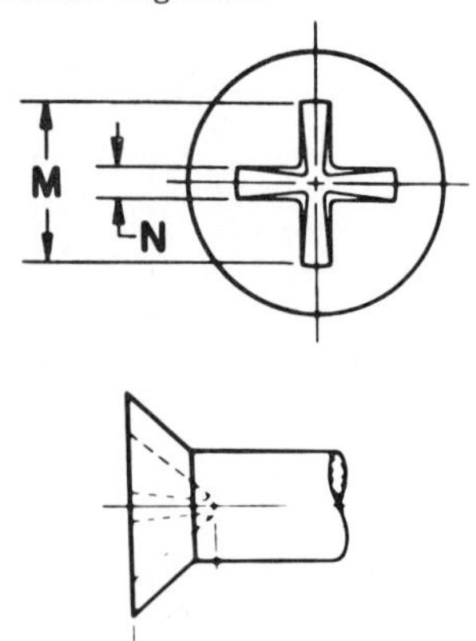

Appendix 39. American Welding Society Standard Welding Symbols

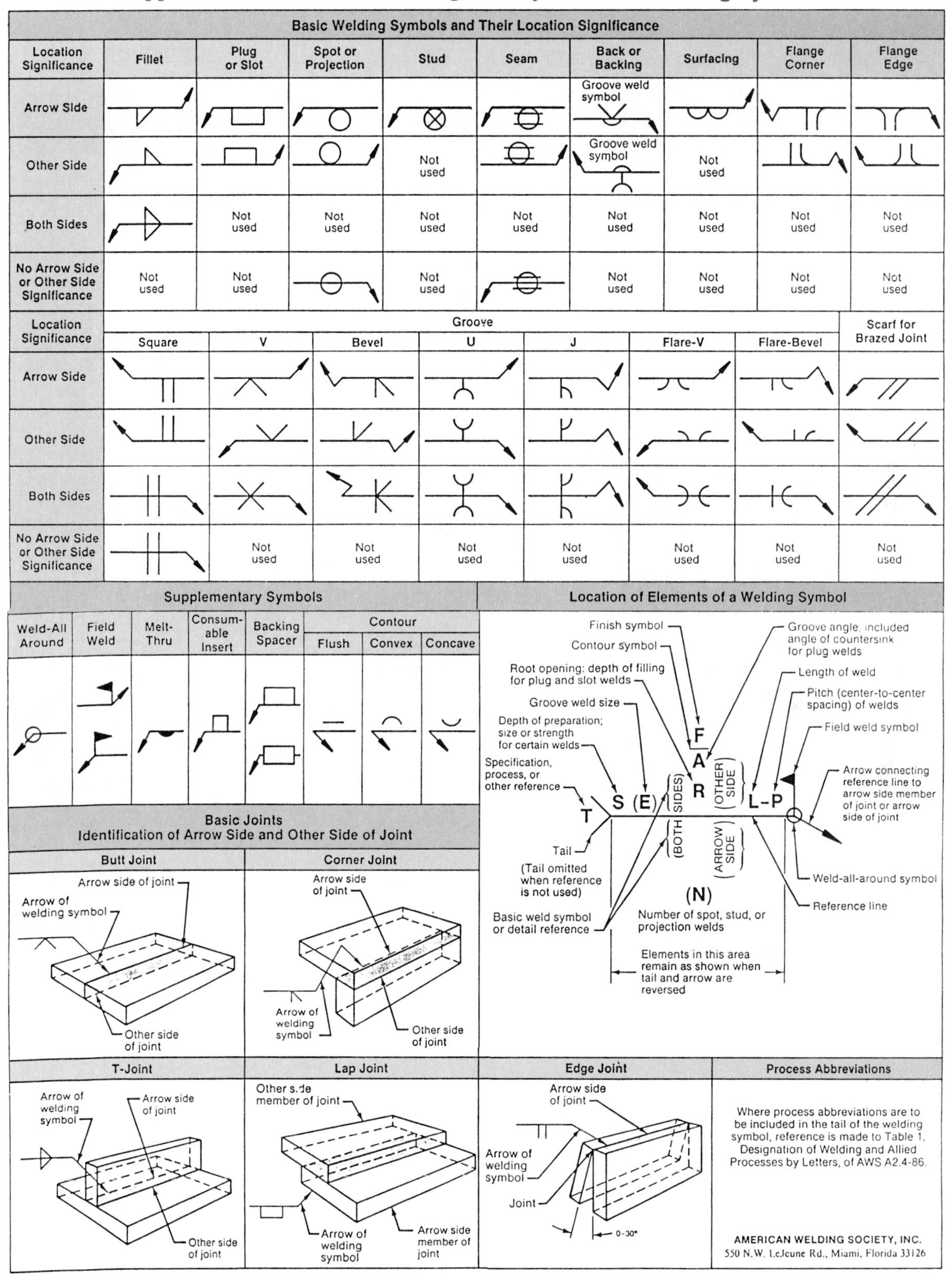

(Continued on next page)

Appendix 39. American Welding Society Standard Welding Symbols* (continued)

Typical Welding Symbols

Double-Fillet Welding Symbol

Weld size
Length
1/4 6
1/16 4
Omission indicates that weld extends between abrupt changes in direction or as dimensioned

Chain Intermittant Fillet Welding Symbol

Pitch (distance between centers) of increments
5/16 2-5
7/16 2-6
Size (length of leg)
Length of increments

Staggered Intermittant Fillet Welding Symbol

Pitch (distance between centers) of increments
1/2 3-5
1/2 3-5
Size (length of leg)
Length of increments

Plug Welding Symbol

Include angle of countersink
Pitch (distance between centers) of welds
Size (diameter of hole at root)
30°
1 3/4 4
Depth of filling in inches (omission indicates filling is complete)

Back Welding Symbol

Back weld
or
2nd operation
1st operation

Backing Welding Symbol

Backing weld
or
1st operation
2nd operation

Spot Welding Symbol

Size or strength
Number of welds
Pitch
(5)
0.25 4
RSW
Process

Stud Welding Symbol

1/2 6
(7)
Pitch
Size
Number of studs

Seam Welding Symbol

Increment length
Pitch
Size or strength
0.30 3-9
RSEW
Process

Square-Groove Welding Symbol

(3/16) 1/4
Weld size
Root opening

Single-V Groove Welding Symbol

Depth of preparation
1/2 (1/2) 1/2
Root opening
60°
Weld size
Groove angle

Double-Bevel-Groove Welding Symbol

Weld size
(1)
(1-1/4)
Weld size
Arrow points toward member to be prepared

Symbol with Backgouging

Depth of preparation
3/8
Back gouge

Flare-V Groove Welding Symbol

(1/4)
Weld size

Flare-Bevel-Groove Welding Symbol

Weld size
(3/4)

Multiple Reference Lines

1st operation on line nearest arrow
2nd operation
3rd operation

Complete Penetration

Indicates complete penetration regardless of type of weld or joint preparation
CJP

Edge Flange Welding Symbol

Radius
3/64 + 1/16
1/16
Weld size
Height above point of tangency

Flash or Upset Welding Symbol

Process reference
FW

Melt-Thru Symbol

1/32

Joint with Backing

R
'R' indicates backing removed after welding

Joint with Spacer

With modified groove weld symbol
Double bevel groove

Flush Contour Symbol

Convex Contour Symbol

G

*It should be understood that these charts are intended only as shop aids. The only complete and official presentation of the standard welding symbols is in A2.4.

Appendix 40. American Standard Running and Sliding Fits
Preferred Limits and Fits for Cylindrical Parts
Basic Hole System, Inch Series

Nominal Size Range Inches	Class RC 1			Class RC 2			Class RC 3			Class RC 4		
	Limits of Clearance	Standard Limits		Limits of Clearance	Standard Limits		Limits of Clearance	Standard Limits		Limits of Clearance	Standard Limits	
Over To		Hole H5	Shaft g4		Hole H6	Shaft g5		Hole H6	Shaft f6		Hole H7	Shaft f7
0.04– 0.12	0.1 0.45	+0.2 0	−0.1 −0.25	0.1 0.55	+0.25 0	−0.1 −0.3	0.3 0.8	+0.25 0	−0.3 −0.55	0.3 1.1	+0.4 0	−0.3 −0.7
0.12– 0.24	0.15 0.5	+0.2 0	−0.15 −0.3	0.15 0.65	+0.3 0	−0.15 −0.35	0.4 1.0	+0.3 0	−0.4 −0.7	0.4 1.4	+0.5 0	−0.4 −0.9
0.24– 0.40	0.2 0.6	+0.25 0	−0.2 −0.35	0.2 0.85	+0.4 0	−0.2 −0.45	0.5 1.3	+0.4 0	−0.5 −0.9	0.5 1.7	+0.6 0	−0.5 −1.1
0.40– 0.71	0.25 0.75	+0.3 0	−0.25 −0.45	0.25 0.95	+0.4 0	−0.25 −0.55	0.6 1.4	+0.4 0	−0.6 −1.0	0.6 2.0	+0.7 0	−0.6 −1.3
0.71– 1.19	0.3 0.95	+0.4 0	−0.3 −0.55	0.3 1.2	+0.5 0	−0.3 −0.7	0.8 1.8	+0.5 0	−0.8 −1.3	0.8 2.4	+0.8 0	−0.8 −1.6
1.19– 1.97	0.4 1.1	+0.4 0	−0.4 −0.7	0.4 1.4	+0.6 0	−0.4 −0.8	1.0 2.2	+0.6 0	−1.0 −1.6	1.0 3.0	+1.0 0	−1.0 −2.0
1.97– 3.15	0.4 1.2	+0.5 0	−0.4 −0.7	0.4 1.6	+0.7 0	−0.4 −0.9	1.2 2.6	+0.7 0	−1.2 −1.9	1.2 3.6	+1.2 0	−1.2 −2.4
3.15– 4.73	0.5 1.5	+0.6 0	−0.5 −0.9	0.5 2.0	+0.9 0	−0.5 −1.1	1.4 3.2	+0.9 0	−1.4 −2.3	1.4 4.2	+1.4 0	−1.4 −2.8
4.73– 7.09	0.6 1.8	+0.7 0	−0.6 −1.1	0.6 2.3	+1.0 0	−0.6 −1.3	1.6 3.6	+1.0 0	−1.6 −2.6	1.6 4.8	+1.6 0	−1.6 −3.2
7.09– 9.85	0.6 2.0	+0.8 0	−0.6 −1.2	0.6 2.6	+1.2 0	−0.6 −1.4	2.0 4.4	+1.2 0	−2.0 −3.2	2.0 5.6	+1.8 0	−2.0 −3.8
9.85–12.41	0.8 2.3	+0.9 0	−0.8 −1.4	0.8 2.9	+1.2 0	−0.8 −1.7	2.5 4.9	+1.2 0	−2.5 −3.7	2.5 6.5	+2.0 0	−2.5 −4.5
12.41–15.75	1.0 2.7	+1.0 0	−1.0 −1.7	1.0 3.4	+1.4 0	−1.0 −2.0	3.0 5.8	+1.4 0	−3.0 −4.4	3.0 7.4	+2.2 0	−3.0 −5.2

(Continued on next page)

Appendix 40. American Standard Running and Sliding Fits
Preferred Limits and Fits for Cylindrical Parts
Basic Hole System, Inch Series (continued)

Nominal Size Range Inches	Class RC 5			Class RC 6			Class RC 7			Class RC 8			Class RC 9		
	Limits of Clearance	Standard Limits		Limits of Clearance	Standard Limits		Limits of Clearance	Standard Limits		Limits of Clearance	Standard Limits		Limits of Clearance	Standard Limits	
Over To		Hole H8	Shaft e7		Hole H9	Shaft e8		Hole H9	Shaft d8		Hole H10	Shaft c9		Hole H11	Shaft
0 – 0.12	0.6 1.6	+0.6 0	−0.6 −1.0	0.6 2.2	+1.0 0	−0.6 −1.2	1.0 2.6	+1.0 0	− 1.0 − 1.6	2.5 5.1	+1.6 0	− 2.5 − 3.5	4.0 8.1	+ 2.5 0	− 4.0 − 5.6
0.12– 0.24	0.8 2.0	+0.7 0	−0.8 −1.3	0.8 2.7	+1.2 0	−0.8 −1.5	1.2 3.1	+1.2 0	− 1.2 − 1.9	2.8 5.8	+1.8 0	− 2.8 − 4.0	4.5 9.0	+ 3.0 0	− 4.5 − 6.0
0.24– 0.40	1.0 2.5	+0.9 0	−1.0 −1.6	1.0 3.3	+1.4 0	−1.0 −1.9	1.6 3.9	+1.4 0	− 1.6 − 2.5	3.0 6.6	+2.2 0	− 3.0 − 4.4	5.0 10.7	+ 3.5 0	− 5.0 − 7.2
0.40– 0.71	1.2 2.9	+1.0 0	−1.2 −1.9	1.2 3.8	+1.6 0	−1.2 −2.2	2.0 4.6	+1.6 0	− 2.0 − 3.0	3.5 7.9	+2.8 0	− 3.5 − 5.1	6.0 12.8	+ 4.0 0	− 6.0 − 8.8
0.71– 1.19	1.6 3.6	+1.2 0	−1.6 −2.4	1.6 4.8	+2.0 0	−1.6 −2.8	2.5 5.7	+2.0 0	− 2.5 − 3.7	4.5 10.0	+3.5 0	− 4.5 − 6.5	7.0 15.5	+ 5.0 0	− 7.0 −10.5
1.19– 1.97	2.0 4.6	+1.6 0	−2.0 −3.0	2.0 6.1	+2.5 0	−2.0 −3.6	3.0 7.1	+2.5 0	− 3.0 − 4.6	5.0 11.5	+4.0 0	− 5.0 − 7.5	8.0 18.0	+ 6.0 0	− 8.0 −12.0
1.97– 3.15	2.5 5.5	+1.8 0	−2.5 −3.7	2.5 7.3	+3.0 0	−2.5 −4.3	4.0 8.8	+3.0 0	− 4.0 − 5.8	6.0 13.5	+4.5 0	− 6.0 − 9.0	9.0 20.5	+ 7.0 0	− 9.0 −13.5
3.15– 4.73	3.0 6.6	+2.2 0	−3.0 −4.4	3.0 8.7	+3.5 0	−3.0 −5.2	5.0 10.7	+3.5 0	− 5.0 − 7.2	7.0 15.5	+5.0 0	− 7.0 −10.5	10.0 24.0	+ 9.0 0	−10.0 −15.0
4.73– 7.09	3.5 7.6	+2.5 0	−3.5 −5.1	3.5 10.0	+4.0 0	−3.5 −6.0	6.0 12.5	+4.0 0	− 6.0 − 8.5	8.0 18.0	+6.0 0	− 8.0 −12.0	12.0 28.0	+10.0 0	−12.0 −18.0
7.09– 9.85	4.0 8.6	+2.8 0	−4.0 −5.8	4.0 11.3	+4.5 0	−4.0 −6.8	7.0 14.3	+4.5 0	− 7.0 − 9.8	10.0 21.5	+7.0 0	−10.0 −14.5	15.0 34.0	+12.0 0	−15.0 −22.0
9.85–12.41	5.0 10.0	+3.0 0	−5.0 −7.0	5.0 13.0	+5.0 0	−5.0 −8.0	8.0 16.0	+5.0 0	− 8.0 −11.0	12.0 25.0	+8.0 0	−12.0 −17.0	18.0 38.0	+12.0 0	−18.0 −26.0
12.41–15.75	6.0 11.7	+3.5 0	−6.0 −8.2	6.0 15.5	+6.0 0	−6.0 −9.5	10.0 19.5	+6.0 0	−10.0 −13.5	14.0 29.0	+9.0 0	−14.0 −20.0	22.0 45.0	+14.0 0	−22.0 −31.0

Extracted from *Preferred Limits and Fits for Cylindrical Parts,* (ANSI B4.1–1967, R1979), with permission of the publisher, The American Society of Mechanical Engineers.

Appendix 41. American Standard Clearance Locational Fits
Preferred Limits and Fits for Cylindrical Parts
Basic Hole System, Inch Series

Nominal Size Range Inches	Class LC 1			Class LC 2			Class LC 3			Class LC 4			Class LC 5		
	Limits of Clearance	Standard Limits		Limits of Clearance	Standard Limits		Limits of Clearance	Standard Limits		Limits of Clearance	Standard Limits		Limits of Clearance	Standard Limits	
Over To		Hole H6	Shaft h5		Hole H7	Shaft h6		Hole H8	Shaft h7		Hole H10	Shaft h9		Hole H7	Shaft g6
0.04 – 0.12	0 0.45	+0.25 − 0	+ 0 − 0.2	0 0.65	+0.4 − 0	+ 0 −0.25	0 1	+0.6 − 0	+ 0 −0.4	0 2.6	+1.6 − 0	+ 0 −1.0	0.1 0.75	+0.4 − 0	−0.1 −0.35
0.12 – 0.24	0 0.5	+0.3 − 0	+ 0 − 0.2	0 0.8	+0.5 − 0	+ 0 −0.3	0 1.2	+0.7 − 0	+ 0 −0.5	0 3.0	+1.8 − 0	+ 0 −1.2	0.15 0.95	+0.5 − 0	−0.15 −0.45
0.24 – 0.40	0 0.65	+0.4 − 0	+ 0 −0.25	0 1.0	+0.6 − 0	+ 0 −0.4	0 1.5	+0.9 − 0	+ 0 −0.6	0 3.6	+2.2 − 0	+ 0 −1.4	0.2 1.2	+0.6 − 0	−0.2 −0.6
0.40 – 0.71	0 0.7	+0.4 − 0	+ 0 −0.3	0 1.1	+0.7 − 0	+ 0 −0.4	0 1.7	+1.0 − 0	+ 0 −0.7	0 4.4	+2.8 − 0	+ 0 −1.6	0.25 1.35	+0.7 − 0	−0.25 −0.65
0.71 – 1.19	0 0.9	+0.5 − 0	+ 0 −0.4	0 1.3	+0.8 − 0	+ 0 −0.5	0 2	+1.2 − 0	+ 0 −0.8	0 5.5	+3.5 − 0	+ 0 −2.0	0.3 1.6	+0.8 − 0	−0.3 −0.8
1.19 – 1.97	0 1.0	+0.6 − 0	+ 0 −0.4	0 1.6	+1.0 − 0	+ 0 −0.6	0 2.6	+1.6 − 0	+ 0 −1	0 6.5	+4.0 − 0	+ 0 −2.5	0.4 2.0	+1.0 − 0	−0.4 −1.0
1.97 – 3.15	0 1.2	+0.7 − 0	+ 0 −0.5	0 1.9	+1.2 − 0	+ 0 −0.7	0 3	+1.8 − 0	+ 0 −1.2	0 7.5	+4.5 − 0	+ 0 −3	0.4 2.3	+1.2 − 0	−0.4 −1.1
3.15 – 4.73	0 1.5	+0.9 − 0	+ 0 −0.6	0 2.3	+1.4 − 0	+ 0 −0.9	0 3.6	+2.2 − 0	+ 0 −1.4	0 8.5	+5.0 − 0	+ 0 −3.5	0.5 2.8	+1.4 − 0	−0.5 −1.4
4.73 – 7.09	0 1.7	+1.0 − 0	+ 0 −0.7	0 2.6	+1.6 − 0	+ 0 −1.0	0 4.1	2.5 − 0	+ 0 −1.6	0 10	+6.0 − 0	+ 0 −4	0.6 3.2	+1.6 − 0	−0.6 −1.6
7.09 – 9.85	0 2.0	+1.2 − 0	+ 0 −0.8	0 3.0	+1.8 − 0	+ 0 −1.2	0 4.6	+2.8 − 0	+ 0 −1.8	0 11.5	+7.0 − 0	+ 0 −4.5	0.6 3.6	+1.8 − 0	−0.6 −1.8
9.85 – 12.41	0 2.1	+1.2 − 0	+ 0 −0.9	0 3.2	+2.0 − 0	+ 0 −1.2	0 5	+3.0 − 0	+ 0 −2.0	0 13	+8.0 − 0	+ 0 −5	0.7 3.9	+2.0 − 0	−0.7 −1.9
12.41 – 15.75	0 2.4	+1.4 − 0	+ 0 −1.0	0 3.6	+2.2 − 0	+ 0 −1.4	0 5.7	+3.5 − 0	+ 0 −2.2	0 15	+9.0 − 0	+ 0 −6	0.7 4.3	+2.2 − 0	−0.7 −2.1

(Continued on next page)

Appendix 41. American Standard Clearance Locational Fits Preferred Limits and Fits for Cylindrical Parts Basic Hole System, Inch Series (continued)

Nominal Size Range Inches	Class LC 6			Class LC 7			Class LC 8			Class LC 9			Class LC 10			Class LC 11		
	Limits of Clearance	Standard Limits		Limits of Clearance	Standard Limits		Limits of Clearance	Standard Limits		Limits of Clearance	Standard Limits		Limits of Clearance	Standard Limits		Limits of Clearance	Standa[rd] Limits	
Over To		Hole H9	Shaft f8		Hole H10	Shaft e9		Hole H10	Shaft d9		Hole H11	Shaft c10		Hole H12	Shaft		Hole H13	Sh
0– 0.12	0.3 1.9	+1.0 −0	−0.3 −0.9	0.6 3.2	+1.6 −0	−0.6 −1.6	1.0 3.6	+1.6 −0	−1.0 −2.0	2.5 6.6	+2.5 −0	−2.5 −4.1	4 12	+4 −0	−4 −8	5 17	+6 −0	
0.12– 0.24	0.4 2.3	+1.2 −0	−0.4 −1.1	0.8 3.8	+1.8 −0	−0.8 −2.0	1.2 4.2	+1.8 −0	−1.2 −2.4	2.8 7.6	+3.0 −0	−2.8 −4.6	4.5 14.5	+5 −0	−4.5 −9.5	6 20	+7 −0	
0.24– 0.40	0.5 2.8	+1.4 −0	−0.5 −1.4	1.0 4.6	+2.2 −0	−1.0 −2.4	1.6 5.2	+2.2 −0	−1.6 −3.0	3.0 8.7	+3.5 −0	−3.0 −5.2	5 17	+6 −0	−5 −11	7 25	+9 −0	
0.40– 0.71	0.6 3.2	+1.6 −0	−0.6 −1.6	1.2 5.6	+2.8 −0	−1.2 −2.8	2.0 6.4	+2.8 −0	−2.0 −3.6	3.5 10.3	+4.0 −0	−3.5 −6.3	6 20	+7 −0	−6 −13	8 28	+10 −0	
0.71– 1.19	0.8 4.0	+2.0 −0	−0.8 −2.0	1.6 7.1	+3.5 −0	−1.6 −3.6	2.5 8.0	+3.5 −0	−2.5 −4.5	4.5 13.0	+5.0 −0	−4.5 −8.0	7 23	+8 −0	−7 −15	10 34	+12 −0	
1.19– 1.97	1.0 5.1	+2.5 −0	−1.0 −2.6	2.0 8.5	+4.0 −0	−2.0 −4.5	3.0 9.5	+4.0 −0	−3.0 −5.5	5 15	+6 −0	−5 −9	8 28	+10 −0	−8 −18	12 44	+16 −0	
1.97– 3.15	1.2 6.0	+3.0 −0	−1.2 −3.0	2.5 10.0	+4.5 −0	−2.5 −5.5	4.0 11.5	+4.5 −0	−4.0 −7.0	6 17.5	+7 −0	−6 −10.5	10 34	+12 −0	−10 −22	14 50	+18 −0	
3.15– 4.73	1.4 7.1	+3.5 −0	−1.4 −3.6	3.0 11.5	+5.0 −0	−3.0 −6.5	5.0 13.5	+5.0 −0	−5.0 −8.5	7 21	+9 −0	−7 −12	11 39	+14 −0	−11 −25	16 60	+22 −0	
4.73– 7.09	1.6 8.1	+4.0 −0	−1.6 −4.1	3.5 13.5	+6.0 −0	−3.5 −7.5	6 16	+6 −0	−6 −10	8 24	+10 −0	−8 −14	12 44	+16 −0	−12 −28	18 68	+25 −0	
7.09– 9.85	2.0 9.3	+4.5 −0	−2.0 −4.8	4.0 15.5	+7.0 −0	−4.0 −8.5	7 18.5	+7 −0	−7 −11.5	10 29	+12 −0	−10 −17	16 52	+18 −0	−16 −34	22 78	+28 −0	
9.85–12.41	2.2 10.2	+5.0 −0	−2.2 −5.2	4.5 17.5	+8.0 −0	−4.5 −9.5	7 20	+8 −0	−7 −12	12 32	+12 −0	−12 −20	20 60	+20 −0	−20 −40	28 88	+30 −0	
12.41–15.75	2.5 12.0	+6.0 −0	−2.5 −6.0	5 20.0	+9.0 −0	−5 −11	8 23	+9 −0	−8 −14	14 37	+14 −0	−14 −23	22 66	+22 −0	−22 −44	30 100	+35 −0	

Extracted from *ANSI B4.1–1967 (R1979) Preferred Limits and Fits for Cylindrical Parts,* with the permission of the publisher, The American Socie[ty of] Mechanical Engineers.

Appendix 42. American Standard Transition Locational Fits Preferred Limits and Fits for Cylindrical Parts Basic Hole System, Inch Series

Nominal Size Range Inches	Class LT 1			Class LT 2			Class LT 3			Class LT 4			Class LT 6		
	Fit	Standard Limits		Fit	Standard Limits		Fit	Standard Limits		Fit	Standard Limits		Fit	Standard Limits	
Over To		Hole H7	Shaft j6		Hole H8	Shaft JS7		Hole H7	Shaft k6		Hole H8	Shaft k7		Hole h7	Shaft n7
0– 0.12	−0.10 +0.50	+0.4 −0	+0.10 −0.10	−0.2 +0.8	+0.6 −0	+0.2 −0.2							−0.65 +0.15	+0.4 −0	−0.65 +0.25
0.12– 0.24	−0.15 +0.65	+0.5 −0	+0.15 −0.15	−0.25 +0.95	+0.7 −0	+0.25 −0.25							−0.8 +0.2	+0.5 −0	+0.8 +0.3
0.24– 0.40	−0.20 +0.8	+0.6 −0	+0.2 −0.2	−0.3 +1.2	+0.9 −0	+0.3 −0.3	−0.5 +0.5	+0.6 −0	+0.5 +0.1	−0.7 +0.8	+0.9 −0	+0.7 +0.1	−1.0 +0.2	+0.6 −0	+1.0 +0.4
0.40– 0.71	−0.2 +0.9	+0.7 −0	+0.2 −0.2	−0.35 +1.35	+1.0 −0	+0.35 −0.35	−0.5 +0.6	+0.7 −0	+0.5 +0.1	−0.8 +0.9	+1.0 −0	+0.8 +0.1	−1.2 +0.2	+0.7 −0	+1.2 +0.5
0.71– 1.19	−0.25 +1.05	+0.8 −0	+0.25 −0.25	−0.4 +1.6	+1.2 −0	+0.4 −0.4	−0.6 +0.7	+0.8 −0	+0.6 +0.1	−0.9 +1.1	+1.2 −0	+0.9 +0.1	−1.7 +0.3	+0.8 −0	+1.4 +0.6
1.19– 1.97	−0.3 +1.3	+1.0 −0	+0.3 −0.3	−0.5 +2.1	+1.6 −0	+0.5 −0.5	−0.7 +0.9	+1.0 −0	+0.7 +0.1	−1.1 +1.5	+1.6 −0	+1.1 +0.1	−2.0 +0.4	+1.0 −0	+1.7 +0.7
1.97– 3.15	−0.3 +1.5	+1.2 −0	+0.3 −0.3	−0.6 +2.4	+1.8 −0	+0.6 −0.6	−0.8 +1.1	+1.2 −0	+0.8 +0.1	−1.3 +1.7	+1.8 −0	+1.3 +0.1	−2.4 +0.4	+1.2 −0	+2.0 +0.8
3.15– 4.73	−0.4 +1.8	+1.4 −0	+0.4 −0.4	−0.7 +2.9	+2.2 −0	+0.7 −0.7	−1.0 +1.3	+1.4 −0	+1.0 +0.1	−1.5 +2.1	+2.2 −0	+1.5 +0.1	−2.8 +0.4	+1.4 −0	+2.4 +1.0
4.73– 7.09	−0.5 +2.1	+1.6 −0	+0.5 −0.5	−0.8 +3.3	+2.5 −0	+0.8 −0.8	−1.1 +1.5	+1.6 −0	+1.1 +0.1	−1.7 +2.4	+2.5 −0	+1.7 +0.1	−3.2 +0.4	+1.6 −0	+2.8 +1.2
7.09– 9.85	−0.6 +2.4	+1.8 −0	+0.6 −0.6	−0.9 +3.7	+2.8 −0	+0.9 −0.9	−1.4 +1.6	+1.8 −0	+1.4 +0.2	−2.0 +2.6	+2.8 −0	+2.0 +0.2	−3.4 +0.6	+1.8 −0	+3.2 +1.4
9.85–12.41	−0.6 +2.6	+2.0 −0	+0.6 −0.6	−1.0 +4.0	+3.0 −0	+1.0 −1.0	−1.4 +1.8	+2.0 −0	+1.4 +0.2	−2.2 +2.8	+3.0 −0	+2.2 +0.2	−3.8 +0.6	+2.0 −0	+3.4 +1.4
12.41–15.75	−0.7 +2.9	+2.2 −0	+0.7 −0.7	−2.2 +4.5	+3.5 −0	+1.0 −1.0	−1.6 +2.0	+2.2 −0	+1.6 +0.2	−2.4 +3.3	+3.5 −0	+2.4 +0.2	−4.3 +0.7	+2.3 −0	+3.8 +1.6

Extracted from *ANSI B4.1–1967 (R 1979) Preferred Limits and Fits for Cylindrical Parts,* with the permission of the publisher, The American Society of Mechanical Engineers.

Appendix 43. American Standard Interference Locational Fits
Preferred Limits and Fits for Cylindrical Parts
Basic Hole System, Inch Series

Nominal Size Range Inches	Class LN 2			Class LN 3			Nominal Size Range Inches	Class LN 2			Class LN 3		
	Limits of Interference	Standard Limits		Limits of Interference	Standard Limits			Limits of Interference	Standard Limits		Limits of Interference	Standard limits	
Over To		Hole H7	Shaft p6		Hole H7	Shaft r6	Over To		Hole H7	Shaft p6		Hole H7	Shaft r6
0.04–0.12	0 0.65	+0.4 − 0	+0.65 +0.4	0.1 0.75	+0.4 − 0	+0.75 +0.5	1.97– 3.15	0.2 2.1	+1.2 − 0	+2.1 +1.4	0.4 2.3	+1.2 − 0	+2.3 +1.6
0.12–0.24	0 0.8	+0.5 − 0	+0.8 +0.5	0.1 0.9	+0.5 − 0	+0.9 +0.6	3.15– 4.73	0.2 2.5	+1.4 − 0	+2.5 +1.6	0.6 2.9	+1.4 − 0	+2.9 +2.0
0.24–0.40	0 1.0	+0.6 − 0	+1.0 +0.6	0.2 1.2	+0.6 − 0	+1.2 +0.8	4.73– 7.09	0.2 2.8	+1.6 − 0	+2.8 +1.8	0.9 3.5	+1.6 − 0	+3.5 +2.5
0.40–0.71	0 1.1	+0.7 − 0	+1.1 +0.7	0.3 1.4	+0.7 − 0	+1.4 +1.0	7.09– 9.85	0.2 3.2	+1.8 − 0	+3.2 +2.0	1.2 4.2	+1.8 − 0	+4.2 +3.0
0.71–1.19	0 1.3	+0.8 − 0	+1.3 +0.8	0.4 1.7	+0.8 − 0	+1.7 +1.2	9.85–12.41	0.2 3.4	+2.0 − 0	+3.4 +2.2	1.5 4.7	+2.0 − 0	+4.7 +3.5
1.19–1.97	0 1.6	+1.0 − 0	+1.6 +1.0	0.4 2.0	+1.0 − 0	+2.0 +1.4	12.41–15.75	0.3 3.9	+2.2 − 0	+3.9 +2.5	2.3 5.9	+2.2 − 0	+5.9 +4.5

Extracted from ANSI B4.1–1967 (R1979) *Preferred Limits and Fits for Cylindrical Parts,* with the permission of the publisher, The American Society of Mechanical Engineers.

Appendix 44. American Standard Force and Shrink Fits
Preferred Limits and Fits for Cylindrical Parts
Basic Hole System, Inch Series

Nominal Size Range Inches	Class FN 1			Class FN 2			Class FN 3			Class FN 4			Class FN 5		
	Limits of Interference	Standard Limits		Limits of Interference	Standard Limits		Limits of Interference	Standard Limits		Limits of Interference	Standard Limits		Limits of Interference	Standard Limits	
Over To		Hole H6	Shaft		Hole H7	Shaft s6		Hole H7	Shaft t6		Hole H7	Shaft u6		Hole H8	Shaft x7
0.04– 0.12	0.05 0.5	+0.25 − 0	+0.5 +0.3	0.2 0.85	+0.4 + 0	+0.85 +0.6				0.3 0.95	+0.4 − 0	+0.95 + 0.7	0.3 1.3	+0.6 − 0	+ 1.3 + 0.9
0.12– 0.24	0.1 0.6	+0.3 − 0	+0.6 +0.4	0.2 1.0	+0.5 − 0	+1.0 +0.7				0.4 1.2	+0.5 − 0	+ 1.2 + 0.9	0.5 1.7	+0.7 − 0	+ 1.7 + 1.2
0.24– 0.40	0.1 0.75	+0.4 − 0	+0.75 +0.5	0.4 1.4	+0.6 − 0	+1.4 +1.0				0.6 1.6	+0.6 − 0	+ 1.6 + 1.2	0.5 2.0	+0.9 − 0	+ 2.0 + 1.4
0.40– 0.56	0.1 0.8	+0.4 − 0	+0.8 +0.5	0.5 1.6	+0.7 − 0	+1.6 +1.2				0.7 1.8	+0.7 − 0	+ 1.8 + 1.4	0.6 2.3	+1.0 − 0	+ 2.3 + 1.6
0.56– 0.71	0.2 0.9	+0.4 − 0	+0.9 +0.6	0.5 1.6	+0.7 − 0	+1.6 +1.2				0.7 1.8	+0.7 − 0	+ 1.8 + 1.4	0.8 2.5	+1.0 − 0	+ 2.5 + 1.8

(Continued on next page)

Appendix 44. American Standard Force and Shrink Fits Preferred Limits and Fits for Cylindrical Parts Basic Hole System, Inch Series (continued)

Nominal Size Range Inches	Class FN 1			Class FN 2			Class FN 3			Class FN 4			Class FN 5		
	Limits of Interference	Standard Limits		Limits of Interference	Standard Limits		Limits of Interference	Standard Limits		Limits of Interference	Standard Limits		Limits of Interference	Standard Limits	
Over To		Hole H6	Shaft		Hole H7	Shaft s6		Hole H7	Shaft t6		Hole H7	Shaft u6		Hole H8	Shaft x7
0.71– 0.95	0.2 1.1	+0.5 −0	+1.1 +0.7	0.6 1.9	+0.8 −0	+1.9 +1.4				0.8 2.1	+0.8 −0	+2.1 +1.6	1.0 3.0	+1.2 −0	+3.0 +2.2
0.95– 1.19	0.3 1.2	+0.5 −0	+1.2 +0.8	0.6 1.9	+0.8 −0	+1.9 +1.4	0.8 2.1	+0.8 −0	+2.1 +1.6	1.0 2.3	+0.8 −0	+2.3 +1.8	1.3 3.3	+1.2 −0	+3.3 +2.5
1.19– 1.58	0.3 1.3	+0.6 −0	+1.3 +0.9	0.8 2.4	+1.0 −0	+2.4 +1.8	1.0 2.6	+1.0 −0	+2.6 +2.0	1.5 3.1	+1.0 −0	+3.1 +2.5	1.4 4.0	+1.6 −0	+4.0 +3.0
1.58– 1.97	0.4 1.4	+0.6 −0	+1.4 +1.0	0.8 2.4	+1.0 −0	+2.4 +1.8	1.2 2.8	+1.0 −0	+2.8 +2.2	1.8 3.4	+1.0 −0	+3.4 +2.8	2.4 5.0	+1.6 −0	+5.0 +4.0
1.97– 2.56	0.6 1.8	+0.7 −0	+1.8 +1.3	0.8 2.7	+1.2 −0	+2.7 +2.0	1.3 3.2	+1.2 −0	+3.2 +2.5	2.3 4.2	+1.2 −0	+4.2 +3.5	3.2 6.2	+1.8 −0	+6.2 +5.0
2.56– 3.15	0.7 1.9	+0.7 −0	+1.9 +1.4	1.0 2.9	+1.2 −0	+2.9 +2.2	1.8 3.7	+1.2 −0	+3.7 +3.0	2.8 4.7	+1.2 −0	+4.7 +4.0	4.2 7.2	+1.8 −0	+7.2 +6.0
3.15– 3.94	0.9 2.4	+0.9 −0	+2.4 +1.8	1.4 3.7	+1.4 −0	+3.7 +2.8	2.1 4.4	+1.4 −0	+4.4 +3.5	3.6 5.9	+1.4 −0	+5.9 +5.0	4.8 8.4	+2.2 −0	+8.4 +7.0
3.94– 4.73	1.1 2.6	+0.9 −0	+2.6 +2.0	1.6 3.9	+1.4 −0	+3.9 +3.0	2.6 4.9	+1.4 −0	+4.9 +4.0	4.6 6.9	+1.4 −0	+6.9 +6.0	5.8 9.4	+2.2 −0	+9.4 +8.0
4.73– 5.52	1.2 2.9	+1.0 −0	+2.9 +2.2	1.9 4.5	+1.6 −0	+4.5 +3.5	3.4 6.0	+1.6 −0	+6.0 +5.0	5.4 8.0	+1.6 −0	+8.0 +7.0	7.6 11.6	+2.5 −0	+11.6 +10.0
5.52– 6.30	1.5 3.2	+1.0 −0	+3.2 +2.5	2.4 5.0	+1.6 −0	+5.0 +4.0	3.4 6.0	+1.6 −0	+6.0 +5.0	5.4 8.0	+1.6 −0	+8.0 +7.0	9.5 13.6	+2.5 −0	+13.6 +12.0
6.30– 7.09	1.8 3.5	+1.0 −0	+3.5 +2.8	2.9 5.5	+1.6 −0	+5.5 +4.5	4.4 7.0	+1.6 −0	+7.0 +6.0	6.4 9.0	+1.6 −0	+9.0 +8.0	9.5 13.6	+2.5 −0	+13.6 +12.0
7.09– 7.88	1.8 3.8	+1.2 −0	+3.8 +3.0	3.2 6.2	+1.8 −0	+6.2 +5.0	5.2 8.2	+1.8 −0	+8.2 +7.0	7.2 10.2	+1.8 −0	+10.2 +9.0	11.2 15.8	+2.8 −0	+15.8 +14.0
7.88– 8.86	2.3 4.3	+1.2 −0	+4.3 +3.5	3.2 6.2	+1.8 −0	+6.2 +5.0	5.2 8.2	+1.8 −0	+8.2 +7.0	8.2 11.2	+1.8 −0	+11.2 +10.0	13.2 17.8	+2.8 −0	+17.8 +16.0
8.86– 9.85	2.3 4.3	+1.2 −0	+4.3 +3.5	4.2 7.2	+1.8 −0	+7.2 +6.0	6.2 9.2	+1.8 −0	+9.2 +8.0	10.2 13.2	+1.8 −0	+13.2 +12.0	13.2 17.8	+2.8 −0	+17.8 +16.0
9.85–11.03	2.8 4.9	+1.2 −0	+4.9 +4.0	4.0 7.2	+2.0 −0	+7.2 +6.0	7.0 10.2	+2.0 −0	+10.2 +9.0	10.2 13.2	+2.0 −0	+13.2 +12.0	15.0 20.0	+3.0 −0	+20.0 +18.0
11.03–12.41	2.8 4.9	+1.2 −0	+4.9 +4.0	5.0 8.2	+2.0 −0	+8.2 +7.0	7.0 10.2	+2.0 −0	+10.2 +9.0	12.0 15.2	+2.0 −0	+15.2 +14.0	17.0 22.0	+3.0 −0	+22.0 +20.0
12.41–13.98	3.1 5.5	+1.4 −0	+5.5 +4.5	5.8 9.4	+2.2 −0	+9.4 +8.0	7.8 11.4	+2.2 −0	+11.4 +10.0	13.8 17.4	+2.2 −0	+17.4 +16.0	18.5 24.2	+3.5 −0	+24.2 +22.0

Appendix 45. American National Standard Preferred Metric Limits and Fits
Preferred Hole Basis Clearance Fits, Cylindrical*

Basic Size		LOOSE RUNNING Hole H11	LOOSE RUNNING Shaft c11	LOOSE RUNNING Fit	FREE RUNNING Hole H9	FREE RUNNING Shaft d9	FREE RUNNING Fit	CLOSE RUNNING Hole H8	CLOSE RUNNING Shaft f7	CLOSE RUNNING Fit	SLIDING Hole H7	SLIDING Shaft g6	SLIDING Fit	LOCATIONAL CLEARANCE Hole H7	LOCATIONAL CLEARANCE Shaft h6	LOCATIONAL CLEARANCE Fit
1	MAX	1.060	0.940	0.180	1.025	0.980	0.070	1.014	0.994	0.030	1.010	0.998	0.018	1.010	1.000	0.016
	MIN	1.000	0.880	0.060	1.000	0.955	0.020	1.000	0.984	0.006	1.000	0.992	0.002	1.000	0.994	0.000
1.2	MAX	1.260	1.140	0.180	1.225	1.180	0.070	1.214	1.194	0.030	1.210	1.198	0.018	1.210	1.200	0.016
	MIN	1.200	1.080	0.060	1.200	1.155	0.020	1.200	1.184	0.006	1.200	1.192	0.002	1.200	1.194	0.000
1.6	MAX	1.660	1.540	0.180	1.625	1.580	0.070	1.614	1.594	0.030	1.610	1.598	0.018	1.610	1.600	0.016
	MIN	1.600	1.480	0.060	1.600	1.555	0.020	1.600	1.584	0.006	1.600	1.592	0.002	1.600	1.594	0.000
2	MAX	2.060	1.940	0.180	2.025	1.980	0.070	2.014	1.994	0.030	2.010	1.998	0.018	2.010	2.000	0.016
	MIN	2.000	1.880	0.060	2.000	1.955	0.020	2.000	1.984	0.006	2.000	1.992	0.002	2.000	1.994	0.000
2.5	MAX	2.560	2.440	0.180	2.525	2.480	0.070	2.514	2.494	0.030	2.510	2.498	0.018	2.510	2.500	0.016
	MIN	2.500	2.380	0.060	2.500	2.455	0.020	2.500	2.484	0.006	2.500	2.492	0.002	2.500	2.494	0.000
3	MAX	3.060	2.940	0.180	3.025	2.980	0.070	3.014	2.994	0.030	3.010	2.998	0.018	3.010	3.000	0.016
	MIN	3.000	2.880	0.060	3.000	2.955	0.020	3.000	2.984	0.006	3.000	2.992	0.002	3.000	2.994	0.000
4	MAX	4.075	3.930	0.220	4.030	3.970	0.090	4.018	3.990	0.040	4.012	3.996	0.024	4.012	4.000	0.020
	MIN	4.000	3.855	0.070	4.000	3.940	0.030	4.000	3.978	0.010	4.000	3.988	0.004	4.000	3.992	0.000
5	MAX	5.075	4.930	0.220	5.030	4.970	0.090	5.018	4.990	0.040	5.012	4.996	0.024	5.012	5.000	0.020
	MIN	5.000	4.855	0.070	5.000	4.940	0.030	5.000	4.978	0.010	5.000	4.988	0.004	5.000	4.992	0.000
6	MAX	6.075	5.930	0.220	6.030	5.970	0.090	6.018	5.990	0.040	6.012	5.996	0.024	6.012	6.000	0.020
	MIN	6.000	5.855	0.070	6.000	5.940	0.030	6.000	5.978	0.010	6.000	5.988	0.004	6.000	5.992	0.000
8	MAX	8.090	7.920	0.260	8.036	7.960	0.112	8.022	7.987	0.050	8.015	7.995	0.029	8.015	8.000	0.024
	MIN	8.000	7.830	0.080	8.000	7.924	0.040	8.000	7.972	0.013	8.000	7.986	0.005	8.000	7.991	0.000
10	MAX	10.090	9.920	0.260	10.036	9.960	0.112	10.022	9.987	0.050	10.015	9.995	0.029	10.015	10.000	0.024
	MIN	10.000	9.830	0.080	10.000	9.924	0.040	10.000	9.972	0.013	10.000	9.986	0.005	10.000	9.991	0.000
12	MAX	12.110	11.905	0.315	12.043	11.950	0.136	12.027	11.984	0.061	12.018	11.994	0.035	12.018	12.000	0.029
	MIN	12.000	11.795	0.095	12.000	11.907	0.050	12.000	11.966	0.016	12.000	11.983	0.006	12.000	11.989	0.000
16	MAX	16.110	15.905	0.315	16.043	15.950	0.136	16.027	15.984	0.061	16.018	15.994	0.035	16.018	16.000	0.029
	MIN	16.000	15.795	0.095	16.000	15.907	0.050	16.000	15.966	0.016	16.000	15.983	0.006	16.000	15.989	0.000
20	MAX	20.130	19.890	0.370	20.052	19.935	0.169	20.033	19.980	0.074	20.021	19.993	0.041	20.021	20.000	0.034
	MIN	20.000	19.760	0.110	20.000	19.883	0.065	20.000	19.959	0.020	20.000	19.980	0.007	20.000	19.987	0.000

BASIC SIZE		LOOSE RUNNING Hole H11	Shaft c11	Fit	FREE RUNNING Hole H9	Shaft d9	Fit	CLOSE RUNNING Hole H8	Shaft f7	Fit	SLIDING Hole H7	Shaft g6	Fit	LOCATIONAL CLEARANCE Hole H7	Shaft h6	Fit
25	MAX	25.130	24.890	0.370	25.052	24.935	0.169	25.033	24.980	0.074	25.021	24.993	0.041	25.021	25.000	0.034
	MIN	25.000	24.760	0.110	25.000	24.883	0.065	25.000	24.959	0.020	25.000	24.980	0.007	25.000	24.987	0.000
30	MAX	30.130	29.890	0.370	30.052	29.935	0.169	30.033	29.980	0.074	30.021	29.993	0.041	30.021	30.000	0.034
	MIN	30.000	29.760	0.110	30.000	29.883	0.065	30.000	29.959	0.020	30.000	29.980	0.007	30.000	29.987	0.000
40	MAX	40.160	39.880	0.440	40.062	39.920	0.204	40.039	39.975	0.089	40.025	39.991	0.050	40.025	40.000	0.041
	MIN	40.000	39.720	0.120	40.000	39.858	0.080	40.000	39.950	0.025	40.000	39.975	0.009	40.000	39.984	0.000
50	MAX	50.160	49.870	0.450	50.062	49.920	0.204	50.039	49.975	0.089	50.025	49.991	0.050	50.025	50.000	0.041
	MIN	50.000	49.710	0.130	50.000	49.858	0.080	50.000	49.950	0.025	50.000	49.975	0.009	50.000	49.984	0.000
60	MAX	60.190	59.860	0.520	60.074	59.900	0.248	60.046	59.970	0.106	60.030	59.990	0.059	60.030	60.000	0.049
	MIN	60.000	59.670	0.140	60.000	59.826	0.100	60.000	59.940	0.030	60.000	59.971	0.010	60.000	59.981	0.000
80	MAX	80.190	79.850	0.530	80.074	79.900	0.248	80.046	79.970	0.106	80.030	79.990	0.059	80.030	80.000	0.049
	MIN	80.000	79.660	0.150	80.000	79.826	0.100	80.000	79.940	0.030	80.000	79.971	0.010	80.000	79.981	0.000
100	MAX	100.220	99.830	0.610	100.087	99.880	0.294	100.054	99.964	0.125	100.035	99.988	0.069	100.035	100.000	0.057
	MIN	100.000	99.610	0.170	100.000	99.793	0.120	100.000	99.929	0.036	100.000	99.966	0.012	100.000	99.978	0.000
120	MAX	120.220	119.820	0.620	120.087	119.880	0.294	120.054	119.964	0.125	120.035	119.988	0.069	120.035	120.000	0.057
	MIN	120.000	119.600	0.180	120.000	119.793	0.120	120.000	119.929	0.036	120.000	119.966	0.012	120.000	119.978	0.000
160	MAX	160.250	159.790	0.710	160.100	159.855	0.345	160.063	159.957	0.146	160.040	159.986	0.079	160.040	160.000	0.065
	MIN	160.000	159.540	0.210	160.000	159.755	0.145	160.000	159.917	0.043	160.000	159.961	0.014	160.000	159.975	0.000
200	MAX	200.290	199.760	0.820	200.115	199.830	0.400	200.072	199.950	0.168	200.046	199.985	0.090	200.046	200.000	0.075
	MIN	200.000	199.470	0.240	200.000	199.715	0.170	200.000	199.904	0.050	200.000	199.956	0.015	200.000	199.971	0.000
250	MAX	250.290	249.720	0.860	250.115	249.830	0.400	250.072	249.950	0.168	250.046	249.985	0.090	250.046	250.000	0.075
	MIN	250.000	249.430	0.280	250.000	249.715	0.170	250.000	249.904	0.050	250.000	249.956	0.015	250.000	249.971	0.000
300	MAX	300.320	299.670	0.970	300.130	299.810	0.450	300.081	299.944	0.189	300.052	299.983	0.101	300.052	300.000	0.084
	MIN	300.000	299.350	0.330	300.000	299.680	0.190	300.000	299.892	0.056	300.000	299.951	0.017	300.000	299.968	0.000
400	MAX	400.360	399.600	1.120	400.140	399.790	0.490	400.089	399.938	0.208	400.057	399.982	0.111	400.057	400.000	0.093
	MIN	400.000	399.240	0.400	400.000	399.650	0.210	400.000	399.881	0.062	400.000	399.946	0.018	400.000	399.964	0.000
500	MAX	500.400	499.520	1.280	500.155	499.770	0.540	500.097	499.932	0.228	500.063	499.980	0.123	500.063	500.000	0.103
	MIN	500.000	499.120	0.480	500.000	499.615	0.230	500.000	499.869	0.068	500.000	499.940	0.020	500.000	499.960	0.000

Dimensions in millimeters.

Extracted from *Preferred Metric Limits and Fits* (ANSI B4.2–1978, R1984), with the permission of the publisher, The American Society of Mechanical Engineers.

Appendix 46. American National Standard Preferred Metric Limits and Fits
Preferred Hole Basis Transition and Interference Fits, Cylindrical*

BASIC SIZE		LOCATIONAL TRANSN. Hole H7	Shaft k6	Fit	LOCATIONAL TRANSN. Hole H7	Shaft n6	Fit	LOCATIONAL INTERF. Hole H7	Shaft p6	Fit	MEDIUM DRIVE Hole H7	Shaft s6	Fit	FORCE Hole H7	Shaft u6	Fit
1	MAX	1.010	1.006	0.010	1.010	1.010	0.006	1.010	1.012	0.004	1.010	1.020	−0.004	1.010	1.024	−0.008
	MIN	1.000	1.000	−0.006	1.000	1.004	−0.010	1.000	1.006	−0.012	1.000	1.014	−0.020	1.000	1.018	−0.024
1.2	MAX	1.210	1.206	0.010	1.210	1.210	0.006	1.210	1.212	0.004	1.210	1.220	−0.004	1.210	1.224	−0.008
	MIN	1.200	1.200	−0.006	1.200	1.204	−0.010	1.200	1.206	−0.012	1.200	1.214	−0.020	1.200	1.218	−0.024
1.6	MAX	1.610	1.606	0.010	1.610	1.610	0.006	1.610	1.612	0.004	1.610	1.620	−0.004	1.610	1.624	−0.008
	MIN	1.600	1.600	−0.006	1.600	1.604	−0.010	1.600	1.606	−0.012	1.600	1.614	−0.020	1.600	1.618	−0.024
2	MAX	2.010	2.006	0.010	2.010	2.010	0.006	2.010	2.012	0.004	2.010	2.020	−0.004	2.010	2.024	−0.008
	MIN	2.000	2.000	−0.006	2.000	2.004	−0.010	2.000	2.006	−0.012	2.000	2.014	−0.020	2.000	2.018	−0.024
2.5	MAX	2.510	2.506	0.010	2.510	2.510	0.006	2.510	2.512	0.004	2.510	2.520	−0.004	2.510	2.524	−0.008
	MIN	2.500	2.500	−0.006	2.500	2.504	−0.010	2.500	2.506	−0.012	2.500	2.514	−0.020	2.500	2.518	−0.024
3	MAX	3.010	3.006	0.010	3.010	3.010	0.006	3.010	3.012	0.004	3.010	3.020	−0.004	3.010	3.024	−0.008
	MIN	3.000	3.000	−0.006	3.000	3.004	−0.010	3.000	3.006	−0.012	3.000	3.014	−0.020	3.000	3.018	−0.024
4	MAX	4.012	4.009	0.011	4.012	4.016	0.004	4.012	4.020	0.000	4.012	4.027	−0.007	4.012	4.031	−0.011
	MIN	4.000	4.001	−0.009	4.000	4.008	−0.016	4.000	4.012	−0.020	4.000	4.019	−0.027	4.000	4.023	−0.031
5	MAX	5.012	5.009	0.011	5.012	5.016	0.004	5.012	5.020	0.000	5.012	5.027	−0.007	5.012	5.031	−0.011
	MIN	5.000	5.001	−0.009	5.000	5.008	−0.016	5.000	5.012	−0.020	5.000	5.019	−0.027	5.000	5.023	−0.031
6	MAX	6.012	6.009	0.011	6.012	6.016	0.004	6.012	6.020	0.000	6.012	6.027	−0.007	6.012	6.031	−0.011
	MIN	6.000	6.001	−0.009	6.000	6.008	−0.016	6.000	6.012	−0.020	6.000	6.019	−0.027	6.000	6.023	−0.031
8	MAX	8.015	8.010	0.014	8.015	8.019	0.005	8.015	8.024	0.000	8.015	8.032	−0.008	8.015	8.037	−0.013
	MIN	8.000	8.001	−0.010	8.000	8.010	−0.019	8.000	8.015	−0.024	8.000	8.023	−0.032	8.000	8.028	−0.037
10	MAX	10.015	10.010	0.014	10.015	10.019	0.005	10.015	10.024	0.000	10.015	10.032	−0.008	10.015	10.037	−0.013
	MIN	10.000	10.001	−0.010	10.000	10.010	−0.019	10.000	10.015	−0.024	10.000	10.023	−0.032	10.000	10.028	−0.037
12	MAX	12.018	12.012	0.017	12.018	12.023	0.006	12.018	12.029	0.000	12.018	12.039	−0.010	12.018	12.044	−0.015
	MIN	12.000	12.001	−0.012	12.000	12.012	−0.023	12.000	12.018	−0.029	12.000	12.028	−0.039	12.000	12.033	−0.044
16	MAX	16.018	16.012	0.017	16.018	16.023	0.006	16.018	16.029	0.000	16.018	16.039	−0.010	16.018	16.044	−0.015
	MIN	16.000	16.001	−0.012	16.000	16.012	−0.023	16.000	16.018	−0.029	16.000	16.028	−0.039	16.000	16.033	−0.044
20	MAX	20.021	20.015	0.019	20.021	20.028	0.006	20.021	20.035	−0.001	20.021	20.048	−0.014	20.021	20.054	−0.020
	MIN	20.000	20.002	−0.015	20.000	20.015	−0.028	20.000	20.022	−0.035	20.000	20.035	−0.048	20.000	20.041	−0.054

BASIC SIZE	LOCATIONAL TRANSN.			LOCATIONAL TRANSN.			LOCATIONAL INTERF.			MEDIUM DRIVE			FORCE		
	Hole H7	Shaft k6	Fit	Hole H7	Shaft n6	Fit	Hole H7	Shaft p6	Fit	Hole H7	Shaft s6	Fit	Hole H7	Shaft u6	Fit
25 MAX	25.021	25.015	0.019	25.021	25.028	0.006	25.021	25.035	−0.001	25.021	25.048	−0.014	25.021	25.061	−0.027
MIN	25.000	25.002	−0.015	25.000	25.015	−0.028	25.000	25.022	−0.035	25.000	25.035	−0.048	25.000	25.048	−0.061
30 MAX	30.021	30.015	0.019	30.021	30.028	0.006	30.021	30.035	−0.001	30.021	30.048	−0.014	30.021	30.061	−0.027
MIN	30.000	30.002	−0.015	30.000	30.015	−0.028	30.000	30.022	−0.035	30.000	30.035	−0.048	30.000	30.048	−0.061
40 MAX	40.025	40.018	0.023	40.025	40.033	0.008	40.025	40.042	−0.001	40.025	40.059	−0.018	40.025	40.076	−0.035
MIN	40.000	40.002	−0.018	40.000	40.017	−0.033	40.000	40.026	−0.042	40.000	40.043	−0.059	40.000	40.060	−0.076
50 MAX	50.025	50.018	0.023	50.025	50.033	0.008	50.025	50.042	−0.001	50.025	50.059	−0.018	50.025	50.086	−0.045
MIN	50.000	50.002	−0.018	50.000	50.017	−0.033	50.000	50.026	−0.042	50.000	50.043	−0.059	50.000	50.070	−0.086
60 MAX	60.030	60.021	0.028	60.030	60.039	0.010	60.030	60.051	−0.002	60.030	60.072	−0.023	60.030	60.106	−0.057
MIN	60.000	60.002	−0.021	60.000	60.020	−0.039	60.000	60.032	−0.051	60.000	60.053	−0.072	60.000	60.087	−0.106
80 MAX	80.030	80.021	0.028	80.030	80.039	0.010	80.030	80.051	−0.002	80.030	80.078	−0.029	80.030	80.121	−0.072
MIN	80.000	80.002	−0.021	80.000	80.020	−0.039	80.000	80.032	−0.051	80.000	80.059	−0.078	80.000	80.102	−0.121
100 MAX	100.035	100.025	0.032	100.035	100.045	0.012	100.035	100.059	−0.002	100.035	100.093	−0.036	100.035	100.146	−0.089
MIN	100.000	100.003	−0.025	100.000	100.023	−0.045	100.000	100.037	−0.059	100.000	100.071	−0.093	100.000	100.124	−0.146
120 MAX	120.035	120.025	0.032	120.035	120.045	0.012	120.035	120.059	−0.002	120.035	120.101	−0.044	120.035	120.166	−0.109
MIN	120.000	120.003	−0.025	120.000	120.023	−0.045	120.000	120.037	−0.059	120.000	120.079	−0.101	120.000	120.144	−0.166
160 MAX	160.040	160.028	0.037	160.040	160.052	0.013	160.040	160.068	−0.003	160.040	160.125	−0.060	160.040	160.215	−0.150
MIN	160.000	160.003	−0.028	160.000	160.027	−0.052	160.000	160.043	−0.068	160.000	160.100	−0.125	160.000	160.190	−0.215
200 MAX	200.046	200.033	0.042	200.046	200.060	0.015	200.046	200.079	−0.004	200.046	200.151	−0.076	200.046	200.265	−0.190
MIN	200.000	200.004	−0.033	200.000	200.031	−0.060	200.000	200.050	−0.079	200.000	200.122	−0.151	200.000	200.236	−0.265
250 MAX	250.046	250.033	0.042	250.046	250.060	0.015	250.046	250.079	−0.004	250.046	250.169	−0.094	250.046	250.313	−0.238
MIN	250.000	250.004	−0.033	250.000	250.031	−0.060	250.000	250.050	−0.079	250.000	250.140	−0.169	250.000	250.284	−0.313
300 MAX	300.052	300.036	0.048	300.052	300.066	0.018	300.052	300.088	−0.004	300.052	300.202	−0.118	300.052	300.382	−0.298
MIN	300.000	300.004	−0.036	300.000	300.034	−0.066	300.000	300.056	−0.088	300.000	300.170	−0.202	300.000	300.350	−0.382
400 MAX	400.057	400.040	0.053	400.057	400.073	0.020	400.057	400.098	−0.005	400.057	400.244	−0.151	400.057	400.471	−0.378
MIN	400.000	400.004	−0.040	400.000	400.037	−0.073	400.000	400.062	−0.098	400.000	400.208	−0.244	400.000	400.435	−0.471
500 MAX	500.063	500.045	0.058	500.063	500.080	0.023	500.063	500.108	−0.005	500.063	500.292	−0.189	500.063	500.580	−0.477
MIN	500.000	500.005	−0.045	500.000	500.040	−0.080	500.000	500.068	−0.108	500.000	500.252	−0.292	500.000	500.540	−0.580

Dimensions in millimeters.

Extracted from *Preferred Metric Limits and Fits* (ANSI B 4.2–1978, R1984), with the permission of the publisher, The American Society of Mechanical Engineers.

Appendix 47. American National Standard Preferred Metric Limits and Fits Preferred Shaft Basis Clearance Fits, Cylindrical*

BASIC SIZE		LOOSE RUNNING			FREE RUNNING			CLOSE RUNNING			SLIDING			LOCATIONAL CLEARANCE		
		Hole C11	Shaft h11	Fit	Hole D9	Shaft h9	Fit	Hole F8	Shaft h7	Fit	Hole G7	Shaft h6	Fit	Hole H7	Shaft h6	Fit
1	MAX	1.120	1.000	0.180	1.045	1.000	0.070	1.020	1.000	0.030	1.012	1.000	0.018	1.010	1.000	0.016
	MIN	1.060	0.940	0.060	1.020	0.975	0.020	1.006	0.990	0.006	1.002	0.994	0.002	1.000	0.994	0.000
1.2	MAX	1.320	1.200	0.180	1.245	1.200	0.070	1.220	1.200	0.030	1.212	1.200	0.018	1.210	1.200	0.016
	MIN	1.260	1.140	0.060	1.220	1.175	0.020	1.206	1.190	0.006	1.202	1.194	0.002	1.200	1.194	0.000
1.6	MAX	1.720	1.600	0.180	1.645	1.600	0.070	1.620	1.600	0.030	1.612	1.600	0.018	1.610	1.600	0.016
	MIN	1.660	1.540	0.060	1.620	1.575	0.020	1.606	1.590	0.006	1.602	1.594	0.002	1.600	1.594	0.000
2	MAX	2.120	2.000	0.180	2.045	2.000	0.070	2.020	2.000	0.030	2.012	2.000	0.018	2.010	2.000	0.016
	MIN	2.060	1.940	0.060	2.020	1.975	0.020	2.006	1.990	0.006	2.002	1.994	0.002	2.000	1.994	0.000
2.5	MAX	2.620	2.500	0.180	2.545	2.500	0.070	2.520	2.500	0.030	2.512	2.500	0.018	2.510	2.500	0.016
	MIN	2.560	2.440	0.060	2.520	2.475	0.020	2.506	2.490	0.006	2.502	2.494	0.002	2.500	2.494	0.000
3	MAX	3.120	3.000	0.180	3.045	3.000	0.070	3.020	3.000	0.030	3.012	3.000	0.018	3.010	3.000	0.016
	MIN	3.060	2.940	0.060	3.020	2.975	0.020	3.006	2.990	0.006	3.002	2.994	0.002	3.000	2.994	0.000
4	MAX	4.145	4.000	0.220	4.060	4.000	0.090	4.028	4.000	0.040	4.016	4.000	0.024	4.012	4.000	0.020
	MIN	4.070	3.925	0.070	4.030	3.970	0.030	4.010	3.988	0.010	4.004	3.992	0.004	4.000	3.992	0.000
5	MAX	5.145	5.000	0.220	5.060	5.000	0.090	5.028	5.000	0.040	5.016	5.000	0.024	5.012	5.000	0.020
	MIN	5.070	4.925	0.070	5.030	4.970	0.030	5.010	4.988	0.010	5.004	4.992	0.004	5.000	4.992	0.000
6	MAX	6.145	6.000	0.220	6.060	6.000	0.090	6.028	6.000	0.040	6.016	6.000	0.024	6.012	6.000	0.020
	MIN	6.070	5.925	0.070	6.030	5.970	0.030	6.010	5.988	0.010	6.004	5.992	0.004	6.000	5.992	0.000
8	MAX	8.170	8.000	0.260	8.076	8.000	0.112	8.035	8.000	0.050	8.020	8.000	0.029	8.015	8.000	0.024
	MIN	8.080	7.910	0.080	8.040	7.964	0.040	8.013	7.985	0.013	8.005	7.991	0.005	8.000	7.991	0.000
10	MAX	10.170	10.000	0.260	10.076	10.000	0.112	10.035	10.000	0.050	10.020	10.000	0.029	10.015	10.000	0.024
	MIN	10.080	9.910	0.080	10.040	9.964	0.040	10.013	9.985	0.013	10.005	9.991	0.005	10.000	9.991	0.000
12	MAX	12.205	12.000	0.315	12.093	12.000	0.136	12.043	12.000	0.061	12.024	12.000	0.035	12.018	12.000	0.029
	MIN	12.095	11.890	0.095	12.050	11.957	0.050	12.016	11.982	0.016	12.006	11.989	0.006	12.000	11.989	0.000
16	MAX	16.205	16.000	0.315	16.093	16.000	0.136	16.043	16.000	0.061	16.024	16.000	0.035	16.018	16.000	0.029
	MIN	16.095	15.890	0.095	16.050	15.957	0.050	16.016	15.982	0.016	16.006	15.989	0.006	16.000	15.989	0.000
20	MAX	20.240	20.000	0.370	20.117	20.000	0.169	20.053	20.000	0.074	20.028	20.000	0.041	20.021	20.000	0.034
	MIN	20.110	19.870	0.110	20.065	19.948	0.065	20.020	19.979	0.020	20.007	19.987	0.007	20.000	19.987	0.000

BASIC SIZE		LOOSE RUNNING			FREE RUNNING			CLOSE RUNNING			SLIDING			LOCATIONAL CLEARANCE		
		Hole C11	Shaft h11	Fit	Hole D9	Shaft h9	Fit	Hole F8	Shaft h7	Fit	Hole G7	Shaft h6	Fit	Hole H7	Shaft h6	Fit
25	MAX	25.240	25.000	0.370	25.117	25.000	0.169	25.053	25.000	0.074	25.028	25.000	0.041	25.021	25.000	0.034
	MIN	25.110	24.870	0.110	25.065	24.948	0.065	25.020	24.979	0.020	25.007	24.987	0.007	25.000	24.987	0.000
30	MAX	30.240	30.000	0.370	30.117	30.000	0.169	30.053	30.000	0.074	30.028	30.000	0.041	30.021	30.000	0.034
	MIN	30.110	29.870	0.110	30.065	29.948	0.065	30.020	29.979	0.020	30.007	29.987	0.007	30.000	29.987	0.000
40	MAX	40.280	40.000	0.440	40.142	40.000	0.204	40.064	40.000	0.089	40.034	40.000	0.050	40.025	40.000	0.041
	MIN	40.120	39.840	0.120	40.080	39.938	0.080	40.025	39.975	0.025	40.009	39.984	0.009	40.000	39.984	0.000
50	MAX	50.290	50.000	0.450	50.142	50.000	0.204	50.064	50.000	0.089	50.034	50.000	0.050	50.025	50.000	0.041
	MIN	50.130	49.840	0.130	50.080	49.938	0.080	50.025	49.975	0.025	50.009	49.984	0.009	50.000	49.984	0.000
60	MAX	60.330	60.000	0.520	60.174	60.000	0.248	60.076	60.000	0.106	60.040	60.000	0.059	60.030	60.000	0.049
	MIN	60.140	59.810	0.140	60.100	59.926	0.100	60.030	59.970	0.030	60.010	59.981	0.010	60.000	59.981	0.000
80	MAX	80.340	80.000	0.530	80.174	80.000	0.248	80.076	80.000	0.106	80.040	80.000	0.059	80.030	80.000	0.049
	MIN	80.150	79.810	0.150	80.100	79.926	0.100	80.030	79.970	0.030	80.010	79.981	0.010	80.000	79.981	0.000
100	MAX	100.390	100.000	0.610	100.207	100.000	0.294	100.090	100.000	0.125	100.047	100.000	0.069	100.035	100.000	0.057
	MIN	100.170	99.780	0.170	100.120	99.913	0.120	100.036	99.965	0.036	100.012	99.978	0.012	100.000	99.978	0.000
120	MAX	120.400	120.000	0.620	120.207	120.000	0.294	120.090	120.000	0.125	120.047	120.000	0.069	120.035	120.000	0.057
	MIN	120.180	119.780	0.180	120.120	119.913	0.120	120.036	119.965	0.036	120.012	119.978	0.012	120.000	119.978	0.000
160	MAX	160.460	160.000	0.710	160.245	160.000	0.345	160.106	160.000	0.146	160.054	160.000	0.079	160.040	160.000	0.065
	MIN	160.210	159.750	0.210	160.145	159.900	0.145	160.043	159.960	0.043	160.014	159.975	0.014	160.000	159.975	0.000
200	MAX	200.530	200.000	0.820	200.285	200.000	0.400	200.122	200.000	0.168	200.061	200.000	0.090	200.046	200.000	0.075
	MIN	200.240	199.710	0.240	200.170	199.885	0.170	200.050	199.954	0.050	200.015	199.971	0.015	200.000	199.971	0.000
250	MAX	250.570	250.000	0.860	250.285	250.000	0.400	250.122	250.000	0.168	250.061	250.000	0.090	250.046	250.000	0.075
	MIN	250.280	249.710	0.280	250.170	249.885	0.170	250.050	249.954	0.050	250.015	249.971	0.015	250.000	249.971	0.000
300	MAX	300.650	300.000	0.970	300.320	300.000	0.450	300.137	300.000	0.189	300.069	300.000	0.101	300.052	300.000	0.084
	MIN	300.330	299.680	0.330	300.190	299.870	0.190	300.056	299.948	0.056	300.017	299.968	0.017	300.000	299.968	0.000
400	MAX	400.760	400.000	1.120	400.350	400.000	0.490	400.151	400.000	0.208	400.075	400.000	0.111	400.057	400.000	0.093
	MIN	400.400	399.640	0.400	400.210	399.860	0.210	400.062	399.943	0.062	400.018	399.964	0.018	400.000	399.964	0.000
500	MAX	500.880	500.000	1.280	500.385	500.000	0.540	500.165	500.000	0.228	500.083	500.000	0.123	500.063	500.000	0.103
	MIN	500.480	499.600	0.480	500.230	499.845	0.230	500.068	499.937	0.068	500.020	499.960	0.020	500.000	499.960	0.000

*Dimensions in millimeters.
Extracted from *Preferred Metric Limits and Fits* (ANSI B4.2–1978, R1984), with the permission of the publisher, The American Society of Mechanical Engineers.

Appendix 48. American National Standard Preferred Metric Limits and Fits Preferred Shaft Basis Transition and Interference Fits, Cylindrical*

BASIC SIZE		LOCATIONAL TRANSN.			LOCATIONAL TRANSN.			LOCATIONAL INTERF.			MEDIUM DRIVE			FORCE		
		Hole K7	Shaft h6	Fit	Hole N7	Shaft h6	Fit	Hole P7	Shaft h6	Fit	Hole S7	Shaft h6	Fit	Hole U7	Shaft h6	Fit
1	MAX	1.000	1.000	0.006	0.996	1.000	0.002	0.994	1.000	0.000	0.986	1.000	−0.008	0.982	1.000	−0.012
	MIN	0.990	0.994	−0.010	0.986	0.994	−0.014	0.984	0.994	−0.016	0.976	0.994	−0.024	0.972	0.994	−0.028
1.2	MAX	1.200	1.200	0.006	1.196	1.200	0.002	1.194	1.200	0.000	1.186	1.200	−0.008	1.182	1.200	−0.012
	MIN	1.190	1.194	−0.010	1.186	1.194	−0.014	1.184	1.194	−0.016	1.176	1.194	−0.024	1.172	1.194	−0.028
1.6	MAX	1.600	1.600	0.006	1.596	1.600	0.002	1.594	1.600	0.000	1.586	1.600	−0.008	1.582	1.600	−0.012
	MIN	1.590	1.594	−0.010	1.586	1.594	−0.014	1.584	1.594	−0.016	1.576	1.594	−0.024	1.572	1.594	−0.028
2	MAX	2.000	2.000	0.006	1.996	2.000	0.002	1.994	2.000	0.000	1.986	2.000	−0.008	1.982	2.000	−0.012
	MIN	1.990	1.994	−0.010	1.986	1.994	−0.014	1.984	1.994	−0.016	1.976	1.994	−0.024	1.972	1.994	−0.028
2.5	MAX	2.500	2.500	0.006	2.496	2.500	0.002	2.494	2.500	0.000	2.486	2.500	−0.008	2.482	2.500	−0.012
	MIN	2.490	2.494	−0.010	2.486	2.494	−0.014	2.484	2.494	−0.016	2.476	2.494	−0.024	2.472	2.494	−0.028
3	MAX	3.000	3.000	0.006	2.996	3.000	0.002	2.994	3.000	0.000	2.986	3.000	−0.008	2.982	3.000	−0.012
	MIN	2.990	2.994	−0.010	2.986	2.994	−0.014	2.984	2.994	−0.016	2.976	2.994	−0.024	2.972	2.994	−0.028
4	MAX	4.003	4.000	0.011	3.996	4.000	0.004	3.992	4.000	0.000	3.985	4.000	−0.007	3.981	4.000	−0.011
	MIN	3.991	3.992	−0.009	3.984	3.992	−0.016	3.980	3.992	−0.020	3.973	3.992	−0.027	3.969	3.992	−0.031
5	MAX	5.003	5.000	0.011	4.996	5.000	0.004	4.992	5.000	0.000	4.985	5.000	−0.007	4.981	5.000	−0.011
	MIN	4.991	4.992	−0.009	4.984	4.992	−0.016	4.980	4.992	−0.020	4.973	4.992	−0.027	4.969	4.992	−0.031
6	MAX	6.003	6.000	0.011	5.996	6.000	0.004	5.992	6.000	0.000	5.985	6.000	−0.007	5.981	6.000	−0.011
	MIN	5.991	5.992	−0.009	5.984	5.992	−0.016	5.980	5.992	−0.020	5.973	5.992	−0.027	5.969	5.992	−0.031
8	MAX	8.005	8.000	0.014	7.996	8.000	0.005	7.991	8.000	0.000	7.983	8.000	−0.008	7.978	8.000	−0.013
	MIN	7.990	7.991	−0.010	7.981	7.991	−0.019	7.976	7.991	−0.024	7.968	7.991	−0.032	7.963	7.991	−0.037
10	MAX	10.005	10.000	0.014	9.996	10.000	0.005	9.991	10.000	0.000	9.983	10.000	−0.008	9.978	10.000	−0.013
	MIN	9.990	9.991	−0.010	9.981	9.991	−0.019	9.976	9.991	−0.024	9.968	9.991	−0.032	9.963	9.991	−0.037
12	MAX	12.006	12.000	0.017	11.995	12.000	0.006	11.989	12.000	0.000	11.979	12.000	−0.010	11.974	12.000	−0.015
	MIN	11.988	11.989	−0.012	11.977	11.989	−0.023	11.971	11.989	−0.029	11.961	11.989	−0.039	11.956	11.989	−0.044
16	MAX	16.006	16.000	0.017	15.995	16.000	0.006	15.989	16.000	0.000	15.979	16.000	−0.010	15.974	16.000	−0.015
	MIN	15.988	15.989	−0.012	15.977	15.989	−0.023	15.971	15.989	−0.029	15.961	15.989	−0.039	15.956	15.989	−0.044
20	MAX	20.006	20.000	0.019	19.993	20.000	0.006	19.986	20.000	−0.001	19.973	20.000	−0.014	19.967	20.000	−0.020
	MIN	19.985	19.987	−0.015	19.972	19.987	−0.028	19.965	19.987	−0.035	19.952	19.987	−0.048	19.946	19.987	−0.054

BASIC SIZE		LOCATIONAL TRANSN. Hole K7	Shaft h6	Fit	LOCATIONAL TRANSN. Hole N7	Shaft h6	Fit	LOCATIONAL INTERF. Hole P7	Shaft h6	Fit	MEDIUM DRIVE Hole S7	Shaft h6	Fit	FORCE Hole U7	Shaft h6	Fit
25	MAX	25.006	25.000	0.019	24.993	25.000	0.006	24.986	25.000	−0.001	24.973	25.000	−0.014	24.960	25.000	−0.027
	MIN	24.985	24.987	−0.015	24.972	24.987	−0.028	24.965	24.987	−0.035	24.952	24.987	−0.048	24.939	24.987	−0.061
30	MAX	30.006	30.000	0.019	29.993	30.000	0.006	29.986	30.000	−0.001	29.973	30.000	−0.014	29.960	30.000	−0.027
	MIN	29.985	29.987	−0.015	29.972	29.987	−0.028	29.965	29.987	−0.035	29.952	29.987	−0.048	29.939	29.987	−0.061
40	MAX	40.007	40.000	0.023	39.992	40.000	0.008	39.983	40.000	−0.001	39.966	40.000	−0.018	39.949	40.000	−0.035
	MIN	39.982	39.984	−0.018	39.967	39.984	−0.033	39.958	39.984	−0.042	39.941	39.984	−0.059	39.924	39.984	−0.076
50	MAX	50.007	50.000	0.023	49.992	50.000	0.008	49.983	50.000	−0.001	49.966	50.000	−0.018	49.939	50.000	−0.045
	MIN	49.982	49.984	−0.018	49.967	49.984	−0.033	49.958	49.984	−0.042	49.941	49.984	−0.059	49.914	49.984	−0.086
60	MAX	60.009	60.000	0.028	59.991	60.000	0.010	59.979	60.000	−0.002	59.958	60.000	−0.023	59.924	60.000	−0.057
	MIN	59.979	59.981	−0.021	59.961	59.981	−0.039	59.949	59.981	−0.051	59.928	59.981	−0.072	59.894	59.981	−0.106
80	MAX	80.009	80.000	0.028	79.991	80.000	0.010	79.979	80.000	−0.002	79.952	80.000	−0.029	79.909	80.000	−0.072
	MIN	79.979	79.981	−0.021	79.961	79.981	−0.039	79.949	79.981	−0.051	79.922	79.981	−0.078	79.879	79.981	−0.121
100	MAX	100.010	100.000	0.032	99.990	100.000	0.012	99.976	100.000	−0.002	99.942	100.000	−0.036	99.889	100.000	−0.089
	MIN	99.975	99.978	−0.025	99.955	99.978	−0.045	99.941	99.978	−0.059	99.907	99.978	−0.093	99.854	99.978	−0.146
120	MAX	120.010	120.000	0.032	119.990	120.000	0.012	119.976	120.000	−0.002	119.934	120.000	−0.044	119.869	120.000	−0.109
	MIN	119.975	119.978	−0.025	119.955	119.978	−0.045	119.941	119.978	−0.059	119.899	119.978	−0.101	119.834	119.978	−0.166
160	MAX	160.012	160.000	0.037	159.988	160.000	0.013	159.972	160.000	−0.003	159.915	160.000	−0.060	159.825	160.000	−0.150
	MIN	159.972	159.975	−0.028	159.948	159.975	−0.052	159.932	159.975	−0.068	159.875	159.975	−0.125	159.785	159.975	−0.215
200	MAX	200.013	200.000	0.042	199.986	200.000	0.015	199.967	200.000	−0.004	199.895	200.000	−0.076	199.781	200.000	−0.190
	MIN	199.967	199.971	−0.033	199.940	199.971	−0.060	199.921	199.971	−0.079	199.849	199.971	−0.151	199.735	199.971	−0.265
250	MAX	250.013	250.000	0.042	249.986	250.000	0.015	249.967	250.000	−0.004	249.877	250.000	−0.094	249.733	250.000	−0.238
	MIN	249.967	249.971	−0.033	249.940	249.971	−0.060	249.921	249.971	−0.079	249.831	249.971	−0.169	249.687	249.971	−0.313
300	MAX	300.016	300.000	0.048	299.986	300.000	0.018	299.964	300.000	−0.004	299.850	300.000	−0.118	299.670	300.000	−0.298
	MIN	299.964	299.968	−0.036	299.934	299.968	−0.066	299.912	299.968	−0.088	299.798	299.968	−0.202	299.618	299.968	−0.382
400	MAX	400.017	400.000	0.053	399.984	400.000	0.020	399.959	400.000	−0.005	399.813	400.000	−0.151	399.586	400.000	−0.378
	MIN	399.960	399.964	−0.040	399.927	399.964	−0.073	399.902	399.964	−0.098	399.756	399.964	−0.244	399.529	399.964	−0.471
500	MAX	500.018	500.000	0.058	499.983	500.000	0.023	499.955	500.000	−0.005	499.771	500.000	−0.189	499.483	500.000	−0.477
	MIN	499.955	499.960	−0.045	499.920	499.960	−0.080	499.892	499.960	−0.108	499.708	499.960	−0.292	499.420	499.960	−0.580

Dimensions in millimeters.

Extracted from *Preferred Metric Limits and Fits* (ANSI B4.2–1978, R1984), with the permission of the publisher, The American Society of Mechanical Engineers.

Appendix 49. Double-Line Symbols for Malleable Iron Screw Fittings

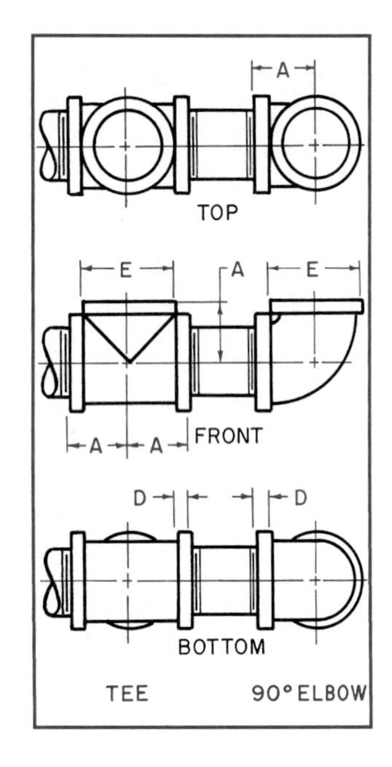

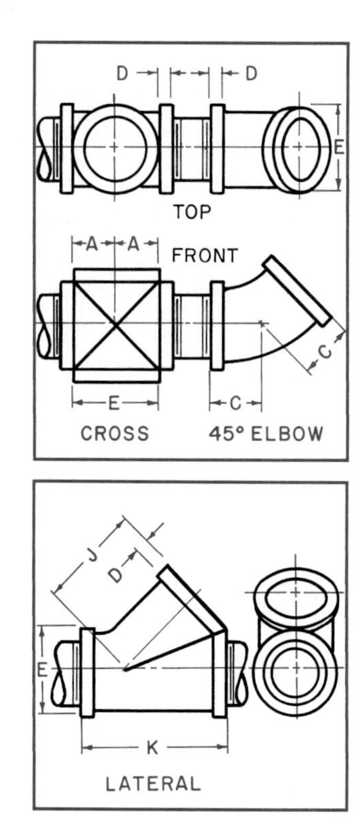

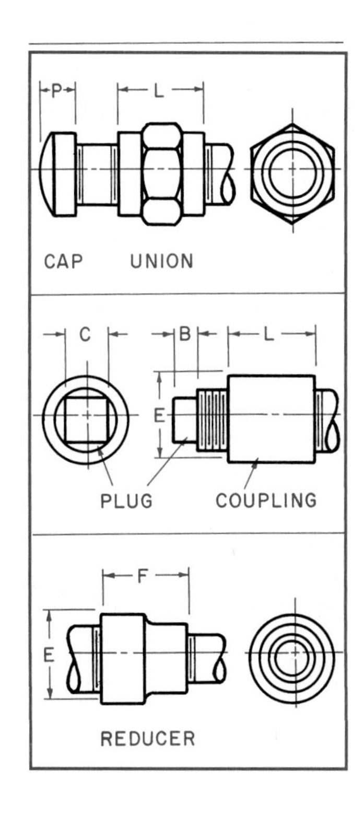

Approximate Malleable Iron Pipe Fitting Sizes (Screw Type)

Pipe Dia.	3/4	1	1 1/2	2	2 1/2
A	1 5/16	1 1/2	1 15/16	2 1/2	2 45/64
B	1/2	37/64	45/64	3/4	59/64
C	1	1 1/8	1 7/16	1 11/16	1 61/64
D	9/32	5/16	3/8	27/64	15/32
E	1 29/64	1 25/32	2 27/64	2 61/64	3 19/32
F	1 7/16	1 11/16	2 5/16	2 13/16	3 1/4
G	1	15/32	1 21/64	1 29/64	1 45/64
J	2 3/64	2 7/16	3 9/32	3 15/16	4 47/64
K	2 25/32	3 9/32	4 3/8	5 11/64	6 1/4
L	1 1/2	1 43/64	2 5/32	2 17/32	2 7/8

Appendix 50. Double-Line Symbols for Steel Butt-Welded Pipe Fittings

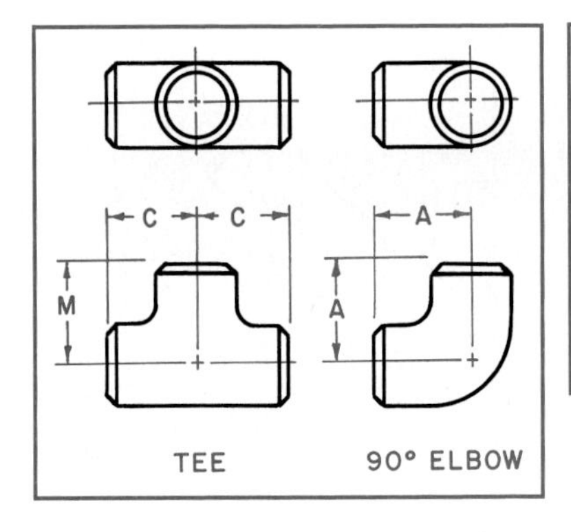

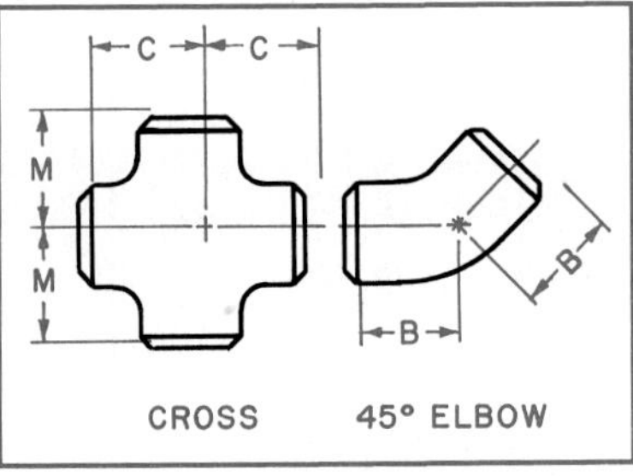

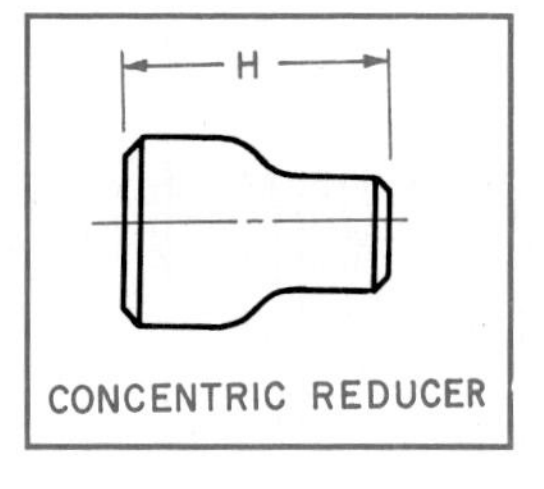

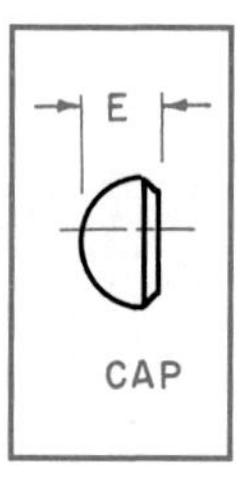

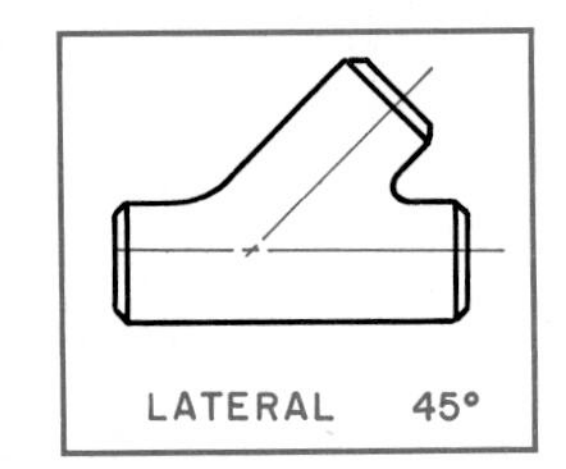

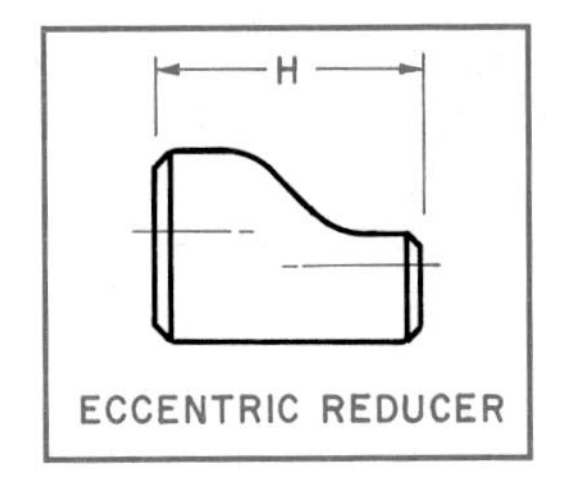

Steel Butt-Welded Reducers*

	Large End	Small End	H
3/4 × 1/2	1.050	0.840	1 1/2
1 × 3/4	1.315	1.050	2
1 1/2 × 1	1.900	1.315	2 1/2
2 × 1 1/2	2.375	1.900	3
2 1/2 × 2	2.875	2.375	3 1/2
3 × 2 1/2	3.500	2.875	3 1/2

*Dimensions in inches.

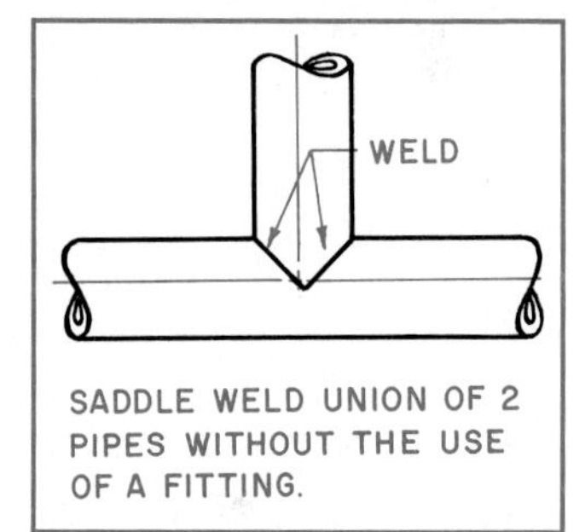

SADDLE WELD UNION OF 2 PIPES WITHOUT THE USE OF A FITTING.

Steel Butt-Welded Fittings

		Elbows*		Tees and Crosses*		Caps*
Nominal Pipe Size	Outside Dia. at Bevel	90° Elbow	45° Elbow	Center to Center		Length E
		A	B	Run C	Outlet M	
3/4	1.050	1 1/8	7/16	1 1/8	1 1/8	1
1	1.315	1 1/2	7/8	1 1/2	1 1/2	1 1/2
1 1/2	1.900	2 1/4	1 1/8	2 1/4	2 1/4	1 1/2
2	2.375	3	1 3/8	2 1/2	2 1/2	1 1/2
2 1/2	2.875	3 3/4	1 3/4	3	3	1 1/2
3	3.500	4 1/2	2	3 3/8	3 3/8	2

*Dimensions are in inches.

Appendix 51. Double-Line Symbols for Cast Iron Flanged Pipe Fittings

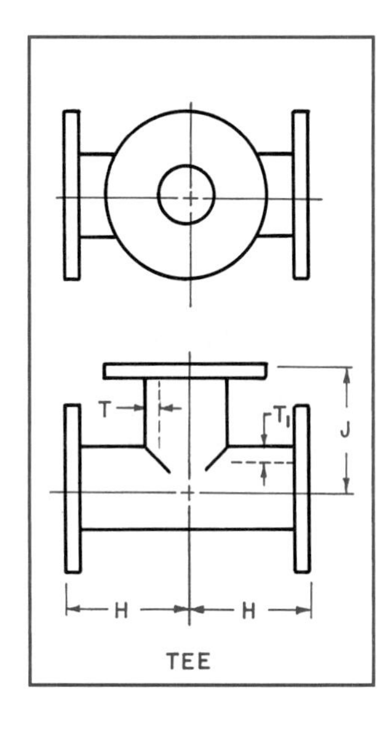

TEE

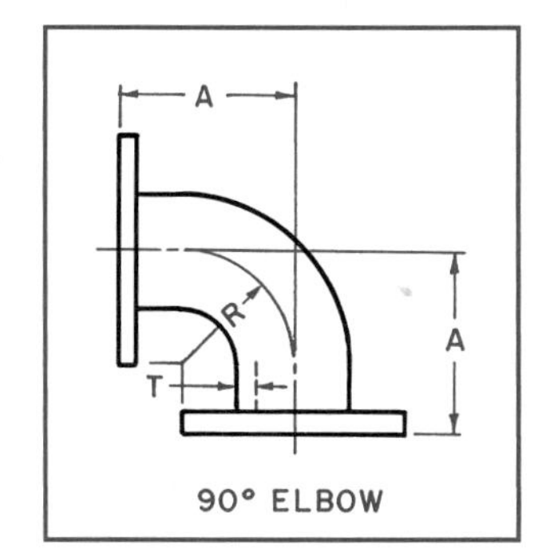

90° ELBOW

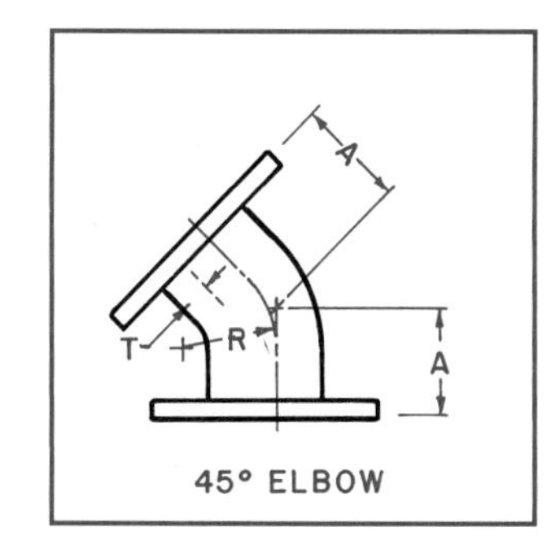

45° ELBOW

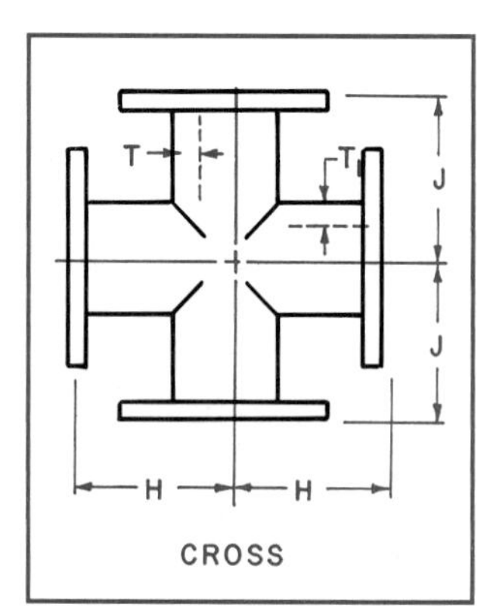

CROSS

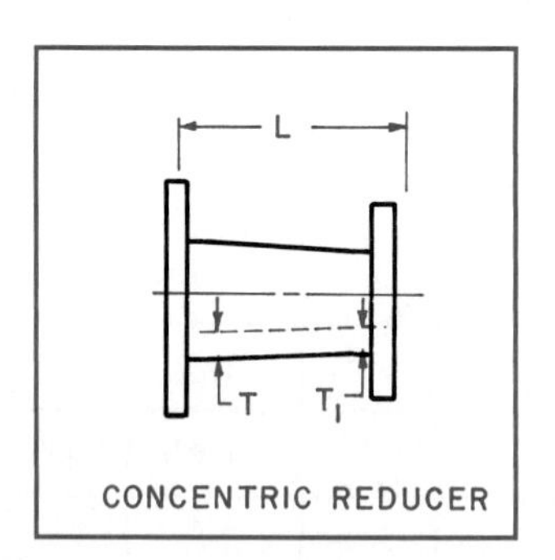

CONCENTRIC REDUCER

Flanged Elbows-Cast Iron*

		90° Elbow		45° Elbow	
Nominal Pipe Size	T	A	R	A	R
2	0.35	4.5	3.0	2.5	2.44
3	0.48	5.5	4.0	3.0	3.62
4	0.52	6.5	4.5	4.0	4.81
6	0.55	8.0	6.0	5.0	7.25

*Dimensions in inches. For flange dimensions, see flange table.

Flanged Tees and Crosses–Cast Iron*

Nominal Pipe Size	T	T_1	H	J
2	0.35	0.35	4.5	4.5
3	0.48	0.35	5.5	5.5
4	0.52	0.35	6.5	6.5
6	0.55	0.35	8.0	8.0

*Dimensions in inches. For flange dimensions, see flange table.

Flange Dimensions for All Types of Flanged Fittings*

Nominal Pipe Size	Outside Diameter	Bolt Hole Circle	Flange Thickness	Bolt Hole Dia.	Bolt Dia. and Length	No. of Bolts
2	6.0	4.75	0.62	¾	⅝ × 2¼	4
3	7.50	6.00	0.75	¾	⅝ × 2¼	4
4	9.00	7.50	0.94	¾	⅝ × 3	8
6	11.00	9.50	1.00	⅞	⅝ × 3¼	8

*Dimensions in inches.

Appendix 52. Double-Line Symbols for Cast Brass Solder-Joint Fittings

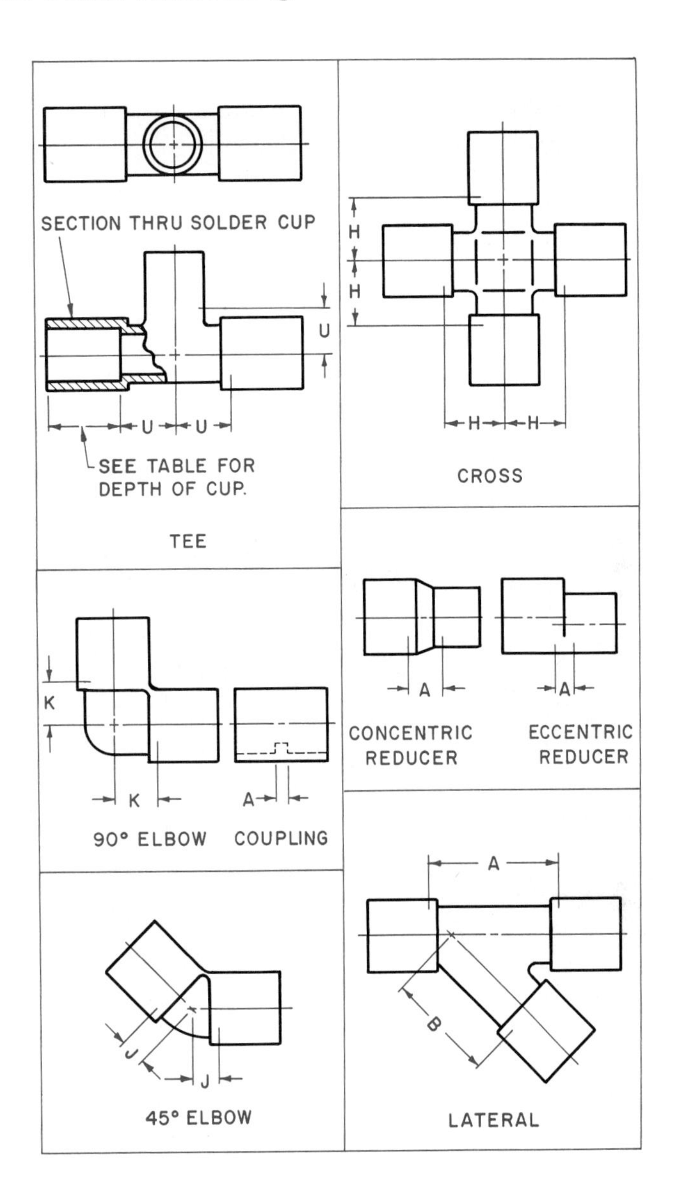

Depth of Solder Cup on Cast Brass Solder-Joint Fittings

Nominal Pipe Size	Solder Cup Depth in Inches
3/4	3/4
1	29/32
1 1/2	1 3/32
2	1 11/32
2 1/2	1 15/32
3	1 21/32

Cast Brass Solder-Joint Fittings

Nominal Pipe Size	Concentric Reducers	Eccentric Reducers
	A	A
3/4 to 1/2	5/16	7/32
1 to 3/4	3/8	1/4
1 1/2 to 1	3/8	3/16
2 to 1 1/2	1/2	9/32
3 to 2	9/16	7/16

Cast Brass Solder-Joint Fittings

Nominal Pipe Size	Tee	90° Elbow	45° Elbow	Coupling	Cross	Lateral	
	U	K	J	A	H	A	B
3/4	9/16	9/16	1/4	1/8	9/16	1 17/32	1 7/32
1	23/32	23/32	5/16	1/8	23/32	1 13/16	1 15/16
1 1/2	1	1	1/2	1/8	1	2 5/8	2 1/8
2	1 1/4	1 1/4	9/16	3/16	1 1/4	3 5/16	2 3/4
2 1/2	1 1/2	1 1/2	5/8	3/16	1 1/2	5	4 1/16
3	1 3/4	1 3/4	3/4	3/16	1 23/32	4 3/4	4 1/32

Appendix 53. Double-Line Symbols for Valves

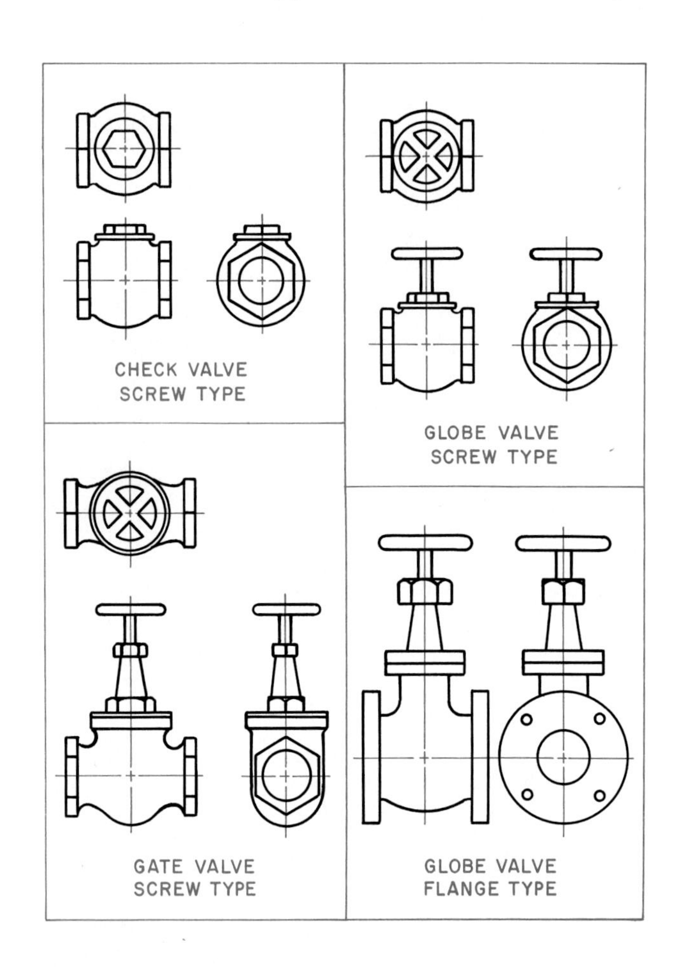

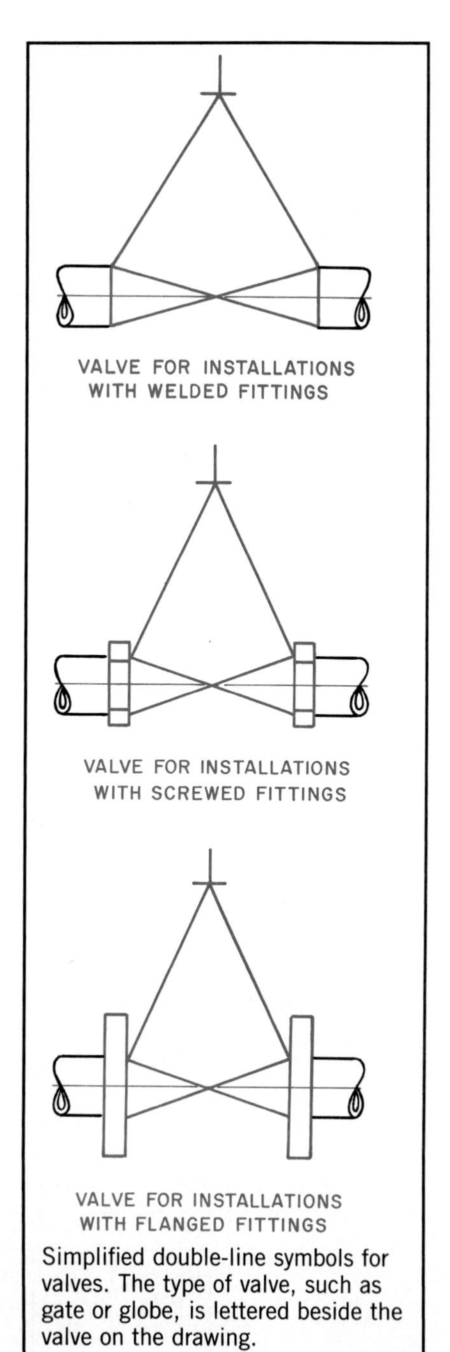

Simplified double-line symbols for valves. The type of valve, such as gate or globe, is lettered beside the valve on the drawing.

Appendix 54A. Preferred Metric Design Sizes

1 mm to 10 mm																	
1st choice	1	1.2	1.6	2	2.5	3	4	5	6	8	10						
2nd choice	1.1	1.4	1.8	2.2	2.8	3.5	4.5	5.5	7	9							
3rd choice	1.3	1.5	1.7	1.9	2.1	2.4	2.6	3.2	3.8	4.2	4.8	5.2	5.8	6.5	7.5	8.5	9.5

10 mm to 100 mm																					
1st choice	10	12	16	20	25	30	40	50	60	80	100										
2nd choice	11	14	18	22	28	32	35	38	42	45	48	52	55	58	65	70	75	85	90	95	
3rd choice	13	15	17	19	21	23	24	26	34	36	44	46	54	56	62	68	72	78	82	88	92	98

100 mm to 1000 mm																									
1st choice	100	110	120	140	160	180	200	220	250	280	300	350	400	450	500	550	600	700	800	900	1000				
2nd choice	130	150	170	190	210	230	240	260	320	380	420	480	520	580	620	650	680	720	750	780	850	950			
3rd choice	105	115	125	135	145	155	165	175	185	195	270	290	340	360	440	460	540	560	640	660	740	760	820	880	920	980

Appendix 54B. Preferred Metric Basic Sizes for Round Shafts and Holes

FIRST CHOICE	SECOND CHOICE	FIRST CHOICE	SECOND CHOICE	FIRST CHOICE	SECOND CHOICE
1		10		100	
	1.1		11		110
1.2		12		120	
	1.4		14		140
1.6		16		160	
	1.8		18		180
2		20		200	
	2.2		22		220
2.5		25		250	
	2.8		28		280
3		30		300	
	3.5		35		350
4		40		400	
	4.5		45		450
5		50		500	
	5.5		55		550
6		60		600	
	7		70		700
8		80		800	
	9		90		900
				1000	

ANSI B4.2–1978

Appendix 54C. Preferred Decimal Inch Basic Sizes for Round Shafts and Holes

0.010	2.00	8.50
0.012	2.20	9.00
0.016	2.40	9.50
0.020	2.60	10.00
0.025	2.80	10.50
0.032	3.00	11.00
0.040	3.20	11.50
0.05	3.40	12.00
0.06	3.60	12.50
0.08	3.80	13.00
0.10	4.00	13.50
0.12	4.20	14.00
0.16	4.40	14.50
0.20	4.60	15.00
0.24	4.80	15.50
0.30	5.00	16.00
0.40	5.20	16.50
0.50	5.40	17.00
0.60	5.60	17.50
0.80	5.80	18.00
1.00	6.00	18.50
1.20	6.50	19.00
1.40	7.00	19.50
1.60	7.50	20.00
1.80	8.00	

Appendix 55A. Common Metric Measures

Metric Measures of Linear Distances

1 kilometer (km) = 1000 meters
1 hectometer (hm) = 100 meters (m)
1 dekameter (dam) = 10 meters (m)
1 decimeter (dm) = one-tenth of a meter (m)
1 centimeter (cm) = one-hundredth of a meter (m)
1 millimeter (mm) = one-thousandth of a meter (m)

1 meter (m) = 10 decimeters (dm)
1 decimeter (dm) = 10 centimeters (cm)
1 centimeter (cm) = 10 millimeters (mm)

Metric Measures of Area

1 square meter (m^2) = 100 square decimeters (dm^2)
1 square meter (m^2) = 10 000 square centimeters (cm^2)
1 square meter (m^2) = 1 000 000 square millimeters (mm^2)

100 square millimeters (mm^2) = 1 square centimeter (cm^2)
100 square centimeters (cm^2) = 1 square decimeter (dm^2)
100 square decimeters (dm^2) = 1 square meter (m^2)

Designations for Large Areas

1 square kilometer (km^2) = 1 000 000 square meters (m^2)
1 hectare (square hectometer) (hm^2) = 10 000 square meters (m^2)

Metric Measures of Volume

1 cubic meter (m^3) = 1000 cubic decimeters (dm^3)
1 cubic meter (m^3) = 1 000 000 cubic centimeters (cm^3)
1 cubic meter (m^3) = 1 000 000 000 cubic millimeters (mm^3)

1000 cubic millimeters (mm^3) = 1 cubic centimeter (cm^3)
1000 cubic centimeters (cm^3) = 1 cubic decimeter (dm^3)
1000 cubic decimeters (dm^3) = 1 cubic meter (m^3)

Metric Measures of Mass

1 kilogram (kg) = 1000 grams (g)
1 hectogram (hg) = 100 grams (g)
1 dekagram (dag) = 10 grams (g)
1 decigram (dg) = one-tenth of a gram (g)
1 centigram (cg) = one-hundredth of a gram (g)
1 milligram (mg) = one-thousandth of a gram (g)

Appendix 55B. Common Conversions

Linear Measures

inches to millimeters—multiply inches by 25.4
millimeters to inches—multiply millimeters by .0394
inches to centimeters—multiply inches by 2.54
centimeters to inches—multiply centimeters by .394
inches to meters—multiply inches by .0254
feet to meters—multiply feet by .3048

Measurement of Area

square inches (in^2) to square millimeters (mm^2)—multiply square inches by 645.4
square inches (in^2) to square centimeters (cm^2)—multiply square inches by 6.454
square centimeters (cm^2) to square inches—multiply square centimeters by 0.155
square feet (ft^2) to square meters (m^2)—multiply square feet by 0.093
square meters (m^2) to square feet—multiply square meters by 10.76
square yards (yd^2) to square meters (m^2)—multiply square yards by 0.836
square meters (m^2) to square yards—multiply square meters by 1.197
acres to square meters (m^2)—acres multiplied by 4046.87
square miles (mi^2) to square meters (m^2)—square miles multiplied by 2 589 998
square miles (mi^2) to square kilometers (km^2)—square miles multiplied by 2.589

Measures of Volume

cubic yards to cubic meters (m^3)—multiply cubic yards by .764
cubic feet to cubic meters (m^3)—multiply cubic feet by 0.028
gallons to cubic meters (m^3)—multiply gallons by 0.0038
quarts to cubic meters (m^3)—multiply quarts by 0.00095
ounces to cubic meters (m^3)—multiply ounces by 0.000029

Measures of Mass

pounds to kilograms—multiply pounds by .454
kilograms to pounds—multiply kilograms by 2.205

Measures of Temperature

Fahrenheit to Celsius = (°F − 32) × .555
Celsius to Fahrenheit = (°C × 1.8) + 32

Appendix 56. Millimeters to Decimal Inches*

mm	In.	mm	In.	mm	In.	mm	In.	mm	In.
1	= 0.0394	21	= 0.8268	41	= 1.6142	61	= 2.4016	81	= 3.1890
2	= 0.0787	22	= 0.8662	42	= 1.6536	62	= 2.4410	82	= 3.2284
3	= 0.1181	23	= 0.9055	43	= 1.6929	63	= 2.4804	83	= 3.2678
4	= 0.1575	24	= 0.9449	44	= 1.7323	64	= 2.5197	84	= 3.3071
5	= 0.1969	25	= 0.9843	45	= 1.7717	65	= 2.5591	85	= 3.3465
6	= 0.2362	26	= 1.0236	46	= 1.8111	66	= 2.5985	86	= 3.3859
7	= 0.2756	27	= 1.0630	47	= 1.8504	67	= 2.6378	87	= 3.4253
8	= 0.3150	28	= 1.1024	48	= 1.8898	68	= 2.6772	88	= 3.4646
9	= 0.3543	29	= 1.1418	49	= 1.9292	69	= 2.7166	89	= 3.5040
10	= 0.3937	30	= 1.1811	50	= 1.9685	70	= 2.7560	90	= 3.5434
11	= 0.4331	31	= 1.2205	51	= 2.0079	71	= 2.7953	91	= 3.5827
12	= 0.4724	32	= 1.2599	52	= 2.0473	72	= 2.8247	92	= 3.6221
13	= 0.5118	33	= 1.2992	53	= 2.0867	73	= 2.8741	93	= 3.6615
14	= 0.5512	34	= 1.3386	54	= 2.1260	74	= 2.9134	94	= 3.7009
15	= 0.5906	35	= 1.3780	55	= 2.1654	75	= 2.9528	95	= 3.7402
16	= 0.6299	36	= 1.4173	56	= 2.2048	76	= 2.9922	96	= 3.7796
17	= 0.6693	37	= 1.4567	57	= 2.2441	77	= 3.0316	97	= 3.8190
18	= 0.7087	38	= 1.4961	58	= 2.2835	78	= 3.0709	98	= 3.8583
19	= 0.7480	39	= 1.5355	59	= 2.3229	79	= 3.1103	99	= 3.8977
20	= 0.7874	40	= 1.5748	60	= 2.3622	80	= 3.1497	100	= 3.9371

*For easy reference, this table has also been printed on the inside back cover of this book.

Appendix 57. Fractional Inches to Millimeters*

In.	mm	In.	mm	In.	mm	In.	mm
1/64	= 0.397	17/64	= 6.747	33/64	= 13.097	49/64	= 19.447
1/32	= 0.794	9/32	= 7.144	17/32	= 13.494	25/32	= 19.844
3/64	= 1.191	19/64	= 7.541	35/64	= 13.890	51/64	= 20.240
1/16	= 1.587	5/16	= 7.937	9/16	= 14.287	13/16	= 20.637
5/64	= 1.984	21/64	= 8.334	37/64	= 14.684	53/64	= 21.034
3/32	= 2.381	11/32	= 8.731	19/32	= 15.081	27/32	= 21.431
7/64	= 2.778	23/64	= 9.128	39/64	= 15.478	55/64	= 21.828
1/8	= 3.175	3/8	= 9.525	5/8	= 15.875	7/8	= 22.225
9/64	= 3.572	25/64	= 9.922	41/64	= 16.272	57/64	= 22.622
5/32	= 3.969	13/32	= 10.319	21/32	= 16.669	29/32	= 23.019
11/64	= 4.366	27/64	= 10.716	43/64	= 17.065	59/64	= 23.415
3/16	= 4.762	7/16	= 11.113	11/16	= 17.462	15/16	= 23.812
13/64	= 5.159	29/64	= 11.509	45/64	= 17.859	61/64	= 24.209
7/32	= 5.556	15/32	= 11.906	23/32	= 18.256	31/32	= 24.606
15/64	= 5.953	31/64	= 12.303	47/64	= 18.653	63/64	= 25.003
1/4	= 6.350	1/2	= 12.700	3/4	= 19.050	1	= 25.400

*For easy reference, this table has also been printed on the inside back cover of this book.

Appendix 58. Metric Equivalents of Two-Place Decimal Inches*

				Millimeter Equivalent					Millimeter Equivalent
		.02		.508			.52		13.208
			.03	.762				.53	13.462
		.04		1.016			.54		13.716
			.05	1.270				.55	13.970
	.06			1.524		.56			14.224
		.08		2.032			.58		14.732
			.09	2.286				.59	14.986
.10				2.540	.60				15.240
	.12			3.048		.62			15.748
		.14		3.556			.64		16.256
			.15	3.810				.65	16.510
		.16		4.064			.66		16.764
	.18			4.572		.68			17.272
			.19	4.826				.69	17.526
.20				5.080	.70				17.780
		.22		5.588			.72		18.288
	.24			6.096		.74			18.796
			.25	6.350				.75	19.050
		.26		6.604			.76		19.304
		.28		7.112			.78		19.812
.30				7.620	.80				20.320
			.31	7.874				.81	20.574
	.32			8.128		.82			20.828
		.34		8.636			.84		21.336
			.35	8.890				.85	21.590
		.36		9.144			.86		21.844
			.37	9.398				.87	22.098
	.38			9.652		.88			22.352
.40				10.160	.90				22.860
			.41	10.414				.91	23.114
		.42		10.668			.92		23.368
	.44			11.176		.94			23.876
			.45	11.430				.95	24.130
		.46		11.684			.96		24.384
			.47	11.938				.97	24.638
		.48		12.192			.98		24.892
.50				12.700	1.00				25.400

*For easy reference, this table has also been printed at the back of this book, on the endpaper.

Appendix 59. Decimal Equivalents of Common Fractions*

4ths	8ths	16ths	32nds	64ths	To 4 Places	To 3 Places	To 2 Places
				1/64	.0156	.016	.02
			1/32		.0312	.031	.03
				3/64	.0469	.047	.05
		1/16			.0625	.062	.06
				5/64	.0781	.078	.08
			3/32		.0938	.094	.09
				7/64	.1094	.109	.11
	1/8				.1250	.125	.12
				9/64	.1406	.141	.14
			5/32		.1562	.156	.16
				11/64	.1719	.172	.17
		3/16			.1875	.188	.19
				13/64	.2031	.203	.20
			7/32		.2188	.219	.22
				15/64	.2344	.234	.23
1/4					.2500	.250	.25
				17/64	.2656	.266	.27
			9/32		.2812	.281	.28
				19/64	.2969	.297	.30
		5/16			.3125	.312	.31
				21/64	.3281	.328	.33
			11/32		.3438	.344	.34
				23/64	.3594	.359	.36
	3/8				.3750	.375	.38
				25/64	.3906	.391	.39
			13/32		.4062	.406	.41
				27/64	.4219	.422	.42
		7/16			.4375	.438	.44
				29/64	.4531	.453	.45
			15/32		.4688	.469	.47
				31/64	.4844	.484	.48
					.5000	.500	.50

4ths	8ths	16ths	32nds	64ths	To 4 Places	To 3 Places	To 2 Places
				33/64	.5156	.516	.52
			17/32		.5312	.531	.53
				35/64	.5469	.547	.55
		9/16			.5625	.562	.56
				37/64	.5781	.578	.58
			19/32		.5938	.594	.59
				39/64	.6094	.609	.61
	5/8				.6250	.625	.62
				41/64	.6406	.641	.64
			21/32		.6562	.656	.66
				43/64	.6719	.672	.67
		11/16			.6875	.688	.69
				45/64	.7031	.703	.70
			23/32		.7188	.719	.72
				47/64	.7344	.734	.73
3/4					.7500	.750	.75
				49/64	.7656	.766	.77
			25/32		.7812	.781	.78
				51/64	.7969	.797	.80
		13/16			.8125	.812	.81
				53/64	.8281	.828	.83
			27/32		.8438	.844	.84
				55/64	.8594	.859	.86
	7/8				.8750	.875	.88
				57/64	.8906	.891	.89
			29/32		.9062	.906	.91
				59/64	.9219	.922	.92
		15/16			.9375	.938	.94
				61/64	.9531	.953	.95
			31/32		.9688	.969	.97
				63/64	.9844	.984	.98
					1.0000	1.000	1.00

Military Standards 8C

*For easy reference, this table has also been printed at the front of this book, on the endpaper.

Index

M

N

Q

R

S

T

U

V

W

X

Y

Z